T0231283

The COMPILER DESIGN Handbook

Optimizations and Machine Code Generation

SECOND EDITION

The COMPILER DESIGN Handbook

Optimizations and Machine Code Generation

SECOND EDITION

Edited by
Y.N. Srikant
Priti Shankar

CRC Press
Taylor & Francis Group
Boca Raton London New York

CRC Press is an imprint of the
Taylor & Francis Group, an **informa** business

CRC Press
Taylor & Francis Group
6000 Broken Sound Parkway NW, Suite 300
Boca Raton, FL 33487-2742

© 2008 by Taylor & Francis Group, LLC
CRC Press is an imprint of Taylor & Francis Group, an Informa business

No claim to original U.S. Government works
Printed in the United States of America on acid-free paper
10 9 8 7 6 5 4 3

International Standard Book Number-13: 978-1-4200-4382-2 (Hardcover)

Library of Congress Cataloging-in-Publication Data

The Compiler design handbook : optimizations and machine code generation / edited by Y.N. Srikant
and Priti Shankar. -- 2nd ed.
    p. cm.
Includes bibliographical references and index.
ISBN 978-1-4200-4382-2 (alk. paper)
    1. Compilers (Computer programs) 2. Code generators. I. Srikant, Y. N. II. Shankar, P. (Priti) III.
Title.

QA76.76.C65C35 2007
005.4'53--dc22   2007018733

Visit the Taylor & Francis Web site at
http://www.taylorandfrancis.com

and the CRC Press Web site at
http://www.crcpress.com

Table of Contents

Preface

Current challenges that face compiler designers have resulted in the amalgamation of ideas from traditional compiler design, other areas of computer science, and intuitive insights for the design of novel techniques. With the proliferation of sensor networks and embedded devices, an important focus of research continues to be the design of effective algorithms for the compiler back end. Further, compiler techniques are now being used extensively for problems that are not specifically concerned with the design of the compiler, for example, timing predictability for hard real-time embedded systems. It has been demonstrated that a compiler can have a significant impact on the power expended by an embedded system. Thus energy aware compilation has become an active area of research. The manual customization of embedded systems puts the onus of exploiting novel architectural features on the compiler. Further, mismatches between the clock speeds of microprocessors and off-chip memories is being bridged by the use of memory hierarchies, and appropriate compiler transformations can reduce the average latency of memory accesses. With the growing number of applications that work across the web, analyzing a program for security holes has become an important issue in the static analysis of programs. The need for maintenance of huge software systems calls for efficient techniques for debugging faulty code. It is a well known fact that compilers are replete with inherent hard problems and complex interactions between optimizations that are often hard to comprehend. Thus inventing new techniques for designing compilers that will adapt to specific programs and target platforms is worthy of investigation. New uses of type systems, for example, ensuring the proper interaction between threads in a concurrent program, have brought type systems closer in terms of both techniques as well as objectives to static analysis.

The second edition of *The Compiler Design Handbook: Optimization and Machine Code Generation* has new chapters addressing each of the topics above in addition to new material on garbage collection and computations on iteration spaces. Further it has revised versions of chapters on transformations on the SSA intermediate form, optimizations for object oriented languages, shape analysis, program slicing, retargetable compilation, register allocation, instruction scheduling, software pipelining and and dynamic compilation.

As in the first edition of the Handbook, the material presented here is addressed to readers having a sound background in traditional compiler design. In particular, we assume that the reader is familiar with traditional front end techniques, elements of data flow analysis, code improvement techniques, and principles of machine code generation. We hope that the material presented here will be useful to graduate students, compiler designers in industry as well as researchers in the area.

Acknowledgments

The patience and efforts of the contributors are the major factors which have made this handbook possible. We thank each one of them for their contribution and for sparing their time to write a chapter. Several of them have responded promptly to our requests and have readily cooperated in reviewing other contributors' chapters. The assistance and continuous cooperation of the editors and other staff of CRC Press have proved to be invaluable in producing this handbook. The partial financial assiatance from the Defence Research and Development Organization of India is gratefully acknowledged.

Editors

Y.N. Srikant is a Professor in the department of Computer Science and Automation at the Indian Institute of Science, Bangalore. He received his Ph.D in Computer Science from the Indian Institute of Science, and is the recipient of Young Scientist medal of the Indian National Science Academy. He joined IISc in 1987 as a faculty member. Since then, he has guided a number of doctoral and master's degree students and has consulted for a large number of industries. His area of interest is compiler design.

He started the Compiler Laboratory in the department Computer Science and Automation of IISc, where several major research projects have been carried out in the last ten years. Some of the research projects currently in progress are: energy-aware instruction scheduling, energy and time estimation of programs, partial flow sensitivity, microarchitecture sensitive empirical models for compiler optimizations, and energy-aware cache reconfiguration.

Priti Shankar is a Professor at the Department of Computer Science and Automation at the Indian Institute of Science (IISc). She received a B.Tech in Electrical Engineering at the Indian Institute of Technology (IIT) Delhi in 1968, and the M.S. and Ph.D. degrees at the University of Maryland, College Park in 1971 and 1972 respectively. She joined the IISc in 1973 and was largely responsible for beginning research activity in compilers in the Department. Her current interests in the area of compilers are in the application of formal methods in the generation of automatic tools for compilers. She is a member of the American Mathematical Society and the Association for Computing Machinery.

Contributors

David I. August
Department of Computer
  Science
Princeton University
  Princeton, NJ
august@cs.princeton.edu

Evelyn Duesterwald
IBM Research
Hawthorne, NY
duester@us.ibm.com

Nikil Dutt
Center for Embedded
  Computer Systems
Donald Bren School of
  Informatics and Computer
  Sciences
University of California
  Irvine, CA
dutt@uci.edu

K. Gopinath
Department of Computer
  Science and Automation
Indian Institute of Science
  Bangalore, India
gopi@csa.iisc.ernet.in

R. Govindarajan
Supercomputer Education &
  Research Centre
Department of Computer
  Science and Automation
Indian Institute of Science
  Bangalore, India
govind@serc.iisc.ernet.in

Gautam
Department of Computer
  Science
Colorado State University
  Fort Collins, CO
ggupta@cs.colostate.edu

Neelam Gupta
Department of Computer
  Sciences
University of Arizona
  Tuscon, AZ
ngupta@cs.arizona.edu

Rajiv Gupta
Department of Computer
  Science and Engineering
University of California
  Riverside, Riverside, CA
gupta@cs.ucr.edu

Nigel Horspool
Department of Computer
  Science
University of Victoria
  Victoria, BC, Canada
nigelh@csr.uvic.ca

Matthew T. Jacob
Department of Computer
  Science and Automation
Indian Institute of Science
  Bangalore, India
mjt@csa.iisc.ernet.in

P. J. Joseph
Freescale Semiconductor
  India Pvt. Ltd.
Noida, India
PJ.Joseph@freescale.com

Uday P. Khedker
Department of Computer
  Science & Engineering
IIT Bombay, Powai
  Mumbai, India
uday@cse.iitb.ac.in

Andreas Krall
Institute fuer
  Computersprachen
Technische Universität Wien
  Wien, Austria
andi@complang.tuwien.ac.at

Sharad Malik
Department of Electrical
  Engineering
Princeton University
  Princeton, NJ
sharad@ee.princeton.edu

Rajib Mall
Department of Computer
  Science and Engineering
Indian Institute of
  Technology, Kharagpur
  India
rajib@cse.iitkgp.ernet.in

Todd Millstein
UCLA Computer Science
　Department
University of California
　Los Angeles, CA
todd@cs.ucla.edu

Tulika Mitra
School of Computing
National University of
　Singapore, Republic of
　Singapore
tulika@comp.nus.edu.sg

G. B. Mund
Department of Computer
　Science and Engineering
Kalinga Institute of Industrial
　Technology, Bhubaneswar,
　India
mundgb@yahoo.com

V. Krishna Nandivada
IBM India Research Labs
New Delhi, India
nvkrishna@in.ibm.com

Jens Palsberg
UCLA Computer Science
　Department
University of California
　Los Angeles, CA
palsberg@cs.ucla.edu

J. Prakash Prabhu
Department of Computer
　Science and Automation
Indian Institute of Science
　Bangalore, India
prakash.prabhu@gmail.com

Wei Qin
Department of Electrical &
　Computer Engineering
Boston University
　Boston, MA
wqin@bu.edu

Subramanian Rajagopalan
Synopsys (India) EDA
　Software Pvt. Ltd.
Bangalore, India
Subramanian.Rajagopalan@
　synopsys.com

Sanjay Rajopadhye
Department of Computer
　Science
Colorado State University
　Fort Collins, CO
Sanjay.Rajopadhye@
　colostate.edu

Easwaran Raman
Department of Computer
　Science
Princeton University
　Princeton, NJ
eraman@cs.princeton.edu

**Lakshminarayanan
Renganarayana**
Department of Computer
　Science
Colorado State University
　Fort Collins, CO
ln@cs.colostate.edu

Thomas Reps
University of
　Wisconsin-Madison
Computer Sciences
　Department, Madison, WI
reps@cs.wisc.edu

Hongbo Rong
Microsoft Corporation
Redmond, WA
hongbor@microsoft.com

Abhik Roychoudhury
School of Computing
National University of
　Singapore, Republic
　of Singapore
abhik@comp.nus.edu.sg

Mooly Sagiv
Department of Computer
　Science
School of Mathematics
　and Science
Tel Aviv University, Tel Aviv
　Israel
sagiv@math.tau.ac.il

Amitabha Sanyal
Department of Computer
　Science & Engineering
IIT Bombay, Powai, Mumbai
　India
as@cse.iitb.ac.in

Priti Shankar
Department of Computer
　Science and Automation
Indian Institute of Science
　Bangalore, India
priti@csa.iisc.ernet.in

Aviral Shrivastava
Department of Computer
　Science and Engineering
School of Computing and
　Informatics
Arizona State University
　Tempe, AZ
Aviral.Shrivastava@asu.edu

Y. N. Srikant
Department of Computer
　Science and Automation
Indian Institute of Science
　Bangalore, India
srikant@csa.iisc.ernet.in

Michelle Mills Strout
Department of Computer
　Science
Colorado State University
　Fort Collins, CO
mstrout@CS.ColoState.edu

K. Ananda Vardhan
Department of Computer
  Science and Automation
Indian Institute of Science
  Bangalore, India
nandu@csa.iisc.ernet.in

Kapil Vaswani
Department of Computer
  Science and Automation
Indian Institute of Science
  Bangalore, India
kapil@csa.iisc.ernet.in

Xiangyu Zhang
Deaprtment of Computer
  Sciences
Purdue University
  West Lafayette, IN
xyzhang@cs.purdue.edu

Reinhard Wilhelm
Fachbereich Informatik
Universitaet des Saarlandes
  Saarbruecken, Germany
wilhelm@cs.uni-sb.de

1

Worst-Case Execution Time and Energy Analysis

Tulika Mitra
and
Abhik Roychoudhury
Department of Computer Science,
School of Computing,
National University of Singapore,
Singapore
tulika@comp.nus.edu.sg and
abhik@comp.nus.edu.sg

1.1 Introduction

Timing predictability is extremely important for hard real-time embedded systems employed in application domains such as automotive electronics and avionics. Schedulability analysis techniques can guarantee the satisfiability of timing constraints for systems consisting of multiple concurrent tasks. One of the key inputs required for the schedulability analysis is the worst-case execution time (WCET) of each of the tasks. WCET of a task on a target processor is defined as its maximum execution time across all possible inputs.

Figure 1.1a and Figure 1.2a show the variation in execution time of a `quick sort` program on a simple and complex processor, respectively. The program sorts a five-element array. The figures show the distribution of execution time (in processor cycles) for all possible permutations of the array elements as inputs. The maximum execution time across all the inputs is the WCET of the program. This simple example illustrates the inherent difficulty of finding the WCET value:

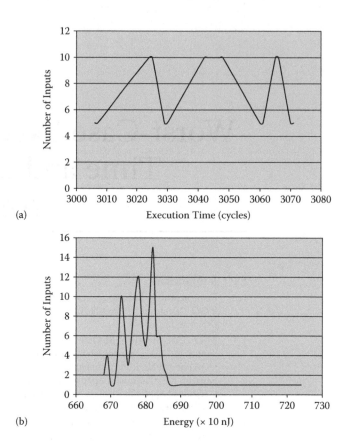

(a)

(b)

FIGURE 1.1   Distribution of time and energy for different inputs of an application on a simple processor.

- Clearly, executing the program for all possible inputs so as to bound its WCET is not feasible. The problem would be trivial if the worst-case input of a program is known *a priori*. Unfortunately, for most programs the worst-case input is unknown and cannot be derived easily.
- Second, the complexity of current micro-architectures implies that the WCET is heavily influenced by the target processor. This is evident from comparing Figure 1.1a with Figure 1.2a. Therefore, the timing effects of micro-architectural components have to be accurately accounted for.

Static analysis methods estimate a bound on the WCET. These analysis techniques are conservative in nature. That is, when in doubt, the analysis assumes the worst-case behavior to guarantee the safety of the estimated value. This may lead to overestimation in some cases. Thus, the goal of static analysis methods is to estimate a *safe* and *tight* WCET value. Figure 1.3 explains the notion of safety and tightness in the context of static WCET analysis. The figure shows the variation in execution time of a task. The *actual WCET* is the maximum possible execution time of the program. The static analysis method generates the *estimated WCET* value such that *estimated WCET* ≥ *actual WCET*. The difference between the estimated and the actual WCET is the overestimation and determines how tight the estimation is. Note that the static analysis methods guarantee that the estimated WCET value can never be less than the actual WCET value. Of course, for a complex task running on a complex processor, the actual WCET value is unknown. Instead, simulation or execution of the program with a subset of possible inputs generates the *observed WCET*, where *observed WCET* ≤ *actual WCET*. In other words, the observed WCET value is not safe, in the sense that it cannot be used to provide absolute timing guarantees for safety-critical systems. A notion related to WCET is the *BCET (best-case execution time)*, which represents the minimum execution time across all possible inputs. In this chapter, we will focus on static analysis

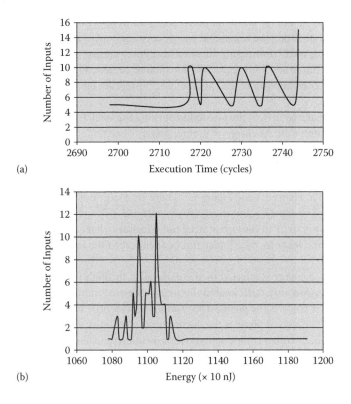

(a)

(b)

FIGURE 1.2 Distribution of time and energy for different inputs of the same application on a complex processor.

techniques to estimate the WCET. However, the same analysis methods can be easily extended to estimate the BCET.

Apart from timing, the proliferation of battery-operated embedded devices has made energy consumption one of the key design constraints. Increasingly, mobile devices are demanding improved functionality and higher performance. Unfortunately, the evolution of battery technology has not been able to keep up with performance requirements. Therefore, designers of mission-critical systems, operating on limited battery life, have to ensure that both the timing and the energy constraints are satisfied under all possible

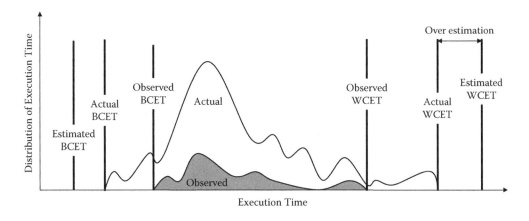

FIGURE 1.3 Definition of observed, actual, and estimated WCET.

scenarios. The battery should never drain out before a task completes its execution. This concern leads to the related problem of estimating the worst-case energy consumption of a task running on a processor for all possible inputs. Unlike WCET, estimating the worst-case energy remains largely unexplored even though it is considered highly important [86], especially for mobile devices. Figure 1.1b and Figure 1.2b show the variation in energy consumption of the quick sort program on a simple and complex processor, respectively.

A natural question that may arise is the possibility of using the WCET path to compute a bound on the worst-case energy consumption. As *energy = average power × execution time*, this may seem like a viable solution and one that can exploit the extensive research in WCET analysis in a direct fashion. Unfortunately, the path corresponding to the WCET may not coincide with the path consuming maximum energy. This is made apparent by comparing the distribution of execution time and energy for the same program and processor pair as shown in Figure 1.1 and Figure 1.2. There are a large number of input pairs $\langle I_1, I_2 \rangle$ in this program, where $time(I_1) < time(I_2)$, but $energy(I_1) > energy(I_2)$. This happens as the energy consumed because of the switching activity in the circuit need not necessarily have a correlation with the execution time. Thus, the input that leads to WCET may not be identical to the input that leads to the worst-case energy.

The execution time or energy is affected by the path taken through the program and the underlying micro-architecture. Consequently, static analysis for worst-case execution time or energy typically consists of three phases. The first phase is the *program path analysis* to identify loop bounds and infeasible flows through the program. The second phase is the *architectural modeling* to determine the effect of pipeline, cache, branch prediction, and other components on the execution time (energy). The last phase, *estimation*, finds an upper bound on the execution time (energy) of the program given the results of the flow analysis and the architectural modeling.

Recently, there has been some work on *measurement-based timing analysis* [6, 17, 92]. This line of work is mainly targeted toward soft real-time systems, such as multimedia applications, that can afford to miss the deadline once in a while. In other words, these application domains do not require absolute timing guarantees. Measurement-based timing analysis methods execute or simulate the program on the target processor for a subset of all possible inputs. They derive the maximum observed execution time (see the definition in Figure 1.3) or the distribution of execution time from these measurements. Measurement-based performance analysis is quite useful for soft real-time applications, but they may underestimate the WCET, which is not acceptable in the context of safety-critical, hard real-time applications. In this article, we only focus on static analysis techniques that provide safe bounds on WCET and worst-case energy. The analysis methods assume uninterrupted program execution on a single processor. Furthermore, the program being analyzed should be free from unbounded loops, unbounded recursion, and dynamic function calls [67].

The rest of the chapter is organized as follows. We proceed with programming-language-level WCET analysis in the next section. This is followed by micro-architectural modeling in Section 1.3. We present a static analysis technique to estimate worst-case energy bound in Section 1.4. A brief description of existing WCET analysis tools appears in Section 1.5, followed by conclusions.

1.2   Programming-Language-Level WCET Analysis

We now proceed to discuss static analysis methods for estimating the WCET of a program. For WCET analysis of a program, the first issue that needs to be determined is the program representation on which the analysis will work. Earlier works [73] have used the *syntax tree* where the (nonleaf) nodes correspond to programming-language-level control structures. The leaves correspond to basic blocks — maximal fragments of code that do not involve any control transfer. Subsequently, almost all work on WCET analysis has used the *control flow graph*. The nodes of a control flow graph (CFG) correspond to basic blocks, and the edges correspond to control transfer between basic blocks. When we construct the CFG of a program, a separate copy of the CFG of a function f is created for every distinct call site of f in the program

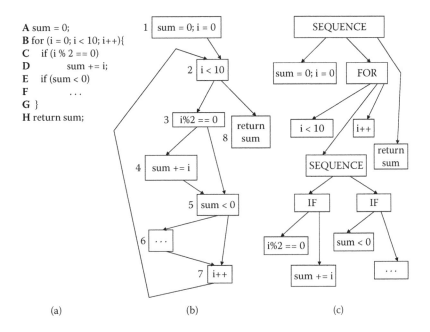

FIGURE 1.4   (a) A code fragment. (b) Control flow graph of the code fragment. (c) Syntax tree of the code fragment.

such that each call transfers control to its corresponding copy of CFG. This is how interprocedural analysis will be handled. Figure 1.4 shows a small code fragment as well as its syntax tree and control flow graph representations.

One important issue needs to be clarified in this regard. The control flow graph of a program can be either at the source code level or at the assembly code level. The difference between the two comes from the compiler optimizations. Our program-level analysis needs to be hooked up with micro-architectural modeling, which accurately estimates the execution time of each instruction while considering the timing effects of underlying microarchitectural features. Hence we always consider the assembly-code-level CFG. However, while showing our examples, we will show CFG at the source code level for ease of exposition.

1.2.1   WCET Calculation

We explain WCET analysis methods in a top-down fashion. Consequently, at the very beginning, we present *WCET calculation* — how to combine the execution time estimates of program fragments to get the execution time estimate of a program. We assume that the *loop bounds* (i.e., the maximum number of iterations for a loop) are known for every program loop; in Section 1.2.2 we outline some methods to estimate loop bounds.

In the following, we outline the three main categories of WCET calculation methods. The path-based and integer linear programming methods operate on the program's control flow graph, while the tree-based methods operate on the program's syntax tree.

1.2.1.1   Tree-Based Methods

One of the earliest works on software timing analysis was the work on *timing schema* [73]. The technique proceeds essentially by a bottom-up pass of the syntax tree. During the traversal, it associates an execution time estimate for each node of the tree. The execution time estimate for a node is obtained from the execution time estimates of its children, by applying the rules in the schema. The schema prescribes rules — one for each control structure of the programming language. Thus, rules corresponding to a sequence of statements, *if-then-else* and *while-loop* constructs, can be described as follows.

- *Time*(S1; S2) = *Time*(S1) + *Time*(S2)
- *Time*(if (B) { S1 } else { S2}) = *Time*(B) + max(*Time*(S1), *Time*(S2))
- *Time* (while (B) { S1}) = $(n + 1)*Time(B) + n*Time(S1)$

Here, n is the loop bound. Clearly, S1, S2 can be complicated code fragments whose execution time estimates need to obtained by applying the schema rules for the control structures appearing in S1, S2. Extensions of the timing schema approach to consider micro-architectural modeling will be discussed in Section 1.3.5.

The biggest advantage of the timing schema approach is its simplicity. It provides an efficient compositional method for estimating the WCET of a program by combining the WCET of its constituent code fragments. Let us consider the following schematic code fragment Pgm. For simplicity of exposition, we will assume that all assignments and condition evaluations take one time unit.

$$i = 0; \text{ while } (i<100) \ \{\text{if } (B') \ S1 \text{ else } S2; \ i++;\}$$

If *Time*(S1) > *Time*(S2), by using the rule for if-then-else statements in the timing schema we get

$$Time(\text{if } (B') \ S1 \text{ else } S2) = Time(B') + Time(S1) = 1 + Time(S1)$$

Now, applying the rule for while-loops in the timing schema, we get the following. The loop bound in this case is 100.

$$
\begin{aligned}
Time(\text{while } (i<100) \ \{\text{if}(B') \ S1 \text{ else } S2\}) &= 101 * Time(i < 100) + \\
&\quad 100 * Time(\text{if } (B') \ S1 \text{ else } S2) \\
&= 101 * 1 + 100 * (1 + Time(S1)) \\
&= 201 + 100 * Time(S1)
\end{aligned}
$$

Finally, using the rule for sequential composition in the timing schema we get

$$
\begin{aligned}
Time(Pgm) &= Time(i = 0) + Time(\text{while } (i<100) \ \{\text{if } (B') \ S1 \text{ else } S2\}) \\
&= 1 + 201 + 100 * Time(S1) = 202 + 100 * Time(S1)
\end{aligned}
$$

The above derivation shows the working of the timing schema. It also exposes one of its major weaknesses. In the timing schema, the timing rules for a program statement are local to the statement; they do not consider the *context* with which the statement is arrived at. Thus, in the preceding we estimated the maximum execution time of if (B') S1 else S2 by taking the execution time for evaluating B and the time for executing S1 (since time for executing S1 is greater than the time for executing S2). As a result, since the if-then-else statement was inside a loop, our maximum execution time estimate for the loop considered the situation where S1 is executed in every loop iteration (i.e., the condition B' is evaluated to true in every loop iteration).

However, in reality S1 may be executed in very few loop iterations for any input; if *Time*(S1) is significantly greater than *Time*(S2), the result returned by timing schema will be a gross overestimate. More importantly, it is difficult to extend or augment the timing schema approach so that it can return tighter estimates in such situations. In other words, even if the user can provide the information that "it is infeasible to execute S1 in every loop iteration of the preceding program fragment Pgm," it is difficult to exploit such information in the timing schema approach. Difficulty in exploiting infeasible program flows information (for returning tighter WCET estimates) remains one of the major weaknesses of the timing schema. We will revisit this issue in Section 1.2.2.

1.2.1.2 Path-Based Methods

The path-based methods perform WCET calculation of a program P via a longest-path search over the control flow graph of P. The loop bounds are used to prevent unbounded unrolling of the loops. The biggest disadvantage of this method is its complexity, as in the worst-case it may amount to enumeration of all program paths that respect the loop bounds. The advantage comes from its ability to handle various kinds of flow information; hence, infeasible path information can be easily integrated with path-based WCET calculation methods.

One approach for restricting the complexity of longest-path searches is to perform symbolic state exploration (as opposed to an explicit path search). Indeed, it is possible to cast the path-based searches for WCET calculation as a (symbolic) model checking problem [56]. However, because model checking is a verification method [13], it requires a temporal property to verify. Thus, to solve WCET analysis using model-checking-based verification, one needs to guess possible WCET estimates and verify that these estimates are indeed WCET estimates. This makes model-checking-based approaches difficult to use (see [94] for more discussion on this topic). The work of Schuele and Schneider [72] employs a *symbolic* exploration of the program's underlying transition system for finding the longest path, without resorting to checking of a temporal property. Moreover, they [72] observe that for finding the WCET there is no need to (even symbolically) maintain data variables that do not affect the program's control flow; these variables are identified via program slicing. This leads to overall complexity reduction of the longest-path search involved in WCET calculation.

A popular path-based WCET calculation approach is to employ an explicit longest-path search, but over a fragment of the control flow graph [31, 76, 79]. Many of these approaches operate on an acyclic fragment of the control flow graph. Path enumeration (often via a breadth-first search) is employed to find the longest path within the acyclic fragment. This could be achieved by a weighted longest-path algorithm (the weights being the execution times of the basic blocks) to find the longest sequence of basic blocks in the control flow graph for a program fragment. The longest-path algorithm can be obtained by a variation of Djikstra's shortest-path algorithm [76]. The longest paths obtained in acyclic control flow graph fragments are then combined with the loop bounds to yield the program's WCET. The path-based approaches can readily exploit any known infeasible flow information. In these methods, the explicit path search is pruned whenever a known infeasible path pattern is encountered.

1.2.1.3 Integer Linear Programming (ILP)

ILP combines the advantages of the tree and path-based approaches. It allows (limited) integration of infeasible path information while (often) being much less expensive than the path-based approaches. Many existing WCET tools such as aiT [1] and Chronos [44] employ ILP for WCET calculation.

The ILP approach operates on the program's control flow graph. Each basic block B in the control flow graph is associated with an integer variable N_B, denoting the *total* execution count of basic block B. The program's WCET is then given by the (linear) objective function

$$\text{maximize} \sum_{B \in \mathcal{B}} N_B * c_B$$

where $\mathcal{B}$ is the set of basic blocks of the program, and c_B is a constant denoting the WCET estimate of basic block B. The linear constraints on N_B are developed from the flow equations based on the control flow graph. Thus, for basic block B,

$$\sum_{B' \to B} E_{B' \to B} = N_B = \sum_{B \to B''} E_{B \to B''}$$

where $E_{B' \to B}$ ($E_{B \to B''}$) is an ILP variable denoting the number of times control flows through the control flow graph edge $B' \to B$ ($B \to B''$). Additional linear constraints are also provided to capture loop bounds and any known infeasible path information.

In the example of Figure 1.4, the control flow equations are given as follows. We use the numbering of the basic blocks 1 to 8 shown in Figure 1.4. Let us examine a few of the control flow equations. For basic block 1, there are no incoming edges, but there is only one outgoing edge $1 \to 2$. This accounts for the constraint $N_1 = E_{1 \to 2}$; that is, the number of executions of basic block 1 is equal to the number of flows from basic block 1 to basic block 2. In other words, whenever basic block 1 is executed, control flows from basic block 1 to basic block 2. Furthermore, since basic block 1 is the entry node, it is executed exactly once; this is captured by the constraint $N_1 = 1$. Now, let us look at the constraints for basic block 2; the inflows to this basic block are the edges $1 \rightarrow 2$ and $7 \rightarrow 2$ and the outflows are the edges $2 \rightarrow 3$ and $2 \rightarrow 8$. This means that whenever block 2 is executed, control must have flown in via either the edge $1 \rightarrow 2$ or the edge $7 \rightarrow 2$; this accounts for the constraint $E_{1 \rightarrow 2} + E_{7 \rightarrow 2} = N_2$. Furthermore, whenever block 2 is executed, control must flow out via the edge $2 \rightarrow 3$ or the edge $2 \rightarrow 8$. This accounts for the constraint $N_2 = E_{2 \rightarrow 3} + E_{2 \rightarrow 8}$. The inflow/outflow constraints for the other basic blocks are obtained in a similar fashion. The full set of inflow/outflow constraints for Figure 1.4 are shown in the following.

$$N_1 = 1 = E_{1 \rightarrow 2}$$
$$E_{1 \rightarrow 2} + E_{7 \rightarrow 2} = N_2 = E_{2 \rightarrow 3} + E_{2 \rightarrow 8}$$
$$E_{2 \rightarrow 3} = N_3 = E_{3 \rightarrow 4} + E_{3 \rightarrow 5}$$
$$E_{3 \rightarrow 4} = N_4 = E_{4 \rightarrow 5}$$
$$E_{4 \rightarrow 5} + E_{3 \rightarrow 5} = N_5 = E_{5 \rightarrow 6} + E_{5 \rightarrow 7}$$
$$E_{5 \rightarrow 6} = N_6 = E_{6 \rightarrow 7}$$
$$E_{6 \rightarrow 7} + E_{5 \rightarrow 7} = N_7 = E_{7 \rightarrow 2}$$
$$E_{2 \rightarrow 8} = N_8 = 1$$

The execution time of the program is given by the following linear function in N_i variables (c_i is a *constant* denoting the WCET of basic block i).

$$\sum_{i=1}^{8} N_i * c_i$$

Now, if we ask the ILP solver to maximize this objective function subject to the inflow/outflow constraints, it will not succeed in producing a time bound for the program. This is because the only loop in the program has not been bounded. The *loop bound* information itself must be provided as linear constraints. In this case, since Figure 1.4 has only one loop, this accounts for the constraint

$$E_{7 \rightarrow 2} \leq 10$$

Using this loop bound, the ILP solver can produce a WCET bound for the program. Of course, the WCET bound can be tightened by providing additional linear constraints capturing infeasible path information; the flow constraints by default assume that all paths in the control flow graph are feasible. It is worthwhile to note that the ILP solver is capable of only *utilizing* the loop bound information and other infeasible path information that is provided to it as linear constraints. Inferring the loop bounds and various infeasible path patterns is a completely different problem that we will discuss next.

Before moving on to infeasible path detection, we note that tight execution time estimates for basic blocks (the constants c_i appearing in the ILP objective function) are obtained by micro-architectural modeling techniques described in Section 1.3. Indeed, this is how the micro-architectural modeling and program path analysis hook up in most existing WCET estimation tools. The program path analysis is done by an ILP solver; infeasible path and loop bound information are integrated with the help of additional linear constraints. The objective function of the ILP contains the WCET estimates of basic blocks as constants. These estimates are provided by micro-architectural modeling, which considers cache, pipeline, and branch prediction behavior to tightly estimate the maximum possible execution time of a basic block B (where B is executed in any possible hardware state and/or control flow context).

1.2.2 Infeasible Path Detection and Exploitation

In the preceding, we have described WCET calculation methods without considering that certain sequences of program fragments may be *infeasible*, that is, not executed on any program input. Our WCET calculation methods only considered the loop bounds to determine a program's WCET estimate. In reality, the WCET calculation needs to consider (and exploit) other information about infeasible program paths. Moreover, the loop bounds also need to be estimated through an off-line analysis. Before proceeding further, we define the notion of an infeasible path.

Definition 1.1

Given a program P, let $\mathcal{B}_P$ be the set of basic blocks of P. Then, an infeasible path of P is a sequence of basic blocks σ over the alphabet $\mathcal{B}_P$, such that σ does not appear in the execution trace corresponding to any input of P.

Clearly, knowledge of infeasible path patterns can tighten WCET estimates. This is simply because the longest path determined by our favorite WCET calculation method may be an infeasible one. Our goal is to efficiently detect *and* exploit infeasible path information for WCET analysis. The general problem of infeasible path detection is NP-complete [2]. Consequently, any approach toward infeasible path detection is an underapproximation — any path determined to be infeasible is indeed infeasible, but not vice versa.

It is important to note that the infeasible path information is often given at the level of source code, whereas the WCET calculation is often performed at the assembly-code-level control flow graph. Because of compiler optimizations, the control flow graph at the assembly code level is not the same as the control flow graph at the source code level. Consequently, infeasible path information that is (automatically) inferred or provided (by the user) at the source code level needs to be converted to a lower level within a WCET estimation tool. This transformation of flow information can be automated and integrated with the compilation process, as demonstrated in [40].

In the following, we discuss methods for infeasible path *detection*. Exploitation of infeasible path information will involve augmenting the WCET calculation methods we discussed earlier. At this stage, it is important to note that infeasible path detection typically involves a smart path search in the program's control flow graph. Therefore, if our WCET calculation proceeds by path-based methods, it is difficult to separate the infeasible path detection and exploitation. In fact, for many path-based methods, the WCET detection and exploitation will be fused into a single step. Consequently, we discuss infeasible path detection methods and along with it exploitation of these in path-based WCET calculation. Later on, we also discuss how the other two WCET calculation approaches (tree-based methods and ILP-based methods) can be augmented to exploit infeasible path information. We note here that the problem of infeasible path detection is a very general one and has implications outside WCET analysis. In the following, we only capture some works as representatives of the different approaches to solving the problem of infeasible path detection.

1.2.2.1 Data Flow Analysis

One of the most common approaches for infeasible path detection is by adapting data flow analysis [21, 27]. In this analysis, each control location in the program is associated with an *environment*. An environment is a mapping of program variables to values, where each program variable is mapped to a *set of values*, instead of a single value. The environment of a control location L captures *all* the possible values that the program variables may assume at L; it captures variable valuations for all possible visits to L. Thus, if x is an integer variable, and at line 70 of the program, the environment at line 70 maps x to [0.5], this means that x is guaranteed to assume an integer value between 0 and 5 when line 70 is visited. An infeasible path is detected when a variable is mapped to the empty set of values at a control location.

Approaches based on data flow analysis are often useful for finding a wide variety of infeasible paths and loop bounds. However, the environments computed at a control location may be too approximate. It is important to note that the environment computed at a control location CL is essentially an *invariant* property — a property that holds for *every* visit to CL. To explain this point, consider the example program in Figure 1.4a. Data flow analysis methods will infer that in line E of the program sum $\in [0..20]$, that is, $0 \leq$ sum ≤ 20. Hence we can infer that execution of lines E, F in Figure 1.4a constitutes an infeasible path. However, by simply keeping track of all possible variable values at each control location we cannot directly infer that line D of Figure 1.4a cannot be executed in consecutive iterations of the loop.

1.2.2.2 Constraint Propagation Methods

The above problem is caused by the merger of environments at any control flow merge point in the control flow graph. The search in data flow analysis is not truly path sensitive — at any control location CL we construct the environment for CL from the environments of all the control locations from which there is an incoming control flow to CL. One way to solve this problem is to perform constraint propagation [7, 71] (or value propagation as in [53]) along paths via *symbolic* execution. Here, instead of assigning possible values to program variables (as in flow analysis), each input variable is given a special value: *unknown*. Thus, if nothing is known about a variable x, we simply represent it as x. The operations on program variables will then have to deal with these symbolic representations of variables. The search then accumulates constraints on x and detects infeasible paths whenever the constraint store becomes unsatisfiable. In the program of Figure 1.4a, by traversing lines C,D we accumulate the constraint i % 2 $\neq$ 0. In the subsequent iteration, we accumulate the constraint i+1 % 2 $\neq$ 0. Note that via symbolic execution we know that the current value of i is one greater than the value in the previous iteration, so the constraint i+1 % 2 $\neq$ 0. We now need to show that the constraint i % 2 $\neq$ 0 $\wedge$ i+1 % 2 $\neq$ 0 is unsatisfiable in order to show that line D in Figure 1.4a cannot be visited in subsequent loop iterations. This will require the help of external constraint solvers or theorem provers such as Simplify [74]. Whether the constraint in question can be solved automatically by the external prover, of course, depends on the prover having appropriate decision procedures to reason about the operators appearing in the constraint (such as the addition [+] and remainder [%] operators appearing in the constraint i % 2 $\neq$ 0 $\wedge$ i+ 1 % 2 $\neq$ 0).

The preceding example shows the plus and minus points of using path-sensitive searches for infeasible path detection. The advantage of using such searches is the precision with which we can detect infeasible program paths. The difficulty in using full-fledged path-sensitive searches (such as model checking) is, of course, the huge number of program paths to consider.[1]

In summary, even though path-sensitive searches are more accurate, they suffer from a huge complexity. Indeed, this has been acknowledged in [53], which accommodates specific heuristics to perform path merging. Consequently, using path-sensitive searches for infeasible path detection does not scale up to large programs. Data flow analysis methods fare better in this regard since they perform merging at control flow merge points in the control flow graph. However, even data flow analysis methods can lead to full-fledged loop unrolling if a variable gets new values in every iteration of a loop (e.g., consider the program while (...){ i++ }).

1.2.2.3 Heuristic Methods

To avoid the cost of loop unrolling, the WCET community has studied techniques that operate on the acyclic graphs representing the control flow of a single loop iteration [31, 76, 79]. These techniques do not detect or exploit infeasible paths that span across multiple loop iterations. The basic idea is to find the weighted longest path in any loop iteration and multiply its cost with the loop bound. Again, the complication arises from the presence of infeasible paths even within a loop iteration. The work of Stappert et al. [76] finds the longest path π in a loop iteration and checks whether it is feasible; if π is infeasible, it employs

---

[1]Furthermore, the data variables of a program typically come from unbounded domains such as integers. Thus, use of a finite-state search method such as model checking will have to either employ data abstractions to construct a finite-state transition system corresponding to a program or work on symbolic state representations representing infinite domains (possibly as constraints), thereby risking nontermination of the search.

graph-theoretic methods to remove π from the control flow graph of the loop. The longest-path calculation is then run again on the modified graph. This process is repeated until a feasible longest path is found. Clearly, this method can be expensive if the feasible paths in a loop have relatively low execution times.

To address this gap, the recent work of Suhendra et al. [79] has proposed a more "infeasible path aware" search of the control flow graph corresponding to a loop body. In this work, the infeasible path detection and exploitation proceeds in two separate steps. In the first step, the work computes "conflict pairs," that is, incompatible (branch, branch) or (assignment, branch) pairs. For example, let us consider the following code fragment, possibly representing the body of a loop.

```
1  if (x > 3)
2      z = z + 1;
3  else
4      x = 1;
5  if (x < 2)
6      z = z/2;
7  else
8      z = z -1;
```

Clearly, the assignment at line 4 conflicts with the branch at line 5 evaluating to false. Similarly, the branch at line 1 evaluating to true conflicts with the branch at line 5 evaluating to true. Such conflicting pairs are detected in a traversal of the control flow directed acyclic graph (DAG) corresponding to the loop body. Subsequently, we traverse the control flow DAG of the loop body from sink to source, always keeping track of the heaviest path. However, if any assignment or branch decision appearing in the heaviest path is involved in a conflict pair, we also keep track of the next heaviest path that is not involved in such a pair. Consequently, we may need to keep track of more than one path at certain points during the traversal; however, redundant tracked paths are removed as soon as conflicts (as defined in the conflict pairs) are resolved during the traversal. This produces a path-based WCET calculation method that detects and exploits infeasible path patterns and still avoids expensive path enumeration or backtracking.

We note that to scale up infeasible path detection and exploitation to large programs, the notion of pairwise conflicts is important. Clearly, this will not allow us to detect that the following is an infeasible path:

$$x = 1; \; y = x; \; \text{if } (y > 2)\{\ldots$$

However, using pairwise conflicts allows us to avoid full-fledged data flow analysis in WCET calculation. The work of Healy and Whalley [31] was the first to use pairwise conflicts for infeasible path detection and exploitation. Apart from pairwise conflicts, this work also detects iteration-based constraints, that is, the behavior of individual branches across loop iterations. Thus, if we have the following program fragment, the technique of Healy and Whalley [31] will infer that the branch inside the loop is true only for the iterations 0..24.

```
for (i = 0; i < 100; i++){
    if (i < 25)
                  { S1; }
    else          { S2;}
}
```

If the time taken to execute S1 is larger than the time taken to execute S2, we can estimate the cost of the loop to be $25 * Time(S1) + 75 * Time(S2)$. Note that in the absence of a framework for using iteration-based constraints, we would have returned the cost of the loop as $100 * Time(S1)$.

In principle, it is possible to combine the efficient control flow graph traversal in [79] with the framework in [31], which combines branch constraints as well as iteration-based constraints. This can result in a path-based WCET calculation that performs powerful infeasible path detection [31] and efficient infeasible path exploitation [79].

```
for (i = 1; i <= 100; i++){
    for (j = i; j <= 100; j++){
        . . .
```

FIGURE 1.5   A nonrectangular loop nest.

1.2.2.4   Loop Bound Inferencing

An important part of infeasible path detection and exploitation is inferencing and usage of loop bounds. Without sophisticated inference of loop bounds, the WCET estimates can be vastly inflated. To see this point, we only need to examine a nested loop of the form shown in Figure 1.5. Here, a naive method will put the loop bound of the inner loop as $100 * 100 = 10,000$, which is a gross overestimate over the actual bound of $1 + 2 + \ldots + 100 = 5050$.

Initial work on loop bounds relied on the programmer to provide manual annotations [61]. These annotations are then used in the WCET calculation. However, giving loop bound annotations is in general an error-prone process. Subsequent work has integrated automated loop bound inferencing as part of infeasible path detection [21]. The work of Liu and Gomez [52] exploits the program structure for high-level languages (such as functional languages) to infer loop bounds. In this work, from the recursive structure of the functions in a functional program, a cost function is constructed automatically. Solving this cost-bound function can then yield bounds on loop executions (often modeled as recursion in functional programs). However, if the program is recursive (as is common for functional programs), the cost bound function is also recursive and does not yield a closed-form solution straightaway. Consequently, this technique [52] (a) performs symbolic evaluation of the cost-bound function using knowledge of program inputs and then (b) transforms the symbolically evaluated function to simplify its recursive structure. This produces the program's loop bounds. The technique is implemented for a subset of the functional language Scheme.[2]

For imperative programs, the work of Healy et al. [30] presents a comprehensive study for inferring loop bounds of various kinds of loops. It handles loops with multiple exits by automatically identifying the conditional branches within a loop body that may affect the number of loop iterations. Subsequently, for each of these branches the range of loop iterations where they can appear is detected; this information is used to compute the loop bounds. Moreover, the work of Healy et al. [30] also presents techniques for automatically inferring bounds on loops where loop exit/entry conditions depend on values of program variables. As an example, let us consider the nonrectangular loop nest shown in Figure 1.5. The technique of Healy et al. [30] will automatically extract the following expression for the bound on the number of executions of the inner loop.

$$N_{inner} = \sum_{i=1}^{100} \sum_{j=i}^{100} 1 = \sum_{i=1}^{100} \left(\sum_{j=1}^{100} 1 - \sum_{j=1}^{i-1} 1 \right) = \sum_{i=1}^{100} (100 - (i - 1))$$

We can then employ techniques for solving summations to obtain N_{inner}.

1.2.2.5   Exploiting Infeasible Path Information in Tree-Based WCET Calculation

So far, we have outlined various methods for detecting infeasible paths in a program's control flow graph. These methods work by traversing the control flow graph and are closer to the path-based methods.

---

[2]Dealing loops as recursive procedures has also been studied in [55] but in a completely different context. This work uses context-sensitive interprocedural analysis to separate out the cache behavior of different executions of the recursive procedure corresponding to a loop, thereby distinguishing, for instance, the cache behavior of the first loop iteration from the remaining loop iterations.

If the WCET calculation is performed by other methods (tree based or ILP), how do we even integrate the infeasible path information into the calculation? In other words, if infeasible path patterns have been detected, how do we let tree-based or ILP-based WCET calculation exploit these patterns to obtain tighter WCET bounds? We first discuss this issue for tree-based methods and then for ILP methods.

One simple way to exploit infeasible path information is to partition the set of program inputs. For each input partition, the program is partially evaluated to remove the statements that are never executed (for inputs in that partition). Timing schema is applied to this partially evaluated program to get its WCET. This process is repeated for every input partition, thereby yielding a WCET estimate for each input partition. The program's WCET is set to the maximum of the WCETs for all the input partitions. To see the benefit of this approach, consider the following schematic program with a boolean input b.

$$If\, Stmt_1 : \quad \texttt{if (b == 0) \{ S1;\} else \{S2;\}}$$
$$If\, Stmt_2 : \quad \texttt{if (b == 1) \{ S3;\} else \{S4;\}}$$

Assume that

$$Time(\texttt{S1}) > Time(\texttt{S2}) \ \text{and} \ Time(\texttt{S3}) > Time(\texttt{S4})$$

Then using the rules of timing schema we have the following. For convenience, we call the first (second) if statement in the preceding schematic program fragment $If\, Stmt_1$ ($If\, Stmt_2$).

$$
\begin{aligned}
Time(If\, Stmt_1) = \quad & Time(\texttt{b == 0}) + Time(\texttt{S1}) \\
Time(If\, Stmt_2) = \quad & Time(\texttt{b == 1}) + Time(\texttt{S3}) \\
Time(If\, Stmt_1; If\, Stmt_2) = \quad & Time(If\, Stmt_1) + Time(If\, Stmt_2) = \\
& Time(\texttt{b == 0}) + Time(\texttt{b == 1}) + Time(\texttt{S1}) + Time(\texttt{S3})
\end{aligned}
$$

We now consider the execution time for the two possible inputs and take their maximum. Let us now consider the program for input b = 0. Since statements S1 and S4 are executed, we have:

$$Time(If\, Stmt_1; If\, Stmt_2)_{b=0} = Time(\texttt{b == 0}) + Time(\texttt{b == 1}) + Time(\texttt{S1}) + Time(\texttt{S4})$$

Similarly, S2 and S3 are executed for b = 1. Thus,

$$Time(If\, Stmt_1; If\, Stmt_2)_{b=1} = Time(\texttt{b == 0}) + Time(\texttt{b == 1}) + Time(\texttt{S2}) + Time(\texttt{S3})$$

The execution time estimate is set to the maximum of $Time(If\, Stmt_1; If\, Stmt_2)_{b=0}$ and $Time(If\, Stmt_1; If\, Stmt_2)_{b=1}$. Both of these quantities are lower than the estimate computed by using the default timing schema rules. Thus, by taking the maximum of these two quantities we will get a tighter estimate than by applying the vanilla timing schema rules.

Partitioning the program inputs and obtaining the WCET for each input partition is a very simple, yet powerful, idea. Even though it has been employed for execution time analysis and energy optimization in the context of timing schema [24, 25], we can plug this idea into other WCET calculation methods as well. The practical difficulty in employing this idea is, of course, computing the input partitions in general. In particular, Gheorghita et al. [25] mention the suitability of the input partitioning approach for multimedia applications performing video and audio decoding and encoding; in these applications there are different computations for different types of input frames being decoded and encoded. However, in general, it is difficult to partition the input space of a program so that inputs with similar execution time estimates get grouped to the same partition. As an example, consider the insertion sort program where the input space consists of the different possible ordering of the input elements in the input array. Thus, in an n-element input array, the input space consists of the different possible permutations of the array element (the permutation $a[1], a[3], a[2]$ denoting the ordering $a[1] < a[3] < a[2]$). First, getting such a partitioning will involve an expensive symbolic execution of the sorting program. Furthermore, even after

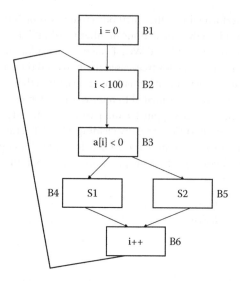

FIGURE 1.6   Example control flow graph.

we obtain the partitioning we still have too many input partitions to work with (the number of partitions for the sorting program is the number of permutations, that is, $n!$). In the worst case, each program input is in a different partition, so the WCET estimation will reduce to exhaustive simulation.

A general approach for exploiting infeasible path information in tree-based WCET calculation has been presented in [61]. In this work, the set of all paths in the control flow graph (taking into account the loop bounds) is described as a regular expression. This is always possible since the set of paths in the control flow graph (taking into account the loop bounds) is finite. Furthermore, all of the infeasible path information given by the user is also converted to regular expressions. Let *Paths* be the set of all paths in the control flow graph and let I_1, I_2 be certain infeasible path information (expressed as a regular expression). We can then safely describe the set of feasible paths as $Paths \cap (\neg I_1) \cap (\neg I_2)$; this is also a regular expression since regular languages are closed under negation and intersection. Timing schema now needs to be employed in these paths, which leads to a practical difficulty. To explain this point, consider the following simple program fragment.

```
for (i=0; i <100; i++){
        if (a[i] < 0) { S1;}
        else { S2;}
}
```

We can draw the control flow graph of this program and present the set of paths in the control flow graph (see Figure 1.6) as a regular expression over basic block occurrences. Thus, the set of paths in the control flow graph fragment of Figure 1.6 is

$$B1(B2B3B4B6 + B2B3B5B6)^{100}$$

Now, suppose we want to feed the information that the block B4 is executed at least in one iteration. If $a[i]$ is an input array, this information can come from our knowledge of the program input. Alternatively, if $a[i]$ was constructed via some computation prior to the loop, this information can come from our understanding of infeasible program paths. In either case, the information can be encoded as the regular expression $\neg B1(B2B3B5B6)^* = \Sigma^* B4\Sigma^*$, where $\Sigma = \{B1, B2, B3, B4, B5, B6\}$ is the set of all basic blocks. The set of paths that the WCET analysis should consider is now given by

$$B1(B2B3B4B6 + B2B3B5B6)^{100} \cap \Sigma^* B4\Sigma^*$$

The timing schema approach will now remove the intersection by unrolling the loop as follows.

$$B1(B2B3B4B6)(B2B3B4B6 + B2B3B5B6)^{99}\cup$$
$$B1(B2B3B4B6 + B2B3B5B6)(B2B3B4B6)(B2B3B4B6 + B2B3B5B6)^{98}\cup$$
$$B1(B2B3B4B6 + B2B3B5B6)^2(B2B3B4B6)(B2B3B4B6 + B2B3B5B6)^{97} \cup \ldots$$

For each of these sets of paths (whose union we represent above) we can employ the conventional timing schema approach. However, there are 100 sets to consider because of unrolling a loop with 100 iterations. This is what makes the exploitation of infeasible paths difficult in the timing schema approach.

1.2.2.6 Exploiting Infeasible Path Information in ILP-Based WCET Calculation

Finally, we discuss how infeasible path information can be exploited in the ILP-based approach for WCET calculation. As mentioned earlier, the ILP-based approach is the most widely employed WCET calculation approach in state-of-the-art WCET estimation tools. The ILP approach reduces the WCET calculation to a problem of optimizing a linear objective function. The objective function represents the execution time of the program, which is maximized subject to flow constraints (in the control flow graph) and loop bound constraints. Note that the variables in the ILP problem correspond to execution counts of control flow graph nodes (i.e., basic blocks and edges).

Clearly, integrating infeasible path information will involve encoding knowledge of infeasible program paths as additional linear constraints [49, 68]. Introducing such constraints will make the WCET estimate (returned by the ILP solver) tighter. The description of infeasible path information as linear constraints has been discussed in several works. Park proposes an information description language (IDL) for describing infeasible path information [62]. This language provides convenient primitives for describing path information through annotations such as *samepath(A,C)*, where A, C can be lines in the program. This essentially means than whenever A is executed, C is executed and vice versa (note that A, C can be executed many times, as they may lie inside a loop). In terms of execution count constraints, such information can be easily encoded as $N_{B_A} = N_{B_C}$, where B_A and B_C are the basic blocks containing A, C, and N_{B_A} and N_{B_C} are the number of executions of B_A and B_C.

Recent work [e.g., 20] provides a systematic way of encoding path constraints as linear constraints on execution counts of control flow graph nodes and edges. In this work, the program's behavior is described in terms of "scopes"; scope boundaries are defined by loop or function call entry and exit. Within each scope, the work provides a systematic syntax for providing path information in terms of linear constraints.

For example, let us consider the control flow graph schematic denoting two if-then-else statements within a loop shown in Figure 1.7. The path information is now given in terms of each/all iterations of the scope (which in this case is the only loop in Figure 1.7). Thus, if we want to give the information that blocks $B2$ and $B6$ are always executed together (which is equivalent to using the *samepath* annotation described earlier) we can state it as $N_{B_2} = N_{B_6}$. On the other hand, if we want to give the information that B2 and B6 are never executed together (in any iteration of the loop), this gets converted to the following format

$$\text{for each iteration } N_{B_2} + N_{B_6} \leq 1$$

Incorporating the number of loop iterations in the above constraints, one can obtain the linear constraint $N_{B_2} + N_{B_6} \leq 100$ (assuming that the loop bound is 100). This constraint is then fed to the ILP solver along with the flow constraints and loop bounds (and any other path information).

In conclusion, we note that the ILP formulation for WCET calculation relies on aggregate execution counts of basic blocks. As any infeasible path information involves sequences of basic blocks, the encoding of infeasible path information as linear constraints over aggregate execution counts can lose information (e.g., it is possible to satisfy $N_{B_2} + N_{B_6} \leq 100$ in a loop with 100 iterations even if B_2 and B_6 are executed together in certain iterations). However, encoding infeasible path information as linear constraints provides a safe and effective way of ruling out a wide variety of infeasible program flows. Consequently, in most existing WCET estimation tools, ILP is the preferred method for WCET calculation.

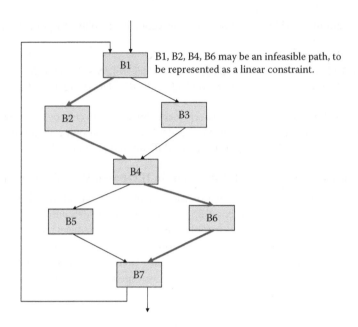

B1, B2, B4, B6 may be an infeasible path, to
be represented as a linear constraint.

FIGURE 1.7   A control flow graph fragment for illustrating infeasible path representation.

1.3   Micro-Architectural Modeling

The execution time of a basic block B in a program executing on a particular processor depends on (a) the number of instructions in B, (b) the execution cycles per instruction in B, and (c) the clock period of the processor. Let a basic block B contain the sequence of instructions $\langle I_1, I_2, \ldots, I_N \rangle$. For a simple micro-controller (e.g., TI MSP430), the execution latency of any instruction type is a constant. Let $latency(I_i)$ be a constant denoting the execution cycles of instruction I_i. Then the execution time of the basic block B can be expressed as

$$time(B) = \left(\sum_{i=1}^{N} latency(I_i) \right) \times period \tag{1.1}$$

where *period* is the clock period of the processor. Thus, for a simple micro-controller, the execution time of a basic block is also a constant and is trivial to compute. For this reason, initial work on timing analysis [67, 73] concentrated mostly on program path analysis and ignored the processor architecture.

However, the increasing computational demand of the embedded systems led to the deployment of processors with complex micro-architectural features. These processors employ aggressive pipelining, caching, branch prediction, and other features [33] at the architectural level to enhance performance. While the increasing architectural complexity significantly improves the average-case performance of an application, it leads to a high degree of timing unpredictability. The execution cycle $latency(I_i)$ of an instruction I_i in Equation 1.1 is no longer a constant; instead it depends on the execution context of the instruction. For example, in the presence of a cache, the execution time of an instruction depends on whether the processor encounters a cache hit or a cache misses while fetching the instruction from the memory hierarchy. Moreover, the large difference between the cache hit and miss latency implies that assuming all memory accesses to be cache misses will lead to overly pessimistic timing estimates. Any effective estimation technique should obtain a safe but tight bound on the number of cache misses.

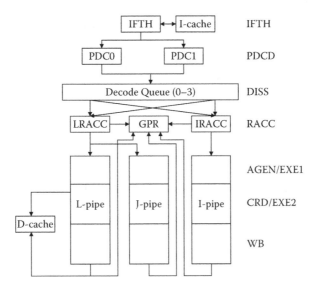

FIGURE 1.8 The IBM PowerPC 440 CPU Pipeline.

1.3.1   Sources of Timing Unpredictability

We first proceed to investigate the sources of timing unpredictability in a modern processor architecture and their implications for timing analysis. Let us use the IBM PowerPC (PPC) 440 embedded core [34] for illustration purposes. The PPC 440 is a 32-bit RISC CPU core optimized for embedded applications. It integrates a superscalar seven-stage pipeline, with support for out-of-order issue of two instructions per clock to multiple execution units, separate instruction and data caches, and dynamic branch prediction.

Figure 1.8 shows the PPC 440 CPU pipeline. The instruction fetch stage (IFTH) reads a cache line (two instructions) into the instruction buffer. The predecode stage (PDCD) partially decodes at most two instructions per cycle. At this stage, the processor employs a combination of static and dynamic branch prediction for conditional branches. The four-entry decode queue accepts up to two instructions per cycle from the predecode stage and completes the decoding. The decode queue always maintains the instructions in program order. An instruction waits in the decode queue until its input operands are ready and the corresponding execution pipeline is available. Up to two instructions can exit the decode queue per cycle and are issued to the register access (RACC) stage. Instruction can be issued out-of-order from the decode queue. After register access, the instructions proceed to the execution pipelines. The PPC 440 contains three execution pipelines: a load/store pipe, a simple integer pipe, and a complex integer pipe. The first execute stage (AGEN/EXE1) completes simple arithmetics and generates load/store addresses. The second execute stage (CRD/EXE2) performs data cache access and completes complex operations. The write back (WB) stage writes back the results into the register file.

Ideally, the PPC 440 pipeline has a throughput of two instructions per cycle. That is, the effective latency of each individual instruction is 0.5 clock cycle. Unfortunately, most programs encounter multiple pipeline hazards during execution that introduce bubbles in the pipeline and thereby reduce the instruction throughput:

Cache miss: Any instruction may encounter a miss in the instruction cache (IFTH stage) and the load/store instructions may encounter a miss in the data cache (CRD/EXE2 stage). The execution of the instruction gets delayed by the cache miss latency.

Data dependency: Data dependency among the instructions may introduce pipeline bubbles. An instruction I dependent on another instruction J for its input operand has to wait in the decode queue until J produces the result.

Control dependency: Control transfer instructions such as conditional branches introduce control dependency in the program. Conditional branch instructions cause pipeline stalls, as the processor does not know which way to go until the branch is resolved. To avoid this delay, dynamic branch prediction in the PPC 440 core predicts the outcome of the conditional branch and then fetches and executes the instructions along the predicted path. If the prediction is correct, the execution proceeds without any delay. However, in the event of a misprediction, the pipeline is flushed and a branch misprediction penalty is incurred.

Resource contention: The issue of an instruction from the decode queue depends on the availability of the corresponding execution pipeline. For example, if we have two consecutive load/store instructions in the decode queue, then only one of them can be issued in any cycle.

Pipeline hazards have significant impact on the timing predictability of a program. Moreover, certain functional units may have variable latency, which is input dependent. For example, the PPC 440 core can be complemented by a floating point unit (FPU) for applications that need hardware support for floating point operations [16]. In that case, the latency of an operation can be data dependent. For example, to mitigate the long latency of the floating point divide (19 cycles for single precision), the PPC 440 FPU employs an iterative algorithm that stops when the remainder is zero or the required target precision has been reached. A similar approach is employed for integer divides in some processors. In general, any unit that complies with the IEEE floating point standard [35] introduces several sources for variable latency (e.g., normalized versus denormalized numbers, exceptions, multi-path adders, etc.).

A static analyzer has to take into account the timing effect of these various architectural features to derive a safe and tight bound on the execution time. This, by itself, is a difficult problem.

1.3.2   Timing Anomaly

The analysis problem becomes even more challenging because of the interaction among the different architectural components. These interactions lead to counterintuitive timing behaviors that essentially preclude any compositional analysis technique to model the components independently.

Timing anomaly is a term introduced to define the counterintuitive timing behavior [54]. Let us assume a sequence of instructions executing on an architecture starting with an initial processor state. The latency of the first instruction is modified by an amount Δt. Let ΔC be the resulting change in the total execution time of the instruction sequence.

Definition 1.2
A timing anomaly is a situation where one the following cases becomes true:

- $\Delta t > 0$ *results in* $(\Delta C > \Delta t)$ *or* $(\Delta C < 0)$
- $\Delta t < 0$ *results in* $(\Delta C < \Delta t)$ *or* $(\Delta C > 0)$

From the perspective of WCET analysis, the cases of concern are the following: (a) The (local) worst-case latency of an instruction does not correspond to the (global) WCET of the program (e.g., $\Delta t > 0$ results in $\Delta C < 0$), and (b) the increase in the global execution time exceeds the increase in the local instruction latency (e.g., $\Delta t > 0$ results in $\Delta C > \Delta t$). Most analysis techniques implicitly assume that the worst-case latency of an instruction will lead to safe WCET estimates. For example, if the cache state is unknown, it is common to assume a cache miss for an instruction. Unfortunately, in the presence of a timing anomaly, assuming a cache miss may lead to underestimation.

1.3.2.1   Examples

An example where the local worst case does not correspond to the global worst case is illustrated in Figure 1.9. In this example, instructions A, E execute on functional unit 1 (FU1), which has variable latency. Instructions B, C, and D execute on FU2, which has a fixed latency. The arrows on the time line show when each instruction becomes ready and starts waiting for the functional unit. The processor

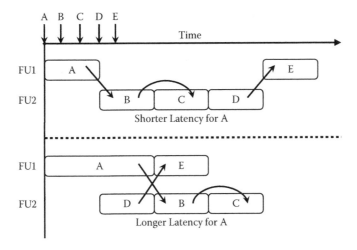

FIGURE 1.9   An example of timing anomaly.

allows out-of-order issue of the ready instructions to the functional units. The dependencies among the instructions are shown in the figure. In the first scenario, instruction A has a shorter latency, but the schedule leads to longer total execution time, as it cannot exploit any parallelism. In the second scenario, A has longer latency, preventing B from starting execution earlier (B is dependent on A). However, this delay opens up the opportunity for D to start execution earlier. This in turn allows E (which is dependent on D) to execute in parallel with B and C. The increased parallelism results in shorter overall execution time for the second scenario even though A has longer latency.

The second example illustrates that the increase in the global execution time may exceed the increase in the local instruction latency. In the PPC 440 pipeline, the branch prediction can indirectly affect instruction cache performance. As the processor caches instructions along the mispredicted path, the instruction cache content changes. This is called *wrong-path instructions prefetching* [63] and can have both constructive and destructive effects on the cache performance. Analyzing each feature individually fails to model this interference and therefore risks missing out on corner cases where branch misprediction introduces additional cache misses.

This is illustrated in Figure 1.10 with an example control flow graph. For simplicity of exposition, let us assume an instruction cache with four lines (blocks) where each basic block maps to a cache block (in reality, a basic block may get mapped to multiple cache blocks or may occupy only part of a cache block). Basic block B1 maps to the first cache block, B4 maps to the third cache block, and B2 and B3 both map to the second cache block (so they can replace each other). Suppose the execution sequence is B1 B2 B4 B1 B2 B4 B1 B2 B4.... That is, the conditional branch at the end of B1 is always taken; however, it is always mispredicted. The conditional branch at the end of B4, on the other hand, is always correctly predicted. If we do not take branch prediction into account, any analysis technique will conclude a cache hit for all the basic blocks for all the iterations except for the first iteration (which encounters cold misses). Unfortunately, this may lead to *underestimation* in the presence of branch prediction. The cache state before the prediction at B1 is shown in Figure 1.10. The branch is mispredicted, leading to instruction fetch along B3. Basic block B3 incurs a cache miss and replaces B2. When the branch is resolved, however, B2 is fetched into the instruction cache after another cache miss. This will result in two additional cache misses per loop iteration. In this case, the total increase in execution time exceeds the branch misprediction penalty because of the additional cache misses. Clearly, separate analysis of instruction caches and branch prediction cannot detect these additional cache misses.

Interested readers can refer to [54] for additional examples of timing anomalies based on a simplified PPC 440 architecture. In particular, [54] presents examples where (a) a cache hit results in worst-case

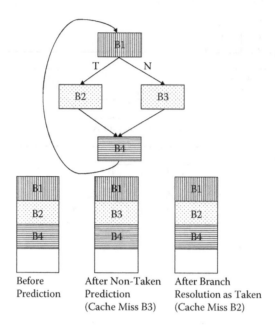

Before   After Non-Taken   After Branch
Prediction   Prediction   Resolution as Taken
                        (Cache Miss B3)   (Cache Miss B2)

FIGURE 1.10   Interference between instruction cache and branch prediction.

timing, (b) a cache miss penalty can be higher than expected, and (c) the impact of a timing anomaly on WCET may not be bounded. The third situation is the most damaging, as a small delay at the beginning of execution may contribute an arbitrarily high penalty to the overall execution time through a domino effect.

Identifying the existence and potential sources of a timing anomaly in a processor architecture remains a hard problem. Lundqvist and Stenström [54] claimed that no timing anomalies can occur if a processor contains only in-order resources, but Wenzel et al. [91] constructed an example of a timing anomaly in an in-order superscalar processor with multiple functional units serving an overlapping set of instruction types. A model-checking-based automated timing anomaly identification method has been proposed [18] for a simplified processor. However, the scalability of this method for complex processors is not obvious.

1.3.2.2 Implications

Timing anomalies have serious implications for static WCET analysis. First, the anomaly caused by scheduling (as shown in Figure 1.9) implies that one has to examine all possible schedules of a code fragment to estimate the longest execution time. A sequence of n instructions, where each instruction can have k possible latency values, generates k^n schedules. Any static analysis technique that examines all possible schedules will have prohibitive computational complexity. On the other hand, most existing analysis methods rely on making safe local decisions at the instruction level and hence run the risk of underestimation.

Second, many analysis techniques adopt a compositional approach to keep the complexity of the modeling architecture under control [29, 81]. These approaches model the timing effects of the different architectural features in separation. Counterintuitive timing interference among the different features (e.g., cache and branch prediction in Figure 1.10 or cache and pipeline) may render the compositional approaches invalid. For example, Healy et al. [29] performed cache analysis followed by pipeline analysis. Whenever a memory block cannot be classified as a cache hit or miss, it is assumed to be a cache miss. This is a conservative decision in the context of cache modeling and works perfectly for the in-order processor pipeline modeled in that work. However, if it is extended to out-of-order pipeline modeling, the cache hit may instead result in worst-case timing, and the decision will not be safe.

Lundqvist and Stenström [54] propose a program modification method that enforces timing predictability and thereby simplifies the analysis. For example, any variable latency instruction can be preceded and succeeded by "synchronization" instructions to force serialization. Similarly, synchronization instructions and/or software-based cache prefetching can be introduced at program path merging points to ensure identical processor states, but this approach has a potentially high performance overhead and requires special hardware support.

An architectural approach to avoid complex analysis due to timing anomalies has been presented in [3]. An application is divided into multiple subtasks with checkpoints to monitor the progress. The checkpoints are inserted based on a timing analysis of a simple processor pipeline (e.g., no out-of-order execution, branch prediction, etc.). The application executes on a complex pipeline unless a subtask fails to complete before its checkpoint (which is rare). At this point, the pipeline is reconfigured to the simple mode so that the unfinished subtasks can complete in a timely fashion. However, this approach requires changes to the underlying processor micro-architecture.

1.3.3 Overview of Modeling Techniques

The micro-architectural modeling techniques can be broadly divided into two groups:

- Separated approaches
- Integrated approaches

The separated approaches work on the control flow graph, estimating the WCET of each basic block by using micro-architectural modeling. These WCET estimates are then fed to the WCET calculation method. Thus, if the WCET calculation proceeds by ILP, only the constants in the ILP problem corresponding to the WCET of the basic blocks are obtained via micro-architectural modeling.

In contrast, the integrated approaches work by augmenting a WCET calculation method with micro-architectural modeling. In the following we see at least two such examples — an augmented ILP modeling method (to capture the timing behavior of caching and branch prediction) and an augmented timing schema approach that incorporates cache/pipeline modeling. Subsequently, we will discuss two examples of separated approaches, one of them using abstract interpretation for the micro-architectural modeling and the other one using a customized fixed-point analysis over the time intervals at which events (changing pipeline state) can occur. In both examples of the separated approach, the program path analysis proceeds by ILP.

In addition, there exist static analysis methods based on *symbolic execution* of the program [53]. This is an integrated method that extends cycle-accurate architectural simulation to perform symbolic execution with partially known operand values. The downside of this approach is the slow simulation speed that can lead to long analysis time.

1.3.4 Integrated Approach Based on ILP

An ILP-based path analysis technique has been described in Section 1.2. Here we present ILP-based modeling of micro-architectural components. In particular, we will focus on ILP-based instruction cache modeling proposed in [50] and dynamic branch prediction modeling proposed in [45]. We will also look at modeling the interaction between the instruction cache and the branch prediction [45] to capture the wrong-path instruction prefetching effect discussed earlier (see Figure 1.10).

The main advantage of ILP-based WCET analysis is the integration of path analysis and micro-architectural modeling. Identifying the WCET path is clearly dependent on the timing of each individual basic block, which is determined by the architectural modeling. On the other hand, behavior of instruction cache and branch prediction depends heavily on the current path. In other words, unlike pipeline, timing effects of cache and branch prediction cannot be modeled in a localized manner. ILP-based WCET analysis techniques provide an elegant solution to this problem of cyclic dependency between path analysis and architectural modeling. The obvious drawback of this method is the long solution time as the modeling complexity increases.

1.3.4.1 Instruction Cache Modeling

Caches are fast on-chip memories that are used to store frequently accessed instructions and data from main memory. Caches are managed under hardware control and are completely transparent to the programmer. Most modern processors employ separate instruction and data caches.

1.3.4.1.1 Cache Terminology

When the processor accesses an address, the address is first looked up in the cache. If the address is present in the cache, then the access is a *cache hit* and the content is returned to the processor. If the address is not present in the cache, then the access is a *cache miss* and the content is loaded from the next level of the memory hierarchy. This new content may replace some old content in the cache. The dynamic nature of the cache implies that it is difficult to statically identify cache hits and misses for an application. Indeed, this is the main problem in deploying caches in real-time systems.

The unit of transfer between different levels of memory hierarchy is called the *block* or *line*. A cache is divided into a number of sets. Let S be the associativity of a cache of size M. Then each cache set contains S cache lines. Alternatively, the cache has S ways. For a direct-mapped cache, $S = 1$. Further, let B be the cache line size. Then the cache contains $N = \frac{M}{S \times B}$ sets. A memory block *Blk* can be mapped to only one cache set given by (*Blk modulo N*).

1.3.4.1.2 Modeling

Li and Malik [50] first model direct-mapped instruction caches. This was later extended to set-associative instruction caches. For simplicity, we will assume a direct-mapped instruction cache here. The starting point of this modeling is again the control flow graph of the program. A basic block B_i is partitioned into n_i l-blocks denoted as $B_{i.1}, B_{i.2}, \ldots, B_{i.n_i}$. A line-block, or l-block, is a sequence of code in a basic block that belongs to the same instruction cache line. Figure 1.11A shows how the basic blocks are partitioned into l-blocks. This example assumes a direct-mapped instruction cache with only two cache lines.

Let $cm_{i.j}$ be the total cache misses for l-block $B_{i.j}$, and cmp be the constant denoting the cache miss penalty. The total execution time of the program is

$$Time = \sum_{i=1}^{N}(cost_i \times v_i + \sum_{j=1}^{n_i} cmp \times cm_{i.j}) \qquad (1.2)$$

where $cost_i$ is the execution time of B_i, assuming a perfect instruction cache, and v_i denotes the number of times B_i is executed. This is the objective function for the ILP formulation that needs to be maximized.

The cache constraints are the linear expressions that bound the feasible values of $cm_{i.j}$. These constraints are generated by constructing a cache conflict graph G_c for each cache line c. The nodes of G_c are the

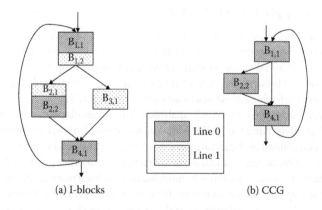

(a) I-blocks   (b) CCG

FIGURE 1.11 l-blocks and cache conflict graph.

l-blocks mapped to cache line c. An edge $B_{i,j} \rightarrow B_{u,v}$ exists in G_c if there exists a path in the control flow graph such that control flows from $B_{i,j}$ to $B_{u,v}$ without going through any other l-block mapped to c. In other words, there is an edge between l-blocks $B_{i,j}$ to $B_{u,v}$ if $B_{i,j}$ can be present in the cache when control reaches $B_{u,v}$. Figure 1.11b shows the cache conflict graph corresponding to cache line 0 for the control flow graph in Figure 1.11a mapped to a cache with two lines.

Let $r_{i,j \rightarrow u,v}$ be the execution count of the edge between l-blocks $B_{i,j}$ and $B_{u,v}$ in a cache conflict graph. Now the execution count of l-block $B_{i,j}$ equals the execution count of basic block B_i. Also, at each node of the cache conflict graph, the inflow equals the outflow and both equal the execution count of the node. Therefore,

$$v_i = \sum_{u,v} r_{i,j \rightarrow u,v} = \sum_{u,v} r_{u,v \rightarrow i,j} \tag{1.3}$$

The cache miss count $cm_{i,j}$ equals the inflow from *conflicting* l-blocks in the cache conflict graph. Any two l-blocks mapped to the same cache block are conflicting if they have different address tags. Two l-blocks mapped to the same cache block do not conflict when the basic block boundary is not aligned with the cache block boundary. For example, l-blocks $B_{1,2}$ and $B_{2,1}$ in Figure 1.11a occupy partial cache blocks and have the same address tags. They do not conflict with each other. Thus, we have

$$cm_{i,j} = \sum_{\substack{u,v \\ B_{u,v} \text{ conflicts } B_{i,j}}} r_{u,v \rightarrow i,j} \tag{1.4}$$

1.3.4.2 Dynamic Branch Prediction Modeling

Modern processors employ branch prediction to avoid performance loss due to control dependency [33]. Branch prediction schemes can be broadly categorized as *static* and *dynamic*. In the static scheme, a branch is predicted in the same direction every time it is executed. Though simple, static schemes are much less accurate than dynamic schemes.

1.3.4.2.1 Branch Terminology

Dynamic schemes predict a branch depending on the execution history. They use a 2^n entry *branch prediction table* to store past branch outcomes. When the processor encounters a conditional branch instruction, this prediction table is looked up using some index, and the indexed entry is used as prediction. When the branch is resolved, the entry is updated with the actual outcome. In practice, two-bit saturating counters are often used for prediction.

Different branch prediction schemes differ in how they compute an n-bit index to access this table. In case of simplest prediction scheme, the index is n lower-order bits of the branch address. More complex schemes use a single shift register called a *branch history register* (BHR) to record the outcomes of the n most recent branches called history π. The prediction table is looked up either using the BHR directly or exclusive or (XOR)-ed with the branch address. Considering the outcome of the neighboring branches exploits the correlation among consecutive branch outcomes.

Engblom [19] investigated the impact of dynamic branch prediction on the predicability of real-time systems. His experiments on a number of commercial processors indicate that dynamic branch prediction leads to high degree of execution time variation even for simple loops. In some cases, executing more iterations of a loop takes less time than executing fewer iterations. These results reaffirm the need to model branch prediction for WCET analysis.

1.3.4.2.2 Modeling

Li et al. [45] model dynamic branch predictors through ILP. The modeling is quite general and can be parameterized with respect to various prediction schemes. Modeling of dynamic branch prediction is somewhat similar to cache modeling. This is because they both use arrays (branch prediction table and cache) to maintain information. However, two crucial differences make branch prediction modeling significantly harder. First, a given branch instruction may use different entries of the prediction table at different points of execution (depending on the outcome of previous branches). However, an l-block always maps to the same cache block. Second, the flow of control between two conflicting l-blocks always implies a cache miss, but the flow of control between two branch instructions mapped to the same entry in the prediction table may lead to correct or incorrect prediction depending on the outcome of the two branches.

To model branch prediction, the objective function given in Equation 1.2 is modified to the following:

$$Time = \sum_{i=1}^{N} (cost_i * v_i + bmp * bm_i) \tag{1.5}$$

where bmp is a constant denoting the penalty for a single branch misprediction, and bm_i is the number of times the branch in B_i is mispredicted. The constraints now need to bound feasible values of bm_i. For simplicity, let us assume that the branch prediction table is looked up using the history π as the index.

First, a terminating least-fixed-point analysis on the control flow graph identifies the possible values of history π for each conditional branch. The flow constraints model the change in history along the control flow graph and thereby derive the upper bound on bm_i^π — the execution count of the conditional branch at the end of basic block B_i with history π. Next, a structure similar to a cache conflict graph is used to bound the quantity $p_{i \to j}^\pi$ denoting the number of times control flows from B_i to B_j such that the πth entry of the prediction table is used for branch prediction at B_i and B_j and is never accessed in between. Finally, the constraints on the number of mispredictions are derived by observing the branch outcomes for consecutive accesses to the same prediction table entry as defined by $p_{i \to j}^\pi$.

1.3.4.3 Interaction between Cache and Branch Prediction

Cache and branch prediction cannot be modeled individually because of the wrong-path instruction prefetching effect (see Figure 1.10). An integrated modeling of these two components through ILP to capture the interaction has been proposed in [45]. First, the objective function is modified to include the timing effect of cache misses as well as branch prediction.

$$Time = \sum_{i=1}^{N} (cost_i * v_i + bmp * bm_i + \sum_{j=1}^{n_i} cmp \times cm_{i,j}) \tag{1.6}$$

If we assume that the processor allows only *one unresolved branch* at any time during execution, then the number of branch mispredictions bm_i is not affected by instruction cache. However, the values of the number of cache misses $cm_{i,j}$ may change because of the instruction fetches along the mispredicted path. The timing effects due to these additional instruction fetches can be categorized as follows:

- An l-block $B_{i,j}$ misses during normal execution since it is displaced by another conflicting l-block $B_{u,v}$ during speculative execution (*destructive effect*).
- An l-block $B_{i,j}$ hits during normal execution, since it is prefetched during speculative execution (*constructive effect*).
- A pending cache miss of $B_{i,j}$ during speculative execution along the wrong path causes the processor to stall when the branch is resolved. How long the stall lasts depends on the portion of cache miss penalty that is masked by the branch misprediction penalty. If the speculative fetching is completely masked by the branch penalty, then there is no delay incurred.

Both the constructive and destructive effects of branch prediction on cache are modeled by modifying the cache conflict graph. The modification adds nodes to the cache conflict graph corresponding to the l-blocks fetched along the mispredicted path. Edges are added among the additional nodes as well as between the additional nodes and the normal nodes depending on the control flow during misprediction. The third factor (delay due to incomplete cache miss when the branch is resolved) is taken care of by introducing an additional delay term in Equation 1.6.

1.3.4.4 Data Cache and Pipeline

So far we have discussed instruction cache and branch prediction modeling using ILP. Data caches are harder to model than instruction caches, as the exact memory addresses accessed by load/store instructions may not be known. A simulation-based analysis technique for data caches has been proposed in [50]. A program is broken into smaller fragments where each fragment has only one execution path. For example, even though there are many possible execution paths in a JPEG decompression algorithm, the execution paths of each computational kernel such as inverse discrete cosine transform (DCT), color transformation, and so on are simple. Each code fragment can therefore be simulated to determine the number of data cache hits and misses. These numbers can be plugged into the ILP framework to estimate the WCET of the whole program. For the processor pipeline, [50] again simulates the execution of a basic block starting with an empty pipeline state. The pipeline state at the end of execution of a basic block is matched against the instructions in subsequent basic blocks to determine the additional pipeline stalls during the overlap. These pipeline stalls are added up to the execution time of the basic block. It should be obvious that this style of modeling for data cache and pipeline may lead to underestimation in the presence of a timing anomaly.

Finally, Ottosson and Sjödin [60] propose a constraint-based WCET estimation technique that extends the ILP-based modeling. This technique takes the context, that is, the history, of execution into account. Each edge in the control flow graph now corresponds to multiple variables each representing a particular program path. This allows accurate representation of the state of the cache and pipeline before a basic block is executed. A constraint-based modeling propagates the cache states across basic blocks.

1.3.5 Integrated Approach Based on Timing Schema

As mentioned in Section 1.2, one of the original works on software timing analysis was based on timing schema [73]. In the original work, each node of the syntax tree is associated with a simple time bound. This simple timing information is not sufficient to accurately model the timing variations due to pipeline hazards, caches, and branch prediction. The timing schema approach has been extended to model a pipeline, instruction cache, and data cache in [51].

1.3.5.1 Pipeline Modeling

The execution time of a program construct depends on the preceding and succeeding instructions on a pipelined processor. A single time bound cannot model this timing variation. Instead a set of *reservation tables* associated with each program construct represents the timing information corresponding to different execution paths. A pruning strategy is used to eliminate the execution paths (and their corresponding reservation tables) that can never become the worst-case execution path of the program construct. The remaining set of reservation tables is called the *worst-case timing abstraction* (WCTA) of the program construct.

The reservation table represents the state of the pipeline at the beginning and end of execution of the program construct. This helps analyze the pipelined execution overlap among consecutive program constructs. The rows of the reservation table represent the pipeline stages and the columns represent time. Each entry in the reservation table specifies whether the corresponding pipeline stage is in use at the given time slot. The execution time of a reservation table is equal to its number of columns. Figure 1.12 shows a reservation table corresponding to a simple five-stage pipeline.

The rules corresponding to the sequence of statements and *if-then-else* and *while-loop* constructs can be extended as follows. The rule for a sequence of statements S: S1; S2 is given by

$$W(S) = W(S1) \oplus W(S2)$$

where W(S), W(S1), and W(S2) are the WCTAs of S, S1, and S2, respectively. The operator $\oplus$ is defined as

$$W_1 \oplus W_2 = \{w_1 \oplus w_2 | w_1 \in W_1, w_2 \in W_2\}$$

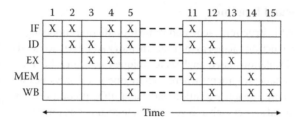

FIGURE 1.12   Reservation table.

where w_1 and w_2 are reservation tables, and $\oplus$ represents the concatenation of two reservation tables following the pipelined execution model. Similarly, the timing schema rule for `S: if (exp) then S1 else S2` is given by

$$W(S) = (W(exp) \oplus (W(S1) \cup W(S2))$$

where $\cup$ is the set union operation. Finally, the rule for the construct `S: while (exp) S1` is given by

$$W(S) = \left(\bigoplus_{i=1}^{N} (W(exp) \oplus W(S1)) \right) \oplus W(exp)$$

where N is the loop bound. In all the cases, a reservation table w can be eliminated from the WCTA W if it can be guaranteed that w will never lead to the WCET of the program. For example, if the worst-case scenario (zero overlap with neighboring instructions) involving $w \in W$ is shorter than the best-case scenario (complete overlap with neighboring instructions) involving $w' \in W$, then w can be safely eliminated from W.

1.3.5.2   Instruction Cache Modeling

To model the instruction cache, the WCTA is extended to maintain the cache state information for a program construct. The main observation is that some of the memory accesses can be resolved locally (within the program construct) as cache hit/miss. Each reservation table should therefore include (a) the first reference to each cache block as its hit or miss depends on the cache content prior to the program construct (*first_reference*) and (b) the last reference to each cache block (*last_reference*). The latter affects the timing of the successor program construct(s).

The timing rules are structurally identical to the pipeline modeling, but the semantics of the $\oplus$ operator is modified. Let us assume a direct-mapped instruction cache. Then $w_1 \oplus w_2$ inherits for a cache block c the first_reference of w_1 except when w_1 does not have any access to c. In that case, $w_1 \oplus w_2$ inherits the first_reference of w_2. Similarly, for a cache block c, $w_1 \oplus w_2$ inherits the last_reference of w_2 except when w_2 does not have a last_reference to c. In this case, the last_reference to c is inherited from w_1. Finally, the number of additional cache hits for w_2 can be determined by comparing the first_references of w_2 with the last_references of w_1. The execution time of $w_1 \oplus w_2$ can be determined by taking into account the pipelined execution across w_1, w_2 and the additional cache hits. As before, a pruning strategy is employed to safely eliminate WCTA elements that can never contribute to the WCET path of the program.

1.3.5.3   Data Cache Modeling

Timing analysis of the data cache is similar to that of the instruction cache. The major difficulty, however, is that the addresses of some data references may not be known at compile time. A global data flow analysis [38] is employed to resolve the data references of load/store instructions as much as possible. A conservative approach is then proposed [38] where two cache miss penalties are assumed for each data reference whose memory address cannot be determined at compile time. The data reference is then ignored in the rest of the analysis. The first penalty accounts for the cache miss possibility of the data reference. The second penalty covers for the possibility that the data reference may replace some memory block (from the cache) that is considered as cache hit in the analysis. Finally, data dependence analysis is utilized to minimize the WCET overestimation resulting from the conservative assumption of two cache misses per unknown reference.

1.3.6 Separated Approach Based on Abstract Interpretation

ILP-based WCET analysis methods can model the architectural components and their interaction in an accurate fashion, thereby yielding tight estimates. However, ILP solution time may increase considerably with complex architectural features. To circumvent this problem, Theiling et al. [82] have proposed a separated approach where abstract interpretation is employed for micro-architectural modeling followed by ILP for path analysis. As there is a dependency between the two steps, micro-architectural modeling has to produce conservative estimates to ensure safety of the result. This overestimation is offset by significantly faster analysis time.

Abstract interpretation [15] is a theory for formally constructing conservative approximations of the semantics of a programming language. A concrete application of abstract interpretation is in static program analysis where a program's computations are performed using *abstract values* in place of concrete values. Abstract interpretation is used in WCET analysis to approximate the "collecting semantics" at a program point. The collecting semantics gives the set of all program states (cache, pipeline, etc.) for a given program point. In general, the collecting semantics is not computable. In abstract interpretation, the goal is to produce an abstract semantics which is less precise but effectively computable. The computation of the abstract semantics involves solving a system of recursive equations/constraints. Given a program, we can associate a variable $[\![p]\!]$ to denote the abstract semantics at program point p. Clearly, $[\![p]\!]$ will depend on the abstract semantics of program points preceding p. Since programs have loops, this will lead to a system of recursive constraints. The system of recursive constraints can be iteratively solved via fixed-point computation. Termination of the fixed-point computation is guaranteed only if (a) the domain of abstract values (which is used to define the abstract program semantics) is free from infinite ascending chains and (b) the iterative estimates of $[\![p]\!]$ grow montonically. The latter is ensured if the semantic functions in the abstract domain, which show the effect of the programming language constructs in the abstract domain and are used to iteratively estimate $[\![p]\!]$, are monotonic.

Once the fixed-point computation terminates, for every program point p, we obtain a stable estimate for $[\![p]\!]$ — the abstract semantics at p. This is an overapproximation of all the concrete states with which p could be reached in program executions. Thus, for cache behavior modeling, $[\![p]\!]$ could be used to denote an overapproximation of the set of concrete cache states with which program point p could be reached in program executions. This abstract semantics is then used to conservatively derive the WCET bounds for the individual basic blocks. Finally, the WCET estimates of basic blocks are combined with ILP-based path analysis to estimate the WCET of the entire program.

1.3.6.1 Cache Modeling

To illustrate AI-based cache modeling, we will assume a fully associative cache with a set of cache lines $L = \{l_1, \ldots, l_n\}$ and least recently used replacement policy. Let $\{s_1, \ldots, s_m\}$ denote the set of memory blocks. The absence of any memory block in a cache line is indicated by a new element I; thus, $S = \{s_1, \ldots, s_m\} \cup \{I\}$.

Let us first define the concrete semantics.

Definition 1.3
A concrete cache state is a function $c : L \rightarrow S$.

If $c(l_x) = s$ for a concrete cache state c, then there are $x - 1$ elements $(c(l_1), \ldots, c(l_{x-1}))$ that are more recently used than s. In other words, x is the relative age of s. C_c denotes the set of all concrete cache states.

Definition 1.4

A cache update function $U : C_c \times S \to C_c$ *describes the new cache state for a given cache state and a referenced memory block.*

Let $s = c(l_x)$ be the referenced memory block. The cache update function shifts the memory blocks $c(l_1), \ldots, c(l_{x-1})$, which have been more recently used than s, by one position to the next cache line. If s was not in the cache, then all the memory blocks are shifted by one position, and the least recently used memory block is evicted from the cache state (if the cache was full). Finally, the update function puts the referenced memory block s in the first position l_1.

The abstract semantics defines the abstract cache states, the abstract cache update function, and the join function.

Definition 1.5

An abstract cache state $\hat{c} : L \to 2^S$ *maps cache lines to sets of memory blocks.*

Let $\hat{C}$ denote the set of all abstract cache states. The abstract cache update function $\hat{U} : \hat{C} \times S \mapsto \hat{C}$ is a straightforward extension of the function U (which works on concrete cache states) to abstract cache states.

Furthermore, at control flow merge points, join functions are used to combined the abstract cache states. That is, join functions approximate the collecting semantics depending on program analysis.

Definition 1.6

A join function $\hat{J} : \hat{C} \times \hat{C} \mapsto \hat{C}$ *combines two abstract cache states.*

Since L is finite and S is finite, clearly the domain of abstract cache states is finite and hence free from any infinite ascending chains. Furthermore, the update and join functions $\hat{U}$ and $\hat{J}$ are monotonic. This ensures termination of a fixed-point computation-based analysis over the above-mentioned abstract domain. We now discuss two such analysis methods.

The program analysis mainly consists of *must analysis* and *may analysis*. The must analysis determines the set of memory blocks that are always in the cache at a given program point. The may analysis determines the memory blocks that may be in the cache at a given program point. The may analysis can be used to determine the memory blocks that are guaranteed to be absent in the cache at a given program point.

The must analysis uses abstract cache states with *upper bounds* on the ages of the memory blocks in the concrete cache states. That is, if $s \in \hat{c}(l_x)$, then s is guaranteed to be in the cache for at least the next $n - x$ memory references (n is the number of cache lines). Therefore, the join function of two abstract cache states $\hat{c}_1$ and $\hat{c}_2$ puts a memory block s in the new cache state if and only if s is present in both $\hat{c}_1$ and $\hat{c}_2$. The new age of s is the maximum of its ages in $\hat{c}_1$ and $\hat{c}_2$. Figure 1.13 shows an example of the join function for must and may analysis.

The may analysis uses abstract cache states with *lower bounds* on the ages of the memory blocks. Therefore, the join function of two abstract cache states $\hat{c}_1$ and $\hat{c}_2$ puts a memory block s in the new cache state if s is present in either $\hat{c}_1$ or $\hat{c}_2$ or both. The new age of s is the minimum of its ages in $\hat{c}_1$ and $\hat{c}_2$.

At a program point, if a memory block s is present in the abstract cache state after must analysis, then a memory reference to s will result in a cache hit (*always hit*). Similarly, if a memory block s is absent in the abstract cache state after may analysis, then a memory reference to s will result in a cache miss (*always miss*). The other memory references cannot be classified as hit or miss. To improve the accuracy, a further persistence analysis can identify memory blocks for which the first reference may result in either hit or miss, but the remaining references will be hits.

These categorization of memory references is used to define the WCET for each basic block. To improve the accuracy, the WCET of a basic block is determined under different calling contexts. Thus, the objective

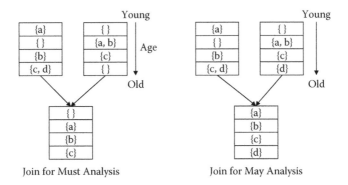

FIGURE 1.13 Join for must and may analysis.

function can be defined as

$$Time = \sum_{i=1}^{N} \sum_{x \in \tau(B_i)} \left(cost_i^x \times v_i^x \right) \tag{1.7}$$

where $\tau(B_i)$ denotes the set of all calling contexts for basic block B_i. The bounds on execution counts v_i^x can be derived by ILP-based path analysis.

An extension for data cache modeling using abstract interpretation has been proposed in [23]. The basic idea is to extend the cache update function such that it can handle cases where not all addresses referenced by a basic block are known.

Another technique for categorizing cache access references into *always hit, always miss, first miss,* and *first hit* has been proposed by the group at Florida State University [4, 57, 93]. They perform categorization through static cache simulation, which is essentially an interprocedural data flow analysis. This categorization is subsequently used during pipeline analysis [29]. Pipeline analysis proceeds by determining the total number of cycles required to execute each path, where a path consists of all the instructions that can be executed during a single iteration of a loop. The data hazards and the structural hazards across paths are determined by maintaining the first and last use of each pipeline stage and register within a path. As mentioned before, this separation of cache analysis from the pipeline analysis may not be safe in the presence of a timing anomaly.

1.3.6.2 Pipeline Modeling

To model a pipeline with abstract interpretation [41], *concrete* execution on a *concrete* pipeline can be viewed as applying a function. This function takes as input a concrete pipeline state s and a sequence of instructions in a basic block b. It produces a sequence of execution states, called a *trace*, and a final concrete state when executing b. The length of the trace determines the number of cycles the execution takes. The concept of trace is similar to the reservation table described in the context of timing-schema-based analysis.

However, in the presence of incomplete information, such as nonclassified cache accesses, the concrete execution is not feasible. Therefore, pipeline analysis employs an *abstract* execution of the sequence of instructions in a basic block starting with an *abstract* pipeline state $\hat{s}$ [41]. This modeling defines an abstract pipeline state as a set of concrete pipeline states, and pipeline states with identical timing behavior are grouped together. Now, suppose that in an abstract pipeline state $\hat{s}$ an event occurs that changes the pipeline states, such as the issue/execution of an instruction I in a basic block. If the latency of this event can be statically determined, $\hat{s}$ has only one successor state. However, if the latency of I's execution cannot be statically determined, a pipeline state will have several successor states resulting from the execution of I corresponding to the various possible latencies of I (thereby causing state space explosion). In this way, reachable pipeline states within a basic block will be enumerated (while grouping together states with identical timing behavior) in order to determine the basic block's WCET.

For a processor without a timing anomaly [41], the abstract execution can be employed to each basic block starting with the empty pipeline state. The abstract execution exploits the memory reference categorization (obtained through cache modeling) to determine memory access delays during pipeline execution. Therefore, abstract execution of a basic block should happen under different contexts. In the presence of a timing anomaly, cache and pipeline analysis cannot be separated [32]. Hence the abstract states now consist of pairs of abstract pipeline states and abstract cache states. Moreover, the final abstract states of a basic block will be passed on to the successor basic block(s) as initial states. Clearly, this can lead to an exponential number of abstract states for complex processor pipelines.

1.3.6.3 Branch Prediction Modeling

Colin and Puaut [14] propose abstract-interpretation-based branch prediction modeling. They assume that the branch prediction table (see Section 1.3.4.2.1) is indexed using the address of the conditional branch instruction. This prediction scheme is simpler and hence easier to model than the BHR-based predictors modeled using ILP [45]. Colin and Puaut use the term *branch target buffer* (BTB) instead of prediction table, as it stores the target address in addition to the branch history. Moreover, each entry in the BTB is tagged with the address of the conditional branch instruction whose history and target address are stored in that entry. When a conditional branch is encountered, if its address is in the BTB, then it is predicted based on the history stored in the BTB. Otherwise, the *default prediction* of the branch not taken is used. The BTB is quite similar to instruction cache and indeed can be organized as direct-mapped or s-way set associative caches.

The abstract execution defines the *abstract buffer state* (ABS) corresponding to the BTB. Each basic block B_i is associated with two ABS: ABS_i^{in} and ABS_i^{out}, representing the BTB state before and after B_i's execution. An ABS indicates for each BTB entry which conditional branch instructions can be in the BTB at that time. At program merge points, a set union operation is carried out. Thus,

$$ABS_i^{in} = \biguplus_{B_j \in Pred\ B_i} ABS_j^{out}$$

where $Pred B_{(i)}$ is the set of basic blocks preceding B_i in the control flow graph. Assuming a set-associative BTB, the union operator $\uplus$ is defined as follows:

$$ABS_i \uplus ABS_j = \forall x \forall y\ \ ABS_i[x,y] \cup ABS_j[x,y]$$

where $ABS_i[s,k]$ is a set containing all the branch instructions that could be in the yth entry of the set x. ABS_i^{out} is derived from ABS_i^{in} by taking into account the conditional branch instruction in B_i.

Given ABS_i^{in}, the conditional branch instruction can be classified as *history predicted* if it is present in the BTB and *default predicted* otherwise. However, a history-predicted instruction does not necessarily lead to correct prediction. Similarly, a default-predicted instruction does not always lead to misprediction. This is taken into account by considering the behavior of the conditional branch instruction. For example, a history-predicted loop instruction is always correctly predicted except for loop exit.

The modeling in [14] was later extended to more complex branch predictors such as bimodal and global-history branch prediction schemes [5, 11]. The semantic context of a branch instruction in the source code is taken into account to classify a branch as easy to predict or hard to predict. Easy-to-predict branches are analyzed, while conservative misprediction penalties are assumed for hard-to-predict branches. The downside of these techniques is that they make a restrictive assumption of each branch instruction mapping to a different branch table entry (i.e., no aliasing).

1.3.7 A Separated Approach That Avoids State Enumeration

The implication of a timing anomaly (see Section 1.3.2) is that all possible schedules of instructions have to be considered to estimate the WCET of even a basic block. Moreover, all possible processor states at the end of the preceding and succeeding basic blocks have to be considered during the analysis of a basic block. This can result in state space explosion for analysis techniques, such as abstract-interpretation-based modeling, that are fairly efficient otherwise [83].

A novel modeling technique [46] obtains safe and tight estimates for processors with timing anomalies without enumerating all possible executions corresponding to variable latency instructions (owing to cache miss, branch misprediction, and variable latency functional units). In particular, [46] models a fairly complex out-of-order superscalar pipeline with instruction cache and branch prediction. First, the problem is formulated as an *execution graph* capturing data dependencies, resource contentions, and degree of superscalarity — the major factors dictating instruction executions. Next, based on the execution graph, the estimation algorithm starts with very coarse yet safe timing estimates for each node of the execution graph and iteratively refines the estimates until a fixed point is reached.

1.3.7.1 Execution Graph

Figure 1.14 shows an example of an execution graph. This graph is constructed from a basic block with five instructions as shown in Figure 1.14a; we assume that the degree of superscalarity is 2. The processor has five pipeline stages: fetch (IF), decode (ID), execute (EX), write back (WB), and commit (CM). A decoded instruction is stored in the re-order buffer. It is issued (possibly out of order) to the corresponding functional unit for execution when the operands are ready and the functional unit is available.

Let $Code_B = I_1 \ldots I_n$ represent the sequence of instructions in a basic block B. Then each node v in the corresponding execution graph is represented by a tuple: an instruction identifier and a pipeline stage denoted as *stage(instruction_id)*. For example, the node $v = IF(I_i)$ represents the fetch stage of the instruction I_i. Each node in the execution graph is associated with the *latency* of the corresponding pipeline stage. For a node u with variable latency $min\_lat_u \sim max\_lat_u$, the node is annotated with an interval $[min\_lat_u, max\_lat_u]$. As some resources (e.g., floating point multiplier) in modern processors are fully pipelined, such resources are annotated with *initiation intervals*. The initiation interval of a resource is defined as the number of cycles that must elapse between issuing two instructions to that resource. For example, a fully pipelined floating point multiplier can have a latency of six clock cycles and an initiation interval of one clock cycle. For a nonpipelined resource, the initiation interval is the same as latency. Also, if there exist multiple copies of the same resource (e.g., two arithmetic logical units (ALUs)), then one needs to define the *multiplicity* of that resource.

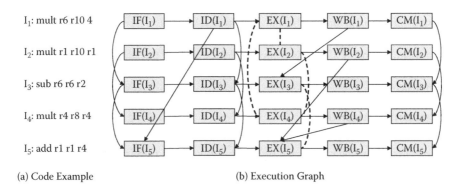

(a) Code Example   (b) Execution Graph

FIGURE 1.14 A basic block and its execution graph. The solid directed edges represent dependencies and the dashed undirected edges represent contention relations. From Li et al. [46] © 2006 Springer.

The dependence relation from node u to node v in the execution graph denotes that v can start execution only after u has completed execution; this is indicated by a solid directed edge from u to v in the execution graph. The analysis models the following dependencies:

- Dependencies among pipeline stages of the same instruction.
- Dependencies due to finite-sized buffers and queues such as I-buffer or ROB. For example, assuming a four-entry I-buffer, there will be no entry available for $IF(I_{i+4})$ before the completion of $ID(I_i)$ (which removes I_i from the I-buffer). Therefore, there should be a dependence edge $ID(I_i) \rightarrow IF(I_{i+4})$.
- Dependencies due to in-order execution in IF, ID, and CM pipeline stages. For example, in a scalar processor (i.e., degree of superscalarity = 1) there will be dependence edges $IF(I_i) \rightarrow IF(I_{i+1})$ because $IF(I_{i+1})$ can only start after $IF(I_i)$ completes. For a superscalar processor with n-way fetch (i.e., degree of superscalarity = n), there are dependence edges $IF(I_i) \rightarrow IF(I_{i+n})$. This captures the fact that I_{i+n} cannot be fetched in the same cycle as I_i.
- Data dependencies among instructions. If instruction I_i produces a result that is used by instruction I_j, then there should be a dependence edge $WB(I_i) \rightarrow EX(I_j)$.

Apart from the dependence relation among the nodes in an execution graph (denoted by solid edges), there also exist contention relations among the execution graph nodes. Contention relations model structural hazards in the pipeline. A contention relation exists between two nodes u and v if they can delay each other by contending for a resource, for example, functional unit or register write port. The contention between u and v is shown as an undirected dashed edge in the execution graph. A contention relation makes it possible for an instruction later in the program order to delay the execution of an earlier instruction.

Finally, a parallelism relation is defined to model superscalarity, for example, multiple issues and multiple decodes. Two nodes u and v participate in a parallelism relation iff (a) nodes u and v denote the same pipeline stage (call it stg) of two different instructions I_i and I_j and (b) instructions I_i and I_j can start execution of this pipeline stage stg in parallel.

1.3.7.2 Problem Definition

Let B be a basic block consisting of a sequence of instructions $Code_B = I_1 \ldots I_n$. Estimating the WCET of B can be formulated as finding the maximum (latest) completion time of the node $CM(I_n)$, assuming that $IF(I_1)$ starts at time zero. Note that this problem is *not* equivalent to finding the longest path from $IF(I_1)$ to $CM(I_n)$ in B's execution graph (taking the maximum latency of each pipeline stage). The execution time of a path in the execution graph is not a summation of the latencies of the individual nodes for two reasons:

- The total time spent in making the transition from $ID(I_i)$ to $EX(I_i)$ is dependent on the contentions from other ready instructions.
- The initiation time of a node is computed as the *max* of the completion times of its immediate predecessors in the execution graph. This models the effect of dependencies, including data dependencies.

1.3.7.3 Estimation Algorithm

The timing effects of the dependencies are accounted for by using a modified longest-path algorithm that traverses the nodes in topologically sorted order. This topological traversal ensures that when a node is visited, the completion times of all its predecessors are known. To model the effect of resource contentions, the algorithms conservatively estimate an upper bound on the delay due to contentions for a functional unit by other instructions. A single pass of the modified longest-path algorithm computes loose bounds on the lifetime of each node. These bounds are used to identify nodes with disjoint lifetimes. These

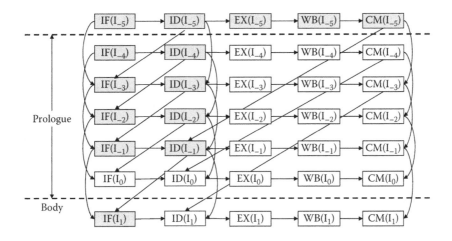

FIGURE 1.15   An example prologue. From Li et al. [46] © 2006 Springer.

nodes are not allowed to contend in the next pass of the longest-path search to get tighter bounds. These two steps repeat until there is no change in the bounds. Termination is guaranteed for the following reasons:

- The algorithm starts with all pairs of instructions in the contention relation (i.e., every instruction can delay every other instruction).
- At every step of the fixed-point computation, pairs are removed from this set — those instruction pairs that are shown to be separated in time.

As the number of instructions in a basic block is finite, the number of pairs initially in the contention relation is also finite. Furthermore, the algorithm removes at least one pair in every step of the fixed-point computation, so the fixed-point computation *must terminate* in finitely many iterations; if the number of instructions in the basic block being estimated is n, the number of fixed-point iterations is bounded by n^2.

1.3.7.3.1 *Basic Block Context*

In the presence of a timing anomaly, a basic block cannot be analyzed in isolation by assuming an empty pipeline at the beginning. The instructions before (after) a basic block B that *directly* affect the execution time of B constitute the contexts of B and are called the *prologue* (*epilogue*) of B. As processor buffer sizes are finite, the prologue and epilogue contain finite number of instructions. Of course, a basic block B may have multiple prologues and epilogues corresponding to the different paths along which B can be entered or exited. To capture the effects of contexts, the analysis technique constructs execution graphs corresponding to all possible combinations of prologues and epilogues. Each execution graph consists of three parts: the prologue, the basic block itself (called the *body*), and the epilogue.

The executions of two or more successive basic blocks have some overlap due to pipelined execution. The overlap δ between a basic block B and its preceding basic block B' is the period during which instructions from both the basic blocks are in the pipeline, that is,

$$\delta = t_{CM(I_0)}^{finish} - t_{IF(I_1)}^{ready} \tag{1.8}$$

where I_0 is the last instruction of block B' (predecessor) and I_1 is the first instruction of block B. To avoid duplicating the overlap in time estimates of successive basic blocks, the execution time t_B of a basic block B is defined as the interval from the time when the instruction immediately preceding B has finished committing to the time when B's last instruction has finished committing, that is,

$$t_B = t_{CM(I_n)}^{finish} - t_{CM(I_0)}^{finish} \qquad (1.9)$$

where I_0 is the instruction immediately prior to B and I_n is the last instruction of B.

The execution time for basic block B is estimated with respect to (w.r.t.) the time at which the first instruction I_1 of B is fetched, i.e., $t_{IF(I_1)}^{ready} = 0$. Thus,

$$t_B = t_{CM(I_n)}^{finish} - \delta \qquad (1.10)$$

t_B can be conservatively estimated by finding the largest value of $t_{CM(I_n)}^{finish}$ and the smallest value of δ.

1.3.7.3.2 Overall Pipeline Analysis

The execution time estimate t_B of a basic block B is obtained *for a specific prologue and a specific epilogue of B*. A basic block B in general has multiple choices of prologues and epilogues. Thus, B's execution time is estimated under all possible combinations of prologues and epilogues. The maximum of these estimates is used as B's WCET c_B. Let $\mathcal{P}$ and $\mathcal{E}$ be the set of prologues and epilogues for B.

$$c_B = \max_{p \in \mathcal{P},\, e \in \mathcal{E}} (t_B \text{ with prologue } p \text{ and epilogue } e)$$

c_B is used in defining the WCET of the program as the following objective function:

$$\text{maximize} \sum_{B \in \mathcal{B}} N_B * c_B$$

The quantity N_B denotes the execution count of basic block B and is a variable. $\mathcal{B}$ is the set of all basic blocks in the program. This objective function is maximized over the constraints on N_B given by ILP-based path analysis.

1.3.7.4 Integrating Cache and Branch Prediction Analysis

The basic idea is to define different scenarios for a basic block corresponding to cache miss and branch misprediction. If these scenarios are defined suitably, then we can estimate a constant that bounds the execution time of a basic block corresponding to each scenario. Finally, the execution frequencies of these scenarios are defined as ILP variables and are bounded by additional linear constraints.

Scenarios corresponding to cache misses are defined as follows. Given a cache configuration, a basic block BB can be partitioned into a fixed number of memory blocks, with instructions in each memory block being mapped to the same cache block (cache accesses of instructions other than the first one in a memory block are always hits). A *cache scenario* of BB is defined as a mapping of hit or miss to each of the memory blocks of BB. The memory blocks are categorized into always hit, always miss, or unknown, using abstract interpretation-based modeling (see Section 1.3.6.1). The upper bounds on the execution time of BB are computed w.r.t. each of the possible cache scenarios. For the first instructions in memory blocks with unknown categorization, the latency of the fetch stage is assumed to be [1, *penalty*] where *penalty* is the cache miss penalty.

Similarly, the scenarios for branch prediction are defined as the two branch outcomes (correct prediction and misprediction) corresponding to each of the predecessor basic blocks. The execution time of the basic block is estimated w.r.t. both the scenarios by adding nodes corresponding to the wrong-path instructions to the execution graph of a basic block.

Considering the possible cache scenarios and correct or wrong prediction of the preceding branch for a basic block, the ILP objective function denoting a program's WCET is now written as follows.

$$\text{Maximize } T = \sum_{i=1}^{N} \sum_{j \to i} \sum_{\omega \in \Omega_i} t_{j \to i}^{c,\omega} * E_{j \to i}^{c,\omega} + t_{j \to i}^{m,\omega} * E_{j \to i}^{m,\omega} \qquad (1.11)$$

TABLE 1.1   Accuracy and analysis time of WCET estimation technique [46]

Benchmark	WCET (cycles)			Analysis time (sec)	
	Estimated	**Observed**	**Ratio**	**ILP formulation**	**ILP solution**
adpcm	139,346	227,134	1.32	3.41	9.90
dhry	275,177	436,610	1.53	1.84	0.10
fdct	15,006	16,956	1.13	0.08	0.01
fft	944,397	1,146,474	1.14	0.44	0.01
fir	77,004	101,333	1.26	0.79	0.31
ludcmp	17,617	23,818	1.28	0.43	0.12
matsum	62,138	62,734	1.01	0.07	0.01
minver	14,221	21,315	1.33	1.28	0.98
qurt	4,114	6,464	1.45	0.85	0.62
whet	760,010	950,818	1.14	1.24	0.01

where $t_{j \to i}^{c,\omega}$ is the WCET of B_i executed under the following context: (a) B_i is reached from a preceding block B_j, (b) the branch prediction at the end of B_j is correct or B_j does not have a conditional branch, and (c) B_i is executed under a cache scenario $\omega \in \Omega_i$. Ω_i is the set of all cache scenarios of block B_i. The bounds on number of scenarios with correct and mispredicted branch instructions are obtained using ILP-based analysis [45] (see Section 1.3.4.2).

Finally, to extend the above approach for modeling data caches, one can adapt the approach of [69]. This work augments the cache miss equation framework of Ghosh et al. [26] to generate accurate hit and miss patterns corresponding to memory references at different loop levels.

1.3.7.5   Accuracy and Scalability

To give the readers a feel of the accuracy and scalability of the WCET analysis techniques, we present in Table 1.1 the experimental results from [46]. The processor configuration used here is fairly sophisticated: a 2-way superscalar out-of-order pipeline with 5 stages containing a 4-entry instruction fetch buffer, an 8-entry re-order buffer, 2 ALUs, variable latency multiplication and floating point units, and 1 load/store unit; perfect data cache; gshare branch predictor with a 128-entry branch history table; a 1-KB 2-way set associative instruction cache with 16 sets, 32 bytes line size, and 30 cycles cache miss penalty. The analysis was run on a 3-GHz Pentium IV PC with 2 GB main memory.

Table 1.1 presents the estimated WCET obtained through static analysis and the observed WCET obtained via simulation (see Figure 1.3 for the terminology). The estimated WCET is quite close to the observed WCET. Also, the total estimation time (ILP formulation + ILP solving) is less than 15 seconds for all the benchmarks.

1.4   Worst-Case Energy Estimation

In this section, we present a static analysis technique to estimate safe and tight bounds for the worst-case energy consumption of a program on a particular processor. The presentation in this section is based on [36].

Traditional power simulators, such as Wattch [9] and SimplePower [96], perform cycle-by-cycle power estimation and then add them up to obtain total energy consumption. Clearly, we cannot use cycle-accurate estimation to compute the worst-case energy bound, as it would essentially require us to simulate all possible scenarios (which is too expensive). The other method [75, 88] is to use fixed per-instruction energy but it fails to capture the effects of cache miss and branch prediction. Instead, worst-case energy analysis is based on the key observation that the energy consumption of a program can be separated out into the following time-dependent and time-independent components:

Instruction-specific energy: The energy that can be attributed to a particular instruction (e.g., energy consumed as a result of the execution of the instruction in the ALU, cache miss, etc.). Instruction-specific energy does not have any relation with the execution time.

Pipeline-specific energy: The energy consumed in the various hardware components (clock network power, leakage power, switch-off power, etc.) that cannot be attributed to any particular instruction. Pipeline-specific energy is roughly proportional to the execution time.

Thus, cycle-accurate simulation is avoided by estimating the two energy components separately. Pipeline-specific energy estimation can exploit the knowledge of WCET. However, switch-off power and clock network power make the energy analysis much more involved — we cannot simply multiply the WCET by a constant power factor. Moreover, cache misses and overlap among basic blocks due to pipelining and branch prediction add significant complexity to the analysis.

1.4.1 Background

Power and *energy* are terms that are often used interchangeably as long as the context is clear. For battery life, however, the important metric is energy rather than power. The energy consumption of a task running on a processor is defined as *Energy* $= P \times t$, where P is the average power and t is the execution time. Energy is measured in Joules, whereas power is measured in Watts (Joules/second). Power consumption consists of two main components: dynamic power and leakage power $P = P_{dynamic} + P_{leakage}$.

Dynamic power is caused by the charging and discharging of the capacitive load on each gate's output due to switching activity. It is defined as $P_{dynamic} = \frac{1}{2} A V_{dd}^2 C f$, where A is the switching activity, V_{dd} is the supply voltage, C is the capacitance, and f is the clock frequency. For a given processor architecture, V_{dd} and f are constants. The capacitance value for each component of the processor can be derived through register-capacitor (RC)-equivalent circuit modeling [9].

Switching activity A is dependent on the particular program being executed. For circuits that charge and discharge every cycle, such as double-ended array bitlines, an activity factor of 1.0 can be used. However, for other circuits (e.g., single-ended bitlines, internal cells of decoders, pipeline latches, etc.), an accurate estimation of the activity factor requires examination of the actual data values. It is difficult, if not impossible, to estimate the activity factors through static analysis. Therefore, an activity factor of 1.0 (i.e., maximum switching) is assumed conservatively for each active processor component.

Modern processors employ clock gating to save power. This involves switching off clock signals to the idle components so they do not consume dynamic power in the unused cycles. Jayaseelan et al. [36] model three different clock gating styles. For simplicity, let us assume a realistic gating style where idle units and ports dissipate 10% of the peak power. A multi-ported structure consumes power proportional to the number of ports accessed in a given cycle. The power consumed in the idle cycles is referred to as *switch-off power*.

A clock distribution network consumes a significant fraction of the total energy. Without clock gating, *clock power* is independent of the characteristics of the applications. However, clock gating results in power savings in the clock distribution network. Whenever the components in a portion of the chip are idle, the clock network in that portion of the chip can be disabled, reducing clock power.

Leakage power captures the power lost from the leakage current irrespective of switching activity. The analysis uses the leakage power model proposed in [98]: $P_{leakage} = V_{dd} \times N \times k_d \times I_{leakage}$, where V_{dd} is the supply voltage and N is the number of transistors. $I_{leakage}$ is a constant specifying the leakage current corresponding to a particular process technology. k_d is an empirically determined design parameter obtained through SPICE simulation corresponding to a particular device.

1.4.2 Analysis Technique

The starting point of the analysis is the control flow graph of the program. The first step of the analysis estimates an upper bound on the energy consumption of each basic block. Once these bounds are known, the worst-case energy of the entire program can be estimated through path analysis.

1.4.2.1 Energy Estimation for a Basic Block

The goal here is to estimate a tight upper bound on the total energy consumption $energy_{BB}$ of a basic block BB through static analysis. From the discussion in Section 1.4.1,

$$energy_{BB} = dynamic_{BB} + switchoff_{BB}$$
$$+ leakage_{BB} + clock_{BB} \qquad (1.12)$$

where $dynamic_{BB}$ is the instruction-specific energy component, that is, the energy consumed as a result of switching activity as an instruction goes through the pipeline stages. $switchoff_{BB}$, $leakage_{BB}$, and $clock_{BB}$ are defined as the energy consumed as a result of the switch-off power, leakage power, and clock power, respectively, during $wcet_{BB}$, where $wcet_{BB}$ is the WCET of the basic block BB. The WCET ($wcet_{BB}$) is estimated using the static analysis techniques. Now we describe how to define bounds for each energy component.

1.4.2.1.1 Dynamic Energy

The instruction-specific energy of a basic block is the dynamic power consumed as a result of the switching activity generated by the instructions in that basic block.

$$dynamic_{BB} = \sum_{instr \in BB} dynamic_{instr} \qquad (1.13)$$

where $dynamic_{instr}$ is the dynamic power consumed by an instruction *instr*. Now, let us analyze the energy consumed by an instruction as it travels through the pipeline:

1. **Fetch and decode:** The energy consumed here is due to fetch, decode, and instruction cache access. This stage needs feedback from cache analysis.
2. **Register access:** The energy consumed for the register file access because of reads/writes can vary from one class of instructions to another. The energy consumption in the register file for an instruction is proportional to the number of register operands.
3. **Branch prediction:** The energy consumption in this stage needs feedback from branch prediction modeling.
4. **Wakeup logic:** When an operation produces a result, the wakeup logic is responsible for making the dependent instructions ready, and the result is written onto the result bus. An instruction places the tag of the result on the wakeup logic and the actual result on the result bus exactly once, and the corresponding energy can be easily accounted for. The enery consumed in the wakeup logic is proportional to the number of output operands.
5. **Selection logic:** Selection logic is interesting from the point of view of energy consumption. The selection logic is responsible for selecting an instruction to execute from a pool of ready instructions. Unlike the other components discussed earlier, an instruction may access the selection logic more than once. This is because an instruction can request a specific functional unit and the request might not be granted, in which case it makes a request in the next cycle. However, we cannot accurately determine the number of times an instruction would access the selection logic. Therefore, it is *conservatively assumed that the selection logic is accessed every cycle.*
6. **Functional units:** The energy consumed by an instruction in the execution stage depends on the functional unit it uses and its latency. For variable latency instructions, one can safely assume the maximum energy consumption. The energy consumption for load/store units depends on data cache modeling.

Now, Equation 1.13, corresponding to dynamic energy consumed in a basic block BB, is redefined as

$$dynamic_{BB} = selection\_power_{cycle} \times wcet_{BB}$$
$$+ \sum_{instr \in BB} dynamic_{instr} \qquad (1.14)$$

where *selection_power*$_{cycle}$ is a constant defining the power consumed in the selection logic per cycle. *wcet*$_{BB}$ is the WCET of *BB*. Note that *dynamic*$_{instr}$ is redefined as the power consumed by *instr* in all the pipeline stages except for selection logic.

As mentioned before, pipeline-specific energy consists of three components: switch-off energy, clock energy, and leakage energy. All three energy components are influenced by the execution time of the basic block.

1.4.2.1.2   Switch-off Energy

The switch-off energy refers to the power consumed in an idle unit when it is disabled through clock gating. Let *access*$_{BB}(C)$ be the total number of accesses to a component *C* by the instructions in basic block *BB*. Let *ports*(*C*) be the maximum number of allowed accesses/ports for component *C* per cycle. Then, switch-off energy for component *C* in basic block *BB* is

$$switchoff_{BB}(C) = \left(wcet_{BB} - \frac{access_{BB}(C)}{ports(C)} \right)$$
$$\times full\_power_{cycle}(C) \times 10\% \qquad (1.15)$$

where *full_power*$_{cycle}(C)$ is the full power consumption per cycle for component *C*. The switch-off energy corresponding to a basic block can now be defined as

$$switchoff_{BB} = \sum_{C \in components} switchoff_{BB}(C) \qquad (1.16)$$

where *components* is the set of all hardware components.

1.4.2.1.3   Clock Network Energy

To estimate the energy consumed in the clock network, clock gating should be taken into account.

$$clock_{BB} = non\_gated\_clock_{BB} \times \left(\frac{circuit_{BB}}{non\_gated\_circuit_{BB}} \right) \qquad (1.17)$$

where *non_gated_clock*$_{BB}$ is the clock energy without gating and can be defined as

$$non\_gated\_clock_{BB} = clock\_power_{cycle} \times wcet_{BB} \qquad (1.18)$$

where *clock_power*$_{cycle}$ is the peak power consumed per cycle in the clock network. *circuit*$_{BB}$ is defined as the power consumed in all the components except clock network in the presence of clock gating. That is,

$$circuit_{BB} = dynamic_{BB} + switchoff_{BB} + leakage_{BB} \qquad (1.19)$$

non_gated_circuit$_{BB}$, however, is the power consumed in all the components except clock network in the absence of clock gating. It is simply defined as

$$non\_gated\_circuit_{BB} = circuit\_power_{cycle} \times wcet_{BB} \qquad (1.20)$$

circuit_power$_{cycle}$ is a constant defining the peak dynamic plus leakage power per cycle excluding the clock network.

1.4.2.1.4   Leakage Energy

The leakage energy is simply defined as *leakage*$_{BB} = P_{leakage} \times wcet_{BB}$, where $P_{leakage}$ is the power lost per processor cycle from the leakage current regardless of the circuit activity. This quantity, as defined in Section 1.4.1, is a constant given a processor architecture. *wcet*$_{BB}$ is, as usual, the WCET of *BB*.

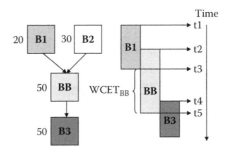

FIGURE 1.16 Illustration of overlap.

1.4.2.2 Estimation for the Whole Program

Given the energy bounds for the basic blocks, we can now estimate the worst-case energy consumption of a program using an ILP formulation. The ILP formulation is similar to the one originally proposed by Li and Malik [50] to estimate the WCET of a program. The execution times of the basic blocks are replaced with the corresponding energy consumptions. Let $energy_{B_i}$ be the upper bound on the energy consumption of a basic block B_i. Then the total energy consumption of the program is given by

$$Total\ energy = \sum_{i=1}^{N} energy_{B_i} \times count_{B_i} \tag{1.21}$$

where the summation is taken over all the basic blocks in the program. The worst-case energy consumption of the program can be derived by maximizing the objective function under the flow constraints through an ILP solver.

1.4.2.3 Basic Block Context

A major difficulty in estimating the worst-case energy arises from the overlapped execution of basic blocks. Let us illustrate the problem with a simple example. Figure 1.16 shows a small portion of the control flow graph. Suppose we are interested in estimating the energy bound for basic block *BB*. The annotation for each basic block indicates the maximum execution count. This is just to show that the execution counts of overlapped basic blocks can be different. As the objective function (defined by Equation 1.21) multiplies each $energy_{BB}$ with its execution count $count_{BB}$, we cannot arbitrarily transfer energy between overlapping basic blocks. Clearly, instruction-specific energy of *BB* should be estimated based on only the energy consumption of its instructions. However, we cannot take such a simplistic view for pipeline-specific energy. Pipeline-specific energy depends critically on $wcet_{BB}$.

If we define $wcet_{BB}$ without considering the overlap, that is, $wcet_{BB} = t_5 - t_2$, then it results in excessive overestimation of the pipeline-specific energy values as the time intervals $t_3 - t_2$ and $t_5 - t_4$ are accounted for multiple times. To avoid this, we can redefine the execution time of *BB* as the time difference between the completion of execution of the predecessor (B_1 in our example) and the completion of execution of *BB*, that is, $wcet_{BB} = t_5 - t_3$. Of course, if *BB* has multiple predecessors, then we need to estimate $wcet_{BB}$ for each predecessor and then take the maximum value among them.

This definition of execution time, however, cannot be used to estimate the pipeline-specific energy of *BB* in a straightforward fashion. This is because switch-off energy and thus clock network energy depend on the idle cycles for hardware ports/units. As we are looking for worst-case energy, we need to estimate an upper bound on idle cycles. Idle cycle estimation (see Equation 1.15) requires an estimate of $access_{BB}(C)$, which is defined as the total number of accesses to a component C by the instructions in basic block *BB*. Now, with the new definition of $wcet_{BB}$ as the interval $t_5 - t_3$, not all these accesses fall within $wcet_{BB}$, and we run the risk of underestimating idle cycles. To avoid this problem, $access_{BB}(C)$ in Equation 1.15 is

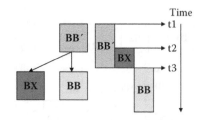

FIGURE 1.17   Illustration of branch misprediction.

replaced with $access_{BB}^{WCET_{BB}}(C)$, which is defined as the total number of accesses to a component C by the instructions in basic block BB that are guaranteed to occur within $wcet_{BB}$.

The number of accesses according to this new definition is estimated during the WCET analysis of a basic block. The energy estimation techniques use the execution-graph-based WCET analysis technique [46] discussed in Section 1.3.7. Let t_3 be the latest commit time of the last instruction of the predecessor node B_1 and let t_5 be the earliest commit time of the last instruction of BB. Then, for each pipeline stage of the different instructions in BB, the algorithm checks whether its earliest or latest start time falls within the interval $t_5 - t_3$. If the answer is yes, then the accesses corresponding to that pipeline stage are guaranteed to occur within $wcet_{BB}$ and are included in $access_{BB}^{WCET_{BB}}(C)$. The pipeline-specific energy is now estimated w.r.t. each of BB's predecessors, and the maximum value is taken.

1.4.2.4   Integrating Cache and Branch Prediction Analysis

Integration of cache and branch prediction modeling is similar to the method described in the context of execution-graph-based WCET analysis (Section 1.3.7). For each cache scenario, the analysis adds the dynamic energy due to cache misses defined as

$$mem\_energy_{BB}^{\omega} = miss_{BB}^{\omega} \times access\_energy \qquad (1.22)$$

where $mem\_energy_{BB}^{\omega}$ is the main memory energy for BB corresponding to cache scenario ω, $miss_{BB}^{\omega}$ is the number of cache misses in BB corresponding to cache scenario ω, and $access\_energy$ is a constant defining the energy consumption per main memory access.

The additional instruction-specific energy due to the execution of speculative instructions is estimated as follows. Let BB be a basic block with BB' as the predecessor (see Figure 1.17). If there is a misprediction for the control flow $BB' \rightarrow BB$, then instructions along the basic block BX will be fetched and executed. The executions along this mispredicted path will continue till the commit of the branch in BB'. Let t_3 be the latest commit time of the mispredicted branch in BB'. For each of the pipeline stages of the instructions along the mispredicted path (i.e., BX), the algorithm checks if its earliest start time is before t_3. If the answer is yes, then the dynamic energy for that pipeline stage is added to the branch misprediction energy of BB. In this fashion, the worst-case energy of a basic block BB corresponding to all possible scenarios can be estimated, where a scenario consists of a preceding basic block BB' and correct/wrong prediction of the conditional branch in BB' and the cache scenario of BB.

1.4.3   Accuracy and Scalability

To give the readers a feel of the accuracy and scalability of the worst-case energy estimation technique, we present in Table 1.2 the experimental results from [36]. The processor configuration used here is as follows: an out-of-order pipeline with five stages containing a 4-entry instruction fetch buffer, an 8-entry re-order buffer, an ALU, variable latency multiplication and floating point units, and a load/store unit; perfect data cache; a gshare branch predictor with a 16-entry branch prediction table; a 4-KB 4-way set associative instruction cache, 32 bytes line size, and a 10-cycle cache miss penalty; 600 MHz clock frequency; and a supply voltage of 2.5 V.

TABLE 1.2   Accuracy and analysis time of worst-case energy estimation technique

Benchmark	Worst-case energy (μJ)			Analysis time (sec)	
	Estimated	Observed	Ratio	ILP formulation	ILP solution
isort	596.93	525.88	1.14	2.88	0.15
fft	13631.21	10260.86	1.33	1.11	0.25
fdct	121.65	105.57	1.15	0.41	0.05
ludcmp	139.75	119.33	1.17	1.37	0.19
matsum	1397.72	1154.31	1.21	0.18	0.01
minver	90.95	80.80	1.13	2.44	1.8
bsearch	3.81	3.07	1.24	0.20	0.01
des	715.58	643.75	1.11	1.37	1.31
matmult	212.94	166.88	1.28	0.15	0.01
qsort	49.84	43.73	1.14	0.85	0.08
qurt	21.95	17.65	1.24	2.14	1.16

Table 1.2 presents the estimated worst-case energy obtained through static analysis and the observed worst-case energy obtained via simulation (Wattch simulator). The estimated values are quite close to the observed values. Moreover, the analysis is quite fast. It takes only $0.15 \sim 2.88$ seconds to formulate the ILP problems for the benchmark programs. The ILP solver (CPLEX) is even faster and completes in under 1.8 seconds for all the benchmarks. All the experiments have been performed on a Pentium IV 1.3 GHz PC with 1 GB of memory.

1.5   Existing WCET Analysis Tools

There are some commercial and research prototype tools for WCET analysis. We discuss them in this section. The most well known and extensively used commercial WCET analyzer is the aiT tool [1] from AbsInt Angewandte Informatik. aiT takes in a code snippet in executable form and computes its WCET. The analyzer uses a two-phased approach where micro-architectural modeling is performed first followed by path analysis. It employs abstract interpretation for cache/pipeline analysis and estimates an upper bound on the execution time of each basic block. These execution time bounds of basic blocks are then combined using ILP to estimate the WCET of the entire program. Versions of aiT are available for various platforms including Motorola PowerPC, Motorola ColdFire, ARM, and so on. The aiT tool is not open-source; so the user cannot change the analyzer code to model timing effects of new processor platforms. The main strength of the aiT tool is its detailed modeling of complex micro-architectures. It is probably the only WCET estimation tool to have a full modeling of the processor micro-architecture for a complex real-life processor like Motorola ColdFire [22] and Motorola PowerPC [32].

Another commercial WCET analyzer is the Bound-T tool [87], which also takes in binary executable programs. It concentrates mainly on program path analysis and does not model cache, complex pipeline, or branch prediction. In path analysis, an important focus of the tool is inferring loop bounds, for which it extensively uses the well-known Omega-calculator [66]. Bound-T has been targeted toward Intel 8051 series micro-controllers, Analog Devices ADSP-21020 DSP, and ATMEL ERC32 SPARC V7-based platforms. Like aiT, Bound-T is not open-source.

The Chronos WCET analyzer [44] incorporates timing models of different micro-architectural features present in modern processors. In particular, it models both in-order and out-of-order pipelines, instruction caches, dynamic branch prediction, and their interactions. The modeling of different architectural features is parameterizable. Chronos is a completely open-source distribution especially suited to the needs of the research community. This allows the researcher to modify and extend the tool for his or her individual needs. Current state-of-the-art WCET analyzers, such as aiT [1], are commercial tools that do not provide the source code. Unlike other WCET analyzers, Chronos is *not* targeted toward one or more commercial embedded processors. Instead, it is built on top of the freely available SimpleScalar simulator infrastructure. SimpleScalar is a widely popular cycle-accurate architectural simulator that allows the user to model a variety of processor platforms in software [10]. Chronos targets its analyzer to processor models supported by SimpleScalar. This choice of platform ensures that the user does not need to purchase a particular embedded platform and its associated compiler, debugger, and other tools (which are often fairly expensive) to conduct research in WCET analysis using Chronos. Also, the flexibility of SimpleScalar enables development and verification of modeling a variety of micro-architectural features for WCET analysis. Thus, Chronos provides a low-overhead, zero-cost, and flexible infrastructure for WCET research. However, it does not support as detailed micro-architectural modeling as is supported by the commercial aiT analyzer; in particular, certain processor features such as data cache are not modeled in Chronos.

Among the research prototypes, HEPTANE [64] is an open-source WCET analyzer. HEPTANE models in-order pipeline, instruction cache, and branch prediction, but it does not include any automated program flow analysis. Symta/P [77] is another research prototype that estimates WCET for C programs. It models caches and simple pipelines but does not support modeling of complex micro-architectural features such as out-of-order pipelines and branch prediction. Cinderella [48] is an ILP-based research prototype developed at Princeton University. The main distinguishing feature of this tool is that it performs both program path analysis and micro-architectural modeling by solving an ILP problem. However, this formulation makes the tool less scalable because the ILP solving time does not always scale up for complex micro-architectures. Also, Cinderella mostly concentrates on program path analysis and cache modeling; it does not analyze timing effects of complex pipelines and branch prediction. The SWEET analyzer from Paderborn, Uppsala, and Malarden Universities focuses mostly on program flow analysis and does not model complex micro-architectures (such as out-of-order pipelines). The program flow analysis proceeds by abstract execution where variable values are abstracted to intervals. However, the abstraction in the flow analysis is limited to data values; the control flow is not abstracted. Consequently, abstract execution in the SWEET tool [27] may resort to a complete unrolling of the program loops.

In addition to the above-mentioned tools, several other research groups have developed their own in-house timing analysis prototypes incorporating certain novel features. One notable effort is by the research group at Florida State University. Their work involves sophisticated flow analysis for inferring infeasible path patterns and loop bounds [31] — features that are not commonly present in many WCET analyzers. However, the tool is currently not available for use or download; it is an in-house research effort.

1.6 Conclusions

In this chapter, we have primarily discussed software timing and energy analysis of an isolated task executing on a target processor without interruption. This is an important problem and forms the building blocks of more complicated performance analysis techniques. As we have seen, the main steps of software timing and energy analysis are (a) program path analysis and (b) micro-architectural modeling. We have also discussed a number of analysis methods that either perform an integrated analysis of the two steps or separate the two steps. It has been observed that integrated analysis methods are not scalable to large programs [94], and hence separated approaches for timing analysis may have a better chance of being integrated into compilers. Finally, we outline here some possible future research directions.

1.6.1 Integration with Schedulability Analysis

The timing and energy analysis methods discussed in this chapter assume *uninterrupted* execution of a program. In reality, a program (or "task," using the terminology of the real-time systems community) may get preempted because of interrupts. The major impact of task preemption is on the performance of the instruction and data caches. Let T_l be a lower-priority task that gets preempted by a higher-priority task T_h. When T_l resumes execution, some of its cache blocks have been replaced by T_h. Clearly, if the WCET analysis does not anticipate this preemption, the resulting timing guarantee will not be safe. *Cache-related preemption delay* [42, 58] analysis derives an upper bound on the number of additional cache misses per preemption. This information is integrated in the schedulability analysis [37] to derive the maximum number of possible preemptions and their effect on the worst-case cache performance.

1.6.2   System-Level Analysis

In a system-on-chip device consisting of multiple processing elements (typically on a bus), a system-wide performance analysis has to be built on top of task-level execution time analysis [70, 85]. Integrating the timing effects of shared bus and complex controllers in the WCET analysis is quite involved. In a recent work, Tanimoto et al. [80] model the shared bus on a system-on-chip device by defining bus scenario as representing a set of possible execution sequences of tasks and bus transfers. They use the definition of bus scenario to automatically derive the deadline and period for each task starting with high-level real-time requirements.

1.6.3   Retargetable WCET Analysis

Retargetability is one of the major issues that needs to be resolved for WCET analysis tools to gain wider acceptability in industry [12]. Developing a complex WCET analyzer for a new platform requires extensive manual effort. Unfortunately, the presence of a large number of platforms available for embedded software development implies that we cannot ignore this problem. The other related problem is the correctness of the abstract processor models used in static timing analysis. The manual abstraction process cannot guarantee the correctness of the models. These two problems can be solved if the static timing analyzer can be generated (semi-)automatically from a formal description of the processor.

One possibility in this direction is to start with the processor specification in some architecture description language (ADL). ADLs precisely describe the instruction-set architecture as well as the micro-architecture of a given processor platform. Certain architectural features are highly parameterizable and hence easy to retarget from a WCET analysis point of view, but other features such as out-of-order pipelines are not easily parameterizable. Li et al. [47] propose an approach to automatically generate static WCET analyzers starting from ADL descriptions for complex processor pipelines. On the other end of the spectrum, we can start with processor specification in hardware description languages (HDLs) such as Verilog or VHDL. The timing models have to be obtained from this HDL specification via simplification and abstraction. Thesing [84] takes this approach for timing models of a system controller. It remains to be seen whether this method scales to complex processor pipelines.

1.6.4   Time-Predictable System Design

The increasing complexity of systems and software leads to reduced timing predictability, which in turn creates serious difficulties for static analysis techniques [86]. An alternative is to design systems and software that are inherently more predictable in terms of timing without incurring significant performance loss. The Virtual Simple Architecture (VISA) approach [3] counters the timing anomaly problem in complex processor pipelines by augmenting the processor micro-architecture with a simpler pipeline. Proposals for predictable memory hierarchy include cache locking [65, 89], cache partitioning [39, 95], as well as replacing cache with scratchpad memory [78, 90] such that WCET analysis is simplified. At the software level, the work in [28, 59] discusses code transformations to reduce the number of program paths considered for WCET analysis. Moreover, Gustafsson et al. [28] also propose WCET-oriented programming to produce code with a very simple control structure that avoids input-data-dependent control flow decisions as far as possible.

1.6.5   WCET-Centric Compiler Optimizations

Traditional compiler optimization techniques guided by profile information focus on improving the average-case performance of a program. In contrast, the metric of importance to real-time systems is the worst-case execution time. Compiler techniques to reduce the WCET of a program have started to receive attention very recently. WCET-centric optimizations are more challenging, as the worst-case path changes as optimizations are applied.

Lee et al. [43] have developed a code generation method for dual-instruction-set ARM processors to simultaneously reduce the WCET and code size. They use a full ARM instruction set along the WCET path to achieve faster execution and at the same time use reduced Thumb instructions along the noncritical paths to reduce code size. Bodin and Puaut [8] designed a customized static branch prediction scheme for reducing a program's WCET. Zhao et al. [99] present a code positioning and transformation method to avoid the penalties associated with conditional and unconditional jumps by placing the basic blocks on WCET paths in contiguous positions whenever possible. Suhendra et al. [78] propose WCET-directed optimal and near-optimal variable allocation strategies to scratchpad memory. Finally, Yu and Mitra [97] exploit application-specific extensions to the base instruction set of a processor for reducing the WCET of real-time tasks. Clearly, there are many other contexts where WCET-guided compiler optimization can play a critical role.

Acknowledgments

Portions of this chapter were excerpted from R. Jayaseelan, T. Mitra, and X. Li, 2006, "Estimating the worst-case energy consumption of embedded software," in *Proceedings of the 12th IEEE Real-Time and Embedded Technology and Applications Symposium (RTAS)*, pages 81–90, and adapted from X. Li, A. Roychoudhury, and T. Mitra, 2006, "Modeling out-of-order processors for WCET analysis," *Real-Time Systems*, 34(3): 195–227.

The authors would like to acknowledge Ramkumar Jayaseelan for preparing the figures in the introduction section.

References

1. AbsInt Angewandte Informatik GmbH. aiT: Worst case execution time analyzer. http://www.absint.com/ait/.
2. P. Altenbernd. 1996. On the false path problem in hard real-time programs. In *Proceedings of the Eighth Euromicro Workshop on Real-Time Systems (ECRTS)*, 102–07.
3. A. Anantaraman, K. Seth, K. Patil, E. Rotenberg, and F. Mueller. 2003. Virtual simple architecture (VISA): Exceeding the complexity limit in safe real-time systems. In *Proceedings of the 30th IEEE/ACM International Symposium on Computer Architecture (ISCA)*, 350–61.
4. R. Arnold, F. Mueller, D. B. Whalley, and M. G. Harmon. 1994. Bounding worst-case instruction cache performance. In *Proceedings of the 15th IEEE Real-Time Systems Symposium (RTSS)*, 172–81.
5. I. Bate and R. Reutemann. 2004. Worst-case execution time analysis for dynamic branch predictors. In *Proceedings of the 16th Euromicro Conference on Real-Time Systems (ECRTS)*, 215–22.
6. G. Bernat, A. Colin, and S. M. Petters. 2002. WCET analysis of probabilistic hard real-time systems. In *Proceedings of the 23rd IEEE Real-Time Systems Symposium (RTSS)*, 279–88.
7. R. Bodik, R. Gupta, and M. L. Soffa. 1997. Refining data flow information using infeasible paths. In *Proceedings of the 6th European Software Engineering Conference* held jointly with the *5th ACM SIGSOFT International Symposium on Foundations of Software Engineering ESEC/FSE*, Vol. 1301 of *Lecture Notes in Computer Science*, 361–77. New York: Springer.
8. F. Bodin and I. Puaut. 2005. A WCET-oriented static branch prediction scheme for real-time systems. In *Proceedings of the 17th Euromicro Conference on Real-Time Systems*, 33–40.
9. D. Brooks, V. Tiwari, and M. Martonosi. 2000. Wattch: A framework for architectural-level power analysis and optimizations. In *Proceedings of the 27th Annual ACM/IEEE International Symposium on Computer Architecture (ISCA)*, 83–94.
10. D. Burger and T. Austin. 1997. The SimpleScalar tool set, version 2.0. Technical Report CS-TR-1997-1342, University of Wisconsin, Madison.
11. C. Burguiére and C. Rochange. 2005. A contribution to branch prediction modeling in WCET analysis. In *Proceedings of the IEEE Design, Automation and Test in Europe Conference and Exposition*, Vol. 1, 612–17.
12. K. Chen, S. Malik, and D. I. August. 2001. Retargetable static timing analysis for embedded software. In *Proceedings of IEEE/ACM International Symposium on System Synthesis (ISSS)*.
13. E. M. Clarke, E. A. Emerson, and A. P. Sistla. 1986. Automatic verification of finite-state concurrent systems using temporal logic specifications. *ACM Transactions on Programming Languages and Systems* 8(2):244–63.

14. A. Colin and I. Puaut. 2000. Worst case execution time analysis for a processor with branch prediction. *Real-Time Systems* 18(2):249–74.

15. P. Cousot and R. Cousot. 1977. Abstract interpretation: A unified lattice model for static analysis of programs by construction or approximation of fixpoints. In *Proceedings of the Fourth Annual ACM Symposium on Principles of Programming Languages (POPL)*, 238–52.

16. K. Dockser. 2001. "Honey, I shrunk the supercomputer!" — The PowerPC 440 FPU brings supercomputing to IBM blue logic library. *IBM MicroNews* 7(4):27–29.

17. S. Edgar and A. Burns. 2001. Statistical analysis of WCET for scheduling. In *Proceedings of the 22nd IEEE Real-Time Systems Symposium (RTSS)*, 215–24.

18. J. Eisinger, I. Polian, B. Becker, A. Metzner, S. Thesing, and R. Wilhelm. 2006. Automatic identification of timing anomalies for cycle-accurate worst-case execution time analysis. In *Proceedings of the Ninth IEEE Workshop on Design and Diagnostics of Electronic Circuits and Systems (DDECS)*, 15–20.

19. J. Engblom. 2003. Analysis of the execution time unpredictability caused by dynamic branch prediction. In *Proceedings of the 9th IEEE Real-Time and Embedded Technology and Applications Symposium (RTAS)*, 152–59.

20. J. Engblom and A. Ermedahl. 2000. Modeling complex flows for worst-case execution time analysis. In *Proceedings of IEEE Real-Time Systems Symposium (RTSS)*.

21. A. Ermedahl and J. Gustafsson. 1997. Deriving annotations for tight calculation of execution time. In *Third International Euro-Par Conference on Parallel Processing (Euro-Par)*. Vol. 1300 of *Lecture Notes in Computer Science*, 1298–307. New York: Springer.

22. C. Ferdinand, R. Heckmann, M. Langenbach, F. Martin, M. Schmidt, H. Theiling, S. Thesing, and R. Wilhelm. 2001. Reliable and precise WCET determination for a real-life processor. In *Proceedings of International Workshop on Embedded Software (EMSOFT)*, 469–85.

23. C. Ferdinand and R. Wilhelm. 1998. On predicting data cache behavior for real-time systems. In *Proceedings of the ACM SIGPLAN Workshop on Languages, Compilers, and Tools for Embedded Systems (LCTES)*, 16–30.

24. S. V. Gheorghita, T. Basten, and H. Corporaal. 2005. Intra-task scenario-aware voltage scheduling. In *International Conference on Compiler, Architectures and Synthesis for Embedded Systems (CASES)*.

25. S. V. Gheorghita, S. Stuijk, T. Basten, and H. Corporaal. 2005. Automatic scenario detection for improved WCET estimation. In *ACM Design Automation Conference (DAC)*.

26. S. Ghosh, M. Martonosi, and S. Malik. 1999. Cache miss equations: A compiler framework for analyzing and tuning memory behavior. *ACM Transactions on Programming Languages and Systems*, 21(4):707–46.

27. J. Gustafsson. 2000. Eliminating annotations by automatic flow analysis of real-time programs. In *Proceedings of the Seventh International Conference on Real-Time Computing Systems and Applications (RTCSA)*, 511–16.

28. J. Gustafsson, B. Lisper, R. Kirner, and P. Puschner. 2006. Code analysis for temporal predictability. *Real-Time Systems* 32:253–77.

29. C. Healy, R. Arnold, F. Mueller, D. Whalley, and M. Harmon. 1999. Bounding pipeline and instruction cache performance. *IEEE Transactions on Computers* 48(1):53–70.

30. C. Healy, M. Sjodin, V. Rustagi, D. Whalley, and R. van Englen. 2000. Supporting timing analysis by automatic bounding of loop iterations. *Real-Time Systems* 18:129–56.

31. C. A. Healy and D. B. Whalley. 2002. Automatic detection and exploitation of branch constraints for timing analysis. *IEEE Transactions on Software Engineering* 28(8):763–81.

32. R. Heckmann, M. Langenbach, S. Thesing, and R. Wilhelm. 2003. The influence of processor architecture on the design and the results of WCET tools. *Proceedings of the IEEE*, 91(7):1038–054.

33. J. L. Hennessy and D. A. Patterson. 2003. *Computer architecture — a quantitative approach*. 3rd ed. San Francisco: Morgan Kaufmann.

34. IBM Microelectronics Division. 1999. *The PowerPC 440 core*.

35. Institute of Electrical and Electronics Engineers. 1985. *IEEE 754: Standard for binary floating-point arithmetic*.

36. R. Jayaseelan, T. Mitra, and X. Li. 2006. Estimating the worst-case energy consumption of embedded software. In *Proceedings of the 12th IEEE Real-Time and Embedded Technology and Applications Symposium (RTAS)*, 81–90.

37. L. Ju, S. Chakraborty, and A. Roychoudhury. 2007. Accounting for cache-related preemption delay in dynamic priority schedulability analysis. In *Proceedings of Design Automation and Test in Europe (DATE)*.

38. S.-K. Kim, S. L. Min, and R. Ha. 1996. Efficient worst case timing analysis of data caching. In *Proceedings of the Second IEEE Real-Time Technology and Applications Symposium (RTAS)*, 230–40.

39. D. B. Kirk. 1989. SMART (strategic memory allocation for real-time) cache design. In *Proceedings of the Real-Time Systems Symposium (RTSS)*, 229–39.

40. R. Kirner. 2003. Extending optimizing compilation to support worst-case execution time analysis. PhD thesis, T. U. Vienna.

41. M. Langenbach, S. Thesing, and R. Heckmann. Pipeline modeling for timing analysis. In *Proceedings of the 9th International Symposium on Static Analysis (SAS)*. Vol. 2477 of *Lecture Notes in Computer Science*, 294–309. New York: Springer.

42. C.-G. Lee, H. Hahn, Y.-M. Seo, S. L. Min, R. Ha, S. Hong, C. Y. Park, M. Lee, and C. S. Kim. 1998. Analysis of cache-related preemption delay in fixed-priority preemptive scheduling. *IEEE Transactions on Computers* 47(6):700–13.

43. S. Lee et al. 2004. A flexible tradeoff between code size and WCET using a dual instruction set processor. In *Proceedings of the 8th International Workshop on Software and Compilers for Embedded Systems (SCOPES)*. Vol. 3199 of *Lecture Notes in Computer Science*, 244–58. New York: Springer.

44. X. Li, Y. Liang, T. Mitra, and A. Roychoudhury. Chronos: A timing analyzer for embedded software. *Science of Computer Programming*, special issue on Experiment Software and Toolkit 2007 (to appear), http://www.comp.nus.edu.sg/~rpembed/chronos/.

45. X. Li, T. Mitra, and A. Roychoudhury. 2005. Modeling control speculation for timing analysis. *Real-Time Systems* 29(1):27–58.

46. X. Li, A. Roychoudhury, and T. Mitra. 2006. Modeling out-of-order processors for WCET analysis. *Real-Time Systems* 34(3):195–227.

47. X. Li, A. Roychoudhury, T. Mitra, P. Mishra, and X. Cheng. 2007. A retargetable software timing analyzer using architecture description language. In *Proceedings of the 12th Asia and South Pacific Design Automation Conference (ASP-DAC)*, 396–401.

48. Y.-T. S. Li. Cinderella 3.0 WCET analyzer. http://www.princeton.edu/~yauli/cinderella-3.0/.

49. Y.-T. S. Li and S. Malik. 1997. Performance analysis of embedded software using implicit path enumeration. *IEEE Transactions on Computer-Aided Design of Integrated Circuits and Systems (TCAD)* 16(12):1477–87.

50. Y.-T. S. Li and S. Malik. 1998. *Performance analysis of real-time embedded software.* New York: Springer.

51. S.-S. Lim, Y. H. Bae, G. T. Jang, B-D. Rhee, S. L. Min, C. Y. Park, H. Shin, K. Park, S.-M. Moon, and C. S. Kim. 1995. An accurate worst case timing analysis for RISC processors. *IEEE Transactions on Software Engineering* 21(7):593–604.

52. Y. A. Liu and G. Gomez. 2001. Automatic accurate cost-bound analysis for high-level languages. *IEEE Transactions on Computers* 50(12):1295–309.

53. T. Lundqvist and P. Stenström. 1999. An integrated path and timing analysis method based on cycle-level symbolic execution. *Real-Time Systems* 17(2/3):183–207.

54. T. Lundqvist and P. Stenström. 1999. Timing anomalies in dynamically scheduled microprocessors. In *Proceedings of the 20th IEEE Real-Time Systems Symposium (RTSS)*, 12–21.

55. F. Martin, M. Alt, R. Wilhelm, and C. Ferdinand. 1998. Analysis of loops. In *Compiler Construction (CC)*. New York: Springer.

56. A. Metzner. 2004. Why model checking can improve WCET analysis. In *Proceedings of the 16th International Conference on Computer Aided Verification (CAV)*. Vol. 3114 of *Lecture Notes in Computer Science*, 361–71. New York: Springer.

57. F. Mueller. 2000. Timing analysis for instruction caches. *Real-Time Systems* 18:217–47.

58. H. S. Negi, T. Mitra, and A. Roychoudhury. 2003. Accurate estimation of cache-related preemption delay. In *Proceedings of the 1st IEEE/ACM/IFIP International Conference on Hardware/Software Codesign and System Synthesis (CODES+ISSS)*, 201–06.

59. H. S. Negi, A. Roychoudhury, and T. Mitra. 2004. Simplifying WCET analysis by code transformations. In *Proceedings of the 4th International Workshop on Worst-Case Execution Time Analysis (WCET)*.

60. G. Ottosson and M. Sjödin. 1997. Worst-case execution time analysis for modern hardware architectures. In *Proceedings of the ACM SIGPLAN Workshop on Languages, Compilers, and Tools for Real-Time Systems (LCT-RTS)*.

61. C. Y. Park. 1993. Predicting program execution times by analyzing static and dynamic program paths. *Real-time Systems* 5(1):31–62.

62. C. Y. Park. 1992. Predicting deterministic execution times of real-time programs. PhD thesis, University of Washington, Seattle.

63. J. Pierce and T. Mudge. 1996. Wrong-path instruction prefetching. In *Proceedings of the 29th Annual IEEE/ACM International Symposium on Microarchitectures (MICRO)*, 165–75.

64. I. Puaut. HEPTANE static WCET analyzer. http://www.irisa.fr/aces/work/heptane-demo/heptane.html.

65. I. Puaut and D. Decotigny. 2002. Low-complexity algorithms for static cache locking in multitasking hard real-time systems. In *Proceedings of the 23rd IEEE Real-Time Systems Symposium (RTSS)*, 114–23.

66. W. Pugh. 1991. The omega test: A fast and practical integer programming algorithm for dependence analysis. In *ACM/IEEE Conference on Supercomputing*.

67. P. Puschner and C. Koza. 1989. Calculating the maximum execution time of real-time programs. *Real-Time Systems* 1(2):159–76.

68. P. Puschner and A. Schedl. 1997. Computing maximum task execution times: A graph based approach. *Real-Time Systems* 13(1):67–91.

69. H. Ramaprasad and F. Mueller. Bounding worst-case data cache behavior by analytically deriving cache reference patterns. In *IEEE Real-Time Technology and Applications Symposium (RTAS)*, 148–57.

70. K. Richter, D. Ziegenbein, M. Jersak, and R. Ernst. 2002. Model composition for scheduling analysis in platform design. In *Proceedings of the 39th Annual ACM/IEEE Design Automation Conference (DAC)*, 287–92.

71. A. Roychoudhury, T. Mitra, and H. S. Negi. 2005. Analyzing loop paths for execution time estimation. In *Lecture Notes in Computer Science*. Vol. 3816, 458–69. New York: Springer.

72. T. Schuele and K. Schneider. 2004. Abstraction of assembler programs for symbolic worst case execution time analysis. In *Proceedings of the 41st ACM/IEEE Design Automation Conference (DAC)*, 107–12.

73. A. C. Shaw. 1989. Reasoning about time in higher-level language software. *IEEE Transactions on Software Engineering* 1(2):875–89.

74. Simplify. Simplify theorem prover. http://www.research.compaq.com/SRC/esc/Simplify.html.

75. A. Sinha and A. P. Chandrakasan. 2001. Jouletrack: A web based tool for software energy profiling. In *Proceedings of the Design Automation Conference (DAC)*.

76. F. Stappert, A. Ermedahl, and J. Engblom. 2001. Efficient longest executable path search for programs with complex flows and pipeline effects. In *Proceedings of the First International Conference on Compilers, Architecture, and Synthesis for Embedded Systems (CASES)*, 132–40.

77. J. Staschulat. Symta/P: Symbolic timing analysis for processes. http://www.ida.ing.tu-bs.de/research/projects/symta/home.e.shtml.

78. V. Suhendra, T. Mitra, A. Roychoudhury, and T. Chen. 2005. WCET centric data allocation to scratchpad memory. In *Proceedings of the 26th IEEE Real-Time Systems Symposium (RTSS)*, 223–32.

79. V. Suhendra, T. Mitra, A. Roychoudhury, and T. Chen. 2006. Efficient detection and exploitation of infeasible paths for software timing analysis. In *Proceedings of the 43rd ACM/IEEE Design Automation Conference (DAC)*, 358–63.

80. T. Tanimoto, S. Yamaguchi, A. Nakata, and T. Higashino. 2006. A real time budgeting method for module-level-pipelined bus based system using bus scenarios. In *Proceedings of the 43rd ACM/IEEE Design Automation Conference (DAC)*, 37–42.

81. H. Theiling and C. Ferdinand. 1998. Combining abstract interpretation and ILP for microarchitecture modelling and program path analysis. In *Proceedings of the 19th IEEE Real-Time Systems Symposium (RTSS)*, 144–53.

82. H. Theiling, C. Ferdinand, and R. Wilhelm. 2000. Fast and precise WCET prediction by separated cache and path analyses. *Real-Time Systems* 18(2/3):157–79.

83. S. Thesing. Safe and precise worst-case execution time prediction by abstract interpretation of pipeline models. PhD thesis, University of Saarland, Germany.

84. S. Thesing. 2006. Modeling a system controller for timing analysis. In *Proceedings of the 6th ACM/IEEE International Conference on Embedded Software (EMSOFT)*, 292–300.

85. L. Thiele, S. Chakraborty, M. Gries, A. Maxiaguine, and J. Greutert. 2001. Embedded software in network processors — models and algorithms. In *Proceedings of the First International Workshop on Embedded Software (EMSOFT)*. Vol. 2211 of *Lecture Notes in Computer Science*, 416–34. New York: Springer.

86. L. Thiele and R. Wilhelm. 2004. Design for timing predictability. *Real-Time Systems*, 28(2/3):157–77.

87. Tidorum Ltd. Bound-T execution time analyzer. http://www.bound-t.com.

88. V. Tiwari, S. Malik, and A. Wolfe. 1994. Power analysis of embedded software: A first step towards software power minimization. *IEEE Transactions of VLSI Systems* 2(4):437–45.

89. X. Vera, B. Lisper, and J. Xue. 2003. Data cache locking for higher program predictability. In *Proceedings of the International Conference on Measurements and Modeling of Computer Systems (SIGMETRICS)*, 272–82.

90. L. Wehmeyer and P. Marwedel. 2005. Influence of memory hierarchies on predictability for time constrained embedded software. In *Proceedings of the Conference on Design, Automation and Test in Europe (DATE)*, 600–605.

91. I. Wenzel, R. Kirner, P. Puschner, and B. Rieder. 2005. Principles of timing anomalies in superscalar processors. In *Proceedings of the Fifth International Conference on Quality Software (QSIC)*, 295–303.

92. I. Wenzel, R. Kirner, B. Rieder, and P. Puschner. Measurement-based worst-case execution time analysis. In *Proceedings of the Third IEEE Workshop on Software Technologies for Future Embedded and Ubiquitous Systems (SEUS)*, 7–10.

93. R. White, F. Mueller, C. Healy, D. Whalley, and M. Harmon. 1997. Timing analysis for data caches and set-associative caches. In *Proceedings of the Third IEEE Real-Time Technology and Applications Symposium (RTAS)*, 192–202.

94. R. Wilhelm. 2004. Why AI + ILP is good for WCET, but MC is not, nor ILP alone. In *Proceedings of the 5th International Conference on Verification, Model Checking, and Abstract Interpretation (VMCAI)*. Vol. 2937 of *Lecture Notes in Computer Science*, 309–22. New York: Springer.

95. A. Wolfe. 1994. Software-based cache partitioning for real-time applications. *Journal of Computer and Software Engineering, Special Issue on Hardware-Software Codesign*, 2(3):315–27.

96. W. Ye et al. 2000. The design and use of simplepower: A cycle-accurate energy estimation tool. In *Proceedings of the ACM/IEEE Design Automation Conference (DAC)*.

97. P. Yu and T. Mitra. 2005. Satisfying real-time constraints with custom instructions. In *Proceedings of the ACM International Conference on Hardware/Software Codesign and System Synthesis (CODES+ISSS)*, 166–71.

98. Y. Zhang et al. 2003. Hotleakage: A temperature-aware model of subthreshold and gate leakage for architects. Technical Report CS-2003-05, University of Virginia.

99. W. Zhao, D. Whalley, C. Healy, and F. Mueller. 2004. WCET code positioning. In *Proceedings of the 25th IEEE Real-Time Systems Symposium (RTSS)*, 81–91.

2

Static Program Analysis for Security

K. Gopinath
*Department of Computer Science
and Automation,
Indian Institute of Science,
Bangalore, India*
gopi@csa.iisc.ernet.in

Abstract

In this chapter, we discuss static analysis of the security of a system. First, we give a brief background on what types of static analysis are feasible in principle and then move on to what is practical. We next discuss static analysis of buffer overflow and mobile code, followed by access control. Finally, we discuss static analysis of information flow expressed in a language that has been annotated with flow policies.

2.1   Introduction

Analyzing a program for security holes is an important part of the current computing landscape, as security has not been an essential ingredient in a program's design for quite some time. With the critical importance of a secure program becoming clearer in the recent past, designs based on explicit security policies are likely to gain prominence.

Static analysis of a program is one technique to detect security holes. Compared to monitoring an execution at runtime (which may not have the required coverage), a static analysis — even if incomplete because of loss of precision — potentially gives an analysis on all runs possible instead of just the ones seen so far.

However, security analysis of an arbitrary program is extremely hard. First, what security means is often unspecified or underspecified. The definition is either too strict and cannot cope with the "commonsense" requirement or too broad and not useful. For example, one definition of security involves the notion of "noninterference" [24]. If it is very strict, even cryptoanalytically strong encryption and decryption does not qualify as secure, as there is information flow from encrypted text to plain text [59]. If it is not very strict, by definition, some flows are not captured that are important in some context for achieving security and hence, again, not secure. For example, if electromagnetic emissions are not taken into account, the key may be easily compromised [50]. A model of what security means is needed, and this is by no means an easy task [8]. Schneider [28] has a very general definition: a security policy defines a binary partition of all (computable) sets of executions — those that satisfy and those that do not. This is general enough to cover access control policies (a program's behavior on an arbitrary individual execution for an arbitrary finite period), availability policies (behavior on an arbitrary individual execution over an infinite period), and information flow (behavior in terms of the set of all executions).

Second, the diagonalization trick is possible, and many analyses are undecidable. For example, there are undecidable results with respect to viruses and malicious logic [16]: it is undecidable whether an arbitrary program contains a computer virus. Similarly, viruses exist for which no error-free detection algorithm exists.

Recently, there have been some interesting results on computability classes for enforcement mechanisms [28] with respect to execution monitors, program rewriting, and static analysis. Execution monitors intervene whenever execution of an untrusted program is about to violate the security policy being enforced. They are typically used in operating systems using structures such as access control lists or used when executing interpreted languages by runtime type checking. It is possible to rewrite a binary so that every access to memory goes through a monitor. While this can introduce overhead, optimization techniques can be used to reduce the overhead. In many cases, the test can be determined to be not necessary and removed by static analysis. For example, if a reference to a memory address m has already been checked, it may not be necessary for a later occurrence.

Program rewriting modifies the untrusted program before execution to make it incapable of violating the security policy. An execution monitor can be viewed as a special case of program rewriting, but Hamlen et al. [28] point out certain subtle cases. Consider a security policy that makes halting a program illegal; an execution monitor cannot enforce this policy by halting, as this would be illegal! There are also classes of policies in certain models of execution monitors that cannot be enforced by any program rewriter.

As is to be expected, the class of statically enforceable policies is the class of recursively decidable properties of programs (class Π_0 of the arithmetic hierarchy): a static analysis has to be necessarily total (i.e., terminate) and return safe or unsafe. If precise analysis is not possible, we can relax the requirement by being conservative in what it returns (i.e., tolerate false positives). Execution monitors are the class of co-recursively enumerable languages (class Π_1 of the arithmetic hierarchy).[1]

A system's security is specified by its security policy (such as access control or information flow model) and implemented by mechanisms such as physical separation or cryptography.[2] Consider access control. An access control system guards access to resources, whereas an information flow model classifies information to prevent disclosure. Access control is a component of security policy, while cryptography is one technical mechanism to effect the security policy. Systems have employed basic access control since timesharing systems began (1960) (e.g., Multics, Unix). Simple access control models for such "stand-alone" machines assumed the universe of users known, resulting in the scale of the model being "small." A set of access control moves can unintentionally leak a right (i.e., give access when it should not). If it is possible to analyze the system and ensure that such a result is not possible, the system can be said to be secure, but theoretical models inspired by these systems (the most well known being the Harrison, Ruzzo, Ullman [HRU] model) showed "surprising" undecidability results [29], chiefly resulting from the unlimited number of subjects and objects possible. Technically, in this model, it is undecidable whether a given state of a given protection

---

[1]A security policy P is co-recursively enumerable if there exists a Turing machine M that takes an arbitrary execution monitor $E M$ as an input and rejects it in finite time if $\sim P(E M)$; otherwise $M(E M)$ loops forever.

[2]While cryptography has rightfully been a significant component in the design of large-scale systems, its relation to security policy, especially its complementarity, has not often been brought out in full. "If you think cryptography is the solution to your problem, you don't know what your problem is" (Roger Needham).

system is safe for a given generic right. However, the need for automated analysis of policies was small in the past, as the scale of the systems was small.

Information flow analysis, in contrast to access control, concerns itself with the downstream use of information once obtained after proper access control. Carelessness in not ensuring proper flow models has resulted recently in the encryption system in HD-DVD and BluRay disks to be compromised (the key protecting movie content is available in memory).[3]

Overt models of information flow specify the policy concerning how data should be used explicitly, whereas covert models use "signalling" of information through "covert" channels such as timing, electromagnetic emissions, and so on. Note that the security weakness in the recent HD-DVD case is due to improper overt information flow. Research on some overt models of information flow such as Bell and LaPadula's [6] was initiated in the 1960s, inspired by the classification of secrets in the military. Since operating systems are the inspiration for work in this area, models of secure operating systems were developed such as C2 and B2 [57]. Work on covert information flow progressed in the 1970s. However, the work on covert models of information flow in proprietary operating systems (e.g., DG-UNIX) was expensive and too late, so late that it could not be used on the by then obsolescent hardware. Showing that Trojan horses did not use covert channels to compromise information was found to be "ultimately unachievable" [27].

Currently, there is a considerable interest in studying overt models of information flow, as the extensive use of distributed systems and recent work with widely available operating systems such as SELinux and OpenSolaris have expanded the scope of research. The scale of the model has become "large," with the need for automated analysis of policies being high.

Access control and information flow policies can both be statically analyzed, with varying effectiveness depending on the problem. Abstract interpretation, slicing, and many other compiler techniques can also be used in conjunction. Most of the static analyses induce a constraint system that needs to be solved to see if security is violated. For example, in one buffer overflow analysis [23], if there is a solution to a constraint system, then there is an attack. In another analysis in language-based security, nonexistence of a solution to the constraints is an indication of a possible leak of information.

However, static analysis is not possible in many cases [43] and has not yet been used on large pieces of software. Hence, exhaustive checking using model checking [14] is increasingly being used when one wants to gain confidence about a piece of code. In this chapter, we consider model checking as a form of static analysis. We will discuss access control on the Internet that uses model checking (Section 2.4.4). We will also discuss this approach in the context of Security-Enhanced Linux (SELinux) where we check if a large application has been given only sufficient rights to get its job done [31].

In spite of the many difficulties in analyzing the security of a system, policy frameworks such as access control and information flow analysis and mechanisms such as cryptography have been used to make systems "secure." However, for any such composite solution, we need to trust certain entities in the system such as the compiler, the BIOS, the (Java) runtime system, the hardware, digital certificates, and so on — essentially the "chain of trust" problem. That this is a tricky problem has been shown in an interesting way by Ken Thompson [53]; we will discuss it below. Hence, we need a "small" trusted computing base (TCB): all protection mechanisms within a system (hardware, software, firmware) for enforcing security policies.

In the past (early 1960s), operating systems were small and compilers larger in comparison. The TCB could justifiably be the operating system, even if uncomfortably larger than one wished. In today's context, the compilers need not be as large as current operating systems (for example, Linux kernel or Windows is many millions of lines of code), and a TCB could profitably be the compiler. Hence, a compiler analysis of security is meaningful nowadays and likely to be the only way large systems (often a distributed system) can be crafted in the future. Using a top-level security policy, it may be possible to automatically partition the program so that the resulting distributed system is secure by design [60].

---

[3]The compromise of the security system DeCSS in DVDs was due to cryptanalysis, but in the case of HD-DVD it was simply improper information flow.

To illustrate the effectiveness of static security analysis, we first discuss a case where static analysis fails completely (Ken Thomson's Trojan horse). We then outline some results on the problem of detecting viruses and then discuss a case study in which static analysis can in principle be very hard (equivalent to cryptanalysis in general) but is actually much simpler because of a critical implementation error. We will also briefly touch upon obfuscation that exploits difficulty of analysis.

We then discuss static analysis of buffer overflows, loading of mobile code, and access control and information flow, illustrating the latter using Jif [35] language on a realistic problem. We conclude with likely future directions in this area of research.

2.1.1   A Dramatic Failure of Static Analysis: Ken Thompson's Trojan Horse

Some techniques to defeat static analysis can be deeply effective; we summarize Ken Thompson's ingenious Trojan horse trick [53] in a compiler that uses self-reproduction, self-learning, and self-application. We follow his exposition closely.

First, self-reproduction is possible; for example, one can construct a self-reproducing program such as ((lambda x. (list x x)) (lambda x. (list x x))). Second, it is possible to teach a compiler (written in its own language) to compile new features (the art of bootstrapping a compiler): this makes self-learning possible. We again give Ken Thompson's example to illustrate this. Let us say that a lexer knows about '\n' (newline) but not '\v' (vertical tab). How does one teach it to compile '\v' also? Let the initial source be:

```
c=next();
if (c !='\\') return(c);
c=next();
if (c !='\\') return('\\')
if (c ='n') return('\n')
```

Adding to the compiler source

```
if (c ='v') return('\v')                //error
```

and using previous compiler binary does not work, as that binary does not know about '\v'. However, the following works:

```
  if (c ='v') return(11)
```

Now a new binary (from the new source using the old compiler binary) knows about '\v' in a portable way. Now we can use it to compile the previously uncompilable statement:

```
  if (c ='v') return('\v')
```

The compiler has "learned." Now, we can introduce a Trojan horse into the login program that has a backdoor to allow a special access to log in as any user:

```
compile(char *s) {(* compiler main that accepts s as input (a program)*)
      if match(s, "pattern_login") {
         (* "Trojan" compile for a login program in the system*)
         compile("bug_login"); return
      }
      ... (* normal compile *)
   }
```

However, this is easily detectable (by examining the source). To make this not possible, we can add a second Trojan horse aimed at the compiler:

```
if match(s, "pattern_compiler") {
   (* "Trojan" compile for a compiler program in the system*)
   compile("bug_compiler"); return
}
```

Now we can code a self-reproducing program that reintroduces both Trojan horses into the compiler with a learning phase where the buggy compiler binary with two Trojan horses now reinserts it into any new compiler binary compiled from a clean source! The detailed scheme is as follows: first a clean compiler binary (A) is built from a clean compiler source (S). Next, as part of the learning phase, a modified compiler source (S') can be built that incorporates the bugs. The logic in S' looks at the source code of any program submitted for compilation and, say by pattern matching, decides whether a program submitted is a login or a compiler program. If it decides that it is one of these special programs, it reproduces[4] one or more Trojan horses (as necessary) when presented with a clean source. Let the program S' be compiled with A. We have a new binary A' that reinserts the two Trojans on any clean source!

The virus exists in the binary but not in the source. This is not possible to discern unless the history of the system (the sequence of compilations and alterations) is kept in mind. Static analysis fails spectacularly!

2.1.2 Detecting Viruses

As discussed earlier, the virus detection problem is undecidable. A successful virus encapsulates itself so that it cannot be detected — the opposite of "self-identifying data." For example, a very clever virus would put logic in I/O routines so that any read of suspected portions of a disk returns the original "correct" information! A polymorphic virus inserts "random" data to vary the signature. More effectively, it can create a random encryption key to encrypt the rest of the virus and store the key with the virus.

2.1.2.1 Cohen's Results

Cohen's impossibility result states that it is impossible for a program to perfectly demarcate a line, enclosing all and only those programs that are infected with some virus: no algorithm can properly detect all possible viruses [16]: $\forall Alg.\exists Virus.Alg$ does not detect *Virus*. For any candidate computer virus detection algorithm A, there is a program p(pgm): if A(pgm) then exit; else spread. Here, *spread* means behave like a virus. We can diagonalize this by setting pgm=p and a contradiction follows immediately as p spreads iff not A(p).

Similarly, Chess and White [11] show there exists a virus that no algorithm perfectly detects, i.e., one with no false positives: $\exists Virus.\forall Alg.Alg$ does not detect *Virus*. Also, there exists a virus that no algorithm loosely detects, i.e., claims to find the virus but is infected with some other virus: $\exists Virus.\forall Alg.Alg$ does not loosely-detect *Virus*. In other words, there exist viruses for which, even with the virus analyzed completely, no program detects only that virus with no false positives. Further, another Chess and White's impossibility result states that there is no program to classify programs (only those) with a virus V and not include any program without any virus.

Furthermore, there are interesting results concerning *polymorphic viruses*: these viruses generate a set of other viruses that are not identical to themselves but are related in some way (for example, are able to reproduce the next one in sequence). If the size of this set is greater than 1, call the set of viruses generated the viral set S.

An algorithm A detects a set of viruses S iff for every program p, $A(p)$ terminates and returns TRUE iff p is infected with some virus V in S. If W is a polyvirus, for any candidate W-detection algorithm C, there is a program $s(pgm)$ that is an instance of W:

---

[4]It is easiest to do so in the binary being produced as the compiled output.

```
if subroutine_one(pgm) then exit, else {
   replace text of subroutine_one with a random program;
   spread; exit; }
subroutine_one: return C(argument)
```

This can be diagonalized by `pgm=s`, resulting in a contradiction: `if C(s) true exit; other-wise polyvirus.`

Hence, it is clear that static analysis has serious limits.

2.1.3 A Case Study for Static Analysis: GSM Security Hole

We will now attempt to situate static analysis with other analyses, using the partitioning attack [50] on GSM hashing on a particular implementation as an example. GSM uses the COMP128 algorithm for encryption. A 16B (128b) key of subscriber (available with the base station along with a 16B challenge from the base station is used to construct a 12B (96b) hash. The first 32b is sent as a response to the challenge, and the remaining 64b is used as the session key. The critical issue is that there should be no leakage of the key in any of the outputs *this includes any electromagnetic (EM) leakages* during the computation of the hash. The formula for computation of the hash it has a butterfly structure) is as follows:

```
//X is 32B (16B key || 16B challenge)
//Tj[r] is a lookuptable of 8-j bits value with r rows:
//alg uses T0[512], T1[256], T2[128], T3[64], T4[32]
for j=0..4
  for k=0..2^j -1
    for l=0..2^(4-j) -1
      m=l+k*2^(5-j)
      n=m+2^(4-j)
      y=(X[m]+2*X[n]) mod 2^(9-j)
      z=(2*X[m]+X[n]) mod 2^(9-j)
      X[m]=Tj[y]
      X[n]=Tj[z]
```

If we expand `y` in the expression for `X[m]`, we have

```
X[m]=Tj[(X[m]+2*X[n]) mod 2^(9-j)]
```

A simple-minded flow analysis will show that there is direct dependence of the EM leakage on the key; hence, this is not information flow secure. However, cryptographers, using the right "confusion" and "diffusion" operators such as the above code, have shown that the inverse can be very difficult to compute. Hence, even if very simple static analysis clearly points out the flow dependence of the EM leakage on the key, it is not good enough to crack the key. However, even if the mapping is cryptanalytically strong, "implementation" bugs can often give away the key. An attack is possible if one does not adhere to the following principle [50] of statistical independence (or more accurately *noninterference*, which will be discussed later): *relevant bits of all intermediate cycles and their values should be statistically independent of the inputs, outputs, and sensitive information.*

Normally, those bits that emit high EM are good candidates for analysis. One set of candidates are the array and index values, as they need to be amplified electrically for addressing memory. They are therefore EM sensitive, whereas other internal values may not be so.

Because of the violation of this principle, a cryptographically strong algorithm may have an implementation that leaks secrets. For example, many implementations use 8b microprocessors, as COMP128 is optimized for them, so the actual implementation for T0 is two tables, T00 and T01 (each 256 entries):

```
//X is 32B (16B key || 16B challenge)
//Tj[r] is a lookuptable of 8-j bits value with r rows:
//alg uses T0[512], T1[256], T2[128], T3[64], T4[32]
for j=0..4
  for k=0..2^j -1
    for l=0..2^(4-j) -1
      m=l+k*2^(5-j)
      n=m+2^(4-j)
      y=(X[m]+2*X[n]) mod 2^(9-j)
      z=(2*X[m]+X[n]) mod 2^(9-j)
      if (y<256) X[m]=Tj0[y] else Tj1[y-256] // old: X[m]=Tj[y]
      if (z<256) X[m]=Tj0[z] else Tj1[z-256] // old: X[n]=Tj[z]
```

If the number of intermediate lookups from tables T00 or T01 have statistical significance, then because of the linearity of the index y with respect to R for the first round, some information can be gleaned about the key. The technique of differential cryptanalysis is based on such observations. In addition, if it is possible to know when access changes from one table (say, T00) to another (T01) by changing R, then the R value where it changes is given by K + 2*R=256, from which X, the first byte of the GSM key, can be determined.

In general, we basically have a set of constraints, such as

$$0 \leq x + 2y_1 \leq 127$$
$$128 \leq x + 2y_2 \leq 256$$

where y_1 and y_2 are two close values that map the index into different arrays (T0 or T1).[5]

If there is a solution to these Diophantine equations, then we have an attack. Otherwise, no. Since the cryptographic confusion and diffusion operations determine the difficulty (especially with devices such as S-boxes in DES), in general the problem is equivalent to the cryptanalysis problem. However, if we assume that the confusion and diffusion operations are linear in subkey and other parameters (as in COMP128), we just need to solve a set of linear Diophantine equations [39].

What we can learn from the above is the following: we need to identify EM-sensitive variables. Other values can be "declassified"; even if we do not take any precautions, we can assume they cannot be observed from an EM perspective. We need to check the flow dependence of the EM-sensitive variables (i.e., externally visible) on secrets that need to be safeguarded.

Recently, work has been done that implies that success or failure of branch prediction presents observable events that can be used to crack encryption keys.

The above suggests the following problems for study:

- Automatic downgrading of "insensitive" variables
- Determination of the minimal declassification to achieve desired flow properties

2.1.4 Obfuscation

Given that static analyses are often hard, some applications use them to good effect. An example is the new area of "obfuscation" of code so that it cannot be reverse-engineered easily. Obfuscation is the attempt to make code "unreadable" or "unintelligible" in the hope that it cannot be used by competitors. This is effected by performing semantic-preserving transformations so that automatic static analysis can reveal nothing useful. In one instance, control flow is intentionally altered so that it is difficult to understand or

---

[5]In general, $l \leq f(a,b) \leq u, \ldots$, with f being an affine function for tractability.

use it by making sure that any analysis that can help in unravelling code is computationally intractable (for example, PSPACE-hard or NP-hard). Another example is intentionally introducing aliases, as certain alias analysis problems are known to be hard (if not undecidable), especially in the interprocedural context. Since it has been shown that obfuscation is, in general, impossible [4], static analysis in principle could be adopted to undo obfuscation unless it is computationally hard.

2.2   Static Analysis of Buffer Overflows

Since the advent of the Morris worm in 1988, buffer overflow techniques to compromise systems have been widely used. Most recently, the SQL slammer worm in 2003, using a small UDP packet (376B), compromised 90% of all target machines worldwide in less than 10 minutes.

Since buffer overflow on a stack can be avoided, for example, by preventing the return address from being overwritten by the "malicious" input string, array bounds checking the input parameters by the callee is one technique. Because of the cost of this check, it is useful to explore compile time approaches that eliminate the check through program analysis. Wagner et al. [56] use static analysis (integer range analysis), but it has false positives due to imprecision in pointer analysis, interprocedural analysis, and so on and a lack of information on dynamically allocated sizes.

CCured [46] uses static analysis to insert runtime checks to create a type-safe version of C program. CCured classifies C pointers into SAFE, SEQ, or WILD pointers. SAFE pointers require only a null check. SEQ pointers require a bounds check, as these are involved in pointer arithmetic, but the pointed object is known statically, while WILD ones require a bounds check as well as a runtime check, as it is not known what type of objects they point to at runtime. For such dynamically typed pointers, we cannot rely on the static type; instead, we need, for example, runtime tags to differentiate pointers from nonpointers.

Ganapathy et al. [23] solve linear programming problems arising out of modeling C string programs as linear programs to identify buffer overruns. Constraints result from buffer declarations, assignments, and function call/returns. C source is first analyzed by a tool that builds a program-dependence graph for each procedure, an interprocedural CFG, ASTs for expressions, along with points-to and side-effect information. A constraint generator then generates four constraints for each pointer to a buffer (between max/min buffer allocation and max/min buffer index used), four constraints on each index assignment (between previous and new values as well as for the highest and lowest values), two for each buffer declaration, and so on. A taint analysis next attempts to identify and remove any uninitialized constraint variables to make it easy for the constraint solvers.

Using LP solvers, the best possible estimate of the number of bytes used and allocated for each buffer in any execution is computed. Based on these values, buffer overruns are inferred. Some false positives are possible because of the flow-insensitive approach followed; these have to be manually resolved. Since infeasible linear programs are possible, they use an algorithm to identify irreducibly inconsistent sets. After such sets of constraints are removed, further processing is done before solvers are employed. This approach also employs techniques to make program analysis context sensitive.

Engler et al. [19] use a "metacompilation" (MC) approach to catch potential security holes. For example, any use of "untrusted input"[6] could be a potential security hole. Since a compiler potentially has information about such input variables, a compiler can statically infer some of the problematic uses and flag them. To avoid hardwiring some of these inferences, the MC approach allows implementers to add rules to the compiler in the form of high-level system-specific checkers. Jaeger et al. [31] use a similar approach to make SELinux aware of two trust levels to make information flow analysis possible; as of now, it is not possible. We will discuss this further in Section 2.4.3.

---

[6]Examples in the Linux kernel code are system call parameters, routines that copy data from user space, and network data.

2.3 Static Analysis of Safety of Mobile Code

The importance of safe executable code embedded in web pages (such as Javascript), applications (as macros in spreadsheets), OS kernel (such as drivers, packet filters [44], and profilers such as DTrace [51]), cell phones, and smartcards is increasing every day. With unsafe code (especially one that is a Trojan), it is possible to get elevated privileges that can ultimately compromise the system. Recently, Google Desktop search [47] could be used to compromise a machine (to make all of its local files available outside, for example) in the presence of a malicious Java applet, as Java allows an applet to connect to its originating host.

The most simple model is the "naive" sandbox model where there are restrictions such as limited access to the local file system for the code, but this is often too restrictive. A better sandbox model is that of executing the code in a virtual machine implemented either as an OS abstraction or as a software isolation layer and using emulation. In the latter solution, the safety property of the programming language and the access checks in the software isolation layer are used to guarantee security.

Since object-oriented (OO) languages such as Java and C# have been designed for making possible "secure" applets, we will consider OO languages here. Checking whether a method has access permissions may not be local. Once we use a programming language with function calls, the call stack has information on the current calling sequence. Depending on this path, a method may or may not have the permissions. Stack inspection can be carried out to protect the callee from the caller by ensuring that the untrusted caller has the right credentials to call a higher-privileged or trusted callee. However, it does not protect the caller from the callee in the case of callback or event-based systems. We need to compute the intersection of permissions of all methods invoked per thread and base access based on this intersection. This protects in both directions.

Static analysis can be carried out to check security loopholes introduced by extensibility in OO languages. Such holes can be introduced through subclassing that overrides methods that check for corner cases important for security. We can detect potential security holes by using a combination of model checking and abstract interpretation: First, compute all the possible execution histories; pushdown systems can be used for representation. Next, use temporal logic to express properties of interest (for example, a method from an applet cannot call a method from another applet). If necessary, use abstract interpretation and model checking to check properties of interest.

Another approach is that of the proof-carrying code (PCC). Here, mobile code is accompanied by a proof that the code follows the security policy. As a detailed description of the above approaches for the safety of mobile code is given in the first edition of this book [36], we will not discuss it here further.

2.4 Static Analysis of Access Control Policies

Lampson [38] introduced access control as a mapping from {entity, resource, op} to {permit, deny} (as commonly used in operating systems). Later models have introduced structure for entities such as roles ("role-based access control") and introduced a noop to handle the ability to model access control modularly by allowing multiple rules to fire: {role, resource, op} to {permit, deny, noop}. Another significant advance is access control with anonymous entities: the subject of trust management, which we discuss in Section 2.4.4.

Starting from the early simple notion, theoretical analysis in the HRU system [29] of access control has the following primitives:

- Create subject s: creates a new row, column in the access control matrix (ACM)
- Create object o: creates a new column in the ACM
- Destroy subject s: deletes a row, column from the ACM
- Destroy object o: deletes a column from the ACM
- Enter r into $A[s, o]$: adds r rights for subject s over object o
- Delete r from $A[s, o]$: removes r rights from subject s over object o

Adding a generic right r where there was not one is "leaking." If a system S, beginning in initial state $s0$, cannot leak right r, it is safe with respect to the right r. With the above primitives, there is no algorithm for determining whether a protection system S with initial state $s0$ is safe with respect to a generic right r. A Turing machine can be simulated by the access control system by the use of the infinite two-dimensional access control matrix to simulate the infinite Turing tape, using the conditions to check the presence of a right to simulate whether a symbol on the Turing tape exists, adding certain rights to keep track of where the end of the corresponding tape is, and so on.

Take grant models [7], in contradistinction to HRU models, are decidable in linear time [7]. Instead of generic analysis, specific graph models of granting and deleting privileges and so on are used. Koch et al. [37] have proposed an approach in which safety is decidable in their graphical model if each graph rule either deletes or adds graph structure but not both. However, the configuration graph is fixed.

Recently, work has been done on understanding and comparing the complexity of discretionary access control (DAC) (Graham–Denning) and HRU models in terms of state transition systems [42]. HRU systems have been shown to be not as expressive as DAC. In the Graham–Denning model, if a subject is deleted, the objects owned are atomically transferred to its parent. In a highly available access control system, however, there is usually more than one parent (a DAG structure rather than a tree), and we need to decide how the "orphaned" objects are to be shared. We need to specify further models (for example, dynamic separation of duty). If a subject is the active entity or leader, further modeling is necessary. The simplest model usually assumes a fixed static alternate leader, but this is inappropriate in many critical designs. The difficulty in handling a more general model is that leader election also requires resources that are subject to access control, but for any access control reconfiguration to take place, authentications and authorizations have to be frozen for a short duration until the reconfiguration is complete. Since leader election itself requires access control decisions, as it requires network, storage, and other resources, we need a special mechanism to keep these outside the purview of the freeze of the access control system. The modeling thus becomes extremely complex. This is an area for investigation.

2.4.1 Case Studies

2.4.1.1 Firewalls

Firewalls are one widely known access control mechanism. A firewall examines each packet that passes through the entry point of a network and decides whether to accept the packet and allow it to proceed or to discard the packet. A firewall is usually designed as a sequence of rules; each rule is of the form <pred> → <decision>, where <pred> is a boolean expression over the different fields of a packet, and the <decision> is either accept or discard. Designing the sequence of rules for a firewall is not an easy task, as it needs to be consistent, complete, and compact. Consistency means the rules are ordered correctly, completeness means every packet satisfies at least one rule in the firewall, and compactness means the firewall has no redundant rules. Gouda and Liu [25] have examined the use of "firewall decision diagrams" for automated analysis and present polynomial algorithms for achieving the above desirable goals.

2.4.1.2 Setuid Analysis

The access control mechanism in Unix-based systems is based critically on the setuid mechanism. This is known to be a source of many privilege escalation attacks if this feature is not used correctly. Since there are many variations of setuid in different Unix versions, the correctness of a particular application using this mechanism is difficult to establish across multiple Unix versions. Static analysis of an application along with the model of the setuid mechanism is one attempt at checking the correctness of an application.

Chen et al. [10] developed a formal model of transitions of the user IDs involved in the setuid mechanism as a finite-state automaton (FSA) and developed techniques for automatic construction of such models. The resulting FSAs are used to uncover problematic uses of the Unix API for uid-setting system calls, to identify differences in the semantics of these calls among various Unix systems, to detect inconsistency in the handling of user IDs within an OS kernel, and to check the proper usage of these calls in programs automatically. As a Unix-based system maintains per-process state (e.g., the real, effective, and saved uids) to track privilege levels, a suitably abstracted FSA (by mapping all user IDs into a single "nonroot" composite ID) can be devised to maintain all such relevant information per state. Each uid-setting system call then leads to a number of possible transitions; FSA transitions are labeled with system calls. Let this FSA be called the setuid-FSA. The application program can also be suitably abstracted and modeled as an FSA (the program FSA) that represents each program point as a state and each statement as a transition. By composing the program FSA with the setuid-FSA, we get a composite FSA. Each state in the composite FSA is a pair of one state from the setuid-FSA (representing a unique combination of the values in the real uid, effective uid, and saved uid) and one state from the program FSA (representing a program point). Using this composite FSA, questions such as the following can be answered:

- Can the setuid system call fail? This is possible if an error state in the setuid-FSA part in a composite state can be reached.
- Can the program fail to drop the privilege? This is possible if a composite state can be reached that has a privileged setuid-FSA state, but the program state should be unprivileged at that program point.
- Which parts of an application run at elevated privileges? By examining all the reachable composite states, this question can be answered easily.

2.4.2 Dynamic Access Control

Recent models of access control are declarative, using rules that encode the traditional matrix model. An access request is evaluated using the rules to decide whether access is to be provided or not. It also helps to separate access control policies from business logic.

Dougherty et al. [18] use Datalog to specify access control policies. At any point in the evolution of the system, facts ("ground terms") interact with the policies ("datalog rules"); the resulting set of deductions is a fixpoint that can be used to answer queries whether an access is to be allowed or not. In many systems, there is also a temporal component to access control decisions. Once an event happens (e.g., a paper is assigned to reviewer), certain accesses get revoked (e.g., the reviewer cannot see the reviews of other reviewers of the same paper until he has submitted his own) or allowed. We can therefore construct a transition system with edges being events that have a bearing on the access control decisions. The goal of analysis is now either safety or availability (a form of liveness): namely, is there some accessible state in the dynamic access model that satisfies some boolean expression over policy facts? These questions can be answered efficiently, as any fixed Datalog query can be computed in polynomial time in the size of the database, and the result of any fixed conjunctive query over a database Q can be computed in space $O(\log|Q|)$ [18, 54].

Analysis of access control by abstract interpretation is another approach. Given a language for access control, we can model leakage of a right as an abstract interpretation problem. Consider a simple language with assignments, conditionals, and sequence (";"). If A is a user, let $[A, stmt]s$ represent whether A can execute *stmt* in state s. Let $r(q,A)s$ mean that A can read q in state s and $w(p,A)s$ means A can write p in state s. Then we have the following interpretation:

$$[A, p = q]s = r(q,A)s \text{ and } w(p,A)s$$
$$[A, \text{if } c \text{ then } p \text{ else } q]s = r(c,A)s \text{ and } ((c \text{ and } [A,p]s) \text{ or } (\text{not } c \text{ and } [A,q]s))$$
$$[A, (a; b)]s = [A, a]s \text{ and } [A, b]s'$$

where s' is the new state after executing a.

Here, A can be a set of users also. Next, if $[A, prog](startstate) = 1$, then A can execute *prog*. The access control problem now becomes: Is there a program P that A can execute and that computes the value of some, forbidden, value and writes it to a location that A can access? With HRU-type models, the set of programs to be examined is essentially unbounded, and we have undecidability. However, if we restrict the programs to be finite, decidability is possible.

It is also possible to model dynamic access control by other methods such as using pushdown systems [34], graph grammars [5], but we will only discuss the access control problem on the Internet that can be modeled using pushdown systems.

2.4.3   Retrofitting Simple MAC Models for SELinux

Since SELinux is easily available, we will use it as an example for discussing access control. In the past, the lack of such operating systems made research difficult; they were either classified or very expensive.

Any machine hosting some services on the net should not get totally compromised if there is a break-in. Can we isolate the breach to those services and not let it affect the rest of the system? It is possible to do so if we can use mandatory access policies (MACs) rather than the standard DACs. In a MAC the system decides how you share your objects, whereas in a DAC you can decide how you share your objects. A break-in into a DAC system has the potential to usurp the entire machine, whereas in a MAC system the kernel or the system still validates each access according to a policy loaded beforehand.

For example, in some recent Linux systems (e.g., Fedora Core 5/6, which is based on SELinux) that employ MAC, there is a "targeted" policy where every access to a resource is allowed implicitly, but deny rules can be used to prevent accesses. By default, most processes run in an "unconfined" domain, but certain daemons or processes[7] (targeted ones) run in "locked-down" domains after starting out as unconfined. If cracker breaks into apache and gets a shell account, it can run only with the privileges of the locked-down daemon, and the rest of the system is usually safe. The rest of the system is not safe only if there is a way to effect a transition into the unconfined domain. With the more stringent "strict policy," also available in Fedora Core 5/6, that implicitly denies everything and "allow" rules are used to enable accesses, it is even more difficult.

Every subject (process) and object (e.g., file, socket, IPC object, etc.) has a security context that is interpreted only by a security server. Policy enforcement code typically handles security identifiers (SIDs); SIDs are nonpersistent and local identifiers. SELinux implements a combination of:

- Type enforcement and (optional) multilevel security: Typed models have been shown to be more tractable for analysis. Type enforcement requires that the type of domains and objects be respected when making transitions to other domains or when acting on objects of a certain type. It also offers some preliminary support for models that have information at different levels of security. The bulk of the rules in most policies in SELinux are for type enforcement.
- Role-based access control (RBAC): Roles for processes. Specifies domains that can be entered by each role and specifies roles that are authorized for each user with an initial domain associated with each user role. It has the ease of management of RBAC with fine granularity of type enforcement.

The security policy is specified through a set of configuration files.

Overt flows transfer data directly; these are often high bandwidth and easily controllable by a policy. Covert flows are indirect (e.g., file existence or CPU usage); these are often low bandwidth and difficult to control in SELinux. Examples of overt flows are:

- Direct information flow: For example, `allow subject_t object_t:file write`. Here, the domain `subject_t` is being given the permission to write to a file of type `object_t`.
- Transitive information flow: For example, `allow subject_a_t object_t:file write; allow subject_b_t object_t:file read`. Here, one domain is writing to an object of type file that is being read by another domain.

---

[7]httpd, dhcpd, mailman, mysqld, named, nscd, ntpd, portmap, postgresql, squid, syslogd, winbind, and snmpd.

However, one downside is that very fine level control is needed. Every major component such as NFS or X needs extensive work on what permissions need to be given before it can do its job. As the default assumption is "deny," there could be as many as 30,000 allow rules. There is a critical need for automated analysis. If the rules are too lax, we can have a security problem. If we have too few rules (too strict), a program can fail at runtime, as it does not have enough permissions to carry out its job. Just as in software testing, we need to do code coverage analysis. In the simplest case, without alias analysis or interprocedural analysis, it is possible to look at the static code and decide what objects are needed to be accessed. Assuming that all the paths are possible, one can use abstract interpretation or program slicing to determine the needed rules. However, these rules will necessarily be conservative. Without proper alias analysis in the presence of aliasing, we will have to be even more imprecise; similar is the case in the context of interprocedural analysis.

Ganapathy et al. [22] discuss automated authorization policy enforcement for user-space servers and the Linux kernel. Here, legacy code is retrofitted with calls to a reference monitor that checks permissions before granting access (MAC). For example, information "cut" from a sensitive window in an X server should not be allowed to be "pasted" into an ordinary one. Since manual placing of these calls is error prone, an automated analysis based on program analysis is useful. First, security-sensitive operations to be checked ("MACed") are identified. Next, for each such operation, the code-level constructs that must be executed are identified by a static analysis as a conjunction of several code-level patterns in terms of their ASTs. Next, locations where these constructs are potentially performed have to be located and, where possible, the "subject" and "object" identified. Next the server or kernel is instrumented with calls to a reference monitor with subject, object, and op triple as the argument, with a jump to the normal code on success or with call to a code that handles the failure case.

We now discuss the static analysis for automatic placement of authorization hooks, given, for example, the kernel code and the reference monitor code [21]. Assuming no recursion, the call graph of the reference monitor code is constructed. For each node in the call graph, a summary is produced. A summary of a function is the set of $(pred, op)$ pairs that denotes the condition under which op can be authorized by the function. For computing the summary, a flow and context-sensitive analysis is used that propagates a predicate through the statements of the function. For example, at a conditional statement with q as the condition, the "if" part is analyzed with $pred \wedge q$, and the "then" part by $pred \wedge \neg q$. At a call site, each pair in the summary of the function is substituted with the actuals of the call, and the propagation of the predicate continues. When it terminates, we have a set of pairs as summary. Another static analysis on the kernel source recovers the set of conceptual operations that may be performed by each kernel function. This is done by searching for combinations of code patterns in each kernel function. For each kernel function, it then searches through a set of idioms for these code patterns to determine if the function performs a conceptual operation; an idiom is a rule that relates a combination of code patterns to conceptual operations.

Once the summary of each function h in the reference monitor code and the set of conceptual operations (S) for each kernel function k is available, finding the set of functions h_i in the monitor code that guards k reduces to finding a cover for the set S using the summary of functions h_i.

Another tractable approach is for a less granular model but finer than non-MAC systems. For example, it is typically the case in a large system that there are definitely forbidden accesses and allowable accesses but also many "gray" areas. "Conflicting access control subspaces" [32] result if assignments of permissions and constraints that prohibit access to a subject or a role conflict. Analyzing these conflicts and resolving them, an iterative procedure, will result in a workable model.

2.4.4 Model Checking Access Control on the Internet: Trust Management

Access control is based on identity. However, on the Internet, there is usually no relationship between requestor and provider prior to request (though cookies are one mechanism used). When users are unknown, we need third-party input so that trust, delegation, and public keys can be negotiated. With public key cryptography, it becomes possible to deal with anonymous users as long as they have a public key: authentication/authorization is now possible with models such as SPKI/SDSI (simple public key infrastructure/simple distributed security infrastructure) [15] or trust management. An issuer authorizes specific permissions to specific principals; these credentials can be signed by the issuer to avoid tampering. We can now have credentials (optionally with delegation) with the assumption that locally generated public keys do not collide with other locally generated public keys elsewhere. This allows us to exploit local namespaces: any local resource controlled by a principal can be given access permissions to others by signing this grant of permission using the public key.

We can now combine access control and cryptography into a larger framework with logics for authentication/authorization and access control. For example, an authorization certificate (K, S, D, T, V) in SPKI/SDSI can be viewed as an ACL entry, where keys or principals represented by the subject S are given permission, by a principal with public key K, to access a "local" resource T in the domain of the principal with public key K. Here, T is the set of authorizations (operations permitted on T), D is the delegation control (whether S can in turn give permissions to others), and V is the duration during which the certificate is valid.

Name certificates define the names available in an issuer's local namespace, whereas authorization certificates grant authorizations or delegate the ability to grant authorizations. A certificate chain provides proof that a client's public key is one of the keys that has been authorized to access a given resource either directly or transitively, via one or more name definition or authorization delegation steps. A set of SPKI/SDSI name and authorization certificates defines a pushdown system [34], and one can "model check" many of the properties in polynomial time. Queries in SPKI/SDSI [15] can be as follows:

- Authorized access: Given resource R and principal K, is K authorized to access R? Given resource R and name N (not necessarily a principal), is N authorized to access R? Given resource R, what names (not necessarily principals) are authorized to access R?
- Shared access: For two given resources R1 and R2, what principals can access both R1 and R2? For two given principals K1 and K2, what resources can be accessed by both K1 and K2?
- Compromisation assessment: Due (solely) to the presence of maliciously or accidentally issued certificate set C0 ⊂ C, what resources could principal K have gained access to? What principals could have gained access to resource R?
- Expiration vulnerability: If certificate set C0 ⊂ C expires, what resources will principal K be prevented from accessing? What principals will be excluded from accessing resource R?
- Universally guarded access: Is it the case that all authorizations that can be issued for a given resource R must involve a cert signed by principal K? Is it the case that all authorizations that grant a given principal K0 access to some resource must involve a cert signed by K?

Other models of trust management such as RBAC-based trust management (RT) [41] are also possible. The following rules are available in the base model RT[]:

- Simple member: $A.r \rightarrow D$. A asserts that D is a member of A's r role.
- Simple inclusion: $A.r \rightarrow B.r1$. This is delegation from A to B.

The model RT[∩] adds to RT[] the following intersection inclusion rule: $A.r \rightarrow B1.r1$ *and* $B2.r2$. This adds partial delegations from A to B1 and to B2. The model RT[⇐] adds to RT[] the following linking inclusion rule: $A.r \rightarrow A.r1.r2$. This adds delegation from A to all the members of the role $A.r1$. RT[∩, ⇐] is all of the above four rules. The kinds of questions we would like to ask are:

- Simple safety (existential): Does a principal have access to some resource in some reachable state?
- Simple availability: In every state, does some principal have access to some resource?
- Bounded safety: In every state, is the number of principals that have access to some resource bounded?
- Liveness (existential): Is there a reachable state in which no principal has access to a given resource?

- Mutual exclusion: In every reachable state, are two given properties (or two given resources) mutually exclusive (i.e., no principal has both properties [or access to both resources] at the same time)?
- Containment: In every reachable state, does every principal that has one property (e.g., has access to a resource) also have another property (e.g., is an employee)? Containment can express safety or availability (e.g., by interchanging the two example properties in the previous sentence).

The complexity of queries such as simple safety, simple availability, bounded safety, liveness, and mutual exclusion analysis for $RT[\cap, \Leftarrow]$ is decidable in poly time in size of state. For containment analysis [41], it is P for $RT[]$, coNP-complete for $RT[\cap]$, PSPACE-complete for $RT[\Leftarrow]$, and decidable in coNEXP for $RT[\cap, \Leftarrow]$.

However, permission-based trust management cannot easily authorize principals with a certain property. For example [41], to give a 20% discount to students of a particular institute, the bookstore can delegate discount permission to the institute key. The institute has to delegate its key to each student with respect to "bookstore" context; this can be too much burden on the institute. Alternatively, the institute can create a new group key for students and delegate it to each student key, but this requires that the institute create a key for each meaningful group; this is also too much burden. One answer to this problem is an attribute-based approach, which combines RBAC and trust management.

The requirements in an attribute-based system [40] are decentralization, provision of delegation of attribute authority, inference, attribute-based delegation of attribute authority, conjunction of attributes, attributes with fields (expiry, age, etc.) with the desirable features of expressive power, declarative semantics, and tractable compliance checking. Logic programming languages such as Prolog or, better, Datalog can be used for a delegation logic for ABAC: this combines logic programming with delegation and possibly with monotonic or nonmonotonic reasoning. With delegation depth and complex principals such as k out of n (static/dynamic) thresholds, many more realistic situations can be addressed.

Related to the idea of attribute-based access control and to allow for better interoperability across administrative boundaries of large systems, an interesting approach is the use of proof-carrying authentication [2]. An access is allowed if a proof can be constructed for an arbitrary access predicate by locating and using pieces of the security policy that have been distributed across arbitrary hosts. It has been implemented as modules that extend a standard web server and web browser to use proof-carrying authorization to control access to web pages. The web browser generates proofs mechanically by iteratively fetching proof components until a proof can be constructed. They provide for iterative authorization, by which a server can require a browser to prove a series of challenges.

2.5 Language-Based Security

As discussed earlier, current operating systems are much bigger than current compilers, so it is worthwhile to make the compiler part of the TCB rather than an OS. If it is possible to express security policies using a programming language that can be statically analyzed, a compiler as part of a TCB makes eminent sense.

The main goal of language-based security is to check the *noninterference* property, that is, to detect all possible leakages of some sensitive information through computation, timing channels, termination channels, I/O channels, and so on. However, the noninterference property is too restrictive to express security policies, since many programs do *leak* some information. For example, sensitive data after encryption can be leaked to the outside world, which is agreeable with respect to security as long as the encryption is effective. Hence, the noninterference property has to be *relaxed* by some mechanisms like *declassification*.

Note that static approaches cannot quantify the leakage of information, as the focus is on whether a program violated some desired property with respect to information flow. It is possible to use a dynamic approach that quantifies the amount of information leaked by a program as the entropy of the program's outputs as a distribution over the possible values of the secret inputs, with the public inputs held constant [43]. Noninterference has an entropy of 0. Such a quantitative approach will often be more useful and flexible than a strict static analysis approach, except that analysis has to be repeated multiple times for coverage.

One approach to static analysis for language-based security has been to use type inference techniques, which we discuss next.

2.5.1   The Type-Based Approach

Type systems establish safety properties (invariants) that hold throughout the program, whereas noninterference requires that two programs give the same output in spite of different input values for their "low" values. Hence, a noninterference proof can be viewed as a bisimulation. For simpler languages (discussed below), a direct proof is possible, but for languages with advanced features such as concurrency and dynamic memory allocation, noninterference proofs are more complex. Before we proceed to discuss the type-based approach, we will briefly describe the lattice model of information flow.

The lattice model of information flow started with the work of Bell and LaPadula [6] and Denning and Denning [17]. Every program variable has a static security class (or label); the security label of each variable can be global (as in early work) or local for each owner, as in the decentralized label model (DLM) developed for Java in Jif [45].

If x and y are variables, and there is (direct) information flow from x to y, it is permissible iff the label of x is less than that of y. Indirect flows arise from control flow such as if (y=1) then x=1 else x=2. If the label of $x \leq$ the label of y, some information of y flows into x (based on whether x is 1 or 2) and should be disallowed. Similarly, if (y=z) then x=1 else w=2, the *lub* of the levels of y and z should be $\leq glb$ of the levels of x and w. To handle this situation, we can assign a label to the program counter (pc). In the above example, we can assign the label of the *lub* to pc just after evaluating the condition; the condition now that needs to be satisfied is that both the arms of the if should have at least the same level as the pc.

Dynamic labels are also possible. A method may take parameters, and the label of the parameter itself could be an another formal. In addition, array elements could have different labels based on index, and hence an expression could have a dynamic label based on the runtime value of its index.

Checking that the static label of an expression is at least as restrictive as the dynamic label of any value it might produce is now one goal of analysis (preferably static). Similarly, in the absence of declassification, we need to check that the static label of a value is at least as restrictive as the dynamic label of any value that might affect it. Because of the limitations of analysis, static checking may need to use conservative approximations for tractability.

Denning and Denning proposed program certification as a lattice-based static analysis method [17] to verify secure information flow. However, soundness of the analysis was not addressed. Later work such as that of Volpano and colleagues [55] showed that a program is secure if it is "typable," with the "types" being labels from a security lattice. Upward flows are handled through subtyping. In addition to checking correctness of flows, it is possible to use type inference to reduce the need to annotate the security levels by the programmer. Type inference computes the type of any expression or program. By introducing type variables, a program can be checked if it can be typed by solving the constraint equations (inequalities) induced by the program. In general, simple type inference is equivalent to first-order unification, whereas in the context of dependent types it is equivalent to higher-order unification.

For example, consider a simple imperative language with the following syntax [55]:

```
(phrases) p::= e | c
(expr)    e::= x | l | n | e arith e' | e relop e'
(cmds)    c::= e:=e' | c; c' | if e then c else c' |
              while e do c | let var x:=e in c
```

Here, l denotes locations (i.e., program counter values), n integers, x variables, and c constants. The types in this system are types of variables, locations, expressions, and commands; these are given by one of the partially ordered security labels of the security system. A cmd has a type t_{cmd} only if it is guaranteed that every assignment in cmd is made to a variable whose security class is t or higher. The type system for security analysis is as follows (L and T are location and type environments, respectively):

$$L; T \vdash n : t \text{ An integer constant can be typed}$$

$$L; T \vdash x : t \text{ if } T(x) = t$$

$$L; T \vdash l : t \text{ if } L(l) = t$$

$$\frac{L; T \vdash e : t \quad L; T \vdash e' : t}{L; T \vdash e \; relop \; e' : t}$$

$$\frac{L; T \vdash x : t \quad L; T \vdash e' : t}{L; T \vdash x := e : t_{cmd}}$$

$$\frac{L; T \vdash c : t_{cmd} \quad L; T \vdash c' : t_{cmd}}{L; T \vdash c; c' : t_{cmd}}$$

$$\frac{L; T \vdash e : t \quad L; T \vdash c : t \quad L; T \vdash c' : t}{L; T \vdash if \; e \; then \; c \; else \; c' : t}$$

$$\frac{L; T \vdash e : t \quad L; T \vdash c : t}{L; T \vdash while \; e \; do \; c : t}$$

Consider the rules for assignment above. In order for information to flow from e' to e, both have to be at the same security level. However, upward flow is allowed, for secrecy, for example, if e is at a higher level and e' is at a lower level. This is handled by extending the partial order by subtyping and coercion: the low level (derived type) is smaller (for secrecy) in this extended order than the high level (base type). Note that the extended relation has to be contravariant in the types of commands t_{cmd}.

It can be proved [55] that if an expression e can be given a type t in the above type system, then, for secrecy, only variables at level t or lower in e will have their contents read when e is evaluated (no read up). For integrity, every variable in e stores information at integrity level t. If a command has the property that every assignment within c is made to a variable whose security class is at least t, then the confinement property for secrecy says that no variable below level t is updated in c (no write down). For integrity, every variable assigned in c can be updated by a type t variable.

Soundness of the type system induces the noninterference property, that is, a high value cannot influence any lower value (or information does not leak from high values to low values).

Smith and Volpano [53] have studied information flow in multithreaded programs. The above type system does not guarantee noninterference; however, by restricting the label of all the while-loops and its guards to low, the property is restored. Abadi has modeled encryption as declassification [1] and presented the resulting type system.

Myers and colleagues have developed static checking for DLM-based Jif language [12, 16], while Pottier and colleagues [49] have developed OCaml-based FlowCAML. We discuss the Jif approach in some detail.

2.5.2   Java Information Flow (Jif) Language

Jif is a Java-based information flow programming language that adds *static analysis* of information flow for improved security assurance. Jif is mainly based on *static type checking*. Jif also performs some *runtime* information flow checks.

Jif is based on decentralized labels. A *label* in Jif defines the security level, represented by a set of *policy expressions* separated by semicolons. A policy expression {*owner* : *reader*$_1$, *reader*$_2$, ...} means the principal *owner* wants to allow labeled information to flow to at most the principal's *reader*$_i$. Unlike the

MAC model, these labels contain fine-grained policies, which have an advantage of being able to represent *decentralized access control*. These labels are called decentralized labels because they enforce security on behalf of the owning principals, not on behalf of an implicitly centralized policy specifier. The policy $\{o_1 : r_1, r_2; o_2 : r_2, r_3\}$ specifies that both o_1 and o_2 own the information with each allowing either r_1, r_2 or r_2, r_3, respectively. For integrity, another notation is adopted.

Information can flow from label L_1 to label L_2 only if $L_1 \sqsubseteq L_2$ (i.e., L_1 is less restrictive than L_2), where $\sqsubseteq$ defines a preorder on labels in which the equivalence classes form a join semilattice. To label an expression (such as $w + x$), a join operator is defined as the *lub* of the labels of the operands, as it has to have a secrecy as strong as any of them. In the context of control flow, such as if cond then x=... else x=..., we also need the *join* operator. To handle implicit flows through control flow, each program visible location is given an implicit label.

A *principal hierarchy* allows one principal to **actfor** another. This helps in simplifying the policy statements in terms of representation of groups or roles. For example, suppose principal *Alice* **actsfor** *Adm* and principal *Bob* **actsfor** *Adm*; thus, in the following code whatever value *Adm* has is also readable by *Alice* and *Bob*.

```
void examplePrincipalHierarchy(){
    int{Alice:} a;
    int{Bob:} b;
    int{Adm:} c = 10;
    a = c; /* is valid stmt */
    b = c; /* is valid stmt */
}
```

The *declassification* mechanism gives the programmer an explicit escape hatch for releasing information whenever necessary. The *declassification* is basically carried out by relaxing the policies of some labels by principals having sufficient *authority*. For example,

```
void exampleDeclassification() where authority (Alice) {
    int{Alice:} x;
    int{Alice:Bob} y;
    int{Bob:} a;
    int{Bob:Alice} b;
     /*Here PC has label {}*/
    if (x > 10) {
       /*Here PC has label {Alice:}*/
       declassify(y = 25, {Alice:Bob});
          /*stmt ``y = 25" is declassified from {Alice:} to {Alice:Bob}*/
          /*valid because has Alice's authority*/
    }
       /*Here PC has again the label {}*/
       b = declassify(a, {Bob:Alice});
       /*invalid because doesn't have Bob's authority*/
}
```

Jif has label polymorphism. This allows the expression of code that is generic with respect to the security class of the data it manipulates. For example,

```
void exampleLP(int{Alice:;Bob:}i){
    ...
    /*Assures a security level upto {Alice:;Bob:}*/
    }
```

To the above function (which assures a security level up to {*Alice:;Bob:*}) one can pass any integer variable having one of the following labels:

- {}
- {*Alice:Bob*}
- {*Alice:*}
- {*Bob:*}
- {*Alice:;Bob:*}

Jif has *automatic label inference*. This makes it unnecessary to write many *type-annotations*. For example, suppose the following function is called (which can only be called from a program point with label at most {*Alice:;Bob:*}) from a valid program point; "*a*" will get a default label of {*Alice:;Bob:*}.

```
void exampleALI{Alice:;Bob:}(int i){
    int a;
    /* Default label of ``a"' is {Alice:;Bob:}*/
    }
```

Runtime label checking and *first-class label values* in Jif make it possible to discover and define new policies at runtime. Runtime checks are statically checked to ensure that information is not leaked by the success or failure of the runtime check itself. Jif provides a mechanism for comparing runtime labels and also a mechanism for comparing runtime principals. For example,

```
void m(int{Alice:pr} i, principal{} pr){
    int{Alice:Bob} x;
    if (Bob actsfor pr) {
      x = i;
       /* OK, since {Alice:pr} <= {Alice:Bob}*/
    }
    else {
      x = 0;
    }
  }
  void n(int{*lbl} i, label{} lbl){

    int{Alice:} x;
    if (lbl <= new label {Alice:}) {
      x = i;
       /* OK, since {*lbl} <= {Alice:Bob}*/
    }
    else {
      x = 0;
    }
}
```

Note that in the above function **n(...)**, **lbl* represents an actual label held by the variable *lbl*, whereas just *lbl* inside a label represents the label of the variable *lbl* (i.e., {} here).

Labels and principals can be used as first-class values represented at runtime. These *dynamic* labels and principals can be used in the specification of other labels and used as the parameters of parameterized classes. Thus, Jif's type system has *dependent types*. For example,

```
class C[label L] { ... }
...
void m() {
   final label lb = new label {Alice: Bob};
   int{*lb; Bob:} x = 4;
   C[lb] foo = null;
     /*Here ``lb" acts as first-class label*/
   C[{*lb}] bar = foo;
}
```

Note that, unlike in Java, method arguments in Jif are always implicitly *final*. Some of the limitations of Jif are that there is no support for *Java Threads, nested classes, initializer blocks*, or *native methods*.

Interaction of Jif with existing Java classes is possible by generating *Jif signatures* for the interface corresponding to these Java classes.

2.5.3   A Case Study: VSR

The context of this section[8] is the security of archival storage. The central objective usually is to guarantee the availability, integrity, and secrecy of a piece of data *at the same time*. Availability is usually achieved through redundancy, which reduces secrecy (it is sufficient for the adversary to get one copy of the secret). Although the requirements of availability and secrecy seem to be in conflict, an information-theoretic secret sharing protocol was proposed by Shamir in 1979 [51], but this algorithm does not provide data integrity. Loss of shares can be tolerated up to a threshold but not to arbitrary modifications of shares.

A series of improvements have therefore been proposed over time to build secret sharing protocols resistant to many kinds of attacks. The first one takes into account the data-integrity requirement and leads to *verifiable* secret sharing (VSS) algorithms [20, 48]. The next step is to take into account mobile adversaries that can corrupt any number of parties given sufficient time. It is difficult to limit the number of corrupted parties on the large timescales over which archival systems are expected to operate. An adversary can corrupt a party, but redistribution can make that party whole again (in practice, this happens, for example, following a system re-installation). Mobile adversaries can be tackled by means of proactive secret sharing (PSS), wherein redistributions are performed periodically. In one approach, the secret is reconstructed and then redistributed. However, this causes extra vulnerability at the node of reconstruction. Therefore, another approach, redistribution without reconstruction, is used [33]. A combination of VSS and PSS is verifiable secret redistribution (VSR); one such protocol is proposed in [58]. In [26] we proposed an improvement of this protocol, relaxing some of the requirements.

Modeling the above protocol using Jif can help us understand the potential and the difficulties of Jif static analysis. We now discuss the design of a simplified VSR [58] protocol:

- **Global values:** This class contains the following variables that are used for generation and verification of shares and subshares during reconstruction and redistribution phases.

   (m,n): m is the threshold number of servers required for reconstruction of a secret, and n is the total number of servers to which shares are distributed.

   p: The prime used for Z_p; r is the prime used for Z_r.

   g: The Diffie–Hellman exponentiator.

   KeyID: Unique for each secret across all clients.

   ClientID: ID of the owner of the secret.

- **Secret:** This class contains ***secret***'s value (i.e., the secret itself), the ***polynomial*** used for distribution of shares, **N[]** — the array of server IDs to which the shares are distributed.

---

[8]This is joint work with a former student, S. Roopesh.

- **Points2D**: This class contains two values **x** and **f(x)**, where *f* is the polynomial used for generating the shares (or subshares). It is used by *Lagrangian interpolator* to reconstruct the secret (or shares from subshares).
- **Share**: This class contains an immutable *original share* value (used to check whether shares are uncorrupted or not) and a *redistributing polynomial* (used for redistribution of this share to a new access structure).
- **SubShare**: This is the same as the Share class, except that the share value actually contains a *subshare* value and no redistributing polynomial.
- **SubShareBox**: This class keeps track of the subshares from a set of valid servers (i.e., B[] servers) in the redistribution phase. This is maintained by all servers belonging to the new *(m′,n′)-access structure* to which new shares are redistributed.
- **Client**: This class contains zero or more secrets and is responsible for initial distribution of shares and reconstruction of secrets from valid sets of servers.
- **Server**: This class maintains zero or more shares from different clients and is responsible for redistribution of shares after the client (who is the owner of the secret corresponding to this share) has approved the redistribution.
- **Node**: Each node contains two units, one *Client* and the other *Server* (having the same IDs as this node). On message reception from the reliable communication (group communication system [GCS]) interface *GCSInterface*, this node extracts the information inside the message and gives it to either the client or the server, based on the message type.
- **GCSInterface**: This communication interface class is responsible for acting as an interface between underlying reliable messaging systems (e.g., the Ensemble GCS [30] system), user requests, and *Client* and *Server*.
- **UserCommand** (Thread): This handles three types of commands from the user:

  Distribution of the user's secret *S* to *(m,n)-access structure*

  Redistribution from *(m,n)-access structure* to *(m′,n′)-access structure*

  Reconstruction of secret *S*

- **SendMessage** (Thread): This class packetizes the *messages* (from Node) (depends on the communication interface), then either sends or multicasts these packets to the destination node(s).
- **Attacker**: This class is responsible for attacking the servers, getting their shares, and corrupting all their valid shares (i.e., changing the *y* values in the *Points2D* class to arbitrary values). This class also keeps all valid shares it got by attacking the servers. It also reconstructs all possible secrets from the shares collected. Without loss of generality, we assume that at most one attacker can be running in the whole system.
- **AttackerNode**: This is similar to *Node*, but instead of *Client* and *Server* instances, it contains only one instance of the *Attacker* class.
- **AttackerGCSInterface**: Similar to the *GCSInterface* class.
- **AttackerUserCommand**: This handles two types of commands from the attacker:

  Attack server S_i

  Construct all possible secrets from collected valid shares

We do not dwell on some internal bookkeeping details during redistribution and reconstruction phases. Figure 2.1 gives the three phases of the simplified VSR protocol. The distribution and reconstruction phases almost remain the same. Only the redistribution phase is slightly modified, where the client acts as the manager of the redistribution process. The following additional assumptions are also made:

- There are no "**Abort**" or "**Commit**" messages from servers.
- In the *redistribution phase*, the *client*, who is the owner of the *secret* corresponding to this *redistribution process*, will send the "**commit**" messages instead of the redistributing servers.

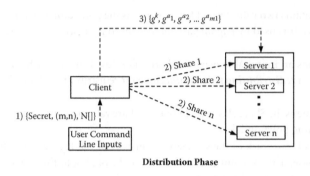

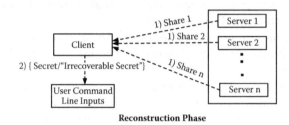

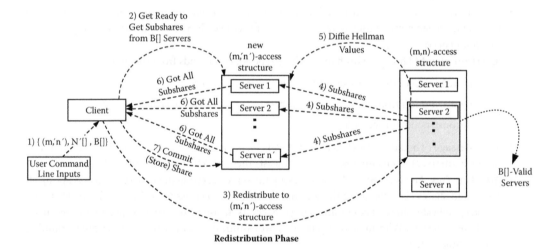

FIGURE 2.1   Simplified VSR protocol.

- Attacker is restricted to only attacking the servers and thereby getting all the original shares, which are held by the servers, and corrupting them.
- There are no "**reply** and **DoS** attacks" messages.

2.5.3.1 Jif Analysis of Simplified VSR

In this section, we discuss an attempt to do a Jif analysis on a simplified VSR implementation and its difficulties. As shown in Figure 2.2, every *Node* (including the *AttackerNode*) runs with "**root**" *authority* (who is above all and can *actfor* all principals). Every message from and out of the network will have an *empty label* (as we rely on underlying Java Ensemble for *cryptographically* perfect end-to-end and multicast communication). The **root** (i.e., *Node*) receives the message from the network. Based on the message type,

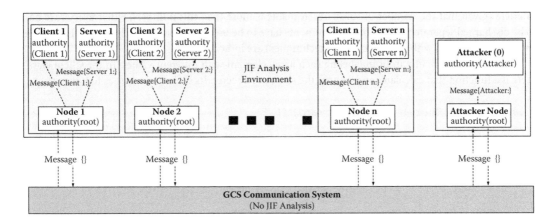

FIGURE 2.2   Jif analysis on VSR.

root will appropriately (*classify* or) *declassify* the contents of the message and handles them by giving them to either the *client* or *server*. Similarly, all outgoing messages from *Client/Server* to communication network would be properly *declassified*.

First, the communication interface (Java Ensemble) uses some *native* methods to contact **ensemble-server**.[9] To do *asynchronous* communication we use *threads* in our VSR implementation. Since Jif does not support **Java Threads** and **native** methods, Jif analysis cannot be done at this level. Hence, we have to restrict the Jif analysis above the communication layer.[10]

Next, let us proceed to do static analysis on the remaining part using the Jif compiler. Consider the *Attacker Node* (see Figure 2.2). The attacker node (running with **root** authority) classifies all content going to the principal Attacker with the label-{$Attacker$:}. If the attacker has compromised some server, it would get all the valid shares belonging to that server. Hence, all shares of a compromised server output from the communication interface (GCS) go to the attacker node first. Since it is running with **root** authority and sees that shares are semi-sensitive, it (de)classifies these shares to {Server:}. For these shares to be read by the attacker, the following property should hold: {**Server:**} ⊑ {**Attacker:**}.

However, since the Attacker principal does not *actsfor* this Server (principal), the above relation does not hold, so the attacker cannot read the shares. The Jif compiler detected this as an error. However, if we add an *actsfor* relation from Attacker (principal) to this Server (principal) (i.e., *Attacker actsfor Server*), it could read the *shares*. In this case, the Jif compiler does not report any error, implying that the information flow from {Server:} to {Attacker:} is valid (since {Server:} ⊑ {Attacker:} condition holds).

VSR has a threshold property: if the number of shares (or subshares) is more than the threshold, the secret can be reconstituted. This implies that any computation knowing less than the threshold number of shares cannot interfere with any computation that has access to more than this threshold. We need a *threshold set noninterference* property. This implies that we can at most declassify less than any set of a threshold number of shares. Since such notions are not expressible in Jif, even if mathematically proved as in [58], more powerful dependent-type systems have to be used in the analysis. Since modern type inference is based on explicit constraint generation and solving, it is possible to model the threshold property through constraints, but solving such constraints will have a high complexity. Fundamentally, we have to express the property that any combination of subshares less than the threshold number cannot

---

[9]*ensemble-server* is a daemon serving group communication.

[10]Because of this separation, we encounter a problem with the Jif analysis even if we want to abstract out the lower GCS layer. Some of the classes (such as the *Message* class) are common to both upper and lower layers, but they are compiled by two different compilers (Jif and Java) that turn out to have incompatible *.class* files for the common classes.

interfere (given that the number of subshares available is more than this number); either every case has to be discharged separately or symmetry arguments have to be used. Approaches such as model checking (possibly augmented with symmetry-based techniques) are indicated.

There is also a difficulty in Jif in labeling each array element with different values. In the Client class, there is a function that calculates and returns the shares of a secret. Its skeleton code is as follows:

```
1 Share[] getShares(int[] ServerID){

2      Share{Client:}[]{Client:} shares = new Share{Client:}[]{Client:};
3      int n = ServerIDs.length;
4      for (int i = 0 ; i < n ; i++){
5        shares[i] = {Client:Server_{ServerID[i]}} polynomial(ServerID[i]);
6      }
7      return shares;
8 }
```

In line 2, "*shares*" is an array of type "*Share*" with a label of {*Client:*} for both array variable and individual elements of this array. Line 5 calculates different shares for different servers based on ServerIDs and assigns a different label of {*Client: Server*$_{ServerID[i]}$} for each of these shares. However, Jif does not allow for different labels for different elements of an array; it only allows a common label for all elements of an array. If this is possible, we should ensure that accessing some array element does not itself leak some information about the index value.

There are currently many other difficulties in using Jif to do analysis such as the need to recode a program in Jif. The basic problem is the (post hoc) analysis after an algorithm has already been designed. What is likely to be more useful[11] is an explicit security policy before the code is developed [60] and then use of Jif or a similar language to express and check these properties where possible. McCamant and Ernst [43] argue that Jif-type static analysis is still relatively rare and no "large" programs have yet been ported to Jif or FlowCaml. They instead use a fine-grained dynamic bit-tracking analysis to measure the information revealed during a particular execution.

2.6   Future Work

Current frameworks such as Jif have not been sufficiently developed. We foresee the following evolution:

- The analysis in Section 2.5.1 assumed that typing can reveal interesting properties. This needs to be extended to a more general static analysis that deals with values in addition to types as well as when the language is extended to include arrays and so on. Essentially, the analysis should be able to handle array types with affine index functions. This goes beyond the analysis possible with type analysis.
- Incorporate pointers and heaps in the analysis. Use dataflow analysis on lattices but with complex lattice values (such as regular expressions, etc.) or use shape analysis techniques.
- Integrate static analysis (such as abstract interpretation and compiler dataflow analysis) with model checking to answer questions such as: What is the least/most privilege a variable should have to satisfy some constraint? This may be coupled with techniques such as counterexample guided abstraction refinement [13]. This, however, requires considerable machinery. Consider, for example, a fully developed system such as the SLAM software model checking [3]. It uses counterexample-driven

---

[11]In the verification area, it is increasingly becoming clear that checking the correctness of finished code is an order of magnitude more difficult than intervening in the early phases of the design.

abstraction of software through "boolean" programs (or abstracted programs) using model creation (c2bp), model checking (bebop), and model refinement (newton). SLAM builds on the OCaml programming language and uses dataflow and pointer analysis, predicate abstraction and symbolic model checking, and tools such as CUDD, a SAT solver, and an SMT theorem prover [3]. Many such analyses have to be developed for static analysis of security properties.

References

1. Martn Abadi. Secrecy by Typing in Security Protocols. TACS 1997.
2. Andrew W. Appel and Edward W. Felten. 1999. Proof-carrying authentication. *Proceedings of the 6th ACM Conference on Computer and Communications Security*, 52–62. New York: ACM Press.
3. Thomas Ball and Sriram K. Rajamani. 2002. The SLAM project: debugging system software via static analysis. In *Proceedings of the 29th ACM SIGPLAN-SIGACT Symposium on Principles of Programming Languages*, 1–3.
4. B. Barak, O. Goldreich, R. Impagliazzo, S. Rudich, A. Sahai, S. Vadhan, and K. Yang. 2001. On the (im)possibility of obfuscating programs. In *Advances in cryptology (CRYPTO'01)*. Vol. 2139 of *Lecture notes in computer science*, 1–18, New York: Springer.
5. Jorg Bauer, Ina Schaefer, Tobe Toben, and Bernd Westphal. 2006. Specification and verification of dynamic communication systems. In *Proceedings of the Sixth International Conference on Application of Concurrency to System Design*, 189–200. Washington, DC: IEEE Computer Society.
6. D. Elliott Bell and Leonard J. LaPadula. 1973. *Secure computer systems: Mathematical foundations.* MITRE Corporation.
7. Matt Bishop. 2002. *Computer security: Art and science.* Reading, MA: Addison-Wesley.
8. R. Canetti. 2001. Universally composable security: A new paradigm for cryptographic protocols. 42nd FOCS, 2001. Revised version (2005). Online: http://eprint.iacr.org/2000/067
9. Bryan M. Cantrill, Michael W. Shapiro, and Adam H. Leventhal. 2004. Dynamic instrumentation of production systems. In *Proceedings of the 2004 USENIX Annual Technical Conference*.
10. Hao Chen, David Wagner, and Drew Dean. 2002. Setuid demystified. In *Proceedings of the 11th USENIX Security Symposium*, 171–90. Berkeley, CA: USENIX Assoc.
11. David M. Chess and Steve R. White. 2000. An undetectable computer virus. Virus Bulletin Conference.
12. S. Chong, A. C. Myers, K. Vikram, and L. Zheng. Jif reference manual. http://www.cs.cornell.edu/jif/doc/jif-3.0.0/manual.html.
13. Edmund M. Clarke, Orna Grumberg, Somesh Jha, Yuan Lu, and Helmut Veith. 2000. Counterexample-guided abstraction refinement. In *Proceedings of the 12th International Conference on Computer Aided Verification*, 154–69. London: Springer Verlag.
14. Edmund M. Clarke, Orna Grumberg, and Doron A. Peled. 2000. *Model checking.* Cambridge, MA: MIT Press.
15. P. Clarke, J. E. Elien, C. Ellison, M. Fridette, A. Morcos, and R. L. Rivest. 2001. Certificate chain discovery in SPKI? SDSI. *J. Comp. Security* 9:285–332.
16. Fred Cohen. 1987. Computer viruses: Theory and experiments. *Comp. Security* 6, Vol 6(1), pp. 22–35.
17. Dorothy E. Denning and Peter J. Denning. 1977. Certification of programs for secure information flow. *Commun. ACM* 20(7):504–13.
18. D. Dougherty, K. Fisler, and S. Krishnamurthi. 2006. Specifying and reasoning about dynamic access-control policies. International Joint Conference on Automated Reasoning (IJCAR), August 2006.
19. Dawson Engler, Benjamin Chelf, Andy Chou, and Seth Hallem. Checking system rules using system-specific, programmer-written compiler extensions. In *Proceedings of the 4th Symposium on Operating System Design and Implementation*, San Diego, CA, October 2000.
20. P. Feldman. 1987. A practical scheme for non-interactive verifiable secret sharing. In *Proceedings of the 28th IEEE Annual Symposium on Foundations of Computer Science*, 427–37.

21. Vinod Ganapathy, Trent Jaeger, and Somesh Jha. 2005. Automatic placement of authorization hooks in the Linux security modules framework. In *Proceedings of the 12th ACM Conference on Computer and Communications Security*, 330–39. New York: ACM Press.

22. Vinod Ganapathy, Trent Jaeger, and Somesh Jha. 2006. Retrofitting legacy code for authorization policy enforcement. *Proceedings of the 2006 Symposium on Security and Privacy*, 214–29. Washington, DC: IEEE Computer Society.

23. V. Ganapathy, S. Jha, D. Chandler, D. Melski, and D. Vitek. 2003. Buffer overrun detection using linear programming and static analysis. In *Proceedings of the 10th ACM Conference on Computer and Communications Security (CCS)*, 345–54. New York: ACM Press.

24. Joseph A. Goguen and Jos Meseguer. 1982. Security policies and security models. In *Proceedings of the 1982 IEEE Symposium on Security and Privacy*, 11–20. Los Alamitos, CA: CS Press.

25. Mohamed G. Gouda and Alex X. Liu. 2004. Firewall design: Consistency, completeness, and compactness. In *Proceedings of the 24th International Conference on Distributed Computing Systems*, 320–27. Washington, DC: IEEE Computer Society.

26. V. H. Gupta and K. Gopinath. 2006. An extended verifiable secret redistribution protocol for archival systems. In *International Conference on Availability, Reliability and Security*, 100–107. Los Alamitos, CA: IEEE Computer Society.

27. Joshua D. Guttman, Amy L. Herzog, and John D. Ramsdel. 2003. Information flow in operating Systems: Eager formal methods. WITS2003.

28. K. W. Hamlen, Greg Morrisett, and F. B. Schneider. 2005. Computability classes for enforcement mechanisms. *ACM TOPLAS* 28:175–205.

29. M. A. Harrison, W. L. Ruzzo, and J. D. Ullman. 1976. Protection in operating systems. *Commun. ACM* 19:461–71.

30. Mark Hayden and Ohad Rodeh. 2004. Ensemble reference manual. Available online: http://dsl.cs. technion.ac.il/projects/Ensemble/doc/ref.pdf

31. Trent Jaeger, Reiner Sailer, and Xiaolan Zhang. 2003. Analyzing integrity protection in the SELinux example policy. In *Proceedings of the 11th USENIX Security Symposium*. USENIX.

32. Trent Jaeger, Xiaolan Zhang, and Fidel Cacheda. 2003. Policy management using access control spaces. *ACM Trans. Inf. Syst. Security*. 6(3):327–64.

33. Sushil Jajodia and Yvo Desmedt. 1997. Redistributing secret shares to new access structures and its applications. Tech. Rep. ISSE TR-97-01, George Mason University.

34. S. Jha and T. Reps. 2002. Analysis of SPKI/SDSI certificates using model checking. In *Proceedings of the 15th IEEE Workshop on Computer Security Foundations*, 129. Washington, DC: IEEE Computer Society.

35. Jif: Java + information flow. www.cs.cornell.edu/jif/.

36. R. B. Keskar and R. Venugopal. 2002. Compiling safe mobile code. In *The compiler design handbook*. Boca Raton, FL: CRC Press.

37. M. Koch, L. V. Mancini, and F. Parisi-Presicce. 2002. Decidability of safety in graph-based models for access control. In *Proceedings of the 7th European Symposium on Research in Computer Security*, 229–43. London: Springer-Verlag.

38. Butler W. Lampson. 1973. A note on the confinement problem. *Commun. ACM* 16(10):613–15.

39. Felix Lazebnik. 1996. On systems of linear Diophantine equations. *Math. Mag.* 69(4).

40. N. Li, B. N. Grosof, and J. Feigenbaum. 2003. Delegation logic: A logic-based approach to distributed authorization. *ACM Trans. Inf. Syst. Security* 6:128–171.

41. Ninghui Li, John C. Mitchell, and William H. Winsborough. 2005. Beyond proof-of-compliance: Security analysis in trust management. *J. ACM* 52:474–514.

42. Ninghui Li and Mahesh V. Tripunitara. 2005. Safety in discretionary access control. In *Proceedings of the 2005 IEEE Symposium on Security and Privacy*, 96–109. Washington, DC: IEEE Computer Society.

43. Stephen McCamant and Michael D. Ernst. 2006. Quantitative information-flow tracking for C and related languages. MIT Computer Science and Artificial Intelligence Laboratory Tech. Rep. MIT-CSAIL-TB-2006-076, Cambridge, MA.

44. Steven McCanne and Van Jacobson. 1993. The BSD packet filter: A new architecture for user-level packet capture. In *Proceedings of the Winter Usenix Technical Conference*, 259–69.

45. Andrew C. Meyers. 1999. Mostly-static decentralized information flow control. MIT Laboratory for Computer Science, Cambridge, MA.

46. George C. Necula, Scott McPeak, and Westley Weimer. 2002. CCured: Type-safe retrofitting of legacy code. In *Annual Symposium on Principles of Programming Languages*, 128–39. New York: ACM Press.

47. Seth Nielson, Seth J. Fogarty, and Dan S. Wallach. 2004. Attacks on Local Searching Tools. Tech. Rep. TR04-445, Department of Computer Science, Rice University, Honston, TX.

48. T. P. Pedersen. 1991. Non-interactive and information theoretic secure verifiable secret sharing. In *Proceedings of CRYPTO 1991, the 11th Annual International Cryptology Conference*, 129–40. London: Springer Verlag.

49. Franois Pottier and Vincent Simonet. 2003. Information flow inference for ML. *ACM Transac. Programming Languages Syst.* 25(1):117–58.

50. J. Rao, P. Rohatgi, H. Scherzer, and S. Tinguely. 2002. Partitioning attacks: Or how to rapidly clone some GSM cards. In *Proceedings of the 2002 IEEE Symposium on Security and Privacy*, 31. Washington, DC: IEEE Computer Society.

51. A. Shamir. 1979. How to share a secret. *Commun. ACM*, 22(11):612–13.

52. Geoffrey Smith and Dennis M. Volpano. 1998. Secure information flow in a multi-threaded imperative language. In *Proceedings of the 25th ACM SIGPLAN-SIGACT Symposium on Principles of Programming Languages*, 355–64. New York: ACM Press.

53. Ken Thompson. 1984. Reflections on trusting trust. *Commun. ACM*, 27(8): 761–63.

54. M. Y. Vardi. 1982. The complexity of relational query languages (extended abstract). In *Proceedings of the 14th Annual Symposium on the Theory of Computing*, 137–46. New York: ACM Press.

55. Dennis M. Volpano, Cynthia E. Irvine, and Geoffrey Smith. 1996. A sound type system for secure flow analysis. *J. Comput. Security* 4(2/3):167–88.

56. D. Wagner, J. Foster, E. Brewer, and A. Aiken. 2000. A first step towards automated detection of buffer overrun vulnerabilities. In *Symposium on Network and Distributed Systems Security* (NDSS00), February 2000, San Diego, CA.

57. Wlkipedia. Trusted Computer System evaluation criteria. http://en.wikipedia.org/wiki/Trusted_Computer_System_Evaluation.

58. Theodore M. Wong, Chenxi Wang, and Jeannette M. Wing. 2002. Verifiable secret redistribution for archival systems. In *Proceedings of the First IEEE Security in Storage Workshop*, 94. Washington, DC: IEEE Computer Society.

59. Steve Zdancewic and Andrew C. Myers. 2001. Robust declassification. In *Proceedings of the 2001 IEEE Computer Security Foundations Workshop*.

60. Steve Zdancewic, Lantian Zheng, Nathaniel Nystrom, and Andrew C. Myers. 2001. Untrusted hosts and confidentiality: Secure program partitioning. In *ACM Symposium on Operating Systems Principles*. New York: ACM Press.

3

Compiler-Aided Design of Embedded Computers

Aviral Shrivastava

Department of Computer Science and Engineering,
School of Computing and Informatics,
Arizona State University, Tempe, AZ
Aviral.Shrivastava@asu.edu

Nikil Dutt

Center for Embedded Computer Systems,
Donald Bren School of Informatics and Computer Sciences,
University of California, Irvine, CA
dutt@uci.edu

3.1   Introduction

Embedded systems are computing platforms that are used inside a product whose main function is different than general-purpose computing. Cell phones and multipoint fuel injection systems in cars are examples of embedded systems. Embedded systems are characterized by application-specific and multidimensional design constraints. While decreasing time to market and the need for frequent upgrades are pushing embedded system designs toward programmable implementations, these stringent requirements demand that designs be *highly customized*. To customize embedded systems, standard design features of general-purpose processors are often omitted, and several new features are introduced in embedded systems to meet all the design constraints simultaneously.

Consequently, software development for embedded systems has become a very challenging task. Traditionally humans used to code for embedded systems directly in assembly language, but now with the software content reaching multimillion lines of code and increasing at the rate of 100 times every decade, compilers have the onus of generating code for embedded systems. With embedded system designs still being manually customized, compilers have a dual responsibility: first to exploit the novel architectural features in the embedded systems, and second to avoid the loss due to missing standard architectural features. Existing compiler technology falls tragically short of these goals.

While the task of the compiler is challenging in embedded systems, it has been shown time and again that whenever possible, a compiler can have a very significant impact on the power, performance, and so on of the embedded system. Given that the compiler can have a very significant impact on the design constraints of embedded systems. Consequently, it is only logical to include the compiler during the design of the embedded system. Existing embedded system design techniques do not include the compiler during the design space exploration. While it is possible to use ad hoc methods to include the compiler's effects during the design of a processor, a systematic methodology to perform compiler-aware embedded systems design is needed. Such design techniques are called *compiler-aided* design techniques.

This chapter introduces our *compiler-in-the-loop* (CIL) design methodology, which systematically includes compiler effects to design embedded processors. The core capability in this methodology is a design space exploration (DSE) compiler. A DSE compiler is different from a normal compiler in that a DSE compiler has heuristics that are parameterized on the architectural parameters of the processor architecture. While typical compilers are built for one microarchitecture, a DSE compiler can generate good-quality code for a range of architectures. A DSE compiler takes the architecture description of the processor as an input, along with the application source code, and generates an optimized executable of the application for the architecture described.

The rest of the chapter is organized as follows. In Section 3.2, we describe our whole approach of using a compiler for processor design. In particular, we attempt to design the popular architectural feature, in embedded processors, called horizontally partitioned cache (HPC), using our CIL design methodology. Processors with HPC have two caches at the same level of memory hierarchy, and wisely partitioning the data between the two caches can achieve significant energy savings. Since there is no existing effective compiler technique to achieve energy reduction using HPCs, in Section 3.4, we first develop a compiler technique to partition data for HPC architectures to achieve energy reduction. The compilation technique is generic, in the sense that it is not for specific HPC parameters but works well across HPC parameters. Being armed with a parametric compilation technique for HPCs, Section 3.5 embarks upon the quest designs an embedded processor by choosing the HPC parameters using inputs from the compiler.

Finally, Section 3.6 summarizes this chapter.

3.2   Compiler-Aided Design of Embedded Systems

The fundamental difference between an embedded system and a general-purpose computer system is in the usage of the system. An embedded system is very *application specific*. Typically a set of applications are installed on an embedded system, and the embedded system continues to execute those applications throughout its lifetime, while general-purpose computing systems are designed to be much more flexible to allow and enable rapid evolution in the application set. For example, the multipoint fuel injection systems in automobiles are controlled by embedded systems, which are manufactured and installed when the car is made. Throughout the life of the car, the embedded system performs no other task than controlling the multipoint fuel injection into the engine. In contrast, a general-purpose computer performs a variety of tasks that change very frequently. We continuously install new games, word processing software, text editing software, movie players, simulation tools, and so on, on our desktop PCs. With the popularity of automatic updating features in PCs, upgrading has become more frequent than ever before. It is the application-specific nature of embedded systems that allows us to perform more aggressive optimizations through customization.

3.2.1   Design Constraints on Embedded Systems

Most design constraints on the embedded systems come from the environment in which the embedded system will operate. Embedded systems are characterized by *application-specific, stringent,* and *multidimensional* design constraints:

> **Application-specific design constraints:** The design constraints on embedded systems differ widely; they are very application specific. For instance, the embedded system used in interplanetary

surveillance apparatus needs to be very robust and should be able to operate in a much wider range of temperatures than the embedded system used to control an mp3 player.

Multidimensional design constraints: Unlike general-purpose computer systems, embedded systems have constraints in multiple design dimensions: power, performance, cost, weight, and even form. A new constraint for handheld devices is the thickness of handheld electronic devices. Vendors only want to develop sleek designs in mp3 players and cell phones.

Stringent design constraints: The constraints on embedded systems are much more stringent than on general-purpose computers. For instance, a handheld has much tighter constraint on weight of the system than a desktop system. This comes from the portability requirements of handhelds such as mp3 players. While people want to carry their mp3 players everywhere with them, desktops are not supposed to be moved very often. Thus, even if a desktop weighs a pound more, it does not matter much, while in an mp3 player every once matters.

3.2.2   Highly Customized Designs of Embedded Systems

Owing to the increasing market pressures of short time to market and frequent upgrading, embedded system designers want to implement their embedded systems using programmable components, which provide faster and easier development and upgrades through software. The stringent, multidimensional, and application-specific constraints on embedded systems force the embedded systems to be *highly customized* to be able to meet all the design constraints simultaneously. The programmable component in the embedded system (or the embedded processor) is designed very much like general-purpose processors but is more specialized and customized to the application domain. For example, even though register renaming increases performance in processors by avoiding false data dependencies, embedded processors may not be able to employ it because of the high power consumption and the complexity of the logic. Therefore, embedded processors might deploy a "trimmed-down" or "light-weight" version of register renaming, which provides the best compromise on the important design parameters.

In addition, designers often implement *irregular design features*, which are not common in general-purpose processors but may lead to significant improvements in some design parameters for the relevant set of applications. For example, several cryptography application processors come with hardware accelerators that implement the complex cryptography algorithm in the hardware. By doing so, the cryptography applications can be made faster and consume less power but may not have any noticeable impact on normal applications. Embedded processor architectures often have such application-specific "idiosyncratic" architectural features.

Last, some design features that are present in general-purpose processors may be entirely missing in embedded processors. For example, support for prefetching is now a standard feature in general-purpose processors, but it may consume too much energy and require too much extra hardware to be appropriate in an embedded processor.

To summarize, embedded systems are characterized by *application-specific, multidimensional,* and *stringent* constraints, which result in the embedded system designs being highly customized to meet all the design constraints simultaneously.

3.2.3   Compilers for Embedded Systems

High levels of customization and the presence of idiosyncratic design features in embedded processors create unique challenges for their compilers. This leaves the compiler for the embedded processor in a very tough spot. Compilation techniques for general-purpose processors may not be suitable for embedded processors for several reasons, some of which are listed below:

Different ISA: Typically, embedded processors have different instruction set architectures (ISAs) than general-purpose processors. While IA32 and PowerPC are the most popular ISAs in the general-purpose processors, ARM and MIPS are the most popular instruction sets in embedded processors.

The primary reason for the difference in ISAs is that embedded processors are often built from the ground up to optimize for their design constraints. For instance, the ARM instruction set has been designed to reduce the code size. The code footprint of an application compiled in ARM instructions is very small.

Different optimization goals: Even if compilers can be modified to compile for a different instruction set, the optimization goals of the compilers for general-purpose processors and embedded processors are different. Most general-purpose compiler technology aims toward high performance and less compile time. However, for many embedded systems, energy consumption and code size may very important goals. For battery-operated handheld devices energy consumption is very important and, due to the limited amount of RAM size in the embedded system, the code size may be very important. In addition, for most embedded systems compile time may not be an issue, since the applications are compiled on a server—somewhere other than the embedded system—and only the binaries are loaded on the embedded system to execute as efficiently as possible.

Limited compiler technology: Even though techniques may be present to exploit the regular design features in general-purpose processors, compiler technology to exploit the "customized" version of the architectural technique may be absent. For example, predication is a standard architectural feature employed in most high-end processors. In predication, the execution of each instruction is conditional on the value of a bit in the processor state register, called the condition bit. The condition bit can be set by some instructions. Predication allows a dynamic decision about whether to execute an instruction. However, because of the architectural overhead of implementing predication, sometimes very limited support for predication is deployed in embedded processors. For example, in the Starcore architecture [36], there is no condition bit, there is just a special conditional move instruction (e.g., *cond_move R1 R2, R3 R4*), whose semantics are: if (R1 > 0) *move R1 R3*, else *move R1 R4*. To achieve the same effect as predication, the computations should be performed locally, and then the conditional instruction can be used to dynamically decide to commit the result or not. In such cases, the existing techniques and heuristics developed for predication do not work. New techniques have to be developed to exploit this "flavor" of predication in the architecture. The first challenge in developing compilers for embedded processors is therefore to enhance the compiler technology to exploit novel and idiosyncratic architectural features present in embedded processors.

Avoid penalty due to missing design features: Several embedded systems simply omit some architectural features that are common in general-purpose processors. For example, the support for prefetching may be absent in an embedded processor. In such cases, the challenge is to minimize the power and performance loss resulting from the missing architectural feature.

To summarize, code generation for embedded processors is extremely challenging because of their nonregular architectures and their stringent multidimensional constraints.

3.2.4   Compiler-Assisted Embedded System Design

While code generation for embedded systems is extremely challenging, a good compiler for an embedded system can significantly improve the power, performance, etc. of the embedded system. For example, a compiler technique to support partial predication can achieve almost the same performance as complete predication [13]. Compiler-aided prefetching in embedded systems with minimal support for prefetching can be almost as effective as a complete hardware solution [37].

3.2.4.1   Compiler as a CAD Tool

Given the significance of the compiler on processor power and performance, it is only logical that the compiler must play an important role in embedded processor design. To be able to use compilers to

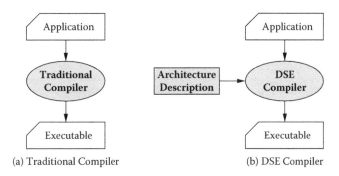

FIGURE 3.1 Traditional and DSE compiler.

design processors, the key capability required is an architecture-sensitive compiler, or what we call a DSE compiler. It should be noted here that the DSE compiler we use and need here is conceptually different than a normal compiler. As depicted in Figure 3.1, a normal compiler is *for a specific processor*; it takes the source code of the application and generates code that is as fast as possible, as low-power consuming as possible, and so on for that specific processor. A DSE compiler is more generic; it is for a range of architectures. A DSE compiler takes the source code of the application, and the processor architecture description as input, and generates code for the processor described. The main difference between a normal compiler and a DSE compiler is in the heuristics used. The heuristics deployed in a traditional compiler may not have a large degree of parameterization. For example, the register allocator in the compiler for a machine that has 32 registers needs to be efficient in allocating just 32 registers, while a DSE compiler should be able to efficiently register allocate using any number of registers. One example is that the instruction scheduling heuristic of a DSE compiler will be parameterized on the processor pipeline description, while in a normal compiler, it can be fixed. Another example is the register allocation heuristic in the compiler. The register allocation algorithm in a compiler for a machine that has 32 registers needs to be efficient in allocating just 32 registers, while a DSE compiler should be able to efficiently register allocate using any number of registers. No doubt, all compilers have some degree of parameterizations that allow some degree of compiler code reuse when developing a compiler for a different architecture. DSE compilers have an extremely high degree of parametrization and allow large-scale compiler code reuse.

Additionally, while a normal compiler can have ad hoc heuristics to generate code, a DSE compiler needs to truthfully and accurately model the architecture and have compilation heuristics that are parameterized on the architecture model. For example, simple *scheduling rules* are often used to generate code for a particular bypass configuration. The scheduling rules, for example, a dependent load instruction should always be separated by two or more cycles after the add instruction, work for the specific bypass configuration. A DSE compiler will have to model the processor pipeline and bypasses as a graph or a grammar and generate code that selects instructions that form a path in the pipeline or a legitimate word in the grammar.

The DSE compiler gets the processor description in Architecture Description Language (ADL). While there is a significant body of research in developing ADLs[1, 4, 5, 8, 9, 20, 21, 38] to serve as golden specification for simulation, verification, synthesis, and so on, here we need an ADL that can describe the processor at an abstraction that the compiler needs. We use the EXPRESSION ADL [10, 25] to parameterize our DSE compiler that we call EXPRESS [13].

3.2.4.2 Traditional Design Space Exploration

Figure 3.2 models the traditional design methodology for exploring processor architectures. In the traditional approach, the application is compiled once to generate an executable. The executable is then

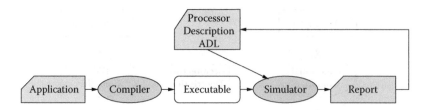

FIGURE 3.2   Traditional simulation-only design space exploration.

simulated over various architectures to choose the best architecture. We call such traditional design method-ology *simulation-only* (SO) DSE. The SO DSE of embedded systems does not incorporate compiler effects in the embedded processor design. However, the compiler effects on the eventual power and performance characteristics can be incorporated in embedded processor design in an ad hoc manner in the existing methodology. For example, the hand-generated code can be used to reflect the code the actual compiler will eventually generate. This hand-generated code can be used to evaluate the architecture. However, such a scheme may be erroneous and result in suboptimal design decisions. A systematic way to incorporate compiler hints while designing the embedded processor is needed.

3.2.4.3   Compiler-in-the-Loop Exploration

Figure 3.3 describes our proposed CIL schema for DSE. In this scheme, for each architectural variation, the application is compiled (using the DSE compiler), and the executable is simulated on a simulator of the architectural variation. Thus, the evaluation of the architecture incorporates the compiler effects in a systematic manner. The overhead CIL DSE is the extra compilation time during each exploration step, but that is insignificant relative to the simulation time.

We have developed various novel compilation techniques to exploit architectural features present in embedded processors and demonstrate the need and usefulness of CIL DSE at several abstractions of processor design, as shown in Figure 3.4: at the processor instruction set design abstraction, at the processor pipeline design abstraction, at the memory design abstraction, and at the processor memory interaction abstraction.

At the processor pipeline design abstraction, we developed a novel compilation technique for generating code for processors with partial bypassing. Partial bypassing is a popular microarchitectural feature present in embedded systems because although full bypassing is the best for performance, it may have significant area, power, and wiring complexity overheads. However, partial bypassing in processors poses a challenge for compilers, as no techniques accurately detect pipeline hazards in partially bypassed processors. Our operation-table-based modeling of the processor allows us to accurately detect all kinds of pipeline hazards and generates up to 20% better performing code than a bypass-insensitive compiler [23, 32, 34].

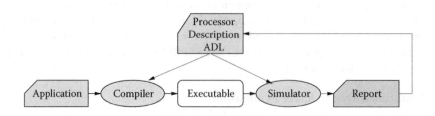

FIGURE 3.3   Compiler-in-the-loop design space exploration.

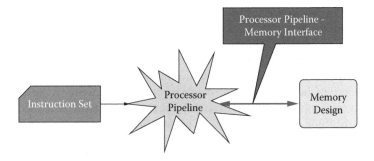

FIGURE 3.4 Processor design abstractions.

During processor design, the decision to add or remove a bypass is typically made by designer's intuition or SO DSE. However, since the compiler has significant impact on the code generated for a bypass configuration, the SO DSE may be significantly inaccurate. The comparison of our CIL with SO DSE demonstrates that not only do these two explorations result in significantly different evaluations of each bypass configuration, but they also exhibit different trends for the goodness of bypass configurations. Consequently, the traditional SO DSE can result in suboptimal design decisions, justifying the need and usefulness of our CIL DSE of bypasses in embedded systems [26, 31].

At the instruction set design abstraction, we first develop a novel compilation technique to generate code to exploit *reduced bit-width instruction set architectures* (rISAs). rISA is a popular architectural feature in which the processor supports two instruction sets. The first instruction set is composed of instructions that are 32 bits wide, and the second is a narrow instruction set composed of 16-bit-wide instructions. rISAs were originally conceived to reduce the code size of the application. If the application can be expressed in the narrow instructions only, then up to 50% code compression can be achieved. However, since the narrow instructions are only 16 bits wide, they implement limited functionality and can access only a small subset of the architectural registers. Our register pressure heuristic consistently achieves 35% code compression as compared to 14% achieved by existing techniques [12, 30].

In addition, we find out that the code compression achieved is very sensitive on the narrow instruction set chosen and the compiler. Therefore, during processor design, the narrow instruction set should be designed very carefully. We employ our CIL DSE technique to design the narrow instruction set. We find that correctly designing the narrow instruction set can double the achievable code compression [9, 29].

At the processor pipeline–memory interface design abstraction, we first develop a compilation technique to aggregate the processor activity and therefore reduce the power consumption when the processor is stalled. Fast and high-bandwidth memory buses, although best for performance, can have very high costs, energy consumption, and design complexity. As a result, embedded processors often employ slow buses. Reducing the speed of the memory bus increases the time a processor is stalled. Since the energy consumption of the processor is lower in the stalled state, the power consumption of the processor decreases. However, there is further scope for power reduction of the processor by switching the processor to IDLE state while it is stalled. However, switching the state of the processor takes 180 processor cycles in the Intel XScale, while the largest stall duration observed in the *qsort* benchmark of the MiBench suite is less than 100 processor cycles. Therefore, it is not possible to switch the processor to a low-power IDLE state during naturally occurring stalls during the application execution. Our technique aggregates the memory stalls of a processor into a large enough stall so that the processor can be switched to the low-power IDLE state. Our technique is able to aggregate up to 50,000 stall cycles, and by switching the processor to the low-power IDLE state, the power consumption of the processor can be reduced by up to 18% [33].

There is a significant difference in the processor power consumption between the SO DSE and CIL DSE. SO DSE can significantly overestimate the processor power consumption for a given memory bus configuration. This bolsters the need and usefulness of including compiler effects during the exploration and therefore highlights the need for CIL DSE.

This chapter uses a very simple architectural feature called horizontally partitioned caches (HPCs) to demonstrate the need and usefulness of CIL exploration design methodology. HPC is a popular memory architectural feature present in embedded systems in which the processors have multiple (typically two) caches at the same level of memory hierarchy. Wisely partitioning data between the caches can result in performance and energy improvements. However, existing techniques target performance improvements and achieve energy reduction only as a by-product. First we will develop energy-oriented data partitioning techniques to achieve high degrees of energy reduction, with a minimal hit on performance [35], and then we show that compared to SO DSE of HPC configurations, CIL DSE results in discovering HPC configurations that result in significantly less energy consumption.

3.3 Horizontally Partitioned Cache

Caches are one of the major contributors of not only system power and performance, but also of the embedded processor area and cost. In the Intel XScale [17], caches comprise approximately 90% of the transistor count and 60% of the area and consume approximately 15% of the processor power [3]. As a result, several hardware, software, and cooperative techniques have been proposed to improve the effectiveness of caches.

Horizontally partitioned caches are one such feature. HPCs were originally proposed in 1995 by Gonzalez et al. [6] for performance improvement. HPCs are a popular microarchitectural feature and have been deployed in several current processors such as the popular Intel StrongArm [16] and the Intel XScale [17]. However, compiler techniques to exploit them are still in their nascent stages.

A horizontally partitioned cache architecture maintains multiple caches at the same level of hierarchy, but each memory address is mapped to exactly one cache. For example, the Intel XScale contains two data caches, a 32KB main cache and a 2KB mini-cache. Each virtual page can be mapped to either of the data caches, depending on the attributes in the page table entry in the data memory management unit. Henceforth in this paper we will call the additional cache the mini-cache and the original cache the main cache.

The original idea behind such cache organization is the observation that array accesses in loops often have low temporal locality. Each value of an array is used for a while and then not used for a long time. Such array accesses sweep the cache and evict the existing data (like frequently accessed stack data) out of the cache. The problem is worse for high-associativity caches that typically employ first-in-first-out page replacement policy. Mapping such array accesses to the small mini-cache reduces the pollution in the main cache and prevents thrashing, leading to performance improvements. Thus, a horizontally partitioned cache is a simple, yet powerful, architectural feature to improve performance. Consequently, most existing approaches for partitioning data between the horizontally partitioned caches aim at improving performance.

In addition to performance improvement, horizontally partitioned caches also result in a reduction in the energy consumption due to two effects. First, reduction in the total number of misses results in reduced energy consumption. Second, since the size of the mini-cache is typically small, the energy consumed per access in the mini-cache is less than that in the large main cache. Therefore, diverting some memory accesses to the mini-cache leads to a decrease in the total energy consumption. Note that the first effect is in line with the performance goal and was therefore targeted by traditional performance improvement optimizations. However, the second effect is orthogonal to performance improvement. Therefore, energy reduction by the second effect was not considered by traditional performance-oriented techniques. As we show in this paper, the second effect (of a smaller mini-cache) can lead to energy improvements even in the presence of slight performance degradation. Note that this is where the goals of performance improvement and energy improvement diverge.

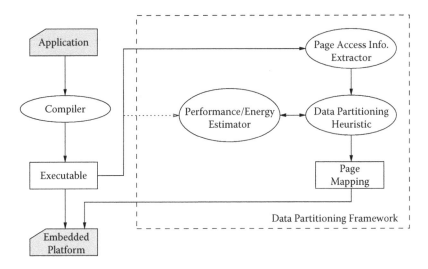

FIGURE 3.5   Compiler framework.

3.4   Compiler for Horizontally Partitioned Cache

3.4.1   HPC Compiler Framework

The problem of energy optimization for HPCs can be translated into a data partitioning problem. The data memory that the program accesses is divided into pages, and each page can be independently and exclusively mapped to exactly one of the caches. The compiler's job is then to find the mapping of the data memory pages to the caches that leads to minimum energy consumption.

As shown in Figure 3.5, we first compile the application and generate the executable. The *page access information extractor* calculates the number of times each page is accessed during the execution of the program. Then it sorts the pages in decreasing order of accesses to the pages. The complexity of simulation used to compute the number of accesses to each page and sorting the pages is $O[n + m\log(m)]$, where n is the number of data memory accesses, and m is the number of pages accessed by the application.

The *data partitioning heuristic* finds the best mapping of pages to the caches that minimizes the energy consumption of the target embedded platform. The *data partitioning heuristic* can be tuned to obtain the best-performing, or minimal energy, data partition by changing the cost function *performance/energy estimator*.

The executable together with the page mapping are then loaded by the operating system of the target platform for optimized execution of the application.

3.4.2   Experimental Framework

We have developed a framework to evaluate data partitioning algorithms to optimize the memory latency or the memory subsystem energy consumption of applications. We have modified *sim-safe* simulator from the SimpleScalar toolset [2] to obtain the number of accesses to each data memory page. This implements our *page access information extractor* in Figure 3.5. To estimate the performance/energy of an application for a given mapping of data memory pages to the main cache and the mini-cache, we have developed performance and energy models of the memory subsystem of a popular PDA, the HP iPAQ h4300 [14].

Figure 3.6 shows the memory subsystem of the iPAQ that we have modeled. The iPAQ uses the Intel PXA255 processor [15] with the XScale core [17], which has a 32KB main cache and 2KB mini-cache. PXA255 also has an on-chip memory controller that communicates with PC100-compatible SDRAMs

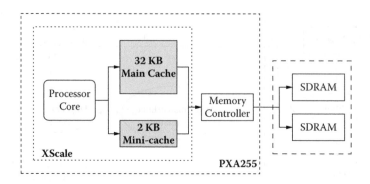

FIGURE 3.6   Modeled memory subsystem.

via an off-chip bus. We have modeled the low-power 32MB Micron MT48V8M32LF [24] SDRAM as the off-chip memory. Since the iPAQ has 64MB of memory, we have modeled two SDRAMs.

We use the memory latency as the performance metric. We estimate the memory latency as $(A_m + A_M) + MP \times (M_m + M_M)$, where A_m and A_M are the number of accesses, and M_m and M_M are the number of misses in the mini-cache and the main cache, respectively. We obtain these numbers using the *sim-cache* simulator [2], modified to model HPCs. The miss penalty MP was estimated as 25 processor cycles, taking into account the processor frequency (400 MHz), the memory bus frequency (100 MHz), the SDRAM access latency in power-down mode (6 memory cycles), and the memory controller delay (1 processor cycle).

We use the memory subsystem energy consumption as the energy metric. Our estimate of memory energy consumption has three components: energy consumed by the caches, energy consumed by off-chip busses, and energy consumed by the main memory (SDRAMs). We compute the energy consumed in the caches using the access and miss statistics from the modified sim-cache results. The energy consumed per access for each of the caches is computed using eCACTI [23]. Compared to CACTI [28], eCACTI provides better energy estimates for high-associativity caches, since it models sense-amps more accurately and scales device widths according to the capacitive loads. We have used linear extrapolation on cache size to estimate energy consumption of the mini-cache, since neither CACTI nor eCACTI model caches with less than eight sets.

We use the Printed Circuit Board (PCB) and layout recommendations of the PXA255 and Intel 440MX chipset [18, 16] and the relation between Z_o, C_o, and ϵ_r [19] to compute the the energy consumed by the external memory bus in a read/write burst as shown in Table 3.1.

We used the parameters shown in Table 3.2 from the MICRON MT48V8M32LF SDRAM to compute the energy consumed by the SDRAM per read/write burst operation (cache line read/write), shown in Table 3.2.

We perform our experiments on applications from the MiBench suite [7] and an implementation of the H.263 encoder [22]. To compile our benchmarks we used GCC with all optimizations turned on.

TABLE 3.1   External Memory Bus Parameters

Input pin capacitance	3.5 pF
Input/output pin capacitance	5 pF
Bus wire length	2.6 in.
PCB characterisitc impedance, Z_o	60 Ω
Relative permitivity of PCB dielectric, ϵ_r	4.4
Capacitance per unit length, C_o	2.34 pF/in.
Capacitance per trace	6.17 pF
Bus energy per burst	9.46 nJ

TABLE 3.2   SDRAM Energy Parameters

SDRAM current I_{dd}	100 mA
SDRAM supply voltage V_{dd}	2.5 V
Memory bus frequency f_{mem}	100 MHz
Number of memory cycles/burst N_{cyc}	13
SDRAM energy per read/write burst E_{mbst}	32.5 nJ

3.4.3   Simple Greedy Heuristics Work Well for Energy Optimization

In this section, we develop and explore several data partitioning heuristics with the aim of reducing the memory subsystem energy consumption.

3.4.3.1   Scope of Energy Reduction

To study the maximum scope of energy reduction achievable by page partitioning, we try all possible page partitions and estimate their energy consumption. Figure 3.7 plots the maximum energy reduction that we achieved by exhaustive exploration of all possible page mappings. We find the page partition that results in the minimum energy consumption by the memory subsystem and plot the reduction obtained compared to the case when all the pages are mapped to the main cache. Since the number of page partitions possible is exponential on the number of pages accessed by the application, it was not possible to complete the simulations for all the benchmarks. Exhaustive exploration was possible only for the first five benchmarks. The plot shows that compared to the case when all pages are mapped to the main cache, the scope of energy reduction is 55% on this set of benchmarks.

Encouraged by the effectiveness of page mapping, we developed several heuristics to partition the pages and see if it is possible to achieve high degrees of energy reduction using much faster techniques.

3.4.3.2   Complex Page Partitioning Heuristic: OM2N

The first technique we developed and examined is the heuristic OM2N, which is a greedy heuristic with one level of backtracking. Figure 3.8 describes the OM2N heuristic. Initially, M (list of pages mapped to the main cache) and m (list of pages mapped to the mini-cache) are empty. All the pages are initially undecided

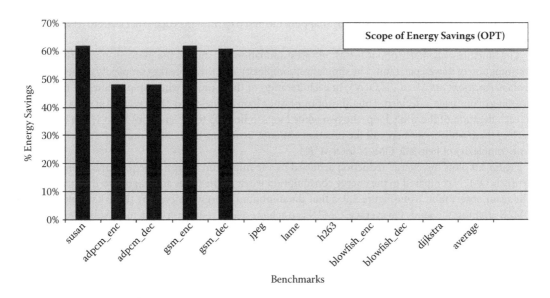

FIGURE 3.7   Maximum scope of energy reduction possible by page mapping.

Heuristic OM2N(Pages P)
```
01: M = φ, m = φ, U = P
02: while (U ≠ φ)
03:    p = U.pop()
04:    M1 = M + p, m1 = m, U1 = U
05:    while (U1 ≠ φ)
06:       p′ = U1.pop()
07:       cost1′ = evaluatePartitionCost(M1 + p′, m1)
08:       cost2′ = evaluatePartitionCost(M1, m1 + p′)
09:          if (cost1′ ≤ cost2′) M1+ = p′ else m1+ = p′
10:    endwhile
11:    M2 = M, m2 = m + p, U2 = U
12:    while (U2 ≠ φ)
13:       p′ = U2.pop()
14:       cost1′ = evaluatePartitionCost(M2 + p′, m2)
15:       cost2′ = evaluatePartitionCost(M2, m2 + p′)
16:          if (cost1′ ≤ cost2′) M2+ = p′ else m2+ = p′
17:    endwhile
18:    cost1 = evaluatePartitionCost(M1, m1)
19:    cost2 = evaluatePartitionCost(M2, m2)
20:       if (cost1 ≤ cost2) M+ = p else m+ = p
21: endwhile
22: return M, m
```

FIGURE 3.8   Heuristic OM2N.

and are in U (line 01). U is a list containing pages sorted in decreasing order of accesses. The heuristic picks the first page in U and tries both the mappings of this page — first to the main cache (line 04) and then to the mini-cache (line 11). In lines 05 to 10, after mapping the first page to the main cache, the while loop tries to map each of the remaining pages one by one into the main cache (line 07) and the mini-cache (line 08) and keeps the best solution. Similarly, it tries to find the best page partition in lines 12 to 17 after assuming that the first page is mapped to the mini-cache and remembers the best solution. In lines 18 to 20 it evaluates the energy reduction achieved by the two assumptions. The algorithm finally decides on the mapping of the first page in line 20 by mapping the first page into the cache that leads to lesser energy consumption.

The function *evaluatePartitionCost(M, m)* uses simulation to estimate the performance or the energy consumption of a given partition. The simulation complexity, and therefore the complexity of the function *evaluatePartitionCost(M, m)*, is $O(N)$. In each iteration of the topmost while loop in lines 02 to 21, the mapping of one page is decided. Thus, the topmost while loop in lines 02 to 21 is executed at most M times. In each iteration of the while loop, the two while loops in lines 05 to 10 and lines 12 to 17 are executed. Each of these while loops may call the function *evaluatePartitionCost(M, m)* at most M times. Thus, the time complexity of heuristic OM2N is $O(M^2 N)$.

Figure 3.9 plots the energy reduction achieved by the minimum energy page partition found by our heuristic OM2N compared to the energy consumption when all the pages are mapped to the main cache. The main observation from Figure 3.9 is that the minimum energy achieved by the exhaustive and the OM2N is almost the same. On average, OM2N can achieve a 52% reduction in memory subsystem energy consumption.

3.4.3.3   Simple Page Partitioning Heuristic: OMN

Encouraged by the fact the algorithm of complexity $O(M^2 N)$ can discover page mappings that result in near-optimal energy reductions, we tried to develop simpler and faster algorithms to partition the pages. Figure 3.10 is a greedy approach for solving the data partitioning problem. The heuristic picks the first

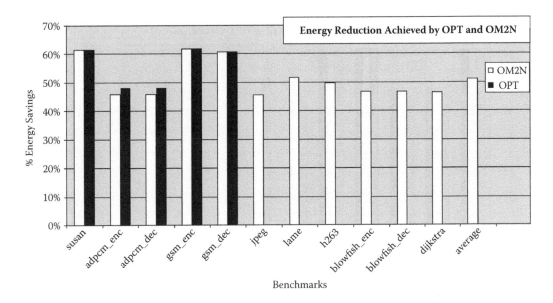

FIGURE 3.9 Energy reduction achieved by OM2N.

page in U and evaluates the cost of the partition when the page is mapped to the main cache (line 04) and when it is mapped to the mini-cache (line 05). The heuristic finally maps a page to the partition that results in minimum cost (line 06). There is only one while loop in this algorithm (lines 02 to 07), and in each step it decides upon the mapping of one page. Thus, it performs at most M simulations, each of complexity O(N). Thus, the complexity of this heuristic OMN is O(MN).

The leftmost bars in Figure 3.11 plot the energy reduction achieved by the minimum energy page partition found by our heuristic OMN compared to the energy consumption when all the pages are mapped to the main cache. On average, OMN can discover page mappings that result in a 50% reduction in memory subsystem energy consumption.

3.4.3.4 Very Simple Page Partitioning Heuristic: ON

Figure 3.12 shows a very simple single-step heuristic. If we define $k = \frac{mini-cache\_size}{page\_size}$, then the first k pages with the maximum number of accesses are mapped to the mini-cache, and the rest are mapped to the main cache. This partition aims to achieve energy reduction while making sure there is no performance loss (for high-associativity mini-caches). Note that for this heuristic we do not need to sort the list of all the pages. Only k pages with the highest number of accesses are required. If the number of pages is m, then the time complexity of selecting the k pages with highest accesses is O(km). Thus, the complexity of the heuristic is only O($n + km$), which can be approximated to O(n), since both k and m are very small compared to n.

Heuristic OMN(Pages P)
01: $M = \phi, m = \phi, U = P$
02: **while** $(U \neq \phi)$
03:　　$p = U.pop()$
04:　　$cost1 = evaluatePartitionCost(M + p, m)$
05:　　$cost2 = evaluatePartitionCost(M, m + p)$
06:　　**if** $(cost1 \leq cost2)$ $M+ = p$ **else** $m+ = p$
07: **endwhile**
08: **return** M, m

FIGURE 3.10 Heuristic OMN.

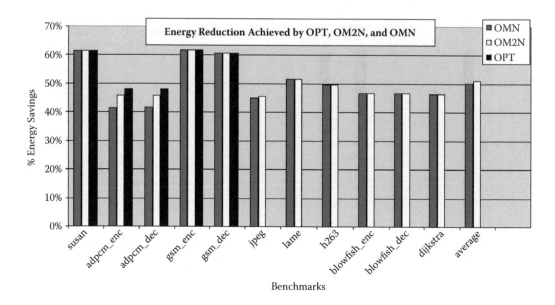

FIGURE 3.11   Energy reduction achieved by OMN.

Figure 3.13 plots the energy reduction achieved by the minimum energy page partition found by our heuristic OMN compared to the energy consumption when all the pages are mapped to the main cache. On average, OMN can discover page mappings that result in a 5% reduction in memory subsystem energy consumption.

The leftmost bars in Figure 3.13 plot the energy reduction obtained by the lowest-energy-consuming page partition discovered by the ON heuristic compared to when all the pages are mapped to the main cache. Figure 3.13 shows that ON could not obtain as impressive results as the previous more complex heuristics. On average, the ON heuristic achieves only a 35% energy reduction in memory subsystem energy consumption.

3.4.3.5   Goodness of Page Partitioning Heuristics

We define the goodness of a heuristic as the energy reduction achieved by it compared to the maximum energy reduction that is possible, that is, $\frac{(E_{Main}-E_{alg})}{(E_{Main}-E_{best})}$, where E_{Main} is the energy consumption when all the pages are mapped to the main cache, E_{alg} is the energy consumption of the best energy partition the heuristic found, and E_{best} is the energy consumption of the best energy partition. Figure 3.14 plots the goodness of the ON and OMN heuristic in obtaining energy reduction. For the last seven benchmarks for which we could not perform the optimal search. We assume the partition found by the heuristic OM2N is the best energy partition. The graph shows that the OMN heuristic could obtain on average 97% of the possible energy reduction, while ON could achieve on average 64% of the possible

Heuristic ON(Pages P)
```
01: M = φ, m = φ
02: for (i = 0; i < mini−cache_size/page_size ; i + +)
03:    m+ = U.pop()
05: endFor
06: M = U
07: return M, m
```

FIGURE 3.12   Heuristic ON.

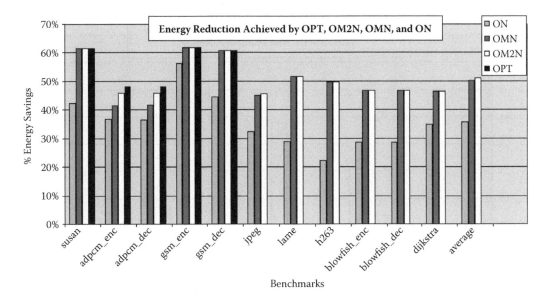

FIGURE 3.13 Energy reduction achieved by ON.

energy reduction. It is important to note that the GCC compiler for XScale does not exploit the mini-cache at all. The ON heuristic provides a simple yet effective way to exploit the mini-cache without incurring any performance penalty (for a high-associativity mini-cache).

3.4.4 Optimizing for Energy Is Different Than Optimizing for Performance

This experiment investigates the difference in optimizing for energy and optimizing for performance. We find the partition that results in the least memory latency and the partition that results in the least energy consumption. Figure 3.15a plots $\frac{E_{br} - E_{be}}{E_{be}}$, where E_{br} is the memory subsystem energy consumption of the

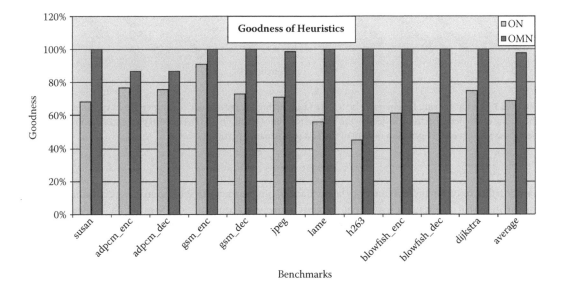

FIGURE 3.14 Goodness of heuristics.

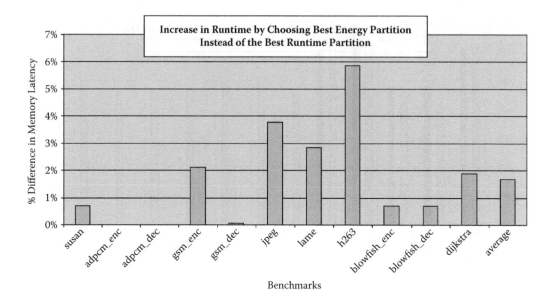

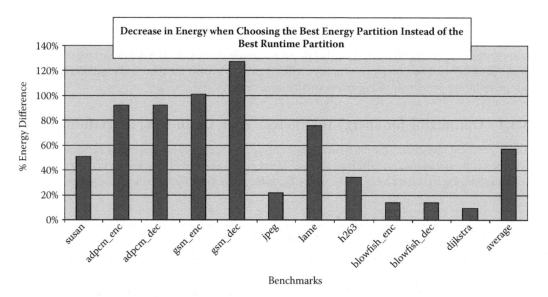

FIGURE 3.15 Optimizing for energy is better than optimizing for performance. (a) Increase in runtime, when we choose the best. energy partition instead of the best performance partition. (b) Decrease in energy consumption, when we choose the best energy partition instead of the best performance partition.

partition that results in the least memory latency, and E_{be} is the memory subsystem energy consumption by the partition that results in the least memory subsystem energy consumption. For the first five benchmarks (*susan to gsm_dec*), the number of pages in the footprint were small, so we could explore all the partitions. For the last seven benchmarks (*jpeg to dijkstra*), we took the partition found by the OM2N heuristic as the best partition, as OM2N gives close-to-optimal results in cases when we were able to search optimally. The graph essentially plots the increase in energy if you choose the best performance partition as your design point. The increase in energy consumption is up to 130% and on average is 58% for this set of benchmarks.

Figure 3.15b plots $\frac{R_{be} - R_{br}}{R_{be}}$, where R_{be} is the memory latency (in cycles) of the best energy partition, and R_{br} is the memory latency of the best-performing partition. This graph shows the increase in memory latency when you choose the best energy partition compared to using the best performance partition. The increase in memory latency is on average 1.7% and is 5.8% in the worst case for this set of benchmarks. Thus, choosing the best energy partition results in significant energy savings at a minimal loss in performance.

3.5   Compiler-in-the-Loop HPC Design

So far we have seen that HPC is a very effective microarchitectural technique to reduce the energy consumption of the processor. The energy savings achieved are very sensitive to the HPC configuration; that is, if we change the HPC configuration, the page partitioning should also change.

In the traditional DSE techniques, for example, SO DSE, the binary and the page mapping are kept the same, and the binary with the page mapping is executed on different HPC configurations. This strategy is not useful for HPC DSE, since it does not make sense to use the same page mapping after changing the HPC parameters. Clearly, the HPC parameters should be explored with the CIL during the exploration. To evaluate HPC parameters, the page mapping should be set to the given HPC configuration.

Our CIL DSE framework to explore HPC parameters is depicted in Figure 3.16. The CIL DSE framework is centered around a textual description of the processor. For our purposes, the processor description contains information about (a) HPC parameters, (b) the memory subsystem energy models, and (c) the processor and memory delay models.

We use the OMN page partitioning heuristic and generate a binary executable along with the page mapping. The page mapping specifies to which cache (main or mini) each data memory page is mapped. The compiler is tuned to generate page mappings that lead to the minimum memory subsystem energy consumption. The executable and the page mapping are both fed into a simulator that estimates the runtime and the energy consumption of the memory subsystem.

The Design Space Walker performs HPC design space exploration by updating the HPC design parameters in the processor description. The mini-cache, which is configured by Design Space Walker, is specified using two attributes: the mini-cache size and the mini-cache associativity. For our experiments, we vary cache size from 256 bytes to 32 KB, in exponents of 2. We explore the whole range of mini-cache associativities, that is, from direct mapped to fully associative. We do not model the mini-cache configurations

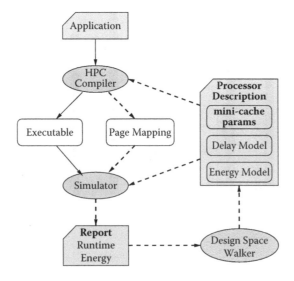

FIGURE 3.16   Compiler-in-the-loop methodology to explore the design space of HPCs.

ExhaustiveExploration()
01: *minEnergy = MAX_ENERGY*
02: **foreach** $(c \in C)$
03:    *energy = estimateEnergy(c)*
04:    **if** *(energy < minEnergy)*
05:       *minEnergy = energy*
06:    **endIf**
07: **endFor**
08: **return** minEnergy

FIGURE 3.17   Exhaustive exploration algorithm.

for which eCACTI [23] does not have a power model. We set the cache line size to be 32 bytes, as in the Intel XScale architecture. In total we explore 33 mini-cache configurations for each benchmark.

3.5.1   Exhaustive Exploration

We first present experiments to estimate the importance of exploration of HPCs. To this end, we perform exhaustive CIL exploration of HPC design space and find the minimum-energy HPC design parameters. Figure 3.17 describes the exhaustive exploration algorithm. The algorithm estimates the energy consumption for each mini-cache configuration (line 02) and keeps track of the minimum energy. The function *estimate_energy* estimates the energy consumption for a given mini-cache size and associativity.

Figure 3.18 compares the energy consumption of the memory subsystem with three cache designs. The leftmost bar represents the energy consumed by the memory subsystem when the system has only a 32KB

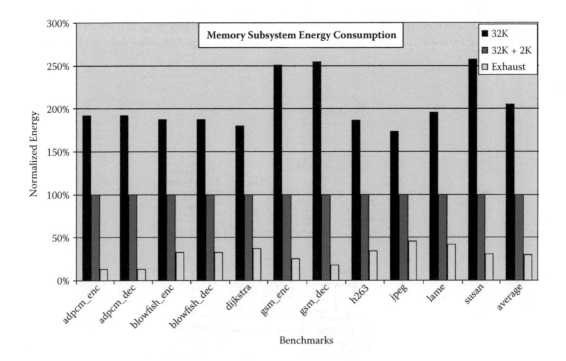

FIGURE 3.18   Energy savings achieved by exploration.

TABLE 3.3   Optimal Mini-cache Parameters

Benchmark	Mini-cache Parameters
adpcm_dec	8K, direct mapped
adpcm_enc	4K, 2-way
dijkstra	8K, 2-way
blowfish_dec	16K, 2-way
blowfish_enc	16K, 2-way
gsm_dec	2K, direct mapped
gsm_enc	2K, direct mapped
h263	8K, 2-way
jpeg	16K, 2-way
lame	8K, 2-way
susan	2K, 4-way

main cache (no mini-cache is present.) The middle bar shows the energy consumed when there is a 2KB mini-cache in parallel with the 32KB cache, and the application is compiled to achieve minimum energy. The rightmost bar represents the energy consumed by the memory subsystem, when the mini-cache parameters (size and associativity) are chosen using exhaustive CIL exploration. All the energy values are normalized to the case when there is a 2KB mini-cache (the Intel XScale configuration). The last set of bars is the average over the applications.

We make two important observations from this graph. The first is that HPC is very effective in reducing the memory subsystem energy consumption. Compared to not using any mini-cache, using a default mini-cache (the default mini-cache is 2KB, 32-way set associative) leads to an average of a 2 times reduction in the energy consumption of the memory subsystem. The second important observation is that the energy reduction obtained using HPCs is very sensitive to the mini-cache parameters. Exhaustive CIL exploration of the mini-cache DSE to find the minimum-energy mini-cache results in an additional 80% energy reduction, thus reducing the energy consumption to just 20% of the case with a 2KB mini-cache.

Furthermore, the performance of the energy-optimal HPC configuration is very close to the performance of the best-performing HPC configuration. The performance degradation was no more than 5% and was 2% on average. Therefore, energy-optimal HPC configuration achieves high energy reductions at minimal performance cost. Table 3.3 shows the energy-optimal mini-cache configuration for each benchmark. The table suggests that low-associativity mini-caches are good candidates to achieve low-energy solutions.

3.5.2   HPC CIL DSE Heuristics

We have demonstrated that CIL DSE of HPC design parameters is very useful and important to achieve significant energy savings. However, since the mini-cache design space is very large, exhaustive exploration may consume a lot of time. In this section we explore heuristics for effective and efficient HPC DSE.

3.5.2.1   Greedy Exploration

The first heuristic we develop for HPC CIL DSE is a pure greedy algorithm, outlined in Figure 3.19. The greedy algorithm first greedily finds the cache size (lines 02 to 04) and then greedily finds the associativity (lines 05 to 07). The function *betterNewConfiguration* tells whether the new mini-cache parameters result in lower energy consumption than the old mini-cache parameters.

Figure 3.20 plots the energy consumption when the mini-cache configuration is chosen by the greedy algorithm compared to when using the default 32KB main cache and 2KB mini-cache configuration. The plot shows that for most applications, greedy exploration is able to achieve good results, but for *blowfish* and *susan*, the greedy exploration is unable to achieve any energy reduction; in fact, the solution it has found consumes even more energy than the base configuration. However, on average, the greedy CIL HPC DSE can reduce the energy consumption of the memory subsystem by 50%.

GreedyExploration()
01: $size = MIN\_SIZE, assoc = MIN\_ASSOC$

// greedily find the size
02: **while** (*betterNewConf*($size \times 2, assoc, size, assoc$))
03:    $size = size \times 2$
04: **endWhile**

// greedily find the assoc
05: **while** (*betterNewConf*($size, assoc \times 2, size, assoc$))
06:    $assoc = assoc \times 2$
07: **endWhile**

08: **return** *estimateEnergy*($size, assoc$)

betterNewConf(size′, assoc′, size, assoc)
01: **if** (!*existsCacheConfig*($size′, assoc′$))
02:    **return** *false*
03: $energy = estimateEnergy(size, assoc)$
04: $energy′ = estimateEnergy(size′, assoc′)$
05: **return** ($energy′ < energy$)

FIGURE 3.19   Greedy exploration.

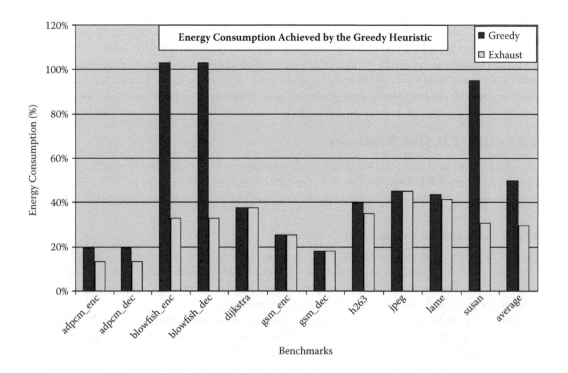

FIGURE 3.20   Energy reduction achieved by the greedy exploration.

HybridExploration()
01: $size = MIN\_SIZE, assoc = MIN\_ASSOC$

// greedily find the size
02: **while** $(betterNewConf(size \times 4, assoc, size, assoc))$
03:    $size = size \times 4$
04: **endWhile**

// search in the neighborhood
05: $done = false$
06: **while** $(!done)$
07:    **if** $(betterNewConf(size \times 2, assoc, size, assoc))$
08:       $size = size \times 2$
09:    **else if** $(betterNewConf(size, assoc \times 2, size, assoc))$
10:       $assoc = assoc \times 2$
11:    **else if** $(betterNewConf(size \div 2, assoc, size, assoc))$
12:       $size = size \div 2$
13:    **else if** $(betterNewConf(size \div 2, assoc \div 2, size, assoc))$
14:       $size = size \div 2, assoc = assoc \div 2$
15:    **else if** $(betterNewConf(size \div 2, assoc \times 2, size, assoc))$
16:       $size = size \div 2, assoc = assoc \times 2$
17:    **else**
18:       $done = true$
19: **endWhile**

08: **return** $estimateEnergy(size, assoc)$

FIGURE 3.21 Hybrid exploration.

3.5.2.2 Hybrid Exploration

To achieve energy consumption close to the optimal configurations, we developed a hybrid algorithm, outlined in Figure 3.21. The hybrid algorithm first greedily searches for the optimal mini-cache size (lines 02 to 04). Note, however, that it tries every alternate mini-cache size. The hybrid algorithm tries mini-cache sizes in exponents of 4, rather than 2 (line 03). Once it has found the optimal mini-cache size, it explores exhaustively in the size-associativity neighborhood (lines 07 to 15) to find a better size-associativity configuration.

The middle bar in Figure 3.22 plots the energy consumption of the optimal configuration compared to the energy consumption when the XScale default 32-way, 2K mini-cache is used and compares the energy reductions achieved with the greedy and exhaustive explorations. The graph shows that the hybrid exploration can always find the optimal HPC configuration for our set of benchmarks.

3.5.2.3 Energy Reduction and Exploration Time Trade-Off

There is a clear trade-off between the energy reductions achieved by the exploration algorithms and the time required for the exploration. The rightmost bar in Figure 3.23 plots the time (in hours) required to explore the design space using the exhaustive algorithm. Although the exhaustive algorithm is able to discover extremely low energy solutions, it may take tens of hours to perform the exploration. The leftmost bar in Figure 3.23 plots the time that greedy exploration requires to explore the design space of the mini-cache. Although the greedy algorithm reduces the exploration time on average by a factor of 5 times, the energy consumption is on average 2 times more than what is achieved by the optimal algorithm.

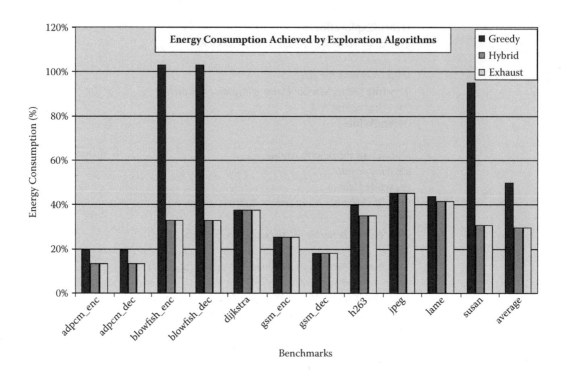

FIGURE 3.22   Relative energy reduction achieved by exploration algorithms.

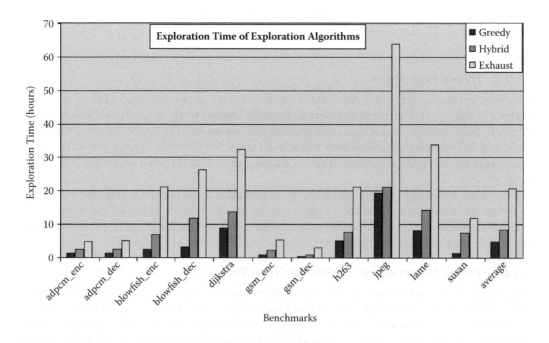

FIGURE 3.23   Exploration time of exploration algorithms.

Finally, the middle bar in Figure 3.23 plots the time required to find the mini-cache configuration when using the hybrid algorithm. Our hybrid algorithm is able to find the optimal mini-cache configuration in all of our benchmarks, while it takes about 3 times less time than the optimal algorithm. Thus, we believe the hybrid exploration is a very effective and efficient exploration technique.

3.5.3   Importance of Compiler-in-the-Loop DSE

Our next set of experiments show that although SO DSE can also find HPC configurations with less memory subsystem energy consumption, it does not do as well as CIL DSE. To this end, we performed SO DSE of HPC design parameters. We compile once for the 32KB/2KB (i.e., the original XScale cache configuration) to obtain an executable and the minimum energy page mapping. While keeping these two the same, we explored all the HPC configurations to find the HPC design parameters that minimize the memory subsystem energy consumption. Figure 3.24 plots the the energy consumption of the HPC configuration found by the SO DSE (middle bar) and CIL DSE (right bar) and the original Intel XScale HPC configuration (left bar) for each benchmark. The rightmost set of bars represents the average over all the benchmarks. All the energy consumption values are normalized to energy consumption of the 32KB/2KB configuration.

It should be noted that the overhead of compilation time in CIL DSE is negligible, because simulation times are several orders of magnitude more than compilation times. The important observation to make from this graph is that although even SO DSE can find HPC configurations that result in, on average, a 57% memory subsystem energy reduction, CIL DSE is much more effective and can uncover HPC configurations that result in a 70% reduction in the memory subsystem energy reduction.

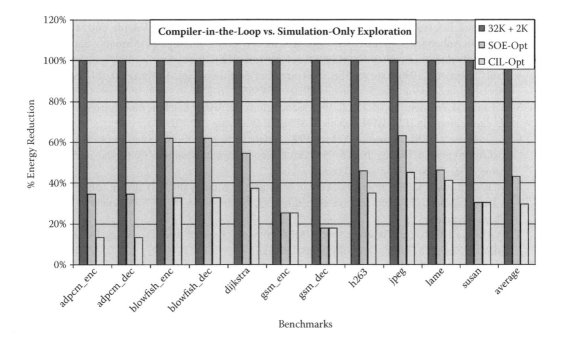

FIGURE 3.24   CIL versus SO exploration.

3.6   Summary

Embedded systems are characterized by stringent, application-specific, multidimensional constraints on their designs. These constraints, along with the shrinking time to market and frequent upgrade needs of embedded systems, are responsible for programmable embedded systems that are *highly customized*. While code generation for these highly customized embedded systems is a challenge, it is also very rewarding in the sense that an architecture-sensitive compilation technique can have significant impact on the system power, performance, and so on. Given the importance of the compiler on the system design parameters, it is reasonable for the compiler to take part in designing the embedded system. While it is possible to use ad hoc methods to include the *compiler effects* while designing an embedded system, a systematic methodology to design embedded processors is needed. This chapter introduced the CIL design methodology, which systematically includes the compiler in the embedded system DSE. Our methodology requires an architecture-sensitive compiler. To evaluate a design point in the embedded system design space, the application code is first compiled for the embedded system design and is then executed on the embedded system model to estimate the various design parameters (e.g., power, performance, etc.). Owing to the lack of compiler technology for embedded systems, most often, first an architecture-sensitive compilation technique needs to be developed, and only then can it be used for CIL design of the embedded processor. In the chapter we first developed a compilation technique for HPCs, which can result in a 50% reduction in the energy consumption of the memory subsystem. When we use this compilation technique in our CIL approach, we can come up with HPC parameters that result in an 80% reduction in the energy consumption by the memory subsystem, demonstrating the need and usefulness of our approach.

References

1. M. R. Barbacci. 1981. Instruction set processor specifications (ISPS): The notation and its applications. *IEEE Transactions on Computing* C-30(1):24–40. New York: IEEE Press.
2. D. Burger and T. M. Austin. 1997. The SimpleScalar tool set, version 2.0. *SIGARCH Computer Architecture News* 25(3):13–25.
3. L. T. Clark, E. J. Hoffman, M. Biyani, Y. Liao, S. Strazdus, M. Morrow, K. E. Velarde, and M. A. Yarch. 2001. An embedded 32-b microprocessor core for low-power and high-performance applications. *IEEE J. Solid State Circuits* 36(11):1599–608. New York: IEEE Press.
4. Paul C. Clements. 1996. A survey of architecture description languages. In *Proceedings of International Workshop on Software Specification and Design (IWSSD)*, 16–25.
5. A. Fauth, M. Freericks, and A. Knoll. 1993. Generation of hardware machine models from instruction set descriptions. In *IEEE Workshop on VLSI Signal Processing*, 242–50.
6. A. Gonzalez, C. Aliagas, and M. Valero. 1995. A data cache with multiple caching strategies tuned to different types of locality. In *ICS '95: Proceedings of the 9th International Conference on Supercomputing*, 338–47. New York: ACM Press.
7. M. R. Guthaus, J. S. Ringenberg, D. Ernst, T. M. Austin, T. Mudge, and R. B. Brown. 2001. MiBench: A free, commercially representative embedded benchmark suite. In *IEEE Workshop in Workload Characterization*.
8. J. Gyllenhaal, B. Rau, and W. Hwu. 1996. HMDES version 2.0 specification. Tech. Rep. IMPACT-96-3, IMPACT Research Group, Univ. of Illinois, Urbana.
9. G. Hadjiyiannis, S. Hanono, and S. Devadas. 1997. ISDL: An instruction set description language for retargetability. In *Proceedings of Design Automation Conference (DAC)*, 299–302. New York: IEEE Press.
10. A. Halambi, P. Grun, V. Ganesh, A. Khare, N. Dutt, and A. Nicolau. 1999. EXPRESSION: A language for architecture exploration through compiler/simulator retargetability. In *Proceedings of Design Automation and Test in Europe*. New York: IEEE Press.
11. A. Halambi, A. Shrivastava, P. Biswas, N. Dutt, and A. Nicolau. 2002. A design space exploration framework for reduced bit-width instruction set architecture (risa) design. In *ISSS '02: Proceedings of the 15th International Symposium on System Synthesis*, 120–25. New York: ACM Press.

12. A. Halambi, A. Shrivastava, P. Biswas, N. Dutt, and A. Nicolau. 2002. An efficient compiler technique for code size reduction using reduced bit-width isas. In *Proceedings of the Conference on Design, Automation and Test in Europe*. New York: IEEE Press.

13. A. Halambi, A. Shrivastava, N. Dutt, and A. Nicolau. 2001. A customizable compiler framework for embedded systems. In *Proceedings of SCOPES*.

14. Hewlett Packard. *HP iPAQ h4000 series-system specifications*. http://www.hp.com.

15. Intel Corporation. *Intel PXA255 processor: Developer's manual*. http://www.intel.com/design/pca/applicationsprocessors/manuals/278693.htm.

16. Intel Corporation. *Intel StrongARM* SA-1110 microprocessor brief datasheet*. http://download.intel.com/design/strong/datashts/27824105.pdf.

17. Intel Corporation. *Intel XScale(R) Core: Developer's manual*. http://www.intel.com/design/intelxscale/273473.htm.

18. Intel Corporation. *LV/ULV Mobile Intel Pentium III Processor-M and LV/ULV Mobile Intel Celeron Processor (0.13u)/Intel 440MX Chipset: Platform design guide*. http://www.intel.com/design/mobile/desguide/251012.htm.

19. *IPC-D-317A: Design guidelines for electronic packaging utilizing high-speed techniques*. 1995. Institute for Interconnecting and Packaging Electronic Circuits.

20. D. Kastner. 2000. TDL: A hardware and assembly description language. Tech. Rep. TDL 1.4, Saarland University, Germany.

21. R. Leupers and P. Marwedel. 1998. Retargetable code generation based on structural processor descriptions. *Design Automation Embedded Syst.* 3(1):75–108. New York: IEEE Press.

22. K. Lillevold et al. 1995. *H.263 test model simulation software*. Telenor R&D.

23. M. Mamidipaka and N. Dutt. 2004. eCACTI: An enhanced power estimation model for on-chip caches. Tech. Rep. TR-04-28, CECS, UCI.

24. Micron Technology, Inc. *MICRON Mobile SDRAM MT48V8M32LF datasheet*. http://www.micron.com/products/dram/mobilesdram/.

25. P. Mishra, A. Shrivastava, and N. Dutt. 2004. Architecture description language (adl)-driven software toolkit generation for architectural exploration of programmable socs. In *DAC '04: Proceedings of the 41st Annual Conference on Design Automation*, 626–58. New York: ACM Press.

26. S. Park, E. Earlie, A. Shrivastava, A. Nicolau, N. Dutt, and Y. Paek. 2006. Automatic generation of operation tables for fast exploration of bypasses in embedded processors. In *DATE '06: Proceedings of the Conference on Design, Automation and Test in Europe*, 1197–202. Leuven, Belgium: European Design and Automation Association. New York: IEEE Press.

27. S. Park, A. Shrivastava, N. Dutt, A. Nicolau, Y. Paek, and E. Earlie. 2006. Bypass aware instruction scheduling for register file power reduction, In *LCTES 2006: Proceedings of the 2006 ACM SIGPLAN/SIGBED conference on language, compilers, and tool support for embedded systems*, 173–81. New York: ACM Press.

28. P. Shivakumar and N. Jouppi. 2001. Cacti 3.0: An integrated cache timing, power, and area model. WRL Technical Report 2001/2.

29. A. Shrivastava, P. Biswas, A. Halambi, N. Dutt, and A. Nicolau. 2006. Compilation framework for code size reduction using reduced bit-width isas (risas). *ACM Transaction on Design Automation of Electronic Systems* 11(1):123–46. New York: ACM Press.

30. A. Shrivastava and N. Dutt. 2004. Energy efficient code generation exploiting reduced bit-width instruction set architecture. In *Proceedings of The Asia Pacific Design Automation Conference (ASPDAC)*. New York: IEEE Press.

31. A. Shrivastava, N. Dutt, A. Nicolau, and E. Earlie. 2005. Pbexplore: A framework for compiler-in-the-loop exploration of partial bypassing in embedded processors. In *DATE '05: Proceedings of the Conference on Design, Automation and Test in Europe*, 1264–69. Washington, DC: IEEE Computer Society.

32. A. Shrivastava, E. Earlie, N. D. Dutt, and A. Nicolau. 2004. Operation tables for scheduling in the presence of incomplete bypassing. In *CODES+ISSS*, 194–99. New York: IEEE Press.

33. A. Shrivastava, E. Earlie, N. Dutt, and A. Nicolau. 2005. Aggregating processor free time for energy reduction. In *CODES+ISSS '05: Proceedings of the 3rd IEEE/ACM/IFIP International Conference on Hardware/Software Codesign and System Synthesis*, 154–59. New York: ACM Press.

34. A. Shrivastava, E. Earlie, N. Dutt, and A. Nicolau. 2006. Retargetable pipeline hazard detection for partially bypassed processors, In *IEEE Transactions on Very Large Scale Integrated Circuits*, 791–801. New York: IEEE Press.

35. A. Shrivastava, I. Issenin, and N. Dutt. 2005. Compilation techniques for energy reduction in horizontally partitioned cache architectures. In *CASES '05: Proceedings of the 2005 International Conference on Compilers, Architectures and Synthesis for Embedded Systems*, 90–96. New York: ACM Press.

36. Starcore LLC. *SC1000-family processor core reference*.

37. S. P. Vanderwiel and D. J. Lilja. 2000. Data prefetch mechanisms. *ACM Computing Survey (CSUR)* 32(2):174–99. New York: ACM Press.

38. V. Zivojnovic, S. Pees, and H. Meyr. 1996. LISA — Machine description language and generic machine model for HW/SW co-design. In *IEEE Workshop on VLSI Signal Processing*, 127–36.

4

Whole Execution Traces and Their Use in Debugging

Xiangyu Zhang
Department of Computer Sciences,
Purdue University,
West Lafayette, IN
xyzhang@cs.purdue.edu

Neelam Gupta
Department of Computer Sciences,
University of Arizona, Tuscon, AZ
ngupta@cs.arizona.edu

Rajiv Gupta
Department of Computer Science
and Engineering,
University of California, Riverside, CA
gupta@cs.ucr.edu

Abstract

Profiling techniques have greatly advanced in recent years. Extensive amounts of dynamic information can be collected (e.g., control flow, address and data values, data, and control dependences), and sophisticated dynamic analysis techniques can be employed to assist in improving the performance and reliability of software. In this chapter we describe a novel representation called *whole execution traces* that can hold a vast amount of dynamic information in a form that provides easy access to this information during dynamic analysis. We demonstrate the use of this representation in locating faulty code in programs through dynamic-slicing- and dynamic-matching-based analysis of dynamic information generated by failing runs of faulty programs.

4.1 Introduction

Program profiles have been analyzed to identify program characteristics that researchers have then exploited to guide the design of superior compilers and architectures. Because of the large amounts of dynamic information generated during a program execution, techniques for space-efficient representation and time-efficient analysis of the information are needed. To limit the memory required to store different types of profiles, lossless compression techniques for several different types of profiles have been developed. Compressed representations of *control flow* traces can be found in [15, 30]. These profiles can be analyzed for the presence of hot program paths or traces [15] that have been exploited for performing path-sensitive optimization and prediction techniques [3, 9, 11, 21]. *Value profiles* have been compressed using value predictors [4] and used to perform code specialization, data compression, and value encoding [5, 16, 20, 31]. *Address profiles* have also been compressed [6] and used for identifying hot data streams that exhibit data locality, which can help in finding cache-conscious data layouts and developing data prefetching mechanisms [7, 13, 17]. *Dependence profiles* have been compressed in [27] and used for computating dynamic slices [27], studying the characteristics of performance-degrading instructions [32], and studying instruction isomorphism [18]. More recently, program profiles are being used as a basis for the debugging of programs. In particular, profiles generated from failing runs of faulty programs are being used to help locate the faulty code in the program.

In this chapter a unified representation, which we call *whole execution traces* (WETs), is described, and its use in assisting faulty code in a program is demonstrated. WETs provide an ability to relate different types of profiles (e.g., for a given execution of a statement, one can easily find the control flow path, data dependences, values, and addresses involved). For ease of analysis of profile information, WET is constructed by labeling a static program representation with profile information such that relevant and related profile information can be directly accessed by analysis algorithms as they traverse the representation. An effective compression strategy has been developed to reduce the memory needed to store WETs.

The remainder of this chapter is organized as follows. In Section 4.2 we introduce the WET representation. We describe the uncompressed form of WETs in detail and then briefly outline the compression strategy used to greatly reduce its memory needs. In Section 4.3 we show how the WETs of failing runs can be analyzed to locate faulty code. Conclusions are given in Section 4.4.

4.2   Whole Execution Traces

WET for a program execution is a comprehensive set of profile data that captures the complete functional execution history of a program run. It includes the following dynamic information:

Control flow profile: The control flow profile captures the complete control flow path taken during a single program run.

Value profile: This profile captures the values that are computed and referenced by each executed statement. Values may correspond to data values or addresses.

Dependence profile: The dependence profile captures the information about data and control dependences exercised during a program run. A data dependence represents the flow of a value from the statement that defines it to the statement that uses it as an operand. A control dependence between two statements indicates that the execution of a statement depends on the branch outcome of a predicate in another statement.

The above information tells what statements were executed and in what order (control flow profile), what operands and addresses were referenced as well as what results were produced during each statement execution (value profile), and the statement executions on which a given statement execution is data and control dependent (dependence profile).

4.2.1   Timestamped WET Representation

WET is essentially a static representation of the program that is labeled with dynamic profile information. This organization provides direct access to all of the relevant profile information associated with every execution instance of every statement. A statement in WET can correspond to a source-level statement, intermediate-level statement, or machine instruction.

To represent profile information of every execution instance of every statement, it is clearly necessary to distinguish between execution instances of statements. The WET representation distinguishes between execution instances of a statement by assigning unique *timestamps* to them [30]. To generate the timestamps a *time* counter is maintained that is initialized to one and each time a basic block is executed, the current value of *time* is assigned as a timestamp to the current execution instances of all the statements within the basic block, and then *time* is incremented by one. Timestamps assigned in this fashion essentially remember the ordering of all statements executed during a program execution. The notion of timestamps is the key to representing and accessing the dynamic information contained in WET.

The WET is essentially a labeled graph, whose form is described next. A label associated with a node or an edge in this graph is an ordered sequence where each element in the sequence represents a subset of profile information associated with an execution instance of a node or edge. The relative ordering of elements in the sequence corresponds to the relative ordering of the execution instances. For ease of presentation it is assumed that each basic block contains one statement, that is, there is one-to-one correspondence between statements and basic blocks. Next we describe the labels used by WET to represent the various kinds of profile information.

4.2.1.1 Whole Control Flow Trace

The whole control flow trace is essentially a sequence of basic block ids that captures the precise order in which they were executed during a program run. Note that the same basic block will appear multiple times in this sequence if it is executed multiple times during a program run. Now let us see how the control flow trace can be represented by appropriately labeling the basic blocks or nodes of the static control flow graph by timestamps.

When a basic block is executed, the timestamp generated for the basic block execution is added as a label to the node representing the basic block. This process is repeated for the entire program execution. The consequence of this process is that eventually each node n in the control flow graph is labeled with a sequence of timestamp values $(t_0, t_1, t_2, \cdots)$ where node n was executed at each time value t_i. Consider the example program and the corresponding control flow graph shown in Figure 4.1. Figure 4.2 shows the representation of the control flow trace corresponding to a program run. The control flow trace for a program run on the given inputs is first given. This trace is essentially a sequence of basic block ids. The subscripts of the basic block ids in the control flow trace represent the corresponding timestamp values. As shown in the control flow graph, each node is labeled with a sequence of timestamps corresponding to its executions during the program run. For example, node 8 is labeled as $(7, 11, 15)$ because node 8 is executed three times during the program run at timestamp values of 7, 11, and 15.

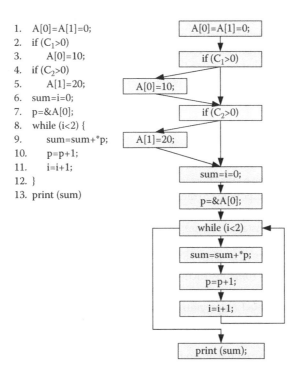

FIGURE 4.1 Example program.

With input $C_1=1$ and $C_2=0$, the execution trace is:
$1_1\, 2_2\, 3_3\, 4_4\, 6_5\, 7_6\, 8_7\, 9_8\, 10_9\, 11_{10}\, 8_{11}\, 9_{12}\, 10_{13}\, 11_{14}\, 8_{15}\, 13_{16}$

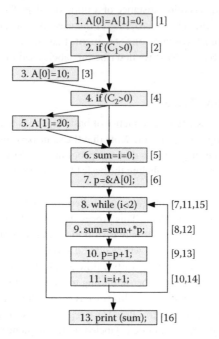

FIGURE 4.2   Timestamped control flow representation.

Let's see how the above timestamped representation captures the complete control flow trace. The path taken by the program can be generated from a labeled control flow graph using the combination of static control flow edges and the sequences of timestamps associated with nodes. If a node n is labeled with timestamp value t, the node that is executed next must be the static control flow successor of n that is labeled with timestamp value $t + 1$. Using this observation, the complete path or part of the program path starting at any execution point can be easily generated.

4.2.1.2   Whole Value Trace

The whole value trace captures all values and addresses computed and referenced by executed statements. Instrumentation code must be introduced for each instruction in the program to collect the value trace for a program run. To represent the control flow trace, with each statement, we already associate a sequence of timestamps $(t_0, t_1, t_2, \cdots)$ corresponding to the statement execution instances. To represent the value trace, we also associate a sequence of values $(v_0, v_1, v_2, \cdots)$ with the statement. These are the values computed by the statement's execution instances. Hence, there is one-to-one correspondence between the sequence of timestamps and the sequence of values.

Two points are worth noting here. First, by capturing values as stated above, we are actually capturing both values and addresses, as some instructions compute data values while others compute addresses. Second, with each statement, we only associate the result values computed by that statement. We do not explicitly associate the values used as operands by the statement. This is because we can access the operand values by traversing the data dependence edges and then retrieving the values from the value traces of statements that produce these values.

Now let us illustrate the above representation by giving the value traces for the program run considered in Figure 4.2. The sequence of values produced by each statement for this program run is shown in Figure 4.3. For example, statement 11 is executed twice and produces values 1 and 2 during these executions.

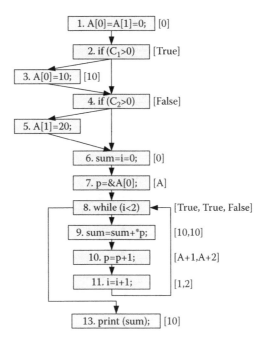

FIGURE 4.3 Value trace representation.

4.2.1.3 Whole Dependence Trace

A dependence occurs between a pair of statements; one is the source of the dependence and the other is the destination. Dependence is represented by an edge from the source to the destination in the static control flow graph. There are two types of dependences:

Static data dependence: A statement d is statically data dependent upon statement s if a value computed by statement s may be used as an operand by statement d in some program execution.

Static control dependence: A statement d is statically control dependent upon a predicate s if the outcome of predicate s can directly determine whether d is executed in some program execution.

The whole data and control dependence trace captures the dynamic occurrences of all static data and control dependences during a program run. A static edge from the source of a dependence to its destination is labeled with dynamic information to capture each dynamic occurrence of a static dependence during the program run. The dynamic information essentially identifies the execution instances of the source and destination statements involved in a dynamic dependence. Since execution instances of statements are identified by their timestamps, each dynamic dependence is represented by a pair of timestamps that identify the execution instances of statements involved in the dynamic dependence. If a static dependence edge $s \rightarrow d$ is exercised multiple times during a program run, it will be labeled by a sequence of timestamp pairs $([t_s^0, t_d^0], [t_s^1, t_d^1], \cdots)$ corresponding to multiple occurrences of the dynamic dependence observed during the program run.

Let us briefly discuss how dynamic dependences are identified during a program run. To identify dynamic data dependences, we need to further process the address trace. For each memory address the execution instance of an instruction that was responsible for the latest write to the address is remembered. When an execution instance of an instruction uses the value at an address, a dynamic data dependence is established between the execution instance of the instruction that performed the latest write to the address and the execution instance of the instruction that used the value at the address. Dynamic control dependences are also identified. An execution instance of an instruction is dynamically control dependent upon the execution instance of the predicate that caused the execution of the instruction. By first computing the static

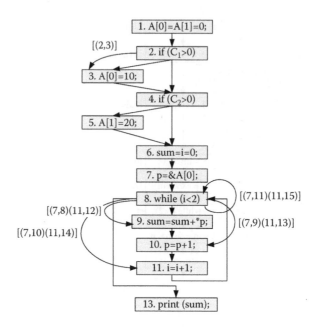

FIGURE 4.4   Timestamped control dependence trace representation.

control predecessors of an instruction, and then detecting which one of these was the last to execute prior to a given execution of the instruction from the control flow trace, dynamic control dependences are identified.

Now let us illustrate the above representation by giving the dynamic data and control dependences for the program run considered in Figure 4.2. First let's consider the dynamic control dependences shown in Figure 4.4. The control dependence edges in this program include $2 \to 3$, $4 \to 5$, $8 \to 9$, $8 \to 10$, and $8 \to 11$. These edges are labeled with timestamp pairs. The edge $2 \to 3$ is labeled with ($[2, 3]$) because this dependence is exercised only once and the timestamps of the execution instances involved are 2 and 3. The edge $4 \to 5$ is not labeled because it is not exercised in the program run. However, edge $8 \to 9$ is labeled with ($[7, 8], [11, 12]$), indicating that this edge is exercised twice. The timestamps in each pair identify the execution instances of statements involved in the dynamic dependences.

Next let us consider the dynamic data dependence edges shown in Figure 4.5. The darker edges correspond to static data dependence edges that are labeled with sequences of timestamp pairs that capture dynamic instances of data dependences encountered during the program run. For example, edge $11 \to 8$ shows the flow of the value of variable i from its definition in statement 11 to its use in statement 8. This edge is labeled ($[10, 11], [14, 15]$) because it is exercised twice in the program run. The timestamps in each pair identify the execution instances of statements involved in the dynamic dependences.

4.2.2 Compressing Whole Execution Traces

Because of the large amount of information contained in WETs, the storage needed to hold the WETs is very large. In this section we briefly outline a two-tier compression strategy for greatly reducing the space requirements.

The first tier of our compression strategy focuses on developing separate compression techniques for each of the three key types of information labeling the WET graph: (a) timestamps labeling the nodes, (b) values labeling the nodes, and (c) timestamp pairs labeling the dependence edges. Let us briefly consider these compression techniques:

Timestamps labeling the nodes: The total number of timestamps generated is equal to the number of basic block executions, and each of the timestamps labels exactly one basic block. We can reduce the

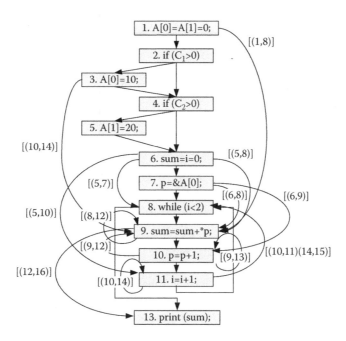

FIGURE 4.5 Timestamped data dependence trace representation.

space taken up by the timestamp node labels as follows. Instead of having nodes that correspond to basic blocks, we create a WET in which nodes can correspond to Ball Larus paths [2] that are composed of multiple basic blocks. Since a unique timestamp value is generated to identify the execution of a node, now fewer timestamps will be generated. In other words, when a Ball Larus path is executed, all nodes in the path share the same timestamp. By reducing the number of timestamps, we save space without having any negative impact on the traversal of WET to extract the control flow trace.

Values labeling the nodes: It is well known that subcomputations within a program are often performed multiple times on the same operand values. This observation is the basis for widely studied techniques for reuse-based redundancy removal [18]. This observation can be exploited in devising a compression scheme for sequence of values associated with statements belonging to a node in the WET. The list of values associated with a statement is transformed such that only a list of unique values produced by it is maintained along with a pattern from which the exact list of values can be generated from the list of unique values. The pattern is often shared across many statements. The above technique yields compression because by storing the pattern only once, we are able to eliminate all repetitions of values in value sequences associated with all statements.

Timestamp pairs labeling the dependence edges: Each dependence edge is labeled with a sequence of timestamp pairs. Next we describe how the space taken by these sequences can be reduced. Our discussion focuses on data dependences; however, similar solutions exist for handling control dependence edges [27]. To describe how timestamp pairs can be reduced, we divide the data dependences into two categories: edges that are *local* to a Ball Larus path and edges that are *nonlocal* as they cross Ball Larus path boundaries.

Let us consider a node n that contains a pair of statements s_1 and s_2 such that a *local* data dependence edge exists due to flow of values from s_1 to s_2. For every timestamp pair (t_{s_1}, t_{s_2}) labeling the edge, it is definitely the case that $t_{s_1} = t_{s_2}$. In addition, if s_2 *always receives* the involved operand value from s_1, then we do not need to label this edge with timestamp pairs. This is because the timestamp pairs that label the edge can be inferred from the labels of node n. If node n is labeled with timestamp t_n, under

TABLE 4.1   WET Sizes

Benchmark	Input	Stmts. Executed (Millions)	Orig. WET (MB)	Comp. WET (MB)	Orig./ Comp.
099.go	training	685.28	10369.32	574.65	18.04
126.gcc	ref/insn-emit.i	364.80	5237.89	89.03	58.84
130.li	ref	739.84	10399.06	203.01	51.22
164.gzip	training	650.46	9687.88	537.72	18.02
181.mcf	testing	715.16	10541.86	416.21	25.33
197.parser	training	615.49	8729.88	188.39	46.34
255.vortex	training/lendian	609.45	8747.64	104.59	83.63
256.bzip2	training	751.26	11921.19	220.70	54.02
300.twolf	training	690.39	10666.19	646.93	16.49
Avg.	n/a	646.90	9588.99	331.25	41.33

the above conditions, the data dependence edge must be labeled with the timestamp pair $< t_n, t_n >$. It should be noted that by creating nodes corresponding to Ball Larus paths, opportunities for elimination of timestamp pair labels increase greatly. This is because many nonlocal edges get converted to local edges.

Let us consider *nonlocal* edges next. Often multiple data dependence edges are introduced between a pair of nodes. It is further often the case that these edges have identical labels. In this case we can save space by creating a representation for a group of edges and save a single copy of the labels.

For the second-tier compression we view the information labeling the WET as consisting of streams of values arising from the following sources: (a) a sequence of $< t, v >$ pairs labeling a node gives rise to two streams, one corresponding to the timestamps (ts) and the other corresponding to the values (vs), and (b) a sequence of $< t_{s_1}, t_{s_2} >$ pairs labeling a dependence edge also gives rise to two streams, one corresponding to the first timestamps (t_{s_1}s) and the other corresponding to the second timestamps (t_{s_2}s). Each of the above streams is compressed using a value-prediction-based algorithm [28].

Table 4.1 lists the benchmarks considered and the lengths of the program runs, which vary from 365 and 751 million intermediate-level statements. WETs could not be collected for complete runs for most benchmarks even though we tried using Trimaran-provided inputs with shorter runs. The effect of our two-tier compression strategy is summarized in Table 4.1. While the average size of the original uncompressed WETs (Orig. WET) is 9589 megabytes, after compression their size (Comp. WET) is reduced to 331 megabytes, which represents a compression ratio (Orig./Comp.) of 41. Therefore, on average, our approach enables saving of the whole execution trace corresponding to a program run of 647 million intermediate statements using 331 megabytes of storage.

4.3   Using WET in Debugging

In this section we consider two debugging scenarios and demonstrate how WET-based analysis can be employed to assist in fault location in both scenarios. In the first scenario we have a program that fails to produce the correct output for a given input, and it is our goal to assist the programmer in locating the faulty code. In the second scenario we are given two versions of a program that should behave the same but do not do so on a given input, and our goal is to help the programmer locate the point at which the behavior of the two versions diverges. The programmer can then use this information to correct one of the versions.

4.3.1   Dynamic Program Slicing

Let us consider the following scenario for fault location. Given a failed run of a program, our goal is to identify a fault candidate set, that is, a small subset of program statements that includes the faulty code whose execution caused the program to fail. Thus, we assume that the fact that the program has failed is known because either the program crashed or it produced an output that the user has determined to be incorrect. Moreover, this failure is due to execution of faulty code and not due to other reasons (e.g., faulty environment variable setting).

The statements executed during the failing run can constitute a first conservative approximation of the fault candidate set. However, since the user has to examine the fault candidate set manually to locate faulty code, smaller fault candidate sets are desirable. Next we describe a number of dynamic-slicing-based techniques that can be used to produce a smaller fault candidate set than the one that includes all executed statements.

4.3.1.1 Backward Dynamic Slicing

Consider a failing run that produces an incorrect output value or crashes because of dereferencing of an illegal memory address. The incorrect output value or the illegal address value is now known to be related to faulty code executed during this failed run. It should be noted that identification of an incorrect output value will require help from the user unless the correct output for the test input being considered is already available to us. The fault candidate set is constructed by computing the backward dynamic slice starting at the incorrect output value or illegal address value. The backward dynamic slice of a value at a point in the execution includes all those executed statements that effect the computation of that value [1, 14]. In other words, statements that directly or indirectly influence the computation of faulty value through chains of *dynamic data and/or control dependences* are included in the backward dynamic slices. Thus, the backward reachable subgraph forms the backward dynamic slice, and all statements that appear at least once in the reachable subgraph are contained in the backward dynamic slice. During debugging, both the statements in the dynamic slice and the dependence edges that connect them provide useful clues to the failure cause.

We illustrate the benefit of backward dynamic slicing with an example of a bug that causes a heap overflow error. In this program, a heap buffer is not allocated to be wide enough, which causes an overflow. The code corresponding to the error is shown in Figure 4.6. The heap array A allocated at line 10 overflows at line 51, causing the program to crash. Therefore, the dynamic slice is computed starting at the address of $A[i]$ that causes the segmentation fault. Since the computation of the address involves $A[]$ and i, both statements at lines 10 and 50 are included in the dynamic slice. By examining statements at lines 10 and 50, the cause of the failure becomes evident to the programmer. It is easy to see that although *a_count* entries have been allocated at line 10, *b_count* entries are accessed according to the loop bounds of the *for* statement at line 50. This is the cause of the heap overflow at line 51. The main benefit of using dynamic slicing is that it focuses the attention of the programmer on the two relevant lines of code (10 and 50), enabling the fault to be located.

We studied the execution times of computing backward dynamic slices using WETs. The results of this study are presented in Figure 4.7. In this graph each point corresponds to the average dynamic slicing time for 25 slices. For each benchmark 25 new slices are computed after an execution interval of 15 million statements. These slices correspond to 25 distinct memory references. Following each execution interval slices are computed for memory addresses that had been defined since the last execution interval. This was done to avoid repeated computation of the same slices during the experiment. The increase in slicing times is linear with respect to the number of statements executed. More importantly, the slicing times are very promising. For 9 out of 10 benchmarks the average slicing time for 25 slices computed at the end of the run is below 18 seconds. The only exception is 300.twolf, for which the average slicing time at the

```
...
10. A = (int *) malloc(a_count * sizeof(int));
...
50. for (i=0; i < b_count; i++)
51.     A[i] = NULL;
...
```

FIGURE 4.6 Understanding a heap overflow bug using backward slice.

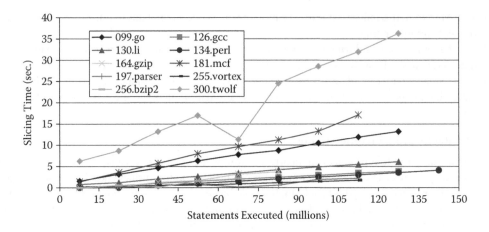

FIGURE 4.7   Dynamic slicing times using WETs.

end of the program run is roughly 36 seconds. We noted that the compression algorithm did not reduce the graph size for this program as much as many of the other benchmarks. Finally, at earlier points during program runs the slicing times are even lower.

4.3.1.2   Forward Dynamic Slicing

Zeller introduced the term *delta debugging* [22] for the process of determining the causes of program behavior by looking at the differences (the deltas) between the old and new configurations of the programs. Hildebrandt and Zeller [10, 23] then applied the delta debugging approach to simplify and isolate the failure-inducing input difference. The basic idea of delta debugging is as follows. Given two program runs r_s and r_f corresponding to the inputs I_s and I_f, respectively, such that the program fails in run r_f and completes execution successfully in run r_s, the delta debugging algorithm can be used to systematically produce a pair of inputs I'_s and I'_f with a minimal difference such that the program fails for I'_f and executes successfully for I'_s. The difference between these two inputs isolates the failure-inducing difference part of the input. These inputs are such that their values play a critical role in distinguishing a successful run from a failing run.

Since the minimal failure-inducing input difference leads to the execution of faulty code and hence causes the program to fail, we can identify a fault candidate set by computing a *forward dynamic slice* starting at this input. In other words, all statements that are influenced by the input value directly or indirectly through a chain of data or control dependences are included in the fault candidate set. Thus, now we have a means for producing a new type of dynamic slice that also represents a fault candidate set. We recognized the role of forward dynamic slices in fault location for the first time in [8].

Let us illustrate the use of forward dynamic slicing using the program in Figure 4.8. In this program if the length of the input is longer than 1,024, the writes to *Buffer[i]* at line 6 overflow the buffer corrupting the pointer stored in *CorruptPointer*. As a result, when we attempt to execute the *free* at line 9, the program crashes.

Let us assume that to test the above program we picked the following two inputs: the first input is *'aaaaa'*, which is a successful input, and the second input is *'a < repeated 2000 times >'*, which is a failing input because the length is larger than 1,024. After applying the *sddmin* algorithm in [23] on them, we have two new inputs: the new successful input is *'a < repeated 1024 times >'* and the new failing input is *'a < repeated 1025 times >'*. The failure-inducing input difference between them is the last character *'a'* in the new failed input.

Now we compute the forward dynamic slice of 1,025th *'a'* in the failing input. The resulting dynamic slice consists of a data dependence chain originating at statement *INPUT[i]* at line 5, leading to the write to

```
1.  char * CorruptPointer;
2.  char Buffer[1024];
3.  CorruptPointer = (char *) malloc( ... );
4.  i = 0;
5.  while (INPUT[i]) {
6.      Buffer[i] = INPUT[i];
7.      i++;
8.  }
9.  free(CorruptPointer);
```

FIGURE 4.8   Understanding a buffer overflow bug using a forward slice.

Buffer[i] at line 6, and then leading to the statement *free(CorruptPointer)* at line 9. When the programmer examines this data dependence chain, it becomes quite clear that there is an unexpected data dependence from *Buffer[i]* at line 6 to *free(CorruptPointer)* at line 9. Therefore, the programmer can conclude that *Buffer[i]* has overflowed. This is the best result one can expect from a fault location algorithm. This is because, other than the input statement, the forward dynamic slice captures exactly two statements. These are the two statements between which the spurious data dependence was established, and hence they must be minimally present in the fault candidate set.

4.3.1.3   Bidirectional Dynamic Slicing

Given an erroneous run of the program, the objective of this method is to explicitly force the control flow of the program along an alternate path at a critical branch predicate such that the program produces the correct output. The basic idea of this approach is inspired by the following observation. Given an input on which the execution of a program fails, a common approach to debugging is to run the program on this input again, interrupt the execution at certain points to make changes to the program state, and then see the impact of changes on the continued execution. If we can discover the changes to the program state that cause the program to terminate correctly, we obtain a good idea of the error that otherwise was causing the program to fail. However, automating the search of state changes is prohibitively expensive and difficult to realize because the search space of potential state changes is extremely large (e.g., even possible changes for the value of a single variable are enormous if the type of the variable is integer or float). However, changing the outcomes of predicate instances greatly reduces the state search space since a branch predicate has only two possible outcomes: true or false. Therefore, we note that through forced *switching* of the outcomes of some predicate instances at runtime, it may be possible to cause the program to produce correct results.

Having identified a critical predicate instance, we compute a fault candidate set as the *bidirectional dynamic slice* of the critical predicate instance. This bidirectional dynamic slice is essentially the union of the backward dynamic slice and the forward dynamic slice of the critical predicate instance. Intuitively, the reason the slice must include both the backward and forward dynamic slice is as follows. Consider the situation in which the effect of executing faulty code is to cause the predicate to evaluate incorrectly. In this case the backward dynamic slice of the critical predicate instance will capture the faulty code. However, it is possible that by changing the outcome of the critical predicate instance we avoid the execution of faulty code, and hence the program terminates normally. In this case the forward dynamic slice of the critical predicate instance will capture the faulty code. Therefore, the faulty code will be in either the backward dynamic slice or the forward dynamic slice. We recognized the role of bidirectional dynamic slices in fault location for the first time in [26], where more details on identification of the critical predicate instance can also be found.

Next we present an example to illustrate the need for bidirectional dynamic slices. We consider a simple program shown in Figure 4.9 that sums up the elements of an array $(A[1] + A[2] + \cdots + A[N])$. While this is the correct version of the program, next we will create three faulty versions of this program. In each of these versions the critical predicate instance can be found. However, the difference in these versions

```
1.  Start = 1;
2.  End = N;
3.  Sum = 0;
4.  i = Start;
5.  while (i <= End) {
6.      Sum = Sum + A[i];
7.          i = i + 1;
8.  }
9.  print(Sum);
```

FIGURE 4.9   Correct version: $Sum = A[1] + A[2] + \cdots + A[N]$.

is where in the bidirectional dynamic slice the faulty code is present, that is, the critical predicate, the backward dynamic slice of the critical predicate, and the forward dynamic slice of the critical predicate:

Fault in the critical predicate: Figure 4.10 shows a faulty version of the program from Figure 4.9. In this faulty version, the error in the predicate of the *while* loop results in the loop executing for one fewer iterations. As a result, the value of $A[N]$ is not added to *Sum*, producing an incorrect output. The critical predicate instance identified for this program run is the last execution instance of the *while* loop predicate. This is because if the outcome of the last execution instance of the *while* loop predicate is switched from false to true, the loop iterates for another iteration and the output produced by the program becomes correct. Given this information, the programmer can ascertain that the error is in the *while* loop predicate, and it can be corrected by modifying the relational operator from < to <=.

Fault in the backward dynamic slice of the critical predicate instance: In the previous faulty version, the critical predicate identified was itself faulty. Next we show that in a slightly altered version of the faulty version, the fault is not present in the critical predicate but rather in the backward dynamic slice of the critical predicate. Figure 4.11 shows this faulty version. The fault is in the initialization of *End* at line 3, and this causes the *while* loop to execute for one fewer iterations. Again, the value of $A[N]$ is not added to *Sum*, producing an incorrect output. The critical predicate instance identified for this program run is the last execution instance of the *while* loop predicate. This is because if the outcome of the last execution instance of the *while* loop predicate is switched from false to true, the loop iterates for another iteration and the output produced by the program becomes correct. However, in this situation the programmer must examine the backward dynamic slice of the critical predicate to locate the faulty initialization of *End* at line 3.

Fault in the forward dynamic slice of the critical predicate instance: Finally, we show a faulty version in which the faulty code is present in the forward dynamic slice of the critical predicate instance. Figure 4.12 shows this faulty version. The fault is at line 6, where reference to $A[i + 1]$ should be

```
1.  Start = 1;
2.  End = N;
3.  Sum = 0;
4.  i = Start;
5.  while (i < End) { – faulty statement
6.      Sum = Sum + A[i];
7.          i = i + 1;
8.  }
9.  print(Sum);
```

FIGURE 4.10   Fault in the critical predicate.

```
1.  Start = 1;
2.  End = N-1; – faulty statement
3.  Sum = 0;
4.  i = Start;
5.  while (i <= End) {
6.      Sum = Sum + A[i];
7.      i = i + 1;
8.  }
9.  print(Sum);
```

FIGURE 4.11   Fault in backward dynamic slice of critical predicate instance.

replaced by reference to $A[i]$. When this faulty version is executed, let us consider the situation in which when the last loop iteration is executed, the reference to $A[N + 1]$ at line 6 causes the program to produce a segmentation fault. The most recent execution instance of the *while* loop predicate is evaluated to true. However, if we switch this evaluation to false, the loop executes for one fewer iterations, causing the program crash to disappear. Note that the output produced is still incorrect because the value of $A[1]$ is not added to *Sum*. However, since the program no longer crashes, the programmer can analyze the program execution to understand why the program crash is avoided. By examining the forward dynamic slice of the critical predicate instance, the programmer can identify the statements, which when not executed avoid the program crash. This leads to the identification of reference to $A[i + 1]$ as the fault.

In the above discussion we have demonstrated that once the critical predicate instance is found, the fault may be present in the critical predicate, its backward dynamic slice, or its forward dynamic slice. Of course, the programmer does not know beforehand where the fault is. Therefore, the programmer must examine the critical predicate, the statements in the backward dynamic slice, and the statements in the forward dynamic slice one by one until the faulty statement is found.

4.3.1.4   Pruning Dynamic Slices

In the preceding discussion we have shown three types of dynamic slices that represent reduce fault candidate sets. In this section we describe two additional techniques for further pruning the sizes of the fault candidate sets:

Coarse-grained pruning: When multiple estimates of fault candidate sets are found using backward, forward, and bidirectional dynamic slices, we can obtain a potentially smaller fault candidate set by intersecting the three slices. We refer to this simple approach as the coarse-grained pruning approach. In [24] we demonstrate the benefits of this approach by applying it to a collection of real bugs reported by users. The results are very encouraging, as in many cases the fault candidate set contains only a handful of statements.

```
1.  Start = 0;
2.  End = N;
3.  Sum = 0;
4.  i = Start;
5.  while (i <= End) {
6.      Sum = Sum + A[i+1]; – faulty statement
7.      i = i + 1;
8.  }
9.  print(Sum);
```

FIGURE 4.12   Fault in forward dynamic slice of critical predicate instance.

Fine-grained pruning: In general, it is not always possible to compute fault candidate sets using backward, forward, and bidirectional dynamic slices. For example, if we fail to identify a minimal failure-inducing input difference and a critical predicate instance, then we cannot compute the forward and bidirectional dynamic slices. As a result, coarse-grained pruning cannot be applied. To perform pruning in such situations we developed a fine-grained pruning technique that reduces the fault candidate set size by eliminating statements in the backward dynamic slice that are expected not to be faulty.

The fine-grained pruning approach is based upon value profiles of executed statements. The main idea behind this approach is to exploit correct outputs produced during a program run before an incorrect output is produced or the program terminates abnormally. The executed statements and their value profiles are examined to find likely correct statements in the backward slice. These statements are such that if they are altered, they will definitely cause at least one correct output produced during the program run to change. All statements that fall in this category are marked as likely correct and thus pruned from the backward dynamic slice. The detailed algorithm can be found in [25] along with experimental data that show that this pruning approach is highly effective in reducing the size of the fault candidate set. It should be noted that this fine-grained pruning technique makes use of both dependence and value traces contained in the WET.

4.3.2   Dynamic Matching of Program Versions

Now we consider a scenario in which we have two versions of a program such that the second version has been derived through application of transformations to the first version. When the two versions are executed on an input, it is found that while the first version runs correctly, the second version does not. In such a situation it is useful to find out the execution point at which the dynamic behavior of the two versions deviates, since this gives us a clue to the cause of differing behaviors.

The above scenario arises in the context of optimizing compilers. Although compile-time optimizations are important for improving the performance of programs, applications are typically developed with the optimizer turned off. Once the program has been sufficiently tested, it is optimized prior to its deployment. However, the optimized program may fail to execute correctly on an input even though the unoptimized program ran successfully on that input. In this situation the fault may have been introduced by the optimizer through the application of an unsafe optimization, or a fault present in the original program may have been exposed by the optimizations. Determining the source and cause of the fault is therefore important.

In [12] a technique called *comparison checking* was proposed to address the above problem. A comparison checker executes the optimized and unoptimized programs and continuously compares the results produced by *corresponding instruction executions* from the two versions. At the earliest point during execution at which the results differ, they are reported to the programmer, who can use this information to isolate the cause of the faulty behavior. It should be noted that not every instruction in one version has a corresponding instruction in the other version because optimizations such as reassociation may lead to instructions that compute different intermediate results. While the above approach can be used to test optimized code thoroughly and assist in locating a fault if one exists, it has one major drawback. For the comparison checker to know which instruction executions in the two versions correspond to each other, the compiler writer must write extra code that determines *mappings* between execution instances of instructions in the two program versions. Not only do we need to develop a mapping for each kind of optimization to capture the effect of that optimization, but we must also be able to compose the mappings for different optimizations to produce the mapping between the unoptimized and fully optimized code. The above task is not only difficult and time consuming, but it must be performed each time a new optimization is added to the compiler.

We have developed a WET-based approach for automatically generating the *mappings*. The basic idea behind our approach is to run the two versions of the programs and regularly compare their execution histories. The goal of this comparison is to find *matches* between the execution history of each instruction in the optimized code with execution histories of one or more instructions in the unoptimized code.

If execution histories match closely, it is extremely likely that they are indeed the corresponding instructions in the two program versions. At each point when executions of the programs are interrupted, their histories are compared with each other. Following the determination of matches, we determine if faulty behavior has already manifested itself, and accordingly potential causes of faulty behavior are reported to the user for inspection. For example, instructions in the optimized program that have been executed numerous times but do not match anything in the unoptimized code can be reported to the user for examination. In addition, instructions that matched each other in an earlier part of execution but later did not match can be reported to the user. This is because the later phase of execution may represent instruction executions after faulty behavior manifests itself. The user can then inspect these instructions to locate the fault(s).

The key problem we must solve to implement the above approach is to develop a matching process that is highly accurate. We have designed a WET-based *matching algorithm* that consists of the following two steps: *signature matching* and *structure matching*. A signature of an instruction is defined in terms of the frequency distributions of the result values produced by the instruction and the addresses referenced by the instruction. If signatures of two instructions are consistent with each other, we consider them to *tentatively* match. In this second step we match the structures of the data dependence graphs produced by the two versions. Two instructions from the two versions are considered to match if there was a tentative signature match between them and the instructions that provided their operands also matched with each other.

In the *Trimaran* system [19] we generated two versions of very long instructional word (VLIW) machine code supported under the Trimaran system by generating an *unoptimized* and an *optimized* version of programs. We ran the two versions on the same input and collected their detailed whole execution traces. The execution histories of corresponding functions were then matched. We found that our matching algorithm was highly accurate and produced the matches in seconds [29]. To study the effectiveness of matching for comparison checking as discussed above, we created another version of the optimized code by manually injecting an error in the optimized code. We plotted the number of distinct instructions for which no match was found as a fraction of distinct executed instructions over time in two situations: when the optimized program had no error and when it contained an error. The resulting plot is shown in Figure 4.13. The points in the graph are also annotated with the actual number of instructions in the optimized code for which no match was found. The interval during which an error point is encountered during execution is marked in the figure.

Compared to the optimized program without error, the number of unmatched instructions increases sharply after the *error interval* point is encountered. The increase is quite sharp — from 3 to 35%. When we look at the actual number of instructions reported immediately before and after the execution interval during which the error is first encountered, the number reported increases by an order of magnitude.

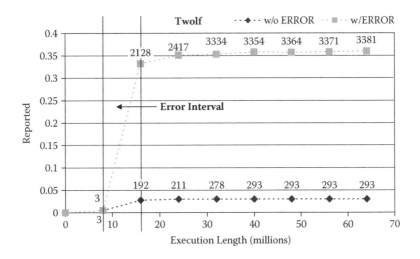

FIGURE 4.13 Unmatched statements over time.

By examining the instructions in the order they are executed, erroneous instructions can be quickly isolated. Other unmatched instructions are merely dependent upon the instructions that are the root causes of the errors. Out of the over 2,000 unmatched instructions at the end of the second interval, we only need to examine the first 15 unmatched instructions in temporal order to find an erroneous instruction.

4.4 Concluding Remarks

The emphasis of earlier research on profiling techniques was separately studying single types of profiles (control flow, address, value, or dependence) and capturing only a subset of profile information of a given kind (e.g., hot control flow paths, hot data streams). However, recent advances in profiling enable us to simultaneously capture and compactly represent complete profiles of all types. In this chapter we described the WET representation that simultaneously captures complete profile information of several types of profile data. We demonstrated how such rich profiling data can serve as the basis of powerful dynamic analysis techniques. In particular, we described how dynamic slicing and dynamic matching can be performed efficiently and used to greatly assist a programmer in locating faulty code under two debugging scenarios.

References

1. H. Agrawal and J. Horgan. 1990. Dynamic program slicing. In *ACM SIGPLAN Conference on Programming Language Design and Implementation*, 246–56. New York, NY: ACM Press.
2. T. Ball and J. Larus. 1996. Efficient path profiling. In *IEEE/ACM International Symposium on Microarchitecture*, 46–57. Washington, DC, IEEE Computer Society.
3. R. Bodik, R. Gupta, and M. L. Soffa. 1998. Complete removal of redundant expressions. In *ACM SIGPLAN Conference on Programming Language Design and Implementation*, 1–14, Montreal, Canada. New York, NY: ACM Press.
4. M. Burtscher and M. Jeeradit. 2003. Compressing extended program traces using value predictors. In *International Conference on Parallel Architectures and Compilation Techniques*, 159–69. Washington, DC, IEEE Computer Society.
5. B. Calder, P. Feller, and A. Eustace. 1997. Value profiling. In *IEEE/ACM International Symposium on Microarchitecture*, 259–69. Washington, DC, IEEE Computer Society.
6. T. M. Chilimbi. 2001. Efficient representations and abstractions for quantifying and exploiting data reference locality. In *ACM SIGPLAN Conference on Programming Language Design and Implementation*, 191–202, Snowbird, UT. New York, NY: ACM Press.
7. T. M. Chilimbi and M. Hirzel. 2002. Dynamic hot data stream prefetching for general-purpose programs. In *ACM SIGPLAN Conference on Programming Language Design and Implementation*, 199–209. New York, NY: ACM Press.
8. N. Gupta, H. He, X. Zhang, and R. Gupta. 2005. Locating faulty code using failure-inducing chops. In *IEEE/ACM International Conference on Automated Software Engineering*, 263–72, Long Beach, CA. New York, NY: ACM Press.
9. R. Gupta, D. Berson, and J. Z. Fang. 1998. Path profile guided partial redundancy elimination using speculation. In *IEEE International Conference on Computer Languages*, 230–39, Chicago, IL. Washington, DC, IEEE Computer Society.
10. R. Hildebrandt and A. Zeller. 2000. Simplifying failure-inducing input. In *International Symposium on Software Testing and Analysis*, 135–45. New York, NY: ACM Press.
11. Q. Jacobson, E. Rotenberg, and J. E. Smith. 1997. Path-based next trace prediction. In *IEEE/ACM International Symposium on Microarchitecture*, 14–23. Washington, DC, IEEE Computer Society.
12. C. Jaramillo, R. Gupta, and M. L. Soffa. 1999. Comparison checking: An approach to avoid debugging of optimized code. In *ACM SIGSOFT 7th Symposium on Foundations of Software Engineering and 8th European Software Engineering Conference. LNCS 1687*, 268–84. Heidelberg, Germany: Springer-Verlag. New York, NY: ACM Press.

13. D. Joseph and D. Grunwald. 1997. Prefetching using Markov predictors. In *International Symposium on Computer Architecture*, 252–63. New York, NY: ACM Press.
14. B. Korel and J. Laski. 1988. Dynamic program slicing. *Information Processing Letters*, 29(3): 155–63. Amsterdam, The Netherlands: Elsevier North-Holland, Inc.
15. J. R. Larus. 1999. Whole program paths. In *ACM SIGPLAN Conference on Programming Language Design and Implementation*, 259–69, Atlanta, GA. New York, NY: ACM Press.
16. M. H. Lipasti and J. P. Shen. 1996. Exceeding the dataflow limit via value prediction. In *IEEE/ACM International Symposium on Microarchitecture*, 226–37. Washington, DC, IEEE Computer Society.
17. S. Rubin, R. Bodik, and T. Chilimbi. 2002. An efficient profile-analysis framework for data layout optimizations. In *ACM SIGPLAN-SIGACT Symposium on Principles of Programming Languages*, 140–53, Portland, OR. New York, NY: ACM Press.
18. Y. Sazeides. 2003. Instruction-isomorphism in program execution. *Journal of Instruction Level Parallelism*, 5:1–22.
19. L. N. Chakrapani, J. Gyllenhaal, W. Hwu, S. A. Mahlke, K. V. Palem, and R. M. Rabbah, Trimaran: An infrastructure for research in instruction-level parallelism, languages, and compliers for high performance computing 2004, *LNCS 8602*, 32–41, Berlin: Springer/Heidlberg.
20. J. Yang and R. Gupta. 2002. Frequent value locality and its applications. *ACM Transactions on Embedded Computing Systems*, 1(1):79–105. New York, NY: ACM Press.
21. C. Young and M. D. Smith. 1998. Better global scheduling using path profiles. In *IEEE/ACM International Symposium on Microarchitecture*, 115–23. Washington, DC, IEEE Computer Society.
22. A. Zeller. 1999. Yesterday, my program worked. Today, it does not. Why? In *ACM SIGSOFT Seventh Symposium on Foundations of Software Engineering and Seventh European Software Engineering Conference*, 253–67. New York, NY: ACM Press.
23. A. Zeller and R. Hildebrandt. 2002. Simplifying and isolating failure-inducing input. *IEEE Transactions on Software Engineering*, 28(2):183–200. Washington, DC, IEEE Computer Society.
24. X. Zhang, N. Gupta, and R. Gupta. Locating Faulty Code by Multiple Points Slicing. *Software - Practice & Experience*, vol 37, Issue 9, pp. 935–961, July 2007. New York, NY: John Wiley & Sons, Inc.
25. X. Zhang, N. Gupta, and R. Gupta. 2006. Pruning dynamic slices with confidence. In *ACM SIGPLAN Conference on Programming Language Design and Implementation*, 169–80, Ottawa, Canada. New York, NY: ACM Press.
26. X. Zhang, N. Gupta, and R. Gupta. 2006. Locating faults through automated predicate switching. In *International Conference on Software Engineering*, 272–81, Shanghai, China. New York, NY: ACM Press.
27. X. Zhang and R. Gupta. 2004. Cost effective dynamic program slicing. In *ACM SIGPLAN Conference on Programming Language Design and Implementation*, 94–106, Washington, DC. New York, NY: ACM Press.
28. X. Zhang and R. Gupta. 2005. Whole execution traces and their applications. *ACM Transactions on Architecture and Code Optimization*, 2(3):301–34. New York, NY: ACM Press.
29. X. Zhang and R. Gupta. 2005. Matching execution histories of program versions. In *Joint 10th European Software Engineering Conference and ACM SIGSOFT 13th Symposium on Foundations of Software Engineering*, 197–206, Lisbon, Portugal. New York, NY: ACM Press.
30. Y. Zhang and R. Gupta. 2001. Timestamped whole program path representation and its applications. In *ACM SIGPLAN Conference on Programming Language Design and Implementation*, 180–90, Snowbird, UT. New York, NY: ACM Press.
31. Y. Zhang and R. Gupta. 2002. Data compression transformations for dynamically allocated data structures. In *International Conference on Compiler Construction*, Grenoble, France. London, UK: Springer–Verlag.
32. C. B. Zilles and G. Sohi. 2000. Understanding the backward slices of performance degrading instructions. In *ACM/IEEE 27th International Symposium on Computer Architecture*, 172–81. New York, NY: ACM Press.

5

Optimizations for Memory Hierarchies

Easwaran Raman
and
David I. August

Department of Computer Science,
Princeton University, Princeton, NJ
eraman@cs.princeton.edu
august@cs.princeton.edu

5.1 Introduction

Since the advent of microprocessors, the clock speeds of CPUs have increased at an exponential rate. While the speed at which off-chip memory can be accessed has also increased exponentially, it has not increased at the same rate. A *memory hierarchy* is used to bridge this gap between the speeds of the CPU and the memory. In a memory hierarchy, the off-chip main memory is at the bottom. Above it, one or more levels of memory reside. Each level is faster than the level below it but stores a smaller amount of data. Sometimes registers are considered to be the topmost level of this hierarchy.

There are two ways in which the intermediate levels of this hierarchy can be organized.

The most popular approach, used in most general-purpose processors, is to use *cache memory*. A cache contains some frequently accessed subset of the data in the main memory. When the processor wants to access a piece of data, it first checks the topmost level of the hierarchy and goes down until the data sought is found. Most current processors have either two or three levels of caches on top of the main memory. Typically, the hardware controls which subset of state of one level is stored in a higher level. In a cache-based memory hierarchy, the average time to access a data item depends upon the probability of finding the data at each level of the hierarchy.

Embedded systems use *scratch pad memory* in their memory subsystems. Scratch pads are smaller and faster *addressable* memory spaces. The compiler or the programmer directs where each data item should reside.

In this chapter, we restrict our attention to cache-based memory hierarchies. This chapter discusses compiler transformations that reduce the average access latency of accessing data and instructions from memory. The average latency of accesses can be reduced in several ways. The first approach is to reduce the number of cache misses. There are three types of cache misses. *Cold misses* are the misses that are seen when a data item is accessed for the first time in the program and hence is not found in the cache. *Capacity misses* are caused when the current working set of the program is greater than the size of the cache. *Conflict misses* happen when more than one data item maps to the same cache line, thereby evicting some data in the current working set from the cache, but the current working set itself fits within the cache. The second approach is to reduce the average number of cycles needed to service a cache miss. The optimizations discussed later in this chapter reduce the average access latency by reducing the cache misses, by minimizing the access latency on a miss, or by a combination of both.

In many processor architectures, the compiler cannot explicitly specify which data items should be placed at each level of the memory hierarchy. Even in architectures that support such directives, the hardware still maintains primary control. In either case, the compiler performs various transformations to reduce the access latency:

1. **Code restructuring optimizations:** The order in which data items are accessed highly influences which set of data items are available in a given level of cache hierarchy at any given time. In many cases, the compiler can restructure the code to change the order in which data items are accessed without altering the semantics of the program. Optimizations in this category include loop interchange, loop blocking or tiling, loop skewing, loop fusion, and loop fission.

2. **Prefetching optimizations:** Some architectures do provide support in the instruction set that allows the compiler some explicit control of what data should be in a particular level of the hierarchy. This is typically in the form of *prefetch* instructions. As the name suggests, a prefetch instruction allows a data item to be brought into the upper levels of the hierarchy before the data is actually used by the program. By suitably inserting such prefetch instructions in the code, the compiler can decrease the average data access latency.

3. **Data layout optimizations:** All data items in a program have to be mapped to addresses in the virtual address space. The mapping function is often irrelevant to the correctness of the program, as long as two different items are with overlapping lifetimes not mapped to overlapping address ranges. However, this can have an effect on which level of the hierarchy a data item is found and hence on the execution time. Optimizations in this category include structure splitting, structure peeling, and stack, heap, and global data object layout.

4. **Code layout optimizations:** The compiler has even more freedom in mapping instructions to address ranges. Code layout optimizations use this freedom to ensure that most of the code is found in the topmost level of the instruction cache.

While the compiler can optimize for the memory hierarchies in many ways, it also faces some significant challenges. The first challenge is in analyzing and understanding memory access patterns of applications. While significant progress has been made in developing better memory analyses, there remain many cases where the compiler has little knowledge of the memory behavior of applications. In particular, irregular access patterns such as traversals of linked data structures still thwart memory optimization efforts on the part of modern compilers. Indeed, most of the optimizations for memory performance are designed to handle regular access patterns such as strided array references. The next challenge comes from the fact that many optimizations for memory hierarchies are highly machine dependent. Incorrect assumptions about the underlying memory hierarchy often reduce the benefits of these optimizations and may even degrade the performance. An optimization that targets a cache with a line size of 64 bytes may not be suitable for processors with a cache line size of 128 bytes. Thus, the compiler needs to model the memory hierarchy as accurately as possible. This is a challenge given the complex memory subsystem found in modern processors. In Intel's Itanium®, for example, the miss latency of first-level and second-level caches depends upon on factors such as the occupancy of the queue that contains the requests to the second-level cache. Despite these difficulties, memory optimizations are worthwhile, given how much memory behavior dictates overall program performance.

The rest of this chapter is organized as follows. Section 5.2 provides some background on the notion of dependence within loops and on locality analysis that is needed to understand some of the optimizations. Sections 5.3, 5.4, and 5.5 discuss some code restructuring optimizations. Section 5.6 deals with data prefetching, and data layout optimizations are discussed in Section 5.7. Optimizations for the instruction cache are covered in Section 5.8. Section 5.9 briefly summarizes the optimizations and discusses some future directions, and Section 5.10 discusses references to various works in this area.

5.2 Background

Code restructuring and data prefetching optimizations typically depend on at least partial regularity in data access patterns. Accesses to arrays within loops are often subscripted by loop indices, resulting in very regular patterns. As a result, most of these techniques operate at the granularity of loops. Hence, we first present some theory on loops that will be used in subsequent sections. A different approach to understanding loop accesses by using the domain of $\mathcal{Z}$-polyhedra to model the loop iteration space is discussed by Rajopadhye et al. [24].

5.2.1 Dependence in Loops

Any compiler transformation has to preserve the semantics of the original code. This is ensured by preserving all true dependences in the original code. Traditional definitions of dependence between instructions are highly imprecise when applied to loops. For example, consider the loop in Figure 5.1 that computes the row sums of a matrix. By the traditional definition of true dependence, the statement row_sum[i] += matrix[i][j] in the inner loop has a true dependence on itself, but this is an imprecise statement since this is true only for statements within the same inner loop. In other words, row_sum[i+1] does not depend on row_sum[i], and both can be computed in parallel. This shows the need for a more precise definition of dependence within a loop. While a detailed discussion of loop dependence is beyond the scope of this chapter, we briefly discuss loop dependence theory and refer readers to Kennedy and Allen [13].

An important drawback of the traditional definition of dependences when applied to loops is that they do not have any notion of loop iteration. As the above example suggests, if a statement is qualified by its loop iteration, the dependence definitions will be more precise. The following definitions help precisely specify a loop iteration:

```
for (i=0; i< m; i++){
    for(j = 0; j< n; j++){
        row_sum[i] += matrix[i][j]
    }
}
```

FIGURE 5.1 Matrix row sum computation.

Normalized iteration number: If a loop index I has a lower bound of L and a step size of S, then the normalized iteration number of an iteration is given by $\frac{I-L}{S} + 1$.

Iteration vector: An iteration vector of a loop nest is a vector of integers representing loop iterations in which the value of the kth component denotes the normalized iteration number of the kth outermost loop. As an example, for the loop shown above, the iteration vector i = (1, 2) denotes the second iteration of the j loop within the first iteration of the i loop.

Using iteration vectors, we can define statement instances. A statement instance S(i) denotes the statement S executed on the iteration specified by the iteration vector **i**. Defining the dependence relations on statement instances, rather than just on the statements, makes the dependence definitions precise. Thus, if S denotes the statement `row_sum[i] += matrix[i][j]`, then there is true dependence between $S(i, j_1)$ and $S(i, j_2)$, where $j_1 < j_2 <= n$. In general, there can be a dependence between two statement instances $S_1(i_1)$ and $S_2(i_2)$ only if the statement S_2 is dependent on S_1 and the iteration vector $\mathbf{i_1}$ is lexicographically less than or equal to $\mathbf{i_2}$. To specify dependence between statement instances, we define two vectors:

Dependence distance vector: If there is a dependence from statement S_i in iteration **i** to statement S_j in iteration **j**, then the dependence distance vector **d(i, j)** is a vector of integers defined as **d(i, j) = j-i**.

Dependence direction vector: If **d(i, j)** is a dependence distance vector, then the corresponding direction vector **D(i, j)** is defined as

$$D(i, j) = (D(i, j)_0, D(i, j)_1, \ldots D(i, j)_n), \text{ where,}$$

$$D(i, j)_k = \begin{cases} <, \text{ if } d(i, j)_k > 0 \\ =, \text{ if } d(i, j)_k = 0 \\ >, \text{ if } d(i, j)_k < 0 \end{cases}$$

In any *valid dependence*, the leftmost non $=$ component of the direction vector must be $<$. The loop transformations described in this chapter are known as *reordering transformations*. A reordering transformation is one that does not add or remove statements from a loop nest and only reorders the execution of the statements that are already in the loop. Since reordering transformations do not add or remove statements, they do not add or remove any new dependences. A reordering transformation is valid if it preserves all existing dependences in the loop.

5.2.2 Reuse and Locality

Most compiler optimizations for memory hierarchies rely on the fact that programs reuse the same data repeatedly. There are two types of data reuse:

Temporal reuse: When a program accesses a memory location more than once, it exhibits temporal reuse.

Spatial reuse: When a program accesses multiple memory locations within the same cache line, it exhibits spatial reuse.

These reuses can be due to the same reference or a group of data references. In the former case, the reuse is known as *self* reuse, and in the latter case, it is known as *group* reuse. In the loop

```
// even_sum and odd_sum are global variables

for (i=0; i<m; i+=2){
    even_sum += a[i];
    odd_sum += a[i+1];


}
```

even_sum and odd_sum exhibit self-temporal reuse. The references to a[i] and a[i+1] show self-spatial reuse individually and exhibit group-spatial reuse together.

Data reuse is beneficial since a reused data item is likely to be found in the caches and hence incur a low average access latency. If the caches are infinitely large, then all reused data would be found in the cache, but since the cache sizes are finite, a data object D in the cache may be replaced by some other data between two uses of D.

5.2.3 Quantifying Reuse and Locality

The compiler must be able to quantify reuse and locality in order to exploit them. Reuse resulting from array accesses in loops whose subscripts are affine functions of the loop indices can be modeled using linear algebra. Consider an n-dimensional array X accessed inside a loop nest $N = (L_1, L_2, \ldots L_m)$ m loops. Let the loop indices be represented by an $n \times 1$ matrix i. Then each access to X can be represented as X$[Ai + C]$, where A is an $m \times n$ matrix that applies a linear transformation on i, and C is an $m \times 1$ constant vector. For example, an access X[0][i-j] in a loop nest with two loops, whose indices are i and j, is represented as X$[Ai + C]$, where

$$A = \begin{pmatrix} 0 & 0 \\ 1 & -1 \end{pmatrix}, i = \begin{pmatrix} i \\ j \end{pmatrix}, \text{ and } C = \begin{pmatrix} 0 \\ 0 \end{pmatrix}$$

This access exhibits self-temporal reuse if it accesses the same memory location in two iterations of the loop. If, in iterations i_1 and i_2, the reference to X accesses the same location, then $Ai_1 + C$ must be equal to $Ai_2 + C$. Let $d = i_1 - i_2$. The vector d is known as the reuse distance vector. If a non-null value of d satisfyies the equation $A \times d = 0$, then the reference exhibits self-temporal reuse. In other words, if the kernel of the vector space given by A is non-null, then the reference exhibits self-temporal reuse. For the above example, there is a non-null d satisfying the equation since $A \times \binom{k}{k} = \binom{0}{0}$. This implies that the location accessed in iteration (i_1, j_1) is also accessed in iteration $(i_1 + k, j_1 + k)$ for any value of k. Group-temporal reuse is identified in a similar manner. If two references $Ai_1 + C_1$ and $Ai_2 + C_2$ to the same array in iterations i_1 and i_2, respectively, access the same location, then $Ai_1 + C_1$ must be equal to $Ai_2 + C_2$. In other words, if a non-null d satisfies $A \times d = (C_1 - C_2)$, the pair of references exhibit group-temporal reuse. If an entry of the d vector is 0, then the corresponding loop carries temporal reuse. The amount of reuse is given by the product of the iteration count of those loops that have a 0 entry in d.

An access exhibits self-spatial reuse if two references access different locations that are in the same cache line. We define the term *contiguous dimension* to mean the dimension of the array along which two adjacent elements of the array are in adjacent locations in memory. In this chapter, all the code examples are in C, where arrays are arranged in row-major order and hence the nth dimension of an n-dimensional array is the contiguous dimension. For simplicity, it is assumed that two references exhibit self-spatial reuse only if they have the same subscripts in all dimensions except the contiguous dimension. This means the first test for spatial reuse is similar to that of temporal reuse ignoring the contiguous dimension subscript. The second test is to ensure that the subscripts of the last contiguous dimension differ by a value that is less than the cache line size. Given an access X$[Ai + C]$, the entries of the last row of matrix A are coefficients of the affine function used as a subscript in the contiguous dimension. We define a new matrix A_s that is obtained by setting all the columns in the last row of A to 0. If the kernel of A_s is non-null and the subscripts of the contiguous dimension differ by a value less than the cache line size, then the reference exhibits self-spatial reuse. Similarly, group-spatial reuse is present between two references $Ai_1 + C_1$ and $Ai_2 + C_2$ if there is a solution to $A_s \times d = (C_1 - C_2)$. Let $ic_1, ic_2, \ldots ic_k$ denote the iteration counts of loops that carry temporal reuse in all but the contiguous dimension. Let s be the stride along the contiguous dimension and cls be the cache line size. Then, the total spatial reuse is given by $\prod_i ic_i \times \frac{cls}{s}$.

As described earlier, reuse does not always translate to locality. The term *data footprint* of a loop invocation refers to the total amount of data accessed by an invocation of the loop. It is difficult to exactly determine the size of data accessed by a set of loops, so it is estimated based on factors such as estimated iteration count. The term *localized iteration space* [21] refers to the subspace of the iteration space whose data footprint is smaller than the cache size. Reuse translates into locality only if the intersection of the reuse distance vector space with the localized iteration vector space is not empty.

5.3   Loop Interchange

Loop interchange changes the order of loops in a perfect loop nest to improve spatial locality. Consider the loop in Figure 5.2. This loop adds two $m \times n$ matrices. Since multidimensional arrays in C are stored in row-major order, two consecutive accesses to the same array are spaced far apart in memory. If all the three matrices do not completely fit within the cache, every access to a, b, and c will miss in the cache, but a compiler can transform the loop into the following equivalent loop:

```
for (i=0; i< m; i++){
    for(j = 0; j< n; j++){
        c[i][j] = a[i][j] + b[i][j];
    }
}
```

The above loop is semantically equivalent to the first loop but results in fewer conflict misses.

As this example illustrates, the order of nesting may be unimportant for correctness and yet may have a significant impact on the performance. The goal of loop interchange is to find a suitable nesting order that reduces memory access latency while retaining the semantics of the original loop. Loop interchange comes under a category of optimizations known as *unimodular transformations*. A loop transformation is called unimodular if it transforms the dependences in a loop by multiplying it with a matrix, whose determinant has a value of either -1 or 1. Other unimodular transformations include loop reversal, skewing, and so on.

5.3.1   Legality of Loop Interchange

It is legal to apply loop interchange on a loop nest if and only if all dependences in the original loop are preserved after the interchange. Whether an interchange preserves dependence can be determined by looking at the direction vectors of the dependences after interchange. Consider a loop nest $N = (L_1, L_2, \ldots L_n)$. Let $\mathbf{D(i, j)}$ be the direction vector of a dependence in this loop nest. If the order of loops in the loop nest is permuted by some transformation, then the direction vector corresponding to the dependence in the permuted loop nest can be obtained by permuting the entries of $\mathbf{D(i, j)}$. To understand this, consider the iteration vectors i and j corresponding to the source and sink of the dependence. A permutation of the loop nest results in a corresponding permutation of the components of i and j. This permutes the distance vector $\mathbf{d(i, j)}$ and hence the direction vector $\mathbf{D(i, j)}$. To determine whether N can be permuted to $N' = (L_{i_1}, L_{i_2}, \ldots L_{i_n})$, the same permutation is applied to the direction vectors of all dependences, and the resultant direction vectors are checked for validity.

```
for (j=0; j< n; j++){
    for(i = 0; i< m; i++){
        c[i][j] = a[i][j] + b[i][j];
    }
}
```

FIGURE 5.2   A suboptimal loop ordering.

As an example, consider the loop

```
for(i=1; i<m; i++){
    for(j=1; j<n; j++){
        a[i][j] = a[i][j-1] + a[i-1][j+1];
    }
}
```

This loop nest has two dependences:

- Dependence from a[i][j-1] to a[i][j], represented by the distance vector $(0, 1)$ and the direction vector $(=, <)$
- Dependence from a[i-1][j+1] to a[i][j], represented by the distance vector $(1, -1)$ and the direction vector $(<, >)$

Loop interchange would permute a direction vector by swapping its two components. While the direction vector $(=, <)$ remains valid after this permutation, the vector $(<, >)$ gets transformed to $(>, <)$, which is an invalid direction vector. Hence, the loop nest above cannot be interchanged.

5.3.2 Dependent Loop Indices

Even if interchange does not violate any dependence, merely swapping the loops may not be possible if the loop indices are dependent on one another. For example, the loop in Figure 5.1 can be rewritten as

```
for (j=0; j<n; j++){
    for(i=j; i<j+m; i++){
        row_sum[i-j] += matrix[i-j][j];
    }
}
```

Except for the change in the bounds of the inner loop index, the loop is semantically identical to the one in Figure 5.1. It has the same dependence vectors, so it must be legal to interchange the loops. However, since the inner loop index is dependent on the outer loop index, the two loops cannot simply be swapped. The indices of the loops need to be adjusted after interchange. This can be done by noting that loop interchange transposes the iteration space of the loop. Figure 5.3 shows the iterations of the original loop for $n = 3$ and $m = 4$. In Figure 5.4, the iteration space is transposed. This corresponds to the following interchanged loop.

```
for(i=0; i < m+n-1 ; i++){
    for(j=max(0, i-m+1); j < min(n, i+1); j++){
        row_sum[i-j] += matrix[i-j][j];
    }
}
```

5.3.3 Profitability of Loop Interchange

Even if loop interchange can be performed on a loop nest, it is not always profitable to do so. In the example shown in Figure 5.2, it is obvious that traversing the matrices first along the rows and then along the columns is optimal, but when the loop body accesses multiple arrays, each of the accesses may prefer a different loop order. For example, in the following loop,

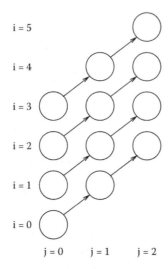

FIGURE 5.3   Iteration space before interchange.

```
for (j=0; j<m; j++){
    for(i=0; i<n; i++){
        sum += (matrix[i][j]*row[j]);
    }
}
```

the loop order is suboptimal for accesses to the `matrix` array. However, this is the best loop order to access the `row` array. In this loop nest, the element `row[j]` shows self-temporal reuse between the iterations of the inner loop. Hence, it needs to be accessed just once per iteration of the outer loop. If the loops are interchanged, this array is accessed once per iteration of the inner loop. If the size of the `row` array is much larger than the size of the cache, then there is no reuse across outer loop iterations.

For this loop nest:

- Number of misses to `matrix` = $n \times m$
- Number of misses to `row` = $\frac{n}{cls}$
- Total number of misses = $n \times m + \frac{n}{cls}$

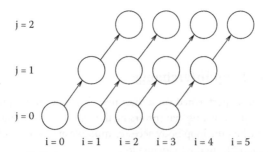

FIGURE 5.4   Iteration space after interchange.

If the loops are interchanged, then

- Number of misses to `matrix` = $\frac{n \times m}{cls}$
- Number of misses to `row` = $\frac{m \times n}{cls}$
- Total number of misses = $2 \times \frac{n \times m}{cls}$

In the above equations, *cls* denotes the size of the cache line in terms of elements of the two arrays. From this set of equations, it is clear that the interchanged loop is better only when the cache line can hold more than two elements of the two arrays.

5.3.4 A Loop Permutation Algorithm

In theory, one could similarly determine the right order for any given loop nest, but when the maximum depth of a perfect loop nest is n, the number of possible loop orderings is $n!$, and computing the total number of misses for those $n!$ loops is prohibitive for a large value of n. Instead, we can use some simple set of heuristics to determine an order that is likely to be profitable.

The algorithm in Figure 5.5 computes a good order of loops for the given loop nest N. For every loop in N, it computes cache_misses(L), which is the estimated number of cache misses when the loop L is made the inner loop. To compute cache_misses(L), the algorithm first computes a set. Each element of this set is either a single array reference or a group of array references that have group reuse. For instance, in the following loop,

```
for (i=0; i<N; i++)
    sum += A[i] + A[i+1];
```

Input: Loop nest $N = (L_1, L_2, \ldots, L_n)$
Output: Good loop order $N' = (L_{g_1}, L_{g_2}, \ldots, L_{g_n})$

for every loop L_i in N **do**
    cache_misses(L_i) = 0
    RGSet = { reuse groups in N }
    for every access group RG in RGSet **do**
        A = leading reference in RG
        if index of L_i is not in A's subscripts **then**
            cache_misses(L_i, A) = 1
        else
          if stride(A, L_i) > *cls* **then**
            cache_misses(L_i, A) = # iterations of L_i
          else
            cache_misses(L_i, A) = # iterations of L_i * $\frac{stride(A,L_i)}{cls}$
          end if
          for L_j in N, $L_i \neq L_j$ **do**
            cache_misses(L_i,A) *= # iterations of L_j
          end for
        end if
        cache_misses(L_i) += cache_misses(L_i, A)
    end for
end for

$N' = \{L_i$ in N sorted by cache_misses(L_i)$\}$

FIGURE 5.5 An algorithm to obtain a good permutation of a loop nest.

array references A[i] and A[i+1] belong to the same reuse group. Only the leading array reference in a reuse group is used for calculating cache_misses. If a reference does not have L_i in the subscripts, its cost is considered as 1. This is because such a reference will correspond to the same memory location for all iterations of L_i and will therefore miss only once during the first access. If a reference has the index of L_i in one of its subscripts, the stride of the reference is first computed. $stride(A, L_i)$ is defined as the difference in address in accesses of A in two consecutive iterations of L_i, where the ith component of the iteration vector differs by 1. As an example, $stride(X[i][j], i)$ is 256 in the following code:

```
char X[128][128];
for (i=0; i<128; i+=2){
    for(j=0; j<128; j++){
        sum += X[i][j];
    }
}
```

If the stride is greater than the cache line size, then every reference to A will miss in the cache, so cache_misses is incremented by the number of iterations of L_i. If the stride is less than cache line size, then only one in every $\frac{cls}{stride}$ accesses will miss in the cache. Hence, the number of iterations of L_i is divided by $\frac{cls}{stride}$ and added to cache_misses(L_i). Then, for each reference, cache_misses(L_i) is multiplied by the iteration counts of other loops whose indices are used as some subscript of A. Finally, the cache_misses for all references are added together.

After cache_misses(L_i) for each loop L_i is computed, the loops are sorted by cache_misses in descending order, and this is to obtain a good loop order. In other words, the loop with the lowest estimated cache misses is a good candidate for the innermost loop, the loop with the next lowest cache misses is a good candidate for the next innermost loop, and so on.

To see how this algorithm works, it is applied to the loop

```
double matrix[m][n], row[m];
for (j=0; j<n; j++){
    for(i=0; i<m; i++){
        sum += (matrix[i][j]*row[j]);
    }
}
```

Let L_i denote the loop with induction variable i and L_j denote the loop with induction variable j. For the loop nest with i as the induction variable:

1. cache_misses(L_i, row) $= n$, since i is not a subscript in row.
2. cache_misses(L_i, matrix) $= m*n$. Here we assume that n is a large enough number so that stride(n, L_i) is greater than the cache line size.
3. cache_misses(L_i) $= n + n*m$.

For the loop L_j:

1. cache_misses(L_j, row) $= m*n*$sizeof(double)/cls
2. cache_misses(L_j, matrix) $= m*n*$sizeof(double)/cls
3. cache_misses(L_j) $= 2*m*n*$sizeof(double)/cls

If the cache line can hold at least two doubles, cache_misses(L_j) is less than cache_misses(L_i), and therefore L_j is the candidate for the inner loop position.

While the algorithm in Figure 5.5 determines a good loop order in terms of profitability, it is not necessarily a valid order, as some of the original dependences may be violated in the new order. The order produced by the algorithm is therefore used as a guide to obtain a good legal order. This can be done in the following manner:

1. Pick the best candidate for the outermost loop among the loops that are not yet in the best legal order.
2. Assign it to the outermost legal position in the best legal order.
3. Repeat steps 1 and 2 until all loops are assigned a place in the best legal order.

Kennedy and Allen [13] show why the resulting order is always a legal order.

5.4 Loop Blocking

A very common operation in many scientific computations is matrix multiplication. The code shown below is a simple implementation of the basic matrix multiplication algorithm.

```
double a[m1][n1], b[n1][n2], c[m1][n2];
for (i=0; i<m1; i++){
    for(j=0; j<n2; j++){
        c[i][j] = 0;
        for (k=0; k < n1; k++){
            c[i][j] += a[i][k]*b[k][j];
        }
    }
}
```

This code has a lot of data reuse. For instance, every row of matrix a is used to compute n2 different elements of the product matrix c. However, this reuse does not translate to locality if the matrices do not fit in the cache. If the matrices do not fit in the cache, the loop also suffers from capacity misses that are not eliminated by loop interchange. In this case, the following transformed loop improves locality:

```
for(i=0; i<m1; i++){
    for(j=0; j<n2; j++){
        c[i][j]=0;
    }
}
for(i1=0; i1<m1; i1 += block_size){
    for(j1=0; j1<n2; j1 += block_size){
        for(k1=0; k1<n1; k1 += block_size){
            for (i=i1; i<min(m1, i1+block_size); i++){
                for(j=j1; j<min(n2, j1+block_size); j++){
                    for (k=k1; k < min(n1, k1+block_size); k++){
                        c[i][j] += a[i][k]*b[k][j];
                    }
                }
            }
        }
    }
}
```

First the initialization of the matrix c is separated out from the rest of the loop. Then, a transformation known as *blocking* or *tiling* is applied. If the value of block_size is chosen carefully, the reuse in the innermost three loops translates to locality. Figures 5.6 and 5.7 show the iteration space of a two-dimensional loop nest before and after blocking. In this example, the original iteration space has been covered by using four nonoverlapping rectangular tiles or blocks. In general, the tiles can take the shape of a *parallelepiped* for an *n*-dimensional iteration space. A detailed discussion of tiling shapes can be found in Rajopadhye [24].

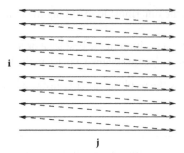

FIGURE 5.6   Iteration space before blocking.

The basic transformation used in blocking is called *strip-mine and interchange.* Strip mining transforms a loop into two nested loops, with the inner loop iterating over a small strip of the original loop and the outer loop iterating across strips. Strip mining the innermost loop in the matrix multiplication example results in the following loop nest:

```
for (i=0; i<m1; i++){
    for(j=0; j<n2; j++){
        for (k1=0; k1 < n1; k1 += block_size){
            for (k=k1; k < min(n1, k1+block_size); k++){
                c[i][j] += a[i][k]*b[k][j];
            }
        }
    }
}
```

Assuming `block_size` is smaller than n1, the innermost loop now executes fewer iterations than before, and there is a new loop with index `k1`. After strip mining, the loop with index `k1` is interchanged with the two outer loops, resulting in blocking along one dimension:

```
for (k1=0; k1 < n1; k1 += block_size){
    for (i=0; i<m1; i++){
        for(j=0; j<n2; j++){
            for (k=k1; k < min(n1, k1+block_size); k++){
                c[i][j] += a[i][k]*b[k][j];
            }
        }
    }
}
```

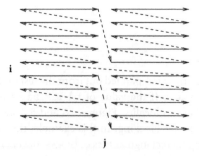

FIGURE 5.7   Iteration space after blocking.

Doing strip-mine and interchange on the i and j loops also results in the blocked matrix multiplication code shown earlier. An alternative approach to understand blocking in terms of clustering and tiling matrices is discussed by Rajopadhye [24].

5.4.1 Legality of Strip-Mine and Interchange

Let L_k be a loop which is to be strip mined into an outer loop L_k' and an inner loop L_k''. Let L_j be the loop with which the loop L_k'' is to be interchanged. Strip mining L_k is always a legal transformation since it does not alter any of the existing dependences and only relabels the iteration space of the loop. Thus, the legality of blocking depends on the legality of the interchange of L_k' and L_j, but determining the legality of this interchange requires strip mining L_k, which necessitates the recomputation of the direction vectors of all the dependences. A faster alternative is to test the legality of interchanging L_j with L_k instead. While the legality of this interchange ensures the correctness of blocking, this test is conservative and may prevent valid strip-mine and interchange in some cases [13].

5.4.2 Profitability of Strip-Mine and Interchange

We now discuss the conditions under which strip-mine and interchange transformation is profitable, given the strip sizes, which determine the block size. For a discussion of optimal block sizes refer to Rajopadhye [24] (Chapter 15 in this text).

Given a loop nest $N = (L_1, L_2 \dots L_n)$, a loop L_k and another loop L_j, where L_k is nested within L_j, the profitability of strip mining L_k and interchanging the by-strip loop with L_j depends on the following:

- The reuse carried by L_j
- The data footprint of all inner loops of L_k
- The cost of strip mining L_k

The goal of strip mining L_k and interchanging the by-strip loop with L_j is to ensure that reuse carried by L_j is translated into locality. For this to happen, L_j must carry reuse between its iterations. This can happen under any of the following circumstances:

- There is some dependence carried by L_j. If L_j carries some dependence, it means an iteration of L_j reuses a location accessed by some earlier iteration of L_j.
- There is an array index that does not have the index of L_j in any of its subscripts.
- The index of L_j is used as a subscript in the contiguous dimension, resulting in spatial reuse.

The data footprint of L_j must be larger than the cache size, as otherwise the reuse carried by L_j lies within the localized iteration space. The data footprint of L_k must also be larger than the cache size, as otherwise it is sufficient to strip-mine some other loop between L_j and L_k. Finally, the benefits of reuse must still outweigh the cost of strip mining L_k. Strip mining can cause a performance penalty in two ways:

- If the strips are not aligned to cache line boundaries, it would reduce the spatial locality of array accesses that have the index of L_k as the subscript in the contiguous dimension.
- Every dependence carried by the loop L_k shows decreased temporal locality.

Typically, the cost of doing strip-mine and interchange is small and is often outweighed by the benefits.

5.4.3 Blocking with Skewing

In loop nests where interchange violates dependences, *loop skewing* can be applied first to enable loop interchange. Consider the loop iteration shown in Figure 5.8. The outer loop of this nest is indexed by **i**, and the inner loop by **j**. Even if it is profitable, the diagonal edges from iteration vector $(\mathbf{i}, \mathbf{j})$ to $(\mathbf{i} + \mathbf{1}, \mathbf{j} - \mathbf{1})$

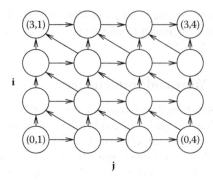

FIGURE 5.8   Dependences before skewing.

prevent loop interchange, but the same loop can be transformed to the one in Figure 5.9, which allows loop interchange. This transformation is known as *loop skewing*.

Given two loops, the inner loop can be skewed with respect to the outer loop by adding the outer loop index to the inner loop index. Consider a dependence with direction vector $(<, >)$. This pair of loops does not permit loop interchange. Consider two iteration vectors $(\mathbf{i1}, \mathbf{j1})$ and $(\mathbf{i2}, \mathbf{j2})$, which are the source and target, respectively, of the dependence. Let $(\mathbf{d1}, \mathbf{d2})$ be the distance vector corresponding to this dependence, where $(\mathbf{d1}, \mathbf{d2}) = (\mathbf{i1} - \mathbf{i2}, \mathbf{j1} - \mathbf{j2})$. The goal is to transform this distance vector into $(\mathbf{d1}, \mathbf{d2} + f \times \mathbf{d1})$, where $f \times \mathbf{d1} >= |\mathbf{d2}|$, so that the direction vector becomes either $(<, <)$ or $(<, =)$, which permits loop interchange. This can be achieved by multiplying the outer loop index and adding the product to the inner loop index. Thus, the two iteration vectors become $(\mathbf{i1}, \mathbf{j1} + \mathbf{f}.\mathbf{i1})$ and $(\mathbf{i2}, \mathbf{j2} + \mathbf{f}.\mathbf{i2})$, and the distance vector becomes $(\mathbf{i1} - \mathbf{i2}, \mathbf{j1} - \mathbf{j2} + \mathbf{f}.(\mathbf{i1} - \mathbf{i2}))$, which is equal to $(\mathbf{d1}, \mathbf{d2} + \mathbf{f}.\mathbf{d1})$.

The loop shown in Figure 5.8 applies an averaging filter m times on an array as shown below:

```
for (i=0; i<m; i++){
    for(j=1; j<n-1; j++){
        a[j] = (a[j-1]+a[j]+a[j+1])/3;
    }
}
```

The dependences in this loop are characterized by three dependence distance vectors $(0, 1), (1, 0)$, and $(1, -1)$, which are depicted in Figure 5.8, for the case where m takes a value of 4 and n takes a value of 6. While the $\mathbf{j}$ loop can be strip mined into two loops with induction variables $\mathbf{j1}$ and $\mathbf{j}$, the $\mathbf{j1}$ loop cannot be interchanged with the $\mathbf{i}$ loop because of the dependence represented by $(-1, 1)$. We skew the inner $\mathbf{j}$

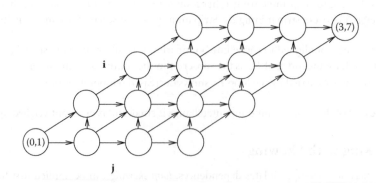

FIGURE 5.9   Dependences after skewing.

loop with respect to the outer **i** loop. Since a distance vector of $(1, 0)$ allows interchange, the value of f is chosen as 1. Figure 5.9 shows the dependences in the loop after applying skewing. This corresponds to the following loop:

```
for(i=0; i<m; i++){
    for(j=i+1; j<i+n-1; j++){
        a[j-i] = (a[j-i-1]+a[j-i]+a[j-i+1])/3;
    }
}
```

Strip mining the loop with index **j** results in the following code:

```
for(i=0; i<m; i++){
    for(j1=i+1; j1<i+n-1; j1+=block_size){
        for(j=j1; j<min(i+n-1, j1+block_size); j++){
            a[j-i] = (a[j-i-1]+a[j-i]+a[j-i+1])/3;
        }
    }
}
```

The outer two loops can now be interchanged after suitably adjusting the bounds. This results in the tiled loop:

```
for(j1=1; j1< m+n-2; j1+= block_size){
    for(i=max(0, j1-n+2); i< min(j1, m-1); i++){
        for(j=j1; j<min(i+n-1, j1+block_size); j++){
            a[j-i] = (a[j-i-1]+a[j-i]+a[j-i+1])/3;
        }
    }
}
```

5.5   Other Loop Transformations

5.5.1   Loop Fusion

When there is reuse of data across two independent loops, the loops can be fused together, provided their indices are compatible. For example, the following set of loops:

```
max = a[0];
for(i=1; i<N; i++){
    if (a[i] > max)
        max = a[i];
}
min = a[0];
for(i=1; i<N; i++){
    if (a[i] < min)
        min = a[i];
}
```

can be fused together into a single loop:

```
max = a[0];
min = a[0];
for(i=1; i<N; i++){
    if (a[i] > max)
        max = a[i];
    if (a[i] < min)
        min = a[i];
}
```

If the value of N is large enough that the array does not fit in the cache, the original loop suffers from conflict misses that are minimized by the fused loop. Loop fusion is often combined with *loop alignment*. Two loops may have compatible loop indices, but the bounds may be different. For example, if the first loop computes the maximum of all N elements and the second loop computes the minimum of only the first $N/2$ elements, the indices have to be aligned. This is done by first splitting the max loop into two, one iterating over the first $N/2$ elements and the other over the rest of the elements. Then the loop computing the minimum can be fused with the first portion of the loop computing the maximum.

5.5.2   Loop Fission

Loop fission or loop distribution is a transformation that does the opposite of loop fusion by transforming a loop into multiple loops such that the the body of the original loop is distributed across those loops. To determine whether the body can be distributed, the program dependence graph (PDG) of the loop body is constructed first. Different strongly connected components of the PDG can be distributed into different loops, but all nodes in the same strongly connected component have to be in the same loop. For example, the following loop,

```
for (i=0; i<n; i++){
    a[i] = a[i-1]+b[i-1];
    b[i] = k*a[i];
    c[i] = c[i-1]+1;
}
```

can be distributed into

```
for (i=0; i<n; i++){
    a[i] = a[i-1]+b[i-1];
    b[i] = k*a[i];
}

for (i=0; i<n; i++){
    c[i] = c[i-1]+1;
}
```

since the statement c[i] = c[i-1]+1 is independent of the other two statements, which are dependent on each other. Loop fission reduces the memory footprint of the original loop. This is likely to reduce the capacity misses in the original loop.

5.6   Data Prefetching

Data prefetching differs from the other loop optimizations discussed above in some significant aspects:

- Instead of transforming the data access pattern of the original code, prefetching introduces additional code that attempts to bring in the cache lines that are likely to be accessed in the near future.

- Data prefetching can reduce all three types of misses. Even when it is unable to avoid a miss, it can reduce the time to service that miss.
- While the techniques described above are applicable only for arrays, data prefetching is a much more general technique.
- Prefetching requires some form of hardware support.

Consider the simple loop

```
for (i=0; i<N; i++){
    sum = sum + A[i];
}
```

This loop computes the sum of the elements of the array A. The cache misses to the array A are all cold misses. To avoid these misses, prefetching code is inserted in the loop to bring into the cache an element that is required some pd iterations later:

```
for (i=0; i<N; i++){
    prefetch(A[i+pd]);
    sum = sum + A[i];
}
```

The value *pd* above is referred to as the *prefetch distance*, which specifies the number of iterations between the prefetching of a data element and its actual access. If the value of *pd* is carefully chosen, then on most iterations of the loop, the array element that is accessed in that iteration would already be available in the cache, thereby minimizing cold and capacity misses.

5.6.1   Hardware Support

In the above example, we have shown `prefetch` as a function routine. In practice, `prefetch` corresponds to some machine instruction. The simplest option is to use the load instruction to do the prefetch. By loading the value to some unused register or a register hardwired to a particular value and making sure the instruction is not removed by dead code elimination, the array element is brought into the cache. While this has the advantage of requiring no additional support in the instruction set architecture (ISA), this approach has some major limitations. First, the load instructions have to be *nonblocking*. A load instruction is nonblocking if the load miss does not stall other instructions in the pipeline that are independent of the load. If the load blocks the pipeline, the prefetches themselves would suffer cache misses and block the pipeline, defeating the very purpose of prefetching. This is not an issue in most modern general-purpose processors, where load instructions are nonblocking. The second, and major, limitation is that since executing loads can cause exceptions, prefetches might introduce new exceptions in the transformed code that were not there in the original code. For instance, the code example shown earlier might cause an exception when it tries to access an array element beyond the length of the array. This can be avoided by making sure an address is prefetched only if it is guaranteed not to cause an exception, but this would severely limit the applicability of prefetching. To avoid such issues, some processors have special prefetch instructions in their instruction set architecture. These instructions do not block the pipeline and silently ignore any exceptions that are raised during their execution. Moreover, they typically do not use any destination register, thereby improving the register usage in code regions with high register pressure.

5.6.2   Profitability of Prefetching

Introducing data prefetching does not affect the correctness of a program, except for the possibility of spurious exceptions discussed above. When there is suitable ISA support to prevent such exceptions, the compiler has to consider only the profitability aspect when inserting prefetches. For a prefetch instruction to be profitable, it has to satisfy two important criteria: *accuracy* and *timeliness*.

Accuracy refers to the fact that a prefetched cache line is actually accessed later by the program. Prefetches may be inaccurate because of the compiler's inability to determine runtime control flow. Consider the following loop:

```
for (i=0; i<N; i++){
    if (A[foo(i)] > 100)
        sum = sum+B[i];
}
```

A prefetch for the array B inserted at iteration i is accurate only if the element B[i+pd] is accessed later, but this depends on A[foo(i+pd)], and it may not always be safe to compute foo(i+pd). Even if it is safe to do so, the cost of invoking foo may far outweigh the potential benefits of prefetching, forcing the compiler to insert prefetches unconditionally for every future iteration. As will be described later, prefetching of a loop traversing a recursive data structure could also be inaccurate. The accuracy of a prefetch is important because useless prefetches might evict useful data — data that will definitely be accessed in the future — from the cache line and increase the memory traffic, thereby degrading the performance.

Accuracy alone does not guarantee that a prefetch is beneficial. If we set the value of pd to be 0, then we can guarantee that the prefetch is accurate by issuing it right before the access, but this obviously does not result in any benefit. A prefetch is *timely* if it is issued at the right time, so that, when the access happens, it finds the data in the cache. The prefetch distance determines the timeliness of a prefetch for a given loop. Let h be the average schedule height of a loop iteration and pd be the prefetch distance. Then the number of cycles separating the prefetch and the actual access is roughly $pd * h$. This value must be greater than the access latency without prefetch for the access to be a hit in the L1 cache. At the same time, if this value is too high, the probability of the prefetched cache line being displaced subsequently, before the actual access, increases, thereby reducing the benefits of the prefetch. Thus, determining the right prefetch distance for a given loop is a crucial step in prefetching implementations.

A third factor that determines the benefits of prefetching is the overhead involved in prefetching. The following example shows how the prefetching overhead could negate the benefits of prefetching. Consider the code fragment

```
char c[MAX];
for(i=0; i<MAX; i++){
    prefetch(c[i+pd]);
    sum += c[i];
}
```

This code is the same as the example shown earlier, except that it sums up an array of chars. Assume that this code is executed on a machine with an issue width of 1, and the size of the L1 cache line is 64 bytes. Let c be the number of cycles required to service a cache miss. Under this scenario,

$$\text{Number of cache misses in the absence of prefetching} = \left\lceil \tfrac{MAX}{64} \right\rceil$$
$$\text{Prefetching overhead} = MAX \text{ cycles}$$

Even if all the misses are prefetched,

$$\text{Cycles saved by prefetching} = \left\lceil \tfrac{MAX}{64} \right\rceil \times c$$

For prefetching to be beneficial,

$$\left\lceil \frac{MAX}{64} \right\rceil \times c > MAX$$
$$\Rightarrow c > 64$$

Input: Loop nest N
Output: Loop nest N with prefetches inserted if profitable

perform locality analysis
identify leading references
for every leading reference lref **do**
    pred(lref) = compute_prefetch_predicate(lref)
end for
decompose loops based on pred(lref)
schedule prefetches

FIGURE 5.10   A loop prefetching algorithm.

If the miss latency for this access without prefetching is <64 cycles, then the overhead of prefetching outweighs its benefits. In such cases, optimizations such as loop splitting or loop unrolling can make prefetching still be profitable. If the value of *MAX* is determined, either statically or using profiling, to be a large number, unrolling the loop 64 times will make prefetching profitable. After unrolling, only one prefetch is issued per iteration of the unrolled loop, and hence the prefetching overhead reduces to $\frac{MAX}{64}$ cycles. For prefetching to be beneficial in this unrolled loop,

$$\left\lceil \frac{MAX}{64} \right\rceil \times c > \frac{MAX}{64} \tag{5.1}$$

$$\Rightarrow c > 1 \tag{5.2}$$

Thus, as long as the cost of a cache miss is more than one cycle, prefetching will be beneficial.

5.6.3   Prefetching Affine Array Accesses

Figure 5.10 outlines an algorithm by Mowry et al. [21] that issues prefetches for array accesses within a loop.

The prefetching algorithm consists of the following steps:

1. Perform locality analysis. The first step of the algorithm is to obtain reuse distance vectors and intersect them with the localized iteration space to determine the locality exhibited by the accesses in the loop.
2. Identify accesses requiring prefetches. All accesses exhibiting self-reuse are candidates for prefetching. When accesses exhibit group reuse, only one of the accesses needs to be prefetched. Among references that exhibit group reuse, the one that is executed first is called the *leading reference*. It is sufficient to prefetch only the leading reference of each group.
3. Compute prefetch predicates. When a reference has spatial locality, multiple instances of that reference access the same cache line, so only one of them needs to be prefetched. To identify this, every instance of an access is associated with a *prefetch predicate*. An instance of a reference is prefetched only if the predicate associated with it is 1. Since all references are affine array accesses, the predicates are some functions of the loop indices. As an example, consider the following loop nest:

```
for (i=0; i<m; i++){
    sum[0] += a[i];
}
```

Assuming that the arrays are aligned to cache line boundaries, the prefetch predicate for sum is (i==0) and for a is (a mod n)==0, where n is the number of elements of a in a cache line.

4. Perform loop decomposition. One way to ensure that a prefetch is inserted only when the predicate is true is to use a conditional statement based on the prefetch predicate or, if the architecture supports it, use predicate registers to guard the execution of the prefetch. This requires computing the predicate during every iteration of the loop, which takes some cycles, and issuing the prefetch on every iteration, which takes up issue slots. Since the prefetch predicates are well-defined functions of the loop indices, a better approach is to decompose the loop into sections such that all iterations in a particular section either satisfy the predicate or do not satisfy the predicate. The code above, with respect to the reference to array a, satisfies the predicate in iterations 0, n, 2n, and so on and does not satisfy the predicates in the rest of the iterations. There are many ways to decompose the loops into such sections. If a prefetch is required only during the first iteration of a loop, then the first iteration can be peeled out of the loop and the prefetch inserted only in the peeled portion. If the prefetch is required once in every n iterations, the loop can be unrolled n times so that in the unrolled iteration space, every iteration satisfies the predicate. However, depending on the unroll factor and the size of the loop body, unrolling might degrade the instruction cache behavior. In those cases, the original iterations that do not satisfy the prefetching predicate can be rerolled back. For instance, if the loop above is transformed into

```
for (i=0;  i<m;  i+=n){
    sum[0] += a[i];
    for(i1 = i+1;  i1 < min(m, i+n);  i1++){
        sum[0] += a[i1];
    }
}
```

all iterations of the inner loop do not satisfy the predicate, while all iterations of the outer loop satisfy the predicate. This process is known as *loop splitting*. This is performed for all distinct prefetch predicates.

5. Schedule the prefetches. A prefetch has to be timely to be effective. If $cycles_{miss}$ denotes the cache miss penalty in terms of cycles, then the prefetch has to be inserted that many cycles before the reference to completely eliminate the miss penalty. This can be achieved by software pipelining the loop, assuming that the latency of the prefetch instruction is $cycles_{miss}$. This will have the effect of moving the prefetch instruction $\frac{cycles_{miss}}{cycles_{loop}}$ iterations ahead of the corresponding access in the software pipelined loop.

Certain architecture-specific features can also be used to enhance the last two steps above. For example, the Intel Itanium® architecture has a feature known as *rotating registers*. This allows registers used in a loop body to be efficiently renamed after every iteration of a counted loop such that the same register name used in the code refers to two different physical registers in two consecutive iterations of the loop. The use of rotating registers to produce better prefetching code is discussed in [8].

5.6.4 Prefetching Other Accesses

Programs often contain other references that are predictable. These include:

- Arrays with subscripts that are not affine functions of the loop indices but still show some predictability of memory accesses
- Traversal of a recursive data structure such as a linked list, where all the nodes in the list are allocated together and hence are placed contiguously
- Traversal of a recursive data structure allocated using a custom allocator that allocates objects of similar size or type together

A compiler cannot statically identify and prefetch such accesses and usually relies on some form of runtime profiling to analyze these accesses. A technique known as *stride profiling* [35] is often used to identify such patterns. The goal of stride profiling is to find whether the addresses produced by a static load/store instruction exhibit any exploitable pattern. Let $a_1, a_2, a_3 \ldots a_n, a_{n+1}$ denote the addresses generated by a static load/store instruction. Then $s_1 = a_2 - a_1, s_2 = a_3 - a_2, \ldots s_n = a_{n+1} - a_n$ denote the access strides. An instruction is said to be *strongly single strided* if most of the s_is are equal to some S. In other words, there is some S such that $\frac{\sum_1^n (s_i == S)}{n}$ is close to 1. This profiled stride S is used in computing the prefetch predicate and scheduling of the prefetch. Some instructions may not have a single dominant stride but still show regularity. For instance, a long sequence of s_i may all be equal to S_1, followed by a long sequence of s_i being all equal to S_2, and so on. In other words, an access has a particular stride in one phase, followed by a different stride in the next phase, and so on. Such accesses are called *strongly multi-strided*. The following example illustrates how a strongly multi-strided access can be prefetched:

```
while (ptr != NULL){
    sum += *ptr;
    ptr = ptr->next;
}
```

If `ptr` is found to be strongly multi-strided, it can be prefetched as follows:

```
while (ptr != NULL){
    sum += *ptr;
    stride = ptr - prev;
    prefetch(ptr+k*stride);
    prev = ptr;
    ptr = ptr->next;

}
```

The loop contains code to dynamically calculate the stride. Assuming that phases with a particular stride last for a long time, the current observed stride is used to prefetch the pointer that is likely to be accessed k iterations later. Since the overhead of this prefetching is high, it must be employed judiciously after considering factors such as the iteration count of the loop, the average access latency of the load that is prefetched, and the length of the phases with a single stride.

Another type of access that is useful to prefetch is indirect access. In the code fragment

```
for (i=0; i<n; i++){
    box[i]->area = box[i]->length * box[i]->breadth;
}
```

the references to the array `box` can be prefetched. However, the references to `area`, `length`, and `breadth` may cause a large number of cache misses if the pointers stored in the `box` array do not point to contiguous memory locations. Profile-based prefetching techniques are also not helpful in this case. The solution is to perform *indirect prefetching*. After applying indirect prefetching, the loop would look like

```
for (i=0; i<n; i++){
    temp = box[i+pd];
    prefetch(temp);
    box[i]->area = box[i]->length * box[i]->breadth;
}
```

The pointer that would be dereferenced pd iterations later is loaded into a temporary variable, and a prefetch is issued for that address. The new loop requires a prefetch to `box[i+pd]`, as otherwise the load to `temp` could result in stalls that may negate any benefits from indirect prefetching.

5.7   Data Layout Transformations

The techniques discussed so far transform the code accessing the data so as to decrease the memory access latency. An orthogonal approach to reduce memory access latency is to transform the layout of data in memory. For example, if two pieces of data are always accessed close to each other, the spatial locality can be improved by placing those two pieces of data in the same cache line. Data layout transformations are a class of optimization techniques that optimize the layout of data to improve spatial locality. These transformations can be done either within a single aggregate data type or across data objects.

5.7.1   Field Layout

The fields of a record type can be classified as *hot* or *cold* depending on their access counts. It is often beneficial to group hot fields together and separate them from cold fields to improve cache utilization. Consider the following record definition:

```
struct S1 {
    char hot[4];
    char cold[60];
};
S1 s1[512];
```

Each instance of S1 occupies a single cache line, and the entire array s1 occupies 512 cache lines. The total size of this array is well above the size of L1 cache in most processors, but only 4 bytes of the above record type are used frequently. If the struct consists of only the field hot, then the entire array fits within an L1 cache.

Structure splitting involves separating a set of fields in a record type into a new type and inserting a pointer to this new type in the original record type. Thus, the above struct can be transformed to:

```
struct S1_cold {
    char cold[60];
};

struct S1 {
    char hot[4];
    struct S1_cold *cold_link;
};

S1 s1[512];
```

After the transformation, the array s1 fits in L1 cache of most processors. While this increases the cost of accessing cold, as it requires one more indirection, this does not hurt much because it is accessed infrequently. However, even this cost can be eliminated for the struct defined above, by transforming S1, as follows:

```
struct S1_cold {
    char cold[60];
};

struct S1_hot {
    char hot[4];
};
S1_hot s1_hot[512];
S1_cold s1_cold[512];
```

The above transformation is referred to as *structure peeling*. While peeling is always better than splitting in terms of access costs, it is sometimes difficult to peel a record type. For instance, if a record type has pointers to itself, then peeling becomes difficult.

While merely grouping the hot fields together is often useful, this method suffers from performance penalties when the size of the record is large. As an example, consider the following record definition and some code fragments that use the record:

```
struct S2 {
    char hot1[32];
    char hot2[32];
    char not_so_hot[32];
};

struct S2 s2[128];
...
for (i = 0; i<128; i++) // Invoked N times
    x += foo (s2[i].hot1);
...
for (i = 0; i<128; i++) // Invoked N times
    x += foo(s2[i].hot2);
...
for (i = 0; i<128; i++) // Invoked N/10 times
    sum3 += foobar(s2[i].not_so_hot) + foo(s2[i].hot1);
...
```

The fields hot1 and hot2 are more frequently accessed than not_so_hot, but not_so_hot is always accessed immediately before hot1. Hence placing not_so_hot together with hot1 improves spatial locality and reduces the misses to hot1. This fact is captured by the notion of *reference affinity*. Two fields have a high reference affinity if they are often accessed close to each other. In the above example, we have considered accesses within the same loop as affine. One could also consider other code granularities such as basic block, procedure, arbitrary loop nest, and so on.

Structure splitting, peeling, and re-layout involve the following steps:

1. Determine if it is safe to split a record type. Some examples of the unsafe behaviors include:

   (a) Implicit assumptions on offset of fields. This typically involves taking the address of a field within the record.

   (b) Pointers passed to external routines or library calls. This can be detected using pointer escape analysis.

   (c) Casting from or to the record type under consideration. Casting from type A to type B implicitly assumes a particular relative ordering of the fields in both A and B.

   If a record type is subjected to any of the above, it is deemed unsafe to transform.

2. Classify the fields of a record type as hot or cold. This involves computing the dynamic access counts of the structure fields. This can be done either using static heuristics or using profiling. Then fields whose access counts are above a certain threshold are labeled as hot and the other fields are labeled as cold.

3. Move the cold fields to a separate record and insert a pointer to this cold record type in the original type.

4. Determine an ordering of the hot fields based on the reference affinity between them. This involves the following steps:

   (a) Compute reference affinity between all pairs of fields using some heuristic.

   (b) Group together fields that have high affinity between them.

5.7.2 Data Object Layout

Data object layout attempts to improve the layout of data objects in stack, heap, or global data space. Some of the techniques for field layout are applicable to data object layout as well. For instance, the local variables in a function can be treated as fields of a record, and techniques for field layout can be applied. Similarly, all global variables can likewise be considered as fields of a single record type.

However, additional aspects of the data object layout problem add to its complexity. One such aspect is the issue of cache line conflicts. Two distinct cache lines containing two different data objects may conflict in the cache. If those two data objects are accessed together frequently, a large number of conflict misses may result. Conflicts are usually not an issue in field layout because the size of structures rarely exceeds the size of the cache, while global, stack, and heap data exceed the cache size more often.

Heap layout is usually more challenging because the objects are allocated dynamically. Placing a particular object at some predetermined position relative to some other object in the heap requires the cooperation of the memory allocator. Thus, all heap layout techniques focus on customized memory allocation that uses runtime profiles to guide allocation. The compiler has little role to play in most of these techniques.

A different approach to customized memory allocation, known as *pool allocation*, has been proposed by Lattner and Adve[14]. Pool allocation identifies the data structure instances used in the program. The allocator then tries to allocate each data structure instance in its own pool. Consider the following code example:

```
struct list;
struct tree;
struct linked_list{
    int n;
    struct linked_list *next;
};
struct tree{
    int n;
    struct tree *left;
    struct tree *right;
};
struct linked_list *l1, *l2;
struct tree *t;
...
```

Assume that after the necessary dynamic allocations, the pointer l1 points to the head of a linked list, l2 points to the head of a different linked list, and t points to the root of a binary tree. The memory for each of these three distinct data structure instances in the program would be allocated in three distinct pools. Thus, if a single data structure instance is traversed repeatedly without accessing the other instances, the cache will not be polluted by unused data, thereby improving the cache utilization. The drawback to this technique is that it does not profile the code to identify the access patterns and hence may cause severe performance degradation when two instances, such as l1 and l2 above, are concurrently accessed.

5.8 Optimizations for Instruction Caches

Modern processors issue multiple instructions per clock cycle. To efficiently utilize this ability to issue multiple instructions per cycle, the memory system must be able to supply the processor with instructions at a high rate. This requires that the miss rate of instructions in the instruction cache be very low. Several compiler optimizations have been proposed to decrease the access latency for instructions in the instruction cache.

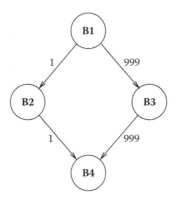

FIGURE 5.11   Control flow graph.

5.8.1   Code Layout in Procedures

A procedure consists of a set of basic blocks. The mapping from the basic blocks to the virtual address space can have a big impact on the instruction cache performance. Consider the section of control flow graph in Figure 5.11 that corresponds to an if-then-else statement. The numbers on the edges denote the frequency of execution of the edges. Thus, most of the time, the basic block B3 is executed after B1, and B2 is seldom executed. It is not desirable to place B2 after B1, as that could fill the cache line containing B1 with infrequently used code. The code layout algorithms use profile data to guide the mapping of basic blocks to the address space.

Figure 5.12 shows an algorithm proposed by Pettis and Hansen [22] to do profile-guided basic block layout. The algorithm sorts the edges of the control flow graph by the frequency of execution. Initially, every basic block is assumed to be in a *chain* containing just itself. A chain is simply a straight line path

> **Input:** control flow graph CFG of procedure P
> **Output:** relative order of basic blocks
>
> sort edges of the CFG by edge weight in descending order
> **for** every edge E in the sorted list **do**
>   B1 = Src(E)
>   B2 = Dest(E)
>   **if** B1 and B2 are not part of any chains **then**
>   form a chain B1-B2
>   **else if** if B1 is the tail of a chain C1 and B2 is the head of a chain C2 **then**
>   append C2 to C1
>   **end if**
> **end for**
> break cycles in CFG and perform a topological sort
> **for** every node B of CFG in topological sort order **do**
>   C = chain containing B
>   **if** basic blocks in C are not laid out **then**
>   lay out basic blocks of C in order
>   **end if**
> **end for**

FIGURE 5.12   Code layout algorithm.

in the control flow graph. The algorithm tries to merge the chains into longer chains by looking at each edge, starting from the one with the highest frequency. If the source of the edge is the tail node of a chain and the destination is the head node of a chain, the chains are merged. This process continues until all edges are processed. Then the algorithm does a topological sort of the control flow graph after breaking cycles. The chains corresponding to the nodes in the topological order are laid out consecutively in memory.

5.8.2   Procedure Splitting

While the above algorithm separates hot blocks from cold blocks within a procedure, a poor instruction cache performance still might result. To see this, consider a program with two procedures P1 and P2. Let the hot basic blocks in each procedure occupy one cache line and the cold basic blocks occupy one cache line. If the size of the cache is two cache lines, using the layout produced by the previous algorithm might result in the hot blocks of the two procedures getting mapped to the same cache line, producing a large number of cache misses.

A simple solution to this problem is to place the hot basic blocks of different procedures close to each other. In the above example, this will avoid conflict between the hot basic blocks of the two procedures. *Procedure splitting* [22] is a technique that splits a procedure into hot and cold sections. Hot sections of all the procedures are placed together, and the cold sections are placed far apart from the hot sections. If the size of the entire program exceeds the cache size, while the hot sections alone fit within the cache, then procedure splitting will result in the cache being occupied by hot code for a large fraction of time. This results in very few misses during the execution of the hot code, resulting in considerable performance improvement.

Procedure splitting simply classifies the basic blocks as hot or cold based on some threshold and moves the cold basic blocks to a distant memory location. For processors that require a special instruction to jump beyond a particular distance, all branches that conditionally jump to a cold basic block are redirected to a stub code, which jumps to the actual cold basic block. Thus, the transitions between hot and cold regions have to be minimal, as they require two control transfer instructions including the costlier special jump instruction.

5.8.3   Cache Line Coloring

Another technique known to improve placement of blocks across procedures is cache line coloring [9]. Cache line coloring attempts to place procedures that call each other in such a way that they do not map to the same set of cache blocks.

Consider the following code fragment:

```
void foo(){
   bar();
   ...
}
int foobar(){
   for (i=0; i<n; i++)
       foo();
   ...
}
```

The procedure `foobar` contains a loop in which `foo` is called and `foo` calls `bar`. The procedures `foo` and `bar` must not map to the same cache line, as that will result in a large number of misses. If it is assumed that the size of the hot code blocks in the program exceeds the size of the instruction cache, procedure splitting might still result in the code blocks of `foo` and `bar` being mapped to the same set of cache lines.

The input to cache line coloring is the call graph of a program with weighted undirected edges, where the weight on the edge connecting P1 and P2 denotes the number of times P1 calls P2 or vice versa. The nodes are labeled with the number of cache lines required for that procedure. The output of the technique is a mapping between procedures to a set of cache lines. The algorithm tries to minimize the cache line conflicts between nodes that are connected by edges with high edge weights. The edges in the call graph are first sorted by their weights, and each edge is processed in the descending order of weights. If both the nodes connecting an edge have not been assigned to any cache lines, they are assigned nonconflicting cache lines. If one of the nodes is unassigned, the algorithm tries to assign nonconflicting cache lines to the unassigned node without changing the assignment of the other node. If nonconflicting colors cannot be found, it is assigned a cache line close to the other node. If both nodes have already been assigned cache lines, the technique tries to reassign colors to one of them based on several heuristics, which try to minimize the edge weight of edges connected to conflicting nodes.

The main drawback of this technique is that it only considers one level of the call depth. If there is a long call chain, the technique does not attempt to minimize conflicts between nodes in this chain that are adjacent to each other. In addition, the technique assumes that the sizes of procedures are multiples of cache line sizes, which may result in holes between procedures.

5.9   Summary and Future Directions

The various optimizations for memory hierarchies described in this chapter are essential components of optimizing compilers targeting modern architectures. While no single technique is a silver bullet for bridging the processor–memory performance gap, many of these optimizations complement each other, and their combination helps a wide range of applications. Table 5.1 summarizes how each of the optimizations achieve improved memory performance.

While these optimizations were motivated by the widening performance gap between the processor and the main memory, the recent trend of stagnant processor clock frequencies may narrow this gap. However, the stagnation of clock frequencies is accompanied by another trend — the prevalence of *chip multiprocessors* (CMPs). CMPs pose a new set of challenges to memory performance and increase the importance of compiler-based memory optimizations. Compiler techniques need to focus on multithreaded applications, as more applications will become multi-threaded to exploit the parallelism offered by CMPs. Compilers also have to efficiently deal with the changes in the memory hierarchy that may have some levels of private caches and some level of caches that are shared among the different cores. The locality in the shared levels of the hierarchy for an application is influenced by applications that are running in the other cores of the CMP.

TABLE 5.1   Classification of optimizations for memory hierarchies

Optimization	Performance improvement
Loop interchange	Fewer conflict misses
Loop blocking	Fewer capacity misses
Loop fusion	Fewer capacity misses
Loop fission	Fewer conflict misses
Data prefetching	Fewer cold, capacity, and conflict misses
	Misses partially hidden
Data layout	Fewer conflict misses, improved cache utilization
Instruction cache optimizations	Fewer conflict misses, improved cache utilization

5.10   References

The work of Abu-Sufah et al. [1] was among the first to look at compiler transformations to improve memory locality. Allen and Kennedy [3] proposed the technique of automatic loop interchange. Loop tiling was proposed by Wolfe [32, 33], who also proposed loop skewing [31]. Several enhancements to tiling including tiling at the register level [11], tiling for imperfectly nested loops [2], and other improvements [12, 27] have been proposed in the literature. Wolf and Lam [29, 30] proposed techniques for combining various unimodular transformations and tiling to improve locality. Detailed discussion of various loop restructuring techniques can be found in textbooks written by Kennedy and Allen [13] and Wolfe [34] and in the dissertations of Porterfield [23] and Wolf [28].

Software prefetching was first proposed by Callahan et al. [5] and Mowry et al. [19–21]. Machine-specific enhancements to software prefetching have been proposed by Santhanam et al. [26] and Doshi et al. [8]. Luk and Mowry [15] proposed some compiler techniques to prefetch recursive data structures that may not have a strided access pattern. Saavedra-Barrera et al. [25] discuss the combined effects of unimodular transformations, tiling, and software prefetching. Mcintosh [17] discusses various compiler-based prefetching strategies and evaluates them.

Hundt et al. [10] developed an automatic compiler technique for structure layout optimizations. Calder et al. [4] and Chilimbi et al. [6, 7] proposed techniques for data object layout that require some level of programmer intervention or library support. Mcintosh et al. [18] describe an interprocedural optimization technique for placement of global values. Lattner and Adve [14] developed compiler analysis and transformation for pool allocation based on the types of data objects.

McFarling [16] first proposed optimizations targeting instruction cache performances. He gave results on optimal performance under certain assumptions. Pettis and Hansen [22] proposed several profile-guided code positioning techniques including basic block ordering, basic block layout, and procedure splitting. Hashemi et al. [9] proposed the coloring-based approach to minimize cache line conflicts in instruction caches.

References

1. W. Abu-Sufah, D. J. Kuck, and D. H. Lawrie. 1979. Automatic program transformations for virtual memory computers. In *Proceedings of the National Computer Conference*, 1979.
2. Nawaaz Ahmed, Nikolay Mateev, and Keshav Pingali. 2000. Tiling imperfectly-nested loop nests. In *Supercomputing '00: Proceedings of the 2000 ACM/IEEE Conference on Supercomputing* 31–34. Washington, DC: IEEE Computer Society.
3. John R. Allen and Ken Kennedy. 1984. Automatic loop interchange. *SIGPLAN Notices* 19(6):233–46, New York, NY: ACM Press.
4. Brad Calder, Chandra Krintz, Simmi John, and Todd Austin. 1998. Cache-conscious data placement. In *ASPLOS-VIII: Proceedings of the Eighth International Conference on Architectural Support for Programming Languages and Operating Systems*, 139–49. New York: ACM Press.
5. David Callahan, Ken Kennedy, and Allan Porterfield. 1991. Software prefetching. *SIGARCH Computer Architecture News* 19(2):40–52. In *ASPLOS-IV: Proceedings of the Fourth International Conference on Architectural Support for Programming Languages and Operating Systems*, New York, NY: ACM Press.
6. Trishul M. Chilimbi, Bob Davidson, and James R. Larus. 1999. Cache-conscious structure definition. In *PLDI '99: Proceedings of the ACM SIGPLAN 1999 Conference on Programming Language Design and Implementation*, Vol. 34, 13–24. New York: ACM Press.
7. Trishul M. Chilimbi, Mark D. Hill, and James R. Larus. 1999. Cache-conscious structure layout. In *PLDI '99: Proceedings of the ACM SIGPLAN 1999 Conference on Programming Language Design and Implementation*, Vol. 34, 1–12. New York: ACM Press.

8. G. Doshi, R. Krishnaiyer, and K. Muthukumar. 2001. Optimizing software data prefetches with rotating registers, 257–67. In *PACT 2001: Proceedings of the Tenth International Conference on Parallel Architectures and Compilation Techniques*, New York, NY: ACM Press.

9. Amir H. Hashemi, David R. Kaeli, and Brad Calder. 1997. Efficient procedure mapping using cache line coloring. In *PLDI '97: Proceedings of the ACM SIGPLAN 1997 Conference on Programming Language Design and Implementation*, 171–82. New York: ACM Press.

10. Robert Hundt, Sandya Mannarswamy, and Dhruva Chakrabarti. 2006. Practical structure layout optimization and advice. In *CGO '06: Proceedings of the International Symposium on Code Generation and Optimization*, 233–44. Washington, DC: IEEE Computer Society.

11. M. Jimnez, J. M. Llabera, A. Fernandez, and E. Morancho. 1998. A general algorithm for tiling the register level. In *ICS '98: Proceedings of the 12th International Conference on Supercomputing*, 133–40. New York: ACM Press.

12. Guohua Jin, John Mellor-Crummey, and Robert Fowler. 2001. Increasing temporal locality with skewing and recursive blocking. In *Supercomputing '01: Proceedings of the 2001 ACM/IEEE Conference on Supercomputing (CDROM)*, 43–43. New York: ACM Press.

13. Ken Kennedy and John R. Allen. 2002. *Optimizing compilers for modern architectures: A dependence-based approach*. San Francisco: Morgan Kaufmann.

14. Chris Lattner and Vikram Adve. 2005. Automatic pool allocation: Improving performance by controlling data structure layout in the heap. In *PLDI '05: Proceedings of the 2005 ACM SIGPLAN Conference on Programming Language Design and Implementation*, Vol. 40, 129–42. New York: ACM Press.

15. Chi-Keung Luk and Todd C. Mowry. 1996. Compiler-based prefetching for recursive data structures. In *ASPLOS-VII: Proceedings of the Seventh International Conference on Architectural Support for Programming Languages and Operating Systems*, 222–33. New York: ACM Press.

16. S. Mcfarling. 1989. Program optimization for instruction caches. In *ASPLOS-III: Proceedings of the Third International Conference on Architectural Support for Programming Languages and Operating Systems*, 183–91. New York: ACM Press.

17. Nathaniel Mcintosh. 1998. Compiler support for software prefetching. PhD thesis Rice University, Houston, TX.

18. Nathaniel Mcintosh, Sandhya Mannarswamy, and Robert Hundt. 2006. Whole-program optimization of global variable layout. In *PACT 2006: Proceedings of the Fifteenth International Conference on Parallel Architectures and Compilation Techniques*, 164–172, New York: ACM Press.

19. Todd C. Mowry. 1995. Tolerating latency through software-controlled data prefetching. PhD thesis, Stanford University, Stanford, CA.

20. Todd Mowry and Anoop Gupta. 1991. Tolerating latency through software-controlled prefetching in shared-memory multiprocessors. *J. Parallel Distrib. Comput.* 12(2):87–106. In *ASPLOS-V: Proceedings of the Fifth International Conference on Architectural Support for Programming Languages and Operating Systems*, 62–73. New York: ACM Press.

21. Todd C. Mowry, Monica S. Lam, and Anoop Gupta. 1992. Design and evaluation of a compiler algorithm for prefetching. *SIGPLAN Notices* 27(9):62–73.

22. Karl Pettis and Robert C. Hansen. 1990. Profile guided code positioning. *SIGPLAN Notices* 25(6):16–27, In *PLDI '90: Proceedings of the 2005 ACM SIGPLAN Conference on Programming Language Design and Implementation*, 16–27, New York: ACM Press.

23. Allan K. Porterfield. 1989. Software methods for improvement of cache performance on supercomputer applications. PhD thesis, Rice University, Houston, TX.

24. Sanjay Rajopadhye, Lakshminarayanan Renganarayana, Gautam Gupta, and Michelle Strout. 2007. Computations on iteration spaces. In *The compiler design handbook: Optimizations and machine code generation*, 2nd ed. Boca Raton, FL: CRC Press.

25. Rafael H. Saavedra-Barrera, Weihua Mao, Daeyeon Park, Jacqueline Chame, and Sungdo Moon. 1996. The combined effectiveness of unimodular transformations, tiling, and software prefetching.

In *IPPS '96: Proceedings of the 10th International Parallel Processing Symposium*, 39–45. Washington, DC: IEEE Computer Society.

26. Vatsa Santhanam, Edward H. Gornish, and Wei-Chung Hsu. 1997. Data prefetching on the hp pa-8000. In *ISCA '97: Proceedings of the 24th Annual International Symposium on Computer Architecture*, 264–73. New York: ACM Press.

27. Yonghong Song and Zhiyuan Li. 1999. New tiling techniques to improve cache temporal locality. In *PLDI '99: Proceedings of the ACM SIGPLAN 1999 Conference on Programming Language Design and Implementation*, 215–28. New York: ACM Press.

28. Michael E. Wolf. 1992. Improving locality and parallelism in nested loops. PhD thesis, Stanford University, Stanford, CA.

29. Michael E. Wolf and M. S. Lam. 1991. A loop transformation theory and an algorithm to maximize parallelism. *IEEE Trans. Parallel Distrib. Syst.* 2(4):452–71, Washington, DC: IEEE Computer Society.

30. Michael E. Wolf and Monica S. Lam. 1991. A data locality optimizing algorithm. In *PLDI '91: Proceedings of the ACM SIGPLAN 1991 Conference on Programming Language Design and Implementation*, 30–44. New York: ACM Press.

31. Michael J. Wolfe. 1986. Loops skewing: The wavefront method revisited. *Int. J. Parallel Programming* 15(4):279–93, Norwell, MA: Kluwer Academic Publishers.

32. Michael J. Wolfe. 1987. Iteration space tiling for memory hierarchies. In *Proceedings of the Third SIAM Conference on Parallel Processing for Scientific Computing*, 357–61. Philadelphia, PA: Society for Industrial and Applied Mathematics.

33. Michael J. Wolfe. 1989. More iteration space tiling. In *Supercomputing '89: Proceedings of the 1989 ACM/IEEE Conference on Supercomputing*, 655–64. New York: ACM Press.

34. Michael J. Wolfe. 1995. *High performance compilers for parallel computing*. Boston: Addison-Wesley Longman Publishing.

35. Youfeng Wu. 2002. Efficient discovery of regular stride patterns in irregular programs and its use in compiler prefetching. In *PLDI '02: Proceedings of the ACM SIGPLAN 2002 Conference on Programming Language Design and Implementation*, 210–21. New York: ACM Press.

6

Garbage Collection Techniques

Amitabha Sanyal
and
Uday P. Khedker
*Department of Computer Sciences
& Engineering,
IIT Bombay, Mumbai, India
as@cse.iitb.ac.in
uday@cse.iitb.ac.in*

This chapter provides a comprehensive discussion of various garbage collection methods. The goal of this chapter is to highlight fundamental concepts rather than cover all the details. After introducing the underlying issues in garbage collection, the basic methods of garbage collection are described. This is followed by their generational, incremental, and concurrent versions.

6.1   Introduction

This section examines the need for garbage collection and introduces the basic concepts related to garbage collection. It also establishes several metrics on the basis of which garbage collectors are evaluated and identifies the distinguishing features of garbage collectors.

6.1.1   The Need for Garbage Collection

A program in execution needs memory to store the data manipulated by it. The data is named by variables in the program. Memory is allocated in various ways that differ from each other in their answers to the following questions: At what point of time is a variable bound to a chunk of memory, and how long does the binding last?

In the earliest form of memory allocation, called *static allocation*, the binding is established at compile time and does not change throughout the program execution. In the case of *stack allocation*, the binding is created during the invocation of the function that has the variable in its scope and lasts for the lifetime of the function. In *heap allocation*, the binding is created explicitly by executing a statement that allocates a chunk of memory and explicitly binds an access expression to it. An access expression is a generalization of a variable and denotes an address. In its general form, it is a reference or a pointer variable followed by a sequence of field names. One of the ways in which the binding can be undone is by disposing of the activation record that contains the pointer beginning the access expression. Then the access expression ceases to have any meaning. The other way is to execute an assignment statement that will bind the access expression to a different memory chunk.

After a binding is changed, the chunk of memory may be inaccessible. An important issue here is the reclamation of such unreachable memory. The reclaimed memory can be subsequently allocated to a different access expression. This is especially important in the cases of stack and heap allocation because it is not possible to place a bound on the memory requirement under these allocation policies. In the case of stack allocation, reclamation takes place at the end of a function invocation by adjusting the stack pointer. This reflects the fact that the space occupied by the local variables of the function is now free. This space is then allocated to the variables of the next function invoked.

In the case of heap allocation, reclamation is easy if it is done explicitly through a deallocation statement. However, a misjudgment by the programmer in inserting such a statement may lead to a reachable memory cell being reclaimed. To free the programmer from concerns of memory management, it is important to use an automatic tool that can detect unreachable memory and reclaim it. In the context of heap allocation, an unreachable memory chunk is called *garbage*, and a tool that detects and collects garbage is called a *garbage collector*. The executing program whose memory requirements are being serviced by the garbage collector is called the *mutator*.

Most garbage collectors base their collection on reachability; that is, they collect memory that cannot be reached by the program in the rest of the execution. However, one can also have memory that is reachable but not live, that is, memory that will not be accessed in further execution of the program. A garbage collector working on the basis of liveness instead of reachability may collect more garbage. Since it is difficult to detect memory that is not live, most collectors approximate liveness by reachability. We shall use the terms *reachable* and *unreachable* instead of the terms *live* and *dead*, which are more popular in garbage collection literature.

Following normal conventions, we shall identify a memory chunk with its contents, which we shall call an *object* from now on. All further discussion will be in terms of objects. For example, the phrase "pointer pointing to an object" will refer to a variable holding the address of the memory chunk containing the object. Similarly, "an object is garbage collected" will mean the memory chunk occupied by the object is garbage collected.

The concepts discussed above are illustrated through a program that creates the graph shown in Figure 6.1b. This graph is represented as the adjacency list shown in Figure 6.1c. The stack and heap configurations are shown in Figure 6.1d. If the node with label 4 is deleted from the graph by deleting the pointers from B to E and G to I and changing the pointer from B to D to point to G, then the objects E, D, F, and I become unreachable. The space occupied by these objects could be reclaimed by a garbage collector and used for further allocation. The pointers in the stack area, static area, and address registers form the entry points from where all the reachable objects in a heap can be accessed by the program. This set of pointers is called the *root set*. Starting from the root set, if we trace the objects and their connectivity, the resulting graph is called a *reachability graph*.

6.1.2 Features of Garbage Collection Methods

Most garbage collection methods can be distinguished by the following features that offer various design choices:

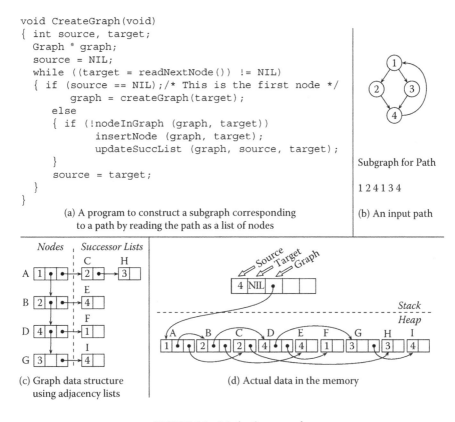

```
void CreateGraph(void)
{ int source, target;
  Graph * graph;
  source = NIL;
  while ((target = readNextNode()) != NIL)
  { if (source == NIL);/* This is the first node */
      graph = createGraph(target);
    else
    { if (!nodeInGraph (graph, target))
          insertNode (graph, target);
          updateSuccList (graph, source, target);
    }
      source = target;
  }
}
```

(a) A program to construct a subgraph corresponding to a path by reading the path as a list of nodes

Subgraph for Path

1 2 4 1 3 4

(b) An input path

(c) Graph data structure using adjacency lists

(d) Actual data in the memory

FIGURE 6.1　Motivating example.

- **Tracing vs. nontracing collectors:** Tracing[1] collectors traverse the data of the mutator and treat the left out objects as unreachable. In the case of nontracing collection, the reachability information is explicitly recorded during mutator execution. This information is encoded within the objects themselves. A nontracing collector updates the reachability information of an object only when the object is manipulated. Thus, it has a local view of the reachability graph. In contrast, a tracing collector traverses the entire reachable data structure in memory. This requires knowledge of the root set, which contains the entry points of the data structures used by the mutator. After tracing, a garbage collector may either directly collect garbage cells or preserve reachable cells, thereby collecting garbage indirectly. The process of preserving reachable cells is also called *scavenging*.
- **Moving vs. nonmoving collectors:** When garbage is reclaimed for allocation, a moving collector reorganizes reachable data to reduce fragmentation and to increase locality. For example, reachable objects are moved toward one end of the memory, freeing the rest of the memory for allocation. The contiguous grouping of reachable data is called *compaction*. Since some objects may remain reachable for a long time, they may be repeatedly moved across partitions. This problem is alleviated by maintaining objects of different ages in separate partitions, so that partitions containing long-lived objects are not garbage collected frequently.

---

[1]When memory is traversed by following the links of a data structure, it is called *tracing*. In contrast, *scanning* is the traversal of contiguous memory by following addresses sequentially.

- **Exhaustive collectors vs. collectors with interleaved detection and collection:** Once garbage collection starts in an exhaustive collector, control is not returned to the mutator until all objects detected as garbage are collected. As a consequence, exhaustive collection methods may be considered disruptive. Often, collectors are used in situations where long pauses due to garbage collection are considered unacceptable. In such situations, collection is interleaved with mutator execution; that is, the mutator may resume execution before the entire garbage is collected. In *incremental collectors* the interleaving is within a single thread of execution, whereas in *concurrent collectors* the mutator and the collector run concurrently as separate threads. Reference counting, which is also incremental, attempts to reduce pause times by disallowing accumulation of large amounts of garbage. This is done by invoking the collector as soon as garbage is created. This is different from all other methods that perform garbage collection only when an allocation request cannot be fulfilled. In effect, reference counting performs garbage collection continuously on very small amounts of data.

6.1.3   Effectiveness of Garbage Collection

The effectiveness of garbage collection can be assessed in terms of the following:

- **Space overheads:** Space is required by garbage collectors to maintain information of various kinds. Most collectors require objects to carry information about types and sizes of objects and tags to distinguish pointer fields from nonpointer fields. Some garbage collection methods such as reference counting require reachability information to be encoded within objects. Moving garbage collection methods (Section 6.1.2) have an overhead of keeping some free space to allow movement of data for compaction, and generational collectors (Section 6.4) have to maintain information regarding intergenerational pointers.
- **Time overheads:** A garbage collector may incur the following time overheads:

  CPU time: This is the actual time spent in garbage collection activity. It also includes the costs of recording old to new generation pointers in generational collectors and various synchronization costs in incremental and concurrent collectors (Section 6.5). The mutator may also have to bear some cost of garbage collection. For example, in incremental collectors, the mutator has to execute some additional code in the form of write-barriers and read-barriers.

  Overheads due to virtual memory: This is the time required to service cache misses or page faults. Cache misses and page faults can effect the execution of the garbage collector as well as the mutator.

- **Pause times:** The performance of garbage collectors is also judged by the lengths of pauses noticed by users. Minimizing pause times to make them imperceptible to users is a key design goal of garbage collection methods. Real-time garbage collectors are even required to offer a guarantee on the upper bound on the lengths of pauses.

6.1.4   The Pragmatics of Garbage Collection

Garbage collectors need to cooperate with not only the compiler that is used to produce the executable version of the mutator but also the underlying operating system. A garbage collector provides an interface between the operating system (or the virtual machine) and the user program for allocation and deallocation of heap memory. All other interactions between the operating system and user programs remain unchanged. Typically, a garbage collector seeks a large chunk of memory from the operating system and services all allocation requests by the mutator.

Garbage collectors also need to have information about the sizes of the objects, the layout of fields within the objects, and whether each field is a pointer. In languages with flow-sensitive and dynamically checked types such as Lisp or Smalltalk, the field of an object can contain both a pointer and a nonpointer at different times during execution. For such languages, each field must be tagged. At any time during execution, this tag identifies whether a field contains a pointer. Other information such as size and layout must also be encoded explicitly within objects.

For languages with flow-insensitive, statically checked types such as Java and Haskell, each object is equipped with a header that has complete information about the object. This header remains unchanged throughout the execution of the program. Tags identifying pointer fields are not necessary. With close cooperation between the compiler and the garbage collector, even the header information can be eliminated. If the garbage collector can be made aware of the types in the mutator, and if the types of the stack variables are known during a collection, then complete information about every object in the heap required for garbage collection can be deduced. However, this detection entails runtime overheads, so explicit headers are preferred.

In some methods, type information is not associated with a single object, but with a page containing objects of the same type. Thus, every page constituting the heap has associated type information. In these methods, every pointer access is processed to find the page containing the pointed object, and the type information associated with it is used for garbage collection.[2]

The real challenge is to design garbage collectors that work in environments in which the compiler is oblivious of the collector and objects do not carry the runtime type information that is required for garbage collection. The Boehm–Demers–Weiser collector [13, 16] is an example of such a collector. These garbage collectors resort to *conservative pointer-finding*, that is, they regard anything that can possibly be a pointer as a pointer. While this strategy may also preserve memory cells that are actually garbage, it is safe for nonmoving collectors. In a moving collector, this technique may result in overwriting an integer by a pointer value during compaction.

Another issue where a garbage collector needs to cooperate with the compiler is in deciding when garbage collection can be performed. Among the possible alternatives, the restrictive strategy is called *safe-points gc* [62], where the compiler designates certain program points as safe for garbage collection. It is ensured that such points occur frequently during execution. The advantage is that in the rest of the program, the compiler is free to perform optimizations of the kind that can make garbage collection unsafe. The unrestricted strategy is called *any-time gc*, where the compiler ensures that garbage collection can be performed any time without affecting the working of the mutator.

In the context of object-oriented languages, objects may represent resources other than heap memory. For example, objects may represent file handles, graphics contexts, network connections, and so on. When such objects are reclaimed, procedures to release the resources must be invoked. The action taken by these procedures is called *finalization*. The mutator registers such finalizable objects with the garbage collector. During tracing, if a finalizable object is marked as reachable, it is scavenged as any other object. Otherwise the procedure for finalization is invoked before reclamation.

6.1.5 Locality of Program Execution

This section defines the concepts related to the effect of memory hierarchy on garbage collection and mutator execution. Every page fault or cache miss during the execution of a program extracts a performance penalty. A program has good locality if the number of page faults or cache misses for a given number of memory accesses is small. There are two notions of locality:

- **Temporal locality:** Temporal locality captures the intuitive idea of clustering in time. If a program has good temporal locality, a data item that has been referenced in the immediate past has a high probability of being referenced in the immediate future.
- **Spatial locality:** Spatial locality captures clustering in space. In a program with good spatial locality, a data item located physically near an item that has been referenced in the immediate past has a high probability of being referenced in the immediate future. An example of data items that are physically close is adjacent nodes in a graph.

---

[2]This technique is called BIBOP (big bag of pages).

If a program has good temporal locality, then, in the best case, the pages containing all the data items referred may be in the main memory, and there may be no page fault. Even if there is a page fault, a recency-based policy for eviction such as least recently used (LRU) works well if the program has good temporal locality. Spatial locality suggests which pages could be loaded together to reduce the number of page faults. Fetch or prefetch policies that bring in pages containing physically adjacent items along with the item requested appeal to spatial locality. In the context of garbage collection, it is desirable that the collector has good locality. Furthermore, the mutator should have good locality even if its data is moved around by the collector. Indeed, some authors [14, 43] suggest that an appropriately designed garbage collector could even improve the locality of the mutator.

6.2 Basic Methods of Garbage Collection

In this section, we describe the early collectors called *reference counting* and *mark-sweep*. Both these collectors were designed for Lisp implementations around 1960. We also describe the *mark-compact* method, which is a later refinement of mark-sweep incorporating compaction.

6.2.1 Reference Counting-Based Collectors

The reference counting method [26, 37] stores in each object a count of the number of pointers pointing to the object. Each pointer assignment also involves manipulation of these counts. When a new object is allocated, the count is initialized to 1. Let pointers p and q hold the addresses of objects A and B before the assignment p = q. When the assignment is executed, both p and q point to B. These assignments are executed through the garbage collector, which decrements the reference count of A by 1 and increments the reference count of B by 1. If the reference count of an object becomes 0, the object is detected as garbage and the space occupied by it is added to a list of free cells. If the reclaimed object contains pointers to other objects, then the reference counts of these objects are also decremented. Clearly, this method is incremental, nontracing, and nonmoving.

This method requires additional space for storing reference counts and additional time for manipulating these counts. In particular, most pointer assignments require two counts to be adjusted rather than one. The limitations of this method are:

- The objects that are a part of a cyclic data structure continue to have nonzero counts even if they are unreachable. Thus, cyclic data structures cannot be garbage collected.
- A chain of reclamations increases pause times, affecting the almost-real-time behavior of the method.

Several variants of reference counting have been proposed to reduce its overheads and overcome its limitations. The space overheads can be overcome by using reference counts saturating at small values. When the maximum value is reached, the counts are neither decremented nor incremented. Such objects can then be garbage collected by a tracing method and their counts reset to 0. By resorting to a tracing method, this approach also overcomes the limitation of not being able to garbage collect cyclic data structures. In the extreme form of this method, one-bit counts can be used. The bit value, in combination with an incoming pointer, captures three states of an object, as illustrated in Figure 6.2. There are variations of this approach in which the bit is stored in the objects themselves [68], or as a tag within pointers [57]. The latter has the advantage that the cell pointed to by a pointer need not be read during garbage collection. This may improve the cache and paging behavior of the mutator.

While most approaches use mark-sweep collectors for garbage collecting shared objects with saturated counts, some use mark-compact collectors [66] or even copying collectors [67]. A further variation tries to extend the range of single bit counts to accurately determine up to two or more pointers over short program fragments [47].

The time overheads can be reduced by *deferred reference counting* [27]. This method avoids adjusting the counts for objects pointed to by parameters and local variables. This is done on the basis that the counts of such objects will eventually have to be decremented at the end of the procedure. For these objects, the

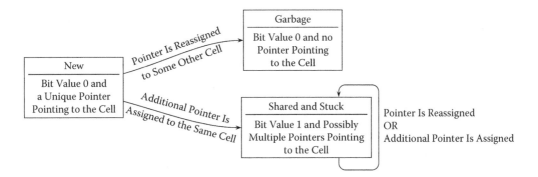

FIGURE 6.2 A single bit reference count captures three states of an object.

counts indicate the number of global pointers only. They may be pointed to by local pointers even when their counts have reduced to zero. Such objects are added to a zero count table (ZCT). If an object that is already in a ZCT is assigned to a global pointer during mutator execution, its count is incremented and it is removed from the ZCT. To collect those objects in the ZCT that are actually unreachable, the following procedure is applied periodically: All objects that are directly pointed to by the stack are marked. The unmarked objects in the ZCT are freed, and the marked objects are unmarked.

The Brownbridge–Salkild [18, 48] variant of reference counting tries to perform garbage collection in the presence of cyclic data structures. In this method, all pointers are characterized as either *weak* or *strong*, based on the following invariant properties:

- Every reachable object is reachable from the root set through a chain of strong pointers.
- There is no cycle consisting only of strong pointers.

For every object, separate counts of the number of incoming strong pointers and weak pointers are maintained. These are called *strong reference count* (SRC) and *weak reference count* (WRC). If the SRC of an object becomes 0, it indicates the possibility that the object *may be* unreachable. In the examples in Figure 6.3, weak pointers are shown as dotted arrows.

A pointer assigned to point to a newly created object is characterized as strong. This makes the new object reachable through a chain of strong pointers, satisfying the first invariant. Moreover, since this pointer cannot create a cycle, the second invariant is also satisfied. Similarly a pointer assigned to point to an old object is characterized as a weak pointer. It is easy to verify that as a result of this characterization, both the invariants are satisfied. In both cases, the WRC and SRC of the relevant objects are suitably updated.

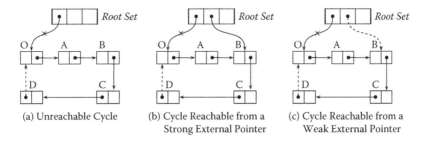

FIGURE 6.3 Different possibilities in Brownbridge–Salkild algorithm. The reference counts of an object are found by simply counting the incoming pointers.

The critical operation is the removal of a pointer pointing to an object, for instance, O. If this is a weak pointer, then WRC(O) is decreased by 1. However, if this is a strong pointer whose removal brings both WRC(O) and SRC(O) down to 0, then O is unreachable and can be reclaimed. The most critical case is when as a result of removing the pointer, O is pointed to by weak pointers only. The different possibilities are:

1. O may be a part of an unreachable cycle. This is illustrated in Figure 6.3a. Because of the two invariants, in this situation only the last pointer in the cycle (from D to O) can be a weak pointer. This situation is detected by a traversal starting with and ending at O. The entire cycle is then reclaimed.

2. O may be part of a reachable cycle, but the external pointer to this cycle points to some other object belonging to the cycle as shown in Figure 6.3b.[3] This situation is detected by a traversal starting at O and ending at B, whose SRC is more than 1. To satisfy the first invariant, the weak pointer D to O is made strong. However, to satisfy the second invariant, the pointer from A to B is made weak.

3. O is reachable but not in a cycle. The action taken is the same as the action in the previous case except that the weakening of the cycle-closing pointer is not required.

The algorithm wrongly detects the cycle in Figure 6.3c as being unreachable, because it relies on the external pointer being strong. Salkild [48] suggests a remedy in which the pointer from A to B is made weak and the pointers from the root to B and D to O are made strong. Furthermore, just as it was done for the object O, a traversal from B is started to ensure that the invariants are satisfied. However, the resulting algorithm does not terminate in some cases [45]. The termination issue was addressed by Pepels et al. [45], but the algorithm that results is complex and inefficient [37]. Axford [5] avoids the problem by assuming that every cycle has a unique entry. This assumption may not be restrictive in certain situations. Implementations of functional programming languages, for example, involve construction of such graphs.

6.2.2   Mark-Sweep Collectors

The reference-counting method has space and time overheads of maintaining counts. In addition, it has the limitation of not being directly able to reclaim cyclic data structures. The space overheads can be reduced by using one-bit counts, and the overheads of maintaining counts at each pointer update can be eliminated by tracing the reachability graph starting from root variables. Conceptually, this results in the mark-sweep method [41]. In this method, detection consists of a *marking phase* in which, starting with root variables, the data structures created by the mutator are traced, and the objects that are reached are marked. Thus, the method requires a bit for each object. Collection consists of a *sweeping phase* in which the entire heap is scanned sequentially, and the garbage is collected by adding the unmarked objects to a free list. Detection and collection are not interleaved — sweeping should begin only after the completion of marking. During tracing, the mutator's execution is suspended, at least until marking is over. Therefore, this is an exhaustive method. Since reclamation adds the garbage cells to the free list in situ, this is a nonmoving collection.

The space overheads of the mark-sweep method are not high except that explicit recursion may cause stack overflows. Additionally, if the memory residency of data is high (i.e., a large part of data remains reachable and very little garbage is created), this method may have to be called very frequently with very little progress in mutator execution between consecutive runs of the collector. This will reduce the effectiveness of the collection still further, as the mutator will not have the opportunity to make a significant number of objects unreachable. This repeated and ineffective execution of the garbage collector is called *thrashing*. Since this method is disruptive, thrashing is particularly undesirable.

The time overheads of the mark-sweep method are:

- If an object is reachable, its mark bit must be set. If mark bits are allocated in objects, setting the mark bit dirties the page containing the object. During eviction, the operating system must write this page back to the disk. This affects the virtual memory performance of the collector.

---

[3]This situation is possible only if B is also a part of another cycle. This is not shown in the figure.

- Although the effort required for detection is proportional to the size of reachable data, collection requires the entire heap to be traversed.
- If tracing is performed using recursive functions, the overheads of function calls can be significant.

The nonmoving nature of this method causes two problems. The first problem is that the collected memory may be fragmented. Therefore, if allocation requests are made for objects of different sizes, it becomes necessary to find a good fit for each object to prevent further fragmentation. This can be time consuming. Moreover, requests for large objects may fail even if the total available memory exceeds the size requested. Second, because of noncontiguous free space, closely related objects may be separated during allocation. This may affect the locality of the mutator. These limitations are reduced by compacting the space occupied by reachable objects. The resulting moving collector is called mark-compact and is described in Section 6.2.3.

Some of the overheads of the mark-sweep method can be reduced as described below:

- The adverse effect on the virtual memory performance due to distributed mark bits can be alleviated by storing the mark bits in a bitmap [16]. Then no page other than the bitmap page needs to be written back during detection. However, the efficiency of this technique requires that the mapping between the bit position and the address that it corresponds to be simple. Furthermore, it is desirable that the bitmap be memory resident. This may require special support from the operating system.
- The effect of traversing the entire heap during collection can be somewhat mitigated by prefetching pages, thanks to a predictable order of traversal. It is also possible to perform lazy collection by distributing it over several allocation requests in the mutator [13, 35, 72]. Thus, after marking is over, any allocation request is served with the explicit involvement of the sweeping phase. The sweeping phase ensures that allocation is performed only from the area that has already been swept. Since the mutator cannot access garbage cells, sweeping is unaffected. Allocation of unswept cells is prohibited, so there is no fear of freshly allocated (and hence unmarked) objects being collected as garbage.
- The overheads of recursive function calls can be reduced by replacing recursion with iteration and by explicit management of the stack required for traversal [11]. Effectively, the stack of activation records is replaced by a stack of marked nodes. Since the amount of data on the stack reduces, both space and time are saved. However, the marking stack may still be deep and require a substantial amount of memory. An elegant solution to the problem of large size of stack is the Deutsch–Schorr–Waite pointer reversal algorithm [31, 50, 58], illustrated in Figure 6.4. This algorithm traverses the heap in depth-first order and simulates the stack of nodes in the heap itself. The stack top is available through the *curr* pointer. The deeper nodes can be accessed through a chain of reversed pointers reachable through the *prev* pointer. There is no pointer between the top element and the element below it. There are three phases of the traversal:

  - When the traversal advances on an edge X → Y, X is the stack top. Its forward pointer to Y is reversed and made to point to the first element of the chain of reversed pointers reachable through *prev*. *prev* is then made to point to X, and a new stack top is created by making *curr* point to Y.
  - When the traversal switches from the left child Y of X to its right child Z, Y is the stack top and X is the previous element. Since the forward pointer in X should now be restored to Y, the reversed pointer in X is copied from its left field to the right field. Z is held in *curr*, the new top.
  - When the traversal retreats from Y to X, Y is the stack top and X is the previous element. The forward pointer from X to Y should now be restored. Hence, *curr* is made to point to X, and the reversed pointer in X is replaced by a forward pointer to Y.

- This method has been extended to general *n*-nary nodes [58] by storing an additional field containing the number of children of a node.
- While the pointer reversal algorithm saves on additional stack space, it requires enough extra bits in each object to store the number of its children. These bits are used to decide the next step

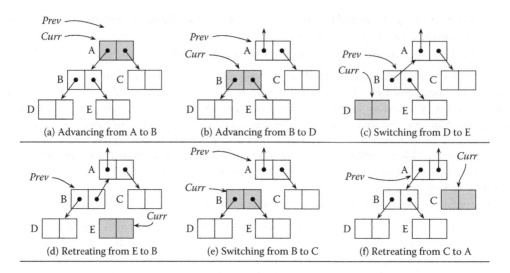

FIGURE 6.4   Depth-first traversal using Deutsch–Schorr–Waite pointer reversal algorithm.

during traversal. Further, unlike the methods that use a stack, every step in this method requires manipulation of three pointers. Hence, this method turns out to be considerably worse in terms of time than the method that maintains an explicit stack of nodes.

6.2.3   Mark-Compact Collectors

Mark-compact is a moving collection technique. It detects reachable objects by tracing data structures, marks objects that are reachable, and then rearranges the reachable objects so that they occupy a contiguous area in the heap. This also makes the free area contiguous. Though the method avoids fragmentation, it may affect locality if the relative order of reachable objects (in terms of addresses) is not preserved.

The compaction may be performed either by "folding" the heap so that reachable objects from one end of the heap are copied into the free cells at the other end or by "sliding" all reachable objects to one side, thereby squeezing out the free cells. The folding-based method was developed by Edwards [39, 49]. It is illustrated in Figure 6.5. In the first pass of this method, reachable objects are marked. In the second pass, two pointers are used to scan reachable objects and free cells. The *free* pointer points to free cells at one end of the heap, and the *live* pointer points to reachable objects at the other end. The reachable objects at one end are copied into the free cells at the other. For each object relocated, a forwarding address is recorded in the original space occupied by the object. The two pointers are advanced toward each other. The *free* pointer is advanced until the next free cell is reached, and the *live* pointer is advanced until the next reachable object is reached. The process is repeated until the two pointers meet. Finally, in a single pass, the reachable objects are scanned and pointers are updated by copying the forwarding addresses recorded in the first pass.

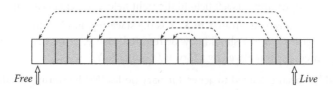

FIGURE 6.5   Two pointer algorithm "folds" the heap memory by relocating reachable objects (indicated by cross-hatching) from one end of the heap into free cells at the other end of the heap.

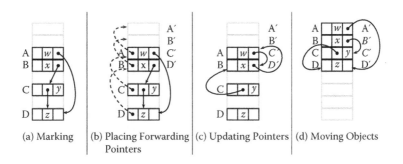

| (a) Marking | (b) Placing Forwarding Pointers | (c) Updating Pointers | (d) Moving Objects |

FIGURE 6.6   Marking followed by three passes of the Lisp 2 algorithm of compaction.

This method does not directly work for variable-sized data. One solution to this problem is to divide the heap into different regions, each region occupying data of a certain size. Another alternative is to allocate variable-sized data in a variable number of fixed-sized chunks [10].

An additional limitation of the two-pointer method is that it disturbs the relative order of the objects, thereby disturbing locality. Sliding algorithms relocate all objects starting from one end. They allow for variable-sized data by first calculating the new position of each reachable object. This is also the forwarding address of the object. Two additional passes are required, the first for updating the pointers in the reachable objects to reflect the forwarding addresses and the other for relocating the objects to their new addresses.

The Lisp 2 algorithm [22, 23, 25] uses an additional pointer space in each memory cell to store the forwarding address. As illustrated in Figure 6.6, after the marking phase is over, the first pass of compaction computes and places the forwarding addresses in each cell. Let the new address of A be denoted by A'. This is placed in A as its forwarding address. In the second pass, all the pointers are updated by their corresponding forwarding addresses. For example, since A points to D, A is made to point to D'. The objects are relocated in the third pass. Note that this movement of objects does not affect the pointers contained, as they move along with the objects. For example, A continues to point to D. Although we have shown all objects of the same size, this method handles variable sizes also.

The main disadvantage of the Lisp 2 algorithm is the space overhead for storing forwarding addresses in the objects. Instead, it is possible to collect the forwarding addresses in a separate table and view the sliding of reachable objects as the relocation performed by a relocating loader while loading a program. The main difference between the two is that a sliding compactor has to deal with a much larger number of entities that need relocation. The relocation table [25, 32] can be created in the first free location discovered. As reachable objects are moved, this table must be slid to the other end of the heap. Since the table should remain sorted, sliding may require special care.

Jonker's algorithm [38] uses a completely different approach based on threading all parents of an object [30]. In the first pass, all reachable objects are marked. In the second pass, the heap is scanned, and when an object is reached, its new address is calculated. Simultaneously, its parents that have already been visited and have been remembered through a thread of pointers are updated to point to the new location of the object. However, if some parents of this object are discovered later in the scan, they are threaded back to the object. In the third pass, all objects are moved to their new locations, and parents in the thread are updated to point to the new location.

The threading performed in the second pass is illustrated in Figure 6.7. When object A is visited, it is discovered to be a parent of D, and D does not have any threaded back pointer. Hence, the data of D is copied in A, thereby making a place for a pointer to A in D. Then B is scanned, the data of its child C is copied in B, and a threading pointer to B is stored in C. Whenever a reachable object is visited, its new address is calculated. When C is visited, its parent node B is reached via the threaded pointer. This allows the data in C to be restored, and the pointer in B is made to point to C'. Note that C is not moved to C' in this pass. Since C has a child D that has a threaded pointer, it is traversed, and the node reached (A) is the node to which C is made to point, whereas D is made to point to C. When D is scanned, all its parents

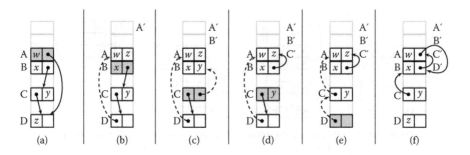

FIGURE 6.7   A trace of the threading used by Jonker's compactor. Objects are scanned in the order A, B, C, D, and E. C and D represent the addresses where X will be moved after compaction.

(A and C) are available through the thread. They are made to point to the new address D', and the data in D, which is available in the last parent in the thread (A), is restored in D.

Note that D may have a parent that may be visited after D is scanned. In such a case a backward thread is created in the second pass (after the situation in Figure 6.7f), which chains such parents of D. If the algorithm could update the pointers in the objects reachable through the forward thread of D (i.e., objects that were scanned before D was reached), why can it not do the same for the objects included in the backward thread? The answer is that to save space, the algorithm *does not record* the new addresses of the objects. The new address D' is calculated only when D is reached. At this point, its parents, which are reachable through the forward thread, can be updated to point to D'. However, when a parent is included in the backward thread, there is no record of the fact that the new address of D is D'. Hence, these parents can be updated in a subsequent pass only. In the third pass, all reachable objects are be moved to their new locations by scanning the heap and calculating the new addresses. Simultaneously, the remaining parents are also updated.

Jonker's method also does not require objects to be of the same size. However, it requires the data to be large enough to be overwritten by a pointer and some way of distinguishing between data and pointers. Its main drawback is that it has the overheads of threading and unthreading pointers.

6.3   Copying Collectors

Similar to mark-compact collectors, copying collectors are also tracing collectors that scavenge and compact reachable objects. However, instead of using the same space for both allocation and compaction, copying collectors use a separate empty space for compaction. This makes compaction easy, since the marked data is just laid contiguously starting from one end of this empty space. Tracing and copying in a copying collector requires a single traversal over all reachable objects. The earliest copying collector with both semi-spaces in memory was proposed by Fenichel and Yochelson [29].[4] However, their formulation of the tracing traversal was recursive and thus consumed stack space. The first practical copying collector was suggested by Cheney [20]. Cheney's formulation is nonrecursive and uses a part of the space taken by the copied objects as a replacement for the stack.

6.3.1   Cheney's Copying Collector

The copying collector divides the heap into two semi-spaces. The mutator uses just one of the two semi-spaces between consecutive garbage collections. The semi-spaces are traditionally called *FromSpace* and *ToSpace* because the garbage collector scavenges reachable data from *FromSpace* and copies them to *ToSpace*.

---

[4]An earlier copying collector proposed by Minsky [43] had one of the semi-spaces on disk.

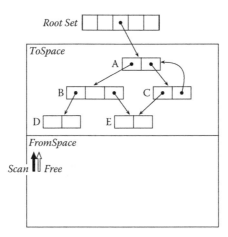

FIGURE 6.8 Heap configuration during program execution. The two spaces are flipped prior to garbage collection.

Assume that the program is running and using memory from *ToSpace* as shown in Figure 6.8. Also assume that there is a single variable in the root set pointing to the object labeled A. When *ToSpace* becomes full, a request for memory by the mutator triggers a garbage collection. At this point, the roles of the two spaces are reversed — *ToSpace* becomes the new *FromSpace* and *FromSpace* becomes the new *ToSpace*. This action is known in the garbage collection literature as a *flip*. Two pointers, *scan* and *free*, are set to the beginning of *ToSpace*. *FromSpace* is now scavenged by starting from each pointer in the root set, collecting all reachable data, and copying them in *ToSpace*. This is explained with the example shown in Figure 6.9.

Starting at the lowest address A′, the object A is first copied into *ToSpace*. We shall use A′ to refer to both the copied object and its address. To redirect future references to A to the copied object, the forwarding address A′ is recorded in the first field of A. The pointer *free* is moved to the first free address after the object A′. During garbage collection, *scan* and *free* are positioned in such a manner that copied objects whose children have also been copied occur between the beginning of *ToSpace* and *scan*. Similarly, the objects that have been copied but all of whose children have not been copied occur between *scan* and *free*. The configuration after A has been copied is shown in Figure 6.9a.

Now the pointers in the fields of A′ are used to copy the objects B and C at B′ and C′. The forwarding addresses are recorded in B and C, and *scan* and *free* are moved to the first addresses after A′ and C′. Now the processing of A is complete. The objects that have been processed completely, in this case A′, have been shaded black in the figure. In contrast, B′ and C′ are gray since their children have not been copied yet. These are copied next, using the pointer fields of B′ and C′. When E is reached a second time from the object C′, the forwarding address indicates that E does not have to be copied, and only the forwarding address is copied into C′. The final configuration is indicated in Figure 6.9d. When *scan* catches up with *free*, it indicates that the entire reachable data starting from the root set has been copied. The control is passed to the mutator, and subsequent allocations are done from *ToSpace* starting at the location pointed to by *free*. Since the mutator is stopped until the current round of garbage collection is over, copying collectors in this form are also called *stop-and-copy* collectors.

6.3.2 Performance of Copying Collectors

We shall first look at the CPU cost of garbage collection using copying collectors. Recollect that the CPU cost only considers the time expended by the CPU and does not take into account the time required for servicing cache misses and page faults. The CPU cost of collection and allocation is low. There is no marking and consequently no bit manipulation as in the case of mark-sweep and mark-compact algorithms. Since copied objects occupy memory from one end of a contiguous address space, they are

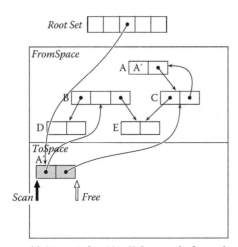

(a) A is copied to A′ in *ToSpace* and a forwarding address is placed in A in *FromSpace*. Its children are yet to be copied.

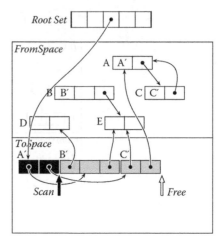

(b) A's children B and C are copied in *ToSpace*. All children of A have been copied but the children of B and C are yet to be copied.

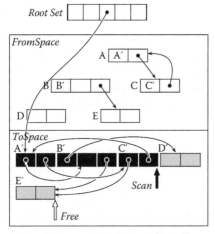

(c) D and E are copied in *ToSpace*. Thus all children of B and C are in *ToSpace*. The children of D and E are yet to be examined.

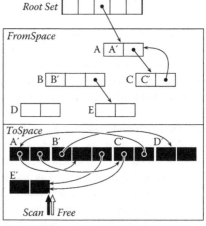

(d) Since D and E have no children, this completes the copying of live objects.

FIGURE 6.9 Snapshots of heap during garbage collection.

automatically compacted. A copying collector thus enjoys all the benefits of compaction — there are no free lists to be maintained for allocation and, since there is no fragmentation, it is not required to find a good fit for variable-sized objects.

One can define the efficiency of a garbage collector as the average cost of scavenging a single object. Appel [1] has argued that increasing the size of the memory results in an increase in the efficiency of the copying collector. If we assume that the number of reachable objects at every garbage collection is roughly the same, then the cost of garbage collection is constant and does not increase with the size of the heap. This is borne out in Appel's experiments. However, increasing the heap size results in fewer garbage collections. Thus, the efficiency of a copying collector increases with an increase in memory size.

The CPU cost of memory allocation is very low. It merely consists of checking whether there is enough memory in *ToSpace* and then advancing the *free* pointer to reflect the allocation. The check for sufficiency of memory can be eliminated by having a write-protected page at the end of *ToSpace* [1]. The resulting page fault raises an exception that is handled by the collector code. After collection, program control is handed back to the instruction that caused the page fault.

Since the copying collector collects reachable objects, there is a situation that can cause a possible rise in CPU costs. If an object is long-living, it gets copied from one semi-space to the other for a large number of garbage collection cycles. A solution is to identify objects that have survived a certain number of cycles and copy them into a separate long-living area. This area is garbage collected through an alternate scheme such as mark-sweep instead of copying collection. However, the objects in the long-living area may have pointers back to the main heap. Such pointers have to be treated as root set pointers while collecting the main heap. A generalization of this scheme is to keep objects in separate spaces according to their ages. These spaces are called generations, and the scheme is called *generational garbage collection* (Section 6.4).

6.3.3 Locality Issues

The memory in Appel's argument [1] refers to physical memory. The argument may not hold for virtual memory because, beyond a certain point, the advantages of increasing the heap size will be nullified by the number of page faults. In general, if the system on which the mutator and the collector are running has a hierarchy of memories, then their performance will be affected by their locality. Poor locality will show up in the form of page faults or cache misses. Page faults can be very costly. Servicing a page fault can cost up to a million cycles.

Some of the locality issues associated with garbage collection that have been identified by Wilson [62] are:

- Locality of the collector itself
- Effects of the collection process on the locality of the mutator

During tracing, all reachable data will be touched very few times, most often only once. Similarly, while copying, the *ToSpace* versions of the objects will be touched only once. Thus, the copying collector has little temporal locality — there are no repeated touches to the same object. However, copying collectors may show significant spatial locality. The tracing process successively touches closely linked objects, and such objects may be laid out closely in memory (see discussion below). Similarly, copying lays out the copied objects successively in *ToSpace*.

Studies by Hayes [33] have shown that objects created by programs show considerable spatial locality; that is, objects closely related in a data structure are also created and therefore reside close to each other in memory. A problem with copying collectors is that while copying, they alter the original layout of the data created by the mutator, thus degrading its spatial locality. To improve the locality of the mutator, several copying strategies have been tried. Moon [43] has shown that depth-first copying results in better spatial locality for the mutator than breadth-first copying. Since Cheney's collector copies objects in a breadth-first manner, Moon has suggested a copying strategy called *approximately depth-first*, which is a modification of the strategy in Cheney's collector. In this strategy, the last partially filled page in *ToSpace* is selected for scanning. Stamos [53, 54] and Blau [12] have shown that both breadth-first and depth-first copying result in better locality than random copying.

According to Wilson [62], the feature of a copying collector that has the most effect on the temporal locality of both the mutator and the collector is the alternate reuse of memory among two large semi-spaces. The size of the semi-spaces cannot be reduced beyond a point, since it will result in frequent collections during which long-living objects will be repeatedly copied from one semi-space to another. Furthermore, because of the alternate use of the semi-spaces, the next page in *ToSpace* that will be accessed during either copying or allocation is likely to be among the least recently used and will probably result in a page fault. As a contrast, consider a scheme in which allocation is done from a single space, which we shall call the *creation space*. When this space is full, objects are scavenged into a separate space called the *old space*, and the creation space can be used once again for allocation. The old space is garbage collected infrequently by a noncopying scheme. The locality of such a system will be much better than a copying collector, especially if the creation space is small in size. However, there is the danger that short-living objects may be scavenged into the old area and die soon after but continue to be resident and occupy heap space for a long time.

Thus, to improve locality of the system, there is a need to distinguish between short-living and long-living objects. The short-living objects should be allocated in an address space that is of relatively small size. Because of frequent reuse, this space will remain resident in physical memory, resulting in fewer page faults. This forms the basis for generational garbage collectors.

6.4 Generational Collectors

There is a mismatch between the demographic behavior of objects and the strategy employed by tracing collectors. If all objects are collected with the same frequency, objects that survive a number of collections will be repeatedly traced. This represents unremunerative work for the garbage collector. Baker [8] has expressed the cost per allocated object of a garbage collection as the *mark/cons*[5] ratio, which should be as low as possible. Generational collectors [40, 43, 59] decrease the *mark/cons* ratio by putting objects of different lifetimes in separate collection spaces.

The basis of the generational collectors is the *weak generational hypothesis*, which states that younger objects have a shorter life expectancy than older objects. This hypothesis seems to be true in practice [33, 56]. Generational collectors take advantage of this by having more than one collection space, which are collected at varying frequencies. The youngest collection space is collected most often. Objects that survive the youngest collection for a certain amount of time are promoted to an older generation, which is collected less often. The object continues in the older generation for some time before it either dies and is garbage collected or is promoted to the next generation. Thus, to recover the same amount of garbage as a nongenerational garbage collector, the generational collector incurs fewer tracing and copying costs.

The central idea of generational garbage collection is to collect different generations at different frequencies. The weak generational hypothesis only makes it profitable to collect younger generations more frequently than older generations. As a theoretical exercise, Baker [8] has considered a model of life expectancy of objects in which the fraction of objects expected to survive a period t is given by $2^{-t/\tau}$, where τ is a constant. In this exponential decay model, the life expectancy of an object does not depend on its age. Using this model, it has been shown [24, 33, 56] that older generations will have fewer survivors than young ones, so it will be profitable to collect older generations more frequently than younger generations.

In summary, the advantages of the generational scheme are the following:

- Because of the decreased *mark/cons* ratio, the time taken by the generational collector over the entire program execution is smaller.
- Since the younger generation has fewer survivors, its collection is faster and the pauses experienced by the mutator are shorter.
- The size of each generation is made considerably smaller than the next. Thus, the size of the youngest generation is quite small. Since most of the activities of the collector are limited to the youngest generation, the locality of the collector and possibly the mutator is improved. Indeed, some authors [64] think this is the real reason for the effectiveness of generational collectors.

6.4.1 The Basic Design

We shall consider a generational extension of a copying collector with two generations. Generalization to more than two generations does not pose any additional conceptual problems. Assume that the state of *ToSpace* after two rounds of garbage collection is as shown in Figure 6.10. We shall assume for now that each object has an extra field to record its age in terms of the number of collections it has survived.

---

[5]*cons* has its origins in the language Lisp, where it is used for allocating a new object. In the context of a copying collector, *mark* will also include the cost of copying.

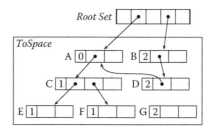

FIGURE 6.10   Young generation before promotion.

After an object has survived a certain number of collections called the *tenuring threshold* (three in this case), it is considered to be a long-living object. It is now promoted to a separate space called the *old generation space*; the earlier space is called the *young generation space*. Since the number of survivors in the young generation is low, the older generation will fill up slowly and will be collected less often than the younger generation. In general, there will be a number of generations of increasing ages and with decreasing frequencies of collection. A collection of the young generation is called a *minor collection*, while that of an old generation is called a *major collection*.

Figure 6.11 shows the state of *ToSpace* of both generations after tenuring. The objects B and D, which have survived three collections, are now tenured into the older generation, while the object G has been collected away. The tenuring ages of the remaining objects in the younger generation have been updated. In addition, the mutator has allocated a new object H, and the sole pointer pointing to H is from B. Note that the tenuring ages of the objects in the oldest generation are irrelevant.

We now mention the issues that could arise in a generational scheme. Some of these issues are later described in detail.

- The size of the younger generation and the tenuring threshold together determine how quickly an object will be promoted. The effectiveness of generational collection depends on whether objects are promoted at the right time. A delay in promoting a long-lived object causes the object to be copied repeatedly during minor collections. However, a short-lived object promoted to the older generation prematurely may become unreachable soon after promotion. Since older generations are collected infrequently, such an object may continue to occupy heap space for a long time. Worse still, objects in the younger generation that are pointed to solely by such objects will also be considered to be reachable and will be repeatedly copied from one semi-space to another during minor collections.

- The size of the younger generation also affects the length of the pause during a minor collection and the localities of the collector and the mutator. A smaller size results in shorter pauses and improved locality.

- The number of generations is also important for generational collectors. The reason for having several generations is that if a short-lived object survives a collection and then becomes unreachable,

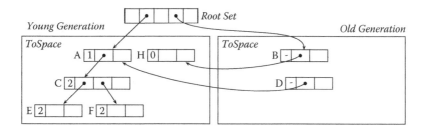

FIGURE 6.11   Young and old generations.

there is still the possibility that its space will be recovered soon enough in a major collection. The *strong generational hypothesis* [24, 33, 59] says that even among generations containing long-lived objects, the relatively younger generations have fewer survivors than older generations. While this might justify the use of several generations, the strong generational hypothesis itself is more an article of faith and is not supported by empirical data [37]. Thus, while there have been collectors with as many as seven generations [19], most generational collectors have two to three generations.

- The recording of the age of objects is another concern. The example described above seems to suggest that an extra field is added to each object to record its age. This is expensive and almost never done. Instead, objects whose ages are close are approximated to the same age. There is a trade-off between expending space to record ages precisely and risking repeated copying or premature promotion of objects due to approximate recording.

- The final issue is that of intergenerational pointers. As can be seen in Figure 6.11, the object H is reachable only through the older generation. To prevent reclamation of H during a minor collection, the pointer from B to H has to be considered as a root set pointer. Pointers can become intergenerational during promotion (D to A) or be created by the mutator through an assignment (B to H). Intergenerational pointers from the older to the younger generation have to be recorded. However, if we collect the young generation whenever the old generation is collected, then the young-to-old pointers do not have to be recorded during creation, because they will be identified during the minor collection.

6.4.2 Tenuring Policy

The tenuring threshold can either be decided a priori or be dynamically decided during garbage collection. In the first case, the tenuring policy is called *fixed*, and in the second, *adaptive*.

6.4.2.1 Fixed Tenuring

The main issue here is to decide the tenuring threshold a priori. For a given environment in which the garbage collector has to operate, the lifetime behavior of objects suggests a tenuring threshold. The original SML/NJ collector [2] had a tenuring threshold of one, since only about 2% of the objects survived a minor collection. However, Wilson and Moher [65] have studied the lifetime behavior of objects in Smalltalk and Lisp systems. They have found that, of the objects allocated at a certain time, the fraction that survive subsequent collections has the characteristics shown in Figure 6.12. The figure shows that a tenure threshold of one will lead to premature tenuring, as there are many survivors (the entire hatched region) after the first scavenge. However, few (the cross-hatched region) survive the second scavenge, and a tenure threshold of two is more appropriate.[6]

6.4.2.2 Adaptive Tenuring

A fine interplay occurs between the size of the new generation, the tenure threshold, and the number of generations on one hand and pause times and effective utilization of the heap space on the other. As a way of changing the tenuring policy dynamically, Zorn [72] has argued against having fixed-sized semi-spaces. If the birth rate of objects becomes more than the death rate, each collection results in a high number of survivors, and collections become frequent. In such situations, instead of waiting for the semi-space to be full, a better policy is to trigger collections only if the volume of data allocated since the last collection reaches a certain threshold and to grow the heap otherwise. This amortizes the cost of the garbage collection.

---

[6]In fact, Wilson and Moher's collector has provisions for tuning the tenure threshold to a value between one and two.

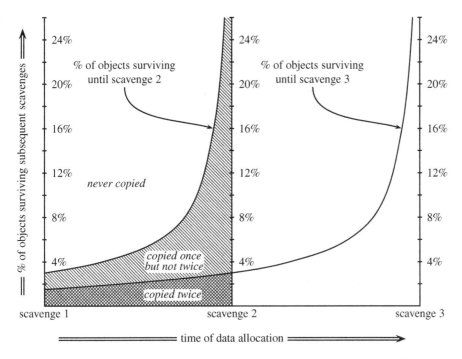

FIGURE 6.12 Fraction of objects surviving subsequent scavenges. (Reproduced with the permission of ACM.)

Ungar and Jackson [60, 61] have extensively studied the garbage collector of a Smalltalk-80 system. Their main goal was to reduce pause times to acceptably low levels. They have identified the limitations of fixed tenuring threshold due to dynamic variations in the lifetime behavior of objects. In particular, if a large number of objects are younger than the threshold, pauses will be long. Furthermore, if very few objects are being scavenged, there is no need to tenure any object, but a fixed-threshold policy will still tenure objects that are old. Since pause times are related to the volume of data surviving a collection,[7] they suggest that the volume of data surviving should decide the tenuring threshold in the next collection:

- If the volume of surviving data is less than a predetermined value, the pause time in the next collection is judged to be acceptable, and the tenuring threshold is set to infinity, so that there is no tenuring in the next collection.
- If the volume of surviving data is greater than the predetermined value, then the tenuring threshold is adjusted to tenure enough objects to bring down the volume of surviving data to acceptable levels. The value of the new tenuring threshold is decided using a table that records the volume of surviving data of each age. This table is updated at each collection.

Other tenuring policies that bring an already tenured object back to the younger generation have also been considered [9].

6.4.3 Recording Ages of Objects

A generational collector has to keep track of the ages of objects to decide when to promote them to the next generation. However, maintaining the ages of individual objects is costly in terms of both space and time. Therefore, age is recorded for groups of objects whose ages are close to each other. The generational

---

[7]For various reasons, the method is actually based not on the number of objects, but on the space consumed by these objects.

space that an object is currently residing in gives an indication of its age. We now discuss how the heap can be organized into different generational spaces to achieve different tenuring thresholds.

In the simplest case, a generation can have only one space (instead of two semi-spaces) called *creation space*. During garbage collection, objects can be promoted en masse to the next generation, thus obviating the need of a *ToSpace*. An advantage of this method is that the creation space is fully used between consecutive collections. Therefore, this method shows better locality than methods that use only one of two semi-spaces at a time. While en masse promotion also has the problem of premature tenuring, it can be alleviated by having more than one generation.

An improvement over this method developed by Ungar [59] is to have a creation space and an aging space that consists of two semi-spaces. At garbage collection time, the creation space and the currently used semi-space of the aging space are scavenged into the other semi-space. This solves the problem of premature tenuring, since a prematurely tenured object can still be collected in the aging space. However, it has the problem of long-living objects being repeatedly garbage collected.

The methods discussed so far have a tenure threshold of one. We can also have a tenure threshold of more than one. In Shaw's method [51], the younger generation consists of a number of buckets. Each bucket is garbage collected with the same frequency. After a predetermined number of collections, survivors of a bucket are promoted to the next bucket, and the oldest bucket is promoted to the next generation. For example, if the number of buckets and the number of collections per bucket are both two, the method guarantees that an object survives at least three collections (at least one in the first bucket and two in the second) before it is promoted to the old generation. By choosing the number of collections appropriately, the collector can be tuned to a particular lifetime characteristic of objects.

Wilson and Moher [65] have combined the single semi-space of Ungar's method with Shaw's greater-than-one tenuring threshold. In their method, the younger generation consists of a creation space and an aging space made up of two semi-spaces. There is also an old generation. At every collection, survivors of the creation space are promoted to the *ToSpace* of the aging space, and the survivors of *FromSpace* in the aging space are promoted to the old generation. Thus, the method in this form has an effective tenure threshold of two.

The method incorporates a modification of what has been described before, to adjust the tenure threshold to a value between one and two. Instead of promoting all the survivors of the creation space to the aging space, only objects below a certain age are transferred. The rest of the objects are directly promoted to the old generation. Detecting all objects below a threshold age is possible because objects are allocated sequentially in the creation space. The threshold gives a handle to tune the tenure threshold to a value between one and two.

6.4.4   Recording of Intergenerational Pointers

It is important to record intergenerational pointers from the old to the young generation. An object in the young generation may only be reachable through an object in the old generation. In such a situation, the young generation object should be detected as being reachable through the intergenerational pointer during a minor collection.

Intergenerational pointers can be discovered by explicitly scanning the older generation during a minor collection. However, this is expensive, and most methods use a write-barrier instead. A write-barrier is a fragment of code that is added to the mutator around pointer assignments. This code checks whether the assignment will create an intergenerational pointer and, if it does, records this information. The issues that arise in handling intergenerational pointers are:

- The overhead to the mutator due to the write-barrier.
- The space overhead for recording the intergenerational pointer.
- The overhead to the collector during scavenge time.

Fortunately, the number of intergenerational pointer stores is a small fraction of the total number of instructions. Of all pointer stores, the stores to registers and the stack can be ignored. The initializing stores

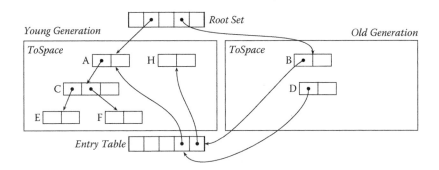

FIGURE 6.13   Entry tables.

(for example, to cons cells in Lisp) can be ignored because they cannot introduce old-to-young generation pointers. The number of old-to-young generation pointer assignments in Lisp or ML that have to be trapped through write-barriers have been estimated as being about 1% of the total number of instructions [37]. However, if each write-barrier consists of 10 instructions, the mutator execution time increases by 10%. While static analysis techniques have been suggested to reduce the number of barriers required [71], there is still a need to optimize recording of intergenerational pointers. We now describe some of the methods used to record intergenerational pointers.

6.4.4.1   Entry Tables

This technique was used as a part of the first generational collector by Lieberman and Hewitt [40]. The method uses a table, called the *entry table*, for each generation except the oldest. If the execution of the mutator results in the potential creation of an old-to-new pointer, then the object in the old generation is made to point instead to a new slot in the entry table for the young generation. The slot contents, in turn, are made to point to the original destination, the object in the younger generation. The scheme for two generations is illustrated in Figure 6.13.

During a collection of the younger generation, the entry table of this generation also forms a part of the root set. Collection under this method is fast; the only cause of inefficiency is possible duplication of pointers from the entry table to the same object. This makes the cost of scanning the entry table proportional to the number of assignments creating intergenerational pointers. However, the mutator may be slowed down considerably, because an original reference from an old object to a new object now involves an indirection through the entry table.

6.4.4.2   Remembered Sets

Unlike entry tables, in this method [59] the old object containing the intergenerational pointer is itself recorded. This is shown in Figure 6.14. Each object has a bit in the header indicating whether it is in the

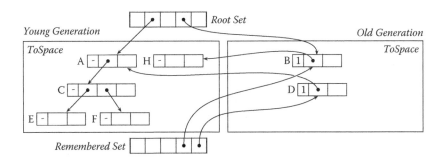

FIGURE 6.14   Remembered set.

remembered set. When an intergenerational pointer is stored in an old object, the object is recorded in the remembered set if the bit in its header is not already set. These objects in the remembered set are treated as part of the root set during a minor collection.

The cost at collection time is proportional to the number of intergenerational pointers and not to the number of stores. However, all the fields of a remembered object have to be scanned. If the object is large, this represents a substantial overhead. The added cost to the mutator is high, since every pointer store involves a checking of whether the pointer is intergenerational and whether the header bit of the object in which the pointer is being stored is set.

Appel's SML/NJ collector [2] has a fast implementation of the remembered set. The remembered set records every pointer store, including those to objects that are already present in the remembered set, and stores of pointers from new generations to old generations. At collection time, the irrelevant pointers in the remembered set are filtered out. This strategy is effective for a language such as ML in which most of the stores are initializing stores and there are very few old-to-young intergenerational pointers. This also works well for collectors that perform en masse promotion. After the en masse promotion, the entire remembered set is cleared.

Approximating the recording of intergenerational pointers results in less overhead for the mutator, but the collector's overhead increases because it has to filter out the wrong kind of pointers. Hudson and Diwan [34] extend Appel's idea by storing intergenerational pointer information at two levels. During mutator execution, addresses of objects that may have intergenerational pointers are quickly recorded in a sequential buffer without any checks. To avoid checking for overflow at every write, the buffer has a write-protected page at the end. During garbage collection, the information in the buffer is transferred to the remembered set after filtering. For example, duplicate entries in the sequential buffer are filtered out by implementing the remembered set as a hash table.

The techniques discussed above mark individual objects. The next few techniques described approximate object marking at larger levels of granularity.

6.4.4.3 Page Marking

Instead of recording individual objects that contain intergenerational pointers, this technique records pages that contain intergenerational pointers. In its simplest form, one can have a bit for every virtual page. Whenever an intergenerational pointer is stored in a page, the event is trapped and the bit corresponding to the page is set. During collection of a younger generation, every page whose bit is set is scanned for the presence of an intergenerational pointer from an older generation to this generation.

The problem with the approach in this form is that pages that have been swapped out may be brought in and then searched to discover that they do not contain intergenerational pointers to the generation concerned. The cost of such an eventuality is very high, so a bitmap is maintained for every swapped out page. This bitmap contains a bit for every generation. Whenever a page is swapped out, it is scanned, and if a generation has a pointer from this page, the corresponding bit is set. During a collection of the ith generation, a swapped out page is brought in only if the ith bit of the bitmap of the page is set.

The Ephemeral Garbage Collector [43] built for the Lisp implementation on top of the Symbolics 3600 machine employed this scheme with minor variations. The garbage collector was aided by the specialized hardware of this machine:

- The write-barrier was implemented in hardware.
- The page size was small, and the presence of an intergenerational pointer was detected by hardware [52]. Therefore, pages could be scanned very quickly.
- The machine had a tagged architecture that allowed the hardware to differentiate between words containing pointers and words containing nonpointers. Because of this, one could scan without taking into account object boundaries and yet not run the risk of interpreting a pointer field as a nonpointer field. This also added to the speed of scanning.

6.4.4.4   Card Marking

Unlike page marking, this technique is designed to work on stock hardware. The memory is divided into cards. The size of the cards is appropriately chosen. If the card is too big, the cost of scanning is large. If the card size is small, the number of cards is large, and the per-card information to be maintained also becomes large. In the design described [52], the card size was 256 words. Thus, the granularity of object marking in this technique is between page marking and word marking.

The card marking scheme works much like the page marking scheme. There is a primary card-mark table that contains a bit for every card. The bit records whether a pointer has been stored in the card. There is also a secondary card mark table that contains a bitmap for every card. The ith bit of this bitmap records whether the card contains a pointer to an object in the ith generation.

During the execution of the mutator, if a pointer is stored in a card, the primary card-mark of the card is set. Garbage collection now proceeds as follows: Suppose the ith generation is being garbage collected. Every card whose bitmap in the secondary card-mark table has the ith bit set is scanned. This involves checking each word to determine whether it contains a pointer to the generation being garbage collected and, in that case, treating the pointer as a root set pointer and performing a scavenging. After a scan, if a card contains a pointer to a younger generation, the secondary card-mark table for that generation is reset. The bit in the primary card-mark table is cleared for the card.

Next, the cards whose bits in the primary card-mark table are set are scanned. These are the cards that did not have any pointers to the generation being collected at the end of the last collection but have had a pointer stored during the last round of mutator execution. These cards are also scanned in the manner described above. At the end of this, the scavenging of the current generation is complete. Clearly, the information in the card-mark tables of the cards in the generation being collected are now meaningless, so they are cleared.

Scanning a card in software requires a solution to the following problem. The header contains the information distinguishing pointer from nonpointer fields. Since a card could begin in the middle of an object, the pointer information of the remaining part of the object is not available. The solution in [52] was to maintain information to determine the nearest preceding card that starts with an object header. Starting with this object, the sequence of objects is followed until the object whose fields begin the current card is reached. This object header is examined to determine whether the words beginning the card are pointers.

6.5   Incremental and Concurrent Collectors

In some applications large pauses cannot be tolerated. Such applications are mostly interactive and in some instances rely on a guarantee of a bound on the length of pauses. While running such applications, it may not be possible to allow the associated garbage collector to run until the end of a round of collection. Hence, one has to use garbage collectors that are incremental, that is, collectors in which a single round of garbage collection is interspersed with mutator execution. The reference counting method described earlier is, in its natural form, incremental. However, reference counting has its own limitations of not being able to handle cycles naturally and of uneven distribution of pause times (Section 6.2.1). Therefore, it becomes necessary to think of incremental versions of other kinds of garbage collectors.

A garbage collector is said to be *incremental* if garbage collection is interleaved with mutator execution in a single thread of execution. It is said to be *concurrent* if the collector and the mutator can run concurrently on multiple threads. In effect, this means the interleaving between the collector and the mutator is well defined in the case of incremental collectors and arbitrary in the case of concurrent collectors. Finally, a garbage collector is said to be *real-time* if it can guarantee a bound on the length of pauses.

The central issue with incremental collectors is that because of the interleaving of mutator and collector, their views of the reachability graph may be different. Instead of trying to make the two views identical, the collector's view of the reachability graph is made an overapproximation of the mutator's view. Thus,

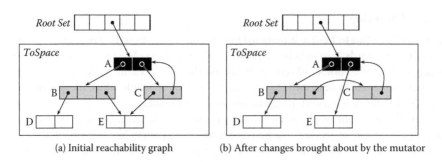

(a) Initial reachability graph   (b) After changes brought about by the mutator

FIGURE 6.15   Situation in which a reachable object can be wrongly reclaimed.

the collector may conservatively preserve an object that is unreachable. However, it will never discard an object that is reachable.

In the context of incremental and concurrent collectors, Dijkstra et al. [28] have introduced a *tricolor abstraction* for indicating the state of tracing of objects. We used this coloring scheme for explaining copying collectors in Figure 6.9. In this scheme, a black object indicates that the collector has traced the object, as well as all its immediate descendants. A gray object indicates that the collector has traced the object but not its descendants. Finally, a white object indicates that the object has not been traced at all. Some incremental collectors explicitly maintain the color of objects. More often, the tricolor abstraction is used as a conceptual mechanism to reason about properties of collectors. It is used in soundness arguments to show that the collector does not wrongly claim any reachable object and in monotonicity arguments to prove the termination of tracing.

Consider the following example, which illustrates conditions under which an incremental or concurrent collector could go wrong. Figure 6.15a shows the state of tracing in the reachability graph recorded by the collector. Assume that the mutator now makes the changes shown in Figure 6.15b. In particular, the link from B to E is severed, and the link from A to C is made to point to E. Since A continues to be labeled black, it will be assumed by the collector that its descendants have been traced. Since there is no path from a gray object to E, it will never be traced and consequently will be collected as garbage. This illustrates the following necessary and sufficient conditions for the garbage collector to fail to preserve a reachable white object O:

- A black object directly points to O.
- There is no path from any gray object to O.

An incremental garbage collector ensures safety by trying to prevent either of the two conditions. This is done by the use of *barriers*. A barrier is an extra fragment of code that is wrapped around certain critical mutator actions. In the incremental versions of nonmoving collectors, a barrier is wrapped around pointer stores, while for moving collectors, it is wrapped around pointer reads. We first discuss the algorithms that use write-barriers.

6.5.1   Incremental and Concurrent Nonmoving Collectors

We first discuss incremental mark-sweep algorithms. In the case of mark-sweep algorithms, the tricolor abstraction during tracing is to be interpreted as follows. Objects with incoming pointers from the mark stack are considered gray. Black objects are those whose children are either black or gray. The rest of the objects are white. Thus, for an object, its mark bit and the presence of an incoming pointer from stack can together be considered as an implicit coding of its color.

There are two kinds of incremental mark-sweep techniques which use the write-barrier in different ways. These are the *snapshot-at-beginning* and *incremental update* techniques. Snapshot-at-beginning techniques ensure correctness by preventing the second condition, whereas incremental update techniques ensure correctness by preventing the first condition.

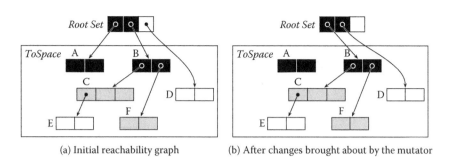

(a) Initial reachability graph   (b) After changes brought about by the mutator

FIGURE 6.16   The need of scanning root set as an atomic action.

6.5.1.1 Snapshot-at-Beginning

We describe a well-known technique proposed by Yuasa [69]. The name *snapshot-at-beginning* derives from the fact that the reachability graph at the beginning of a collection phase is retained in spite of mutator interference. This algorithm prevents the second condition as follows: During a collection phase, if the mutator tries to change[8] a pointer to a white object, a write-barrier is invoked and the pointer is pushed on the mark-stack. For example, in Figure 6.15a, when the pointer from B is changed from E to C, it is first pushed on the mark-stack before the mutator is allowed to continue its normal actions. Since objects pointed by the mark stack will be taken out and traced later by the collector, they will not be lost. Also, every newly allocated object during the collection phase is colored black by setting its mark bit.

The collection starts with a scanning of the root set as an atomic action. The reason scanning needs to be atomic is shown in Figure 6.16. Assume that scanning is incremental, the portion of the root set that has already been scanned is shown in black in Figure 6.16a. If the mutator now takes over and makes the changes as shown in Figure 6.16b, the object D, which is pointed from an already scanned stack slot, will not be marked. Though reachable, the cell will be collected away. The cell could have been preserved if a write-barrier had been used for stack entries. However, this would have resulted in a prohibitively large overhead. The method developed by Yuasa et al. [70] suggests a possible solution to this problem.

The snapshot-at-beginning algorithm is conservative. In Figure 6.15, if the only pointer to E had been from B, then changing this pointer would have resulted in E becoming garbage. However, the pointer to E would have still been recorded on the mark stack and prevented E from being collected in this round. In fact, all the objects that become unreachable during a round of collection will be collected only during the next round. Also note that while snapshot-at-beginning algorithm prevents the second condition, they may allow the first condition to arise. For example, the scenario in Figure 6.15b is allowed.

6.5.1.2 Incremental Update Methods

Incremental update methods ensure correctness by preventing any pointer to a white object from being stored in a black object. Dijkstra et al.'s method [28] conservatively approximates this idea by using a write-barrier to trap the store of a pointer to a white object in an object of any color. The white object is colored gray and thus cannot be garbage collected. Because of the store of the pointer to E in A as shown in Figure 6.15b, the object E will be colored gray. This is shown in Figure 6.17a.

Steele's [55] write-barrier, however, changes the color of the object into which the pointer to the white object was being stored to gray. In Figure 6.17b, object A turns from black to gray. Since this object, which was originally black, has now turned gray, there will be a second round of tracing starting with this object. During this tracing the white object will be marked and thus not be collected.

After changing the color of E to gray under Dijkstra's method, even if the mutator deletes the pointer from A to E, making E unreachable, it will not be reclaimed during this collection. Under Steele's method,

---

[8]A change includes deletion.

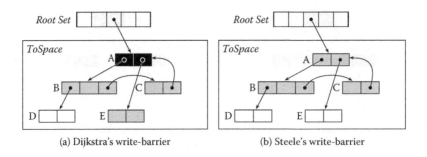

(a) Dijkstra's write-barrier   (b) Steele's write-barrier

FIGURE 6.17   Write-barriers.

however, if the mutator deletes the pointer from A to E while E is still white, it will be reclaimed in the current round of garbage collection. Thus, Dijkstra's method is more conservative than Steele's method. Note that both these methods assume a fixed root set, unlike the Doligez–Leroy–Gonthier method (Section 6.5.2). Moreover, in both these collectors, the mutator and the collector are largely independent. Therefore, they can be viewed as concurrent collectors.

The collectors described so far are incremental or concurrent versions of mark-sweep. We now describe a different kind of incremental nonmoving collector.

6.5.1.3   Baker's Treadmill Noncopying Collector

Many of the drawbacks of copying collectors arise because objects are copied from one semi-space to another. As a result of copying, read-barriers, which are costly, have to be used in incremental versions of copying collectors. The noncopying collector suggested by Baker [7] retains the major advantage of copying collectors, that is, the collection effort is proportional to the number of reachable objects. Additionally, since it does not move objects, there are no forwarding pointers, and there is only one version of an object. Thus, the mutator does not have to be protected from the changes brought about by the collector, as happens in moving collectors (see Section 6.5.2).

Any incremental tracing collector has to distinguish between four sets of objects: black objects, gray objects, white objects, and free objects. New objects are allocated by the mutator from the set of free objects. The incremental versions of mark-sweep and copying collectors maintain and identify these sets as shown in Figure 6.18.

As shown in Figure 6.19, Baker's noncopying collector explicitly maintains the four sets by chaining them in a circular doubly linked list. Within the list, the sets are demarcated by the four pointers *bottom*, *top*, *scan*, and *allocate*. Free objects are hatched. As explained below, this method only needs to distinguish between white and nonwhite objects, and a single bit is sufficient for this purpose. The other colors are merely conceptual and have been used for purposes of exposition.

When the mutator allocates a new object, the *allocate* pointer is advanced clockwise and the allocated object is colored black. The method is practical only when all objects are of the same size. The mutator

	Mark-sweep	*Copying*
White objects	Mark bit of the object is clear and object is not in stack	Object is in *FromSpace*
Black objects	Mark bit of the object is set and the object is not in stack	Object is between one end of *ToSpace* and the *scan* pointer
Gray objects	Mark bit of object is set and object is in stack	Object is between *scan* and *free* pointers in *ToSpace*
Free objects	Object is in free list	Object is at the other end of *ToSpace*

FIGURE 6.18   Identification of objects of different colors.

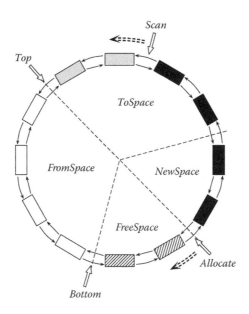

FIGURE 6.19 Baker's treadmill noncopying collector.

can also interfere with the collector by storing a pointer to a white object in a black object. Though Baker suggests the use of a read-barrier, it is enough to use a write-barrier to trap such writes and convert the white object to a gray object. The collector scans by repeatedly selecting the gray object next to the *scan* pointer. If a child of this object is black or gray, then nothing is done. If it is white, it is unlinked from its place and linked at the beginning of the region of gray objects. This is the only step that requires explicit color information. An object is colored black after all its children are processed.

When *scan* meets *top*, the current collection round is complete. At this point there are just black and white objects, and the roles of the two spaces between the *bottom* and the *top* pointers are reversed by swapping the two pointers. The black segment can now be seen as white and the white segment as the hatched segment consisting of free objects.

This method avoids the use of a read-barrier. Furthermore, such nonmoving collectors can work in noncooperative environments such as compilers that do not identify pointer variables [36, 62]. Baker also claims that the space required for the extra links is compensated for small objects by the absence of a mark stack or separate *FromSpace* and *ToSpace* versions of reachable objects. For large objects, the space required by the noncopying collector compares favorably with noncopying versions, and the time required for unlinking and linking is also significantly less than for copying the object.

The main problem with this method is regarding allocation of new objects when the sizes of objects are different. Then the allocation process has to search the free memory region for a chunk of memory that fits the object size. Wilson and Johnstone [63] suggest a variation of the method where the free memory region consists of a number of lists of objects segregated by their sizes. An object of a certain size is quickly approximated to the list of the right size, and a chunk of free memory is allocated from the list.

6.5.2 Incremental and Concurrent Moving Collectors

We have seen that in a mark-sweep collector, changes made by the mutator can affect the collector's view of the reachability graph. A moving collector has an additional concern — the collector actions can also affect the mutator's view of the reachability graph. In the context of a copying collector, for example, when the mutator tries to read the field of an object, the object may be just a copy in *FromSpace*, and the field may contain a forwarding address to the actual object that has been copied into *ToSpace*. Thus, even read

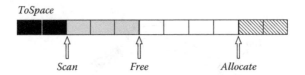

FIGURE 6.20   *ToSpace* in Baker's copying method.

actions of the mutator should be trapped to ensure the validity of the data that is read. We now look at the better-known incremental moving collectors.

6.5.2.1   Baker's Copying Method

Like Cheney's method, Baker's copying method [6] also consists of two semi-spaces — *FromSpace* and *ToSpace*. Garbage collection begins with flipping the two semi-spaces, following which reachable objects are copied from *FromSpace* to *ToSpace*. However, since the collector can be interrupted by the mutator before collection is over, the space from which new cells are to be allocated by the mutator is an issue. If objects are copied at one end of *ToSpace*, allocation is done at the other end as shown in Figure 6.20.

Correctness is ensured by using read-barriers to trap the mutator when it tries to read objects through pointers. The read-barrier checks whether the object that is being read through the pointer has already been copied into *ToSpace*. In this case, the forwarding address stored in the object is returned. Otherwise, the object is copied into *ToSpace* in the manner of a Cheney collector, and the address at which it is copied is returned. In effect, if the mutator tries to read a white object, it is made gray.[9] Thus, the mutator is protected from the changes made by the collector, since it only reads the *ToSpace* version of any object. The collector is also protected from the mutator. Since the mutator cannot read a pointer to a white object, it cannot store such a pointer in a black object. Thus, the second condition is prevented, and the collector does not mistakenly collect a reachable object.

Another issue has to be handled by Baker's copying collector. Since the mutator may interrupt the collector, the mutator may run out of space for allocation before collection is completed. The system then comes to a standstill. To prevent such a situation, the amount of work done by the collector is tied to the amount of allocation of new space. Whenever new space is allocated by the mutator, control is transferred to the collector. The collector does a predetermined amount of copying and scanning and passes control back to the mutator.

Baker's method is conservative. Objects allocated during a collection round are assumed to be black and are not collected, even if they die during the current round. Similarly, an object that has already been traversed will not be collected during the current round of collection, even if it is made unreachable after the traversal.

One of the drawbacks of Baker's method is the cost of the read-barrier, in terms of both space and time. Zorn [73] reports that pointer reads from about 15% of the total number of instructions. Inlining the read-barrier would cause the mutator code size to become unacceptably large. Zorn also reports a 20% time overhead for read-barriers implemented in software.

6.5.2.2   Brooks' Variation of Baker's Copying Method

Brooks [17] introduced a write-barrier variation of Baker's scheme. The idea is to eliminate the check in Baker's read-barrier that determines whether an object needs to be forwarded to *ToSpace*. In this method, every object comes armed with an indirection pointer. If a *FromSpace* object has already been copied, the indirection pointer points to the *ToSpace* version as usual; otherwise the indirection pointer points to

---

[9]For the interpretation of colors in a copying collector, refer to Figure 6.18.

itself. Thus, the method does unconditional single-level dereferencing, which, on average, is assumed to be cheaper than checking. The indirection can also be regarded as a simpler form of read-barrier.

Unlike Baker's method, Brooks' method allows the mutator to see both white and gray objects. To ensure correctness of the collector, the method uses a write-barrier to detect the setting of a black-to-white pointer and moves the white object to *ToSpace*. This is similar to the snapshot-at-beginning algorithms.

6.5.2.3 Appel–Ellis–Li Collector

Without hardware support, Baker's copying method is very inefficient. Moreover, it is inherently sequential. The Appel–Ellis–Li [3] collector is efficient and concurrent. Instead of using a read-barrier to trap access of individual objects through pointers, the method uses virtual-memory page protection to detect access of pages in *FromSpace*. The method makes use of the fact that most operating systems provide two modes of execution — user mode and kernel mode. Page protection applies only to user mode.

The *ToSpace* is organized as in Baker's copying collector. The only difference is that the page occupied by a gray object is write protected. When the mutator tries to access a gray object, a trap is raised. In response to this, the collector runs a thread in the kernel mode, scans the entire page, and converts every gray object in the page to a black object by bringing its children into *ToSpace*. The write protection of the page is removed after scanning. The mutator and the collector then run concurrently, with the collector scanning pages containing gray objects. Since the mutator does not see a pointer to a white object, it cannot store one in a black object. Thus, safety of garbage collection is ensured. New objects are allocated at the other end of *ToSpace*.

Thus, the virtual memory hardware forms an efficient medium-grained synchronization between the collector and the mutator. One of the reasons for the large latency in Baker's collector was that following a flip, the root set, which could be very large, was scanned atomically. The Appel–Ellis–Li collector, however, write protects the root set and then scans it incrementally, thus decreasing the latency. However, it has been claimed [38, 65] that both Baker's and the Appel–Ellis–Li collector fail to provide the bounds on latency that are required of real-time collectors.

The methods described earlier restrict mutator access to *ToSpace* objects only. The *replication methods*, however, do quite the opposite. They restrict the mutator access to *FromSpace* objects. The collector replicates the latest version of *FromSpace* objects into *ToSpace*. When this replication is complete, *FromSpace* is discarded and the mutator switches over to *ToSpace*.

6.5.2.4 Nettles' Replicating Collector

In this method [44], collection starts by copying objects into *ToSpace*, but the root set continues to point to the objects in *FromSpace*. For the mutator to be able to execute by seeing *FromSpace* objects only, forwarding of objects by the collector must be nondestructive, that is, the forwarding address must not overwrite the data in the object as in a normal copying collector. This is done by having a separate space in the object for the forwarding address. Therefore, until the *FromSpace* objects are discarded at the end of the collection, the mutator sees a correct and updated view of the reachability graph.

If the mutator changes an object after it is copied, its *ToSpace* version has to be updated. To do this, a write-barrier is used to log all mutator changes during collection. The collector periodically uses the mutator log to update the *ToSpace* objects. Since the updated objects could be pointing to new objects in *FromSpace*, they are rescanned so that the new objects are also copied. At the end of the collection phase, the root set is updated to point directly to *ToSpace* objects. As in other incremental methods, the amount of work done during each incremental round of collection is linked to the amount of allocation of new cells during the previous mutator round. This ensures that the mutator does not run out of memory before copying is complete.

Nettles' collector can also be made concurrent. Notice that one of the ways in which the collector and mutator may interfere with each other is that an object that is being read by the collector can be written by the mutator. However, this write will appear in the mutator log and will be updated in the future to its correct value. Thus, the mutator and the collector are largely independent. However synchronizations

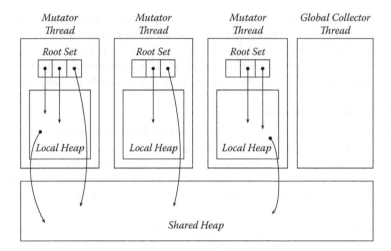

FIGURE 6.21   Outline of the Doligez–Leroy–Gonthier collector. Each mutator thread also includes a collector for the local heap.

are required around the accesses of the mutator log and the root set, since they are written by both the mutator and the collector.

Trapping each mutator write and updating *ToSpace* objects through the mutator log is expensive. Therefore, the method is effective for functional or near-functional languages, in which the number of mutator writes are very few.

6.5.2.5   The Doligez–Leroy–Gonthier Collector

This is a near-real-time concurrent garbage collector for Concurrent Caml Light. The collector is shown in Figure 6.21. Each mutator thread has a local heap, which is also its young generation space. A shared heap forms a common old generation space for all threads. Every mutator thread also includes a fast asynchronous copying collector to collect its young generation. The shared heap is collected using a mark-sweep collector activated as a concurrent thread.

Each thread allocates immutable objects in its local heap and mutable objects in the shared heap. Distinguishing immutable and mutable objects is easy in Caml because they have different types. When the local heap of a thread becomes full, the mutator is stopped and a copying collector scavenges survivors in the local heap into the shared heap. It is ensured that there are no pointers from the shared heap to the local heaps or from one local heap to another. If an assignment causes an object in the shared heap to point to an object O in the local heap, then O and all objects in the local heap reachable from O are copied to the shared heap. Thus, the local root set of a thread is enough to scavenge the local heap.

The Doligez–Leroy–Gonthier collector uses a variant of Dijkstra's method. It performs minor collections along with a major collection. A fundamental assumption in the design of the collector is that a thread can track its own root set only. Therefore, the global collector thread does not have access to any root variable. Hence, it requests each mutator to shade objects directly reachable from its roots as gray. Following this, the collector repeatedly scans and marks the local heaps and global heaps, as long as there are gray objects in the heaps. At the beginning of the sweep phase, there are no gray objects — only black and white ones. The sweep phase returns the white objects to a free list and converts the black objects to white. This ends a round of garbage collection.

If an object is newly allocated by the mutator, it is colored black as usual. To prevent black objects from pointing to white ones, when the mutator changes a pointer from A to B to A to C, the object C is marked as in Dijkstra's method. Where this method differs from Dijkstra's method is that because of the absence of the assumption of a fixed root set, the object B also has to be marked. The reason for this is shown in Figure 6.22. Let us say a marking request by the collector has resulted in the marking of A and C. Now the

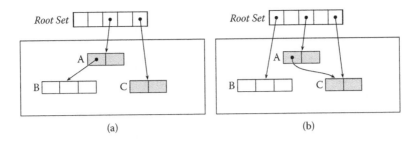

FIGURE 6.22   Old objects need to be marked during a change of pointer.

mutator expands its root set by writing the value in the first field of A to a register and then changes the pointer from A to point to C. The old object B will remain white and will be incorrectly reclaimed.

Most incremental copying collectors copy an object as an atomic action. If the object is large, this can result in large pauses. Ungar and Jackson [61] suggest ways to avoid copying large objects by separating the objects into a short header and a body. The header is treated like any other object, whereas the body is allocated space in a separate large object area. Very often, such large objects are strings and bitmaps of images and do not contain pointers. In such cases, the problem of handling intergenerational pointers also does not arise.

6.6   Conclusions

We have reviewed some basic garbage collection strategies along with their generational and incremental versions. Zorn's [74] empirical studies with the Boehm–Weiser collector have shown that, in some aspects, the performance of programs with even a basic collector such as mark-sweep is comparable to that of programs with explicit memory management. For example, the CPU costs of execution under the Boehm–Weiser collector are often comparable to and sometimes better than the costs of execution under explicitly managed memory. However, their memory requirements are high. One of the reasons for this is the conservative nature of the Boehm–Weiser collector. The page and cache faults are also higher compared to programs with explicit memory management.

An important shortcoming of the basic collectors is their large pause times. This is a serious concern in interactive and multimedia applications. Both incremental and generational techniques reduce pause times. Generational collectors also result in better locality of references. Zorn [72] has found the time overheads of generational collection over explicit memory management to be within 20%. While generational collectors seem to perform well on the whole, our review suggests that the choice of a collector is highly dependent on the context of its usage.

We now describe the directions of current work on garbage collection. Some of these developments are attempts to improve existing ideas, while others are new developments.

- **Garbage collection in uncooperative environments:** Garbage collectors that do not depend on information from the compiler are useful, because they can be made to work with already existing systems without modifications. However, because of incomplete information, these collectors have to be conservative and therefore preserve significant amounts of garbage. Thus, their memory requirement is higher than that of collectors that work in more cooperative environments. Boehm [13, 15] describes methods that increase precision by avoiding misidentification of pointers.
- **Parallel garbage collection:** One of the ways of using a multiprocessor environment effectively is to use a concurrent collector, that is, a collector that consists of several mutator threads running concurrently with a single collector thread. However, even in such configurations, the collector may become a bottleneck. To increase the throughput of the collector, it should be parallelized and run on more than one thread. It is not difficult to identify collector activities that can be done in

parallel. Tracing, for example, is easily parallelizable. However, the challenges are in load balancing and minimizing synchronization overheads. Attanasio et al. [4] have evaluated the performance of five parallel garbage collectors. Most of the collectors scale linearly up to eight processors, showing the effectiveness of parallelization.

- **Distributed garbage collection:** An interesting extension of basic garbage collection is to distribute it across multiple networked computers. Just as distributed systems allow transparent placement and invocation of local and remote objects, the idea behind distributed garbage collectors is transparent management of object spaces. One of the difficulties is to provide a consistent view of object references that keep changing in different local spaces. The distributed collector should also be able to tolerate failures of the components of the distributed system and the communication links. The challenge is to build a scalable and fault-tolerant system overcoming these difficulties. Plainfosse and Shapiro [46] have surveyed different techniques for distributed systems.

- **Cache-conscious garbage collectors:** Modern architectures organize their memory in a hierarchy of increasing access speeds. Since the fastest of these is the cache, it is necessary that the garbage collector and the mutator find most of the data in the cache. The penalty of a cache miss runs into hundreds of cycles. Chilimbi and Larus [21] show how a generational collector can be used to organize the data layout to improve the cache behavior of the mutator. Boehm [14], however, suggests methods of improving the cache behavior of the collector itself.

- **Garbage collection for persistent object stores:** Objects are called persistent when they outlive the execution of the program that created them. Typically, such objects are written into nonvolatile memory such as disks by tedious and error-prone input–output routines. Persistent object stores are a natural extension of the memory hierarchy to include such nonvolatile memory. Thus, persistent objects can be written into and read from nonvolatile memory in a transparent manner, as if such memory were part of the memory hierarchy. This also means that persistent object stores have to be garbage collected periodically; otherwise they tend to accumulate unreclaimed garbage.

On the whole, garbage collection is an extremely successful technology. Implementations of almost all declarative languages have been accompanied by garbage collectors, as have implementations of Smalltalk and scripting languages such as Perl, Ruby, and Python. Furthermore, even implementations of procedural languages such as C and C++, which do not have automatic memory management as a part of their language definitions, are being supported by external garbage collectors. In the modern era, the Java Virtual Machine and the .NET common language runtime, the virtual machines for Java and C#, both support garbage collectors. It is expected that the future will see more programming environments where garbage collectors will free programmers from concerns of memory management.

References

1. Andrew W. Appel. 1987. Garbage collection can be faster than stack allocation. *Information Processing Letters* 25(4):275–79.
2. Andrew W. Appel. 1989. Simple generational garbage collection and fast allocation. *Software Practice and Experience* 19(2):171–83.
3. Andrew W. Appel, John R. Ellis, and Kai Li. 1988. Real-time concurrent collection on stock multi-processors. *ACM SIGPLAN Notices* 23(7):11–20.
4. Clement R. Attanasio, David F. Bacon, Anthony Cocchi, and Stephen Smith. 2001. A comparative evaluation of parallel garbage collectors. In *Fourteenth Annual Workshop on Languages and Compilers for Parallel Computing*. Lecture Notes in Computer Science. Heidelberg, Germany: Springer-Verlag.
5. Thomas H. Axford. 1990. Reference counting of cyclic graphs for functional programs. *Computer Journal* 33(5):466–70.
6. Henry G. Baker. 1978. List processing in real-time on a serial computer. *Communications of the ACM* 21(4):280–94.

7. Henry G. Baker. 1992. The treadmill, real-time garbage collection without motion sickness. *ACM SIGPLAN Notices* 27(3):66–70.

8. Henry G. Baker. 1993. 'Infant mortality' and generational garbage collection. *ACM SIGPLAN Notices* 28(4):55–57.

9. David A. Barrett and Benjamin G. Zorn. Using lifetime predictors to improve memory allocation performance. In *Proceedings of the ACM SIGPLAN Conference on Programming Language Design and Implementation*, 187–96. New York: ACM Press.

10. Joel F. Bartlett. 1989. SCHEME->C: A portable scheme-to-C compiler. Technical report, DEC Western Research Laboratory, Palo Alto, CA.

11. F. L. Bauer and H. Wössner. 1982. *Algorithmic language and program development*. Heidelberg, Germany: Springer-Verlag.

12. Ricki Blau. 1983. Paging on an object-oriented personal computer for Smalltalk. In *ACM SIG-METRICS Conference on Measurement and Modeling of Computer Systems, Minneapolis.* New York: ACM Press.

13. Hans-Juergen Boehm. Space efficient conservative garbage collection. In *Proceedings of the ACM SIGPLAN Conference on Programming Language Design and Implementation*, 197–206. New York: ACM Press.

14. Hans-Juergen Boehm. 2001. Reducing garbage collector cache misses. *ACM SIGPLAN Notices* 36(1):59–64.

15. Hans-Juergen Boehm. 2002. Bounding space usage of conservative garbage collectors. *ACM SIGPLAN Notices* 37(1):93–100.

16. Hans-Juergen Boehm and Mark Weiser. 1988. Garbage collection in an uncooperative environment. *Software Practice and Experience* 18(9):807–20.

17. Rodney A. Brooks. Trading data space for reduced time and code space in real-time garbage collection on stock hardware. In *Conference record of the 1984 ACM Symposium on Lisp and Functional Programming*, 256–62. New York: ACM Press.

18. David R. Brownbridge. 1985. Cyclic reference counting for combinator machines. In *Record of the 1985 Conference on Functional Programming and Computer Architecture*, ed. Jean-Pierre Jouannaud. Vol. 201 of Lecture Notes in Computer Science. Heidelberg, Germany: Springer-Verlag.

19. Patrick J. Caudill and Allen Wirfs-Brock. 1986. A third-generation Smalltalk-80 implementation. *ACM SIGPLAN Notices* 21(11):119–30.

20. C. J. Cheney. 1970. A nonrecursive list compacting algorithm. *Communications of the ACM* 13(11):677–78.

21. Trishul M. Chilimbi and James R. Larus. 1999. Using generational garbage collection to implement cache-conscious data placement. *ACM SIGPLAN Notices* 34(3):37–48.

22. Douglas W. Clark. 1979. Measurements of dynamic list structure in Lisp. *ACM Transactions on Software Engineering* 5(1):51–59.

23. Douglas W. Clark and C. Cordell Green. 1977. An empirical study of list structure in Lisp. *Communications of the ACM* 20(2):78–86.

24. William D. Clinger and Lars T. Hansen. 1997. Generational garbage collection and the radioactive decay model. *ACM SIGPLAN Notices* 32(5):97–108.

25. Jacques Cohen and Alexandru Nicolau. 1983. Comparison of compacting algorithms for garbage collection. *ACM Transactions on Programming Languages and Systems* 5(4):532–53.

26. George E. Collins. 1960. A method for overlapping and erasure of lists. *Communications of the ACM* 3(12):655–57.

27. L. Peter Deutsch and Daniel G. Bobrow. 1976. An efficient incremental automatic garbage collector. *Communications of the ACM* 19(9):522–26.

28. Edsgar W. Dijkstra, Leslie Lamport, A. J. Martin, C. S. Scholten, and E. F. M. Steffens. 1978. On-the-fly garbage collection: An exercise in cooperation. *Communications of the ACM* 21(11):965–75.

29. Robert R. Fenichel and Jerome C. Yochelson. 1969. A Lisp garbage collector for virtual memory computer systems. *Communications of the ACM* 12(11):611–12.

30. David A. Fisher. 1974. Bounded workspace garbage collection in an address order preserving list processing environment. *Information Processing Letters* 3(1):25–32.

31. David Gries. 1979. The Schorr–Waite graph marking algorithm. *Acta Informatica* 11(3):223–32.

32. B. K. Haddon and W. M. Waite. 1967. A compaction procedure for variable length storage elements. *Computer Journal* 10:162–65.

33. Barry Hayes. 1991. Using key objects opportunism to collect old objects. *ACM SIGPLAN Notices* 26(11):33–46.

34. Richard L. Hudson and Amer Diwan. 1990. Adaptive garbage collection for Modula-3 and Smalltalk. In *OOPSLA/ECOOP '90 Workshop on Garbage Collection in Object-Oriented Systems*, ed. Eric Jul and Niels-Christian Juul.

35. R. John M. Hughes. 1982. A semi-incremental garbage collection algorithm. *Software Practice and Experience* 12(11):1081–84.

36. Mark S. Johnstone. 1997. Noncompacting memory allocation and real-time garbage collection. PhD thesis, University of Texas at Austin.

37. Richard Jones and Rafael Lins. 1996. *Garbage collection algorithms for automatic memory management*, 1st ed. New York: John Wiley & Sons.

38. H. B. M. Jonkers. 1979. A fast garbage compaction algorithm. *Information Processing Letters* 9(1):25–30.

39. Donald E. Knuth. 1973. *The art of computer programming*, Vol. I, *Fundamental algorithms*, 2nd ed., chapter 2. Reading, MA: Addison-Wesley.

40. Henry Lieberman and Carl E. Hewitt. 1983. A real-time garbage collector based on the lifetimes of objects. *Communications of the ACM* 26(6):419–29.

41. John McCarthy. 1960. Recursive functions of symbolic expressions and their computation by machine. *Communications of the ACM* 3:184–95.

42. Marvin L. Minsky. 1963. A Lisp garbage collector algorithm using serial secondary storage. Technical Report Memo 58 (rev.), Project MAC, MIT, Cambridge, MA.

43. David A. Moon. 1984. Garbage collection in a large LISP system. In *Conference Record of the 1984 Symposium on Lisp and Functional Programming*, 235–45. New York: ACM Press.

44. Scott M. Nettles, James W. O'Toole, David Pierce, and Nicholas Haines. 1992. Replication-based incremental copying collection. In *Proceedings of International Workshop on Memory Management*, ed. Yves Bekkers and Jacques Cohen. Vol. 637 of Lecture Notes in Computer Science. Heidelberg, Germany: Springer-Verlag.

45. E. J. H. Pepels, M. C. J. D. van Eekelen, and M. J. Plasmeijer. 1988. A cyclic reference counting algorithm and its proof. Technical Report 88–10, Computing Science Department, University of Nijmegen, The Netherlands.

46. David Plainfossé and Marc Shapiro. 1991. Distributed garbage collection in the system is good. In *International Workshop on Object Orientation in Operating Systems*, ed. Luis-Felipe Cabrera, Vincent Russo, and Marc Shapiro, 94–99. Washington, DC: IEEE Press.

47. David J. Roth and David S. Wise. 1998. One-bit counts between unique and sticky. *ACM SIGPLAN Notices* 34(3):49–56.

48. Jon D. Salkild. 1987. Implementation and analysis of two reference counting algorithms. Master's thesis, University College, London.

49. Robert A. Saunders. 1974. The LISP system for the Q–32 computer. In E. C. Berkeley. *The programming language LISP: Its operation and applications*, ed. E. C. Berkeley and Daniel G. Bobrow, 220–31. Cambridge, MA: Information International.

50. H. Schorr and W. Waite. 1967. An efficient machine independent procedure for garbage collection in various list structures. *Communications of the ACM* 10(8):501–506.

51. Robert A. Shaw. 1988. Empirical analysis of a Lisp system. Technical Report CSL–TR–88–351, PhD thesis, Stanford University, Stanford, CA.

52. Patrick Sobalvarro. 1988. A lifetime-based garbage collector for Lisp systems on general-purpose computers. Technical Report AITR-1417, Bachelor of Science thesis, MIT AI Lab, Cambridge, MA.

53. James W. Stamos. 1982. A large object-oriented virtual memory: Grouping strategies, measurements, and performance. Technical Report SCG-82-2, Xerox PARC, Palo Alto, CA.

54. James W. Stamos. 1984. Static grouping of small objects to enhance performance of a paged virtual memory. *ACM Transactions on Computer Systems* 2(3):155–80.

55. Guy L. Steele. 1975. Multiprocessing compactifying garbage collection. *Communications of the ACM* 18(9):495–508.

56. Darko Stefanović and J. Eliot B. Moss. 1994. Characterisation of object behaviour in Standard ML of New Jersey. In *Conference Record of the 1994 ACM Symposium on Lisp and Functional Programming*, 43–54. New York: ACM Press.

57. Will R. Stoye, T. J. W. Clarke, and Arthur C. Norman. 1984. Some practical methods for rapid combinator reduction. In *Conference Record of the 1984 Symposium on Lisp and Functional Programming*, ed. Guy L. Steele, 159–66. New York: ACM Press.

58. Lars-Erik Thorelli. Marking algorithms. *BIT* 12(4):555–68.

59. David M. Ungar. 1984. Generation scavenging: A nondisruptive high performance storage reclamation algorithm. *ACM SIGPLAN Notices* 19(5):157–67.

60. David M. Ungar and Frank Jackson. 1988. Tenuring policies for generation-based storage reclamation. *ACM SIGPLAN Notices* 23(11):1–17.

61. David M. Ungar and Frank Jackson. 1992. An adaptive tenuring policy for generation scavengers. *ACM Transactions on Programming Languages and Systems* 14(1):1–27.

62. Paul R. Wilson. 1994. Uniprocessor garbage collection techniques. Technical report, University of Texas.

63. Paul R. Wilson and Mark S. Johnstone. 1993. Truly real-time noncopying garbage collection. In *OOPSLA/ECOOP '93 Workshop on Garbage Collection in Object-Oriented Systems*, ed. Eliot Moss, Paul R. Wilson, and Benjamin Zorn, New York: ACM Press.

64. Paul R. Wilson, Mark S. Johnstone, Michael Neely, and David Boles. 1995. Dynamic storage allocation: A survey and critical review. In *Proceedings of International Workshop on Memory Management*, ed. Henry Baker. Vol. 986 of Lecture Notes in Computer Science. Heidelberg, Germany: Springer-Verlag.

65. Paul R. Wilson and Thomas G. Moher. 1989. Design of the opportunistic garbage collector. *ACM SIGPLAN Notices*, 24(10):23–35.

66. David S. Wise. 1979. Morris' garbage compaction algorithm restores reference counts. *ACM Transactions on Programming Languages and Systems* 1:115–20.

67. David S. Wise. 1993. Stop-and-copy and one-bit reference counting. Technical Report 360, Indiana University, Computer Science Department.

68. David S. Wise and Daniel P. Friedman. 1977. The one-bit reference count. *BIT* 17(3):351–59.

69. Taichi Yuasa. 1990. Real-time garbage collection on general-purpose machines. *Journal of Software and Systems* 11(3):181–98.

70. Taichi Yuasa, Yuichiro Nakagawa, Tsuneyasu Komiya, and Masahiro Yasugi. 2002. Return barrier: Incremental stack scanning for snapshot real-time GC. In *International Lisp Conference, San Francisco, California*, Heidelberg: Springer Verlag.

71. Karen Zee and Martin Rinard. Write barrier removal by static analysis. *ACM SIGPLAN Notices* 37(11):191–210.

72. Benjamin G. Zorn. 1989. Comparative performance evaluation of garbage collection algorithms. Technical Report UCB/CSD 89/544, PhD thesis, University of California at Berkeley.

73. Benjamin Zorn. 1990. Barrier methods for garbage collection. Technical Report CU-CS-494-90, University of Colorado, Boulder.

74. Benjamin Zorn. 1993. The measured cost of conservative garbage collection. *Software Practice and Experience* 23:733–56.

7

Energy-Aware Compiler Optimizations

Y. N. Srikant and
K. Ananda Vardhan

*Department of Computer Science
and Automation,
Indian Institute of Science,
Bangalore, India*
srikant@csa.iisc.ernet.in
nandu@csa.iisc.ernet.in

7.1 Introduction

The demands of modern society on computing technology are very severe. Its productivity is closely linked to availability of inexpensive and energy-efficient computing devices. While battery technology is developing at a slow pace, the demands of sophisticated mobile computing systems that require to be powered for several hours by batteries without recharging them require novel techniques to be devised to utilize the available energy efficiently. Similarly, large data centers that support complex Web searches and online purchases house thousands of servers that guzzle power. Any energy saved by the microprocessors will lower electricity bills. Under these circumstances, it is no surprise that modern microprocessor design considers energy efficiency as a first-class design constraint [31].

Apart from the above extremely important reasons, there are other technical reasons for considering power dissipation seriously. According to Pollack [35], the power densities in microprocessors are rapidly approaching that of a nuclear reactor. High power dissipation produces excessive heat and lowers reliability and lifetimes of chips. It increases production cost due to complex cooling and packaging requirements, impacts our environment, and may endanger the human body.

There are many ways to save power and energy in a computing system. Some are hardware-based and the others are software-based techniques. The major subsystems of a computing system that can benefit from power- and energy-aware designs are the disk, the memory, the CPU, and the communication subsystems. Energy-saving schemes could be built into the hardware, the operating system, or the device drivers or inserted by the compiler into the application program itself. Various proposals have been made to reduce power consumption from both the micro-architecture community as well as the compiler community. As in performance-oriented optimizations, both techniques follow a similar process of analyzing and understanding the program's behavior, either empirically or analytically, converting this behavior into requirements and mapping these requirements onto the processor's capabilities, thus enabling the program to run with more power efficiency. Even though the two techniques have a similar goal and a similar process, they adopt different approaches to achieve it. The difference is due to the levels at which they analyze the program and, more importantly, how the analysis information is related to power consumption. Micro-architectural techniques normally use simple program properties within a narrow window of time, such as usage of a function unit or cache line in the last few cycles. Compiler techniques consider global program properties, such as profile information, data flow analysis information, and so on and therefore can do better in several cases.

A compiler writer's interpretation of the program behavior from the point of view of power consumption is different from that of a person designing a compiler that would produce time-efficient code. This involves mapping program properties such as resource usage, memory accesses, maximum Instruction Level Parallelism (ILP), availability of slack, and so on onto a power model and estimating the power consumption. This requires a very intricate and detailed analysis. For example, from the memory perspective, once the sequence of blocks being accessed is analyzed, designing a fast prefetching algorithm is not sufficient. The requirement to control power consumption brings a need to understand information such as at what rate the prefetching should be done, because this value directly impacts the power consumption. Even though analysis at such a detailed level is not feasible at the micro-architecture level, controlling power consumption dynamically using the first hand knowledge available at runtime is possible in this case and is quite attractive in certain cases. Using the performance counters in Intel processors for dynamic voltage scaling in just-in-time compilers, as in Section 7.3.4, is such an example.

This chapter discusses a few energy-saving optimizations that can be implemented in a compiler. The discussion will be limited to CPU and memory subsystems. The major advantage of using the compilation approach is that the techniques can be automated. It should be emphasized here that hardware-based and operating system–based techniques increase the effectiveness of the compiler optimizations.

The chapter begins with a brief discussion of power models used in compilers [8, 17, 24], followed by a discussion of the various optimization techniques. Among the major optimization techniques that an energy-aware compiler can implement, *dynamic voltage scaling* of the CPU [21, 52, 53], *instruction scheduling for leakage energy reduction* [46], and *scratchpad memory allocation* [2, 41] are the most important ones, and sufficient attention will be paid to these techniques in this chapter. For a more extensive discussion on power models, the reader is referred to [7, 8, 11, 12, 17, 23, 26, 31, 35, 47, 50]. Other techniques for dynamic voltage scaling are discussed in [3, 36, 38, 45]. Additional instruction scheduling techniques are reported in [27, 32–34, 37, 51, 54–56]. A variety of other scratchpad allocation strategies are available in [1, 5, 16, 28, 48].

Apart from the optimizations discussed in this chapter, other compiler optimization techniques for saving energy have been reported in the literature. Allocation of variables to memory banks is discussed in [6, 13–15, 19, 25, 29, 42], and compiler-controlled cache management is discussed in [39, 57, 58].

7.2   Power and Energy Models

It is essential to understand the models for power and energy consumption in CMOS chips in order to design *any* energy-saving optimization in compilers or operating systems. Since energy is the product of power and time, it may seem that optimizing one will automatically optimize the other. This is not true. For example, the power consumed by a computer system can be reduced by halving the clock frequency, but if the computer takes twice as long to complete the same task, the total energy consumed may not be lower. Energy optimization seems more important in mobile systems because it makes the battery run longer. Servers, however, need power optimization so that instantaneous power consumption and hence temperature are within limits, with energy consumption taking the back seat.

Several simple equations that govern the behavior of power dissipation in a CMOS device, as supply voltage and/or frequency are varied, are well known in the literature [8, 17, 24]. Three of these equations are presented below. The relation governing frequency is

$$f \propto (V - V_{th})^{\alpha} / V \tag{7.1}$$

where V is the device's supply voltage, V_{th} is its threshold or switching voltage, f is its operating frequency, and α is a technology-dependent constant, which has been experimentally determined to be about 1.3. As voltage is reduced to the level of V_{th}, f decreases to zero. This equation can be approximated by a linear relationship:

$$V_n = \beta + (1 - \beta) * f_n \tag{7.2}$$

where $V_n = V/V_{max}$, $f_n = f/f_{max}$, and $\beta = 0.3$ for today's technology. V_{max} and f_{max} are the maximum possible device supply voltage and operating frequency, respectively. Substituting for β in Equation 7.2, assuming $V >> V_{th}$, $V_{max} >> V_{th}$, and simplifying, Equation 7.2 becomes

$$f = kV \tag{7.3}$$

where $k = f_{max}/V_{max}$ is a constant within the operating range of voltage and frequency. Power dissipation in a device is given by

$$P_{device} = \frac{1}{2}CVV_{swing}af + I_{leakage}V + I_{short}V \tag{7.4}$$

where P_{device} is the power dissipation of the device, C is the output capacitance, a is the activity factor ($0 < a < 1$), V_{swing} is the voltage swing across C, $I_{leakage}$ is the leakage current in the device, and I_{short} is the average short-circuit current when the device switches, short-circuiting the drain and the source. The three terms in Equation 7.4 correspond to switching power loss (dynamic power loss), leakage power loss, and short-circuit power loss, respectively. With today's technology, switching power loss is the dominating term in Equation 7.4, $V_{swing} \approx V$. Assuming C to be the average over all devices in the chip, and summing over all the blocks on the chip, an equation for P_{chip}, the power dissipation of the chip, is obtained:

$$P_{chip} = \frac{1}{2} \sum_{all\ blocks\ i} CV_i^2 a_i f_i \tag{7.5}$$

Making a first-order approximation that $a_i = 1$, f_i and V_i are the same for all the blocks i on the chip, and using Equation 7.3, Equation 7.5 simplifies to

$$P_{chip} = K_v V^3 = K_f f^3 \tag{7.6}$$

where K_v and K_f are constants. Equation 7.6 is the so-called *cube root rule*. It is clear from Equation 7.6 that changing the chip voltage or frequency is one of the most effective and powerful methods of reducing dynamic power and energy consumption. Leakage power and energy loss cannot be reduced by dynamic voltage scaling, but can be controlled by voltage gating of function units, memory, and cache. Short-circuit power loss cannot be controlled by software, but only by superior VLSI technology.

7.3 Dynamic Voltage Scaling (DVS)

It is well known that Dynamic Random Access Memory (DRAM) is slower than the CPU by at least a factor of 3. Caches are used to offset this incompatibility. However, in the case of programs that use memory heavily and whose data do not fit into the cache, the CPU stalls to let memory catch up and to ensure that data dependencies are honored. It is easy to come to the conclusion that the CPU can use a lower clock frequency in certain parts of the program that are memory bound and thereby save power and energy.

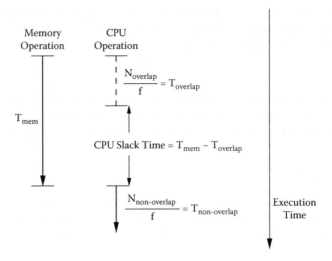

FIGURE 7.1   Illustration of CPU slack time.

This may increase the execution time of the program if not done carefully. Figure 7.1 (adapted from [52]) helps illustrate the concept of DVS more clearly.

The basic assumption here is that the memory is asynchronous, has a clock and power supply that is independent of the CPU, and is much slower than the CPU. Therefore, the CPU stalls in memory-bound applications, and *CPU slack time* can identified as shown in Figure 7.1. In Figure 7.1, T_{mem} is the time for memory operations, f is the CPU clock frequency, $N_{overlap}$ is the number of cycles of CPU operation that can be overlapped with the memory operation without requiring the result of memory operations, and $N_{nonoverlap}$ is the number of cycles of CPU operation that require the result of pending memory operations. The CPU is idle in the CPU slack time period. Therefore, it is possible to reduce the operating voltage and frequency of the CPU in the overlap region such that $T_{overlap}$ stretches and reduces the CPU slack time close to zero. The CPU voltage and frequency may have to be changed again for the nonoverlap region so that there is no performance penalty.

DVS is a hardware-based technique that allows a CPU to change its operating voltage (and thereby frequency) using a program instruction. Currently a number of processors from Intel, AMD, Transmeta, TI, and so on used in laptops and handheld devices provide this facility. Voltage-frequency settings are available as a set of pairs of discrete values, with a particular voltage setting tied to a fixed frequency. Furthermore, switching from the current voltage-frequency setting to a new one requires time equivalent to approximately 100 memory transactions (typically, 20 μs). A good DVS algorithm should consider all these factors.

Program analysis can be used to determine *when* to change the current CPU voltage-frequency setting and *which* setting to use, so that considerable savings in energy can be obtained with not much impact on performance. Such analysis uses program profile information and other program properties to make these decisions. Compiler analysis is only one of the ways in which DVS can be implemented. Interval-based, intertask, and intratask approaches are the other techniques, and a survey of these techniques is found in [47]. The next two sections (7.3.1 and 7.3.3) contain descriptions of a region-based algorithm [21] and an integer-linear-programming–based algorithm [53], respectively, both of which can be incorporated into a compiler. Both these algorithms are *static* in nature and cannot be used in dynamic compilers. A DVS algorithm suitable for use in dynamic compilers [52] is presented in Section 7.3.4.

7.3.1   Region-Based Compiler Algorithm for DVS

This algorithm works well only with memory-bound programs. It partitions a given program into two regions, one of them running at (V_{max}, f_{max}), the maximum voltage-frequency pair for the CPU, and the other running at a lower voltage-frequency pair, for instance, (V, f). The constraints on this partitioning process are on the overall performance and the size of the region running at (V, f). Formally, the problem can be stated as follows.

$$\min_{R, f} \{W_f \cdot \tau(R, f) + W_{f_{max}} \cdot \tau(P - R, f_{max}) + W_{trans} \cdot 2 \cdot n(R)\} \tag{7.7}$$

subject to

$$\tau(R, f) + \tau(P - R, f_{max}) + \tau_{trans} \cdot 2 \cdot n(R) \leq (1 + r) \cdot \tau(P, f_{max}) \tag{7.8}$$

and

$$\tau(R, f_{max})/\tau(P, f_{max}) \geq \rho \tag{7.9}$$

where

P: Given program
R: Region within P that is to be found
$P - R$: Part of P, leaving out R
f_{max}: Maximum frequency of operation of the CPU
f: Operating frequency of region R
$W_{f_{max}}$: Power dissipation of the whole system at frequency f_{max}
W_f: Power dissipation of the whole system at frequency f
W_{trans}: Power dissipation during a single voltage-frequency switch
$\tau(R, f)$: Total execution time of region R running at frequency f
$\tau(P - R, f_{max})$: Total execution time of region $P - R$ running at frequency f_{max}
$\tau(P, f_{max})$: Total execution time of program P running at frequency f_{max}
τ_{trans}: Performance (time) overhead during a single voltage-frequency switch
$n(R)$: Number of times region R is executed
r: Percentage of performance loss (due to DVS) that is tolerated
ρ: Compiler design parameter, controlling the size of R

In words, the problem statement can be explained thus. Given a program P, carve out a region R from it and find a frequency f, such that when R is run at the frequency f and the rest of the program, $P - R$, is run at the frequency f_{max}, the total energy usage is minimized, and the total execution time, including the switching overhead, is not increased more than $r\%$ of the original execution time (P running at f_{max}). The structure of the region R and its determination are explained in Section 7.3.2. It suffices to say at this point that R is a single-entry, single-exit program structure. It is necessary to put a lower bound on the size of region R (see Equation 7.9), to ensure that the execution time of region R is more than the time overhead of a single voltage-frequency switch. This is controlled by the factor ρ and can be set by experimentation. Its value is usually 0.2 to 0.3.

After the region R is found, the compiler introduces two DVS calls, one just before entry into R, to change the frequency to f, and another just after exit from R, to change the frequency back to f_{max}.

7.3.2 Finding Regions

A program region R is a single-entry and single-exit program structure. Loop nests, function and procedure call sites, procedures and functions, sequences of statements, or even the whole program, can all be considered as regions. This definition ensures that all the top-level statements inside a region are executed the same number of times. This in turn implies that the number of times the program executes DVS calls (voltage-frequency change settings) is precisely $2 \cdot N(R)$. Regions are classified as basic regions and composite regions. Tightly nested loops, and call sites of functions and procedures, are basic regions. Composite regions are formed from basic regions by the rules of composition given in Figure 7.2. Any other type of statement, such as an assignment (with no function call in it) or an if-then-else (with no

if-then-else statement:

R: **if** () R_1; **else** R_2

$$\tau(R, f) = \tau(R_1, f) + \tau(R_2, f)$$
$$n(R) = n(R_1) + n(R_2)$$

Either one of R_1 or R_2 is definitely executed; hence addition

Loop structure with imperfect nesting:

R: **for** () R_1

$$\tau(R, f) = \tau(R_1, f)$$
$$n(R) \text{ is obtained by profiling}$$

$\tau(R_1, f)$ takes care of the count of the for-loop as well

While-loop is similar to the above for-loop

Function call site:

R: $F()$

$$\tau(R, f) = \tau(F, f) \cdot n(R)/n(F)$$
$$n(R) \text{ is obtained by profiling}$$

Sequence of regions:

R: $\{R_1; R_2; \cdots R_n\}$

$$\tau(R, f) = \sum_{1 \le i \le n} \tau(R_i, f)$$
$$n(R_1) = n(R_2) = \cdots = n(R_n)$$

Function:

F: $< type\ of\ F > F()\ R$

$$\tau(F, f) = \tau(R, f)$$
$$n(F) = \sum_{R_i \text{ is a call site to } F()} n(R_i)$$

FIGURE 7.2   Rules for combining basic regions to form composite regions. Syntax of constructs is as in the language C.

call site or loop within), is treated as a part of the region following it. An example of regions is given in Figure 7.3.

In Figure 7.3, C1, C2, C3, L2, L3 (together with L4), L5, and L6 (together with L7) are all basic regions. S1 is treated as a part of the region L1. Region L1 is treated as a composite region consisting of regions C2 and L2. Similarly, I1 consists of L3-L4 and C3; F1 consists of C1, S1-L1, and I1; F2 consists of L5; and F3 consists of L6.

The DVS algorithm consists of the following steps:

1. Identify the basic and composite regions in the source code. An abstract syntax tree (AST) representation of the source code will be useful here.
2. Instrument the source code by inserting book-keeping code at the beginning of each region (basic and composite). This book-keeping code is in the form of a system call that takes an index for the region R, for which it carries out book-keeping. At the time of profiling, it counts $n(R)$ for each basic region R and for imperfect loops as shown in Figure 7.2. It also records $\tau(R, f)$ for a loop-nest basic region during profiling.
3. Profile the instrumented code and compute $n(R)$ and $\tau(R, f)$ for basic regions and a few other regions, for all the frequencies, f, that are permitted in the target system. The power dissipation of the system should be measured either with a digital power meter as in [21] or with a simulator such as Wattch [8]. Suitable values for τ_{trans} and W_{trans} must be chosen ([21] used values of 20 μs and 0 W, respectively, for these two quantities).
4. Compute $n(R)$ and $\tau(R, f)$ for the composite regions using the rules shown in Figure 7.2. This step requires interprocedural analysis. A call graph is needed along with the AST. Only regions at

```
F1:  main(){
C1:    init();
S1:    i=1; y=10;
L1:    while (i<100){
C2:      x=func1();
L2:      for(int j=0; j<M; j++) { ··· }
I1:      if(x>y)
L3:        for(int j=0; j<N; j++)
L4:          for(int k=0; k<P; k++) {···}
C3:      else y=func1();
       }
     }
F2:  void init(){
L5:    for(int i=0; i<20; i++){···}
     }
F3:  int func1(){
L6:    while(a>b)
L7:      for(int j=100; j>5; j−−){···}
     }
```

FIGURE 7.3   An example of regions.

the same level of nesting can be considered for composition. For example, C1 cannot be composed with C2, and L2 cannot be composed with L3.

5. Solve the minimization problem stated in Equation 7.7 by considering all possible candidate regions (basic and composite), which can be considered as R and $P - R$, and find R and f. This step requires the values of r and ρ to be specified ([21] used values of 5 and 20%, respectively, for these two quantities).

6. Insert DVS calls at the entry and exit points of the region R to change the voltage and frequency. For example, in Figure 7.3, if region L3-L4 is chosen as R during the minimization phase, and f is the associated frequency, L3-L4 will operate at f, and the rest of the program will operate at f_{max}. To facilitate this, a DVS call is inserted just before the region L3-L4 to change the frequency to f (since the program would have started to execute with the frequency f_{max}), and another DVS call is inserted just after the region L3-L4 to change the frequency back to f_{max}.

7.3.2.1 An Example for Region Computation

Only one of the many regions in a program can be active at any time. Code instrumentation can be implemented through a system call that carries a region identifier as a parameter. For each execution, this system call increments $n(R)$ once for its parameter region R (as per Figure 7.2). Thus, $n(R)$ accumulates the required count of the number of times the region R is executed. This system call also records the time taken to execute basic loop-nest regions. This is done by recording the value of a high-precision timer in $\tau(R, f)$ for the basic loop-nest region R and updating it once another such system call for a different region is executed.

For example, in Figure 7.4, once the execution of the instrumentation code for the region L2 has begun, $\tau(L2, f)$ records the timer value at that time. As soon as the instrumentation code for region L3-L4 has begun, not only is the timer value recorded for L3-L4, but $\tau(L2, f)$ is updated with the difference in timer values (current value and old value in $\tau[L2, f]$). This difference is the time spent in the execution of region L2. The basic assumption here is that straight line code that is executed only once consumes much less time than loops and hence can be either neglected or assigned a small constant value as its time of

```
F1: main(){
        instrumentation code for C1
C1:     init();
        instrumentation code for S1-L1
S1:     i=1; y=10;
L1:     while (i<100){
        instrumentation code for C2
C2:         x=func1();
        instrumentation code for L2
L2:         for(int j=0; j<M; j++) { ··· }
I1:         if(x>y)
            instrumentation code for L3-L4
L3:             for(int j=0; j<N; j++)
L4:                 for(int k=0; k<P; k++) {···}
            instrumentation code for C3
C3:         else y=func1();
        }
    }
F2: void init(){
        instrumentation code for L5
L5:     for(int i=0; i<20; i++){···}
    }
F3: int func1(){
        instrumentation code for L6-L7
L6:     while(a>b)
L7:         for(int j=100; j>5; j--){···}
    }
```

FIGURE 7.4   An example of code instrumentation for regions.

execution. Recursive functions are handled correctly by the above procedure. $n(R)$ gets incremented for each call (as in a loop), and timer value will get updated correctly on termination of the recursive call. The final recorded value of $\tau(R, f)$ will be the time of execution of the call that terminates recursion, even though its intermediate values are not meaningful. All the vaues are averaged over several sets of inputs (as is usual in profiling).

After the profiling operation is complete, $n(R)$ and $\tau(R, f)$ for all the composite regions are computed using the equations shown in Figure 7.2. Sample values of $n(R)$ and $\tau(R, f)$ for the regions in Figure 7.3 are shown in Figure 7.5. The computations are carried out during a traversal of the call graph in reverse topological order. The members of a mutually recursive set of functions can be traversed in any order. The ASTs of each function and procedure are traversed in a bottom-up fashion. The entire set of operations is repeated for the various frequencies available in the target system. Examples of computing $n(R)$ and $\tau(R, f)$ are provided in Figure 7.6 (for some of the entries in Figure 7.5).

7.3.3   Integer Linear Program–Based Compiler Algorithm for DVS

Section 7.3.1 described a DVS algorithm that minimizes total energy consumption. It uses a clever region composition heuristic, which works very well in practice. It also uses coarse-grain profiling (at the level of loop nests and function calls) to enable accurate power and time measurements during executions of

R (basic)	$n(R)$	$\tau(R, f)$
C1	1^+	30
C2	99^+	262.8
C3	14^+	37.2
L2	99^+	8^+
L3-L4	85^+	200^+
L5	1^+	30^+
L6-L7	113^+	300^+
R (composite)	$n(R)$	$\tau(R, f)$
S1-L1	1^+	508
I1	99	237.2
F2=init	1	30
F3=func1	113	300
F1=main	1	538

FIGURE 7.5 Sample values of $n(R)$ and $\tau(R, f)$ for the regions in Figure 7.3. Numbers flagged with $+$ indicate profiled values.

basic regions. It is also possible to use an energy simulator such as Wattch [9] to perform cycle-accurate simulation to get these figures for the purposes of DVS. However, such simulations do not consider power dissipation in the complete computer system. They consider only the CPU and memory and leave out input–output (I/O) controllers, disks, and so on from the simulation. The figures obtained from such simulations are still useful in performing DVS, since present-day I/O controllers and disks are power-aware and minimize power dissipation independently on their own.

The advantage in using such simulators to get the execution time and energy dissipation figures is that these figures can be obtained at a very fine grain level (for example, basic blocks). Therefore, it is possible to perform DVS in a better fashion with finer control. However, at this grain size, voltage transition time and energy become significant when compared to the power dissipation and time consumption figures for basic blocks and cannot be ignored. Trying all possible region combinations by enumerating them as in Section 7.3.1 can be extremely time consuming when the number of regions is very large (as is the case when a region is a basic block). In such cases, the problem of DVS can also be formulated using *mixed integer linear programming* (MILP) to produce better results [36, 53]. MILP formulations produce optimal results, and their accuracy is limited only by the accuracy of profiling and the time available to solve the problem. Several commercial packages such as CPLEX (from ILOG) are available to solve MILP problems, and these produce results in a very reasonable amount of time. An MILP formulation of the DVS problem adapted from [53] is presented in Section 7.3.3.1.

$$
\begin{aligned}
\tau(F2, f) &= \tau(L5, f) = 30 \\
n(F2) &= 1 \text{ (only one call site)} \\
\tau(F3, f) &= \tau(L6 - L7, f) = 300 \\
n(F3) &= n(C2) + n(C3) = 99 + 14 = 113 \\
\tau(C1, f) &= \tau(F2, f) \cdot n(C1)/n(F2) = 30 \times 1/1 = 30 \\
\tau(C2, f) &= \tau(F3, f) \cdot n(C2)/n(F3) = 300 \times 99/113 = 262.8 \\
\tau(C3, f) &= \tau(F3, f) \cdot n(C3)/n(F3) = 300 \times 14/113 = 37.2 \\
\tau(S1 - L1, f) &= \tau(C2, f) + \tau(L2, f) + \tau(I1, f) = 262.8 + 8 + 237.2 = 508 \\
\tau(I1, f) &= \tau(L3 - L4, f) + \tau(C3, f) = 200 + 37.2 = 237.2 \\
n(I1) &= n(L3 - L4) + n(C3) = 85 + 14 = 99 \\
\tau(F1, f) &= \tau(C1, f) + \tau(S1 - L1, f) = 30 + 508 = 538 \\
n(F1) &= 1 \text{ (main is called only once)}
\end{aligned}
$$

FIGURE 7.6 Examples of computing $\tau(R, f)$ and $n(R)$ for the values shown in Figure 7.5.

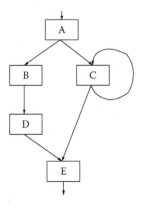

FIGURE 7.7   Control flow graph and DVS call placement.

7.3.3.1   An MILP Formulation of the DVS Problem

This algorithm requires the control flow graph of the program along with profiling information on the number of times each edge is executed. The energy consumption and execution time for each basic block are also needed. The algorithm places DVS calls to change the voltage along chosen edges, rather than at the beginning and end of regions. Consider Figure 7.7. Depending on the execution time of the blocks and the criticality of paths, the voltage setting for block E may depend on the edge through which it is entered, viz., CE or DE. It uses a similar target architectural model as the algorithm in Section 7.3.1.

Burd and Broderson [10] have given precise equations to account for the transition time and energy for a transition from voltage V_i to voltage V_j. The larger the voltage change, the larger are the transition time and energy values.

$$\tau_{trans} = \frac{2 \cdot C}{I_{max}} \cdot |V_i - V_j| \tag{7.10}$$

$$E_{trans} = (1 - \eta) \cdot C \cdot \left| V_i^2 - V_j^2 \right| \tag{7.11}$$

where

   τ_{trans}: Time for transition from voltage V_i to V_j
   E_{trans}: Energy consumed in transition from voltage V_i to V_j
   C: Voltage regulator capacitance
   I_{max}: Maximum allowed output current of the voltage regulator

Typical values are $C = 10 \ \mu F$, $\tau_{trans} = 12 \ \mu s$, and, $E_{trans} = 1.2 \ \mu J$ for a voltage change from $V_i = 1.3$ V, to $V_j = 0.7$ V. The optimization problem can be stated as follows.
   Minimize

$$\sum_{i=1}^{r} \sum_{j=1}^{r} \sum_{k=1}^{v} n_{ij} E_{jk} s_{ijk} + \sum_{p=1}^{r} \sum_{i=1}^{r} \sum_{j=1}^{r} n_{pij} e_{pij} C_e \tag{7.12}$$

subject to the constraints

$$\sum_{i=1}^{r} \sum_{j=1}^{r} \sum_{k=1}^{v} n_{ij} T_{jk} s_{ijk} + \sum_{p=1}^{r} \sum_{i=1}^{r} \sum_{j=1}^{r} n_{pij} t_{pij} C_t \le (1 + \alpha)\tau \tag{7.13}$$

$$\sum_{k=1}^{v} s_{ijk} = 1 \tag{7.14}$$

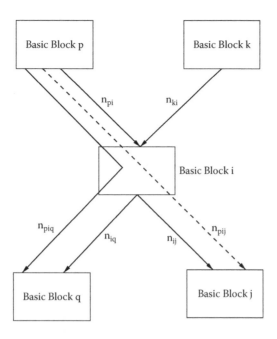

FIGURE 7.8   Edge graph to illustrate definitions in MILP formulation of DVS.

$$-e_{pij} \leq \sum_{k=1}^{v} \left(s_{pik} V_k^2 - s_{ijk} V_k^2 \right) \leq e_{pij} \tag{7.15}$$

$$-t_{pij} \leq \sum_{k=1}^{v} (s_{pik} V_k - s_{ijk} V_k) \leq t_{pij} \tag{7.16}$$

where (see Figure 7.8 for illustrations)

r: Number of regions in the program (i.e., basic blocks)
v: Number of available levels of voltage change
n_{ij}: Number of times region j is entered through edge (i, j)
E_{jk}: Energy consumed by region j at voltage level k
e_{pij}: Transition energy consumed by a voltage change (if any) when passing from edge (p, i) to (i, j)
    during execution
T_{jk}: Time for execution of region j at voltage level k
t_{pij}: Transition time for a voltage change (if any) when passing from edge (p, i) to (i, j) during execution
s_{ijk}: 0-1 integer variable; $s_{ijk} = 1$ if a DVS call sets the voltage level along edge (i, j) to k; otherwise 0
n_{pij}: Number of times region i is entered through edge (p, i) and exited through edge (i, j)
C_e: Equals $C \cdot (1 - \eta)$, a constant for a given voltage regulator
C_t: Equals $\frac{2 \cdot C}{I_{max}}$, a constant for a given voltage regulator
τ: Time of execution of the whole program at maximum permitted voltage with no DVS
α: Tolerance in reduction of performance (time)
V_k: Supply voltage at voltage level k

The two terms in the objective function in Equation 7.12 correspond to the total energy consumed by all the regions in the program and the total energy cost for voltage level changes, respectively. Similarly, the two terms on the left side of the inequality in Equation 7.13 represent the total execution time of the regions and the total time for voltage transitions, respectively. The variables e_{pij} and t_{pij} have been introduced in the constraints to eliminate the absolute value terms that would otherwise be introduced into the function to be minimized, viz., Equation 7.12. It must be noted that the variables s_{pik} and s_{ijk} can have a value of 1 for at most one value of k (by definition). After finding the values of variables s_{ijk} for all the edges (i, j) in the control flow graph, a pass over the control flow graph inserts appropriate DVS calls on the edges by breaking the edge and introducing the instruction in between. More details are available in [53]. It is not yet known whether the MILP-based DVS scheme offers substantial improvement over the region-based scheme presented earlier in real-life applications.

7.3.4 DVS Algorithm for a Dynamic Compiler

The two DVS algorithms presented in Sections 7.3.1 and 7.3.3 are both static. They do not adapt to different program conditions, such as different input sets, or different target machine parameters. For example, the memory requirements of several programs go up as the input size increases [52]. This results in more cache misses and more memory boundedness for programs and more opportunities for DVS to change over the CPU to a much lower voltage and save more energy. In other words, for smaller input sizes, the program may have to run at a higher voltage compared to that for larger input sizes, so as to have acceptable performance penalties due to DVS. Similarly, machines with more cache memory and/or with faster DRAM will have fewer opportunities for DVS compared to machines with less cache and/or slower memory. Such variations cannot be accommodated with a static DVS scheme that inserts DVS calls into code at compile time, but can be handled by dynamic compilers [18] very well.

A dynamic compiler is a much more generalized software system than a just-in-time compiler. It recompiles, modifies, and optimizes parts or all of a program as it runs. It refers to techniques for runtime generation of executable code. A dynamic compiler uses runtime information of a program to exploit optimization opportunities not available to a static compiler. Two good examples are loop unrolling based on runtime information regarding loop limits and optimizations performed across dynamic binding. It also presents a good opportunity to enable legacy executable code produced using outdated compiler technology and currently running on outdated hardware to run on present-day hardware with good speedup. The most important disadvantage of a dynamic compiler is that all the time spent in carrying out optimizations is added to the execution time. Hence, all optimizations and transformations must be very lightweight in nature.

A DVS scheme built into a dynamic compiler measures the slack available because of the difference in the speed of memory and CPU and makes decisions regarding DVS. Such measurements are made based on the frequency of execution of code regions and whether the regions are memory bound. They use the hardware performance counters (HPCs) available in several modern-day processors and very little computation to estimate the slack. Such a scheme, based on [52], is explained in the next few sections.

7.3.4.1 An Overview of the Scheme

The first step in the DVS algorithm is to identify the so-called *hot regions*. Since it is profitable to optimize only long-running code, loops and functions qualify as code regions to be monitored for hotness. The lightweight profiling mechanisms available in dynamic compilers can be used for this purpose [18]. No code regions will be hot at program start time. Once hot regions are identified, the usual optimizations of the dynamic compiler are applied on them, and segments of test code to collect runtime information from HPCs, such as the number of memory bus transactions, are inserted at the entry and exit points of a hot region. After collecting sufficient information, the DVS decision algorithm decides the appropriate voltage/frequency setting for the hot region, removes the test code, and inserts the required DVS code at the entry and exit points of the hot region. While permitting multiple DVS regions may sound natural in a dynamic compilation framework, permitting nested DVS regions may increase the overheads beyond the permitted limit. Furthermore, if a region changes its hotness, then the applied DVS will have to be changed after a reevaluation.

7.3.4.2 The DVS Decision Algorithm

As in any DVS scheme, the assumptions made in Section 7.3 hold here, and the terminology of Figure 7.1 will be used in this section. A new term, *relative CPU slack time*, can be defined as

$$\text{relative CPU slack time} = \frac{T_{mem} - T_{overlap}}{T_{total}} \tag{7.17}$$

where T_{total} is defined as $T_{mem} + T_{nonoverlap}$. If βf is the new CPU frequency, $(1 - \beta)$ is the reduction in frequency. It can be observed from Figure 7.1 that the larger the CPU slack time (and relative CPU slack time), the larger is the frequency reduction (i.e., $[1 - \beta]$) that can be achieved. If η is the tolerated performance loss, then

$$(1 - \beta) = c_0 \cdot \eta \cdot \left(\frac{T_{mem}}{T_{total}} - \frac{T_{overlap}}{T_{total}} \right) \tag{7.18}$$

where c_0 is a constant. Simplifying Equation 7.18, an equation for β is obtained:

$$\beta = 1 - c_1 \frac{T_{mem}}{T_{total}} + c_2 \frac{T_{overlap}}{T_{total}} \tag{7.19}$$

The second and third terms in Equation 7.19 can be expressed using information available in HPC events.

$$\frac{T_{mem}}{T_{total}} \simeq k_1 \frac{\#mem\_bus\_transactions}{\#\mu ops\_retired} \tag{7.20}$$

$$\frac{T_{overlap}}{T_{total}} \simeq k_2 \frac{\#FP\_INT\_instructions}{\#\mu ops\_retired} \tag{7.21}$$

For an x86 processor, the quantities in Equations 7.20 and 7.21 are directly available through HPC events, and hence β in Equation 7.19 is inexpensive to compute. More details regarding HPC events and performance monitoring are available in [22, 40, 52]. However, this method may not be very accurate in practice. More precise computation of β in Equation 7.19 enables better DVS but will perhaps be more expensive in time and may affect performance considerably. According to [52], DVS in a dynamic compilation environment may offer more than two times the energy savings compared to DVS in a static compilation environment.

7.4 Optimizations for Leakage Energy Reduction

Power consumption in processor cores has become an important concern in architectural design as well in compiler construction. In older technologies, as indicated in Equation 7.4 and the associated discussion, dynamic energy dissipation was the dominating factor. The leakage energy consumption corresponding to the inactive state of the circuits was negligible. This assumption will no longer hold in the near future.

Static energy dissipation results from leakage current (see Equation 7.4), which in turn is very sensitive to increases in threshold voltage, V_t. As integrated circuit technology improves and scales to smaller dimensions, supply voltages are reduced, and this in turn decreases V_t. With the threshold voltages reaching low values, the leakage energy is estimated to be on par with the dynamic switching energy in all the units of the processor, with the forthcoming 70-nm technology. Studies show that with the technology trend, leakage power consumption is going to increase linearly, whereas dynamic power consumption is going to remain almost constant [42], the former thereby constituting a significant portion of the total power consumption. In the literature, the terms *static energy* and *leakage energy* are used interchangeably, and they will be used similarly in this chapter. Leakage power consumption can be reduced if the number of idle cycles is increased. This can be done automatically by circuit-level switching techniques or by explicitly

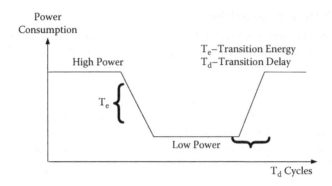

FIGURE 7.9   Behavior of a functional unit. This shows the transition from a high-power mode to a low-power mode.

gating the supply voltage of devices or function units through program instructions, thereby placing the function units in low-power mode. Static power consumption in low-power mode is negligible compared to that in the active mode of the unit.

7.4.1   Transition-Aware Scheduling

Figure 7.9 shows the behavior of a unit with a circuit-level switching technique. In [17, 20, 43] it is shown that transition energy, depending on the controlling technique, varies between 1 and 10% of the active energy. Moreover, the transition delay ranges from 3 to 30 cycles in 70-nm technology [20, 43]. The greater the transition delay, the greater is the performance penalty, but a smaller transition delay (sharp transition) requires more transition energy. If the transition delay is critical and is made very small, the transition energy loss can be up to 50% of the dynamic energy consumption. The break-even time period is calculated as the minimum time that a circuit should be in the idle state so that the energy overhead of transition is nullified. The break-even time is estimated to be anywhere between 10 and 30 cycles.

If the number of transitions between active and idle modes can be reduced, the saved transition delay can be added to the idle period of the functional unit. This reduces the frequent discharging and charging of the capacitance in the circuits of functional units, thereby reducing the power consumed due to transitions as well. It also increases the chances of the function unit staying in an idle period or in an active period continuously for longer durations, thus using the resources more efficiently, from both the performance and power point of view. Circuit-level switching techniques on their own do not take any program parameters into account and hence can cause a certain amount of performance impact too. *Transition-aware scheduling* [46], an instruction scheduling technique, can aid such hardware mechanisms by reordering instructions such that the function units that are in active mode can continue to be in active mode, and once they become idle, they remain so for several cycles. The number of transitions of the function units from active to low power and vice versa is reduced by this intelligent scheduling. This reduction not only saves power, but also improves performance. The presentation in the next few sections is based on [46].

7.4.1.1   An Example

Let us take the code segment in Figure 7.10, possibly generated by a traditional scheduler. The figure also shows how different functional units, ALU, MUL, and memory Read Ports (RP) are used. It must be noted that the functional unit usage in each line is the cumulative demand of the previous instruction(s) and the current instruction. The load instruction scheduled in the cycle i will use an ALU in the first cycle and a read port RP in the next cycle. The initial stages of the pipeline that are common to all the instructions are not modeled. A switch from 1 to 0 and 0 to 1 indicates a transition from the active mode to the idle mode and vice versa, respectively. The *add* instruction scheduled in the $i+1$ cycle and the *mul* instruction scheduled in $i+2$ force a transition in the ALU. Figure 7.11 shows how to rearrange these instructions such

Cycle	Instruction	ALU	MUL	RP
i	ld r1, 10[r2]	1	0	0
i+1	add r1,r1,r4	1	0	1
i+2	mul r3,r1,r3	0	1	0
i+3	ld r5,10[r2]	1	0	0
i+4	add r4,r1,0x04	1	0	1
i+5	mul r4,r1,r4	0	1	0
i+6	add r1,r5,r3	1	0	0

FIGURE 7.10 This is a valid schedule generated by a traditional scheduler.

that the number of transitions decreases and the continuous idle period in each functional unit increases. The schedule in Figure 7.11 has been generated by a transition-aware scheduler. The heuristics adopted by a traditional scheduler suit well if the aim is to just avoid stalls and improve the total execution time. However, in addition to these requirements, the new transition-aware scheduler schedules instructions that use the same set of functional units.

7.4.1.2 Dynamic Resource Usage

Figures 7.12 and 7.13 show plots of continuous idle periods (Y-axis) in ALU with and without transition-aware scheduling applied. The figures are the snapshots of the continuous idle periods in a window of 100,000 transitions (X-axis). The total number of execution cycles with and without optimization differ by only 0.6%. This indicates that in the window that has been captured, both programs are executing almost the same part of the program. Short durations and fewer idle cycles are observed in Figure 7.12. The behavior has been transformed, for the same number of transitions, into longer and more numerous continous idle cycles after applying the optimization, as shown in Figure 7.13. This indicates that there is an opportunity to create extra idle periods in functional units.

7.4.1.3 The Target Machine Architecture

The different functional units considered to be a part of the target machine are ALU, MUL, FPU, and a memory port. Single-issue processors, such as most of the present-day embedded processors, and a single instance of each functional unit, are assumed. These units are exposed to the compiler through a machine description of the target machine. The other units, which are implicitly affected by accessing these units, are the write buffer, the memory buses, and the buses that connect MAC and FPU co-processors. *A transition from active to idle mode can take place if there is an idle period of more than one cycle.* The reverse transition from idle to active mode takes place when the unit is accessed for use. Both transitions are handled by the hardware. The reason for a functional unit being idle can be a cache miss, a branch misprediction, functional unit unavailability being ahead of it, or a dependent instruction being blocked in the pipeline. The machine description associates an instruction with a set of functional units and the time for which each functional unit is used. Thus, an instruction can access any subset of these components during any cycle of its operation in the pipeline. A transition-aware scheduler is a modified form of the standard

Cycle	Instruction	ALU	MUL	RP
i	ld r1, 10[r2]	1	0	0
i+1	ld r5,10[r2]	1	0	1
i+2	add r1,r1,r4	1	0	1
i+3	add r4,r1,0x04	1	0	0
i+4	add r1,r5,r4	1	0	0
i+5	mul r3,r1,r3	0	1	0
i+6	mul r4,r1,r4	0	1	0

FIGURE 7.11 Scheduling instructions with a transition-aware scheduler.

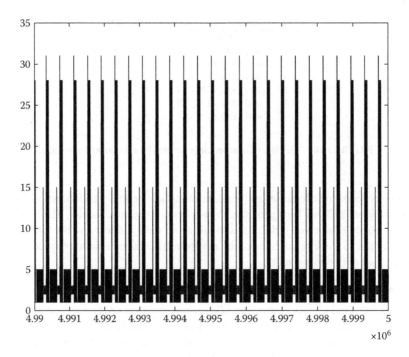

FIGURE 7.12   A snapshot of continuous idle periods (Y-axis) without transition-aware scheduling. Note the short and small number of continuous idle periods (sparse).

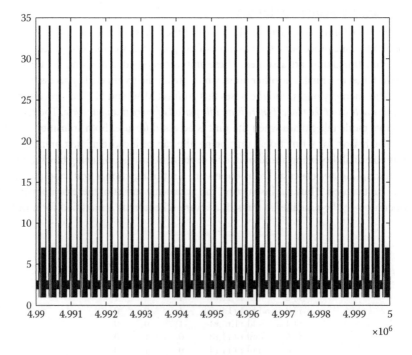

FIGURE 7.13   A snapshot of continuous idle periods (Y-axis) with transition-aware scheduling. Note the long and large number of continuous idle periods (dense).

automaton-based instruction scheduler [4] that, while scheduling a basic block, schedules instructions that use similar resources together, so that the resources need not be switched on and off frequently.

7.4.1.4   The Energy Model

An energy model aiming to capture the total power consumption in functional units should take into consideration the following parameters:

1. Number of active cycles and idle cycles of different functional units
2. Static idle energy, transition energy, and dynamic energy that vary with the circuit-level control techniques used
3. The number of transitions in the functional unit

In addition to functional units, the energy leakage in the read ports and write ports of the data cache can be modeled, and their usage information can be provided to the compiler. The power model appropriate to this discussion is derived from [17]. The model estimates the energy dissipation using different policies of a circuit-level leakage reduction scheme with various design- and technology-dependent constants. The model captures the dynamic behavior of a functional unit through the *state* of each functional unit and the changes in the access pattern of the unit. Energy dissipations in the various states of a functional unit, viz., E_{total}, $E_{dynamic}$, $E_{leakage}$, $E_{transition}$, and E_{sleep}, are given by the following equations:

$$E_{total} = E_{dynamic} + E_{leakage} + E_{transition} + E_{sleep} \tag{7.22}$$

$$E_{dynamic} = N_{active}(\alpha + (1 - D)p) \tag{7.23}$$

$$E_{leakage} = (N_{active} \cdot D + N_{uncontr\_idle})(\alpha \cdot s \cdot p + (1 - \alpha)p) \tag{7.24}$$

$$E_{transition} = N_{transitions}\left((1 - \alpha) + \frac{E_{sl\_signal}}{E_{active}}\right) \tag{7.25}$$

$$E_{sleep} = N_{sleep} \cdot s \cdot p \tag{7.26}$$

$$N = N_{actual} + N_{sleep} + N_{transitions} \tag{7.27}$$

where

N: Total number of cycles in which the unit is used.

N_{active}: Number of cycles in which the unit is in active mode.

N_{uncont_idle}: Number of cycles in which the unit is in active mode but idle.

$N_{transitions}$: Number of transitions of the unit.

N_{sleep}: Number of cycles in which the unit is in low-power sleep mode.

α: Activity factor; that is, the fraction of cells discharged in the functional unit. This factor is dependent on the application.

D: Duty cycle of the clock.

p: Leakage energy factor; fraction of static leakage in active but idle mode to the static leakage in the active mode.

s: Leakage energy factor; fraction of the energy dissipated in sleep mode to the static leakage in active but idle mode.

E_{sl_signal}: Energy consumed in putting a unit into sleep mode.

E_{active}: Energy consumed in the active mode.

The parameters p, s, E_{sleep}, and E_{active} are dependent on the technology and design of the logic. Typical values for the various parameters are $\alpha = 0.25 - 0.75$, $p = 0.05$, $s = 0.01$, and $E_{sleep}/E_{active} = 0.001$. With a value of 1, $N_{uncontr\_idle}$ can be ignored when compared to $(N_{active} \cdot D)$. More discussion on this topic is available in [17].

```
Input:A Basic block in the form of a data
dependency graph
Output: Scheduled sequence of instructions

begin
init(Readylist);schedtime=0;
while(!all insns in BB have been scheduled)
    do
        Global=GetGlobal (PresentState)
      while (!ReadyList)
          do
              for(each element in ReadyList)
                do
                    op=Element of ReadyList;
                    Closeness=Close(op,Global);
                    ChangePriority(op,Closeness);
                done
                    insn=ChooseInsn(ReadyList)
              if(insn)
                    ScheduleInstruction(schedtime,insn)
            schedtime=schedtime+1;
            update(Readylist);
        done
    done
end
```

FIGURE 7.14   The list scheduler is modified to incorporate the information of global resource usage and compute the closeness of an instruction with respect to the global usage.

7.4.1.5   The Scheduling Algorithm

The list scheduling algorithm for basic blocks achieves reduction in the number of transitions in functional units. The modified problem of the scheduling can be stated as follows:

> Given an instruction sequence as a data dependence graph and the resource usage information in the form of bit vectors (resource reservation tables), minimize the number of bit transitions across the usage vectors of the scheduled instruction sequence generated.

The algorithm shown in Figure 7.14 provides a mechanism for fetching the global resource usage vector by querying the scheduling automaton [4] that is constructed from the machine description provided. Then it computes the "closeness" between the vectors and assigns priorities to each instruction. The ready list is sorted based on the priorities of the instructions. The scheduler uses the scheduling automaton to choose suitable instructions (no resource conflicts), one at a time from the ready list, and schedules them in successive time slots. If no operation is ready, then no-ops are introduced until some instruction becomes ready. The scheduler tries to avoid introducing no-ops as in a traditional scheduler.

7.4.1.6   Global Resource Usage Vector

An important part of the algorithm is to obtain the global resource usage vector (GRV), which contains information on the usage of each of the functional units during the present scheduling cycle. Some instructions scheduled in the previous cycles continue to consume different resources during the current cycle, thus changing the global usage in every cycle. To obtain this vector, the pipeline description model that is used to detect structural hazards during scheduling has been modified. The pipeline description model can be in any form, viz., a global resource reservation table or an automaton-based model. An automaton

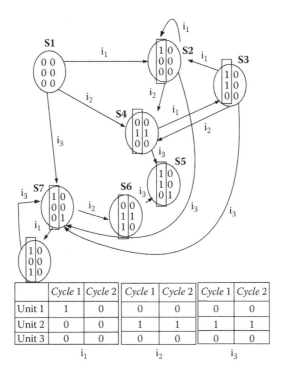

FIGURE 7.15 A scheduling automaton.

model has been used here. This model implicitly captures the global resource usage that is needed during the construction of the automaton, and this reduces the scheduling time. It is also easy to specify new resources other than function units that are used by a traditional scheduler. For example, memory port has been added as another resource component. The number of states increases with the addition of new components to the description. Every state has a GRV, and it is updated during the building of the automaton. The modified algorithm to build the automaton is described in the paragraph below.

The two-column matrix in each state is the resource usage matrix in that state (for two cycles). The first column of this matrix is the GRV for the current cycle. Since our machine model is a single-issue model, each state is a cycle-advancing state, so the second column is inspected for any resource conflicts while deciding which instruction arcs to add to the state during the automaton building process. If there are no conflicts, the state matrix is shifted left by one column and OR-ed with the reservation table of an instruction to generate the matrix for the new state. The rest of the procedure is the same as in [4]. Figure 7.15 shows a snapshot of an automaton to model a pipeline. The figure shows three instruction classes i_1, i_2, and i_3, with their reservation tables and the automaton built using these classes.

7.4.1.7 Closeness Heuristic

The next part of the algorithm calculates the closeness between the instructions in the ready list and the GRV. Each instruction class is associated with a resource usage vector (RUV). The RUV is a weighted vector in which each element corresponds to the number of cycles for which a particular unit is used by an instruction in the associated class. It is computed by summing up all the entries in each row of the reservation table for the corresponding instruction. For example, in Figure 7.15, the RUVs are i_1: 100, i_2: 020, i_3: 101. The closeness of the RUVs of two instructions x and y indicates the number of transitions that may occur if the instructions x and y are scheduled in consecutive cycles. During every iteration of the scheduling process, the GRV is OR-ed with the RUV of the last scheduled instruction. In this operation, a bit version of an RUV (BRUV) that contains 1's for its nonzero elements is used. By OR-ing the two vectors, the GRV, and the BRUV, the current as well as the future usage of the last scheduled instruction can be captured.

Let two vectors be $< i_1, i_2, i_3 >$ and $< j_1, j_2, j_3 >$. The *closeness heuristic* is based on the *Euclidean distance* between these two vectors, which is calculated as $\sqrt{(i_1 - j_1)^2 + (i_2 - j_2)^2 + (i_3 - j_3)^2}$. The *Euclidean distance* between a GRV and an RRV gives an estimate of how close these two vectors are. If this value for an instruction in the ready list is minimal, then that instruction needs to be scheduled as early as possible because it will cause fewer transitions. In a traditional scheduler, the priority that is defined for each instruction before scheduling is computed using the maximum length of the path from the instruction node to a leaf in the dependence graph. In this algorithm, an additional priority, the *power consumption priority*, which depends on the closeness factor, is introduced. The lower the closeness value of an instruction in the ready list is, the higher its priority will be. The ready list is first sorted based on the power consumption priority and then on the dependence chain length priority.

7.4.1.8   A Detailed Example

Consider the data dependence graph (DDG) shown in Figure 7.16. The nodes are annotated with the instruction id and the functional units they use. *alu,mp* means the instruction accesses the *ALU* in the first cycle and the memory port in the second cycle. *mul*2* means the multiplication unit, MUL, is used for two cycles. The issue latency is therefore two cycles in each of these two cases. A single-cycle latency is assumed for both *alu* and *memory port*. The weight on an edge indicates the latency in producing the result. The scheduler output is given in Figure 7.17. The second column indicates the dependence chain priority. The last column indicates the closeness values for the instructions in the ready list in the same order as they appear in the eighth column.

With a traditional list scheduler, instruction C gets a higher priority than E because of its dependence chain length, thereby scheduling C earlier than E, but when the closeness heuristic is used, E gets a higher priority than C, and thus E gets scheduled earlier and is closer to A and B. Similarly, D and E get scheduled, in that order (after C), with the traditional heuristic, whereas with the closeness criterion, instructions C, D, and G are scheduled in that order. If the GRV is observed (shown in the sixth and ninth columns), it can

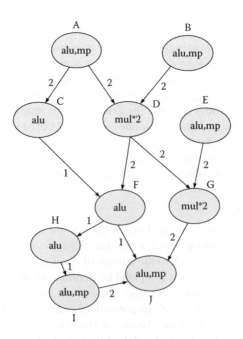

FIGURE 7.16   An example DDG.

Insn	Priority	Cycle	Insn	List	GRV	Insn	List	GRV	Closeness
A	5	i	A	B,E	100	A	B,E	100	0,0
B	5	i+1	B	C,E	101	B	E,C	101	0.1
C	4	i+2	C	D,E	101	E	C,D	101	1, $\sqrt{3}$
D	4	i+3	D	E	010	C	D	101	$\sqrt{3}$
E	2	i+4	E	F	110	D		010	
F	3	i+5	F	H,G	100		G	010	0
G	1	i+6	H	G,I	100	G	F	010	$\sqrt{2}$
H	2	i+7	I	G,J	100	F	H	110	0
I	1	i+8	G	J	010	H	I	100	1
J	0	i+10	J		110	I	J	100	0
		i+11			001	J		101	
		i+12			001			101	

FIGURE 7.17 Scheduler output using traditional heuristics and closeness heuristics.

be seen that the continuous idle cycles of ALU, MP, and MUL have increased. The number of transitions in ALU and MUL has been reduced from four to two, and in the memory port from five to three. However, with the closeness heuristic there is a no-op in the sequence after D (in column 7). This is because no other instructions in the ready list can be scheduled at that point in time. This is the performance penalty (execution cycles) that is paid when the closeness heuristic is used. However, simulation studies show that this impact is not so significant when compared to the power consumption benefits.

Experimental results and more details are available in [46]. Modifications to the above scheme and application to clustered VLIW architectures are described in [30, 31].

7.5   Compile-Time Allocation for Scratchpad Memory

It is well known that off-chip memory is the slowest and most energy consuming type of memory in the memory hierarchy. Cache memories are much smaller than off-chip main memories and are much faster. They are used to keep frequently used code blocks or variables and thereby reduce the number of main memory accesses. Caches are well known and are used in most processors. Caches have a tag memory to store valid addresses and extra hardware for a fast comparison of addresses with the contents of tag memory. These are needed to detect hits and misses, but they consume energy continuously since comparisons are made during each memory access. Caches integrate very well with software and are transparent to the software in their operation. This has made caches very popular with hardware designers. However, cache hits and misses bring a certain amount of nondeterminism to the execution time and its estimation. Most *worst-case execution time* (WCET) estimation software ignores the presence of caches and regards the time of execution of each instruction assuming a cache miss. WCET estimation in the presence of cache is a hot research topic [30].

Scratchpad memories have been proposed not only to make WCET estimation more accurate, but also to reduce the energy consumption in memory hierarchies. Scratchpad memories are as fast as cache memories but do not have the tag memory and address comparison hardware. Therefore, they consume less energy than caches but require the services of a compiler or the programmer to fill it with code and/or variables. Letting programmers handle scratchpad memories is not desirable since the allocation can become quite complex and may lead to inefficient allocation and possible program errors. Compile-time algorithms based on integer linear programming have been proposed for scratchpad memory allocation. The discussion in the sections that follow is based on such algorithms described in [2, 41, 44, 49].

7.5.1   The Static Allocation Algorithm

A typical embedded system contains several types of memories, and each type has its own advantages and disadvantages. Cache, on-chip scratchpad Static Random Access Memory (SRAM), off-chip SRAM, on-chip DRAM, off-chip DRAM, ROM, and external flash memory are among the most common ones. The speeds of these memories decrease approximately in the same order as mentioned above. It is possible to consider the entire memory hierarchy in the allocation algorithm, but for clarity, only off-chip DRAM and on-chip scratchpad SRAM are considered here. Avissar et al. [2] present such an extension.

The allocation algorithm should consider the energy benefits obtained by placing frequently accessed data variables and frequently executed code segments in the scratchpad. The size of scratchpad memory is much smaller than that of main memory and therefore cannot accommodate all the variables and code segments. Therefore, it is necessary to profile the program for which scratchpad allocation is to be performed and determine the counts of execution of basic blocks and the number of accesses to each variable. While entire functions and basic blocks are both considered as program memory object candidates for scratchpad allocation, individual scalar and nonscalar global variables are considered as data memory object candidates for the same purpose. Local variables (stack variables) and dynamically allocated objects are not considered for allocation on scratchpad. The allocation is purely static, and there is no dynamic reloading of objects. These extensions are discussed in the literature suggested in Section 7.1. The following discussion considers the savings or gain in energy consumption obtained by allocating an object to the scratchpad. For example, if E_m and E_s are the energy consumptions for a data access to the main memory and scratchpad, respectively, $(E_m - E_s)$ is the savings in energy.

7.5.1.1   Energy Savings for a Complete Function

It is assumed that functions are single-entry, single-exit entities. Therefore, moving the complete code for a function into scratchpad requires changing none of the instructions in the function body or the corresponding function calls. The savings or gain in energy consumption of a function f can be computed using the following equation:

$$\mathcal{E}^g(f) = \sum_i n_i \cdot \mathcal{E}^g_{instr-fetch} \tag{7.28}$$

where

$\mathcal{E}^g(f)$: Gain in energy consumed by the function f
n_i^f: Number of times instruction i in the function f is executed
$\mathcal{E}^g_{instr-fetch}$: Gain in energy consumed for a single instruction fetch

n_i^f can be computed easily using profiling information for basic blocks.

7.5.1.2   Energy Savings for a Basic Block

Similarly, the energy consumed by a single basic block can also be computed. When a basic block is moved to scratchpad, jump instructions to jump from a basic block in regular memory to the basic block in scratchpad and perhaps back are required. These constitute an extra overhead and can be expensive in the presence of small basic blocks. It is preferable to move consecutive basic blocks to minimize the overhead. The gain in energy consumption of a basic block b can be computed using the following equation:

$$\mathcal{E}^g(b) = I_b \cdot n_b \cdot \mathcal{E}^g_{instr-fetch} - J_b \cdot \mathcal{E}^g_{jump} \tag{7.29}$$

where

$\mathcal{E}^g(b)$: Gain in energy consumed by the basic block b
n_b: Number of times the basic block b is executed (from profiling)
I_b: Number of instructions in the basic block b
$\mathcal{E}^g_{instr-fetch}$: Gain in energy consumed for a single instruction fetch
J_b: Number of jumps from main memory to block b in scratchpad or vice versa
$\mathcal{E}^g_{jump}$: Energy consumed for a single jump instruction

7.5.1.3 Energy Savings for Data Memory Objects

Each global scalar and nonscalar variable is considered a data memory object. Local variables are not considered here. The number of times a global variable v is accessed (including both reads and writes) and the total energy consumed by all the accesses are computed as follows.

$$n_v = \sum_b ref_b(v) \cdot n_b \tag{7.30}$$

$$\mathcal{E}^g(v) = n_v \cdot \mathcal{E}^g_{ls} \tag{7.31}$$

where

n_v: Total number of accesses to variable v
$ref_b(v)$: Number of static references to variable v in basic block b
n_b: Number of times the basic block b is executed (from profiling)
$\mathcal{E}^g(v)$: Total gain in energy consumed due to variable v
$\mathcal{E}^g_{ls}$: Average gain in energy consumed by a memory access (read or write)

7.5.1.4 The Integer Linear Programming Formulation

Maximize

$$\mathcal{E}^g_{total} = \sum_{f \in F} x(f) \cdot \mathcal{E}^g(f) + \sum_{b \in B} x(b) \cdot \mathcal{E}^g(b) + \sum_{v \in V} x(v) \cdot \mathcal{E}^g(v) \tag{7.32}$$

where F, B, and V are the total number of functions, basic blocks, and global variables, in the program, respectively. $x(i)$ is the integer programming 0-1 variable, taking a value of 1 if the object i (function, basic block, or variable) is allocated to the scratchpad, and 0 otherwise. The size constraint can be stated as

$$\sum_{f \in F} x(f) \cdot size(f) + \sum_{b \in B} x(b) \cdot size(b) + \sum_{v \in V} x(v) \cdot size(v) \leq scratchpadsize \tag{7.33}$$

where $size(i)$ is the size of object i. The size of a basic block includes the size of the jump instruction as well. Minor optimizations to this scheme to avoid counting jumps to consecutive basic blocks that are both moved scratchpad are described in [41].

7.6 Conclusions and Future Directions

This chapter considered a few energy-saving compiler optimizations in detail. Dynamic voltage scaling, which reduces the dynamic energy consumption in processors, was formulated as a minimization problem, and transition-aware scheduling, which reduces leakage energy consumption, was formulated as an instruction scheduling problem. Scratchpad memories, which are an alternative to cache memories in embedded processors, require the compiler to perform the allocation of code and data objects onto them, and this problem was formulated as an integer linear programming problem.

The last word on energy-aware compiler optimizations has hardly been said. The processors used in sensor networks and other embedded systems need aggressive energy optimizations. No professional compilers incorporate all the energy-aware optimizations such as DVS, instruction scheduling, memory bank allocation, scratchpad allocation, cache management, and communication reduction (in clusters), and building such a compiler is an interesting research project. Furthermore, the interaction of various energy-saving optimizations with performance-improving optimizations has not been studied thoroughly. Integration of optmizations possible at the operating system level into a compiler framework and design of a programming framework to develop power- and energy-aware applications would be other research possibilities. In addition, innovations in architecture are introducing novel features into hardware to save power and energy, and future compilers would be challenged to use such features effectively. The problem of reducing power consumption can be more effectively solved by designing hybrid algorithms that combine micro-architectural techniques and compiler techniques such that they complement each other. For example, rather than controlling cache configurations dynamically, the same can be achieved through annotations inserted into programs after a compiler analysis. This would reduce the burden on the hardware at runtime.

References

1. Absar, M. J., and Catthoor, F. 2005. Compiler-based approach for exploiting scratchpad in presence of irregular array access. In *ACM Conference on Design, Automation and Test in Europe*, 1162–67.
2. Avissar, O., Barua, R., and Steware, D. 2002. An optimal memory allocation scheme for scratchpad-based embedded systems. *ACM TECS* 1:6–26.
3. Azvedo, A., Issenin, I., Cornea, R., Gupta, R., Dutt, N., Veidenbaum, A., and Nicolau, A. 2002. Profile-based dynamic voltage scheduling using program checkpoints. In *ACM Conference on Design, Automation and Test in Europe*, 168–75.
4. Bala, V., and Rubin, N. 1995. Efficient instruction scheduling using finite state automata. In *ACM International Symposium on Microarchitecture (MICRO-29)*, 46–56.
5. Banakar, R., Steinke, S., Lee, B.-S., Balakrishnan, M., and Marwedel, P. 2002. Scratchpad memory: A design alternative for cache on-chip memory in embedded systems. In *ACM International Conference on Hardware Software Codesign (CODES)*, 73–78.
6. Benini, L., Macii, A., and Poncino, M. 2000. A recursive algorithm for low-power memory partitioning. In *ACM International Symposium on Low Power Electronics and Design (ISLPED)*, 78–83.
7. Borkar, S. 1999. Design challenges of technology scaling. *IEEE Micro* 19(4):23–29.
8. Brooks, D. M., Bose, P., Schuster, S. E., Jacobson, H., Kudva, P. N., Buyuktosunoglu, A., Wellman, J.-D., Zyuban, V., Gupta, M., and Cook, P. W. 2000. Power-aware microarchitecture: Design and modelling challenges for next-generation microprocessors. *IEEE Micro* 20(6):26–44.
9. Brooks, D. M., Tiwari, V., and Martonosi, M. 2000. Wattch: A framework for architectural-level power analysis and optimizations. In *ACM International Conference on Computer Architecture* (ISCA), 83–94.
10. Burd, T., and Broderson, R. 2000. Design issues for dynamic voltage scaling. In *ACM International Symposium on Low Power Electronics and Design (ISLPED)*, 9–14.
11. Butts, J. A., and Sohi, G. S. 2000. A static power model for architects. In *ACM/IEEE International Symposium on Microarchitecture (MICRO-33)*, 191–201.
12. De, V., and Borkar, S. 1999. Technology and design challenges for low power and high performance. In *ACM International Symposium on Low Power Electronics and Design (ISLPED '99)*, 163–68.
13. De La Luz, V., Kandemir, M., Vijayakrishnan, N., Sivasubramaniam, A., and Irwin, M. J. 2001. Hardware and software techniques for controlling DRAM power modes. *IEEE Trans. Comput.* 50(11):1154–73.
14. De La Luz, V., Kandemir, M., and Kolcu, I. 2002. Automatic data migration for reducing energy consumption in multi-bank memory systems. In *ACM/IEEE Conference on Design Automation (DAC)*, 213–18.
15. De La Luz, V., Sivasubramaniam, A., Kandemir, M., Vijayakrishnan, N., and Irwin, M. J. 2002. Scheduler-based DRAM energy management. In *ACM/IEEE Conference on Design Automation (DAC)*, 697–702.
16. Dominguez, A., Udayakumaran, S., and Barua, R. 2005. Heap data allocation to scratchpad memory in embedded systems. *J. Embedded Comput.* 1(4):1–17.
17. Dropsho, S., Kursun, V., Albonesi, D. H., Dwarkadas, S., and Friedman, E. G. 2002. Managing static leakage energy in microprocessor functional units. In *ACM International Symposium on Microarchitecture (MICRO-35)*, 321–32.

18. Duesterwald, E. 2007. Dynamic compilation. In *The compiler design handbook: Optimizations and machine code generation*, ed. Srikant, Y. N. and Shankar, P. Boca Raton, FL: CRC Press.

19. Falk, H., and Verma, M. 2004. Combined data partitioning and loop nest splitting for energy consumption minimization. In *International Workshop on Software and Compilers for Embedded Systems (SCOPES)*, 137–51. Heidelberg, Germany: Springer Verlag.

20. Heo, S., Barr, K., Hampton, M., and Arisovic, K. 2002. Dynamic fine-grain leakage reduction using leakage-biased bitlines. In *ACM International Conference on Computer Architecture (ISCA)*, 137–47.

21. Hsu, C.-H., and Kremer, U. 2003. The design, implementation, and evaluation of a compiler algorithm for CPU energy reduction. In *ACM Conference on Programming Language Design and Implementation (PLDI)*, 38–48.

22. Isci, C., and Martonosi, M. 2003. Runtime power monitoring in high-end processors: Methodology and empirical data. In *ACM International Symposium on Microarchitecture (MICRO-36)*, 93–104.

23. Iyer, A., and Marculescu, D. 2002. Microarchitecture-level power management. *IEEE Trans. VLSI Syst.* 10(3):230–39.

24. Kim, N. S., Austin, T., Blaauw, D., Mudge, T., Flautner, K., Hu, J. S., Irwin, M. J., Kandemir, M., and Narayanan, V. 2003. Leakage current: Moore's law meets static power. *IEEE Comput.* 36(12):68–75.

25. Koc, H., Ozturk, M., Kandemir, M., Narayanan, S. H. K., and Ercanli, E. 2006. Minimizing energy consumpton of banked memories using data recomputation. In *ACM International Symposium on Low Power Electronics and Design (ISLPED)*, 358–62.

26. Kursun, V., and Friedman, E. G., 2002. Low swing dual threshold voltage domino logic. *ACM Great Lakes Symposium GLVLSI*, 47–52.

27. Lee, C., Lee, J. K., and Hwang, T. 2003. Compiler optimization on VLIW instruction scheduling for low power. *ACM TODAES* 8(2):252–68.

28. Li, L., Gao, L., and Xue, J. 2005. Memory coloring: A compiler approach for scratchpad memory management. In *ACM Conference on Parallal Architectures and Compilation Techniques (PACT)*, 329–38.

29. Lyuh, C.-G., and Kim, T. 2004. Memory access scheduling and binding considering energy minimization in multi-bank memory systems. In *ACM/IEEE Conference on Design Automation (DAC)*, 81–86.

30. Mitra, T., and Roychoudhury, A. 2007. Worst-case execution time and energy analysis. In *The compiler design handbook: Optimizations and machine code generation*, ed. Srikant, Y. N. and Shankar, P. Boca Raton FL: CRC Press.

31. Mudge, T. 2001. Power: A first-class architectural design constraint. *IEEE Comput.* 34(4):52–58.

32. Nagpal, R., and Srikant, Y. N. 2006. Compiler-assisted leakage energy optimization for clustered VLIW architectures. In *ACM International Conference on Embedded Software (EMSOFT)*, 233–41.

33. Nagpal, R., and Srikant, Y. N. 2006. Exploring energy-performance tradeoffs for heterogeneous interconnect clustered VLIW processors. In *International Conference on High Performance Computing (HiPC)*, 497–508, Springer-Verlag.

34. Pokam, G., Rochecouste, O., Seznec, A., and Bodin, F. 2004. Speculative software management of datapath-width for energy optimization. In *ACM International Conference on Language, Compiler and Tool Support for Embedded Systems (LCTES)*, 78–87.

35. Pollack, F. 1999. New microarchitecture challenges in the coming generations of CMOS process technologies. Keynote at the *ACM International Symposium of Microarchitecture (MICRO-32)*. http://www.intel.com/research/mrl/Library/micro32Keynote.pdf.

36. Saputra, H., Kandemir, M., Vijaykrishnan, N., Irwin, M. J., Hu, J. S., Hsu, C.-H., and Kremer, U. 2002. Energy-conscious compilation based on voltage scaling. In *ACM International Conference on LCTES/SCOPES*, 2–11.

37. Sharkey, J. J., Ponomarev, D. V., Ghose, K., and Ergin, O. 2006. Instruction packing: Toward fast and energy-efficient instruction scheduling. *ACM TACO* 3(2):158–81.

38. Shin, D., Kim, J., and Lee, S. 2001. Low-energy intra-task voltage scheduling using static timing analysis. In *ACM International Conference on Design Automation (DAC)*, 438–43.

39. Shrivastava, A., Issenin, I., and Dutt, N. 2005. Compilation techniques for energy reduction in horizontally partitioned cache architectures. In *ACM International Conference on Compilers, Architecture and Synthesis for Embedded Systems (CASES)*, 90–96.

40. Sprunt, B. 2002. Pentium 4 performance-monitoring features. *IEEE Micro* 22(4):72–82.

41. Steinke, S., Wehmeyer, L., Lee, B.-S., and Marwedel, P. 2002. Assigning program and data objects to scratchpad for energy reduction. In *ACM Conference on Design, Automation and Test in Europe*, 409–17.

42. Sudarsanam, A., and Malik, S. 2000. Simultaneous reference allocation in code generation for dual memory bank ASIPs. *ACM TODAES* 5:242–64.

43. Tsai, Y. F., Duarte, D., Vijayakrishnan, N., and Irwin, M. J. 2004. Impact of process scaling on efficacy of leakage reduction schemes. In *IEEE International Conference on IC Design and Technology (ICICDT)*, 3–11.

44. Udayakumaran, S., Dominguez, A., and Barua, R. 2006. Dynamic allocation for scratch-pad memory using compile-time decisions. *ACM TECS* 5(2):472–511.

45. Unnikrishnan, P., Chen, G., Kandemir, M., and Mudgett, D. R. 2002. Dynamic compilation for energy adaptation. In *IEEE Conference on Computer-Aided Design*, 158–63.

46. Vardhan, K. A., and Srikant, Y. N. 2005. Transition aware scheduling: Increasing continuous idle periods in resource units. In *ACM International Conference on Computing Frontiers*, 189–98.

47. Venkatachalam, V., and Franz, M. 2005. Power reduction techniques for microprocessor systems. *ACM Comput. Surveys* 37(3):195–237.

48. Verma, M., Wehmeyer, L., and Marwedel, P. 2004. Cache-aware scratchpad allocation algorithm. In *ACM Conference on Design, Automation and Test in Europe*, 21264–69.

49. Verma, M., and Marwedel, P. 2006. Overlay techniques for scratchpad memories in low power embedded systems. *IEEE Trans. VLSI Syst.* 14(8):802–15.

50. Vijayakrishnan, N., Kandemir, M., Irwin, M. J., Kim, H. S., and Yei, W. 2000. Energy-driven integrated hardware-software optimizations using simpower. In *International Conference on Computer Architecture (ISCA)*, 95–106.

51. Wang, Z., and Hu, X. S. 2005. Energy-aware variable partitioning and instruction scheduling for multibank memory architectures. *ACM TODAES* 10(2):369–88.

52. Wu, Q., Martonosi, M., Clark, D. W., Reddi, V. J., Connors, D., Wu, Y., Lee, J., and Brooks, D. 2005. A dynamic compilation framework for controlling microprocessor energy and performance. In *ACM International Conference on Microarchitecture (MICRO-38)*, 271–82.

53. Xie, F., Martonosi, M., and Malik, S. 2004. Intraprogram dynamic voltage scaling: Bounding opportunities with analytic modeling. *ACM TACO* 1(3):323–67.

54. Yang, H., Govindarajan, R., Gao, G. R., Cai, G., and Hu, Z. 2002. Exploiting schedule slacks for rate-optimal power-minimum software pipelining. In *Workshop on Compilers and Operating Systems for Low Power (COLP)*, COLP Workshop Proceedings.

55. You, Y.-P., Lee, C., and Lee, J. K. 2006. Compilers for leakage power reduction. *ACM TODAES* 11(1):147–64.

56. Zhang, W., Vijaykrishnan, N., Kandemir, M., Irwin, M. J., Duarte, D., and Tsai, Y.-F. 2001. Exploiting VLIW schedule slacks for dynamic and leakage energy reduction. In *ACM International Symposium on Microarchitecture (MICRO-34)*, 102–13.

57. Zhang, W., Hu, J. S., Degalahal, V., Kandemir, M., Vijayakrishnan, N., and Irwin, M. J. 2002. Compiler-directed instruction cache leakage optimization. In *ACM International Conference on Microarchitecture (MICRO-35)*, 208–18.

58. Zhang, W., Karakoy, M., Kandemir, M., and Chen, G. 2003. A compiler approach for reducing data cache energy. In *ACM International Conference on Supercomputing (ICS)*, 76–85.

8

Statistical and Machine Learning Techniques in Compiler Design

P.J. Joseph
Freescale Semiconductor India Pvt. Ltd.,
Noida, India
PJ.Joseph@freescale.com

Matthew T. Jacob,
Y.N. Srikant, and
Kapil Vaswani
*Department of Computer Science
and Automation,
Indian Institute of Science,
Bangalore, India*
mjt@csa.iisc.ernet.in
srikant@csa.iisc.ernet.in
kapil@csa.iisc.ernet.in

8.1 Introduction

In computer science, the word *optimize* refers to *the process of modifying to achieve maximum efficiency in storage capacity or time or cost*. True to their name, *optimizing compilers* transform computer programs into semantically equivalent programs with the objective of maximizing performance. The transformation is achieved using a set of optimizations. Each optimization usually consists of an analysis component that identifies specific performance anomalies and a transformation component that eliminates the anomalies via a series of code transformations.

Despite significant advances in compilation techniques and numerous compiler optimizations that have been proposed over the years, the goal of building optimizing compilers that deliver optimal or even near-optimal performance[1] is far from being achieved. For instance, it has been observed that performance of compiler-generated code falls far short of hand-optimized code in certain application domains [1]. One might argue that the superiority of hand-optimized code is due to domain knowledge or special insights into the functioning of code that programmers may have, which allows them to make

---

[1] Refer to Appendix A for a formal description of the notion of optimality in program compilation.

program/hardware-specific transformations. Even if we discount such transformations, optimizing compilers have been shown to fall short in several other ways. For instance, recent research has shown that the sequence in which compiler optimizations are applied to a program has a significant bearing on its execution time and that the performance of the default optimization sequence in most compilers is far below that of the best program-specific optimization sequence [2]. It has also been observed that most programs perform better if some of the optimizations are disabled [3, 4]. These observations suggest that compilers may not be exploiting the power of existing optimizations to the fullest.

In this chapter, we examine the reasons behind these apparent shortcomings of compilers. Not surprisingly, we find that compilers are replete with inherently hard problems and complex interactions between optimizations that are difficult for compiler writers to comprehend. We show that the key to solving these problems and building high-performance compilers lies in *adaptability*, that is, the ability to reconfigure parts of the compiler to the requirements of the program being compiled and the target platform. We discuss how statistical and machine learning techniques can help address some of the complex issues that arise in building adaptive compilers. In particular, we show that problems such as identifying compiler interactions, design of efficient heuristics, finding program-specific optimization sequences, and selection of program-specific optimization flags can be formulated as problems in learning and statistical inference, and near-optimal solutions to these problems can be found efficiently and automatically. This automated approach greatly simplifies the design of performance-critical components of an optimizing compiler, letting compiler writers focus on other important aspects such as correctness and retargetability.

8.1.1   Motivation

Let us start by examining some of the reasons compilers fail to deliver optimal performance.

8.1.1.1   Inherently Hard Problems

Compiler optimizations are the source of some of the hardest and most challenging problems in computer science. This is specially the case with architecture-dependent optimizations such as register allocation, instruction scheduling, and data layout. Since finding optimal solutions to these problems is provably hard, compiler writers are forced to develop heuristics to find approximate solutions to these problems. Consider register allocation, for instance. It has been shown that optimal register allocation can be reduced to the problem of determining whether a graph can be k-colored [5] (refer to [6] for details). Therefore, a coloring-based register allocator uses heuristics to select variables for spilling when the interference graph cannot be k-colored (where k is the number of unallocated registers). For some compiler optimizations, heuristics can be designed to guarantee solutions that are within a constant factor of the optimal solution, but more often than not, heuristics are designed for the average case, and there are no guarantees of how well the heuristics perform for a given program and target platform.

8.1.1.2   Interactions between Optimizations

It is well known that most compiler optimizations *interact*, that is, the effect of any given optimization on the performance may depend on the presence or absence of other optimizations in the compilation sequence. Furthermore, the nature of the interaction may also depend on the order in which the optimizations are applied. For example, loop unrolling is known to interact with global common subexpression elimination, in other words, (GCSE). This implies that, the benefits of unrolling depend on whether GCSE is applied and on the order in which unrolling and GCSE are applied.

Interactions between optimizations can be *positive* or *negative*. A positive interaction usually occurs when one optimization enables or creates opportunities for another optimization. Negative interactions occur when one optimization restricts the applicability of another optimization. A good example of a negative interaction is the one between register allocation and instruction scheduling, which we discuss in detail later in this section.

Apart from interacting with other optimizations, compiler optimizations also interact with the hardware. This is inherently true for architecture-dependent optimizations such as loop optimizations, prefetching, basic block reordering, and instruction scheduling. Despite years of research, the nature of these interactions is only partially understood. Consequently, few interactions are taken into account while designing optimizations.

8.1.1.3 Multiple Objective Functions

As if optimizing programs for performance was not hard enough, modern compilers are routinely expected to generate code that is simultaneously optimized for multiple objective functions. For example, in embedded systems such as mobile devices or sensor networks, the compiler must satisfy additional constraints on metrics such as code size and power consumption. In several cases, some of the objective functions may be at odds with each other, that is, optimizations that improve the quality of code with respect to one objective function may adversely influence the other. Designing optimizations that meet these stringent requirements is a complex task.

8.1.2 Classical Solutions

Having examined some of the problems that plague optimizing compilers, let us review some of the strategies that existing compiler implementations use to deal with these problems. We primarily focus on techniques that deal with interactions between compiler optimizations and interactions with the architecture. The general strategy adopted by compiler writers can be summarized as follows:

- **Identify and characterize specific interactions:** The presence of interactions is usually detected by manually inspecting the semantics of optimizations and architectural specifications. Alternatively, some simple interactions can be automatically detected if the compiler writer is willing to specify the semantics of optimizations in specification languages such as Gospel [7–9].
- **Adapt the optimizations involved:** Compiler optimizations may require restructuring if they are involved in strong interactions. In the simplest of cases, the heuristics that drive the optimization are changed to deal with the interaction [10, 11]. In other cases, two or more optimizations may be combined and performed together [12–15].

We illustrate this process using two specific instances, namely, the interaction between instruction scheduling and register allocation, and the interactions between various loop transformations and the hardware.

8.1.2.1 Interaction between Instruction Scheduling and Register Allocation

Instruction scheduling is a compiler optimization that rearranges instructions in a given code sequence with the goal of reducing the execution latency of the sequence. If sufficient parallelism is available in the code sequence, instruction scheduling can hide functional unit latencies, memory latencies, and other delays such as pipeline interlocks. However, a deeper analysis of the effects of instruction scheduling shows the presence of a complex interaction with register allocation [14]. The nature of this interaction depends on the order in which scheduling and register allocation are performed. If scheduling is performed before register allocation (known as the *prepass* strategy), the scheduler can utilize full parallelism in the code sequence and generate a compact schedule by interleaving the execution of independent instructions. However, in this process, the scheduler can potentially increase the maximum number of live variables and cause the register allocator to generate more spill code. If register allocation is performed before scheduling (known as the *postpass* strategy), the scheduler may not see the full parallelism available in code because of false dependencies induced by the reuse of registers. This may limit the extent to which the scheduler can reorder instructions and result in a suboptimal schedule with unnecessary stalls. The following example illustrates this interaction.

8.1.2.2 Example

Consider the code snippet shown in Figure 8.1 (adapted from [14]). The figure also shows the intermediate code before register allocation and scheduling and the corresponding dependence graph. Assume that the architecture has two registers, $r1$ and $r2$. Figure 8.2 shows the schedule and live ranges of virtual variables if instruction scheduling is performed before register allocation. For this code sequence, the scheduler

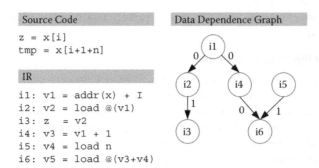

FIGURE 8.1   Example illustrating the interaction between instruction scheduling and register allocation.

generates an optimal schedule that executes in six cycles with no stalls. The scheduler hides latencies of the loads $i2$ and $i5$ by separating them from their respective uses. Although this schedule is optimal, the separation of dependent instructions causes an increase in the number of simultaneously live ranges. In this case, the register allocator is forced to spill one of the variables because there are three live ranges at cycles 3 and 4 and only two registers available for allocation.

Now consider the scenario where register allocation is performed before scheduling. The live ranges seen by the allocator are shown in Figure 8.3. Observe that at no time in this code sequence are more than two variables live, and the register allocator is able to allocate registers to all variables ($v1 \rightarrow r1, v2 \rightarrow r2, v3 \rightarrow r2, v4 \rightarrow r1$) without spilling. However, the false dependencies introduced because of the reuse of registers (both $r1$ and $r2$) limit the parallelism seen by the scheduler. Consequently, the scheduler is unable to hide load latencies and is forced to introduce stalls in the final schedule, which now runs into eight cycles.

Both these scenarios illustrate the presence of a *negative* interaction between instruction scheduling and register allocation, irrespective of the order in which the optimizations are applied. An interaction with similar characteristics also occurs between partial redundancy elimination (PRE) and register allocation [9, 16]. More precisely, the code-hoisting step in PRE might increase the live ranges of variables involved in the redundant expression, resulting in increased register pressure and possible spilling.

Once an interaction is identified, the optimizations involved in the interaction may be reformulated to ensure that the interaction does not result in suboptimal code. Several such formulations for register allocation and instruction scheduling have been proposed. While the specific details may vary, the formulations seek to make the optimizations *aware* of their influence on each other. This is achieved indirectly by introducing some means of communication between the optimizations [9, 17] or directly by integrating the optimizations into a single pass [18]. For instance, Motwani et al. [14] formulate the combined scheduling and register allocation as a single optimization problem, show that the combined problem is NP-hard but approximable, and propose a combined parameterized heuristic that can be used to weigh the conflicting considerations of register pressure and parallelism.

Interestingly, while most of these formulations have been proposed for in-order, VLIW processors, it has been observed that for out-of-order (OoO) superscalar processors, register pressure is often a greater concern than exploiting parallelism and creating better schedules [19, 20]. This is due to several features supported by OoO processors such as register renaming, which eliminates false dependencies

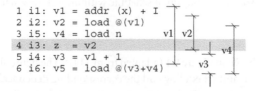

FIGURE 8.2   The schedule generated by a prepass scheduler and the live ranges of virtual registers.

```
i1: v1 = addr(x) + I          1 i1: r1 = addr(x) + I
i2: v2 = load @(v1)           2 i2: r2 = load @(r1)
i3: z  = v2                   3
i4: v3 = v1 + 1               4 i3: z  = r2
i5: v4 = load n               5 i4: r2 = r1 + 1
i6: v5 = load @(v3+v4)        6 i5: r1 = load n
                              7
                              8 i6: r1 = load @(r2+r1)
```

FIGURE 8.3 The live ranges seen by the register allocator and the schedule generated by the postpass scheduler.

introduced by register reuse, and dynamic scheduling with large instruction windows and speculation, which allow out-of-order processors to exploit instruction-level parallelism across large code sequences. Consequently, formulations that assign a higher weight to the register allocator (e.g., postpass strategy) are likely to perform better on OoO processors. This observation shows that interactions may involve more optimizations than anticipated (hardware optimizations in this case) and may be more complex than what meets the eye. Therefore, it is critical for compiler writers to characterize these interactions in their entirety and ensure that all factors that can influence the nature of the interactions are accounted for. The absence of a comprehensive characterization may lead to wrong conclusions and adversely affect performance.

8.1.2.3 Interactions between Loop Transformations and Caches

High-level loop transformations such as unrolling, tiling, fission, fusion, and interchange have individually proved to be some of the most effective compiler optimizations, especially in loop-dominated scientific, media, and graphics programs. Apart from exposing instruction-level parallelism, reducing loop overheads, and increasing functional-unit utilization, these transformations are especially useful in improving the loop's memory behavior. The potential of these optimizations is best illustrated using an example (adapted from [15]). Consider the MATLAB code in Listing 8.1, which contains a simple loop that multiplies the matrix *b* with an array *c* and stores the results in an array *a*.

Listing 8.1: MATLAB code for multiplying an array with a matrix
```
% a, c are arrays of size 1000
% b is a matrix of size 1000x1000
for i=1:1000,
  for j=1:1000,
    a(i) = a(i) + b(i,j)*c(i)
  end
end
```

If the matrix *b* were stored in column major order (which is the case in MATLAB), the accesses to the elements of matrix *b* would incur a cache miss in each iteration of the loop (because the matrix is accessed row-wise). Loop interchange [21] can be used to avoid this performance anomaly. Assuming a cache block size of 32 bytes, the interchanged loop shown in Listing 8.2 incurs a cache miss once every four iterations.

Listing 8.2: Source code after loop interchange has been applied
```
for j=1:1000,
  for i=1:1000,
    a(i) = a(i) + b(i,j)*c(i)
  end
end
```

However, this transformation comes at a cost. In the original loop, the computation in each iteration can be expressed using one *mutliply-and-add* instruction (supported on several current processors such as the PowerPC and Itanium), whereas in the transformed loop, two extra loads (for elements of arrays *a* and *c*) and an extra store (to array *a*) are required in the inner loop. Depending on the cache miss latency, loop interchange may either improve or hurt performance. Hence, cache latency must be factored into the heuristic that evaluates the benefit of loop interchange.

This interaction with cache latency is just one of the many interactions that affect loop interchange. In general, loop transformations are known to interact strongly with each other. Consider the loop in Listing 8.3, which is obtained after applying loop unrolling and interchange to the loop in the previous example. Unrolling this loop amortizes the cost of the extra load and store instructions inserted because of loop interchange. Elements of arrays a and c are now loaded once in every iteration of the inner loop, and their values are reused across four computations. This version of the loop enjoys the benefits of interchange (i.e., reduced cache misses) while minimizing the runtime cost of additional instructions. This example shows that loop unrolling can interact positively with loop interchange, assuming that other side effects of unrolling, such as the increase in code size and register pressure, are ignored.

Listing 8.3: Source code after loop unrolling and interchange has been applied

```
for j =1:1000:4,
   for i=1:1000,
      a(i) = a(i) + b(i,j)*c(i)
      a(i) = a(i) + b(i,j+1)*c(i)
      a(i) = a(i) + b(i,j+2)*c(i)
      a(i) = a(i) + b(i,j+3)*c(i)
   end
end
```

The story about interactions in loop transformations does not quite end here. In our example, we only considered the effect of the transformations on the cache. The transformations may affect other aspects of code such as register pressure, functional unit utilization, and code size. Adding other transformations such as loop tiling, fission, and fusion to the mix further complicates matters. Finding the right sequence in which to apply these transformations and selecting the right parameters for each transformation are difficult tasks, and a method that determines the most effective loop optimization strategy for any given program and platform continues to elude researchers.

Several researchers have explored the possibility of combining two or more loop transformations [15, 22, 23] to deal with interactions. Wolf et al. [15] propose an algorithm that combines several loop transformations. The goal of their algorithm is to find a loop transformation strategy (combination of the order in which loop transformations are applied and the manner in which each transformation is used) that exploits positive interactions between these transformations. Since the number of potential loop transformation strategies is large, their approach uses heuristics to narrow the search space. Within this space, each strategy is evaluated using simple static cost models for caches, registers, and computational resources. Interestingly, their experimental evaluation shows that the combined analysis and transformation results in significant gains over a baseline implementation for a number of programs, with one exception, where the combined analysis performs significantly worse. In this program, some of the loops generated by the combined loop transformations were too complex for the software prefetching optimization to analyze. Hence, no prefetch instructions are generated, leading to a significant performance loss. This observation illustrates the complexity of identifying interactions in compilers and the pitfalls of ignoring key interactions while designing optimizations.

8.1.2.4   Discussion

Our discussion in this section can be summarized as follows. We find that hard problems and complex interactions are pervasive in compilers. More importantly, solutions to these problems are dependent on the program and the intricacies of the target platform. A solution that is best for one program or platform is not necessarily the best for the other. This leads us to the conclusion that compilers must be flexible enough to adapt to the requirements of specific programs and hardware configurations. Aspects of compilation such as the compilation sequence and optimization heuristics must be adaptable by design if high performance is to be achieved.

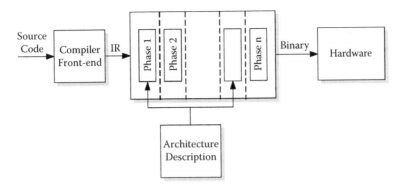

FIGURE 8.4   The structure of a typical optimizing compiler.

8.2   Adaptive Compilers

Let us first review the structure of existing optimizing compilers and see how they fare in terms of flexibility and adaptability. Figure 8.4 illustrates the architecture of a typical optimizing compiler. The back-end of the compiler consists of a set of optimizations or *phases*. These phases are applied to an intermediate representation of the program in a predetermined sequence, irrespective of the program being optimized. Each phase is almost always designed independently, and interactions between optimizations are often ignored. Coarse models of the underlying hardware are sometimes employed to guide optimizations, with the aim of reducing the influence of negative interactions between the optimizations and the hardware. It is easy to see that the structure of a modern optimizing compiler is more or less rigid. This lack of flexibility is one of the key reasons that optimizing compilers fail to deliver optimal performance.

In light of these shortcomings of conventional compilers, many in the compiler community have favored the idea of building compilers that adapt to the program and the hardware [24]. Although the idea of an adaptive compiler sounds compelling, several concerns need to be addressed before an actual implementation can be envisaged. For example, what are the necessary building blocks of an adaptive compiler? How is the compiler structured? What are the compile time requirements of the compiler? What performance guarantees can the compiler make? In the rest of this section, we attempt to answer some of these questions.

Figure 8.5 illustrates the proposed architecture of an adaptive compiler [24]. This architecture differs from the traditional compiler on several counts. An adaptive compiler consists of some form an *adaptation unit*, which controls the entire optimization process. The adaption unit uses a combination of

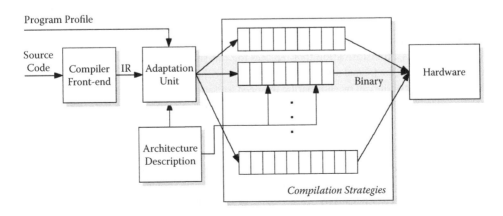

FIGURE 8.5   The architecture of an adaptive optimizing compiler.

program analysis and abstract architecture descriptions to evaluate different compilation strategies and select a strategy that is best suited to the program and a target platform. More precisely, the adaption unit selects (a) a custom compilation sequence and (b) custom heuristics for each optimization. Additionally, the adaptation unit may use a representative program profile provided by the programmer to tune the compilation strategy to the program and a specific class of inputs. This can be seen as an advanced form of profile-guided optimization that involves restructuring the compiler itself. Once a compilation strategy is chosen, compilation proceeds as normal and an optimized binary is generated. We now outline some of the challenges that must be addressed by any implementation that adheres to this conceptual design:

1. Identify and characterize compiler interactions and design synergetic optimization phases.
2. Design an adaptation unit that can automatically:

- Explore the space of compilation sequences and determine the best compilation sequence for a given program and architecture. This is known as the *phase ordering* problem.
- For each optimization in the compilation sequence that is driven by a heuristic, select the best heuristic for the given program and platform. We refer to this problem as one of *heuristic selection*.

In the rest of this chapter, we consider these problems in detail. In addition, we consider the problem of selecting the best compiler optimization flags and parameters for a given program. *Optimization flag selection* is relevent for conventional compilers where the phase ordering is fixed but the compiler allows user control over individual optimization flags and parameters. We show how some of these hard problems are naturally expressed as problems in inference and learning, and well-known statistical and machine learning techniques can be employed to find effective solutions to these problems automatically.

8.3   Characterizing Compiler Interactions

We have already seen that compiler interactions play a crucial role in determining the eventual performance of a compiler. Understanding these interactions is critical to problems such as phase ordering and the design of interaction-sensitive compiler optimizations and heuristics. Traditionally, the task identifying and characterizing key interactions has been performed manually by compiler writers and researchers. Given a set of compiler optimizations, compiler writers are expected to analyze the optimizations and determine if (a) two or more optimizations interact and (b) any of the optimizations interact with the architecture. However, the manual nature of this task leaves open the possibility of oversights. It is possible that the compiler writer may have misread or even overlooked some important interactions. We have already seen an example of an unexpected and complex interaction register allocation, instruction scheduling, and OoO processors that are usually overlooked by compiler writers [20]. The following example further illustrates this possibility.

Function inlining is a transformation that replaces the call-site with the body of the callee. The transformation also replaces the formal arguments in the callee with the actual arguments. This optimization reduces the overheads of the call by eliminating the need for setting up a stack frame and allows the callee to be optimized in the context of the caller. However, inlining leads to an increase in code size and a possible increase in register pressure. It is therefore natural to expect inlining to interact positively with classical optimizations such as common subexpression elimination, PRE, and so on and interact negatively with register allocation and the instruction cache. Hence, virtually all optimizing compilers use heuristics to limit the amount of code growth due to inlining, and some compilers also consider the impact on register pressure before making inlining decisions. However, it was recently shown [4] that on OoO superscalar processors, inlining interacts negatively with the size of the reorder buffer. In other words, the efficacy of inlining is reduced on OoO superscalar processors with large reorder buffers. If this interaction is not considered by the inlining heuristic, excessive inlining may result, which may even hurt performance.

The reasons behind this interaction are easily understood. Modern OoO processors with branch prediction and large reorder buffers are capable of predicting targets of call-sites and fetching instructions across call-sites. This enables the processor to exploit parallelism across call boundaries and hide the latency of instructions that set up the call stack. It would be unfair to expect compiler writers to detect, characterize, and model these intricate interactions. In fact, the difficulty of this task can only increase with the increasing number of compiler optimizations and the complexity of modern processors.

To address these problems, we require a method that automatically detects and quantifies the degree of interactions between any given set of optimizations. It turns out that the problem of detecting and quantifying interactions is not unique to compiler optimizations. Interactions between components exist in most natural and artificial systems. Consequently, there exist a number of automatic techniques to study and quantify interactions. One such class of techniques, known as *empirical regression models*, has been well studied in the statistical and machine learning literature and has found widespread use. In the rest of this section, we describe various regression modeling techniques and evaluate the applicability of these techniques to compiler optimizations.

8.3.1 Empirical Regression Models

Regression modeling refers to a class of methods used for building *black box* models of systems. Say we are given a system defined by a set of input parameters and an output variable (also known as the *response*). Assume that the system is described by some mathematical function, which represents the relationship between the inputs and the response. In regression modeling, we want to find an approximation of this function solely by observing the response of the system for different inputs.

Let us state this notion formally. We are given a system S with n input parameters $\{x_i \mid 1 \leq i \leq n\}$ and a response variable y. We are also given a set of observations $\{(X_1, y_1), (X_2, y_2), \ldots, (X_m, y_m)\}$, where each X_i is an n-dimensional vector and represents an assignment of values to the input parameters, and y_i is the response of the system for the input X_i. Also assume that the system is characterized by the following equation:

$$y = f(x_1, x_2, \ldots, x_n)$$

In regression modeling, we use the observations to *learn* a function $\hat{f}$ that satisfactorily *approximates* f. Formally,

$$y = f(X) = \hat{f}(X) + \epsilon$$

where ϵ represents the approximation error. One can think of regression modeling as the process of searching for a function that best explains the observations. However, since there are infinitely many functions, the search for the closest approximation may never terminate. Regression modeling techniques restrict the search space by making some assumptions about the structure and form of the approximation function $\hat{f}$. We now describe two techniques, linear regression models [25] and multivariate adaptive regression splines (MARS) [26], that are commonly used for building regression models.

8.3.1.1 Linear Regression Models

Linear regression models are the simplest and perhaps the most commonly used regression modeling tools. In linear modeling, we want to find the closest *linear* approximation of the relationship between the inputs and the response. In the simplest case, we assume that the approximation function $\hat{f}$ takes the following form:

$$y = \beta_0 + \sum_{i=1}^{n} \beta_i x_i + \epsilon$$

Here, the approximation function is a weighted sum of the input parameters plus a constant and the approximation error. The unknown weights $\{\beta_i \mid 1 \leq i \leq n\}$ are known as *partial regression coefficients*. These coefficients represent the expected change in the response per unit change in the corresponding input. In situations where we anticipate the presence of interactions between input variables, we can extend linear models to incorporate interactions. For example, the following equation represents a linear model for a system with two input parameters and an interaction term.

$$y = \beta_0 + \beta_1 x_1 + \beta_2 x_2 + \beta_{12} x_1 x_2 + \epsilon$$

For generality, we can also represent this model as follows:

$$y = \beta_0 + \beta_1 x_1 + \beta_2 x_2 + \beta_3 x_3 + \epsilon$$

where $x_3 = x_1 x_2$ is an interaction term. Similarly, linear models can be extended to model interactions between three or more variables. In general, a *complete* linear model for a system with n input variables has 2^n terms and $2^n + 1$ unknown partial regression coefficients, the values for which must be determined from the observations.

8.3.1.1.1   Building Linear Models

Because of their simplicity, building linear models from a set of observations is relatively straightforward. Let $\{(X_1, y_1), (X_2, y_2), \ldots, (X_m, y_m)\}$ be a given set of observations. Let us assume that we are interested in building a linear model of the form shown in Equation 8.1.

$$y = \beta_0 + \sum_{i=1}^{k} \beta_i x_i + \epsilon \tag{8.1}$$

Here, the model consists of k terms. Each term x_i represents an input variable or an interaction between two or more input variables. We now wish to determine the values of the partial regression coefficients that minimize the approximation error. First, we substitute the m observations in Equation 8.1 and represent the resulting m equations in a matrix form as follows:

$$\mathbf{y} = \mathbf{X}\beta + \epsilon \tag{8.2}$$

Here, the vector $\mathbf{y} = \{y_1, y_2, \ldots, y_m\}$ is a vector of responses from the observations, the matrix $\mathbf{X} = \{X_1; X_2; \ldots; X_m\}$ represents the set of inputs (also known as the *design or model* matrix), and β is a vector representing the unknown partial regression coefficients. Since there are many solutions to this set of equations, a unique solution is obtained by defining an error metric and finding partial regression coefficients that minimize the specified error metric. For instance, the *sum of squares error* defined as follows is a commonly used error metric.

$$SSE = \sum_{i=1}^{n} (y_i - \hat{f}(\mathbf{x}_i))^2 \tag{8.3}$$

This metric is the cumulative sum of squared differences between the observed response values and the values predicted by the linear model. Partial regression coefficients that minimize this metric are easily computed as follows:

$$\hat{\beta} = (\mathbf{X}'\mathbf{X})^{-1}\mathbf{X}'\mathbf{y} \tag{8.4}$$

For obvious reasons, this estimate of the partial regression coefficients is known as the *least squares estimate*. These values of $\hat{\beta}$ can be substituted in Equation 8.1 to obtain the linear model $\hat{f}(X)$.

8.3.1.1.1.1   Example

The utility of linear models in analyzing performance of complex systems is best illustrated using an example. Consider the well-known matrix multiplication program shown in Listing 8.4. If the cumulative size of the matrices a, b, and c is large, the working set of the program may not fit into the cache, causing cache misses. Loop tiling is a loop transformation that increases the efficiency of loops by partitioning the loops working set into smaller chunks or *blocks* such that each block stays in the cache while it is being used. Listing 8.5 shows the same program after tiling has been performed.

Listing 8.4: Matrix multiply
```
1  float  a[SIZE][SIZE], b[SIZE][SIZE], c[SIZE][SIZE];
2  for(i=0; i < SIZE; i++)
3    for(j=0; i < SIZE; j++)
4      for(k=0; k < SIZE; k++)
5        c[i][j] = c[i][j] + a[i][k] x b[k][j];
```

Listing 8.5: Matrix multiply after loop tiling has been applied
```
1  float  a[SIZE][SIZE], b[SIZE][SIZE], c[SIZE][SIZE];
2  for( ii =0; ii < SIZE; ii = ii + BLOCK)
3    for( jj =0; jj < SIZE; jj = jj + BLOCK)
4      for(kk=0; kk < SIZE; kk = kk + BLOCK)
5        for(i = 0; i < MIN(SIZE, ii + BLOCK − 1); i++)
6          for(j = 0; j < MIN(SIZE, jj + BLOCK − 1); j++)
7            for(k = 0; k < MIN(SIZE, kk + BLOCK −1); k++)
8              c[i][j] = c[i][j] + a[i][k] * b[k][j];
```

Deciding the best blocking factor for arbitrary loops is hard because it requires reasoning about the size of the loop's working set and knowledge about the target platform's cache size hierarchy. An idea of how critical this decision is can be gauged from Figure 8.6a, which shows the variation in execution time (in seconds) of this program (SIZE = 1,024, measured using a cycle accurate simulator) for various blocking factors ranging from 8 to 128 and L1 data cache sizes ranging from 8 to 128KB. The plot shows that small blocking factors are preferable, irrespective of the cache size. A sharp increase in execution time is observed for high blocking factors, especially on platforms with small caches.

Based on these observations, we can build a linear model that approximately represents the relationship between the execution time of the program, the blocking factor, and the cache size. Equation 8.5 represents the linear model we obtain. The model suggests that the blocking factor has a higher influence on performance than the size of the data cache. As expected, the high value of the coefficient associated with the product terms indicates the presence of a strong interaction between the blocking factor and the cache size. The signs of the coefficients also reveal that increasing the blocking factor hurts performance, whereas the converse is true for the cache size. The sign of the interaction coefficient suggests that performance improves if both the blocking factor and cache size are increased together.

$$\hat{y} = 6.20 + 4.46 * blocking\_factor \quad (8.5)$$
$$- 1.79 * cache\_size$$
$$- 1.85 * blocking\_factor * cache\_size$$

Figure 8.6b graphically illustrates the execution time predicted by the model. Although the linear model does not exactly *fit* all observations (especially the performance variations for large blocking factors and small cache sizes), it captures the overall performance trends.

Apart from the model itself, most linear modeling tools also provide as output a number of statistics that estimate the accuracy of the linear model. These include the mean squared error, R^2 statistic, and residuals [25]. If these statistics suggest that the linear model is not a reasonable approximation of the underlying system (because of nonlinear variations), other modeling techniques such as MARS may be employed.

8.3.1.2 Multivariate Adaptive Regression Splines (MARS)

MARS [26] belongs to a class of *recursive-partitioning*-based regression techniques that use a divide-and-conquer strategy to find functions that approximate arbitrarily complex relationships. Instead of attempting to find one global function that explains the response (like linear models), MARS recursively partitions the

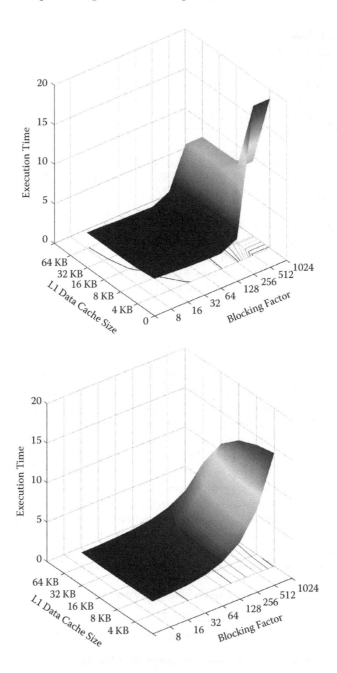

FIGURE 8.6   (a) Variation in execution time of matrix multiply for different blocking factors and L1 data cache sizes. (b) A linear model for the execution time of matrix multiply with blocking factor and L1 data cache size as input parameters. The linear model captures the high-level trends in execution time but is unable to capture sharp, nonlinear variations in the space defined by small cache sizes and high blocking factors.

domain of input variables into disjoint *regions* until the response in each region can be accurately described using simple functions. The final MARS model is simply a linear combination of these functions.

Let us formally state the basic ideas behind MARS models. Assume that the input variables $\{x_i \mid 1 \leq i \leq n\} \in D$, where the domain D defines the input space. The MARS algorithm starts by considering the whole domain as one region and uses a simple function (typically a simple spline function [26]) to model the response in the region. If this simple model is not accurate enough (results in a high approximation error), the algorithm partitions the current region (D) into two disjoint regions, for example, R_1 and R_2, along the axis of one of the input variables. It then tries to model the regions R_1 and R_2 using simple functions. This process repeats until the response in each region $\{R_i | 1 \leq i \leq m\}$ can be adequately modeled using simple functions. If $B_m(x)$ denotes the function the describes the response in region R_m, then the overall model is represented as

$$\text{if } \mathbf{x} \in R_m, \text{ then } \hat{f}(\mathbf{x}) = w_0 + w_m B_m(\mathbf{x}) \tag{8.6}$$

Here, the functions B_m are known as the *basis* functions and w_m are the regression coefficients. The model can also be represented as

$$\hat{f}(\mathbf{x}) = w_0 + \sum_{m=1}^{M} w_m B_m(\mathbf{x}) I_m(\mathbf{x}) \tag{8.7}$$

where $I_m(\mathbf{x})$ is an indicator function that assumes a value 1 if $\mathbf{x} \in \mathbf{R_m}$ and 0 otherwise.

The key aspect of the MARS algorithm is that all parameters of the model, that is, the regions R_m, unknown parameters of the basis functions B_m, and the regression coefficients w_m, are determined using experimental data. Furthermore, a MARS model (Equation 8.7) can be rewritten in a form in which the terms corresponding to input variables and their interactions are associated with their own coefficients (much like Equation 8.1). Because of these features, models produced by MARS are not only more accurate than simple linear models but are also interpretable (partial regression coefficients obtained using the MARS model also indicate the influence of the corresponding term on the response).

8.3.1.3 Case Study: Detecting Interactions in the *gcc* Compiler

Vaswani et al. [4] consider the problem of identifying interactions in the *gcc* compiler using regression models. The system they model (Figure 8.7) is a combination of the compiler and the target platform. The compiler is parameterized by a set of optimization flags (one for each optimization) that control whether the corresponding optimization is applied. Also included as inputs are the settings on heuristics such as the maximum unroll factor and the maximum code growth due to function inlining. The platform is modeled as a generic OoO superscalar processor parameterized by a micro-architectural configuration, which includes variables such as the issue width, reorder buffer size, cache sizes, memory latency, branch predictor configuration, and the number of functional units.

The modeling technique proceeds as follows. For each program/input pair, a set of observations is collected at carefully selected points in the input space. Each observation is a pair (X_i, y_i), where the input vector X_i is an assignment of values to compiler optimization flags, heuristics, and the micro-architectural parameters. Each y_i is the execution time (in cycles) of the program generated using the flag and heuristic settings specified by X_i and executed on a target machine whose configuration is specified by the micro-architectural parameter settings in X_i. The execution time is measured using a detailed cycle-accurate

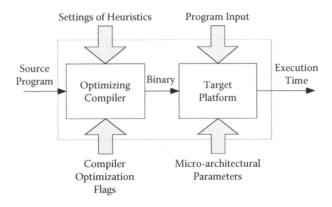

FIGURE 8.7 Components and parameters of the system considered for regression modeling.

processor simulator. A MARS model is built based on these observations. The model relates the execution time of the program to the optimization flags, heuristics, micro-architectural parameters, and interactions between these variables. The partial regression coefficients of these models tell us how significant each of these variables and their interactions are.

An analysis of the regression coefficients obtained for a set of benchmark programs yields some interesting insights. The coefficients reveal that performance is by far determined by the micro-architectural configuration and that conventional compiler optimizations have a relatively smaller role to play. Among compiler optimizations, function inlining and omitting the frame pointer are found to be the most effective across benchmarks. More importantly, the model identifies several interesting positive and negative interactions. For instance, the reorder buffer size, a parameter that is hard to analytically model and hence is ignored by most static modeling techniques, interacts negatively with function inlining but positively with the maximum unroll factor heuristic. The latter interaction, coupled with the observation that increasing the maximum unroll factor tends to hurt performance, suggests that loops should be aggressively unrolled on machines with larger reorder buffers. These important insights into the inner workings of compiler optimizations would have been hard to derive without the assistance of empirical models.

8.4   Heuristic Selection

It is common for a compiler optimization to find itself in situations where it must choose from one of the several ways it can affect a code transformation (which may include the option of not affecting the transformation). Ideally, we would like a model of the system that accurately computes the costs and benefits of applying each transformation. The compiler would then be able to evaluate each of the options and pick the optimal transformation to apply. However, this approach is generally not feasible for two reasons. (a) The number of possible transformations may be too large to enumerate. (b) Building models that accurately compute the costs and benefits of transformations on modern platforms is difficult. For these reasons, compiler optimizations are forced to rely on heuristics to guide decision making. Optimizations may use heuristics to narrow the space of transformations to evaluate and estimate the costs and benefits of each transformation. In essence, heuristics make compiler optimizations tractable at the cost of optimal decision making.

This leads us to the following important question: How do we design good heuristics? This question has plagued compiler writers for several years. The difficulty in designing good heuristics is best illustrated by an example. Consider function inlining and the problem of determining whether to inline a given call-site. The costs and benefits of inlining depend on a host of factors including:

- Static parameters such as the function sizes (both caller and callee), number and type of arguments, depth in the call tree, available parallelism, dependence relationships, etc.
- The target platform configuration, which includes parameters such as issue width, cache sizes, reorder buffer size, number of functional units, branch predictor configuration, etc.
- Other optimizations in the compilation sequence that may interact with inlining

Our goal is to derive an inlining heuristic that considers all these factors and determines whether the call-site in question should be inlined. We require that the heuristic make the right decisions across different programs and different target platforms. Designing such a heuristic is a challenge. Not surprisingly, the approach taken by compiler writers is one of *trial and error*. Compiler writers are usually adept at identifying program features that should be considered in the heuristic; deriving the heuristic itself is, however, based on experience and repeated experiments.

In this section, we consider the problem of *automatically* deriving good heuristics. It has been shown that in many cases, heuristic selection can be reduced to well-known problems in machine learning such as supervised classification [27, 28], function approximation [29, 30], and function selection [31]. The key idea behind the learning-based approach is to consider a heuristic as a function that takes as input various factors that influence the effectiveness of an optimization and outputs a metric that reflects the *goodness* of each candidate transformation. This reduction from heuristics to functions enables the use of standard machine learning techniques such as support vector machines (SVMs) [32] and genetic programming [33] to automatically *learn* heuristics using training data generated by running a set of representative programs on the target platform. We illustrate this approach with two specific examples, namely, the problem of determining the best unroll factors for loops and a more general problem of learning good priority functions for optimizations such as function inlining and instruction scheduling.

8.4.1 Deciding Best Loop Unroll Factors

Loop unrolling is a loop transformation that replicates the body of a loop several times. Listing 8.3 shows an unrolled version of the loop in Listing 8.1. Loop unrolling reduces the overheads of instructions that check for loop termination. Other important benefits of unrolling include (a) the increase in integer linear programming (ILP) that may help generate better schedules, (b) positive interactions with classical optimizations and other loop transformations [15], and (c) exposing regular memory access patterns [34, 35]. However, loop unrolling has several undesired side effects such as increase in code size and register pressure, which can neutralize the benefits of unrolling and even hurt performance. The task of a compiler writer is to design a heuristic function that considers these factors and determines if a loop should be unrolled and the best unroll factor for the loop.

Monsifrot et al. [27] proposed the use of supervised classification to infer unroll heuristic functions for a given target platform. Their approach focuses on the problem of finding a heuristic that determines if a given loop should be unrolled. Stephenson and Amarasinghe [28] extended this approach to find heuristics that predict the best unroll factor for a loop. Before discussing these approaches, we introduce some of the basic concepts of supervised classification.

8.4.1.1 Supervised Classification

Consider a set of objects, each characterized by a finite number of measurable quantities or *features*. Also assume that there exists an equivalence relation that partitions these objects into a finite number of equivalence classes. Supervised classification [32] is a machine learning technique that automatically finds an approximation to this equivalence relation from training data. The training data consists of a subset of the objects (represented using *feature vectors*), each associated with an equivalence class (represented using *class labels*). If the approximate equivalence relation inferred from the training data resembles the actual relation, it will classify "unseen" objects (objects not part of the training data) correctly with high probability. This approximate relation can then be used as a proxy for the actual equivalence relation.

8.4.1.2 Loop Unrolling and Supervised Classification

Let us formulate the problem of finding loop unrolling heuristics for a given target platform in the supervised classification framework. Consider loops as the objects of interest and assume that there exists an equivalence relation that partitions all loops into equivalence classes such that two loops L_1 and L_2 belong to the same equivalence class if they have the same best loop unroll factor. Since this equivalence relation is unknown, we use supervised classification to learn an approximate relation, which will be our unrolling heuristic. In other words, if the heuristic we learn is a close approximation of the unknown equivalence relation, the result of classifying an unseen loop using this heuristic is likely to give the best unroll factor for that loop.

The procedure for using supervised classification to find the best loop unrolling heuristic consists of the following steps:

Feature extraction: Identify a set of static loop features that are likely to determine the costs and benefits of loop unrolling. Stephenson and Amarasinghe [28] propose to use 38 loop features in their classification scheme; a subset of these 38 features is listed in Table 8.1.

Generating training data: The training data consist of pairs $\{(L_i, u_i) \mid 1 \le i \le n\}$, where L_i is the feature vector and u_i is the best unroll factor for loop i. The training data is generated by first

TABLE 8.1   Subset of loop features considered for
supervised classification

Loop characteristics
Loop nest level
Number of operands
Number of operations
Number of floating point operations
Number of memory operations
Number of branches
Instruction fan-in in DAG
Live range size
Critical path length
Known trip count
Number of parallel computations in loop
The min. memory to memory loop carried dependency
The language (C or Fortran)
Number of indirect references in loop body
The max dependence height of computations

selecting a set of benchmark programs. The compiler then identifies loops in these programs and statically analyzes the loop bodies to generate feature vectors. Finding the best unroll factor for each loop in the training programs is also relatively straightforward. For each loop in the training data set, the execution time of the loop is measured on the target platform for all unroll factors (below a reasonable upper bound), and the loop is labeled with the unroll factor that results in the smallest execution time. This assignment partitions the training data into as many classes as the number of unroll factors considered for evaluation.

Feature selection: Based on the training data, feature selection techniques [32] such as *mutual information score* and *greedy feature selection* are used to reduce the dimensionality of the feature space. These techniques identify and eliminate redundant features and only retain the most informative ones. Using smaller feature vectors for training is recommended, both for accuracy and efficiency reasons.

Select a classification technique: The choice of the classification technique is typically governed by factors such as accuracy, efficiency, and interpretability. Monsifrot et al. [27] use decision trees [36] for classification for their efficiency and interpretability. A decision-tree-based approach recursively partitions the feature space into regions, much like MARS. The nodes of the decision tree are simple conditions over one or more features, which can be easily read and interpreted by compiler writers. However, Stephenson and Amarasinghe [28] prefer more accurate but less interpretable methods such as nearest-neighbor classification and support vector machines.

The result of this exercise is an unrolling heuristic customized for a given target platform. Stephenson and Amarasinghe [28] found that the heuristic learned using supervised classification outperforms a highly tuned, manually developed heuristic in several cases, and unlike the manual heuristic that took several years to develop, the classifier-based heuristic was obtained in a matter of days, including the time to generate the training data. Interestingly, they also found that the loop unroll factors suggested by the classifier for unseen loops are almost always the same as the optimal unroll factors obtained using an oracle.

8.4.2   Learning Priority Functions

In several optimizations, compiler heuristics are structured as *priority functions* [31, 37]. A priority function is used when a compiler must choose one from among several code transformations it can effect. The priority function considers several program characteristics and assigns a *weight* to each candidate transformation, which reflects the potential gains of applying the transformation. The nature and importance of priority functions is best illustrated using examples:

- **Function inlining:** Inlining a function at a call-site can improve performance by eliminating the overheads of setting up a stack frame and by allowing the callee's code to be optimized in the context of the caller. However, unconstrained inlining and the resulting increase in code size and register pressure can adversely affect performance. Therefore, it is not uncommon for compilers to control the increase in code size using a tunable threshold. Given this threshold, inliners typically use a priority function to choose the most profitable call-sites in the program. An example priority function is shown below.

$$P(c) = \frac{(level(c) + 1) * largestSize}{size(c)} \qquad (8.8)$$

  Here c refers to a callee function being considered for inlining, $level(c)$ is the call depth of the callee, $size(c)$ is the size of the callee (typically measured in the number of IR (intermediate) instructions), $largestSize$ represents the size of the largest function in the program, and $P(c)$ is the priority value assigned to the callee c. This priority function assigns a higher priority to callees that have relatively small code size and are deeper in the call-graph.

- **List scheduling:** A list scheduler first creates a dependence graph of instructions in the optimization unit (typically a basic block). In each subsequent step, the scheduler identifies a set of *ready* instructions, that is, instructions whose input dependencies have been resolved and micro-architectural constraints (such as the availability of functional units) have been satisfied. The scheduler then selects one of the ready instructions for scheduling. The selection is typically based on a priority function, which estimates the potential benefit of scheduling each of the ready instructions next. Equation 8.9 shows a typical list scheduling priority function $P(i)$ for instruction i that prioritizes instructions that head long-latency dependence chains. Here, $latency(i)$ is the estimated latency of the instruction (in cycles).

$$P(i) = \begin{array}{ll} latency(i) & : \text{if } i \text{ is independent} \\ max_{i \text{ depends on } j} latency(i) + P(j) & : \text{otherwise} \end{array} \qquad (8.9)$$

These two examples are by no means exhaustive; priority functions are integral to several other optimizations such as register allocation [10, 38], hyperblock formation [39, 40], software pipelining [11], and data prefetching [31, 41]. It has been shown that priority functions play a critical role in determining the eventual effectiveness of an optimization.

Unfortunately, effective priority functions are hard to engineer. In the absence of a systematic process for deriving good priority functions, compiler writers manually design and evaluate priority functions via ad hoc experimentation. For some optimizations, priority functions may be found using an approach based on supervised learning (Section 8.4.1). However, for optimizations like instruction scheduling and register allocation, where the difference in execution time between good and bad *individual* choices may not be discernible, supervised learning may not be the best alternative (although efforts along those lines have been made [29]). Stephenson et al. [31] propose an alternate and rather interesting approach to solving this problem. They show that effective priority functions may be found using an unsupervised learning technique known as *genetic programming* [33]. In this formulation, the best priority function is found by *searching* through the space of priority functions and evaluating each candidate function by executing one or more representative programs on the target platform and measuring the execution time. Let us try to understand how genetic programming can search this potentially infinite space automatically and efficiently.

8.4.2.1 Genetic Programming

Genetic programming (GP) [33] is a machine learning technique that attempts to find a program that optimally solves a given problem by mimicking the evolutionary behavior of a natural population. Just as a population consists of a set of individuals, a GP population consists of a set of programs, represented

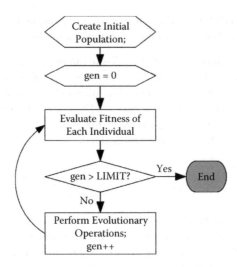

FIGURE 8.8   The sequence of operations in a genetic programming-based algorithm.

using *parse trees*. Each program is also associated with a *fitness function*, which is an estimate of how well the program solves the given problem. GP algorithms start with an initial population of programs (usually randomly generated). Each step in the algorithm (illustrated in Figure 8.8) corresponds to a *generation* in the evolutionary process of the natural population. During each generation, programs *evolve* via mechanisms analogous to *natural selection, reproduction (cross-over)*, and *mutation* amongst others, while following the principle of *survival of the fittest*. In a mechanism analogous to natural selection, a predetermined number of programs are randomly chosen to *die* in each generation. The probability of a program being selected for dying is inversely proportional to its fitness. Hence, programs that are relatively fitter (i.e., solve the problem better) are more likely to survive and remain a part of the population after selection. In cross-over, a predetermined number of *offsprings* are produced by a cross-over of random individuals in the current population. Again, the likelihood of a program being chosen for cross-over is proportional to its fitness. This strategy ensures that features of good programs survive through the generations, while features that cause programs to perform poorly die out. Programs may also mutate by randomly changing into other programs. Mutation ensures that an algorithm does not get stuck with locally optimal programs and that the algorithm *eventually* finds the globally optimal program for the given problem. The algorithm usually terminates when the number of generations reaches a user-specified limit. At this point, the fittest function in the current population is returned as the best solution.

8.4.2.2   Priority Functions and GP

Priority functions can be thought of as programs that assign weights to code transformations. We now outline a GP-based technique [31] that exploits this correspondence between priority functions and programs and finds effective priority functions for a given target platform. The technique relies on genetic programming for searching through the space of priority functions. To use GP, we must resolve the following issues: (a) identify the set of program features to be included in the priority function, (b) decide how priority functions can be represented, (c) define the space of valid priority functions that the genetic algorithm is allowed to explore, (d) define a fitness function, and (e) define evolutionary operations over priority functions. Let's take a closer look at each of these tasks:

> **Feature extraction:** Given an optimization, we must first identify a set of measurable code characteristics that are likely to influence the costs and benefits of the optimization. This task is similar to the feature extraction procedure discussed in Section 8.4.1.

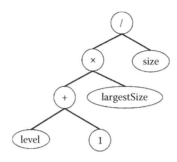

FIGURE 8.9　Parse tree representation of the inlining priority function.

Representing priority functions: Priority functions can be represented using *parse trees* [33]. For example, Figure 8.9 shows the parse tree corresponding to the inlining priority function (Equation 8.8). The inner nodes of the parse tree are operations, and the leaves correspond to features.

Priority function space: The priority function space represents all functions that GP is allowed to generate and explore. A function space is defined by the set of operators that may be used to generate expressions. A simple set of operators over real-valued and boolean-valued features used by Stephenson et al. [31] are listed in Table 8.2. In general, a larger number of operators allows the GP to explore a larger function space and enhances the GP's ability to express good priority functions. However, allowing a larger number of operators may adversely affect the time taken by GP to converge to a locally optimal priority function.

Fitness functions: We require a fitness function that estimates a priority function's ability to recognize and prioritize profitable transformations over unprofitable ones. One way of assessing the fitness of a priority function is to build the function into the optimization, compile a set of benchmark programs, run the programs on the target platform, and measure the total execution time. The best priority function obtained by GP using this fitness function will be tuned for the class of applications that the benchmarks represent. Alternatively, if the goal is to obtain a customized fitness function for a given program (and perhaps a given input), the execution time of that program on the target platform may be used as the fitness value. The latter can also be seen as a profile-driven optimization.

Evolutionary operations: As shown in Figure 8.8, the GP uses one or more evolutionary operations during each generation to evolve the population. We define these operations over parse trees. Figure 8.10 shows the priority functions that may be produced after a cross-over is performed between two priority functions. Here, two subtrees of the parents are randomly selected and swapped to produce new individuals. Similarly, mutation may be defined as an operation in which a subtree of an individual is randomly selected and replaced with another randomly generated expression (Figure 8.11).

TABLE 8.2　Real-valued and Boolean-valued operators used for constructing priority functions

Real-valued operators	Boolean-valued operators
$Real_1 + Real_2$	$Bool_1$ and $Bool_2$
$Real_1 - Real_2$	$Bool_1$ or $Bool_2$
$Real_1 \times Real_2$	not $Bool_1$
$Real_1 / Real_2$ if $Real_2 \neq 0$, 0 otherwise	$Real_1 < Real_2$
$\sqrt{Real_1}$	$Real_1 > Real_2$
$Real_1$: if $Bool_1$	$Real_1 = Real_2$
$Real_2$: if not $Bool_1$	const $Bool_1$
const $Real_1$	

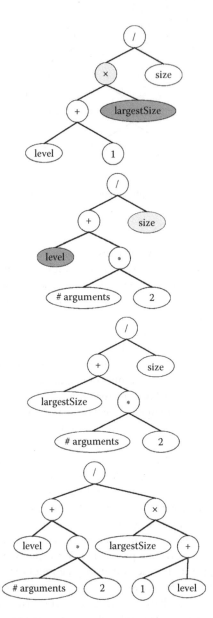

FIGURE 8.10   Figure illustrating the cross-over operation. If the subtrees rooted at gray-color-nodes in expressions (a) and (b) are selected for cross-over and swapped, one of the resulting offspring is shown in (c). If the subtrees rooted at the blue-colored nodes are chosen and swapped, the offspring shown in (d) may result.

Once the problem of finding an effective priority function is reduced to the GP framework, one of the many GP implementations may be used. The result of running a GP over the program feature space is a parse tree, which can be embedded in a compiler and used as a priority function.

8.4.2.3   Implementation Issues

Having established the connection between priority functions and GP, we can use one of the many readily available GP implementations to find effective priority functions. However, depending on the optimization at hand and the GP implementation used, one or more of the following implementation issues may arise.

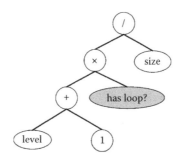

FIGURE 8.11 Priority function produced after mutating the function in Figure 8.10(a). The subtree rooted at node *largestSize* has been replaced with a randomly generated subtree representing the expression *has loop?*.

First, note that expressions created via the evolutionary operations are not bounded in size. If left to its own devices, GP may create arbitrarily large expressions. However, if one assumes that Occam's razor holds and most priority functions are likely to be simple, a GP implementation may place some form of external bounds on the size of the expressions. For instance, larger priority functions could be penalized by using the function size in computing the expression's fitness. Such a fitness function would ensure that large expressions are not likely to survive unless they are significantly better. Stephenson et al. [31] use a simpler version of this strategy.

The efficacy and running time of genetic programming has been shown to depend significantly on the degree to which each evolutionary operator is applied. Most GP implementations take input parameters such as the population size, number of generations, number of expressions replaced in each generation, mutation rate, and so on. Policies like *elitism*, which ensures that the fittest member of each generation is not replaced, may also be used. Finding the best values for these parameters will usually require some experimentation. Finally, note that in the specific instance of finding priority functions, the running time of the GP is dominated by the time spent in evaluating the fitness of priority functions in each generation. This comprises the time spent in compiling the benchmark program(s) using each priority function and executing the program(s) on the target platform. However, this process can be entirely automated, eliminating the manual search process that compiler writers currently employ.

8.5 Phase Ordering

In discussions leading up this section, we observed that the nature of interactions between optimizations depends on the program being compiled *and* the order of applying the optimizations. We also observed that applying optimizations in the *right* order is likely to yield significant benefits. These observations naturally lead us to the problem of finding the sequence of applying optimizations that results in "optimal" code. This problem, also known as the *phase ordering* problem, has intrigued researchers for several years.

The complex nature of the phase ordering problem has been recognized for several years. Although few in number, attempts have been made to analytically infer phase orderings for specific classes of optimizations. For instance, Whitfield and Soffa [7, 8] proposed a framework for specifying optimizations. The formal specifications are used to automatically identify potential interactions, which in turn could be used to select local phase orderings. Long and Fursin [42] consider the special class of iteration reordering loop transformations and show that the phase ordering problem for this class of optimizations can be transformed to one of searching over a polyhedral space. These methods are unfortunately restricted to a select set of optimizations, restricting their use in practice.

Because of these limitations of analytical methods, researchers have tended to favor empirical techniques such as *iterative compilation* [43] for finding effective solutions to the phase ordering problem. Unlike analytical methods, empirical techniques [2, 44–48] evaluate different phase orderings by executing programs on the target platform. Since each phase ordering is evaluated based on its execution time on the target platform, all intricacies of the hardware are automatically accounted for, eliminating the need for complex hardware models. Hence, the empirical approach produces high-quality phase orderings, tuned for a given program(s) and target platform. However, empirical methods are inherently associated with significantly high compilation costs. The compilation time, which now includes the time spent compiling programs with each phase ordering and executing the programs on the target platform, depends on the number of phase orderings evaluated before a satisfactory solution is found. This in turn depends on the technique used to search the phase ordering space and the characteristics of the space itself. In the rest of this section, we take a closer look at these two aspects vital to the feasibility of empirical approaches.

8.5.1 Characterizing Phase Ordering Space

The phase ordering space is a discrete space consisting of all possible phase orderings. This space is obviously infinite because each optimization may be occur infinitely many times in a phase ordering.[2] For the purpose of characterizing and searching, it is usual practice to restrict the space by imposing constraints, such as imposing an upper bound on the length of the phase orderings. However, even a constrained phase ordering space can be huge if a nontrivial number of optimizations are considered. A compiler with 16 optimizations and a maximum length of 10 results in a space of 16^{10} phase orderings [2]. Therefore, most studies that characterize phase orderings are limited to a much smaller phase ordering space.

Interestingly, Kulkarni et al. [50, 51] consider the problem of finding optimal phase orderings for individual procedures. Their studies show that for simple procedures and a typical set of optimizations, the effective phase ordering space is finite in practice. They argue that for most procedures, the phase ordering space can be reduced to a finite set of phase orderings such that each phase ordering in the original infinite space is functionally equivalent (i.e., produces the same optimized procedure) to at least one phase ordering in the finite set. If a finite phase ordering space does exist, an exhaustive characterization and search may be possible [52].

We now discuss conclusions from two separate attempts at characterizing the phase ordering space [2, 52]. These conclusions are critical to our understanding of the phase ordering space and to the design of efficient search techniques employed for finding the optimal phase ordering.

Almagor et al. [2] experimentally characterized a 10-of-5 space (sequences of length 5 drawn from 10 optimizations) with close to 10 million points for a set of small programs. The programs were compiled for each of these 10 million points, and the number of instructions executed was used as the fitness metric. Two points in this space are said to be adjacent if the Hamming distance between them is 1. An analysis of the results shows that the phase ordering space is all but smooth. The surface of the fitness metric in this space resembles a *heavily cracked glacier*. The surface does not reveal any relationships or correlations between the optimizations. Also, the difference between the best- and the worst-performing phase ordering was sometimes found to be 100% or larger. This confirms the importance of finding the right phase ordering for programs. A more detailed analysis of the surface reveals the existence of several local minima (all adjacent phase orderings perform worse) and at times several global minima. Furthermore, the distance between a randomly chosen point in the space and the closest local minima tends to be small (a Hamming distance of 16 or less in most cases). This property was shown to hold for a larger 10-of-16 phase ordering space as well. However, the analysis also finds that both *good* and *bad* local minima exist. In other words, some of the local minima are fairly close to the global optimal (within 5%), wheres other local minima are farther away (20% or more). Furthermore, the distribution of the local minima in space is heavily dependent on the program and the set of transformations considered. These observations suggest that simple search techniques like hill climbing [53] may reach a local minima in a small number of steps and better solutions may be obtained using adaptive techniques like genetic algorithms [54].

While Almagor et al. [2] characterized spaces with a fixed phase ordering length, Kulkarni et al. [52] evaluated phase ordering spaces without a bound on the length. Their study reveals a correlation between

---

[2]Touati and Barthou [49] formalize this notion and prove that the phase ordering problem is undecidable in general.

the length of the phase orderings and performance. They observe that as the length of the phase orderings considered increases, the average difference between the local minima for that length and the global minima decreases. However, increasing the length beyond a threshold does not result in proportional benefits. Since searching in spaces with longer phase orderings is expensive, efficient search algorithms must strike a balance between the time taken for searching and the quality of the search results.

8.5.2 Search Techniques

We now describe two general search techniques that have been shown to be very effective in finding close-to-optimal phase orderings:

Hill climbing: Hill climbing [53] is a simple search technique that attempts to find a local optimum of a function over a discrete space. There are several versions of this search technique. In the simplest version of hill climbing, the search starts at an initial point (usually randomly selected) and in each subsequent step moves to a neighboring point with a better fitness value (lower execution time in our case). The search terminates at a point where no neighboring point has a better fitness value, that is, a locally optimal point is reached. In steepest descent hill climbing, the algorithm evaluates the fitness of all neighboring points and moves to the point with the best fitness value. An *impatient* version of steepest descent restricts the number of neighbors it evaluates before making a decision to move. The version of hill climbing commonly used for finding effective phase orderings is random research hill climbing. Here, one of the basic versions of hill climbing is performed multiple times with different randomly selected initial points. Most studies [2, 52] indicate that random research hill climbing finds phase orderings that are very close to the global optimum within a small number of steps. This result follows from some of the characteristics of the search space.

Genetic algorithms: A genetic algorithm (GA) [54] is a search technique that belongs to the class of evolutionary algorithms. GAs are a generalized form of GP. In GAs, the input space is an abstract representation called *chromosomes* of possible solutions. Traditionally, the solutions or chromosomes are represented as binary strings. However, many other representations of the solution space are also possible. Furthermore, evolutionary operators such as cross-over and mutation are also defined over chromosomes. Much like GPs, the randomness in the evolutionary operators ensures that the GA does not get stuck in locally optimal solutions and reaches the global optimal if run for a sufficient number of generations. In our problem, a phase ordering naturally maps to the notion of a chromosome and is easily represented as a string over the alphabet of compiler optimizations. Cooper et al. [44] first proposed the use of GAs for finding phase orderings that reduce the size of optimized code. Kulkarni et al. [52] show that GAs are very effective in finding phase orderings close to the global optimum within a reasonable number of generations. They also find that the maximum length of the phase orderings considered is an important parameter in the performance of the GA. As the length of the phase ordering is increased, the difference between the GA solution and the global optimum decreases.

8.5.3 Focused Search Techniques

The search techniques discussed above address the problem of finding good phase orderings automatically. However, the techniques have found limited use in practice because of the large amounts of time they incur in finding good phase orderings. Even with the most efficient search techniques, finding an effective phase ordering could take several hours, if not days [47, 48]. For instance, Kulkarni et al. [47] reported that finding phase orderings for a *jpeg* implementation using a conventional genetic algorithm requires over 20 hours. Similarly, Agakov et al. [48] found that over 100 phase ordering evaluations are required to find a locally optimal phase ordering using random search. The search time is dominated by the time spent in compiling and evaluating each phase ordering. While this effort may be justified in some settings, faster searching algorithms are essential for a wider acceptance and use of compilers that adapt phase orderings.

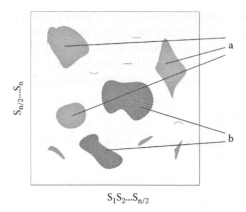

FIGURE 8.12   Graphical illustration of a phase ordering space. The shaded areas represent near-optimal phase orderings for two programs.

There are several ways of improving the efficiency of the search process. For instance, Cooper et al. [55] proposed methods for reducing the amount of time spent in evaluating each phase ordering. The speed of searching could also be improved by caching the results of phase orderings as they are evaluated and reusing the results from the cache when a *redundant* phase ordering is detected. A phase ordering is said to be redundant if (a) it is syntactically the same as a phase ordering that has previously been searched and evaluated or (b) it is functionally equivalent to the previously evaluated phase orderings. This may happen because of the presence of dormant optimizations or two different phase orderings producing the same code or two phase orderings producing *equivalent* code [47]. An orthogonal approach is to reduce the number of phase orderings considered for evaluation by focusing the search techniques to regions of the phase ordering space where the optimal points are more likely to occur [45, 47, 48]. For illustrative purposes, let us consider a focused search technique based on predictive modeling [48].

8.5.3.1   Focused Search Using Predictive Models

The search techniques we have discussed so far are generic techniques that do not require any prior knowledge of the programs or optimizations under consideration. As a result, these techniques apply to a wide variety of programs and optimizations. However, the generic nature of these techniques comes at the cost of increased search times. This is because in the absence of any prior information about the nature of the search space, generic search techniques must start by considering all phase orderings as potential candidates and slowly narrow down the search to regions of the phase ordering space that yield the most benefit.

In light of this shortcoming, researchers have considered the possibility of using more specialized search techniques for increasing the efficiency of the search process. Agakov et al. [48] propose one such specialized search technique based on machine learning. The technique relies on the assumption that "similar" programs are likely to respond in a similar fashion to different phase orderings. In other words, a phase ordering that is beneficial for one program is likely to be beneficial for all similar programs. This assumption also implies that the best phase orderings for similar programs are likely to occur in the same regions of the phase ordering space. Therefore, if the regions that contain effective phase orderings for one program are known a priori, the process of finding the best phase orderings for similar programs can be speeded up by biasing the search toward these regions.

Figure 8.12 illustrates the basis of this search technique graphically. Consider a set Γ of m optimizations $\{o_1, \ldots, o_m\}$. The bounded area in Figure 8.12 represents a constrained phase ordering space in which each phase ordering is a sequence $s_1 s_2 \ldots s_n$ of length n, where each $s_i \in \Gamma$. To simplify the visualization, the phase ordering space is flattened out into two dimensions. The x-axis represents the first half of the sequence $s_1 s_2 \ldots s_{n/2}$, and the y-axis represents the second half $s_{n/2} \ldots s_n$. The figure also shows regions (shaded with plain gray) that contain phase orderings that result in a significant speedup for program a and other regions (shaded in hatched gray) that benefit program b. The best phase orderings for another program c similar to b are more likely to be found in regions that benefit b.

We now describe the search technique in more detail. The proposed methodology involves an offline training step, in which a mapping between a set of training programs and regions of the phase ordering space that yield significant benefits is learned. Then, given a new program, the technique finds the closest program in the training set and initiates a conventional search from regions that are known to benefit the closest training program. Let us take a closer look at each of these steps.

8.5.3.1.1 Offline Training

In this offline step, we want to learn a mapping between a set of training programs and regions in the phase ordering space that yield significant performance benefits for these programs. There are several ways of learning such a mapping. In the simplest of strategies, we could run the training programs for a randomly selected set of phase orderings and associate each training program with its own best phase ordering(s). Then an effective phase ordering for a new program may be found simply by finding a program in the training set that is most similar to the new program, evaluating its best phase ordering(s) and picking the best ordering among them. While this focused search technique is extremely quick, it constrains the search space to a very small set of phase orderings. As a result, it lacks the ability to *generalize* beyond the set of programs in the training data set and may not yield the best phase orderings for sufficiently different programs.

Let us now consider an alternate strategy that generalizes to a larger class of programs while retaining much of the performance advantages of the previous strategy. The key to the alternate approach lies in realizing that the previous strategy assigns an equal, nonzero probability to phase orderings that were effective for the most similar program in the training set and a probability of 0 to all other phase orderings. In the alternate approach, we try to assign high probabilities to phase orderings that were effective for the program in the training set and progressively lower (nonzero) probabilities to other phase orderings. In effect, such a probability assignment ensures that given a new program, the search procedure evaluates phase orderings that were beneficial for similar programs in the training set and hence are more likely to benefit the new program, before evaluating other phase orderings.

We are now left with the task of assigning probabilities to phase orderings for each of the programs in the training data set. Ideally, the probability assigned to each phase ordering should reflect the likelihood of the phase ordering being effective for similar programs. There are several ways to compute such an assignment. For instance, if we assume that the individual optimizations are independent, each phase ordering could be assigned a probability equal to the product of the probabilities of the optimizations in that sequence being effective (Equation 8.10).

$$P(s_1 s_2 \ldots s_n) = \prod_{i=1}^{n} P(s_i) \tag{8.10}$$

Here, $P(s_i)$ is the probability that the optimization s_i occurs in effective phase orderings. This probability is easily computed by running the training program for a randomly generated set of phase orderings and counting the number of times the optimization appears in an effective phase ordering. Here, an effective phase ordering is defined as a phase ordering that achieves at least a certain fraction of the performance achieved by the best phase ordering.

This simple method of assigning probabilities to phase orderings ignores the presence of interactions. As we have already seen, interactions between optimizations are pervasive. For example, some optimizations are most effective in the presence of other optimizations. In such scenarios, a simple product of the individual probabilities is not a good measure of the effectiveness of the phase ordering. To account for

TABLE 8.3   Some of the features used
to classify programs

Features
loop nest depth
are loop bounds constant?
number of array references within loops
number of load/store instructions in loops
number of integer variables in loops
number of floating point variables in loops
number of call instructions in loops
number of branch instructions in loops

interactions between optimizations, phase orderings can be modeled as Markov chains [56]. A Markov chain for phase orderings is defined as follows:

$$P(s_1 s_2 \ldots s_n) = P(s_1) \prod_{i=2}^{n} P(s_i \mid s_{i-1}) \tag{8.11}$$

This richer characterization of phase orderings assumes that the effectiveness of an optimization depends only on the previous optimization in the phase ordering. Since this model accounts for some of the interactions between optimizations, the resulting probability assignment more accurately reflects the potential benefits of phase orderings. Consequently, a search guided by this model is likely to find effective phase orderings quickly.

8.5.3.1.2   *Program Classification*

The focused search technique we have discussed hinges on the notion of similarity of programs. Given a new program, we want to identify a program in the training set that responds in the same fashion to changes in phase orderings. This problem can be cast as an unsupervised classification problem. Here, each program is represented using a set of measurable *features* that characterize the way the program responds to different phase orderings. We first collect the feature vectors of all training programs. Given the feature vector of a new program, any unsupervised classification technique like nearest neighbors can be used to find the program most similar to the new program. The subset of features used by Agakov et al. is listed in Table 8.3; a detailed listing can be found elsewhere [48].

8.5.3.1.3   *Focused Search*

After identifying the training program most similar to the given program, we drive one of the many conventional search techniques using the probability model associated with the training program. For instance, in random search, phase orderings are selected based on the probability model of the training program instead of the original method of selecting phase orderings at random. In case of a GA, the initial population of phase orderings is selected using the probability model of the training program instead of a randomly generated set of phase orderings.

The focused search technique we have discussed has shown a lot of promise [48]. These techniques arrive at the best phase orderings using a fraction of the number of evaluations incurred by the corresponding conventional search techniques. However, just like other learning techniques, it has been found that the effectiveness of focused search depends on the quality and the amount of training data, which in this case equates to the number of different programs included in the training data and the number of phase orderings evaluated for learning the probability models. The latter is especially important if we use richer models such as Markov chains. Here, the number of unknown probabilities that must be learned from training data is $O(m^2)$, where m is the number of optimizations. It has been found that the simpler product of probabilities model may outperform Markov chains if the amount of training data available is not sufficient to accurately learn all the unknown probabilities in the chain.

8.6 Optimization Flags and Parameter Selection

The selection of optimization flags is probably the only form of (manual) adaptation built into a conventional compiler. Most compilers allow users to choose their own optimization settings by providing flags that can be used to enable or disable optimizations individually. However, developers and compilation tools rarely exercise this option and rely on the default settings instead (typically the -Ox settings, where x is the level of optimization). Recent research has shown that the use of default settings often leads to suboptimal performance [3, 4]. This is because compiler writers usually choose the default settings based on the average performance measured over a number of benchmark programs on a limited number of platforms. Therefore, these settings are not guaranteed to deliver the best performance for a given program on a given target platform. For instance, it is not uncommon for some of the *aggressive* optimizations like loop unrolling to be turned off by default [57] because the optimization was found to hurt average performance, probably due to negative interactions and/or ineffective heuristics. These optimizations are not used (unless explicitly turned on by the user), even when the program being compiled may have benefited from the optimizations.

Ideally, we would like the compiler to automatically select the best set of optimization flags for each program, probably by performing a cost-benefit analysis based on an accurate model of all optimizations and the target platform. Such analysis is beyond the capabilities of existing analytical modeling techniques. As an alternate approach, researchers have proposed a number of automatic techniques based on statistics and machine learning to address this problem [3, 4, 58, 59]. Let us consider the technique based on inferential statistics [3] in some detail.

8.6.1 Optimization Flag Selection Using Inferential Statistics

In this approach, compiler optimizations are considered as *factors* that influence the performance of the given program. We design a set of *experiments* to test whether each of the factors has a *significant positive* effect on performance. Here, each experiment involves measuring the execution time of the program compiled using one of the 2^k possible settings of the optimization flags. For the moment, assume that we are interested in determining whether one of the optimizations A is effective for the program. We can design a set of experiments G, which can be split into two equal groups G_0 and G_1 with N experiments each. The group G_0 consists of experiments in which the optimization A is turned off, and G_1 consists of experiments where A is turned on. All other optimization flags are randomly selected. A statistical test known as the Mann–Whitney test [60] can be used to analyze the data collected by measuring the execution times of the given program for each of these experiments and determine whether the optimization A is effective. Informally, the test is based on the observation that if optimization A is effective, the performance of experiments in group G_1 should be better than the performance of experiments in group G_0. Note that simply comparing the average execution times of experiments in G_0 with the average of experiments in G_1 may lead to false conclusions because this comparison does not take the variance into account.

The Mann–Whitney test works as follows. The test *ranks* each of the experiments in G based on the measured execution time. The experiment with the smallest execution time is assigned rank 1, and the experiment with the highest time is assigned rank $2N$. Then the test computes statistics S_0 and S_1, where S_0 and S_1 are the sums of ranks of all experiments in groups G_0 and G_1, respectively. Consider the statistic S_1. This statistic can have a minimum value $(1 + 2 + \cdots + N)$ and a maximum value $([N+1] + \cdots + 2N)$. The first value occurs when all the experiments in group G_1 have execution times smaller than the experiments in group G_0, and the second values occurs in the opposite case. It can be shown that if the optimization A is not effective for the program, the statistic S_1 is normally distributed with mean

$$\mu = \frac{N(N+1)}{2} \tag{8.12}$$

and standard deviation

$$\sigma = \sqrt{\frac{N^2(2N+1)}{12}} \tag{8.13}$$

Since the statistic S_1 is normally distributed, one can use ordinary statistics to determine the probability of a specific value of S_1 as originating from this distribution. If this probability is small (less than a threshold, typically 5%), we can conclude that the optimization A is effective for the program.

So far, we have discussed the scenario where we are interested in testing for the significance of one optimization. Although this test can be independently repeated for each optimization, it would require a large number of experiments. An experimental design technique known as *orthogonal arrays* [61] may be used to design a single set of experiments that can be reused for all optimizations. Using orthogonal arrays, Haneda et al. [3] show that a set of 48 experiments is sufficient to select the best optimization flag settings for 23 optimizations in *gcc*.

Several other techniques may be employed to select optimization flags. For instance, Cavazos et al. [58] use logistic regression [62] for selecting method-specific optimization flags for compilation in a Java runtime system. Vaswani et al. [4] use neural network models to select both optimization flags and optimization parameters and find that performance can be improved by as much as 20% by using program-specific optimization settings. Cavazos and O'Boyle [59] propose the use of regression models to select optimization flags from data obtained using hardware performance counters. It is important to note that these techniques are statisitcally rigorous, completely automatic, and require no prior knowledge of the optimizations involved.

8.7   Conclusions

Optimizing compilers were originally conceived as tools that would free programmers from the burden of manually optimizing and tuning programs for performance, allowing them to focus on writing correct and modular code instead. It is therefore ironic that compiler writers find themselves in a similar situation today. Writing high-performance optimizing compilers is becoming an increasingly difficult and tedious task because of the sheer number and nature of compiler optimizations, complex interactions, multiple (often conflicting) objective functions, and the complexity of modern hardware. Therefore, tools and techniques that automatically address these concerns and allow compiler writers to focus on the task of writing correct and retargetable optimizations are desirable.

In this chapter, we observe that the key to high performance lies in flexible and adaptable compilation strategies, a feature missing in conventional compilers. Through a series of examples, we show that effective compilation strategies can be automatically learned using statistical and machine learning techniques. In this model of compilation (illustrated in Figure 8.13), the compiler *learns* a mapping between different

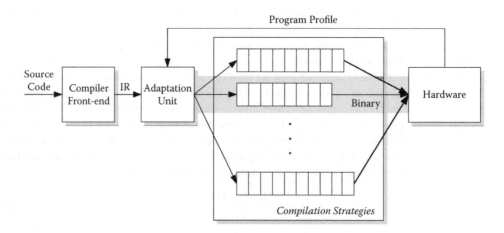

FIGURE 8.13   An adaptive compilation model that uses feedback from the target platform to search and evaluate compilation strategies.

types of programs and effective compilation strategies, which includes aspects such as phase ordering, optimization flags, and heuristics. The learning process is based entirely on experimental data collected by observing the behavior of compiler optimizations for a set of representative programs on the target platform. Note that this model of adaptive compilation differs slightly from the model shown in Figure 8.5 in the use of an architecture description for driving optimizations. Researchers have preferred using the *target platform as the hardware model* rather than relying on an abstract model of the hardware. This is a direct consequence of the fact that existing techniques for modeling the hardware [9] are not accurate or flexible enough to drive adaptation. The design of techniques that can quickly and accurately predict the performance implications of compiler optimizations continues to be an area of active research.

This *black-box* approach to the design of compiler optimizations has several advantages over traditional analytical approaches. The approach does not rely on a compiler writer's knowledge of the optimizations or on the availability of detailed hardware models; the approach is targeted at complex systems like compilers where human understanding is often limited. For instance, empirical modeling techniques offer insights into the inner workings of compiler optimizations that are hard to infer otherwise. Furthermore, some statistical and learning techniques often find optimal or close-to-optimal solutions to hard problems. Finally, the empirical approach also automates tedious tasks such as heuristic selection, often resulting in significant savings in design and development time.

We conclude our discussion with a brief mention of some of the shortcomings of statistical and machine learning techniques. First, these techniques rely on the availability of adequate training data, which must be collected by executing programs on the target platform or via simulation. This results in a significant, often intolerable increase in compilation time. Consequently, the use of learning-based adaptive compilation techniques has been restricted to environments such as embedded systems, where large compilation times are acceptable. Second, most learning techniques are heavily parameterized and cannot be used out of the box. For instance, the efficiency of a GA relies heavily on a number of tunable parameters such as the initial population size, the fraction of population selected for mutation and cross-over, and the maximum number of generations the GA is allowed to run. Tuning these parameters often requires a certain amount of trial and error. Addressing these shortcomings is critical for the feasibility and wider acceptability of adaptive compilers.

Appendix A: Optimality in Program Compilation

In the absolute sense, a program P_{abs} is said to be *optimal* if there exists no semantically equivalent program P' that executes faster than P_{abs} for any input of the program. Therefore, given an arbitrary program, we would ideally like an optimizing compiler to generate the corresponding optimal program. It turns out that this notion of optimality is only of theoretical interest and cannot be realized in practice. In fact, there are simple programs for which an equivalent optimal program does not exist [63]. For instance, consider programs that contain loops with branches conditioned on input variables. The execution paths of these programs clearly depend on the input data. For such programs, one requires the notion of an unbounded speculative window to express or execute an equivalent optimal program.

Since this notion of optimality is not practical, it is relaxed by restricting our reasoning to the performance of the program on a given input [49]. Formally, a program P_I can be considered optimal if there exists no semantically equivalent program P' that executes faster for a given input I. Note that although optimality is defined in terms of the program's performance over a specific input, the optimal program is still required to execute correctly for all other inputs. It turns out that even this notion of optimality is hard to achieve because it requires reasoning over a large (potentially infinite) number of possible compiler optimizations (those that exist or may exist in the future). Therefore, it makes practical sense to restrict the notion of optimality to a given set of compiler optimizations. Thus, from the perspective of program compilation, a program P_{MI} is considered optimal [49] if there exists no semantically equivalent program that executes faster than P_{MI} for a given input data set I, and P_{MI} is obtained using compiler optimizations from some finite set M.

References

1. Kamen Yotov, Xiaoming Li, Gang Ren, Michael Cibulskis, Gerald DeJong, Maria Garzaran, David Padua, Keshav Pingali, Paul Stodghill, and Peng Wu. 2003. A comparison of empirical and model-driven optimization. In *Proceedings of the Conference on Programming Languages Design and Implementation.*

2. L. Almagor, Keith D. Cooper, Alexander Grosul, Timothy J. Harvey, Steven W. Reeves, Devika Subramanian, Linda Torczon, and Todd Waterman. 2004. Finding effective compilation sequences. In *Proceedings of the Symposium on Languages, Compilers and Tool Support for Embedded Systems.*

3. M. Haneda, P. M. W. Knijnenburg, and H. A. G. Wijshoff. 2005. Automatic selection of compiler options using non-parametric inferential statistics. In *Proceedings of the International Conference on Parallel Architectures and Compilation Techniques.*

4. Kapil Vaswani, P. J. Joseph, Matthew J. Thazhuthaveetil, and Y. N. Srikant. 2007. Microarchitecture sensitive empirical models for compiler optimizations. In *Proceedings of the International Symposium on Code Generation and Optimization.*

5. G. J. Chaitin, M. A. Auslander, A. K. Chandra, J. Cocke, M. E. Hopkins, and P. W. Markstein. 1981. Register allocation via coloring. *Computer Languages* 6:47–57.

6. Florent Bouchez, Alain Darte, Fabrice Rastello, and Christophe Guillon. 2006. Register allocation: What does the np-completeness proof of Chaitin et al. really prove? In *Workshop on Duplicating, Deconstructing, and Debunking.*

7. D. Whitfield and M. L. Soffa. 1990. An approach to ordering optimizing transformations. In *Proceedings of the ACM SIGPLAN Symposium on Principles & Practice of Parallel Programming.*

8. Deborah L. Whitfield and Mary Lou Soffa. 1997. An approach for exploring code improving transformations. *ACM Transactions on Programming Languages and Systems* 19(6):1053–84.

9. B. R. Childers, M. L. Soffa, and M. Zhao. 2005. A model-based framework: An approach for profit-driven optimization. In *Proceedings of the International Symposium on Code Generation and Optimization.*

10. D. Bernstein and D. Goldin. 1989. Spill code minimization techniques for optimizing compilers. In *Proceedings of the Conference on Programming Language Design and Implementation,* 258–63.

11. B. R. Rau. 1994. Iterative modulo scheduling: An algorithm for software pipelining loops. In *Proceedings of the International Symposium on Microarchitecture.*

12. Cliff Click and Keith D. Cooper. 1995. Combining analyses, combining optimizations. *ACM Transactions on Programing Languages and Systems* 17(2):181–96.

13. Sorin Lerner, David Grove, and Craig Chambers. 2002. Composing dataflow analyses and transformations. In *Proceedings of the International Symposium on Principles of Programming Languages,* 270–82.

14. Rajeev Motwani, Krishna V. Palem, Vivek Sarkar, and Salem Reyen. 1995. Combining register allocation and instruction scheduling. Technical Report TR 698, Courant Institute.

15. Michael E. Wolf, Dror E. Maydan, and Ding-Kai Chen. 1996. Combining loop transformations considering caches and scheduling. In *Proceedings of the 29th Annual ACM/IEEE International Symposium on Microarchitecture,* 274–86.

16. Rajiv Gupta and Rastislav Bodik. 1999. Register pressure sensitive redundancy elimination. In *Proceedings of the International Conference on Compiler Construction,* 107–21. Heidelberg, Germany: Springer-Verlag.

17. Shlomit S. Pinter. 1993. Register allocation with instruction scheduling: A new approach. In *ACM SIGPLAN Conference on Programming Language Design and Implementation,* 248–57.

18. David G. Bradlee, Susan J. Eggers, and Robert R. Henry. 1991. Integrating register allocation and instruction scheduling for RISCS. *ACM SIGOPS Operating Systems Review* 25, 122–131.

19. R. Silvera, J. Wang, G. Gao, and R. Govindarajan. 1997. A register pressure sensitive instruction scheduler for dynamic issue processors. In *Proceedings of the International Conference on Parallel Architectures and Compilation Techniques,* 78–89.

20. Madhavi Gopal Valluri and R. Govindrajan. 1999. Evaluating register allocation and instruction scheduling techniques in out-of-order issue processors. In *Proceedings of the International Conference on Parallel Architectures and Compilation Techniques*.

21. Ken Kennedy and Randy Allen. 2001. *Optimizing compilers for modern architectures: A dependence-based approach*. San Francisco: Morgan Kaufmann.

22. Stephanie Coleman and Kathryn S. McKinley. 1995. Tile size selection using cache organization and data layout. In *Proceedings of the Conference on Programming Language Design and Implementation*, 279–90.

23. Apan Qasem and Ken Kennedy. 2006. Profitable loop fusion and tiling using model-driven empirical search. In *Proceedings of International Conference on Supercomputing*.

24. Keith D. Cooper, Devika Subramaniam, and Linda Torczon. 2002. Adaptive compilers of the 21st century. *Journal of Supercomputing* 23:7–22.

25. D. C. Montgomery. 2001. *Design and analysis of experiments*, 5th ed. New York: Wiley.

26. J. H. Friedman. 1991. Multivariate adaptive regression splines. *Annals of Statistics* 19:1–141.

27. Antoine Monsifrot, Francois Bodin, and Rene Quiniou. 2002. A machine learning approach to the automatic production of compiler heuristics. In *Lecture Notes on Artificial Intellegence*, 41–50, Springer-Verlag.

28. Mark Stephenson and Saman Amarasinghe. 2005. Predicting unroll factors using supervised classification. In *Proceedings of the Symposium on Code Generation and Optimization*.

29. E. Moss, P. Utgoff, J. Cavazos, D. Precup, D. Stefanovi, C. Brodley, and D. Scheeff. 1997. Learning to schedule straight line code. In *Proceedings of Neural Information Processing Symposium*.

30. A. McGovern and E. Moss. 1998. Scheduling straight-line code using reinforcement learning and rollouts. In *Proceedings of Neural Information Processing Symposium*.

31. Mark Stephenson, Saman Amarasinghe, Martin Martin, and Una-May O'Reilly. 2003. Meta optimization: Improving compiler heuristics with machine learning. In *Proceedings of the Symposium on Programming Languages Design and Implementation*.

32. R. Duda, P. Hart, and D. Stork. 2001. *Pattern classification*. New York: Wiley.

33. J. Koza. 1992. *Genetic programming: On the programming of computers by means of natural selection*. Cambridge, MA: MIT Press.

34. J. W. Davidson and S. Jinturkar. 1994. Memory access coalescing: A technique for eliminating redundant memory accesses. In *Proceedings of the Conference on Programming Language Design and Implementation*, 186–95.

35. S. Larsen and S. Amarasinghe. 2000. Exploiting superword level parallelism with multimedia instruction sets. In *Proceedings of the Conference on Programming Language Design and Implementation*, 145–56.

36. J. R. Quinlan. 1987. Simplifying decision trees. *International Journal of Man-Machine Studies* 27(3):221–34.

37. E. Ozer, S. Banerjia, and T. Conte. 1998. Unified assign and schedule: A new approach to scheduling for clustered register file microarchitectures. In *Proceedings of the 27th Annual International Symposium on Microarchitecture*, 308–15.

38. F. C. Chow and J. L. Hennessey (1990). The priority-based coloring approach to register allocation. *ACM Transactions on Programming Languages and Systems*, 501–36.

39. S. A. Mahlke. 1996. Exploiting instruction level parallelism in the presence of branches. PhD thesis, Department of Electrical and Computer Engineering, University of Illinois at Urbana-Champaign.

40. S. A. Mahlke, D. Lin, W. Chen, R. Hank, and R. Bringmann. 1992. Effective compiler support for predicated execution using the hyperblock. In *Proceedings of the International Symposium on Microarchitecture*, Vol. 25, 45–54.

41. Open research compiler. http://ipf-orc.sourceforge.net.

42. Shun Long and Grigori Fursin. 2005. A heuristic search algorithm based on unified transformation framework. In *Workshop on High Performance Scientific and Engineering Computing*.

43. T. Kisuki, P. M. W. Knijnenburg, M. F. P. Boyle, and H. A. G. Wijshoff. 2000. Iterative compilation in program optimization. In *Workshop on Compilers for Parallel Computers*.

44. Keith D. Cooper, Philip J. Schielke, and Devika Subrarnanian. 1999. Optimizing for reduced code space using genetic algorithms. In *Proceedings of the Symposium on Languages, Compilers and Tool Support for Embedded Systems*.

45. Spyridon Triantafyllis, Manish Vachharajani, Neil Vachharajani, and David I. August. 2003. Compiler optimization-space exploration. In *Proceedings of the International Symposium on Code Generation and Optimization*.

46. Prasad Kulkarni, Wankang Zhao, Hwashin Moon, Kyunghwan Cho, David Whalley, Jack Davidson, Mark Bailey, Yunheung Paek, and Kyle Gallivan. Finding effective optimization phase sequences. In *Proceedings of the Symposium on Languages, Compilers and Tool Support for Embedded Systems*.

47. Prasad Kulkarni, Stephen Hines, Jason Hiser, David Whalley, Jack Davidson, and Douglas Jones. 2004. Fast searches for effective optimization phase sequences. In *Proceedings of the Conference on Programming Language Design and Implementation*, 171–82.

48. F. Agakov, E. Bonilla, J. Cavazos, B. Franke, G. Fursin, M. F. P. O'Boyle, J. Thomson, M. Toussaint, and C. K. I. Williams. 2006. Using machine learning to focus iterative compilation. In *Proceedings of the International Symposium on Code Generation and Optimization*.

49. Sid-Ahmed-Ali Touati and Denis Barthou. 2006. On the decidability of phase ordering problem in optimizing compilation. In *Proceedings of the International Conference on Computing Frontiers*.

50. Prasad Kulkarni, David Whalley, Gary Tyson, and Jack Davidson. 2006. In search of near-optimal optimization phase orderings. In *Proceedings of the Conference on Language, Compilers and Tool Support for Embedded Systems*.

51. P. Kulkarni, D. Whalley, G. Tyson, and J. Davidson. 2006. Exhaustive optimization phase order space exploration. In *Proceedings of the International Symposium on Code Generation and Optimization*.

52. Prasad A. Kulkarni, David B. Whalley, Gary S. Tyson, and Jack W. Davidson. 2007. Evaluating heuristic optimization phase order search algorithms. In *Proceedings of the International Symposium on Code Generation and Optimization*.

53. Stuart Russell and Peter Norvig. 1995. *Artificial intelligence: A modern approach*, chapter 4. New York: Prentice Hall.

54. David E. Goldberg. 1989. *Genetic algorithms in search, optimization and machine learning*. Dordrecht, The Netherlands: Kluwer Academic.

55. K. D. Cooper, A. Grosul, T. J. Harvey, S. Reeves, D. Subramanian, L. Torczon, and T. Waterman. 2005. Acme: Adaptive compilation made efficient. In *Proceedings of the Conference on Language, Compilers and Tool Support for Embedded Systems*.

56. J. L. Doob. 1953. *Stochastic processes*. New York: John Wiley & Sons.

57. GNU gcc compiler. http://gcc.gnu.org.

58. J. Cavazos and Michael F. P. O'Boyle. 2006. Method specific dynamic compilation using logistic regression. In *Proceedings of the International Conference on Object Oriented Programming Systems, Languages and Architecture*.

59. John Cavazos, Grigori Fursin, Felix Agakov, Edwin Bonilla, Michael F. P. O'Boyle, and Olivier Temam. 2007. Rapidly selecting good compiler optimizations using performance counters. In *Proceedings of the International Symposium on Code Generation and Optimization*.

60. Myles Hollander and Douglas A. Wolfe. 1999. *Nonparametric statistical methods*. Wiley Series in Probability and Statistics. New York: Wiley.

61. A. S. Hedayat, N. J. A. Sloane, and J. Stufken. 1999. *Orthogonal arrays: The theory and applications*. Springer-Verlag Series in Statistics. Heidelberg, Germany: Springer-Verlag.

62. C. Bishop. 2005. *Neural networks for pattern recognition*. Oxford: Oxford University Press.

63. U. Schwiegelshohn, F. Gasperoni, and K. Ebcioglu. 1991. On optimal parallelization of arbitrary loops. *Journal of Parallel and Distributed Computing* 11(2):130–34.

9

Type Systems: Advances and Applications

Jens Palsberg
and
Todd Millstein

UCLA Computer Science Department,
University of California, Los Angeles, CA
palsberg@cs.ucla.edu
todd@cs.ucla.edu

9.1 Introduction

This chapter is about the convergence of type systems and static analysis. Historically, these two approaches to reasoning about programs have had different purposes. Type systems are developed to catch common kinds of programming errors early in the software development cycle. In contrast, static analyses were developed to automatically optimize the code generated by a compiler. The two fields also have different theoretical foundations: type systems are typically formalized as logical inference systems [61], while static analyses are typically formalized as abstract program executions [20, 46].

Recently, however, there has been a convergence of the objectives and techniques underlying type systems and static analysis [42, 55, 57, 58]. On the one hand, static analysis is increasingly being used for program understanding and error detection, rather than purely for code optimization. For example, the LCLint tool [30] uses static analysis to detect null-pointer dereferences and other common errors in C programs, and it relies on type-system-like program annotations for efficiency and precision. As another example, the Error Detection via Scalable Program Analysis (ESP) tool [21] uses static analysis to detect violations of Application Programming Interface (API) usage protocols, for example, that a file can only be read or written after it has been opened.

On the other hand, type systems have become a mature and widely accepted technology. Programmers write most new software in languages such as C [45], C++ [29], Java [39], and C# [49], which all feature varying degrees of static type checking. For example, the Java type system guarantees that if a program calls a method on some object, at runtime the object will actually have a method of that name, expecting the proper number and kind of arguments. Types are also used in the intermediate languages of compilers and even in assembly languages [51], such as the typed assembly language for x86 called TALx86 [50].

With this success, researchers have been motivated to explore the potential to extend traditional type systems to detect a variety of interesting classes of program errors. This exploration has shown type systems to be a robust approach to static reasoning about programs and their properties. For example, type systems have been used recently to ensure the safety of manual memory management (e.g., [40, 53, 65]), to track and restrict the aliasing relationships among pointers (e.g., [1, 11, 17, 31]), and to ensure the proper interaction of threads in concurrent programs (e.g., [9, 32, 33]).

These new uses of type systems have brought type systems closer to the domain of static analysis, in terms of both objectives and techniques. For example, reasoning about aliasing is traditionally done via a static analysis to compute the set of *may-aliases*, rather than via a type system. As another example, some sophisticated uses of type systems have required making types *flow sensitive* [23, 37], whereby the type of an expression can change at each program point (e.g., a file's type might denote that the file is open at one point but closed at another point). This style of type system has a natural relationship to traditional static analysis, where the set of "flow facts" can change at each program point.

In this chapter, we describe two type systems that have a strong relationship to static analysis. Each of the type systems is a *refinement* of an existing and well-understood type system: the first refines a subset of the Java type system, while the second refines a system of simple types for the lambda calculus. The refinements are done via annotations that refine existing types to specify and check finer-grained properties. Many of the sophisticated type systems mentioned above can be viewed as refinements of existing types and type systems. Such type systems are examples of *type-based analyses* [56]; that is, they assume and leverage the existing type system and they provide information only for programs that type check with the existing type system.

In the following section we describe a type system that ensures a strong form of encapsulation in object-oriented languages. Namely, the analysis guarantees that an object of a class declared `confined` will never dynamically escape the class's scope. Object confinement goes well beyond the guarantees of traditional privacy modifiers such as `protected` and `private`, and it bears a strong relationship to standard static analyses.

Language designers cannot anticipate all of the refinements that will be useful for programmers or all of the ways in which these refinements can be used to practically check programs. Therefore, it is desirable to provide a framework that allows programmers to easily augment a language's type system with new refinements of interest for their applications. In Section 9.3 we describe a representative framework of this kind, supporting programmer-defined *type qualifiers*. A type qualifier is a simple but useful kind of type refinement consisting solely of an uninterpreted "tag." For example, C's `const` qualifier refines an existing type to further indicate that values of this type are not modifiable, and a `nonnull` qualifier could refine a pointer type to further indicate that pointers of this type are never null.

9.2   Types for Confinement

In this section we will use types to ensure that an object cannot escape the scope of its class. Our presentation is based on results from three papers on *confined types* [8, 41, 72].

9.2.1   Background

Object-oriented languages such as Java provide a way of protecting the name of a field but not the contents of a field. Consider the following example.

```
package p;

public class Table {
    private Bucket[] buckets;
    public Object[] get(Object key) { return buckets; }
}

class Bucket {
    Bucket next;
    Object key, val;
}
```

The hash table class `Table` is a public class that uses a package-scoped class `Bucket` as part of its implementation. The programmer has declared the field `buckets` to be private and intends the hash-table-bucket objects to be internal data structures that should not escape the scope of the `Bucket` class. The declaration of `Bucket` as packaged scoped ensures that the `Bucket` class is not visible outside the package p. However, even the combination of a private field and a package-scoped class does *not* prevent `Bucket` objects from being accessible outside the scope of the `Bucket` class. To see why, notice that the public `get` method in class `Table` has body `return buckets;` that provides an array of bucket objects to any client, including clients outside the package p. Any client can now update the array and thereby change the behavior of the hash table.

The example shows how an object reference can leak out of a package. Such leakage is a problem because (a) the object may represent private information such as a private key and (b) code outside the package may update the object, making it more difficult for programmers to reason about the program. The problem stems from a combination of aliasing and side effects. Aliasing occurs when an object is accessible through different access paths. In the above example, code outside the package can access bucket objects and update them.

How can we ensure that an object cannot escape the scope of its class? We will briefly discuss how one can solve the problem using static analysis and then proceed to show a type-based solution.

9.2.2 Static Analysis

Static analysis can be used to determine whether an object can escape the scope of its class. We will explain a whole-program analysis, that is, an approach that requires access to all the code in the application and its libraries.

Assuming that we have the whole program, let U be the set of class names in the program. The basic idea is to statically compute, for each expression e in the program, a subset of U that conservatively approximates the possible values of e. We will call that set the *flow set* for e. For example, if the flow set for e is the set $\{A, B, C\}$, that means the expression e will evaluate to either an A-object, a B-object, or a C-object. Notice that we allow the set to be a conservative approximation; for example, e might never evaluate to a C-object. All we require is that if e evaluates to an X-object, then X is a member of the flow set for e.

Researchers have published many approaches to statically computing flow sets for expressions in object-oriented programs; see, for example, [2, 22, 59, 64, 67] for some prominent and efficient whole-program analyses. For the purposes of this discussion, all we rely on is that flow sets can be computed statically.

Once we have computed flow sets, for each package-scoped class C we can determine whether C ever appears in the flow set for an expression *outside* the package of C. For each class that never appears in flow sets outside its package, we know its objects do not escape its package in this particular program.

The whole-program-analysis approach has several drawbacks:

Bug finding after the program is done: The approach finds bugs *after* the whole program is done. While that is useful, we would like to help the programmer find bugs *while* he or she is writing the program.

No enforcement of discipline: The static analysis does not enforce any discipline on the programmer. A programmer can write crazy code, and the static analysis may then simply report that every object can escape the scope of its class. While that should be a red flag for the programmer, we would like to help the programmer determine which lines of code to fix to avoid some of the problems.

Fragility: The static analysis tends to be sensitive to small changes in the program text. For one version of a program, a static analysis may find no problems with escaping objects, and then after a few lines of changes, suddenly the static analysis finds problems all over the place. We would like to help the programmer build software in a modular way such that changes in one part of the program do not affect other parts of the program.

The type-based approach in the next section has none of these three drawbacks.

The static-analysis approach in this section is one among many static analyses that solve the same or similar problems. For example, researchers have published powerful escape analyses [5–7, 27] some of which can be adapted to the problem we consider in this chapter.

9.2.3 Confined Types

We can use types to ensure that an object cannot escape the scope of its class. We will show an approach for Java that extends Java with the notions of *confined type* and *anonymous method*. The idea is that if we declare a class to be confined, the type system will enforce rules that ensure that an object of the class cannot escape the scope of the class. If a program type checks in the extended type system, an object cannot escape the scope of its class.

Confinement can be enforced using two sets of constraints. The first set of constraints, *confinement rules*, applies to the classes defined in the same package as the confined class. These rules track values of confined types and ensure that they are neither exposed in public members nor widened to nonconfined types. The second kind of constraints, *anonymity rules*, applies to methods inherited by the confined classes, potentially including library code, and ensures that these methods do not leak a reference to the distinguished variable this, which may refer to an object of confined type.

We will discuss the confinement and anonymity rules next and later show how to formalize the rules and integrate them into the Java type system.

9.2.3.1 Confinement Rules

The following confinement rules must hold for all classes of a package containing confined types.

- $C1$: A confined type must not appear in the type of a public (or protected) field or the return type of a public (or protected) method.
- $C2$: A confined type must not be public.
- $C3$: Methods invoked on an expression of confined type must either be defined in a confined class or be anonymous.
- $C4$: Subtypes of a confined type must be confined.
- $C5$: Confined types can be widened only to other confined types.
- $C6$: Overriding must preserve anonymity of methods.

Rule $C1$ prevents exposure of confined types in the public interface of the package, as client code could break confinement by accessing values of confined types through a type's public interface. Rule $C2$ is needed to ensure that client code cannot instantiate a confined class. It also prevents client code from declaring field or variables of confined types. The latter restriction is needed so that code in a confining package will not mistakenly assign objects of confined types to the fields or variables outside that package. Rule $C3$ ensures that methods invoked on an object enforce confinement. In the case of methods defined in the confining package, this ensues from the other confinement rules. Inherited methods defined in another package do not have access to any confined fields, since those are package scoped (rule $C1$). However, an inherited method of confined class may leak the this reference, which is implicitly widened to the method's declaring class. To prevent this, rule $C3$ requires these methods to be anonymous (as explained below). Rule $C4$ prevents the declaration of a public subclass of a confined type. This prevents *spoofing* leaks, where a public subtype defined outside of the confined package is used to access private fields [18]. Rule $C5$ prevents code within confining packages from assigning values of confined types to fields or variables of public types. Finally, rule $C6$ allows us to statically verify the anonymity of the methods that are invoked on expressions of confined types.

9.2.3.2 Anonymity Rule

The anonymity rule applies to inherited methods that may reside in classes outside of the enclosing package. This rule prevents a method from leaking the this reference. A method is *anonymous* if it has the following property.

- $\mathcal{A}1$: The this reference is used only to select fields and as the receiver in the invocation of other anonymous methods.

This prevents an inherited method from storing or returning this as well as using it as an argument to a call. Selecting a field is always safe, as it cannot break confinement because only the fields visible in the current class can be accessed. Method invocation (on this) is restricted to other methods that are anonymous as well. Note that we check this constraint assuming the static type of this, and rule $\mathcal{C}6$ ensures that the actual method invoked on this will also be anonymous. Thus, rule $\mathcal{C}6$ ensures that the anonymity of a method is independent of the result of method lookup.

Rule $\mathcal{C}6$ could be weakened to apply only to methods inherited by confined classes. For instance, if an anonymous method m of class A is overridden in both class B and C, and B is extended by a confined class while C is not, then the method m in B must be anonymous while m of C need not be. The reason is that the method m of C will never be invoked on confined objects, so there is no need for it to be anonymous.

9.2.3.3 Confined Featherweight Java

Confined Featherweight Java (ConfinedFJ) is a minimal core calculus for modeling confinement for a Java-like object-oriented language. ConfinedFJ extends Featherweight Java (FJ), which was designed by Igarashi et al. [43] to model the Java type system. It is a core calculus, as it limits itself to a subset of the Java language with the following five basic expressions: object construction, method invocation, field access, casts, and local variable access. This spartan setting has proved appealing to researchers. ConfinedFJ stays true to the spirit of FJ. The surface differences lie in the presence of class- and method-level visibility annotations. In ConfinedFJ, classes can be declared to be either public or confined, and methods can optionally be declared as anonymous. One further difference is that ConfinedFJ class names are pairs of identifiers bundling a package name and a class name just as in Java.

9.2.3.4 Syntax

Let metavariable L range over class declarations, C, D, E range over a denumerable set of class identifiers, K, M range over constructor and method declarations, respectively, and f and x range over field names and variables (including parameters and the pseudo-variable this), respectively. Let e, d range over expressions and u, v, w range over values.

We adopt FJ notational idiosyncrasies and use an overbar to represent a finite (possibly empty) sequence. We write $\bar{f}$ to denote the sequence $f_1, \ldots, f_n$ and similarly for $\bar{e}$ and $\bar{v}$. We write $\overline{C f}$ to denote $C_1 f_1, \ldots C_n f_n$, $\bar{C} <: \bar{D}$ to denote $C_1 <: D_1, \ldots, C_n <: D_n$, and finally $\text{this}.\bar{f} = \bar{f}$ to denote $\text{this}.f_1 = f_1, \ldots, \text{this}.f_n = f_n$.

The syntax of ConfinedFJ is given in Figure 9.1. An expression e can be either one of a variable x (including this), a field access $e.f$, a method invocation $e.m(\bar{e})$, a cast $(C)\, e$, or an object $\text{new}\, C(\bar{e})$. Since ConfinedFJ has a call-by-value semantics, it is expedient to add a special syntactic form for fully evaluated objects, denoted $\text{new}\, C(\bar{v})$.

Class identifiers are pairs $p.q$ such that p and q range over denumerable disjoint sets of names. For ConfinedFJ class name $p.q$, p is interpreted as a *package name* and q as a *class name*. In ConfinedFJ, class identifiers are fully qualified. For a class identifier C, *packof*(C) denotes the identifier's package prefix, so, for example, the value of *packof*$(p.0)$ is p.

$$
\begin{aligned}
C &::= \text{p.q} \\
L &::= [\text{public}|\text{conf}]\, \text{class}\ C \lhd D\ \{\overline{Cf};\ K\ \bar{M}\} \\
K &::= C(\overline{C\,f})\ \{\text{super}(\bar{f});\ \text{this}.\bar{f} = \bar{f};\ \} \\
M &::= [\text{anon}]\ C\ m(\overline{C\ x})\ \{\text{return};\ \} \\
e &::= x\ |\ e.f\ |\ e.m(\bar{e})\ |\ (C)\ e\ |\ \text{new}\, C(\bar{e}) \\
v &::= \text{new}\, C(\bar{v})
\end{aligned}
$$

FIGURE 9.1 ConfinedFJ: Syntax.

Each class declaration is annotated with one of the visibility modifiers `public`, `conf`, or none; a public class is declared by `public class C ◁ D {...}`, a package-scoped, confined class is `conf class C ◁ D{...}`, and a package-scoped, nonconfined class is `class C ◁ D{...}`. Methods can be annotated with the optional `anon` modifier to denote anonymity.

We will not formalize the dynamic semantics of ConfinedFJ (for full details, see [75]). We assume a class table *CT* that stores the definitions of all classes of the ConfinedFJ program such that *CT*(C) is the definition of class C. The subtyping relation C <: D denotes that class C is a subtype of class D; <: is the smallest reflexive and transitive class ordering that has the property that if C extends D, then C <: D. Every class is a subtype of `1.Object`. The function *fields*(C) returns the list of all fields of the class C including inherited ones; *methods*(C) returns the list of all methods in the class C; *mdef*(m) returns the identifier of defining class for the method m.

9.2.3.5　Type Rules

Figure 9.2 defines relations used in the static semantics. The predicate *conf*(C) holds if the class table maps C to a class declared as confined. Similarly, the predicate *public*(C) holds if the class table maps C to a class declared as public. The function *mtype*(m, C) yields the type signature of a method. The predicate *override*(m, C, D) holds if m is a valid, anonymity-preserving redefinition of an inherited method or if this is the method's original definition. Class visibility, written *visible*(C, D), states that a class C is visible from D if C is public or if both classes are in the same package.

The *safe subtyping* relation, written C ⪯ D, is a confinement-preserving restriction of the subtyping relation <:. A class C is a safe subtype of D if C is a subtype of D and either C is public or D is confined. This relation is used in the typing rules to prevent widening a confined type to a public type; confinement-preserving widening requires safe subtyping to hold. The type system further constrains subtyping by enforcing that all subclasses of a confined class must belong to the same package (see the T-CLASS rule and the definition of visibility in Figure 9.4). Notice that safe subtyping is reflexive and transitive.

Figure 9.3 defines constraints imposed on anonymous methods. A method m is anonymous in class C, written *anon*(m, C), if its declaration is annotated with the `anon` modifier. The following syntactic restrictions are imposed on the body of an anonymous method. An expression e is anonymous in class C, written *anon*(e, C), if the pseudo-variable `this` is used solely for field selection and anonymous method invocation. (C) e is anonymous if e is anonymous. `new C(ē)` and `e.m(ē)` are anonymous if e ≠ `this` and

Confined types, type visibility, and safe subtyping:

$$\frac{CT(\text{C}) = \text{conf class C} \triangleleft \text{D}\{\ldots\}}{conf\,(\text{C})}$$

$$\frac{public\,(\text{C})}{visible\,(\text{C}, \text{D})} \qquad \frac{packof\,(\text{C}) = packof(\text{D})}{visible(\text{C}, \text{D})}$$

$$\frac{\text{C} <: \text{D}\ conf(\text{C}) \Rightarrow conf\,(\text{D})}{\text{C} \preceq \text{D}}$$

Method type lookup:

$$\frac{mdef\,(\text{m}, \text{C}) = \text{D}\ [\text{anon}]\ \text{B m}(\overline{\text{B x}}\,\{\text{returne; }\}) \in methods\,(\text{D})}{mtype\,(\text{m}, \text{C}) = \overline{\text{B}} \rightarrow \text{B}}$$

Valid method overriding:

$$\frac{\text{either m is not defined in D or any of its parents, or}}{mtype\,(\text{M}, \text{C}) = \overline{\text{C}} \rightarrow \text{C}_0 \quad mtype\,(\text{m}, \text{D}) = \overline{\text{C}} \rightarrow \text{C}_0\ (anon(\text{m}, \text{D}) \Rightarrow anon\,(\text{m}, \text{C}))}{override\,(\text{m}, \text{C}, \text{D})}$$

FIGURE 9.2　ConfinedFJ: Auxiliary definitions.

Anonymous method:

$$\frac{mdef\,(m, C_0) = C_0'\quad \text{anon } C\ m\ (\overline{C\,x}\ \{\ldots\}) \in methods\,(C_0')}{anon(m, C_0)}$$

Anonymity constraints:

$$\frac{anon\,(e, C)}{anon\,((C')\,e, C)}\quad \frac{anon\,(\overline{e}, C)}{anon\,(\text{new } C'\,(\overline{e}),\, C)}\quad \frac{x \neq \text{this}}{anon\,(x, C)}$$

$$\frac{anon\,(e, C)}{anon\,(e, f, C)}\quad \frac{anon\,(e, C)\ \ anon\,(\overline{e}, C)}{anon\,(e.m(\overline{e}), C)}$$

$$\frac{}{anon(\text{this.f}, C)}\quad \frac{anon(m, C)\ \ anon\,(\overline{e}, C)}{anon\,(\text{this.m}(\overline{e}), C)}$$

FIGURE 9.3 ConfinedFJ: Syntactic anonymity constraints.

$e, \overline{e}$ are anonymous. With the exception of `this` all variables are anonymous. `this.f` is always anonymous, and `this.m`$(\overline{e})$ is anonymous in C if m is anonymous in C and $\overline{e}$ is anonymous. We write $anon(\overline{e}, C)$ to denote that all expressions in $\overline{e}$ are anonymous.

9.2.3.6 Expression Typing Rules

The typing rules for ConfinedFJ are given in Figure 9.4, where type judgments have the form $\Gamma \vdash e : C$, in which Γ is an environment that maps variables to their types. The main difference with FJ is that these rules disallow unsafe widening of types. This is captured by conditions of the form $C \preceq D$ that enforce safe subtyping:

- Rules T-VAR and T-FIELD are standard.
- Rule T-NEW prevents instantiating an object if any of the object's fields with a public type is given a confined argument. That is, for fields with declared types $\overline{D}$ and argument types $\overline{C}$, relation $\overline{C} \preceq \overline{D}$ must hold. By definition of $C_i \preceq D_i$, if C_i is confined, then D_i is confined as well.
- Rule T-INVK prevents widening of confined arguments to public parameters by enforcing safe subtyping of argument types with respect to parameter types. To prevent implicit widening of the receiver, we consider two cases. Assume that the receiver has type C_0 and the method m is defined in D_0; then it must be the case either that C_0 is a safe subtype of D_0 or that m has been declared anonymous in D_0.
- Rule T-UCAST prevents casting a confined type to a public type. Notice that a down cast preserves confinement because by rule T-CLASS a confined class can only have confined subclasses.

9.2.3.7 Typing Rules for Methods and Classes

Figure 9.4 also gives rules for typing methods and classes:

- Rule T-METHOD places the following constraints on a method m defined in class C_0 with body e. The type D of e must be a safe subtype of the method's declared type C. The method must preserve anonymity declarations. If m is declared anonymous, e must comply with the corresponding restrictions. The most interesting constraint is the visibility enforced on the body by $\Gamma \vdash visible(e, C_0)$, which is defined recursively over the structure of terms. It ensures that the types of all subexpressions of e are visible from the defining class C_0. In particular, the method parameters used in the method body e must have types visible in C_0.
- Rule T-CLASS requires that if class C extends D, then D must be visible in C, and if D is confined, then so is C. Rule T-CLASS allows the fields of a class C to have types not visible in C, but the constraint of $\Gamma \vdash visible(e, C)$ in rule T-METHOD prohibits the method of C from accessing such fields.

Expression typing:

$$\Gamma \vdash \mathtt{x} : \Gamma(\mathtt{x}) \qquad\qquad \text{(T-VAR)}$$

$$\frac{\Gamma \vdash \mathtt{e} : \mathtt{C} \; \mathit{fields}(\mathtt{C}) = (\overline{\mathtt{C}\,\mathtt{f}})}{T \vdash \mathtt{e.f_i} : \mathtt{C_i}} \qquad\qquad \text{(T-FIELD)}$$

$$\frac{\begin{array}{c} \Gamma \vdash \mathtt{e} : \mathtt{C_0} \quad \Gamma \vdash \overline{\mathtt{e}} : \overline{\mathtt{C}} \quad \mathit{mtype}(\mathtt{m,C_0}) = \overline{\mathtt{D}} \to \mathtt{C} \quad \overline{\mathtt{C}} \prec \overline{\mathtt{D}} \\ \mathit{mdef}(\mathtt{m,C_0}) = \mathtt{D_0} \; (\mathtt{C_0} \preceq \mathtt{D_0} \vee \mathit{anon}(\mathtt{m,D_0})) \end{array}}{\Gamma \vdash \mathtt{e.m}(\overline{\mathtt{e}}) : \mathtt{C}} \qquad\qquad \text{(T-INVK)}$$

$$\frac{\mathit{fields}(\mathtt{C}) = (\overline{\mathtt{D}\,\mathtt{f}}) \quad \Gamma \vdash \overline{\mathtt{e}} : \overline{\mathtt{C}} \quad \overline{\mathtt{C}} \prec \overline{\mathtt{D}}}{\Gamma \vdash \mathtt{new\,C}(\overline{\mathtt{e}}) : \mathtt{C}} \qquad\qquad \text{(T-NEW)}$$

$$\frac{\Gamma \vdash \mathtt{e} : \mathtt{D} \quad \mathit{conf}(\mathtt{D}) \Rightarrow \mathit{conf}(\mathtt{C})}{\Gamma \vdash \mathtt{(C)\,e} : \mathtt{C}} \qquad\qquad \text{(T-UCAST)}$$

Method typing:

$$\frac{\begin{array}{c} \overline{\mathtt{x}} : \overline{\mathtt{C}}, \mathtt{this} : \mathtt{C_0} \vdash \mathtt{e} : \mathtt{D} \quad \mathtt{D} \preceq \mathtt{C} \quad \mathit{override}(\mathtt{m,C_0,D_0}) \\ \overline{\mathtt{x}} : \overline{\mathtt{C}}, \mathtt{this} : \mathtt{C_0} \vdash \mathit{visible}(\mathtt{e,C_0}) \; (\mathit{anon}(\mathtt{m,C_0}) \Rightarrow \mathit{anon}(\mathtt{e,C_0})) \end{array}}{[\mathtt{anon}]\,\mathtt{C}\,\mathtt{m}(\overline{\mathtt{C}\,\mathtt{x}})\,\{\mathtt{returne;}\}\,\text{OK IN } \mathtt{C_0} \lhd \mathtt{D_0}} \qquad \text{(T-METHOD)}$$

Class typing:

$$\frac{\begin{array}{c} \mathit{fields}(\mathtt{D}) = (\overline{\mathtt{D}\,\mathtt{g}}) \quad \mathtt{K} = \mathtt{C}(\overline{\mathtt{D}\,\mathtt{g}}, \; \overline{\mathtt{C}\,\mathtt{f}})\,\{\mathtt{super}(\overline{\mathtt{g}}); \mathtt{this.\overline{f}} = \overline{\mathtt{f}};\} \\ \mathit{visible}(\mathtt{D,C}) \; (\mathit{conf}(\mathtt{D}) \Rightarrow \mathit{conf}(\mathtt{C})) \; \overline{\mathtt{M}} \, \text{OK IN } \mathtt{C} \lhd \mathtt{D} \end{array}}{[\mathtt{public}|\mathtt{conf}]\,\mathtt{class}\,\mathtt{C} \lhd \mathtt{D}\,\{\overline{\mathtt{C}\,\mathtt{f}}; \mathtt{K}\,\overline{\mathtt{M}}\}\,\text{OK}} \qquad \text{(T-CLASS)}$$

Static expression visibility:

$$\frac{\mathit{visible}(\Gamma(\mathtt{x}),\mathtt{C})}{\Gamma \vdash \mathit{visible}(\mathtt{x,C})} \quad \frac{\Gamma \vdash \mathtt{e.f_i} : \mathtt{C'} \; \mathit{visible}(\mathtt{C',C}) \quad \Gamma \vdash \mathit{visible}(\mathtt{e,C})}{\Gamma \vdash \mathit{visible}(\mathtt{e.f_i,C})}$$

$$\frac{\mathit{visible}(\mathtt{C',C}) \quad \Gamma \vdash \mathit{visible}(\mathtt{e,C})}{\Gamma \vdash \mathit{visible}((\mathtt{C'})\,\mathtt{e,C})} \quad \frac{\mathit{visible}(\mathtt{C',C}) \quad \forall i, \Gamma \vdash \mathit{visible}(e_i,C)}{\Gamma \vdash \mathit{visible}(\mathtt{new\,C'}(\overline{\mathtt{e}}),\mathtt{C})}$$

$$\frac{\Gamma \vdash \mathtt{e.m}(\overline{\mathtt{e}}) : \mathtt{C'} \quad \mathit{visible}(\mathtt{C',C}) \quad \Gamma \vdash \mathit{visible}(\mathtt{e,C}) \quad \forall i, \Gamma \vdash \mathit{visible}(\mathtt{e_i,C})}{\Gamma \vdash \mathit{visible}(\mathtt{e.m}(\overline{\mathtt{e}}),\mathtt{C})}$$

FIGURE 9.4 ConfinedFJ: Typing rules.

The class table *CT* is well-typed if all classes in *CT* are well-typed. For the rest of this paper, we assume *CT* to be well-typed.

9.2.3.8 Relation to the Informal Rules

We now relate the confinement and anonymity rules with the ConfinedFJ type system. The effect of rule $\mathcal{C}1$, which limits the visibility of fields if their type is confined, is obtained as a side effect of the visibility constraint as it prevents code defined in another package from accessing a confined field. ConfinedFJ could be extended with a field and method access modifier without significantly changing the type system. The expression typing rules enforce confinement rules $\mathcal{C}3$ and $\mathcal{C}5$ by ensuring that methods invoked on an object of confined type are either anonymous or defined in a confined class and that widening is confinement preserving. Rule $\mathcal{C}2$ uses access modifiers to limit the use of confined types, and the same effect is achieved by the visibility constraint $\Gamma \vdash \mathit{visible}(\mathtt{e,C})$ on the expression part of T-METHOD. Rule

$C4$, which states that subclassing is confinement preserving, is enforced by T-CLASS. Rule $C6$, which states that overriding is anonymity preserving, is enforced by T-METHOD. Finally, the anonymity constraint of rule $A1$ is obtained by the *anon* predicate in the antecedent of T-METHOD.

9.2.3.9 Two ConfinedFJ Examples

Consider the following stripped-down version of a hash table class written in ConfinedFJ. The hash table is represented by a class p.Table defined in some package p that holds a single bucket of class p.Buck. The bucket can be obtained by calling the method get() on a table, and the bucket's data can then be obtained by calling getData(). In this example, buckets are confined, but they extend a public class p.Cell. The interface of p.Table.get() specifies that the method's return type is p.Cell; this is valid, as that class is public. In this example a factory class, named p.Factory, is needed to create instances of p.Table. because the table's constructor expects a bucket and since buckets are confined, they cannot be instantiated outside of their defining package.

```
class p.Table ◁ l.Object {
    p.Buck buck;
    Table(p.Buck buck) { super(); this.buck = buck; }
    p.Cell get() { return this.buck; }
}

class p.Cell ◁ l.Object {
    l.Object data;
    l.Object getData() { return this.data; }
}

conf class p.Buck ◁ p.Cell {
    p.Buck() { super(); }
}

class p.Factory ◁ l.Object {
    p.Factory() { super(); } }
    p.Table table() { return new p.Table( new p.Buck() ); }
}
```

This program does not preserve confinement as the body of the p.Table.get() method returns an instance of a confined class in violation of the widening rule. The breach can be exhibited by constructing a class o.Breach in package o that creates a new table and retrieves its bucket.

```
class o.Breach ◁ l.Object {
    l.Object main () { return new p.Factory().table().get(); }
}
```

The expression new o.Breach().main() eventually evaluates to new p.Buck(), exposing the confined class to code defined in another package. This example is not typable in the ConfinedFJ type system. The method p.Table.get() does not type-check because rule T-METHOD requires the type of the expression returned by the method to be a safe subtype of the method's declared return type. The expression has the confined type p.Buck, while the declared return type is the public type p.Cell.

In another prototypical breach of confinement, consider the following situation in which the confined class p.Self extends a Broken parent class that resides in package o. Assume further that the class inherits its parent's code for the reveal() method.

```
conf class p.Self ⊲ o.Broken {
    p.Self() { super(); }
}

class p.Main ⊲ l.Object {
    p.Main() { super(); }
    l.Object get() { return new p.Self().reveal(); }
}
```

Inspection of this code does not reveal any breach of confinement, but if we widen the scope of our analysis to the o.Broken class, we may see

```
class o.Broken ⊲ l.Object {
    o.Broken() { super(); }
    l.Object reveal() { return this; }
}
```

Invoking reveal() on an instance of p.Self will return a reference to the object itself. This does not type-check because the invocation of reveal() in p.Main.get() violates the rule T-INVK (because the non-anonymous method reveal(), inherited from a public class o.broken, is invoked on an object of a confined type p.Self). The method reveal() cannot be declared anonymous, as the method returns this directly.

9.2.3.10   Type Soundness

Zhao et al. [72] presented a small-step operational semantics of ConfinedFJ, which is a computation-step relation $P \to P'$ on program states P, P'. They define that a program state *satisfies confinement* if every object is in the scope of its defining class. They proceed to prove the following type soundness result (for a version of ConfinedFJ without downcast).

Theorem (confinement) [72]: If P is well-typed, satisfies confinement, and $P \to^* P'$, then P' satisfies confinement.

The confinement theorem states that a well-typed program that initially satisfies confinement preserves confinement. Intuitively, this means that during the execution of a well-typed program, all the objects that are accessed within the body of a method are visible from the method's defining package. The only exception is for anonymous methods, as they may have access to this, which can evaluate to an instance of a class confined in another package, and if this occurs the use of this is restricted to be a receiver object.

Confined types have none of the three drawbacks of whole-program static analysis: we can type-check fragments of code well before the entire program is done, the type system enforces a discipline that can help make many types confined, and a change to a line of code only affects types locally.

9.2.3.11   Confinement Inference

Every type-correct FJ program can be transformed into a type-correct ConfinedFJ program by putting all the classes into the same package. Conversely, every type-correct ConfinedFJ program can be transformed into a type-correct Java program by removing all occurrences of the modifiers conf and anon. (The original version of FJ does not have packages.)

The modifiers conf and anon can help enforce more discipline than Java does. If we begin with a program in FJ extended with packages and want to enforce the stricter discipline of ConfinedFJ, we face what we call the *confinement inference* problem.

> **The confinement inference problem:** Given a Java program, find a subset of the package-scoped classes that we can make confined and find a subset of the methods that we can make anonymous.

The confinement inference problem has a trivial solution: make no classes confined and make no method anonymous. In practice we may want the largest subsets we can get.

Grothoff et al. [41] studied confinement inference for a variant of the confinement and anonymity rules in this chapter. They used a constraint-based program analysis to infer confinement and method anonymity. Their constraint-based analysis proceeds in two steps: (a) generate a system of constraints from program text and then (b) solve the constraint system. The constraints are of the following six forms:

$$A ::= \text{not-anon(methodId)}$$
$$T ::= \text{not-conf(classId)}$$
$$C ::= A \mid T \mid T \Rightarrow A \mid A \Rightarrow A \mid A \Rightarrow T \mid T \Rightarrow T$$

A constraint not-anon(methodId) asserts that the method methodId is *not* anonymous; similarly, not-conf(classId) asserts that the class classId is *not* confined. The remaining four forms of constraints denote logical implications. For example, not-anon(A.m()) $\Rightarrow$ not-conf(C) is read "if method m in class A is not anonymous, then class C will not be confined."

From each expression in a program, we generate one or more constraints. For example, for a type cast expression (C) e for which the static Java type of e is D, we generate the constraint not-conf(C) $\Rightarrow$ not-conf(D), which represents the condition from the T-UCAST rule that *conf*(D) $\Rightarrow$ *conf*(C).

All the constraints are ground Horn clauses. The solution procedure computes the set of clauses not-conf(classId) that are either immediate facts or derivable via logical implication. This computation can be done in linear time [28] in the number of constraints, which, in turn, is linear in the size of the program.

A solution represents a set of classes that cannot be confined and a set of methods that are not anonymous. The complements of those sets represent a maximal solution to the confinement inference problem.

Grothoff et al. [41] presented an implementation of their constraint-based analysis. They gathered a suite of 46,000 Java classes and analyzed them for confinement. The average time to analyze a class file is less than 8 milliseconds. The results show that, without any change to the source, 24% of the package-scoped classes (exactly 3,804 classes, or 8% of all classes) are confined. Furthermore, they found that by using generic container types, the number of confined types could be increased by close to 1,000 additional classes. Finally, with appropriate tool support to tighten access modifiers, the number of confined classes can be well over 14,500 (or over 29% of all classes) for that same benchmark suite.

9.2.4 Related Work on Alias Control

The type-based approach in this chapter is one among many type-based approaches that solve the same or similar problems. For example, a popular approach is to use a notion of ownership type [1, 3, 4, 9, 10, 11, 16, 19, 26, 52, 62] for controlling aliasing. The basic idea of ownership types is to use the concept of *domination* on the dynamic object graph. (In a graph with a starting vertex s, a vertex u dominates another vertex v if every path from s to v must pass through u.) In a dynamic object graph, we may have an object we think of as *owning* several *representation* objects. The goal of ownership types is to ensure that the owner object dominates the representation objects. The dominance relation guarantees that the only way we can access a representation object is via the owner. An ownership type system has type rules that are quite different than the rules for confined types.

9.3   Type Qualifiers

In this section we will use types to allow programmers to easily specify and check desired properties of their applications. This is achieved by allowing programmers to introduce new *qualifiers* that refine existing types. For example, the type `nonzero int` is a refinement of the type `int` that intuitively denotes the subset of integers other than zero.

9.3.1   Background

Static type systems are useful for catching common programming errors early in the software development cycle. For example, type systems can ensure that an integer is never accidentally used as a string and that a function is always passed the right number and kinds of arguments. Unfortunately, language designers cannot anticipate all of the program errors that programmers will want to statically detect, nor can they anticipate all of the practical ways in which such errors can be detected.

As a simple example, while most type systems in mainstream programming languages can distinguish integers from strings and ensure that each kind of data is used in appropriate ways, these type systems typically cannot distinguish positive from negative integers. Such an ability would enable stronger assurances about a program, for example, that it never attempts to take the square root of a negative number. As another example, most type systems cannot distinguish between data that originated from one source and data that originated from a different source within the program. Such an ability could be useful to track a form of *value flow*, for example, to ensure that a string that was originally input from the user is treated as *tainted* and therefore given restricted capabilities (e.g., such a string should be disallowed as the format-string argument to C's `printf` function, since a bad format string can cause program crashes and worse).

Without static checking for these and other kinds of errors, programmers have little recourse. They can use `assert` statements, which catch errors, but only as they occur in a running system. They can specify desired program properties in comments, which are useful documentation but need have no relation to the actual program behavior. In the worst case, programmers simply leave the desired program properties completely implicit, making these properties easy to misunderstand or forget entirely.

9.3.2   Static Analysis

Static analysis could be used to ensure desired program properties and thereby guarantee the absence of classes of program errors. Indeed, generic techniques exist for performing static analyses of programs (e.g., [20, 46]), which could be applied to the properties of interest to programmers. As with confinement, one standard approach is to compute a flow set for each expression e in the program, which conservatively overapproximates the possible values of e. However, instead of using class names as the elements of a flow set, each static analysis defines its own domain of *flow facts*.

For example, to track positive and negative integers, a static analysis could use a domain of signs [20], consisting of the three elements $+$, 0, and $-$ with the obvious interpretations. If the flow set computed for an expression e contains only the element $+$, we can be sure that e will evaluate to a positive integer. In our format-string example, a static analysis could use a domain consisting of the elements `tainted` and `untainted`, representing, respectively, data that do and do not come from the user. If the flow set computed for an expression e contains only the element `untainted`, we can be sure that e does not come from the user.

While this approach is general, it suffers from the drawbacks discussed in Section 9.2.2. First, whole-program analysis is typically required for precision, so errors are only caught once the entire program has been implemented. Second, the static analysis is *descriptive*, reporting the properties that are true of

$$\tau ::= \texttt{int} \mid \tau \to \tau$$
$$e ::= n \mid e_1 + e_2 \mid x \mid \lambda x : \tau.e \mid e_1 e_2$$

FIGURE 9.5   The syntax of the simply typed lambda calculus.

a given program, rather than *prescriptive*, providing a discipline to help programmers achieve the desired properties. Finally, the results of a static analysis can be sensitive to small changes in the program.

The type-based approach described next is less precise than some static analyses but has none of the above drawbacks.

9.3.3   A Type System for Qualifiers

We now develop a type system that supports programmer-defined type qualifiers. After a brief review of the simply typed lambda calculus, types are augmented with user-defined *tags* and language support for *tag checking*. A notion of subtyping for tagged types provides a natural form of type qualifiers. Finally, more expressiveness is achieved by allowing users to provide specialized typing rules for qualifier checking.

9.3.3.1   Simply Typed Lambda Calculus

We assume familiarity with the simply typed lambda calculus and briefly review the portions that are relevant for the rest of the section. Many other sources contain fuller descriptions of the simply typed lambda calculus, for example, the text by Pierce [61].

Figure 9.5 shows the syntax for the simply typed lambda calculus augmented with integers and integer addition. The metavariable τ ranges over types, and e ranges over expressions. The syntax $\tau_1 \to \tau_2$ denotes the type of functions with argument type τ_1 and result type τ_2. The metavariable n ranges over integer constants, and x ranges over variable names. The syntax $\lambda x : \tau.e$ represents a function with formal parameter x (of type τ) and body e, and the syntax $e_1 e_2$ represents application of the function expression e_1 to the actual argument e_2.

Figure 9.6 presents static typechecking rules for the simply typed lambda calculus. The rules define a judgment of the form $\Gamma \vdash e : \tau$. The metavariable Γ ranges over *type environments*, which are finite mappings from variables to types. Informally, the judgment $\Gamma \vdash e : \tau$ says that expression e is *well-typed* with type τ under the assumption that free variables in e have the types associated with them in Γ. The rules in Figure 9.6 are completely standard.

Static type-checking enforces a notion of well-formedness on programs at compile time, thereby preventing some common kinds of runtime errors. For example, the rules in Figure 9.6 ensure that a well-typed expression (with no free variables) will never attempt to add an integer to a function at runtime. A type system's notion of well-formedness is formalized by a type soundness theorem, which specifies the properties

$$\frac{\Gamma \vdash e : \tau}{\Gamma \vdash n : \texttt{int}} \quad \text{(T-Int)}$$

$$\frac{\Gamma \vdash e_1 : \texttt{int} \quad \Gamma \vdash e_2 : \texttt{int}}{\Gamma \vdash e_1 + e_2 : \texttt{int}} \quad \text{(T-Plus)}$$

$$\frac{\Gamma(x) = \tau}{\Gamma \vdash x : \tau} \quad \text{(T-Var)}$$

$$\frac{\Gamma, x : \tau_1 \vdash e : \tau_2}{\Gamma \vdash \lambda x : \tau_1.e : \tau_1 \to \tau_2} \quad \text{(T-Abs)}$$

$$\frac{\Gamma \vdash e_1 : \tau_2 \to \tau \quad \Gamma \vdash e_2 : \tau_2}{\Gamma \vdash e_1 e_2 : \tau} \quad \text{(T-App)}$$

FIGURE 9.6   Static type-checking for the simply typed lambda calculus.

$$\tau ::= q\,\nu$$
$$\nu ::= \texttt{int}\,|\,\tau \rightarrow \tau$$
$$e ::= \ldots\,|\,\texttt{annot}(e,q)\,|\,\texttt{assert}\,(e,q)$$

FIGURE 9.7   Adding user-defined tags to the syntax.

of well-typed programs. Intuitively, type soundness for the simply typed lambda calculus says that the evaluation of well-typed expressions will not "get stuck," which happens when an operation is attempted with operand values of the wrong types.

A type soundness theorem relies on a formalization of a language's evaluation semantics. There are many styles of formally specifying language semantics and of proving type soundness, and common practice today is well described by others [61, 68]. These topics are beyond the scope of this chapter.

9.3.3.2   Tag Checking

One way to allow programmers to easily extend their type system is to augment the syntax for types with a notion of programmer-defined *type tags* (or simply *tags*). The new syntax of types is shown in Figure 9.7. The metavariable q ranges over an infinite set of programmer-definable type tags. Each type is now augmented with a tag. For example, `positive int` could be a type, where `positive` is a programmer-defined tag denoting positive integers. Function types include a top-level tag as well as tags for the argument and result types.

For programmers to convey the intent of a type tag, the language is augmented with two new expression forms, as shown in Figure 9.7. Our presentation follows that of Foster et al. [35, 36]. The expression $\texttt{annot}(e,q)$ evaluates e and tags the resulting value with q. For example, if the expression e evaluates to a string input by the user, one can use the expression `annot(e, tainted)` to declare the intention to consider e's value as tainted [54, 63]. The expression $\texttt{assert}(e,q)$ evaluates e and checks that the resulting value is tagged with q. For example, the expression `assert(e, untainted)` ensures that e's value does not originate from the user and is therefore an appropriate format-string argument to `printf`. A failed `assert` causes the program to terminate erroneously.

Just as our base type system in Figure 9.6 statically tracks the type of each expression, so does our augmented type system, using the augmented syntax of types. The rules are shown in Figure 9.8. For simplicity, the rules are set up so that each runtime value created during the program's execution will have

$$\boxed{\Gamma \vdash e\,:\,\nu}$$

$$\frac{}{\Gamma \vdash n\,:\,\texttt{int}} \qquad \text{(Q-INT)}$$

$$\frac{\Gamma \vdash e_1\,:\,q_1\,\texttt{int} \quad \Gamma \vdash e_2\,:\,q_2\,\texttt{int}}{\Gamma \vdash e_1 + e_2\,:\,\texttt{int}} \qquad \text{(Q-PLUS)}$$

$$\frac{\Gamma, x:\tau_1 \vdash e\,:\,\tau_2}{\Gamma \vdash \lambda x:\tau_1.e\,:\,\tau_1 \rightarrow \tau_2} \qquad \text{(Q-ABS)}$$

$$\boxed{\Gamma \vdash e\,:\,\tau}$$

$$\frac{\Gamma(x) = \tau}{\Gamma \vdash x\,:\,\tau} \qquad \text{(Q-VAR)}$$

$$\frac{\Gamma \vdash e_1\,:\,\tau_2 \rightarrow \tau \quad \Gamma \vdash e_2\,:\,\tau_2}{\Gamma \vdash e_1 e_2\,:\,\tau} \qquad \text{(Q-APP)}$$

$$\frac{\Gamma \vdash e\,:\,\nu}{\Gamma \vdash \texttt{annot}\,(e,q)\,:\,q\,\nu} \qquad \text{(Q-ANNOT)}$$

$$\frac{\Gamma \vdash e\,:\,q\,\nu}{\Gamma \vdash \texttt{assert}\,(e,q)\,:\,q\,\nu} \qquad \text{(Q-ASSERT)}$$

FIGURE 9.8   Adding user-defined tags to the type system.

exactly one tag (a conceptually untagged value can be modeled by tagging it with a distinguished `notag` tag). This invariant is achieved via two interrelated typing judgments. The judgment $\Gamma \vdash e : \nu$ determines an untagged type for a given expression. This judgment is only defined for *constructor expressions*, which are expressions that dynamically create new values. The judgment $\Gamma \vdash e : \tau$ is the top-level type-checking judgment. It is defined for all other kinds of expressions. The Q-ANNOT rule provides a bridge between the two judgments, requiring each constructor expression to be tagged in order to be given a complete type τ.

Intuitively, the type system conservatively ensures that if $\Gamma \vdash e : q\,\nu$ holds, the value of e at run time will be tagged with q. The rules for $\text{annot}(e, q)$ and $\text{assert}(e, q)$ are straightforward: Q-ANNOT includes q as the tag on the type of e, while Q-ASSERT requires that e's type already includes the tag q. The rest of the rules are unchanged from the original simply typed lambda calculus, except that the premises of Q-PLUS allow for the tags on the types of the operands. Nonetheless, these unchanged rules have exactly the desired effect. For example, Q-APP requires the actual argument's type in a function application to match the formal argument type, thereby ensuring that the function only ever receives values tagged with the expected tag.

Together the rules in Figure 9.8 provide a simple form of value-flow analysis, statically ensuring that values of a given tag will flow at runtime only to places where values of that tag are expected. For example, a programmer can define a square-root function of the form

$$\lambda x : \texttt{positive int}.e$$

and the type system guarantees that only values explicitly tagged as `positive` will be passed to the function. As another example, the programmer can statically detect possible division-by-zero errors by replacing each divisor expression e (assuming our language included integer division) with the expression $\text{assert}(e, \texttt{nonzero})$. Finally, the type of the following function has type `tainted int`→`untainted int` which ensures that although the function accepts a tainted integer as an argument, this integer does not flow to the return value:

$$\lambda x : \texttt{tainted int}.\text{annot}(0, \texttt{untainted})$$

However, the following function, which returns the given tainted argument, is forced to record this fact in its type, `tainted int`→`tainted int`:

$$\lambda x : \texttt{tainted int}.x$$

9.3.3.2.1 Type Soundness

The notion of type soundness in the presence of tags is a natural extension of that of the simply typed lambda calculus. Type soundness still ensures that well-typed expressions will not get stuck, but the notion of stuckness now includes failed `assert`s. This definition of stuckness formalizes the idea that tagged values will only flow where they are expected. Type soundness can be proven using standard techniques.

9.3.3.2.2 Tag Inference

It is possible to consider *tag inference* for our language. Constructor expressions are no longer explicitly annotated via `annot`, and formal argument types no longer include tags. Tag inference automatically determines the tag of each constructor expression and the tags on each formal argument or determines that the program cannot be typed. Programmers still must employ `assert` explicitly to specify constraints on where values of particular tags are expected.

As with confinement inference, a constraint-based program analysis can be used for tag inference. Conceptually, each subexpression in the program is given its own *tag variable*, and the analysis then generates equality constraints based on each kind of expression. For example, in a function application, the tag of the actual argument is constrained to match the tag of the formal argument type. The simple equality constraints generated by tag inference can be solved in linear time [60, 66]. Furthermore, if the constraints have a solution, there exists a *principal solution*, which is more general than every other solution. Intuitively, this is the solution that produces the largest number of tags.

For example, consider the following function:

$$\lambda x : \mathtt{int}.\lambda y : \mathtt{int}.\mathtt{assert}(x, \mathtt{tainted})$$

One possible typing for the function gives both x and y the type tainted int. However, a more precise typing gives y's type a fresh tag q_y, since the function's constraints do not require it to have the tag tainted. This new typing encodes that fact as well as the fact that x and y flow to disjoint places in the program. Finally, the following program generates constraints that have no solution, since x is required to be both tainted and untainted:

$$(\lambda x : \mathtt{int}.\mathtt{assert}(x, \mathtt{tainted}))\, \mathtt{assert}(e, \mathtt{untainted})$$

9.3.3.3 Qualifier Checking

While the type system in the previous subsection allows programmers to specify and check new properties of interest via tags, its expressiveness is limited because tags are completely uninterpreted. For example, the type system does not "know" the intent of tags such as positive, nonzero, tainted, and untainted; it only knows that these tags are not equivalent to one another. However, tags often have natural relationships to one another. For example, intuitively it should be safe to pass a positive int where a nonzero int is expected, since a positive integer is also nonzero. Similarly, we may want to allow untainted data to be passed where tainted data is expected, since allowing that cannot cause tainted data to be improperly used. The type system of the previous section does not permit such flexibility.

Foster et al. observed that this expressiveness can be naturally achieved by allowing programmers to specify a partial order $\sqsubseteq$ on type tags [35, 36]. Intuitively, if $q_1 \sqsubseteq q_2$, then q_1 denotes a stronger constraint than q_2. The programmer can now declare positive $\sqsubseteq$ nonzero and, similarly, untainted $\sqsubseteq$ tainted, where untainted denotes the set of values that are definitely untainted, and tainted now denotes the set of values that are *possibly* tainted. The programmer-defined partial order naturally induces a subtyping relation among tagged types. For example, given the above partial order, positive int would be considered a subtype of nonzero int, which therefore allows a value of the former type to be passed where a value of the latter type is expected.

With this added expressiveness, type tags can be considered full-fledged *type qualifiers*. For example, a canonical example of a type qualifier is C's const annotation, which indicates that the associated value can be initialized but not later updated. C allows a value of type int* to be passed where a (const int) * is expected. This is safe because it simply imposes an extra constraint on the given pointer value, namely, that its contents are never updated. However, a value of type (const int) * cannot safely be passed where an int* is expected, since this would allow the pointer value's constness to be forgotten, allowing its contents to be modified. Another useful example qualifier is nonnull for pointers, whereby it is safe to pass a nonnull pointer where an arbitrary pointer is expected, but not vice versa.

Figure 9.9 shows the extension of the rules in the previous subsection to support qualifiers, adapted from [36]. Q-Sub is a *subsumption* rule, which allows an expression's type to be promoted to any supertype.

$$\boxed{\tau \leq \tau'}$$

$$\frac{q \sqsubseteq q'}{q\,\mathtt{int} \leq q'\,\mathtt{int}} \quad \text{(S-INT)}$$

$$\frac{q \sqsubseteq q' \quad \tau_1' \leq \tau_1 \quad \tau_2 \leq \tau_2'}{q(\tau_1 \to \tau_2) \leq q'(\tau_1' \to \tau_2')} \quad \text{(S-FUN)}$$

$$\boxed{\Gamma \vdash e : \tau}$$

$$\frac{\Gamma \vdash e : \tau' \quad \tau' \leq \tau}{\Gamma \vdash e : \tau} \quad \text{(Q-SUB)}$$

FIGURE 9.9   Adding subtyping to the type system.

The subtyping relation $\leq$ depends on the partial order $\sqsubseteq$ among qualifiers in a straightforward way. As usual, subtyping is *contravariant* on function argument types for soundness [13].

As an example of this type system in action, consider an expression e of type `positive int`. Assuming that the programmer specifies `positive` $\sqsubseteq$ `nonzero`, then by S-INT we have `positive int` $\leq$ `nonzero int` and by Q-SUB e we have `nonzero int`. Therefore, by the Q-APP rule from Figure 9.8, e may be passed to a function expecting an argument of type `nonzero int`.

As an aside, the addition of subtyping makes our formal system expressive enough to encode multiple qualifiers per type. For example, to encode a type like `untainted positive int`, one can define a new qualifier, `untainted_positive`, along with the partial-order `untainted_positive` $\sqsubseteq$ `untainted` and `untainted_positive` $\sqsubseteq$ `positive`. Then the subtyping and subsumption rules allow an `untainted_positive` value to be treated as being both `untainted` and `positive`, as desired.

As before, type soundness says that the type system guarantees that all `assert`s will succeed at runtime, where the runtime assertion check now requires a value's associated qualifier to be "less than" the specified qualifier, according to the declared partial order. The type soundness proof again uses standard techniques. It is also possible to generalize tag inference to support *qualifier inference*. The approach is similar to that described above, although the generated constraints are now *subtype constraints* instead of equality constraints.

Foster's thesis discusses type soundness and qualifier inference in detail [34]. It also discusses CQUAL, an implementation of programmer-defined type qualifiers that adapts the described theory to the C language. CQUAL has been used successfully for a variety of applications, including inference of `const`ness [36], detection of format-string vulnerabilities [63], detection of user/kernel pointer errors [44], validation of the placement of authorization hooks in the Linux kernel [71], and the removal of sensitive information from crash reports [12].

9.3.3.4 Qualifier-Specific Typing Rules

The $\sqsubseteq$ partial order allows programmers to specify more information about each qualifier, making the overall type system more flexible. However, most of the intent of a qualifier must still be conveyed indirectly via `annot`s, which is tedious and error prone. For example, the programmer must use `annot` to explicitly annotate each constructor expression that evaluates to a positive integer as being `positive`, or else it will not be considered as such by the type system. Therefore, the programmer has the burden of manually figuring out which expressions are positive and which are not. Furthermore, if the programmer accidentally annotates an expression such as `-34 + 5` as `positive`, the type system will happily allow this expression to be passed to a square-root function expecting a `positive int`, even though that will likely cause a runtime error.

Qualifier inference avoids the need for explicit annotations using `annot`. However, qualifier inference simply determines which expressions must be treated as `positive` to satisfy a program's `assert`s. There is no guarantee that these expressions actually evaluate to positive integers, and many expressions that do evaluate to positive integers will not be found to be `positive` by the inferencer.

To address the burden and fragility of qualifier annotations, we consider an alternate approach to expressing a qualifier's intent. Instead of relying on program annotations, we require qualifier designers to specify a *programming discipline* for each qualifier, which indicates when an expression may be given that qualifier. For example, a programming discipline for `positive` might say that all positive constants can be considered `positive` and that an expression of the form $e_1 + e_2$ can be considered `positive` if each operand expression can itself be considered `positive` according to the discipline. In this way, the discipline declaratively expresses the fact that `34 + 5` can be considered `positive`, while `-34 + 5` cannot.

The approach described is used by the Clarity framework for programmer-defined type qualifiers in C [14]. Clarity provides a declarative language for specifying programming disciplines. For example, Figure 9.10 shows how the discipline informally described above for `positive` would be specified in Clarity. The figure declares a new qualifier named `positive`, which refines the type `int`. It then uses

```
qualifier positive(int Expr E)
   case E of
      decl int Const C:
         C, where C>0
    | decl int Expr E1, E2:
         E1 + E2, where positive(E1) && positive(E2)
```

FIGURE 9.10   A programming discipline for `positive` in Clarity.

pattern matching to specify two ways in which an expression E can be given the qualifier `positive`. The Clarity framework includes an *extensible type-checker*, which employs user-defined disciplines to automatically type-check programs.

Formally, consider the type system consisting of the rules in Figures 9.8 and 9.9. We remove all the rules of the form $\Gamma \vdash e : v$, which perform type-checking on constructor expressions, and we remove the `annot` expression form along with its type-checking rule Q-ANNOT. When a programmer introduces a new qualifier, he or she must also augment the type system with new inference rules indicating the conditions under which each constructor expression may be given this qualifier. For example, the rules in Figure 9.10 are formally represented by adding the following two rules to the type system:

$$\frac{n > 0}{\Gamma \vdash n : \texttt{positive int}} \qquad \text{[P-INT]}$$

$$\frac{\Gamma \vdash e_1 : \texttt{positive int} \quad \Gamma \vdash e_2 : \texttt{positive int}}{\Gamma \vdash e_1 + e_2 : \texttt{positive int}} \qquad \text{(P-PLUS)}$$

Assuming that the programmer also declares `positive` $\sqsubseteq$ `nonzero`, the subtyping and subsumption rules in Figure 9.9 allow the above rules to be used to give the qualifier `nonzero` to an expression as well.

Not all qualifiers have natural rules associated with them. For example, the programming disciplines associated with qualifiers such as `tainted` and `untainted` could be program dependent and/or quite complicated. Therefore, in practice both the Clarity and CQUAL approaches are useful.

9.3.3.4.1   Type Soundness

A type soundness theorem analogous to that for traditional type qualifiers, which guarantees that `asserts` succeed at runtime, can be proven in this setting. In addition, it is possible to prove a stronger notion of type soundness. Clarity allows the programmer to optionally specify the set of values associated with a particular qualifier. For example, the programmer could associate the set of positive integers with the `positive` qualifier. Given this information, type soundness says that a well-typed expression with the qualifier `positive` will evaluate to a member of the specified set.

To ensure this form of type soundness, Clarity generates one *proof obligation* per programmer-defined rule. For example, the second rule for `positive` above requires proving that the sum of two integers greater than zero is also an integer greater than zero. Clarity discharges proof obligations automatically using off-the-shelf decision procedures [25], but in general these may need to be manually proven by the qualifier designer.

This form of type soundness validates the programmer-defined rules. For example, if the second rule for `positive` above were erroneously defined for subtraction rather than addition, the error would be caught because the associated proof obligation is not valid: the difference between two positive integers is not necessarily positive. In this way, programmers obtain a measure of confidence that their qualifiers and associated inference rules are behaving as intended.

9.3.3.4.2   Qualifier Inference

Qualifier inference is also possible in this setting and is implemented in Clarity, allowing the qualifiers for variables to be inferred rather than declared by the programmer. Similar to qualifier inference in the previous subsection, a set of subtype constraints is generated and solved. However, handling programmer-defined inference rules requires a form of *conditional* subtype constraints to be solved [15].

9.3.4   Related Work on Type Refinements

Work on *refinement types* for the ML language allows programmers to create subtypes of data type definitions [38], each denoting a subset of the values of the data type. For example, a standard list data type could be refined to define a type of nonempty lists. The language for specifying these refinements is analogous to the language for programmer-defined inference rules in Clarity.

Other work has shown how to make refinement types and type qualifiers *flow sensitive* [23, 24, 37, 47], which allows the refinement of an expression to change over time. For example, a file pointer could have the qualifier closed upon creation and the qualifier open after it has been opened. In this way, type refinements can be used to track *temporal protocols*, for example, that a file must be opened before it can be read or written.

Finally, others have explored type refinements through the notion of *dependent types* [48], in which types can depend on program expressions. An instance of this approach is Dependent ML [69, 70], which allows types to be refined through their dependence on linear arithmetic expressions. For example, the type int list(5) represents integer lists of length 5, and a function that adds an element to an integer list would be declared to have the argument type int list(n) for some integer n and to return a value of type int list(n+1). These kinds of refinements are targeted at qualitatively different kinds of program properties from those targeted by type qualifiers.

References

1. Jonathan Aldrich, Valentin Kostadinov, and Craig Chambers. 2002. Alias annotations for program understanding. In *Proceedings of the 17th ACM SIGPLAN Conference on Object-Oriented Programming, Systems, Languages, and Applications*, 311–30. New York: ACM Press.

2. David F. Bacon and Peter F. Sweeney. 1996. Fast static analysis of C++ virtual function calls. *SIGPLAN Notices* 31(10): 324–41.

3. Anindya Banerjee and David A. Naumann. 2002. Representation independence, confinement and access control. In *Proceedings of POPL'02, SIGPLAN–SIGACT Symposium on Principles of Programming Languages*, 166–77.

4. Mike Barnett, Robert DeLine, Manuel Fähndrich, K. Rustan M. Leino, and Wolfram Schulte. 2003. Verification of object-oriented programs with invariants. In *Fifth Workshop on Formal Techniques for Java-Like Programs*.

5. Bruno Blanchet. 1999. Escape analysis for object oriented languages. Application to Java. *SIGPLAN Notices* 34(10):20–34.

6. Bruno Blanchet. 2003. Escape analysis for Java: Theory and practice. *ACM Transactions on Programming Languages and Systems* 25(6):713–75.

7. Jeff Bogda and Urs Hölzle. 1999. Removing unnecessary synchronization in Java. *SIGPLAN Notices* 34(10): 35–46.

8. Boris Bokowski and Jan Vitek. 1999. Confined types. In *Proceedings of the Fourteenth Annual Conference on Object-Oriented Programming Systems, Languages, and Applications (OOPSLA'99)*, 82–96.

9. Chandrasekhar Boyapati, Robert Lee, and Martin Rinard. 2002. Ownership types for safe programming: Preventing data races and deadlocks. In *Proceedings of the 17th ACM SIGPLAN Conference on Object-Oriented Programming, Systems, Languages, and Applications*, 211–30. New York: ACM Press.

10. Chandrasekhar Boyapati, Alexandru Salcianu, William Beebee, and Martin Rinard. 2003. Ownership types for safe region-based memory management in real-time Java. In *ACM Conference on Programming Language Design and Implementation*, 324–37.

11. John Boyland. 2001. Alias burying: Unique variables without destructive reads. *Software Practice and Experience*, 31(6):533–53.

12. Pete Broadwell, Matt Harren, and Naveen Sastry. 2003. Scrash: A system for generating secure crash information. In *USENIX Security Symposium*.

13. Luca Cardelli. 1988. A semantics of multiple inheritance. *Information and Computation* 76(2/3): 138–64.

14. Brian Chin, Shane Markstrum, and Todd Millstein. 2005. Semantic type qualifiers. In *PLDI '05: Proceedings of the 2005 ACM SIGPLAN Conference on Programming Language Design and Implementation*, 85–95. New York: ACM Press.

15. Brian Chin, Shane Markstrum, Todd Millstein, and Jens Palsberg. 2006. Inference of user-defined type qualifiers and qualifier rules. In *European Symposium on Programming*.

16. David Clarke. 2001. Object ownership and containment. PhD thesis, School of Computer Science and Engineering, University of New South Wales, Sydney, Australia.

17. David G. Clarke, John M. Potter, and James Noble. 1998. Ownership types for flexible alias protection. In *Proceedings of the 13th ACM SIGPLAN Conference on Object-Oriented Programming, Systems, Languages, and Applications*, 48–64. New York: ACM Press.

18. Dave Clarke, Michael Richmond, and James Noble. 2003. Saving the world from bad beans: Deployment-time confinement checking. In *Proceedings of the ACM Conference on Object-Oriented Programming, Systems, Languages, and Applications (OOPSLA)*, 374–87.

19. David Clarke and Tobias Wrigstad. 2003. External uniqueness. In *10th Workshop on Foundations of Object-Oriented Languages (FOOL)*.

20. Patrick Cousot and Radhia Cousot. 1977. Abstract interpretation: A unified lattice model for static analysis of programs by construction or approximation of fixpoints. In *Fourth ACM Symposium on Principles of Programming Languages*, 238–52.

21. Manuvir Das, Sorin Lerner, and Mark Seigle. 2002. Esp: Path-sensitive program verification in polynomial time. In *PLDI '02: Proceedings of the ACM SIGPLAN 2002 Conference on Programming Language Design and Implementation*, 57–68. New York: ACM Press.

22. J. Dean, D. Grove, and C. Chambers. 1995. Optimization of object-oriented programs using static class hierarchy analysis. In *Proceedings of the Ninth European Conference on Object-Oriented Programming (ECOOP'95)*, ed. W. Olthoff, 77–101. Aarhus, Denmark: Springer-Verlag.

23. Robert DeLine and Manuel Fahndrich. 2001. Enforcing high-level protocols in low-level software. In *Proceedings of the ACM SIGPLAN 2001 Conference on Programming Language Design and Implementation*, 59–69. New York: ACM Press.

24. Robert DeLine and Manuel Fahndrich. 2004. Typestates for objects. In *Proceedings of the 2004 European Conference on Object-Oriented Programming*, LNCS 3086. Heidelberg, Germany: Springer-Verlag.

25. David Detlefs, Greg Nelson, and James B. Saxe. 2005. Simplify: A theorem prover for program checking. *Journal of the ACM* 52(3):365–473.

26. David Detlefs, K. Rustan, M. Leino, and Greg Nelson. 1996. Wrestling with rep exposure. Technical report, Digital Equipment Corporation Systems Research Center.

27. Alain Deutsch. 1995. Semantic models and abstract interpretation techniques for inductive data structures and pointers. In *Proceedings of the ACM SIGPLAN Symposium on Partial Evaluation and Semantics-Based Program Manipulation*, 226–229.

28. William F. Dowling and Jean H. Gallier. 1984. Linear-time algorithms for testing the satisfiability of propositional horn formulae. *Journal of Logic Programming* 1(3):267–84.

29. Margaret A. Ellis and Bjarne Stroustrup. 1990. *The annotated C++ reference manual*. Reading, MA: Addison-Wesley.

30. David Evans. 1996. Static detection of dynamic memory errors. In *PLDI '96: Proceedings of the ACM SIGPLAN 1996 Conference on Programming Language Design and Implementation*, 44–53. New York: ACM Press.

31. Manuel Fahndrich, K. Rustan, and M. Leino. 2003. Declaring and checking non-null types in an object-oriented language. In *Proceedings of the 18th ACM SIGPLAN Conference on Object-Oriented Programing, Systems, Languages, and Applications*, 302–12. New York: ACM Press.

32. Cormac Flanagan and Stephen N. Freund. 2000. Type-based race detection for Java. In *Proceedings of the ACM SIGPLAN 2000 Conference on Programming Language Design and Implementation*, 219–32. New York: ACM Press.

33. Cormac Flanagan and Shaz Qadeer. 2003. A type and effect system for atomicity. In *Proceedings of the ACM SIGPLAN 2003 Conference on Programming Language Design and Implementation*, 338–49. New York: ACM Press.

34. Jeffrey S. Foster. 2002. Type qualifiers: Lightweight specifications to improve software quality. PhD dissertation, University of California, Berkeley.

35. Jeffrey S. Foster, Manuel Fähndrich, and Alexander Aiken. 1999. A theory of type qualifiers. In *Proceedings of the 1999 ACM SIGPLAN Conference on Programming Language Design and Implementation*, 192–203. New York: ACM Press.

36. Jeffrey S. Foster, Robert Johnson, John Kodumal, and Alex Aiken. 2006. Flow-insensitive type qualifiers. *ACM Transactions on Programming Languages and Systems* 28(6):1035–87.

37. Jeffrey S. Foster, Tachio Terauchi, and Alex Aiken. 2002. Flow-sensitive type qualifiers. In *Proceedings of the ACM SIGPLAN 2002 Conference on Programming Language Design and Implementation*, 1–12. New York: ACM Press.

38. Tim Freeman and Frank Pfenning. 1991. Refinement types for ML. In *PLDI '91: Proceedings of the ACM SIGPLAN 1991 Conference on Programming Language Design and Implementation*, 268–77. New York: ACM Press.

39. James Gosling, Bill Joy, and Guy Steele. 1996. *The Java language specification*. Reading, MA: Addison-Wesley.

40. Dan Grossman, Greg Morrisett, Trevor Jim, Michael Hicks, Yanling Wang, and James Cheney. 2002. Region-based memory management in Cyclone. In *Proceedings of the ACM SIGPLAN 2002 Conference on Programming Language Design and Implementation*, 282–93. New York: ACM Press.

41. Christian Grothoff, Jens Palsberg, and Jan Vitek. 2001. Encapsulating objects with confined types. In *ACM Transactions on Programming Languages and Systems. Proceedings of OOPSLA'01, ACM SIGPLAN Conference on Object-Oriented Programming Systems, Languages and Applications*, 241–53 (to appear in 2007).

42. Nevin Heintze. 1995. Control-flow analysis and type systems. In *Proceedings of SAS'95, International Static Analysis Symposium*, 189–206. Heidelberg, Germany: Springer-Verlag.

43. Atsushi Igarashi, Benjamin C. Pierce, and Philip Wadler. 2001. Featherweight Java: a minimal core calculus for Java and GJ. *ACM Transactions on Programming Languages and Systems* 23(3):396–450.

44. Rob Johnson and David Wagner. 2004. Finding user/kernel pointer bugs with type inference. In *Proceedings of the 13th USENIX Security Symposium*, 119–34.

45. Brian W. Kernighan and Dennis M. Ritchie. 1978. *The C programming language*. New York: Prentice-Hall.

46. Gary A. Kildall. 1973. A unified approach to global program optimization. In *Conference Record of the ACM Symposium on Principles of Programming Languages*, 194–206.

47. Yitzhak Mandelbaum, David Walker, and Robert Harper. 2003. An effective theory of type refinements. In *Proceedings of the Eighth ACM SIGPLAN International Conference on Functional Programming*, 213–25. New York: ACM Press.

48. Per Martin-Löf. 1982. Constructive mathematics and computer programming. In *Sixth International Congress for Logic, Methodology, and Philosophy of Science*, 153–75. Amsterdam: North-Holland.

49. Microsoft. Microsoft Visual C#. http://msdn.microsoft.com/vcsharp.

50. Greg Morrisett, Karl Crary, Neal Glew, Dan Grossman, Richard Samuels, Frederick Smith, David Walker, Stephanie Weirich, and Steve Zdancewic. 1999. Talx86: A realistic typed assembly language. Presented at 1999 ACM Workshop on Compiler Support for System Software, May 1999.

51. Greg Morrisett, David Walker, Karl Crary, and Neal Glew. 1998. From system F to typed assembly language. In *Proceedings of POPL'98, 25th Annual SIGPLAN–SIGACT Symposium on Principles of Programming Languages*, 85–97.

52. Peter Müller and Arnd Poetzsch-Heffter. 1999. Universes: A type system for controlling representation exposure. In *Programming Languages and Fundamentals of Programming*, ed. A. Poetzsch-Heffter and J. Meyer. Fernuniversität Hagen.

53. George C. Necula, Scott McPeak, and Westley Weimer. 2002. CCured: Type-safe retrofitting of legacy code. In *Proceedings of the 29th ACM SIGPLAN-SIGACT Symposium on Principles of Programming Languages*, 128–39. New York: ACM Press.

54. Peter Ørbæk and Jens Palsberg. 1995. Trust in the λ-calculus. *Journal of Functional Programming* 7(6):557–91.

55. Jens Palsberg. 1998. Equality-based flow analysis versus recursive types. *ACM Transactions on Programming Languages and Systems* 20(6):1251–64.

56. Jens Palsberg. 2001. Type-based analysis and applications. In *Proceedings of PASTE'01, ACM SIGPLAN/SIGSOFT Workshop on Program Analysis for Software Tools and Engineering*, 20–27.

57. Jens Palsberg and Patrick M. O'Keefe. 1995. A type system equivalent to flow analysis. *ACM Transactions on Programming Languages and Systems* 17(4):576–99.

58. Jens Palsberg and Christina Pavlopoulou. 2001. From polyvariant flow information to intersection and union types. *Journal of Functional Programming*, 11(3):263–17.

59. Jens Palsberg and Michael I. Schwartzbach. 1991. Object-oriented type inference. In *Proceedings of OOPSLA'91, ACM SIGPLAN Sixth Annual Conference on Object-Oriented Programming Systems, Languages and Applications*, 146–61.

60. M. S. Paterson and M. N. Wegman. 1978. Linear unification. *Journal of Computer and System Sciences* 16:158–67.

61. Benjamin C. Pierce. 2002. *Types and programming languages*. Cambridge MA: MIT Press.

62. K. Rustan, M. Leino, and Peter Müller. 2004. Object invariants in dynamic contexts. In *Proceedings of ECOOP'04, 16th European Conference on Object-Oriented Programming*, 491–516.

63. Umesh Shankar, Kunal Talwar, Jeffrey S. Foster, and David Wagner. 2001. Detecting format string vulnerabilities with type qualifiers. In *Proceedings of the 10th Usenix Security Symposium*.

64. Frank Tip and Jens Palsberg. 2000. Scalable propagation-based call graph construction algorithms. In *Proceedings of OOPSLA'00, ACM SIGPLAN Conference on Object-Oriented Programming Systems, Languages and Applications*, 281–93.

65. Mads Tofte and Jean-Pierre Talpin. 1994. Implementation of the typed call-by-value λ-calculus using a stack of regions. In *Proceedings of the 21st ACM SIGPLAN-SIGACT Symposium on Principles of Programming Languages*, 188–201. New York: ACM Press.

66. Mitchell Wand. 1987. A simple algorithm and proof for type inference. *Fundamentae Informaticae* X:115–22.

67. John Whaley and Monica Lam. 2004. Cloning-based context-sensitive pointer alias analysis using binary decision diagrams. In *Proceedings of PLDI'04, ACM SIGPLAN Conference on Programming Language Design and Implementation*.

68. Andrew K. Wright and Matthias Felleisen. 1994. A syntactic approach to type soundness. *Information and Computation* 115(1):38–94.

69. Hongwei Xi and Frank Pfenning. 1998. Eliminating array bound checking through dependent types. In *Proceedings of ACM SIGPLAN Conference on Programming Language Design and Implementation*, 249–57.

70. Hongwei Xi and Frank Pfenning. 1999. Dependent types in practical programming. In *Proceedings of the 26th ACM SIGPLAN Symposium on Principles of Programming Languages*, 214–27.

71. Xiaolan Zhang, Antony Edwards, and Trent Jaeger. 2002. Using cqual for static analysis of authorization hook placement. In *USENIX Security Symposium*, ed. Dan Boneh, 33–48.

72. Tian Zhao, Jens Palsberg, and Jan Vitek. 2006. Type-based confinement. *Journal of Functional Programming* 16(1):83–128.

10

Dynamic Compilation

Evelyn Duesterwald
IBM T.J. Watson Research Center,
Yorktown Heights, NY
duester@us.ibm.com

10.1 Introduction

The term *dynamic compilation* refers to techniques for runtime generation of executable code. The idea of compiling parts or all the application code while the program is executing challenges our intuition about overheads involved in such an endeavor, yet recently a number of approaches have evolved that effectively manage this challenging task.

The ability to dynamically adapt executing code addresses many of the existing problems with traditional static compilation approaches. One such problem is the difficulty for a static compiler to fully exploit the performance potential of advanced architectures. In the drive for greater performance, today's microprocessors provide capabilities for the compiler to take on a greater role in performance delivery, ranging from predicated and speculative execution (e.g., for the Intel Itanium processor) to various power consumption control models. To exploit these architectural features, the static compiler usually has to rely on profile information about the dynamic execution behavior of a program. However, collecting valid execution profiles ahead of time may not always be feasible or practical. Moreover, the risk of performance degradation that may result from missing or outdated profile information is high.

Current trends in software technology create additional obstacles to static compilation. These are exemplified by the widespread use of object-oriented programming languages and the trend toward shipping software binaries as collections of dynamically linked libraries instead of monolithic binaries. Unfortunately, the increased degree of runtime binding can seriously limit the effectiveness of traditional static compiler optimization, because static compilers operate on the statically bound scope of the program.

Finally, the emerging Internet and mobile communications marketplace creates the need for the compiler to produce portable code that can efficiently execute on a variety of machines. In an environment of networked devices, where code can be downloaded and executed on the fly, static compilation at the target device is usually not an option. However, if static compilers can only be used to generate platform-independent intermediate code, their role as a performance delivery vehicle becomes questionable.

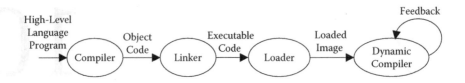

FIGURE 10.1   Dynamic compilation pipeline.

This chapter discusses dynamic compilation, a radically different approach to compilation that addresses and overcomes many of the preceding challenges to effective software implementation. Dynamic compilation extends our traditional notion of compilation and code generation by adding a new dynamic stage to the classical pipeline of compiling, linking, and loading code. The extended dynamic compilation pipeline is depicted in Figure 10.1.

A dynamic compiler can take advantage of runtime information to exploit optimization opportunities not available to a static compiler. For example, it can customize the running program according to information about actual program values or actual control flow. Optimization may be performed across dynamic binding, such as optimization across dynamically linked libraries. Dynamic compilation avoids the limitations of profile-based approaches by directly utilizing runtime information. Furthermore, with a dynamic compiler, the same code region can be optimized multiple times should its execution environment change. Another unique opportunity of dynamic compilation is the potential to speed up the execution of legacy code that was produced using outdated compilation and optimization technology.

Dynamic compilation provides an important vehicle to efficiently implement the "write-once-run-anywhere" execution paradigm that has recently gained a lot of popularity with the Java programming language [22]. In this paradigm, the code image is encoded in a mobile platform-independent format (e.g., Java bytecode). Final code generation that produces native code takes place at runtime as part of the dynamic compilation stage.

In addition to addressing static compilation obstacles, the presence of a dynamic compilation stage can create entirely new opportunities that go beyond code compilation. Dynamic compilation can be used to transparently migrate software from one architecture to a different host architecture. Such a translation is achieved by dynamically retargeting the loaded nonnative guest image to the host machine native format. Even for machines within the same architectural family, a dynamic compiler may be used to upgrade software to exploit additional features of the newer generation.

As indicated in Figure 10.1, the dynamic compilation stage may also include a feedback loop. With such a feedback loop, dynamic information, including the dynamically compiled code itself, may be saved at runtime to be restored and utilized in future runs of the program. For example, the FX!32 system for emulating x86 code on an Alpha platform [27] saves runtime information about executed code, which is then used to produce translations offline that can be incorporated in future runs of the program. It should be noted that FX!32 is not strictly a dynamic compilation system, in that translations are produced between executions of the program instead of online during execution.

Along with its numerous opportunities, dynamic compilation also introduces a unique set of challenges. One such challenge is to amortize the dynamic compilation overhead. If dynamic compilation is sequentially interleaved with program execution, the dynamic compilation time directly contributes to the overall execution time of the program. Such interleaving greatly changes the cost-benefit compilation trade-off that we have grown accustomed to in static compilation. Although in a static compiler increased optimization effort usually results in higher performance, increasing the dynamic compilation time may actually diminish some or all of the performance improvements that were gained by the optimization in the first place. If dynamic compilation takes place in parallel with program execution on a multiprocessor system, the dynamic compilation overhead is less important, because the dynamic compiler cannot directly slow down the program. It does, however, divert resources that could have been devoted to execution. Moreover, long dynamic compilation times can still adversely affect performance. Spending too much time on compilation can delay the employment of the dynamically compiled code and diminish the benefits. To maximize the benefits, dynamic compilation time should therefore always be kept to a minimum.

To address the heightened pressure for minimizing overhead, dynamic compilers often follow an adaptive approach [23]. Initially, the code is optimized with little or no optimization. Aggressive optimizations are considered only later, when more evidence has been found that added optimization effort is likely to be of use.

A dynamic compilation stage, if not designed carefully, can also significantly increase the space requirement for running a program. Controlling additional space requirements is crucial in environments where code size is important, such as embedded or mobile systems. The total space requirements of execution with a dynamic compiler include not only the loaded input image but also the dynamic compiler itself, plus the dynamically compiled code. Thus, care must be taken to control both the footprint of the dynamic compiler and the size of the currently maintained dynamically compiled code.

10.2 Approaches to Dynamic Compilation

A number of approaches to dynamic compilation have been developed. These approaches differ in several aspects, including the degree of transparency, the extent and scope of dynamic compilation, and the assumed encoding format of the loaded image. On the highest level, dynamic compilation systems can be divided into transparent and nontransparent systems. In a transparent system, the remainder of the compilation pipeline is oblivious to the fact that a dynamic compilation stage has been added. The executable produced by the linker and loader is not specially prepared for dynamic optimization, and it may execute with or without a dynamic compilation stage. Figure 10.2 shows a classification of the various approaches to transparent and nontransparent dynamic compilation.

Transparent dynamic compilation systems can further be divided into systems that operate on binary executable code (binary dynamic compilation) and systems that operate on an intermediate platform-independent encoding (just-in-time [JIT] compilation). A binary dynamic compiler starts out with a loaded fully executable binary. In one scenario, the binary dynamic compiler recompiles the binary code to incorporate native-to-native optimizing transformations. These recompilation systems are also referred to as *dynamic optimizers* [3, 5, 7, 15, 36]. During recompilation, the binary is optimized by customizing the code with respect to specific runtime control and data flow values. In dynamic binary translation, the

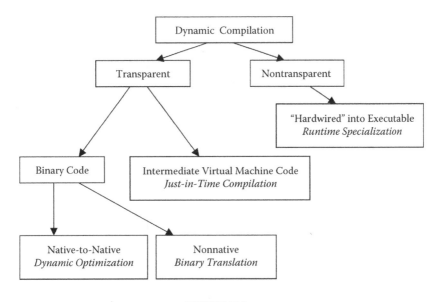

FIGURE 10.2

loaded input binary is in a nonnative format, and dynamic compilation is used to retarget the code to a different host architecture [19, 35, 39]. The dynamic code translation may also include optimization.

JIT compilers present a different class of transparent dynamic compilers [11, 12, 18, 28, 29]. The input to a JIT compiler is not a native program binary; instead, it is code in an intermediate, platform-independent representation that targets a virtual machine. The JIT compiler serves as an enhancement to the virtual machine to produce native code by compiling the intermediate input program at runtime, instead of executing it in an interpreter. Typically, semantic information is attached to the code, such as symbol tables or constant pools, which facilitates the compilation.

The alternative to transparent dynamic compilation is the nontransparent approach, which integrates the dynamic compilation stage explicitly within the earlier compilation stages. The static compiler cooperates with the dynamic compiler by delaying certain parts of the compilation to runtime, if their compilation can benefit from runtime values. A dynamic compilation agent is compiled (i.e., hardwired) into the executable to fill and link in a prepared code template for the delayed compilation region. Typically, the programmer indicates adequate candidate regions for dynamic compilation via annotations or compiler directives. Several techniques have been developed to perform runtime specialization of a program in this manner [9, 23, 31, 33].

Runtime specialization techniques are tightly integrated with the static compiler, whereas transparent dynamic compilation techniques are generally independent of the static compiler. However, transparent dynamic compilation can still benefit from information that the static compiler passes down. Semantic information, such as a symbol table, is an example of compiler information that is beneficial for dynamic compilation. If the static compiler is made aware of the dynamic compilation stage, more targeted information may be communicated to the dynamic compiler in the form of code annotations to the binary [30].

The remainder of this chapter discusses the various dynamic compilation approaches shown in Figure 10.2. We first discuss transparent binary dynamic optimization as a representative dynamic compilation system. We discuss the mechanics of dynamic optimization systems and their major components, along with their specific opportunities and challenges. We then discuss systems in each of the remaining dynamic compilation classes and point out their unique characteristics.

Also, a number of hardware approaches are available to dynamically manipulate the code of a running program, such as the hardware in out-of-order superscalar processors or hardware dynamic optimization in trace cache processors [21]. However, in this chapter, we limit the discussion to software dynamic compilation.

10.3   Transparent Binary Dynamic Optimization

A number of binary dynamic compilation systems have been developed that operate as an optional dynamic stage [3, 5, 7, 15, 35]. An important characteristic of these systems is that they take full control of the execution of the program. Recall that in the transparent approach, the input program is not specially prepared for dynamic compilation. Therefore, if the dynamic compiler does not maintain full control over the execution, the program may escape and simply continue executing natively, effectively bypassing dynamic compilation altogether. The dynamic compiler can afford to relinquish control only if it can guarantee that it will regain control later, for example, via a timer interrupt.

Binary dynamic compilation systems share the general architecture shown in Figure 10.3. Input to the dynamic compiler is the loaded application image as produced by the compiler and linker. Two main components of a dynamic compiler are the compiled code cache that holds the dynamically compiled code fragments and the dynamic compilation engine. At any point in time, execution takes place either in the dynamic compilation engine or in the compiled code cache. Correspondingly, the dynamic compilation engine maintains two distinct execution contexts: the context of the dynamic compilation engine itself and the context of the application code.

Execution of the loaded image starts under control of the dynamic compilation engine. The dynamic compiler determines the address of the next instruction to execute. It then consults a lookup table to determine whether a dynamically compiled code fragment starting at that address already exists in the

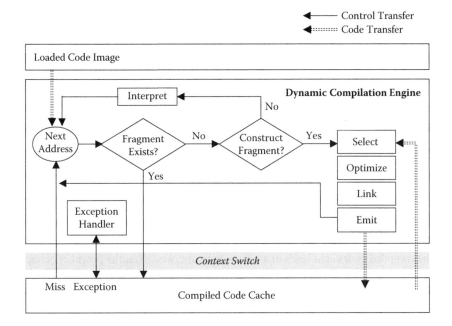

FIGURE 10.3

code cache. If so, a context switch is performed to load the application context and to continue execution in the compiled code cache until a code cache miss occurs. A code cache miss indicates that no compiled fragment exists for the next instruction. The cache miss triggers a context switch to reload the dynamic compiler's context and reenter the dynamic compilation engine.

The dynamic compiler decides whether a new fragment should be compiled starting at the next address. If so, a code fragment is constructed based on certain fragment selection policies, which are discussed in the next section. The fragment may optionally be optimized and linked with other previously compiled fragments before it is emitted into the compiled code cache.

The dynamic compilation engine may include an instruction interpreter component. With an interpreter component, the dynamic compiler can choose to delay the compilation of a fragment and instead interpret the code until it has executed a number of times. During interpretation, the dynamic compiler can profile the code to focus its compilation efforts on only the most profitable code fragments [4]. Without an interpreter, every portion of the program that is executed during the current run can be compiled into the compiled code cache.

Figure 10.3 shows a code transfer arrow from the compiled code cache to the fragment selection component. This arrow indicates that the dynamic compiler may choose to select new code fragments from previously created code in the compiled code cache. Such fragment reformation may be performed to improve fragment shape and extent. For example, several existing code fragments may be combined to form a single new fragment. The dynamic compiler may also reselect an existing fragment for more aggressive optimization. Reoptimization of a fragment may be indicated if profiling of the compiled code reveals that it is a hot (i.e., frequently executing) fragment.

In the following sections, we discuss the major components of the dynamic compiler in detail: fragment selection, fragment optimization, fragment linking, management of the compiled code cache, and exception handling.

10.3.1 Fragment Selection

The fragment selector proceeds by extracting code regions and passing them to the fragment optimizer for optimization and eventual placement in the compiled code cache. The arrangement of the extracted code

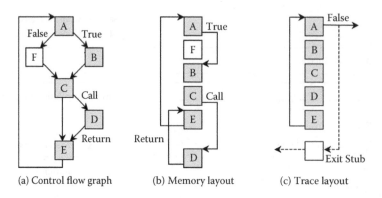

(a) Control flow graph   (b) Memory layout   (c) Trace layout

FIGURE 10.4

regions in the compiled code cache leads to a new code layout, which has the potential of improving the performance of dynamically compiled code. Furthermore, by passing isolated code regions to the optimizer, the fragment selector dictates the scope and kind of runtime optimization that may be performed. Thus, the goal of fragment selection is twofold: to produce an improved code layout and to expose dynamic optimization opportunities.

New optimization opportunities or improvements in code layout are unlikely if the fragment selector merely copies static regions from the loaded image into the code cache. Regions such as basic blocks or entire procedures are among the static regions of the original program and have been already exposed to, and possibly optimized by, the static compiler. New optimization opportunities are more likely to be found in the dynamic scope of the executing program. Thus, it is crucial to incorporate dynamic control flow into the selected code regions.

Because of the availability of dynamic information, the fragment selector has an advantage over a static compiler in selecting the most beneficial regions to optimize. At the same time, the fragment selector is more limited because high-level semantic information about code constructs is no longer available. For example, without information about procedure boundaries and the layout of switch statements, it is generally impossible to discover the complete control flow of a procedure body in a loaded binary image.

In the presence of these limitations, the units of code commonly used in a binary dynamic compilation system are a partial execution trace, or trace for short BD [4, 7]. A trace is a dynamic sequence of consecutively executing basic blocks. The sequence may not be contiguous in memory; it may even be interprocedural, spanning several procedure boundaries, including dynamically linked modules. Thus, traces are likely to offer opportunities for improved code layout and optimization. Furthermore, traces do not need to be computed; they can be inferred simply by observing the runtime behavior of the program.

Figure 10.4 illustrates the effects of selecting dynamic execution traces. The graph in Figure 10.4a shows a control flow graph representation of a trace, consisting of blocks A, B, C, D, and E that form a loop containing a procedure call. The graph in Figure 10.4b shows the same trace in a possible noncontiguous memory layout of the original loaded program image. The graph in Figure 10.4c shows a possible improved layout of the looping trace in the compiled code cache as a contiguous straight-line sequence of blocks. The straight-line layout reduces branching during execution and offers better code locality for the loop.

10.3.1.1 Adaptive Fragment Selection

The dynamic compiler may select fragments of varying shapes. It may also stage the fragment selection in a progressive fashion. For example, the fragment selector may initially select only basic block fragments. Larger composite fragments, such as traces, are selected as secondary fragments by stringing together frequently executing block fragments [4]. Progressively larger regions, such as tree regions, may then be constructed by combining individual traces [19]. Building composite code regions can result in potentially large amounts of code duplication because code that is common across several composite regions is replicated in each region. Uncontrolled code duplication can quickly result in excessive cache size requirements, the so-called *code explosion* problem. Thus, a dynamic compiler has to employ some form of execution profiling to limit composite region construction to only the (potentially) most profitable candidates.

10.3.1.2 Online Profiling

Profiling the execution behavior of the loaded code image to identify the most frequently executing regions is an integral part of dynamic compilation. Information about the hot spots in the code is used in fragment selection and for managing the compiled code cache space. Hot spots must be detected online as they are becoming hot, which is in contrast to conventional profiling techniques that operate offline and do not establish relative execution frequencies until after execution. Furthermore, to be of use in a dynamic compiler, the profiling techniques must have very low space and time overheads.

A number of offline profiling techniques have been developed for use in feedback systems, such as profile-based optimization. A separate profile run of the program is conducted to accumulate profile information that is then fed back to the compiler. Two major approaches to offline profiling are statistical PC sampling and binary instrumentation for the purpose of branch or path profiling. Statistical PC sampling [1, 10, 40] is an inexpensive technique for identifying hot code blocks by recording program counter hits. Although PC sampling is efficient for detecting individual hot blocks, it provides little help in finding larger hot code regions. One could construct a hot trace by stringing together the hottest code blocks. However, such a trace may never execute from start to finish because the individual blocks may have been hot along disjoint execution paths. The problem is that individually collected branch frequencies do not account for branch correlations, which occur if the outcome of one branch can influence the outcome of a subsequent branch.

Another problem with statistical PC sampling is that it introduces nondeterminism into the dynamic compilation process. Nondeterministic behavior is undesirable because it greatly complicates development and debugging of the dynamic compiler.

Profiling techniques based on binary instrumentation record information at every execution instance. They are more costly than statistical sampling, but can also provide more fine-grained frequency information. Like statistical sampling, branch profiling techniques suffer the same problem of not adequately addressing branch correlations. Path-profiling techniques overcome the correlation problem by directly determining hot traces in the program [6]. The program binary is instrumented to collect entire path (i.e., trace) frequency information at runtime in an efficient manner.

A dynamic compiler could adopt these techniques by inserting instrumentation in first-level code fragments to build larger composite secondary fragments. The drawback of adapting offline techniques is the large amount of profile information that is collected and the overhead required to process it. Existing dynamic compilation systems have employed more efficient, but also more approximate, profiling schemes that collect a small amount of profiling information, either during interpretation [5] or by instrumenting first-level fragments [19]. Ephemeral instrumentation is a hybrid profiling technique [37] based on the ability to efficiently enable and disable instrumentation code.

10.3.1.3 Online Profiling in the Dynamo System

As an example of a profiling scheme used in a dynamic compiler, we consider the next executing tail (NET) scheme used in the Dynamo system [16]. The objective of the NET scheme is to significantly reduce profiling overhead while still providing effective hot path predictions. A path is divided into a path head (i.e., the path starting point) and a path tail, which is the remainder of the path following the starting point. For example, in path ABCDE in Figure 10.4a, block A is the path head and BCDE is the path tail. The NET scheme reduces profiling cost by using speculation to predict path tails, while maintaining full profiling support to predict hot path heads. The rationale behind this scheme is that a hot path head indicates that the program is currently executing in a hot region, and the next executing path tail is likely to be part of that region.

Accordingly, execution counts are maintained only for potential path heads, which are the targets of backward taken branches or the targets of cache exiting branches. For example, in Figure 10.4a, one profiling count is maintained for the entire loop at the single path head at the start of block A. Once the counter at block A has exceeded a certain threshold, the next executing path is selected as the hot path for the loop.

10.3.2   Fragment Optimization

After a fragment has been selected, it is translated into a self-contained location-independent intermediate representation (IR). The IR of a fragment serves as a temporary vehicle to transform the original instruction stream into an optimized form and to prepare it for placement and layout in the compiled code cache. To enable fast translation between the binary code and the IR, the abstraction level of the IR is kept close to the binary instruction level. Abstraction is introduced only when needed, such as to provide location independence through symbolic labels and to facilitate code motion and code transformations through the use of virtual registers.

After the fragment is translated into its intermediate form, it can be passed to the optimizer. A dynamic optimizer is not intended to duplicate or replace conventional static compiler optimization. On the contrary, a dynamic optimizer can complement a static compiler by exploiting optimization opportunities that present themselves only at runtime, such as value-based optimization or optimization across the boundaries of dynamically linked libraries. The dynamic optimizer can also apply path-specific optimization that would be too expensive to apply indiscriminately over all paths during static compilation. On a given path, any number of standard compiler optimizations may be performed, such as constant and copy propagation, dead code elimination, value numbering, and redundancy elimination [4, 15]. However, unlike in static compiler optimization, the optimization algorithm must be optimized for efficiency instead of generality and power. A traditional static optimizer performs an initial analysis phase over the code to collect all necessary data flow information that is followed by the actual optimization phase. The cost of performing multiple passes over the code is likely to be prohibitive in a runtime setting. Thus, a dynamic optimizer typically combines analysis and optimization into a single pass over the code [4]. During the combined pass all necessary data flow information is gathered on demand and discarded immediately if it is no longer relevant for current optimization [17].

10.3.2.1   Control Specialization

The dynamic compiler implicitly performs a form of control specialization of the code by producing a new layout of the running program inside the compiled code cache. Control specialization describes optimizations whose benefits are based on the execution taking specific control paths. Another example of control specialization is code sinking [4], also referred to as hot–cold optimization [13]. The objective of code sinking is to move instructions from the main fragment execution path into fragment exits to reduce the number of instructions executed on the path. An instruction can be sunk into a fragment exit block if it is not live within the fragment. Although an instruction appears dead on the fragment, it cannot be removed entirely because it is not known whether it is also dead after exiting the fragment.

An example of code sinking is illustrated in Figure 10.5. The assignment X: = Y in the first block in fragment 1 is not live within fragment 1 because it is overwritten by the read instruction in the next block. To avoid useless execution of the assignment when control remains within fragment 1, the assignment can be moved out of the fragment and into a so-called compensation block at every fragment exit at which the assigned variable may still be live, as shown in Figure 10.5i. Once the exit block is linked to a target fragment (fragment 2 in Figure 10.5) the code inside the target fragment can be inspected to determine whether the moved assignment becomes dead after linking. If it does, the moved assignment in the compensation block can safely be removed, as shown in Figure 10.5ii.

Another optimization is prefetching, which involves the placement of prefetch instructions along execution paths prior to the actual usage of the respective data to improve the memory behavior of the dynamically compiled code. If the dynamic compiler can monitor data cache latency, it can easily identify candidates for prefetching. A suitable placement of the corresponding prefetch instructions can be determined by consulting collected profile information.

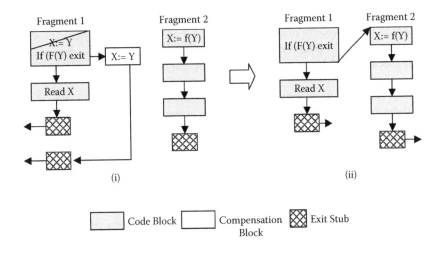

FIGURE 10.5

10.3.2.2 Value Specialization

Value specialization refers to an optimization that customizes the code according to specific runtime values of selected specialization variables. The specialization of a code fragment proceeds like a general form of constant propagation and attempts to simplify the code as much as possible.

Unless it can be established for certain that the specialization variable is always constant, the execution of the specialized code must be guarded by a runtime test. To handle specialization variables that take on multiple values at runtime, the same region of code may be specialized multiple times. Several techniques, such as polymorphic in-line caches [25], have been developed to efficiently select among multiple specialization versions at runtime.

A number of runtime techniques have been developed that automatically specialize code at runtime, given a specification of the specialization variables [9, 23, 31]. In code generated from object-oriented languages, virtual method calls can be specialized for a common receiver class [25]. In principle, any code region can be specialized with respect to any number of values. For example, traces may be specialized according to the entry values of certain registers. In the most extreme case, one can specialize individual instructions, such as complex floating point instructions, with respect to selected fixed-input register values [34].

The major challenge in value specialization is to decide when and what to specialize. Overspecialization of the code can quickly result in code explosion and may severely degrade performance. In techniques that specialize entire functions, the programmer typically indicates the functions to specialize through code annotations prior to execution [9, 23]. Once the specialization regions are determined, the dynamic specializer monitors the respective register values at runtime to trigger the specialization. Runtime specialization is the primary optimization technique employed by nontransparent dynamic compilation systems. We revisit runtime specialization in the context of nontransparent dynamic compilation in Section 10.6.

10.3.2.3 Binary Optimization

The tasks of code optimization and transformation are complicated by having to operate on executable binary code instead of a higher-level intermediate format. The input code to the dynamic optimizer has previously been exposed to register allocation and possibly also to static optimization. Valuable semantic information that is usually incorporated into compilation and optimization, such as type information and information about high-level constructs (i.e., data structures), is no longer available and is generally difficult to reconstruct.

An example of an optimization that is relatively easy to perform on intermediate code but difficult on the binary level is procedure inlining. To completely inline a procedure body, the dynamic compiler has to reverse engineer the implemented calling convention and stack frame layout. Doing this may be difficult, if not impossible, in the presence of memory references that cannot be disambiguated from stack frame references. Thus, the dynamic optimizer may not be able to recognize and entirely eliminate instructions for stack frame allocation and deallocation or instructions that implement caller and callee register saves and restores.

The limitations that result from operating on binary code can be partially lifted by making certain assumptions about compiler conventions. For example, assumptions about certain calling or register usage conventions help in the procedure inlining problem. Also, if it can be assumed that the stack is only accessed via a dedicated stack pointer register, stack references can be disambiguated from other memory references. Enhanced memory disambiguation may then in turn enable more aggressive optimization.

10.3.3   Fragment Linking

Fragment linking is the mechanism by which control is transferred among fragments without exiting the compiled code cache. An important performance benefit of linking is the elimination of unnecessary context switches that are needed to exit and reenter the code cache.

The fragment-linking mechanism may be implemented via exit stubs that are initially inserted at every fragment exiting branch, as illustrated in Figure 10.6. Prior to linking, the exit stubs direct control to the context switch routine to transfer control back to the dynamic compilation engine. If a target fragment for the original exit branch already exists in the code cache, the dynamic compiler can patch the exiting branch to jump directly to its target inside the cache. For example, in Figure 10.6, the branches A to E and G to A have been directly linked, leaving their original exit stubs inactive. To patch exiting branches, some information about the branch must be communicated to the dynamic compiler. For example, to determine the target fragment of a link, the dynamic compiler must know the original target address of the exiting branch. This kind of branch information may be stored in a link record data structure, and a pointer to it can be embedded in the exit stub associated with the branch [4].

The linking of an indirect computed branch is more complicated. If the fragment selector has collected a preferred target for the indirect branch, it can be inlined directly into the fragment code. The indirect target is inlined by converting the indirect branch into a conditional branch that tests whether the current target is equal to the preferred target. If the test succeeds, control falls through to the preferred target inside the fragment. Otherwise, control can be directed to a special lookup routine that is permanently resident in the compiled code cache. This routine implements a lookup to determine whether a fragment for the indirect branch target is currently resident in the cache. If so, control can be directed to the target fragment without having to exit the code cache [4].

Although its advantages are obvious, linking also has some disadvantages that need to be kept in balance when designing the linker. For example, linking complicates the effective management of the code cache,

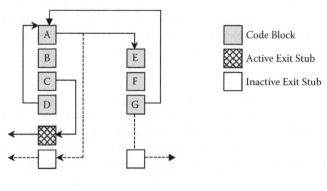

FIGURE 10.6

which may require the periodic removal or relocation of individual fragments. The removal of a fragment may be necessary to make room for new fragments, and fragment relocation may be needed to periodically defragment the code cache. Linking complicates both the removal and relocation of individual fragments because all incoming fragment links have to be unlinked first. Another problem with linking is that it makes it more difficult to limit the latency of asynchronous exception handling. Asynchronous exceptions arise from events such as keyboard interrupts and timer expiration. Exception handling is discussed in more detail in Section 10.3.5.

Linking may be performed on either a demand basis or a preemptive basis. With on-demand linking, fragments are initially placed in the cache with all their exiting branches targeting an exit stub. Individual links are inserted as needed each time control exits the compiled code cache via an exit stub. With preemptive linking, all possible links are established when a fragment is first placed in the code cache. Preemptive linking may result in unnecessary work when links are introduced that are never executed. On the other hand, demand-based linking causes additional context switches and interruptions of cache execution each time a delayed link is established.

10.3.4 Code Cache Management

The code cache holds the dynamically compiled code and may be organized as one large contiguous area of memory, or it may be divided into a set of smaller partitions. Managing the cache space is a crucial task in the dynamic compilation system. Space consumption is primarily controlled by a cache allocation and deallocation strategy. However, it can also be influenced by the fragment selection strategy. Cache space requirements increase with the amount of code duplication among the fragments. In the most conservative case, the dynamic compiler selects only basic block fragments, which avoids code duplication altogether. However, the code quality and layout in the cache is likely to be unimproved over the original binary. A dynamic compiler may use an adaptive strategy that permits unlimited duplication if sufficient space is available but moves toward shorter, more conservatively selected fragments as the available space in the cache diminishes. Even with an adaptive strategy, the cache may eventually run out of space, and the deallocation of code fragments may be necessary to make room for future fragments.

10.3.4.1 Fragment Deallocation

A fragment deallocation strategy is characterized by three parameters: the granularity, the timing, and the replacement policy that triggers deallocation. The granularity of fragment deallocation may range from an individual fragment deallocation to an entire cache flush. Various performance tradeoffs are to be considered in choosing the deallocation granularity. Individual fragment deallocation is costly in the presence of linking because each fragment exit and entry has to be individually unlinked. To reduce the frequency of cache management events, one might choose to deallocate a group of fragments at a time. A complete flush of one of the cache partitions is considerably cheaper because individual exit and entry links do not have to be processed. Moreover, complete flushing does not incur fragmentation problems. However, uncontrolled flushing may result in loss of useful code fragments that may be costly to reacquire.

The timing of a deallocation can be demand or preemptive based. A demand-based deallocation occurs simply in reaction to an out-of-space condition of the cache space. A preemptive strategy is used in the Dynamo system for cache flushing [4]. The idea is to time a cache flush so that the likelihood of losing valuable cache contents is minimized. The Dynamo system triggers a cache flush when it detects a phase change in the program behavior. When a new program phase is entered, a new working set of fragments is built, and it is likely that most of the previously active code fragments are no longer relevant. Dynamo predicts phase changes by monitoring the fragment creation rate. A phase change is signaled if a sudden increase in the creation rate is detected.

Finally, the cache manager has to implement a replacement policy. A replacement policy is particularly important if individual fragments are deallocated. However, even if an entire cache partition is flushed, a decision has to be made as to which partition to free. The cache manager can borrow simple common replacement policies from memory paging systems, such as first-in, first-out (FIFO) or least recently used (LRU). Alternatively, more advanced garbage collection strategies, such as generational garbage collection strategies, can be adopted to manage the dynamic compilation cache.

Besides space allocation and deallocation, an important JIT code cache service is the fast lookup of fragments that are currently resident in the code cache. Fragment lookups are needed throughout the dynamic compilation system and even during the execution of cached code fragments when it is necessary to look up an indirect branch target. Thus, fast implementation of fragment lookups, for example, via hash tables, is crucial.

10.3.4.2   Multiple Threads

The presence of multithreading can greatly complicate the cache manager. Most of the complication from multithreading can simply be avoided by using thread-private caches. With thread-private caches, each thread uses its own compiled code cache, and no dynamically compiled code is shared among threads. However, the lack of code sharing with thread-private caches has several disadvantages. The total code cache size requirements are increased by the need to replicate thread-shared code in each private cache. Besides additional space requirements, the lack of code sharing can also cause redundant work to be carried out when the same thread-shared code is repeatedly compiled.

To implement shared code caches, every code cache access that deletes or adds fragment code must be synchronized. Operating systems usually provide support for thread synchronization. To what extent threads actually share code and, correspondingly, to what extent shared code caches are beneficial are highly dependent on the application behavior.

Another requirement for handling multiple threads is the provision of thread-private state. Storage for thread-private state is needed for various tasks in the dynamic compiler. For example, during fragment selection a buffer is needed to hold the currently collected fragment code. This buffer must be thread private to avoid corrupting the fragment because multiple threads may be simultaneously in the process of creating fragments.

10.3.5   Handling Exceptions

The occurrence of exceptions while executing in the compiled code cache creates a difficult issue for a dynamic compiler. This is true for both user-level exceptions, such as those defined in the Java language, and system-level exceptions, such as memory faults. An exception has to be serviced as if the original program is executing natively. To ensure proper exception handling, the dynamic compiler has to intercept all exceptions delivered to the program. Otherwise, the appropriate exception handler may be directly invoked, and the dynamic compiler may lose control over the program. Losing control implies that the program has escaped and can run natively for the remainder of the execution.

The original program may have installed an exception handler that examines or even modifies the execution state passed to it. In binary dynamic compilation, the execution state includes the contents of machine registers and the program counter. In JIT compilation, the execution state depends on the underlying virtual machine. For example, in Java, the execution state includes the contents of the Java runtime stack.

If an exception is raised when control is inside the compiled code cache, the execution state may not correspond to any valid state in the original program. The exception handler may fail or operate inadequately when an execution state has been passed to it that was in some way modified through dynamic compilation. The situation is further complicated if the dynamic compiler has performed optimizations on the dynamically compiled code.

Exceptions can be classified as asynchronous or synchronous. Synchronous exceptions are associated with a specific faulting instruction and must be handled immediately before execution can proceed. Examples of synchronous exceptions are memory or hardware faults. Asynchronous exceptions do not require immediate handling, and their processing can be delayed. Examples of asynchronous exceptions include external interrupts (e.g., keyboard interrupts) and timer expiration.

A dynamic compiler can deal with asynchronous exceptions by delaying their handling until a safe execution point is reached. A safe point describes a state at which the precise execution state of the original program is known. In the absence of dynamic code optimization, a safe point is usually reached when control is inside the dynamic compilation engine. When control exits the code cache, the original execution state is saved by the context switch routine prior to reentering the dynamic compilation engine. Thus, the saved context state can be restored before executing the exception handler.

If control resides inside the code cache at the time of the exception, the dynamic compiler can delay handling the exception until the next code cache exit. Because the handling of the exception must not be delayed indefinitely, the dynamic compiler may have to force a code cache exit. To force a cache exit, the fragment that has control at the time of the exception is identified, and all its exit branches are unlinked. Unlinking the exit branches prevents control from spinning within the code cache for an arbitrarily long period of time before the dynamic compiler can process the pending exception.

10.3.5.1 Deoptimization

Unfortunately, postponing the handling of an exception until a safe point is reached is not an option for synchronous exceptions. Synchronous exceptions must be handled immediately, even if control is at a point in the compiled code cache. The original execution state must be recovered as if the original program had executed unmodified. Thus, at the very least, the program counter address, currently a cache address, has to be set to its corresponding address in the original code image.

The situation is more complicated if the dynamic compiler has applied optimizations that change the execution state. This includes optimizations that eliminate code, remap registers, or reorder instructions. In Java JIT compilation, this also includes the promotion of Java stack locations to machine registers. To reestablish the original execution state, the fragment code has to be deoptimized. This problem of deoptimization is similar to one that arises with debugging optimized code, where the original unoptimized user state has to be presented to the programmer when a break point is reached.

Deoptimization techniques for runtime compilation have previously been discussed for JIT compilation [26] and binary translation [24]. Each optimization requires its own deoptimization strategy, and not all optimizations are deoptimizable. For example, the reordering of two memory load operations cannot be undone once the reordered earlier load has executed and raised an exception. To deoptimize a transformation, such as dead code elimination, several approaches can be followed. The dynamic compiler can store sufficient information at every optimization point in the dynamically compiled code. When an exception arises, the stored information is consulted to determine the compensation code that is needed to undo the optimization and reproduce the original execution state. For dead code elimination, the compensation code may be as simple as executing the eliminated instruction. Although this approach enables fast state recovery at exception time, it can require substantial storage for deoptimization information.

An alternative approach, which is better suited if exceptions are rare events, is to retrieve the necessary deoptimization information by recompiling the fragment at exception time. During the initial dynamic compilation of a fragment, no deoptimization information is stored. This information is recorded only during a recompilation that takes place in response to an exception.

It may not always be feasible to determine and store appropriate deoptimization information, for example, for optimizations that exploit specific register values. To be exception-safe and to faithfully reproduce original program behavior, a dynamic compiler may have to suppress optimizations that cannot be deoptimized if an exception were to arise.

10.3.6 Challenges

The previous sections have discussed some of the challenges in designing a dynamic optimization system. A number of other difficult issues still must be dealt with in specific scenarios.

10.3.6.1 Self-Modifying and Self-Referential Code

One such issue is the presence of self-modifying or self-referential code. For example, self-referential code may be inserted for a program to compute a check sum on its binary image. To ensure that self-referential behavior is preserved, the loaded program image should remain untouched, which is the case if the dynamic compiler follows the design illustrated in Figure 10.3.

Self-modifying code is more difficult to handle properly. The major difficulty lies in the detection of code modification. Once code modification has been detected, the proper reaction is to invalidate all fragments currently resident in the cache that contain copies of the modified code. Architectural support can make the detection of self-modifying code easy. If the underlying machine architecture provides page–write protection, the pages that hold the loaded program image can simply be write protected. A page protection violation can then indicate the occurrence of code modification and can trigger the corresponding fragment invalidations in the compiled code cache. Without such architectural support, every store to memory must be intercepted to test for self-modifying stores.

10.3.6.2 Transparency

A truly transparent dynamic compilation system can handle any loaded executable. Thus, to qualify as transparent a dynamic compiler must not assume special preparation of the binary, such as explicit relinking or recompilation with dynamic compilation code. To operate fully transparently, a dynamic compiler should be able to handle even legacy code. In a more restrictive setting, a dynamic compiler may be allowed to make certain assumptions about the loaded code. For example, an assumption may be made that the loaded program was generated by a compiler that obeys certain software conventions. Another assumption could be that it is equipped with symbol table information or stack unwinding information, each of which may provide additional insights into the code that can be valuable during optimization.

10.3.6.3 Reliability

Reliability and robustness present another set of challenges. If the dynamic compiler acts as an optional transparent runtime stage, robust operation is of even greater importance than in static compilation stages. Ideally, the dynamic compilation system should reach hardware levels of robustness, though it is not clear how this can be achieved with a piece of software.

10.3.6.4 Real-Time Constraints

Handling real-time constraints in a dynamic compiler has not been sufficiently studied. The execution speed of a program that runs under the control of a dynamic compiler may experience large variations. Initially, when the code cache is nearly empty, dynamic compilation overhead is high and execution progress is correspondingly slow. Over time, as a program working set materializes in the code cache, the dynamic compilation overhead diminishes and execution speed picks up. In general, performance progress is highly unpredictable because it depends on the code reuse rate of the program. Thus, it is not clear how any kind of real-time guarantees can be provided if the program is dynamically compiled.

10.4   Dynamic Binary Translation

The previous sections have described dynamic compilation in the context of code transformation for performance optimization. Another motivation for employing a dynamic compiler is software migration. In this case, the loaded image is native to a guest architecture that is different from the host architecture, which runs the dynamic compiler. The binary translation model of dynamic compilation is illustrated in Figure 10.7. Caching instruction set simulators [8] and dynamic binary translation systems [19, 35, 39] are examples of systems that use dynamic compilation to translate nonnative guest code to a native host architecture.

An interesting aspect of dynamic binary translation is achieving separation of the running software from the underlying hardware. In principle, a dynamic compiler can provide a software implementation

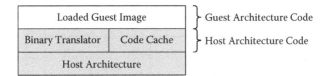

FIGURE 10.7

of an arbitrary guest architecture. With the dynamic compilation layer acting as a bridge, software and hardware may evolve independently. Architectural advances can be hidden and remain transparent to the user. This potential of dynamic binary translation has recently been commercially exploited by Transmeta's code morphing software [14] and Transitive's emulation software layer [38].

The high-level design of a dynamic compiler, if used for binary translation, remains the same as illustrated in Figure 10.3, with the addition of a translation module. This additional module translates fragments selected from guest architecture code into fragments for the host architecture, as illustrated in Figure 10.8.

To produce a translation from one native code format to another, the dynamic compiler may choose to first translate the guest architecture code into an intermediate format and then generate the final host architecture instructions. Going through an intermediate format is especially helpful if the differences in host and guest architecture are large. To facilitate the translation of instructions, it is useful to establish a fixed mapping between guest and host architecture resources, such as machine registers [19].

Although the functionality of the major components in the dynamic compilation stage, such as fragment selection and code cache management, is similar to the case of native dynamic optimization, a number of important challenges are unique to binary translation.

If the binary translation system translates code not only across different architectures but also across different operating systems, it is called full system translation. The Daisy binary translation system that translates from code for the PowerPC under IBM's UNIX system, AIX, to a customized very long instruction word (VLIW) architecture is an example of full system translation [19]. Full system translation may be further complicated by the presence of a virtual address space in the guest system. The entire virtual memory address translation mechanism has to be faithfully emulated during the translation, which includes the handling of such events as page faults. Furthermore, low-level boot code sequences must also be translated. Building a dynamic compiler for full system translation requires in-depth knowledge of both the guest and host architectures and operating systems.

10.5 Just-in-Time Compilation

JIT compilation refers to the runtime compilation of intermediate virtual machine code. Thus, unlike binary dynamic compilation, the process does not start out with already compiled executable code. JIT compilation was introduced for Smalltalk-80 [18] but has recently been widely popularized with the introduction of the Java programming language and its intermediate bytecode format [22].

The virtual machine environment for a loaded intermediate program is illustrated in Figure 10.9. As in binary dynamic compilation, the virtual machine includes a compilation module and a compiled code cache. Another core component of the virtual machine is the runtime system that provides various system services that are needed for the execution of the code. The loaded intermediate code image is inherently

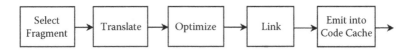

FIGURE 10.8   Dynamic translation pipeline.

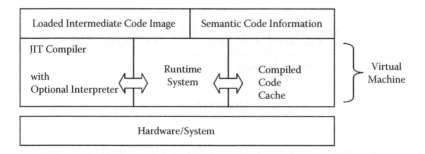

FIGURE 10.9

tied to, and does not execute outside, the virtual machine. Virtual machines are an attractive model to implement a "write-once-run-anywhere" programming paradigm. The program is statically compiled to the virtual machine language. In principle, the same statically compiled program may run on any hardware environment, as long as the environment provides an appropriate virtual machine. During execution in the virtual machine, the program may be further (JIT) compiled to the particular underlying machine architecture. A virtual machine with a JIT compiler may or may not include a virtual machine language interpreter.

JIT compilation and binary dynamic compilation share a number of important characteristics. In both cases, the management of the compiled code cache is crucial. Just like a binary dynamic compiler, the JIT compiler may employ profiling to stage the compilation and optimization effort into several modes, from a quick base compilation mode with no optimization to an aggressively optimized mode.

Some important differences between JIT and binary dynamic compilation are due to the different levels of abstraction in their input. To facilitate execution in the virtual machine, the intermediate code is typically equipped with semantic information, such as symbol tables or constant pools. A JIT compiler can take advantage of the available semantic information. Thus, JIT compilation more closely resembles the process of static compilation than does binary recompilation.

The virtual machine code that the JIT compiler operates on is typically location independent, and information about program components, such as procedures or methods, is available. In contrast, binary dynamic compilers operate on fully linked binary code and usually face a code recovery problem. To recognize control flow, code layout decisions that were made when producing the binary have to be reverse engineered, and full code recovery is in general not possible. Because of the code recovery problem, binary dynamic compilers are more limited in their choice of compilation unit. They typically choose simple code units, such as straight-line code blocks, traces, or tree-shaped regions. JIT compilers, on the other hand, can recognize higher-level code constructs and global control flow. They typically choose whole methods or procedures as the compilation unit, just as a static compiler would do. However, recently it has been recognized that there are other advantages to considering compilation units at a different granularity than whole procedures, such as reduced compiled code sizes [2].

The availability of semantic information in a JIT compiler also allows for a larger optimization repertoire. Except for overhead concerns, a JIT compiler is just as capable of optimizing the code as a static compiler. JIT compilers can even go beyond the capabilities of a static compiler by taking advantage of dynamic information about the code. In contrast, a binary dynamic optimizer is more constrained by the low-level representation and the lack of a global view of the program. The aliasing problem is worse in binary dynamic compilation because the higher-level-type information that may help disambiguate memory references is not available. Furthermore, the lack of a global view of the program forces the binary dynamic compiler to make worst-case assumptions at entry and exit points of the currently processed code fragment, which may preclude otherwise safe optimizations.

The differences in JIT compilation and binary dynamic compilation are summarized in Table 10.1. A JIT compiler is clearly more able to produce highly optimized code than a binary compiler. However,

TABLE 10.1 Differences in JIT Compilation and Binary Dynamic Compilation

JIT Compilation	Dynamic Binary Compilation
Semantic information available	Lack of semantic information
Full code recovery	Limited code recovery; limited choice in compilation unit
Full optimization repertoire	Limited optimization potential

consider a scenario where the objective is not code quality but compilation speed. Under these conditions, it is no longer clear that the JIT compiler has an advantage. A number of compilation and code generation decisions, such as register allocation and instruction selection, have already been made in the binary code and can often be reused during dynamic compilation. For example, binary translators typically construct a fixed mapping between guest and host system machine registers. Consider the situation where the guest architecture has fewer registers, for instance, 32, than the host architecture, for instance, 64, so that the 32 guest registers can be mapped to the first 32 host registers. When translating an instruction *opcode*, *op1,op2*, the translator can use the fixed mapping to directly translate the operands from guest to host machine registers. In this fashion, the translator can produce code with globally allocated registers without any analysis, simply by reusing register allocation decisions from the guest code.

In contrast, a JIT compiler that operates on intermediated code has to perform a potentially costly global analysis to achieve the same level of register allocation. Thus, what appears to be a limitation may prove to have its virtues depending on the compilation scenario.

10.6 Nontransparent Approach: Runtime Specialization

A common characteristic among the dynamic compilation systems discussed so far is transparency. The dynamic compiler operates in complete independence from static compilation stages and does not make assumptions about, or require changes to, the static compiler.

A different, nontransparent approach to dynamic compilation has been followed by staged runtime specialization techniques [9, 31, 33]. The objective of these techniques is to prepare for dynamic compilation as much as possible at static compilation time. One type of optimization that has been supported in this fashion is value-specific code specialization. Code specialization is an optimization that produces an optimized version by customizing the code to specific values of selected specialization variables.

Consider the code example shown in Figure 10.10. Figure 10.5i shows a dot product function that is called from within a loop in the main program, such that two parameters are fixed ($n = 3$ and $row = [5, 0, 3]$) and only a third parameter (*col*) may still vary. A more efficient implementation can be achieved by specializing the dot function for the two fixed parameters. The resulting function *spec doc*, which retains only the one varying parameter, is shown in Figure 10.10ii.

In principle, functions that are specialized at runtime, such as *spec dot*, could be produced in a JIT compiler. However, code specialization requires extensive analysis and is too costly to be performed fully at runtime. If the functions and the parameters for specialization are fixed at compile time, the static compiler can prepare the runtime specialization and perform all the required code analyses. Based on the analysis results, the compiler constructs code templates for the specialized procedure. The code templates for *spec dot* are shown in Figure 10.11ii in C notation. The templates may be parameterized with respect to missing runtime values. Parameterized templates contain holes that are filled in at runtime with the respective values. For example, template T2 in Figure 10.11ii contains two holes for the runtime parameters $row[0] \ldots row[2]$ (hole h1) and the values $0, \ldots, (n-1)$ (hole h2).

By moving most of the work to static compile time, the runtime overhead is reduced to initialization and linking of the prepared code templates. In the example from Figure 10.10, the program is statically compiled so that in place of the call to routine *dot*, a call to a specialized dynamic code generation agent is inserted. The specialized code generation agent for the example from Figure 10.10, *make spec dot*, is shown in Figure 10.11i. When invoked at runtime, the specialized dynamic compiler looks up the appropriate

```
dot (inf n, int row[ ], int           spec_dot (int col [ ]){
col [ ])                                  int sum = 0;
{                                         sum += 5* col [0];
   int i, sum;                            sum += 3* col [2];
   sum = 0;                               return sum;
 for (i = 0; i<n; i++)                 }
     sum + = row (i)* col (i);
   return sum;                        main {} {
}                                        read (&n, row);
                                         make_spec_dot (n,
 main ( ) {                            row);
   read (&n, row);
   . . .                                 . . .
   while (. . .) {                       while (. . . ) {
     /* n=3, row = {5, 0, 3} */            spec_dot (col);
     dot (n, row, col);                     . . .
     . . .                               }
   }                                   }
}
```

                (i)   (ii)

FIGURE 10.10   Code specialization example. A dot–product function before (i) and after (ii) specialization.

code templates for *spec dot*, fills in the holes for parameters *n* and *row* with their runtime values, and patches the original main routine to link in the new specialized code.

The required compiler support renders these runtime specialization techniques less flexible than transparent dynamic compilation systems. The kind, scope, and timing of dynamic code generation are fixed at compile time and hardwired into the code. Furthermore, runtime code specialization techniques usually require programmer assistance to choose the specialization regions and variables (e.g., via code annotations or compiler directives). Because overspecialization can easily result in code explosion and performance degradation, the selection of beneficial specialization candidates is likely to follow an interactive approach,

```
make_spec_dot (int n, int row [ ] ) {
  buf = alloc ( );                    /* allocate buffer space for spec_dot */

  copy_temp (buf, T1);                /* copy template T1 into buffer */
  for (i=0; i<n; i++) {
      copy_temp (buf, T2);                     /* copy template T2 */
      fill_hole (buf, h1, row [i]);            /* fill hole h1 in T2 */
      fill_hole (buf, h2, i);                  /* file hole h2 in T2 */
  }
  copy_temp (buf, T3);                         /* copy template T3 */
  return buf;
}
```
                              (i)

```
Template T1:        spec_dot (int col []) { int sum = 0;
Template T2:        sum += {hole h1} * col [ {hole h2}];
Template T3:        return sum; }
```
                              (ii)

FIGURE 10.11   Runtime code generation function (i) and code templates (ii) for specializing function dot from Figure 10.10.

where the programmer explores various specialization opportunities. Recently, a system has been developed toward automating the placement of compiler directives for dynamic code specialization [32].

The preceding techniques for runtime specialization are classified as declarative. Based on the programmer declaration, templates are produced automatically by the static compiler. An alternative approach is imperative code specialization. In an imperative approach, the programmer explicitly encodes the runtime templates. C is an extension of the C languages that allows the programmer to specify dynamic code templates [33]. The static compiler compiles these programmer specifications into code templates that are initiated at runtime in a similar way to the declarative approach. Imperative runtime specialization is more general because it can support a broader range of runtime code generation techniques. However, it also requires deeper programmer involvement and is more error prone, because of the difficulty of specifying the dynamic code templates.

10.7 Summary

Dynamic compilation is a growing research field fueled by the desire to go beyond the traditional compilation model that views a compiled binary as a static immutable object. The ability to manipulate and transform code at runtime provides the necessary instruments to implement novel execution services. This chapter discussed the mechanisms of dynamic compilation systems in the context of two applications: dynamic performance optimization and transparent software migration. However, the capabilities of dynamic compilation systems can go further and enable such services as dynamic decompression and decryption or the implementation of security policies and safety checks.

Dynamic compilation should not be viewed as a technique that competes with static compilation. Dynamic compilation complements static compilation, and together they make it possible to move toward a truly write-once-run-anywhere paradigm of software implementation.

Although dynamic compilation research has advanced substantially in recent years, numerous challenges remain. Little progress has been made in providing effective development and debugging support for dynamic compilation systems. Developing and debugging a dynamic compilation system is particularly difficult because the source of program bugs may be inside transient dynamically generated code. Break points cannot be placed in code that has not yet materialized, and symbolic debugging of dynamically generated code is not an option. The lack of effective debugging support is one of the reasons the engineering of dynamic compilation systems is such a difficult task. Another area that needs further attention is code validation. Techniques are needed to assess the correctness of dynamically generated code. Unless dynamic compilation systems can guarantee high levels of robustness, they are not likely to achieve widespread adoption.

This chapter surveys and discusses the major approaches to dynamic compilation with a focus on transparent binary dynamic compilation. For more information on the dynamic compilation systems that have been discussed, we encourage the reader to explore the sources cited in the References section.

References

1. L. Anderson, M. Berc, J. Dean, M. Ghemawat, S. Henzinger, S. Leung, L. Sites, M. Vandervoorde, C. Waldspurger, and W. Weihl. 1977. Continuous profiling: Where have all the cycles gone? In *Proceedings of the 16th ACM Symposium of Operating Systems Principles*, 14.
2. D. Bruening and E. Duesterwald. 2000. Exploring optimal compilation unit shapes for an embedded just-in-time compiler. In *Proceedings of the 3rd Workshop on Feedback-Directed and Dynamic Optimization*.
3. D. Bruening, E. Duesterwald, and S. Amarasinghe. 2001. Design and implementation of a dynamic optimization framework for Windows. In *Proceedings of the 4th Workshop on Feedback-Directed and Dynamic Optimization*.
4. V. Bala, E. Duesterwald, and S. Banerjia. 1999. Transparent dynamic optimization: The design and implementation of Dynamo. Hewlett-Packard Laboratories Technical Report HPL-1999–78.

5. V. Bala, E. Duesterwald, and S. Banerjia. 2000. Dynamo: A transparent runtime optimization system. In *Proceedings of the SIGPLAN '00 Conference on Programming Language Design and Implementation*, 1–12.

6. T. Ball and J. Larus. 1996. Efficient path profiling. In *Proceedings of the 29th Annual International Symposium on Microarchitecture (MICRO-29)*, 46–57.

7. W. Chen, S. Lerner, R. Chaiken, and D. Gillies. 2000. Mojo: A dynamic optimization system. In *Proceedings of the 3rd Workshop on Feedback-Directed and Dynamic Optimization*.

8. R. F. Cmelik and D. Keppel. 1993. Shade: A fast instruction set simulator for execution profiling. Technical Report UWCSE-93-06-06, Department of Computer Science and Engineering, University of Washington, Seattle.

9. C. Consel and F. Noel. 1996. A general approach for run-time specialization and its application to C. In *Proceedings of the 23rd Annual Symposium on Principles of Programming Languages*, 145–56.

10. T. Conte, B. Patel, K. Menezes, and J. Cox. 1996. Hardware-based profiling: an effective technique for profile-driven optimization. *Int. J. Parallel Programming* 24:187–206.

11. C. Chambers and D. Ungar. 1989. Customization: Optimizing compiler technology for SELF, a dynamically-typed object-oriented programming language. In *Proceedings of the SIGPLAN '89 Conference on Programming Language Design and Implementation*, 146–60.

12. Y. C. Chung and Y. Byung-Sun. The Latte Java Virtual Machine. Mass Laboratory, Seoul National University, Korea. latte.snu.ac.kr/manual/html mono/latte.html.

13. R. Cohn and G. Lowney. 1996. Hot cold optimization of large Windows/NT applications. In *Proceedings of the 29th Annual International Symposium on Microarchitecture*, 80–89.

14. D. Ditzel. 2000. Transmeta's Crusoe: Cool chips for mobile computing. In *Proceedings of Hot Chips 12*, Stanford University, Stanford, CA.

15. D. Deaver, R. Gorton, and N. Rubin. 1999. Wiggins/Redstone: An online program specializer. In *Proceedings of Hot Chips 11*, Palo Alto, CA.

16. E. Duesterwald and V. Bala. 2000. Software profiling for hot path prediction: Less is more. In *Proceedings of 9th International Conference on Architectural Support for Programming Languages and Operating Systems*, 202–211.

17. E. Duesterwald, R. Gupta, and M. L. Soffa. 1995. Demand-driven computation of interprocedural data flow. In *Proceedings of the 22nd ACM Symposium on Principles on Programming Languages.* 37–48.

18. L. P. Deutsch and A. M. Schiffman. 1994. Efficient implementation of the Smalltalk-80 system. In *Conference Record of the 11th Annual ACM Symposium on Principles of Programming Languages*, 297–302.

19. K. Ebcioglu and E. Altman. 1997. DAISY: Dynamic compilation for 100% architectural compatibility. In *Proceedings of the 24th Annual International Symposium on Computer Architecture*, 26–37.

20. D. R. Engler. 1996. VCODE: A retargetable, extensible, very fast dynamic code generation system. In *Proceedings of the SIGPLAN '96 Conference on Programming Language Design and Implementation (PLDI '96)*, 160–70.

21. D. H. Friendly, S. J. Patel, and Y. N. Patt. 1998. Putting the fill unit to work: Dynamic optimizations for trace cache microprocessors. In *Proceedings of the 31st Annual International Symposium on Microarchitecture (MICRO-31)*, 173–81.

22. J. Gosling, B. Joy, and G. Steele. 1999. *The Java language specification*. Reading, MA: Addison-Wesley.

23. B. Grant, M. Philipose, M. Mock, C. Chambers, and S. Eggers. 1999. An evaluation of staged run-time optimizations in DyC. In *Proceedings of the SIGPLAN '99 Conference on Programming Language Design and Implementation*, 293–303.

24. M. Gschwind and E. Altman. 2000. Optimization and precise exceptions in dynamic compilation. In *Proceedings of Workshop on Binary Translation*.

25. U. Hoelzle, C. Chambers, and D. Ungar. 1991. Optimizing dynamically-typed object-oriented languages with polymorphic inline caches. In *Proceedings of ECOOP 4th European Conference on Object-Oriented Programming*, 21–38.

26. U. Hoelzle, C. Chambers, and D. Ungar. 1992. Debugging optimized code with dynamic deoptimization. In *Proceedings of the SIGPLAN '92 Conference on Programming Language Design and Implementation*, 32–43.

27. R. J. Hookway and M. A. Herdeg. 1997. FX!32: Combining emulation and binary translation. *Digital Tech. J.* 0(1):3–12.

28. IBM Research. The IBM Jalapeno Project. www.research.ibm.com/jalapeno/.

29. Intel Microprocessor Research Lab. Open Runtime Platform. www.intel.com/research/mrl/orp/.

30. C. Krintz and B. Calder. 2001. Using annotations to reduce dynamic optimization time. In *Proceedings of the SIGPLAN '01 Conference on Programming Language Design and Implementation*, 156–67.

31. M. Leone and P. Lee. 1996. Optimizing ML with run-time code generation. In *Proceedings of the SIGPLAN '96 Conference on Programming Language Design and Implementation*, 137–48.

32. M. Mock, M. Berryman, C. Chambers, and S. Eggers. 1999. Calpa: A tool for automatic dynamic compilation. In *Proceedings of the 2nd Workshop on Feedback-Directed and Dynamic Optimization*.

33. M. Poletta, D. R. Engler, and M. F. Kaashoek. 1997. TCC: A system for fast flexible, and high-level dynamic code generation. In *Proceedings of the SIGPLAN '97 Conference on Programming Language Design and Implementation*, 109–21.

34. S. Richardson. 1993. Exploiting trivial and redundant computation. In *Proceedings of the 11th Symposium on Computer Arithmetic*.

35. K. Scott and J. Davidson. 2001. Strata: A software dynamic translation infrastructure. In *Proceedings of the 2001 Workshop on Binary Translation*.

36. A. Srivastava, A. Edwards, and H. Vo. 2001. Vulcan: Binary translation in a distributed environment. Technical Report MSR-TR-2001-50, Microsoft Research.

37. O. Taub, S. Schechter, and M. D. Smith. 2000. Ephemeral instrumentation for lightweight program profiling. Technical report, Harvard University.

38. Transitive Technologies. www.transitives.com/.

39. D. Ung and C. Cifuentes. 2000. Machine-adaptable dynamic binary translation. *ACM Sigplan Notices* 35(7):41–51.

40. X. Zhang, Z. Wang, N. Gloy, J. Chen, and M. Smith. 1997. System support for automatic profiling and optimization. In *Proceedings of the 16th ACM Symposium on Operating Systems Principles*, 15–26.

11

The Static Single Assignment Form: Construction and Application to Program Optimization

J. Prakash Prabhu[1],
Priti Shankar,
and
Y. N. Srikant

*Department of Computer Science
and Automation,
Indian Institute of Science,
Bangalore, India
prakash.prabhu@gmail.com
priti@csa.iisc.ernet.in
srikant@csa.iisc.ernet.in*

11.1 Introduction

The emergence of the *static single assignment* (SSA) form as an important intermediate representation for compilers has resulted in considerable research in algorithms to compute this form efficiently. The SSA representation is an example of a *sparse* representation where definition sites are directly associated with use sites. Analysis of sparse representations has the advantage of being able to directly access points where relevant data flow information is available. Therefore, one can profitably use this property in improving algorithms for optimizations carried out on older, more traditional intermediate representations.

---

[1] Presently at Bell Laboratories, Bangalore.

```
Read A, B, C
if (A > B)
    if (A > C) max = A
    else max = C
else if (B > C) max = B
    else max = C
Print max
```

FIGURE 11.1   Program in non-SSA form.

In this chapter we present the original construction [18], some recent improvements [19, 51], and applications to some specific optimizations. We discuss conditional constant propagation [44, 45, 55], global value numbering [4, 10, 16, 44, 45], and partial redundancy elimination [33, 38, 43–45, 48]. Several other algorithms for optimizations such as dead code elimination [6, 18, 44], strength reduction [17, 20, 34, 36, 44, 46, 58], array bound check elimination [6, 7, 27, 39], and liveness analysis [5] on sparse representations have not been included because of space constraints.

Section 11.2 discusses the SSA form and its construction. Some variants of the SSA form proposed by Briggs et al. [9] are described in Section 11.3. Section 11.4 contains a detailed presentation of conditional constant propagation algorithms on the the SSA representation. This is followed by a description of value-numbering algorithms in Section 11.5 and partial redundancy elimination algorithms in Section 11.6. Some issues involved in back translation of the SSA form to executable form are discussed in Section 11.7. Section 11.8 concludes this chapter with a discussion. We assume that the reader is familiar with the notion of control flow graphs, basic blocks, paths, and so forth. These definitions are available in [3, 6, 18, 44].

11.2   The Static Single Assignment Form

A program is in SSA form if each of its variables has exactly one definition, which implies that each use of a variable is reached by exactly one definition. The control flow remains the same as in a traditional (non-SSA) program. A special merge operator, denoted ϕ, is used for the selection of values in join nodes. The SSA form is usually augmented with use–definition and/or definition–use chains in its data structure representation to facilitate design of faster algorithms.

Figures 11.1 and 11.4 show two non-SSA form programs, Figures 11.2 and 11.5 show their SSA forms, and Figures 11.3 and 11.6 show the flowcharts of the SSA form.

The program in Figure 11.1 is not in SSA form because there are several assignments to the variable *max*. In the program in Figure 11.2 (with the accompanying flowchart in Figure 11.6), each assignment is made to a different variable, max_i. The variable max_5 is assigned the correct value by the ϕ-function, which takes the value max_i, if the control reaches it via the ith incoming branch from left to right.

The ϕ-functions in the two blocks $B1$ and $B5$ in Figure 11.6 are meant to choose the appropriate value based on the control flow. For example, the ϕ-assignment to RSR_5 in the block $B5$ in Figure 11.6 selects one of RSR_3, RSR_2, or RSR_4 based on the execution following the arc $B2 \rightarrow B5$, $B3 \rightarrow B5$, or $B4 \rightarrow B5$, respectively.

```
Read A, B, C
if (A > B)
    if (A > C) max₁ = A
    else max₂ = C
else if (B > C) max₃ = B
    else max₄ = C
max₅ = φ(max₁, max₂, max₃, max₄)
Print max₅
```

FIGURE 11.2   Program of Figure 11.1 in SSA form.

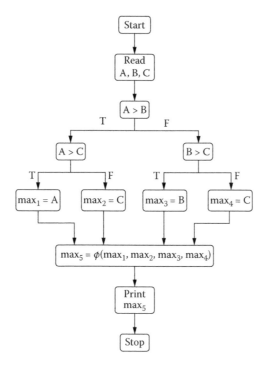

FIGURE 11.3 Flowchart of program in Figure 11.2.

Usually, compilers construct a control flow graph representation of a program first and then convert it to SSA form. The conversion process involves the introduction of statements with assignment to ϕ-functions in several join nodes and renaming of variables that are targets of more than one definition. Of course, the usages of such variables will also be changed appropriately. Not every join node needs a ϕ-function for every variable in the program. Algorithms for the suitable placement of ϕ-functions to ensure a minimal SSA form are described in the next section.

A reader unfamiliar with the SSA form may wonder about the role of ϕ-functions in the final machine code of the program. Since there is no direct translation of a ϕ-function to machine code, a copy instruction needs to be inserted at the end of each predecessor block of the block containing a ϕ-function. This introduces some inefficiency into machine code, which can be compensated for, to some extent, by good register allocation [18]. Carrying out dead code elimination before ϕ-conversion is also required to remove redundant assignment statements.

```
Read A; LSR = 1; RSR = A; LSR = (LSR + RSR)/2;
Repeat
        T = SR * SR;
        if (T > A) RSR = SR
        else if (T < A) LSR = SR
                else begin LSR = SR; RSR = SR; end
        SR = (LSR + RSR)/2;
until (LSR ≠ RSR)
Print SR
```

FIGURE 11.4 Another program in non-SSA form.

Read A; $LSR_1 = 1$; $RSR_1 = A$; $SR_1 = (LSR_1 + RSR_1)/2$;
Repeat
       $LSR_2 = \phi(LSR_5, LSR_1)$;
       $RSR_2 = \phi(RSR_5, RSR_1)$;
       $SR_2 = \phi(SR_3, SR_1)$;
       $T = SR_2 * SR_2$;
       if $(T > A)$ $RSR_3 = SR_2$
       else if $(T < A)$ $LSR_3 = SR_2$
           else begin $LSR_4 = SR_2$; $RSR_4 = SR_2$; end
       $LSR_5 = \phi(LSR_2, LSR_3, LSR_4)$;
       $RSR_5 = \phi(RSR_3, RSR_2, RSR_4)$;
       $SR_3 = (LSR_5 + RSR_5)/2$;
until $(LSR_5 \neq RSR_5)$
Print SR_3

FIGURE 11.5   Program in Figure 11.4 in SSA form.

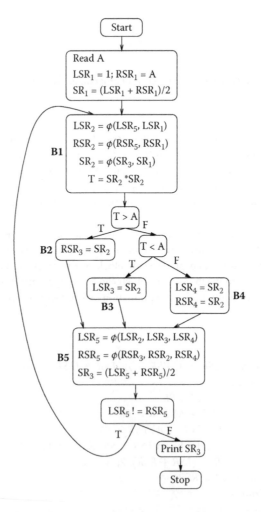

FIGURE 11.6   Flowchart of program in Figure 11.5.

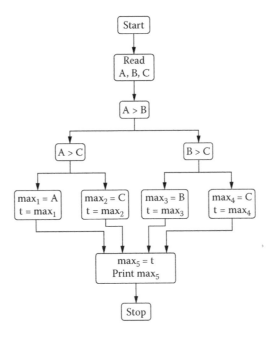

FIGURE 11.7 ϕ-Conversion on program in Figure 11.3.

Figure 11.7 shows a simple example of the effect of ϕ-conversion. During register allocation, the same register would be assigned to t, max_1, max_2, max_3, max_4, and max_5, and thereby copying is eliminated.

11.2.1 Construction of the SSA Form

We now discuss the construction of an SSA form from a flow graph. We consider only scalars and refer the reader to [18, 35] for details on how structures, pointers, and arrays are handled. Our presentation is based on the material in [18].

11.2.1.1 Conditions on the SSA Form

After a program has been translated to SSA form, the new form should satisfy the following conditions for every variable v in the original program:

1. If two paths from nodes with a definition of v converge at a node p, then p contains a trivial ϕ-function of the form $v = \phi(v, v, \ldots, v)$, with the number of arguments equal to the in-degree of v.
2. Each appearance of v in the original program or a ϕ-function in the new program has been replaced by a new variable v_i, leaving the new program in SSA form.
3. Any use of a variable v along any control path in the original program and the corresponding use of v_i in the new program yield the same value for both v and v_i.

Condition 1 above is recursive. It implies that ϕ-assignments introduced by the translation procedure will also qualify as assignments to v, and this in turn may lead to introduction of more ϕ-assignments at other nodes.

11.2.1.2 The Join Set and ϕ-Nodes

Given a set $\mathcal{S}$ of flow graph nodes, we define the set $JOIN(\mathcal{S})$ of nodes from the flow graph to be the set of all nodes n, such that there are at least two nonnull paths in the flow graph that start at two distinct nodes in $\mathcal{S}$ and converge at n. The *iterated join set*, $JOIN^+(\mathcal{S})$ is the limit of the monotonic nondecreasing

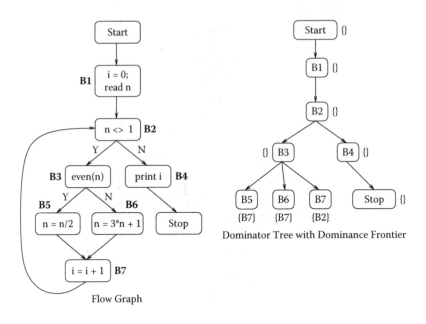

FIGURE 11.8   Example flow graph for SSA form construction.

sequence of sets of nodes

$$JOIN^{(1)}(\mathcal{S}) = JOIN(\mathcal{S})$$
$$JOIN^{(i+1)}(\mathcal{S}) = JOIN(\mathcal{S} \cup JOIN^{(i)}(\mathcal{S}))$$

If $\mathcal{S}$ is defined to be the set of assignment nodes for a variable v, then $JOIN^+(\mathcal{S})$ is precisely the set of flow graph nodes, where ϕ-functions are needed (for v). See [18] for proofs. The set $JOIN^+(\mathcal{S})$ is termed the *iterated dominance frontier*, $DF^+(\mathcal{S})$, which can be computed efficiently in a manner to be described shortly.

11.2.1.3   Dominator Tree

Given two nodes x and y in a flow graph, x *dominates* y if x appears in all paths from the *Start* node to y. The node x *strictly dominates* y if x dominates y and $x \neq y$. x is the *immediate dominator* of y (denoted $idom[y]$) if x is the closest strict dominator of y. A *dominator tree* shows all the immediate dominator relationships. Figure 11.8 shows a flow graph and its dominator tree. Dominator trees can be constructed in time almost linear in the number of edges of a flow graph [42] (see [29] for a linear time but more complicated algorithm).

11.2.1.4   Dominance Frontier

For a flow graph node x, the set of all flow graph nodes y, such that x dominates a predecessor of y but does not strictly dominate y, is called the *dominance frontier* of x and is denoted by $DF(x)$. The following redefinition of $DF(x)$ makes it simple to compute in linear time.

$$DF(x) = DF_{local}(x) \cup \bigcup_{z \, \in \, children(x)} DF_{up}(z)$$
$$DF_{local}(x) = \{y \ \in \ successor(x) \mid idom(y) \neq x\}$$
$$DF_{up}(x) = \{y \ \in \ DF(x) \mid idom(y) \neq parent(x)\}$$

```
function dominance_frontier(n) // n is a node in the dominator tree
begin
    for all children c of n in the dominator tree do
        dominance_frontier(n.c);
    end for
    DF(n) = ∅;
    // DF_local computation
    for all successors of n in the flow graph do
        if (idom(s) ≠ n) then DF(n) = DF(n) ∪ {s};
    end for
    // DF_up computation
    for all children c of n in the dominator tree do
        for all p ∈ DF(c) do
            if (idom(p) ≠ n) then DF(n) = DF(n) ∪ {p};
        end for
    end for
end
```

FIGURE 11.9 Dominance frontier computation.

Here, $children(x)$ and $parent(x)$ are defined over the dominator tree and $successor(x)$ is defined over the flow graph. The algorithm in Figure 11.9 computes the dominance frontier based on the definitions given above. It is called on the root of the dominator tree, and the tree is traversed in postorder. Figure 11.8 shows the DF sets as annotations on the nodes of the dominator tree.

We now extend the definition of DF to act on sets and define the *iterated dominance frontier* on lines similar to $JOIN^+(\mathcal{S})$.

$$DF(\mathcal{S}) = \cup_{x \in \mathcal{S}} DF(x)$$
$$DF^{(1)}(\mathcal{S}) = DF(\mathcal{S})$$
$$DF^{(i+1)}(\mathcal{S}) = DF(\mathcal{S} \cup DF^{(i)}(\mathcal{S}))$$

For each variable v, the set of flow graph nodes that need ϕ-functions is $DF^+(\mathcal{S})$, where $\mathcal{S}$ is the set of nodes containing assignments to v. We do not construct $DF^+(\mathcal{S})$ explicitly. It is computed implicitly during the placement of ϕ-functions.

11.2.1.5 Minimal SSA Form Construction

There are three steps in the construction of the minimal SSA form:

1. Compute DF sets for each node of the flow graph using the algorithm in Figure 11.9.
2. Place trivial ϕ-functions for each variable in the nodes of the flow graph using the algorithm in Figure 11.10.
3. Rename variables using the algorithm in Figure 11.11.

The function *place-phi-function(v)* is called once for each variable v. It can be made more efficient by using integer flags instead of boolean flags as described in [18]. Our presentation uses boolean flags to make the algorithm simpler to understand.

The renaming algorithm in Figure 11.11 performs a top-down traversal of the dominator tree. It maintains a version stack V, whose top element is always the version to be used for a variable usage encountered (in the appropriate range, of course). A counter v is used to generate a new version number. It is possible to use a separate stack for each variable as in [18], so as to reduce the overheads a bit. However, we believe our presentation is simpler to comprehend.

```
function Place-phi-function(v) // v is a variable
// This function is executed once for each variable in the flow graph
begin
    // has-phi(B) is true if a φ-function has already
    // been placed in B
    // processed(B) is true if B has already been processed once
    // for variable v
    for all nodes B in the flow graph do
        has-phi(B) = false; processed(B) = false;
    end for
    W = ∅; // W is the work list
    // Assignment-nodes(v) is the set of nodes containing
    // statements assigning to v
    for all nodes B ∈ Assignment-nodes(v) do
        processed(B) = true; Add(W, B);
    end for
    while W ≠ ∅ do
    begin
        B = Remove(W);
        for all nodes y ∈ DF(B) do
            if (not has-phi(y)) then
            begin
                place < v = φ(v, v, ..., v) > in y;
                has-phi(y) = true;
                if (not processed(y)) then
                begin processed(y) = true; Add(W, y); end
            end
        end for
    end
end
```

FIGURE 11.10   Minimal SSA construction.

Let us trace the steps of the SSA construction algorithm with the help of the example in Figure 11.12. Let us concentrate on the variable n. Blocks B_1, B_5, and B_6 have assignments to n. (Read n is also considered as an assignment.) The dominance frontier of B_1 being null, no ϕ-function is introduced while processing it. A ϕ-function is introduced for n in B_7 while processing B_5 (this will not be repeated while processing B_6). B_2 gets a ϕ-function for n when B_7 is handled.

Let us now understand how different instances of n are renamed. The instruction read n in B_1 becomes read n_0 while processing B_1. At the same time, the second parameter of the ϕ-function for n in block B_2 is changed to n_0. Processing B_2 in the top-down order results in changing the statement $n = \phi(n, n_0)$ to $n_1 = \phi(n, n_0)$, and the new version number 1 is pushed onto the version stack V. This results in changing the comparisons $n \neq 1$ to $n_1 \neq 1$ in block B_2 and $even(n)$ to $even(n_1)$ in block B_3. The expressions $n/2$ in block B_5 and $3 * n + 1$ in block B_6 also change to $n_1/2$ and $3 * n_1 + 1$, respectively. A new version of n, viz., n_2 (or n_3, respectively) is created while processing the assignment in block B_5 (or B_6, respectively). The version number 2 (or 3, respectively) is pushed onto V. This results in changing the parameters of the ϕ-function in block B_7 as shown in Figure 11.12. After finishing with B_5 (or B_6, respectively), V is popped to remove 2 (or 3, respectively), before processing B_7. A new version n_4 is created while processing the ϕ-statement in B_7, which in turn changes the first parameter of the ϕ-function for n in block B_2, from n to n_4. V is then popped and recursion unwinds. The variable i is treated in a similar manner.

```
function Rename-variables(x, B) // x is a variable and B is a block
begin
    v_e = Top(V); // V is the version stack of x
    for all statements s ∈ B do
        if s is a non-φ statement then
            replace all uses of x in the RHS(s) with Top(V);
        if s defines x then
        begin
            replace x with x_v in its definition; push x_v onto V;
            // x_v is the renamed version of x in this definition
            v = v + 1; // v is the version number counter
        end
    end for
    for all successors s of B in the flow graph do
        j = predecessor index of B with respect to s
        for all φ-functions f in s which define x do
            replace the j^th operand of f with Top(V);
        end for
    end for
    for all children c of B in the dominator tree do
        Rename-variables(x, c);
    end for
    repeat Pop(V); until (Top(V) == v_e);
end

begin // calling program
    for all variables x in the flow graph do
        V = ∅; v = 1; push 0 onto V; // end-of-stack marker
        Rename-variables(x, Start);
    end for
end
```

FIGURE 11.11   Renaming variables.

11.2.1.6   Complexity of SSA Graph Construction

We define R, the size of a flow graph, as follows.

$$R = max\{N_f, E_f, A, M\},$$

where

> N_f is the number of nodes in the flow graph.
>
> E_f is the number of edges in the flow graph.
>
> A is the number of assignments in the flow graph.
>
> M is the number of uses of variables in the flow graph.

The construction of the dominance frontier and the SSA form in theory take $O(R^2)$ and $O(R^3)$ time, respectively. However, according to [18], measurements on programs show that the size of dominance frontiers in practice is small, and hence the entire construction, including the construction of dominance frontiers, has complexity $O(R)$ in practice.

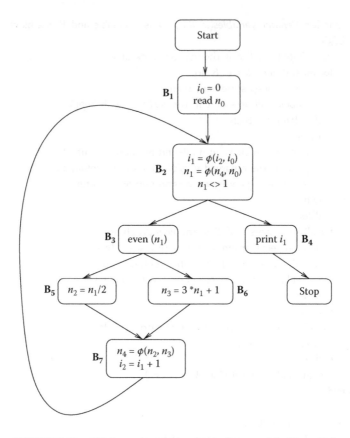

FIGURE 11.12   SSA form construction for the flow graph in Figure 11.8.

11.2.1.7   A Note on the Size of SSA Graphs

An SSA form is usually augmented with links from every unique definition of a variable to its uses (corresponding to d-u information). Some algorithms need SSA forms augmented with links that go from every use of a variable to its unique definition (corresponding to u-d information). If there are n definitions in a program, and each of these could reach n uses, then both d-u and u-d chains will have $O(n^2)$ links. However, an SSA graph with d-u or u-d information will have only $O(n)$ links because of the factoring carried out by ϕ-functions. This is illustrated in Figures 11.13 and 11.14. In Figure 11.13, the d-u chain of each definition of i contains all three uses of i in the second switch statement. However, in Figure 11.14, because of the factoring introduced by the ϕ-function, each definition reaches only one use.

```
switch(j)
    case 1: i=10; break;
    case 2: i=20; break;
    case 3: i=30; break;
end
switch(k)
    case 1: x = i * 3 + 5; break;
    case 2: y = i * 6 + 15; break;
    case 3: z = i * 9 + 25; break;
end
```

FIGURE 11.13   Program in non-SSA form.

```
switch( j )
    case 1: i₁=10; break;
    case 2: i₂=20; break;
    case 3: i₃=30; break;
end
i₄ = φ(i₁, i₂, i₃);
switch(k)
    case 1: x = i₄ * 3 + 5; break;
    case 2: y = i₄ * 6 + 15; break;
    case 3: z = i₄ * 9 + 25; break;
end
```

FIGURE 11.14   Program in Figure 11.13 in SSA form.

The renaming algorithm of Figure 11.11 can be augmented to establish *d-u* and *u-d* links. This requires every statement to be a separate node in the flow graph. It also requires keeping a pointer to the node defining a variable on the version stack along with the name of the variable. The rest of the process is simple.

The SSA form increases the number of variables. If there are n variables in a program and each of these has k definitions, then the SSA form would have nk variables to take care of the nk definitions. It may not be possible to map these nk variables back to n variables during machine code generation because of code movements that could have taken place as a result of optimization.

11.2.2   A Linear Time Algorithm for Placing ϕ-Nodes

In this section, we describe a simple linear time algorithm for computing ϕ-nodes developed by Sreedhar and Gao [51]. This algorithm is based on a new data structure, called the DJ-graph, which is basically the dominator tree augmented with edges called "join" edges, which may lead to nodes where data flow information is merged. Cytron's original algorithm [18] for computing ϕ-nodes precomputes the dominance frontier for every node in the flow graph. In some cases this precomputation takes time quadratic in the number of nodes of the flow graph. For example, consider the "ladder graph" shown in Figure 11.15a. To compute $DF(2)$, we use the expression

$$DF(2) = DF_{up}(4) \cup \{3\}$$

Computing the set $DF_{up}(4)$ involves first computing $DF(4)$ and then selecting those nodes in $DF(4)$ that are not dominated by node 2 and thus require time linear in the size of $DF(4)$. Similarly, computing $DF(4)$ requires time linear in the size of $DF(6)$:

$$DF(4) = DF_{up}(6) \cup \{5\}$$
$$DF(6) = DF_{up}(8) \cup \{7\}$$
$$DF(8) = \{9\}$$

Every step in the dominance frontier computation adds a new node to the dominance frontier (DF_{local}) and examines all the nodes in the dominance frontier of the node(s) it immediately dominates. In general, a ladder graph with n nodes, computing the dominance frontier for a node placed similarly to node 2 above, takes time

$$1 + 2 + \cdots + \frac{n}{2} = O(n^2)$$

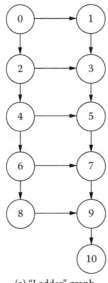

(a) "Ladder" graph

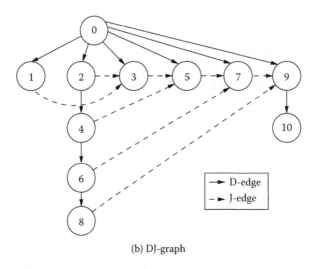

(b) DJ-graph

FIGURE 11.15   The "ladder" graph and its DJ-graph.

However, as can be seen from Table 11.1, the maximum number of ϕ-nodes required for this ladder graph (for a single variable) is linear in the number of nodes of the graph. While the example above may seem contrived, there are examples of nested *repeat-until* loops for which the dominance frontier computation takes quadratic time [18]. In the following section, we will see a data structure called the DJ-graph that is used to compute the iterated dominance frontier without precomputing the dominance frontier of every node.

11.2.2.1   The DJ-Graph and Its Properties

The DJ-graph is a directed graph whose nodes are the nodes of the control flow graph. There are two kinds of edges between the nodes, called the D-edges (dominator edges) and J-edges (join edges). The construction of the DJ-graph is performed in two steps:

TABLE 11.1 Dominance
frontiers for the "ladder"
graph shown in Figure 11.15

n	$DF(n)$
0	{}
1	{3}
2	{3,5,7,9}
3	{5}
4	{5,7,9}
5	{7}
6	{7,9}
7	{9}
8	{9}
9	{}
10	{}

1. Build the dominator tree of the control flow graph. The edges in the dominator tree constitute the D-edges of the DJ-graph.

2. Add the J-edges to the dominator tree to obtain the complete DJ-graph. A J-edge is an edge $x \rightarrow y$ from the control flow graph such that x does not strictly dominate y. In other words, y is a control flow join point and x is a predecessor of y. The time complexity of adding the J-edges to the DJ-graph from the control flow graph is O(E), where E is the number of edges in the control flow graph.

The total time for constructing the DJ-graph is linear in the size of the control flow graph since the dominator tree can be contructed in linear time [29]. The DJ-graph for the ladder graph of Figure 11.15a is shown in Figure 11.15b. The solid edges represent the D-edges, while dotted ones represent J-edges. Any DJ-graph has the following three key properties:

1. The number of edges in the DJ-graph is less than the total number of nodes and edges in the corresponding control flow graph. This can be shown as

$$\text{Number of edges in DJ-graph} = |\text{D-edges}| + |\text{J-edges}|$$
$$\leq N_f - 1 + E_f$$
$$< N_f + E_f$$

2. If a node y has to be in the $DF(x)$, then there should be a J-edge from some node $z \in SubTree(x)$ to y. $Subtree(x)$ is subtree rooted at x in the dominator tree and hence is the set of all the nodes dominated by x.
   Proof: Let z be any node in $SubTree(x)$ and $y \in DF(x)$. Suppose there exists no J-edge $z \rightarrow y$.
   Case 1: There is a D-edge $z \rightarrow y$. Then $(x \text{ dom } z \wedge z \text{ dom } y) \Rightarrow (x \text{ dom } y) \Rightarrow [y \notin DF(x)]$, giving a contradiction.
   Case 2: There is no D-edge $z \rightarrow y$. Since $y \in DF(x)$, x dominates a predecessor of y but does not strictly dominate y. Let $pred(y)$ be z. Since $(x \text{ dom } z)$, $[z \in Subtree(x)]$. Now there are two cases: (a) $(z \text{ dom } y) \Rightarrow (x \text{ sdom } y) \Rightarrow y \notin DF(x)$, giving a contradiction, and (b) $(z \text{ !dom } y)$. By definition of a J-edge, there should exist a J-edge from z to y.

   Thus to compute $DF(x)$, we do not need to look beyond those nodes y, which are connected by means of a J-edge to a node z in the $SubTree(x)$.

3. Let W be the set of nodes y such that there exists a J-edge from a node $z \in SubTree(x)$ to y. The $DF(x)$ is composed of *exactly* those set of those nodes $y \in W$ such that $level(y) \leq level(x)$. The *level* of a node is the depth of the node in the dominator tree with the *level* of root being defined as 0.

For every $z \in SubTree(x)$
if $((z \to y$ is a J-edge) $\wedge$ $(level(y) \le level(x)))$
DF(x) = DF(x) $\cup$ {y}

FIGURE 11.16 Computing the dominance frontier of single node x.

Proof: Let $w \in W$ and $z \to w$ be a J-edge. There are two cases to consider with respect to the levels of w and x:

Case 1: $level(w) \le level(x)$. Now $[z \in pred(w)]$ and $(x$ dom $z)$. Also $[level(w) \le level(x)] \Rightarrow (x$!sdom $w)$. From this, it follows that x dominates a predecessor of w and does not strictly dominate w. Hence $w \in DF(x)$.

Case 2: $level(w) > level(x)$. Now there are two possibilities: (a) $[w \in SubTree(x)] \Rightarrow (x$ sdom $w)$ $\Rightarrow [w \notin DF(x)]$. (b) $[w \notin SubTree(x)]$. Let $u = idom(w)$. Now since $level(w) > level(x)$, $level(u) \ge level(x) \Rightarrow (u$!dom $x)$. This means there exists a path (in the flow graph) from start node to w through x and z that does not pass through u, which implies $(u$!dom $w)$, which is a contradiction. Thus, $[w \in SubTree(x)]$, which reduces to case 1.

11.2.2.2 Dominance Frontier of a Single Node Using a DJ-Graph

The algorithm shown in Figure 11.16 computes $DF(x)$ for a given node x. It directly follows from properties 2 and 3 where two conditions are checked for every node in the $SubTree(x)$ to determine membership in $DF(x)$. For example, to compute $DF(2)$, only the join edges $8 \to 9, 6 \to 7, 4 \to 5$, and $2 \to 3$ are considered. Since the heads of all these J-edges are at the same level as 2, $DF(2) = \{3, 5, 7, 9\}$.

To compute the dominance frontier of a set of nodes, we can call the above algorithm for every node in the set and take the union of the resulting dominance frontiers. However, this could result in redundant computation and lead to quadratic behavior. For example, to compute $DF(\{4, 2\})$, since $DF(\{4, 2\}) = DF(4) \cup DF(2)$, we can compute $DF(4)$ and $DF(2)$ individually and take their union. Assume we call $DF(4)$ first and then $DF(2)$. While computing $DF(2)$, we would visit the subtree rooted at node 4 again, even though it has been processed by the first call to $DF(4)$. To prevent this reprocessing, the nodes have to be properly ordered by their levels, and state has to be maintained about the visits that have already been made.

11.2.2.3 Dominance Frontier for a Set of Nodes N_α

The algorithm given in Figures 11.17, 11.18, and 11.19 computes the dominance frontier of a set of nodes N_α. The algorithm is based on two observations:

1. Before computing $DF(x)$, if the $DF(z)$ for some node $z \in SubTree(x)$ has already been computed, it is not recomputed while finding $DF(x)$.

Compute_IDF(N_α)
   For every $x \in N_\alpha$,
      $next(x) = DAT[level(x)]$
      $DAT[level(x)] = x$
   endfor
   while ($x =$ **Next_Bottom_Up_Node**() != NULL)
      $CurrentRoot = x$
      Mark x as *visited*
      Top_Down_Visit(x)
   endwhile

FIGURE 11.17 Dominance frontier computation of set of nodes N_α.

Top_Down_Visit(*x*)

    For each *y* which is a successor of *x* in the DJ-graph

        if (*x* → *y*) is a J-edge

          if (*level*(*y*) ≤ *level*(*x*))

            if (φ has not been placed at *y*)

              Place a φ at *y*

              IDF = *IDF* ∪ {*y*}

              if (*y* ∉ *N*_α)

                next(*y*) = *DAT*[*level*(*y*)]

                DAT[*level*(*y*)] = *y*

              endif

            endif

          endif

        else

          if (*y* has not been marked as visited)

            Mark *y* as visited

            Top_Down_Visit(*y*)

          endif

        endif

    endfor

FIGURE 11.18 Dominance frontier computation of set of nodes N_α.

2. To compute $DF(x)$, we need to look at only those J-edges $z \rightarrow y$, where $z \in SubTree(x)$ and $level(y) \leq level(x)$.

Let the set of nodes for which the dominance frontier needs to computed be N_α. The algorithm can be summarized by the following sequence of steps:

1. It orders the set of nodes in N_α in bottom-up fashion based on their level numbers in the dominator tree. It uses this ordering to call the dominance frontier computation (described in step 2). A direct access table (DAT), indexed by level numbers, is used to maintain this ordering. It allows for dynamic updates when new nodes are added to the iterated dominance frontier while making sure the newly added nodes do not violate the bottom-up processing of nodes.

2. During the computation of $DF(x)$, the nodes in $SubTree(x)$ are visited top-down (using the D-edges), without processing those nodes that are marked, that is, those that have already been visited, thus avoiding entire subtrees that have been processed once.

3. During the top-down visit, the J-edges are used to identify the nodes to be added to the iterated dominance frontier (IDF) (based on their level numbers). These new nodes are added to the DAT if not already present. These nodes along with remaining nodes in the DAT then carry forward the recursion.

4. No node is added more than once to any list of bottom-up ordered nodes maintained in the DAT.

The algorithm never processes a D-edge or a J-edge more than once. Also, each node is inserted and removed from the DAT not more than once. Thus, it takes time linear in the size of the DJ-graph:

$$
\begin{aligned}
\text{Time complexity} &= O(N_{DJ} + E_{DJ}) \\
&= O(N_f + E_{DJ}) \\
&= O(N_f + E_f) \text{ (by Property 1 of DJ-graph)} \\
&= O(E)
\end{aligned}
$$

Next_Bottom_Up_Node()
    if $(DAT[CurrentLevel] != NULL)$
      $x = DAT[CurrentLevel]$
      $DAT[CurrentLevel]=next(x)$
      return x
    endif
    for $i = CurrentLevel$ downto 1 do
      if $(DAT[i]! = NULL)$
        $CurrentLevel = i$
        $x = DAT[i]$
        $DAT[i] = next(x)$
        return x
      endif
    endfor

FIGURE 11.19   Dominance frontier computation of set of nodes N_α.

The algorithm thus runs in time linear in the number of edges of the control flow graph. The key to this linear time complexity is the direct access table. Each element of the table is a linked list that contains nodes that are at the same level (the corresponding index into the table). Insertion and deletion of a node from the DAT take constant time.

An application of the above algorithm on the DJ-graph of Figure 11.15b for $N_\alpha = \{0, 2\}$ is traced in Figure 11.20. It shows the direct access table at various points in the algorithm. Initially, the two nodes are inserted into *DAT* and aligned in bottom-up order. A call to *Top_Down_Visit*(2) is placed that first examines the J-edge $2 \rightarrow 3$, adds 3 to the iterated dominance frontier *IDF* and to *DAT*. It goes on to examine edges $4 \rightarrow 5, 6 \rightarrow 7$, and $8 \rightarrow 9$ while adding 5, 7, and 9 to *IDF* and *DAT*. Note that this top-down traversal visits the D-edges of *SubTree*(2) exactly once. The process is then carried out with *Top_Down_Visit*(9), *Top_Down_Visit*(7), and *Top_Down_Visit*(5), none of them contributing anything more to *IDF*. The final call to *Top_Down_Visit*(0) does not visit the subtrees rooted at 2, 3, 5, 7, and 9 since they have already been visited once. This avoids the recomputation that leads to quadratic behavior in Cytron's algorithm. The J-edges correspond to the DF_{local} relation used in Cytron's dominance frontier computation. In the DJ-graph, the DF_{local} relation is explicitly represented with join edges that help in getting a linear time algorithm for placing ϕ-nodes.

The algorithms for ϕ-function placement seen in this section were recursive in nature. Das and Ramakrishna [19] proposed an efficient iterative algorithm for ϕ-function computation based on the computation of *merge set* of the nodes in the control flow graph [8]. The merge set of a node n is the set of nodes where ϕ-functions may have to be placed if a variable definition is placed in n. The algorithm precomputes the merge set for every node in the control flow graph and later uses them for actual ϕ-function placement. The merge sets are computed using the DJ-graph structure. This algorithm is especially advantageous when the basic blocks are very *dense*, that is, when they have a lot of variable definitions. The earlier approaches to finding ϕ-functions computed the ϕ-points for every variable definition. This implies that if two or more variable definitions are in the same basic block, although the ϕ-function placement may be at the same location, it is still recomputed. However, in the new approach, the ϕ-function is computed with respect to a basic block instead of a variable definition. Once the merge sets have been precomputed, to find the nodes at which a ϕ-function has to be placed because of variable definitions in a set of nodes N_α, it is sufficient to take the union of merge sets of nodes in N_α. Details of the algorithm can be found in [19].

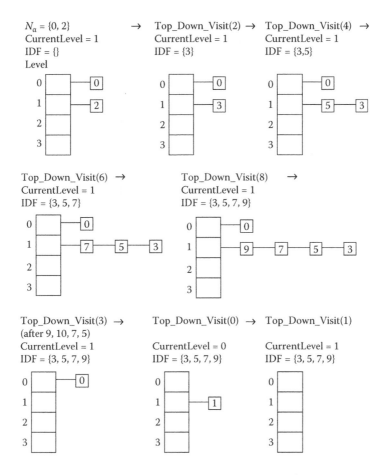

FIGURE 11.20 Trace of *DAT* during IDF computation of $N_\alpha = \{0, 2\}$.

11.3 Variants of the SSA Form

The dominance frontier algorithm developed by Cytron et al. [18], called the "minimal" SSA form, can insert a ϕ-function for a variable at a program flow point where the variable is not live; that is, it can insert a ϕ-function at a point even if the corresponding variable is not used later in the program.

Consider the example program skeleton shown in Figure 11.21a. Of the three variables x, y, and z defined and used in the program, x is defined and used wholly in blocks B1 and B2; y is defined in blocks B1 and B2 and used only in blocks B3 and B4; z is defined in blocks B1 and B2 and used only in block B5. As shown in Figure 11.21b, the minimal SSA form would insert a ϕ for both x and y in B5 even though they are no longer used after B5.

11.3.1 Pruned SSA

Choi et al. [12] proposed a variant form of SSA called pruned SSA. In this form, once the dominance frontiers have been computed, a live variable analysis is done to find the set of variables that are live at the entry of a basic block. Pruned SSA would then insert a ϕ-function for a variable v in a basic block b only if v is live at the entry of b. The pruned SSA form for the example program of Figure 11.21a is shown in Figure 11.22a. While pruned SSA may insert fewer ϕ-functions than minimal SSA, it is expensive because of the need for a global data flow analysis.

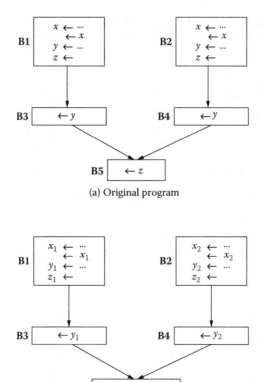

(a) Original program

(b) SSA form of program

FIGURE 11.21   Example program and its minimal SSA form.

11.3.2   Semi-Pruned SSA

Briggs et al. [9] suggested another variant of SSA called semi-pruned SSA based on the observation that many temporary variables have very short lifetimes, that is, they are defined and used *only* within a single basic block. These variables would not require ϕ-functions at the join points. Semi-pruned SSA computes that set of variables that are live on entry to *some* basic block. These variables are named *non-locals*, that is, those that have definitions outside the current basic block. Semi-pruned SSA computes the iterated dominance frontier only for the definitions in *non-locals*. Thus, the number of ϕ-functions inserted would be between that for minimal and pruned SSA. The cost of computing *non-locals* is less than that of a full-blown live-variable analysis since it does not involve any iteration or elimination. Figure 11.23 gives the algorithm for computing *non-locals*. The semi-pruned SSA form for the example program of Figure 11.21a is shown in Figure 11.22b. Since x is not live at the entry of any of the blocks (it is defined and used wholly within B1 and B2), a ϕ is not placed at B5 for x, whereas a ϕ is placed for y, as it is live across a basic block.

Briggs et al. [9] also noted that different SSA-based optimizations may require different flavors of SSA. For example, the extra ϕ-functions of minimal SSA may help value numbering discover congruences that are usually not found in their absence [18]. However, if SSA is used for finding live ranges during register

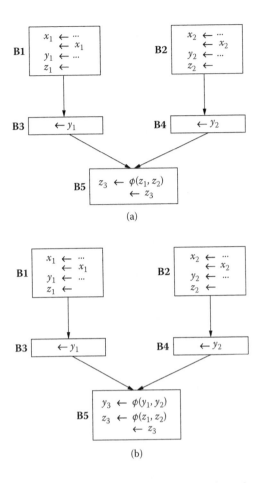

FIGURE 11.22 Pruned and semi-pruned SSA forms. (a) Pruned SSA form for example in Figure 11.21(a). (b) Semi-pruned SSA form for example in Figure 11.21(a).

allocation, pruned and semi-pruned SSA forms are more useful since the extra ϕ can cause the lifetime of its argument variables to be extended unnecessarily. Results of constant propagation are not affected by any specific flavor of SSA.

We next discuss some optimizations that can be carried out on the SSA form.

Compute_Non_Locals
    For every block B
        $killed = \phi$
        For each instruction $v \leftarrow x \; op \; y$ in B
            if $x \notin killed$ then
                $non\_locals \leftarrow non\_locals \cup \{x\}$
            if $y \notin killed$ then
                $non\_locals \leftarrow non\_locals \cup \{y\}$
            $killed \leftarrow killed \cup \{v\}$

FIGURE 11.23 Algorithm for computing the "non-locals" set.

11.4   Conditional Constant Propagation

Constant propagation is a well-known compiler optimization during which variables that can be determined to contain only constant values at runtime in all possible executions of the program are discovered. These values are propagated throughout the program, and expressions whose operands can be determined to be only constants are also discovered and evaluated. In effect, there is a symbolic execution of the program with the limited purpose of discovering constant values. The following examples help explain the intricacies of constant propagation.

The simplest case is that of straight line code without any branches, as in a basic block. This requires only one pass through the code, with forward substitutions and no iteration. However, such a one-pass strategy cannot discover constants in conditionals and loops. For such cases we need to carry out data flow analysis involving work-lists and iterations. A simple algorithm of this kind adds successor nodes to the work-list as symbolic execution proceeds. Nodes are removed one at a time from the work-list and executed. If the new value at a node is different from the old value, all the successors of the node are added to the work-list. The algorithm stops when the work-list becomes empty. Such an algorithm can catch constants in programs such as program A but not in programs such as program B in Figure 11.24.

Conditional constant propagation handles programs such as program B in Figure 11.24 by evaluating all conditional branches with only constant operands. This uses a work-list of edges (instead of nodes) from the flow graph. Furthermore, neither successor edge of a branch node is added to the work-list when the branch condition evaluates to a constant value (true or false); only the relevant successor (true or false) is added.

Conditional constant propagation algorithms on SSA graphs are faster and more efficient than the ones on flow graphs. They find at least as many constants as the algorithms on flow graphs (even with d-u chains, this efficiency cannot be achieved with flow graphs; see [55]). In the following sections, we illustrate the conditional constant propagation algorithms on flow graphs by means of an example and follow it up with a detailed explanation of the algorithm on SSA graphs. Our description is based on the algorithms presented in [55]. We assume that each node contains exactly one instruction and that expressions at nodes can contain at most one operator and two operands (except for ϕ-functions). The graphs are supposed to contain N nodes, E_f flow edges, E_s SSA edges (corresponding to d-u information), and V variables.

Program A	Program A after simple CP
$a = 10$;	$a = 10$;
if $(i > j)$ $b = a$ else $c = a$;	if $(i > j)$ $b = 10$ else $c = 10$;
Program B	**Program B after simple CP**
$a = 10$; $b = 20$;	$a = 10$; $b = 20$;
if $(a == 10)$ $x = b$ else $x = a$;	if $(a == 10)$ $x = 20$ else $x = 10$;
	Program B after CCP with flow graph
	$a = 10$; $b = 20$;
	$x = 10$;
Program C	**Program C after CCP with flow graph**
$a = 10; b = 20$;	$a = 10; b = 20$;
if $(b == 20)$ $a = 30$;	$a = 30$;
$d = a$;	$d = a$; (a cannot be determined to be a constant)
	Program C after CCP with SSA graph
	$a = 10$; // unused code, removable
	$b = 20$;
	$a = 30$;
	$d = 30$; (a has been determined to be a constant)

FIGURE 11.24   Limitations of various CP algorithms.

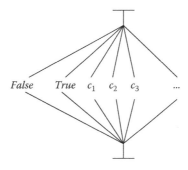

FIGURE 11.25   The lattice of constants.

$$\text{any} \sqcap \top = \text{any}$$
$$\text{any} \sqcap \bot = \bot$$
$$C_i \sqcap C_j = \bot, \text{ if } i \neq j$$
$$C_i \sqcap C_j = C_i, \text{ if } i = j$$

FIGURE 11.26   The meet operator.

11.4.1   The Lattice of Constants

The infinite lattice of constants used in the constant propagation algorithms is a flat lattice as shown in Figure 11.25. $\top$ stands for an as yet undetermined constant, $\bot$ for a nonconstant, and C_i for a constant value. The meet operator $\sqcap$ is defined as in Figure 11.26. The symbol any stands for any lattice value in Figure 11.26.

11.4.2   The CCP Algorithm: Flow Graph Version

All variables are supposed to be used only after they are defined and are initialized by the Conditional Constant Propagation (CCP) algorithm to $\top$ at every node. This is a special feature of the Wegman–Zadeck algorithm, which enables it to find more constants than other algorithms (in programs with loops). For details of this effect, the reader is referred to [55]. Each node is supposed to have two lattice cells per variable, one to store the incoming value and the other to store the computed value (exit value). There are also two lattice cells at each node to store the old and new values of the expression. All the edges going out of the start node are initially put on the work-list and marked as *executable*. The algorithm does not process any edges that are not so marked. A marked edge is removed from the work-list, and the node at the target end of the edge is symbolically executed. This execution involves computing the lattice value of all variables (not just the assigned variable). The incoming value of a variable x at a node y is computed as the *meet* of the exit values of x at the preceding nodes of y, and the values are stored in the incoming lattice cells of the respective variables at the node.

Now the expression at the node is evaluated according to the rules given in Figure 11.28. If the node contains an assignment statement, and if the new value of the expression is lower (in lattice value) than the existing value of the assignment target variable, then all outgoing edges of the node are marked as executable and added to the work-list. The new value is also stored in the exit lattice cell of the assignment target variable.

If the node contains a branch condition, and if the new value of the expression is lower than its existing value, then one of the following two actions is taken. If the new value is $\bot$, *both* the successor edges of the node are marked and added to the work-list. If the new value is a constant (*true or false*), *only* the corresponding successor edge is marked and added to the work-list. This step enables elimination of *unreachable code*. The new value of the expression is also stored in a lattice cell at the node.

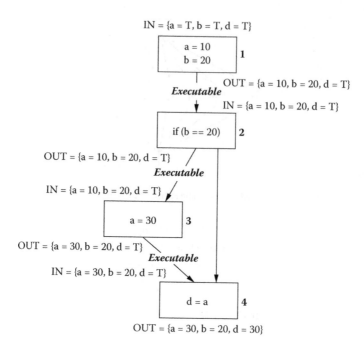

FIGURE 11.27   Example illustrating CCP on the flow graph.

In both these cases, note that no action is taken if the value of the expression does not change from one visit to the next. There is also a copying of all incoming lattice cells apart from the target variable of the assignment to their exit cells to enable the successor nodes to pick up values of variables. Figure 11.27 gives the trace of this algorithm on the example seen earlier in Figure 11.24. All edges except $2 \rightarrow 4$ are marked executable. This helps in determining that d is always assigned a value of 30 and helps eliminate the *if* statement.

However, this algorithm is too slow in practice and hence is not used as it is in any compiler. It is used with *d-u* information that helps in eliminating most of the lattice cell copy operations. The major source of inefficiency is that it uses the flow graph edges for both propagation of values and tracking reachable code. The SSA-based algorithm does this more efficiently because it uses different types of edges for each task. We present this version now.

$$\text{val}(a \ op \ b) = \bot, \text{ if either or both of } a \text{ and } b \text{ is } \bot$$
$$= C_i \ op \ C_j, \text{ if val}(a) \text{ and val}(b) \text{ are constants, } C_i \text{ and } C_j, \text{ respectively}$$
$$= \top, \text{ otherwise}$$

Special rules for $\vee$ and $\wedge$
any stands for an element of $\{\text{true,false}, \bot, \top\}$
These rules indicate that if one of the operands is known in the shown cases, then the other operand is irrelevant

$$\text{any} \vee \text{true} = \text{true}$$
$$\text{true} \vee \text{any} = \text{true}$$
$$\text{any} \wedge \text{false} = \text{false}$$
$$\text{false} \wedge \text{any} = \text{false}$$

FIGURE 11.28   Expression evaluation.

```
// G = (N,E_f,E_s) is the SSA graph, with flow edges and SSA edges,
// and V is the set of variables used in the SSA graph
begin
    Flowpile = {(Start → n) | (Start → n) ∈ E_f };
    SSApile = ∅;
    for all e ∈ E_f do e.executable = false; end for
    v.cell is the lattice cell associated with the variable v
    for all v ∈ V do v.cell = ⊤; end for
    // y.oldval and y.newval store the lattice values of expressions at node y
    for all y ∈ N do
        y.oldval = ⊤; y.newval = ⊤;
    end for
    while (Flowpile ≠ ∅) or (SSApile ≠ ∅) do
    begin
        if (Flowpile ≠ ∅) then
        begin
            (x, y) = remove(Flowpile);
            if (not (x, y).executable) then
            begin
                (x, y).executable = true;
                if (φ-present(y)) then visit-φ(y)
                    else if (first-time-visit(y)) then visit-expr(y);
                // visit-expr is called on y only on the first visit
                // to y through a flow edge; subsequently, it is called
                // on y on visits through SSA edges only
                if (flow-outdegree(y) == 1) then
                    // Only one successor flow edge for y
                    Flowpile = Flowpile ∪ {(y, z) | (y, z) ∈ E_f};
        end
        // if the edge is already marked, then do nothing
    end
```

FIGURE 11.29   CCP algorithm with SSA graphs.

11.4.3 The CCP Algorithm: SSA Version

SSA forms along with extra edges (SSA edges) corresponding to d-u information enable efficient algorithms for constant propagation of the flow graph. We add an edge from every definition to each of its uses in the SSA form. The new algorithm uses both flow graph and SSA edges and maintains two work-lists, one for each. Flow graph edges are used to keep track of reachable code, and SSA edges help in propagation of values. Flow graph edges are added to the flow work-list whenever a branch node is symbolically executed or whenever an assignment node has a single successor (all this is subject to value changes as before). SSA edges coming out of a node are added to the SSA work-list whenever there is a change in the value of the assigned variable at the node. This ensures that all *uses* of a definition are processed whenever a definition changes its lattice value. This algorithm needs only one lattice cell per *variable* (globally, not on a per node basis) and two lattice cells per node to store expression values. Conditional expressions at branch nodes are handled as before. However, at any join node, the *meet* operation considers only those predecessors that are marked *executable*. The SSA-based algorithm is presented in Figures 11.29, 11.30, 11.31, and 11.32.

```
        if (SSApile ≠ Ø) then
        begin
            (x, y) = remove(SSApile);
            if (φ-present(y)) then visit-φ(y)
                else if (already-visited(y)) then visit-expr(y);
                // A false returned by already-visited implies that y
                // is not yet reachable through flow edges
        end
    end // Both piles are empty
end


function φ-present(y) // y ∈ 𝒩
begin
    if y is a φ-node then return true
        else return false
end


function visit-φ(y) // y ∈ 𝒩
begin
    y.newval = ⊤;
    // ‖y.instruction.inputs‖ is the number of parameters of the φ-instruction at node y
    for i = 1 to ‖y.instruction.inputs‖ do
        Let p_i be the i^{th} predecessor of y ;
        if ((p_i, y).executable) then
        begin
            Let a_i = y.instruction.inputs[i];
            // a_i is the i^{th} input and a_i.cell is the lattice cell associated with that variable
            y.newval = y.newval ⊓ a_i.cell;
        end
    end for
    if (y.newval < y.instruction.output.cell) then
    begin
        y.instruction.output.cell = y.newval;
        SSApile = SSApile ∪ {(y, z) | (y, z) ∈ ℰ_s };
    end
end
```

FIGURE 11.30   CCP algorithm with SSA graphs cont'd.

11.4.3.1 An Example

Consider program C in Figure 11.24. The flow graph and the SSA graph for this program are shown in and Figures 11.33 and 11.34. It is clear that at node $B5$, a cannot be determined to be a constant by the CCP algorithm using a flow graph because of the two definitions of a reaching $B5$. The problem is due to the *meet* operation executed at $B5$ using both its predecessors while determining the lattice value of a. This problem is avoided by the CCP algorithm using the SSA graph. It uses only those predecessors that have edges to the current node marked as *executable*. In this example, only $C4$ would be considered while performing the *meet* operation to compute the new value for a_3, because the other edge $(C3, C5)$ is not marked *executable*. As shown in Figure 11.24, the SSA version of the algorithm determines a_3 to be a constant (of the same value as a_2), and assigns its value to d.

```
function visit-expr(y) // y ∈ 𝒩
begin
    Let input₁ = y.instruction.inputs[1];
    Let input₂ = y.instruction.inputs[2];
    if (input₁.cell == ⊥ or input₂.cell == ⊥) then
        y.newval = ⊥
    else if (input₁.cell == ⊤ or input₂.cell == ⊤) then
            y.newval = ⊤
        else // evaluate expression at y as in Figure 11.20
            y.newval = evaluate(y);
            It is easy to handle instructions with one operand
    if y is an assignment node then
        if (y.newval < y.instruction.output.cell) then
        begin
            y.instruction.output.cell = y.newval;
            SSApile = SSApile ∪ {(y,z) | (y,z) ∈ ℰₛ };
        end
    else if y is a branch node then
        begin
            if (y.newval < y.oldval) then
            begin
                y.oldval = y.newval;
                switch(y.newval)
                    case ⊥: // Both true and false branches are equally likely
                        Flowpile = Flowpile ∪ {(y,z) | (y,z) ∈ ℰ_f };
                    case true: Flowpile = Flowpile ∪ {(y,z) | (y,z) ∈ ℰ_f and
                                        (y,z) is the true branch edge at y };
                    case false: Flowpile = Flowpile ∪ {(y,z) | (y,z) ∈ ℰ_f and
                                        (y,z) is the false branch edge at y };
                end switch
            end
        end
end
```

FIGURE 11.31   CCP algorithm with SSA graphs cont'd.

11.4.3.2 Asymptotic Complexity

Each SSA edge will be examined at least once and at most twice because the lattice value of each variable can be lowered only twice. Each flow graph edge will be examined only once. During a visit to a node, all operations take constant time. As before, the time for adding an edge (either flow or SSA) to a pile is charged to the edge removal operation. Thus, the total time taken by the algorithm is $O(E_f + E_s)$. Theoretically, E_s can be as large as $O(max\{E_f, N\})^2$ [18], and thus, this algorithm can become $O(E_f^2)$. However, in practice, E_s is usually $O(max\{E_f, N\})$, so the time for constant propagation with the SSA graph is practically linear in the size of the program ($max\{E_f, N\}$).

11.5   Value Numbering

Value numbering is one of the oldest and still a very effective technique that is used for performing several optimizations in compilers [3, 4, 10, 15, 16, 46, 49]. The central idea of this method is to assign numbers (called *value numbers*) to expressions in such a way that two expressions receive the same number if

function *first-time-visit*(y) // $y \in \mathcal{N}$
// This function is called when processing a flow graph edge
begin // Check in-coming flow graph edges of y
   for all $e \in \{(x, y) \mid (x, y) \in \mathcal{E}_f\}$
     if *e.executable* is true for more than one edge e
       then return *false* else return *true*
   end for
   // At least one in-coming edge will have *executable* true
   // because the edge through which node y is entered is
   // marked as *executable* before calling this function
end

function *already-visited*(y) // $y \in \mathcal{N}$
// This function is called when processing an SSA edge
begin // Check in-coming flow graph edges of y
   for all $e \in \{(x, y) \mid (x, y) \in \mathcal{E}_f\}$
     if *e.executable* is true for at least one edge e
       then return *true* else return *false*
   end for
end

FIGURE 11.32   CCP algorithm with SSA graphs cont'd.

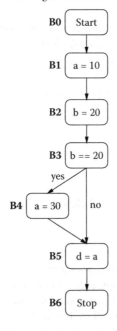

FIGURE 11.33   Flow graph of program C in Figure 11.24.

the compiler can prove that they are equal for all possible program inputs. This technique is useful in finding redundant computations and folding constants. Even though it was originally proposed as a local optimization technique applicable to basic blocks, it can be modified to operate globally. We assume that expressions in basic blocks have at most one operator (except for ϕ-functions in SSA forms). There are two principal techniques to prove equivalence of expressions: *hashing* and *partitioning*.

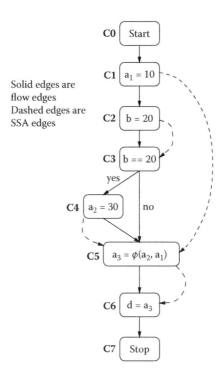

Solid edges are
flow edges
Dashed edges are
SSA edges

FIGURE 11.34   SSA graph of program C in Figure 11.24.

The hashing scheme is simple and easy to understand. It uses a hashing function that combines the operator and the value numbers of operands of an expression (assume non-ϕ, for the present) contained in the instruction and produces a unique value number for the expression. If this number is already contained in the hash table, the name corresponding to this existing value number refers to an earlier computation in the block. This name holds the same value as the expression for all inputs, so the expression can be replaced by the name. Any operator with known constant values is evaluated, and the resulting constant value is used to replace any subsequent references. It is easy to adapt this algorithm to apply commutativity and simple algebraic identities without increasing its complexity.

The partitioning method operates on SSA forms and uses the technique of Deterministic Finite Automaton (DFA) minimization [1] to partition the values into congruence classes. Two expressions or computations are *congruent* to each other if their operators are identical and their operands are congruent to each other. Two constants are congruent to each other if their values are the same. Two variables are congruent to each other if the computations defining them are congruent to each other. The process starts by putting all expressions with the same operator into the same class and then refining the classes based on the equivalence of operands. The partitioning-based technique is described in [4, 10, 14, 45]. Partitioning-based methods are global techniques that operate on the whole SSA graph, unlike hash-based methods. We provide a sample of the partitioning method through an example (see Figures 11.35 and 11.36).

The initial and final value partitions are as shown in Figure 11.35. To begin with, variables that have been assigned constants with the same value are put in the same partition (P_1 and P_5), and so are variables assigned expressions with the same operator (P_3). ϕ-instructions of the same block are also bundled together (P_2 and P_4). Now we start examining instructions in each partition pairwise and split the partition if the operands of the instructions are not in the same partition. For example, we split P_3 into Q_3 and Q_4, because $x_2 + 2$ is not equivalent to $x_2 + 3$ (constants 2 and 3 are not equivalent). The partitioning technique splits partitions only when necessary, whereas the hashing technique combines partitions whenever equivalence is found. Continuing the example, we do not split x_2 and y_2 into

Initial Partitions	Final Partitions
$P_1 = \{x_1, y_1\}$	$Q_1 = \{x_1, y_1\}$
$P_2 = \{x_2, y_2\}$	$Q_2 = \{x_2, y_2\}$
$P_3 = \{x_5, y_5, x_6, y_6\}$	$Q_3 = \{x_5, y_5\}$
$P_4 = \{x_4, y_4\}$	$Q_4 = \{x_6, y_6\}$
$P_5 = \{x_3, y_3\}$	$Q_5 = \{x_4, y_4\}$
	$Q_6 = \{x_3, y_3\}$

FIGURE 11.35   Initial and final value partitions for the SSA graph in Figure 11.36

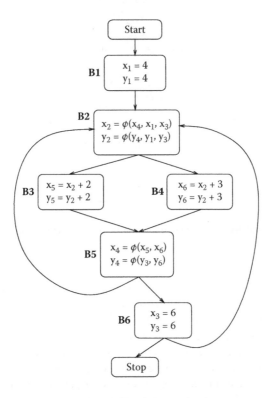

FIGURE 11.36   SSA graph for value numbering with the partitioning technique.

different classes because their corresponding inputs, viz., (x_1, y_1), (x_3, y_3), and (x_4, y_4), belong to identical equivalent classes, P_1, P_5, and P_4, respectively. A similar argument holds for x_4 and y_4 as well. After partitioning, this information may be used for several optimizations such as common subexpression elimination, invariant code motion, and so on as explained in [4].

Even though there are programs on which partitioning is more effective than hashing (and vice-versa), partitioning is not as effective as hashing on real-life programs [10]. Hashing is the method of choice for most compilers. We informally describe the local hashing-based value numbering on basic blocks in the next section and then go on to describe the global algorithm on SSA graphs in detail in the following sections. Our presentation is based on the techniques described in [10, 15, 45, 46].

11.5.1   Hashing-Based Value Numbering: Basic Block Version

We briefly describe the data structures used in the basic block version of value numbering using hashing and illustrate the algorithm with an example. The basic block version uses three tables: *HashTable*, *ValnumTable*, and *NameTable*. *HashTable* is indexed using the output of a hashing function that takes an expression (with

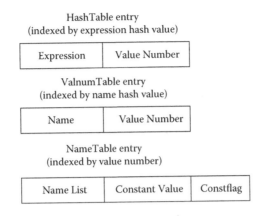

FIGURE 11.37 Data structures for value numbering with basic blocks.

value numbers of its operands replacing the operands themselves) as a parameter. It stores expressions and their value numbers. *ValnumTable* is indexed using another hashing function that takes a variable name (a string) as a parameter. It stores variable names and their value numbers. *NameTable* is indexed using a value number, and it stores the variable names (a list) corresponding to this value number and the constant value associated with this variable, if any, indicated by the field *Constflag*. The first name in the name list always corresponds to the first definition for the corresponding value number. A counter is used to generate new value numbers by simple incrementing. The array B, of size n, stores the n instructions of a basic block. Any good hashing scheme presented in books on data structures and algorithms can be used for the hashing functions [2, 53]. The data structures used in value numbering are shown in Figure 11.37.

We illustrate the hashing algorithm on flow graphs by means of an example. Source-level language statements and the corresponding intermediate code in the form of quadruples are shown in Figure 11.38, along with the final transformed code. Let us run through the instructions in the basic block one at a time and observe how value numbering works on them. Processing the first instruction, $a = 10$, results in a being entered into *ValnumTable* and *NameTable* with a value number of, for instance, 1, along with its constant value. When handling instruction 2, a is looked up in the *ValnumTable*, and its constant value is fetched from *NameTable*. Constant folding results in b getting a constant value of 40. The name b is entered into *ValnumTable* and *NameTable* with a new value number of, for instance, 2, and its constant value of 40.

HLL Program	Quadruples before Value Numbering	Quadruples after Value Numbering
$a = 10$	1. $a = 10$	1. $a = 10$
$b = 4*a$	2. $b = 4*a$	2. $b = 40$
$c = i*j+b$	3. $t1 = i*j$	3. $t1 = i*j$
$d = 15*a*c$	4. $c = t1+b$	4. $c = t1+40$
$e = i$	5. $t2 = 15*a$	5. $t2 = 150$
$c = e*j+i*a$	6. $d = t2*c$	6. $d = 150*c$
	7. $e = i$	7. $e = i$
	8. $t3 = e*j$	8. $t3 = i*j$
	9. $t4 = i*a$	9. $t4 = i*10$
	10. $c = t3+t4$	10. $c = t1+t4$
		(Instructions 5 and 8 can be deleted)

FIGURE 11.38 Example of value numbering.

ValNumTable

Name	Value Number
a	1
b	2
i	3
j	4
t1	5
c	10
t2	7
d	8
e	3
t3	5
t4	9

HashTable

Expression	Value Number
$i * j$	5
$t1 + 40$	6
$150 * c$	8
$i * 10$	9
$t1 + t4$	10

FIGURE 11.39   The *HashTable* and *ValNumTable* for the example in Figure 11.38.

The names i and j are entered into the two tables with new value numbers when processing instruction 3. The expression $i * j$ is searched in *HashTable* (with the value numbers of i and j used during hashing). Since it is not found there, it is entered into *HashTable* with a new value number. The name $t1$ is also entered into *ValnumTable* and *NameTable* with the same value number. The constant value of b, viz., 40, replaces b in instruction 4 and is inserted into *HashTable* with a new value number, and c enters the tables with this value number. Instructions 5, 6, and 7 are processed in a similar way. The name $t2$ becomes a constant of value 150, and this value of 150 replaces $t2$ in instruction 6. During the processing of instruction 8, e is replaced by i, and the expression $i * j$ is found in *HashTable*. *NameTable* gives the name of the first definition corresponding to it as $t1$. Therefore, $t1$ replaces $t3$ in instruction 10. Contents of the three tables are shown in Figures 11.39 and 11.40. Instructions 5 and 8 can be deleted, since the holding variables, $t2$ and $t3$, have already been replaced by 150 and $t1$, respectively, in instructions referencing them.

11.5.2   Value Numbering with Hashing and SSA Forms

In this section, we describe the value numbering algorithm on the SSA form of a program using hashing. The local value numbering algorithm for basic blocks can be applied on the SSA form of a program to obtain a global value numbering algorithm. We use a reverse postorder over the SSA flow graph to guide traversals over the dominator tree. One of the reasons for choosing a reverse postorder over the SSA flow graph is to ensure that all the predecessors of a block are processed before the block is processed (this is obviously not possible in the presence of loops, and the effect of loops will be explained later). The dominator tree for a flow graph is shown in Figure 11.43, along with a depth-first tree for the same flow

NameTable

Name	ConstantValue	Constant Flag
a	10	T
b	40	T
e, i		
j		
t3, t1		
t2	150	T
d		
c		

FIGURE 11.40   The *NameTable* for the example in Figure 11.38.

```
function SSA_Value_Numbering(B) // B is a basic block
begin
  scope = new HashTable_Scope
  Push scope onto the Scope Stack
  for i = 1 to n // n = number of statements in B
  begin
    Let s be the statement at B[i]
    if s is an assignment of the form x = y op z in B
      expr = ValNum(y) op ValNum(z)
      // Apply algebraic laws here to simplify expr
      v = Hash(expr) // Apply the hash function to expr
      if Present_Hash(v)
        w = Get_Defining_Var(v)
        Insert_ValNum(x, v, w)
        Remove s from B
      else
        Insert_ValNum(x, v, x)
        Insert_Hash(expr, v, x)
      endif
    endif
    if s is an assignment of the form x = y in B
      w = Get_Replacing_Var(y)
      v = ValNum(y)
      Insert_ValNum(x, v, w)
    endif
    if s is φ-function of the form x ← φ(y₁, y₂, ...)
      if s is meaningless
        Insert_ValNum(x, ValNum(y₁), y₁)
        Remove s from B
      else if s is redundant
        Let w be the defining variable of the earlier non-redundant φ
        Insert_ValNum(x, ValNum(w), w)
        Remove s from B
      else
        v = Hash(φ(y₁, y₂, ...))
        Insert_ValNum(x, v, x)
        Insert_Hash(φ(y₁, y₂, ...), v, x)
      endif
    endif
  endfor
  for i = 1 to | succ(B) |
    Let Cᵢ be a successor of B at index i
    Update the φ function inputs in Cᵢ
  endfor
  // Note that in the following for loop, the recursive calls on
  // the children of B are performed in reverse postorder
  for i = 1 to | Children(B) | in the Dominator Tree
    Let Cᵢ be a child of B at index i
    SSA_Value_Numbering(Cᵢ)
  endfor
  Pop the top of the Scope Stack
end
```

FIGURE 11.41   Value numbering with SSA form.

function *ValNum*(x)
>   Returns the Value Number corresponding to variable x from the *ValumTable*

function *Hash*(expr)
>   Applies the hashing function on expr and returns the result

function *Present_Hash*(v)
>   Returns *true* is an entry indexed by v exists in *Hashtable* else returns *false*

function *Get_Defining_Var*(v)
>   Returns the defining variable corresponding to an entry indexed by v in *HashTable*

function *Insert_Valnum*(x, v, y)
>   Inserts an entry in *ValnumTable* at index given by value number v for variable x
>   and sets the replacing variable as y

function *Insert_Hash*(expr, v, x)
>   Inserts an entry in *Hashtable* at index given by value number v for an expression expr
>   and sets the defining variable as x

function *Get_Replacing_Var*(v)
>   Returns the replacing variable corresponding to an entry indexed by v in *Valnumtable*

FIGURE 11.42   Value numbering with SSA form cont'd.

graph and traversal orders on these. The algorithm requires a scoped *HashTable* similar to one used in the processing of symbol tables for lexically scoped languages. The scope of the tables extends along the paths in the dominator tree. Every time a basic block is visited during the traversal of the dominator tree, new table entries corresponding to the block's scope are added to the table. These entries are removed when a basic block has completed processing. The SSA form used here does not need any additional edges in the form of *u-d* or *d-u* information. The structure of the tables used in this algorithm is shown in Figure 11.44.

The SSA form introduces some specialties in instruction processing. There will be no "hanging expressions" (expressions not attached to any variable), because no definition is killed in the SSA form.

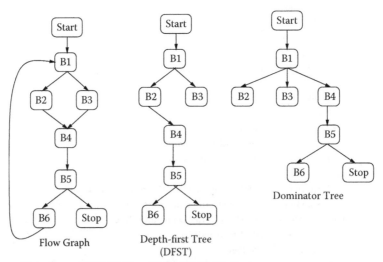

Postorder on DFST: B6, Stop, B5, B4, B2, B3, B1, Start
Reverse postorder on DFST: Start, B1, B3, B2, B4, B5, Stop, B6
Visit order on dominator tree is same as reverse postorder

FIGURE 11.43   Flowchart, DFST, and dominator tree.

HashTable entry
(indexed by expression hash value)

Expression	Value Number	Parameters for ϕ-function	Defining Variable

ValnumTable
(indexed by name hash value)

Variable Name	Value Number	Constant Value	Replacing Variable

FIGURE 11.44   Data structures for value numbering with SSA forms.

This also means there is no need to restore any old entries in tables when the processing of a block is completed. It is enough if the new entries are deleted. This reduces the overheads in managing scoped tables. We do not need the *NameTable* here. The SSA names of the variables themselves can be used as value numbers.

A computation C (e.g., of the form $a = b\ op\ c$) becomes redundant if $b\ op\ c$ has already been computed in one of the nodes dominating the node containing C. If the defining variable of the dominating computation is x, then C can be deleted, and all occurrences of a can be replaced by x. This is recorded in the *ValnumTable* by entering a and the value number of x into it and setting the field *replacing-variable* to x. From now on, whenever an expression involving a is to be processed, we search for a in the *ValnumTable* and get its *replacing-variable* field (which contains x). This replaces a in the expression being processed. While processing an instruction of the form $p = q$, we take the *replacing-variable* of q (e.g., r) and enter it along with p in the *ValnumTable*. This ensures that any future references of p are also replaced by r.

We maintain a global *ValnumTable* and a scoped *HashTable* as before, but over the dominator tree (*ValnumTable* is not scoped). For example, in Figure 11.43, a computation in block $B5$ can be replaced by a computation in block $B1$ or $B4$, since the tables for $B1$, $B4$, and $B5$ will be together while processing $B5$. It is possible that no such previous computations are found in the *HashTable*, in which case we generate a new value number and store the expression in the computation along with the new value number in the *HashTable*. The defining variable of the computation is also stored in the global *ValnumTable* along with the new value number. A global table is needed for *ValnumTable* while processing ϕ-instructions.

Processing ϕ-instructions is slightly more complicated. A ϕ-instruction receives inputs from several variables along different predecessors of a block. The inputs need not be defined in the immediate predecessors or dominators of the current block. They may be defined in any block that has a control path to the current block. For example, in Figure 11.45, while processing block $B9$, we need definitions of a_2, a_6, and so on, which are not in the same scope as $B9$ (over the dominator tree). However, each incoming arc corresponds to exactly one input parameter of the ϕ-instruction. This global nature of inputs requires a global *ValnumTable*, containing all the variable names in the SSA graph.

During the processing of a ϕ-instruction, it is possible that one or more of its inputs are not yet defined because the corresponding definitions reach the block through back arcs. Such entries will not be found in the *ValnumTable*. In such a case, we simply assign a new value number to the ϕ-expression and record the defining variable of the ϕ-instruction along with this new value number in the global *ValnumTable*. The ϕ-expression is also stored with the new value number in the scoped *HashTable*. It may not be out of place to mention here that value numbering based on *partitioning* can handle some of the cases where definitions reach through back arcs. For details, refer to [10, 45] and the example discussed later in this section.

If all the input variables are found in the global *ValnumTable*, then we first replace the inputs of the ϕ-instruction by the entries found in the *ValnumTable* and then go on to check whether the ϕ-expression is either *meaningless* or *redundant*. If neither of these is true, a new value number is generated, and the simplified ϕ-expression and its defining variable are entered into the tables as explained before.

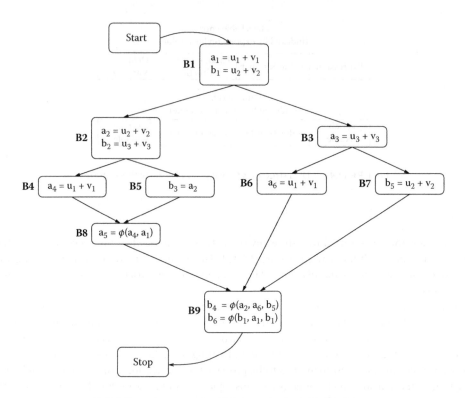

FIGURE 11.45   Example of value numbering with SSA forms.

A ϕ-expression is *meaningless* if all its inputs are identical. In such a case, the corresponding ϕ-instruction can be deleted, and all occurrences of the defining variable of the ϕ-instruction can be replaced by the input parameter. This is recorded in the global *ValnumTable* along with the value number of the input parameter. For example, the instruction $u = \phi(a, b, c)$ may become $u = \phi(x, x, x)$ if a, b, and c are all equivalent to x, as determined from the entries in *ValnumTable*. In such a case, we delete the instruction and record u in *ValnumTable* along with x and its value number, so that future occurrences of u can be replaced by x.

A ϕ-expression is *redundant* if there is another ϕ-expression in the *same basic block* with exactly the same parameters. Note that we cannot use another ϕ-expression from a dominating block here because the control conditions for the blocks may be different. For example, the blocks $B1$ and $B4$ in Figure 11.43 may have the same ϕ-expression, but they may yield different values at runtime depending on the control flow. *HashTable* can be used to check the redundancy of a ϕ-expression in the block. If a ϕ-expression is indeed redundant, then the corresponding ϕ-instruction can be deleted and all occurrences of the defining variable in the redundant ϕ-instruction can be replaced by the earlier nonredundant one. This information is recorded in the tables. The complete algorithm is shown in Figures 11.41 and 11.42.

Figures 11.45 and 11.46 show an SSA graph before and after value numbering. Figure 11.47 shows the dominator tree and a reverse postorder for the same SSA graph. Block $B8$ has a meaningless ϕ-instruction, and block $B9$ has a redundant ϕ-instruction. Instructions such as $a_2 = b_1$ in block $B2$ can perhaps be deleted but are shown in Figure 11.46 to explain the functioning of the algorithm. The SSA graph in Figure 11.45 has not been obtained by translation from a flow graph; it has been constructed to demonstrate the features of the algorithm.

As another example, consider the SSA graph shown in Figure 11.36. Hashing-based techniques discover fewer equivalences as shown below.

$$\{x_1, y_1\}, \{x_3, y_3\}, \{x_2\}, \{x_4\}, \{x_5\}, \{x_6\}, \{y_2\}, \{y_4\}, \{y_5\}, \{y_6\}$$

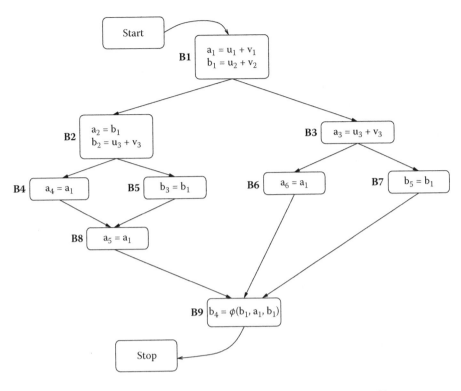

FIGURE 11.46 Example of value numbering with SSA forms cont'd.

This is partly because of the back arcs. x_2 and y_2 will always be assigned different value numbers, because the values of x_3, x_4, y_3, and y_4 reach the block $B2$ through back edges, and their corresponding instructions would not have been processed (present in blocks $B5$ and $B6$) while block $B2$ was being processed. x_5 and y_5 are not assigned the same value number because x_2 and y_2 do not have the same value number. The same is true of x_6 and y_6.

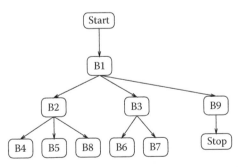

Reverse postorder on the SSA graph that is used with the dominator tree above:

Start, B1, B3, B7, B6, B2, B5, B4, B8, B9, Stop

FIGURE 11.47 Dominator tree and reverse postorder for Figure 11.45.

11.6   Partial Redundancy Elimination

Partial redundancy elimination (PRE) is a powerful compiler optimization that subsumes global common subexpression elimination and loop-invariant code motion and can be modified to perform additional code improvements such as strength reduction as well. PRE was originally proposed by Morel and Renvoise [43]. They showed that elimination of redundant computations and the moving of invariant computations out of loops can be combined by solving a more general problem, that is, the elimination of computations performed twice on a given execution path. Such computations were termed partially redundant. PRE performs insertions and deletions of computations on a flow graph in such a way that after the transformation, each path, in general, contains fewer occurrences of such computations than before. Most compilers today perform PRE. It is regarded as one of the most important optimizations, and it has generated substantial interest in the research community [13, 20–24, 33, 34, 36–38, 45, 46, 48].

In spite of its benefits, Morel and Renvoise's algorithm has some shortcomings. It is not optimal in the sense that it does not eliminate all partial redundancies that exist in a program, and it performs redundant code motion. It involves performing bidirectional data flow analysis, which, some claim, is in general more complex than unidirectional analysis [38]. Knoop et al. decomposed the bidirectional structure of the PRE algorithm into a sequence of unidirectional analyses and proposed an optimal solution to the problem with no redundant code motion [37, 38].

In this section, we informally describe a simple algorithm for partial redundancy elimination for a program not in the SSA form [48]. It is based on well-known concepts, namely, availability, anticipability, partial availability, and partial anticipability. The algorithm is computationally and lifetime optimal. Its essential feature is the integration of the notion of safety into the definition of partial availability and partial anticipability. It requires four unidirectional bit vector analyses. A special feature of this algorithm is that it does not require edge-splitting transformation to be done before application of the algorithm.

An informal description of the idea behind the algorithm follows. We say an expression is *available* at a point if it has been computed along all paths reaching this point with no changes to its operands since the computation. An expression is said to be *anticipable* at a point if every path from this point has a computation of that expression with no changes to its operands in between. *Partial availability* and *partial anticipability* are weaker properties with the requirement of a computation along "at least one path" as against "all paths" in the case of availability and anticipability.

We say a point is *safe* for an expression if it is either available or anticipable at that point. *Safe partial availability (anticipability)* at a point differs from partial availability (anticipability) in that it requires all points on the path along which the computation is partially available (anticipable) to be safe. In the example given in Figure 11.48, partial availability at the entry of node 4 is true, but safe partial availability at that point is false, because the entry and exit points of node 3 are not safe. In Figure 11.49, safe partial availability at the entry of node 4 is true. We say a computation is *safe partially redundant* in a node if it is locally anticipable and is safe partially available at the entry of the node. In Figure 11.49, the computation in node 4 is safe partially redundant.

The basis of the algorithm is to identify safe partially redundant computations and make them totally redundant by the insertion of new computations at proper points. The totally redundant computations after the insertions are then replaced. If $a + b$ is the expression of interest, then by insertion we mean insertion of the computation $h = a + b$, where h is a new variable; replacement means substitution of a computation, such as $x = a + b$, by $x = h$.

Given a control flow graph, we compute availability, anticipability, safety, safe partial availability, and safe partial anticipability at the entry and exit points of all the nodes in the graph. We then mark all points that satisfy both safe partial availability and safe partial anticipability. Now we note that the points of insertion for the transformation are the entry points of all nodes containing the computation whose exit point is marked but whose entry point is not, as well as all edges whose heads are marked but whose tails are not. We also note that replacement points are the nodes containing the computation whose entry or exit point is marked.

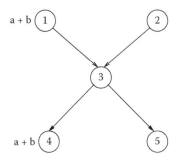

FIGURE 11.48 Node 4: a + b not safe partially available.

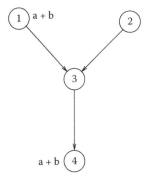

FIGURE 11.49 Node 4: a + b is safe partially redundant.

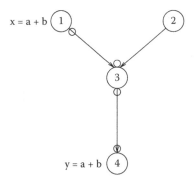

FIGURE 11.50 Before PRE.

Alternatively, if we consider the paths formed by connecting all the adjacent points that are marked, we observe that the points of insertion are the nodes corresponding to the starting points of such paths and the edges that enter *junction* nodes on these paths. The computations to be replaced are the ones appearing on these paths. For the example in Figure 11.50, small circles correspond to marked points. Based on the above observation, we see that node 1 and edge (2, 3) are the points of insertion, and nodes 1 and 4 are the points of replacement. The graph after the transformation is shown in Figure 11.51.

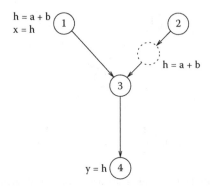

FIGURE 11.51   After PRE.

In the next section, we will describe in detail an algorithm for performing PRE on the SSA form of a program [33].

11.6.1   Partial Redundancy Elimination on the SSA Form

Most optimizations that use the SSA form are based on variables and do not act on program expressions. In this section, we describe SSAPRE [33], an algorithm for performing PRE based entirely on the SSA form. In the absence of SSAPRE, to do PRE on any program in SSA form, the program first needs to be translated out of SSA, and then a traditional bit-vector-based PRE is performed on the program and the result back-converted into SSA form to continue performing other SSA-based optimizations. SSAPRE avoids this conversion back and forth between SSA and non-SSA forms.

A distinctive feature of any SSA-based optimization is that it does not require an iterative data flow analysis, unlike traditional optimizations that operate on the control flow graph. This is because information can be represented only at those points where it changes in the SSA graph and hence can be propagated faster. Also, SSA-based optimizations do not need to distinguish between global and local versions of problems (inter- and intra-basic block), which is necessary in case of bit-vector problems on the control flow graph. Both these features are present in SSAPRE.

The main challenge in performing PRE on the SSA form is to identify expressions that may have different instances of variables as operands but that are potentially redundant. For example, consider the example in Figure 11.52. Although $a_3 + b_1$ seems to be computationally different from $a_1 + b_1$, it is actually partially redundant with respect to $a_1 + b_1$.

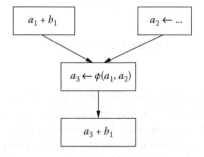

FIGURE 11.52   Partially redundant expressions having different SSA variable instances.

11.6.2 The SSAPRE Algorithm

The SSAPRE algorithm works by representing the computation of an expression *exp* by a hypothetical temporary *h*, for instance, the *redundancy class variable* (RCV). A definition of *exp* corresponds to a store of *exp* into *h*. The use of *exp* corresponds to a load from *h*. It then applies the analysis of PRE on the SSA form of *h* to introduce full redundancies. Finally, compiler temporaries are introduced to remove these redundancies by saving expressions into temporaries and reusing them. The program in its final representation will be in SSA form.

The algorithm assumes that all the critical edges in the control flow graph have been split. A critical edge is an edge from a node with more than one successor (for instance, *a*) to a node with more than one predecessor (for instance, *b*). Usually these edges are points that act as bottlenecks for code hoisting (out of nodes like *b*). By splitting this edge (and inserting a node, for instance, *c*, on this edge), code can still be hoisted out of *b* and placed at the new node *c*, thereby eliminating a partial redundancy.

The SSAPRE algorithm is performed in six steps. Before applying these six steps, it performs a single pass on the SSA program to identify *lexically* identical expressions. These are expressions whose operators and operands are the same, ignoring the SSA versions. For example, $a_3 + b_1$ and $a_1 + b_2$ are lexically identical. The algorithm uses a Φ-operator similar to ϕ of SSA computation for representing the merge of the redundancy class variables.

The first two steps of the algorithm correspond to the SSA computation algorithm of [18], the only difference being that here both these steps work on the expressions (i.e., the RCVs) instead of the original variables of the program. The steps involved in SSAPRE, in order, are:

1. **Φ-insertion:** Inserts a Φ-operator for the RCVs at specific merge points in the program. The Φ insertions are not necessarily minimal.
2. **Rename:** Gives SSA versions to the occurrences of RCVs in the program. After this step, all occurrences of an expression $a + b$ represented by the same SSA version of RCV compute the same value. It also renames the operands of the Φ-function. Some Φ-operands correspond to paths along which $a + b$ does not occur but that are merge points for operands of the expression. These operands are represented by $\perp$. After this step, the SSA graph is very dense because of introduction of RCV instances and their corresponding Φ-operators.
3. **Down-safety:** Identifies Φs that are down-safe, that is, Φ-blocks at which the expression $a + b$ represented by RCV *h* is fully anticipable. This is done by a backward analysis on the SSA graph.
4. **WillBeAvail:** Identifies Φs at which the expression $a + b$ is available assuming PRE insertions will be performed on appropriate incoming edges to the Φ-block.
5. **Finalize:** Uses the results of the previous step to insert computations of $a + b$ into *h* on an incoming edge to the Φ-block to ensure $a + b$ is available at Φ. It then links up uses of *h* to the definitions that have been newly inserted. Finally, it removes extraneous Φs and gets the SSA form of *h* in minimal form.
6. **Code Motion:** Introduces real temporaries to eliminate redundant computations of $a+b$ by walking over the SSA graph of *h*, introducing the store of $a + b$ into *t* and giving each *t* its unique SSA version. The Φs for *h* are translated into ϕ for the temporaries.

Figure 11.53 shows the candidate SSA program that we will use to illustrate the application of the SSAPRE algorithm. The final result after application of SSAPRE on this program is shown in Figure 11.58, below. In the following subsections, we give a detailed description of each of the six steps. Before the algorithm begins, it assumes that the critical edges have already been split and that the dominance tree and dominance frontier relations have been computed.

11.6.2.1 Φ-Insertion

A Φ for an expression is required whenever two different values of an expression merge at a common point in the program. There are two different conditions for the placement of Φ:

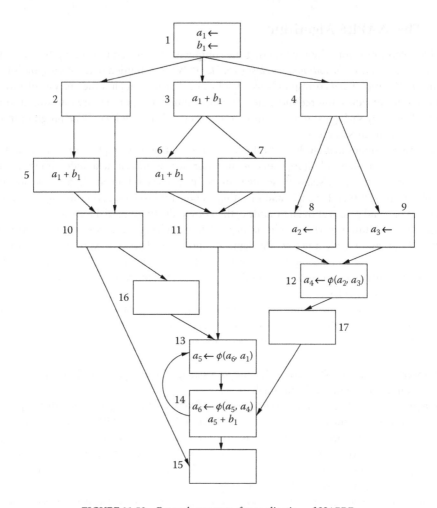

FIGURE 11.53　Example program for application of SSAPRE.

- Wherever an expression *exp* occurs, a Φ has to be inserted at its iterated dominance frontier. This is similar to a definition for *h*, its RCV.
- Wherever there is a ϕ for a variable that is an operand of *exp*, a corresponding Φ has to be placed there. This is because a ϕ indicates a changed value of *exp* reaching the block because of different values of the operand along its incoming edges. Also, a Φ is inserted only if *exp* is used after that (i.e., it is not dead). This may require precomputation of points of last use for *exp* using a dead variable analysis.

The Φs are placed by a single pass through the program, during which both the IDF of the nodes where *exp* is defined and the IDF of the definition points of variables that are operands of *exp* (with RCV *h*) are taken into account. Figure 11.54 shows the program after Φ-insertion. Here *h* is the RCV for the expression $a + b$. The Φs placed in nodes 10 and 11 are due to these nodes being in the IDF of nodes 5 and 6, where the expression $a + b$ is defined, while the Φ placed in node 12 is due to the presence of ϕ for *a* in that node, which is an operand in $a + b$. A Φ-operator is not inserted at node 15 since the expression is dead at this node.

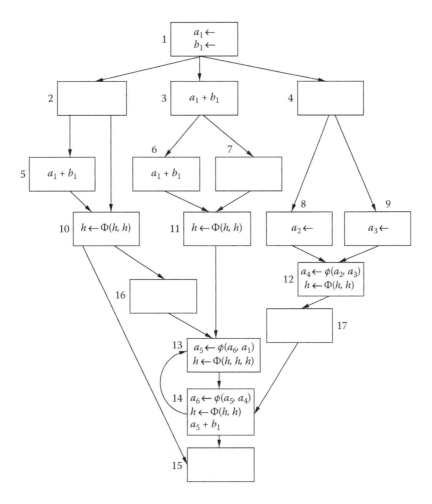

FIGURE 11.54   Program after Φ insertion.

11.6.2.2   Rename

The rename step assigns SSA versions to the RCV h. After this step, the SSA instances of h have two properties:

1. Identical SSA instances of h represent identical expression values for the expression exp that the RCV represents.
2. A control flow path with two different instances of h has to cross either an assignment to an operand of exp or a Φ of h.

Renaming is achieved by a preorder traversal on the dominator tree, similar to the scalar SSA variable renaming step [18]. Here a renaming stack for every expression exp is maintained along with the renaming stack for each variable at the same time. Entries are popped off the expression stack when backing out of the current basic block. At every block, three kinds of expressions can be encountered:

1. Real occurrences, that is, exp's that were present in the original program.
2. Operators Φ inserted in the insertion step.
3. Operands of a Φ-operator.

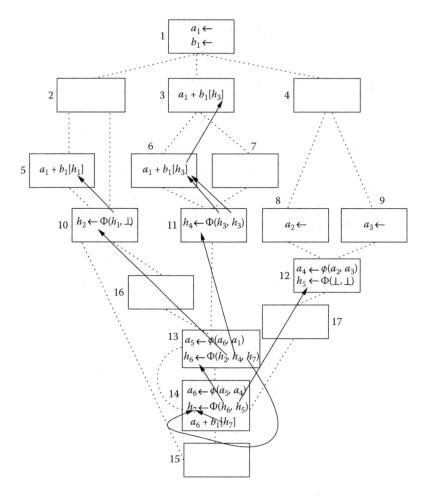

FIGURE 11.55   SSA graph of program after rename.

When an expression occurrence q is encountered:

- If $q = h$ is the result of the Φ-function, a new version is assigned to h.
- In case of a real occurrence $q = exp$, the versions of occurrence of every variable v in exp on the top of v's stack are compared with the corresponding version of v on the top of exp's stack:

  - If there is a match for all such vs in exp, $q = exp$ is assigned the same version as the version exp on the top of exp's stack.
  - If there is a mismatch for at least one variable v, assign a new version to exp (i.e., the corresponding RCV).

- In case of a Φ-operand, the same process of trying to match the versions of the variables on the top of renaming stacks of the variables and the expression is carried out. The only difference here is that if there is a mismatch, a special version $\bot$ is assigned to the Φ-operand. This indicates that there is no valid computation of exp along that incoming edge.

Every time a new version is assigned, it is pushed onto the expression renaming stack. After the preorder traversal is completed, all dead Φs are eliminated.

Figure 11.55 shows the SSA graph of the program after the renaming step. The results of all the Φ-operators in nodes 10, 11, 12, 13, and 14 get new SSA versions since the Φ might represent a merge point where two or more instances of $a + b$ meet. The two real occurrences of $a + b$ in nodes 3 and 6 get assigned the same SSA version for *h* since there is no change in either of the operands between the two nodes. One of the Φ-operands for the Φ in node 10 is assigned the $\bot$ element since there is no valid instance of $a + b$ along the corresponding incoming edge. The same holds for both operands of the Φ in node 12.

11.6.2.3 Down-Safety

Whenever an expression is inserted at a program point, it has to be down-safe, that is, it should not introduce a new computation along a path on which it was not originally present. In SSAPRE, the expression insertions are required only at the Φ-operands. Since there are no critical edges, down-safety at a Φ-operand is equivalent to down-safety at the Φ itself [33]. A Φ-block is *not* down safe if:

1. There is a path to exit from the Φ-block along which the Φ result is not used.
2. There is a path to exit from the Φ-block along which the only use of Φ is as an operand of another Φ that itself is not down-safe.

Down-safety is computed as follows:

1. Initially, all Φs from which there is a path to exit, along which the Φ result is not used, are marked as not down-safe. The remaining Φs are marked as down-safe. This can be done during the rename step itself. (When the exit block is reached during the pre-order traversal, the instance of *exp* on top of the expression stack is examined. If it is a Φ, then that Φ is marked as not down-safe.) Also, all Φ-operands that are real occurrences are marked with the flag *has_real_use* (this can also be done during the rename step [33]). The rest of the Φ-operands are either $\bot$ or are Φ themselves.
2. All Φs from which there is a path to exit, along which the only use of the Φ is as an operand of another Φ that itself is not down-safe, are marked as not down-safe. This is achieved by backward propagation along the use-def edges of the SSA graph: if the operand of a Φ (which is already marked as not down-safe) is not a real occurrence and not a $\bot$, then the operand (which is a Φ) is marked as not down-safe, and this procedure recursed.

Figure 11.56 shows the result of the down-safety step on the SSA graph of the candidate program. The only node marked as not being down-safe is 10 since there is a direct path from 10 to the exit where h_2, the result of the Φ, is not used. The rest of the Φs are marked as down-safe.

11.6.2.4 WillBeAvail

This step has two forward propagation passes, performed one after the other, to compute the set of points where an expression *exp* can be inserted to make the partially redundant expression fully redundant.

The first pass computes a predicate called *can_be_avail*: At the end of this computation, all Φs for which *can_be_avail* is true are: (a) those that are down-safe for insertion of *exp* along with (b) those that are not down-safe but where *exp* is fully available. The first pass begins with the set of Φ-nodes that are not down-safe and for which one of the operands is $\bot$. It sets the *can_be_avail* of such Φs to false. It then forward propagates this value along the chain of def-use edges to all Φs for which the former Φ is an operand (excluding edges along which *has_real_use* is set to true). Once such Φs are found, the operand corresponding to the former Φ is set to $\bot$, the *can_be_avail* of this Φ is set to false, and the procedure recursed.

The second pass computes a predicate called *later*: Φ-nodes at which *later* is true belong to the set of Φs beyond which insertions cannot be postponed without introducing a new unnecessary redundancy. It works within the region computed by the first pass. Initially, *later* is set to true wherever *can_be_avail* is true. This pass then assigns a false value for Φs with at least one operand with *has_real_use* set to true and forward propagates this value to other Φ-nodes similar to *can_be_avail*.

At the end of the two passes, the *will_be_avail* predicate is computed:

$$will\_be\_avail = can\_be\_avail \wedge \neg later$$

The *can_be_avail* portion of the predicate represents the computational optimality condition of PRE, while the *later* predicate represents the lifetime optimality criterion.

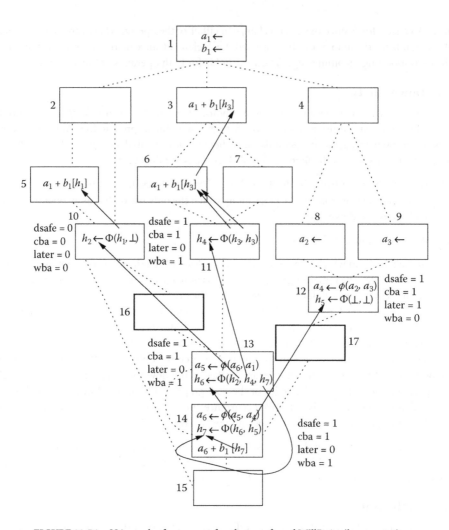

FIGURE 11.56 SSA graph of program after down-safe and WillBeAvail computations.

At the end of this phase, the insertion points are computed. These insertion points are indicated by a predicate called *insert*. Insertions are done for Φ-operands that satisfy *insert*, that is, along the predecessor edge corresponding to the Φ-operand. The predicate *insert* holds for a Φ-operand if:

1. *will_be_avail*(Φ) = true and
2. (Φ operand = ⊥) ∨ (*has_real_use* [Φ-operand] = false ∧ operand is defined by a Φ that does not satisfy *will_be_avail*)

Figure 11.56 shows the values of *can_be_avail* and *later* computed for the different Φ-nodes in the SSA graph. The points of insertion are shown in Figure 11.57 indicated by the predicate *insert* being set to *true*.

11.6.2.5 Finalize

The finalize phase transforms the program with RCVs into a valid SSA form that is suitable for insertions of expressions and in which no Φ-operand is a ⊥. This phase does the following:

- For every real occurrence of an expression *exp*, finalize decides whether it should be computed and stored into a temporary *t* (*saved*) or whether it should be *reloaded* from an already saved temporary *t*.

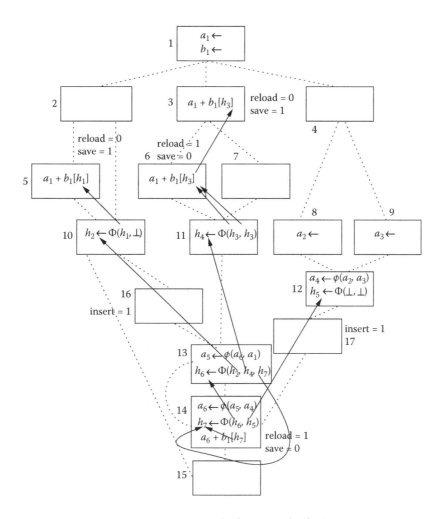

FIGURE 11.57 SSA graph of program after finalize.

- For every Φ for which *will_be_avail* is true, an insertion of *exp* is performed on every incoming edge that corresponds to a Φ-operand at which the *exp* is not available (whose *insert* flag has been set to true).
- All the Φs for which *will_be_avail* is true become ϕs of the temporary t of the SSA form. The other Φs for which *will_be_avail* is false are not a part of the SSA form for t.
- All extraneous Φs are removed.

The finalize phase performs a preorder traversal on the dominator tree of the control flow graph. It maintains a table *AvailDef* indexed by the SSA subscript of the RCV h_x. The entry *AvailDef*[x] points to the defining occurrence of the expression exp_i. This can either be a real occurrence of the expression *exp* or a Φ for which *will_be_avail* is true.

During the preorder traversal, three kinds of expressions may be visited:

- Φ: If the *will_be_avail* flag has not been set, the Φ is ignored and is not a part of the final SSA form. Otherwise, *AvailTable*[x] is set to this Φ where h_x is the left hand side (lhs) variable of the Φ.
- Real occurrence of exp_i: Assuming exp_i is represented by the RCV h_x, if *AvailTable*[x] is set to $\bot$ or some expression that does not dominate exp_i, then *AvailTable*[x] is set to exp_i. Otherwise, the *save* flag of entry pointed to by *AvailTable*[x] and the *reload* flag of exp_i is set to true.

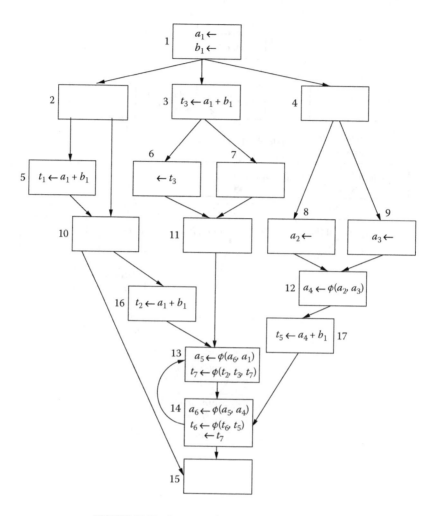

FIGURE 11.58   Program after the code motion phase.

- Operand of a Φ that is in a successor block: Three cases arise depending on the flags *will_be_avail* of Φ and the *insert* flag of the Φ-operand:

  - If the *will_be_avail* of the Φ is set to false, nothing is done.
  - If the *will_be_avail* of the Φ is set to true and the *insert* flag of the Φ-operand is set to true, a computation of exp_i is inserted at the exit of the current block.
  - If the *will_be_avail* of the Φ is set to true and the *insert* flag of the Φ-operand is set to false, the *save* flag of the entry pointed to by *AvailTable*$[x]$ is set to true and the Φ-operand is updated to refer to that entry.

The extraneous Φs do not affect the correctness of the algorithm but do consume some space. The minimization of Φs thus improves the efficiency of SSA-based optimizations that are done after SSAPRE. Removal of extraneous Φs involves one more pass over the program [33].

Figure 11.57 shows the program after the finalize phase along with the *reload* and *save* flags.

11.6.2.6   Code Motion

The final phase of the SSAPRE algorithm is code motion. It involves introducing the temporary t to eliminate redundant computations. It is done by the following steps:

- At a real occurrence of an expression *exp*, if the *save* flag is set, the result of the computation is stored into a new version of the temporary *t*. However, if the *reload* flag is set, the computation of *exp* is replaced by a use of the current version of *t*.
- At an inserted occurrence of expression, the value of the inserted computation is stored into a new version of *t*.
- At a Φ, a corresponding ϕ is generated for *t*.

The final result of the algorithm after application of all six steps is given in Figure 11.58. The partially redundant computations of $a+b$ from nodes 6 and 14 have been replaced by loads from a saved temporary.

11.7   Translation from SSA Form to Executable Form

Once optimizations have been performed on the SSA form, the SSA form of code has to be converted back to the executable form, replacing the hypothetical ϕ-functions by commonly implemented instructions. One way of removing the ϕ-functions while still preserving program semantics is to insert a *copy* statement at the end of each predecessor block of the ϕ-node, corresponding to each ϕ-node's argument. Figure 11.59b shows the result of copy insertion applied to the SSA program of Figure 11.59a.

An optimization that is frequently used to reduce the number of SSA variable instances is copy folding. It is usually coupled with the SSA renaming step. At a copy or assign statement $y \leftarrow x$, instead of pushing the new version of y (for instance, y_i) onto the renaming stack of y, copy folding can be achieved by pushing the current version of x (for instance, x_j) onto the renaming stack of y. This step would make sure that subsequent uses of y_i are directly replaced by x_j since they would see x_j on the top of y's stack instead of y_i. Figure 11.60 illustrates the effect of copy folding y with x.

Copy folding can result in the live ranges of different instances of a variable to overlap with each other. In the example shown in Figure 11.60, the live ranges of x_2 get extended because of copy folding and now interfere with the live range of x_3. Briggs et al. [9] point out two problems that might arise in the back-translated code when the original algorithm of inserting copies for back-translation is performed in the presence of copy folding.

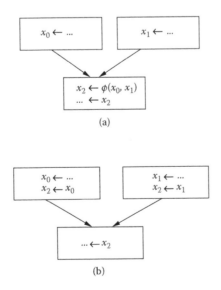

FIGURE 11.59   Back-translation and copy insertion. (a) Candidate program for back-translation. (b) Result of back-translation using copy insertion.

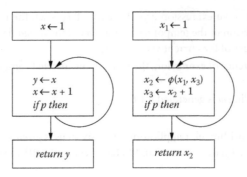

FIGURE 11.60   Copy folding during SSA renaming.

11.7.1   The Lost Copy Problem

This problem is illustrated in Figure 11.61a, which shows the result of back-translation of the program in Figure 11.60. A naive insertion of copy instructions in the predecessor blocks results in a lost copy. For program correctness, the value of x_2 should be returned before it is overwritten by x_3 in the last iteration. The main cause of this problem is the use of x_2 beyond the scope of the ϕ-function that defined it. It can be solved by splitting the critical edge as shown in Figure 11.62 or by saving x_2 into a temporary t before it is overwritten by x_3 and then replacing all subsequent references to x_2 by t (Figure 11.61b).

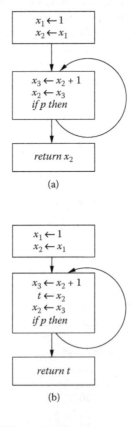

FIGURE 11.61   Lost copy problem and a solution. (a) The lost copy problem. (b) Solving the lost copy problem by introduction of a temporary solution.

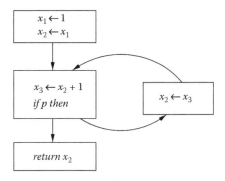

FIGURE 11.62   Solving the lost copy problem by critical edge splitting.

11.7.2   The Swap Problem

Another problem that arises when back-translation is performed on SSA code that has been copy folded is called the swap problem. For example, when copy folding is done during SSA conversion on the program snippet shown in Figure 11.63a, it results in the code shown in Figure 11.63b. In the translated code, there is now a cyclic dependency between the results of ϕ-nodes; each of a_2 and b_2 is used as an argument to the ϕ-function that defines the other variable. A naive back-translation would result in the code shown in Figure 11.64a. This code is incorrect since on the first iteration a_2 is overwritten by b_2 when actually it

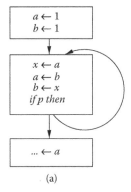

(a)

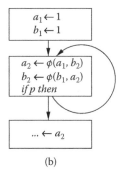

(b)

FIGURE 11.63   The swap problem. (a) Program to illustrate the swap problem. (b) Program segment after SSA conversion with copy folding.

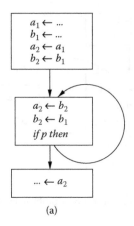

(a)

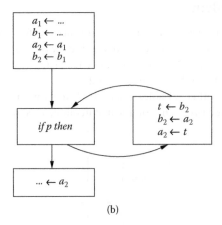

(b)

FIGURE 11.64   The swap problem. (a) Result of naive translation by inserting copy instructions. (b) One solution to the swap problem.

should have been assigned the value of a_1. One way of solving of this problem is by splitting the critical edge and using a temporary as shown in Figure 11.64b.

Briggs et al. give an algorithm for translation of SSA back to normal code without splitting the critical edge. More details can be found in [9].

11.8   Discussion and Conclusions

In this chapter, we have discussed in detail several algorithms on the traditional flow graph and on the static single assignment form. These algorithms demonstrate how the features of SSA graphs such as ϕ-functions, single assignment to variables, and SSA edges (corresponding to d-u information) facilitate faster operation of algorithms for some global optimizations. For example, conditional constant propagation operates much faster on SSA graphs and discovers at least the same set of constants as the algorithm on the flow graph. In the case of value numbering, the SSA version is not only faster and simpler, but also global, because of the single assignment property. For partial redundancy elimination the benefits stem from enabling PRE to be seamlessly integrated into a global optimizer that uses SSA as its internal representation.

As a consequence of its practical advantages, several well-known compilers have adopted the SSA form for code optimizations. The latest version of the GNU compiler collection (GCC) uses a form of SSA based on its internal tree intermediate representation, called the Tree SSA [47]. Tree SSA is designed to be language and target machine independent so that high-level transformations that cannot be performed over the register transfer language (GCC's low-level representation) can be easily done.

The Jalapeno virtual machine for Java, now known as Jikes RVM (Research Virtual Machine), is another compiler that uses the SSA form [11]. It makes use of a variant of SSA for carrying out optimizations on object fields and arrays, called the heap array SSA [26]. Some optimizations that are performed are intraprocedural flow-sensitive optimizations, anticipatory SSAPRE [54], conditional constant propagation for scalars and arrays, global value numbering, redundant load elimination, and dead store elimination [26].

The SUIF 2 Compiler System [57] constructs the SSA form using the techniques of [9]. It provides options to construct semi-pruned, minimal, or pruned SSA forms and construct use-def and def-use chains [32]. The relevant optimizations that are supported include eliminating copy instructions and useless ϕ-nodes and dead code elimination [44].

Investigations on whether optimizations can be combined on the lines of [14] and on alternative SSA forms [31, 56] on which more optimization algorithms can operate are needed. The issue of using dependence information in SSA forms also arises while tackling parallelization transformations. Knobe and Sarkar [35] discuss the array SSA form and its use in parallelization. Lapowski and Hendren [40] extend SSA to pointers. Hasti and Horwitz [30] show that SSA helps convert flow-insensitive analysis to flow-sensitive analysis. Concurrent SSA has been addressed by Srinivisan et al. [52], Lee et al. [41], and Novillo et al. [47].

The static single information (SSI) form is a recently proposed extension of SSA [5] that has yet to be widely adopted by the static analysis community. Some comparisons of SSI and SSA have been presented in [50] along with algorithms for some analyses. Applications of SSA to compiler back-end problems such as register allocation and code generation have been attempted. Hack et al. [28] use the SSA form for register allocation, and Eckstein et al. [25] use SSA graphs for instruction selection. Thus, back-end problems also can be profitably addressed using SSA.

References

1. Alfred V. Aho and John E. Hopcroft. 1974. *The design and analysis of computer algorithms*. Boston, MA: Addison-Wesley Longman.
2. Alfred V. Aho, John E. Hopcroft, and Jeffrey Ullman. 1983. *Data structures and algorithms*. Boston, MA: Addison-Wesley Longman.
3. Alfred V. Aho, Ravi Sethi, and Jeffrey D. Ullman. 1986. *Compilers: Principles, techniques, and tools*. Boston, MA: Addison-Wesley Longman.
4. B. Alpern, M. N. Wegman, and F. K. Zadeck. 1988. Detecting equality of variables in programs. In *POPL '88: Proceedings of the 15th ACM SIGPLAN-SIGACT Symposium on Principles of Programming Languages*, 1–11. New York: ACM Press.
5. C. Ananian. 1999. The static single information form. Master's thesis, Tech. Rep. MIT-LCS-TR-801, MIT, Cambridge, MA.
6. Andrew W. Appel. 1997. *Modern compiler implementation in Java: Basic techniques*. New York: Cambridge University Press.
7. Jonathan M. Asuru. 1992. Optimization of array subscript range checks. *ACM Lett. Program. Lang. Syst.* 1(2):109–18.
8. Gianfranco Bilardi and Keshav Pingali. 2003. Algorithms for computing the static single assignment form. *J. ACM* 50(3):375–425.
9. Preston Briggs, Keith D. Cooper, Timothy J. Harvey, and L. Taylor Simpson. 1998. Practical improvements to the construction and destruction of static single assignment form. *Softw. Pract. Exper.* 28(8):859–81.
10. Preston Briggs, Keith D. Cooper, and L. Taylor Simpson. 1997. Value numbering. *Softw. Pract. Exper.* 27(6):701–24.

11. Michael G. Burke, Jong-Deok Choi, Stephen Fink, David Grove, Michael Hind, Vivek Sarkar, Mauricio J. Serrano, V. C. Sreedhar, Harini Srinivasan, and John Whaley. 1999. The Jalapeno Dynamic Optimizing Compiler for Java. In *JAVA '99: Proceedings of the ACM 1999 Conference on Java Grande*, 129–41. New York: ACM Press.

12. Jong-Deok Choi, Ron Cytron, and Jeanne Ferrante. 1991. Automatic construction of sparse data flow evaluation graphs. In *POPL '91: Proceedings of the 18th ACM SIGPLAN-SIGACT Symposium on Principles of Programming Languages*, 55–66. New York: ACM Press.

13. F. Chow. A portable machine independent optimizer — Design and implementation. PhD thesis, Dept. of Electrical Engineering, Stanford University, Stanford, CA.

14. Cliff Click and Keith D. Cooper. 1995. Combining analyses, combining optimizations. *ACM Trans. Program. Lang. Syst.* 17(2):181–96.

15. John Cocke. 1969. *Programming languages and their compilers: Preliminary notes*. Courant Institute of Mathematical Sciences, New York University.

16. K. Cooper and Taylor Simpson. 1995. SCC-based value numbering. Tech. Rep. CRPC-TR95636-S, Rice University, Houston, TX.

17. Keith D. Cooper, L. Taylor Simpson, and Christopher A. Vick. 2001. Operator strength reduction. *ACM Trans. Program. Lang. Syst.* 23(5):603–25.

18. Ron Cytron, Jeanne Ferrante, Barry K. Rosen, Mark N. Wegman, and F. Kenneth Zadeck. 1991. Efficiently computing static single assignment form and the control dependence graph. *ACM Trans. Program. Lang. Syst.* 13(4):451–90.

19. Dibyendu Das and U. Ramakrishna. 2005. A practical and fast iterative algorithm for ϕ function computation using DJ graphs. *ACM Trans. Program. Lang. Syst.* 27(3):426–40.

20. D. Dhamdhere. 1989. A new algorithm for composite hoisting and strength reduction. *Int. J. Comput. Math.* 27:1–14.

21. D. M. Dhamdhere. 1988. A fast algorithm for code movement optimisation. *SIGPLAN Notices* 23(10):172–80.

22. D. M. Dhamdhere. 1991. Practical adaption of the global optimization algorithm of Morel and Renvoise. *ACM Trans. Program. Lang. Syst.* 13(2):291–94.

23. V. M. Dhaneshwar and D. M. Dhamdhere. 1995. Strength reduction of large expressions. *J. Program. Lang.* 3:95–120.

24. Karl-Heinz Drechsler and Manfred P. Stadel. 1993. A variation of Knoop, Rüthing, and Steffen's lazy code motion. *SIGPLAN Notices* 28(5):29–38.

25. Erik Eckstein, Oliver König, and Bernhard Scholz. 2003. Code instruction selection based on SSA-graphs. In *Software and Compilers for Embedded Systems*, 49–65. Heidelberg, Germany: Springer-Verlag.

26. Stephen J. Fink, Kathleen Knobe, and Vivek Sarkar. 2000. Unified analysis of array and object references in strongly typed languages. In *SAS '00: Proceedings of the Ninth International Static Analysis Symposium*, 155–74. Heidlberg, Germany: Springer-Verlag.

27. Rajiv Gupta. 1993. Optimizing array bound checks using flow analysis. *ACM Lett. Program. Lang. Syst.* 2(1–4):135–50.

28. Sebastian Hack, Daniel Grund, and Gerhard Goos. 2006. Register allocation for programs in ssa-form. In *CC '06: Proceedings of the Fifteenth International Conference on Compiler Construction*, 247–62.

29. D. Harel. 1985. A linear algorithm for finding dominators in flow graphs and related problems. In *STOC '85: Proceedings of the Seventeenth Annual ACM Symposium on Theory of Computing*, 185–94. New York: ACM Press.

30. Rebecca Hasti and Susan Horwitz. 1998. Using static single assignment form to improve flow-insensitive pointer analysis. In *PLDI '98: Proceedings of the ACM SIGPLAN 1998 Conference on Programming Language Design and Implementation*, 97–105. New York: ACM Press.

31. Paul Havlak. 1994. Construction of thinned gated single-assignment form. In *Proceedings of the 6th International Workshop on Languages and Compilers for Parallel Computing*, 477–99. London: Springer-Verlag.

32. Glenn Holloway. 2001. The Machine-SUIF static single assignment library. http://www.eecs.harvard.edu/hobe/software/nei/ssa.pdf

33. Robert Kennedy, Sun Chan, Shin-Ming Liu, Raymond Lo, Peng Tu, and Fred Chow. 1999. Partial redundancy elimination in SSA form. *ACM Trans. Program. Lang. Syst.* 21(3): 627–76.

34. Robert Kennedy, Fred C. Chow, Peter Dahl, Shin-Ming Liu, Raymond Lo, and Mark Streich. 1998. Strength reduction via ssapre. In *CC '98: Proceedings of the 7th International Conference on Compiler Construction*, 144–58. London: Springer-Verlag.

35. Kathleen Knobe and Vivek Sarkar. 1998. Array SSA form and its use in parallelization. In *POPL '98: Proceedings of the 25th ACM SIGPLAN-SIGACT Symposium on Principles of Programming Languages*, 107–20. New York: ACM Press.

36. J. Knoop, O. Ruthing, and B. Steffen. 1993. Lazy strength reduction. *J. Program. Lang.* 1:71–91.

37. Jens Knoop, Oliver Rüthing, and Bernhard Steffen. 1992. Lazy code motion. In *PLDI '92: Proceedings of the ACM SIGPLAN 1992 Conference on Programming Language Design and Implementation*, 224–34. New York: ACM Press.

38. Jens Knoop, Oliver Rüthing, and Bernhard Steffen. 1994. Optimal code motion: Theory and practice. *ACM Trans. Program. Lang. Syst.* 16(4):1117–55.

39. Priyadarshan Kolte and Michael Wolfe. 1995. Elimination of redundant array subscript range checks. In *PLDI '95: Proceedings of the ACM SIGPLAN 1995 Conference on Programming Language Design and Implementation*, 270–78. New York: ACM Press.

40. Christopher Lapkowski and Laurie J. Hendren. 1996. Extended SSA numbering: Introducing SSA properties to languages with multi-level pointers. In *CASCON '96: Proceedings of the 1996 Conference of the Centre for Advanced Studies on Collaborative Research*, Toronto, Ontario, Canada, 23. IBM Press.

41. Jaejin Lee, Samuel P. Midkiff, and David A. Padua. 1997. Concurrent static single assignment form and constant propagation for explicitly parallel programs. In *Proceedings of the 10th International Workshop on Languages and Compilers for Parallel Computing*. Vol. 1366 of Lecture Notes in Computer Science, 114–30. New York: Springer.

42. Thomas Lengauer and Robert Endre Tarjan. 1979. A fast algorithm for finding dominators in a flowgraph. *ACM Trans. Program. Lang. Syst.* 1(1):121–41.

43. E. Morel and C. Renvoise. 1979. Global optimization by suppression of partial redundancies. *Commun. ACM* 22(2):96–103.

44. Robert Morgan. 1998. *Building an optimizing compiler*. Newton, MA: Digital Press.

45. Steven S. Muchnick. 1997. *Advanced compiler design and implementation*. San Francisco: Morgan Kaufmann.

46. Steven S. Muchnick and Neil D. Jones. 1981. *Program flow analysis: Theory and application*. New York: Prentice Hall Professional Technical Reference.

47. Diego Novillo, Ronald C. Unrau, and Jonathan Schaeffer. 1998. Concurrent SSA form in the presence of mutual exclusion. In *ICPP '98: Proceedings of the 1998 International Conference on Parallel Processing*, 356. Washington, DC: IEEE Computer Society.

48. V. K. Paleri, Y. N. Srikant, and P. Shankar. 2003. Partial redundancy elimination: a simple, pragmatic, and provably correct algorithm. *Sci. Comput. Program.* 48(1):1–20.

49. B. K. Rosen, M. N. Wegman, and F. K. Zadeck. 1988. Global value numbers and redundant computations. In *POPL '88: Proceedings of the 15th ACM SIGPLAN-SIGACT Symposium on Principles of Programming Languages*, 12–27. New York: ACM Press.

50. Jeremy Singer. 2005. Static program analysis based on virtual register renaming. PhD thesis, Christ's College, University of Cambridge, Cambridge, UK.

51. Vugranam C. Sreedhar and Guang R. Gao. 1995. A linear time algorithm for placing ϕ-nodes. In *POPL '95: Proceedings of the 22nd ACM SIGPLAN-SIGACT Symposium on Principles of Programming Languages*, 62–73. New York: ACM Press.

52. Harini Srinivasan, James Hook, and Michael Wolfe. 1993. Static single assignment for explicitly parallel programs. In *Proceedings of the 20th ACM SIGPLAN-SIGACT Symposium on Principles of Programming Languages*, 260–72.

53. Ronald L. Rivest, Thomas Cormen, and Charles E. Leiserson. 1990. *Introduction to algorithms.* New York: McGraw-Hill.

54. Thomas VanDrunen and Antony L. Hosking. 2004. Anticipation-based partial redundancy elimination for static single assignment form. *Softw. Pract. Exper.* 34(15):1413–39.

55. Mark N. Wegman and F. Kenneth Zadeck. 1991. Constant propagation with conditional branches. *ACM Trans. Program. Lang. Syst.* 13(2):181–210.

56. Daniel Weise, Roger F. Crew, Michael Ernst, and Bjarne Steensgaard. 1994. Value dependence graphs: representation without taxation. In *POPL '94: Proceedings of the 21st ACM SIGPLAN-SIGACT Symposium on Principles of Programming Languages*, 297–310. New York: ACM Press.

57. Robert P. Wilson, Robert S. French, Christopher S. Wilson, Saman P. Amarasinghe, Jennifer-Ann M. Anderson, Steven W. K. Tjiang, Shih-Wei Liao, Chau-Wen Tseng, Mary W. Hall, Monica S. Lam, and John L. Hennessy. 1994. SUIF: An infrastructure for research on parallelizing and optimizing compilers. *SIGPLAN Notices* 29(12):31–37.

58. Michael Wolfe. 1992. Beyond induction variables. In *PLDI '92: Proceedings of the ACM SIGPLAN 1992 Conference on Programming Language Design and Implementation*, 162–174. New York: ACM Press.

12

Shape Analysis and Applications[1]

Thomas Reps[2]
Computer Sciences Department,
University of Wisconsin-Madison, WI
reps@cs.wisc.edu

Mooly Sagiv
Department of Computer Science,
School of Mathematics and Science,
Tel Aviv University, Tel Aviv, Israel
Sagiv@math.tau.ac.il

Reinhard Wilhelm
Fachbereich Informatik,
Universitaet des Saarlandes,
Saarbruecken, Germany
Wilhelm@cs.uni-sb.de

Abstract

A shape-analysis algorithm statically analyzes a program to determine information about the heap-allocated data structures that the program manipulates. The results can be used to understand programs or to verify properties of programs. Shape analysis also recovers information that is valuable for debugging, compile-time garbage collection, instruction scheduling, and parallelization.

[1] Portions of this paper were adapted from [65] (© Springer-Verlag) and excerpted from [58] (© ACM).

[2] Supported in part by NSF Grants CCR-9619219, CCR-9986308, CCF-0540955, and CCF-0524051; by ONR Grants N00014-01-1-0796 and N00014-01-1-0708; by the Alexander von Humboldt Foundation; and by the John Simon Guggenheim Memorial Foundation. Address: Comp. Sci. Dept.; Univ. of Wisconsin; 1210 W. Dayton St.; Madison, WI 53706.

[3] Address: School of Comp. Sci.; Tel Aviv Univ.; Tel Aviv 69978; Israel.

[4] Address: Fachrichtung Informatik, Univ. des Saarlandes; 66123 Saarbrücken; Germany.

12.1   Introduction

Pointers and heap-allocated storage are features of all modern imperative programming languages. However, they are ignored in most formal treatments of the semantics of imperative programming languages because their inclusion complicates the semantics of assignment statements: an assignment through a pointer variable (or through a pointer-valued component of a record) may have far-reaching side effects. Works that have treated the semantics of pointers include [5, 42, 43, 45].

These far-reaching side effects also make program dependence analysis harder, because they make it difficult to compute the aliasing relationships among different pointer expressions in a program. Having less precise program dependence information decreases the opportunities for automatic parallelization and for instruction scheduling.

The usage of pointers is error prone. Dereferencing NULL pointers and accessing previously deallocated storage are two common programming mistakes. The usage of pointers in programs is thus an obstacle for program understanding, debugging, and optimization. These activities need answers to many questions about the structure of the heap contents and the pointer variables pointing into the heap.

By *shapes*, we mean descriptors of heap contents. *Shape analysis* is a generic term denoting static program-analysis techniques that attempt to determine properties of the heap contents relevant for the applications mentioned above.

12.1.1   Structure of the Chapter

Section 12.2 lists a number of questions about the contents of the heap. Figure 12.1 presents a program that will be used as a running example, which inserts an element into a singly linked list. Section 12.2.3 shows how shape analysis would answer the questions about the heap contents produced by this program. Section 12.3 then informally presents a parametric shape-analysis framework along the lines of [58], which provides a generative way to design and implement shape-analysis algorithms. The "shape semantics" — plus some additional properties that individual storage elements may or may not possess — are specified in logic, and the shape-analysis algorithm is automatically generated from such a specification. Section 12.4 shows how the informal treatment from Section 12.3 can be made precise by basing it on predicate logic. In particular, it is shown how a 2-valued interpretation and a 3-valued interpretation of the same set of

```
                              / * insert.c */
                              #include ''list.h''
                              void insert (List x, int d) {
                                List y, t, e;
                                assert(acyclic _list (x) && x != NULL);
                                y = x;
                                while (y->n ! = NULL && ...) {
                                  y = y->n;
                                }
                                t = malloc( );
/* list.h */                    t->data = d;
typedef struct node {           e = y->n;
  struct node *n;               t->n = e;
  int data;                     y->n = t;
} *List;                        }
(a)                           (b)
```

FIGURE 12.1   (a) Declaration of a linked-list data type in C. (b) A C function that searches a list pointed to by parameter x, and splices in a new element.

formulas can be used to define the concrete and abstract semantics, respectively, of pointer-manipulating statements. Section 12.5 lists some applications of shape analysis. Section 12.6 briefly describes several extensions of the shape-analysis framework that have been investigated. Section 12.7 discusses related work. Section 12.8 presents some conclusions.

12.2 Questions about the Heap Contents

Shape analysis has a somewhat constrained view of programs. It is not concerned with numeric or string values that programs compute, but exclusively with the linked data structures they build in the heap and the pointers into the heap from the stack, from global memory, or from cells in the heap.[5] We will therefore use the term *execution state* to mean the set of cells in the heap, the connections between them (via pointer components of heap cells), and the values of pointer variables in the store.

12.2.1 Traditional Compiler Analyses

We list some questions about execution states that a compiler might ask at points in a program, together with (potential) actions enabled by the respective answers:

NULL pointers: Does a pointer variable or a pointer component of a heap cell contain NULL at the entry to a statement that dereferences the pointer or component?

> **Yes (for every state):** Issue an error message.
> **No (for every state):** Eliminate a check for NULL.
> **Maybe:** Warn about the potential NULL dereference.

Alias: Do two pointer expressions reference the same heap cell?

> **Yes (for every state):** Trigger a prefetch to improve cache performance, predict a cache hit to improve cache-behavior prediction, or increase the sets of uses and definitions for an improved liveness analysis.
> **No (for every state):** Disambiguate memory references and improve program dependence information [11, 55].[6]

Sharing: Is a heap cell shared?[7]

> **Yes (for some state):** Warn about explicit deallocation, because the memory manager may run into an inconsistent state.
> **No (for every state):** Explicitly deallocate the heap cell when the last pointer to it ceases to exist.

Reachability: Is a heap cell reachable from a specific variable or from any pointer variable?

> **Yes (for every state):** Use this information for program verification.
> **No (for every state):** Insert code at compile time that collects unreachable cells at runtime.

Disjointness: Do two data structures pointed to by two distinct pointer variables ever have common elements?

> **No (for every state):** Distribute disjoint data structures and their computations to different processors [24].

---

[5] However, the shape-analysis techniques presented in Sections 12.3 and 12.4 can be extended to account for both numeric values and heap-allocated objects. See Section 12.6.4 and [20, 21, 28].

[6] The answer "yes (for some state)" indicates the case of a may-alias. This answer prevents reordering or parallelizing transformations from being applied.

[7] Later in the chapter, the sharing property that is formalized indicates whether a cell is "heap-shared," that is, pointed to by two or more pointer components of heap cells. Sharing due to two pointer variables or one pointer variable and one heap cell component pointing to the same heap cell is also deducible from the results of shape analysis.

Cyclicity: Is a heap cell part of a cycle?

> **No (for every state):** Perform garbage collection of data structures by reference counting. Process all elements in an acyclic linked list in a *doall*–parallel fashion.

Memory leak: Does a procedure or a program leave behind unreachable heap cells when it returns?

> **Yes (in some state):** Issue a warning.

The questions in this list are ones for which several traditional compiler analyses have been designed, motivated by the goal of improving optimization and parallelization methods. The may-alias-analysis problem, which seeks to find out whether the answer to the alias question is "yes (in some state)" is of particular importance in compiling. The goal of providing better may-alias information was the motivation for our work that grew into shape analysis.

Alias, sharing, and disjointness properties are related but different. To appreciate the difference, it suffices to see that they are defined on different domains and used in different types of compiler tasks. *Alias* relations concern pairs of pointer expressions; they are relevant for disambiguating memory references. *Sharing* properties concern the organization of neighboring heap cells; they are relevant for compile-time memory management. *Disjointness* relations concern pairs of data structures; they are relevant for determining whether traversals of two data structures can be parallelized. The relations between these properties are as follows:

Disjointness-aliasing: Two data structures D_1 and D_2 are disjoint in every state if there exist no two pointer expressions e_1, referring to D_1, and e_2, referring to D_2, that may be aliased in any state.

Disjointness-sharing: If two data structures D_1 and D_2 are not disjoint in some state, at least one of the common elements of D_1 and D_2 is shared in this state.

Aliasing-sharing: If two different pointer expressions e_1 and e_2 reference the same heap cell in some state, then this cell or one of its "predecessors" must be shared in this state. However, the opposite need not hold because not all heap cells are necessarily reachable from a variable.

Some of the other questions in the list given earlier concern memory-cleanness properties [14], for example, no NULL-dereferences, no deallocation of shared cells, and no memory leaks.

12.2.1.1 Memory Disambiguation

Many compiler transformations and their enabling analyses are based on information about the independence of program statements. Such information is used extensively in compiler optimizations, automatic program parallelizations, code scheduling for instruction-level parallel machines, and in software-engineering tools such as code slicers. The concept of *program dependence* is based on the notions of *definition* and *use of resources*. Such analyses can be performed at the source-language level, where resources are mostly program variables, as well as at the machine-language level, where resources are registers, memory cells, status flags, and so on. For source-level analysis, these notions have been generalized from scalar variables to array components. Definitions and uses, in the form of indexed array names, now denote resources that are subsections of an array. Definitions and uses, which were uniquely determining resources in the case of scalar variables, turn into *potential* definitions (respectively uses) of sets of resources. Using these sets in the computation of dependences may induce spurious dependences. Many alias tests have been developed to ascertain whether two sets of potentially referenced resources are actually disjoint, that is, whether two given references to the same array never access the same element [66].

The same is overdue for references to the heap through pointer expressions. However, pointer expressions may refer to an unbounded amount of storage that is located in the heap. Appropriate analyses of pointer expressions should find information about:

Must-aliases: Two pointer expressions refer to the same heap cell on all executions that reach a given program point.

May-aliases: Two pointer expressions may refer to the same heap cell on an execution that reaches a given program point.

Approaches that attempt to identify may-aliases and must-aliases have traditionally used path expressions [27]. In Section 12.5.1 we provide a new approach based on shape analysis, which yields very precise results.

12.2.2 Analyzing Programs for Shapes

Several of the properties listed above can be combined to formulate more complex properties of heap contents:

Shape: What is the "shape" of (some part of) the contents of the heap? Shapes (or, more precisely, shape descriptors) characterize data structures. A shape descriptor could indicate whether the heap contains a singly linked list, potentially with (or definitely without) a cycle, a doubly linked list, a binary tree, and so on. The need to track many of the properties listed above, for example, sharing, cyclicity, reachability, and disjointness, is an important aspect of many shape-analysis algorithms. Shape analysis can be understood as an extended type analysis; its results can be used as an aid in program understanding and debugging [13].

Nonstructural properties: In addition to the shape of some portions of the contents of the heap, what properties hold among the value components of a data structure? These combined properties can be used to prove the partial correctness of programs [35].

History properties: These track where a heap cell was allocated and what kinds of operations have been performed on it. This kind of information can be used to identify dependences between points in the program (see Section 12.5.2).

12.2.2.1 Shape Descriptors and Data Structures

We claimed above that shape descriptors can characterize data structures. The constituents of shape descriptors that can be used to characterize a data structure include:

i. Root pointer variables, that is, information about which pointer variables point from the stack or from the static memory area into a data structure stored in the heap
ii. The types of the data-structure elements and, in particular, which fields hold pointers
iii. Connectivity properties, such as:

Whether all elements of the data structure are reachable from a root pointer variable
Whether any data-structure elements are shared
Whether there are cycles in the data structure
Whether an element v pointed to by a "forward" pointer of another element v' has its "backward" pointer pointing to v'

iv. Other properties, for instance, whether an element of an ordered list is in the correct position

Each data structure can be characterized by a certain set of such properties.

Most semantics track the values of pointer variables and pointer-valued fields using a pair of functions, often called the *environment* and the *store*. Constituents *i* and *ii* above are parts of any such semantics; consequently, we refer to them as *core* properties.

Connectivity and other properties, such as those mentioned in *iii* and *iv*, are usually not explicitly part of the semantics of pointers in a language but instead are properties derived from this core semantics. They are essential ingredients in program verification, however, as well as in our approach to shape analysis of programs. Noncore properties will be called *instrumentation* properties (for reasons that will become clear shortly).

Let us start by taking a Platonic view, namely that ideas exist without regard to their physical realization. Concepts such as "is shared," "lies on a cycle," and "is reachable" can be defined either in graph-theoretic terms, using properties of paths, or in terms of the programming-language concept of pointers. The definitions of these concepts can be stated in a way that is independent of any particular data structure; for instance:

Example 12.1

A heap cell is *heap-shared* if it is the target of two pointers — either from two different heap cells or from two different pointer components of the same heap cell.

Data structures can now be characterized using sets of such properties, where "data structure" is still independent of a particular implementation; for instance:

Example 12.2

An *acyclic singly linked list* is a set of objects, each with one pointer field. The objects are *reachable from a root pointer* either directly or by following pointer fields. No object *lies on a cycle*, that is, is reachable from itself by following pointer fields.

To address the problem of verifying or analyzing a particular program that uses a certain data structure, we have to leave the Platonic realm and formulate shape invariants in terms of the pointer variables and data-type declarations from that program.

Example 12.3

Figure 12.1a, above, shows the declaration of a linked-list data type in C, and Figure 12.1b shows a C program that searches a list and splices a new element into the list. The characterization of an acyclic singly linked list in terms of the properties "is reachable from a root pointer" and "lies on a cycle" can now be specialized for that data-type declaration and that program as follows:

- "Is reachable from a root pointer" means "is reachable from x, or is reachable from y, or is reachable from t, or is reachable from e."
- "Lies on a cycle" means "is reachable from itself following one or more n-fields."

This chapter deals with analyses that attempt to determine the shapes of all data structures in the heap. To obtain shape descriptors, these analyses track many of the properties that have been discussed above. Looking at things in the other direction, however, once such shape descriptors have been obtained, answers to many of the above questions can merely be "read off" of the shape descriptors.

12.2.3   Answers as Given by Shape Analysis

This section discusses the results obtained by analyzing `insert` using a particular shape-analysis algorithm designed to analyze programs that manipulate singly linked lists. In this case, the analysis of `insert` has been carried out under the assumption that the inputs to `insert` are a nonempty, acyclic singly linked list and an integer. The former requirement is captured by the shape descriptors shown in Figure 12.2, which are provided as input to the shape-analysis algorithm.

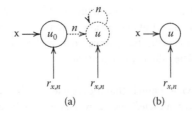

(a)   (b)

FIGURE 12.2   Shape descriptors that describe the input to `insert`. (a) Represents acyclic lists of length at least 2. (b) Represents acyclic lists of length 1.

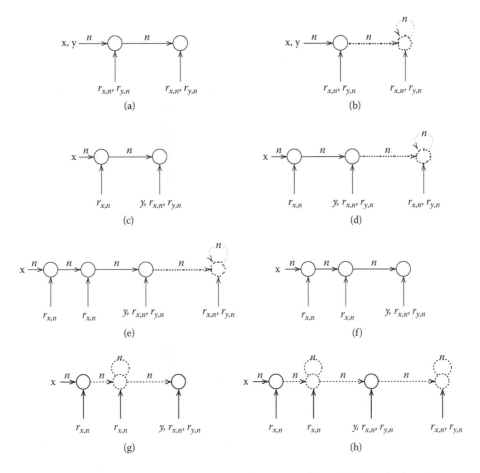

FIGURE 12.3 The eight shape graphs that arise at the beginning of the while-loop body in the program of Figure 12.1.

The shape-analysis algorithm produces information for each program point that describes the lists that can arise there. At the entry to the while-loop body some of the properties are:

- Pointer variables x and y point into the same list: x always points to the head; y points to either the head of the x-list or some tail of the x-list.
- All other pointer variables of the program have the value NULL.
- The list is acyclic.
- No memory leaks occur.

In addition, the information obtained by the shape-analysis algorithm shows that no attempt to dereference a NULL-valued pointer is ever made. Figure 12.3 shows the eight shape graphs produced by the analysis for the program point at the entry to the loop body.

Each shape graph represents a set of concrete memory configurations. In `insert`, the loop body is executed when the argument list is of length 2 or greater, and it advances variable y along the list that is pointed to by x. The shape-analysis algorithm is able to discover eight shape graphs that represent all such memory configurations. The graphs represent lists of various lengths, with various numbers of list cells between the list cells pointed to by x and y: Figure 12.3a and 12.3b represent lists in which x and y point to the same list cell; Figure 12.3c and 12.3d represent lists in which x and y point to list cells that are one apart; Figure 12.3e and 12.3f represent lists in which x and y point to list cells that are two apart; Figure 12.3g and 12.3h represent lists in which x and y point to list cells that are three or more apart.

Heap cells and their properties in the represented heaps can be read off from a shape graph in the following way:

- The name p represents pointer variable p. For instance, two of the pointer variables of program insert, namely x and y, appear in the shape graphs in Figures 12.2 and 12.3. The absence of the name p in a shape graph means that, in the stores represented by the shape graph, program variable p definitely has the value NULL. In Figure 12.3a, the absence of the name t means that t definitely has the value NULL in the stores that the shape graph represents.

- Circles stand for abstract nodes. A solid circle stands for an abstract node that represents exactly one heap cell. In Figure 12.2b, the circle u represents the one cell of an input list of length 1. Solid circles could be viewed as abstract nodes with the property "uniquely representing." (This is the complement of the "summary" property sm that is introduced later on.)

- A dotted circle stands for an abstract node that may represent one or more heap cells; in Figure 12.2a, the dotted circle u represents the cells in the tail of the input list.

- A solid edge labeled c between abstract nodes m and m' represents the fact that the c-field of the heap cell represented by m points to the heap cell represented by m'. Figure 12.3a indicates that the n-field of the first list cell points to the second list cell.

- A dotted edge labeled c between abstract nodes m and m' tells us that the c-field of one of the heap cells represented by m may point to one of the heap cells represented by m'. When m and m' are the same abstract nodes, this edge may or may not represent a cycle. In Figure 12.3b, the dotted self-cycle on the dotted circle represents n-fields of heap cells represented by this abstract node possibly pointing to other heap cells represented by the dotted circle. Additional information about noncyclicity (see below) implies that, in this case, the dotted self-cycle does not represent a cycle in the heap.

- A unary property q that holds for all heap cells represented by an abstract node is represented in the graph by having a solid arrow from the property name q to that node. (These names are typically subscripted, such as $r_{x,n}$ or c_n.) For example, the property "reachable-from-x-via-n," denoted in the graph by $r_{x,n}$, means that the heap cells represented by the corresponding abstract nodes are (transitively) reachable from pointer variable x via n-fields. Both nodes in Figure 12.3b are the targets of a solid edge from an instance of property name $r_{x,n}$. This means the concrete cell represented by the first abstract node and all concrete cells represented by the second abstract node are reachable from x via n-fields.

- A dotted arrow from a property name p to an abstract node represents the fact that p may be true for some of the heap cells represented by the abstract node and may be false for others. The absence of an arrow from p to an abstract node means that none of the represented heap cells has property p. (Examples with dotted edges are given in Section 12.3.4.)

In summary, the shape graphs portray information of three kinds:

- **Solid**, meaning "always holds" for properties (including "uniquely representing")
- **Absent**, meaning "never holds" for properties
- **Dotted**, meaning "don't know" for properties (including "uniquely representing")

Shape analysis associates sets of shape graphs with each program point. They describe (a superset of) all the execution states that can occur whenever execution reaches that program point. To determine whether a property always (ever) holds at a given program point, we must check that it holds for all (some) of the shape graphs for that point.

With this interpretation in mind, all of the claims about the properties of the heap contents at the entry to the while-loop body listed at the beginning of this subsection can be checked by verifying that they hold for all of the graphs shown in Figure 12.3.

12.3   Shape Analysis

The example program `insert` works for lists of arbitrary lengths. However, as described in the preceding section (at least for one program point), the description of the lists that occur during execution is finite. As shown in Figure 12.3, eight shape graphs are sufficient to describe all of the execution states that occur at the entry of the loop body in `insert`. This is a general requirement for shape analysis. Although the data structures that a program builds or manipulates are in general of unbounded size, the shape descriptors, manipulated by a shape-analysis algorithm, have to have *bounded size*.

This representation of the heap contents has to be *conservative* in the sense that whoever asks for properties of the heap contents — for example, a compiler, a debugger, or a program-understanding system — receives a reliable answer. The claim that "pointer variable p or pointer field p->c never has the value NULL at this program point" may only be made if this is indeed the case for all executions of the program and all program paths leading to the program point. It may still be the case that in no program execution p (respectively p->c) will be NULL at this point but that the analysis will be unable to derive this information. In the field of program analysis, we say that program analysis is allowed to (only) err on the safe side.

In short, shape analysis computes for a given program and each point in the program:

> a finite, conservative representation of the heap-allocated data structures that could arise when a path to this program point is executed.

12.3.1   Summarization

The constraint that we must work with a bounded representation implies a loss of information about the heap contents. Size information, such as the lengths of lists or the depths of trees, will in general be lost. However, structural information may also be lost because of the chosen representation. Thus, a part of the execution state (or some of its properties) is exactly represented, and some part of the execution state (or some of its properties) is only approximately represented. The process leading to the latter is called *summarization*. Summarization intuitively means the following:

- Some heap cells will lose their identity, that is, will be represented together with other heap cells by one abstract node.
- The connectivity among those jointly represented heap cells will be represented conservatively; that is, each pointer in the heap will be represented, but several such pointers (or the absence of such pointers) may be represented jointly.
- Properties of these heap cells will also be represented conservatively. This means the following:
  - A property that holds for all (for none of the) summarized cells will be found to hold (not to hold) for their summary node.
  - A property that holds for some but not all of the summarized cells will have the value "don't know" for the summary node.

12.3.2   Parametric Shape Analysis

Shape analysis is a generic term representing a whole class of algorithms of varying power and complexity that try to answer questions about the structure of heap-allocated storage. In our setting, a particular shape-analysis algorithm is determined by a set of properties that heap cells may have and by relations that may or may not hold between heap cells.

First, there are the aforementioned *core properties*, for example, the "pointed-to-by-p" property for each program pointer variable p, and the property "connected-through-c," which pairs of heap cells (l_1, l_2) possess if the c-field of l_1 points to l_2 (see Table 12.1). These properties are part of any pointer semantics. The core properties in the particular shape analysis of the `insert` program are

TABLE 12.1   Predicates used for representing the stores manipulated by programs that use the `List` data-type declaration from Figure 12.1(a)

Predicate	Intended Meaning
$q(v)$	Does pointer variable q point to cells v?
$n(v_1, v_2)$	Does the n-field of v_1 point to v_2?

"pointed-to-by-x," denoted by x, "pointed-to-by-y," denoted by y, "pointed-to-by-t," denoted by t, "pointed-to-by-e," denoted by e, and "connected-through-n," denoted by $n(\cdot, \cdot)$.

The *instrumentation properties* [58], denoted by $\mathcal{I}$, together with the core properties determine what the analysis is capable of observing. These are expressed in terms of the core properties. Our example analysis is designed to identify properties of programs that manipulate acyclic singly linked lists. Reachability properties from specific pointer variables have the effect of keeping disjoint sublists summarized separately. This is particularly important when analyzing a program in which two pointers are advanced along disjoint sublists.

Therefore, the instrumentation properties in our example analysis are "is-on-an-n-cycle," denoted by c_n, "reachable-from-x-via-n," denoted by $r_{x,n}$, "reachable-from-y-via-n," denoted by $r_{y,n}$, and "reachable-from-t-via-n," denoted by $r_{t,n}$. For technical reasons, a property that is part of every shape analysis is "summary," denoted by $sm(\cdot)$.

12.3.3 Abstraction Functions

The abstraction function of a particular shape analysis is determined by a distinguished subset of the set of all unary properties, the so-called *abstraction properties*, $\mathcal{A}$. Given a set $\mathcal{A}$ of abstraction properties, the corresponding abstraction function will be called $\mathcal{A}$-*abstraction function* (and the act of applying it, $\mathcal{A}$-*abstraction*). If the set $\mathcal{W} = \mathcal{I} - \mathcal{A}$ is not empty, that is, if there are instrumentation predicates that are not used as abstraction predicates, we will call the abstraction $\mathcal{A}$-*abstraction with* $\mathcal{W}$.

The principle of abstraction is that heap cells that have the same definite values for the abstraction properties are summarized to the same abstract node. Thus, if we view the set of abstraction properties as our means of observing the contents of the heap, the heap cells summarized by one summary node have no observable difference.

All concrete heap cells represented by the same abstract heap cells agree on their abstraction properties; that is, either they all have these abstraction properties, or none of them have them. Thus, summary nodes inherit the values of the abstraction properties from the nodes they represent. For nonabstraction properties, their values are computed in the following way: if all summarized cells agree on this property — that is, they have the same value — the summary node receives this value. If not all summarized cells agree on a property, their summary node will receive the value "don't know." The values of binary properties are computed the same way.

From what has been said above, it is clear that there is a need for three values: two definite values, representing 0 (false) and 1 (true), and an additional value, 1/2, representing uncertainty. This abstraction process is called *truth-blurring embedding* (see also Section 12.4.4).

Example 12.4

The shape graphs in Figure 12.2 and the ones in Figure 12.3 are obtained using the $\{x, y, t, e, r_{x,n}, r_{y,n}, r_{t,n}, r_{e,n}, c_n\}$-abstraction function. In Figure 12.2a, all the cells in the tail of an input list of length at least 2 are summarized by the abstract node u, because they all have the property $r_{x,n}$ and do not have the properties $x, y, t, e, r_{y,n}, r_{t,n}, r_{e,n}$, and c_n. The abstract node u_0 represents exactly one cell — the first cell of the input list. It has the properties x and $r_{x,n}$ and none of the other properties.

Now consider how the value of the property n is computed for the summary node u. The different list cells that are summarized by u do not have the same values for n, because at any one time a pointer field

Name	Graphical Representation
$S_0^\natural$	
$S_1^\natural$	$x \to (u_1)$
$S_2^\natural$	$x \to (u_1) \xrightarrow{n} (u_2)$
$S_3^\natural$	$x \to (u_1) \xrightarrow{n} (u_2) \xrightarrow{n} (u_3)$
$S_4^\natural$	$x \to (u_1) \xrightarrow{n} (u_2) \xrightarrow{n} (u_3) \xrightarrow{n} (u_4)$

FIGURE 12.4 Concrete lists pointed to by x of length ≤ 4.

may point to at most one heap cell. Thus, the connected-by-n-field properties of the resulting summary nodes have the value 1/2.

12.3.4 Designing a Shape Abstraction

This section presents a sequence of example shape abstractions to demonstrate how the precision of a shape abstraction can be changed by changing the properties used — both abstraction and nonabstraction properties. Here *precision* refers to the set of concrete heap structures that each abstract shape descriptor represents; a more precise shape descriptor represents a smaller set of concrete structures. One abstraction is more precise than another if it yields more precise shape descriptors. All examples treat singly linked lists of the type declared in Figure 12.1. The core properties are *x*, later also *y*, and *n*.

Example 12.5

Consider the case of $\{x\}$-abstraction; that is, the only abstraction property is *x*. Figure 12.4 depicts four lists of length 1 to 4 pointed to by x and the empty list. Figure 12.5 shows the shape graphs obtained by applying $\{x\}$-abstraction to the concrete lists of Figure 12.4. In addition to the lists of length 3 and 4 from Figure 12.4 (i.e., $S_3^\natural$ and $S_4^\natural$), the shape graph S_3 also represents:

- The acyclic lists of length 5, 6, and so on that are pointed to by x
- The cyclic lists of length 3 or more that are pointed to by x, such that the backpointer is not to the head of the list, but to the second, third, or later element

Thus, S_3 is a finite shape graph that captures an infinite set of (possibly cyclic) concrete lists. The example shows that a "weak" abstraction may lose valuable information: even when only acyclic lists are abstracted, the result of the abstraction is a shape graph that also represents cyclic lists.

Name	Graphical Representation
S_0	
S_1	$x \to (u_1)$
S_2	$x \to (u_1) \xrightarrow{n} (u)$
S_3	$x \to (u_1) \xrightarrow{n} (u) \circlearrowleft^{n}$

FIGURE 12.5 Shape graphs that are obtained by applying $\{x\}$-abstraction to the concrete lists that appear in Figure 12.4.

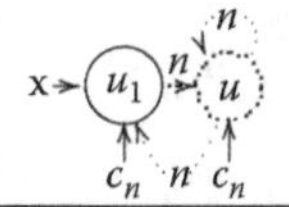

FIGURE 12.6   The shape graphs that are obtained by applying $\{x\}$-abstraction with $\{c_n\}$ to acyclic lists (top) and cyclic lists (bottom).

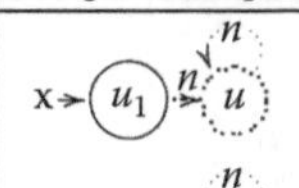

Example 12.6

The next example uses $\{x\}$-abstraction with $\{c_n\}$; that is, it is uses cyclicity properties in addition to the abstraction property x. Figure 12.6 shows two shape graphs: $S_{acyclic}$, the result of applying this abstraction to acyclic lists, and S_{cyclic}, the result of applying it to cyclic lists. Although $S_{acyclic}$, which is obtained by $\{x\}$-abstraction with $\{c_n\}$, looks just like S_3 in Figure 12.5, which is obtained just by $\{x\}$-abstraction (without $\{c_n\}$), $S_{acyclic}$ describes a smaller set of lists than S_3, namely only acyclic lists of length at least 3. The absence of a c_n-arrow to u_1 expresses the fact that none of the heap cells summarized by u_1 lie on a cycle.

In contrast, S_{cyclic} describes lists in which the heap cells represented both by u_1 and by u definitely lie on a cycle. These are lists in which the last list element has a backpointer to the head of the list.

Example 12.7

This example shows what it means to make an instrumentation property an abstraction property. $\{x, c_n\}$-abstraction and $\{x\}$-abstraction with $\{c_n\}$ are applied to cyclic lists, that is, lists that have a backpointer into the middle of the list. Figure 12.7 shows how the additional abstraction property c_n causes there to be two different summary nodes.

Instrumentation properties that track information about *reachability from pointer variables* are particularly important for avoiding a loss of precision, because they permit the abstract representations of data structures — and different parts of the same data structure — that are disjoint in the concrete world to be kept separate [57, p. 38]. A reachability property $r_{q,n}(v)$ captures whether a heap cell v is (transitively) reachable from pointer variable q along n-fields.

Example 12.8

The power of reachability information is illustrated in our next example. Figures 12.8 and 12.9 show how a concrete list in which x points to the head and y points into the middle is mapped to two different shape graphs, depending on whether $\{x, y, r_{x,n}, r_{y,n}\}$-abstraction or just $\{x, y\}$-abstraction is used.

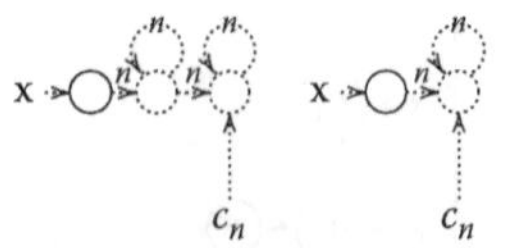

FIGURE 12.7   (a) $\{x, c_n\}$-abstraction and (b) $\{x\}$-abstraction with $\{c_n\}$. The two abstractions have been applied to a list of length at least 5, with a backpointer into the middle of the list. The ≥ 5 elements of the lists represented by shape graph (a) are distributed as follows: at least three of them form the acyclic prefix of the list, and at least two of them form the cycle.

Name	Graphical Representation
$S_6^\natural$	$x \rightarrow (u_1) \xrightarrow{n} (u_3) \xrightarrow{n} (u_4) \xrightarrow{n} (u_2) \xrightarrow{n} (u_5) \xrightarrow{n} (u_6)$ y

FIGURE 12.8   A concrete list pointed to by x, where y points into the middle of the list.

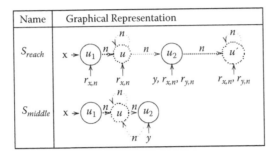

Name	Graphical Representation
S_{reach}	$x \rightarrow (u_1) \xrightarrow{n} (u) \xrightarrow{n} (u_2) \xrightarrow{n} (u)$ $r_{x,n} \quad r_{x,n} \quad y, r_{x,n}, r_{y,n} \quad r_{x,n}, r_{y,n}$
S_{middle}	$x \rightarrow (u_1) \xrightarrow{n} (u) \xrightarrow{n} (u_2)$ $n \quad y$

FIGURE 12.9   The shape graphs that are obtained by applying $\{x, y, r_{x,n}, r_{y,n}\}$-abstraction and $\{x, y\}$-abstraction, respectively, to the list $S_6^\natural$ from Figure 12.8.

Note that the situation depicted in Figure 12.8 occurs in `insert` as y is advanced down the list; the reachability abstraction properties play a crucial role in developing a shape-analysis algorithm that is capable of obtaining precise shape information for `insert`.

12.4   An Overview of a Shape-Analysis Framework

This section provides an overview of the formal underpinnings of the shape-analysis framework presented in [58]. The framework is *parametric*; that is, it can be instantiated in different ways to create a variety of specific shape-analysis algorithms. The framework is based on 3-valued logic. In this paper, the presentation is at a semi-technical level; for a more detailed treatment of this material, as well as several elaborations on the ideas covered here, the reader should refer to [58].

To be able to perform shape analysis, the following concepts need to be formalized:

- An encoding (or representation) of stores, so that we can talk precisely about store elements and the relationships among them.
- A language in which to state properties that store elements may or may not possess.
- A way to extract the properties of stores and store elements.
- A definition of the concrete semantics of the programming language, in particular, one that makes it possible to track how properties change as the execution of a program statement changes the store.
- A technique for creating abstractions of stores so that abstract interpretation can be applied.

In our approach, the formalization of each of these concepts is based on predicate logic.

12.4.1   Representing Stores via 2-Valued and 3-Valued Logical Structures

To represent stores, we work with what logicians call *logical structures*. A logical structure is associated with a *vocabulary* of predicate symbols (with given arities). So far we have talked about *properties* of different classes, that is, core, instrumentation, and abstraction properties. Properties in our specification language, predicate logic, correspond to predicates.

Each logical structure S, denoted by $\langle U^S, \iota^S \rangle$, has a universe of *individuals* U^S. In a 2-valued logical structure, ι^S maps each arity-k predicate symbol p and possible k-tuple of individuals $(u_1, \ldots, u_k)$, where $u_i \in U^S$, to the value 0 or 1 (i.e., *false* and *true*, respectively). In a 3-valued logical structure, ι^S maps p and $(u_1, \ldots, u_k)$ to the value 0, 1, or 1/2 (i.e., *false*, *true*, and *unknown*, respectively).

2-valued logical structures will be used to encode concrete stores; 3-valued logical structures will be used to encode abstract stores; members of these two families of structures will be related by "truth-blurring embeddings" (explained in Section 12.4.4).

2-valued logical structures are used to encode concrete stores as follows: individuals represent memory locations in the heap; pointers from the stack into the heap are represented by unary "pointed-to-by-variable-q" predicates; and pointer-valued fields of data structures are represented by binary predicates.

Example 12.9

Table 12.1 lists the predicates used for representing the stores manipulated by programs that use the List data-type declaration from Figure 12.1a. In the case of insert, the unary predicates x, y, t, and e correspond to the program variables x, y, t, and e, respectively. The binary predicate n corresponds to the n-fields of List elements.

Figure 12.10 illustrates the 2-valued logical structures that represent lists of length ≤ 4 that are pointed to by program variable x. Column 3 of Figure 12.10 gives a graphical rendering of these 2-valued logical structures; note that these graphs are identical to those depicted in Figure 12.4:

- Individuals of the universe are represented by circles with names inside.
- A unary predicate p is represented in the graph by having a solid arrow from the predicate name p to node u for each individual u for which $\iota(p)(u) = 1$ and no arrow from predicate name p to node u' for each individual u' for which $\iota(p)(u') = 0$. (If $\iota(p)$ is 0 for all individuals, the predicate name p will not be shown.)
- A binary predicate q is represented in the graph by a solid arrow labeled q between each pair of individuals u_i and u_j for which $\iota(q)(u_i, u_j) = 1$ and no arrow between pairs u'_i and u'_j for which $\iota(q)(u'_i, u'_j) = 0$.

$S_0^\natural$

unary preds. indiv.	x	y	t	e		binary preds. n

Graphical Representation: (empty)

$S_1^\natural$

unary preds. indiv.	x	y	t	e		binary preds. n	u_1
u_1	1	0	0	0		u_1	0

Graphical Representation: $x \to (u_1)$

$S_2^\natural$

unary preds. indiv.	x	y	t	e		binary preds. n	u_1	u_2
u_1	1	0	0	0		u_1	0	1
u_2	0	0	0	0		u_2	0	0

Graphical Representation: $x \to (u_1) \overset{n}{\to} (u_2)$

$S_3^\natural$

unary preds. indiv.	x	y	t	e		binary preds. n	u_1	u_2	u_3
u_1	1	0	0	0		u_1	0	1	0
u_2	0	0	0	0		u_2	0	0	1
u_3	0	0	0	0		u_3	0	0	0

Graphical Representation: $x \to (u_1) \overset{n}{\to} (u_2) \overset{n}{\to} (u_3)$

$S_4^\natural$

unary preds. indiv.	x	y	t	e		binary preds. n	u_1	u_2	u_3	u_4
u_1	1	0	0	0		u_1	0	1	0	0
u_2	0	0	0	0		u_2	0	0	1	0
u_3	0	0	0	0		u_3	0	0	0	1
u_4	0	0	0	0		u_4	0	0	0	0

Graphical Representation: $x \to (u_1) \overset{n}{\to} (u_2) \overset{n}{\to} (u_3) \overset{n}{\to} (u_4)$

FIGURE 12.10 The 2-valued logical structures that represent lists of length ≤ 4.

Thus, in structure $S_2^\natural$, pointer variable x points to individual u_1, whose n-field points to individual u_2. The n-field of u_2 does not point to any individual (i.e., u_2 represents a heap cell whose n-field has the value NULL).

12.4.2   Extraction of Store Properties

2-valued structures offer a systematic way to answer questions about properties of the concrete stores they encode. For example, consider the formula

$$\varphi_{is}(v) \stackrel{\text{def}}{=} \exists v_1, v_2 : n(v_1, v) \wedge n(v_2, v) \wedge v_1 \neq v_2 \tag{12.1}$$

which expresses the "is-shared" property. Do two or more different heap cells point to heap cell v via their n-fields? For instance, $\varphi_{is}(v)$ evaluates to 0 in $S_2^\natural$ for the assignment $[v \mapsto u_2]$, because there is no assignment of the form $[v_1 \mapsto u_i, v_2 \mapsto u_j]$ such that $\iota^{S_2^\natural}(n)(u_i, u_2)$, $\iota^{S_2^\natural}(n)(u_j, u_2)$, and $u_i \neq u_j$ all hold.

As a second example, consider the formula

$$\varphi_{c_n}(v) \stackrel{\text{def}}{=} n^+(v, v) \tag{12.2}$$

which expresses the property of whether a heap cell v appears on a directed n-cycle. Here n^+ denotes the transitive closure of the n-relation. Formula $\varphi_{c_n}(v)$ evaluates to 0 in $S_2^\natural$ for the assignment $[v \mapsto u_2]$, because the transitive closure of the relation $\iota^{S_2^\natural}(n)$ does not contain the pair (u_2, u_2).

The preceding discussion can be summarized as the following principle:

Observation 12.1   (Property-Extraction Principle). *By encoding stores as logical structures, questions about properties of stores can be answered by evaluating formulas. The property holds or does not hold, depending on whether the formula evaluates to 1 or 0, respectively, in the logical structure.*

The language in which queries are posed is standard first-order logic with a transitive-closure operator. The notion of evaluating a formula φ in logical structure S with respect to assignment Z (where Z assigns individuals to the free variables of φ) is completely standard (e.g., see [17, 58]). We use the notation $[[\varphi]]_2^S(Z)$ to denote the value of φ in S with respect to Z.

12.4.3   Expressing the Semantics of Program Statements

Our tool for expressing the semantics of program statements is also based on evaluating formulas:

Observation 12.2   (Expressing the Semantics of Statements via Logical Formulas). *Suppose that σ is a store that arises before statement st, that σ' is the store that arises after st is evaluated on σ, and that S is the logical structure that encodes σ. A collection of predicate-update formulas — one for each predicate p in the vocabulary of S — allows one to obtain the structure S' that encodes σ'. When evaluated in structure S, the predicate-update formula for a predicate p indicates what the value of p should be in S'.*

In other words, the set of predicate-update formulas captures the concrete semantics of st.

This process is illustrated in Figure 12.11 for the statement y = y->n, where the initial structure $S_a^\natural$ represents a list of length 4 that is pointed to by both x and y. Figure 12.11 shows the predicate-update formulas for the five predicates of the vocabulary used in conjunction with insert: x, y, t, e, and n; the symbols x', y', t', e', and n' denote the values of the corresponding predicates in the structure that arises after execution of y = y->n. Predicates x', t', e', and n' are unchanged in value by y = y->n. The predicate-update formula $y'(v) = \exists v_1 : y(v_1) \wedge n(v_1, v)$ expresses the advancement of program variable y down the list.

Structure Before

unary preds.					binary preds.				
indiv.	x	y	t	e	n	u_1	u_2	u_3	u_4
u_1	1	1	0	0	u_1	0	1	0	0
u_2	0	0	0	0	u_2	0	0	1	0
u_3	0	0	0	0	u_3	0	0	0	1
u_4	0	0	0	0	u_4	0	0	0	0

$$x \rightarrow \boxed{u_1} \overset{n}{\rightarrow} \boxed{u_2} \overset{n}{\rightarrow} \boxed{u_3} \overset{n}{\rightarrow} \boxed{u_4}$$
$$S_a^{\natural} \qquad y$$

Statement

$$y = y\, \text{->}\, n$$

Predicate Update Formulae

$$x'(v) = x(v)$$
$$y'(v) = \exists v_1 : y(v_1) \wedge n(v_1, v)$$
$$t'(v) = t(v)$$
$$e'(v) = e(v)$$
$$n'(v_1, v_2) = n(v_1, v_2)$$

Structure After

unary preds.					binary preds.				
indiv.	x	y	t	e	n	u_1	u_2	u_3	u_4
u_1	1	0	0	0	u_1	0	1	0	0
u_2	0	1	0	0	u_2	0	0	1	0
u_3	0	0	0	0	u_3	0	0	0	1
u_4	0	0	0	0	u_4	0	0	0	0

$$x \rightarrow \boxed{u_1} \overset{n}{\rightarrow} \boxed{u_2} \overset{n}{\rightarrow} \boxed{u_3} \overset{n}{\rightarrow} \boxed{u_4}$$
$$S_b^{\natural} \qquad y$$

FIGURE 12.11   The given predicate-update formulas express a transformation on logical structures that corresponds to the semantics of $y\ =\ y\text{->}n$.

12.4.4   Abstraction via Truth-Blurring Embeddings

The abstract stores used for shape analysis are 3-valued logical structures that, by the construction discussed below, are a priori of bounded size. In general, each 3-valued logical structure corresponds to a (possibly infinite) set of 2-valued logical structures. Members of these two families of structures are related by *truth-blurring embeddings*.

The principle behind truth-blurring embedding is illustrated in Figure 12.12, which shows how 2-valued structure $S_a^{\natural}$ is abstracted to 3-valued structure S_a when we use $\{x, y, t, e\}$-abstraction. Abstraction is driven by the values of the "vector" of unary predicate values that each individual u has — that is, for $S_a^{\natural}$, by the values $\iota(x)(u)$, $\iota(y)(u)$, $\iota(t)(u)$, and $\iota(e)(u)$ — and, in particular, by the equivalence

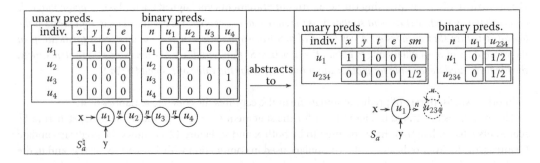

FIGURE 12.12   The abstraction of 2-valued structure $S_a^{\natural}$ to 3-valued structure S_a when we use $\{x, y, t, e\}$-abstraction. The boxes in the tables of unary predicates indicate how individuals are grouped into equivalence classes; the boxes in the tables for predicate n indicate how the quotient of n with respect to these equivalence classes is performed.

TABLE 12.2 Kleene's 3-valued interpretation of the propositional operators

$\wedge$	0	1	1/2	$\vee$	0	1	1/2	$\neg$	
0	0	0	0	0	0	1	1/2	0	1
1	0	1	1/2	1	1	1	1	1	0
1/2	0	1/2	1/2	1/2	1/2	1	1/2	1/2	1/2

classes formed from the individuals that have the same vector for their unary predicate values. In $S_a^\natural$, there are two such equivalence classes: (a) $\{u_1\}$, for which x, y, t, and e are 1, 1, 0, and 0, respectively, and (b) $\{u_2, u_3, u_4\}$, for which x, y, t, and e are all 0. (The boxes in the table of unary predicates for $S_a^\natural$ show how individuals of $S_a^\natural$ are grouped into two equivalence classes.)

All members of such equivalence classes are mapped to the same individual of the 3-valued structure. Thus, all members of $\{u_2, u_3, u_4\}$ from $S_a^\natural$ are mapped to the same individual in S_a, called u_{234};[8] similarly, all members of $\{u_1\}$ from $S_a^\natural$ are mapped to the same individual in S_a, called u_1.

For each non-unary predicate of the 2-valued structure, the corresponding predicate in the 3-valued structure is formed by a *truth-blurring quotient*. For instance:

- In $S_a^\natural$, $\iota^{S_a^\natural}(n)$ evaluates to 0 for the only pair of individuals in $\{u_1\} \times \{u_1\}$. Therefore, in S_a the value of $\iota^{S_a}(n)(u_1, u_1)$ is 0.
- In $S_a^\natural$, $\iota^{S_a^\natural}(n)$ evaluates to 0 for all pairs from $\{u_2, u_3, u_4\} \times \{u_1\}$. Therefore, in S_a the value of $\iota^{S_a}(n)(u_{234}, u_1)$ is 0.
- In $S_a^\natural$, $\iota^{S_a^\natural}(n)$ evaluates to 0 for two of the pairs from $\{u_1\} \times \{u_2, u_3, u_4\}$ (i.e., $\iota^{S_a^\natural}(n)(u_1, u_3) = 0$ and $\iota^{S_a^\natural}(n)(u_1, u_4) = 0$), whereas $\iota^{S_a^\natural}(n)$ evaluates to 1 for the other pair (i.e., $\iota^{S_a^\natural}(n)(u_1, u_2) = 1$); therefore, in S_a the value of $\iota^{S_a}(n)(u_1, u_{234})$ is 1/2.
- In $S_a^\natural$, $\iota^{S_a^\natural}(n)$ evaluates to 0 for some pairs from $\{u_2, u_3, u_4\} \times \{u_2, u_3, u_4\}$ (e.g., $\iota^{S_a^\natural}(n)(u_2, u_4) = 0$), whereas $\iota^{S_a^\natural}(n)$ evaluates to 1 for other pairs (e.g., $\iota^{S_a^\natural}(n)(u_2, u_3) = 1$); therefore, in S_a the value of $\iota^{S_a}(n)(u_{234}, u_{234})$ is 1/2.

In Figure 12.12, the boxes in the tables for predicate n indicate these four groupings of values.

An additional unary predicate, called *sm* (standing for "summary"), is added to the 3-valued structure to capture whether individuals of the 3-valued structure represent more than one concrete individual. For instance, $\iota^{S_a}(sm)(u_1) = 0$ because u_1 in S_a represents a single individual of $S_a^\natural$. However, u_{234} represents three individuals of $S_a^\natural$. For technical reasons, *sm* can be 0 or 1/2, but never 1; therefore, $\iota^{S_a}(sm)(u_{234}) = 1/2$.

12.4.5 Conservative Extraction of Store Properties

Questions about properties of 3-valued structures can be answered by evaluating formulas using Kleene's semantics of 3-valued logic (see [58]). The value of a formula is obtained in almost exactly the same way that it is obtained in ordinary 2-valued logic, except that the propositional operators are given the interpretations shown in Table 12.2. (The evaluation rules for $\exists$, $\forall$, and transitive closure are adjusted accordingly; that is, $\exists$ and $\forall$ are treated as indexed-$\vee$ and indexed-$\wedge$ operators, respectively.) We use the notation $[\![\varphi]\!]_3^S(Z)$ to denote the value of φ in 3-valued logical structure S with respect to 3-valued assignment Z.

We define a partial order $\sqsubseteq$ on truth values to reflect their degree of definiteness (or *information content*): $l_1 \sqsubseteq l_2$ denotes that l_1 is at least as definite as l_2.

---

[8]The reader should bear in mind that the names of individuals are completely arbitrary. u_{234} could have been called u_{17} or u_{99} and so on; in particular, the subscript "234" is used here only to remind the reader that, in this example, u_{234} of S_a is the individual that represents $\{u_2, u_3, u_4\}$ of $S_a^\natural$. (In many subsequent examples, u_{234} will be named u.)

Name	Logical Structure		Graphical Representation
S_0	**unary preds.** indiv. | x | y | t | e | sm	**binary preds.** n	
S_1	**unary preds.** indiv. | x | y | t | e | sm u_1 | 1 | 0 | 0 | 0 | 0	**binary preds.** n | u_1 u_1 | 0	$x \to (u_1)$
S_2	**unary preds.** indiv. | x | y | t | e | sm u_1 | 1 | 0 | 0 | 0 | 0 u | 0 | 0 | 0 | 0 | 0	**binary preds.** n | u_1 | u u_1 | 0 | 1 u | 0 | 0	$x \to (u_1) \overset{n}{\to} (u)$
S_3	**unary preds.** indiv. | x | y | t | e | sm u_1 | 1 | 0 | 0 | 0 | 0 u | 0 | 0 | 0 | 0 | 1/2	**binary preds.** n | u_1 | u u_1 | 0 | 1/2 u | 0 | 1/2	$x \to (u_1) \overset{n}{\to} (u)$ $\overset{n}{\curvearrowleft}$

FIGURE 12.13   The 3-valued logical structures that are obtained by applying truth-blurring embedding to the 2-valued structures that appear in Figure 12.10.

Definition 12.1   (Information Order). *For $l_1, l_2 \in \{0, 1/2, 1\}$, we define the **information order** on truth values as follows: $l_1 \sqsubseteq l_2$ if $l_1 = l_2$ or $l_2 = 1/2$. The symbol $\sqcup$ denotes the least-upper-bound operation with respect to $\sqsubseteq$:*

$\sqcup$	0	1/2	1
0	0	1/2	1/2
1/2	1/2	1/2	1/2
1	1/2	1/2	1

The 3-valued semantics is monotonic in the information order (see Table 12.2).

In [58] the *embedding theorem* states that the 3-valued Kleene interpretation in S of every formula is consistent with (i.e., $\sqsupseteq$) the formula's 2-valued interpretation in every concrete store $S^\natural$ that S represents. Consequently, questions about properties of stores can be answered by evaluating formulas using Kleene's semantics of 3-valued logic:

- If a formula evaluates to 1, then the formula holds in every store represented by the 3-valued structure S.
- If a formula evaluates to 0, then the formula does not hold in any store represented by S.
- If a formula evaluates to 1/2, then we do not know if this formula holds in all stores, does not hold in any store, or holds in some stores and does not hold in some other stores represented by S.

Consider the formula $\varphi_{c_n}(v)$ defined in Equation 12.2. (Does heap cell v appear on a directed cycle of n-fields?) Formula $\varphi_{c_n}(v)$ evaluates to 0 in structure S_3 from Figure 12.13 for the assignment $[v \mapsto u_1]$, because $n^+(u_1, u_1)$ evaluates to 0 in Kleene's semantics.

Formula $\varphi_{c_n}(v)$ evaluates to 1/2 in S_3 for the assignment $[v \mapsto u]$, because $\iota^{S_3}(n)(u, u) = 1/2$, and thus $n^+(u, u)$ evaluates to 1/2 in Kleene's semantics. Because of this, we do not know whether S_3 represents a concrete store that has a cycle; this uncertainty implies that (the tail of) the list pointed to by x *might* be cyclic.

In many situations, however, we are interested in analyzing the behavior of a program under the assumption, for example, that the program's input is an acyclic list. If an abstraction is not capable of expressing the distinction between cyclic and acyclic lists, an analysis algorithm based on that abstraction will usually be able to recover only very imprecise information about the actions of the program.

For this reason, we are interested in having our parametric framework support abstractions in which, for instance, the acyclic lists are distinguished from the cyclic lists. Our framework supports such distinctions by using *instrumentation predicates*.

The preceding discussion illustrates the following principle:

Observation 12.3 (**Instrumentation Principle**). *Suppose that S is a 3-valued structure that represents the 2-valued structure $S^\natural$. By explicitly "storing" in S the values that a formula φ has in $S^\natural$, it is sometimes possible to extract more precise information from S than can be obtained just by evaluating φ in S.*

In our experience, we have found three kinds of instrumentation predicates to be useful:

- Nullary predicates record Boolean information (and are similar to the "predicates" in predicate abstraction [3, 22]). For example, to distinguish between cyclic and acyclic lists, we can define an instrumentation predicate c_0 by the formula

$$\varphi_{c_0} \stackrel{\text{def}}{=} \exists v : n^+(v,v) \tag{12.3}$$

  which expresses the property that some heap cell v lies on a directed n-cycle. Thus, when $\iota^S(c_0)$ is 0, we know that S does not represent any memory configurations that contain cyclic data structures.

- Unary instrumentation predicates record information for unbounded sets of objects. Examples of some unary instrumentation predicates are given in Section 12.3.4. Notice that the unary cyclicity predicate c_n (defined by an open formula [see Equation 12.2]) allows finer distinctions than are possible with the nullary cyclicity predicate (defined by a closed formula [see Equation 12.3]). Unary cyclicity predicate c_n records information about the cyclicity properties of individual nodes — namely, $c_n(v)$ records whether node v lies on a cycle; nullary cyclicity predicate c_0 records a property of the heap as a whole — namely, whether the heap contains *any* cycle.

- Binary instrumentation predicates record relationships between unbounded sets of objects. For example, the instrumentation predicate $t[n](v_1, v_2) \stackrel{\text{def}}{=} n^+(v_1, v_2)$ records the existence of n-paths from v_1 to v_2.

Moreover, instrumentation predicates that are unary can also be used as abstraction predicates.

In Section 12.3.4, we saw how it is possible to change the shape abstraction in use by changing the set of instrumentation predicates in use and/or by changing which unary instrumentation predicates are used as abstraction predicates. By using the right collection of instrumentation predicates and abstraction predicates, shape-analysis algorithms can be created that, in many cases, determine precise shape information for programs that manipulate several (possibly cyclic) data structures simultaneously. The information obtained is more precise than that obtained from previous work on shape analysis.

In Section 12.5, several other instrumentation predicates are introduced that augment shape descriptors with auxiliary information that permits flow-dependence information to be read off from the results of shape analysis.

12.4.6 Abstract Interpretation of Program Statements

The goal of a shape-analysis algorithm is to associate with each vertex v of control-flow graph G, a finite set of 3-valued structures that "describes" all of the 2-valued structures that can arise at v (and possibly more). The abstract semantics can be expressed as the least fixed point (in terms of set inclusion) of a system of equations over variables that correspond to vertices in the program. The right-hand side of each equation is a transformer that represents the abstract semantics for an individual statement in the program.

The most complex issue we face is the definition of the abstract semantics of program statements. This abstract semantics has to (a) be conservative, that is, must account for every possible runtime situation, and (b) should not yield too many "unknown" values.

The fact that the concrete semantics of statements can be expressed via logical formulas (Observation 12.2), together with the fact that the evaluation of a formula φ in a 3-valued structure S is guaranteed to be safe with respect to the evaluation of φ in any 2-valued structure that S represents (the embedding theorem), means that one abstract semantics falls out automatically from the concrete semantics. One merely has to evaluate the predicate-update formulas of the concrete semantics on 3-valued structures.

Observation 12.4 (**Reinterpretation Principle**). *Evaluation of the predicate-update formulas for a statement st in 2-valued logic captures the transfer function for st of the concrete semantics. Evaluation of the same formulas in 3-valued logic captures a sound transfer function for st of the abstract semantics.*

If st is a statement, $[\![st]\!]_3$ denotes the transformation on 3-valued structures that is defined by evaluating in 3-valued logic the predicate-update formulas that represent the concrete semantics of st.

Figure 12.14 combines Figures 12.11 and 12.12 (see column 2 and row 1, respectively, of Figure 12.14). Column 4 of Figure 12.14 illustrates how the predicate-update formulas that express the concrete semantics for y = y->n also express a transformation on 3-valued logical structures — that is, an abstract semantics — that is safe with respect to the concrete semantics (cf. $S_a^\natural \to S_b^\natural$ versus $S_a \to S_b$).[9]

As we will see, this approach has a number of good properties:

- Because the number of elements in the 3-valued structures that we work with is bounded, the abstract-interpretation process always terminates.
- The embedding theorem implies that the results obtained are conservative.
- By defining appropriate instrumentation predicates, it is possible to emulate some previous shape-analysis algorithms (e.g., [8, 25, 30, 33]).[10]

Unfortunately, there is also bad news: the method described above and illustrated in Figure 12.14 can be very imprecise. For instance, the statement y = y->n illustrated in Figure 12.14 sets y to the value of y->n; that is, it makes y point to the next element in the list. In the abstract semantics, the evaluation in structure S_a of the predicate-update formula $y'(v) = \exists v_1 : y(v_1) \wedge n(v_1, v)$ causes $\iota^{S_b}(y)(u_{234})$ to be set to $1/2$. When $\exists v_1 : y(v_1) \wedge n(v_1, v)$ is evaluated in S_a, we have $\iota^{S_a}(y)(u_1) \wedge \iota^{S_a}(n)(u_1, u_{234}) = 1 \wedge 1/2 = 1/2$.

---

[9]The abstraction of $S_b^\natural$, as described in Section 12.4.4, is S_c. Figure 12.14 illustrates that in the abstract semantics we also work with structures that are even further "blurred." We say that S_c *embeds into* S_b; u_1 in S_c maps to u_1 in S_b; u_2 and u_{34} in S_c both map to u_{234} in S_b; the n predicate of S_b is the truth-blurring quotient of n in S_c under this mapping.

Our notion of the 2-valued structures that a 3-valued structure represents is based on this more general notion of embedding [58]. Note that in Figure 12.13, S_2 can be embedded into S_3; thus, structure S_3 also represents the acyclic lists of length 2 that are pointed to by x.

[10]The discussion above ignores the fact that for every statement and condition in the program, we also need to define how to update each instrumentation predicate p. That is, if p is defined by φ_p, an update formula is needed for transformation $[\![st]\!]_3(S)$ to produce an appropriate set of values for predicate p.

The simplest way is to reevaluate φ_p on the core predicates produced by $[\![st]\!]_3(S)$. In practice, however, this approach does not work very well because information will be lost under abstraction. As observed elsewhere [58], when working in 3-valued logic, Observation 12.3 implies that it is usually possible to retain more precision by defining a special *instrumentation-predicate maintenance formula*, $\mu_{p,st}(v_1, \ldots, v_k)$, and evaluating $\mu_{p,st}(v_1, \ldots, v_k)$ in structure S.

In [37, 50] algorithms are given that create an alternative predicate-maintenance formula $\mu_{p,st}$ for $p \in \mathcal{I}$ in terms of two *finite-differencing operators*, denoted by $\Delta_{st}^-[\cdot]$ and $\Delta_{st}^+[\cdot]$, which capture the negative and positive changes, respectively, that execution of statement st induces in an instrumentation predicate's value. The formula $\mu_{p,st}$ is created by combining p with $\Delta_{st}^-[\varphi_p]$ and $\Delta_{st}^+[\varphi_p]$ as follows: $\mu_{p,st} = p ? \neg \Delta_{st}^-[\varphi_p] : \Delta_{st}^+[\varphi_p]$.

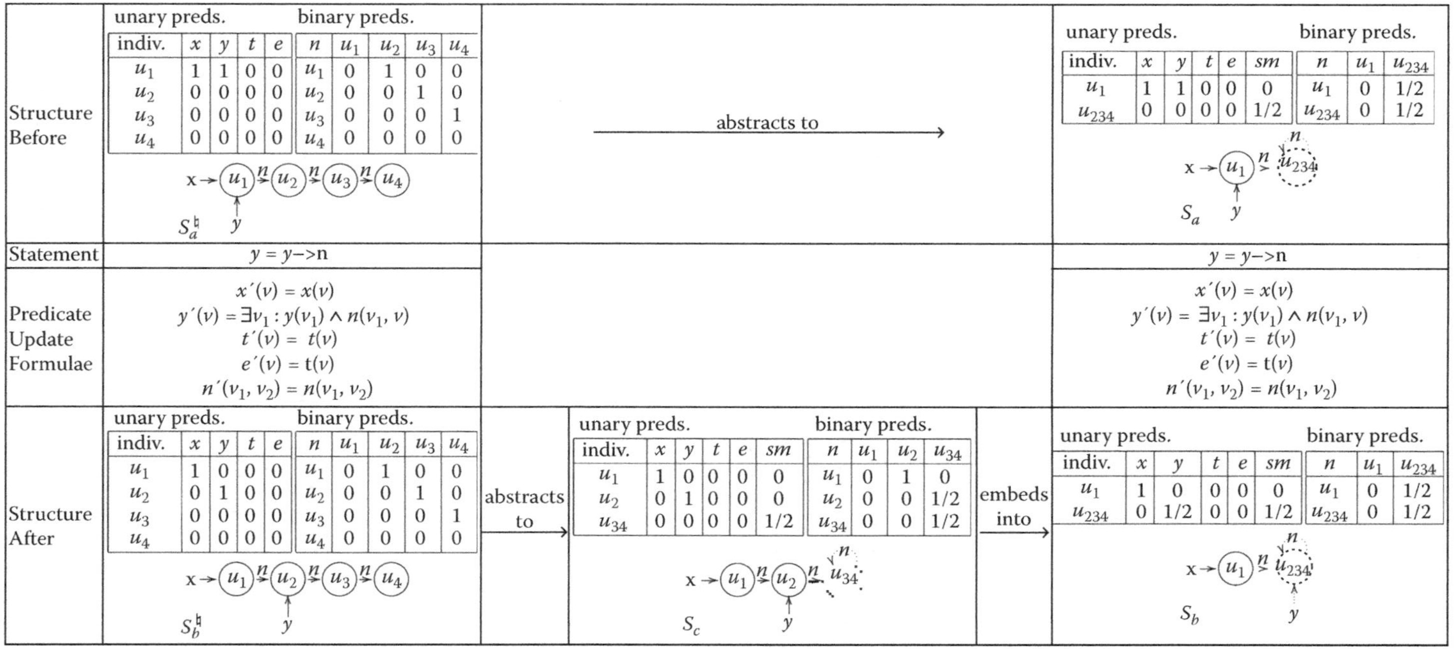

FIGURE 12.14 Commutative diagram that illustrates the relationships among (i) the transformation on 2-valued structures (defined by predicate-update formulas) that represents the concrete semantics for $y = y\text{->}n$, (ii) abstraction, and (iii) the transformation on 3-valued structures (defined by the same predicate-update formulas) that represents the simple abstract semantics for $y = y\text{->}n$ obtained via the reinterpretation principle (Observation 12.4). (In this example, $\{x, y, t, e\}$-abstraction is used.)

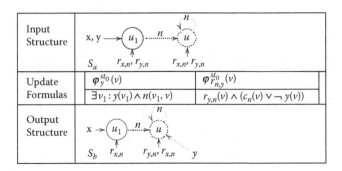

FIGURE 12.15   An application of the simplified abstract transformer for statement st_0: y = y->n in insert.

Consequently, all we can surmise after the execution of y = y->n is that y may point to one of the heap cells that summary node u_{234} represents (see S_b).

In contrast, the truth-blurring embedding of $S_b^\natural$ is S_c; thus, column 4 and row 4 of Figure 12.14 show that the abstract semantics obtained via Observation 12.4 can lead to a structure that is not as precise as what the abstract domain is capable of representing (cf. structures S_c and S_b).

As mentioned in Example 12.8, the use of reachability information is very important for retaining precision during shape analysis. However, even this mechanism is not sufficiently powerful to fix the problem. The same problem still occurs even if we use $\{x, y, t, e, is, r_{x,n}, r_{y,n}, r_{t,n}, r_{e,n}\}$-abstraction with $\{c_n\}$. Figure 12.15 shows the result of applying the abstract semantics of the statement $st_0 : y = y->n$ to structure S_a — one of the 3-valued structures that arises in the analysis of insert just before y is advanced down the list by statement st_0. Similar to what was illustrated in Figure 12.14, the resulting structure S_b shown in Figure 12.15 is not as precise as what the abstract domain is capable of representing. For instance, S_b does not contain a node that is definitely pointed to by y.

This imprecision leads to problems when a destructive update is performed. In particular, the first column in Table 12.3 shows what happens when the abstract transformers for the five statements that follow the search loop in insert are applied to S_b. Because $y(v)$ evaluates to 1/2 for the summary node, we eventually reach the situation shown in the fourth row of structures, in which y, e, r_x, r_y, r_e, r_t, and is are all 1/2 for the summary node. As a result, with the approach that has been described thus far, the abstract transformer for y->n = t sets the value of c_n for the summary node to 1/2. Consequently, the analysis fails to determine that the structure returned by insert is an acyclic list.

In contrast, the analysis that uses the techniques described in the remainder of this section is able to determine that at the end of insert the following properties always hold: (a) x points to an acyclic list that has no shared elements, (b) y points into the tail of the x-list, and (c) the value of e and y->n are equal.

It is worthwhile to note that the precision problem becomes even more acute for shape-analysis algorithms that, like [8], do not explicitly track reachability properties. The reason is that, without reachability, S_b represents situations in which y points to an element that is not even part of the x-list.

12.4.6.1 Mechanisms for an Improved Abstract Semantics

The remainder of this section describes the main ideas behind two mechanisms that provide a more precise way of defining the abstract semantics of program statements. In particular, these mechanisms are able to "materialize" new nonsummary nodes from summary nodes as data structures are traversed. As we will see, these improvements allow us to determine more precise shape descriptors for the data structures that arise in the insert program.

In formulating an improved approach, our goal is to retain the property that the transformer for a program statement falls out automatically from the predicate-update formulas of the concrete semantics and the predicate-update formulas supplied for the instrumentation predicates. Thus, the main idea

TABLE 12.3 Selective applications of the abstract transformers using the one-stage and the multi-stage approaches, for the statements in `insert` that come after the search loop. (For brevity, r_z is used in place of $r_{z,n}$ for all variables z, and node names are not shown.)

One-Stage	Multi-Stage		
x, r_x $\;\;y, r_x$ $\;\;$ y	x, r_x $\;\;y, r_x,$ r_y	x, r_x $\;\;y, r_x,$ r_y $\;\;r_x, r_y$	
$t = \texttt{malloc}();\; t\text{->}data = d;$			
t, r_t $\;\;x, r_x$ $\;\;y, r_x$ $\;\;$ y	t, r_t $\;\;x, r_x$ $\;\;y, r_x,$ r_y	t, r_t $\;\;x, r_x$ $\;\;y, r_x,$ r_y $\;\;r_x, r_y$	
$e = y\text{->}n$			
t, r_t $\;\;x, r_x$ $\;\;y, r_x$ $\;\;y, e,$ r_e	t, r_t $\;\;x, r_x$ $\;\;y, r_x,$ r_y	t, r_t $\;\;x, r_x$ $\;\;y, r_x,$ r_y $\;\;e, r_e,$ r_x, r_y	t, r_t $\;\;x, r_x$ $\;\;y, r_x,$ r_y $\;\;e, r_e,$ r_x, r_y $\;\;r_e, r_x,$ r_y
$t\text{->}n = e;$			
t, r_t $\;\;x, r_x$ $\;\;y, r_x$ $\;\;y, e,$ $r_e, r_t,$ is	t, r_t $\;\;x$ $\;\;y, r_y,$ r_x	t, r_t $\;\;x, r_x$ $\;\;y, r_y,$ r_x $\;\;e, r_e,$ $r_x, r_y,$ r_t, is	t, r_t $\;\;x, r_x$ $\;\;y, r_y,$ r_x $\;\;e, r_x,$ $r_y, r_t,$ is $\;\;r_e, r_x,$ r_y, r_t
$y\text{->}n = t;$			
$c_n, r_y;$ r_x, r_e $\;\;t, r_t$ $\;\;y, e,$ $r_y, r_x,$ $r_e, r_t,$ is, c_n $\;\;x, r_x$	t, r_t $\;\;r_y, r_x$ $\;\;x$ $\;\;y, r_y,$ r_x	t, r_t $\;\;r_y, r_x$ $\;\;x, r_x$ $\;\;e, r_e,$ $r_t, r_y,$ r_x $\;\;y, r_y,$ r_x	t $\;\;r_y, r_x$ $\;\;x$ $\;\;y, r_x$ $\;\;e, r_t,$ r_x, r_y $\;\;r_e, r_t,$ r_x, r_y

behind the improved approach is to decompose the transformer for *st* into a composition of several functions, as depicted in Figure 12.16 and explained below, each of which falls out automatically from the predicate-update formulas of the concrete semantics and the predicate-update formulas supplied for the instrumentation predicates:

- The operation *focus* refines 3-valued structures so that the formulas that define the meaning of *st* evaluate to definite values. The *focus* operation thus brings these formulas "into focus."
- The simple abstract meaning function for statement *st*, $[[st]]_3$, is then applied.
- The operation *coerce* converts a 3-valued structure into a more precise 3-valued structure by removing certain kinds of inconsistencies.

(The 10 structures referred to in Figure 12.16 are depicted in Figure 12.17. Figure 12.17 will be used to explain the improved mechanisms that are presented in Sections 12.4.6.2 and 12.4.6.3.)

It is worth noting that both *focus* and *coerce* are *semantic-reduction* operations (a concept originally introduced in [12]). That is, they convert a set of 3-valued structures into a more precise set of 3-valued structures that describe the same set of stores. This property, together with the correctness of the structure transformer $[[st]]_3$, guarantees that the overall multi-stage semantics is correct. In the context of a

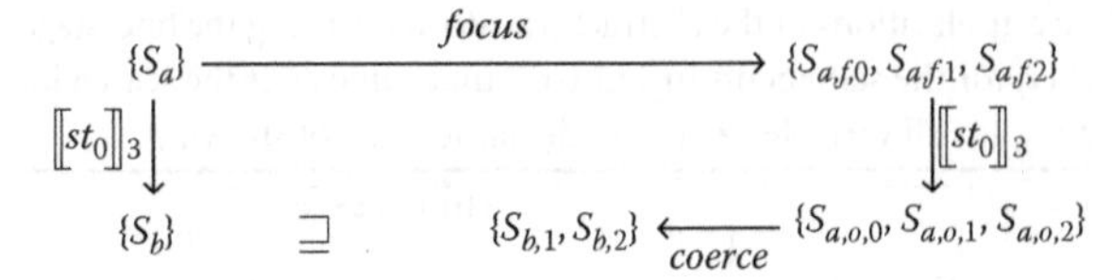

$$\{S_a\} \xrightarrow{\quad focus \quad} \{S_{a,f,0}, S_{a,f,1}, S_{a,f,2}\}$$

$$[\![st_0]\!]_3 \Big\downarrow \qquad\qquad\qquad\qquad \Big\downarrow [\![st_0]\!]_3$$

$$\{S_b\} \qquad \sqsupseteq \qquad \{S_{b,1}, S_{b,2}\} \xleftarrow{\quad coerce \quad} \{S_{a,0,0}, S_{a,0,1}, S_{a,0,2}\}$$

FIGURE 12.16 One-stage vs. multi-stage abstract semantics for statement $st_0 : \text{y = y->n}$.

parametric framework for abstract interpretation, semantic reductions are valuable because they allow the transformers of the abstract semantics to be defined in the modular fashion shown in Figure 12.16.

12.4.6.2 The Focus Operation

The operation $focus_F$ generates a set of structures on which a given set of formulas F have definite values for all assignments. (This operation will be denoted by $focus$ when F is clear from the context or when we are referring to a focus operation for F in the generic sense.) The focus formulas used in shape analysis are determined from the left-hand side (as an L-value) and right-hand side (as an R-value) of each kind of statement in the programming language. These are illustrated in the following example.

Example 12.10

For the statement $st_0: \text{y = y->n}$ in procedure `insert`, we focus on the formula

$$\varphi_0(v) \overset{\text{def}}{=} \exists v_1 : y(v_1) \wedge n(v_1, v) \tag{12.4}$$

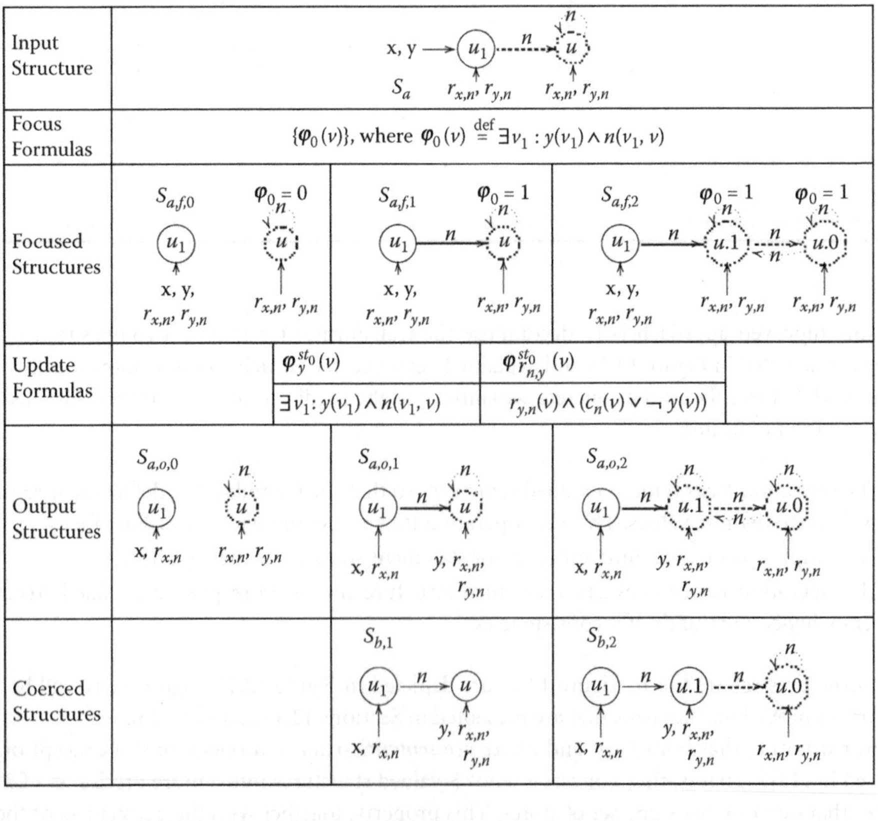

FIGURE 12.17 The first application of the improved transformer for statement $st_0: \text{y = y->n}$ in `insert`.

which corresponds to the R-value of the right-hand side of st_0 (the heap cell pointed to by y->n). The upper part of Figure 12.17 illustrates the application of $focus_{\{\varphi_0\}}(S_a)$, where S_a is the structure shown in Figure 12.15 that occurs in insert just before the first application of statement st_0: y = y->n. This results in three structures: $S_{a,f,0}$, $S_{a,f,1}$, and $S_{a,f,2}$:

- In $S_{a,f,0}$, $[\![\varphi_0]\!]_3^{S_{a,f,0}}([v \mapsto u])$ equals 0. This structure represents a situation in which the concrete list that x and y point to has only one element, but the store also contains garbage cells, represented by summary node u. (As we will see later, this structure is inconsistent because of the values of the $r_{x,n}$ and $r_{y,n}$ instrumentation predicates and will be eliminated from consideration by *coerce*.)

- In $S_{a,f,1}$, $[\![\varphi_0]\!]_3^{S_{a,f,1}}([v \mapsto u])$ equals 1. This covers the case where the list that x and y point to has exactly two elements. For all of the concrete cells that summary node u represents, φ_0 must evaluate to 1, so u must represent just a single list node.

- In $S_{a,f,2}$, $[\![\varphi_0]\!]_3^{S_{a,f,2}}([v \mapsto u.0])$ equals 0 and $[\![\varphi_0]\!]_3^{S_{a,f,2}}([v \mapsto u.1])$ equals 1. This covers the case where the list that x and y point to is a list of three or more elements. For all of the concrete cells that $u.0$ represents, φ_0 must evaluate to 0, and for all of the cells that $u.1$ represents, φ_0 must evaluate to 1. This case captures the essence of node materialization as described in [57]: individual u is bifurcated into two individuals.

The structures shown in Figure 12.17 are constructed by $focus_{\{\varphi_0\}}(S_a)$ by considering the reasons why $[\![\varphi_0]\!]_3^{S_a}(Z)$ evaluates to 1/2 for various assignments Z. In some cases, $[\![\varphi_0]\!]_3^{S_a}(Z)$ already has a definite value; for instance, $[\![\varphi_0]\!]_3^{S_a}([v \mapsto u_1])$ equals 0, and therefore φ_0 is already in focus at u_1. In contrast, $[\![\varphi_0]\!]_3^{S_a}([v \mapsto u])$ equals 1/2. We can construct three (maximal) structures S from S_a in which $[\![\varphi_0]\!]_3^{S}([v \mapsto u])$ has a definite value:

- $S_{a,f,0}$, in which $\iota^{S_{a,f,0}}(n)(u_1, u)$ is set to 0, and thus $[\![\varphi_0]\!]_3^{S_{a,f,0}}([v \mapsto u])$ equals 0.
- $S_{a,f,1}$, in which $\iota^{S_{a,f,1}}(n)(u_1, u)$ is set to 1, and thus $[\![\varphi_0]\!]_3^{S_{a,f,1}}([v \mapsto u])$ equals 1.
- $S_{a,f,2}$, in which u has been bifurcated into two different individuals, $u.0$ and $u.1$. In $S_{a,f,2}$, $\iota^{S_{a,f,2}}(n)$ $(u_1, u.0)$ is set to 0, and thus $[\![\varphi_0]\!]_3^{S_{a,f,2}}([v \mapsto u.0])$ equals 0, whereas $\iota^{S_{a,f,2}}(n)(u_1, u.1)$ is set to 1, and thus $[\![\varphi_0]\!]_3^{S_{a,f,2}}([v \mapsto u.1])$ equals 1.

An algorithm for *focus* that is based on these ideas is given in [58].

The greater the number of formulas on which we focus, the greater the number of distinctions that the shape-analysis algorithm can make, leading to improved precision. However, using a larger number of focus formulas can increase the number of structures that arise, thereby increasing the cost of analysis. Our preliminary experience indicates that in shape analysis there is a simple way to define the formulas on which to focus that guarantees that the number of structures generated grows only by a constant factor. The main idea is that in a statement of the form *lhs* = *rhs*, we only focus on formulas that define the heap cells for the L-value of *lhs* and the R-value of *rhs*. Focusing on L-values and R-values ensures that the application of the abstract transformer does not set to 1/2 the entries of core predicates that correspond to pointer variables and fields that are updated by the statement. This approach extends naturally to program conditions and to statements that manipulate multiple L-values and R-values.

For our simplified language and type List, the target formulas on which to focus can be defined as shown in Table 12.4. Let us examine a few of the cases from Table 12.4:

- For the statement x = NULL, the set of target formulas is the empty set because neither the *lhs* L-value nor the *rhs* R-value is a heap cell.
- For the statement x = t->n, the set of target formulas is the singleton set $\{\exists v_1 : t(v_1) \wedge n(v_1, v)\}$ because the *lhs* L-value cannot be a heap cell, and the *rhs* R-value is the cell pointed to by t->n.
- For the statement x->n = t, the set of target formulas is the set $\{x(v), t(v)\}$ because the *lhs* L-value is the heap cell pointed to by x, and the *rhs* R-value is the heap cell pointed to by t.
- For the condition x == t, the set of target formulas is the set $\{x(v), t(v)\}$; the R-values of the two sides of the conditional expression are the heap cells pointed to by x and t.

TABLE 12.4   The target formulas for *focus*, for statements
and conditions of a program that uses type List

st	Focus Formulae
x = NULL	$\emptyset$
x = t	$\{t(v)\}$
x = t->n	$\{\exists v_1 : t(v_1) \wedge n(v_1, v)\}$
x->n = t	$\{x(v), t(v)\}$
x = malloc()	$\emptyset$
x == NULL	$\{x(v)\}$
x != NULL	$\{x(v)\}$
x == t	$\{x(v), t(v)\}$
x != t	$\{x(v), t(v)\}$
UninterpretedCondition	$\emptyset$

12.4.6.3   The Coerce Operation

The operation *coerce* converts a 3-valued structure into a more precise 3-valued structure by removing certain kinds of inconsistencies. The need for *coerce* can be motivated by the following example:

Example 12.11

After *focus*, the simple transformer $[[st]]_3$ is applied to each of the structures produced. For instance, in Example 12.10, $[[st_0]]_3$ is applied to structures $S_{a,f,0}$, $S_{a,f,1}$, and $S_{a,f,2}$ to obtain structures $S_{a,o,0}$, $S_{a,o,1}$, and $S_{a,o,2}$, respectively (see Figure 12.17).

However, this process can produce structures that are not as precise as we would like. The intuitive reason for this state of affairs is that there can be interdependences between different properties stored in a structure, and these interdependences are not necessarily incorporated in the definitions of the predicate-update formulas. In particular, consider structure $S_{a,o,2}$. In this structure, the n-field of $u.0$ can point to $u.1$, which suggests that y may be pointing to a heap-shared cell. However, this is incompatible with the fact that $\iota(is)(u.1) = 0$ (i.e., $u.1$ cannot represent a heap-shared cell) and the fact that $\iota(n)(u_1, u.1) = 1$ (i.e., it is known that $u.1$ definitely has an incoming n-edge from a cell other than $u.0$).

Also, the structure $S_{a,o,0}$ describes an impossible situation: $\iota(r_{y,n})(u) = 1$ and yet u is not reachable — or even potentially reachable — from a heap cell that is pointed to by y.

The *coerce* mechanism is a systematic method that captures interdependences among the properties stored in 3-valued structures; *coerce* removes indefinite values that violate certain consistency rules, thereby "sharpening" the structures that arise during shape analysis. This remedies the imprecision illustrated in Example 12.11. In particular, when the sharpening process is applied to structure $S_{a,o,2}$ from Figure 12.17, the structure that results is $S_{b,2}$. In this case, the sharpening process discovers that (a) two of the n-edges with value 1/2 can be removed from $S_{a,o,2}$ and (b) individual $u.1$ can only ever represent a single individual in each of the structures that $S_{a,o,2}$ represents, and hence $u.1$ should not be labeled as a summary node. These facts are not something that the mechanisms that have been described in earlier sections are capable of discovering. Also, the structure $S_{a,o,0}$ is discarded by the sharpening process.

The sharpening mechanism that *coerce* provides is crucial to the success of the improved shape-analysis framework because it allows a more accurate job of materialization to be performed than would otherwise be possible. For instance, note how the sharpened structure, $S_{b,2}$, clearly represents an unshared list of length 3 or more that is pointed to by x and whose second element is pointed to by y. In fact, in the domain of $\{x, y, t, e, is, r_{x,n}, r_{y,n}, r_{t,n}, r_{e,n}\}$-abstraction with $\{c_n\}$, $S_{b,2}$ is the most precise representation possible for the family of unshared lists of length 3 or more that are pointed to by x and whose second element is pointed to by y. Without the sharpening mechanism, instantiations of the framework would rarely be able to determine such things as "The data structure being manipulated by a certain list-manipulation program is actually a list."

The *coerce* operation is based on the observation that 3-valued structures obey certain consistency rules that are a consequence of truth-blurring embedding. These consistency rules can be formalized as a system of "compatibility constraints." Moreover, the constraint system can be obtained automatically from formulas that express certain global invariants on concrete stores.

Example 12.12

Consider a 2-valued structure $S^\natural$ that can be embedded in a 3-valued structure S, and suppose that the formula φ_{is} for "inferring" whether an individual u is shared evaluates to 1 in S (i.e., $[[\varphi_{is}(v)]]_3^S([v \mapsto u]) = 1$). By the embedding theorem, $\iota^{S^\natural}(is)(u^\natural)$ must be 1 for any individual $u^\natural \in U^{S^\natural}$ that the embedding function maps to u.

Now consider a structure S' that is equal to S except that $\iota^{S'}(is)(u)$ is 1/2. $S^\natural$ can also be embedded in S'. However, the embedding of $S^\natural$ in S is a "better" embedding — one that preserves more definite values. This has operational significance: it is needlessly imprecise to work with structure S' in which $\iota^{S'}(is)(u)$ has the value 1/2; instead, we should discard S' and work with S. In general, the "stored predicate" *is* should be at least as precise as its inferred value; consequently, if it happens that φ_{is} evaluates to a definite value (1 or 0) in a 3-valued structure, we can sharpen the stored predicate *is*.

Similar reasoning allows us to determine, in some cases, that a structure is inconsistent. In $S_{a,o,0}$, for instance, $\varphi_{r_{y,n}}(u) = 0$, whereas the value stored in S for $r_{y,n}$, namely $\iota^{S_{a,o,0}}(r_{y,n})(u)$, is 1; consequently, $S_{a,o,0}$ is a 3-valued structure that does not represent any concrete structures at all. Structure $S_{a,o,0}$ can therefore be eliminated from further consideration by the shape-analysis algorithm.

This reasoning applies to all instrumentation predicates, not just *is* and $r_{y,n}$, and to both of the definite values, 0 and 1.

The reasoning used in Example 12.12 can be summarized as the following principle:

Observation 12.5 (**The Sharpening Principle**). *In any structure S, the value stored for $\iota^S(p)(u_1, \ldots, u_k)$ should be at least as precise as the value of p's defining formula, φ_p, evaluated at $u_1, \ldots, u_k$ (i.e., $[[\varphi_p]]_3^S([v_1 \mapsto u_1, \ldots, v_k \mapsto u_k])$). Furthermore, if $\iota^S(p)(u_1, \ldots, u_k)$ has a definite value and φ_p evaluates to an incomparable definite value, then S is a 3-valued structure that does not represent any concrete structures at all.*

This observation can be formalized in terms of *compatibility constraints*, defined as follows:

Definition 12.2 A **compatibility constraint** is a term of the form $\varphi_1 \triangleright \varphi_2$, where φ_1 is an arbitrary 3-valued formula, and φ_2 is either an atomic formula or the negation of an atomic formula over distinct logical variables. We say that a 3-valued structure S and an assignment Z **satisfy** $\varphi_1 \triangleright \varphi_2$ if, whenever Z is an assignment such that $[[\varphi_1]]_3^S(Z) = 1$, we also have $[[\varphi_2]]_3^S(Z) = 1$. (If $[[\varphi_1]]_3^S(Z)$ equals 0 or 1/2, S and Z satisfy $\varphi_1 \triangleright \varphi_2$, regardless of the value of $[[\varphi_2]]_3^S(Z)$.)

The compatibility constraint that captures the reasoning used in Example 12.12 is $\varphi_{is}(v) \triangleright is(v)$. That is, when φ_{is} evaluates to 1 at u, then *is* must evaluate to 1 at u to satisfy the constraint. The compatibility constraint used to capture the similar case of sharpening $\iota(is)(u)$ from 1/2 to 0 is $\neg\varphi_{is}(v) \triangleright \neg is(v)$.

Compatibility constraints can be generated automatically from formulas that express certain global invariants on concrete stores. We call such formulas *compatibility formulas*. There are two sources of compatibility formulas:

- The formulas that define the instrumentation predicates
- Additional formulas that formalize the properties of stores that are compatible with the semantics of C (i.e., with our encoding of C stores as 2-valued logical structures)

The following definition supplies a way to convert formulas into compatibility constraints:

Definition 12.3   *Let φ be a closed formula and a be an atomic formula such that (a) a contains no repetitions of logical variables, and (b) $a \not\equiv sm(v)$. Then the **compatibility constraint generated from** φ is defined as follows:*

$$\varphi_1 \triangleright a \qquad \text{if } \varphi \equiv \forall v_1, \ldots v_k : (\varphi_1 \Rightarrow a) \tag{12.5}$$

$$\varphi_1 \triangleright \neg a \qquad \text{if } \varphi \equiv \forall v_1, \ldots v_k : (\varphi_1 \Rightarrow \neg a) \tag{12.6}$$

The intuition behind Equations 12.5 and 12.6 is that for an atomic predicate, a truth-blurring embedding is forced to yield $1/2$ only in cases in which a evaluates to 1 on one tuple of values for $v_1, \ldots, v_k$ but evaluates to 0 on a different tuple of values. In this case, the left-hand side will evaluate to $1/2$ as well.

Our first source of compatibility formulas is the set of formulas that define the instrumentation predicates. For every instrumentation predicate $p \in \mathcal{I}$ defined by a formula $\varphi_p(v_1, \ldots, v_k)$, we generate a compatibility formula of the following form:

$$\forall v_1, \ldots, v_k : \varphi_p(v_1, \ldots, v_k) \Leftrightarrow p(v_1, \ldots, v_k) \tag{12.7}$$

So that we can apply Definition 12.3, this is then broken into two implications:

$$\forall v_1, \ldots, v_k : \varphi_p(v_1, \ldots, v_k) \Rightarrow p(v_1, \ldots, v_k) \tag{12.8}$$

$$\forall v_1, \ldots, v_k : \neg \varphi_p(v_1, \ldots, v_k) \Rightarrow \neg p(v_1, \ldots, v_k) \tag{12.9}$$

For instance, for each program variable x, we have the defining formula of instrumentation predicate $r_{x,n}$:

$$\varphi_{r_{x,n}}(v) \stackrel{\text{def}}{=} x(v) \vee \exists v_1 : x(v_1) \wedge n^+(v_1, v) \tag{12.10}$$

and thus

$$\forall v : x(v) \vee \exists v_1 : x(v_1) \wedge n^+(v_1, v) \Leftrightarrow r_{x,n}(v) \tag{12.11}$$

which is then broken into

$$\forall v : x(v) \vee \exists v_1 : x(v_1) \wedge n^+(v_1, v) \Rightarrow r_{x,n}(v) \tag{12.12}$$

$$\forall v : \neg(x(v) \vee \exists v_1 : x(v_1) \wedge n^+(v_1, v)) \Rightarrow \neg r_{x,n}(v) \tag{12.13}$$

We then use Definition 12.3 to generate the following compatibility constraints:

$$x(v) \vee \exists v_1 : x(v_1) \wedge n^+(v_1, v) \,\triangleright r_{x,n}(v) \tag{12.14}$$

$$\neg(x(v) \vee \exists v_1 : x(v_1) \wedge n^+(v_1, v)) \triangleright \neg r_{x,n}(v) \tag{12.15}$$

The constraint-generation rules defined in Definition 12.3 generate interesting constraints only for certain specific syntactic forms, namely implications with exactly one (possibly negated) predicate symbol on the right-hand side. Thus, when we generate compatibility constraints from formulas written as implications (such as Equations 12.12 and 12.13 and those in Table 12.5), the set of constraints generated depends on the form in which the compatibility formulas are written. However, not all of the many equivalent forms possible for a given compatibility formula lead to useful constraints. For instance, when Definition 12.3 is applied to the formula $\forall v_1, \ldots v_k : (\varphi_1 \Rightarrow a)$, it generates the constraint $\varphi_1 \triangleright a$; however, Definition 12.3 does not generate a constraint for the equivalent formula $\forall v_1, \ldots v_k : (\neg \varphi_1 \vee a)$.

This phenomenon can prevent an instantiation of the shape-analysis framework from having a suitable compatibility constraint at its disposal that would otherwise allow it to sharpen or discard a structure that arises during the analysis — and hence can lead to a shape-analysis algorithm that is more conservative than we would like.

TABLE 12.5 The formulas listed above the line are compatibility formulas for structures that represent a store of a C program that operates on values of the type List defined in Figure 12.1(a). The corresponding compatibility constraints are listed below the line.

		$\neg \exists v : sm(v)$	(12.22)
for each $x \in P\,Var, \forall v_1, v_2 : x(v_1) \land x(v_2)$	$\Rightarrow$	$v_1 = v_2$	(12.23)
$\forall v_1, v_2 : (\exists v_3 : n(v_3, v_1) \land n(v_3, v_2))$	$\Rightarrow$	$v_1 = v_2$	(12.24)
$(\exists v : sm(v))$	$\triangleright$	0	(12.25)
for each $x \in P\,Var, x(v_1) \land x(v_2)$	$\triangleright$	$v_1 = v_2$	(12.26)
$(\exists v_3 : n(v_3, v_1) \land n(v_3, v_2))$	$\triangleright$	$v_1 = v_2$	(12.27)

The way around this difficulty is to augment the constraint-generation process to generate constraints for some of the logical consequences of each compatibility formula:

Example 12.13

The defining formula for instrumentation predicate *is* is

$$\varphi_{is}(v) \stackrel{\text{def}}{=} \exists v_1, v_2 : n(v_1, v) \land n(v_2, v) \land v_1 \neq v_2 \tag{12.16}$$

We obtain the following formula from Equation 12.16:

$$\forall v : (\exists v_1, v_2 : n(v_1, v) \land n(v_2, v) \land v_1 \neq v_2) \Leftrightarrow is(v) \tag{12.17}$$

which is broken into the two formulas

$$\forall v : (\exists v_1, v_2 : n(v_1, v) \land n(v_2, v) \land v_1 \neq v_2) \Rightarrow is(v) \tag{12.18}$$

$$\forall v : \neg(\exists v_1, v_2 : n(v_1, v) \land n(v_2, v) \land v_1 \neq v_2) \Rightarrow \neg is(v) \tag{12.19}$$

By rewriting the implication in Equation 12.18 as a disjunction and then applying De Morgan's laws, we have

$$\forall v, v_1, v_2 : \neg n(v_1, v) \lor \neg n(v_2, v) \lor v_1 = v_2 \lor is(v) \tag{12.20}$$

One of the logical consequences of Equation 12.20 is

$$\forall v, v_2 : (\exists v_1 : n(v_1, v) \land v_1 \neq v_2 \land \neg is(v)) \Rightarrow \neg n(v_2, v) \tag{12.21}$$

from which we obtain the following compatibility constraint:

$$(\exists v_1 : n(v_1, v) \land v_1 \neq v_2 \land \neg is(v)) \triangleright \neg n(v_2, v) \tag{12.22}$$

(In addition to Equation 12.22, we obtain a number of other compatibility constraints from other logical consequences of Equation 12.20 [58].)

As we will see shortly, Equation 12.22 allows a more accurate job of materialization to be performed than would otherwise be possible: When $is(u)$ is 0 and one incoming n-edge to u is 1, to satisfy Equation 12.22 a second incoming n-edge to u cannot have the value 1/2. It must have the value 0; that is, the latter edge cannot exist (cf. Examples 12.11 and 12.15). This allows edges to be removed (safely) that a more naive materialization process would retain (cf. structures $S_{a,o,2}$ and $S_{b,2}$ in Figure 12.17), and permits the improved shape-analysis algorithm to generate more precise structures for insert than the ones generated by the simple shape-analysis algorithm sketched at the beginning of Section 12.4.6.

12.4.6.3.1 Compatibility Constraints from Hygiene Conditions

Our second source of compatibility formulas stems from the fact that not all structures $S^\natural \in$ STRUCT[$\mathcal{P}$] represent stores that are compatible with the semantics of C. For example, stores have the property that each pointer variable points to at most one element in heap-allocated storage.

Example 12.14

The set of formulas listed above the line in Table 12.5 is a set of compatibility formulas that must be satisfied for a structure to represent a store of a C program that operates on values of the type `List` defined in Figure 12.1a. Equation 12.23 captures the condition that concrete stores never contain any summary nodes. Equation 12.24 captures the fact that every program variable points to at most one list element. Equation 12.25 captures a similar property of the n-fields of `List` structures: whenever the n-field of a list element is non-`NULL`, it points to at most one list element. The corresponding compatibility constraints generated according to Definition 12.3 are listed below the line.

12.4.6.3.2 An Example of Coerce in Action

We are now ready to show how the *coerce* operation uses these compatibility constraints to either sharpen or discard a 3-valued logical structure. The *coerce* operation is a constraint-satisfaction procedure that repeatedly searches a structure S for assignments Z that fail to satisfy $\varphi_1 \triangleright \varphi_2$ (i.e., $[[\varphi_1]]_3^S(Z) = 1$ but $[[\varphi_2]]_3^S(Z) \neq 1$). This is used to improve the precision of shape analysis by (a) sharpening the values of predicates stored in S when the constraint violation is repairable, and (b) eliminating S from further consideration when the constraint violation is irreparable. (An algorithm for this process is given in [58].)

Example 12.15

The application of *coerce* to the structures $S_{a,o,0}$, $S_{a,o,1}$, and $S_{a,o,2}$ yields $S_{b,1}$ and $S_{b,2}$, as shown in the bottom block of Figure 12.17:

- The structure $S_{a,o,0}$ is discarded because the violation of Equation 12.15 is irreparable.
- The structure $S_{b,1}$ was obtained from $S_{a,o,1}$ by removing incompatibilities as follows:
  - Consider the assignment $[v \mapsto u, v_1 \mapsto u_1, v_2 \mapsto u]$. Because $\iota(n)(u_1, u) = 1$, $u_1 \neq u$, and $\iota(is)(u) = 0$, Equation 12.22 implies that $\iota(n)(u, u)$ must equal 0. Thus, in $S_{b,1}$ the (indefinite) n-edge from u to u has been removed.
  - Consider the assignment $[v_1 \mapsto u, v_2 \mapsto u]$. Because $\iota(y)(u) = 1$, Equation 12.28 implies that $[[v_1 = v_2]]_3^{S_{b,1}}([v_1 \mapsto u, v_2 \mapsto u])$ must equal 1, which in turn means that $\iota^{S_{b,1}}(sm)(u)$ must equal 0. Thus, in $S_{b,1}$ u is no longer a summary node.
- The structure $S_{b,2}$ was obtained from $S_{a,o,2}$ by removing incompatibilities as follows:
  - Consider the assignment $[v \mapsto u.1, v_1 \mapsto u_1, v_2 \mapsto u.0]$. Because $\iota(n)(u_1, u.1) = 1$, $u_1 \neq u.0$, and $\iota(is)(u.1) = 0$, Equation 12.22 implies that $\iota^{S_{b,2}}(n)(u.0, u.1)$ must equal 0. Thus, in $S_{b,2}$ the (indefinite) n-edge from $u.0$ to $u.1$ has been removed.
  - Consider the assignment $[v \mapsto u.1, v_1 \mapsto u_1, v_2 \mapsto u.1]$. Because $\iota(n)(u_1, u.1) = 1$, $u_1 \neq u.1$, and $\iota(is)(u.1) = 0$, Equation 12.22 implies that $\iota^{S_{b,2}}(n)(u.1, u.1)$ must equal 0. Thus, in $S_{b,2}$ the (indefinite) n-edge from $u.1$ to $u.1$ has been removed.
  - Consider the assignment $[v_1 \mapsto u.1, v_2 \mapsto u.1]$. Because $\iota(y)(u.1) = 1$, Equation 12.28 implies that $[[v_1 = v_2]]_3^{S_{b,2}}([v_1 \mapsto u.1, v_2 \mapsto u.1])$ must equal 1, which in turn means that $\iota^{S_{b,2}}(sm)(u.1)$ must equal 0. Thus, in $S_{b,2}$ $u.1$ is no longer a summary node.

Important differences between the structures $S_{b,1}$ and $S_{b,2}$ result from applying the multi-stage abstract transformer for statement $st_0 : \text{y = y->n}$, compared with the structure S_b that results from applying the one-stage abstract transformer (see Figure 12.15). For instance, y points to a summary node in S_b, whereas y does not point to a summary node in either $S_{b,1}$ or $S_{b,2}$; as noted earlier, in the domain of $\{x, y, t, e, is, r_{x,n}, r_{y,n}, r_{t,n}, r_{e,n}\}$-abstraction with $\{c_n\}$, $S_{b,2}$ is the most precise representation possible for

TABLE 12.6 The structures that occur before and after successive applications of the multi-stage abstract transformer for the statement y = y->n during the abstract interpretation of `insert`. (For brevity, node names are shown.)

Iter.	Structure Before	Structures After
1	x, y → ○ ⋯ⁿ⋯ ; S_a $r_{x,n}, r_{y,n}$ $r_{x,n}, r_{y,n}$	x → ○ —ⁿ→ ○ ; $S_{b,1}$ $r_{x,n}$ y, $r_{x,n}, r_{y,n}$; x → ○ —ⁿ→ ○ ⋯ⁿ⋯; $S_{b,2}$ $r_{x,n}$ y, $r_{x,n}, r_{y,n}$ $r_{x,n}, r_{y,n}$
2	x → ○ —ⁿ→ ○ ; $S_{b,1}$ $r_{x,n}$ y, $r_{x,n}, r_{y,n}$	x → ○ —ⁿ→ ○ ; S_c $r_{x,n}$ $r_{x,n}$
3	x → ○ —ⁿ→ ○ ⋯ⁿ⋯; $S_{b,2}$ $r_{x,n}$ y, $r_{x,n}, r_{y,n}$ $r_{x,n}, r_{y,n}$	x → ○ —ⁿ→ ○ —ⁿ→ ○ ; $S_{d,1}$ $r_{x,n}$ $r_{x,n}$ y, $r_{x,n}, r_{y,n}$; x → ○ —ⁿ→ ○ —ⁿ→ ○ ⋯ⁿ⋯; $S_{d,2}$ $r_{x,n}$ $r_{x,n}$ y, $r_{y,n}, r_{x,n}$ $r_{y,n}, r_{x,n}$; = S_c
4	x → ○ —ⁿ→ ○ ; S_c $r_{x,n}$ $r_{x,n}$	x → ○ ⋯ⁿ⋯; S_e $r_{x,n}$ $r_{x,n}$
5	x → ○ —ⁿ→ ○ —ⁿ→ ○ ; $S_{d,1}$ $r_{x,n}$ $r_{x,n}$ y, $r_{x,n}, r_{y,n}$	x → ○ ⋯ⁿ⋯; S_e $r_{x,n}$ $r_{x,n}$
6	x → ○ —ⁿ→ ○ —ⁿ→ ○ ⋯ⁿ⋯; $S_{d,2}$ $r_{x,n}$ $r_{x,n}$ y, $r_{y,n}, r_{x,n}$ $r_{y,n}, r_{x,n}$	x → ○ ⋯ⁿ⋯ ○; $S_{f,1}$ $r_{x,n}$ $r_{x,n}$ y, $r_{y,n}, r_{x,n}$; x → ○ ⋯ⁿ⋯ ○ —ⁿ→; $S_{f,2}$ $r_{x,n}$ $r_{x,n}$ y, $r_{y,n}, r_{x,n}$ $r_{y,n}, r_{x,n}$
7	x → ○ ⋯ⁿ⋯; S_e $r_{x,n}$ $r_{x,n}$	= S_e
8	x → ○ ⋯ⁿ⋯ ○; $S_{f,1}$ $r_{x,n}$ $r_{x,n}$ y, $r_{y,n}, r_{x,n}$	= S_e
9	x → ○ ⋯ⁿ⋯ ○ —ⁿ→; $S_{f,2}$ $r_{x,n}$ $r_{x,n}$ y, $r_{y,n}, r_{x,n}$ $r_{y,n}, r_{x,n}$	= $S_{f,1}$; = $S_{f,2}$

the family of unshared lists of length 3 or more that are pointed to by x and whose second element is pointed to by y.

Example 12.16

Table 12.6 shows the 3-valued structures that occur before and after applications of the abstract transformer for the statement y = y->n during the abstract interpretation of `insert`.

The material in Table 12.3 that appears under the heading "Multi-Stage" shows the application of the abstract transformers for the five statements that follow the search loop in `insert` to $S_{b,1}$ and $S_{b,2}$. For space reasons, we do not show the abstract execution of these statements on the other structures shown in Table 12.6; however, the analysis is able to determine that at the end of `insert` the following properties always hold: (a) x points to an acyclic list that has no shared elements, (b) y points into the tail of the x–list, and (c) the value of e and y->n are equal. The identification of the latter condition is rather remarkable: the analysis is capable of showing that e and y->n are must-aliases at the end of `insert` (see also Section 12.5.1).

12.5   Applications

The algorithm sketched in Section 12.4 produces a set of 3-valued structures for each program point *pt*. This set provides a conservative representation, that is, it describes a superset of the set of concrete stores that can possibly occur in any execution of the program that ends at *pt*. Therefore, questions about the stores at *pt* can be answered (conservatively) by posing queries against the set of 3-valued structures that the shape-analysis algorithm associates with *pt*. The answers to these questions can be utilized in an optimizing compiler, as explained in Section 12.2. Furthermore, the fact that the shape-analysis framework is based on logic allows queries to be specified in a uniform way using logical formulas.

In this section, we discuss several kinds of questions. Section 12.5.1 discusses how instantiations of the parametric shape-analysis framework that have been described in previous sections can be applied to the problem of identifying may- and must-aliases. Section 12.5.2 shows that the shape-analysis framework can be instantiated to produce flow-dependence information for programs that manipulate linked data structures. Finally, Section 12.5.3 sketches some other applications for the results of shape analysis.

12.5.1   Identifying May- and Must-Aliases

We say that two pointer access paths, e_1 and e_2, are *may-aliases* at a program point *pt* if there exists an execution sequence ending at *pt* that produces a store in which both e_1 and e_2 point to the same heap cell. We say that e_1 and e_2 are *must-aliases* at *pt* if, for every execution sequence ending at *pt*, e_1 and e_2 point to the same heap cell.[11]

Consider the access paths $e_1 \equiv$ x->f_1->$\cdots$->f_n and $e_2 \equiv$ x->g_1->$\cdots$->g_m. To extract aliasing information, we use the formula

$$al[e_1, e_2] \stackrel{\text{def}}{=} \begin{aligned} \exists v_0, \ldots, v_n, w_0, \ldots, w_m \ : \ & x(v_0) \wedge f_1(v_0, v_1) \wedge \cdots \wedge f_n(v_{n-1}, v_n) \\ & \wedge \ y(w_0) \wedge g_1(w_0, w_1) \wedge \cdots \wedge g_m(w_{m-1}, w_m) \\ & \wedge \ v_n = w_m \end{aligned} \qquad (12.23)$$

If Equation 12.23 evaluates to 0 in every 3-valued structure that the shape-analysis algorithm associates with program point *pt*, we know that e_1 and e_2 are not may-aliases at *pt*. Similarly, when $al[e_1, e_2]$ evaluates to 1 in every such structure, we know that e_1 and e_2 are must-aliases at *pt*. In all other cases, e_1 and e_2 are considered may-aliases.

Note that in some cases, $al[e_1, e_2]$ may evaluate to 1/2, in which case e_1 and e_2 are considered may-aliases; this is a conservative result.

The answer can sometimes be improved by first applying *focus* with Equation 12.23. This will produce a set of structures in which $al[e_1, e_2]$ does not evaluate to an indefinite value. Finally, one can run *coerce* on the 3-valued structures produced by *focus* to eliminate infeasible 3-valued structures.

Example 12.17

Consider the 3-valued structure at the bottom right corner of Table 12.3. The formula $al[$y->n->n,e$]$ evaluates to 1 in this structure and in all of the other structures arising after y->n = t; thus, y->n->n

---

[11]Variants of these definitions can be defined that account for the case when e_1 or e_2 has the value NULL.

int y;
List p, q;
 	
q = (List) malloc();
p = q;
 	
l_1: p->data = 5;
$l_{1.5}$:
l_2: y = q->data;
	int y;
	List p, q, t;
	
	q = (List) malloc();
	p = q;
	t = p;
	l_1: p->data = 5;
	$l_{1.5}$: t->data = 7;
	l_2: y = q->data;
		int y;
		List p, q;
		
		q = (List) malloc();
		p = q;
		
		l_1: p->data = 5;
		$l_{1.5}$: p = (List) malloc();
		l_2: y = q->data;

(a)	(b)	(c)

FIGURE 12.18   A motivating example to demonstrate the differences between may-aliases and flow dependences.

and e are must-aliases at this point. Also, $al[$e->n,y->n$]$ evaluates to 0 in this structure and in all of the other structures; thus, e->n and y->n are not may-aliases.

However, the formula $al[$e->n->n,e->n$]$ evaluates to 1/2 in this structure. If we focus on the $al[$e->n->n,e->n$]$ formula, we obtain several structures; in one of them, e->n points to a nonsummary node that has a definite n-edge to itself. This structure is eliminated by *coerce*. In all of the remaining structures, the formula $al[$e->n->n,e->n$]$ evaluates to 0, which shows that e->n->n and e->n are not may-aliases.

12.5.2 Constructing Program Dependences

This section shows how to use information obtained from shape analysis to construct program dependence graphs [19, 31, 46]. To see why the problem of computing flow dependences is nontrivial, consider the example program fragments shown in Figure 12.18. A formal definition of flow dependence is given in Definition 12.4; for the purposes of this discussion, a statement l_b *depends on* l_a if the value written to a resource in l_a is directly used at l_b, that is, without intervening writes to this resource.

A naive (and unsafe) criterion that one might use to identify flow dependences in Figure 12.18a would be to say that l_2 depends on l_1 if p and q can refer to the same location at l_2 (i.e., if p and q are may-aliases at l_2). In Figure 12.18a, this would correctly identify the flow dependence from l_1 to l_2. The naive criterion sometimes identifies more flow dependences than we might like. In Figure 12.18b it would say there is a flow dependence from l_1 to l_2, even though $l_{1.5}$ overwrites the location that p points to. The naive criterion is unsafe because it may miss dependences. In Figure 12.18c, there is a flow dependence from l_1 to l_2; this would be missed because statement $l_{1.5}$ overwrites p, and thus p and q are never may-aliases at l_2.

One safe way to identify dependences in a program that uses heap-allocated storage is to introduce an abstract variable for each allocation site, use the results of a flow-insensitive points-to analysis [1, 18, 23, 61, 62] to determine a safe approximation of the variables that are possibly defined and possibly used at each program point, and then use a traditional algorithm for reaching definitions (where each allocation site is treated as a use of its associated variable).

In this section, we utilize the parametric shape-analysis framework to define an alternative, and much more precise, algorithm. This algorithm is based on an idea developed by Horwitz et al. [25].[12] They introduced an augmented semantics for the programming language; in addition to all of the normal

---

[12]An alternative approach would have been to use the Ross–Sagiv construction [56], which reduces the problem of computing program dependences to the problem of computing may-aliases, and then to apply the method of Section 12.5.1. The method presented in this section is a more direct construction for identifying program dependences and thereby provides a better demonstration of the utility of the parametric shape-analysis framework for this problem.

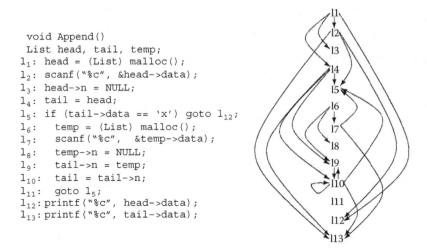

```
void Append()
 List head, tail, temp;
l₁: head = (List) malloc();
l₂: scanf("%c", &head->data);
l₃: head->n = NULL;
l₄: tail = head;
l₅: if (tail->data == 'x') goto l₁₂;
l₆:    temp = (List) malloc();
l₇:    scanf("%c",  &temp->data);
l₈:    temp->n = NULL;
l₉:    tail->n = temp;
l₁₀:   tail = tail->n;
l₁₁:   goto l₅;
l₁₂: printf("%c", head->data);
l₁₃: printf("%c", tail->data);
```

FIGURE 12.19   A program that builds a list by appending elements to `tail`, and its flow-dependence graph.

aspects of the language's semantics, the augmented semantics also records information about the history of resource usage — in this case, "last-write" information — for each location in the store. As we will see, it is natural to record this extra information using additional core predicates. As discussed in more detail below, this instantiation of the shape-analysis framework creates an algorithm from which conservative dependence information can then be extracted.

The resulting algorithm is the most precise algorithm known for identifying the data dependences of programs that manipulate heap-allocated storage. In addition, it does not need the artificial concept of introducing an abstract variable for each allocation site. For example, Figure 12.19 shows a program that builds a list by destructively appending elements to `tail`, together with a graph that shows the flow dependences that the algorithm identifies.

The rest of this subsection is organized as follows: Dependences are discussed in Section 12.5.2.1. Predicates for recording history information are introduced in Section 12.5.2.2, which also illustrates the results obtained via this dependence-analysis method.

12.5.2.1   Program Dependences

Program dependences can be grouped into flow dependences (def-use), output dependences (def-def), and anti-dependences (use-def) [19, 31]. In this section, we focus on flow dependences between program statements. Other types can be handled in a similar fashion.

We allow programs to explicitly modify the store via assignments through pointers. Because of this, we phrase the definition of flow dependence in terms of memory locations rather than program variables [25].

Definition 12.4   (**Flow Dependence**). *Consider labeled statements* l_i : st_i *and* l_j : st_j. *We say that* l_i *has a flow dependence on* l_j *if there is an execution path along which* st_j *writes into a memory location, loc, that* st_i *reads, and there is no intervening write into loc.*

Example 12.18

In the program fragment shown in Figure 12.18b, statement l_2 does not depend on l_1 because statement l_2 reads from a location that is last written at statement $l_{1.5}$. In Figure 12.18c, $l_{1.5}$ does not interrupt the dependence between l_2 and l_1 because it does not write into a location that is read by l_2.

TABLE 12.7 Predicates for recording history information

Predicate	Intended Meaning
$lst\_w\_v[l,z]$	Program variable z was last written into by the statement at label l.
$lst\_w\_f[l,n](v)$	The n-field of list element v was last written into by the statement at label l.
$lst\_w\_f[l,d](v)$	The data-field of list element v was last written into by the statement at label l.

Example 12.19

Consider the program and graph of flow dependences shown in Figure 12.19. Notice that l_{12} is flow dependent only on l_1 and l_2, while l_{13} is flow dependent on l_2, l_4, l_7, and l_{10}. This information could be used by a slicing tool to determine that the loop need not be executed in order to print head->data in l_{12}, or by an instruction scheduler to reschedule l_{12} to be executed any time after l_2. Also, l_3, l_8, and l_{11} have no statements that are dependent on them, making them candidates for elimination.

Thus, even in this simple example, knowing the flow dependences would allow several code transformations.

12.5.2.2 Recording History Using Predicates

Table 12.7 shows the predicates that are introduced to implement the augmented semantics à la Horwitz et al. [25]. As indicated in the column labeled "Intended Meaning," the intention is that these predicates will record the label of the program statement that last writes into a given memory location.

The predicate $lst\_w\_v[l,z]$ is similar to the one used in reaching-definitions analysis; it records that program variable z was last written into by the statement at label l. The other two predicates record, for each field of each list element, which statement last wrote into that location.

Table 12.8 shows the predicate-update formulas for recording which statement last wrote into a location. The definitions given in Table 12.8 would be used to augment the instantiation of the shape-analysis

TABLE 12.8 Predicate-update formulae for recording last-write information. Here rhs denotes an arbitrary expression.

Statement	Cond.	Predicate
l_1: x = rhs		$\varphi^{st}_{lst\_w\_v[l_1,x]} = 1$
	$l \not\equiv l_1$	$\varphi^{st}_{lst\_w\_v[l,x]} = 0$
	$l \not\equiv l_1, z \not\equiv x$	$\varphi^{st}_{lst\_w\_v[l,z]} = lst\_w\_v[l,z]$
		$\varphi^{st}_{lst\_w\_f[l,n]}(v) = lst\_w\_f[l,n](v)$
		$\varphi^{st}_{lst\_w\_f[l,d]}(v) = lst\_w\_f[l,d](v)$
l_1: x->n = rhs		$\varphi^{st}_{lst\_w\_v[l,z]} = lst\_w\_v[l,z]$
		$\varphi^{st}_{lst\_w\_f[l_1,n]}(v) = (lst\_w\_f[l_1,n](v) \wedge \neg x(v)) \vee x(v)$
	$l \not\equiv l_1$	$\varphi^{st}_{lst\_w\_f[l,n]}(v) = lst\_w\_f[l,n](v) \wedge \neg x(v)$
		$\varphi^{st}_{lst\_w\_f[l,d]}(v) = lst\_w\_f[l,d](v)$
l_1: x->data =rhs		$\varphi^{st}_{lst\_w\_v[l,z]} = lst\_w\_v[l,z]$
		$\varphi^{st}_{lst\_w\_f[l,n]}(v) = lst\_w\_f[l,n](v)$
		$\varphi^{st}_{lst\_w\_f[l_1,d]}(v) = lst\_w\_f[l_1,d](v) \wedge \neg x(v)) \vee x(v)$
	$l \not\equiv l_1$	$\varphi^{st}_{lst\_w\_f[l,d]}(v) = lst\_w\_f[l,d](v) \wedge \neg x(v)$

TABLE 12.9   Formulas that use the last-write information in the structures associated with statement at l_2 to identify flow dependences from statement l_1. (In this table, c stands for any constant.)

Statement	Formula
l_2: x = NULL	0
l_2: x = malloc()	0
l_2: x = y	$lst\_w\_v[l_1, y]$
l_2: x = y->n	$lst\_w\_v[l_1, y] \vee \exists v : y(v) \wedge lst\_w\_f[l_1, n](v)$
l_2: x = y->data	$lst\_w\_v[l_1, y] \vee \exists v : y(v) \wedge lst\_w\_f[l_1, d](v)$
l_2: x->f = NULL	$lst\_w\_v[l_1, x]$
l_2: x->f = y	$lst\_w\_v[l_1, x] \vee lst\_w\_v[l_1, y]$
l_2: x->data = c	$lst\_w\_v[l_1, x]$

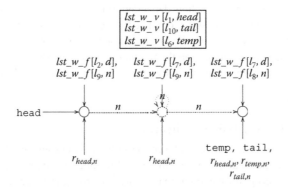

FIGURE 12.20   The most complex structure that the analysis yields at l_{12}.

framework that was described in Section 12.4. When the shape-analysis algorithm that we obtain in this way is applied to a program, it produces a set of 3-valued structures for each program point. Then, for each statement l_2, to determine whether there is a flow dependence from l_1 to l_2, each of the structures associated with l_2 is checked by evaluating the formulas from the appropriate line of Table 12.9.[13] These formulas use the last-write information in the structures associated with l_2 to determine whether there is flow dependence from a statement l_1. The idea behind Table 12.9 is that for each location accessed in the evaluation of l_2's left-hand side (as an L-value) and l_2's right-hand side (as an R-value), we need to check which statement last wrote into that location. If the formula is potentially satisfied by some 3-valued structure at l_2, there is a flow dependence from l_1 to l_2.

Example 12.20

Figure 12.20 shows one of the 3-valued structures that occurs at l_{12} when the program shown in Figure 12.19 is analyzed. (Three other structures arise at l_{12}; these correspond to simpler configurations of memory than the one depicted.) The formula $lst\_w\_v[l_2, head] \vee \exists v : head(v) \wedge lst\_w\_f[l_2, d](v)$ evaluates to 1, which indicates that l_{12} depends on l_2.

In contrast, the formula $lst\_w\_v[l_7, head] \vee \exists v : head(v) \wedge lst\_w\_f[l_7, d](v)$ evaluates to 0 (in this structure and in all of the other structures that arise at l_{12}). This allows us to conclude that l_{12} does not depend on l_7.

---

[13]It is straightforward to provide similar formulas to extract flow dependences from the structures that the shape-analysis algorithm associates with program conditions.

12.5.3   Other Applications

12.5.3.1   Cleanness of Programs That Manipulate Linked Lists

Java programs are more predictable than C programs because the language assures that certain types of error conditions will be trapped, such as NULL-dereference errors. However, most Java compilers and Java Virtual Machine (JVM) implementations check these conditions at runtime, which slows the program down and does not provide any feedback to the programmer at compile time. In [14], it is shown how shape analysis can be used to detect the absence (and presence) of NULL dereferences and memory leaks in many C programs. It is also shown there that shape analysis yields much more precise information than what is obtained by a flow-sensitive points-to analysis, such as the one developed in [15]. If such methods could be implemented in Java compilers and/or a JVM, some runtime checks could be avoided.

12.5.3.2   Correctness of Sorting Implementations

In [35] a shape-analysis abstraction is developed to analyze programs that sort linked lists. It is shown that the analysis is precise enough to discover that (correct versions of) bubble-sort and insertion-sort procedures always produce correctly sorted lists as outputs and that the invariant "is-sorted" is maintained by list-manipulation operations such as merge. In addition, it is shown that when the analysis is applied to erroneous versions of bubble-sort and insertion-sort procedures, it is able to discover the error. In [37, 38] a novel abstraction-refinement method is defined, based on inductive logic programming, and successfully used to derive abstractions *automatically* that establish the partial correctness of several sorting algorithms. The derived abstractions are also used to establish that the algorithms possess additional properties, such as stability and antistability.

12.5.3.3   Conformance to API Specifications

[48] shows how to verify that client programs using a library conform to the library's API specifications. In particular, an analysis is provided for verifying the absence of concurrent-modification exceptions in Java programs that use Java collections and iterators. In [68] separation and heterogeneous abstraction are used to scale the verification algorithms and to allow verification of larger programs that use libraries such as the Java Database Connectivity (JDBC) API.

12.5.3.4   Checking Multithreaded Systems

In [67] it is shown how to apply 3-valued logic to the problem of checking properties of multithreaded systems. In particular, [67] addresses the problem of state-space exploration for languages, such as Java, that allow (a) dynamic creation and destruction of an unbounded number of threads, (b) dynamic allocation and freeing of an unbounded number of storage cells from the heap, and (c) destructive updating of structure fields. This combination of features creates considerable difficulties for any method that tries to check program properties.

In this chapter, the problem of program analysis is expressed as a problem of annotating a control-flow graph with sets of 3-valued structures; in contrast, the analysis algorithm given in [67] builds and explores a 3-valued transition system on-the-fly.

In [67] problems (a) and (b) are handled essentially via the techniques developed in this chapter; problem (c) is addressed by reducing it to problem (b). Threads are modeled by individuals, which are abstracted using truth-blurring embedding — in this case, with respect to the collection of unary thread properties that hold for a given thread. This naming scheme automatically discovers commonalities in the state space, but without relying on explicitly supplied symmetry properties, as in, for example, [10, 16].

Unary core predicates are used to represent the program counter of each thread object; *focus* implements the interleaving of threads. The analysis described in [67] is capable of proving the absence of deadlock in a dining-philosophers program that permits there to be an unbounded number of philosophers.

In [70] this approach was applied to verify partial correctness of concurrent-queue implementations.

In [69] the above approach was extended to provide a general framework for proving temporal properties of programs by representing program traces as logical structures. A more efficient technique for proving local temporal properties is presented in [60] and applied to compile-time garbage collection in Javacard programs. (While the aforementioned technique was developed as a verification technique, it can also be utilized to reduce synchronization overhead, e.g., [64].)

12.6 Extensions

The approach to shape analysis presented in this chapter has been implemented by T. Lev-Ami in a system called *TVLA* (three-valued-logic analyzer) [34, 36]. TVLA provides a language in which the user can specify (a) an operational semantics (via predicates and predicate-update formulas), (b) a control-flow graph for a program, and (c) a set of 3-valued structures that describe the program's input. Using this specification, TVLA builds the corresponding equation system and finds its least fixed point.

The experience gained from building TVLA led to a number of improvements to, and extensions of, the methods described in this chapter (and in [58]). The enhancements that TVLA incorporates include:

- The ability to declare that certain binary predicates specify *functional properties*.
- The ability to specify that structures should be stored only at nodes of the control-flow graph that are targets of backedges.
- An enhanced version of *coerce* that incorporates some methods that are similar in flavor to relational-database query-optimization techniques (cf. [63]).
- An enhanced *focus* algorithm that generalizes the methods of Section 12.4.6.2 to handle focusing on arbitrary formulas.[14] In addition, this version of *focus* takes advantage of the properties of predicates that are specified to be functions.
- The ability to specify criteria for merging together structures associated with a program point. This feature is motivated by the idea that when the number of structures that arise at a given program point is too large, it may be better to create a smaller number of structures that represent at least the same set of 2-valued structures. In particular, nullary predicates (i.e., predicates of 0-arity) are used to specify which structures are to be merged together. For example, for linked lists, the "x-is-not-null" predicate, defined by the formula $nn[x]() = \exists v : x(v)$, discriminates between structures in which x points to a list element, and structures in which it does not. By using $nn[x]()$ as the criterion for whether to merge structures, the structures in which x is NULL are kept separate from those in which x points to an allocated memory cell.

Further details about these features can be found in [34].

The remainder of this section describes several other extensions of our parametric logic-based analysis framework that have been investigated; many of these extensions have also been incorporated into TVLA.

12.6.1 Interprocedural Analysis

Several papers have investigated interprocedural shape analysis. In [53] procedures are handled by explicitly representing stacks of activation records as linked lists, allowing rather precise analysis of (possibly recursive) procedures. In [29] procedures are handled by automatically creating summaries of their behavior. Abstractions of *two-vocabulary structures* are used to capture an over-approximation of the relation that describes the transformation effected by a procedure. In [52] a new concrete semantics for programs that manipulate heap-allocated storage is presented, which only passes "local" heaps to procedures. A simplified version of this semantics is used in [54] to perform more modular summarization by only representing reachable parts of the heap.

---

[14]The enhanced *focus* algorithm may not always succeed.

12.6.2 Computing Intersections of Abstractions

Arnold et al. [2] considers the problem of computing the intersection (meet) of heap abstractions, namely the greatest lower bound of two sets of 3-valued structures. This problem turns out to have many applications in program analysis, such as interpreting program conditions, refining abstract configurations, reasoning about procedures [29], and proving temporal properties of heap-manipulating programs, either via greatest-fixed-point approximation over trace semantics or in a staged manner over the collecting semantics. [2] describes a constructive formulation of meet that is based on finding certain relations between abstract heap objects. The enumeration of those relations is reduced to finding constrained matchings over bipartite graphs.

12.6.3 Efficient Heap Abstractions and Representations

Manevich et al. [39] addresses the problem of space consumption in first-order state representations by describing and evaluating two new representation techniques for logical structures. One technique uses ordered binary decision diagrams (OBDDs) [7]; the other uses a variant of a functional map data structure [44, 51]. The results show that both the OBDD and functional implementations reduce space consumption in TVLA by a factor of 4 to 10 relative to the original TVLA state representation, without compromising analysis time.

Manevich et al. [40] present a new heap abstraction that works by merging shape descriptors according to a partial isomorphism similarity criterion, resulting in a partially disjunctive abstraction. There it is also shown that on the existing TVLA examples, the abstract interpretation using this abstraction is drastically faster than the powerset heap abstraction, practically without a significant loss of precision.

Manevich et al. [41] provide a family of simple abstractions for potentially cyclic linked lists. In particular, it provides a relatively efficient predicate abstraction that allows verification of programs that manipulate potentially cyclic linked lists.

12.6.4 Abstracting Numeric Values

In this chapter, we ignore numeric values in programs, so the analysis would be imprecise for programs that perform numeric computations. [20] presents a generic solution for combining abstractions of numeric values and heap-allocated storage. This solution has been integrated into a version of TVLA. In [21] a new abstraction of numeric values is presented, which like canonical abstraction, tracks correlations between aggregates and not just indices. For example, it can identify loops that perform array-kills (i.e., assign values to an entire array). In [28] this approach has been generalized to define a family of abstractions (for relations as well as numeric quantities) that is more precise than pure canonical abstraction and allows the basic idea from [20] to be applied more widely.

12.6.5 Abstraction Refinement

The model-checking community has had much success with the notion of *automatic abstraction refinement*, in which an analyzer is started with a crude abstraction, and the results of analysis runs that fail to establish a definite answer (about whether the property of interest does or does not hold) are used as feedback about how the abstraction should be refined [4, 9, 32]. However, the abstract domains used in shape analysis are based on first-order logic, whereas model-checking tools that use abstraction refinement are based on predicate-abstraction domains [3, 22].

Abstraction-refinement methods suitable for use with shape-analysis domains were investigated in [37, 38]. The methods are based on inductive logic programming [47], which is a machine-learning technique for identifying general rules from a set of observed instances. In [37, 38] this was used to identify appropriate formulas that define new instrumentation relations (and thereby change the abstraction in use).

12.7   Related Work

The shape-analysis problem was first investigated by Reynolds [49], who studied it in the context of a Lisp-like language with no destructive updating. Reynolds treated the problem as one of simplifying a collection of set equations. A similar shape-analysis problem, but for an imperative language supporting nondestructive manipulation of heap-allocated objects, was formulated independently by Jones and Muchnick, who treated the problem as one of solving (i.e., finding the least fixed point of) a collection of equations using regular tree grammars [30].

Jones and Muchnick [30] also began the study of shape analysis for languages *with* destructive updating. To handle such languages, they formulated an analysis method that associates program points with sets of finite shape-graphs.[15] To guarantee that the analysis terminates for programs containing loops, the Jones–Muchnick approach limits the length of acyclic selector paths by some chosen parameter k. All nodes beyond the "k-horizon" are clustered into a summary node. The Jones–Muchnick formulation has two drawbacks:

- The analysis yields poor results for programs that manipulate cons-cells beyond the k-horizon. For example, in the list-reversal program of Figure 12.1, little useful information is obtained. The analysis algorithm must model what happens when the program is applied to lists of length greater than k. However, the tail of such a list is treated conservatively, as an arbitrary, and possibly cyclic, data structure.

- The analysis may be extremely costly because the number of possible shape-graphs is doubly exponential in k.

In addition to Jones and Muchnick's work, k-limiting has also been used in a number of subsequent papers (e.g., Horwitz et al. [25]).

Another well-known shape-analysis algorithm, developed by Chase et al. [8], is based on the following ideas:

- Sharing information, in the form of abstract heap reference counts (0, 1, and ∞), is used to characterize shape-graphs that represent list structures.[16]

- Several heuristics are introduced to allow several shape-nodes to be maintained for each allocation site.

- For an assignment to x->n, when the shape-node that x points to represents only concrete elements that will definitely be overwritten, the n-field of the shape-node that x points to can be overwritten (a so-called strong update).

The Chase–Wegman–Zadeck algorithm is able to identify list-preservation properties in some cases; for instance, it can determine that a program that appends a list to a list preserves "listness." However, as noted in [8], allocation-site information alone is insufficient to determine interesting facts in many programs. For example, it cannot determine that "listness" is preserved for either the list-insert program or a list-reversal program that uses destructive-update operations. In particular, in the list-reversal program, the Chase–Wegman–Zadeck algorithm reports that a possibly cyclic structure may arise, and that the two lists used by the program might share cells in common (when in fact the two lists are always disjoint).

The parametric framework presented in this paper can be instantiated to implement the Chase–Wegman–Zadeck algorithm, as well as other shape-analysis algorithms. Furthermore, in Section 12.5.2, we presented a new algorithm for computing flow dependences using our parametric approach.

For additional discussion of related work, the reader is referred to [57, 58].

---

[15]In this section, we use the term *shape-graph* in the generic sense, meaning any finite graph structure used to approximate the shapes of runtime data structures.

[16]The idea of augmenting shape-graphs with sharing information also appears in the earlier work of Jones and Muchnick [30].

12.8 Conclusions

Many of the classical data-flow-analysis algorithms use bit vectors to represent the characteristic functions of set-valued data-flow values. This corresponds to a logical interpretation (in the abstract semantics) that uses two values. It is *definite* on one of the bit values and *conservative* on the other. That is, either "false" means "false" and "true" means "may be true/may be false," or "true" means "true" and "false" means "may be true/may be false." Many other static-analysis algorithms have a similar character.

Most static analyses have such a one-sided bias; exceptions include data-flow analyses that simultaneously track "may" and "must" information, for example, [6, 59].

The material presented in this chapter shows that while *indefiniteness* is inherent (i.e., a static analysis is unable, in general, to give a definite answer), one-sidedness is not. By basing the abstract semantics on 3-valued logic, definite truth and definite falseness can both be tracked, with the third value, 1/2, capturing indefiniteness.

This outlook provides some insight into the true nature of the values that arise in other work on static analysis:

- A one-sided analysis that is precise with respect to "false" and conservative with respect to "true" is really a 3-valued analysis over 0, 1, and 1/2 that conflates 1 and 1/2 (and uses "true" in place of 1/2).

- Likewise, an analysis that is precise with respect to "true" and conservative with respect to "false" is really a 3-valued analysis over 0, 1, and 1/2 that conflates 0 and 1/2 (and uses "false" in place of 1/2).

In contrast, the analyses developed in this chapter are unbiased: They are precise with respect to both 0 and 1 and use 1/2 to capture indefiniteness. Other work that uses 3-valued logic to develop unbiased analyses includes [26].

We hope the ideas presented in this chapter (and the TVLA system, which embodies these ideas) will help readers implement new static-analysis algorithms that identify interesting properties of programs that make use of heap-allocated data structures.

References

1. L. O. Andersen. 1994. Program analysis and specialization for the C programming language. Ph.D. dissertation, DIKU, University of Copenhagen (DIKU report 94/19).
2. G. Arnold, R. Manevich, M. Sagiv, and R. Shaham. 2006. Combining shape analyses by intersecting abstractions. In *Verification, Model Checking, and Abstract Interpretation*.
3. T. Ball, R. Majumdar, T. Millstein, and S. K. Rajamani. 1998. Automatic predicate abstraction of C programs. In *SIGPLAN Conference on Programming Language Design and Implementation*.
4. T. Ball and S. K. Rajamani. 2000. Bebop: A symbolic model checker for Boolean programs. In *Proceedings of the 7th International SPIN Workshop on SPIN Model Checking and Software Verification*, 113–30. London: Springer-Verlag.
5. A. Bijlsma. 1999. A semantics for pointers. Talk at the Dagstuhl-Seminar on Program Analysis.
6. R. Bodik, R. Gupta, and M. L. Soffa. 1998. Complete removal of redundant computations. In *SIGPLAN Conference on Programming Language Design and Implementation*.
7. R. E. Bryant. 1986. Graph-based algorithms for Boolean function manipulation. In *IEEE Trans. Comput.* 6:677–91.
8. D. R. Chase, M. Wegman, and F. Zadeck. 1990. Analysis of pointers and structures. In *SIGPLAN Conference on Programming Language Design and Implementation*.
9. E. M. Clarke, O. Grumberg, S. Jha, Y. Lu, and H. Veith. 2000. Counterexample-guided abstraction refinement. In *Computer Aided Verification*.
10. E. M. Clarke and S. Jha. 1995. Symmetry and induction in model checking. In *Computer Science Today: Recent Trends and Developments*, New York: Springer-Verlag.

11. F. Corbera, R. Asenjo, and E. L. Zapata. 1999. New shape analysis techniques for automatic paral-lelization of C codes. In *ACM International Conference on Supercomputing*, 220–27.

12. P. Cousot and R. Cousot. 1979. Systematic design of program analysis frameworks. In *ACM SIGPLAN-SIGACT Symposium on Principles of Programming Language*.

13. N. Dor, M. Rodeh, and M. Sagiv. 1998. Detecting memory errors via static pointer analysis. In *Proceedings of the ACM SIGPLAN-SIGSOFT Workshop on Program Analysis for Software Tools and Engineering*.

14. N. Dor, M. Rodeh, and M. Sagiv. 2000. Checking cleanness in linked lists. In *Static Analysis Symposium*.

15. M. Emami, R. Ghiya, and L. Hendren. 1994. Context-sensitive interprocedural points-to analysis in the presence of function pointers. In *SIGPLAN Conference on Programming Language Design and Implementation*.

16. E. Emerson and A. P. Sistla. 1993. Symmetry and model checking. In *Computer Aided Verification*.

17. H. B. Enderton. 1972. *A mathematical introduction to logic*. New York: Academic Press.

18. M. Fähndrich, J. Foster, Z. Su, and A. Aiken. 1998. Partial online cycle elimination in inclusion constraint graphs. In *SIGPLAN Conference on Programming Language Design and Implementation*.

19. J. Ferrante, K. Ottenstein, and J. Warren. 1987. The program dependence graph and its use in optimization. *ACM Trans. Program. Lang. Syst.* 3(9):319–49.

20. D. Gopan, F. DiMaio, N. Dor, T. Reps, and M. Sagiv. 2004. Numeric domains with summarized dimensions. In *Tools and Algorithms for the Construction and Analysis of Systems*.

21. D. Gopan, T. Reps, and M. Sagiv. 2005. A framework for numeric analysis of array operations. In *Symposium on Principles of Programming Language*.

22. S. Graf and H. Saïdi. 1997. Construction of abstract state graphs with PVS. In *Computer Aided Verification*.

23. N. Heintze and O. Tardieu. 2001. Ultra-fast aliasing analysis using CLA: A million lines of C code in a second. In *SIGPLAN Conference on Programming Language Design and Implementation*.

24. L. Hendren, J. Hummel, and A. Nicolau. 1992. Abstractions for recursive pointer data structures: Improving the analysis and the transformation of imperative programs. In *SIGPLAN Conference on Programming Language Design and Implementation*.

25. S. Horwitz, P. Pfeiffer, and T. Reps. 1989. Dependence analysis for pointer variables. In *SIGPLAN Conference on Programming Language Design and Implementation*.

26. M. Huth, R. Jagadeesan, and D. A. Schmidt. 2001. Modal transition systems: A foundation for three-valued program analysis. In *European Symposium on Programming*.

27. Y. S. Hwang and J. Saltz. 1997. Identifying def/use information of statements that construct and traverse dynamic recursive data structures. In *Languages and Compilers for Parallel Computing*, 131–45.

28. B. Jeannet, D. Gopan, and T. Reps. 2005. A relational abstraction for functions. In *Static Analysis Symposium*.

29. B. Jeannet, A. Loginov, T. Reps, and M. Sagiv. 2004. A relational approach to interprocedural shape analysis. In *Static Analysis Symposium*.

30. N. D. Jones and S. S. Muchnick. 1981. Flow analysis and optimization of Lisp-like structures. In *Program flow analysis: Theory and applications*, ed. S. S. Muchnick and N. D. Jones, 102–31. Englewood Cliffs, NJ: Prentice-Hall.

31. D. J. Kuck, R. H. Kuhn, B. Leasure, D. A. Padua, and M. Wolfe. 1981. Dependence graphs and com-piler optimizations. In *ACM SIGPLAN-SIGACT Symposium on Principles of Programming Language*.

32. R. Kurshan. 1994. *Computer-Aided Verification of Coordinating Processes*. Princeton, NJ: Princeton University Press.

33. J. R. Larus and P. N. Hilfinger. 1988. Detecting conflicts between structure accesses. In *SIGPLAN Conference on Programming Language Design and Implementation*.

34. T. Lev-Ami. 2000. TVLA: A framework for Kleene based static analysis. Master's thesis, Tel-Aviv University, Tel-Aviv, Israel.

35. T. Lev-Ami, T. Reps, M. Sagiv, and R. Wilhelm. 2000. Putting static analysis to work for verification: A case study. In *International Symposium on Software Testing and Analysis*, 26–38.

36. T. Lev-Ami and M. Sagiv. 2000. TVLA: A system for implementing static analyses. In *Static Analysis Symposium*, 280–301.

37. A. Loginov. 2006. Refinement-based program verification via three-valued-logic analysis. Ph.D. dissertation and Tech. Rep. TR-1574, Computer Science Department, University of Wisconsin, Madison.

38. A. Loginov, T. Reps, and M. Sagiv. 2005. Abstraction refinement via inductive learning. In *Computer Aided Verification*.

39. R. Manevich, G. Ramalingam, J. Field, D. Goyal, and M. Sagiv. 2002. Compactly representing first-order structures for static analysis. In *Static Analysis Symposium*.

40. R. Manevich, M. Sagiv, G. Ramalingam, and J. Field. 2004. Partially disjunctive heap abstraction. In *Static Analysis Symposium*.

41. R. Manevich, E. Yahav, G. Ramalingam, and M. Sagiv. 2005. Predicate abstraction and canonical abstraction for singly-linked lists. In *Verification, Model Checking, and Abstract Interpretation*.

42. B. Möller. 1999. Calculating with acyclic and cyclic lists. *Inf. Sci.* 119(3-4):135–54.

43. J. M. Morris. 1982. Assignment and linked data structures. In *Theoretical foundations of programming methodology*, ed. M. Broy and G. Schmidt, 35–41. Boston: D. Reidel.

44. E. W. Myers. 1984. Efficient applicative data types. In *ACM SIGPLAN-SIGACT Symposium on Principles of Programming Language*, 66–75.

45. G. Nelson. 1983. Verifying reachability invariants of linked structures. In *ACM SIGPLAN-SIGACT Symposium on Principles of Programming Language*.

46. K. J. Ottenstein and L. M. Ottenstein. 1984. The program dependence graph in a software development environment. In *Proceedings of the ACM SIGSOFT/SIGPLAN Software Engineering Symposium on Practical Software Development Environments*.

47. J. R. Quinlan. 1990. Learning logical definitions from relations. *Machine Learn.* 5:239–66.

48. G. Ramalingam, A. Warshavsky, J. Field, D. Goyal, and M. Sagiv. 2002. Deriving specialized program analyses for certifying component-client conformance. In *SIGPLAN Conference on Programming Language Design and Implementation*.

49. J. C. Reynolds. 1968. Automatic computation of data set definitions. In *Information Processing 68: Proceedings of the IFIP Congress*, 456–61. New York: North-Holland.

50. T. Reps, M. Sagiv, and A. Loginov. 2003. Finite differencing of logical formulas for static analysis. In *European Symposium on Programming*.

51. T. Reps, T. Teitelbaum, and A. Demers. 1983. Incremental context-dependent analysis for language-based editors. *ACM Trans. Program Lang. Syst.* 5(3):449–77.

52. N. Rinetzky, J. Bauer, T. Reps, M. Sagiv, and R. Wilhelm. 2005. A semantics for procedure local heaps and its abstractions. In *Symposium on Principles of Programming Language*.

53. N. Rinetzky and M. Sagiv. 2001. Interprocedural shape analysis for recursive programs. In *Proceedings of the 10th International Conference on Compiler Construction*, 133–49. London: Springer-Verlag.

54. N. Rinetzky, M. Sagiv, and E. Yahav. 2005. Interprocedural shape analysis for cutpoint-free programs. In *Static Analysis Symposium*.

55. J. L. Ross and M. Sagiv. 1998. Building a bridge between pointer aliases and program dependences. In *European Symposium on Programming*.

56. J. L. Ross and M. Sagiv. 1998. Building a bridge between pointer aliases and program dependences. *Nordic J. Comput.* 8:361–86.

57. M. Sagiv, T. Reps, and R. Wilhelm. 1998. Solving shape-analysis problems in languages with destructive updating. *ACM Trans. Program Lang. Syst.* 20(1):1–50.

58. M. Sagiv, T. Reps, and R. Wilhelm. 2002. Parametric shape analysis via 3-valued logic. *ACM Trans. Program Lang. Syst.* 24(3):217–98.

59. S. Sagiv, N. Francez, M. Rodeh, and R. Wilhelm. 1998. A logic-based approach to data flow analysis problems. *Acta Inf.* 35(6):457–504.

60. R. Shaham, E. Yahav, E. K. Kolodner, and M. Sagiv. 2003. Establishing local temporal heap safety properties with applications to compile-time memory management. In *Static Analysis Symposium*.
61. M. Shapiro and S. Horwitz. 1997. Fast and accurate flow-insensitive points-to analysis. In *ACM SIGPLAN-SIGACT Symposium Principles of Programming Language*, 1–14.
62. B. Steensgaard. 1996. Points-to analysis in almost-linear time. In *ACM SIGPLAN-SIGACT Symposium on Principles of Programming Language*, 32–41.
63. J. D. Ullman. 1989. *Principles of database and knowledge-base systems*, Vol. II, *The new technologies*. Rockville, MD: Computer Science Press.
64. C. Ungureanu and S. Jagannathan. 2000. Concurrency analysis for Java. In *Static Analysis Symposium*.
65. R. Wilhelm, M. Sagiv, and T. Reps. 2000. Shape analysis. In *Compiler Construction*.
66. M. Wolfe and U. Banerjee. 1987. Data dependence and its application to parallel processing. *Int. J. Parallel Program.* 16(2):137–78.
67. E. Yahav. 2001. Verifying safety properties of concurrent Java programs using 3-valued logic. In *Symposium on Principles of Programming Language*.
68. E. Yahav and G. Ramalingam. 2004. Verifying safety properties using separation and heterogeneous abstractions. In *SIGPLAN Conference on Programming Language Design and Implementation*.
69. E. Yahav, T. Reps, M. Sagiv, and R. Wilhelm. 2003. Verifying temporal heap properties specified via evolution logic. In *European Symposium on Programming*.
70. E. Yahav and M. Sagiv. 2003. Automatically verifying concurrent queue algorithms. In *Workshop on Software Model Checking*.

13

Optimizations for Object-Oriented Languages

Andreas Krall
Institute fuer Computersprachen,
Technische Universität Wien,
Wein, Austria
andi@complang.tuwien.ac.at

Nigel Horspool
Department of Computer Science,
University of Victoria,
Victoria, BC, Canada
nigelh@csr.uvic.ca

13.1 Introduction

This chapter introduces optimization techniques appropriate for object-oriented languages. The topics covered include object and class layout, method invocation, efficient runtime-type checks, devirtualization with type analysis techniques, and escape analyses. Object allocation and garbage collection techniques are also very important for the efficiency of object-oriented programming languages. However, because of their complexity and limited space, this important topic must unfortunately be omitted. A good reference is the book by Jones and Lins [18].

Optimization issues relevant to a variety of programming languages, including C++, Java, Eiffel, Smalltalk, and Theta, are discussed. However, to ensure consistent treatment, all examples have been converted to Java syntax. When necessary, some liberties with Java syntax, such as true multiple inheritance, have been made.

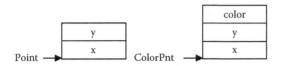

FIGURE 13.1   Single inheritance layout.

13.2   Object Layout and Method Invocation

The memory layout of an object and how the layout supports dynamic dispatch are crucial factors in the performance of object-oriented programming languages. For single inheritance there are only a few efficient techniques: dispatch using a virtual dispatch table and direct calling guarded by a type test. For multiple inheritance many techniques with different compromises are available: embedding superclasses, trampolines, and table compression.

13.2.1   Single Inheritance

In the case of single inheritance, the layout of a superclass is a prefix of the layout of the subclass. Figure 13.1 shows the layouts of an example class and subclass. Access to instance variables requires just one load or store instruction. Adding new instance variables in subclasses is simple.

Invocation of virtual methods can be implemented by a method pointer table (virtual function table, *vtbl*). Each object contains a pointer to the virtual method table. The *vtbl* of a subclass is an extension of the superclass. If the implementation of a method of the superclass is used by the subclass, the pointer in the *vtbl* of the subclass is the same as the pointer in the superclass.

Figure 13.2 shows the virtual tables for the classes `Point` and `ColorPnt` defined as follows:

```
class Point {
      int x, y;
      void move(int x, int y) {...}
      void draw() {...}
      }

class ColorPnt extends Point {
      int color;
      void draw() {...}
      void setcolor(int c) {...}
      }
```

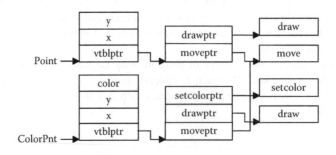

FIGURE 13.2   Single inheritance layout with virtual method table.

FIGURE 13.3 Multiple inheritance layout.

Each method is assigned a fixed offset in the virtual method table. Method invocation is just three machine code instructions (two load instructions and one indirect jump):

```
LD vtblptr,(obj)           ;    load vtbl pointer
LD mptr,method(vtblptr)    ;    load method pointer
JSR (mptr)                 ;    call method
```

Dynamic dispatching using a virtual method table has the advantage that it is fast and executes in constant time. It is possible to both add new methods and override old ones. One extra word of memory is needed for every object. On modern architectures, load instructions and indirect jumps are expensive. Therefore, Rose [25] suggested fat dispatch tables, where the method code is directly placed in the virtual method table eliminating one indirection. The problem with fat dispatch tables is that the offsets for different method implementations must be the same. Either memory is wasted or large methods must branch to overflow code.

13.2.2 Multiple Inheritance

While designing multiple inheritance for C++, Stroustrup also proposed different implementation strategies [29]. Extending the superclasses as in single inheritance does not work anymore. The fields of the superclass are embedded as a contiguous block. Figure 13.3 demonstrates embedding for the class ColorPnt, which is defined as follows:

```
class Point {
    int x, y;
    }

class Colored {
    int color;
    }

class ColorPnt extends Point, Colored {}
```

Embedding allows fast access to instance variables exactly as in single inheritance. The object pointer is adjusted to the embedded object whenever explicit or implicit pointer casting occurs (assignments, type casts, parameter and result passing). Pointer adjustment has to be suppressed for casts of null pointers:

```
Colored col;
Colorpoint cp;
col = cp; // col=cp; if (cp!=null) col=(Colored)((int*)cp+2)
```

In C++ the pointer adjustments break if type casts outside the class hierarchy are used (e.g., casting to void*). Garbage collection becomes more complex because pointers also point into the middle of objects.

Dynamic dispatching also can be solved for embedding. For every superclass, virtual method tables have to be created, and multiple *vtbl* pointers are included in the object. A problem occurs with implicit casts

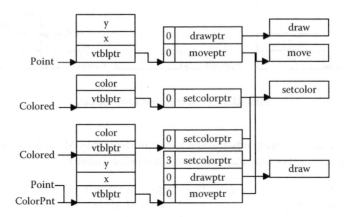

FIGURE 13.4   Multiple inheritance layout with virtual method table.

from the actual receiver to the formal receiver. The caller does not know the type of the formal receiver in the callee, and the callee does not know the type of the actual receiver of the caller. Therefore, this type information has to be stored as an adjustment offset in the virtual method table. Given the following definition for the class `ColorPnt`, the virtual tables are organized as shown in Figure 13.4.

```
class Point {
    int x, y;
    void move(int x, int y) {...}
    void draw() {...}
    }

class Colored {
    int color;
    void setcolor(int c) {...}
    }

class ColorPnt extends Point, Colored {
    void draw() {...}
    }
```

Method invocation now requires four to five machine instructions, depending on the computer architecture:

```
LD    vtblptr,(obj)              ; load vtbl pointer
LD    mptr,method_ptr(vtblptr)   ; load method pointer
LD    off,method_off(vtblptr)    ; load adjustment offset
ADD   obj,off,obj                ; adjust receiver
JSR   (mptr)                     ; call method
```

This overhead in table space and program code is even necessary when multiple inheritance is not used. Furthermore, adjustments to the remaining parameters and the result are not possible. A solution that eliminates much of the overhead is to insert a small piece of code called a *trampoline* that performs the pointer adjustments and then jumps to the original method. The advantages are a smaller table (no storing of an offset) and fast method invocation when multiple inheritance is not used (the same dispatch code as

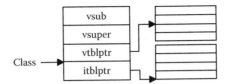

FIGURE 13.5 Bidirectional object layout with dispatch tables.

in single inheritance). In the example of Figure 13.4, the `setcolorptr` method pointer in the virtual method table of `Colorpoint` would point to code that adds three to the receiver before jumping to the code of method `setcolor`:

```
ADD obj,3,obj    ;    adjust receiver
BR setcolor      ;    call method
```

When instance variables of common superclasses need to be shared, the offset of each such variable has to be stored in the virtual method table. Each access to a shared variable then incurs an additional penalty of loading and adding the appropriate offset.

13.2.3 Bidirectional Object Layout

The programming language Theta [21], like Java, uses single inheritance with multiple subtyping. For this language, Myers proposed a bidirectional object layout and showed how the bidirectional layout rules can be extended to support true multiple implementation inheritance [22]. The idea behind bidirectional layout is that data structures can be extended in two directions and information can be shared in a way that leads to less indirection and smaller memory usage. Both the object and virtual method table extend in both directions. The object contains the instance variable fields, the pointer to the object's virtual method table, with and, at negative offsets of, the pointer to the interface method tables. The method dispatch tables also extend in both directions. The superclass information fields are in the middle of the tables, and subclass fields are at both ends of the tables. Figure 13.5 shows the object and method table layout scheme.

The key to the efficiency of the bidirectional layout is the merging of interface and class headers. Determining an optimal layout is not feasible; therefore, a heuristic is used. The details of the algorithm can be found in the original article by Myers [22].

Given classes C1 and C2 defined as follows, the bidirectional layout scheme would be as shown in Figure 13.6.

```
interface A {                    interface B extends A {
    int a1() {...}                   int b1() {...}
    int a2() {...}                   }
    }


class C1 implements A {           class C2 implements B {
    int v1;                          int v2;
    int a1() {...}                   int a2() {...}
    int a2() {...}                   int b1() {...}
    int c1() {...}                   int c2() {...}
    }                                }
```

As Figure 13.6 shows, the bidirectional layout reduces object and *vtbl* sizes. No additional dispatch headers and method tables are needed.

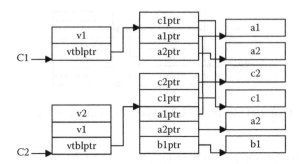

FIGURE 13.6   Bidirectional object layout with virtual method table.

In a large C++ library with 426 classes, bidirectional layout only needs additional dispatch headers in 27 classes with a maximum of 4 dispatch headers compared with 54 classes with more than 7 dispatch headers.

Zibin and Gil [34] describe a space optimal two-dimensional bidirectional object layout scheme for languages that support multiple inheritance. Their scheme has similarities to the work of Myers (above) and Pugh and Weddell [23]. Instead of using pointer adjustments as presented in Section 13.2.2, the fields of an object are accessed using an additional indirection. The fields are organized in a set of bidirectional layers. In the canonical format, all layers are accessed via the layer dispatch table, which contains pointers to all layers. In the compact format, the first layer is accessed directly. The first layer contains a pointer to the layer dispatch table. Here the dispatch table contains offsets to the layers, which makes it possible to share the same dispatch table for all objects of the same type. Experiments show that the field access efficiency is always better than with the standard C++ layout model.

13.2.4   Dispatch Table Compression

Invoking a method in an object-oriented language requires looking up the address of the block of code that implements that method and passing control to it. In some cases, the lookup may be performed at compile time. Perhaps there is only one implementation of the method in the class and its subclasses; perhaps the language has provided a declaration that forces the call to be nonvirtual; perhaps the compiler has performed a static analysis that can determine that a unique implementation is always called at a particular call site. In other cases, a runtime lookup is required.

In principle, the lookup can be implemented as indexing a two-dimensional table. Each method name in the program can be given a number, and each class in the program can be given a number. Then the method call:

```
result = obj.m(a1,a2);
```

can be implemented by these three actions:

- Fetch a pointer to the appropriate row of the dispatch table from the `obj` object.
- Index the dispatch table row with the method number.
- Transfer control to the address obtained.

Note that if two classes C1 and C2 are not related by an ancestor relationship or do not have a subclass in common (because of multiple inheritance), and if the language is strongly typed (as with C++ and Java), then the numbers assigned to the methods of C1 do not have to be disjoint from the numbers used for the methods of C2.

The standard implementation of method dispatch in a strongly typed language such as C++ using a virtual table (*vtbl*) may be seen to be equivalent. Each virtual table implements one row of the two-dimensional dispatch table.

The size of the dispatch table (or of all the virtual tables summed together) can be appreciable. As reported in [31], the dispatch table for the ObjectWorks Smalltalk class library would require approximately 16 MB if no compression were to be performed. In this library, there are 776 classes and 5,325 selectors (i.e., method names).

For a statically typed language, virtual tables provide an effective way to compress the dispatch table because all entries in each virtual table are used. For a dynamically typed language, such as Smalltalk, any object can in principle be invoked with any method name. Most methods are not implemented, and therefore most entries in the dispatch table can be filled in with the address of the "not implemented" error reporting routine. The table for dynamically typed languages therefore tends to be quite sparse, and that is a property that can be exploited to achieve table compression.

A second property possessed by the dispatch tables for both statically and dynamically typed languages is that many rows tend to be similar. When a subclass is defined, only some of the methods in the parent class are normally redefined. Therefore, the rows for these two classes would have identical entries for all except the few redefined methods.

Both properties are also possessed by LR parser tables, and there has been considerable research into compressing such tables. A comprehensive survey of parse table compression is provided by Dencker et al. [12]. It should not be surprising that two of the most effective techniques for compressing static dispatch tables have also been used for parse table compression.

13.2.4.1 Virtual Tables

Virtual tables provide an effective implementation of the dispatch table for statically typed languages. Because methods can be numbered compactly for each class hierarchy to leave no unused entries in each virtual table, a good degree of compression is automatically obtained. For the ObjectWorks example used in [31], if virtual tables could be used (they cannot), the total size of the dispatch tables would be reduced from 16 MB to 868 KB.

13.2.4.2 Selector Coloring Compression

This is a compression method based on graph coloring. Two rows of the dispatch table can be merged if no column contains different method addresses for the two classes. (An unimplemented method corresponds to an empty entry in the table.) The graph is constructed with one node per class, and an edge connects two nodes if the corresponding classes provide different implementations for the same method name. A heuristic algorithm may then be used to assign colors to the nodes so that no two adjacent nodes have the same color, and the minimal number of distinct colors is used. (Heuristics must be used in practice because graph coloring is an NP-complete problem.) Each color corresponds to the index for a row in the compressed table.

Note: A second graph coloring pass may be used to combine columns of the table.

Implementation of the method invocation code does not need to change from that given earlier. Each object contains a reference to a possibly shared row of the dispatch table. However, if two classes C1 and C2 share the same row and C1 implements method m whereas C2 does not, then the code for m should begin with a check that the control was reached via dispatching on an object of type C1. This extra check is the performance penalty for using table compression.

For the ObjectWorks example, the size of the dispatch table would be reduced from 16 to 1.15 MB. Of course, an increase occurs in code size to implement the checks that verify the correct class of the object. That increase is estimated as close to 400 KB for the ObjectWorks library.

13.2.4.3 Row Displacement Compression

The idea is to combine all the rows of the dispatch table into a single very large vector. If the rows are simply placed one after the other, this is exactly equivalent to using the two-dimensional table. However, it is possible to have two or more rows overlapping in memory as long as an entry in one row corresponds to empty entries in the other rows.

A simply greedy algorithm works well when constructing the vector. The first row is placed at the beginning of the vector; then the second row is aligned on top of the first row and tested to see if a conflicting entry exists. If there is a conflict, the row is shifted one position to the right and the test is repeated and so on until a nonconflicting alignment is found.

Implementation of the method invocation code is again unchanged. As before, a test to verify the class of the current object must be placed at the beginning of any method that can be accessed via more than one row of the dispatch table. For the ObjectWorks example, the size of the dispatch table would be reduced from 16 MB to 819 KB, with the same 400 KB penalty for implementing checks in methods to verify the class.

13.2.4.4 Partitioning

It is pointed out in [31] that good compression combined with fast access code can be achieved by breaking dispatch tables into pieces. If two classes have the same implementations for 50 methods, for instance, and different implementations for just 5 methods, we could have a single shared dispatch table for the common methods plus two separate but small tables for the 5 methods where implementations differ. The partitioning approach generalizes this insight by allowing any number of partitions to be created.

For each method, the compiler must predetermine its partition number within the owning class and its offset within a partition table. The method lookup code requires indexing the class top-level table with the method partition number to obtain a pointer to the partition table and then indexing that partition table with the method offset.

To keep the access code as efficient as possible, the partitioning should be regular. That is, each class must have the same number of partitions, and all partitions accessed via the same offset in each class table must have the same size.

The partitioning approach advocated by Vitek and Horspool proceeds in three steps [31]. First, the compiler divides the method selectors into two sets: one set contains the conflict selectors that are implemented by two classes unrelated by inheritance, and the other set contains all other method selectors. Two separate dispatch tables are created for the two sets of methods.

Second, two columns in a table may be combined if no two classes provide different implementations for the two methods. Merging columns may be performed using graph coloring heuristics to achieve the best results.

Third, and finally, the two tables are divided into equal-sized partitions, and any two partitions that are discovered to have identical contents are shared. It is possible to use two partition sizes for splitting the two tables. Although a clever reordering of the columns might increase the opportunities for partition table sharing, good results are achieved without that extra work.

Vitek and Horspool report that a partition size of 14 entries gave good results with the ObjectWorks library [31]. The total size of all the tables came to 221 KB for the library, with a penalty for increased code size of less than 300 KB.

13.2.5 Java Class Layout and Method Invocation

Java and the programming language Theta do not implement multiple inheritance, but implement single inheritance with multiple subtyping. This important difference makes object layout and method dispatch more efficient. Although the bidirectional layout was designed for a language with multiple subtyping, it has the problem that more than one *vtbl* pointer has to be included in objects. The CACAO JIT compiler [19] moves the dispatch header of the bidirectional layout into the class information with negative offsets from the *vtbl*. For each implemented interface a distinct interface *vtbl* exists. Unimplemented interfaces are represented by null pointers. An example of the layout used by CACAO is shown in Figure 13.7.

To call a virtual method, two memory access instructions are necessary (load the class pointer, load the method pointer) followed by the call instruction. Calling an interface method needs an additional indirection.

In the faster scheme, we store interface methods in an additional table at negative offsets from the *vtbl*, as shown in Figure 13.8. Segregating the interface virtual function table keeps the standard virtual

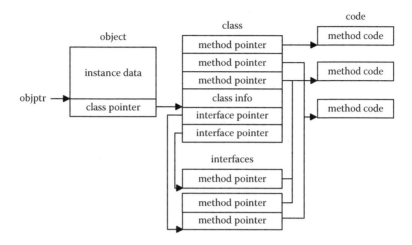

FIGURE 13.7 CACAO object and compact class descriptor layout.

function table small and allows interface methods to be called with just two memory accesses. The memory consumption of virtual function tables containing interface and class methods would be *number of (classes + interfaces) × number of distinct methods*. The memory consumption of the interface tables is only *number of classes that implement interfaces × number of interface methods*. Coloring can be used to reduce the number of distinct offsets for interface methods further but complicates dynamic class loading, leading to renumbering and code patching.

The Jalapeno virtual machine (VM) implements an interface method invocation similar to the fast class descriptor layout of CACAO; however, instead of coloring, hashing of the method indices is used [4]. The table for the interface method pointers is allocated with a fixed size much smaller than the number of interface methods. When two method indices hash to the same offset, a conflict resolving stub is called instead of the interface methods directly. For conflict resolution the stub is passed to the method index in a scratch register as additional argument. An interface method invocation can be executed with the following four machine instructions:

```
LD    vtblptr,(obj)                      ;   load vtbl pointer
LD    mptr,hash(method_ptr)(vtblptr)     ;   load method pointer
MV    mindex,idreg                       ;   load method index
JSR   (mptr)                    ; call method (or conflict stub)
```

The number of machine instructions is the same as in the compact class descriptor layout of CACAO, but the indirection is eliminated, which reduces the number of cycles needed for the execution of this

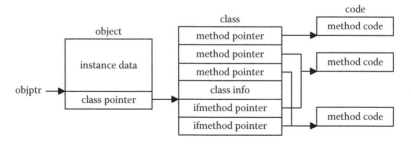

FIGURE 13.8 CACAO object and fast class descriptor layout.

instruction sequence on a pipelined architecture. Compared with the old interface method invocation in the Jalapeno VM, which searched the superclass hierarchy for a matching method signature, the new method yields runtimes that range from a 1% reduction in speed to a 51% speedup.

13.2.6 Dispatch without Virtual Method Tables

Virtual method tables are the most common way to implement dynamic dispatch. Despite the use of branch target caches and similar techniques, indirect branches are expensive on modern architectures. The SmallEiffel compiler [33] uses a technique similar to polymorphic inline caches [16]. The indirect branch is replaced by a binary search for the type of the receiver and a direct branch to the method or the inlined code of the method.

Usually an object contains a pointer to the class information and the virtual method table. SmallEiffel replaces the pointer by an integer representing the type of the object. This type identifier is used in a dispatch function that searches for the type of the receiver. SmallEiffel uses a binary search, but a linear search weighted by the frequency of the receiver type would also be possible. The dispatch functions are shared between calls with the same statically determined set of concrete types. Assuming that the type identifiers *TA*, *TB*, *TC*, and *TD* are sorted by increasing number, the dispatch code for calling x.f is:

```
if idx ≤ TB then
if idx ≤ TA then fA(x)
else fB(x)
else if idx ≤ TC then fC(x)
else fD(x)
```

Obviously, the method calls are inlined when the code is reasonably small. An empirical evaluation showed that for a method invocation with three concrete types, dispatching with binary search is between 10 and 48% faster than dispatching with a virtual method table. For a megamorphic call with 50 concrete types, the performance of the two dispatch techniques is about the same. Dispatch without virtual method calls cannot be used easily in a language with dynamic class loading (e.g., Java). Either the code has to be patched at runtime, or some escape mechanism is necessary.

13.3 Fast Type Inclusion Tests

In a statically typed language, an assignment may require a runtime test to verify correctness. For example, if class B is a subclass of A, the assignment to b in the following Java code:

```
A a = new B();
... // intervening code omitted
B b = a;
```

needs validation to ensure that a holds a value of type B (or some subclass of B) instead of type A. Usually that validation can be a runtime test.

Java also has an explicit instanceof test to check whether an object has the same type or is a subtype of a given type. Other object-oriented languages have similar tests. Static analysis is not very effective in eliminating these tests [13]. Therefore, efficient runtime type checking is very important.

The obvious implementation of a type inclusion test is for a representation of the class hierarchy graph to be held in memory and that graph to be traversed, searching to see whether one node is an ancestor of the other node. The traversal is straightforward to implement for a language with single inheritance and less so for multiple inheritance. However, the defect with this approach is that the execution time of the test increases with the depth of the class hierarchy. Small improvements are possible if one or two supertypes are cached [28].

13.3.1   Binary Matrix Implementation

Given two types, $c1$ and $c2$, it is straightforward to precompute a binary matrix that is indexed by numbers associated with the two types $BM[c1, c2]$ to determine whether one type is a subtype of the other.

Accessing an entry in the binary matrix requires only a few machine instructions, but the matrix can be inconveniently large, perhaps hundreds of kilobytes in size. Compaction of the binary matrix is possible, but that makes the access code more complicated.

13.3.2   Cohen's Algorithm

Cohen [10] adapted the notion of a display table (used for finding frames in block structured programming languages). Cohen's idea applies to languages with single inheritance, so that the class hierarchy graph is a tree.

Each type has a unique type identifier, *tid*, which is simply a number. A runtime data structure records the complete path of each type to the root of the class as a vector of type identifiers. The *tid* in, for instance, position three of that vector would identify the ancestor at level 3 in the class hierarchy.

If the compiler has to implement the test:

```
if (obj instanceof C) ...
```

then the level and type identifier for class C are both constants, C_level and C_tid, determined by the compiler. The steps needed to implement the test are simply

```
level := obj.level;
if level < C_level then
    result := false
else
    result := (obj.display[C_level] = C_tid)
```

Cohen's algorithm is easy to implement, requires constant time for lookups, and uses little storage. However, it works only for single inheritance hierarchies. Extending the algorithm to work with multiple inheritance hierarchies is not trivial.

13.3.3   Relative Numbering

A well-known algorithm for testing whether one node is an ancestor of another in a tree works by associating a pair of integers with each node. An example tree and the numbering are shown in Figure 13.9.

To test whether node $n1$ is a descendant of $n2$ (or equal), the test is implemented as

```
n1.left ≤ n2.left and n1.right ≥ n2.right
```

where the two numbers associated with a node are named *left* and *right*.

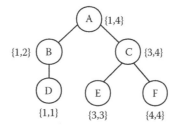

FIGURE 13.9   Relative numbering for a hierarchy.

The scheme is simple and efficient. It is probably the method of choice for implementing the subtype test for single inheritance hierarchies. Extension of the idea to multiple inheritance hierarchies was a long time coming, but has now been achieved by Gil and Zibin. Their scheme is known as PQ-encoding [14] and is explained in a following section.

13.3.4   Hierarchical Encoding

With hierarchical encoding, a bit vector is associated with each type. Each bit vector implements a compact set of integers. The test of whether type t1 is a subclass of t2 is implemented as the subset test t2.vector $\subseteq$ t1.vector. The approach works with multiple inheritance hierarchies.

A simple way to implement the bit vectors is to number all the nodes in the class hierarchy graph. The bit vector for node n represents the set of all the integers associated with n and the ancestors of n. That yields correct but unnecessarily large vectors.

Caseau [8], Aït-Kaci et al. [3], and Krall et al. [20] provide algorithms for constructing much smaller bit vectors. The algorithms are, however, computationally expensive and would need to be reexecuted after even a small change to the class hierarchy.

13.3.5   More Algorithms

Vitek et al. [32] describe three more type test algorithms that they call *packed encoding*, *bit-packed encoding*, and *compact encoding*. All three perform worse than hierarchical encoding if the total size of the data tables is used as the only criterion. However, these algorithms are much faster and are therefore more suitable for an environment where the class hierarchy may be dynamically updated, as with Java or Smalltalk, for example.

13.3.6   Partitioning the Class Hierarchy

Because type tests for trees are more (space) efficient than type tests for direct acyclic graphs (DAGs), a possible solution is to split a DAG into a tree part and the remaining graph. For languages with single inheritance and multiple subtyping, this partitioning of the class hierarchy is already done in the language itself.

CACAO uses a subtype test based on relative numbering for classes and a kind of matrix implementation for interfaces. Two numbers *low* and *high* are stored for each class in the class hierarchy. A depth-first traversal of the hierarchy increments a counter for each class and assigns the counter to the low field when the class is first encountered and assigns the counter to the high field when the traversal leaves the class. In languages with dynamic class loading, a renumbering of the hierarchy is needed whenever a class is added. A class *sub* is a subtype of another class *super* if *super.low* $\leq$ *sub.low* $\leq$ *super.high*. Because a range check is implemented more efficiently by an unsigned comparison, CACAO stores the difference between the low and high values and compares it against the difference between the low values of both classes. The code for instanceof looks similar to:

```
return (unsigned) (sub->vftbl->baseval - super->vftbl->baseval)
        <= (unsigned) (super->vftbl->diffval);
```

Figure 13.10 shows an example hierarchy using baseval and diffval pairs. For leaf nodes in the class hierarchy the diffval is 0, which results in a faster test (a simple comparison of the baseval fields of the sub- and superclass). In general, a just-in-time (JIT) compiler can generate the faster test only for final classes. An ahead-of-time (AOT) compiler or a JIT compiler that does patching of the already generated machine code may also replace both the baseval and the diffval of the superclass by a constant. Currently, CACAO uses constants only when dynamic class loading is not used.

CACAO stores an interface table at negative offsets from the virtual method table (as seen in Figure 13.7). This table is needed for the invocation of interface methods. The same table is also used

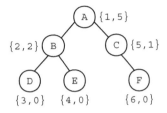

FIGURE 13.10 Relative numbering with baseval and diffval pairs.

by the subtype test for interfaces. If the table is empty for the index of the superclass, the subtype test fails. The code for `instanceof` looks similar to

return (sub->vftbl->interfacetable[-super->index] != NULL);

Both subtype tests can be implemented by just a few machine code instructions without using branches that are expensive on modern processors.

13.3.7 PQ-Encoding

An observation made by Gil and Zibin [14] is that a minor generalization to the relative numbering scheme enables it to handle many multiple inheritance hierarchies. The example from their paper is shown in Figure 13.11. In this hierarchy, class C has two parents, A and B, and so on.

Each class in the hierarchy has three associated integers: L, R, and id. The L and R numbers represent a range [L, R], analogous to the relative numbering scheme. If we wish to test whether class C1 is a subtype of class C2, we would perform the following test:

$$C1.id \geq C2.L \textbf{ and } C1.id \leq C2.R$$

Like the relative numbering scheme, this test requires a small fixed number of instructions.

Unfortunately, it is not always possible to find L, R, and id numbers that work for an arbitrary multiple inheritance hierarchy. However, Gil and Zibin recognized that the PQ-encoding scheme could be used for most nodes in the hierarchy and that the exceptional cases would be relatively uncommon. Their solution solves the problem as a number of *slices*. The first slice is a PQ-encoding solution, like that shown in Figure 13.11, where as many nodes as possible in the hierarchy are handled. The second slice is another PQ-encoding solution that handles as many nodes as possible of those that were not handled by the first slice, and so on. The number of slices needed for even the most complicated multiple inheritance hierarchies found in practice is small.

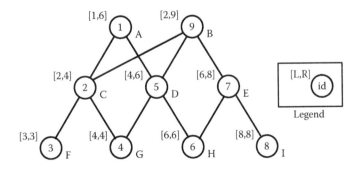

FIGURE 13.11 PQ-encoding for a multiple inheritance hierarchy.

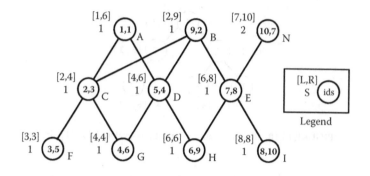

FIGURE 13.12   A two-slice PQ-encoding for a multiple inheritance hierarchy.

Figure 13.12 shows a one-node addition to the previous hierarchy, which makes it impossible to find a correct PQ-encoding.

The PQ-encoding solution for this hierarchy requires two slices. Each class in the hierarchy now has three integers plus a list of integers associated with it. The first two integers, L and R, represent a range as before. The third integer, S, represents the slice to which that class belongs. Finally, ids is a list of integers, providing an id number for each slice. The test to determine whether class C1 is a subclass of C2 can now be implemented as follows:

$$\texttt{C1.ids[C2.S]} \geq \texttt{C2.L} \ \textbf{and} \ \texttt{C1.ids[C2.S]} \leq \texttt{C2.R}$$

In this code, `C2.S` specifies the slice to which node `C2` belongs; then the expression `C1.ids[C2.S]` gives the id number for node `C1` in that slice.

As Gil and Zibin admit, their algorithm for constructing the slices and determining the L, R, and id numbers in each slice is not incremental. If the hierarchy changes because of dynamic class loading, it is necessary to rerun the algorithm from scratch. However, the same is true of most of the type inclusion test implementations described in this chapter. The major advantage of the multislice PQ-encoding scheme is that it requires the least amount of data compared to all the other approaches, while being a (very fast) constant-time test.

13.4   Devirtualization

Devirtualization is a technique to reduce the overhead of virtual method invocation in object-oriented languages. The aim of this technique is to statically determine which methods can be invoked by virtual method calls. If exactly one method is resolved for a method call, the method can be inlined or the virtual method call can be replaced by a static method call. The analyses necessary for devirtualization also improve the accuracy of the call graph and the accuracy of subsequent interprocedural analyses. We first discuss different type analysis algorithms, comparing their precision and complexity. Then different solutions for devirtualization of extensible class hierarchies and similar problems are presented.

13.4.1   Class Hierarchy Analysis

The simplest devirtualization technique is class hierarchy analysis (CHA), which determines the class hierarchy used in a program. A Java class file contains information about all referenced classes. This information can be used to create a conservative approximation of the class hierarchy. The approximation is formed by computing the transitive closure of all classes referenced by the class containing the main method. A more accurate hierarchy can be constructed by computing the call graph [11]. CHA uses the declared types for the receiver of a virtual method call for determining all possible receivers.

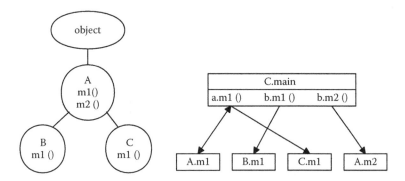

FIGURE 13.13 Class hierarchy and call graph.

As an example, Figure 13.13 shows the class hierarchy and call graph that correspond to the following fragment of Java code:

```
class A extends Object {
    void m1() {...}
    void m2() {...}
    }

class B extends A {
    void m1() {...}
    }

class C extends A {
    void m1() {...}
    public static void main(...) {
        A a = new A();
        B b = new B();

        a.m1(); b.m1(); b.m2();
        }
    }
```

Informally, CHA collects all methods in the call graph in a work list of methods. This work list is initialized with the *main* method. To this work list is added every method that is inherited by a subtype of the declared type of a virtual method call in the body of each method in the work list. The algorithm given in Figure 13.14 gives the computation of the class hierarchy and call graph in more detail.

13.4.2 Rapid Type Analysis

A more accurate class hierarchy and call graph can be computed if the type of the receiver can be determined more precisely than the declared type specifies. Rapid type analysis (RTA) uses the fact that a method m of a class c can be invoked only if an object of type c is created during the execution of a program [5, 6]. RTA refines the class hierarchy by only including classes for which objects can be created at runtime. The pessimistic algorithm includes all classes in the class hierarchy for which instantiations occur in methods of the call graph from CHA.

The optimistic algorithm initially assumes that no methods besides main are called and that no objects are instantiated. It traverses the call graph, initially ignoring virtual calls (only marking them in a class

```
/* Notation:
   main              -- the main method in a program
   x ( )             -- call of static method x
   type (x)          -- the declared type of the expression x
   x. y ()           -- call of virtual method y in expression x
   subtype (x)       -- x and all classes which are subtypes of x
   method (x, y)     -- the method y which is defined for class x
*/
callgraph := main
hierarchy := { }
```
for each m **in** *callgraph* **do**
    for each m_{stat} ( ) **occurring in** m **do**
    if m_{stat} **not in** *callgraph* **then**
        add m_{stat} **to** *callgraph*
    for each $e.m_{vir}$ ( ) **occurring in** m **do**
    for each c **in** *subtype* (*type* (*e*)) **do**
      m_{def} := *method* (c, m_{vir})
     if m_{def} **not in** *callgraph* **then**
    add m_{def} **to** *callgraph*
    add c **to** *hierarchy*

FIGURE 13.14   Class hierarchy analysis.

mapping as a potential call) and following static calls only. When the instantiation of an object is found during the analysis, all virtual methods of the corresponding class that were left out previously are then traversed as well. The live part of the call graph and the set of instantiated classes grow interleaved as the algorithm proceeds.

Figure 13.15 shows the rapid type analysis algorithm.

13.4.3 Other Fast Precise-Type Analysis Algorithms

Tip and Palsberg [30] evaluated different algorithms that are more precise but are on average only a factor of five slower than RTA. The different type analysis algorithms differ primarily in the number of sets of types used. RTA uses one set for the whole program. 0-Control flow analysis (CFA) [15, 26] uses one set per expression. k–l-CFA makes separate analyses for k levels of method invocations and uses more than one set per expression. These algorithms have high computational complexity and only work for small programs. Therefore, Tip and Palsberg evaluated the design space between RTA and 0-CFA [30] by proposing and comparing three analysis algorithms which they name XTA, MTA, and FTA.

XTA uses a distinct set for each field and method of a class, fast type analysis (FTA) uses one set for all fields and a distinct set for every method of a class, and MTA uses one set for all methods and a distinct set for every field of a class. Arrays are modeled as classes with one instance field. All algorithms use iterative propagation-based flow analysis to compute the results. Three work lists are associated with each method or field that keeps track of processed types. *New* types are propagated to the method or field in the current iteration and can be propagated onward in the next iteration. Current types can be propagated onward in the current iteration. Processed types have been propagated onward in previous iterations.

Tip and Palsberg efficiently implemented the type sets using a combination of array-based and hash-based data structures to allow efficient membership tests, element additions, and iterations over all elements [30]. Type inclusion is implemented by relative numbering. These techniques are necessary because the type sets are filtered by the types of fields, method parameters, and method return types. Additionally, type casts restricting return types are used for filtering.

```
/* Notation:
  main              -- the main method in a program
  new x             -- instantiation of an object of class x
  marked (x)        -- marked methods of class x
  x ( )             -- call of static method x
  type (x)          -- the declared type of the expression x
  x.y ( )           -- call of virtual method y in expression x
  subtype (x)       -- x and all classes which are subtypes of x
  method (x, y)     -- the method y which is defined for class x
  mark (m, x)       -- mark method m in class x
*/
```
callgraph := main
hierarchy := { }
for each *m* **in** *callgraph* **do**
   for each new *c* **occurring in** *m* **do**
      if *c* **not in** *hierarchy* **then**
        add *c* **to** *hierarchy*
        for each m_{mark} **in** *marked (c)* **do**
          add m_{mark} **to** *callgraph*
   for each m_{stat} *( )* **occurring in** *m* **do**
      if m_{stat} **not in** *callgraph* **then**
        add m_{stat} **to** *callgraph*
   for each $e.m_{vir}$ *( )* **occurring in** *body (m)* **do**
      for each *c* **in** *subtype (type (e))* **do**
        m_{def} *:= method c, m_{vir})*
        if m_{def} **not in** *callgraph* **then**
          if *c* **not in** *hierarchy* **then**
            mark (m_{def}, c)
          else
            add *mdef* **do** *callgraph*

FIGURE 13.15 Rapid type analysis.

All these algorithms are more precise than RTA. On the range of Java benchmark programs (benchmark code only), MTA computes call graphs with 0.6% fewer methods and 1.6% fewer edges than RTA. FTA computes call graphs with 1.4% fewer methods and 6.6% fewer edges than RTA. XTA computes call graphs with 1.6% fewer methods and 7.2% fewer edges than RTA. All algorithms are about five times as slow as RTA. Therefore, XTA has the best precision and performance trade-off of the three algorithms compared.

13.4.4 Variable Type Analysis

RTA is imprecise because every type that is instantiated somewhere in the program and is a subtype of the declared type can be the receiver of a method invocation. Variable type analysis (VTA) is more precise because it computes reaching type information, taking into consideration chains of assignments between instantiations and method invocations [27], but it does ignore type casts. It is a flow-insensitive algorithm that avoids iterations over the program. The analysis works by constructing a directed type propagation graph where nodes represent variables and edges represent assignments. Reaching type information is initialized by object instantiations and propagated along the edges of the type propagation graph.

The type propagation graph is constructed from the classes and methods contained in the conservative call graph. For every class c and for every field f of c that has a reference type, a node *c.f* is created. Additionally for every method *c.m* and:

- For every formal parameter *p* (including *this*) of *c.m* that has a reference type, create a node *c.m.p*.
- For every local variable *l* of *c.m* that has a reference type, create a node *c.m.l*.
- If *c*.m returns a reference type, create a node *c.m.ret*.

After the nodes are created, for every assignment of reference types an edge is added to the graph. Assignments are either explicit assignments of local variables or object fields or assignments resulting from passing of parameters or returning a reference value. Native methods are handled by summarizing their effects on the analysis.

To avoid alias analysis, all variable references and all their aliases are represented by exactly one node. In Java, locals and parameters cannot be aliased. All instances of a field of a class are represented by one node. Arrays can introduce aliasing. Different variables can point to the same array. Therefore, if both sides of an assignment are of type Object or if at least one side is an array type, edges in both directions are added.

Type propagation is accomplished in two phases. The first phase detects strongly connected components in the type propagation graph. All nodes of a strongly connected component are collapsed into one supernode. The type of this supernode is the union of the types of all its subnodes. The remaining graph is a DAG. Types are propagated in a topological order where a node is processed after all its predecessors have been processed. The complexity of both strongly connected component detection and type propagation is linear in the maximum of the number of edges and nodes. The most expensive operation is the union of type sets.

The algorithm does no killing of types on casts or declared types. An algorithm using declared type information would be more precise, but collapsing of strongly connected components would not be possible anymore. Impossible types are filtered after type propagation has been finished. Over the set of Java benchmarks (the benchmark code only), VTA computes call graphs with 0.1 to 6.6% fewer methods and 1.1 to 18% fewer edges than RTA. For the set of Java applications (including the libraries), VTA computes call graphs with 2.1 to 20% fewer methods and 7.7 to 27% fewer edges than RTA. The implementation is untuned and written in Java. The performance numbers indicate that the algorithm scales linearly with the size of the program (54 sec for 27,570 instructions, 102 sec for 55,468 instructions).

13.4.5   Cartesian Product Algorithm

Agesen [2] developed a type analysis algorithm called the Cartesian product algorithm (CPA), in which a method is analyzed separately for every combination of argument types (Cartesian product of the argument types). For example, a method with two arguments, where the first argument can be of type Int and Long and the second argument can be of type Float and Double, can result in four different analyses with argument types (Int, Float), (Long, Float), (Int, Double), and (Long, Double). For better precision, CPA computes the needed analyses lazily. Only argument type combinations that occur during program analysis are analyzed. The return type of a method is the union of the return types of the different analyses for a specific method invocation.

CPA is precise in the sense that it can analyze arbitrary deep call chains without loss of precision. CPA is efficient, because redundant analysis is avoided. However, megamorphic call sites, where the method has many arguments and an argument has a high number of different types, can lead to long analysis times. Therefore, Agesen restricted the number of different types for an argument and combined the analyses if the number exceeded a small constant. The bounded CPA scales well for bigger programs too.

13.4.6   Comparisons and Related Work

Grove et al. investigated the precision and complexity of a set of type analysis algorithms [15]. They implemented a framework where different algorithms can be evaluated. They evaluated RTA, the bounded CPA, and different levels of k–l-CFA. CPA gives more accuracy than 0-CFA with reasonable computation times. The higher levels of k–l-CFA cannot be used for large applications.

CHA and RTA have the same time complexity, but RTA always produces more accurate results. The results for XTA and VTA cannot be compared directly, since different programs are used for benchmarking, and the implementation of VTA was not done for performance. It can be estimated that the running time for the algorithms will be similar, but VTA may produce more accurate results. 0-CFA is more accurate than the other algorithms at slightly higher analysis costs.

Qian and Hendren [24] did a limit study and compared dynamic CHA and VTA with an ideal type analysis based on efficient call graph profiling using the Jikes Research Virtual Machine (JikesRVM) as a testbed. The ideal type analysis logs the call targets of each virtual method call during the first benchmark run and determines all monomorphic method invocations. The logged information is used in subsequent benchmark runs to compare the inlining results of other analyses against the ideal one.

Qian and Hendren developed a common dynamic type analysis framework for method inlining in the JikesRVM. The framework supports speculative inlining with invalidations. Since the framework is written in Java, the data structures impose more work on the garbage collector. VTA analysis increases the live data by 60% for the javac benchmark program.

Already the simple CHA comes close to the ideal type analysis for virtual method calls but misses some inlining opportunities for interface calls. Dynamic VTA is as effective as the ideal type analysis for interface calls, leading to speedups between 0.6 and 2.1% over CHA.

13.4.7   Inlining and Devirtualization Techniques

Ishizaki et al. [17] point out that straightforward devirtualization may have little effect on the performance of Java programs. Because Java is strongly typed, a *vtbl* can be used for dispatching. Devirtualizing simply removes the lookup in the *vtbl*, and that is not significant compared with the other costs of calling a method. Significant performance gains only arise if the devirtualized method is inlined at the call site; many opportunities for devirtualization are lost in any case because Java has dynamic class loading.

Ishizaki et al. [17] propose a technique based on code patching that allows methods to be inlined and inlining to be removed if dynamic class loading subsequently requires the method call to be implemented by the normal *vtbl* dispatching again. Their code patching technique avoids any need to recompile the code of the caller.

An example can make the idea clear. Suppose that the program to be compiled contains the following Java statements:

```
i = i + 1;
obj.meth(i,j);
j = j-1;
```

The compiler would normally generate the following pattern of code for those statements:

```
// code for i = i + 1
// code to load arguments i and j
// dispatch code to lookup address of methodmeth
// ... and pass obj and arguments to that method
// code for j = j-1
```

Now suppose analysis of the program shows that only one possible implementation of method *meth* can be invoked at this call site. If the body of that method is reasonably small, it can be inlined. The generated code corresponds to the following pattern:

```
          // code for i = i + 1
          // inlined code for method meth parameterized
          // ... by the arguments obj, i, and j
   L2:    // code for j = j - 1

   ...    // much omitted code

   L1:    // code to load arguments i and j
          // dispatch code to lookup address of method meth
          // ... and pass arguments and obj to that method
          goto L2;
```

Label L1 is not reached with this version of the code.

Now suppose dynamic class loading causes an alternative implementation of method *meth* to be loaded. The runtime environment now uninlines the method call by patching the code. It overwrites the first word of the inlined method with a branch instruction, so that the patched code corresponds to the following pattern:

```
          // code for i = i + 1
          goto L1
          // remainder of code of inlined method, which
          // is now unreachable.
   L2:    // code for j = j - 1

   ...    // much omitted code

   L1:    // code to load arguments i and j
          // dispatch code to lookup address of method meth
          // ... and pass arguments and obj to that method
          goto L2;
```

This patched code contains two more branches than the original unoptimized program and would therefore run more slowly; it also contains unreachable code that incurs a modest space penalty. The assumption is that dynamic class loading is rare and that methods rarely need to be uninlined.

Experiments with a JIT compiler showed that the number of dynamic method calls is reduced by 9 to 97% on their test suite. The effect on execution speed ranged from a small 1% worsening of performance to an impressive 133% improvement in performance, with a geometric mean speedup of 16%.

Ishizaki et al. [17] also point out that a similar technique is applicable to a method invoked via an interface. If only one class implements an interface class, we can generate code that assumes this class is actually used and we can inline methods of that class that are not overridden by any of its subclasses. If the assumption is later broken by dynamically loading a new class that also implements the interface or that overrides the method, we can patch the code to revert to the original full scheme of looking up the class and looking up the method.

13.5   Escape Analysis

In general, instances of classes are dynamically allocated. Storage for these instances is normally allocated on the heap. In a language such as C++, where the programmer is responsible for allocating and deallocating memory for objects on the heap, the program should free the memory for a class instance when it is no longer needed. Other languages, such as Java, provide automatic garbage collection. At periodic intervals, the garbage collector is invoked to perform the computationally intensive task of tracing through the references between objects and determining which objects can no longer be referenced. The storage for these objects is reclaimed.

The goal of escape analysis is to determine which objects have lifetimes that do not stretch outside the lifetime of their immediately enclosing scopes. The storage for such objects can be safely allocated as part of the current stack frame; that is, their storage can be allocated on the runtime stack. (For C programmers who use the `gcc` C compiler, the transformation is equivalent to replacing a use of the `malloc` function with the `alloca` function.) This optimization is valuable for Java programs. The transformation also improves the data locality of the program and, depending on the computer cache, can significantly reduce execution time.

Another benefit of escape analysis is that objects with lifetimes that are confined to within a single scope cannot be shared between two threads. Therefore, any synchronization actions for these objects can be eliminated. Escape analysis does not capture all possibilities for synchronization removal, however. If this is deemed to be an important optimization, a separate analysis to uncover unnecessary synchronization operations should be performed.

Algorithms for escape analysis are based on abstract interpretation techniques [1]. Different algorithms make different trade-offs between the precision of the analysis and the length of time the analysis takes. The better the precision, the more opportunities for optimization that should be found.

The reported speedup of Java programs can range up to 44% [7], but that figure includes savings due to synchronization removal and due to inlining of small methods. (Blanchet [7] reports an average speedup of 21% in his experiments.) Inlining significantly increases the number of opportunities for finding objects that do not escape from their enclosing scope, especially because many methods allocate a new object that is returned as the result of the method call.

13.5.1 Escape Analysis by Abstract Interpretation

A prototype implementation of escape analysis was included in the IBM High Performance Compiler for Java. This implementation is based on straightforward abstract interpretation techniques and has been selected for presentation in this text because it is relatively easy to understand. Further details of the algorithm may be found in the paper published by Choi et al. [9].

The approach of Choi et al. [9] attempts to determine two properties for each allocated object—whether the object escapes from a method (i.e., from the scope where it is allocated), and whether the object escapes from the thread that created it. It is possible that an object escapes the method but does not escape from the thread, and thus synchronization code may be removed. The converse is not possible; if an object does not escape a method, then it cannot escape its thread. The analysis therefore uses the very simple lattice of three values shown in Figure 13.16. If analysis determines that an object status is *NoEscape*, the object definitely does not escape from its method or from its thread; if the status is *ArgScape*, the object may escape from a method via its arguments but definitely does not escape the thread; finally, *GlobalEscape* means the object may escape from both the method and the thread.

The two versions of the analysis are a fast flow-insensitive version that yields imprecise results and a slower flow-sensitive version that gives better results. *Imprecise* means the analysis can be overly conservative, reporting many objects as having *GlobalEscape* status when a more accurate analysis might have shown the status as one of the other two possibilities, or reporting *ArgEscape* instead of *NoEscape*. Imprecision in this manner does not cause incorrect optimizations to be made; some opportunities for optimization can simply be missed. We give only the more precise flow-sensitive version of the analysis in this chapter.

13.5.1.1 Connection Graphs

As its name suggests, abstract interpretation involves interpretive execution of the program. With this form of execution, the contents of variables (or fields of classes when analyzing a Java program) are

FIGURE 13.16   Lattice elements for escape analysis.

tracked. However, we do not attempt to determine the contents of the variables for normal execution of the program—we would of course simply execute the program to do that. To perform escape analysis, we are interested only in following an object O from its point of allocation, knowing which variables reference O and which other objects are referenced by O fields. The abstraction implied in the name *abstract interpretation* is to abstract out just the referencing information, using a graph structure where nodes represent variables and objects and directed edges represent object references and containment of fields inside objects. Choi et al. [9] call this graph a *connection graph*.

A sample connection graph is shown in Figure 13.17; it shows the program state after executing the code:

```
A a = new A();    //  line L1
a.b1 = new B();   //  line L2
a.b2 = a.b1;
```

where we assume that the only fields of AC are b1 and b2, and BC has only fields with intrinsic types (i.e., the types int and char).

The notational conventions used in the connection graph are as follows. A circle node represents a variable (i.e., a field of a class or a formal parameter of a method); a square node represents an object instance. An edge from a circle to a square represents a reference; an edge from a square to a circle represents ownership of fields.

The graph shown in Figure 13.17 on the left is a simplification of that used by Choi et al. [9]. Their more complicated graph structure has two kinds of edges. An edge drawn as a dotted arrow is called a *deferred edge*. When there is an assignment from copies of one object reference to another, which copies an object reference from one variable to another, such as

```
p = q;   //  p and q have class types
```

then the effect of the assignment is shown as a deferred edge from the node for p to the node for q. In Figure 13.17, the graph on the right uses a deferred edge to show the effect of an assignment from one variable to another.

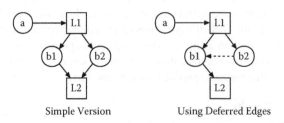

Simple Version   Using Deferred Edges

FIGURE 13.17   Sample connection graph.

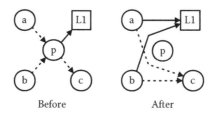

Before   After

FIGURE 13.18   Effect of bypass operation.

Each node in a connection graph has an associated escape state, chosen from the three possibilities given in Figure 13.16. If a connection graph has been constructed for a method *M* and if O is an object node in *M*, then if O can be reached from any node in the graph whose escape state is other than *NoEscape*, then O may escape from *M*. A similar property holds for objects escaping from a thread.

13.5.1.2   Intraprocedural Abstract Interpretation

The abstract interpretation is performed on a low-level version of the code where only one action at a time is performed. The Java bytecode is adequate for this purpose, as are other intermediate representation formats used in a typical compiler. Assuming that the code has been suitably simplified, abstract interpretation of the code within a method steps through the code and performs an action appropriate for each low-level operation. This interpretive execution involves following control flow edges, as explained in more detail later.

The actions for assignment statements and instantiations of new object instances are shown next. Each action involves an update to the connection graph. An assignment to a variable p *kills* any value that the variable previously had. The kill part of an assignment to p is implemented by an update to the connection graph that is called *ByPass(p)*. The *ByPass(p)* operation redirects or removes deferred edges as illustrated in Figure 13.18. Note also that compound operations, such as p.a.b.c, are assumed to be decomposed into simpler steps that dereference only one level at a time. The bytecode form of the program automatically possesses this property:

```
p = new C();   //  line L
```

If the connection graph does not already contain an object node labeled L, one is created and added to the graph. If the node needs to be created, then nodes for the fields of C that have nonintrinsic types are also created and are connected by edges pointing from the object node. Any outgoing edges from the node for p are deleted by applying the *ByPass(p)* operation, and then a new edge from p to the object node for L is added:

```
p=q;
```

The *ByPass(p)* operation is applied to the graph. Then a new deferred edge from p to q is created:

```
p.f=q;
```

If p does not point to any object nodes in the connection graph, a new object node (with the appropriate fields for the datatype of p) is created and an edge from p to the new object node is added to the graph. (Choi et al. [9] call this new object node a *phantom node*. Two reasons phantom nodes may arise are (a) the original program may contain an error and p would actually be null, referencing no object, when this statement is reached and (b) p may reference an object outside the current method—and that situation can be covered by the interprocedural analysis.) Then, for each object node that is connected to p by an edge, an assignment to the f field of that object is performed. That assignment is implemented by adding a deferred edge from the f node to the q node. Note that no *ByPass* operation is performed (to kill the previous value of f) because there is not necessarily a unique object that p references, and we cannot therefore be sure that the assignment kills all previous values for f:

```
p = q.f;
```

As before, if q does not point at any object nodes, a phantom node is created, and an edge from q to the new node is added to the graph. Then *ByPass(p)* is applied, and deferred edges are added from p to all the f nodes that q is connected to by field edges.

In principle, a different connection graph represents the state of the program at each statement in the method. Thus, when the abstract interpretation action modifies the graph, it is modifying a copy of the graph. When analyzing a sequence of statements in a basic block, the analysis proceeds sequentially through the statements in order. At a point where control flow diverges, such as at a conditional statement, each successor statement of the conditional is analyzed using a separate copy of the connection graph. At a point where two or more control paths converge, the connection graphs from each predecessor statement are merged.

A small example is given to make the process clearer. Suppose the code inside some method is as follows, with the declarations of classes A, B1, and B2 omitted:

```
A a = new A();          // line L1
if (i > 0)
    a.f1 = new B1();    // line L3
else
    a.f1 = new B2();    // line L5
a.f2 = a.f1;            // line L6
```

The connection graphs that are constructed by the abstract interpretation actions are shown in Figure 13.19. Diagram 1 in the figure shows the state after executing line L1; diagrams 2 and 3 show

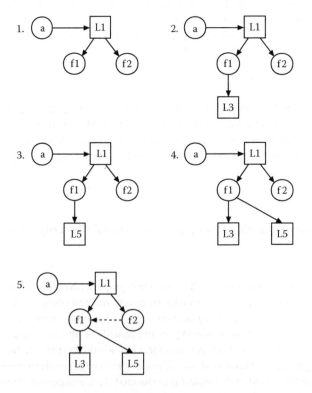

FIGURE 13.19   Sequence of connection graphs.

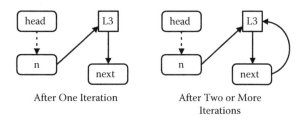

After One Iteration   After Two or More
                                   Iterations

FIGURE 13.20   Connection graph for a loop.

the states after lines L3 and L5, respectively. Note that diagrams 2 and 3 are obtained by applying the effects of lines L3 and L5 to the state in diagram 1. After the if statement, the two control flow paths merge; the graph in diagram 4 is the result of merging diagrams 2 and 3. Finally, diagram 5 shows the effect of applying line L6.

If the program contains a loop, abstract interpretation is performed repeatedly until the connection graphs converge. Convergence is guaranteed because the maximum number of nodes in the graph is a finite number that is proportional to the number of occurrences of new in the source code, and the number of edges that can be added between the nodes is also finite. If, for example, the source code to be analyzed is

```
Node head = null;
for( int cnt = 0; cnt < 1000; cnt++ ) {
    Node n = new Node(); // Line L3
    n.next = head;
    n.data = cnt;
    head = n;
}
```

then analysis gives the connection graphs shown in Figure 13.20. After analyzing the loop body once, the graph has the structure shown on the left; after analyzing a second time, the graph has converged to the diagram on the right. Even though the actual program allocates 1,000 instances of the Node class, only one new operation is in the code, and therefore only one object node is in the graph. The fact that one graph node represents 1,000 objects in the program is one of the approximations inherent in the graph structure invented by Choi et al. [9].

13.5.1.3 Interprocedural Abstract Interpretation

A call to a method M is equivalent to copying the actual parameters (i.e., the arguments are passed in the method call) to the formal parameters, then executing the body of M, and finally copying any value returned by M as its result back to the caller. If M has already been analyzed intraprocedurally following the approach described earlier, the effect of M can be summarized with a connection graph. That summary information eliminates the need to reanalyze M for each call site in the program.

It is necessary to analyze each method in the reverse of the order implied by the *call graph*. If method A may call methods B and C, then B and C should be analyzed before A. Recursive edges in the call graph are ignored when determining the order. Java has virtual method calls. At a method call site where it is not known which method implementation is invoked, the analysis must assume that all the possible implementations are called, combining the effects from all the possibilities. The interprocedural analysis iterates over all the methods in the call graph until the results converge. The extra actions needed to create the summary information for a method M follow:

Entry to a method M: If M has $n - 1$ formal parameters, fi, . . . fn, then n object nodes a1 . . . an are created. These correspond to the actual parameters (or arguments). The extra parameter corresponds to the implicit this parameter. Because Java has call-by-value semantics, an implicit assignment exists from each actual parameter to the corresponding formal parameter at method invocation. These assignments are modeled by creating a deferred edge from fi to ai for each parameter. The escape state initially associated with an fi node is *NoEscape*, and the state initially associated with an ai node is *ArgEscape*. By having created the first object nodes for the parameters, the body of the method is analyzed using the approach described for intraprocedural analysis.

Exit from method M: A `return` statement is processed by creating a dummy variable named *return*, to which the returned result is assigned. If there are multiple `return` statements in the method, the different result values are merged into the connection graph by adding deferred edges from the *return* node to the different results.

When the whole method body has been processed, the connection graph that was created after the `return` statement represents the summary information for the method. In particular, after the *ByPass* function has been used to eliminate all deferred edges, the connection graph can be partitioned into three subgraphs:

Global escape nodes: All nodes that are reachable from a node whose associated state is *GlobalEscape* are themselves nodes that are considered to be global escape nodes and form the first subgraph. The nodes initially marked as *GlobalEscape* are the static fields of any classes and instances of any class that implements the `Runnable` interface.

Argument escape nodes: All nodes reachable from a mode whose associated state is *ArgEscape*, but are not reachable from a *Global Escape* node, are in the second subgraph. The nodes initially marked as *ArgEscape* are the argument nodes a1 ... an.

No escape nodes: All other nodes in the connection graph form the third subgraph and have *NoEscape* status.

All objects created within a method M and that have the *NoEscape* status after the three subgraphs are determined can be safely allocated on the stack. The third subgraph represents the summary information for the method because it shows which objects can be reached via the arguments passed to the method. The part that remains to be covered is how to process a method call when it is encountered in the body of the method analyzed.

Suppose that while analyzing some method m1, we reach a method call:

```
result = obj.m2(p1,p2);
```

and that we have previously analyzed method m2 in the class to which `obj` belongs. The analysis algorithm creates three formal parameter nodes $\hat{a}_1$, $\hat{a}_2$, and $\hat{a}_3$ and processes three assignments:

$$\hat{a}_1 = obj; \quad \hat{a}_2 = p1; \quad \hat{a}_3 = p2;$$

The nodes $\hat{a}_1$ ... correspond to the argument nodes a1 ... in the connection graph that summarizes method m2, whereas the values `obj`, `p1`, and `p2` are nodes within the connection graph constructed for method m1.

The summary connection graph for m2 is used like a macro. Connections from a_1 ... are copied, becoming edge connections from `obj`. Field nodes that are children of object values that a_i nodes reference are matched against field nodes that are children of the `obj`, `p1`, and `p2` nodes and so on recursively. During this process, many of the phantom nodes that were introduced into the connection graph of m2 can be matched against nodes of m1. The algorithm for matching and duplicating the nodes and edges is omitted for space reasons.

One more issue needs to be covered. If the method call graph contains cycles, as occurs in the presence of direct or indirect recursion, then it appears to be necessary to use a method summary connection graph before it has been created. For this situation, a special bottom graph is used in place of the connection graph for the unanalyzed method. The bottom graph represents a worst-case (or conservative) scenario for escape of objects from any method in the program. The bottom graph has one node for every class. A points-to edge is from the node for class C1 to the node for class C2 if C1 contains a field of type C2; a deferred edge is from the node for C1 to the node for C2 if C2 is a subtype of C1. This graph can be used to make a conservative estimate for the escape of objects passed to a method by matching the argument types against the nodes in the bottom graph.

13.5.2 Other Approaches

Blanchet [7] developed a different analysis technique. His approach also uses abstract interpretation, but three significant differences exist.

First, the Java bytecode is directly interpreted so that an abstract representation of the values on the Java runtime stack is managed. During abstract interpretation, each bytecode operation has an effect on that representation of the stack.

Second, information is propagated both forward and backward. Forward propagation occurs when instructions are analyzed following the normal flow of control — as in the approach of Choi et al. [9]. Backward propagation is performed by interpretively executing the bytecode instructions along the reverse of control flow paths. The reverse execution mode has, of course, different semantics for the abstract meaning of each instruction (e.g., an instruction that pushes a value when interpreted forward becomes one that pops a value when analyzed backward). The combination of forward and backward analysis passes enables much more precise results to be obtained, especially when used in conjunction with the program analyzed containing no errors. (This is a common assumption for code optimization.)

Third, Blanchet [7] uses a quite different domain of values to represent escape of objects. He represents each class type by an integer, which is the context for what may escape from an instance of that class. His abstract values are equations (or context transformers) that map from the contexts of the arguments and result of a method to the escape contexts of concrete values. Instead of manipulating a collection of graphs, like Choi et al. [9], Blanchet [7] manipulates sets of equations.

No comparison has been made between the approaches of Choi et al. and Blanchet. Blanchet states that the Choi et al. analysis is more time consuming, and that is almost certainly true. Blanchet also claims bigger speedups for his set of sample programs, but Blanchet also performs extensive inlining of small methods.

13.6 Conclusion

We presented the most important optimizations for object-oriented programming languages that should give language implementors good choices for their work. Method invocation can be efficiently solved by different kinds of dispatch tables and inlining. Inlining and specialization greatly improve the performance but need precise and efficient type analysis algorithms. Escape analysis computes the information necessary to allocate objects on the runtime stack.

References

1. S. Abramsky and C. Hankin. 1987. *Abstract interpretation of declarative languages.* New York: Ellis Horwood.
2. O. Agesen. 1995. The Cartesian product algorithm: Simple and precise type inference of parametric polymorphism. In *ECOOP '95 — Object-Oriented Programming, 9th European Conference*, ed. W. G. Olthoff, 2–26. Lecture Notes in Computer Science, Vol. 952. New York: Springer-Verlag.
3. H. Aït-Kaci, R. Boyer, P. Lincoln, and R. Nasr. 1989. Efficient implementation of lattice operations. *ACM Trans. Program. Lang. Syst.* 11(1):115–46.
4. B. Alpern, A. Cocchi, D. Grove, and D. Lieber. 2001. Efficient dispatch of Java interface methods. In *HPCN '01: Java in High Performance Computing*, ed. V. Getov and G. K. Thiruvathukal, 621–28. Lecture Notes in Computer Science, Vol. 2110. New York: Springer-Verlag.
5. D. F. Bacon. 1997. Fast and effective optimization of statically typed object-oriented languages. Ph.D. thesis, University of California, Berkeley.

6. D. F. Bacon and P. F. Sweeney. 1996. Fast static analysis of C++ virtual function calls. In *Conference on Object-Oriented Programming Systems, Languages & Applications (OOPSLA '96)*, 324–41. New York: ACM Press.

7. B. Blanchet. 2000. Escape analysis for object oriented languages: Application to java. In *Proceedings of OOPSLA '99*, 20–34. New York: ACM Press.

8. Y. Caseau. 1993. Efficient handling of multiple inheritance hierarchies. In *Proceedings of OOPSLA '93*, 271–87. New York: ACM Press.

9. J.-D. Choi, M. Gupta, M. J. Serrano, V. C. Sreedhar, and S. P. Midkiff. 2003. Stack allocation and synchronization optimizations for Java using escape analysis. *ACM Trans. Program. Lang. Syst.* 25(6):876–910.

10. N. J. Cohen. 1991. Type-extension type tests can be performed in constant time. *ACM Trans. Program. Lang. Syst.* 13(4):626–29.

11. J. Dean, D. Grove, and C. Chambers. 1995. Optimization of object-oriented programs using static class hierarchy analysis. In *Proceedings of the 9th European Conference on Object-Oriented Programming (ECOOP '95)*, 77–101. Lecture Notes in Computer Science, Vol. 952. New York: Springer-Verlag.

12. P. Dencker, K. Dürre, and J. Heuft. 1984. Optimization of parser tables for portable compilers. *ACM Trans. Program. Lang. Syst.* 6(4):546–72.

13. S. Gehmawat, K. H. Randall, and D. J. Scales. 2000. Field analysis: getting useful and low-cost inter-procedural information. In *Conference on Programming Language Design and Implementation*, 334–44. New York: ACM Press.

14. J. Gil and Y. Zibin. 2005. Efficient subtyping tests with PQ-encoding. *ACM Trans. Program. Lang. Syst.* 27(5):819–56.

15. D. Grove, G. DeFouw, J. Dean, and C. Chambers. 1997. Call graph construction in object-oriented languages. In *Proceedings of the ACM SIGPLAN Conference on Object-Oriented Programming Systems, Languages and Applications (OOPSLA-97)*, 108–24. New York: ACM Press.

16. U. Hölzle, C. Chambers, and D. Ungar. 1991. Optimizing dynamically-typed object-oriented programming languages with polymorphic inline caches. In *Proceedings of the European Conference on Object-Oriented Programming (ECOOP '91)*, 21–38. Lecture Notes in Computer Science, Vol. 512. New York: Springer-Verlag.

17. K. Ishizaki, M. Kawahito, T. Yasue, H. Komatsu, and T. Nakatani. 2000. A study of devirtualization techniques for a JavaTM just-in-time compiler. In *Proceedings of the Conference on Object-Oriented Programming, Systems, Languages and Application (OOPSLA-00)*, 294–310. New York: ACM Press.

18. R. Jones and R. Lins. 1996. *Garbage collection*. New York: John Wiley & Sons.

19. A. Krall and R. Grafl. 1997. CACAO — A 64 bit JavaVM just-in-time compiler. *Concurrency Practice Exper.* 9(11):1017–30.

20. A. Krall, J. Vitek, and N. Horspool. 1997. Near optimal hierarchical encoding of types. In *11th European Conference on Object Oriented Programming (ECOOP '97)*, ed. M. Aksit and S. Matsuoka, 128–45. Lecture Notes in Computer Science, Vol. 1241. New York: Springer-Verlag.

21. B. Liskov, D. Curtis, M. Day, S. Ghemawat, R. Gruber, P. Johnson, and A. C. Myers. 1995. Theta reference manual. In *Technical Report Programming Methodology Group Memo 88*. Laboratory for Computer Science, Massachusetts Institute of Technology, Cambridge, MA.

22. A. C. Myers. 1995. Bidirectional object layout for separate compilation. In *OOPSLA '95 Conference Proceedings: Object-Oriented Programming Systems, Languages, and Applications*, 124–39. New York: ACM Press.

23. W. Pugh and G. Weddell. 1990. Two-directional record layout for multiple inheritance. In *Proceedings of the ACM SIGPLAN '90 Conference on Programming Language Design and Implementation (PLDI '90)*, 85–91. New York: ACM Press.

24. F. Qian and L. J. Hendren. 2005. A study in type analysis for speculative method inlining in a JIT environment. In *Proceedings of International Conference on Compiler Construction (CC 2005)*, 255–70, Lecture Notes in Computer Science, Vol. 3443. New York: Springer-Verlag.

25. J. R. Rose. 1991. Fast dispatch mechanisms for stock hardware. In *OOPSLA '88: Object-Oriented Programming Systems, Languages and Applications: Conference Proceedings*, ed. N. Meyrowitz, 27–35. New York: ACM Press.

26. O. Shivers. 1991. Control-flow analysis of higher-order languages. Ph.D. thesis, Carnegie-Mellon University, Pittsburgh, PA.

27. V. Sundaresan, L. J. Hendren, C. Razafimahefa, R. Vallée-Rai, P. Lam, E. Gagnon, and C. Godin. 2000. Practical virtual method call resolution for Java. In *Proceedings of the Conference on Object-Oriented Programming, Systems, Languages and Application (OOPSLA-00)*, 264–80. New York: ACM Press.

28. T. Suganuma, T. Ogasawara, M. T. Yasuekeuchi, M. Kawahito, K. Ishizaki, and H. Komatsuatani. 2000. Overview of the IBM Java just-in-time compiler. *IBM Syst. J.* 39(1):175–93.

29. B. Stroustrup. 1989. Multiple inheritance for C++. *Comput. Syst.* 2(4):367–95.

30. F. Tip and J. Palsberg. 2000. Scalable propagation-based call graph construction algorithms. In *Proceedings of the Conference on Object-Oriented Programming Systems, Languages and Application (OOPSLA-00)*, 281–93. New York: ACM Press.

31. J. Vitek and R. N. Horspool. 1996. Compact dispatch tables for dynamically typed object oriented languages. In *Proceedings of International Conference on Compiler Construction (CC '96)*, 307–25, Lecture Notes in Computer Science, Vol. 1060. New York: Springer-Verlag.

32. J. Vitek, N. Horspool, and A. Krall. 1997. Efficient type inclusion tests. In *Conference on Object-Oriented Programming Systems, Languages and Applications (OOPSLA '97)*, ed. T. Bloom, 142–57. New York: ACM Press.

33. O. Zendra, D. Colnet, and S. Collin. 1997. Efficient dynamic dispatch without virtual function tables: The Small Eiffel compiler. In *Proceedings of the ACM SIGPLAN Conference on Object-Oriented Programming Systems, Languages and Applications (OOPSLA-97)*, 125–41. New York: ACM Press.

34. Y. Zibin and J. Gil. 2002. Fast algorithm for creating space efficient dispatching tables with application to multi-dispatching. In *Proceedings of the 17th ACM SIGPLAN Conference on Object-Oriented Programming, Systems, Languages and Applications (OOPSLA-02)*, 142–60. New York: ACM Press.

14

Program Slicing

G. B. Mund
*Department of Computer Science and
Engineering,
Kalinga Institute of Industrial
Technology, Bhubaneswar, India,
mundgb@yahoo.com*

Rajib Mall
*Department of Computer Science and
Engineering,
Indian Institute of Technology,
Kharagpur, India,
rajib@cse.iitkgp.ernet.in*

14.1 Introduction

Program slicing is a program analysis technique. It can be used to extract the statements of a program that are relevant to a given computation. The concept of program slicing was introduced by Weiser [1–3]. A program can be sliced with respect to a *slicing criterion*. A slicing criterion is a pair $<p, V>$, where p is a program point of interest and V is a subset of the program's variables. If we attach integer labels to all the statements of a program, a program point of interest could be an integer i representing the label associated with a statement of the program. *A slice of a program P with respect to a slicing criterion $<s, V>$ is the*

```
        int i, sum, prd;
1.  read(i);
2.  prd = 1;
3.  sum = 0;
4.  while(i<10);
5.      sum=sum+i;
6.      prd = prd * i;
7.      i = i + 1;
8.  write(sum);
9.  write(prd);
```

FIGURE 14.1   An example program.

set of all the statements of the program P that might affect the slicing criterion for every possible input to the program. The program slicing technique introduced by Weiser [1–3] is now called static backward slicing. It is static in the sense that the slice is independent of the input values to the program. It is backward because the control flow of the program is considered in reverse while constructing the slice.

14.1.1   Static and Dynamic Slicing

Static slicing techniques perform analysis of a static intermediate representation of the source code to derive slices. The source code of the program is analyzed and slices are computed that hold well for all possible input values [3]. A static slice contains all the statements that may affect the value of a variable at a program point for every possible input. Therefore, we need to make conservative assumptions that often lead to relatively larger slices. That is, a static slice may contain some statements that might not be executed during an actual run of the program.

Korel and Laski [4] introduced the concept of dynamic program slicing. *Dynamic slicing* makes use of the information about a particular execution of a program. *A dynamic slice with respect to a slicing criterion <p, V>, for a particular execution, contains statements that affect the slicing criterion in the particular execution.* Therefore, dynamic slices are usually smaller than static slices and are more useful in interactive applications such as program debugging and testing. A comprehensive survey on the existing dynamic program slicing algorithms is reported in Korel and Rilling [5].

Consider the example program given in Figure 14.1. The static slice with respect to the slicing criterion $<8, sum>$ is $\{1, 3, 4, 5, 7\}$. Consider a particular execution of the program with the input value $i = 15$. The dynamic slice with respect to the slicing criterion $<8, sum>$ for the particular execution of the program is $\{3\}$.

14.1.2   Backward and Forward Slicing

A backward slice contains all parts of the program that might directly or indirectly affect the slicing criterion. Thus, a static backward slice provides the answer to the question: Which statements affect the slicing criterion?

A *forward slice* with respect to a slicing criterion $<p, V>$ contains all parts of the program that might be affected by the variables in V used or defined at the program point p [6]. A forward slice provides the answer to the question: Which statements will be affected by the slicing criterion? Unless otherwise specified, we consider backward slices throughout the discussion in this chapter.

14.1.3   Organization of the Chapter

The remainder of this chapter is organized as follows. In Section 14.2, we discuss the applications of program slicing. In Section 14.3, we discuss some basic concepts, notations, and terminologies associated with intermediate representations of sequential programs. Section 14.4 presents some basic slicing algorithms for sequential programs. Section 14.5 deals with intermediate representations and slicing of concurrent and distributed programs. In Section 14.6, we discuss parallel slicing of sequential and concurrent programs. In Section 14.7, we discuss intermediate representations and slicing of object-oriented programs. In Section 14.8, we discuss slicing of concurrent and distributed object-oriented programs. We present our conclusions in Section 14.9.

14.2 Applications of Program Slicing

The program slicing technique was originally developed to realize automated static code decomposition tools. The primary objective of those tools was to aid program debugging [1–3]. From this modest beginning, program slicing techniques have now ramified into a powerful set of tools for use in such diverse applications as program understanding, program verification, debugging, software maintenance and testing, functional cohesion metric computation, dead code elimination, reverse engineering, parallelization of sequential programs, software portability analysis, reusable component generation, program integration, tuning of compilers, compiler optimization, determining uninitialized variables, Y2K problems, and so on [7–65]. Excellent surveys of existing slicing algorithms and their applications are reported in [5, 41, 44, 63, 64, 66, 67]. In the following, we briefly discuss some of these applications of program slicing.

14.2.1 Debugging

Realization of automated tools to help effective program debugging was the original motivation for the development of the static slicing technique. In his doctoral thesis, Weiser provided experimental evidence that programmers unconsciously use a mental form of slicing during program debugging [1]. Locating a bug can be a difficult task when one is confronted with a large program. In such cases, program slicing is useful because it can enable one to ignore many statements while attempting to localize the bug. If a program computes an erroneous value for a variable x, only those statements in its slice would contain the bug; all statements not in the slice can safely be ignored.

The control and data dependences existing in a program are determined during slice computation. A program slicer integrated into a symbolic debugger can help in visualizing control and data dependences. Variants of the basic program slicing technique have been developed to further assist the programmer during debugging; *program dicing* [10] identifies statements that are likely to contain bugs by using information that some variables fail some tests while others pass all tests at some program point. Consider a slice with respect to an incorrectly computed variable at a particular statement. Now consider a correctly computed variable at some program point. The bug is likely to be associated with the slice on the incorrectly computed variable minus the slice on the correctly computed variable. This dicing heuristic can be used iteratively to locate a program bug. Several slices may be combined with each other in different ways. The intersection of two slices contains all statements that lead to an error in both test cases. The union of two slices contains all statements that lead to an error in at least one of the test cases. The symmetric difference of two slices contains all statements that lead to an error in exactly one of the test cases.

Another variant of program slicing is *program chopping* [68, 69]. It identifies statements that lie between two points a and b in the program and are affected by a change at a. Debugging in such a situation should be focused only on statements between a and b that transmit the change of a to b.

14.2.2 Software Maintenance and Testing

Software maintainers often have to perform *regression testing*, that is, retesting a software product after any modifications are carried out to ensure that no new bugs have been introduced [59]. Even after a small change, extensive tests may be necessary, requiring running of many test cases. Suppose a program modification requires only changing a statement that defines a variable x at a program point p. If the forward slice with respect to the slicing criterion $<p, x>$ is disjoint from the coverage of a regression test

t, then it is not necessary to rerun the test t. Let us consider another situation. Suppose a coverage tool reveals that a use of variable x at some program point p has not been tested. What input data is required to cover p? The answer to this question can be provided by examining the backward slice with respect to the slicing criterion $<p, x>$. Work has also been reported concerning testing incrementally through an application of program slicing [27]. These applications are discussed in detail in [47, 70].

Software testers have to locate the parts of code that might affect a safety-critical computation and to ascertain its proper functioning throughout the system. Program slicing techniques can be used to locate all the code parts that influence the values of variables that might be part of a safety-critical computation, but these variables that are part of the safety-critical computation have to be determined beforehand by domain experts.

One way to assure high quality is to incorporate redundancy into the system. If some output values are critical, they should be computed independently. For doing this, one has to ensure that the computation of these values should not depend on the same internal functions, since an error might manifest in both output values in the same way, thereby causing both parts to fail. An example of such a technique is *functional diversity* [43]. In this technique, multiple algorithms are used for the same purpose. Thus, the same critical output values are computed using different internal functions. Program slicing can be used to determine the logical independence of the slices with respect to the output values computing the same result.

14.2.3   Program Integration

Programmers often face the problem of integrating several variants of a base program. To achieve integration, the first step may be to look for textual differences between the variants. Semantics-based program integration is a technique that attempts to create an integrated program that incorporates the changed computations of the variants as well as the computations of the base program that are preserved in all variants [13].

Consider a program *Base*. Let A and B be two variants of Base created by modifying separate copies of Base. The set of preserved components consists of those components of Base that are affected in neither A nor B. This set precisely consists of the components having the same slices in Base, A, and B. Horwitz et al. presented an algorithm for semantics-based program integration that creates the integrated program by merging the program Base and its variants A and B [13]. The integrated program is produced through the following steps: (a) building dependence graphs $D1$, $D2$, and $D3$, which represent *Base*, A, and B, respectively; (b) obtaining a dependence graph of the merged program by taking the graph union of the symmetric difference of $D1$ and $D2$, the symmetric difference of $D1$ and $D3$, and the induced graph on the preserved components; (c) testing the merged graph for certain interference criteria; and (d) reconstructing a program from the merged graph.

14.2.4   Functional Cohesion-Metric Computation

The cohesion metric measures the relatedness of the different parts of some component. A highly cohesive software module is a module that has one function and is indivisible. For developing an effective functional cohesion metric, Beiman and Ott define *data slices* that consist of data tokens (instead of statements) [36]. Data tokens may be variables and constant definitions and references. Data slices are computed for each output of a procedure (e.g., output to a file, output parameter, assignment to a global variable). The tokens that are common to more than one data slice are the connections between the slices; they are the *glue* that binds the slices together. The tokens that are present in every data slice of a function are called *super-glue*. *Strong functional cohesion* can be expressed as the ratio of super-glue tokens to the total number of tokens in the slice, whereas *weak functional cohesion* is the ratio of glue tokens to the total number of tokens. The *adhesiveness* of a token is another measure expressing how many slices are glued together by that token.

14.2.5 Parallelization

Program slicing can be used to decompose a conventional program into substantially independent slices for assignment to separate processors as a way to parallelize the program. A goal of such parallelization is to determine slices with almost no overlap. Assuming that a combination slicer–compiler could produce a sliced executable code suitable for a parallel machine, an issue of some complexity is the problem of reconstructing the original behavior by "splicing" the results of the separate outputs of different slices. Such a technique is investigated in [7].

14.2.6 Other Applications of Program Slicing

Program slicing methods have been used in several other applications such as tuning of compilers, compiler optimizations, detecting dead code, determining uninitialized variables, software portability analysis, program understanding, reverse engineering, program specialization and reuse, program verification, and so on. These applications are discussed in some detail in [41, 44, 63–67, 71].

14.3 Some Basic Concepts and Definitions

In this section we present a few basic concepts, notations, and terminologies that are used later in this chapter. The existing program slicing literature shows a wide variation in and disagreement about the notations used in program slicing. We explain our usage here because of this lack of consensus. The usage presented here does not come from any single source but rather is a personal blending of ideas from many sources.

Definition 14.1 **Directed graph or graph:** *A directed graph (or graph) G is a pair* (N, E)*, where N is a finite non-empty set of* nodes, *and* $E \subseteq N \times N$ *is a set of directed edges between the nodes.*

Let $G = (N, E)$ be a graph. If (x, y) is an edge of G, then x is called a *predecessor* of y, and y is called a *successor* of x. The number of predecessors of a node is its *in-degree*, and the number of successors of the node is its *out-degree*. A *directed path* (or *path*) from a node x_1 to a node x_k in a graph $G = (N, E)$ is a sequence of nodes $(x_1, x_2, ..., x_k)$ such that $(x_i, x_{i+1}) \in E$ for every i, $1 \le i \le k - 1$.

Definition 14.2 **Flow graph:** *A flow graph is a quadruple (N, E, Start, Stop), where (N, E) is a graph,* Start $\in$ N *is a distinguished node of in-degree 0 called the* start *node,* Stop $\in$ N *is a distinguished node of out-degree 0 called the* stop *node, there is a path from Start to every other node in the graph, and there is a path from every other node in the graph to Stop.*

Definition 14.3 **Dominance:** *If x and y are two nodes in a flow graph, then x* dominates *y iff every path from* Start *to y passes through x. y* post-dominates *x iff every path from x to* Stop *passes through y.*

Let x and y be nodes in a flow graph G. Node x is said to be the *immediate post-dominator* of node y iff x is a post-dominator of y, $x \ne y$, and each post-dominator $z \ne x$ of y post-dominates x. The *post-dominator tree* of a flow graph G is the tree that consists of the nodes of G, has the root *Stop*, and has an edge (x, y) iff x is the immediate post-dominator of y.

Consider the flow graph shown in Figure 14.3. In the flow graph, each of the nodes 1, 2, and 3 dominates the node 4. Node 5 does not dominate node 7. Node 7 post-dominates nodes 1, 2, 3, 4, 5, and 6. Node 6 post-dominates node 5. Node 6 post-dominates none of the nodes 1, 2, 3, 4, 7, 8, and 9. Node 3 is the immediate post-dominator of node 2. Node 7 is the immediate post-dominator of node 4.

```
                              int a, b, sum;
                         1.  read(a);
                         2.  read(b);
                         3.  sum = 0;
                         4.  while(a < 8);
                         5.      sum = sum + b;
                         6.      a = a + 1;
                         7.  write(sum);
                         8.  sum = b;
                         9.  write(sum);
```

FIGURE 14.2   An example program.

14.3.1   Intermediate Program Representation

To compute a slice, it is first required to transform the program code into a suitable intermediate representation. In the following, we present a few basic concepts, notations, and terminologies associated with intermediate program representation that are used later in this chapter. A common cornerstone for most of the slicing algorithms is that programs are represented by a directed graph, which captures the notion of data dependence and control dependence.

14.3.1.1   Control Flow Graph

The *control flow graph* (CFG) is an intermediate representation for imperative programs that is useful for data flow analysis and for many optimizing code transformations such as common subexpression elimination, copy propagation, and loop-invariant code motion [11, 72].

Definition 14.4   **Control flow graph:** *Let the set N represent the set of statements of a program P. The control flow graph of the program P is the flow graph $G = (N_1, E)$, where $N_1 = N \cup \{Start, Stop\}$. An edge $(m, n) \in E$ indicates the possible flow of control from the node m to the node n.*

Note that the existence of an edge (x, y) in the control flow graph does not mean control *must* transfer from x to y during program execution. Figure 14.3 represents the CFG of the example program given in Figure 14.2. The CFG of a program P models the branching structures of the program, and it can be built while parsing the source code using algorithms that have linear time complexity in the size of the program [35].

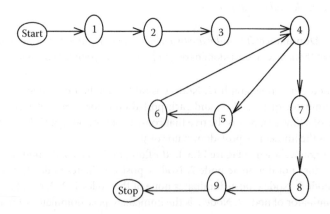

FIGURE 14.3   The CFG of the example program shown in Figure 14.2.

14.3.1.2 Data Dependence Graph

Several types of data dependences can be distinguished, such as flow dependence, output dependence, and anti-dependence [11]. However, for the purpose of slicing, only flow dependence is relevant [41].

The CFG of a program represents the flow of control through the program. However, the concept that is often more useful in program analysis is the flow of data through a program. Data flow describes the flow of the values of variables from the points of their definitions to the points where their values are used.

Definition 14.5   Data dependence: *Let G be the CFG of a program P. A node n is said to be* data dependent *on a node m if there exists a variable var of the program P such that the following hold:*

- *The node m defines var.*
- *The node n uses var.*
- *There exists a directed path from m to n along which there is no intervening definition of var.*

Consider the example program given in Figure 14.2 and its CFG of Figure 14.3. The node 5 has data dependence on each of the nodes 2, 3, and 5. The node 8 has data dependence on node 2. Note that node 8 is data dependent on none of the nodes 3 and 5.

Aho et al. [72] use the term *reaching definition* to mean that a value defined at a node may be used at another node. That is, node x is a reaching definition for a node y iff y is data dependent on x. A data dependence from node x to node y indicates that a value computated at x may be used at y under some path through the control flow graph. A dependence from x to y is a conservative approximation that says that under some conditions a value computed at x may be used at y.

Definition 14.6   Data dependence graph: *The data dependence graph of a program P is the graph $G = (N, E)$, where each node $n \in N$ represents a statement of the program P and $(x, y) \in E$ iff x is data dependent on y.*

14.3.1.3 Control Dependence Graph

Ferrante et al. [11] introduced the notion of *control dependences* to represent the relations between program entities arising because of control flow.

Definition 14.7   Control dependence: *Let G be the control flow graph of a program P. Let x and y be nodes in G. Node y is control dependent on node x if the following hold:*

- *There exists a directed path Q from x to y.*
- *y post-dominates every z in Q (excluding x and y).*
- *y does not post-dominate x.*

Let x and y be two nodes in the CFG G of a program P. If y is control dependent on x, then x must have multiple successors in G. Conversely, if x has multiple successors, at least one of its successors must be control dependent on it.

Consider the example program of Figure 14.2 and its CFG given in Figure 14.3. Each of the nodes 5 and 6 is control dependent on node 4. Note that node 4 has two successor nodes, 5 and 7, and node 5 is control dependent on node 4.

Definition 14.8   Control dependence graph (CDG): *The control dependence graph of a program P is the graph $G = (N, E)$, where each node $n \in N$ represents a statement of the program P and $(x, y) \in E$ iff x is control dependent on y.*

14.3.1.4 Program Dependence Graph

Ferrante et al. [11] presented a new mechanism of program representation called a *program dependence graph* (PDG). Unlike the flow graphs, an important feature of the PDG is that it explicitly represents

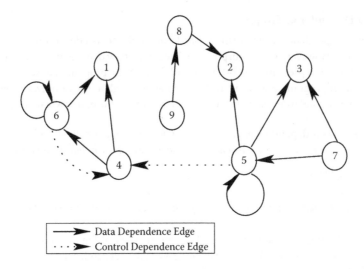

FIGURE 14.4   PDG of the example program shown in Figure 14.2.

both control and data dependences in a single program representation. A PDG models a program as a graph in which the nodes represent the statements and the edges represent interstatement data or control dependences.

Definition 14.9   Program dependence graph (PDG): *The* program dependence graph *G of a program P is the graph* $G = (N, E)$*, where each node* $n \in N$ *represents a statement of the program P. The graph contains two kinds of directed edges:* control dependence edges *and* data dependence edges*. A control (or data) dependence edge* (n, m) *indicates that n is control (or data) dependent on m.*

Note that the PDG of a program P is the union of a pair of graphs: the data dependence graph of P and the control dependence graph of P. Consider the example program given in Figure 14.2. Its PDG is given in Figure 14.4. We consider the PDG representation with the direction of dependences inverse to the traditional representation. In such a representation, graph traversal becomes more natural, as we can traverse along the edges directly without reversing their directions.

The program dependence graph of a program P can be built from its control flow graph in $O(n^2)$ time, where n is the number of nodes in the control flow graph [11].

14.3.1.5   System Dependence Graph

The PDG of a program combines the control dependences and data dependences into a common framework. The PDG has been found to be suitable for intraprocedural slicing. However, it cannot handle procedure calls. Horwitz et al. enhanced the PDG representation to facilitate interprocedural slicing [73]. They introduced the *system dependence graph* (SDG) representation, which models the main program together with all nonnested procedures. The SDG, an extention of the PDG, models programs in a language with the following properties [73]:

- A complete program consists of a main program and a collection of auxiliary procedures.
- Procedures end with return statements. A return statement does not include a list of variables.
- Parameters are passed by value-results.

The SDG is very similar to the PDG. Indeed, a PDG of the main program is a subgraph of the SDG. In other words, for a program without procedure calls, the PDG and the SDG are identical. The technique for constructing an SDG consists of first constructing a PDG for every procedure, including the main procedure, and then adding auxiliary dependence edges that link the various subgraphs together. This

```
main()
int s, i;
{
    s = 0;                  void add(int a, int b)   void inc(int z)
    i = 1;                  {                        {
    while (i < 10) do           a = a + b;               add(z, 1);
        add(s, i);              return;                  return;
        inc(i);             }                        }
    write(s);
}
```

FIGURE 14.5 An example program consisting of a main program and two procedures.

results in a program representation that includes the information necessary for slicing across procedure boundaries.

An SDG includes several types of nodes to model procedure calls and parameter passing:

- *Call-site nodes* represent the procedure call statements in a program.
- *Actual-in* and *actual-out nodes* represent the input and output parameters at the call sites. They are control dependent on the call-site nodes.
- *Formal-in* and *formal-out nodes* represent the input and output parameters at the called procedure. They are control dependent on the procedure's entry node.

Control dependence edges and data dependence edges are used to link the individual PDGs in an SDG. The additional edges used to link the PDGs together are as follows:

- *Call edges* link the call-site nodes with the procedure entry nodes.
- *Parameter-in edges* link the actual-in nodes with the formal-in nodes.
- *Parameter-out edges* link the formal-out nodes with the actual-out nodes.

Finally, *summary edges* are used to represent the transitive dependences that arise from calls. A summary edge is added from an actual-in node A to an actual-out node B, if there exists a path of control, data, and summary edges in the called procedure from the corresponding formal-in node A' to the formal-out node B'. Figure 14.6 represents the SDG of the example program shown in Figure 14.5.

14.3.2 Precision and Correctness of Slices

Let P be a program and S be a static slice of P with respect to a slicing criterion C. In Weiser's original definition [3], the reduced program S is required to be an executable program, and its behavior with respect to the slicing criterion C must be the same as the original program P. A slice S of P with respect to a slicing criterion C is *statement-minimal* if no other slice of P with respect to the slicing criterion has fewer statements than S. Weiser [3] has shown that the problem of computing statement-minimal slices is undecidable.

Another common definition of a static slice is the following: a slice S of a program P with respect to a slicing criterion C is a subset of the statements of the program that directly or indirectly affect the slicing criterion [11, 73, 74]. Note that such a slice need not be executable. Unless specified otherwise, we follow this definition of a slice throughout the discussion in the chapter.

Let G_C be the CFG of a program P. In all the existing program slicing frameworks, for each statement s in the program P, a set *UseSet*(s) of variable names used at s and a set *DefSet*(s) of variable names defined at s are maintained. The interstatement dependences in the program P are captured using the CFG G_C and the variable names in the sets *UseSet*(s) and *DefSet*(s) for each statement s [2, 3, 11, 73, 74].

Note that statement 4 of the example program shown in Figure 14.7 uses the variable m. Though node 4 assigns the value $0 = (m - m)$ to the variable z, it has dependence on node 1 in the program slicing frameworks since node 1 is a reaching definition of the variable m for node 4.

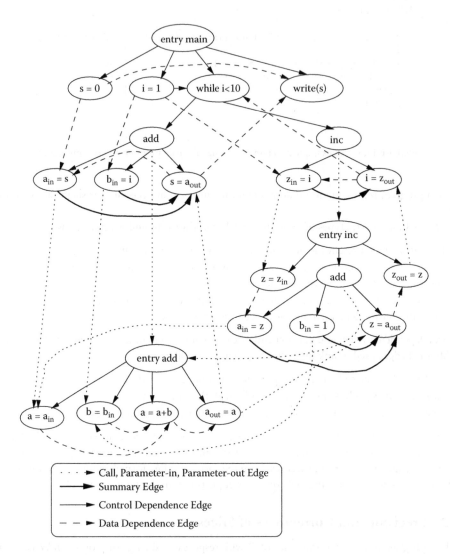

FIGURE 14.6   The *SDG* of the example program shown in Figure 14.5. (An edge (x, y) in the SDG indicates that the node y has dependence on the node x.)

```
      integer m, i, x, z;
1.  read(m);
2.  i = 1;
3.  x = 4;
4.  z = m - m;
5.  write(z)
```

FIGURE 14.7   An example program.

The precision of a dynamic slice in the existing program slicing frameworks is defined as follows. *A dynamic slice is said to be precise if it contains only those statements that actually affect the slicing criterion in the particular execution.*

Note that a precise dynamic slice need not be a statement-minimal slice. Consider the example program in Figure 14.7. For any input value of the variable m, the statement-minimal slice with respect to the slicing criterion $<4, z>$ is {4}, as z is always assigned the value 0 ($= m - m$). In the existing program slicing frameworks, the precise slice for the slicing criterion is certainly {1, 4}, as node 1 is a reaching definition of the variable m for the 4.

A slice is said to be *correct* if it contains all the nodes that affect the slicing criterion. A slice is said to be *incorrect* if it fails to contain at least one statement that affects the slicing criterion. Note that the whole program is always a correct slice of any slicing criterion. A correct slice is *imprecise* if and only if it contains at least one statement that does not affect the slicing criterion.

14.4 Basic Slicing Algorithms: An Overview

This section presents an overview of the basic program slicing techniques and includes a brief history of their development. We do not aim to give a comprehensive review of the related work. Such an attempt would be extremely difficult because of the numerous publications in this area and the diverse theory and techniques used by researchers over the years. Instead, we briefly review the work relevant to our discussion in this chapter. We start with the original approach of Weiser [1], where slicing is considered as a data flow analysis problem, and then examine the slicing techniques where slicing is seen as a graph-reachability problem.

14.4.1 Slicing Using Data Flow Analysis

Weiser introduced the concept of program slicing in his doctoral dissertation [1]. His original motivation was to develop the slicing technique as an aid to program debugging. His ideas were inspired by the abstraction mechanisms used by programmers while analyzing existing programs. In his slicing technique, the units of abstraction were called *slices*. The slices abstract a program based on the behavior of the program with respect to a specified set of data components, for example, variables.

14.4.1.1 Static Slicing

The first slicing algorithm was presented by Weiser [1]. His static slicing method used a CFG as an intermediate representation of the program to be sliced and was based on iteratively solving data and control flow equations representing interstatement influences. Let v be a variable and n be a statement (node) of a program P and S be the slice with respect to the slicing criterion $<n, v>$. Consider a node $m \in S$. Weiser defined the set of *relevant variables* for the node m, *relevant*(m), as the set of variables of the program P whose values (transitively) affect the computation of the value of the variable v at the node n. Consider the example program shown in Figure 14.1a and the slice with respect to the slicing criterion $<9, prd>$. For this example, *relevant*(8) = {*prd*}, *relevant*(7) = {*prd*}, and *relevant*(6) = {*prd, i*}. In Weiser's methodology, computing a slice from a CFG requires computation of the data flow information about the set of *relevant variables* at each node. That is, slices can be computed by solving a set of data and control flow equations derived directly from the CFG of the program being sliced.

Weiser's algorithms [1, 2] for computing static slices did not handle programs with multiple procedures. Later [3], Weiser presented algorithms for interprocedural slicing.

In Weiser's approaches, every slice is computed from scratch. That is, no information obtained during any previous computation of slices is used. This is a serious disadvantage of his algorithm. It has been shown that computation of static slices using his algorithm requires $O(n^2 e)$ time, where n is the number of nodes and e is the number of edges in the CFG [1].

```
        int i, n, z, x, y;
1.  read(i);
2.  n = 3;
3.  z = 1;
4.  while (i < n) do
5.      read(x);
6.      if (x < 0) then
7.          y = x + 2;
        else
8.          y = x + 8;
9.      z = y + 7;
10.     i = i +1;
11. write(z)
```

FIGURE 14.8   An example program.

14.4.1.2   Dynamic Slicing

Korel and Laski [4] introduced a new form of slicing. This new form of slicing is dependent on input data and is generated during execution-time analysis as opposed to Weiser's static-analysis slicing and is therefore called *dynamic slicing*. Similar to the major objective of static slicing, dynamic slicing was specifically designed as an aid to debugging and can be used to help in the search for offending statements that caused the program error.

Consider the example program in Figure 14.8. The static slice of the program with respect to the criterion $<11, z>$ includes the whole program. Suppose the program is executed with the input value $i = 3$. The dynamic slice with respect to the slicing criterion $<11, z>$ for this execution of the program is {3}. If the value of z computed at the end of the program is not as expected, we can infer that the program contains a bug.

The dynamic slice identifies statements that contribute to the value of the variable z when the input $i = 3$ is supplied to the program. Locating the bug using the dynamic slice is thus easier than examining the original program or the corresponding static slice, as the number of statements included in the dynamic slice is normally much less.

Korel and Laski extended Weiser's CFG-based static slicing algorithm to compute dynamic slices [4]. They computed dynamic slices by solving the associated data flow equations. Their method needs $O(N)$ space to store the execution history and $O(N^2)$ space to store the dynamic flow data, where N is the number of statements executed (length of execution) during the run of the program. To compute a dynamic slice using the stored execution history, their method requires $O(N)$ time. Note that for programs containing loops, N may be unbounded. That is, N may not be bounded in the size of the program. This is a major shortcoming of their method. Furthermore, the dynamic slices computed by the algorithm of Korel and Laski [4] may be imprecise [41, 75]. That is, the computed dynamic slice may contain some statements that do not affect the slicing criterion.

Other relevant dynamic slicing algorithms based on execution history are reported in [76–82]. These algorithms use execution history along with other relevant information and compute dynamic slices. Note that these dynamic slicing algorithms have essentially the same space complexity and time complexity as the basic algorithm of Korel and Laski [4].

14.4.2   Slicing Using Graph-Reachability Analysis

Ottenstein and Ottenstein [74] defined slicing as a reachability problem in the dependence graph representation of a program. A directed graph is used as an intermediate representation of the program. This directed graph models the control or data dependences among the program entities. Slices can be computed by traversing along the dependence edges of this intermediate representation. An important advantage of this approach is that data flow analysis has to be performed only once and that the information can be used for computing all slices.

14.4.2.1 Static Slicing Using Dependence Graphs

Ottenstein and Ottenstein introduced the PDG as an intermediate program representation [74]. They demonstrated how the PDG could be used as the basis of a new slicing algorithm. Their algorithm produced smaller slices than Weiser's algorithm. This method differed from Weiser's in an important way: it used a single reachability pass of a PDG compared to Weiser's incremental flow analysis. Ottenstein and Ottenstein presented a linear time solution for intraprocedural static slicing in terms of graph reachability in the PDG [74]. The construction of the PDG of a program requires $O(n^2)$ time, where n is the number of statements in the program. Once the PDG is constructed, the slice with respect to a slicing criterion can be computed in $O(n + e)$ time, where n is the number of nodes and e is the number of edges in the PDG. The process of building the PDG of a program involves computation and storage of most of the information needed for generating slices of the program.

As described earlier, the notion of the PDG was extended by Horwitz et al. into a *system dependence graph* (SDG) to represent multiprocedure programs [73]. Interprocedural slicing can be implemented as a reachability problem over the SDG. Horwitz et al. developed a two-phase algorithm that computes precise interprocedural slices [73]. To compute a slice with respect to a node n in a procedure P requires two phases of computations that perform the following:

- In the first phase, all edges except the parameter-out edges are followed backward starting with node n in procedure P. All nodes that reach n and are either in P itself or in procedures that (transitively) call P are marked. That is, the traversal ascends from procedure P upward to the procedures that called P. Since parameter-out edges are not followed, phase 1 does not "descend" into procedures called by P. The effects of such procedures are not ignored. Summary edges from actual-in nodes to actual-out nodes cause nodes to be included in the slice that would only be reached through the procedure call, though the graph traversal does not actually descend into the called procedure. The marked nodes represent all nodes that are part of the calling context of P and may influence n.

- In the second phase, all edges except parameter-in and call edges are followed backward starting from all nodes that have been marked during phase 1. Because parameter-in edges and call edges are not followed, the traversal does not "ascend" into calling procedures. Again, the summary edges simulate the effects of the calling procedures. The marked nodes represent all nodes in the called procedures that induce summary edges.

14.4.2.2 Dynamic Slicing Using Dependence Graphs

Agrawal and Horgan [75] were the first to present algorithms for finding dynamic program slices using the PDG as the intermediate program representation. In [75] they proposed four intraprocedural dynamic slicing algorithms.

Their first approach to compute dynamic slices uses the PDG as the intermediate program representation and marks the nodes of this graph as the corresponding parts of the program are executed for a given set of input values. A dynamic slice is computed by applying the static slicing algorithm of Ottenstein and Ottenstein [74] to the subgraph of the PDG induced by the marked nodes. This approach is imprecise because it does not consider situations where there exits an edge in the PDG from a marked node u to a marked node v but the definition at v is not used at u.

We illustrate this imprecision through an example. Consider the program of Figure 14.9 and its PDG in Figure 14.10. Let the input value of the variable m be 2 and the input values of x in the first and second iterations of the while loop be 0 and 2, respectively. In the first iteration of the while loop, statement 8 defines a value for y. In the second iteration of the loop, statement 9 defines a value for y without using its previous value and destroys the previous definition of y. Therefore, the dynamic slice for the slicing criterion $<10, w>$ in the second iteration of the while loop should contain the statement 9 and should not contain the statement 8. Let us find the dynamic slice using the first approach of Agrawal and Horgan [75]. We mark node 8 in the first iteration of the loop and node 9 in the second iteration. As node 10 has outgoing dependence edges to both the nodes 8 and 9 in the PDG, both statements 8 and 9 get included in the dynamic slice, which is clearly imprecise.

```
       int m, p, i, q, x, y, w
    1. read(m);
    2. p = 20;
    3. i = 1;
    4. q = 0;
    5. while (i <= m)
    6.     read(x);
    7.     if (x <= 0) then
    8.         y = x + 5;
           else
    9.         y = x - 5;
    10.    w = y + 4;
    11.    if (w > 0) then
    12.        p = p * w;
           else
    13.        q = p - 4;
    14.    i = i + 1;
    15. write(p);
    16. write(q)
```

FIGURE 14.9 An example program.

The second approach of Agrawal and Horgan [75] marks the edges of the PDG as and when the corresponding dependences arise during program execution. A dynamic slice is computed by applying the static slicing algorithm of Ottenstein and Ottenstein [74] and traversing the PDG only along the marked edges. This approach finds precise dynamic slices of programs with no loops. In the presence of loops, the slices may sometimes include more statements than necessary because this approach does not consider

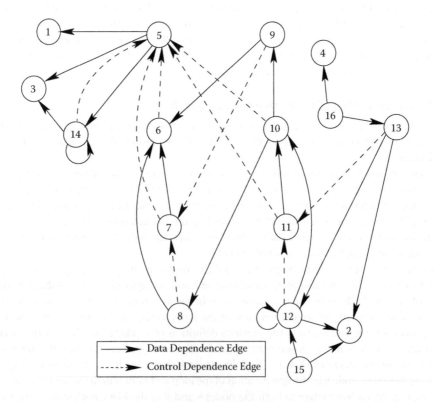

FIGURE 14.10 The PDG of the example program shown in Figure 14.9.

that execution of the same statement during different iterations of a loop may be (transitively) dependent on different sets of statements.

To illustrate this point, consider the example program in Figure 14.9 and its PDG in Figure 14.10, with the input values $m = 2$ and $x = 0$ in the first iteration of the while loop and $x = 2$ in the second iteration. The dynamic slice for the slicing criterion $<10, w>$ in the second iteration of the while loop should contain the statement 9 and should not contain the statement 8. Using the second approach of Agrawal and Horgan [75], we mark the outgoing dependence edge (10, 8) in the first iteration of the while loop. In the second iteration, we mark the outgoing dependence edge (10, 9). Traversing along the marked edges, we include both the statements 8 and 9 in the dynamic slice, which clearly results in an imprecise slice.

Agrawal and Horgan [75] pointed out that their second approach to compute dynamic slices produces results identical to those produced by the algorithm of Korel and Laski [4]. Note that the PDG of a program with n statements requires only $O(n^2)$ space. Thus, the space requirement of Agrawal and Horgan's second algorithm [75] is $O(n^2)$, but the algorithm of Korel and Laski [4] may use unbounded space in the worst case.

The shortcomings of the second approach of Agrawal and Horgan [75] motivated their third approach: construct a *dynamic dependence graph* (DDG) creating a new node for each occurrence of a statement in the execution history along with the associated dependence edges. A dynamic slicing criterion is identified with a node in the DDG, and a dynamic slice is computed by determining all the nodes of the DDG that can be reached from the slicing criterion.

Clearly, the space complexity and time complexity of the DDG-based algorithm are linear in the number of nodes in the DDG. The disadvantage of using the DDG is that the number of nodes in a DDG is equal to the number of executed statements (length of execution), which may be unbounded for programs with loops.

In their fourth approach, Agrawal and Horgan [75] proposed to reduce the number of nodes in the DDG by merging nodes whose transitive dependences map to the same set of statements. In other words, a new node is introduced only if it can create a new dynamic slice. This check incurs $O(d_u n)$ runtime overhead, where n is the number of statements in the program under consideration and d_u is the number of stored (different) dynamic slices corresponding to node u prior to its present execution in the particular execution of the program. The resulting reduced graph is called the *reduced dynamic dependence graph* (RDDG). The size of the RDDG is proportional to the number of dynamic slices that can arise during execution of the program. Note that in the worst case, the number of dynamic slices of a program having n statements is $O(2^n)$ [41].

Other relevant dependence graph–based dynamic slicing methods that use DDGs are reported in [83–86]. The interprocedural dynamic slicing algorithms of Agrawal et al. [83] and Kamkar et al. [84–86] use DDGs with distinct vertices for different occurrences of a statement in the execution history. Therefore, these algorithms have essentially the same space complexity and time complexity as the basic DDG-based algorithm of Agrawal and Horgan [75].

In [87] Mund et al. introduced the concepts of *stable and unstable edges*. They modified the PDG and used the *modified program dependence graph* (MPDG) as the intermediate program representation for their algorithm. Their algorithm [87] is based on marking and unmarking the unstable edges of the MPDG appropriately as and when the dependences arise and cease during execution of the program. The space complexity of their algorithm [87] is quadratic in the number of statements in the program. The dynamic slicing algorithm requires at most $O(n^2)$ time to compute a dynamic slice after execution of a statement of interest.

Goswami and Mall [88] proposed a dynamic slicing algorithm based on a *compact dynamic dependence graph* (CDDG). In their approach, they construct the CDDG at runtime. During execution of a *loop containing conditionals* (LCC), their method examines the path (sequence of the nodes executed) taken in every iteration. If the path is not taken in any previous iterations of the LCC, then the path along with its iteration number is stored. If the path is already taken in some previous executions, then its associated iteration number is updated with the current iteration number. After execution of the LCC, the paths are arranged in ascending order of their associated iteration numbers, and control and data dependence edges are added. The control and data dependences among nodes of the arranged paths are obtained by traversing backward along the sequence of paths.

Mund et al. [89] have shown that the dynamic slicing algorithm of Goswami and Mall [88] may compute incorrect dynamic slices. That is, it may fail to include some statements that actually affect a slicing criterion.

The *edge-marking* and *node-marking* dynamic slicing algorithms of Mund et al. [89] use the PDG as the intermediate program representation and compute precise dynamic slices. The space complexity of each of these algorithms is quadratic in the number of statements in the program. Each of the dynamic slicing algorithms reported in [89] requires at most $O(n)$ time to compute and store a dynamic slice after execution of a statement of interest, where n is the number of statements in the program.

Zhang et al. [90] presented three precise dynamic slicing algorithms. The algorithms differ in the degree of preprocessing they carry out prior to computation of dynamic slices. They have reported that their limited preprocessing (LP) algorithm is better than the other two algorithms. Their LP algorithm augments the execution trace with summary information for faster traversal of the trace. During slicing it performs demand-driven analysis to capture dynamic dependences from the compacted execution trace. The space complexity of the algorithm is proportional to the length of the execution trace. The time required to compute a dynamic slice after execution of a statement of interest is also proportional to the length of the execution history.

Recently, Mund and Mall [91] presented two precise dynamic slicing algorithms named the *IntraSlice* algorithm and the *InterSlice* algorithm. The intraprocedural dynamic slicing algorithm (*IntraSlice algorithm*) uses a CDG as the intermediate program representation. The *IntraSlice* algorithm is based on computing and updating the *data and control dependences* at runtime. The interprocedural dynamic slicing algorithm (*InterSlice algorithm*) uses a collection of CDGs (one for each function of the program) as the intermediate program representation. The *InterSlice* algorithm is similar to the *IntraSlice* algorithm. However, it uses additional data structures to handle the interprocedural dependences caused by the calling nodes, the paramater passing mechanisms, and the *RETURN* mechanisms.

Mund and Mall [91] have shown that the space complexity of each of their algorithms is quadratic in the number of statements in the program. The time required to compute and store a dynamic slice after execution of a statement of interest is at most linear in the number of statements in the program.

14.4.3   Slicing in the Presence of Composite Data Types and Pointers

Lyle [8] proposed a conservative solution to the problem of static slicing in the presence of arrays. Essentially, any update to an element of an array is regarded as an update and a reference of the entire array. In the presence of pointers, situations may occur where two or more variables refer to the same memory location; this phenomenon is commonly called *aliasing*.

Slicing in the presence of aliasing may require a generalization of the notion of data dependence to take potential aliases into account. Horwitz et al. [92] defined this notion of data dependence in terms of potential definitions and uses of abstract memory locations. The DDG-based slicing algorithm of Agrawal et al. [83] implements a similar idea to deal with both composite data types and pointers.

Jiang et al. [93] presented an algorithm for slicing C programs with pointers and arrays. Tip [41] has shown that the algorithm of Jiang et al. [93] may compute incorrect slices. Other relevant methods to compute static slices in the presence of pointer variables are reported in [94–98].

Korel and Laski [4, 76] considered dynamic slicing in the presence of arrays by regarding each element of an array as a distinct variable. Dynamic data structures are treated as two distinct entities: the element pointed to and the pointer itself [76]. Agrawal et al. [83] presented a DDG-based algorithm for dynamic slicing in the presence of arrays and pointers. To handle arrays and pointers, their algorithm uses the actual memory location of the variables provided by the compiler. The space complexity and time complexity of these dynamic slicing algorithms remain essentially the same as the basic algorithm of Korel and Laski [4].

In [91] Mund and Mall considered dynamic slicing in the presence of arrays, structures, and pointers. They follow the approach of Korel and Laski [4, 76] and regard each element of an array as distinct. Let *arr* be an array of k elements in a program P. Consider each array element $arr[1], \ldots, arr[k]$ as distinct. Let a statement s use or define $arr[i]$. During an actual run of the program, the index value i is known prior to execution of the statement s. As the sets of array elements used or defined at s corresponding to its different executions may be different, they consider computing and storing these sets dynamically. Let *ptr* be a pointer variable in the program P. Mund and Mall [91] consider *ptr* and *∗ptr* as two distinct variables. Following an approach similar to their approach of handling arrays, they have proposed efficient handling of pointers and structures.

14.4.4 Slicing of Unstructured Programs

Computation of slices becomes complicated in the presence of unstructured constructs (e.g., `goto`). It is difficult to capture the interstatement dependences caused by the `goto` statements or their restricted forms such as `return`, `exit`, `break`, and `continue` in a program [41, 44].

Some methods to compute static and dynamic slices for unstructured programs have been reported [8, 16, 20, 35, 79–81, 93, 99–104]. Lyle [8] reported that the static slicing algorithms of Weiser [2, 3] may compute incorrect slices in the presence of unstructured control flow. He proposed a solution to slicing in the presence of unstructured control flows. Gallagher [16] and Gallagher and Lyle [20] used a variation of Weiser's methods. They proposed including a `goto` statement in the slice if it can jump to a label of an included statement.

Jiang et al. [93] proposed extending Weiser's static slicing algorithms to C programs by introducing a number of rules to include the unstructured control flow statements in a slice. However, they have not given any formal justification for proper handling of the unstructured flow constructs.

Agrawal [99] presented an algorithm to compute static slices of unstructured programs and showed that the algorithm of Gallagher and Lyle [20] may produce incorrect slices. Agrawal's algorithm [99] uses the standard PDG as the intermediate representation and computes slices as a graph-reachability problem. Finally, affecting jump statements are added to the slices.

Ball and Horwitz [100] and Choi and Ferrante [101] discovered independently that conventional PDG-based slicing algorithms may produce incorrect slices in the presence of unstructured control flow. The algorithms presented in Ball and Horwitz [100] and Choi and Ferrante [101] use an augmented CFG (ACFG) to build a dependence graph. Augmented PDGs (APDG) are constructed using the ACFG. The control dependence edges in the APDGs are different from those in the PDGs.

Harman and Daninic [102] presented an algorithm for static slicing of unstructured programs. Their algorithm is an extension of Agrawal [99], and it produces smaller slices by using a refined rule for adding jump statements.

Sinha et al. [103] proposed a system dependence graph–based static slicing algorithm to handle arbitrary interprocedural control flow. Their algorithm handles interprocedural control flows in which a *return* may not occur, and if it does, it may not be to the call site.

Kumar and Horwitz [104] presented a static slicing algorithm for programs with jumps and switches. Their algorithm uses ACFG to build a dependence graph called a pseudo-predicate PDG (PPDG) and computes slices as a restricted graph-reachability problem in the *PPDG*. Finally, a label L is included in the static slice iff a `goto L` statement is already in the slice.

Korel [79] presented a dynamic slicing algorithm for an unstructured program. He employs the notion of removable blocks to find dynamic program slices. In his approach, data dependencies are used to compute the parts of the program that affect the value of a variable of interest. The removable blocks are used to identify parts of the noncontributing computations.

Huynh and Song [80] presented a dynamic slicing algorithm for programs with structured jump statements. Their algorithm also employs the notion of removable blocks to find dynamic program slices. It can handle unstructured programs that have only structured jumps.

Beszédes et al. [81] presented a dynamic slicing algorithm. Their program representation records two lists of variable names for each statement of the program: a list of variables defined and a list of variables used. During execution of the program they keep track of the last definitions of the variables and predicates in the execution history. Using the last definitions of variables and predicates, they compute and store the dynamic slices corresponding to each execution of a statement in the execution history.

Mund et al. [89] presented a node-marking algorithm to compute precise dynamic slices of unstructured programs. They introduced the concept of *jump dependence*. Based on the notion of jump dependence, they built an *unstructured program dependence graph* (UPDG) as an intermediate representation of the unstructured program to be sliced. Their algorithm is based on marking and unmarking the nodes of the UPDG as and when the corresponding dependences arise and cease during runtime.

14.5   Slicing of Concurrent and Distributed Programs: An Overview

In this section, we first discuss some basic issues associated with concurrent programming. Later, we discuss how these issues have been addressed in computation of slices of concurrent and distributed programs.

The basic "unit" of concurrent programming is the *process* (also called *task* in the literature). A process is an execution of a program or a section of a program. Multiple processes can execute the same program (or a section of the program) simultaneously. A set of processes can execute on one or more processors. In the limiting case of a single processor, all processes are interleaved or time-shared on this processor. *Concurrent program* is a generic term that is used to describe any program involving potential parallel behavior. Parallel and distributed programs are subclasses of concurrent programs that are designed for execution in specific parallel processing environments.

14.5.1   Nondeterminism

A sequential program imposes a total ordering on the actions it performs. In a concurrent program, there is an uncertainty over the precise order of occurrence of some events. This property of a concurrent program is referred to as *nondeterminism*. A consequence of nondeterminism is that when a concurrent program is executed repeatedly, it may take different execution paths even when operating on the same input data.

14.5.2   Process Interaction

A concurrent program normally involves process interaction. This occurs for two main reasons:

- Processes compete for exclusive access to shared resources, such as physical devices or data, and therefore need to coordinate access to the resource.
- Processes communicate to exchange data.

In both the above cases, it is necessary for the processes concerned to *synchronize* their execution, either to avoid conflict, when acquiring resources, or to make contact, when exchanging data. Processes can interact in one of two ways: through *shared variables* or by *message passing*. Process interaction may be *explicit* within a program description or may occur *implicitly* when the program is executed.

A process needing to use a *shared resource* must first *acquire* the resource, that is, obtain permission to access it. When the resource is no longer required, it is *released*. If a process is unable to acquire a resource, its execution is usually suspended until that resource is available. Resources should be administered so that no process is delayed unduly.

14.5.3   A Coding View

The main concerns in the representation of concurrent programs are:

- The representation of processes.
- The representation of process interactions.

Concurrent behavior may be expressed directly in a programming notation or implemented by system calls. In a programming notation, a process is usually described in a program block, and process instances are created through declaration or invocation references to that block. Process interaction is achieved via shared variables or by message passing from one process to another.

14.5.4 Interaction via Shared Variables

A commonly used mechanism for enforcing mutual exclusion is the use of semaphores. Entry to and exit from a critical region are controlled by using P and V operations, respectively. This notation was proposed by Dijkstra [105], and the operations can be read as "wait if necessary" and "signal" (the letters actually represent Dutch words meaning "pass" and "release"). Some semaphores are defined to give access to competing processes based on their arrival order. The original definition, however, does not stipulate an order. The less strict definition gives greater flexibility to the implementor but forces the program designer to find other means of managing queues of waiting processes.

14.5.5 Interaction by Message Passing

Process interaction through message passing is very popular. This model has been adopted by the major concurrent programming languages [106]. It is also a model amenable to implementation in a distributed environment. Within the general scheme of message passing, there are two alternatives:

- *Synchronous,* in which the sending process is blocked until the receiving process has accepted the message (implicitly or by some explicit operation).
- *Asynchronous,* in which the sender does not wait for the message to be received but continues immediately. This is sometimes called a nonblocking or no-wait send.

Synchronous message passing, by definition, involves a synchronization as well as a communication operation. Since the sender process is blocked while awaiting receipt of the message, there can be at most one pending message from a given sender to a given receiver, with no ordering relation assumed between messages sent by different processes. The buffering problem is simple because the number of sending messages is bounded.

In asynchronous message passing, the pending messages are buffered transparently, leading to potential unreliability in case of a full buffer. For most applications, synchronous message passing is thought to be the easier method to understand and use and is more reliable as well. Asynchronous message passing allows a higher degree of concurrency.

14.5.6 Concurrency at the Operating System Level

In this section we confine our attention to a discussion of the Unix Operating System. Unix [107, 108] is not a single operating system but an entire family of operating systems. The discussion here is based chiefly on the POSIX standard, which describes a common, portable, Unix programmer's interface.

Unix uses a pair of system calls, *fork* and *exec*, for process creation and activation. The fork call creates a copy of the forking process with its own address space. The exec call is invoked by either the original or a copied process to replace its own virtual memory space with the new program, which is loaded into memory, destroying the memory image of the calling process. The parent process of a process terminating by using the *exit* system call can wait on the termination event of its child process by using the *wait* system call. Process synchronization is implemented using *semaphores*. Interprocess communication is achieved through shared memory and message passing mechanisms [109]. A shared memory segment is created using the *shmget* function. It returns an identifier to the segment. The system call *shmat* is used to map a segment to the address space of a particular process. Message queues are created by using the *msgget* function. Messages are sent by using the *msgsnd* function, and these messages get stored in the message queue. The *msgrcv* function is used by a process to receive a message addressed to it from the message queue.

14.5.7   Slicing Concurrent and Distributed Programs

Research in slicing of concurrent programs is scarcely reported in the literature. In the following, we review the reported work in static and dynamic slicing of concurrent and distributed programs.

14.5.7.1   Static Slicing

Cheng [110] generalized the notion of a CFG and a PDG to a *nondeterministic parallel control flow net* and a *program dependence net* (PDN), respectively. In addition to edges for data dependence and control dependence, a *PDN* may also contain edges for selection dependences, synchronization dependences, and communication dependences. Selection dependence is similar to control dependence but involves non-deterministic selection statements, such as the ALT statement of Occam-2. Synchronization dependence reflects the fact that the start or termination of the execution of a statement depends on the start or termination of the execution of another statement. Communication dependence corresponds to a situation where a value computed at one point in the program influences the value computed at another point through interprocess communication. Static slices are computed by solving for the reachability problem in a PDN. However, Cheng did not precisely define the semantics of synchronization and communication dependences, nor did he state or prove any property of the slices computed by his algorithm [110].

Goswami et al. [111] presented an algorithm for computing static slices of concurrent programs in a Unix process environment. They introduced the concept of a *concurrent program dependence graph* (CPDG) and constructed the graph representation of a concurrent program through three hierarchical levels: *process graph*, *concurrency graph*, and *CPDG*. A process graph captures the basic process structure of a concurrent program and represents process creation, termination, and joining of processes. A *process node* consists of a sequence of statements of a concurrent program that would be executed by a process.

Krinke [112] proposed a method for slicing threaded programs. Krinke's work extends the structures of the CFG and PDG for threaded programs with *interference*. She defines *interference* as data flow that is introduced through the use of variables common to parallel executing statements. In [112] she proposed a slicing algorithm to compute slices from the new constructs for threaded programs called *threaded-PDG* and *threaded-CFG*. Nanda and Ramesh [113] later pointed out some inaccuracies in Krinke's slicing algorithm and proposed some improvements over it. Their algorithm has a worst-case complexity of $O(N^t)$, where N is the number of nodes in the graph and t is the number of threads in the program. They also proposed three optimizations to reduce this exponential complexity.

A process graph captures only the basic process structure of a program. This has been extended to capture other Unix programming mechanisms such as interprocess communication and synchronization. A *concurrency* graph is a refinement of a process graph where the process nodes of the process graph containing message passing statements are split up into three different kinds of nodes: *send node, receive node,* and *statement node*. The significance of these nodes and their construction procedure are explained in the following:

- **Send node**: A send node consists of a sequence of statements that ends with a msgsend statement.
- **Receive node**: A receive node consists of a sequence of statements that begins with a msgrecv statement.
- **Statement node**: A statement node consists of a sequence of statements without any message passing statement.

Each node of the concurrency graph is called a concurrent component. A concurrency graph captures the dependencies among different components arising from message passing communications among them. However, components may also interact through other forms of communication such as shared variables. Access to shared variables may be either unsynchronized or synchronized using semaphores. Furthermore, to compute a slice, in addition to representing concurrency and interprocess communication aspects, one needs to represent all traditional (sequential) program instructions. To achieve this, Goswami and Mall [111] extended the concurrency graph to construct a third-level graph called a *concurrent program dependence graph* (CPDG). Consider the example program given in Figure 14.11a. Its process graph,

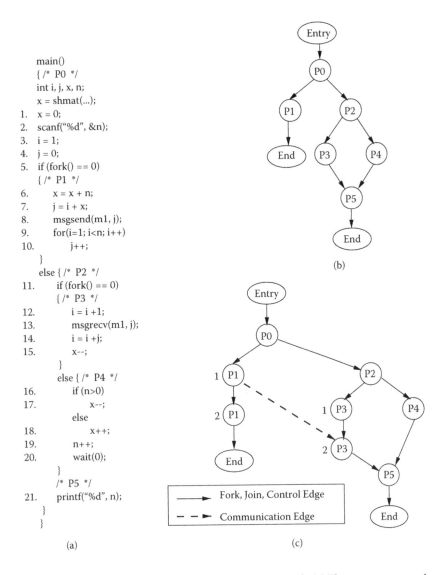

```
      main()
      { /* P0 */
      int i, j, x, n;
      x = shmat(...);
1.    x = 0;
2.    scanf("%d", &n);
3.    i = 1;
4.    j = 0;
5.    if (fork() == 0)
      { /* P1 */
6.        x = x + n;
7.        j = i + x;
8.        msgsend(m1, j);
9.        for(i=1; i<n; i++)
10.           j++;
      }
      else { /* P2 */
11.       if (fork() == 0)
          { /* P3 */
12.           i = i + 1;
13.           msgrecv(m1, j);
14.           i = i + j;
15.           x--;
          }
          else { /* P4 */
16.           if (n>0)
17.               x--;
              else
18.               x++;
19.           n++;
20.           wait(0);
          }
          /* P5 */
21.       printf("%d", n);
      }
      }
```

(a)

(b)

(c)

FIGURE 14.11 (a) An example program. (b) The process graph. (c) The concurrency graph.

concurrency graph, and CPDG are shown in Figure 14.11b, 14.11c, and Figure 14.12, respectively. Once the CPDG is constructed, slices can be computed through simple graph-reachability analysis.

Goswami et al. [111] implemented a static slicing tool that supports an option to view slices of programs at different levels, that is, process level, concurrent component level, or code level. They reported on the basis of implementation experience that their approach of hierarchical presentation of the slicing information helps users get a better understanding of the behaviors of concurrent programs.

Krinke [114] proposed a context-sensitive method to slice concurrent recursive programs. She used extended CFG and extended PDG to represent concurrent programs with interference. Her technique does not require serialization or inlining of called procedures.

14.5.7.2 Dynamic Slicing

Korel and Ferguson extended the dynamic slicing method of [4, 76] to distributed programs with Ada-type rendezvous communication [115]. For a distributed program, the execution history is formalized as

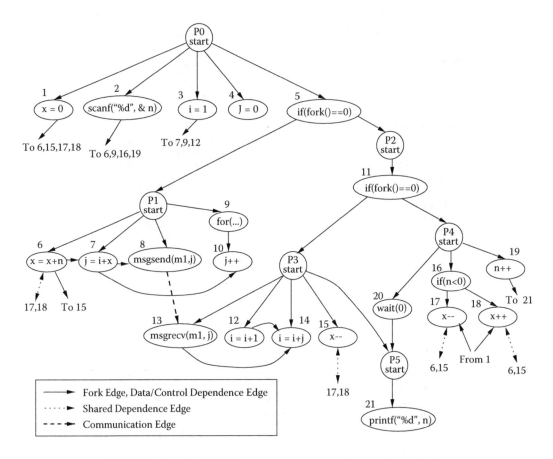

FIGURE 14.12   The CPDG for the example program shown in Figure 14.11a.

a *distributed program path* that, for each task, comprises of: (a) the sequences of statements (trajectory) executed by it and (b) a sequence of triples (A, C, B) identifying each rendezvous in which the task is involved.

A dynamic slicing criterion of a distributed program specifies: (a) the inputs to each task, (b) a distributed program path P, (c) a task W, (d) a statement occurrence q in the trajectory of w, and (e) a variable v. A dynamic slice with respect to such a criterion is an executable projection of the program that is obtained by deleting statements from it. However, the computed slice is only guaranteed to preserve the behavior of the program if the rendezvous in the slice occurs in the same relative order as in the program.

Duesterwald et al. presented a dependence graph–based algorithm for computing dynamic slices of distributed programs [116]. They introduced a DDG for representing distributed programs. A DDG contains a single vertex for each statement and predicate in a program. Control dependences between statements are determined statically, prior to execution. Edges for data and communication dependences are added to the graph at runtime. Slices are computed in the usual way by determining the set of DDG vertices for which the vertices specified in the criterion can be reached. Both the construction of the DDG and the computation of slices are performed in a distributed manner. Thus, a separate DDG construction process and slicing process are assigned to each process P_i in the program. The different processes communicate when a *send* or *receive* statement is encountered, but, because a single vertex is used for all occurrences of a statement in the execution history, inaccurate slices may be computed in the presence of loops.

Cheng presented an alternative graph-based algorithm for computing dynamic slices of distributed and concurrent programs [110]. Cheng's algorithm is basically a generalization of the initial approach proposed by Agrawal and Horgan in [75]; the PDN vertices corresponding to executed statements are marked, and the static slicing algorithm is applied to the PDN subgraph induced by the marked vertices. This, however, yields inaccurate slices in the presence of loops.

In [117] Goswami and Mall extended their static slicing framework [111] to compute dynamic slices of concurrent programs. They introduced the notion of the *dynamic program dependence graph (DPDG)* to represent various intra- and inter-process dependences of concurrent programs. They constructed the DPDG of a concurrent program through three hierarchical stages. At compile time, a *dynamic process graph* and a *static program dependence graph* (SPDG) are constructed. A dynamic process graph is the most abstract representation of a concurrent program. It captures the basic information about processes that is obtained through a static analysis of the program code. The SPDG of a concurrent program represents the static part of data and control dependences of the program. Trace files are generated at runtime to record the information regarding the relevant events that occur during the execution of concurrent programs. Using the information stored in the trace files, the dynamic process graph is refined to realize a *dynamic concurrency graph*. The SPDG, the information stored in the trace files, and the dynamic concurrency graph are then used to construct the DPDG. The DPDG of a concurrent program represents dynamic information regarding fork/join, semaphore/shared dependences and communication dependences due to message passing in addition to data and control dependences. After construction of the DPDG of a concurrent program, dynamic slices can be computed using some simple graph-reachability algorithm.

The dynamic slicing algorithm of Goswami and Mall [117] can handle both shared memory and message passing constructs. They have shown that their dynamic slicing algorithm computes more precise dynamic slices than the dynamic slicing algorithms of Duesterwald et al. and Soffa [116] and Cheng [110].

Rilling et al. [118, 119] proposed predicate-based dynamic slicing algorithms for message passing programs. They considered slicing criteria to focus on parts of the program that affect the predicates. Their dynamic predicate slices capture all statements that are relevant to some global requirements or suspected error properties of the distributed program.

14.6 Parallelization of Slicing

Parallel algorithms have the potential of being faster than their sequential counterparts since the computation work can be shared by many computing agents all executing at the same time. Also, for large programs, sequential algorithms become very slow. Slicing algorithms for concurrent programs are highly compute-intensive, as the graphs required for intermediate representations of the programs often become very large for practical problems. Therefore, parallelization of slicing algorithms seems to be an attractive option to improve efficiency. In the following, we review the research results in parallelization of slicing algorithms for sequential and concurrent programs.

14.6.1 Parallel Slicing of Sequential Programs

Harman et al. presented a *parallel slicing algorithm* to compute intraprocedural slices for sequential programs. In their method, a *process network* is constructed from the program to be sliced. A process network is a network of concurrent processes. It is represented as a directed graph in which nodes represent processes and edges represent communication channels among processes.

The process network is constructed using the CFG of the program. The *reverse control flow graph* (RCFG) is constructed by reversing the direction of every edge in the CFG. The topology of the process network is obtained from the RCFG, with one process for each of its nodes and with communication channels corresponding to its edges. The edges entering a node i represent input to process i, and the edges leaving node i represent outputs from process i.

To compute a slice for the slicing criterion $<n, V>$, where V is a set of variables of the program and n is a node of the CFG of the program, network communication is initiated by outputting the message V from the process n of the process network. Messages will then be generated and passed around the network until it eventually stabilizes, that is, when no new message arrives from any node. The algorithm computes the slice of a program by including the set of nodes whose identifiers are input to the *entry* node of the process network. The parallel slicing algorithm has been shown to be correct and finitely terminating [120]. Implementation details of the algorithm were not reported in [120].

14.6.2 Parallel Slicing of Concurrent Programs

Goswami et al. [121–123] extended the parallel static slicing algorithm of Harman et al. [120] for sequential programs to concurrent programs. They introduced the concept of the *concurrent control flow graph* (CCFG). The CCFG of a concurrent program consists of the CFGs of all the processes, with nodes and edges added to represent interprocess communications. Note that fork edges in a process graph represent flow of control among processes in a concurrent program. When a process forks, it creates a child process and executes concurrently with the child, so a fork edge in a process can be used to represent parallel flow of control. Process graphs and concurrency graphs are constructed as already discussed in the context of static slicing of concurrent programs. For every node x of the process graph, a CFG is constructed from the process represented by node x. The CCFG is then constructed by interconnecting the individual CFGs.

The algorithm of Goswami et al. [121, 122] first constructs the process network of a given concurrent program. The topology of the process network is given by the *reverse concurrent control flow graph* (RCCFG). The RCCFG is constructed from the CCFG by reversing the direction of all the edges. Every node of the RCCFG represents a process, and the edges represent communication channels among processes. Consider a slicing criterion $<P, s, V>$ of a concurrent program, where P is a process, s is a statement in the process P, and V is a set of variables of the program. To compute a slice with respect to this criterion, the process network is first initiated. Let n be the node in the CCFG corresponding to the statement s in the process P and m be the process in the process network corresponding to the CCFG node m. The process network is initiated by transmitting the message $\{m, V\}$ on all output channels of m. Each process in the process network repeatedly sends and receives messages until the network stabilizes. The network stabilizes when no messages are generated in the whole network. The set of all node identifiers that reach the *Entry* node gives the required static slice. Goswami et al. [121, 122] proved that their algorithm is correct and finite terminating. The steps for computation of slices are summarized below:

1. Construct the hierarchical CCFG for the concurrent program.
2. Reverse the CCFG.
3. Compile the RCCFG into a process network.
4. Initiate network communication by outputting the message $\{s, v\}$ from the process in the process network representing statement s in the process P, where $<P, s, v>$ is the slicing criterion.
5. Continue the process of message generation until no new messages are generated in the network.
6. Add to the slice all statements whose node identifiers have reached the *entry* node of the CCFG.

14.6.2.1 Implementation Results

Goswami et al. [121, 122] implemented their parallel algorithm for computing dynamic slices of concurrent programs in a Digital–Unix environment. They considered a subset of C language with Unix primitives for process creation and interprocess communications. Standard Unix tools *Lex* and *Yacc* have been used for lexical analysis and parsing of the source code.

A major aim of Goswami et al.'s implementation was to investigate the achieved speed-up in computing slices. They examined their algorithm with several input concurrent programs. The lengths of the input programs were between 30 to 100 lines.

They reported the following encouraging results of the implementation. The speed-up achieved for different programs in a two-processor environment is between 1.13 and 1.56; in a three-processor environment it is between 1.175 and 1.81; in a four-processor environment it is between 1.255 and 2.08. For the same number of processors used, speed-up varies for different program samples. It is shown that speed-up is more for larger programs. This may be because the number of nodes in the process network for larger programs is higher compared to smaller programs, leading to higher utilization of processors. Goswami et al.'s implemetation supported up to four-processor environment and considered small input programs (up to 100 lines).

14.7 Slicing of Object-Oriented Programs

Object-oriented programming languages have become very popular during the last decade. The concepts of classes, inheritance, polymorphism, and dynamic binding are the basic strengths of object-oriented programming languages. However, these concepts raise new challenges for program slicing. Intermediate representations for object-oriented programs need to model classes, objects, inheritance, scoping, persistence, polymorphism, and dynamic binding effectively. In the literature, research efforts to slice object-oriented programs are rarely reported. In the following, we briefly review the reported work on static and dynamic slicing of object-oriented programs.

14.7.1 Static Slicing of Object-Oriented Programs

Several researchers have extended the concepts of intermediate procedural program representation to intermediate object-oriented program representation. Kung et al. [124–126] presented a representation for object-oriented software. Their model consists of an *object relation diagram* and a *block branch diagram*. The object relation diagram of an object-oriented program provides static structural information on the relationships existing between objects. It models the relationship that exists between classes such as inheritance, aggregation, association, and so on. The block branch diagram of an object-oriented program contains the CFG of each of the class methods and presents a static implementation view of the program. Harrold and Rothermel [127] presented the concept of the *call graph*. A call graph provides a static view of the relationship between object classes. A call graph is an interprocedural program representation in which nodes represent individual methods and edges represent call sites. However, a call graph does not represent important object-oriented concepts such as inheritance, polymorphism, and dynamic binding.

Krishnaswamy [128] introduced the concept of the *object-oriented program dependence graph* (OPDG). The OPDG of an object-oriented program represents control flow, data dependences, and control dependences. The OPDG representation of an object-oriented program is constructed in three layers: *class hierarchy subgrach* (CHS), *control dependence subgraph* (CDS), and *data dependence subgraph* (DDS). The CHS represents the inheritance relationship between classes and the composition of methods into a class. A CHS contains a single *class header node* and a *method header node* for each method that is defined in the class. Inheritance relationships are represented by edges connecting class headers. Every method header is connected to the class header by a *membership edge*. Subclass representations do not repeat representations of methods that are already defined in the superclasses. *Inheritance edges* of a CHS connect the class header node of a derived class to the class header nodes of its superclasses. *Inherited membership edges* connect the class header node of the derived class to the method header nodes of the methods that it inherits. A CDS represents the static control dependence relationships that exist within and among the different methods of a class. The DDS represents the data dependence relationship among the statements and predicates of the program. The OPDG of an object-oriented program is the union of the three subgraphs: CHS, CDS, and DDS. Slices can be computed using an OPDG as a graph-reachability problem.

The OPDG of an object-oriented program is constructed as the classes are compiled, and hence it captures the complete class representations. The main advantage of OPDG representation over other representations is that the representation has to be generated only once during the entire life of the class. It does not need to be changed as long as the class definition remains unchanged. Figure 14.13b represents the CHS of the example program of Figure 14.13a.

Larsen and Harrold [129] extended the concept of the *system dependence graph* (SDG) to represent some of the features of object-oriented programs. They introduced the notions of the *class dependence graph*, *class call graph*, *class control flow graph*, and *interclass dependence graph*. A class dependence graph captures the control and data dependence among statements in a single class hierarchy. It connects individual PDGs

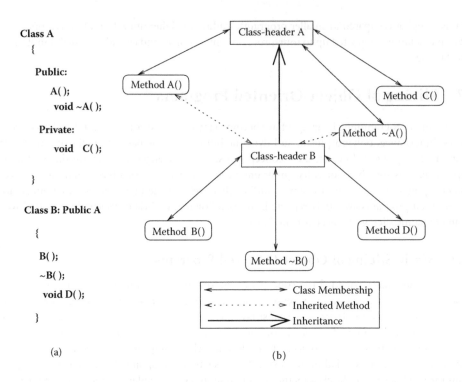

FIGURE 14.13   (a) An object-oriented program and (b) its CHS.

for methods that are members of the class. A class dependence graph uses a polymorphic choice node to represent the dynamic choice among the possible destinations. Such a node has all edges incident to subgraphs representing calls to each possible destination. A class call graph captures calling relationships among methods in a class hierarchy. Like the CHS, it contains class header nodes, method header nodes, virtual method header nodes, membership edges, inheritance edges, and inherited membership edges. A class call graph also includes edges that represent method calls. A class control flow graph captures the static control flow relationships that exist within and among methods of the class. It consists of a class call graph in which each method header node is replaced by the control flow graph for its associated method. An interclass dependence graph captures the control and data dependences for interacting classes that are not in the same hierarchy. In object-oriented programs, a composite class may instantiate its component class either through a declaration or by using an operation, such as *new*. Larsen and Harrold [129] constructed SDGs for these individual classes, groups of interacting classes, and finally for the complete object-oriented program. Slices are computed using a graph-reachability algorithm.

14.7.2   Dynamic Slicing of Object-Oriented Programs

Zhao [130] presented an algorithm for dynamic slicing of object-oriented programs. He adopted the following concepts for slicing of object-oriented programs:

- A slicing criterion for an object-oriented program is of the form (s, v, t, i), where s is a statement in the program, v is a variable used at s, and t is an execution trace of the program with input i.
- A dynamic slice of an object-oriented program on a given slicing criterion (s, v, t, i) consists of all statements in the program that affected the value of the variable v at the statement s.

Zhao [130] introduced the concept of the *dynamic object-oriented dependence graph* (DODG). Construction of DODG involves creating a new node for each occurrence of a statement in the execution history and creating all the dependence edges associated with the occurrence at runtime. Zhao's method of construction of the DODG of an object-oriented program is based on performing dynamic analysis of data and control flow of the program. This is similar to the methods of Agrawal and Horgan [75] and Agrawal et al. [83] for constructing DDGs of procedural programs. Computation of dynamic slices using the DODG is carried out as a graph-reachability problem. Other relevant dynamic slicing methods for object-oriented programs have been reported in [131–136].

Mohapatra et al. [134–136] proposed dynamic slicing algorithms for object-oriented programs. They used the basic concepts of the edge-marking and node-marking algorithms of Mund et al. [65, 89]. Their edge-marking dynamic slicing algorithm [135] and node-marking dynamic slicing algorithm [134] for object-oriented programs use an extended system dependence graph as the intermediate program representation. Mohapatra et al. [134–136] have shown that each of their algorithms is more space and time efficient than the existing dynamic slicing algorithms for object-oriented programs.

14.8 Slicing of Concurrent and Distributed Object-Oriented Programs

The nondeterministic nature of concurrent and distributed programs, unsynchronized interactions among objects, lack of global states, and dynamically varying numbers of objects make the understanding and debugging of concurrent and distributed object-oriented programs difficult. An increasing amount of resources are being spent in testing and maintaining these software products. Slicing techniques promise to come in handy at this point. However, research reports dealing with intermediate representations and slicing of concurrent and distributed object-oriented programs are scarce in the literature [57, 67, 131, 137–148].

14.8.1 Static Slicing

Zhao et al. [57] introduced an intermediate program representation called *system dependence net* (SDN) to capture various dependence relations in a concurrent object-oriented program. To represent interprocess communications between different methods in a class of a concurrent object-oriented program, they introduced a new type of dependence edge called an *external communication dependence edge*. An SDN captures the object-oriented features as well as the concurrency issues in a concurrent object-oriented program. Using SDN as the intermediate program representation, Zhao et al. [57] presented a static slicing algorithm for concurrent object-oriented programs.

Zhao [138, 139] introduced the concept of a *multithreaded dependence graph* (MDG) as an intermediate representation of a concurrent Java program. The MDG consists of a collection of *thread dependence graphs* (TDGs). A TDG represents a single thread of the program. Zhao's algorithm [139] uses an MDG as the intermediate program representation to compute static slices of the concurrent Java program.

Chen and Xu [140] proposed a static slicing algorithm for concurrent Java programs. Their algorithm uses a CPDG as the intermediate program representation and computes more precise slices than Zhao's algorithm [139].

14.8.2 Dynamic Slicing

Recently, Mohapatra et al. [136, 146, 147] presented algorithms for dynamic slicing of concurrent Java programs. They use a *concurrent system dependence graph* (CSDG) as the intermediate program representation. A CSDG contains control dependence, data dependence, synchronization dependence, and communication dependence edges. Their edge-marking algorithm is based on marking and unmarking the edges of the CSDG as and when the corresponding dependences arise and cease during an actual run of the program. The space complexity of their algorithm is $O(n^2)$, and the time complexity is $O(n^2 S)$, where n is the number of statements in the program and S is the length of the execution trace of the program.

Lallchandani and Mall [149] proposed a novel dynamic slicing technique for object-oriented concurrent programs. They introduced the notion of the *object-oriented concurrent program dependence graph* (OOCPDG). Their dynamic slicing algorithm [149] uses an OOCPDG as the intermediate program representation and is based on marking and unmarking the dependence edges as and when the dependences arise and cease at runtime. Their technique encompasses different aspects of an object-oriented programming paradigm, namely inheritance and polymorphism from the slicing arena. Their approch handles dynamically created object-based processes and process interactions through shared memory and message passing. They have also reported a dynamic slicing tool called the concurrent dynamic slicer for object-oriented concurrent programs, which implements their dynamic slicing technique.

Mohapatra et al. [145] were the first to present an algorithm for dynamic slicing of distributed object-oriented programs. They introduced the notion of the DPDG. Based on the DPDG, Mohapatra et al. [136, 145] presented an algorithm for dynamic slicing of distributed object-oriented programs. Their algorithm addresses the concurrency issues of the object-oriented program while computing the dynamic slices. The algorithm also handles the communication dependences arising because of objects shared among processes on the same machine and because of message passing among processes on different machines. The space complexity of their dynamic slicing algorithm is $O(n^2)$, and the time complexity is $O(n^2 S)$, where n is the total number of statements in the distributed object-oriented program and S is the total length of execution of the program. Mohapatra et al. [136, 148] also presented an efficient dynamic slicing algorithm for distributed Java programs.

14.9 Conclusions

We started with a discussion on the basic concepts and terminologies used in the area of program slicing. We also reviewed recent work in the area of program slicing, including slicing of sequential, concurrent, and object-oriented programs. Starting with the basic sequential program constructs, researchers have now started to address various issues of slicing distributed object-oriented programs. Also, slicing algorithms are being extended to architectural slicing and slicing of low-level programs such as hardware description languages. Since modern software products often require programs with millions of lines of code, development of parallel algorithms for slicing has assumed importance to reduce the slicing time.

References

1. M. Weiser. 1979. Program slices: Formal, psychological, and practical investigations of an automatic program abstraction method. PhD thesis, University of Michigan, Ann Arbor.
2. M. Weiser. 1982. Programmers use slices when debugging. *Communications of the ACM* 25(7): 446–52.
3. M. Weiser. 1984. Program slicing. *IEEE Transactions on Software Engineering* 10(4):352–57.
4. B. Korel and J. Laski. 1988. Dynamic program slicing. *Information Processing Letters* 29(3):155–63.
5. B. Korel and J. Rilling. 1998. Dynamic program slicing methods. *Information and Software Technology* 40:647–49.
6. J.-F. Bergeretti and B. Carré. 1985. Information-flow and data-flow analysis of while-programs. *ACM Transactions on Programming Languages and Systems* 7(1):37–61.
7. M. Weiser. 1983. Reconstructing sequential behavior from parallel behavior projections. *Information Processing Letters* 17(10):129–35.
8. J. Lyle. Evaluating variations on program slicing for debugging. PhD thesis, University of Maryland, College Park.
9. S. Rapps and E. J. Weyuker. 1985. Selecting software test data using data flow information. *IEEE Transactions on Software Engineering* SE-11(4):367–75.
10. J. R. Lyle and M. D. Weiser. 1987. Automatic program bug location by program slicing. *In Proceedings of the Second International Conference on Computers and Applications, Peking, China*, 877–82.

11. J. Ferrante, K. Ottenstein, and J. Warren. 1987. The program dependence graph and its use in optimization. *ACM Transactions on Programming Languages and Systems* 9(3):319–49.

12. J.-D. Choi. 1989. Parallel program debugging with flowback analysis. PhD thesis, University of Wisconsin–Madison.

13. S. Horwitz, J. Prins, and T. Reps. 1989. Integrating non-interfering versions of programs. *ACM Transactions on Programming Languages and Systems* 11(3):345–87.

14. A. Podgurski and L. A. Clarke. 1990. A formal model of program dependences and its implications for software testing, debugging, and maintenance. *IEEE Transactions on Software Engineering* 16(9): 965–79.

15. N. Shahmehri. 1991. Generalized algorithmic debugging. PhD thesis, Linkoping University, Sweden.

16. K. Gallagher. 1989. Using program slicing in software maintenance. PhD thesis, University of Maryland, Baltimore.

17. K. B. Gallagher. 1990. Surgeon's assistant limits side effects. *IEEE Software* 7:64.

18. K. B. Gallagher. 1991. Using program slicing to eliminate the need for regression testing. In *Eighth International Conference on Testing Computer Software*.

19. S. Horwitz. 1990. Identifying the semantic and textual differences between two versions of a program. *SIGPLAN Notices* 25(6):234–45.

20. K. B. Gallagher and J. Lyle. 1991. Using program slicing in software maintenance. *IEEE Transactions on Software Engineering* SE-17(8):751–61.

21. J.-D. Choi, B. Miller, and R. Netzer. 1991. Techniques for debugging parallel programs with flowback analysis. *ACM Transactions on Programming Languages and Systems* 13(4):491–530.

22. H. Agrawal. 1992. Towards automatic debugging of computer programs. PhD thesis, Purdue University, West Lafayette, IN.

23. T. Shimomura. 1992. The program slicing technique and its application to testing, debugging, and maintenance. *Journal of IPS of Japan* 9(9):1078–86.

24. J. Laski and W. Szermer. 1992. Identification of program modifications and its application in software maintenance. In *Proceedings of the IEEE Conference on Software Maintenance*, 282–90. Washington, DC: IEEE Computer Society Press.

25. L. M. Ott and J. J. Thuss. 1992. Effects of software changes in module cohesion. In *Proceedings of the IEEE Conference on Software Maintenance*, 345–53. Washington, DC: IEEE Computer Society Press.

26. D. Binkley. 1992. Using semantic differencing to reduce the cost of regression testing. In *Proceedings of the IEEE Conference on Software Maintenance*, 41–50. Washington, DC: IEEE Computer Society Press.

27. S. Bates and S. Horwitz. 1993. Incremental program testing using program dependence graphs. In *Conference Record of the Twentieth ACM Symposium on Principles of Programming Languages*, 384–96. New York: ACM Press.

28. F. Lanubile and G. Visaggio. 1993. Function recovery based on program slicing. In *Proceedings of the Conference on Software Maintenance*, 396–404. Los Alamitos, CA: IEEE Computer Society Press.

29. L. M. Ott and J. J. Thuss. 1993. Slice based metrics for estimating cohesion. In *Proceedings of the IEEE-CS International Metrics Symposium*, 71–81. Los Alamitos, CA: IEEE Computer Society Press.

30. H. Agrawal, R. A. DeMillo, and E. H. Spafford. 1993. Debugging with dynamic slicing and backtracking. *Software Practice and Experience* 23(6):589–616.

31. M. Kamkar. 1993. Interprocedural dynamic slicing with applications to debugging and testing. PhD thesis, Linkoping University, Sweden.

32. J. A. Beck. 1993. Interface slicing: A static program analysis tool for software engineering. PhD thesis, West Virginia University, Morgantown.

33. J. Beck and D. Eichmann. 1993. Program and interface slicing for reverse engineering. In *Proceedings of the IEEE/ACM Fifteenth International Conference on Software Engineering, ICSE '93*, 509–18.

34. K. B. Gallagher and J. R. Lyle. 1993. Program slicing and software safety. In *Proceedings of the 8th Annual Conference on Computer Assurance, COMPASS '93*, 71–80.

35. T. Ball. 1993. The use of control flow and control dependence in software tools. PhD thesis, Computer Science Department, University of Wisconsin–Madison.

36. J. M. Beiman and L. M. Ott. 1994. Measuring functional cohesion. *IEEE Transactions on Software Engineering* 20(8):644–57.

37. G. Rothermel and M. J. Harrold 1994. Selecting tests and identifying test coverage requirements for modified software. In *Proceedings of the ACM SIGSOFT International Symposium on Software Testing and Analysis*, 169–84. New York: ACM Press.

38. J. Q. Ning, A. Engberts, and W. Kozaczynski. 1994. Automated support for legacy code understanding. *Communications of the ACM* 37(5):50–57.

39. M. Harman and S. Daninic. 1995. Using program slicing to simplify testing. *Journal of Software Testing, Verification and Reliability* 5(3):143–62.

40. M. Harman, S. Daninic, and Y. Sivagurunathan. 1995. Program comprehension assisted by slicing and transformation. In *Proceedings of the First UK Workshop on Program Comprehension*, ed. M. Munro.

41. F. Tip. 1995. A survey of program slicing techniques. *Journal of Programming Languages* 3(3): 121–89.

42. D. Binkley, S. Horwitz, and T. Reps. 1995. Program integration for languages with procedure calls. *ACM Transactions on Software Engineering and Methodology* 4(1):3–31.

43. J. R. Lyle, D. R. Wallace, J. R. Graham, K. B. Gallagher, J. E. Poole, and D. W. Binkley. 1995. A case tool to evaluate funcational diversity in high integrity software. U.S. Department of Commerce, Technology Administration, National Institute of Standards and Technology, Computer Systems Laboratory, Gaithersburg, MD.

44. D. Binkley and K. B. Gallagher. 1996. Program slicing. *Advances in computers*, vol. 43, ed. M. Zelkowitz, 1–50. San Diego, CA: Academic Press.

45. A. De Lucia, A. R. Fasolino, and M. Munro. 1996. Understanding function behaviors through program slicing. In *Proceedings of the Fourth IEEE Workshop on Program Comprehension*, 9–18. Los Alamitos, CA: IEEE Computer Society Press.

46. A. Cimitile, A. De Lucia, and M. Munro. 1996. A specification driven slicing process for identifying reusable functions. *Software Maintenance: Research and Practice* 8:145–78.

47. R. Gupta, M. J. Harrold, and M. L. Soffa. 1996. Program slicing-based regression testing techniques. *Journal of Software Testing, Verification and Reliability* 6(2).

48. F. Lanubile and G. Visaggio. 1997. Extracting reusable functions by flow graph-based program slicing. *IEEE Transactions on Software Engineering* 23(4):246–59.

49. J. Cheng. 1997. Dependence analysis of parallel and distributed programs and its applications. In *Proceedings of the International Conference on Advances in Parallel and Distributed Computing*, 370–77, Los Alamitos, CA: IEEE Computer Society Press.

50. B. Korel and J. Rilling. 1997. Dynamic program slicing in understanding of program execution. In *Proceedings of the 5th International Workshop on Program Comprehension*, 80–90, Los Alamitos, CA: IEEE Computer Society Press.

51. J. Krinke and G. Snelting. 1998. Validation of measurement software as an application of slicing and constraint solving. *Information and Software Technology* 40:661–76.

52. M. Harman and K. B. Gallagher. 1998. Program slicing. *Information and Software Technology* 40: 577–81.

53. D. Binkley. 1998. The application of program slicing to regression testing. *Information and Software Technology* 40:583–94.

54. M. Kamkar. 1998. Application of program slicing in algorithmic debugging. *Information and Software Technology* 40:637–45.

55. L. M. Ott and J. M. Bieman. 1998. Program slices as an abstraction for cohesion measurement. *Information and Software Technology* 40:691–99.

56. T. Reps. 1998. Program analysis via graph reachability. *Information and Software Technology* 40: 701–26.

57. J. Zhao, J. Cheng, and K. Ushijima. 1998. A dependence-based representation for concurrent object-oriented software maintenance. In *Proceedings of the Second Euromicro Conference on Software Maintenance and Re-engineering*, 60–66. Los Alamitos, CA: IEEE Computer Society Press.

58. A. Lakhotia and J. C. Deprez. 1998. Restructuring programs by tucking statements into functions. *Information and Software Technology* 40:677–91.

59. R. Mall. 1999. *Fundamentals of software engineering*. New Delhi: Prentice Hall of India.

60. R. M. Hierons, M. Harman, and S. Daninic. 1999. Using program slicing to assist in the detection of equivalent mutants. *Journal of Software Testing, Verification and Reliability* 9(4):233–62.

61. J. Zhao. 2000. A slicing based approach to extracting reusable software architectures. In *Proceedings of the 4th European Conference on Software Maintenance and Reengineering*, 215–23.

62. R. Mall and A. Chakraborty. An approach to prioritize test cases for object-oriented programs. In *Proceedings of the International Workshop on Software Design and Architecture (SoDA 2002)*, Chennai, India, 21–32.

63. A. De Lucia. 2001. Program slicing: Methods and applications. In *IEEE Proceedings of Workshop on Source Code Analysis and Manipulation*, 142–49. Washington, DC: IEEE Computer Society Press.

64. M. Harman and R. M. Hierons. 2001. An overview of program slicing. *Software Focus* 2(3):85–92.

65. G. B. Mund. 2003. Efficient dynamic slicing of programs. PhD thesis, Department of Computer Science and Engineering, Indian Institute of Technology, Kharagpur.

66. B. Xu, J. Qian, X. Zhang, Z. Wu, and L. Chen. 2005. A brief survey of program slicing. *ACM SIGSOFT Software Engineering Notes* 30(2):1–36.

67. D. P. Mohapatra, R. Mall, and R. Kumar. 2006. An overview of slicing techniques for object-oriented programs. *Informatica* 30(2):253–77.

68. D. Jackson and E. J. Rollins. 1994. A new model of program dependences for reverse engineering. In *Proceedings of the Second ACM SIGSOFT Symposium on the Foundations of Software Engineering*, 2–10. New York: ACM Press.

69. T. Reps and G. Rosay. 1995. Precise interprocedural chopping. In *Proceedings of the Third ACM Symposium on the Foundations of Software Engineering*, 41–52. New York: ACM Press.

70. I. Forgacs and A. Bertolino. 1997. Feasible test path selection by principal slicing. In *Proceedings of the Sixth European Software Engineering Conference (ESEC/FSE97)*. LNCS 1301. Heidelberg, Germany: Springer-Verlag.

71. J. Larus and S. Chandra. 1993. Using tracing and dynamic slicing to tune compilers. Computer Sciences Technical Report 1174, University of Wisconsin–Madison.

72. A. V. Aho, R. Sethi, and J. D. Ullman. 1986. *Compilers: Principles, techniques and tools*. Reading, MA: Addison-Wesley.

73. S. Horwitz, T. Reps, and D. Binkley. 1990. Interprocedural slicing using dependence graphs. *ACM Transactions on Programming Languages and Systems* 12(1):26–61.

74. K. Ottenstein and L. Ottenstein. 1984. The program dependence graph in software development environment. *SIGPLAN Notices* 19(5):177–84.

75. H. Agrawal and J. Horgan. 1990. Dynamic program slicing. *SIGPLAN Notices, Analysis and Verification* 25(6):246–56.

76. B. Korel and J. Laski. 1990. Dynamic slicing of computer programs. *Journal of Systems and Software* 13:187–95.

77. R. Gopal. 1991. Dynamic program slicing based on dependence relations. In *Proceedings of the IEEE Conference on Software Maintenance*, 191–200. Washington, DC: IEEE Computer Society Press.

78. B. Korel and S. Yalamanchili. 1994. Forward computation of dynamic program slices. In *Proceedings of the International Symposium on Software Testing and Analysis (ISSTA)*, 66–79. New York: ACM Press.

79. B. Korel. 1997. Computation of dynamic program slices for unstructured programs. *IEEE Transactions on Software Engineering* 23(1):17–34.

80. D. Huynh and Y. Song. 1997. Forward computation of dynamic slices in the presence of structured jump statements. In *Proceedings of International Symposium on Applied Corporate Computing (ISACC)'97*, 73–81. Los Alamitos, CA: IEEE Computer Society Press.

81. A. Beszédes, T. Gergely, Z. M. Szabó, J. Csirik, and T. Gyimóthy. 2001. Dynamic slicing method for maintenance of large C programs. In *Proceedings of the 5th European Conference on Software Maintenance and Reengineering*, 105–113. Los Alamitos, CA: IEEE Computer Society Press.

82. D. M. Dhamdhere, K. Gururaja, and P. G. Ganu. 2003. A compact execution history for dynamic slicing. *Information Processing Letters* 85(3):145–52.

83. H. Agrawal, R. A. Demillo, and E. H. Spafford. 1991. Dynamic slicing in the presence of unconstrained pointers. In *Proceedings of the ACM Fourth Symposium on Testing, Analysis, and Verification (TAV4)*, 60–73. New York: ACM Press.

84. M. Kamkar, N. Shahmehri, and P. Fritzson. 1992. Interprocedural dynamic slicing. In *Proceeding of the Fourth International Conference on Programming Language Implementation and Logic Programming*, ed. M. Bruynooghe and M. Wirsing, 370–84. LNCS 631. Heidelberg, Germany: Springer-Verlag.

85. M. Kamkar, N. Shahmehri, and P. Fritzson. 1993. Three approaches to interprocedural dynamic slicing. *Microprocessing and Microprogramming* 38:625–36.

86. M. Kamkar, P. Fritzson, and N. Shahmehri. 1993. Interprocedural dynamic slicing applied to interprocedural data-flow testing. In *Proceedings of the Conference on Software Maintenance*, 386–95. Los Alamitos, CA: IEEE Computer Society Press.

87. G. B. Mund, R. Mall, and S. Sarkar. 2002. An efficient dynamic program slicing technique. *Information and Software Technology* 44(2):123–32.

88. D. Goswami and R. Mall. 2002. An efficient method for computing program slices. *Information Processing Letters* 81(2):111–17.

89. G. B. Mund, R. Mall, and S. Sarkar. 2003. Computation of intraprocedural dynamic program slices. *Information and Software Technology* 45(8):499–512.

90. X. Zhang, R. Gupta, and Y. Zhang. 2003. Precise dynamic slicing. In *Proceedings of the 25th International Conference on Software Engineering*, 319–29. Los Alamitos, CA: IEEE Computer Society Press.

91. G. B. Mund and R. Mall. 2006. An efficient interprocedural dynamic slicing technique. *Journal of Systems and Software* 79(6):791–806.

92. S. Horwitz, P. Pfiffer, and T. Reps. 1989. Dependence analysis for pointer variables. *SIGPLAN Notices* 24(7):28–40.

93. J. Jiang, X. Zhou, and D. Robson. 1991. Program slicing for C — The problems in implementation. In *Proceedings of the Conference on Software Maintenance*, 182–90. Los Alamitos, CA: IEEE Computer Society Press.

94. J. R. Lyle and D. Binkley. 1993. Program slicing in the presence of pointers. In *Proceedings of the Third Annual Software Engineering Research Forum*. Los Alamitos, CA: IEEE Computer Society Press.

95. J.-D. Choi, B. Miller, and P. Carini. 1993. Efficient flow-sensitive interprocedural computation of pointer induced aliases and side effects. In *Conference Records of the Twentieth ACM Symposium on Principles of Programming Languages*, 232–45. New York: ACM Press.

96. P. E. Livadas and A. Rosenstein. 1994. Slicing in the presence of pointer variables. Technical Report SERC-TR-74-F, Computer Science and Information Services Department, University of Florida, Gainesville.

97. D. Liang and M. J. Harrold. 1999. Reuse-driven interprocedural slicing in the presence of pointers and recursion. In *Proceedings of the International Conference on Software Maintenance*, 421–32. Los Alamitos, CA: IEEE Computer Society Press.

98. A. Orso, S. Sinha, and M. J. Harrold. 2000. Effects of pointers on data dependences and program slicing. Technical Report GIT-CC-00-33, Georgia Institute of Technology.

99. H. Agrawal. 1994. On slicing programs with jump statements. *SIGPLAN Notices* 29(6):302–12.

100. T. Ball and S. Horwitz. 1993. Slicing programs with arbitrary control flow. In *Proceedings of the First International Workshop on Automated and Algorithmic Debugging*, ed. P. Fritzson, 206–22. Lecture Notes in Computer Science, vol. 749. Heidelberg, Germany: Springer-Verlag.

101. J. D. Choi and J. Ferrante. 1994. Static slicing in the presence of goto statements. *ACM Transactions on Programming Languages and Systems* 16(4):1097–113.

102. M. Harman and S. Daninic. 1998. A new algorithm for slicing unstructured programs. *Journal of Software Maintenance: Research and Practice* 10(6):415–41.

103. S. Sinha, M. J. Harrold, and G. Rothermel. 1999. System dependence graph based slicing of programs with arbitrary interprocedural control flow. In *Proceedings of the 21st International Conference on Software Engineering*, 432–41. Los Alamitos, CA: IEEE Computer Society Press.

104. S. Kumar and S. Horwitz. 2002. Better slicing of programs with jump and switches. In *Proceedings of FASE 2002*. Los Alamitos, CA: IEEE Computer Society Press.

105. E. W. Dijkstra. 1968. Cooperating sequential processes. In *Programming languages*, ed. F. Genuys, 43–112. New York: Academic Press.

106. A. Burns. 1985. *Concurrent programming in Ada.* Cambridge, U.K.: Cambridge University Press.

107. B. W. Kernighan and R. Pike. 1984. *The UNIX programming environment.* Englewood Cliffs, NJ: Prentice Hall.

108. M. Rochkind. 1985. *Advanced UNIX programming.* Englewood Cliffs, NJ: Prentice Hall.

109. M. J. Bach. 1986. *The design of the Unix operating system.* New Delhi: Prentice Hall India.

110. J. Cheng. 1993. Slicing concurrent programs — A graph theoretical approach. In *Proceedings of the First International Workshop on Automated and Algorithmic Debugging*, ed. P. Fritzson, 223–40. Lecture Notes in Computer Science, vol. 749. Heidelberg, Germany: Springer-Verlag.

111. D. Goswami, R. Mall, and P. Chatterjee. 2000. Static slicing in Unix process environment. *Software Practice and Experience* 30(1):17–36.

112. J. Krinke. 1998. Static slicing of threaded programs. In *Program Analysis for Software Tools and Engineering (PASTE'98), ACMSOFT*, 35–42. New York: ACM Press.

113. M. G. Nanda and S. Ramesh. 2000. Slicing concurrent programs. In *Proceedings of the ACM International Symposium on Software Testing and Analysis, ISSTA-2000.* New York: ACM Press.

114. J. Krinke. 2003. Context sensitive slicing of concurrent programs. *ACM SIGSOFT Software Engineering Notes.* New York: ACM Press.

115. B. Korel and R. Ferguson. 1992. Dynamic slicing of distributed programs. *Applied Mathematics and Computer Science* 2(2):199–215.

116. E. Duesterwald, R. Gupta, and M. L. Soffa. 1992. Distributed slicing and partial re-execution for distributed programs. In *Proceedings of the Fifth Workshop on Languages and Compilers for Parallel Computing*, 329–37. LNCS 757. Heidelberg, Germany: Springer-Verlag.

117. D. Goswami and R. Mall. 2000. Dynamic slicing of concurrent programs. In *Proceedings of Seventh International Conference on High Performance Computing*, 15–26. LNCS 1970. Heidelberg, Germany: Springer-Verlag.

118. J. Rilling, H. F. Li, and D. Goswami. 2002. Predicate-based dynamic slicing of message passing programs. In *Proceedings of the Second IEEE International Workshop on Source Code Analysis and Manipulation (SCAM2002)*, 133–42. Washington, DC: IEEE Computer Society Press.

119. H. F. Li, J. Rilling, and D. Goswami. 2004. Granularity-driven dynamic predicate slicing for message passing systems. *Automated Software Engineering Journal* 11(1): 63–89.

120. M. Harman, S. Daninic, and Y. Sivagurunathan. 1996. A parallel algorithm for static program slicing. *Information Processing Letters* 56(6):307–13.

121. M. Haldar, D. Goswami, and R. Mall. 1998. Static slicing of shared memory parallel programs. In *Proceedings of the Seventh International Conference on Advanced Computing.* Los Alamitos, CA: IEEE Computer Society Press.

122. D. Goswami. 2001. Slicing of parallel and distributed programs. PhD thesis, Department of Computer Science and Engineering, Indian Institute of Technology, Kharagpur.

123. D. Goswami and R. Mall. 2004. A parallel algorithm for static slicing of concurrent programs. *Concurrency Practice and Experience* 16(8):751–69.

124. D. Kung, J. Gao, P. Hsia, Y. Toyoshima, and C. Chen. 1993. Design recovery for software testing of object-oriented programs. In *Working Conference on Reverse Engineering*, 202–11. Los Alamitos, CA: IEEE Computer Society Press.

125. D. Kung, J. Gao, P. Hsia, Y. Toyoshima, F. Wen, and C. Chen. 1994. Change impact identification in object-oriented software maintenance. In *International Conference on Software Maintenance*, 202–11. Los Alamitos, CA: IEEE Computer Society Press.

126. D. Kung, J. Gao, P. Hsia, Y. Toyoshima, F. Wen, and C. Chen. 1994. Firewall regression testing and software maintenance of object-oriented systems. *Journal of Object-Oriented Programing*.

127. M. J. Harrold and G. Rothermel. 1994. Performing data flow testing on classes. *Second ACM SIGSOFT Symposium on the Foundation of Software Engineering*, 154–63. New York: ACM Press.

128. A. Krishnaswamy. 1994. Program slicing: An application of object-oriented program dependency graphs. Technical Report TR-94-108, Department of Computer Science, Clemson University, Clemson, SC.

129. L. Larsen and M. J. Harrold. 1996. Slicing object-oriented software. In *Proceedings of the 18th International Conference on Software Engineering*, 495–505. Los Alamitos, CA: IEEE Computer Society Press.

130. J. Zhao. 1998. Dynamic slicing of object-oriented programs. Technical Report SE-98-119, Information Processing Society of Japan, 17–23.

131. Z. Chen and B. Xu. 2001. Slicing objected-oriented Java programs. *ACM SIGPLAN Notices* 36:33–40.

132. F. Ohata, K. Hirose, M. Fuji, and K. Inoue. 2001. A slicing method for object-oriented programs using dynamic light weight information. In *Proceedings of the Eighth Asia Pacific Software Engineering Conference*. Los Alamitos, CA: IEEE Computer Society Press.

133. T. Wang and A. Roychoudhury. 2004. Using compressed bytecode traces for slicing Java programs. In *Proceedings of the IEEE International Conference on Software Engineering*, 512–21. Washington, DC: IEEE Computer Society Press.

134. D. P. Mohapatra, R. Mall, and R. Kumar. 2004. A node marking dynamic slicing technique for object-oriented programs. In *Proceedings of the Workshop on Software Development and Architecture (SoDA)*, 1–15.

135. D. P. Mohapatra, R. Mall, and R. Kumar. 2004. An edge marking technique for dynamic slicing of object-oriented programs. In *Proceedings of the COMPSAC 2004*, 60–65.

136. D. P. Mohapatra. 2005. Dynamic slicing of object-oriented programs. PhD thesis, Department of Computer Science and Engineering, Indian Institute of Technology, Kharagpur.

137. J. Zhao, J. Cheng, and K. Ushijima. 1996. Static slicing of concurrent object-oriented programs. In *Proceedings of the 20th IEEE Annual International Computer Software and Application Conference*, 312–20. Washington, DC: IEEE Computer Society Press.

138. J. Zhao. 1999. Multithreaded dependence graphs for concurrent Java programs. In *Proceedings of the 1999 International Symposium on Software Engineering for Parallel and Distributed Systems (PDSE'99)*, 13–23. Los Alamitos, CA: IEEE Computer Society Press.

139. J. Zhao. 1999. Slicing concurrent Java programs. In *Proceedings of the IEEE International Workshop on Program Compression*, 126–33. Washington, DC: IEEE Computer Society Press.

140. Z. Chen and B. Xu. 2001. Slicing concurrent Java programs. *ACM SIGPLAN Notices* 36:41–47.

141. V. K. Garg and N. Mittal. 2001. On slicing distributed computation. In *Proceedings of the 21st IEEE International Conference on Distributed Computing Systems (ICDCS2001)*, 322–29. Washington, DC: IEEE Computer Society Press.

142. N. Mittal and V. K. Garg. 2001. Computation slicing: Techniques and theory. In *Proceedings of the 15th International Symposium on Distributed Computing (DISC2001)*, 78–92.

143. Z. Chen, B. Xu, and J. Zhao. 2002. An overview of methods for dependence analysis of concurrent programs. *ACM SIGPLAN Notices* 37(8):45–52. New York: Springer-Verlag.

144. J. Zhao and B. Li. 2004. Dependence based representation for concurrent Java programs and its application to slicing. In *Proceedings of ISFST*, 105–12.

145. D. P. Mohapatra, R. Mall, and R. Kumar. 2004. A novel approach for dynamic slicing of distributed object-oriented programs. In *Proceedings of the ICDCIT 2004*, 304–309.

146. D. P. Mohapatra, R. Mall, and R. Kumar. 2004. An efficient technique for dynamic slicing of concurrent Java programs. In *Proceedings of the AACC 2004*, 255–62.

147. D. P. Mohapatra, R. Mall, and R. Kumar. 2005. Computing dynamic slices of concurrent object-oriented programs. *Information and Software Technology* 47(12):805–17.

148. D. P. Mohapatra, R. Kumar, R. Mall, D. S. Kumar, and M. Bhasin. 2007. Distributed dynamic slicing of Java programs. *Journal of Systems and Software*. (2006).

149. J. T. Lallchandani and R. Mall. 2005. Computation of dynamic slices for object-oriented concurrent programs. In *Proceedings of the Asia-Pacific Software Engineering Conference (APSEC 2005)*, 341–50. Los Alamitos, CA: IEEE Computer Society Press.

15

Computations on Iteration Spaces

Sanjay Rajopadhye,
Lakshminarayanan
Renganarayana,
Gautam,
and
Michelle Mills Strout

Department of Computer Science,
Colorado State University,
Fort Collins, CO
Sanjay.Rajopadhye@colostate.edu
ln@cs.colostate.edu
ggupta@cs.colostate.edu
mstrout@cs.colostate.edu

15.1 Introduction

This chapter consists of two independent parts. The first deals with programs involving indexed data sets such as dense arrays and indexed computations such as loops. Our position is that high-level mathematical equations are the most natural way to express a large class of such computations, and furthermore, such equations are amenable to powerful static analyses that would enable a compiler to derive very efficient code, possibly significantly better than what a human would write. We illustrate this by describing a simple equational language and its semantic foundations and by illustrating the analyses we can perform, including one that allows the compiler to reduce the degree of the polynomial complexity of the algorithm embodied in the program.

The second part of this chapter deals with tiling, an important program reordering transformation applicable to imperative loop programs. It can be used for many different purposes. On sequential machines tiling can improve the locality of programs by exploiting reuse, so that the caches are used more effectively. On parallel machines it can also be used to improve the granularity of programs so that the communication and computation "units" are balanced.

We describe the tiling transformation, an optimization problem for selecting tile sizes, and how to generate tiled code for codes with regular or affine dependences between loop iterations. We also discuss approaches for reordering iterations, parallelizing loops, and tiling sparse computations that have irregular dependences.

15.2 The $\mathcal{Z}$-Polyhedral Model and Some Static Analyses

It has been widely accepted that the single most important attribute of a programming language is programmer productivity. Moreover, the shift to multi-core consumer systems, with the number of cores expected to double every year, necessitates the shift to parallel programs. This emphasizes the need for productivity even further, since parallel programming is substantially harder than writing unithreaded code. Even the field of high-end computing, typically focused exclusively on performance, is becoming concerned with the unacceptably high cost per megaflop of current high-end systems resulting from the required programming expertise. The current initiative is to increase programmability, portability, and robustness. DARPA's High Productivity Computing Systems (HPCSs) program aims to reevaluate and redesign the computing paradigms for high-performance applications from architectural models to programming abstractions.

We focus on compute- and data-intensive computations. Many data-parallel models and languages have been developed for the analysis and transformation of such computations. These models essentially abstract programs through (a) variables representing collections of values, (b) pointwise operations on the elements in the collections, and (c) collection-level operations. The parallelism may either be specified explicitly or derived automatically by the compiler. Parallelism detection involves analyzing the dependence between computations. Computations that are independent may be executed in parallel.

We present high-level mathematical equations to describe data-parallel computations succinctly and precisely. Equations describe the kernels of many applications. Moreover, most scientific and mathematical computations, for example, matrix multiplication, LU-decomposition, Cholesky factorization, Kalman filtering, as well as many algorithms arising in RNA secondary structure prediction, dynamic programming, and so on, are naturally expressed as equations.

It is also widely known that high-level programming languages increase programmer productivity and software life cycles. The cost of this convenience comes in the form of a performance penalty compared to lower-level implementations. With the subsequent improvement of compilation technology to accommodate these higher-level constructs, this performance gap narrows. For example, most programmers never use assembly language today. As a compilation challenge, the advantages of programmability offered by equations need to be supplemented by performance. After our presentation of an equational language, we will present automatic analyses and transformations to reduce the asymptotic complexity and to parallelize our specifications. Finally, we will present a brief description of the generation of imperative code from optimized equational specifications. The efficiency of the generated imperative code is comparable to hand-optimized implementations.

For an example of an equational specification and its automatic simplification, consider the following:

$$Y_i = \sum_{j=0}^{i} \sum_{k=0}^{i} A_{i,j+k} \times B_{k,j} \quad \text{for} \quad 0 \leq i \leq n \tag{15.1}$$

Here, the variable Y is defined over a line segment, and the variables A and B, over a triangle and a square, respectively. These are the previously mentioned collections of values and are also called the *domains* of the respective variables. The dependence in the given computation is such that the value of Y at i requires the value of A at $[i, j + k]$ and the value of B at $[k, j]$ for all valid values of j and k.

An imperative code segment that implements this equation is given in Figure 15.1. The triply nested loop (with linear bounds) indicates a $\Theta(n^3)$ asymptotic complexity for such an implementation. However, a $\Theta(n^2)$ implementation of Equation 15.1 exists and can be derived automatically. The code for this "simplified" specification is provided as well in Figure 15.1. The required sequence of transformations required to optimize the initial specification is given in Section 15.2.4. These transformations have been developed at the level of equations.

The equations presented so far have been of a very special form. It is primarily this special form that enables the development of sophisticated analyses and transformations. Analyses on general equations are often impossible. The class of equations that we consider consist of (a) variables defined on $\mathcal{Z}$-polyhedral domains with (b) dependences in the form of affine functions. These restrictions enable us to use linear algebraic theory and techniques. In Section 15.2.1, we present $\mathcal{Z}$-polyhedra and associated mathematical objects in detail that abstract the iteration domains of loop nests. Then we show the advantages of manipulating $\mathcal{Z}$-polyhedra over integer polyhedra. A language to express equations over $\mathcal{Z}$-polyhedral domains is presented in Section 15.2.3. The latter half of this section presents transformations to automatically simplify and parallelize equations. Finally, we provide a brief explanation of the transformations in the backend and code generation.

```
for i = 0 to n {
    Y [i] = 0;
    for j = 0 to i
        for k = 0 to i
            Y[i] += A[i, j + k] * B[k, j];
}
```

```
// Evaluation of temporary variable X
for i = 1 to n {
    l = i - 1;
    X[i, l] = 0;
    for k = 0 to l
        X[i, l] += B[k, l-k];
    for l = 0 to i -2
        X[i, l] = X[i-1, l];
}
// Evaluation of temporary variable W
for i = 0 to n {
    for l = 2i-1 to 2i { // Constant trip count
        W[i, l] = 0;
        for k = l-i to i
            W[l, l] += B[k, l-k];
    }
    for l = i to 2i-2
        W[i, l] = W[i-1, l] + B [l-i, i] + B[i, l-i];
}
// Evaluation of temporary variable Z
for i = 0 to n {
    for l = 0 to i-1
        Z [i, l] = X [i, l];
    for l = i to 2i
        Z [i, l] =W [i, l];
}
// Evaluation of Y
for i = 0 to n {
    Y[i] = 0;
    for l = 0 to 2i
        Y[i] += A[i, l]*Z [i, l];
}
```

FIGURE 15.1   A $\Theta(n^3)$ loop nest for Equation 15.1 and an equivalent $\Theta(n^2)$ loop nest.

15.2.1   Mathematical Background[1]

First, we review some mathematical background on matrices and decribe terminology. As a convention, we denote matrices with upper-case letters and vectors with lower-case. All our matrices and vectors have integer elements. We denote the identity matrix by I. Syntactically, the different elements of a vector v will be written as a list.

We use the following concepts and properties of matrices:

- The kernel of a matrix T, written as ker(T), is the set of all vectors z such that $Tz = 0$.
- The column (respectively row) *rank* of a matrix T is the maximal number of linearly independent columns (respectively rows) of T.
- A matrix is *unimodular* if it is square and its determinant is either 1 or -1.

---

[1]Parts of this section are adapted from [37], © 2007, Association for Computing Machinery, Inc., included by permission.

- Two matrices L and L' are said to be *column equivalent* or *right equivalent* if there exists a unimodular matrix U such that $L = L'U$.
- A unique representative element in each set of matrices that are column equivalent is the one in *Hermite normal form* (HNF).

Definition 15.1

An $n \times m$ matrix H with column rank d is in HNF if:

1. *For columns $2, \ldots, d$, the first nonzero element is positive and is below the first positive element for the previous column.*

$$\exists i_1, \ldots, i_d, 1 \leq i_1 < \ldots < i_d \leq n : H_{i_j, j} > 0$$

2. *In the first d columns, all elements above the first positive element are zero.*

$$\forall 1 \leq j \leq d, 1 \leq i < i_j : H_{i,j} = 0$$

3. *The first positive entry in columns $1, \ldots, d$ is the maximal entry on its row. All elements are nonnegative in this row.*

$$\forall 1 \leq l < j \leq d : 0 \leq H_{i_j, l} < H_{i_j, j}$$

4. *Columns $d + 1, \ldots m$ are zero columns.*

$$\forall d + 1 \leq j \leq m, 1 \leq i \leq n : H_{i,j} = 0$$

$$\begin{pmatrix} \blacktriangledown & 0 & 0 & 0 & 0 \\ \star & 0 & 0 & 0 & 0 \\ \blacktriangle & \blacktriangledown & 0 & 0 & 0 \\ \star & \star & 0 & 0 & 0 \\ \blacktriangle & \blacktriangle & \blacktriangledown & 0 & 0 \\ \star & \star & \star & 0 & 0 \\ \star & \star & \star & 0 & 0 \\ \blacktriangle & \blacktriangle & \blacktriangle & \blacktriangledown & 0 \\ \star & \star & \star & \star & 0 \\ \star & \star & \star & \star & 0 \end{pmatrix}$$

A template of a matrix in HNF is provided above. In the template, $\blacktriangledown$ denotes the maximum element in the corresponding row, $\blacktriangle$ denotes elements that are not the maximum element, and $\star$ denotes any integer. Both $\blacktriangledown$ and $\blacktriangle$ are nonnegative elements.

For every matrix A, there exists a unique matrix H that is in HNF and column equivalent to A, that is, there exists a unimodular matrix U such that $A = HU$. Note that the provided definition of the HNF does not require the matrix A to have full row rank.

15.2.1.1 Integer Polyhedra

An *integer polyhedron*, $\mathcal{P}$, is a subset of $\mathbb{Z}^n$ that can be defined by a finite number of affine inequalities (also called affine constraints or just constraints when there is no ambiguity) with integer coefficients. We follow the convention that the affine constraint c_i is given as $(a_i^T z + \alpha_i \geq 0)$, where $z, a_i \in \mathbb{Z}^n, \alpha_i \in \mathbb{Z}$. The integer polyhedron, $\mathcal{P}$, satisfying the set of constraints $C = \{c_1, \ldots, c_b\}$, is often written as $\{z \in \mathbb{Z}^n | Qz + q \geq 0\}$, where $Q = (a_1 \ldots a_b)^T$ is an $b \times n$ matrix and $q = (\alpha_1 \ldots \alpha_b)^T$ is a b-vector.

Example 15.1

Consider the equation

$$Y_i = \sum_{j=0}^{i} \sum_{k=0}^{i} A_{i,j+k} \times B_{k,j}, 0 \leq i \leq 10$$

The domains of the variables Y, A, and B are, respectively, the sets $\{i | 0 \leq i \leq 10\}$, $\{i, l | 0 \leq i \leq 10, 0 \leq l \leq 2i\}$, and $\{k, j | 0 \leq k \leq 10, 0 \leq j \leq 10\}$. These sets are polyhedra, and deriving the aforementioned representation simply requires us to obtain, through elementary algebra, all affine constraints of the correct form, yielding $\{i | i \geq 0, -i + 10 \geq 0\}$, $\{i, l | i \geq 0, -i + 10 \geq 0, l \geq 0, 2i - l \geq 0\}$, and $\{k, j | k \geq 0, -k + 10 \geq 0, j \geq 0, -j + 10 \geq 0\}$, respectively. Nevertheless, these are less intuitive, and in our presentation, we will not conform to the formalities of representation.

A subtle point to note here is that elements of polyhedral sets are tuples of integers. The index variables i, j, k, and l are simply place holders and can be substituted by other unused names. The domain of B can also be specified by the set $\{i, j | 0 \leq i \leq 10, 0 \leq j \leq 2i\}$.

We shall use the following properties and notation of integer polyhedra and affine constraints:

- For any two coefficients β and β', where $\beta, \beta' \geq 0$ and $\beta + \beta' = 1$, $\beta z + \beta' z'$ is said to be a *convex combination* of z and z'. If z and z' are two iteration points in an integer polyhedron, $\mathcal{P}$, then any convex combination of z and z' that has all integer elements is also in $\mathcal{P}$.
- The constraint $c \equiv (a^T z + \alpha \geq 0)$ of $\mathcal{P}$ is said to be *saturated iff* $(a^T z + \alpha = 0) \cap \mathcal{P} = \mathcal{P}$.
- The *lineality space* of $\mathcal{P}$ is defined as the linear part of the largest affine subspace contained in $\mathcal{P}$. It is given by $\ker(Q)$.
- The *context* of $\mathcal{P}$ is defined as the linear part of the smallest affine subspace that contains $\mathcal{P}$. If the saturated constraints in $\mathcal{C}$ are the rows of $\{Q_0 z + q_0 \geq 0\}$, then the context of $\mathcal{P}$ is $\ker(Q_0)$.

15.2.1.1.1 Parameterized Integer Polyhedra

Recall Equation 15.1. The domain of Y is given by the set $\{i | 0 \leq i \leq n\}$. Intuitively, the variable n is seen as a size parameter that indicates the problem instance under consideration. If we associate every iteration point in the domain of Y with the appropriate problem instance, the domain of Y would be described by the set $\{n, i | 0 \leq i \leq n\}$. Thus, a parameterized integer polyhedron is an integer polyhedron where some indices are interpreted as size parameters.

An equivalence relation is defined on the set of iteration points in a parameterized polyhedron such that two iteration points are equivalent if they have identical values of size parameters. By this relation, a parameterized polyhedron is partitioned into a set of equivalence classes, each of which is identified by the vector of size parameters. Equivalence classes correspond to program instances and are, thus, called instances of the parameterized polyhedron. We identify size parameters by omitting them from the index list in the set notation of a domain.

15.2.1.2 Affine Images of Integer Polyhedra

An (standard) *affine function*, f, is a function from iteration points to iteration points. It is of the form $(z \rightarrow Tz + t)$, where T is an $n \times m$ matrix and t is an n-vector.

Consider the integer polyhedron $\mathcal{P} = \{z \in \mathbb{Z}^m | Qz + q \geq 0\}$ and the standard affine function given above. The image of $\mathcal{P}$ under f is of the form $\{Tz + t | Qz + q \geq 0, z \in \mathbb{Z}^m\}$. These are the so-called *linearly bound lattices* (LBLs). The family of LBLs is a strict superset of the family of integer polyhedra. Clearly, every integer polyhedra is an LBL with $T = I$ and $t = 0$. However, for an example of an LBL that is not an integer polyhedron refer to Figure 15.2.

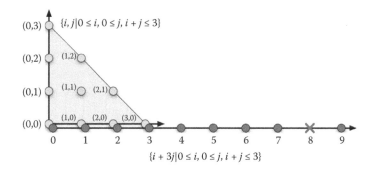

FIGURE 15.2 The LBL corresponding to the image of the polyhedron $\{i, j | 0 \leq i, 0 \leq j, i + j \leq 3\}$ by the affine function $(i, j \rightarrow i + 3j)$ does not contain the iteration point 8 but contains 7 and 9. Since 8 is a convex combination of 7 and 9, the set is not an integer polyhedron. Adapted from [37], © 2007, Association for Computing Machinery, Inc., included by permission.

```
for k = 1 to m {
   start = 0; // int to hold the starting point for j
   forall i = 0 to n {
      forall j = start to n step 2
         C[i, j, k] = 0.25 * (C[i-1, j, k-1] + C[i+1, j, k-1] +
                              C[i, j-1, k-1] + C[i, j+1, k-1] );
      start = 1 - start;
   }
   start = 1;
   forall i = 0 to n {
      forall j = start to n step 2
         C[i, j, k] = 0.25 * (C [i-1, j, k] + C[i+1, j, k] +
                             C [i, j-1, k] + C[i, j+1, k] );
      start = 1 - start;
   }
}
```

FIGURE 15.3   Imperative loop implementing the red–black SOR.

15.2.1.3   Affine Lattices

Often, the domain over which an equation is specified, or the iteration space of a loop program, does not contain every integer point that satisfies a set of affine constraints.

Example 15.2

Consider the red–black SOR for the iterative computation of partial differential equations. Iterations in the (i, j) plane are divided into "red" points and "black" points, similar to the layout of squares in a chess board. First, black points (at even $i + j$) are computed using the four neighboring red points (at odd $i + j$), and then the red points are computed using its four neighboring black points. These two phases are repeated until convergence. Introducing an additional dimension, k to denote the iterative application of the two phases, we get the following equation:

$$C_{i,j,k} = \begin{cases} i + j \text{ even} : \frac{1}{4}(C_{i-1,j,k-1} + C_{i+1,j,k-1} + C_{i,j-1,k-1} + C_{i,j+1,k-1}) & \text{// black} \\ i + j \text{ odd} : \frac{1}{4}(C_{i-1,j,k} + C_{i+1,j,k} + C_{i,j-1,k} + C_{i,j+1,k}) & \text{// red} \end{cases} \quad (15.2)$$

where the domain of C is $\{i, j, k | 0 \le i \le n, 0 \le j \le n, 0 \le k \le m\}$, n and m are size parameters, and $C[i, j, 0]$ is given as input. The imperative loop nest that implements this equation is given in Figure 15.3.

We see that the first (respectively second) branch of the equation is not defined over all iteration points that satisfy a set of affine constraints, namely, $\{0 \le i \le n, 0 \le j \le n, 0 \le k \le m\}$, but over points that additionally satisfy $(i + j) \bmod 2 = 0$ (respectively $[i + j] \bmod 2 = 1$). This additional constraint in the first branch of the equation is satisfied precisely by the iteration points that can be expressed as an integer linear combination of the vectors $\{\binom{1}{1}, \binom{0}{2}\}$. The vectors $\binom{1}{1}$ and $\binom{0}{2}$ are the generators of the lattice on which these iteration points lie.

The additional constraint in the second branch of the equation is satisfied precisely by iteration points that can be expressed as the following affine combination:

$$\binom{1}{1} i' + \binom{0}{2} j' + \binom{0}{1} = \begin{pmatrix} 1 & 0 \\ 1 & 2 \end{pmatrix} \binom{i'}{j'} + \binom{0}{1}$$

Formally, the lattice generated by a matrix L is the set of all integer linear combinations of the columns of L. If the columns of a matrix are linearly independent, they constitute a *basis* of the generated lattice. The lattices generated by two-dimensionally identical matrices are equal *iff* the matrices are column equivalent. In general, the lattices generated by two arbitrary matrices are equal *iff* the submatrices corresponding to the nonzero columns in their HNF are equal.

As seen in the previous example, we need a generalization of the lattices generated by a matrix, additionally allowing offsets by constant vectors. These are called *affine lattices*. An affine lattice is a subset of

$\mathbb{Z}^n$ of the form $\{Lz + l | z \in \mathbb{Z}^m\}$, where L and l are an $n \times m$ matrix and n-vector, respectively. We call z the coordinates of the affine lattice.

The affine lattices $\{Lz + l | z \in \mathbb{Z}^m\}$ and $\{L'z' + l' | z' \in \mathbb{Z}^{m'}\}$ are equal *iff* the lattices generated by L and L' are equal and $l' = Lz_0 + l$ for some constant vector $z_0 \in \mathbb{Z}^m$. The latter requirement basically enforces that the offset of one lattice lies on the other lattice.

15.2.1.4 $\mathcal{Z}$-Polyhedra

A $\mathcal{Z}$-polyhedron is the intersection of an integer polyhedron and an affine lattice. Recall the set of iteration points defined by either branch of the equation for the red–black SOR. As we saw above, these iteration points lie on an affine lattice in addition to satisfying a set of affine constraints. Thus, the set of these iteration points is precisely a $\mathcal{Z}$-polyhedron. When the affine lattice is the canonical lattice, $\mathbb{Z}^n$, a $\mathcal{Z}$-polyhedron is also an integer polyhedron. We adopt the following representation for $\mathcal{Z}$-polyhedra:

$$\{Lz + l | Qz + q \geq 0, z \in \mathbb{Z}^m\} \tag{15.3}$$

where L has full column rank, and the polyhedron $\mathcal{P}^c = \{z | Qz + q \geq 0, z \in \mathbb{Z}^m\}$ has a context that is the universe, $\mathbb{Z}^m$. $\mathcal{P}^c$ is called the coordinate polyhedron of the $\mathcal{Z}$-polyhedron. The $\mathcal{Z}$-polyhedron for which L has no columns has a coordinate polyhedron in $\mathbb{Z}^0$.

We see that every $\mathcal{Z}$-polyhedron is an LBL simply by observing that the representation for a $\mathcal{Z}$-polyhedron is in the form of an affine image of an integer polyhedron. However, the LBL in Figure 15.2 is clearly not a $\mathcal{Z}$-polyhedron. There does not exist any lattice with which we can intersect the integer polyhedron $\{i | 0 \leq i \leq 9\}$ to get the set of iteration points of the LBL. Thus, the family of LBLs is a strict superset of the family of $\mathcal{Z}$-polyhedra.

Our representation for $\mathcal{Z}$-polyhedra as affine images of integer polyhedra is specialized through the restriction to L and $\mathcal{P}^c$. We may interpret the $\mathcal{Z}$-polyhedral representation in Equation 15.3 as follows. It is said to be *based on* the affine lattice given by $\{Lz + l | z \in \mathbb{Z}^m\}$. Iteration points of the $\mathcal{Z}$-polyhedral domain are points of the affine lattice corresponding to valid coordinates. The set of valid coordinates is given by the coordinate polyhedron.

15.2.1.4.1 *Parameterized $\mathcal{Z}$-Polyhedra*

A parameterized $\mathcal{Z}$-polyhedron is a $\mathcal{Z}$-polyhedron where some rows of its corresponding affine lattice are interpreted as size parameters. An equivalence relation is defined on the set of iteration points in a parameterized $\mathcal{Z}$-polyhedron such that two iteration points are equivalent if they have identical value of size parameters. By this relation, a parameterized $\mathcal{Z}$-polyhedron is partitioned into a set of equivalence classes, each of which is identified by the vector of size parameters. Equivalence classes correspond to program instances and are, thus, called instances of the parameterized $\mathcal{Z}$-polyhedron.

For the sake of explanation, and without loss of generality, we may impose that the rows that denote size parameters are before all non-parameter rows. The equivalent $\mathcal{Z}$-polyhedron based on the HNF of such a lattice has the important property that all points of the coordinate polyhedron with identical values of the first few indices belong to the same instance of the parameterized $\mathcal{Z}$-polyhedron.

Example 15.3

Consider the $\mathcal{Z}$-polyhedron given by the intersection of the polyhedron $\{p, i | 0 \leq i \leq p\}$ and the lattice $\{j + k, j - k\}$.[2] It may be written as

$$\{j + k, j - k | 0 \leq j - k \leq j + k\} = \{j + k, j - k | 0 \leq k \leq j\}$$

Now, suppose the first index, p, in the polyhedron is the size parameter. As a result, the first row in the lattice $\{j + k, j - k\}$ corresponding to the $\mathcal{Z}$-polyhedron is the size parameter. The HNF of this lattice is

---

[2]For both the polyhedron and the affine lattice, the specification of the coordinate space $\mathbb{Z}^2$ is redundant. It can be derived from the number of indices and is therefore dropped for the sake of brevity.

$\{j', j' + 2k'\}$. The equivalent $\mathcal{Z}$-polyhedron is

$$\{j', j' + 2k'|k' \leq 0 \leq j' + 2k'\}$$

The iterations of this $\mathcal{Z}$-polyhedron belong to the same program instance *iff* they have the same coordinate index j'. Note that valid values of the parameter row trivially have a one-to-one correspondence with values of j', identity being the required bijection. In the general case, however, this is not the case. Nevertheless, the required property remains invariant. For example, consider the following $\mathcal{Z}$-polyhedron with the first two rows considered as size parameters:

$$\{m, m + 2n, i + m, j + n|0 \leq i \leq m; 0 \leq j \leq n\}$$

Here, valid values of the parameter rows have a one-to-one correspondence with the values of m and n, but it is impossible to obtain identity as the required bijection.

15.2.1.5 Affine Lattice Functions

Affine lattice functions are of the form $(Kz + k \rightarrow Rz + r)$, where K has full column rank. Such functions provide a mapping from the iteration $Kz + k$ to the iteration $Rz + r$. We have imposed that K have full column rank to guarantee that $(Kz + k \rightarrow Rz + r)$ be a function and not a relation, mapping any point in its domain to a unique point in its range. All standard affine functions are also affine lattice functions.

The mathematical objects introduced here are used to abstract the iteration domains and dependences between computations. In the next two sections, we will show the advantages of manipulating equations with $\mathcal{Z}$-polyhedral domains instead of polyhedral domains and present a language for the specification of such equations.

15.2.2 The $\mathcal{Z}$-Polyhedral Model

We will now develop the $\mathcal{Z}$-polyhedral model that enables the specification, analysis, and transformation of equations described over $\mathcal{Z}$-polyhedral domains. It has its origins in the polyhedral model that has been developed for over a quarter century. The polyhedral model has been used in a variety of contexts, namely, automatic parallelization of loop programs, locality, hardware generation, verification, and, more recently, automatic reduction of asymptotic computational complexity. However, the prime limitation of the polyhedral model lay in its requirement for dense iteration domains. This motivated the extension to $\mathcal{Z}$-polyhedral domains. As we have seen in the red–black SOR, $\mathcal{Z}$-polyhedral domains describe the iterations of a regular loop with non-unit stride.

In addition to allowing more general specifications, the $\mathcal{Z}$-polyhedral model enables more sophisticated analyses and transformations by providing greater information in the specifications, namely, pertaining to lattices. The example below demonstrates the advantages of manipulating $\mathcal{Z}$-polyhedral domains. The variable X is defined over the domain $\{i|1 \leq i \leq n\}$.[3]

```
for i = 1 to n
    if ((i%2==0) || (i%3==0)) X[i] = X[i-1];
```

In the loop, only iteration points that are a multiple of 2 or 3 execute the statement $X[i] = X[i - 1]$. The iteration at $i = 5$ may be excluded from the loop nest. Generalizing, any iteration that can be written in the form $6j + 5$ may be excluded from the loop nest. The same argument applies to iterations that can be written in the form $6j + 1$. As result of these "holes," all iterations at $6j + 2$ may be executed in parallel at the first time step. The iterations at $6j + 6$ may also be executed in parallel at the first time step. At the

---

[3]Code fragments in this section are adapted from [37], ©2007, Association for Computing Machinery, Inc., included by permission.

next time step, we may execute iterations at $6j+3$ and finally at iterations $6j+4$. The length of the longest dependence chain is 3. Thus, the loop nest can be parallelized to execute in constant time as follows:

```
forall i = 2 to n step 6
    X[i] = X[i-1];

forall i = 6 to n step 6
    X[i] = X[i-1];

forall i = 3 to n step 6
    X[i] = X[i-1];

forall i = 4 to n step 6
    X[i] = X[i-1];
```

However, our derivation of parallel code requires to manipulate $\mathcal{Z}$-polyhedral domains. A polyhedral approximation of the problem would be unable to result in such a parallelization.

Finally, the $\mathcal{Z}$-polyhedral model allows specifications with a more general dependence pattern than the specifications in the polyhedral model. Consider the following equation that cannot be expressed in the polyhedral model.

$$A[i] = \begin{cases} i \text{ even}: A[i/2] \\ i \text{ odd}: 0 \end{cases} \tag{15.4}$$

where $1 \leq i \leq n$ and the corresponding loop is

```
for i = 1 to N
    A[i] = (i%2==0 ? A[i/2] : 0);
```

This program exhibits a dependence pattern that is richer than the affine dependences of the polyhedral model. In other words, it is impossible to write an equivalent program in the polyhedral model, that is, without the use of the mod operator or non-unit stride loops, that can perform the required computation. One may consider replacing the variable A with two variables X and Y corresponding to the even and odd points of A such that $A[2i] = X[i]$ and $A[2i-1] = Y[i]$. However, the definition of X now requires the mod operator, because $X[2i] = X[i]$ and $X[2i-1] = Y[i]$.

Thus, the $\mathcal{Z}$-polyhedral model is a strict generalization of the polyhedral model and enables more powerful optimizations.

15.2.3 Equational Language

In our presentation of the red–black SOR in Section 15.2.1.3, we studied the domains of the two branches of Equation 15.2. More specifically, these are the branches of the *case* expression in the right-hand side (*rhs*) of the equation. In general, our techniques require the analysis and transformation of the subexpressions that constitute the *rhs* of equations, treating expressions as *first-class* objects.

For example, consider the simplification of Equation 15.1. As written, the simplification transforms the accumulation expression in the *rhs* of the equation. However, one would expect the technique to be able to decrease the asymptotic complexity of the following equation as well.

$$Y_i = 5 + \sum_{j=0}^{i} \sum_{k=0}^{i} A_{i,j+k} \times B_{k,j} \quad \text{for} \quad 0 \leq i \leq n \tag{15.5}$$

Generalizing, one would reasonably expect the existence of a technique to reduce the complexity of evaluation of the accumulation subexpression $(\sum_{j=0}^{i} \sum_{k=0}^{i} A_{i,j+k} \times B_{k,j})$, irrespective of its "level." This motivates a homogeneous treatment of the subexpression at any level. At the lowest level of specification,

TABLE 15.1   Expressions: Syntax and Domains

Expression	Syntax	Domain
Constants	Constant name or symbol	$\mathcal{D}_C$
Variables	V	D_V
Operators	op(Expr$_1$,...,Expr$_M$)	$\bigcap_{i=1}^{M} \mathcal{D}_{\text{Expr}_i}$
Case	caseExpr$_1$;...;Expr$_M$esac	$\biguplus_{i=1}^{M} \mathcal{D}_{\text{Expr}_i}$
Restriction	$\mathcal{D}'$: Expr	$\mathcal{D}' \cap \mathcal{D}_{\text{Expr}}$
Dependence	Expr. f	$f^{-1}(\mathcal{D}_{\text{Expr}})$
Reductions	reduce $(\oplus, f, Expr)$	$f(\mathcal{D}_{\text{Expr}})$

The domain of an expression A is denoted by $\mathcal{D}_A$. The operator, op, may be written in infix notation if it is binary. $\uplus$ denotes disjoint union and f^{-1} denotes relational inverse. Adapted from [38], ©2006, Association for Computing Machinery, Inc., included by permission.

a subexpression is a variable (or a constant) associated with a domain. Generalizing, we associate domains to arbitrary subexpressions.

The treatment of expressions as first-class objects leads to the design of a functional language where programs are a finite list of (mutually recursive) equations of the form Var = Expr, where both Var and Expr denote mappings from their respective domains to a set of values (similar to multidimensional arrays). A variable is defined by at most one equation. Expressions are constructed by the rules given in Table 15.1, column 2. The domains of all variables are declared, the domains of constants are either declared or defined over $\mathbb{Z}^0$ by default, and the domains of expressions are derived by the rules given in Table 15.1, column 3. The function specified in a dependence expression is called the *dependence function* (or simply a *dependence*), and the function specified in a reduction is called the *projection function* (or simply a *projection*).

In this language, Equation 15.5 would be a syntactically sugared version of the following concrete problem.

```
Y=5.(i->) + reduce(+,(i,j,k->i),A.(i,j,k->i,j+k)*B.(i,j,k->k,j))
```

In the equation above, 5 is a constant expression defined over $\{i|0 \leq i \leq n\}$ and Y, A, and B are variables. In addition to the equation, the domains of Y, A, and B would be declared as the sets $\{i|0 \leq i \leq n\}$, $\{i,l|0 \leq i \leq n, 0 \leq l \leq 2i\}$, and $\{k,j|0 \leq k \leq n, 0 \leq j \leq n\}$, respectively. The reduction expression is the accumulation in Equation 15.5. Summation is expressed by the reduction operator + (other possible reduction operators are *, max, min, or, and, etc.). The projection function (i,j,k->i) specifies that the accumulation is over the space spanned by j and k resulting in values in the one-dimensional space spanned by i. A subtle and important detail is that the expression that is accumulated is defined over a domain in three-dimensional space spanned by i, j, and k (this information is implicit in standard mathematical specifications as in Equation 15.5). This is an operator expression equal to the product of the value of A at $[i, j + k]$ and B at $[k, j]$. In the space spanned by i, j, and k, the required dependences on A and B are expressed through dependence expressions A.(i,j,k->i,j+k) and B.(i,j,k->k,j), respectively. The equation does not contain any case or restrict constructs. For an example of these two constructs, refer back to Equation 15.4. In our equational specification, the equation would be written as

$$A = \text{case}$$
$$\{2j|1 \leq 2j \leq n\}:\text{A}.(2j\text{->}j)$$
$$\{2j - 1|1 \leq 2j - 1 \leq n\}:0$$
$$\text{esac}$$

where the domains of A and the constant 0 are $\{i|1 \leq i \leq n\}$ and $\{2i - 1|1 \leq 2i - 1 \leq n\}$, respectively. There are two branches of the case expression, each of which is a restriction expression. We have not provided domains of any of the subexpressions mentioned above for the sake of brevity. These can be computed using the rules given in Table 15.1, column 3.

15.2.3.1 Semantics[4]

At this point, we intuitively understand the semantics of expressions. Here, we provide the formal semantics of expressions over their domains of definition. At the iteration point z in its domain, the value of:

- A constant expression is the associated constant.
- A variable is either provided as input or given by an equation; in the latter case, it is the value, at z, of the expression on its *rhs*.
- An operator expression is the result of applying op on the values, at z, of its expression arguments. op is an arbitrary, iteration-wise, single valued function.
- A case expression is the value at z of that alternative, to whose domain z belongs. Alternatives of a case expression are defined over disjoint domains to ensure that the case expression is not under- or overdefined.
- A restriction over E is the value of E at z.
- The dependence expression E. f is the value of E at $f(z)$. For the affine lattice function $(Lz + l \rightarrow Rz + r)$, the value of the (sub)expression E. $(Lz + l \rightarrow Rz + r)$ at $Lz + l$ equals the value of E at $Rz + r$.
- reduce($\oplus, f, E$) is the application of $\oplus$ on the values of E at all iteration points in $\mathcal{D}_E$ that map to z by f. Since $\oplus$ is an associative and commutative binary operator, we may choose any order of its application.

It is often convenient to have a variable defined either entirely as input or only by an equation. The former is called an *input variable* and the latter is a *computed variable*. So far, all our variables have been of these two kinds only. Computed variables are just names for valid expressions.

15.2.3.2 The Family of Domains

Variables (and expressions) are defined over $\mathcal{Z}$-polyhedral domains. Let us study the compositional constructs in Table 15.1 to get a more precise understanding of the family of $\mathcal{Z}$-polyhedral domains.

For compound expressions to be defined over the same family of domains as their subexpressions, the family should be closed under intersection (operator expressions, restrictions), union (case expression), and preimage (dependence expressions) and image (reduction expressions) by the family of functions. With closure, we mean that a (valid) domain operation on two elements of the family of domains should result in an element that also belongs to the family. For example, the family of integer polyhedra is closed under intersection but not under images, as demonstrated by the LBL in Figure 15.2. The family of integer polyhedra is closed under intersection since the intersection of two integer polyhedra that lie in the same dimensional space results in an integer polyhedron that satisfies the constraints of both the integer polyhedra.

In addition to intersection, union, and preimage and image by the family of functions, most analyses and transformations (e.g., simplification, code generation, etc.) require closure under the difference of two domains. With closure under the difference of domains, we may always transform any specification to have only input and computed variables.

The family of $\mathcal{Z}$-polyhedral domains should be closed under the domain operations mentioned above. This constraint is unsatisfied if the elements of this family are $\mathcal{Z}$-polyhedra. For example, the union of two $\mathcal{Z}$-polyhedra is not a $\mathcal{Z}$-polyhedron. Also, the LBL in Figure 15.2 shows that the image of a $\mathcal{Z}$-polyhedron is not a $\mathcal{Z}$-polyhedron. However, if the elements of the family of $\mathcal{Z}$-polyhedral domains are unions of $\mathcal{Z}$-polyhedra, then all the domain operations mentioned above maintain closure.

15.2.3.3 Parameterized Specifications

Extending the concept of parameterized $\mathcal{Z}$-polyhedra, it is possible to parameterize the domains of variables and expressions with size parameters. This leads to parameterized equational specifications. Instances of the parameterized $\mathcal{Z}$-polyhedra correspond to program instances.

---

[4]Parts of this section are adapted from [38], © 2006, Association for Computing Machinery, Inc., included by permission.

TABLE 15.2   Normalization Rules

#	Input	Output
1	case $E_1,\ldots,E_{k-1}$, case, $E'_1,\ldots,E'_m$ esac, $E_{k+1},\ldots,E_n$ esac	case $E_1,\ldots,E_{k-1},E'_1,\ldots,E'_m,$ $E_{k+1},\;\ldots\;,E_n$ esac
2	$\mathcal{D}'$:case $E_1,\ldots,E_n$ esac	case $\mathcal{D}'\;:\;E_1,\;\ldots\;,\;\mathcal{D}':E_n$ esac
3	$op(E_1,\ldots,E_{k-1},$ case $E'_1,\ldots,E'_m$ esac, $E_{k+1},\ldots,E_n)$	case $op(E_1,\ldots,E_{k-1},E'_1,E_{k+1},\ldots,E_n),\ldots,$ $op(E_1,\ldots,E_{k-1},E'_m,E_{k+1},\ldots,E_n)$ esac
4	(case $E_1,\ldots,E_n$ esac).f	case $E_1.f,\ldots,E_n.f$ esac
5	$\mathcal{D}_1:\mathcal{D}_2:E$	$(\mathcal{D}_1\cap\mathcal{D}_2):E$
6	$op(E_1,\ldots,E_{k-1},\mathcal{D}':E,E_{k+1},\ldots,E_n)$	$\mathcal{D}':op(E_1,\ldots,E_{k-1},E,E_{k+1},\ldots,E_n)$
7	$(\mathcal{D}':E).f$	$f^{-1}(\mathcal{D}'):E$
8	$(op(E_1,\ldots,E_n)).f$	$op(E_1.f,\ldots,E_n.f)$
9	$(E.f_1).f_2$	$E.(f_1\circ f_2)$

Every program instance in a parameterized specification is independent, so all functions should map consumer iterations to producer iterations within the same program instance.

15.2.3.4 Normalization

For most analyses and transformations (e.g., the simplification of reductions, scheduling, etc.), we need equations in a special normal form. Normalization is a transformation of an equational specification to obtain an equivalent specification containing only equations of the canonic forms $V = E$ or $V = \mathtt{reduce}(\oplus, f_p, E)$, where the expression E is of the form

$$\mathtt{case}\;\ldots,\mathcal{D}_{V,i}:op(\ldots,U.f,\ldots),\ldots\;\mathtt{esac} \tag{15.6}$$

and U is a variable or a constant.

Such a normalization transformation[5] first introduces an equation for every `reduce` expression, replacing its occurrence with the corresponding local variable. As a result, we get equations of the forms $V = E$ or $V = \mathtt{reduce}(\oplus, f_p, E)$, where the expression E does not contain any `reduce` subexpressions. Subsequently, these expressions are processed by a rewrite engine to obtain equivalent expressions of the form specified in Equation 15.6. The rules for the rewrite engine are given in Table 15.2. Rules 1 to 4 "bubble" a single `case` expression to the outermost level, rules 5 to 7 then "bubble" a single restrict subexpression to the second level, rule 8 gets the operator to the penultimate level, and rule 9 is a dependence composition to obtain a single dependence at the innermost level.

The validity of these rules, in the context of obtaining a valid specification of the language, relies on the closure properties of the family of unions of $\mathcal{Z}$-polyhedra.

15.2.4 Simplification of Reductions[6]

We now provide a deeper study of reductions. Reductions, commonly called accumulations, are the application of an associative and commutative operator to a collection of values to produce a collection of results.

Our use of equations was motivated by the simplification of asymptotic complexity of an equation involving reductions. We first present the required steps for the simplification. Then we will provide an intuitive explanation of the algorithm for simplification. For the sake of intuition, we use the standard mathematical notation for accumualations rather than the reduce expression.

---

[5]More sophisticated normalization rules may be applied, expressing the interaction of `reduce` expressions with other subexpressions. However, these are unnecessary in the scope of this chapter.

[6]Parts of this section are adapted from [38], © 2006, Association for Computing Machinery, Inc., included by permission.

Our initial specification was

$$Y_i = \sum_{j=0}^{i} \sum_{k=0}^{i} A_{i,j+k} \times B_{k,j} \quad \text{for} \quad 0 \leq i \leq n \tag{15.7}$$

The loop nest corresponding to this equation has a $\Theta(n^3)$ complexity. The cubic complexity for this equation can also be directly deduced from the equational specification. Parameterized by n, there are three independent[7] indices within the summation. The following steps are involved in the derivation of the $\Theta(n^2)$ equivalent specification.

1. Introduce the index variable $l = j + k$ and replace every occurrence of j with l. This is a change of basis of the three-dimensional space containing the domain of the expression that is reduced.

$$Y_i = \sum_{l=0}^{2i} \sum_{k=\max(0,l-i)}^{\min(i,l)} A_{i,l} \times B_{k,l-k} \quad \text{for} \quad 0 \leq i \leq n$$

The change in the order of summation is legal under our assumption that the reduction operator is associative and commutative.

2. Distribute multiplication over the summation since $A_{i,l}$ is independent of k, the index of the inner summation.

$$Y_i = \sum_{l=0}^{2i} \left(A_{i,l} \times \sum_{k=\max(0,l-i)}^{\min(i,l)} B_{k,l-k} \right) \quad \text{for} \quad 0 \leq i \leq n$$

3. Introduce variable $Z_{i,l}$ to hold the result of the inner summation $\sum_{k=\max(0,l-i)}^{\min(i,l)} B_{k,l-k}$.

$$Y_i = \sum_{l=0}^{2i} A_{i,l} \times Z_{i,l} \quad \text{for} \quad 0 \leq i \leq n$$

$$Z_{i,l} = \sum_{k=\max(0,l-i)}^{\min(i,l)} B_{k,l-k} \quad \text{for} \quad 0 \leq i \leq n, 0 \leq l \leq 2i$$

Note that the complexity of evaluating Y is now quadratic. However, we still have an equational specification that has cubic complexity (for the evaluation of Z).

4. Separate the summation over k to remove min and max operators in the equation for $Z_{i,l}$.

$$Z_{i,l} = \begin{cases} 0 \leq l < i : \sum_{k=0}^{l} B_{k,l-k} \\ i \leq l \leq 2i : \sum_{k=l-i}^{i} B_{k,l-k} \end{cases} \quad \text{for} \quad 0 \leq i \leq n$$

5. Introduce variables $X_{i,l}$ and $W_{i,l}$ for each branch of the equation defining $Z_{i,l}$.

$$X_{i,l} = \sum_{k=0}^{l} B_{k,l-k} \quad \text{for} \quad 1 \leq i \leq n, 0 \leq l < i$$

$$W_{i,l} = \sum_{k=l-i}^{i} B_{k,l-k} \quad \text{for} \quad 0 \leq i \leq n, i \leq l \leq 2i$$

Both the equations given above have cubic complexity.

---

[7]With *independent*, we mean that there are no equalities between indices.

6. Reuse. The complexity for the evaluation of X can be decreased by identifying that the expression on the *rhs* is independent of i. We may evaluate each result once (for instance, at a boundary) and then pipeline along i as follows.

$$X_{i,l} = \begin{cases} 1 \leq i \leq n, l = i - 1 : & \sum_{k=0}^{l} B_{k,l-k} \\ 1 \leq i \leq n, 0 \leq l < i - 1 : & X_{i-1,l} \end{cases}$$

The initialization takes quadratic time since there are a linear number of results to evaluate and each evaluation takes linear time. Then the pipelining of results over an area requires quadratic time. This decreases the overall complexity of evaluating X to quadratic time.

7. Scan detection. The simplification of $W_{i,l}$ occurs when we identify

$$W_{i,l} = B_{l-i,i} + B_{i,l-i} + \sum_{k=l-i+1}^{i-1} B_{k,l-k} \quad \text{for} \quad 0 \leq i \leq n, i \leq l \leq 2i - 2$$

$$= B_{l-i,i} + B_{i,l-i} + W_{i-1,l} \quad \text{for} \quad 0 \leq i \leq n, i \leq l \leq 2i - 2$$

The values are, once again, initialized in quadratic time at a boundary (here, $0 \leq i \leq n, 2i - 1 \leq l \leq 2i$). The scan takes constant time per iteration over an area and can be performed in quadratic time as well, thereby decreasing the complexity for the evaluation of W to quadratic time.

8. Summarizing, we have the following system of equations:

$$Y_i = \sum_{l=0}^{2i} A_{i,l} \times Z_{i,l}, 0 \leq i \leq n$$

$$Z_{i,l} = \begin{cases} 0 \leq i \leq n, 0 \leq l < i : X_{i,l} \\ 0 \leq i \leq n, i \leq l \leq 2i : W_{i,l} \end{cases}$$

$$X_{i,l} = \begin{cases} 1 \leq i \leq n, l = i - 1 : \sum_{k=0}^{l} B_{k,l-k} \\ 1 \leq i \leq n, 0 \leq l < i - 1 : X_{i-1,l} \end{cases}$$

$$W_{i,l} = \begin{cases} 0 \leq i \leq n, 2i - 1 \leq l \leq 2i : \sum_{k=l-i}^{i} B_{k,l-k} \\ 0 \leq i \leq n, i \leq l \leq 2i - 2 : B_{l-i,i} + B_{i,l-i} + W_{i-1,l} \end{cases}$$

These equations directly correspond to the optimized loop nest given in Figure 15.1. We have not optimized these equations or the loop nest any further for the sake for clarity, and moreso, because we only want to show an asymptotic decrease in complexity. However, a constant-fold improvement in the asymptotic complexity (as well as the memory requirement) can be obtained by eliminating the variable Z (or, alternatively, the two variables X and W).

We now provide an intuitive explanation of the algorithm for simplification. Consider the reduction

$$Y = \texttt{reduce}(\oplus, f_p, E)$$

where E is defined over the domain $\mathcal{D}_E$. The *accumulation space* of the above equation is characterized by $\ker(f_p)$. Any two points z and z' that contribute to the same element of the result, Y, satisfy $z - z' \in \ker(f_p)$. To aid intuition, we may also write this reduction as

$$Y[f_p(z)] = \oplus E[z] \tag{15.8}$$

$Y[f_p(z)]$ is the "accumulation," using the $\oplus$ operator, of the values of E at all points $z \in \mathcal{D}_E$ that have the same image $f_p(z)$. Now, if E has a distinct value at all points in its domain, they must all be computed, and no optimization is possible. However, consider the case where the expression E exhibits *reuse*: its value is the same at many points in $\mathcal{D}_E$. Reuse is characterized by $\ker(f_r)$, the kernel of a many-to-one affine

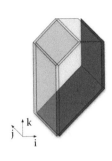

The reuse space is spanned by $\{(1, 0, 0)^T\}$ and the accumulation space is spanned by $\{(0, 0, 1)^T\}$. Translating D by $(1, 0, 0)^T$ exposes the left and top diagonal boundaries (light), and *exceeds* the right and bottom diagonal boundaries (dark). Moreover, left and right boundaries contain the accumulation space. Each boundary is thus treated differently

- The left boundary is the *initialization* domain.
- The top diagonal boundary is the *addition* domain.
- The bottom diagonal boundary is the *subtraction* domain.
- The right boundary is ignored (the projection along the accumulation space is outside the domain of the answer).
- The front, back, and top horizontal boundaries are also ignored since translation is along these boundaries.

FIGURE 15.4   Illustration of the core algorithm for $Y_{i,j} = \sum_{k=i+1}^{\min(i+n,3n/2)} X_{j,k}$ for $1 \leq i, j \leq n$. Adapted from [38], © 2006, Association for Computing Machinery, Inc., included by permission.

function, f_r; the value of E at any two points in $\mathcal{D}_E$ is the same if their difference belongs to $\ker(f_r)$. We can denote this reuse by $E[z] = X[f_r(z)]$, where X is a variable with domain $\mathcal{D}_X$. In our language, this would be expressed by the dependence expression $X.(z \to f_r[z])$. The canonical equation that we analyze is

$$Y[f_p(z)] = \oplus X[f_r(z)] \tag{15.9}$$

Its nominal complexity is the cardinality of the domain $\mathcal{D}_E$ of E. The main idea behind our method is based on analyzing (a) $\mathcal{D}_E$, the domain of the expression E inside the reduction, (b) its reuse space, and (c) the accumulation space.

15.2.4.1   Core Algorithm

Consider two adjacent instances of the answer variable, Y_z and Y_{z-r_Y} along $r_Y = f_p(r_E)$, where r_E is a vector in the reuse space of E. The set of values that contribute to Y_z and Y_{z-r_Y} overlap. This would enable us to express Y_z in terms of Y_{z-r_Y}. Of course, there would be residual accumulations on values outside the intersection that must be "added" or "subtracted" accordingly. We may repeat this for other values of Y along r_Y. The simplification results from replacing the original accumulation by a recurrence on Y_{z-r_Y} and residual accumulations. For example, in the simple scan, $Y_i = \sum_{j=1}^{i} X_j$, the expression inside the summation $F_{i,j} = X_j$ has reuse along $(1, 0)^T$. Taking $r_Y = (1)$, we get $Y_i = Y_{i-1} + F_{i,i} = Y_{i-1} + X_i$.

The geometric interpretation of the above reasoning is that we translate $\mathcal{D}_E$ by a vector in the reuse space of E. Let us call the translated domain $\mathcal{D}_{E'}$. The intersection of $\mathcal{D}_E$ and $\mathcal{D}_{E'}$ is precisely the domain of values, the accumulation over which can be avoided. Their differences account for the residual accumulations. In the simple scan explained above, the residual domain to be added is $\{i, j | 1 \leq i = j \leq n\}$, and the domain to be subtracted is empty. The residual accumulations to be added or subtracted are determined only by $\mathcal{D}_E$ and r_E.

This leads to Algorithm 15.1 (also see Figure 15.4).

Algorithm 15.1   Intuition of the Core Algorithm[8]

1. Choose r_E, a vector in $\ker(f_r)$, along which the reuse of E is to be exploited. In general, $\ker(f_r)$ is multidimensional and therefore there may be infinitely many choices.
2. Determine the domains D^0, D^-, and D^+ corresponding to the domain of initialization, and the residual domains to subtract and to add, respectively. The choice of r_E is made such that the cardinalities of these three domains are polynomials whose degree is strictly less than that for the original accumulation. This leads to simplification of the complexity.

---

[8]Adapted from [38], © 2006, Association for Computing Machinery, Inc., included by permission.

3. For these three domains, D^0, D^-, and D^+, define the three expressions, E^0, E^-, and E^+, consisting of the original expression E, restricted to the appropriate subdomain.
4. Replace the original equation by the following recurrence:

$$Y[f_p(z)] = \begin{cases} f_p(D^0) & : & \oplus E^0 \\ D_Y - f_p(D^0) & : & Y[z - r_E] \oplus \\ & & (\oplus E^+) \ominus (\oplus E^-) \end{cases}$$

5. Apply steps 1 to 4 recursively on the residual reductions over E^0, E^-, or E^+ if they exhibit further reuse.

Note that Algorithm 15.1 assumes that the reduction operation admits an inverse; that is, "subtraction" is defined. If this is not the case, we need to impose constraints on the direction of reuse to exploit: essentially, we require that the domain $\mathcal{D}^-$ is empty. This leads to a *feasible space* of exploitable reuse.

15.2.4.2 Multidimensional Reuse

When the reuse space as well as the accumulation space are multidimensional, there are some interesting interactions. Consider the equation $Y_i = \sum_{j=1}^{i-1} \sum_{k=1}^{i-j} X_j$ for $i \geq 2$. It has two-dimensional reuse (in the $\{i, k\}$ plane), and the accumulation is also two-dimensional (in the $\{j, k\}$ plane). Note that the two subspaces intersect, and this means that in the k direction, not only do all points have identical values, but they also all contribute to the same answer. From the bounds on the k summation we see that there are exactly $i - j$ such values, so the inner summation is just $(i - j) \times X_j$, because multiplication is a *higher-order operator* for repeated addition of identical values (similar situations arise with other operators, e.g., power for multiplication, identity for the idempotent operator max, etc.). We have thus optimized the $\Theta(n^3)$ computation to $\Theta(n^2)$. However, our original equation had two dimensions of reuse, and we may wonder whether further optimization is possible. In the new equation $Y_i = \sum_{j=1}^{i-1} (i - j) \times X_j$, the body is the product of two subexpressions, $(i - j)$ and X_j. They both have one dimension of reuse, in the $(1, 1)^T$ and $(1, 0)^T$ directions, respectively, but their product does not. No further optimization is possible for this equation.

However, if we had first exploited reuse along i, we would have obtained the simplified equation $Y_i = Y_{i-1} + \sum_{j=1}^{i-1} X_j$, initialized with $Y_2 = X_1$. The residual reduction here is itself a scan, and we may recurse the algorithm to obtain $Y_i = Y_{i-1} + Z_i$ and $Z_i = Z_{i-1} + X_{i-1}$ initialized with $Z_2 = X_1$. Thus, our equation can be computed in linear time. This shows how the choice of reuse vectors to exploit, and their order, affects the final simplification.

15.2.4.3 Decomposition of Accumulation

Consider the equation $Y_i = \bigoplus_{k=1}^{n-i} \bigoplus_{j=1}^{i} X_{j,k}$ for $1 \leq i \leq n - 1$. The one-dimensional reuse space is along $\{i\}$, and $\{j, k\}$ is the two-dimensional accumulation space. The set of points that contribute to the ith result lie in an $i \times (n - i)$ rectangle of the two-dimensional input array X. Comparing successive rectangles, we see that as the width decreases from one to the other, the height increases (Figure 15.5). If the operator $\oplus$ does not have an inverse, it seems that we may not be able to simplify this equation. This is not true: we can see that for each k we have an independent scan. The inner reduction $Z_{i,k} = \bigoplus_{j=1}^{i} X_{j,k}$ is just a family of scans, which can be done in quadratic time with $Z_{i,k} = Z_{i-1,k} + X_{i,k}$ initialized with $Z_{1,k} = X_{1,k}$. The outer reduction just accumulates columns of $Z_{i,k}$, which is also quadratic.

What we did in this example was to decompose the original reduction that was along the $\{j, k\}$ space into two reductions, the inner along the $\{j\}$ space yielding partial answers along the $\{i, k\}$ plane and the outer that combined these partial answers along the $\{k\}$ space. Although the default choice of the decomposition — the innermost accumulation direction — of the $\{k, j\}$ space worked for this example, in general this is not the case. It is possible that the optimal solution may require a nonobvious decomposition, for instance,

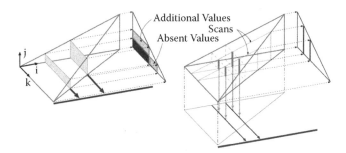

FIGURE 15.5 Geometric interpretation for $Y_i = \oplus_{k=1}^{n-i} \oplus_{j=1}^{i} X_{j,k}$. Adapted from [38], © 2006, Association for Computing Machinery, Inc., included by permission.

along some diagonal. We encourage the reader to simplify[9] the following equation:

$$Y_i = \bigoplus_{j=i}^{n+i} \bigoplus_{k=i}^{n+2i-j} X_{j,k}$$

15.2.4.4 Distributivity and Accumulation Decomposition

Returning to the simplification of Equation 15.7, we see that the methods presented so far do not apply, since the body of the summation, $A_{i,j+k} \times B_{k,j}$, has no reuse. The expression has a distinct value at each point in the three-dimensional space spanned by i, j, and k. However, the expression is composed of two subexpressions, which individually have reuse and are combined with the multiplication operator that distributes over the reduction operator, addition.

We may be able to distribute a subexpression outside a reduction if it has a constant value at all the points that map to the same answer. This was ensured by a change in basis of the three-dimensional space to i, l, and k, followed by a decomposition to summations over k and then l. Values of A were constant for different iterations of the accumulation over k. After distribution, the body of the inner summation exhibited reuse that was exploited for the simplification of complexity.

15.2.5 Scheduling[10]

Scheduling is assigning an execution time to each computation so that precedence constraints are satisfied. It is one of the most important and widely studied problems. We present the scheduling analysis for programs in the $\mathcal{Z}$-polyhedral model. The resultant schedule can be subsequently used to construct a space–time transformation leading to the generation of sequential or parallel code. The application of this schedule is made possible as a result of closure of the family of $\mathcal{Z}$-polyhedral domains under image by the constructed transformation. We showed the advantages of scheduling programs in the $\mathcal{Z}$-polyhedral model in Section 15.2.2. The general problem of scheduling programs with reductions is beyond the scope of this chapter. We will restrict our analysis to $\mathcal{Z}$-polyhedral programs without reductions.

The steps involved in the scheduling analysis are (a) deriving precedence (causality) constraints for programs written in the $\mathcal{Z}$-polyhedral model and (b) formulation of an *integer linear program* to obtain the schedule.

---

[9]Solution: The inner reduction would map all points for which $j + k = c$, for a given constant c, to the same partial answer.

[10]Adapted from [36] © 2007 IEEE, included by permission.

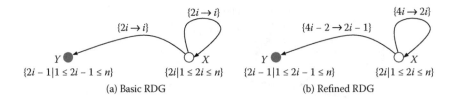

FIGURE 15.6 Basic and refined reduced dependence graphs for Example 15.4. Adapted from [36], © 2007 IEEE, included by permission.

The precedence constraints between variables are derived from the reduced dependence graph (RDG). We will now provide some refinements of the RDG.

15.2.5.1   Basic and Refined RDG

Equations in the $\mathcal{Z}$-polyhedral model can be defined over an infinite iteration domain. For any dependence analysis on an infinite graph, we need a compact representation. A directed multi-graph, the RDG precisely describes the dependences between iterations of variables. It is based on the normalized form of a specification and defined as follows:

- For every variable in the normalized specification, there is a vertex in the RDG labeled by the variable name and annotated by its domain. We will refer to vertices and variables interchangeably.
- For every dependence of the variable V on U, there is an edge from V to U, annotated by the corresponding dependence function. We will refer to edges and dependences interchangeably.

At a finer granularity, every branch of an equation in a normalized specification dictates the dependences between computations. A precise analysis requires that dependences be expressed separately for every branch. Again, for reasons of precision, we may express dependences of a variable separately for every $\mathcal{Z}$-polyhedron in the $\mathcal{Z}$-polyhedral domain of the corresponding branch of its equation. To enable these, we replace a variable by a set of new variables as elaborated below. Remember, our equations are of the form

$$\texttt{case}\ldots,\mathcal{D}_{V,i} : \texttt{op}(\ldots,\texttt{U}.f,\ldots),\ldots \ \ \texttt{esac} \tag{15.10}$$

Let $\mathcal{D}_{V,i}$ be written as a disjoint union of $\mathcal{Z}$-polyhedra given by $\uplus_j \mathcal{Z}_j$. The variable V in the domain $\mathcal{Z}_j$ is replaced by a new variable, for instance, X_j. Similarly, let U be replaced by new variables given as Y_k. The dependence of V in $\mathcal{D}_{V,i}$ on U is replaced by dependences from all X_j on all Y_k. An edge from X_j to Y_k may be omitted if there are no iterations in X_j that map to Y_k (mathematically, if the preimage of Y_k by the dependence function does not intersect with X_j). A naive construction following these rules results in the *basic reduced dependence graph*.

Figure 15.6a gives the basic RDG for Equation 15.4, which is repeated here for convenience.

$$A = \begin{cases} \{2i | 1 \leq 2i \leq n\} : & A.(2i \to i) \quad \text{(denoted by } X) \\ \{2i - 1 | 1 \leq 2i - 1 \leq n\} : 0 & \quad \text{(denoted by } Y) \end{cases}$$

The domains of A and the constant 0 are $\{i | 1 \leq i \leq n\}$ and $\{2i - 1 | 1 \leq 2i - 1 \leq n\}$, respectively. Next, we will study a refinement on this RDG.

In the RDG for the generic equation given in Equation 15.10, let X be a variable derived from V and defined on $\mathcal{Z}_X \in \mathcal{D}_{V,i}$, and let Y be a variable derived from U defined on $\mathcal{Z}_Y \in \mathcal{D}_U$, where $\mathcal{Z}_X$ and $\mathcal{Z}_Y$ are given as follows:

$$\mathcal{Z}_X = \{L_X z_X + l_X | z_X \in \mathcal{P}_X^c\}$$
$$\mathcal{Z}_Y = \{L_Y z_Y + l_Y | z_Y \in \mathcal{P}_Y^c\}$$

A dependence of the form $(Lz + l \rightarrow Rz + r)$ is directed from X to Y. X at $(Lz + l) \in \mathcal{Z}_X$ cannot be evaluated before Y at $(Rz + r) \in \mathcal{Z}_Y$. The affine lattice $\{Lz + l | z \in \mathbb{Z}^n\}$ may contain points that do not lie in the affine lattice $\{L_X z_X + l_X | z_X \in \mathbb{Z}^{n_X}\}$. Similarly, the affine lattice $\{Rz + r | z \in \mathbb{Z}^n\}$ may contain points that do not lie in the affine lattice $\{L_Y z_Y + l_Y | z_Y \in \mathbb{Z}^{n_Y}\}$. As a result, the dependence may be specified on a finer lattice than necessary and may safely be replaced by a dependence of the form $(L'z' + l' \rightarrow R'z' + r')$, where

$$
\begin{aligned}
L' &= L_X S, l' = L_X s + l_X \\
R' &= L_Y S', r' = L_Y s' + l_Y
\end{aligned}
\tag{15.11}
$$

where S and S' are matrices and s and s' are integer vectors. The refined RDG is a refinement of the basic RDG where every dependence has been replaced by a dependence satisfying Equation 15.11. Figure 15.6b gives the refined RDG for Equation 15.4.

15.2.5.2 Causality Constraints

Dependences between the different iterations of variables impose an ordering on their evaluation. A valid schedule of the evaluation of these iterations is the assignment of an execution time to each computation so that precedence (causality) constraints are satisfied.

Let X and Y be two variables in the refined RDG defined on $\{L_X z_X + l_X | z_X \in \mathcal{P}_X^c\}$ and $\{L_Y z_Y + l_Y | z_Y \in \mathcal{P}_Y^c\}$, respectively. We seek to find schedules on X and Y of the following form:

$$
\begin{aligned}
\lambda'_X(z_X) &= (L_X z_X + l_X \rightarrow \lambda_X(z_X)) \\
\lambda'_Y(z_Y) &= (L_Y z_Y + l_Y \rightarrow \lambda_Y(z_Y))
\end{aligned}
\tag{15.12}
$$

where λ_X and λ_Y are affine functions on z_X and z_Y, respectively. Our motivation for such schedules is that all vectors and matrices are composed of integer scalars. If we seek schedules of the form $\lambda'(z')$, where λ' is an affine function and z' is an iteration in the domain of a variable, then we may potentially assign execution times to "holes," or computations that do not exist.

We will now formulate causality constraints using the refined RDG. Consider dependences from X to Y. All such dependences can be written as

$$
(L_X(Sz + s) + l_X \rightarrow L_Y(S'z + s') + l_Y)
$$

where S and S' are matrices and s and s' are vectors. The execution time for Y at $L_Y(S'z + s') + l_Y$ should precede the execution time for X at $L_X(Sz + s) + l_X$. With the nature of the schedules presented in Equation 15.12, our causality constraint becomes

$$
\lambda_X(Sz + s) - \lambda_Y(S'z + s') \geq 1
\tag{15.13}
$$

with the assumption that op is atomic and takes a single time step to evaluate.

From these constraints, we may derive an integer linear program to obtain schedules of the form $\lambda'(z) = (Lz + l \rightarrow \lambda[z])$, where $\{Lz + l | z \in \mathbb{Z}^n\}$ is the lattice corresponding to the $\mathcal{Z}$-polyhedron and $\lambda(z)$ is the affine function (composed of integer scalars) on the coordinates of this lattice. An important feature of this formulation is that the resultant schedule can then be used to construct a space–time transformation.

15.2.6 Backend

After optimization of the equational specification and obtaining a schedule, the following steps are performed to generate (parallel) imperative code.

Analogous to the schedule that assigns a date to every operation, a second key aspect of the parallelization is to assign a processor to each operation. This is done by means of a *processor allocation function*. As with schedules, we confine ourselves to affine lattice functions. Since there are no causality constraints for choosing an allocation function, there is considerable freedom in choosing it. However, in the search for processor allocation functions, we need to ensure that two iteration points that are scheduled at the same time are not mapped to the same processing element.

The final key aspect in the static analysis of our equations is the allocation of operations to memory locations. As with the schedule and processor allocation function, the *memory allocation* is also an affine lattice function. The memory allocation function is, in general, a many-to-one mapping with most values overwritten as the computation proceeds. The validity condition for memory allocation functions is that no value is overwritten before all the computations that depend on it are themselves executed.

Once we have the three sets of functions, namely, schedule, processor allocation, and memory allocation, we are left with the problem of code generation. Given the above three functions, how do we produce parallel code that "implements" these choices? Code generation may produce either sequential or parallel code for programmable processors, or even descriptions of application-specific or nonprogrammable hardware (in appropriate hardware description language) that implements the computation specified by the equation.

Current techniques in code generation produce extremely efficient implementations comparable to hand-optimized imperative programs. With this knowledge, we return to our motivation for the use of equations to specify computations. An imperative loop nest that corresponds to an equation contains more information than required to specify the computation. There is an order (corresponding to the schedule) in the evaluation of values of a variable at different iteration points, namely, the lexicographic order of the loop indices. A loop nest also specifies the order of evaluation of the partial results of accumulations. A memory mapping has been chosen to associate values to memory locations. Finally, in the case of parallel code, a loop nest also specifies a processor allocation. Any analysis or transformation of loop nests that is equivalent to analysis or transformations on equations has to deconstruct these attributes and, thus, becomes unnecessarily complex.

15.2.7   Bibliographic Notes[11]

Our presentation of the equational language and the various analyses and transformations is based on the the ALPHA language [59, 69] and the MMALPHA framework for manipulating ALPHA programs, which relies on a library for manipulating polyhedra [107].

Although the presentation in this section has focused on equational specifications, the impact of the presented work is equally directed toward loop optimizations. In fact, many advances in the development of the polyhedral model were motivated by the loop parallelization and hardware synthesis communities.

To overcome the limitations of the polyhedral model in its requirement of dense iteration spaces, Teich and Thiele proposed LBLs [104]. $\mathcal{Z}$-polyhedra were originally proposed by Ancourt [6]. Le Verge [60] argued for the extension of the polyhedral model to $\mathcal{Z}$-polyhedral domains. Le Verge also developed normalization rules for programs with reductions [59].

The first work that proposed the extension to a language based on unions of $\mathcal{Z}$-polyhedra was by Quinton and Van Dongen [81]. However, they did not have a unique canonic representation. Also, they could not establish the equivalence between identical $\mathcal{Z}$-polyhedra nor provide the difference of two $\mathcal{Z}$-polyhedra or their image under affine functions. Closure of unions of $\mathcal{Z}$-polyhedra under all the required domain operations was proved in [37] as a result of a novel representation for $\mathcal{Z}$-polyhedra and the associated family of dependences. One of the consequences of their results on closure was the equivalence of the family of $\mathcal{Z}$-polyhedral domains and unions of LBLs.

Liu et al. [67] described how incrementalization can be used to optimize polyhedral loop computations involving reductions, possibly improving asymptotic complexity. However, they did not have a cost model and, therefore, could not claim optimality. They exploited reuse only along the indices of the accumulation loops and would not be able to simplify an equation like $Y_{i,j} = \sum_k X_{i-j,k}$. Other limitations were the requirement of an inverse operator. Also, they did not consider reduction decompositions and algebraic

---

[11] Parts of this section are adapted from [36] ©2007 IEEE and [37, 38], ©2007, 2006 Association for Computing Machinery, Inc., included by permission.

transformations and do not handle the case when there is reuse of values that contribute to the same answer. These limitations were resolved in [38], which presented a precise characterization of the complexity of equations in the polyhedral model and an algorithm for the automatic and optimal application of program simplifications.

The scheduling problem on recurrence equations with uniform (constant-sized) dependences was originally presented by Karp et al. [52]. A similar problem was posed by Lamport [56] for programs with uniform dependences. Shang and Fortes [97] and Lisper [66] presented optimal linear schedules for uniform dependence algorithms. Rao [87] first presented affine by variable schedules for uniform dependences (Darte et al. [21] showed that these results could have been interpreted from [52]). The first result of scheduling programs with affine dependences was solved by Rajopadhye et al. [83] and independently by Quinton and Van Dongen [82]. These results were generalized to variable dependent schedules by Mauras et al. [70]. Feautrier [31] and Darte and Robert [23] independently presented the optimal solution to the affine scheduling problem (by variable). Feautrier also provided the extension to multidimensional time [32]. The extension of these techniques to programs in the $\mathcal{Z}$-polyhedral model was presented in [36]. Their problem formulation searched for schedules that could directly be used to perform appropriate program transformations. The problem of scheduling reductions was initially solved by Redon and Feautrier [90]. They had assumed a Concurrent Read, Concurrent Write Parallel Random Access Machine (CRCW PRAM) such that each reduction took constant time. The problem of scheduling reductions on a Concurrent Read, Exclusive Write (CREW) PRAM was presented in [39]. The scheduling problem was studied along with the objective for minimizing communication by Lim et al. [63]. The problem of memory optimization, too, has been studied extensively [22, 26, 57, 64, 79, 105].

The generation of efficient imperative code for programs in the polyhedral model was presented by Quilleré et al. [80] and later extended by Bastoul [9]. Algorithms to generate code, both sequential and parallel, after applying nonunimodular transformations to nested loop programs has been studied extensively [33, 62, 85, 111]. However, these results were all restricted to single, perfectly nested loop nests, with the same transformation applied to all the statements in the loop body. The code generation problem thus reduced to scanning the image, by a nonunimodular function, of a single polyhedron. The general problem of generating loop nests for a union of $\mathcal{Z}$-polyhedra was solved by Bastoul in [11].

Lenders and Rajopadhye [58] proposed a technique for designing multirate VLSI arrays, which are regular arrays of processing elements, but where different registers are clocked at different rates. Their formulation was based on equations defined over $\mathcal{Z}$-polyhedral domains.

Feautrier [30] showed that an important class of conventional imperative loop programs called *affine control loops* (ACLs) can be transformed to programs in the polyhedral model. Pugh [78] extended Feautrier's results. The detection of scans in imperative loop programs was presented by Redon and Feautrier in [89]. Bastoul et al. [10] showed that significant parts of the SpecFP and PerfectClub benchmarks are ACLs.

15.3 Iteration Space Tiling

This section describes an important class of reordering transformations called tiling. Tiling is crucial to exploit locality on a single proccessor, as well as for adapting the granularity of a parallel program. We first describe tiling for dense iteration spaces and data sets and then consider irregular iteration spaces and sparse data sets. Next, we briefly summarize the steps involved in tiling and conclude with bibliographic notes.

15.3.1 Tiling for Dense Iteration Spaces

Tiling is a loop transformation used for adjusting the granularity of the computation so that its characteristics match those of the execution environment. Intuitively, tiling partitions the iterations of a loop into groups called *tiles*. The tile sizes determine the granularity.

In this section, we will study three aspects related to tiling. First, we will introduce tiling as a loop transformation and derive conditions under which it can be applied. Second, we present a constrained optimization approach for formulating and finding the optimal tile sizes. We then discuss techniques for generating tiled code.

15.3.1.1 Tiling as a Loop Transformation

Stencil computations occur frequently in many numerical solvers, and we use them to illustrate the concepts and techniques related to tiling. Consider the typical Gauss–Seidel style stencil computation shown in Figure 15.7 as a running example. The loop specifies a particular order in which the values are computed. An *iteration reordering transformation* specifies a new order for computation. Obviously not every reordering of the iterations is legal, that is, *semantics preserving*. The notion of semantics preserving can be formalized using the concept of *dependence*. A dependence is a relation between a producer and consumer of a value. A dependence is said to be preserved after the reordering transformation if the iteration that produces a value is computed before the consumer iteration. Legal iteration reorderings are those that preserve all the dependences in a given computation.

Array data dependence analyses determine data dependences between values stored in arrays. The relationship can be either *memory-based* or *value-based*. Memory-based dependencies are induced by write to and read from the same memory location. Value-based dependencies are induced by production and consumption of values. Once can view memory-based dependences as a relation between memory locations and valued-based dependences as a relation between values produced and consumed. For computations represented by loop nests, the values produced and consumed can be uniquely associated with an iteration. Hence, dependences can be viewed as a relation between iterations.

Dependence analyses summarize these dependence relationships with a suitable representation. Different dependence representations can be used. Here, we introduce and use *distance vectors* that can represent a particular kind of dependence and discuss legality of tiling with respect to them. More general representations such as direction vectors, dependence polyhedra, and cones can be used to capture general dependence relationships. Legality of tiling transformations can be naturally extended to these representations, and a discussion of them is beyond the scope of this article.

We consider perfect loop nests. Since, through array expansion, memory-based dependences can be automatically transformed to value-based dependences, we consider only the later. For an n-deep perfect

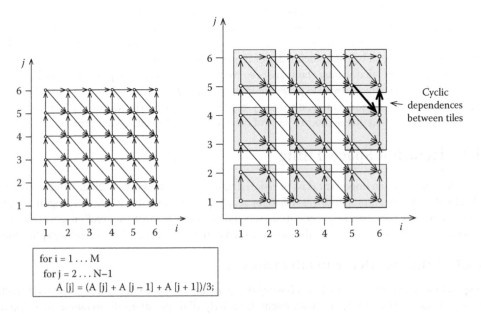

FIGURE 15.7 A stencil computation is shown on the bottom-left and a geometric view of its data flow is shown above it. Shown on the right is an *illegal* rectangular tiling of the iteration space with 2×2 tiles. This is due to the cyclic dependences between the tiles. An instance of this is highlighted.

loop nest, the iterations can be represented as integer n-vectors. A *dependence vector* for an n-dimensional perfect loop nest is an n-vector $d = (d_1, \ldots, d_n)$, where the kth component corresponds to the kth loop (counted from outermost to innermost). A *distance vector* is a dependence vector $d \in \mathbb{Z}^n$, that is, every component d_k is an integer. A *dependence distance* is the distance between the iteration that produces a value and the iteration that consumes a value. Distance vectors represent this information. The dependence distance vector for a value produced at iteration $p = (p_1, \ldots, p_n)$ and consumed at a later iteration $c = (c_1, \ldots, c_n)$ is $d = (c - p) = (c_1 - p_1, \ldots, c_n - p_n)$. The stencil computation given in Figure 15.7 has three dependences. The values consumed at an iteration (i, j) are produced at iterations $(i, j - 1)$, $(i - 1, j)$, and $(i - 1, j + 1)$. The corresponding three dependence vectors are $\binom{0}{1}$, $\binom{1}{0}$, and $\binom{1}{-1}$.

Tiling is an iteration reordering transformation. Tiling reorders the iterations to be executed in a *block-by-block* or *tile-by-tile* fashion. Consider the tiled iteration space shown in Figure 15.7 and the following execution order. Both the tiles and the points within a tile are executed in the lexicographic order. The tiles are also executed in an atomic fashion, that is, all the iterations in a tile are executed before any iteration of another tile. It is very instructive to pause for a moment and ask whether this tiled execution order preserves all the dependences of the original computation. One can observe that the dependence $\binom{1}{-1}$ is not preserved, and hence the tiling is illegal. There exists a nice geometric way of checking the legality of a tiling. A given tiling is *illegal* if there exist cyclic dependences between tiles. An instance of this cyclic dependence is highlighted in Figure 15.7.

The legality of tiling is determined not by the dependences alone, but also by the shape of the tiles.[12] We saw (Figure 15.7) that tiling the stencil computation with rectangles is illegal. However, one might wonder whether there are other tile shapes for which tiling is legal. Yes, tiling with parallelograms is legal as shown in Figure 15.8. Note how the change in the tile shape has avoided the cyclic dependences that were present in the rectangular tiling. Instead of considering nonrectangular shapes that make tiling legal, one could also consider transforming the data dependences so that rectangular tiling becomes legal. Often, one can easily find such transformations. A commonly used transformation is *skewing*. The skewed iteration space is shown in Figure 15.8 together with a rectangular tiling. Compare the dependences between tiles in this tiling with those in the illegal rectangular tiling shown in Figure 15.7. One could also think of more complicated tile shapes, such as hexagons or octagons. Because of complexity of tiled code generation such tile shapes are not used.

A given tiling can be characterized by the shape and size of the tiles, both of which can be concisely specified with a matrix. Two matrices, *clustering* and *tiling*, are used to characterize a given tiling. The clustering matrix has a straightforward geometric interpretation, and the tiling matrix is its inverse and is useful in defining legality conditions. A parallelogram (or a rectangle) has four vertices and four edges. Let us pick one of the vertices to be the origin. Now we have two edges or two vertices adjoining the origin. The shape and size of the tiles can be specified by characterizing these edges or vertices. We can easily generalize these concepts to higher dimensions. In general, an n-dimensional parallelepiped has 2^n vertices and $2n$ facets (higher-dimensional edges), out of which n facets and n vertices adjoin the origin. A *clustering matrix* is an $n \times n$ square matrix whose columns correspond to the facets that determine a tile. The clustering matrix has the property that the absolute value of its determinant is equal to the tile volume.

The clustering matrices of the parallelogram and rectangular tilings shown in Figure 15.8 are

$$G_{ppd} = \begin{pmatrix} 0 & 2 \\ 2 & -2 \end{pmatrix} \text{ and } G_{rect} = \begin{pmatrix} 2 & 0 \\ 0 & 2 \end{pmatrix}$$

---

[12]Legality of tiling also depends on the shape of the iteration space. However, for practical applications, we can check the legality with the shape of the tiles and dependence information alone.

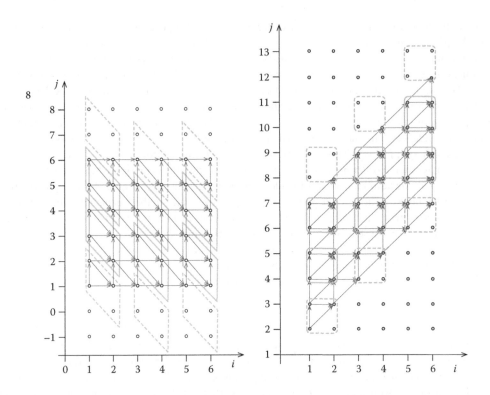

FIGURE 15.8   Shown on the left is a parallelogram tiling. A rectangular tiling of the transformed iteration space is shown on the right. A skewing transformation is required to make the rectangular tiling valid.

In G_{ppd}, the first column represents the horizontal edge, and the second represents the oblique edge. In G_{rect}, the first column represents the horizontal edge, and the second represents the vertical edge.

The *tiling matrix* is the inverse of the clustering matrix. The tiling matrices of the parallelogram and the rectangular tilings shown in Figure 15.8 are

$$H_{ppd} = G_{ppd}^{-1} = \begin{pmatrix} \frac{1}{2} & \frac{1}{2} \\ \frac{1}{2} & 0 \end{pmatrix} \text{ and } H_{rect} = G_{rect}^{-1} = \begin{pmatrix} \frac{1}{2} & 0 \\ 0 & \frac{1}{2} \end{pmatrix}$$

For rectangular tiling the edges are always along the canonical axes, and hence, there is no loss of generality in assuming that the tiling and clustering matrices are diagonal. The tiling is completely described by just the so-called *tile size vector, s* $= (s_1, \ldots, s_n)$, where each s_i denotes the tile size along the ith dimension. The clustering and tiling matrices are simply $G_{rect} = \text{diag}(s_1, \ldots, s_n)$ and $H_{rect} = \text{diag}(\frac{1}{s_1}, \ldots, \frac{1}{s_n})$, where $\text{diag}(x_1, \ldots, x_n)$ denotes the $n \times n$ diagonal matrix with $x_1, \ldots, x_n$ as the diagonal entries.

A geometric way of checking the legality of a given tiling was discussed earlier. One can derive formal legality conditions based on the shape and size of the tiles and the dependences. Let D be the set of dependence distance vectors. A vector $x = (x_1, \ldots, x_n)$ is lexicographically nonnegative if the leading nonzero component of the x is positive, that is, $x_1 > 0$ or both $x_1 = 0$ and $(x_2, \ldots, x_n) \succeq 0$. The floor operator $\lfloor . \rfloor$ when used on vectors is applied component-wise, that is, $\lfloor x \rfloor = (\lfloor x_1 \rfloor, \ldots, \lfloor x_n \rfloor)$. The legality condition for a given (rectangular or parallelepiped) tiling specified by the tiling matrix H and dependence set D is

$$\forall d \in D : \lfloor Hd \rfloor \succeq 0$$

The above condition formally captures the presence or absence of cycles between tiles.

We can now apply this legality condition to the stencil computation example. Let $D_{orig} = \{\binom{0}{1}, \binom{1}{0}, \binom{1}{-1}\}$, the set of dependence vectors from the original or given stencil computation, and $D_{skew} = \{\binom{0}{1}, \binom{1}{1}, \binom{1}{0}\}$, the dependence vectors obtained after applying the skewing transformation $\left(\begin{smallmatrix} 1 & 0 \\ 1 & 1 \end{smallmatrix}\right)$ to the original dependences. Earlier we showed that rectangular tiling of the original iteration space is not legal based on the existence of cycles between tiles (cf. Figure 15.7). This can also be verified by observing that the condition for validity, $\forall d \in D_{orig} : \lfloor H_{rect} d \rfloor \succeq 0$, is *not* satisfied, since, for the dependence vector $\binom{1}{-1}$ in D_{orig}, we have $\lfloor H_{rect}(\binom{1}{-1}) \rfloor = \binom{0}{-1}$. However, for the same dependences, D_{orig}, as shown in Figure 15.8, a parallelogram tiling is valid. This validity is confirmed by the satisfaction of the constraint $\forall d \in D_{orig} : \lfloor H_{ppd} d \rfloor \succeq 0$. We also showed that a skewing transformation of the iteration space can make rectangular tiling valid. This can also be verified by observing the satisfaction of $\forall d \in D_{skew} : \lfloor H_{rect} d \rfloor \succeq 0$. In the case of rectangular tiling the legality condition can be simplified by using the fact that the tiling can be completely specified by the tile size vector $s = (s_1, \ldots, s_n)$. The legality condition for rectangular tiling specified by the tile size vector s for a loop nest with a set of dependences D is

$$\forall d \in D : \begin{pmatrix} \lfloor \frac{d_1}{s_1} \rfloor \\ \vdots \\ \lfloor \frac{d_n}{s_n} \rfloor \end{pmatrix} \succeq 0$$

A rectangular tiling can also be viewed as a sequence of two transformations: strip mining and loop interchange. This view is presented by Raman and August in this text [84].

15.3.1.2 Optimal Tiling

Selecting the tile shape and selecting the sizes are two important tasks in using loop tiling. If rectangular tiling is valid or could be made valid by appropriate loop transformation, then it should be preferred over parallelepipeds. This preference is motivated by the simplicity and efficiency in tiled code generation as well as tile size selection methods. For many practical applications we can transform the loop so that rectangular tiling is valid. We discuss rectangular tiling only. Having fixed the shape of tiles to (hyper-)rectangles, we address the problem of choosing the "best" tile sizes.

Tile size selection methods vary widely depending on the purpose of tiling. For example, when tiling for multi-processor parallelism, analytical models are often used to pick the best tile sizes. However, when tiling for caches or registers, empirical search is often the best choice. Though the methods vary widely, they can be treated in the single unifying formulation of *constrained optimization problems*. This approach is used in the next section to formulate the optimal tile size selection problem.

15.3.1.2.1 Optimal Tile Size Selection Problem

The *optimal tile size selection* problem involves selecting the *best* tile sizes from a set of *valid* tile sizes. What makes a tile size *valid* and what makes it the *best* can be specified in a number of ways. Constrained optimization provides this unified approach. Validity is specified with a set of constraints, and an objective function is used to pick the *best* tile sizes. A constrained optimization problem has the following form:

$$\text{minimize } f(s)$$
$$\text{subject to } g_i(s) \leq 0 \text{ for } i = 1 \ldots m \qquad (15.14)$$

where s is the variable, $f(s)$ is the objective function, and $g_i(s) \leq 0$ are constraints on s. The solution to such an optimization problem has two components: (a) the *minimum value* of f over all valid s and (b) a *minimizer* s^*, which is a value of s at which f attains the minimum value.

All the optimal tile size selection problems can be formulated as a constrained optimization problem with appropriate choice of the f and g_is. Furthermore, the structure of f and g_is determines the techniques that can be used to solve the optimization problem. For example, consider the problem of tiling for data locality, where we seek to pick a tile size that minimizes the number of cache misses. This can be cast into an optimization problem, where the objective function is the number of misses, and the constraints are the positivity constraints on the tile sizes and possibly upper bounds on the tile sizes based on program size parameters as well as cache capacity. In the next two sections, we will present an optimization-based approach to optimal tile size selection in the context of two problems: (a) tiling for data locality and (b) tiling for parallelism. The optimization problems resulting from optimal tiling formulations can be cast into a particular form of numerical convex optimization problems called *geometric programs*, for which powerful and efficient tools are widely available. We first introduce this class of convex optimization problems in the next section and use them in the later sections.

15.3.1.2.2 Geometric Programs

In this section we introduce the class of numerical optimization problems called geometric programs, which will be used in later sections to formulate optimal tile size selection problems.

Let x denote the vector $(x_1, x_2, \ldots, x_n)$ of n real, positive variables. A function f is called a *posynomial* function of x if it has the form

$$f(x_1, x_2, \ldots, x_n) = \sum_{k=1}^{t} c_k x_1^{\alpha_{1k}} x_2^{\alpha_{2k}} \cdots x_n^{\alpha_{nk}}$$

where $c_j \geq 0$ and $\alpha_{ij} \in \mathbb{R}$. Note that the coefficients c_k must be nonnegative, but the exponents α_{ij} can be arbitrary real numbers, including negative or fractional. When there is exactly one nonzero term in the sum, that is, $t = 1$ and $c_1 > 0$, we call f a *monomial* function. For example, $0.7 + 2x_1/x_3^2 + x_2^{0.3}$ is a posynomial (but not monomial), $2.3(x_1/x_2)^{1.5}$ is a monomial (and, hence, a posynomial), while $2x_1/x_3^2 - x_2^{0.3}$ is neither. Note that posynomials are closed under addition, multiplication, and nonnegative scaling. Monomials are closed under multiplication and division.

A *geometric program* (GP) is an optimization problem of the form

$$\begin{aligned}
\text{mimimize} \quad & f_0(x) \\
\text{subject to} \quad & f_i(x) \leq 1, i = 1, \ldots, m \\
& g_i(x) = 1, i = 1, \ldots, p \\
& x_i > 0, \quad i = 1, \ldots, n
\end{aligned} \tag{15.15}$$

where $f_0, \ldots, f_m$ are posynomial functions and $g_1, \ldots, g_p$ are monomial functions. If $\forall i = 1 \ldots n : x_i \in \mathbb{Z}$, we call the GP an integer geometric program (IGP).

15.3.1.2.2.1 Solving IGPs

GPs can be transformed into convex optimization problems using a variable substitution and solved efficiently using polynomial time interior point methods. Integer solutions can be found by using a branch-and-bound algorithm. Tools such as YALMIP provide a high-level symbolic interface (in MATLAB) that can be used to define and solve IGPs. The number of (tile) variables of our IGPs are related to number of dimensions tiled and hence are often small. In our experience with solving IGPs related to tiling, the integer solutions were found in a few (fewer than 10) iterations of the branch-and-bound algorithm. The (wall clock) running time of this algorithm was just a few seconds, even with the overhead of using the symbolic MATLAB interface.

15.3.1.2.3 Tiling for Parallelism

We consider a distributed memory parallel machine as the execution target. Message passing is a widely used interprocess communication mechanism for such parallel machines. The cost of communication in such systems is significant. Programs with fine-grained parallelism require frequent communications and are not suited for message-passing-style parallel execution. We need to increase the granularity of parallel computation and make the communications less frequent. Tiling can be used to increase the granularity of parallelism from fine to coarse. Instead of executing individual iterations in parallel, we can execute tiles in parallel, and instead of communicating after every iteration, we communicate between tiles. The tile sizes determine how much computation is done between communications.

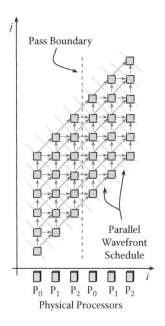

Tile graph for a tiling with $s_i = s_j$ (square tiles) is shown. Each column of tiles is mapped to a single processor resulting in a set of virtual processors. The virtual processors execute the tiles in a parallel wavefront style schedule. Often the number of physical processors available is far less than the virtual processors. The physical processors simulate the virtual processors by executing them in multiple passes. Three physical processors, simulating the execution of six virtual processors in two passes, are shown.

FIGURE 15.9 Tile graph and a parallelization of it.

Consider the 2×2 rectangular tiling shown in Figure 15.8. We seek to execute the tiles in parallel. To do this we need (a) a *processor mapping* that maps tiles to processors and (b) a *schedule* that specifies a (parallel) time stamp for each tile. A *parallelization* of a tiled iteration space involves derivation of a processor mapping and a schedule. A better abstraction of the tiled iteration space useful in comparing and analyzing different parallelizations is the *tile graph*. A tile graph consists of nodes that are tiles and edges representing dependences between them. Figure 15.9 shows the tile graph of the 2×2 tiling shown in Figure 15.8. The dependences between tiles are induced by the dependences between the iterations and the tiles they are grouped into. The shape of the tile graph is determined by the shape of the tiled iteration space as well as the tile sizes. The shapes of the tile graph (Figure 15.9) and the rectangular tiled iteration space (Figure 15.8) are the same because the tile sizes $s_i = s_j = 2$ are the same. It is useful to pause for a moment and think of the tile graph shape when $s_i = 2$ and $s_j = 4$.

To parallelize the tile graph we need to find a parallel schedule and a processor mapping. As shown in Figure 15.9, the wavefronts corresponding to the $i + j = c$ lines define a parallel schedule — all the tiles on a wavefront can be executed in parallel. We can verify that this schedule is valid by observing that any given tile is scheduled after all the tiles it depends on are executed. A processor mapping is valid if it does not map two tiles scheduled to execute at the same time to the same processor. There are many valid processor mappings possible for this schedule. For example, one can easily verify that the following three mappings are valid: (a) map each column of tiles to a processor, (b) map each row of tiles to a processor, and (c) map all the tiles along the $i = j$ diagonal line to a single processor. Though all of them are valid, they have very different properties: the first (column-wise) and the last (diagonal) map the same number of tiles to each processor, whereas the second (row-wise) maps a different number of tiles to different processors. For a load-balanced allocation one would prefer the column-wise or the diagonal mappings. However, for simplicity of the resulting parallel program, one would prefer the column-wise over the diagonal mapping.

Typically the number of processors that results from a processor mapping is far greater than the number of available processors. We call the former the *virtual processors* and the latter the *physical processors*. Fewer physical processors simulate the execution of the virtual processors in multiple passes. For example, the column-wise mapping in Figure 15.9 results in six virtual processors, and they are simulated by three

```
foreach tile t ∈ tiles (p)
    Receive tile-inputs for t from neighboring processors
    compute t
    Send tile-outputs for t to neighboring processors
```

FIGURE 15.10   Steps performed by each processor p for computing the tiles allocated to it. `tiles(p)` gives the tiles allocated to a processor p.

physical processors in two passes. Tiles are executed in an atomic fashion; all the iterations in a tile are executed before any iteration of another tile. The parallel execution proceeds in a wavefront style.

We call a parallelization *idle-free* if it has the property that once a processor starts execution it will never be idle until it finishes all the computations assigned to it. We call a parallelization *load-balanced* if it has the property that all the processors are assigned an (almost) equal amount of work. For example, the column-wise and diagonal mappings are load-balanced, whereas the row-wise mapping is not. Furthermore, within a given pass, the wavefront schedule is *idle-free*. Across multiple passes, it will be idle-free if by the time the first processor finishes its first column of tiles the last processor finishes at least one tile. This will be true whenever the number of tiles in a column is more than the number of physical processors.

15.3.1.2.4 An Execution Time Model

After selecting a schedule and a processor mapping, the next step is to pick the tile size. We want to pick the tile size that minimizes the execution time. We will develop an analytical model for the total running time of the parallel program and then use it to formulate a constrained optimization problem, whose solution will yield the optimal tile sizes.

We choose the wave front schedule and the column-wise processor mapping discussed earlier. Recall that the column-wise mapping is load-balanced, and within a pass the wave front schedule is idle-free. To ensure that the schedule is also idle-free across passes, we will characterize and enforce the constraint that the number of tiles in a column is greater than the number of physical processors. Furthermore, we consider the following *receive–compute–send* execution pattern (shown in Figure 15.10): every processor first receives all the inputs it needs to execute a tile and then executes the tiles and then sends the tile outputs to other processors. The total execution time is the time elapsed between the start of the first tile and the completion of the last tile. Let us assume that all the passes are full, that is, in each pass all the processors have a column of tiles to execute. Now, the last tile will be executed by the last processor, and its completion time will give the total execution time. Given that the parallelization is idle-free, the total time taken by any processor is equal to the initial latency (time it waits to get started) and the time it takes to compute all the tiles allocated to it. Hence, the sum of the latency and the computation time of the last processor is equal to the total execution time. Based on this reasoning, the total execution time is

$$T = L + (\mathsf{TPP} \times \mathsf{TET})$$

where L denotes the latency last processor to start, TPP denotes the number of tiles allocated per processor, and TET is the time to execute a tile (sequentially) by a single processor. Here, the term $\mathsf{TPP} \times \mathsf{TET}$ denotes the time any processor takes to execute all the tiles allocated to it. Given that we have a load-balanced processor mapping, this term is same for all processors. In the following derivations, P is the number of physical processors, N_i and N_j denote the size of the iteration space along i and j, respectively, and s_i and s_j are the tile sizes along i and j, respectively.

Let us now derive closed-form expressions for the terms discussed above. The time to execute a tile, TET, is the sum of the computation and communication time. The computation time is proportional to the area of the rectangular tile and is given by $s_i \times s_j \times \alpha$. The constant α denotes the average time to execute one iteration. The communication time is modeled as an affine function of the message size. Every processor receives the left edge of the tile from its left neighbor and sends its right edge to the right neighbor. This results in two communications with messages of size s_j, the length of the vertical edge of a tile. The cost of sending a message of size x is modeled by $\tau x + \beta$, where τ and β are constants that denote the transmission cost per byte and the start-up cost of a communication call, respectively. The cost of the two communications performed for each tile is $2(\tau s_j + \beta)$. The total time to execute a tile is now TET $= s_i s_j \alpha + 2(\tau s_j + \beta)$.

The number of tiles allocated to a processor is equal to the number of columns allocated to a processor times the number of tiles per column. The number of columns is equal to the number of passes, which is $\frac{N_i}{s_i P}$. The tiles per column is equal to $\frac{N_j + s_i}{s_j}$, which makes TPP $= \frac{N_i}{s_i P} \times \frac{N_j + s_i}{s_j}$.

The dependences in the tile graph induce the delay between the start of the processors. The slope $\sigma = \frac{s_j}{s_i}$, known as the *rise*, plays a fundamental role in determining the latency. The last processor can start as soon as the processor before it completes the execution of its first two tiles. Formally, the last processor can start its first tile only after $(P - 1) \times (\sigma + 1)$. For example, in Figure 15.9 the last processor can start only after four time steps since $\sigma = \frac{2}{2} = 1$ and $P = 3$, yielding $(3 - 1) \times (1 + 1) = 4$. Since at each time step a processor computes a tile, $(P - 1) \times (\sigma + 1) \times$ TET gives the time after which the last processor can start, that is, $L = (P - 1) \times (\sigma + 1) \times$ TET.

To ensure that there is no idle time between passes, we need to constrain the tile sizes such that by the time the first processor finishes its column of tiles, the last processor must have finished its first tile. The time the first processor takes to complete a column of tiles is equal to $\frac{N_j + s_i}{s_j} \times$ TET, and the time by which the last processor would finish its first tile is $[(P - 1 + 1) \times (\sigma + 1)] \times$ TET. The no idle time between passes constraint is $\frac{N_j + s_i}{s_j} \times$ TET $\geq [P \times (\sigma + 1)] \times$ TET.

Using the terms derived above we can now formulate an optimization problem to pick the optimal tile size.

$$\text{minimize } T = \left[((P - 1)(\sigma + 1)) + \left(\frac{N_i}{s_i P} \times \frac{N_j + s_i}{s_j} \right) \right] \times (\alpha s_i s_j + 2(\tau s_j + \beta)) \quad (15.16)$$

$$\text{subject to} \quad \frac{N_j + s_i}{s_j} \geq P(\sigma + 1)$$

The solution to the above optimization problem yields the optimal tile sizes, that is, the tile sizes that minimize the total execution time of the parallel program, subject to the constraint that there is no idle time between passes.

The optimization problem in Equation 15.16 can be transformed into a GP. The objective function T is directly a posynomial. With the approximation of $N_j + s_i \approx N_j$, the constraint transforms into

$$\frac{P(\sigma + 1)s_j}{N_j} \leq 1$$

which is the required form for a GP constraint. Adding to it the obvious constraints that tile sizes are integers and positive, that is, $s_i, s_j \in \mathbb{Z}$, $s_i \geq 1$, and $s_j \geq 1$, we get an IGP that can be solved efficiently as discussed above. The solution to this IGP will yield the optimal tile sizes.

15.3.1.2.4.1 *Generality of Approach*

The analysis techniques presented above can be directly extended to higher-dimensional rectangular or parallelepiped iteration spaces. For example, stencil computations with two-dimensional or three-dimensional data grids, after skewing to make rectangular tiling valid, have parallelepiped iteration spaces, and the techniques described above can be directly applied to them. The GP-based modeling approach is quite general. Because of the fundamental positivity property of tile sizes, often the functions used in modeling parallel execution time or communication or computation volumes are posynomials. This naturally leads to optimization problems that are GPs.

15.3.1.2.5 *Tiling for Data Locality*

Consider the stencil computation shown in Figure 15.7. Every value, A[j], computed at an iteration (i, j) is used by three other computations as illustrated in the geometric view shown in Figure 15.7 (left). The

```
for (iT = 1; iT < = N; iT += Si)
  for (jT = 2; jT < = N+M; jT += Sj)
    for (i= max(iT,1); i<=min(iT+Si-1, N); i++)
      for (j= max(jT, i+1); i<=min(jT+Sj-1, i+N); j++)
          /* Loop Body */
```

FIGURE 15.11   Tiled loop nest with tile sizes as parameters.

three uses are in iterations $(i, j + 1), (i + 1, j)$, and $(i + 1, j - 1)$. Consider the case when the size of A is much larger than the cache size. On the first use at iteration $(i, j + 1)$, the value will be cache. However, for the other two uses, $(i + 1, j)$ and $(i + 1, j - 1)$, the value may not be in cache, resulting in a *cache miss*. This cache miss can be avoided if we can keep the computed values in the cache until their last use. One way to achieve this is by changing the order of the computations such that the iterations that use a value are computed "as soon as" the value itself is computed. Tiling is widely used to achieve such reorderings that improve data locality. Furthermore, the question of how soon the iterations that use a value should be computed is dependent on the size of the cache and processor architecture. This aspect can be captured by appropriately picking the tile sizes.

Consider the rectangular tiling of the skewed iteration space shown in Figure 15.8 (right). Figure 15.11 shows the tiled loop nest of the skewed iteration space, with tile sizes as parameters. The new execution order after tiling is as follows: both the tiles and the points within a tile are executed in column-major order. Observe how the new execution order brings the consumers of a value closer to the producer, thereby decreasing the chances of a cache miss. Figure 15.8 (right) shows a tiling with tiles of sizes 2×2. In general, the sizes are picked so that the volume data touched by a tile, known as its *footprint*, fits in the cache, and some metric such as number of misses or total execution time is minimized. A discussion of other loop transformations (e.g., loop fusion, fission, etc.) aimed at memory hierarchy optimization can be found in the chapter by Raman and August [84] in the same text.

15.3.1.2.5.1   *Tile Size Selection Approaches*

A straightforward approach for picking the best tile size is empirical search. The tiled loop nest is executed for a range of tile sizes, and the one that has the minimum execution time is selected. This search method has the advantage of being accurate, that is, the minimum execution time tile is *the best* for the machine on which it is obtained. However, such a search may not be feasible because of the huge space of tile sizes that needs to be explored. Often, some heuristic model is used to narrow down this space. In spite of the disadvantages, such an empirical search is the popular and widely used approach for picking the best tile sizes. For the obvious reason of huge search time, such an approach is not suitable for a compiler.

Compilers trade off accuracy for search time required to find the best tile size. They use approximate cache behavior models and high-level execution time models. Efficiency is achieved by specializing the tile size search algorithm to the chosen cache and execution time models. However, such specialization of search algorithms makes it difficult to change or refine the models.

Designing a good model for the cache behavior of loop programs is hard, but even harder is the task of designing a model that would keep up with the advancements in processor and cache architectures. Thus, cache models used by compilers are often outdated and inaccurate. In fact, the performance of a tiled loop nest generated by a state-of-the-art optimizing compiler could be a few factors poorer than the one hand-tuned with an empirical search for best tile sizes. This has led to the development of the so-called *auto-tuners*, which automatically generate loop kernels that are highly tuned to a given architecture. Tile sizes are an important parameter tuned by auto-tuners. They use a model-driven empirical search to pick the tile sizes. Essentially they do an empirical search for the best tile size over a space of tile sizes and use an approximate model to prune the search space.

15.3.1.2.5.2 Constrained Optimization Approach

Instead of discussing specialized algorithms, we present a GP-based framework that can be used to develop models, formulate optimal tile size selection problems, and obtain the best tile sizes by using the efficient numerical solvers. We illustrate the use of the GP framework by developing an execution time model for the tiled stencil computation and formulating a GP whose solution will yield the optimal tile sizes. Though we restrict our discussion to this illustration-based presentation, the GP framework is quite general and can be used with several other models. For example, the models used in the IBM production compiler or the one used by the auto-tuner ATLAS can be transformed into the GP framework.

The generality and wide applicability of the GP framework stems from a fundamental property of the models used for optimal tile size selection. The key property is based on the following: *tile sizes are always positive and all these models use metrics and constraints that are functions of the tile sizes.* These functions of tile size variables often turn out to be posynomials. Furthermore, the closure properties of posynomials provide the ability to compose models. We illustrate these in the following sections.

15.3.1.2.5.3 An Analytical Model

We will first derive closed-form characterizations of several basic components related to the execution of a tiled loop and then use them to develop an analytical model for the total execution time. We will use the following parameters in the modeling. Some of them are features of processor memory hierarchy and others are a combination of processor and loop body features:

- α: The average cost (in cycles) of computing an iteration assuming that the accessed data values are in the lowest level of cache. This can be determined by executing the loop for a small number of iterations, such that the data arrays fit in the cache, and taking the average.
- μ: The cost (in cycles) for moving a word from main memory to the lowest level of cache. This can be determined by the miss penalties associated with caches, translation look aside buffers, and so on.
- λ: The average cost (in cycles) to compute and check loop bounds. This can be determined by executing the loops without any body and taking the average.
- *C* **and** *L*: The capacity and line size, in words, of the lowest level of cache. These two can be directly determined from the architecture manual.

15.3.1.2.5.4 Computation Cost

The number of iterations computed by a tile is given by the tile area $s_i \times s_j$. If the data values are present in the lowest level of cache, then the cost of computing the iterations of a tile, denoted by $\Theta(s)$, is $\alpha s_i s_j$, where α is the average cost to compute an iteration.

15.3.1.2.5.5 Loop Overhead Cost

Tiling (all the loops of) a loop nest of depth d results in $2d$ loops of which the outer d loops enumerate the tiles and the inner d loops enumerate points in a tile. We refer to the cost for computing and testing loop bounds as the *loop overhead* cost. In general, the loop overhead is significant for tiled loops and needs to be accounted for in modeling the execution time. The loop overhead cost of a given loop is proportional to the number of iterations it enumerates. In general, λ, the loops bounds check cost, is dependent on the complexity of the loop bounds of a given loop. However, for notational and modeling simplicity we will use the same λ for all the loops. Now the product of λ with the number iterations of a loop gives the loop overhead of that loop.

Consider the tiled loop nest of the skewed iteration space shown in Figure 15.11. The total number of iterations enumerated by the tile-loops (iT and jT loops) is $\frac{N \times (N+M)}{s_j} + \frac{N}{s_i}$. The loop overhead of the tile-loops is $\lambda \frac{N \times (N+M)}{s_j} + \frac{N}{s_i}$. With the small overapproximation of partial tiles by full tiles, the number of iterations enumerated by the point-loops (i and j loops), for any given iteration of the outer tile-loops, is $s_i \times (s_i \times s_j)$. The loop overhead of the point-loops is $\lambda(s_i^2 s_j)$. The total loop overhead of the tiled loop nest is denoted by $\Lambda(s) = \lambda s_i^2 s_j \frac{N \times (N+M)}{s_j} + \frac{N}{s_i}$.

15.3.1.2.5.6   Footprint of a Tile

The *footprint* of a tile is the number of distinct array elements touched by a tile. Let us denote the footprint of a tile of size s by $F(s)$. Deriving closed-form descriptions of $F(s)$ for loop nests with an arbitrary loop body is hard. However, for the case when the loop body consists of references to arrays and the dependences are distance vectors, we can derive closed-form descriptions of $F(s)$. However, for the case when the loop body contains affine references, deriving closed-form expressions for $F(s)$ is complicated. We illustrate the steps involved in deriving $F(s)$ for dependence distance vectors with our stencil computation example.

Consider the tiled stencil computation. Let $s = (s_i, s_j)$ be the tile size vector, where s_i represents the tile size along i and s_j along j. Each (nonboundary, full) tile executes $s_i \times s_j$ iterations updating the values of the one-dimensional array A. The number of distinct values of A touched by a tile is proportional to one of its edges, namely, s_j. One might have to store some intermediate values during the tiled execution, and these require an additional array of size s_i. Adding these two together, we get $F(s) = s_i + s_j$. Note that $F(s)$ takes into account the reuse of values. Loops with good data locality (i.e., with at least one dimension of reuse) have the following property: the footprint is proportional to the (weighted) sum of the facets of the tile. Note that our stencil computation has this property, and hence $F(s)$ is the sum of the facets (here just edges) s_i and s_j. To maximize the benefits of data locality, we should make sure that the footprint $F(s)$ fits in the cache.

15.3.1.2.5.7   Load Store Cost of a Tile

Earlier during the calculation of the computation cost, we assumed that the values are available in the lowest level of the cache. Now we will model the cost of moving the values between main memory and the lowest level of cache. To derive this cost we need a model of the cache. We will assume a fully associative cache of capacity C words with cache lines of size L words. μ is the cost of getting a word from the main memory to the cache. Ignoring the reuse of cache lines across tiles, $F(s)$ provides a good estimated number of values accessed by a tile during its execution. Let $F_L(s)$ be the number of cache lines needed for $F(s)$. We have $F_L(s) = \lceil \frac{F(s)}{L} \rceil$, where L is the cache line size. Then the load store cost of a tile, denoted by $\Delta(s)$, is $F_L(s) \times \mu$.

15.3.1.2.5.8   Total Execution Time of a Tiled Loop Nest

The total execution time of the tiled loop nest is the sum of the time it takes to execute the tiles and the loop overhead. The time to execute the tiles can be modeled as the product of time to execute a tile times the number of tiles. For our stencil computation the iteration space is a parallelogram, and calculating the number of $s_i \times s_j$ rectangles that cover it is a hard problem. However, we can use the reasonable approximation of $\frac{N \times M}{s_i \times s_j}$ to model the number of tiles, denoted by ntiles(s). The total execution time T is given by

$$T = \text{ntiles} \times [\Theta(s) + \Delta(s)] + \Lambda(s) \tag{15.17}$$

where ntiles(s) is the number of tiles, $\Theta(s)$ is the cost of executing a tile, $\Delta(s)$ is the load store cost, and $\Lambda(s)$ is the loop overhead.

15.3.1.2.5.9   Optimal Tile Size Selection Problem Formulation

Using the quantities derived above, we can now formulate an optimization problem whose solution will yield the optimal tile size — one that minimizes the total execution time. Recall that the function T (Equation 15.17) derived above models the execution time under the assumption that the data accessed by a tile fits in the cache. We model this assumption by translating it into a constraint in the optimization problem. Recall that $F(s)$ measures the data accessed by a tile, and $F_L(s)$ gives the number of cache lines needed for $F(s)$. The constraint $F_L(s) \leq C$, where C is the cache capacity, ensures that all the data touched by a tile fits in the cache. Now we can formulate the optimization problem to find the tile size that minimizes T_{base} as follows:

$$\text{minimize} \quad T_{base} = \text{ntiles} \times [\Theta(s) + \Delta(s)] + \Lambda(s) \qquad (15.18)$$
$$\text{subject to} \ F_L(s) \le C$$
$$s_i, s_j \ge 1$$
$$s_i, s_j \in \mathbb{Z}$$

where the last two constraints ensure that s_i and s_j are positive and are integers.

15.3.1.2.5.10 Optimal Tiling Problem Is an IGP

The constrained optimization problem formulated above (Equation 15.18) can be directly cast into an IGP (integer geometric program) of the form of Equation 15.15. The constraints are already in the required form. The objective function T is a posynomial. This can be easily verified by observing that the terms used in the construction of T_{base}, namely, ntiles, $\Theta(s)$, $\Delta(s)$, and $\Lambda(s)$, are all posynomials, and posynomials are closed under addition — the sum of posynomials is a posynomial. Based on the above reasoning, the optimization problem Equation 15.18 is an IGP.

15.3.1.2.5.11 A Sophisticated Execution Time Model

One can also consider a sophisticated execution time model that captures several hardware and compiler optimizations. For example, modern processor architectures support nonblocking caches, out-of-order issue, hardware prefetching, and so on, and compilers can also do latency hiding optimizations such as software prefetching and instruction reordering. As a result of these hardware and compiler optimizations, one can almost completely hide the load–store cost. In such a case, the cost of a tile is not the sum of the computation and communication cost, but the maximum of them. We model this sophisticated execution time with the function T_{opt} as follows:

$$T_{opt} = \text{ntiles} \times \max\left(\Theta(s), \Delta(s)\right) + \Lambda(s)$$

Thanks to our unified view of the optimization problem approach, we can substitute T_{base} with T_{opt} in the optimization problem Equation 15.18 and solve for the optimal tile sizes. However, T_{opt} must be a posynomial for this substitution to yield a GP. We can easily transform T_{opt} to a posynomial by introducing new variables to eliminate the max (.) operator.

15.3.1.3 Tiled Code Generation

An important step in applying the tiling transformation is the generation of the tiled code. This step involves generation of tiled loops and the transformed loop body. Since tiling can be used for a variety of purposes, depending on the purpose, the loop body generation can be simple and straightforward to complicated. For example, loop body generation is simple when tiling is used to improve data cache locality, whereas, in the context of register tiling, loop body generation involves loop unrolling followed by scalar replacement, and in the context of tiling for parallelism, loop body generation involves generation of communication and synchronization. There exist a variety of techniques for loop body generation, and a discussion of them is beyond the scope of this article. We will present techniques that can be used for tiled loop generation both when the tile sizes are fixed and when they are left as symbolic parameters.

15.3.1.3.1 Tiled Loop Generation

We will first introduce the structure of tiled loops and develop an intuition for the concepts involved in generating them. Consider the iteration space of a two-dimensional parallelogram such as the one shown in Figure 15.12, which is the skewed version of the stencil computation. Figure 15.13 shows a geometric view of the iteration space superimposed with a 2×2 rectangular tiling. There are three types of tiles: *full* (which are completely contained in the iteration space), *partial* (which are not completely contained but have a nonempty intersection with the iteration space), and *empty* (which do not intersect the iteration

```
for (i=1; i<=M; i++)
    for (j=i+1; j<=N+i-1; j++)
        A[j-1]=(A[j-1]+A[j-i-1]+A[j-i+1])/3;
```

FIGURE 15.12   Skewed iteration space of the stencil computation.

space). The lexicographically earliest point in a tile is called its *origin*. The goal is to generate a set of loops that *scans* (i.e., visits) each integer point in the original iteration space based on the tiling transformation, where the tiles are visited lexicographically and then the points within each tile are visited lexicographically. We can view the four loops that scan the tiled iteration space as two sets of two loops each, where the first set of two loops enumerate the tile origins and the next set of two loops visit every point within a tile. We call the loops that enumerate the tile origins the *tile-loops* and those that enumerate the points within a tile the *point-loops*.

15.3.1.3.2   *Bounding Box Method*

One solution for generating the tile-loops is to have them enumerate every tile origin in the bounding box of the iteration space and push the responsibility of checking whether a tile contains a valid iteration to the point-loops. The tiled loop nest generated with this bounding box scheme is shown in Figure 15.11. The first two loops (iT and jT) enumerate all the tile origins in the bounding box of size $N \times (N + M)$, and the two inner loops (i and j) scan the points within a tile. A closer look at the point-loop bounds reveals its simple structure. One set of bounds is from what we refer to as the *tile box bounds,* which restrict the loop variable to points within a tile. The other set of bounds restricts the loop variable to points within

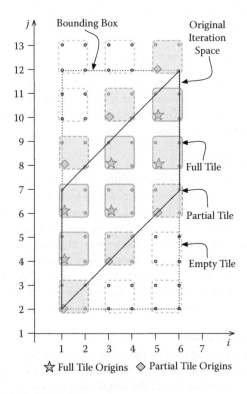

FIGURE 15.13   A 2 × 2 rectangular tiling of the 2D stencil iteration space is shown. The bounding box of the iteration space together with full, partial, and empty tiles and their origins are also shown. Adapted from [92], ©2007 Association for Computing Machinery, Inc., included by permission.

```
for (iT = 0; iT < = ⌊(N/2)⌋; iT++)
  for (jT = max (1, iT); jT <=min (⌊ (2*iT+M+1)/2⌋, (⌊N+M/2⌋); jT++)
    for (i = max (max (1,2*iT), 2*jT-M); i <=min (min (2*iT+1, 2*jT), N); i++)
      for (j = max (2*jT, i+1); j <=min (2* jT+1, i+M); j++)
              /* Loop Body */
```

FIGURE 15.14 Tiled loops generated for fixed tile sizes using the classic scheme.

the iteration space. Combining these two sets of bounds, we get the point-loops that scan points within the iteration space and tiles. Geometrically, the point-loop bounds correspond to the intersection of the tile box (or rectangle) and the iteration space, here the parallelogram in Figure 15.13.

The bounding box scheme provides a couple of important insights into the tiled-loop generation problem. First, the problem can be decomposed into the generation of tile-loops and the generation of point-loops. Such a decomposition leads to efficient loop generation, since the time and space complexity of loop generation techniques is a doubly exponential function of the number of bounds. The second insight is the scheme of combining the tile box bounds and iteration space bounds to generate point-loops. An important feature of the bounding box scheme is that tile sizes need not be fixed at loop generation time, but can be left as symbolic parameters. This feature enables generation of *parameterized tiled loops*, which is useful in iterative compilation, auto-tuners, and runtime tile size adaptation. However, the empty tiles enumerated by tile-loops can become a source of inefficiency, particularly for small tile sizes.

15.3.1.3.3 *When Tile Sizes Are Fixed*
When the tile sizes can be fixed at the loop generation time, an *exact* tiled loop nest can be generated. The tile-loops are exact in the sense that they do not enumerate any empty tile origins. When the tile sizes are fixed, the tiled iteration space can be described as a set of linear constraints. Tools such as OMEGA and CLOOG provide standard techniques to generate loops that scan the integer points in sets described by linear constraints. These tools can be used to generate the tiled loop nest. The exact tiled loop nest for the two-dimensional stencil example is shown in Figure 15.14. Note that the efficiency due to the exactness of the tile-loops has come at the cost of fixing the tile sizes at generation time. Such loops are called *fixed tiled loops*.

The classic schemes for tiled loop generation take as input all the constraints that describe the bounds of the $2d$ loops of the tiled iteration space, where d is the depth of the original loop nest. Since the time–space complexity of the method is doubly exponential on the number of constraints, an increase in the number (from d to $2d$) of constraints might lead to situations where the loop generation time becomes prohibitively expensive. Similar to the bounding box technique, tiled loop generation for fixed tile sizes can also be decomposed into generating tile-loops and point-loops separately. Such a decomposition will reduce the number of constraints considered into each step by half and will improve the scalability of the tiled loop generation method.

15.3.2 Tiling Irregular Applications

Applications that make heavy use of sparse data structures are difficult to parallelize and reschedule for improved data locality. Examples of such applications include mesh-quality optimization, nonlinear equation solvers, linear equation solvers, finite element analysis, N-body problems, and molecular dynamics simulations. Sparse data structures introduce irregular memory references in the form of indirect array accesses (e.g., A[B[i]]), which inhibit compile-time, performance-improving transformations such as tiling. For example, in Figure 15.15, the array A is referenced with two different indirect array accesses, p[i] and q[i].

The flow, memory-based data dependences within the loop in Figure 15.15 can be described with the dependence relation $(i' \rightarrow i)$, where iteration i' depends on the value generated in iteration i.

$$\{(i' \rightarrow i) \mid q(i') = p(i) \text{ and } (1 \leq i, i' \leq N) \text{ and } i' > i\}$$

```
for i = 1 .. N
   A[p[i]]=...
        ...=A[q[i]]
```

FIGURE 15.15   Example loop with irregular memory references.

The uninterpreted functions $p()$ and $q()$ are static place holders for quantities that are unknown until runtime. It is not possible to parallelize or tile the loop in Figure 15.15 without moving some of the required analysis to the runtime.

To address this problem, inspector and executor strategies have been developed where the inspector dynamically analyzes memory reference patterns and reorganizes computation and data, and the executor executes the irregular computation in a different order to improve data locality or exploit parallelism. The ideal role for the compiler in applying inspector and executor strategies is performing program analysis to determine where such techniques are applicable and inserting inspector code and transforming the original code to form the executor. This section summarizes how inspector and executor strategies are currently applied to various loop patterns. The section culminates with the description of a technique called full sparse tiling being applied to irregular Gauss–Seidel.

15.3.2.1 Terminology

Irregular memory references are those that cannot be described with a closed-form, static function. Irregular array references often occur as a result of *indirect array references* where an access to an index array is used to reference a data array (e.g., A[p[i]] and A[q[i]] in Figure 15.15). A *data array* is an array that holds data for the computation. An *index array* is an array of integers, where the integers indicate indices into a data array or another index array.

This section assumes that data dependence analysis has been performed on the loops under consideration. The dependences are represented as relations between integer tuples with contraints specified using Presburger arithmetic including uninterpreted function symbols. *Presburger arithmetic* includes the universal operator $\forall$, existential operator $\exists$, conjunction $\wedge$, disjunction $\vee$, negation $\neg$, integer addition $+$, and multiplication by a constant.

The dependence relations are divided into flow dependences, anti dependences, and output dependences. *Flow dependence* relations are specified as a set of iteration pairs where the iteration in which a read occurs depends on the iteration where a write occurs. The flow dependence relation for Figure 15.15 is as follows:

$$\{(i' \rightarrow i) \mid q(i') = p(i) \wedge (1 \le i, i' \le N) \wedge i' > i\}$$

An *anti dependence* is when a write must happen after a read because of variable reuse. The anti dependence relation for the example in Figure 15.15 is

$$\{(i' \rightarrow i) \mid p(i') = q(i) \wedge (1 \le i, i' \le N) \wedge i' > i\}$$

An *output dependence* is a dependence between two writes to the same memory location. The output dependence relation for the example in Figure 15.15 is

$$\{(i' \rightarrow i) \mid p(i') = p(i) \wedge (1 \le i, i' \le N) \wedge i' > i\}$$

A *reduction loop* has no loop-carried dependences except for statements of the form

```
X = X op expression
```

where X is a reference to a scalar or an array that is the same on the left- and right-hand side of the assignment, there are no references to the variable being referenced by X in the expression on the right-hand side, and op is an associative operator (e.g., addition, max, min). Since associative statements

```
for i = 1 .. n
    for j = 1 .. m
        x[j] = ...
    for k = 1 .. p
        y[i][k] = x[pos[k]]
```

FIGURE 15.16   Compile-time analysis can determine that the x array is privatizable.

may be executed in any order, the loop may be parallelized as long as accesses to X are surrounded with a lock.

15.3.2.2   Detecting Parallelism

In some situations, static analysis algorithms are capable of detecting when array privatization and loop parallelization are possible in loops involving indirect array accesses. Figure 15.16 shows an example where compile-time program analysis can determine that the array x can be privatized, and therefore the i loop can be parallelized. The approach is to analyze the possible range of values that pos[k] might have and verify that it is a subset of the range [1..m], which is the portion of x being defined in the j loop.

If compile-time parallelism detection is not possible, then it is sometimes possible to detect parallelism at runtime. Figures 15.17 and 15.19 show loops where runtime tests might prove that the loop is in fact parallelizable. For the example in Figure 15.17, there are possible sets of flow and anti dependences between the write to A[p[i]] and the read of A[i]. If a runtime inspector determines that for all i, $p(i)$ is greater than N, then the loop is parallelizable. Figure 15.18 shows an inspector that implements the runtime check and an executor that selects between the original loop and a parallel version of the loop.

To show an overall performance improvement, the overhead of the runtime inspector must be amortized over multiple executions of the loop. Therefore, one underlying assumption is that an outer loop encloses the loop to be parallelized. Another assumption needed for correctness is that the indirection arrays p and q are not modified within the loops. Figure 15.19 has an example where the inspection required might be overly cumbersome. In Figure 15.19, there are possible sets of flow and anti dependences between the write to A[p[i]] and the read of A[q[i]]. If it can be shown that for all i and j such that $1 \leq i, j \leq N$, $p(i)$ is not equal to $q(j)$, then there are no dependences in the loop. Notice in Figure 15.20 that for this example, the inspector that determines whether there is a dependence requires $O(N^2)$ time, thus making it quite difficult to amortize such parallelization detection for this example.

15.3.2.3   Runtime Reordering for Data Locality and Parallelism

Many runtime data reordering and iteration reordering heuristics for loops with no dependences or only reduction dependences have been developed. Such runtime reordering transformations inspect data mappings (the mapping of iterations to data) to determine the best data and iteration reordering within a parallelizable loop.

In molecular dynamics simulations there is typically a list of interactions between molecules, and each interaction is visited to modify the position, velocity, and acceleration of each molecule. Figure 15.21 outlines the main loop within the molecular dynamics benchmark moldyn. An outer time-stepping loop makes amortization of inspector overhead possible. The j loop calculates the forces on the molecules using the left and right index arrays, which indicate interaction pairs. In the j loop are two reduction

```
for i = 1 .. N
    A[p[i]] = ... A[i] ...
```

FIGURE 15.17   A runtime inspector can determine this loop is parallelizable if for i = 1 to N, p[i] is greater than N.

```
// inspector
dep = false
for i = 1 . . N
    if p [i] <=N then
        dep = true

for s = . . .
    // executor
    if dep then // serial version
        for i = 1 to N
            A[p[i]] = . . . A[i] . . .

    else      // parallel version
        for all i = 1 to N
            A [p[i]]= . . . A [i] . . .
```

FIGURE 15.18   An inspector checks if any of the writes to array A will occur in the same range as the reads. If not, it is possible to execute the loop from Figure 15.17 in parallel.

```
for i=1 .. N
    A[p[i]] = ... A[q[i]]
```

FIGURE 15.19   To disprove loop-carried dependence in this loop, a runtime inspector would need to show that for all i and j such that $1 \leq i, j \leq N$, $p(i)$ is not equal to $q(j)$.

```
// inspector
dep = false
for i = 1 . . N
    for j = 1 . . N
        if p [i] = = p[j] then
            dep = true

for s = . . .
    // executor
    if dep then // serial version
        for i = 1 to N
            A[p[i]] = . . . A[q[i]]

    else      // parallel version
        for all i = 1 to N
            A [p[i]]= . . . A[q[i]]
```

FIGURE 15.20   An inspector checks if there are any possible conflicts between the reads and writes for the loop in Figure 15.19.

```
for s = 1 .. num_steps
    for i = 1 .. num_nodes
S1    x [i] = x[i] + vx[i] + fx[i]

    for j = 1 .. num_inter
S2    fx[left [j]] += g( x [left [j]], x [right [j]] )
S3    fx[right [j]] -= g( x [left [j]], x [right [j]] )

    for k = 1 .. num_nodes
S4    vx [k] += fx [k]
```

FIGURE 15.21   Simplified moldyn example. Adapted from [102], © 2003 ACM, Inc., included with permission.

```
for s = 1 .. M
  forall i in wavefront(s)
    A[p[i]] = ...
    ... = A[q[i]]
```

FIGURE 15.22   Executor for a loop that has been dynamically scheduled into parallel wavefronts of iterations.

statements where the x-coordinate of the force fx for a molecule is updated as a function of the original x position for that molecule and the x position of some neighboring molecule right[i]. The j loop indirectly accesses the data arrays x and fx with the index arrays left and right.

Runtime data and iteration reorderings are legal for the j loop, because it only involves loop-carried dependences due to reductions. The data and iteration reordering inspectors can be inserted before the s loop, because the index arrays left and right are not modified within s (in some implementations of moldyn the index arrays are modified every 10 to 20 iterations, at which point the reorderings would need to be updated as well). The inspector can use various heuristics to inspect the index arrays and reorder the data arrays x and fx including: packing data items in the order they will be accessed in the loop, ordering data items based on graph partitioning, and sorting iterations based on the indices of the data items accessed. As part of the data reordering, the index arrays should be updated using a technique called pointer update. Iteration reordering is implemented through a reordering of the entries in the index array. Of course in this example, the left and right arrays must be reordered identically since entries left[i] and right[i] indicate an interacting pair of molecules. The executor is the original computation, which uses the reordered data and index arrays.

A significant amount of work has been done to parallelize irregular reduction loops on distributed memory machines. The data and computations are distributed among the processors in some fashion. Often the data is distributed using graph partitioning, where the graph arises from a physical mesh or list of interactions between entities. A common way to distribute the computations is called "owner computes," where all updates to a data element are performed by the processor where the data is allocated. Inspector and executor strategies were originally developed to determine a communication schedule for each processor so that data that is read in the loop is gathered before executing the loop, and at the end of the loop results that other processors will need in the next iteration are scattered. In iterative computations, an owner-computes approach typically involves communication between processors with neighboring data at each outermost iteration of the computation. The inspector must be inserted into the code after the index arrays have been initialized, but preferably outside of a loop enclosing the target loop. The executor is the original loop with gather and scatter sections inserted before and after.

For irregular loops with loop-carried dependences, an inspector must determine the dependences at runtime before rescheduling the loop. The goal is to dynamically schedule iterations into wavefronts such that all of the iterations within one wavefront may be executed in parallel. As an example, consider the loop in Figure 15.15. The flow, anti, and output dependences for the loop are given in Section 15.3.2.1. An inspector for detecting partial parallelism inspects all the dependences for a loop and places iterations into wavefronts. The original loop is transformed into an executor similar to the one in Figure 15.22, where the newly inserted s loop iterates over wavefronts, and all iterations within a wavefront can be executed in parallel.

15.3.2.4   Tiling Irregular Loops with Dependences[13]

The partial parallelism techniques described in Section 15.3.2.3 dynamically discover fine-grained parallelism within a loop. *Sparse tiling* techniques can dynamically schedule between loops or across outermost loops and can create course-grain parallel schedules. Two application domains where sparse tiling techniques have been found useful are iterative computations over interaction lists (e.g., molecular dynamics simulations) and iterative computations over sparse matrices. This section describes *full sparse tiling*, which

---

[13]Parts of this section are adapted from [102], ©ACM, Inc., included with permission, and from [101], with kind permission of Springer Science and Business Media © 2002.

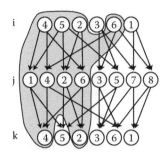

FIGURE 15.23   We highlight the iterations of one sparse tile for the code in Figure 15.21. The j loop has been blocked to provide a seed partitioning. In the full sparse-tiled executor code, the iterations within a tile are executed atomically. Adapted from [102], © 2003, Association for Computing Machinery, Inc., included with permission.

has been used to tile sparse computations across loops in a molecular dynamics benchmark and across the outermost loop of iterative computations.

15.3.2.4.1 *Full Sparse Tiling for Molecular Dynamics Simulations*

The runtime data and iteration reordering transformations described in Section 15.3.2.3 may be applied to the loop j in the molecular dynamics code shown in Figure 15.21. Reordering the data and iterations within the j loop is legal since the j loop is a reduction. Full sparse tiling is capable of scheduling subsets of iterations across the i, j, and k loops in the same example. The full sparse tiling inspector starts with a seed partitioning of iterations in one of the loops (or in one iteration of an outer loop). If other data and iteration reordering transformations have been applied to the loop being partitioned, then consecutive iterations in the loop have good locality, and a simple block partitioning of the iterations is sufficient to obtain an effective seed partitioning. Tiles are *grown* from the seed partitioning to the other loops involved in the sparse tiling by a traversal of the data dependences between loops (or between iterations of an outer loop). Growing from the seed partition to an earlier loop entails including in the tile all iterations in the previous loop that are sources for data dependences ending in the current seed partition and that have not yet been placed in a tile. Growth to a later loop is limited to iterations in the later loop whose dependences have been satisfied by the current seed partition and any previously scheduled tiles.

For the simplified moldyn example, Figure 15.23 shows one possible instance of the data dependences between iterations of the i, j, and k loops after applying various data and iteration reorderings to each of the loops. A full sparse tiling iteration reordering causes subsets of all three loops to be executed atomically as sparse tiles. Figure 15.23 highlights one such sparse tile where the j loop has been blocked to create a seed partitioning. The highlighted iterations that make up the first tile execute in the following order: iterations 4, 5, 2, and 6 in loop i; iterations 1, 4, 2, and 6 in loop j; and iterations 4 and 2 in loop k. The second tile executes the remaining iterations. Figure 15.24 shows the executor that maintains the outer loop over time steps, iterates over tiles, and then within the i, j, and k loops executes the iterations belonging

```
for s = 1 . . num_steps
   for t = 1 . . num_tiles
      for i in sched (t, 1)
         x′ [i] = x′[i] + vx′ [i] + fx′[i]

      for j in sched (t, 2)
         fx′ [left′ [j]] += g(x′ [left′ [j]], x′ [right′ [j]])
         fx′ [right′ [j]] += g(x′ [left′ [j]], x′ [right′ [j]])
      for k in sched (t, 2)
         vx′ [k] += fx′ [k]
```

FIGURE 15.24   Sparse-tiled executor when the composed inspector performs CPACK, lexGroup, full sparse tiling, and tilePack.

```
    for iter =1 .. T
       for i = 0 . . (R-1)
S1       u [i] = f[i]
         for p = ia[i] . . ia [i+1]-1
           if (ja[p] ! = i)
S2             u[i] = u[i] - a [p] * u [ja[p]]
           else
S3             diag[i] = a[p]
S4       u[i] = u[i] / diag[i]
```

FIGURE 15.25 Typical Gauss–Seidel for CSR assuming an initial reordering of sparse matrix rows and columns and entries in vectors **u** and **f**.

to each tile as specified by the schedule data structure. Since iterations within all three loops touch the same or adjacent data locations, locality between the loops is improved in the new schedule.

Full sparse tiling can dynamically parallelize irregular loops by executing the directed, acyclic dependence graph between the sparse tiles in parallel using a master–worker strategy. The small example shown in Figure 15.23 only contains two tiles, where one tile must be executed before the other to satisfy dependences between the i, j, and k loops. In a typical computation where the seed partitions are ordered via a graph coloring, more parallelism between tiles is possible.

15.3.2.4.2 *Full Sparse Tiling for Iterative Computations Over Sparse Matrices*
Full sparse tiling can also be used to improve the temporal locality and parallelize the Gauss–Seidel computation. Gauss–Seidel is an iterative computation commonly used alone or as a smoother within multigrid methods for solving systems of linear equations of the form $Au = f$, where A is a matrix, u is a vector of unknowns, and f is a known right-hand side. Figure 15.25 contains a linear Gauss–Seidel computation written for the compressed sparse row (CSR) sparse matrix format. We refer to iterations of the outer loop as *convergence iterations*. The iteration space graph in Figure 15.26 visually represents an instance of the linear Gauss–Seidel computation. Each iteration point $\langle iter, i \rangle$ in the iteration space represents all the computation for the unknown u_i at convergence iteration *iter* (one instance of S1 and S4 and multiple instances of S2 and S3). The *iter* axis shows three *convergence iterations*. The dark arrows show the data dependences between iteration points for one unknown u_i in the three convergence iterations. The unknowns are indexed by a single variable i, but the computations are displayed in a two-dimensional plane parallel to the x and y axes to exhibit the relationships between iterations. At each convergence iteration *iter* the relationships between the unknowns are shown by the lightly shaded *matrix graph*. Specifically, for each nonzero in the sparse matrix A, $a_{ij} \neq 0$, there is an edge $\langle i, j \rangle$ in the matrix graph. The original

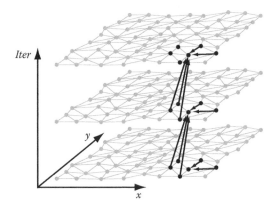

FIGURE 15.26 The arrows show the data dependences for one unknown u_i. The relationships between the iteration points are shown with a matrix graph.

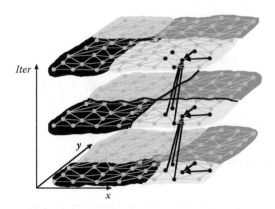

FIGURE 15.27   Full sparse-tiled Gauss–Seidel iteration space. Adapted from [101], with kind permission of Springer Science and Business Media, © 2002.

order, $v = 0, 1, \ldots, (R - 1)$, given to the unknowns and corresponding matrix rows and columns is often arbitrary and can be changed without affecting the convergence properties of Gauss–Seidel. Therefore, if the unknowns are mapped to another order before performing Gauss–Seidel, the final numerical result will vary somewhat, but the convergence properties still hold.

In linear Gauss–Seidel, the data dependences arise from the nonzero structure of the sparse matrix A. Each iteration point $\langle iter, i \rangle$ depends on the iteration points of its neighbors in the matrix graph from either the current or the previous convergence iteration, depending on whether the neighbor's index j is ordered before or after i. The dependences between iteration points within the same convergence iteration make parallelization of Gauss–Seidel especially difficult. Approaches to parallelizing Gauss–Seidel that maintain the same pattern of Gauss–Seidel data dependences use the fact that it is possible to apply an a priori reordering to the unknowns and the corresponding rows of the sparse matrix A. This domain-specific knowledge is impossible to analyze with a compiler, so while automating full sparse tiling, it is necessary to provide some mechanism for a domain expert to communicate such information to the program analysis tool.

Figure 15.27 illustrates how the full sparse tiling inspector divides the Gauss–Seidel iteration space into *tiles*. The process starts by performing a seed partitioning on the matrix graph. In Figure 15.27, the seed-partitioned matrix graph logically sits at the second convergence iteration, and tiles are grown to the first and third convergence iterations.[14] The tile growth must satisfy the dependences. For Gauss–Seidel, that involves creating and maintaining a new data order during tile growth. The full sparse tiling executor is a transformed version of the original Gauss–Seidel computation that executes each tile atomically (see Figure 15.28).

At runtime, the full sparse tiling inspector generates a *data reordering function* for reordering the rows and columns in the matrix, $\sigma(v) : V \rightarrow \{0, \ldots, (R - 1)\}$, and a *tiling function*, $\theta(iter, v) : \{1, .., T\} \times V \rightarrow \{0, \ldots, (k - 1)\}$. The tiling function maps iteration points $\langle iter, v \rangle$ to tiles. From this tiling function, the inspector creates a *schedule function*, $sched(tileID, iter) : \{0, \ldots, (k - 1)\} \times \{1, \ldots, T\} \rightarrow 2^{\{0, \ldots, (R - 1)\}}$. The schedule function specifies for each tile and convergence iteration the subset of the reordered unknowns that must be updated. The transformed code shown in Figure 15.28 performs a tile-by-tile execution of the iteration points using the schedule function, which is created by the inspector to satisfy the following:

$$sched(tileID, iter) = \{\sigma(v) \mid \theta(iter, v) = tileID\}$$

---

[14]The number of iterations for tile growth is usually small (i.e., two to five), and the full sparse tiling pattern can be repeated multiple times if necessary. The tile growth is started from a middle iteration to keep the size of the tiles as small as possible.

```
        for tileID = 0 .. (k-1)
          for iter = 1 .. T
            for i in sched (tileID, iter)
  S1          u' [i] = f' [i]
              for p = ia [i] .. ia[i+1]-1
                if (ja [p] ! = i)
  S2              u' [i] = u' [i] - a [p] * u'[ja[p]]
                else
  S3                diag [i] = a[p]
  S4          u'[i] = u'[i] / diag [i]
```

FIGURE 15.28   Code that performs a serial execution of sparse-tiled Gauss–Seidel for compressed sparse row (CSR).

A matrix graph partitioning serves as a seed partitioning from which tiles can be grown. The seed partitioning determines the tiling at a particular convergence iteration, $iter_s$. Specifically at $iter_s$, where $1 \leq iter_s \leq T$, the tiling function is set to the partition function, $\theta(iter_s, v) = part(v)$. To determine the tiling at other convergence iterations, the tile growth algorithm adds or deletes nodes from the seed partition as needed to ensure that atomic execution of each tile does not violate any data dependences.

The FULLSPARSENAIVE_GSCSR algorithm, shown in Figure 15.29, generates the tiling function θ for the Gauss–Seidel computation in Figure 15.25. While generating the tiling function, ordering constraints between nodes in the matrix graph are maintained in the relation *NodeOrd*. The first two statements in

ALGORITHM FULLSPARSENAIVE_GSCSR $(G(V, E), part(\ ), T, iter_s)$

1: foreach vertex $v \in V, \theta(iter_s, v)$ $\leftarrow part(v)$
2: $NodeOrd \leftarrow \{\langle v, w \rangle | \theta(iter_s, v) < \theta(iter_s, w)$
                          and $(\langle v, w \rangle \in E$ or $\langle w, v \rangle \in E)\}$

Downward tile growth
3: for $iter = (iter_s - 1)$ downto 1
4:     foreach vertex $v \in V, \theta(iter, v) \leftarrow \theta(iter + 1, v)$
5:     do while θ changes
6:         foreach $\langle v, w \rangle \in NodeOrd$
7:             $\theta(iter, w) \leftarrow min(\theta(iter, w), \theta(iter + 1, v))$
8:             $\theta(iter, v) \leftarrow min(\theta(iter, v), \theta(iter, w))$
9:         end foreach
10:    end do while
11:    $NodeOrd \leftarrow NodeOrd \bigcup$
                $\{\langle v, w \rangle | \theta(iter, v) < \theta(iter, w)$
                          and $(\langle v, w \rangle \in E$ or $\langle w, v \rangle \in E)\}$
12: end for

Upward tile growth
13: for $iter = (iter_s + 1)$ to T
14:    foreach vertex $v \in V, \theta(iter, v) \leftarrow \theta(iter - 1, v)$
15:    do while θ changes
16:        foreach $\langle v, w \rangle \in NodeOrd$
17:            $\theta(iter, v) \leftarrow max(\theta(iter, v), \theta(iter, -1, w))$
18:            $\theta(iter, w) \leftarrow max(\theta(iter,, w), \theta(iter, v))$
19:        end foreach
20:    end do while
21:    $NodeOrd \leftarrow NodeOrd \bigcup$
                $\{\langle v, w \rangle | \theta(iter, v) < \theta(iter, w)$
                          and $(\langle v, w \rangle \in E$ or $\langle w, v \rangle \in E)\}$
22: end for

FIGURE 15.29   FullSparseNaive_GSCSR algorithm.

the algorithm initialize the *NodeOrd* relation and all of the tiling function values for the convergence iteration *iter$_s$*. The algorithm then loops through the convergence iterations *iter* such that *iter* < *iter$_s$*, setting θ at each iteration point $\langle iter, v \rangle$ to $\langle iter+1, v \rangle$. Finally, it visits the edges that have endpoints in two different partition cells, adjusting the tiling function θ to ensure that the data dependences are satisfied. The process is repeated for the convergence iterations between *iter$_s$* and *T* in the upward tile growth. Once neighboring nodes, $\langle v, w \rangle \in E$, are put into two different tiles at any iteration *iter*, the relative order between these two nodes must be maintained. The *NodeOrd* relation maintains that relative order. For example, if $\theta(iter, v) < \theta(iter, w)$, then $\langle v, w \rangle \in NodeOrd$.

The running time of this algorithm is $O(Tk|V||E|)$, where *T* is the number of convergence iterations, *k* is the number of tiles, $|V|$ is the number of nodes in the matrix graph, and $|E|$ is the number of edges in the matrix graph. The $k|V||E|$ term is due to the while loops that begin at lines 5 and 15. In the worst case, the while loop will execute $k|V|$ times, with only one $\theta(iter, v)$ value decreasing (or increasing in the forward tile growth) each time through the while loop. Each $\theta(iter, v)$ can take on values between 1 and *k*, where *k* is the number of tiles. In practice, the algorithm runs much faster than this bound.

To exploit parallelism, the inspector creates a tile dependence graph, and the executor for the full sparse-tiled computation executes sets of independent tiles in parallel. The tile dependence graph is used by a master–worker implementation that is part of the executor. The master puts tiles whose data dependences are satisfied on a ready queue. The workers execute tiles from the ready queue and notify the master upon completion. The following is an outline of the full sparse tiling process for parallelism:

- **Partition** the matrix graph to create the seed partitioning.
- **Choose a numbering** on the cells of the seed partition. The numbering dictates the order in which tiles are grown and affects the resulting parallelism in the tile dependence graph (TDG). A numbering that is based on a coloring of a partition interaction graph results in much improved TDG parallelism.
- **Grow tiles** from each cell of the seed partitioning in turn, based on the numbering, to create the tiling function θ that assigns each iteration point to a tile. The tile growth algorithm will also generate constraints on the data reordering function.
- **Reorder** the data using a reordering function that satisfies the constraints generated during tile growth.
- **Reschedule** by creating a schedule function based on the tiling function θ. The schedule function provides a list of iteration points to execute for each tile at each convergence iteration.
- **Generate a TDG** identifying which tiles may be executed in parallel.

15.3.3   Bibliographic Notes

As early as 1969, McKellar and Coffman [71] studied how to match the organization of matrices and their operations to paged memory systems. Early studies of such matching, in the context of program transformation, were done by Abu-Sufah et al. [2] and Wolfe and coworkers [55, 109]. Irigoin and Triolet [49] in their seminal work give validity conditions for arbitrary parallelepiped tiling. These conditions were further refined by Xue [113].

Tiling for memory hierarchy is a well-studied problem, and so is the problem of modeling the cache behavior of a loop nest. Several analytical models measure the number of cache misses for a given class of loop nests. These models can be classified into precise models that use sophisticated (computationally costly) methods and approximate models that provide a closed form with simple analysis. In the precise category, we have the cache miss equations [40] and the refinement by Chatterjee et al. [17], which use Ehrhart polynomials [18] and Presburger formulae to describe the number of cache misses. Harper et al. [44] propose an analytical model of set-associative caches and Cascaval and Padua [15] give a compile-time technique to estimate cache misses using stack distances. In the approximate category, Ferrante et al. [34] present techniques to estimate the number of distinct cache lines touched by a given loop nest.

Sarkar [94] presents a refinement of this model. Although the precise models can be used for selecting the optimal tile sizes, only Abella et al. [1] have proposed a *near-optimal* loop tiling using cache miss equations and genetic algorithms. Sarkar and Megiddo [95] have proposed an algorithm that uses an approximate model [34] and finds the optimal tile sizes for loops of depth up to three.

Several algorithms [16, 19, 47, 54] have been proposed for single-level tile size selection (see Hsu and Kremer [47] for a good comparison). The majority of them use an indirect cost function such as the number of capacity misses or conflict misses, and not a direct metric such as overall execution time. Mitchell et al. [74] illustrate how such local cost functions may not lead to globally optimal performance.

Mitchell et al. [74] were the first to quantify the multilevel interactions of tiling. They clearly point out the importance of using a global metric such as execution time rather than local metrics such as number of misses. Furthermore, they also show, through examples, the interactions between different levels of tiling and hence the need for a framework in which the tile sizes at all the levels are chosen simultaneously with respect to a global cost function. Other results that show the application and importance of multilevel tiling include [14, 50, 75]. Auto-tuners such as PHiPHAC [12] and ATLAS [106] use a model-driven empirical approach to choose the optimal tile sizes.

The description of optimal tiling literature presented above and the GP-based approach presented in this chapter is based on the work of Renganarayanan and Rajopadhye [91][15], who present a general technique for optimal multilevel tiling of rectangular iteration spaces with uniform dependences. YALMIP [68] is a tool that provides a symbolic interface to many optimization solvers. In particular, it provides an interface for defining and solving IGPs.

15.3.3.1 Tiled Loop Generation[16]

Ancourt and Irigoin proposed a technique [6] for scanning a single polyhedron, based on Fourier–Motzkin elimination over inequality constraints. Le Verge et al. [61] proposed an algorithm that exploits the dual representation of polyhedra with vertices and rays in addition to constraints. The general code generation problem for affine control loops requires scanning *unions* of polyhedra. Kelley et al. [53] solved this by extending the Ancourt–Irigoin technique, and together with a number of sophisticated optimizations, developed the widely distributed Omega library [78]. The SUIF [108] tool has a similar algorithm. Quilleré et al. proposed a dual-representation algorithm [80] for scanning the union of polyhedra, and this algorithm is implemented in the CLooG code generator [11] and its derivative Wloog used in the WRaP-IT project.

Code generation for fixed tile sizes can also benefit from the above techniques, thanks to Irigoin and Triolet's proof that the tiled iteration space is a polyhedron if the tile sizes are constants [49]. Either of the above tools may be used (in fact, most of them can generate such tiled code). However, it is well known that since the worst-case complexity of Fourier–Motzkin elimination is doubly exponential in the number of dimensions, this may be inefficient. Goumas et al. [41] decompose the generation into two subproblems, one to scan the tile origins, and the other to scan points within a tile, thus obtaining significant reduction of the worst-case complexity. They propose a technique to generate code for *fixed-sized, parallelogram* tiles.

There has been relatively little work for the case when tile sizes are symbolic parameters, except for the very simple case of *orthogonal* tiling: either rectangular loops tiled with rectangular tiles or loops that can be easily transformed to this. For the more general case, the standard solution, as described in Xue's text [114], has been to simply *extend* the iteration space to a rectangular one (i.e., to consider its bounding box) and apply the orthogonal technique with appropriate guards to avoid computations outside the original iteration space.

Amarasinghe and Lam [4, 5] implemented, in the SUIF tool set, a version of Fourier-Motzkin Elimination (FME) that can deal with a limited class of symbolic coefficients (parameters and block sizes), but the full details have not been made available. Größlinger et al. [42] have proposed an extension to the polyhedral model, in which they allow arbitrary rational polynomials as coefficients in the linear constraints that define the iteration space. Their generosity comes at the price of requiring computationally very expensive

---

[15]Portions reprinted, with permission, from [91], © 2004 IEEE.

[16]Parts of this section are based on [92], © 2007, Association for Computing Machinery, Inc., included with permission.

machinery such as quantifier elimination in polynomials over the real algebra to simplify constraints that arise during loop generations. Because of this their method does not scale with the number of dimensions and the number of nonlinear parameters.

Jiménez et al. [51] develop code generation techniques for register tiling of nonrectangular iteration spaces. They generate code that traverses the bounding box of the tile iteration space to enable parameterized tile sizes. They apply index-set splitting to tiled code to traverse parts of the tile space that include only full tiles. Their approach involves less overhead in the loop nest that visits the full tiles; however, the resulting code experiences significant code expansion.

15.3.3.2   Tiling for Parallelism

Communication-minimal tiling refers to the problem of choosing the tile sizes such that the communication volume is minimized. Schriber and Dongarra [96] are perhaps the first to study communication-minimal tilings. Boulet et al. [13] are the first to solve the communication-minimal tiling optimally. Xue [112] gives a detailed comparison of various communication-minimal tilings.

Hogstedt et al. [45] studied the idle time associated with parallelepiped tiling. They characterize the time processor's wait for data from other processors. Desprez et al. [27] present simpler proofs to the solution presented by Hogstedt et al.

Several researchers [46, 76, 86, 115] have studied the problem of picking the tile sizes that minimize the parallel execution time. Andonov et al. [7, 8] have proposed optimal tile size selection algorithms for n-dimensional rectangular and two-dimensional parallelogram iteration spaces. Our formulation of the optimal tiling problem for parallelism is very similar to theirs. They derive closed-form optimal solutions for both cases. We presented a GP-based framework that can be used to solve their formulation directly. Xue [114] gives a good overview of loop tiling for parallelism.

15.3.3.3   Tiling for Sparse Computations

Irregular memory references are also prevalent in popular games such as Unreal, which was analyzed as having 90% of its integer variables within index arrays such as B [103].

In Section 15.3.2.2, the static analysis techniques described were developed by Lin and Padua [65]. Pugh and Wonnacott [77] and Rus et al. [92] have developed techniques for extending static data dependence analysis with runtime checks, as discussed in Section 15.3.2.2. In [77] constraints for disproving dependences are generated at compile time, with the possibility of evaluating such constraints at runtime. Rus et al. [92] developed an interprocedural hybrid (static and dynamic) analysis framework, where it is possible to disprove all data dependences at runtime, if necessary.

Examples of runtime reordering transformation for data locality include [3, 20, 24, 28, 35, 43, 48, 72, 73, 99]. The pointer update optimization was presented by Ding and Kennedy [28].

Saltz et al. [93] originally developed inspector–executor strategies for the parallelization of irregular programs. Initially such transformations were incorporated into applications manually for parallelism [24]. Next, libraries with runtime transformation primitives were developed so that a programmer or compiler could insert calls to such primitives [25, 98].

Rauchwerger [88] surveys various techniques for dynamically scheduling iterations into wavefronts such that all of the iterations within one wavefront may be executed in parallel. Rauchwerger also discusses many issues such as load balancing, parallelizing the inspector, finding the optimal schedule, and removing anti and output dependences.

Strout et al. developed full sparse tiling [100–102]. Cache blocking of unstructured grids is another sparse tiling transformation, which was developed by Douglas et al. [29]. Wu [110] shows that reordering the unknowns in Gauss–Seidel does not affect the convergence properties.

References

1. J. Abella, A. Gonzalez, J. Llosa, and X. Vera. 2002. Near-optimal loop tiling by mean of cache miss equations and genetic algorithms. In *Proceedings of International Conference on Parallel Processing Workshops*.

2. W. Abu-Sufah, D. Kuck, and D. Lawrie. 1981. On the performance enhancememt of paging systems through program analysis and transformations. *IEEE Trans. Comput.* 30(5):341–56.

3. I. Al-Furaih and S. Ranka. 1998. Memory hierarchy management for iterative graph structures. In *Proceedings of the 1st Merged International Parallel Processing Symposium and Symposium on Parallel and Distributed Processing*, 298–302.

4. S. P. Amarasinghe. 1997. Parallelizing compiler techniques based on linear inequalities. PhD thesis, Computer Science Department, Stanford University, Stanford, CA.

5. Saman P. Amarasinghe and Monica S. Lam. 1993. Communication optimization and code generation for distributed memory machines. In *PLDI '93: Proceedings of the ACM SIGPLAN 1993 Conference on Programming Language Design and Implementation*, 126–38. New York: ACM Press.

6. C. Ancourt. 1991. Génération automatique de codes de transfert pour multiprocesseurs à mémoires locales. PhD thesis, Université de Paris VI.

7. Rumen Andonov, Stephan Balev, Sanjay V. Rajopadhye, and Nicola Yanev. 2003. Optimal semi-oblique tiling. *IEEE Trans. Parallel Distrib. Syst.* 14(9):944–60.

8. Rumen Andonov, Sanjay V. Rajopadhye, and Nicola Yanev. 1998. Optimal orthogonal tiling. In *Euro-Par*, 480–90.

9. C. Bastoul. 2002. Generating loops for scanning polyhedra. Technical Report 2002/23, PRiSM, Versailles University.

10. C. Bastoul, A. Cohen, A. Girbal, S. Sharma, and O. Temam. 2000. Putting polyhedral loop transformations to work. In *Languages and compilers for parallel computers*, 209–25.

11. Cédric Bastoul. 2004. Code generation in the polyhedral model is easier than you think. In *IEEE PACT*, 7–16.

12. Jeff Bilmes, Krste Asanovic, Chee-Whye Chin, and Jim Demmel. 1997. Optimizing matrix multiply using phipac: A portable, high-performance, ANSI C coding methodology. In *Proceedings of the 11th International Conference on Supercomputing*, 340–47. New York: ACM Press.

13. Pierre Boulet, Alain Darte, Tanguy Risset, and Yves Robert. (Pen)-ultimate tiling? *Integr. VLSI J.* 17(1):33–51.

14. L. Carter, J. Ferrante, F. Hummel, B. Alpern, and K. S. Gatlin. 1996. Hierarchical tiling: A methodology for high performance. Technical Report CS96-508, University of California at San Diego.

15. Calin Cascaval and David A. Padua. 2003. Estimating cache misses and locality using stack distances. In *Proceedings of the 17th Annual International Conference on Supercomputing*, 150–59. New York: ACM Press.

16. Jacqueline Chame and Sungdo Moon. 1999. A tile selection algorithm for data locality and cache interference. In *Proceedings of the 13th International Conference on Supercomputing*, 492–99. New York: ACM Press.

17. Siddhartha Chatterjee, Erin Parker, Philip J. Hanlon, and Alvin R. Lebeck. 2001. Exact analysis of the cache behavior of nested loops. In *Proceedings of the ACM SIGPLAN 2001 Conference on Programming Language Design and Implementation*, 286–97. New York: ACM Press.

18. Philippe Clauss. 1996. Counting solutions to linear and nonlinear constraints through Ehrhart polynomials: Applications to analyze and transform scientific programs. In *Proceedings of the 10th International Conference on Supercomputing*, 278–85. New York: ACM Press.

19. Stephanie Coleman and Kathryn S. McKinley. 1995. Tile size selection using cache organization and data layout. In *Proceedings of the ACM SIGPLAN 1995 Conference on Programming Language Design and Implementation*, 279–90. New York: ACM Press.

20. E. Cuthill and J. McKee. 1969. Reducing the bandwidth of sparse symmetric matrices. In *Proceedings of the 24th National Conference ACM*, 157–72.

21. A. Darte, Y. Robert, and F. Vivien. 2000. *Scheduling and automatic parallelization*. Basel, Switzerland: Birkhäuser.

22. A. Darte, R. Schreiber, and G. Villard. 2005. Lattice-based memory allocation. *IEEE Trans. Comput.* 54(10):1242–57.

23. Alain Darte and Yves Robert. 1995. Affine-by-statement scheduling of uniform and affine loop nests over parametric. *J. Parallel Distrib. Comput.* 29(1):43–59.

24. R. Das, D. Mavriplis, J. Saltz, S. Gupta, and R. Ponnusamy. 1992. The design and implementation of a parallel unstructured Euler solver using software primitives. *AIAA J.* 32:489–96.

25. R. Das, M. Uysal, J. Saltz, and Yuan-Shin S. Hwang. 1994. Communication optimizations for irregular scientific computations on distributed memory architectures. *J. Parallel Distrib. Comput.* 22(3):462–78.

26. E. De Greef, F. Catthoor, and H. De Man. 1997. Memory size reduction through storage order optimization for embedded parallel multimedia applications. In *Parallel processing and multimedia*. Geneva, Switzerland. Amsterdam, Netherlands: Elsevier Science.

27. Frederic Desprez, Jack Dongarra, Fabrice Rastello, and Yves Robert. 1997. Determining the idle time of a tiling: New results. In *PACT '97: Proceedings of the 1997 International Conference on Parallel Architectures and Compilation Techniques*, 307. Washington, DC: IEEE Computer Society.

28. C. Ding and K. Kennedy. 1999. Improving cache performance in dynamic applications through data and computation reorganization at run time. In *Proceedings of the 1999 ACM SIGPLAN Conference on Programming Language Design and Implementation (PLDI)*, 229–41. New York: ACM Press.

29. C. C. Douglas, J. Hu, M. Kowarschik, U. Rüde, and C. Weiss. 2000. Cache optimization for structured and unstructured grid multigrid. *Electron. Trans. Numerical Anal.* 10:21–40.

30. P. Feautrier. 1991. Dataflow analysis of array and scalar references. *Int. J. Parallel Program.* 20(1):23–53.

31. P. Feautrier. 1992. Some efficient solutions to the affine scheduling problem. Part I. One-dimensional time. *Int. J. Parallel Program.* 21(5):313–48.

32. P. Feautrier. 1992. Some efficient solutions to the affine scheduling problem. Part II. Multi-dimensional time. *Int. J. Parallel Program.* 21(6):389–420.

33. Agustin Fernández, José M. Llabería, and Miguel Valero-García. 1995. Loop transformation using nonunimodular matrices. *IEEE Trans. Parallel Distrib. Syst.* 6(8):832–40.

34. J. Ferrante, V. Sarkar, and W. Thrash. 1991. On estimating and enhancing cache effectiveness. In *Fourth International Workshop on Languages and Compilers for Parallel Computing*, ed. U. Banerjee, D. Gelernter, A. Nicolau, and D. Padua, 328–43. Vol. 589 of Lecture Notes on Computer Science. Heidelberg, Germany: Springer-Verlag.

35. Jinghua Fu, Alex Pothen, Dimitri Mavriplis, and Shengnian Ye. 2001. On the memory system performance of sparse algorithms. In *Eighth International Workshop on Solving Irregularly Structured Problems in Parallel*.

36. Gautam, DaeGon Kim, and S. Rajopadhye. Scheduling in the $\mathcal{Z}$-polyhedral model. In *Proceedings of the IEEE International Symposium on Parallel and Distributed Systems* (Long Beach, CA, USA, March 26–30, 2007). IPDPS '07. IEEE Press, 1–10.

37. Gautam and S. Rajopadhye. 2007. The $\mathcal{Z}$-polyhedral model. *PPoPP 2007: ACM Symposium on Principles and Practice of Parallel Programming*. In *Proceedings of the 12th ACM SIGPLAN Symposium on Principles and Practice of Parallel Programming* (San Jose, CA, USA, March 14–17, 2007). PPoPP '07. New York, NY: ACM Press, 237–248.

38. Gautam and S. Rajopadhye. 2006. Simplifying reductions. In *POPL '06: Symposium on Principles of Programming Languages*, 30–41. New York: ACM Press.

39. Gautam, S. Rajopadhye, and P. Quinton. 2002. Scheduling reductions on realistic machines. In *SPAA '02: Symposium on Parallel Algorithms and Architectures*, 117–26.

40. Somnath Ghosh, Margaret Martonosi, and Sharad Malik. 1999. Cache miss equations: A compiler framework for analyzing and tuning memory behavior. *ACM Trans. Program. Lang. Syst.* 21(4):703–46.

41. Georgios Goumas, Maria Athanasaki, and Nectarios Koziris. 2003. An efficient code generation technique for tiled iteration spaces. *IEEE Trans. Parallel Distrib. Syst.* 14(10):1021–34.

42. Armin Größlinger, Martin Griebl, and Christian Lengauer. 2004. Introducing non-linear parameters to the polyhedron model. In *Proceedings of the 11th Workshop on Compilers for Parallel Computers*

(CPC 2004), ed. Michael Gerndt and Edmond Kereku, 1–12. Research Report Series. LRR-TUM, Technische Universität München.

43. H. Han and C. Tseng. 2000. A comparison of locality transformations for irregular codes. In *Proceedings of the 5th International Workshop on Languages, Compilers, and Run-time Systems for Scalable Computers*. Vol. 1915 of Lecture Notes in Computer Science. Heidelberg, Germany: Springer.

44. John S. Harper, Darren J. Kerbyson, and Graham R. Nudd. 1999. Analytical modeling of set-associative cache behavior. *IEEE Trans. Comput.* 48(10):1009–24.

45. Karin Hogstedt, Larry Carter, and Jeanne Ferrante. 1997. Determining the idle time of a tiling. In *POPL '97: Proceedings of the 24th ACM SIGPLAN-SIGACT Symposium on Principles of Programming Languages*, 160–73. New York: ACM Press.

46. Karin Hogstedt, Larry Carter, and Jeanne Ferrante. 2003. On the parallel execution time of tiled loops. *IEEE Trans. Parallel Distrib. Syst.* 14(3):307–21.

47. C. Hsu and U. Kremer. 1999. Tile selection algorithms and their performance models. Technical Report DCS-TR-401, Computer Science Department, Rutgers University, New Brunswick, NJ.

48. E. Im and K. Yelick. 2001. Optimizing sparse matrix computations for register reuse in sparsity. In *Computational Science — ICCS 2001*, ed. V. N. Alexandrov, J. J. Dongarra, B. A. Juliano, R. S. Renner, and C. J. K. Tan, 127–36. Vol. 2073 of Lecture Notes in Computer Science. Heidelberg, Germany: Springer-Verlag.

49. F. Irigoin and R. Triolet. 1988. Supernode partitioning. In *15th ACM Symposium on Principles of Programming Languages*, 319–28. New York: ACM Press.

50. M. Jimenez, J. M. Llaberia, and A. Fernandez. 2003. A cost-effective implementation of multilevel tiling. *IEEE Trans. Parallel Distrib. Comput.* 14(10):1006–20.

51. Marta Jiménez, José M. Llabería, and Agustín Fernández. 2002. Register tiling in nonrectangular iteration spaces. *ACM Trans. Program. Lang. Syst.* 24(4):409–53.

52. R. M. Karp, R. E. Miller, and S. V. Winograd. 1967. The organization of computations for uniform recurrence equations. *J. ACM* 14(3):563–90.

53. W. Kelly, W. Pugh, and E. Rosser. 1995. Code generation for multiple mappings. In *Frontiers '95: The 5th Symposium on the Frontiers of Massively Parallel Computation*.

54. Monica D. Lam, Edward E. Rothberg, and Michael E. Wolf. 1991. The cache performance and optimizations of blocked algorithms. In *Proceedings of the Fourth International Conference on Architectural Support for Programming Languages and Operating Systems*, 63–74. New York: ACM Press.

55. Monica S. Lam and Michael E. Wolf. 1991. A data locality optimizing algorithm (with retrospective). In *Best of PLDI*, 442–59.

56. Leslie Lamport. 1974. The parallel execution of DO loops. *Commun. ACM* 17(2) 83–93.

57. V. Lefebvre and P. Feautrier. 1997. Optimizing storage size for static control programs in automatic parallelizers. In *Euro-Par'97*, ed. C. Lengauer, M. Griebl, and S. Gorlatch. Vol. 1300 of Lecture Notes in Computer Science. Heidelberg, Germany: Springer-Verlag.

58. P. Lenders and S. V. Rajopadhye. 1994. Multirate VLSI arrays and their synthesis. Technical Report 94-70-01, Oregon State University.

59. H. Le Verge. 1992. Un environnement de transformations de programmmes pour la synthèse d'architectures régulières. PhD thesis, L'Université de Rennes I, IRISA, Campus de Beaulieu, Rennes, France.

60. H. Le Verge. 1995. Recurrences on lattice polyhedra and their applications. Based on a manuscript written by H. Le Verge just before his untimely death.

61. H. Le Verge, V. Van Dongen, and D. Wilde. 1994. Loop nest synthesis using the polyhedral library. Technical Report PI 830, IRISA, Rennes, France. Also published as INRIA Research Report 2288.

62. Wei Li and Keshav Pingali. 1994. A singular loop transformation framework based on non-singular matrices. *Int. J. Parallel Program.* 22(2):183–205.

63. Amy W. Lim, Gerald I. Cheong, and Monica S. Lam. 1999. An affine partitioning algorithm to maximize parallelism and minimize communication. In *International Conference on Supercomputing*, 228–37.

64. Amy W. Lim, Shih-Wei Liao, and Monica S. Lam. 2001. Blocking and array contraction across arbitrarily nested loops using affine partitioning. In *PPoPP '01: Proceedings of the Eighth ACM SIGPLAN Symposium on Principles and Practices of Parallel Programming*, 103–12. New York: ACM Press.

65. Yuan Lin and David Padua. 2000. Compiler analysis of irregular memory accesses. In *Proceedings of the ACM SIGPLAN Conference on Programming Language Design and Implementation*, 157–68. Vol. 35.

66. B. Lisper. 1990. Linear programming methods for minimizing execution time of indexed computations. In *International Workshop on Compilers for Parallel Computers*.

67. Yanhong A. Liu, Scott D. Stoller, Ning Li, and Tom Rothamel. 2005. Optimizing aggregate array computations in loops. *ACM Trans. Program. Lang. Syst.* 27(1):91–125.

68. J. Löfberg. 2004. YALMIP: A toolbox for modeling and optimization in MATLAB. In *Proceedings of the CACSD Conference*. http://control.ee.ethz.ch/˜joloef/yalmip.php.

69. C. Mauras. 1989. ALPHA: Un langage équationnel pour la conception et la programmation d'architectures parallèles synchrones. PhD thesis, L'Université de Rennes I, Rennes, France.

70. C. Mauras, P. Quinton, S. Rajopadhye, and Y. Saouter. 1990. Scheduling affine parametrized recurrences by means of variable dependent timing functions. In *International Conference on Application Specific Array Processing*, 100–10.

71. A. C. McKellar and E. G. Coffman, Jr. 1969. Organizing matrices and matrix operations for paged memory systems. *Commun. ACM* 12(3):153–65.

72. J. Mellor-Crummey, D. Whalley, and K. Kennedy. 1999. Improving memory hierarchy performance for irregular applications. In *Proceedings of the 1999 ACM SIGARCH International Conference on Supercomputing (ICS)*, 425–33.

73. N. Mitchell, L. Carter, and J. Ferrante. 1999. Localizing non-affine array references. In *Proceedings of the 1999 International Conference on Parallel Architectures and Compilation Techniques*, 192–202.

74. N. Mitchell, N. Hogstedt, L. Carter, and J. Ferrante. 1998. Quantifying the multi-level nature of tiling interactions. *Int. J. Parallel Program.* 26(6):641–70.

75. Juan J. Navarro, Toni Juan, and Toms Lang. 1994. Mob forms: A class of multilevel block algorithms for dense linear algebra operations. In *Proceedings of the 8th International Conference on Supercomputing*, 354–63. New York: ACM Press.

76. Hiroshi Ohta, Yasuhiko Saito, Masahiro Kainaga, and Hiroyuki Ono. 1995. Optimal tile size adjustment in compiling general doacross loop nests. In *ICS '95: Proceedings of the 9th International Conference on Supercomputing*, 270–79. New York: ACM Press.

77. B. Pugh and D. Wonnacott. 1994. Nonlinear array dependence analysis. Technical Report CS-TR-3372, Department of Computer Science, University of Maryland, College Park.

78. W. Pugh. 1992. A practical algorithm for exact array dependence analysis. *Commun. ACM* 35(8):102–14.

79. Fabien Quilleré and Sanjay Rajopadhye. 2000. Optimizing memory usage in the polyhedral model. *ACM Trans. Program. Lang. Syst.* 22(5):773–815.

80. Fabien Quilleré, Sanjay Rajopadhye, and Doran Wilde. 2000. Generation of efficient nested loops from polyhedra. *Int. J. Parallel Program.* 28(5):469–98.

81. P. Quinton, S. Rajopadhye, and T. Risset. 1996. Extension of the alpha language to recurrences on sparse periodic domains. In *ASAP '96*, 391.

82. Patrice Quinton and Vincent Van Dongen. 1989. The mapping of linear recurrence equations on regular arrays. *J. VLSI Signal Process.* 1(2):95–113.

83. S. V. Rajopadhye, S. Purushothaman, and R. M. Fujimoto. 1986. On synthesizing systolic arrays from recurrence equations with linear dependencies. In *Foundations of software technology and theoretical computer science*, 488–503.

84. Easwaran Raman and David August. 2007. Optimizations for memory hierarchy. In *The compiler design handbook: Optimization and machine code generation*. Boca Raton, FL: CRC Press.

85. J. Ramanujam. 1995. Beyond unimodular transformations. *J. Supercomput.* 9(4):365–89.

86. J. Ramanujam and P. Sadayappan. 1991. Tiling multidimensional iteration spaces for nonshared memory machines. In *Supercomputing '91: Proceedings of the 1991 ACM/IEEE Conference on Supercomputing*, 111–20. New York: ACM Press.

87. S. K. Rao. 1985. Regular iterative algorithms and their implementations on processor arrays. PhD thesis, Information Systems Laboratory, Stanford University, Stanford, CA.

88. Lawrence Rauchwerger. 1998. Run-time parallelization: Its time has come. *Parallel Comput.* 24(3–4):527–56.

89. Xavier Redon and Paul Feautrier. 1993. Detection of recurrences in sequential programs with loops. In *PARLE '93: Parallel Architectures and Languages Europe*, 132–45.

90. Xavier Redon and Paul Feautrier. 1994. Scheduling reductions. In *International Conference on Supercomputing*, 117–25.

91. Lakshminarayanan Renganarayana and Sanjay Rajopadhye. 2004. A geometric programming framework for optimal multi-level tiling. In *SC '04: Proceedings of the 2004 ACM/IEEE Conference on Supercomputing*, 18. Washington, DC: IEEE Computer Society.

92. S. Rus, L. Rauchwerger, and J. Hoeflinger. 2002. Hybrid analysis: Static & dynamic memory reference analysis. In *Proceedings of the 16th Annual ACM International Conference on Supercomputing (ICS)*.

93. Joel H. Salz, Ravi Mirchandaney, and Kay Crowley. 1991. Run-time parallelization and scheduling of loops. *IEEE Trans. Comput.* 40(5):603–12.

94. V. Sarkar. 1997. Automatic selection of high-order transformations in the IBM XL Fortran compilers. *IBM J. Res. Dev.* 41(3):233–64.

95. V. Sarkar and N. Megiddo. 2000. An analytical model for loop tiling and its solution. In *Proceedings of ISPASS*.

96. R. Schreiber and J. Dongarra. 1990. Automatic blocking of nested loops. Technical Report 90.38, RIACS, NASA Ames Research Center, Moffett Field, CA.

97. W. Shang and J. Fortes. 1991. Time optimal linear schedules for algorithms with uniform dependencies. *IEEE Trans. Comput.* 40(6):723–42.

98. Shamik D. Sharma, Ravi Ponnusamy, Bongki Moon, Yuan-Shin Hwang, Raja Das, and Joel Saltz. 1994. Run-time and compile-time support for adaptive irregular problems. In *Supercomputing '94*. Washington, DC: IEEE Computer Society.

99. J. P. Singh, C. Holt, T. Totsuka, A. Gupta, and J. Hennessy. 1995. Load balancing and data locality in adaptive hierarchical *N*-body methods: Barnes-Hut, fast multipole, and radiosity. *J. Parallel Distrib. Comput.* 27(2):118–41.

100. M. M. Strout, L. Carter, and J. Ferrante. 2001. Rescheduling for locality in sparse matrix computations. In *Computational Science — ICCS 2001*, ed. V. N. Alexandrov, J. J. Dongarra, B. A. Juliano, R. S. Renner, and C. J. K. Tan. Vol. 2073 of Lecture Notes in Computer Science. Heidelberg, Germany: Springer-Verlag.

101. M. M. Strout, L. Carter, J. Ferrante, J. Freeman, and B. Kreaseck. 2002. Combining performance aspects of irregular Gauss-Seidel via sparse tiling. In *Proceedings of the 15th Workshop on Languages and Compilers for Parallel Computing (LCPC)*.

102. Michelle Mills Strout, Larry Carter, and Jeanne Ferrante. 2003. Compile-time composition of run-time data and iteration reorderings. In *Proceedings of the 2003 ACM SIGPLAN Conference on Programming Language Design and Implementation (PLDI)*.

103. Tim Sweeney. 2006. The next mainstream programming language: A game developer's perspective. Invited talk at *ACM SIGPLAN Conference on Principles of Programming Languages (POPL)*. Charleston, SC, USA.

104. J. Teich and L. Thiele. 1993. Partitioning of processor arrays: A piecewise regular approach. *INTEGRATION: VLSI J.* 14(3):297–332.

105. William Thies, Frédéric Vivien, Jeffrey Sheldon, and Saman P. Amarasinghe. 2001. A unified framework for schedule and storage optimization. In *Proceedings of the 2001 ACM SIGPLAN Conference on Programming Language Design and Implementation*, 232–42, Snowbird, Utah, USA.

106. R. Clint Whaley and Jack J. Dongarra. 1998. Automatically tuned linear algebra software. In *Proceedings of the 1998 ACM/IEEE Conference on Supercomputing (CDROM)*, 1–27. Washington, DC: IEEE Computer Society.

107. D. Wilde. 1993. A library for doing polyhedral operations. Technical Report PI 785, IRISA, Rennes, France.

108. R. P. Wilson, Robert S. French, Christopher S. Wilson, S. P. Amarasinghe, J. M. Anderson, S. W. K. Tjiang, S.-W. Liao, C.-W. Tseng, M. W. Hall, M. S. Lam, and J. L. Hennessy. 1994. SUIF: An infrastructure for research on parallelizing and optimizing compilers. *SIGPLAN Notices* 29(12):31–37.

109. Michael Wolfe. 1989. Iteration space tiling for memory hierarchies. In *Proceedings of the Third SIAM Conference on Parallel Processing for Scientific Computing*, 357–61, Philadelphia, PA: Society for Industrial and Applied Mathematics.

110. C. H. Wu. 1990. A multicolour SOR method for the finite-element method. *J. Comput. Applied Math.* 30(3):283–94.

111. Jingling Xue. 1994. Automating non-unimodular loop transformations for massive parallelism. *Parallel Comput.* 20(5):711–28.

112. Jingling Xue. 1997. Communication-minimal tiling of uniform dependence loops. *J. Parallel Distrib. Comput.* 42(1):42–59.

113. Jingling Xue. 1997. On tiling as a loop transformation. *Parallel Process. Lett.* 7(4):409–24.

114. Jingling Xue. 2000. *Loop tiling for parallelism*. Dordrecht, The Netherlands: Kluwer Academic Publishers.

115. Jingling Xue and Wentong Cai. 2002. Time-minimal tiling when rise is larger than zero. *Parallel Comput.* 28(6):915–39.

16

Architecture Description Languages for Retargetable Compilation

Wei Qin

Department of Electrical & Computer Engineering,
Boston University, Boston, MA
wqin@bu.edu

Sharad Malik

Department of Electrical Engineering,
Princeton University, Princeton, MA
sharad@princeton.edu

16.1 Introduction

Retargetable compilation has posed many challenges to researchers. The ultimate goal of the field is to develop a universal compiler that generates high-quality code for all known architectures. Among all existing compilers, GNU Compiler Collection (GCC) comes closest to this goal by using a myriad of formal algorithms and engineering hacks. Currently, GCC officially supports more than 30 architectures commonly used around the world [12]. Still, for many less known architectures, there is no GCC support. Many recent application-specific instruction set processors (ASIPs) are designed with such specialized architectural features that it is not a straightforward task to create customized compilers for them, let alone retarget GCC. Despite the difficulties, good retargetable compilers are highly desirable for these architectures, especially during their development stages in which many candidate domain-specific features need to be evaluated. If a retargetable compiler is capable of utilizing these candidate features and generating optimized code accordingly, their effectiveness can be quickly evaluated by running the resulting code on an equally retargetable simulator. In this regard, retargetable compilation and retargetable simulation are closely related problems, but retargetable compilation is a much harder one.

The need for retargetability largely comes from the back-ends of compilers. Compiler back-ends may be classified into the following three types: customized, semi-retargetable, and retargetable. Customized back-ends have little retargetability. They are usually written for high-quality proprietary compilers or developed for special architectures requiring nonstandard code generation flow. In contrast, semi-retargetable and retargetable compilers reuse at least part of the compiler back-end by factoring out target-dependent information into machine description systems. The difference of customization and retargetability is illustrated in Figure 16.1.

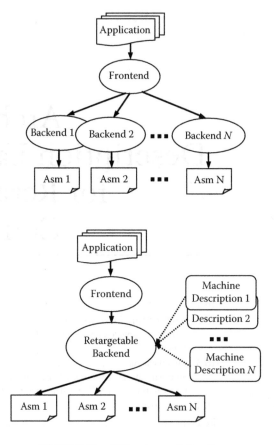

FIGURE 16.1   Different compilation flows.

Semi-retargetable compilers, for example, lcc [11] and GCC [12], share a significant amount of code across the back-ends for different targets. The sharing reduces the time to retarget the compilers to a new processor. Nevertheless, the retargeting effort is still substantial because of target-specific architectural features or function calling conventions. In a semi-retargetable compiler implementation, the instruction set of the target is usually described in tree patterns as is required by popular code generation algorithms [3]. General-purpose register files and usage of the registers are either described as some data structures or simply hard-coded in algorithms. This type of description system serves as the interface between the machine-independent part and the machine-dependent part of the compiler. It typically involves a mixture of the pattern descriptions and C interface functions or macros. It is the primitive form of architecture description languages.

A fully retargetable compiler minimizes coding effort for various target architectures by providing a more friendly machine description interface. Retargetable compilers are important for the development of ASIPs (including digital signal processors [DSPs]). These cost-sensitive processors are usually designed for specialized application domains and have a narrow market window. During their development process, numerous design choices need to be evaluated with benchmarks. It is impractical for designers to manually tune a general-purpose semi-retargetable compiler for each design choice to quantitatively verify its efficacy. A fully retargetable compiler, which can be configured through an architecture description language (ADL), is desirable in such situations.

An ideal ASIP design flow involving an ADL is illustrated as the Y-chart [30] in Figure 16.2. In such a flow, a designer or possibly an automated tool tunes the architecture description. The retargetable compiler reads the description, configures itself, and compiles the applications (usually a set of benchmark programs)

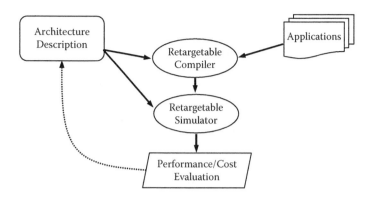

FIGURE 16.2 Design space exploration of ASIP.

into target binary code. The retargetable simulator also reads the description, configures itself, executes the compiled code, and produces performance metrics such as cycle count and power consumption. Based on these metrics, the designer can then tune the architecture and the corresponding architecture description. The updated description will initiate another round of compilation and simulation. The process iterates until a satisfactory trade-off is found. Since the process may iterate for dozens of times, fully retargetable tools driven by a common ADL description are indispensable.

In the past three decades, many interesting ADLs have been introduced together with their related retargetable tools. These ADLs include MIMOLA [33], UDL/I [4], nML [9], ISDL [15], CSDL [48], Maril [7], HMDES [14], TDL [28], LISA [39, 40], RADL [50], EXPRESSION [17], PRMDL [52], and MADL [44]. The common retargetable tools driven by an ADL include a retargetable compiler, an assembler, a disassembler, and a functional simulator. In some rare cases, a cycle-accurate simulator and even hardware synthesis tools are included. This chapter starts with a survey of the existing ADLs. It will then analyze and compare the ADLs and highlight their relative strengths and weaknesses. This will be followed by a study of the necessary elements and the organization of ADLs. The chapter then discusses the major challenges faced by ADL designers and ADL users. In the end, this chapter points out the future architecture trends of ASIPs and their implications for future ADLs.

16.2 Architecture Description Languages

Traditionally, ADLs are classified into three categories: structural, behavioral, and mixed. This classification is based on the nature of the information provided by the language. Structural ADLs describe the microarchitecture organization of the processor. Behavioral ADLs focus on the instruction set architecture (ISA) of the processor, and mixed ADLs contain both microarchitecture and architecture information.

16.2.1 Structural Languages

Structural ADLs are similar to hardware description languages (HDLs). They describe the actual structural organization of microprocessors at the register-transfer level (RT-level). At the RT-level, structural ADLs specify data transformation and transfer among storage units at each clock cycle. It is sufficiently general to model the behavior of arbitrary digital logic, but without the tedious details of gates or transistors. The structural ADLs include MIMOLA [33] and UDL/I [4].

It is worth noting that the RT-level abstraction is not the same as the notion of register-transfer lists (RT-lists) used by computer scientists, though both are often abbreviated as RTL. The latter is a higher abstraction level used for specifying instruction semantics. The major difference between the two is the notion of time. Cycle time is an essential element in an RT-level description. All RT-level events are associated with a specific time stamp and are fully ordered. In contrast, for RT-lists time is not a concern. Only causality is of interest, and events are only partially ordered based on data dependency.

16.2.1.1 MIMOLA

MIMOLA [33] is an interesting combination of a computer hardware description language and a high-level programming language developed at the University of Dortmund, Germany. It was originally proposed for microarchitecture implementation. Over the years, MIMOLA has undergone many revisions, and a number of tools have been developed based on it. These tools cover the areas of hardware synthesis, simulation, test generation, and code generation. They include the MSSH hardware synthesizer, the MSSQ code generator, the MSST self-test program compiler, the MSSB functional simulator, and the MSSU RT-level simulator. MIMOLA was also used by the RECORD compiler as its ADL [32].

MIMOLA contains two parts: the hardware part in which microarchitectures are specified in the form of a component netlist and the software part in which application programs are written using a PASCAL-like syntax. Hardware structures in MIMOLA are described as a network of modules at the RT-level. A simple arithmetic unit module, slightly modified from an example in [33], is shown below.

```
MODULE Alu
  (IN i1, i2: (15:0); OUT outp: (15:0);
   IN ctr: (1:0));
CONBEGIN
  outp <- CASE ctr OF
    0: i1 + i2 ;
    1: i1 - i2 ;
    2: i1 AND i2 ;
    3: i1 ;
    END AFTER 1;
CONEND;
```

The syntax of the hardware part of MIMOLA is similar to the Verilog Hardware Description Language [21]. The first line of the above example declares the module named "Alu." The following two lines describe the names, directions, and bit-widths of the module's ports. The name of the ports can be referred in the rest part of the description, for example, the behavior statements, for their signal values. If multiple statements exist between "CONBEGIN" and "CONEND," they are evaluated concurrently. The "AFTER" statement in the above example describes timing information. It means that the output is produced one cycle after the input is received. For a network of module instances, connections need to be declared to connect the ports of module instances. Below is an example connection statement from [33].

```
CONNECTIONS Alu.outp -> Accu.inp
            Accu.outp -> alu.i1
```

The MSSQ code generator extracts instruction set information from the network of modules. It transforms the RT-level hardware structure into a connection operation graph (COG). The nodes in a COG represent hardware operators or module ports, while the edges represent data flow through the nodes. An instruction tree (I-tree) is also derived from the netlist and the decoder module. It records the encoding information of instructions. During code generation, a MIMOLA application program is translated to an intermediate representation. Then pattern matching is performed to map the intermediate code to COGs. After that, register allocation is performed on the COGs, and binary code is emitted. For the MSSQ compiler to identify important hardware modules such as the register file, the instruction memory, and the program counter, linkage information is provided in MIMOLA as hints. For example, the program counter and instruction memory location can be specified as in the example below [33].

```
LOCATION_FOR_PROGRAMCOUNTER  PCReg;
LOCATION_FOR_INSTRUCTIONS  IM[0..1023];
```

The above statements point out that the "PCReg" is the program counter and the "IM" is the instruction memory. Despite the linkage information, it is a very difficult task to extract the COG and I-trees from the MIMOLA description in general. Extra constraints are imposed for the MSSQ code generator to work properly. The constraints limit the architecture scope of MSSQ to micro-programmable controllers, in which all control signals must originate directly from the instruction word [33].

MIMOLA's software programming part is an extension of PASCAL. It is different from other high-level programming languages in that it allows programmers to designate variables to physical registers and to refer to hardware as procedures calls. For example, to use an operation performed by a module called "Simd," programmers can write the following:

```
x: = Simd(y,z);
```

This special feature helps programmers control code generation and map intrinsics effectively. Compiler intrinsics are assembly instructions disguised as library functions. They help programmers utilize complex machine instructions while avoiding writing inline assembly.

16.2.1.2 UDL/I

A second RT-level ADL used for compiler generation is UDL/I developed at Kyushu University in Japan [4]. ADL descriptions in UDL/I serve as the input to an automated ASIP design system named COACH. From the structural description of a processor, the COACH system will extract the instruction set following the hints given by the designer. Similar to MIMOLA, restrictions must be imposed on the type of processors for successful extraction. According to [4], superscalar and very long instruction word (VLIW) architectures are not supported because of their complicated structure.

In general, RT-level ADLs enable flexible and precise microarchitecture descriptions. An RT-level ADL description can be used by a range of electronic design automation (EDA) tools including logic synthesizers and automatic test generators as well as software tools such as retargetable compilers and simulators. However, for users only interested in retargetable compilers, describing a processor at the RT-level is an overly arduous process. The resulting instruction set information is buried under enormous microarchitecture details that only hardware designers care about. Additionally, the extraction of instruction set information is very difficult in general. Restrictions on the processor microarchitecture must be imposed. Consequently, the generality provided by the RT-level abstraction is lost in the support for code generation.

16.2.2 Behavioral Languages

Behavioral ADLs directly specify the ISAs, avoiding the difficulty of extracting the instruction set. In behavioral ADLs, instruction semantics are directly specified in the form of RT-lists, while detailed hardware structures are ignored. A typical behavioral ADL description is close in its form to an instruction set reference manual. ADLs in this category include nML [9], ISDL [15], and CSDL [48].

16.2.2.1 nML

nML is a simple and elegant formalism for instruction set modeling. It was proposed at the Technical University of Berlin, Germany. nML was used by the retargetable code generator CBC [8] and the instruction set simulator Sigh/Sim [34]. It has also been used by the CHESS [31] code generator and the CHECKERS instruction set simulator at IMEC. CHESS and CHECKERS were later commercialized [51].

Designers of nML observed that in any real-world processor, many instructions share common properties. Factorization based on these common properties can yield a simple and compact representation. Consequently, nML uses a hierarchical scheme to factorize instructions. The root elements of the hierarchy are instructions, and the intermediate elements are partial instructions (PIs). Two composition rules can be used to specify the relationships between the elements: the AND-rule groups several PIs into a larger

PI, and the OR-rule enumerates a set of alternatives for one PI. Thus, instruction definitions in nML can be in the form of an AND–OR tree. Each possible derivation of the tree corresponds to an actual instruction.

nML utilizes attribute grammars [38] to simplify the description of instruction properties. Each element in nML defines a few attributes such as semantic behavior and instruction encoding. The attribute values of a non-leaf element can be derived based on the attribute values of its children. Attribute grammar was later adopted by Instruction Set Description Language (ISDL) and several other ADLs.

An example nML instruction semantics specification is provided below:

```
op num_instruction(a:num_action, src:SRC, dst:DST)
    action {
        temp_src = src;
        temp_dst = dst;
        a.action;
        dst = temp_dst;
    }
    op num_action = add | sub | mul | div
    op add()
    action = {
        temp_dst = temp_dst + temp_src
    }
```

The "num_instruction" definition above combines three PIs using the AND-rule. The first PI, "num_action," is formed through an OR-rule. Any of "add," "sub," "mul," and "div" is a valid option for "num_action." The set of instructions covered by the "num_instruction" is the Cartesian product of the sets of "num_action," "SRC," and "DST." The common behavior of all these instructions is defined in the action attribute of "num_instruction." Each option of "num_action" should have its own action attribute defined as its functional behavior, which is referred to via the "a.action" statement in the "num_instruction" module. Besides the action attribute shown in the example, two additional attributes, image and syntax, can be specified in the same hierarchical manner. Image represents the binary encoding of instructions, and syntax is their assembly mnemonic.

Although generally classified as a behavioral language, nML is not completely free of structural information. In addition to instruction properties, nML also specifies the storage units including RAM, register, and transitory storage. Transitory storage refers to machine states that are retained only for a limited number of cycles, for instance, values on busses and latches. nML assumes a simple timing model: computation has no delay; only storage units have delay. Instruction delay slots can be modeled by using storage units as pipeline registers and by propagating the result through the registers in the behavior specification [9].

nML models constraints between operations by enumerating all valid combinations. For instance, consider a VLIW architecture with three issue slots and two operation types: *A* and *B*. If because of some resource contention at most one type *A* operation can be issued at each cycle, then the user needs to enumerate all the possible issue combinations including *ABB*, *BAB*, *BBA*, and *BBB*. The enumeration of valid cases can make nML descriptions lengthy. More complicated constraints, which often appear in DSPs with irregular instruction level parallelism (ILP) constraints or VLIW processors with many issue slots, are hard to model with nML. For example, nML cannot model the temporal constraint that operation *X* cannot directly follow operation *Y*.

16.2.2.2 ISDL

ISDL tackles the problem of constraint modeling with its explicit constraint specification. ISDL was developed at MIT and used by the Aviv compiler [19]. It was also used by the retargetable simulator generation system GENSIM [16]. ISDL was designed to assist hardware-software codesign for embedded systems. Its target scope is VLIW ASIPs.

ISDL mainly contains five sections:

- Storage resources
- Instruction word format
- Global definition
- Instruction set
- Constraints

Similar to nML, ISDL also contains storage resources. In ISDL, the register files, the program counter, and the instruction memory must be defined for each architecture. An instruction word format section describes the composing fields of the instruction word. ISDL assumes a simple tree-like VLIW instruction model; an instruction word contains a list of fields, and each field contains a list of subfields. Each field corresponds to one operation. For VLIW architectures, an ISDL operation is equivalent to an nML PI, and an ISDL instruction is equivalent to an nML instruction.

The global definition section describes the addressing modes of operations. Production rules for tokens and non-terminals can be defined in this section. Tokens are the primitive operands of instructions. For each token, assembly format and binary encoding information must be defined. An example token definition of a register operand is

```
Token "RA"[0..15] RA {[0..15];}
```

In this example, following the keyword "Token" is the assembly format of the operand. Here any one of "RA0" to "RA15" is a valid choice. Next, "RA" defines the name of the token. The second "[0..15]" defines the range of its value. The value will be used during the definition of behavioral action, binary encoding, and assembly format. The non-terminal is a mechanism provided to exploit commonalities among operations. It is often used to describe complex addressing modes. A non-terminal typically groups a few tokens or other child non-terminals as its arguments, whose assembly formats and return values are referred to when defining the properties of the non-terminal itself. An example non-terminal specification is provided below.

```
Non_Terminal SRC:
    RA {$$ = 0x00000 | RA;} {RFA[RA]} {} {} {} |
    RB {$$ = 0x10000 | RB;} {RFB[RB]} {} {} {};
```

The above example shows a non-terminal named "SRC." It defines an addressing mode with two options: "RA" and "RB." The first pair of braces in each line defines the binary encoding of the operand. In this case it is the result of a bitwise-OR expression. The "$$" symbol indicates the return value of the non-terminal, a usage likely borrowed from Yacc [25]. The second pair of braces contains the semantic action. Here the "SRC" operand refers to the data value in the register indexed by the return value of token "RA" as its action. The following three empty pairs of braces specify side effects, cost, and time, which are all nonexistent for this simple operand. These three fields exist in the instruction set section as well.

The instruction set section of an ISDL describes the instruction set in terms of its fields. For each field, a list of alternative operations can be described. Similar to non-terminals, an operation contains a name, a list of tokens or non-terminals as arguments, a binary encoding definition, an action definition, side effects, and costs. Side effects refer to behaviors such as the setting or clearing of a machine flag. Three types of cost can be specified: execution cycles, encoding size, and stall. Stall models the cycle number of pipeline stalls if the next instruction uses the result of the current instruction. The timing model of ISDL contains two parameters: latency and usage. Latency is the number of fetch cycles before the result of the current operation becomes available. Usage is the cycle count that the operation spends in its slot. The difference between the latency and the stall cost is not clear from the available publications of ISDL. The most interesting part of ISDL is its explicit constraint specification. In contrast to nML, which enumerates all valid combinations, ISDL defines invalid combinations in the form of Boolean expressions. This often results in a much simpler constraint specification. It also enables ISDL to capture more irregular ILP constraints. Recall the temporal constraint that instruction X cannot directly follow Y, which cannot be modeled by nML. ISDL can describe the constraint as follows [15]:

```
~(Y *) & (([1] X *, *)
```

The above "[1]" indicates a one-cycle delay between the fetch bundle (instruction) containing Y and the one containing X. The " " is an operator for logic complement and "&" for logical and. The detailed grammar of the Boolean constraints syntax is available in [19]. The way ISDL models constraints affects the code generation process; a constraint checker is needed to check if the selected instructions meet the constraint. In case of a checking failure, alternative code must be generated.

Overall, ISDL provides the means for compact, hierarchical specification of instruction sets. Its Boolean constraint specification models irregular ILP constraints effectively. A shortcoming of ISDL is that its simple tree-like instruction format forbids the description of instruction sets with multiple encoding formats.

16.2.2.3 CSDL

CSDL (Computer System Description Languages) [48] is a family of machine description languages used by the Zephyr compiler infrastructure. The CSDL family currently includes CCL [5], a function-calling convention-specification language; SLED [49], a formalism describing instruction assembly format and a specification language for functions calling conventions, binary encoding, and λ-RTL [47], an RT-list description language for instruction semantics.

SLED (Specification Language for Encoding and Decoding) was developed as part of the New Jersey Machine-Code Toolkit, which helps programmers build binary editing tools. The retargetable linker **mld** [10] and the retargetable debugger **ldb** [46] were developed based on the toolkit. Similar to ISDL, SLED uses a hierarchical model for machine instructions. A SLED instruction is composed of one or more tokens, each of which is composed of one or more fields. The *pattern* construct groups the fields together and binds them to binary encoding values. The *directive* construct concatenates the fields to form instruction words. A detailed description of the syntax can be found in [49]. SLED does not assume a single format of the instruction set. Therefore, it is sufficiently flexible to describe the encoding and the assembly format of both reduced instruction set computers (RISCs) and complex instruction set computers (CISCs). Like nML, SLED enumerates valid combinations of fields. There is neither a notion of hardware resources nor explicit constraint descriptions. Therefore, in its current form, SLED is not suitable for describing VLIW instruction sets.

The **vpo** (very portable optimizer) in the Zephyr system is capable of instruction selection, instruction scheduling, and global optimization. Instruction sets are represented as RT-lists in **vpo**. Since the raw RT-list form is verbose, λ-RTL was developed to improve description efficiency. A λ-RTL description is translated into RT-lists for the use of **vpo**. According to its developers [47], λ-RTL is "a high order, strongly typed, polymorphic, and pure functional language" based on Standard ML [35]. It has many high-level language features such as name space (through the *module* and *import* directives) and function. Users can even overload basic operators to introduce new semantics and precedence.

The CSDL family of languages does not specify any timing information. Therefore, CSDL does not contain enough information for supporting code generation and scheduling of VLIW processors. It is more suitable for modeling conventional general-purpose processors.

In summary, the behavioral languages share one common feature: hierarchical description of the instruction set based on attribute grammars [38]. This feature greatly simplifies the description of instruction sets by sharing common parts among operations. However, the lack of solid pipeline and timing information prevents these languages from being useful to modern optimizing schedulers. Also, the behavioral ADLs cannot be used to generate a cycle-accurate simulator without assumptions on the control behavior of the architecture. In other words, underlying architecture templates must be used to interpret the full behavior of the ADL descriptions.

16.2.3 Mixed Languages

Similar to behavioral languages, mixed ADLs often use RT-lists for specifying instruction semantics. They also include abstract hardware resources and temporal behavior of the instructions.

16.2.3.1 Maril

Maril is a mixed ADL used by the retargetable compiler Marion [7], which targets RISC-style processors. Maril contains both instruction set information and coarse-grained structural information. Its structural information contains storage units as well as highly abstracted pipeline units. Compiler back-ends for the Motorola 88000 [36], the MIPS R2000 [26], and the Intel i860 [22] architectures were developed based on Maril descriptions.

Maril contains the following three sections:

- **Declaration:** The declaration section describes structural information, such as register files, memory, and abstract hardware resources. Abstract hardware resources include pipeline stages and data buses. They are useful for specifying reservation tables, which are essentially temporal mapping relationships between instructions and architecture resources. A reservation table captures the resource usage of an instruction at every cycle starting with the fetch. It is often a one-to-many mapping since there may be multiple alternative paths for an instruction. Besides hardware structures, the declaration section also contains information such as the range of immediate operands or relative branch offset. This is necessary for the correctness of generated code.

- **Runtime model:** The runtime model section specifies the conventions of the architecture, mostly the function calling conventions. However, it is not intended to be a general framework for specifying these, but instead, is a parameter system for configuring the Marion compiler. Specifying the function calling conventions is a complicated subject and falls beyond the scope of this chapter. Interested readers can refer to [5] for more information.

- **Instruction:** The instruction section describes the instruction set. Each instruction is specified in five parts. The first part is the instruction mnemonics and operands, which specify the assembly format of instructions. The optional second part declares data type constraints of the operation for code selection use. The third part describes an expression that can be used by the Marion code generator as the tree-pattern of the instruction. Only one assignment can occur in the expression since the tree-pattern can have only one root node (and an assignment operator is always a root). As a result, this part cannot specify instruction behaviors such as side effects on machine flags and post-increment memory addressing modes because they all involve additional assignments. This limitation can be worked around for code generation purposes since code generators require simplified semantics specification anyway. But for simulators, the Maril description seems insufficient. The fourth part of the instruction description is the reservation table of the instruction. The abstract resources used in each cycle can be described here, starting from the instruction fetching cycle. The fifth and last part of the instruction description is a triple of (*cost, latency, slot*). The cost is used to distinguish actual instructions from the so-called dummy instructions, which are used for type conversion. The latency is the number of cycles before the result of the instruction can be used by other operations. The slot specifies the delay slot count. Instruction encoding information is not provided in Maril. An example definition integer "Add" instruction is given below [7].

```
%instr  Add r, r, r  (int)
{$3 = $1+$2;}
{IF; ID; EX; MEM; WB} (1,1,0)
```

The operands of the "Add" instruction are all general-purpose registers, as denoted by the "r"s in the first line. The ";" in the reservation table specification delimits cycle boundaries. The "Add" instruction will go through five pipeline stages in five cycles. It has a cost of one, takes one cycle, and has no delay slot.

In general, Maril is designed for code generation and scheduling of RISC architectures. Maril does not describe VLIW instruction sets, nor does it provide enough information for cycle-accurate simulation. Some information in Maril is not part of the target architecture, but hints to Marion. Maril does not utilize a hierarchical instruction set description scheme, as do most behavioral languages. Nonetheless, compared with the behavioral languages, it carries more structural information than just storage units. Its

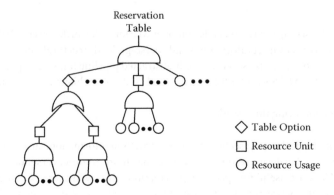

FIGURE 16.3   HMDES reservation table hierarchy.

description of reservation tables enables resource-based instruction scheduling, which can bring notable performance improvement for deeply pipelined processors.

16.2.3.2   HMDES

HMDES is an ADL emphasizing scheduling support [14]. It was developed at the University of Illinois at Urbana-Champaign for the IMPACT research compiler. Because IMPACT focuses on exploring ILP, the reservation tables of instructions become the major content of HMDES. There is no explicit notion of instruction or operation bundle in HMDES. Instead, it represents VLIW issue slots as artificial resources that are consumed by operations. IMPACT is interested in wide-issue architectures in which a single instruction can have numerous scheduling alternatives. For example, if in an eight-issue processor an "Add" instruction can go from any of the eight decoding slots to any of the eight function units, there are in total 64 scheduling alternatives. To avoid laboriously enumerating the alternatives, an AND–OR tree structure is used in HMDES to compress reservation tables. Figure 16.3 shows the description hierarchy used in HMDES. The leaf nodes are called resource usages, which are tuples of (*resource, time*).

A special feature of HMDES is its preprocessing constructs. It performs C-like preprocessing tasks such as conditional file inclusion and macro expansion. More complex preprocessing constructs such as loops are also supported. The preprocessing does not provide extra description power, but helps keep the description compact and readable. Instruction semantics, assembly format, and binary encoding information are not part of HMDES since IMPACT is not designed to be a fully retargetable compiler. Instead, it uses manually written code generation back-ends. After generating machine code, the back-ends query the machine database based on HMDES to do ILP scheduling. A detailed description of the interface between the machine description system and the compiler back-end can be found in the documentation of Elcor [1].

16.2.3.3   TDL

TDL (Target Description Language) was developed at Saarland University in Germany. The language is used in a retargetable post-pass assembly-based code optimization system called PROPAN [27]. In PROPAN, a TDL description is transformed into a parser for target assembly code and a set of ANSI C files containing data structures and functions. The C files are included in the application as a means of generic access to architecture information. A TDL description consists of four parts:

- **Resource section:** Storage units such as register files and memory have built-in syntax support in this section. Moreover, TDL allows the description of the cache memory, which partially exposes the memory hierarchy to the compiler and allows for more optimization opportunities. Function units are also described in this section. TDL also provides flexible syntax support for the user to define generic function units that do not belong to any predefined category.

- **Instruction set section:** Similar to the behavioral languages, TDL utilizes the attribute grammar [38]. TDL supports VLIW architectures explicitly, so it distinguishes the notions of operation and instruction. However, no binary encoding information is provided by TDL. An adapted example description from [28] for an integer add operation is shown below.

```
DefineOp IAdd "%s=%s+%s"
    {dst1="$1" in {gpr}, src1="$2" in {gpr}, src2="$3" in {gpr}},
    {ALU1(latency=1, slots=0, exectime=1) |
     ALU2(latency=1, slots=0, exectime=2);
     WbBus(latency=1)},
    {dst1 := src1+src2;}.
```

The above operation definition contains a name, an assembly format, a list of predefined attributes such as the locations and the types of the source and destination operands, the reservation table, and the semantics in RT-lists form. In this example, the destination operand "$1" corresponds to the first "%s" in the assembly format and is from the register file named "gpr." The operation can be scheduled on either function unit "ALU1" or "ALU2." It will also occupy the result write back bus ("WbBus") resource. The operation performs addition. The detailed TDL syntax rules for describing operation semantics can be found in [28]. This section also contains an operation class definition that groups operations into groups for ease of reference. The format of an instruction word can be defined based on the operation classes. For instance, a declaration "InstructionFormat ifo2 [opclass3, opclass4];" means the instruction format "ifo2" contains one operation from opclass3 and one from opclass4. Similar to ISDL, TDL also provides a non-terminal syntax construct to capture common components between operations.

- **Constraints section:** The Boolean expression used in ISDL is based on lexical elements in the assembly instruction. TDL also uses Boolean expressions for constraint modeling, but its expressions are based on explicitly declared operation attributes defined in the other sections. A constraint definition includes a premise part followed by a rule part, separated by a colon. An example definition from [28] is given below.

```
op1 in {MultiAluFixed} & op2 {MultiMulFixed} :
    !(op1 && op2) | op1.src1 in {iregC} & op1.src2 in {iregD}
                  & op2.src1 in {iregA} & op2.src2 in {iregB}
```

The example specifies the constraint for two operations from the "MultiAluFixed" class and the "MultiMulFixed" class, respectively. Either they do not coexist in one instruction word or their operands have to be in the designated register files. The constraint specification is equally powerful to that of ISDL, but in a cleaner syntax. The Boolean expression will be transformed to integer linear programming constraints to be used in the PROPAN system.

- **Assembly section:** This section describes the lexical elements in assembly code including comment syntax, operation and instruction delimiters, and assembly directives.

Overall, TDL provides a well-organized formalism for describing the assembly code of VLIW DSPs. It supports preprocessing and description of hardware resources such as caches. It also meticulously defines the rules for describing RT-lists. It uses a hierarchical scheme to exploit the commonality among operations and provides both resource-based (function units) and explicit Boolean constraints for constraint modeling. However, its timing model and reservation table model seem to be restricted by the syntax. There is no way for users to flexibly specify operand latency — the cycle time when operands are read and written. These restrictions prevent the use of TDL for accurate RISC architecture modeling. Furthermore, register ports and data transfer paths are not explicitly modeled in the resource section. Both are often resource bottlenecks in VLIW DSPs.

A related but more restrictive machine description formalism for assembly code analysis and transformation applications can be found in the SALTO framework [6]. Its organization is similar to that of TDL.

16.2.3.4 EXPRESSION

The explicit reservation table description in HMDES is not natural and intuitive. Users have to manually translate pipeline structures to abstract resources, which is an annoying task during design space exploration. The EXPRESSION [17] ADL eliminates such manual effort. In EXPRESSION, one can describe a graph of pipeline stages and storage units. The EXPRESSION compiler can automatically generate reservation tables based on the graph. In contrast to MIMOLA's fine-grained netlist, EXPRESSION uses a much coarser representation. Its graph style specification is more friendly to computer architects.

EXPRESSION was developed at the University of California at Irvine. It was used by the research compiler EXPRESS [18] and the research simulator SIMPRESS [29], developed at the same university. A graphical user interface (GUI) front-end for EXPRESSION was also reported [29]. EXPRESSION takes a balanced view of behavioral and structural descriptions. It contains a behavioral section and a structural section. The behavioral section is similar to that of ISDL in that it distinguishes instructions and operations, but it does not cover assembly format and binary encoding. Nor does it use any hierarchical structure for describing operation properties.

The behavioral section contains three subsections: operation, instruction, and operation mapping. The operation subsection is in the form of RT-lists, but detailed rules for semantics description are not publicly available. Similar to TDL, EXPRESSION groups similar operations together for ease of reference. Operation addressing modes are also defined in the operation subsection. The instruction subsection describes the bundling of operations that can be issued in parallel. Instruction width, issuing widths, and the function units bound to the issuing slots are in this section. Such information is essential for VLIW modeling. The operation mapping subsection specifies the translation rules when generating code. Two types of mapping can be specified: mapping from a compiler's intermediate generic operations to target operations and mapping from some target operations to other target-specific operations. The first type of mapping is used for code generation. The second type is required for code optimization purposes. Predicates can be specified for conditional mappings. The mapping subsection in EXPRESSION makes code generation much easier, but it also makes the EXPRESSION language dependent on the code generation algorithm, which is based on direct mapping.

The most interesting part of EXPRESSION is its structural specification in a graph. This part contains three subsections: component declaration, path declaration, and memory subsystem declaration. In the component subsection, coarse-grained structural units such as pipeline stages, memory controllers, memory banks, register files, and latches are specified. Linkage resources such as ports and connections can also be declared in this part. A pipeline stage declaration for the example architecture in Figure 16.4 is shown below.

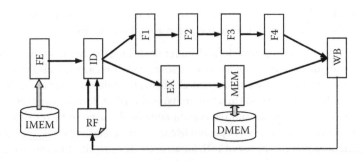

FIGURE 16.4   Example DLX pipeline.

```
(DECODEUnit  ID
   (LATCHES  decodeLatch)
   (PORTS    ID_srcport1  ID_srcport2)
   (CAPACITY 1)
   (OPCODES  all)
   (TIMING   all  1)
)
```

The above example describes an instruction decoding stage. The "LATCHES" statement refers to the output pipeline register of the unit. The "PORTS" statement refers to the abstract data ports of the unit. Here the "ID" unit has two register read ports. "CAPACITY" describes the number of instructions that the pipeline stage can hold at the same time. Normal function units have a capacity of one, while the fetching and decoding stages of VLIW or superscalar processors can have a capacity as large as the issue width of the processor. "OPCODES" describes the operations that can go through this stage. The "all" refers to a predefined operation group containing all operations. "TIMING" is the cycle count that operations spend in the unit. In this example, each operation takes one cycle. Timing can also be specified on a per-operation basis.

In the path subsection of the structural description, the pipeline paths and the data paths between pipeline stages and storage units are specified. This subsection connects the components into a graph. The pipeline path declaration stitches together the pipeline stages to form a directed acyclic pipeline graph. An example path declaration for the simple DLX architecture [20] in Figure 16.4 is shown below.

```
(PIPELINE FE ID EX MEM WB)
(PIPELINE FE ID F1 F2 F3 F4 WB)
```

Recall that the "OPCODES" attribute of pipeline stages declares the operations that can go through each stage, so the paths that an operation may traverse are those whose pipeline stages can all accommodate the operation. Since the time spent by each operation in each stage is specified in the "TIMING" declaration in the component subsection, a reservation table can be generated for the operation. In addition to pipeline stage resources, the usage of register ports and data transfer paths by an operation can also be inferred from its operands and the connections between the pipeline stages and the storage units. The ports and data transfer paths are modeled as resources in the reservation tables too. Compared to the explicit description style of reservation tables in HMDES and Maril, EXPRESSION's graph style is more convenient and intuitive. However, the expressiveness of EXPRESSION in this regard is limited for the same reason. For example, it cannot describe the situation in which an operation should occupy two pipeline stages at the same cycle, which often occurs in floating point pipelines or when artificial resources [45] need to be introduced for irregular ILP constraint modeling.

The last structural subsection is the memory model. EXPRESSION adopts a parameterized memory hierarchy model. Description of memory hierarchy is useful for optimizing compilers to improve the cache behavior of the generated code. EXPRESSION and TDL are the only ADLs that address memory hierarchy.

In general, EXPRESSION captures the data path information in the processor. As with all the aforementioned mixed languages, the control path is not explicitly modeled. Thus, it does not contain complete information for cycle-accurate simulation. An underlying architecture template is necessary to provide the missing information of the control path for simulation purposes. The behavioral model of EXPRESSION does not utilize hierarchical techniques such as the AND–OR tree. This makes it tedious to specify a complete instruction set. The VLIW instruction composition model is simple. It is not clear if interoperation constraints such as sharing of common fields can be modeled. Such constraints can be modeled in ISDL through cross-field encoding assignment.

16.2.3.5 LISA

The emphasis of the behavioral languages is on instruction set specification. Mixed languages look beyond that and provide coarse-grained data path information. However, the description of control paths is largely ignored by the aforementioned mixed languages. This is probably due to the lack of convenient formalism for modeling control paths. Complex control-related behaviors such as speculation and zero-overhead loops are very difficult to model cleanly. Nevertheless, modeling of control behaviors is valuable for code generation and simulation. As the pipeline structures of high-end processors become increasingly complicated, control behaviors such as branching and speculation can have a significant effect on the performance of a processor.

The LISA (Language of Instruction Set Architecture) [40] ADL is capable of accurately specifying the control-related behaviors of a processor. It was developed at Aachen University of Technology in Germany. LISA's development accompanies that of a production-quality cycle-accurate simulator [39]. LISA has many features of an imperative programming language. Using a C-like syntax, it can specify control-related behaviors such as pipeline flush or stall. The flexibility of the imperative-style description allows LISA to specify a wide range of processors.

LISA contains mainly two types of top-level declarations: resource and operation. Resource declarations cover hardware resources including register files, memories, program counters, and pipeline stages. A pipeline definition in LISA enumerates all possible pipeline paths that the operations can go through. A pipeline description corresponding to the same example in Figure 16.4 is as follows:

```
PIPELINE pipe = {FE; ID; EX; MEM; WB}
PIPELINE pipe_fp = {FE; ID; F1; F2; F3; F4; WB}
```

Similar to Maril, the ";"s above are used to delimit the cycle boundary. Machine operations are described with respect to the pipeline stages in LISA. Its basic description unit is an OPERATION, which specifies the behavior, the encoding, and the assembly format of similar instructions at one pipeline stage. The OPERATIONs form an AND–OR tree structure in LISA, greatly reducing the amount of code for a processor. In a slightly modified example from [39], the decoding behavior for arithmetic operations in the DLX ID stage can be described as below.

```
OPERATION arithmetic IN pipe.ID {
  DECLARE {
    GROUP opcode={ADD || ADDU || SUB || SUBU}
    GROUP rs1, rs2, rd = {fix_register};
  }
  CODING {opcode rs1 rs2 rd}
  SYNTAX {opcode rd "," rs1 "," rs2}
  BEHAVIOR {
    reg_a = rs1;
    reg_b = rs2;
    cond = 0;
  }
  ACTIVATION {opcode, writeback}
}
```

The above example captures the common behavior of the arithmetic instructions "ADD," "ADDU," "SUB," and "SUBU" in the decoding stage, as well as their assembly format and binary encoding. In general, an "OPERATION" declaration can contain several parts:

- DECLARE, where local identifiers are specified.
- CODING, where the binary encoding of the operation is described.
- SYNTAX, where the assembly format of the operation is declared.
- BEHAVIOR, where the exact instruction semantics including side effects are specified in a C-like syntax.
- ACTIVATION, where the follow-up operations are activated.

In LISA, one OPERATION can activate another OPERATION. The firing time of the activated OPERATION is dependent on the distance between the two OPERATIONs in the pipeline path. The above

LISA example activates one of the opcodes, which are OPERATIONs declared with respect to the "EX" stage. Since "EX" directly follows "ID," the activated opcode will execute in the following cycle. In the same example [39], an opcode "ADD" is declared as follows:

```
OPERATION ADD IN pipe.EX {
  CODING {0b001000}
  SYNTAX {"ADD"}
  BEHAVIOR {alu = reg_a + reg_b;}
}
```

Each LISA description contains a special main OPERATION, which will be activated at every cycle. It serves as a kernel loop in simulation. Built-in pipeline control functions such as "shift," "stall," and "flush" can be invoked in the main loop. One advantage of LISA is that the user can describe detailed control path information with the activation model. Control path description is important in generating a real cycle-accurate simulator. A fast and accurate simulator for Texas Instrument's TMS320C6201 VLIW DSP [53] based on LISA was reported [39].

To use a LISA description for code generation, the instruction set needs to be extracted. First, the AND–OR tree formed by the OPERATIONs can be analyzed and expanded. Each derivation represents one instruction. It contains architecture information such as the semantics of the instruction, as well as microarchitecture information such as timing and bypassing. Second, semantics information can be extracted in the form of RT-lists. However, for complicated processors, the extraction task is not straight-forward. Consequently, LISA supports an optional SEMANTICS subsection for users to specify simplified instruction semantics directly. The use of the SEMANTICS subsection introduces redundancy into the LISA description since its content overlaps with the BEHAVIOR subsection. Similar redundancy can also be found in EXPRESSION, whose operation subsection and mapping subsection are related to the semantics specification.

16.2.3.6 MADL

The MESCAL Architecture Description Language (MADL) was developed at Princeton University as part of the MESCAL project [44]. It provides support to tools including the cycle-accurate simulator, the functional simulator, the compiler, the assembler, and the disassembler. MADL is an open-source ADL. Its reference manual and the MADL compiler are available for download from [41].

MADL is based on a flexible computation model called the Operation State Machine (OSM) [43]. The OSM model adopts a two-layer view of microprocessors: an operation layer that defines instruction semantics and resource consumption and an abstract hardware layer that controls the resource tokens. The operation layer contains a set of finite state machines (FSMs), each representing an instruction in the pipeline. The hardware layer contains a set of token managers, each of which controls a set of resource tokens. The FSMs communicate with the token managers through token transactions. The model is very flexible in modeling all types of processor architectures, including superscalar and VLIW.

MADL recognizes that there may be both tool-independent and tool-dependent information in descriptions. The former represents the architecture and can be shared among different tools, while the latter serve only individual tools using the ADL. To separate these two types of information, MADL is designed with a two-level structure, the core level and the annotation level. The core level of MADL is a description language for the operation layer of the OSM model. This level defines the FSMs and their interaction with the token managers. Based on the well-defined semantics of the OSM model, the core description is executable. The annotation level of MADL is a generic description language for specifying tool-dependent information. For example, hints for the compiler can be placed in this level. The hints help the compiler extract useful information for its optimizers.

Overall, MADL is designed for balanced support of simulation and compilation. The fast cycle-accurate simulator in the popular SimIt-ARM [42] package was generated from an MADL description of the StrongARM processor. An MADL description is also highly analyzable. For retargetable compilation, it is convenient to extract reservation tables from MADL descriptions owing to the resource-based abstraction used in OSM. MADL also allows users to embed instruction properties into FSMs, such as semantics and assembly encoding. It supports the AND–OR tree description style for both FSMs and instruction properties, allowing the descriptions to be very compact. In theory, it is possible to extract instruction semantics information from MADL for code generation purposes. However, because of the complexity of implementing a retargetable code generator, it remains a work in progress.

16.2.3.7   Other ADLs

An ADL similar to LISA is RADL (Rockwell Architecture Description Language) [50]. It was developed as a follow-up of some earlier work on LISA. The only purpose of RADL is to support cycle-accurate simulation. Control-related library functions such as "stall" and "kill" may be used in RADL descriptions to control the pipeline. There is no available information on the simulator utilizing RADL. Besides RADL, another Architecture Description Language developed in the industry is PRMDL (Philips Research Machine Description Language) [52]. Its goal is to cover a range of VLIW architectures, mostly clustered VLIW architectures with incomplete bypassing networks and shared function units among different issuing slots. A PRMDL description separates the software view (the virtual machine) and the hardware view (the physical machine) of the processor. The separation ensures code portability by keeping application programs independent of changes in hardware. However, no information is available on the mapping from the virtual machine to the physical machine.

16.3   Discussions

16.3.1   Summary of ADLs

ADL as a research area is far from being mature. In contrast to other computer languages such as programming languages and HDLs, there is no common or de facto standard in the ADL field. New ADLs continue to show up and there seems no sign of convergence.

An obvious reason for this situation is the lack of mathematical formalism for general modeling of computer architecture. Modern computer architectures range from simple DSP/ASIPs with small register files and little control logic to complex out-of-order superscalar machines [20] with deep memory hierarchy and heavy control and data speculation. As semiconductor processes evolve, more and more transistors are being squeezed into a single chip. The growing thirst for higher performance and lower power led to the birth of even more sophisticated architectures such as multithreaded and multicore processors. Without a solid underlying mathematical model, it is extremely hard for an ADL to simply use description techniques to cover the vast range of computer architectures.

A second reason for the absence of convergence is the diverging purposes of ADLs. Some ADLs were originally designed as hardware description languages, for example, MIMOLA and UDL/I. The main goal of those languages is accurate hardware modeling. Cycle-accurate simulators and hardware synthesizers are natural products of such languages. However, it is nontrivial to extract the ISA from them for use by a compiler. Some other ADLs were initially developed to be high-level processor modeling languages, such as nML, LISA, EXPRESSION, and MADL. The main goals of these ADLs are general coverage over a wide architecture range and general support for both compilers and simulators. However, because of the diverging needs of compilers and simulators, these ADLs tend to have notable limitations for their support for one type of tools. Furthermore, many ADLs were developed as configuration systems for their associated tools, such as Maril, HMDES, TDL, and PRMDL. In some sense those languages can be viewed as a by-product of the software tools. They are valuable only in the context of their associated tools.

In summary, ADLs are designed for different purposes, at different abstraction levels, with emphasis on different architectures of interest, and with support for different tools. Table 16.1 compares the important features of various ADLs. A few entries in the table were left empty, because information is either not available (no related publication) or not applicable. The parentheses in some entries indicate support with significant limitation. Without a flexible and solid mathematical model for processors, it is hard to unify

TABLE 16.1 Comparison between ADLs

	MIMOLA	UDL/I	nML	ISDL	SLED/λ-RTL	Maril	HMDES	TDL	EXPRESSION	LISA	MADL
category	HDL	HDL	behavioral	behavioral	behavioral	mixed	mixed	mixed	mixed	mixed	mixed
compiler	MSSQ, Record	COACH	CBC, CHESS	Aviv	Zephyr	Marion	IMPACT	PROPAN	EXPRESS		
simulator	MSSB, MSSU	COACH	Sign/Sim Checkers	GENSIM					SIMPRESS	LISA	SimIt
behavioral representation	RT-level	RT-level	RT-lists	RT-lists	RT-lists	RT-lists		RT-lists	RT-lists, mapping	RT-lists	RT-lists
hierarchical behavior			yes	yes	yes	no	yes	yes	no	yes	yes
structural representation	netlist	netlist				resource	resource	resource	netlist	pipeline	token manager
ILP compiler support			yes	yes		yes	yes	yes	yes	yes	yes
cycle simulation support	yes	yes	(yes)	(yes)	no	no	(yes)	no	yes	yes	yes
control path	yes	yes	no	no	no	no	no	no	no	yes	yes
constraint model				Boolean		resource	resource	resource, Boolean	resource		resource
other features							preprocessing	preprocessing	memory	interrupt	annotation

all the ADL efforts. However, this situation may change in the near future. As programmable processors play an increasingly important role in the development of systems, the ADL field is attracting the attention of more researchers and engineers. Some excellent ADLs will eventually stand out.

16.3.2   Essential Elements of an ADL

Modern retargetable compiler back-ends normally consist of three phases: code selection, register allocation, and operation scheduling. Although the phases and the additional optimization steps [37] may sometimes be performed in a different order and in different combinations, the first two phases are indispensable to generating working code. They require operation semantics and addressing mode information. The last phase, operation scheduling, is traditionally viewed as an optimization phase. It minimizes potential pipeline hazards during the execution of the resulting code. Scheduling also packs operations into instruction words for VLIW processors and minimizes the number of wasted slots. As deeper pipelines and wider instruction words are gaining popularity in processors, scheduling is increasingly important to the quality of generated code. A good ILP scheduler can improve performance as well as reduce code size notably. Therefore, we assume in this chapter that it is also an essential part of the compiler back-end.

The data model of a typical operation scheduler contains two basic elements: operand latency and reservation table [14]. The first element models data hazards and is used to build a directed acyclic dependency graph of the operations in a program. The second element models structural hazards between operations. For VLIW architectures, the instruction word packing constraints can also be modeled as reservation tables. The key idea is to convert the packing constraints into artificial resources [45]. For example, if two types of operation cannot be scheduled in the same instruction, an artificial resource is created. Each type of operation is required to obtain the artificial resource to be issued. Therefore, only one operation can be packed into the instruction at the same time. By augmenting the reservation table with artificial resources, the scheduler unifies the operation scheduling problem and the instruction packing problem.

Each machine operation performs some state transition in the processor. A precise description of the state transition should include three elements: what, where, and when. Correspondingly, information required by a compiler back-end contains three basic elements: behavior, resource, and time. Here, behavior means semantic actions including reading of source operands, calculation, and writing of destination operands. Resource refers to abstracted hardware resources used to model structural hazards or artificial resources used to model instruction packing. Common hardware resources include pipeline stages, register file ports, memory ports, and data transfer paths. The last element, time, is the cycle number when the behavior occurs and when resources are occupied. It is usually relative to the operation fetching time or issuing time. In some cases, phase (subcycle) number can be used for even more accurate modeling. With the three basic elements, we can easily represent each machine operation in a list of triples. Each triple is in the form of (behavior, resource, time). For example, an integer "Add" operation in the pipeline depicted by Figure 16.4 can be described as

```
(read operand reg[src1], through register file port a,
    at the 2nd cycle since fetch);
(read operand reg[src2], through register file port b,
    at the 2nd cycle since fetch);
(perform addition, in alu, at the 3rd cycle since fetch);
(write operand reg[dst], through register file write port,
    at the 5th cycle since fetch).
```

From the triple list, a compiler can extract the operation semantics and addressing mode by combining the first elements in the description order. It can also extract operand latency and reservation table information for the scheduler. Operand latency is used to avoid data hazards, while reservation table can be used to avoid structural hazards. The triple list is a simple and general way of operation description. In practice, some triples may sometimes be simplified into tuples. For instance, most architectures contain sufficient register ports so that no resource hazard can ever occur because of those. As a result, the scheduler does not need to take the ports into consideration. In the above example, one can simplify the first two triples by omitting the resource elements. However, when there are resources that may cause contention, but there is no visible behavior associated with the resources, the behavior can be omitted from the triple. In the same integer "Add" example, if the MEM stage becomes a source of conflict, one may model it as a tuple of (MEM stage, at the 4th cycle since fetch).

The triple/tuple list can be found in all mixed languages in some form. For instance, the HMDES language is composed of these two types of tuples: the (behavior, time) tuple for operand latency and the (resource, time) tuple for reservation table. In LISA, operations were described in terms of pipeline stages. Each pipeline stage has an associated time stamp according to its position in the path. Therefore, a LISA description contains the triple list. The pieces of architectural information left out by the triple list are the assembly format and the binary encoding. These two can be viewed as attributes attached to the operation behavior. It is a relatively straightforward task to describe them.

16.3.3 Organization of an ADL

It is possible to directly use the triple list form as an ADL. However, describing a processor based on such an ADL can be a tedious task for a human. A typical RISC processor contains 50 to 100 operations. Each operation has multiple issuing alternatives, proportional to the issuing width of the processor. Each issuing alternative corresponds to one triple list, whose length is approximately the depth of the pipeline. Consequently, the total number of tuples is the product of the operation count, the issuing width, and the depth of the pipeline, which is typically at the order of thousands. Moreover, artificial resources can be used to model instruction packing constraints. A raw triple list representation requires that the user manually perform the artificial resource formation prior to the description. This process can be laborious and error-prone for a human.

The task of an ADL design is to find a simple, concise, and intuitive organization to capture the required information. Conceptually, to capture all relevant architecture and microarchitecture information, an ADL should contain three parts:

- **Behavioral part:** This part contains operation addressing modes, operation semantics, assembly mnemonics, and binary encoding. These correspond to the first element in the triple. Behavior information can be found directly in architecture reference manuals. For typical processors, operations in the same category share common properties. For instance, usually all three-operand arithmetic operations and logic operations share the same addressing modes and binary encoding formats. They differ only in their opcode encoding and semantics. Exploiting the commonalities among them can make the description compact, as has been demonstrated by at least nML and ISDL. Both ADLs adopt hierarchical description schemes, under which common behaviors are described at the roots of the hierarchy while specifics are kept in the leaves. Besides the sharing of common properties, another powerful description scheme is factorization. A single machine operation can be viewed as the combination of several suboperations, each of which has a few alternatives. Take, for example, the "Load" operation of the TMS320C6200 DSP family [53]. The operation contains two suboperations: the load action and the addressing mode. Five versions of load action exist: load byte, load unsigned byte, load half word, load unsigned half word, and load word. The addressing mode can further be decomposed into two parts: the offset mode and the address calculation mode. Two offset modes exist: register offset and constant offset. In the address calculation mode, six options exist: plus-offset, minus-offset, pre-increment, pre-decrement, post-increment, and post-decrement. Overall, the single "Load" operation contains $5 * 2 * 6 = 60$ versions. Under a flat description scheme, it would be a laborious task to enumerate all 60 versions. To reduce the description effort, a hierarchical description scheme can factorize suboperations. It first describes the set of load actions and the set of addressing modes separately and then combines them into a set of operations through the Cartesian product of the suboperation sets. The resulting compact description is much easier to verify and modify.

- **Structural part:** This part describes the hardware resources corresponding to the second element in the triple. Artificial resources may also be included in this part. The level of abstraction in this part can vary from fine-grained RT-level to coarse-grained pipeline level. Two schemes exist for coarse-grained structural descriptions: resource based and netlist based. Maril, TDL, and HMDES utilize the resource-based description scheme. The main advantage of using resources is flexibility. When creating a resource, the description writer does not need to be concerned about its corresponding physical entity and its interconnection with other entities. It may be just an artificial resource used for constraint modeling. In contrast, EXPRESSION and PRMDL utilize the netlist-based scheme. The advantage of a netlist-based description is its intuitiveness. Netlists are more familiar to computer architects. The netlist description scheme also enables the use of friendly GUIs. A reservation table can be extracted from the netlist through some simple automated translation. The disadvantage of netlist is its limited modeling capability. Architectures with complex dynamic pipeline behaviors are hard to model as a simple coarse-grained netlist. Also, netlists of commercial superscalar architectures may not be available to researchers. In summary, the resource-based scheme seems more suitable for complex high-end architectures, while the netlist-based scheme is better suited as an accurate model for simple ASIP/DSP designs whose netlists are available and whose control logic is minimal.
- **Linkage part:** This part completes the triple. The linkage information maps operation behavior to resources and time. It is usually an implicit part of ADLs. In Maril, TDL, and HMDES, linkage is described in the form of an explicit reservation table. For each operation, the resources that it uses and the time of each use are enumerated. HMDES further exploits the commonality of resource usages through a hierarchical description scheme. In EXPRESSION, the linkage information is expressed in multiple ways: operations are mapped to pipeline stages by the OPCODES attribute associated with the pipeline stages and are mapped to data transfer resources according to the interconnection of the pipeline stages and the storage units. Grouping of operations helps simplify the mapping in EXPRESSION.

In summary, the desirable features of an ADL include simplicity, conciseness, and generality. These features may be contradictory to each other. A major task of the ADL design process is to find a good trade-off among the three for the architecture family of interest. A good ADL design should also distinguish true architecture information and artificial information useful only for individual tool implementations.

16.3.4   Challenges

ADL designers are constantly coping with the trade-off of generality and efficiency. A low abstraction level brings more generality but at the cost of efficiency. A high abstraction level makes ADL descriptions more concise but less general in supporting different architectures. Regarding the retargetable tools utilizing ADLs, compilers prefer simplified high-level abstraction, while simulators prefer concrete low-level models. It is very difficult to find a clean and elegant representation to satisfy both. In addition, idiosyncrasies of various architectures make the design of a general ADL an agonizing process. When a new processor needs to be described, designers need to first reevaluate the generality of the ADL. If the ADL is not general enough to cover the new processor, they need to extend the ADL but without seriously disrupting all the existing descriptions based on the ADL. If this is not possible, the entire language must be revised, the parser should be rewritten, and all existing architecture descriptions need to be updated. The constant emergence of new "weird" architectural features keeps designers struggling with such reevaluations, revising and rewriting tasks. Consequently, most ADLs give up generality to some degree but focus on a fixed range of architectures that are interesting to the associated tools.

Below we enumerate a few common challenges faced by ADL designers and users. There seems to be no clean solution for most of the challenges. Trade-offs have to be made depending on the specific needs of each case.

16.3.4.1 Ambiguity

The most common challenge when designing an ADL is to control its ambiguity, or the inability to precisely describe architectural features. Ambiguity is a by-product of abstraction. It exists in most ADLs and especially manifests itself in the description of control-related behaviors. Take, for example, the common behavior of pipeline interlocking. RISC architectures have the capability to detect data and control hazards and stall the pipeline when a hazard exists. Many VLIW architectures, however, do not have the interlocking mechanism. The difference between the two will obviously result in different requirements for the operation scheduler. For RISC processors, the scheduler needs to remove as many read-after-write (RAW) data hazards as possible, while for VLIWs, the scheduler must ensure the correctness of the code by completely removing all data hazards. For these two very different architecture types, it is expected that they result in very different ADL descriptions. However, in most ADLs, there is no difference between the two because of their inability to precisely specify control paths. Among the mixed ADLs, only LISA and RADL can model the interlocking mechanism since they focus on accurate simulator generation.

Another example is the instruction packing of VLIW processors. Some architectures allow only one issue bundle in an instruction word; that is, only operations scheduled to issue at the same cycle can appear in the same instruction word, while other architectures allow multiple issue bundles in one instruction word. The latter architectures will dispatch the issue bundles at different cycles according to the stop bits encoded in the instruction words [24] or according to the instruction templates. Among the mixed ADLs, few can capture such instruction packing details. The code compression [2, 54] feature in some processors is more difficult than simple bundling and is not addressed by ADLs at all.

The amount of ambiguity in an ADL is directly affected by its abstraction level. RT-level languages have the least amount of ambiguity, while behavioral-level ADLs have the most. Among mixed ADLs, executable simulator-oriented ADLs are less ambiguous than descriptive compiler-oriented ADLs. A good ADL involves clever abstraction with minimal ambiguity. In practice, ambiguity can be resolved by using an underlying architecture template; that is, the compiler or simulator presumes some basic architecture information that is not explicitly described by the ADL. This strategy has been adopted by the tool-specific ADLs. A more general ADL may resolve the ambiguity while preserving generality by using multiple architecture templates.

16.3.4.2 Variable Latency

In many processors, operations or their operands may have variable latency. Many compiler-oriented ADLs can only describe a fixed latency rather than the accurate variable latency. Consider again the example of the integer "Add" in Figure 16.4. By default, the source operands will be read at the "ID" stage and the destination written at the "WB" stage. Therefore, the source operand has a latency of one, while the destination operand's latency is four, both relative to the fetching time. However, if there exists a forwarding path, which allows the processor to forward computation results from the "MEM" stage to "EX," then its destination operand latency is equivalent to three. It is sufficient for the compiler if we provide the equivalent latency. Now consider a multi-issue version of the same architecture in which interpipeline forwarding is forbidden, shown in Figure 16.5. Inside each pipeline, forwarding can occur, resulting in an equivalent latency of three. Between pipelines, no forwarding is allowed, so the equivalent latency is four. Here we see a variable latency for the same operand. It may be three or four, depending on whether the forwarding occurs within the same pipeline and between pipelines. Such a type of variable latency is hard to describe explicitly in the triple list, unless the forwarding path and its implication become part of the ADL. ADLs based on reservation tables normally do not capture forwarding paths. The common practice is to inform the scheduler about the worst-case latency, which means there is no distinction to the compiler between the above intrapipeline and interpipeline forwarding case. Consequently, some optimization opportunities are lost.

Another type of variable latency resides in the operations themselves. Operations such as floating point division or square root can take a variable number of execution cycles depending on the value of the source operands. In the triple list model, a variable latency means the time element should be a function, which would complicate the ADL greatly. To save the complexity, usually a worst-case latency is used in the

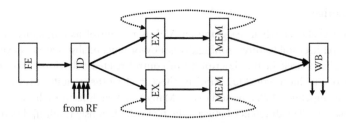

FIGURE 16.5   Two-issue integer pipeline with forwarding.

description. Memory operations can also introduce variable latencies due to the presence of probabilistic components such as TLBs and caches. Most ADLs ignore the memory hierarchy by specifying an optimistic latency for load and store operations since their emphasis is only on the processors. Memory hierarchy modeling is nontrivial because of different types of memory technologies, different hierarchical structures, and different memory management policies. Research has indicated that memory-aware code optimization may yield an average performance gain of 24% [13]. As the speed gap between digital logic and memory continues to enlarge, it is increasingly important that the compiler be aware of the memory hierarchy. So far only EXPRESSION and TDL support a simple parameterized memory hierarchy model.

16.3.4.3   Irregular Constraints

Constraints exist in computer architectures because of the limitation of resources, such as the function units in a pipeline and the bit fields in an instruction word. Common constraints include the range of constant operands and the number of issue slots. These constraints are familiar to compiler developers and can be handled with standard code generation and scheduling techniques. The recent trend of ASIP brings about many extra-irregular constraints. Most ASIP processors are extremely optimized for cost and power consumption. They often use features such as clustered register files, incomplete data transfer paths, and irregular instruction encoding. Irregular instruction encoding helps reduce the size of individual instruction words and therefore the size of the instruction memory. Instruction encoding constraints include intraoperation constraints and interoperation constraints. An intraoperation constraint example from a real proprietary DSP design is that if in operation "ADD D1, S0, S1," "D1" is "AX," then "S0" and "S1" must be chosen from "B0" and "B1." If "D1" is not "AX," there is no constraint on "S0" and "S1." Interoperation constraints may involve operands in different operations. These operand-related constraints are difficult to convert to artificial resource-based constraints. Special compiler techniques are necessary to handle them. For example, the Aviv compiler uses an extra constraint-checking phase [19], and the Propane system uses integer linear programming to solve the constraints [27]. The irregular constraints create new challenges for ADLs as well as compilers. Most existing ADLs cannot model irregular constraints. The exceptions are ISDL and TDL. Both utilize Boolean expressions for constraint modeling, which are effective supplements to the triple list.

16.3.4.4   Operation Semantics

Both the code generator and the instruction set simulator need operation semantics information. For code generation, simple tree-like semantics is most desirable because of the popularity of tree-pattern-based code-generation algorithms [3]. Side effects such as the setting of machine flags cannot be expressed directly as a part of the tree-patterns and are often omitted. Similarly, operations unused for code generation are often omitted too. In contrast, for simulation, precise operation semantics of all operations should be provided. It is not a straightforward task to unify the two types of semantic specification into one, while at the same time making it convenient to access by both the compiler and simulator. Consequently, ADLs like LISA and EXPRESSION separate the two types of semantic specification at the cost of redundancy. Redundancy leads to additional verification challenges to verify the consistency between the two parts of the description.

For compilers, two description schemes for operation semantics have been used in ADLs. The first is a simple mapping from machine-independent intermediate operations to machine-dependent operations. Both EXPRESSION and PRMDL use this scheme. The mapping scheme makes code generation a simple table lookup. The disadvantage of this practical scheme is that the description cannot be reused for simulation purposes, and the description is dependent on the implementation of the compiler. A different compiler may not be able to use the mapping if it uses a different set of intermediate operations. The second description scheme defines for each operation one or more statements based on a predefined set of primitive operators. nML, ISDL, and Maril use this approach. The statements can be directly used for simulation. For most operations, the compiler can derive tree-patterns for its use. It will then be able to apply the pattern matching algorithm for code generation [3]. However, occasionally the compiler's intermediate representations (IR) may fail to match any single operation. For example, in lcc's IR, a comparison operation and a branch operation are unified in the same tree node. Its operator "EQ" compares its two operands and jumps to the associated label if they are equal. But many processors do not have a single operation performing this task. Take the IA32 family, for example [23]. A comparison followed by a conditional branch is used to perform this task. The former sets machine flags. The latter reads the flags and updates the program counter conditionally. The IA32 floating point comparison and branch is even more complicated; the flags set by the comparison need to be moved through an integer register to the machine flags before the branch operation can read them. As a result, a total of four operations is needed to map the single "EQ" node. It is a complicated task for the compiler to understand the meanings of the operations and the flags. Some hints to the compiler should be provided in such cases. Because of these, Maril [7] contains the "glue" and "*func" mechanisms. "Glue" allows users to transform the IR tree for easier pattern matching, and "*func" maps an IR node to multiple operations. However, such mechanisms expose the internals of the compiler to the ADL and make ADL descriptions dependent on the compiler.

16.3.4.5 Debugging of ADL Descriptions

Debugging support is an important feature for any programming language since code written by humans invariably contains bugs. Retargetable compilers and simulators are difficult to write and debug themselves. The bugs inside an ADL description simply make the development process harder. An initial ADL description may contain thousands of lines of code and may have hundreds of errors. Among the errors, syntax ones can be easily detected by the ADL compiler. Some others can be detected by careful semantic analysis in the ADL compiler. However, many functional errors can pass both tests and remain in the software for a long time. The experience of LISA developers shows that it takes longer to debug a machine description than to write the description itself [39]. So far there is no formal methodology for ADL debugging. An ADL description itself does not execute on the host machine. It is usually interpreted or transformed into executable C code. In the former case, debugging the ADL description is intertwined with the task of debugging the tool that utilizes the description. In the latter case, debugging can be performed on the generated C code, which probably looks familiar only to the developers of the tool.

The task of debugging may be easier if there exists a golden reference compiler and simulator for the architecture to be described. Comparison of emitted assembly code or simulation traces helps detect errors. Unfortunately, this is impossible for the descriptions of new architectures. The ADL users invariably spend months to go through a trial-and-error process before getting a working description. Though it takes long to get the right description, it is still worthwhile since it would take even longer to customize a compiler.

16.3.4.6 Miscellaneous

Other challenges in designing an ADL include the handling of register overlapping and the mapping of intrinsics. These challenges have been addressed by several existing ADLs, but they are still well worth the attention of ADL designers. Register overlapping, or alias, is common to many architectures. For example, in the IA32 architecture, the lower 16 bits of the 32-bit register "EAX" can be used independently as "AX." The lower and upper half of "AX" can also be used as 8-bit registers "AL" and "AH." Register overlapping also commonly exists in floating point register files in which a single-precision register occupies half of a double-precision register. An ADL must provide proper syntax constructs to handle such cases.

Most architectures contain a few operations that cannot be expressed by predefined RT-list operators. For example, the TI's TMS320C6200 family implements some bit manipulation operations such as the bit-reverse "BITR." Such operations are useful to special application families, but no high-level language operators resemble such operations. To utilize them, the programmers may either write assembly code or resort to intrinsics, which are essentially assembly-level code in C syntax.

16.3.5   Future Directions

As both chip density and size increase, traditional board-level functionality can now be performed on a single chip. Processors are no longer privileged central components in electronic systems. Sophisticated modern chips contain multiple processors, busses, memories, and peripheral circuits. As EDA tools start to address system-on-chip (SoC) design issues, system-level modeling becomes a very active research area. It is desirable that an ADL that models processors can be extended to model a complete system. RT-level ADLs such as MIMOLA may naturally be used for this purpose. However, because of the large size of a typical SoC, describing the entire system at the RT-level inevitably suffers from poor efficiency. Using mixed abstraction levels seems a more practical approach.

To reuse the existing tools and ADLs for system-level description, a convenient approach is to equip the ADLs with the capability to model communication interfaces, which include bus drivers, memory controllers, interrupt interfaces, and so on. The communication interfaces are useful for the processor simulators to interact with other components in the system-level simulator. They are also important for advanced optimizing compilers. For instance, a system-level compiler may partition tasks and assign them to multiple processors according to their computation power and their communication latency. It can also schedule individual communication transactions to avoid congestion. Among the existing mixed ADLs, LISA is capable of modeling interrupts, and EXPRESSION has a parameterized memory hierarchy model. Both have made important first steps toward the specification of system-level communication interfaces, but overall, the systematic modeling of communication interfaces still faces a lot of challenges.

16.4   Conclusion

Architecture Description Languages provide machine models for retargetable software tools including compilers and simulators. They are crucial to the design space exploration of ASIPs. To effectively support an optimizing compiler, an ADL should contain several pieces of information:

- Behavioral information in the form of RT-lists. Hierarchical behavioral models based on attribute grammars are common and effective means. For VLIW architectures, instruction formats should be also modeled.
- Structural information in the form of a reservation table or coarse-grained netlists. Essential information provided by this part includes abstracted resources such as pipeline stages and data transfer paths.
- Mapping between behavioral and structural information. Information in this part includes the time when semantic actions take place and the resources used by the actions.

In addition, modeling of irregular ILP constraints is useful for ADLs targeting ASIPs. In summary, the desirable features of an ADL include simplicity for comprehending, conciseness for efficiency, generality for supporting wide architecture range, flexibility for supporting wide tool range, minimal ambiguity, and minimal redundancy. It is also helpful to separate the description of true architecture information and artificial information used only for individual tool implementations. It is extremely difficult to design an ADL with these features. In practice, trade-offs are always necessary and helpful. As ASIPs gain popularity in the SoC era, ADLs as well as retargetable compilers will become important additions to electronic design automation tools. Eventually, they should encompass not just single processors, but complete systems.

References

1. S. Aditya, V. Kathail, and B. Rau. 1993. Elcors machine description system: Version 3.0. Technical report, Hewlett-Packard.
2. S. Aditya, S. A. Mahlke, and B. R. Rau. 2000. Code size minimization and retargetable assembly for custom EPIC and VLIW instruction formats. *ACM Transactions on Design Automation of Electronic Systems* 5(4):752–73.
3. A. V. Aho, M. Ganapathi, and S. W. K. Tjiang. 1989. Code generation using tree matching and dynamic programming. *ACM Transactions on Programming Languages and Systems* 11(4):491–516.
4. H. Akaboshi. 1996. A study on design support for computer architecture design. PhD thesis, Department of Information Systems, Kyushu University, Japan.
5. M. W. Bailey and J. W. Davidson. 1995. A formal model and specification language for procedure calling conventions. In *Proceedings of the 22nd ACM SIGPLAN-SIGACT Symposium on Principles of Programming Languages (POPL 95)*, San Francisco, CA, 298–310. New York: ACM Press.
6. F. Bodin, E. Rohou, and A. Seznec. 1996. SALTO: System for assembly-language transformation and optimization. In *Proceedings of the 6th Workshop on Compilers for Parallel Computers*.
7. D. G. Bradlee, R. R. Henry, and S. J. Eggers. 1991. The Marion system for retargetable instruction scheduling. In *Proceedings of the Conference on Programming Language Design and Implementation*, Toronto, Ontario, Canada. New York: ACM Press.
8. A. Fauth and A. Knoll. 1993. Automatic generation of DSP program development tools using a machine description formalism. In *Proceedings of International Conference on Acoustics, Speech and Signal Processing (ICASSP)*, Minneapolis, MN, 457–60. New York: IEEE Press.
9. A. Fauth, J. V. Praet, and M. Freericks. 1995. Describing instructions set processors using nML. In *Proceedings of the 1995 European Conference on Design and Test*, 503–7. Washington, DC: IEEE Computer Society.
10. M. F. Fernandez. 1995. Simple and effective link-time optimization of modula-3 programs. In *Proceedings of the Conference on Programming Language Design and Implementation*, La Jolla, CA, 103–15. New York: ACM Press.
11. C. Fraser and D. Hanson. 1995. *A retargetable C compiler: Design and implementation*. Reading, MA: Addison-Wesley.
12. Free Software Foundation. Using the GNU compiler collection (GCC). http://gcc.gnu.org/onlinedocs/gcc.
13. P. Grun, N. Dutt, and A. Nicolau. 2000. Memory aware compilation through accurate timing extraction. In *Proceedings of Design Automation Conference*, Los Angeles, CA, 316–21. New York: ACM Press.
14. J. C. Gyllenhaal, W. Hwu, and B. R. Rao. 1996. HMDES version 2.0 specification. Technical Report IMPACT-96-3, University of Illinois at Urbana-Champaign.
15. G. Hadjiyiannis, S. Hanono, and S. Devadas. 1997. ISDL: An instruction set description language for retargetability. In *Proceedings of Design Automation Conference*, Anaheim, CA, 299–302. New York: ACM Press.
16. G. Hadjiyiannis, P. Russo, and S. Devadas. 1999. A methodology for accurate performance evaluation in architecture exploration. In *Proceedings of Design Automation Conference*, New Orleans, LA, 927–32. New York: ACM Press.
17. A. Halambi, P. Grun, V. Ganesh, A. Khare, N. Dutt, and A. Nicolau. 1999. EXPRESSION: A language for architecture exploration through compiler/simulator retargetability. In *Proceedings of Conference on Design Automation and Test in Europe*, Munich, Germany, 485–90. New York: ACM Press.
18. A. Halambi, A. Shrivastava, N. Dutt, and A. Nicolau. 2001. A customizable compiler framework for embedded systems. In *5th International Workshop on Software and Compilers for Embedded Systems (SCOPES)*, Schloss Rheinfels, St. Goar, Germany.
19. S. Z. Hanono. 1999. Aviv: A retargetable code generator for embedded processors. PhD thesis, Massachusetts Institute of Technology, Cambridge, MA.

20. J. Hennessy and D. Patterson. 2006. *Computer architecture: A quantitative approach*, 4th ed. San Francisco: Morgan Kaufmann.
21. IEEE. 2001. *IEEE Standard Hardware Description Language based on the Verilog Hardware Description Language (1364-2001)*. New York: IEEE.
22. Intel Corporation. 1989. *i860 64-bit microprocessor programmers reference manual*.
23. Intel Corporation. 1997. *Intel architecture software developer's manual. Vol. 2. Instruction set reference manual*.
24. Intel Corporation. 2000. *Intel IA-64 software developers manual. Vol. 3. Instruction set reference*.
25. S. C. Johnson. Yacc: Yet another compiler-compiler. http://dinosaur.compilertools.net/yacc/index.html.
26. G. Kane. 1988. *MIPS R2000 RISC Architecture*. Chicago: Longman Higher Education.
27. D. Kastner. 2000. Retargetable postpass optimization by integer linear programming. PhD thesis, Saarland University, Germany.
28. D. Kastner. 2000. TDL: A hardware and assembly description languages. Technical report, Saarland University, Germany.
29. A. Khare, N. Savoiu, A. Halambi, P. Grun, N. Dutt, and A. Nicolau. 1999. V-SAT: A visual specification and analysis tool for system-on-chip exploration. In *Proceedings of EUROMICRO*. New York: Elsevier North-Holland, Inc.
30. B. Kienhuis. 1999. Design space exploration of stream-based dataflow architecture. PhD thesis, Delft University of Technology, The Netherlands.
31. D. Lanneer. 1995. Chess: Retargetable code generation for embedded DSP processors. In *Code generation for embedded processors*. Dordrecht, The Netherlands: Kluwer Academic.
32. R. Leupers and P. Marwedel. 1997. Retargetable generation of code selectors from HDL processor models. In *Proceedings of Conference on Design Automation and Test in Europe*, 140–44. Washington, DC: IEEE Computer Society.
33. R. Leupers and P. Marwedel. 1998. Retargetable code generation based on structural processor descriptions. *Design Automation for Embedded Systems* 3(1):1–36.
34. F. Lohr, A. Fauth, and M. Freericks. 1993. SIGH/SIM: An environment for retargetable instruction set simulation. Technical Report 43, Technical University of Berlin, Germany.
35. R. Milner, M. Tofte, and D. Macqueen. 1997. *The definition of Standard ML*. Cambridge, MA: MIT Press.
36. Motorola, Inc. 1988. *MC88100 RISC microprocessor user's manual*.
37. S. S. Muchnick. 1997. *Advanced compiler design and implementation*. San Francisco: Morgan Kaufmann.
38. J. Paakki. 1995. Attribute grammar paradigms — A high-level methodology in language implementation. *ACM Computing Surveys* 27(2):196–255.
39. S. Pees, A. Hoffmann, and H. Meyr. 2000. Retargetable compiled simulation of embedded processors using a machine description language. *ACM Transactions on Design Automation of Electronic Systems* 5(4):815–34.
40. S. Pees, A. Hoffmann, V. Zivojnovic, and H. Meyr. 1999. LISA — Machine description language for cycle-accurate models of programmable DSP architectures. In *Proceedings of Design Automation Conference*, New Orleans, LA, 933–38. New York: ACM Press.
41. W. Qin. http://people.bu.edu/wqin/pub/madl-1.0.tar.gz.
42. W. Qin. http://sourceforge.net/projects/simit-arm/.
43. W. Qin and S. Malik. 2003. Flexible and formal modeling of microprocessors with application to retargetable simulation. In *Proceedings of Conference on Design Automation and Test in Europe*, 556–61. Washington, DC: IEEE Computer Society.
44. W. Qin, S. Rajagopalan, and S. Malik. 2004. A formal concurrency model based architecture description language for synthesis of software development tools. In *Proceedings of the ACM SIGPLAN/SIGBED 2004 Conference on Languages, Compilers, and Tools for Embedded Systems (LCTES'04)*. New York: ACM Press.

45. S. Rajagopalan, M. Vachharajani, and S. Malik. 2000. Handling irregular ILP within conventional VLIW schedulers using artificial resource constraints. In *Proceedings of the International Conference on Compilers, Architecture, and Synthesis for Embedded Systems*, San Jose, CA, 157–64. New York: ACM Press.

46. N. Ramsey. 1992. A retargetable debugger. PhD thesis, Princeton University, Princeton, NJ.

47. N. Ramsey and J. Davidson. 1999. Specifying instructions semantics using λ-RTL. Technical report, University of Virginia.

48. N. Ramsey and J. W. Davidson. 1998. Machine descriptions to build tools for embedded systems. In *Proceedings of the ACM SIGPLAN Workshop on Languages, Compilers, and Tools for Embedded Systems*, 176–92. London: Springer-Verlag.

49. N. Ramsey and M. F. Fernandez. 1997. Specifying representations of machine instructions. *ACM Transactions on Programming Languages and Systems* 19(3):492–524.

50. C. Siska. 1998. A processor description language supporting retargetable multi-pipeline DSP program development tools. In *Proceedings of the International Symposium on System Synthesis*, Hsinchu, Taiwan, 31–36. Washington, DC: IEEE Computer Society.

51. Target Compiler Technologies N.V. http://www.retarget.com.

52. A. Terechko, E. Pol, and J. van Eijndhoven. 2001. PRMDL: A machine description language for clustered VLIW architectures. In *Proceedings of Conference on Design Automation and Test in Europe*, Munich, Germany, 821. Piscataway, NJ: IEEE Press.

53. Texas Instruments Inc. *TMS320C6000 CPU and instructions, set reference guide.*

54. A. Wolfe and A. Chanin. 1992. Executing compressed programs on an embedded RISC architecture. In *Proceedings of International Symposium on Microarchitecture*, Portland, Oregon, 81–91. Los Alamitos, CA: IEEE Computer Society Press.

17

Instruction Selection Using Tree Parsing

Priti Shankar

*Department of Computer Science
and Automation,
Indian Institute of Science,
Bangalore, India*
priti@csa.iisc.ernet.in

17.1 Introduction and Background

17.1.1 Introduction

One of the final phases in a typical compiler is the *instruction selection* phase. This traverses an intermediate representation of the source code and selects a sequence of target machine instructions that implement the code. There are two aspects to this task. The first has to do with finding efficient algorithms for generating an optimal instruction sequence with reference to some measure of optimality. The second has to do with the automatic generation of instruction selection programs from precise specifications of machine instructions. Achieving the second aim is a first step toward retargetabiltiy of code generators. We confine our attention to instruction selection for basic blocks. An optimal sequence of instructions for a basic block is called *locally optimal code*.

Early techniques in code generation centered around interpretive approaches where code is produced for a virtual machine and then expanded into real machine instructions. The interpretive approach suffers from the drawback of having to change the code generator for each machine. The idea of code generation by tree parsing replaced the strategy of virtual machine interpretation. The intermediate representation (IR) of the source program is in the form of a tree, and the target machine instructions are represented as productions of a *regular tree grammar* augmented with semantic actions and costs. The code generator parses the input subject tree, and on each reduction, outputs target code. This is illustrated in Figure 17.1 for a subject tree generated by the grammar of Example 17.1.

Example 17.1

Consider the tree grammar below where the right-hand sides of productions represent trees using the usual list notation. Each production is associated with a cost and a semantic action enclosed in braces. The operator := is the assignment operator; deref is the deferencing operator that dereferences an address to refer to its contents. Subscripts on nonterminals are used to indicate attributes. For example, c_j indicates a constant with value j. Costs are assumed to be additive.

$$
\begin{array}{llll}
S \rightarrow := (+(c_j, G_k), R_i) & [3] & \{\text{emit}("\texttt{store R\%s, \%s[\%s]}"), i, j, k)\} \\
R \rightarrow +(\text{deref}(+(c_j, G_k), R_i) & [3] & \{\text{emit}("\texttt{add \%s[\%s], R\%s}"), j, k, i\} \\
R_i \rightarrow c_j & [2] & \{i = \text{allocreg}(); \text{emit}("\texttt{mov \#\%s, R\%s}"), j, i\} \\
G_i \rightarrow sp & [0] & \{i = sp\}
\end{array}
$$

Consider the source level statement $b := b + 1$, where b is a local variable stored in the current frame pointed at by the stack pointer sp. The IR tree is shown as the first tree in Figure 17.1, with nodes numbered from 1 to 10. The IR tree is processed using the replacements shown in Figure 17.1. Each nonterminal replacing a subtree has its cost shown alongside.

The tree is said to have been *reduced* to S by a tree-parsing process, which implicitly constructs the derivation tree shown in Figure 17.2 for the subject tree. The set of productions used is a *cover* for the tree. In general, there are several covers, given a set of productions, and we aim to obtain the best one according to some measure of optimality. The semantic actions also include a call to a routine to allocate a register. This can go hand in hand with the tree parsing, and the selection of the register is independent of the parsing process. For the sequence of replacements shown, the code emitted is

```
move #1, R0
add b[sp], R0
store R0, b[sp]
```

Fraser [19] and Cattell [12] employed tree-pattern matching along with heuristic search for code generation. Fraser used knowledge-based rules to direct pattern matching, whereas Cattell suggested a goal-directed heuristic search. In 1978, Graham and Glanville [23] opened up new directions in the area of retargetable code generation. They showed that if the intermediate code tree is linearized and the target machine instructions are represented as context-free grammar productions, then bottom-up parsing techniques could be used to generate parsers that parse the linearized intermediate code tree and emit machine instructions while performing reductions. This was a purely syntactic approach to the problem of instruction selection and suffered from the drawback that the effective derivation has a left bias, in that the code for the subtree corresponding to the left operand is selected without considering the right operand. As a result, the code generated is suboptimal in many instances. A second problem with the Graham–Glanville approach is that many architectural restrictions have to be taken into account when generating code, such as register restrictions on addressing modes and so on. A purely syntactic approach to such semantics yields a very large number of productions in the specifications. Several implementations of the Graham–Glanville technique have been described, and the technique has been applied to practical compilers [24, 30]. Ganapathi and Fischer [21] suggested using attribute grammars instead of context-free grammars to handle the problem of semantics. Attributes are used to track multiple instruction results, for example, the setting of condition codes. Furthermore, predicates are used to specify architectural restrictions on the programming model. Instruction selection is therefore performed by attributed parsing. While this solves the problem of handling of semantic attributes, it is still not able to overcome the problem of left bias in the mode of instruction selection. A good survey of early work in this area is [22].

The seminal work of Hoffman and O'Donnell (HOD) [28] provided new approaches that could be adopted for retargetable code generation. They considered the general problem of pattern matching in trees with operators of fixed arity and presented algorithms for both top-down and bottom-up tree

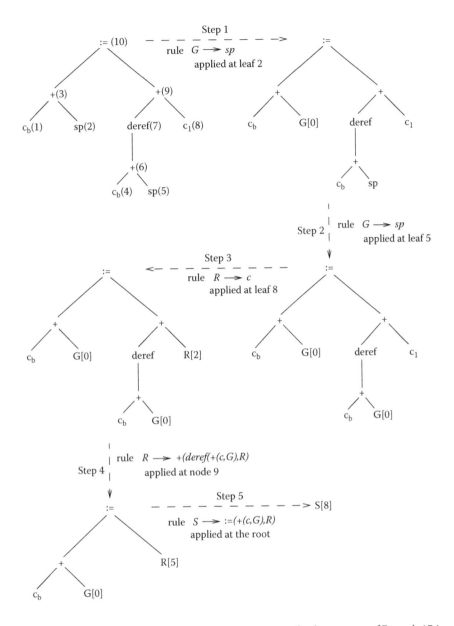

FIGURE 17.1 A sequence of tree replacements for an IR tree for the grammar of Example 17.1.

pattern matching. The fact that choices for patterns that are available can be stored, and decisions can be deferred until an optimal choice can be made, overcomes the problem of left bias in the Graham–Glanville approach. The basic idea here is that a tree automaton can be constructed from a specification of the machine instructions in terms of tree patterns during a "preprocessing" phase, and this can be used to traverse an intermediate code tree during a "matching" phase to find all matches and finally generate object code. Hoffmann and O'Donnell showed that if tables encoding the automaton could be precomputed, matching could be achieved in linear time. The size of the tables precomputed for bottom-up tree-pattern-matching automata can be exponential in the product of the arity and the number of sub-patterns. Chase [13] showed that table compression techniques that could be applied as the tables were being constructed could greatly reduce auxiliary space requirements while performing pattern matching. This important observation made the HOD technique practically useful. Several bottom-up tools for

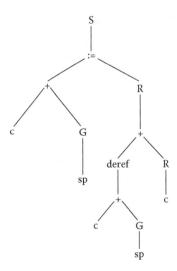

FIGURE 17.2   Derivation tree for the IR tree of Figure 17.1.

generating retargetable code generators were designed. Some of these are "hard coded" in that their control structure mirrors the underlying regular tree grammar [17, 20]. Such techniques have been employed in the tools BEG [17] and iburg [20], where the dynamic programming strategy first proposed by Aho and Johnson [4] is used in conjunction with tree parsing to obtain locally optimal code.

Aho and Ganapathi [2] showed that top-down tree parsing combined with dynamic programming can be used for generating locally optimal code. Their technique is implemented in the code-generator generators twig [41] and olive [42]. The advantage of a top-down technique is that the tables describing the tree-parsing automaton are small. The disadvantage stems from the fact that cost computations associated with dynamic programming are performed at code generation time, thus slowing down the code generator. Dynamic programming using a top-down traversal was also used by Christopher et al. [14]. Appel [7] has generated code generators for the VAX and Motorola 68020 using twig. Weisgerber and Wilhelm [44] describe top-down and bottom-up techniques to generate code. Henry and Damron [27] carried out an extensive comparison of Graham–Glanville style code generators and those based on tree parsing. Hatcher and Christopher [26] showed that costs could be included in the states of the finite-state tree-pattern-matching automaton so that optimal instruction selection could be performed without incurring the extra overhead of the cost computations on-line. Static cost analysis exemplified in the approach of Hatcher and Christopher makes the code-generator generator more complex and involves large space overheads. However, the resultant code generator is simple and fast, which implies faster compilation. The technique of Hatcher and Christopher does not guarantee that the statically selected code will always be optimal and requires interaction from the user.

Pelegri-Llopart and Graham [37] combined static cost analysis with table compression techniques from Chase [13] and used term rewrite systems rather than tree patterns to develop a bottom-up rewrite system (BURS). A BURS is more powerful than a bottom-up pattern-matching system, as it can incorporate algebraic properties of terms into the code generation system. However, systems based on BURS are generally more complex than those based on tree parsing. Balachandran et al. [9] used an extension of the work of Chase [13] to perform static cost analysis and produce optimal code. Proebsting [38] used a simple and efficient algorithm for generating tables with static costs, in which new techniques called triangle trimming and chain rule trimming are used for state reduction. This technique is used in the bottom-up tool burg. Ferdinand et al. [18] reformulated the static bottom-up tree-parsing algorithm based on finite tree automata. This generalized the work of Chase to work for regular tree grammars and included table compression techniques. More recently, Nymeyer and Katoen [36] described an implementation of an algorithm based on BURS theory, which computes all pattern matches and does a search that results in optimal code. Heuristics are used to cut down the search space. Shankar et al. [40] constructed an LR-like parsing algorithm for regular tree parsing, which can be used for code generation with dynamic cost computation. The static cost computation technique of Balachandran et al. and the LR-like parsing approach of Shankar et al. have been combined into a technique for locally optimal code generation in [34]. A treatment of tree parsing for instruction selection is given in [45].

The bottom-up techniques mentioned above all require at least two passes over the intermediate code tree, one for labeling the tree with matched patterns and costs and the next for selecting the least-cost parse based on the information collected during the first pass. Thus, an explicit IR tree needs to be built. A technique that avoids the building of an explicit IR tree is proposed by Proebsting and Whaley [39]. The tool wburg generates parsers that can find an optimal parse in a single pass. An IR tree is not built explicitly, as the tree structure is mirrored in the sequence of procedure invocations necessary to build the tree in a bottom-up fashion. The class of grammars handled by this technique is a proper subset of the grammars that the two pass systems can handle. However, Proebsting and Whaley have claimed that optimal code can be generated for most major instruction sets including the SPARC, the MIPS R3000, and the x86.

We restrict our attention in this chapter to instruction selection techniques based on tree parsing. The techniques based on term rewriting systems [15, 36, 37] are more powerful but not as practical.

17.1.2 Dynamic Programming

Aho and Johnnson [4] used dynamic programming to generate code for expression trees. The algorithm they presented generates optimal code for a machine with r interchangeable registers and instructions of the form $R_i := E$. R_i is one of the registers, and E is any expression involving operators, registers, and memory locations. The dynamic programming algorithm generates optimal code for evaluation of an expression "contiguously." If T is an expression tree with op at its root and T_1 and T_2 as its subtrees, then a program is said to evaluate the tree contiguously if it first evaluates the subtrees of T that need to be computed into memory and then evaluates the remainder of the tree either in the order T_1, T_2, and then the root, or T_2, T_1, and then the root. Aho and Johnson proved that for a uniform register machine, optimal code would always be generated by their algorithm. The implication of the property is that for any expression tree there is always an optimal program that consists of optimal programs for subtrees of the root followed by an instruction to evaluate the root. The original dynamic programming algorithm uses three phases. In the first bottom-up phase it computes a vector of costs for each node n of the expression tree, in which the ith component of the vector is the cost of computing the subtree at that node into a register, assuming i registers are available for the computation $0 \le i \le r$. The zeroth component of the vector is the minimal cost of computing the subtree into memory. In the second phase the algorithm traverses the tree top-down to determine which subtrees should be computed into memory. In the third phase the algorithm traverses each tree using the cost vectors to generate the optimal code.

A simplified form of the dynamic programming algorithm is used in most code generator tools where what is computed at each node is a set of (rule, scalar cost) pairs. Register allocation is not part of the instruction selection algorithm, though it can be carried out concurrently. The cost associated with a subtree is computed either at compile time (i.e., dynamically), by using cost rules provided in the grammar specification, or by simply adding the costs of the children to the cost of the operation at the root or at compiler generation time (i.e., statically) by precomputing differential costs and storing them along with the instructions that match, as part of the state information of a tree-pattern-matching automaton. How exactly this is done will become clear in the following sections.

17.2 Regular Tree Grammars and Tree Parsing

Let A be a finite alphabet consisting of a set of operators OP and a set of terminals T. Each operator op in OP is associated with an *arity*, $arity(op)$. Elements of T have arity 0. The set $TREES(A)$ consists of all trees with internal nodes labeled with elements of OP and leaves with labels from T. Such trees are called *subject trees* in this chapter. The number of children of a node labeled *op* is *arity(op)*. Special symbols called *wildcards* are assumed to have arity 0. If N is a set of wildcards, the set $TREES(A \cup N)$ is the set of all trees with wildcards also allowed as labels of leaves.

We begin with a few definitions drawn from [9] and [18].

Definition 17.1   *A regular cost-augmented tree grammar G is a four tuple (N, A, P, S) where:*

1. *N is a finite set of nonterminal symbols.*
2. *$A = T \cup OP$ is a ranked alphabet, with the ranking function denoted by arity. T is the set of terminal symbols, and OP is the set of operators.*
3. *P is a finite set of production rules of the form $X \rightarrow t$ $[c]$, where $X \in N$ and t is an encoding of a tree in $TREES(A \cup N)$, and c is a cost, which is a nonnegative integer.*
4. *S is the start symbol of the grammar.*

A *tree pattern* is thus represented by the right-hand side of a production of P in the grammar above. A production of P is called a *chain rule* if it is of the form $A \rightarrow B$, where both A and B are nonterminals.

Definition 17.2   *A production is said to be in normal form if it is in one of the three forms below.*

1. *$A \rightarrow op(B_1, B_2 \ldots , B_k)[c]$, where $A, B_i, i = 1, 2, \ldots k$ are all nonterminals, and op has arity k.*
2. *$A \rightarrow B$ $[c]$, where A and B are nonterminals. Such a production is called a* chain rule.
3. *$B \rightarrow b$ $[c]$, where b is a terminal.*

A grammar is in normal form if all its productions are in normal form.

Any regular tree grammar can be put into normal form by the introduction of extra nonterminals and zero-cost rules.

Below is an example of a cost-augmented regular tree grammar in normal form. Arities of symbols in the alphabet are shown in parentheses next to the symbol.

Example 17.2

$$G = (\{V, B, G\}, \{a(2), b(0)\}, P, V)$$

$$P :$$
$$V \rightarrow a(V, B)[0]$$
$$V \rightarrow a(G, V)[1]$$
$$V \rightarrow G \quad [1]$$
$$G \rightarrow B \quad [1]$$
$$V \rightarrow b \quad [7]$$
$$B \rightarrow b \quad [4]$$

Definition 17.3   *For $t, t' \in TREES(A \cup N)$, t directly derives t', written as $t \Rightarrow t'$, if t' can be obtained from t by replacement of a leaf of t labeled X by a tree p where $X \rightarrow p \in P$. We write $\Rightarrow_r$ if we want to specify that rule r is used in a derivation step. The relations $\Rightarrow^+$ and $\Rightarrow^*$ are the transitive closure and reflexive-transitive closure, respectively, of $\Rightarrow$.*

An X-derivation tree, D_X, for G has the following properties:

- The root of the tree has label X.
- If X is an internal node, then the subtree rooted at X is one of the following three types (for describing trees we use the usual list notation).

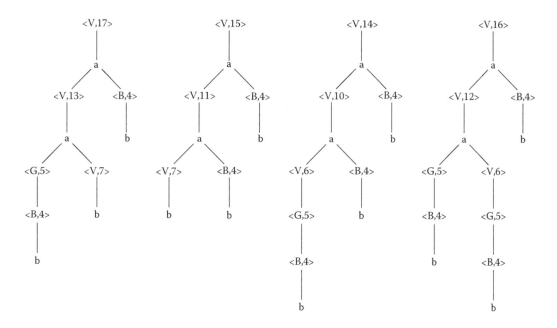

FIGURE 17.3 Four cost-augmented derivation trees for the subject tree $a(a(b,b),b)$ in the grammar of Example 17.2.

- $X(D_Y)$ if $X \rightarrow Y$ is a chain rule and D_Y is a derivation tree rooted at Y.
- $X(a)$ if $X \rightarrow a, a \in T$ is a production of P.
- $X(op(D_{X_1}, D_{X_2}, \ldots D_{X_k}))$ if $X \rightarrow op(X_1, X_2 \ldots X_k)$ is an element of P.

The language defined by the grammar is the set

$$L(G) = \{t \mid t \in \mathit{TREES}(A), \text{ and } S \Longrightarrow^* t\}$$

With each derivation tree is an associated cost, namely, the sum of the costs of all the productions used in constructing the derivation tree. We label each nonterminal in the derivation tree with the cost of the subtree below it. Four cost-augmented derivation trees for the subject tree $a(a(b,b),b)$ in the language generated by the regular tree grammar of Example 17.2 above are displayed in Figure 17.3.

Definition 17.4   *A rule* $r : X \rightarrow p$ *matches a tree* t *if there exists a derivation* $X \Rightarrow_r p \Rightarrow^* t$.

Definition 17.5   *A nonterminal* X *matches a tree* t *if there exists a rule of the form* $X \rightarrow p$ *that matches* t.

Definition 17.6   *A rule or nonterminal matches a tree* t *at node* n *if the rule or nonterminal matches the subtree rooted at the node* n.

Each derivation tree for a subject tree thus defines a set of matching rules at each node in the subject tree (a *set* because there may be chain rules that also match at the node).

Example 17.3

For all the derivation trees of Figure 17.3 the rule $V \rightarrow a(V, B)$ matches at the root.

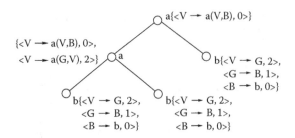

FIGURE 17.4   Subject tree of Figure 17.3 shown with <matching rule, relative cost> pairs.

For a rule $r : X \rightarrow p$ matching a tree t at node n, where t_1 is the subtree rooted at node n, we define:

- The cost of rule r matching t at node n. It is the minimum of the cost of all possible derivations of the form $X \Rightarrow_r p \Rightarrow^* t_1$.
- The cost of nonterminal X matching t at node n is the minimum of the cost of all rules r of the form $X \rightarrow p$ that match t_1.

Typically, any algorithm that does dynamic cost computations compares the costs of all possible derivation trees and selects one with minimal costs while computing matches. To do this it has to compute for each nonterminal that matches at a node the minimal cost of reducing to that nonterminal (or equivalently, deriving the portion of the subject tree rooted at that node from the nonterminal.) In contrast, algorithms that perform static cost computations precompute relative costs, and store differential costs for nonterminals. Thus, the cost associated with a rule r at a particular node in a subject tree is the difference between the minimum cost of deriving the subtree of the subject tree rooted at that node using rule r at the first step and the minimum cost of deriving it using any other rule at the first step. Figure 17.4 shows the matching rules with relative costs at the nodes of the subject tree for which derivation trees are displayed in Figure 17.3. Assuming such differences are bounded for all possible derivation trees of the grammar, they can be stored as part of the information in the states of a finite-state tree-parsing automaton. Thus, no cost analysis need be done at matching time. Clearly, tables encoding the tree automaton with static costs tend to be larger than those without cost information in the states.

The *tree-parsing problem* we will address in this chapter is:

Given a regular tree grammar $G = (N, T, P, S)$, and a subject tree t in TREES(A), find (a representation of) all S-derivation trees for t.

The problem of computing an optimal derivation tree has to take into account costs as well. We will be discussing top-down as well as bottom-up strategies for solving this problem. All the algorithms we will present will solve the following problem, which we will call the *optimal tree-parsing problem*:

Given a cost-augmented regular tree grammar G and a subject tree t in TREES(A), find a representation of a cheapest derivation tree for t in G.

Given a specification of the target machine by a regular tree grammar at the semantic level of a target machine and an IR tree, we distinguish between the following two times when we solve the optimal tree-parsing problem for the IR tree:

- **Preprocessing time:** This is the time required to process the input grammar, independent of the IR tree. It typically includes the time taken to build the matching automaton or the tables.
- **Matching time:** This involves all IR tree–dependent operations and captures the time taken by the driver to match a given IR tree using the tables created during the preprocessing phase.

For the application of code generation, minimizing matching time is important since it adds to compile time, whereas preprocessing is done only once at compiler generation time.

17.3   A Top-Down Tree-Parsing Approach

The key idea here is to reduce tree-pattern matching to string-pattern matching. Each root-to-leaf path in a tree is regarded as a string in which the symbols in the alphabet are interleaved with numbers indicating which branch from father to son has been followed. This effectively generates a set of strings. The well-known Aho–Corasick multiple-keyword pattern-matching algorithm [1] is then adapted to generate a top-down tree-pattern-matching algorithm. The Aho–Corasick algorithm converts the set of keywords into a trie; the trie is then converted into a string-pattern-matching automaton that performs a parallel search for keywords in the input string. If K is the set of keywords, then each keyword has a root leaf path in the trie, whose branch labels spell out the keyword. This trie is then converted into a string-pattern-matching automaton as follows. The states of the automaton are the nodes of the trie, with the root being the start state. All states that correspond to complete keywords are final states. The transitions are just the branches of the trie with the labels representing the input symbols on which the transition is made. There is a transition from the start state to itself on every symbol that does not begin a keyword. The pattern-matching automaton has a *failure function* for every state other than the start state. For a state reached on input w, this is a pointer to the state reached on the longest prefix of a keyword that is a proper suffix of w. The construction of the trie as well as the pattern-matching automaton has complexity linear in the sum of the sizes of the keywords. Moreover, matches of the keywords in an input string w are found in time linearly proportional to the length of w. Thus, the string-pattern-matching problem can be solved in time $O(|K| + |w|)$, where K is the sum of the lengths of the keywords and w is the length of the input string [1].

Hoffman and O'Donnell generalized this algorithm for tree-pattern matching by noting that a tree can be defined by its root-to-leaf paths. A root-to-leaf path contains, alternately, root labels and branch numbers according to a left-to-right ordering. Consider the tree patterns on the right-hand sides of the regular tree grammar in Example 17.4. Arities of various terminals and operators are given in parentheses next to the operators and terminals and rules for computing costs shown along with the productions. Actions to be carried out at each reduction are omitted.

Example 17.4

$G = (\{S, R\}, \{:= (2), +(2), deref(1), sp(0), c(0)\}, P, S)$, where P consists of the following productions:

- $S \rightarrow := (deref(sp), R)$ $cost = 3 + cost(R)$
- $R \rightarrow deref(sp)$ $cost = 2$
- $R \rightarrow +(R, c)$ $cost = 1 + cost(R)$
- $R \rightarrow +(c, R)$ $cost = 1 + cost(R)$
- $R \rightarrow c$ $cost = 1$

Thus, the patterns on the right-hand sides of productions in Example 17.4 are associated with the following set of path strings:

1. $:= 1$ *deref* 1 *sp*
2. $:= 2$ *R*
3. *deref* 1 *sp*
4. $+ 1$ *R*
5. $+ 2$ *c*
6. $+ 1$ *c*
7. $+ 2$ *R*
8. *c*

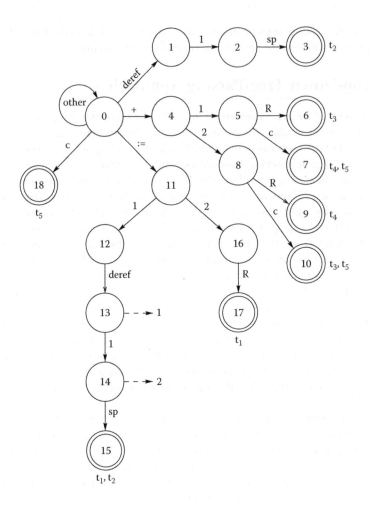

FIGURE 17.5   The tree-pattern-matching automaton for the tree grammar of Example 17.2.

Using the Aho–Corasick algorithm, we can construct the pattern-matching automaton shown in Figure 17.5. The final states are enclosed in double circles, and the failure function pointers that do not point to state 0 are shown as dashed lines. Path strings that match at final states are indicated by specifying the trees they belong to next to the state, t_i, indicating the right-hand side of rule i. Once the pattern-matching automaton is constructed, the subject tree is traversed in preorder, computing automaton states as we visit nodes and traverse edges.

The top-down tree-pattern-matching algorithm was adapted to the problem of tree parsing by Aho and Ganapathi, and the presentation that follows is based on [3]. First, the subject tree is traversed in depth-first order using the routine *MarkStates(n)*, and the automaton state reached at each node is kept track of. This is displayed in Figure 17.6, where δ is the transition function of the underlying string-pattern-matching automaton. The routine also determines the matching-string patterns. The scheme described by Hoffman and O'Donnell [28] using bit vectors to decide whether there is a match at a node is used here. With each node of the subject tree, a bit string b_i is associated with every right-hand side pattern t_i, $1 \leq i \leq m$, where m is the total number of patterns. At any node n of the subject tree, bit j of the bit string b_i is 1 iff every path from the ancestor of n at distance j through n to every descendant of n has a prefix that is a path string of the pattern we want to match. The bit string need not be longer than the height of the corresponding pattern. To find a cover of the intermediate code tree, it is necessary to keep track of reductions that are applicable at a node. The routine *Reduce* shown in Figure 17.7 does this. Since we are

```
procedure MarkStates(n)
if n is the root then
    n.state = δ (0, n.symbol)
else
    n.state = δ(δ(n.parent.state, k), n.symbol)
    where n is the kth child of n.parent
end if
for every child c of n do
    MarkStates(c)
end for
n.bᵢ = 0
if n.state is accepting then
    for each path string of tᵢ of length 2 j + 1 recognized at n.state do
        n.bᵢ = n.bᵢ or 2ʲ
    end for
end if
for every righthandside tree pattern tᵢ do do
    n.bᵢ = n.bᵢ or ∏_{c∈C(n)} right_shift (c.bᵢ)
    where C(n) is the set of all children of node n
end for
Reduce(n)
end procedure
```

FIGURE 17.6   The procedure for visiting nodes and computing states.

looking for an optimal cover, there is the need to maintain a cost for each tree t_i that matches at a node n. The implementation in [3] allows general cost computation rules to be used in place of simple additive costs. The function $cost(t_i, n)$ computes this cost for each node n. For each node n, there is an array $n.cost$ of dimension equal to the number of nonterminals. The entry corresponding to a nonterminal is the cost of the cheapest match of a rule with that nonterminal on the left-hand side. The index of that rule is

```
procedure Reduce(n)
list = set of productions lᵢ → tᵢ such that the zeroth bit of n.bᵢ is 1
while list ≠ ∅ do
    choose and remove next element lᵢ → tᵢ from list
    if cost (tᵢ, n) < n.cost [lᵢ] then
        n.cost [lᵢ] = cost (tᵢ, n)
        n.match [lᵢ]= i
        if n is the root then
            q = δ (0, lᵢ)
        else
            q = δ(δ (n.parent.state, k), lᵢ)
            where n is the kth child or n.parent
        end if
        if q is an accepting state then
            for each path string of tₖ of length 2 j + 1 recognized at q do
                n.bₖ = n.bₖ or 2ʲ
                if the zeroth bit of n.bₖ is a 1 then
                    add lₖ  →  tₖ to list
                end if
            end for
        end if
    end if
end while
end procedure
```

FIGURE 17.7   Procedure for reducing IR trees.

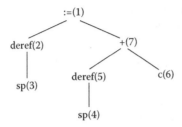

FIGURE 17.8   A subject tree generated by the tree grammar of Example 17.2.

stored in the array *n.match*, which also has dimension equal to the number of nonterminals. Thus, after the parsing is over, the least-cost covers at each node along with their associated costs are available. The result of applying *MarkStates* to the root of the subject tree shown in Figure 17.8 is shown in Figure 17.9. We trace through the first few steps of this computation.

Example 17.5

We use the subject tree of Figure 17.8 with the nodes numbered as shown and the automaton of Figure 17.5 whose transition function is denoted by δ.

1. The start state is 0.
2. $node(1).state = \delta(0, :=) = 11$.
3. $node(2).state = \delta(\delta(11, 1), deref) = 13$.
4. $node(3).state = \delta(\delta(13, 1), sp) = 15$.
5. Path strings corresponding to patterns t_1 and t_3 are matched at $node(3)$. Bit strings $node(3).b_1 = 100$, $node(3).b_2 = 10$, and all the other bit strings are 0.
6. The routine *Reduce*($node(3)$) does nothing, as no reductions are called for.
7. We now return to $node(2)$ and update the bit string $node(2).b_1 = 10$ and $node(2).b_2 = 1$ to reflect the fact that we have moved one level up in the tree.
8. The call to *Reduce*($node(2)$) notes that the zeroth bit of $node(2).b_2 = 1$. Thus, a reduction by the rule $R \rightarrow deref(sp)$ is called for. The cost of this rule is 2 and the rule number is 2; thus, $Cost(R) = 2$ and $Match(R) = 2$ at $node(2)$.
9. We return to $node(1)$ and call *MarkState*($node(7)$); $node(7)$ is the second child of its parent, the failure function is invoked at state 16 on which a transition is made to state 0 and then from state 0 to state 4 on the symbol $+$. Thus, $node(7).state = 4$.
10. $node(5).state = \delta(\delta(4, 1), deref) = 1$. (Note: The failure function is invoked here again at *state* 5.)
11. $node(4).state = \delta(\delta(1, 1), sp) = 3$.
12. A path string corresponding to pattern t_2 is matched at $node(4)$. The bit string $node(4).b_2 = 10$, and all other bit strings are 0.
13. The routine *Reduce*($node(4)$) does nothing, as no reductions are called for.
14. We return to $node(5)$; the bit string $node(5).b_2 = 1$ to reflect that we have moved one level up in the tree.
15. The call to *Reduce*($node(5)$) notes that the zeroth bit of $node(5).b_2$ is a 1. Thus, a reduction by the rule $R \rightarrow deref(sp)$ is called for. The cost of this rule is 2, and the number of the rule is 2. Thus $Cost(R) = 2$ and $Match(R) = 2$ at $node(5)$.
16. The variable q at $node(5)$ is updated to reflect the state after reduction. Thus, $q = \delta(\delta(4, 1), R) = 6$. The state 6 is an accepting state, which matches a string pattern of t_3. Therefore, $node(5).b_3 = 10$.
17. Continuing in this manner, we obtain the labels in Figure 17.9.

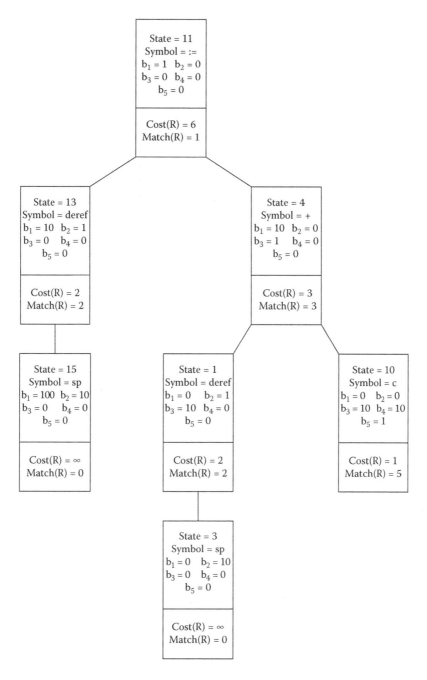

FIGURE 17.9 The information at each node after *MarkStates* is applied to the subject tree of Figure 17.8.

Once a cover has been found, the reductions are performed, during which time the action parts of the productions constituting the cover are executed. As we observed earlier, a reduction may be viewed as replacing a subtree corresponding to the right-hand side of a production with the left-hand side nonterminal. In addition, the action part of the rule is also executed. Usually actions associated with reductions are carried out in depth-first order.

17.4   Bottom-Up Tree-Parsing Approaches

We begin with cost-augmented regular tree grammars in normal form. We first describe a strategy for the generation of tables representing a tree automaton whose states do not encode cost information. Using such an automaton, cost computations for generating locally optimal code will have to be performed at code generation time, that is, dynamically.

Our aim is to find at each node n in the subject tree, minimal cost rules for each nonterminal that matches at the node. We call such a set of nonterminals *matching nonterminals* at node n. If the intermediate code tree is in the language generated by the tree grammar, we expect one of the nonterminals that matches at the root to be the start symbol of the regular tree grammar. The information about rules and nonterminals that match as we go up the tree can be computed with the help of a bottom-up tree-pattern-matching automaton built from the specifications. Thus, during the matching or code generation phase, we traverse the intermediate code tree bottom-up, computing states and costs as we go along. For a given nonterminal in a set, we retain only the minimal cost rule associated with that nonterminal. Finally, when we reach the root of the tree, we have associated with the start symbol the minimal cost of deriving the tree from the start nonterminal. Next, in a top-down pass we select the nonterminals that yield the minimal-cost tree and generate code as specified in the translation scheme.

We present an iterative algorithm as well as a worklist-based algorithm drawing from the work of Balachandran et al. [9] and Proebsting [38].

17.4.1   The Iterative Bottom-Up Preprocessing Algorithm

Let $G = (N, A, P, S)$ be a regular tree grammar in normal form. Before describing the algorithm, we describe some functions that we will be using in the computation. Let *maxarity* be the maximum arity of an operator in A. Let I_l be the set of positive integers of magnitude at most *maxarity*.

$$rules : T \cup OP \mapsto 2^P$$
$$\text{For } a \in T rules(a) = \{r \,|\, r : n \rightarrow a \in P\}$$
$$\text{For } op \in OP rules(op) = \{r : |r : n \rightarrow op(n_1, n_2, \dots, n_k) \in P\}$$

The set *rules(a)* contains all production rules of the grammar whose right-hand side tree patterns are rooted at a.

$$child\_rules : N \times I_l \mapsto 2^P$$
$$child\_rules(n, i) = \{r \,|\, r : n_l \rightarrow op(n_1, n_2, \dots, n_k) \text{ and } n_i = n\}$$

The set $child\_rules(n, i)$ contains all those productions such that the ith nonterminal on the right-hand side is n. The function can be extended to a set of nonterminals N_1 as follows:

$$child\_rules(N_1, i) = \cup_{n \in N_l} child\_rules(n, i)$$

$$child\_NT : OP \times I_l \mapsto 2^N$$
$$child\_NT(op, j) = \{n_j \,|\, r : n_l \rightarrow op \, n_1, n_2, \dots, n_j, \dots, n_k \in P\}$$

In other words, $child\_NT(op, j)$ is the set of all nonterminals that can appear in the jth position in the sequence of nonterminals on the right-hand side of a production for operator op. (If j exceeds $arity[op]$, the function is not defined.)

$$nt : P \mapsto 2^N$$
$$nt(r) = \{n \,|\, r : n \rightarrow \alpha \in P\}$$

```
function Match(t)
begin
if t = a ∈ T then
    match_rules = rules(a);
    match_NT = nt(match_rules);
    Match = match_rules ∪ chain_rule_closure (match_NT);
else
    let t = op(t₁, t₂, . . . tₖ) where arity (op) = k
    for i = 1 to k do
        mrᵢ = Match(tᵢ)
        ntᵢ = nt(mrᵢ)
    end for
    match_rules = rules(op) ∩ child_rules (nt₁, 1) ∩ child_rules (nt₂, 2) . . . ∩
    child_rules (ntₖ, k)
    match_NT = nt(match_rules);
    Match = match_rules ∪ chain_rule_closure (match_NT)
end if
end function
```

FIGURE 17.10   Function that computes matching rules and nonterminals.

The set $nt(r)$ contains the left-hand side nonterminal of the production. The function can be extended to a set R of rules as follows:

$$nt(R) = \cup_{r \in R} nt(r)$$

$$chain\_rule\_closure : N \mapsto 2^P$$

$$chain\_rule\_closure(n) = \{r \, | \, r \in P, r : n_1 \to n_2, n_1 \Rightarrow n_2 \Rightarrow^* n\}$$

The set *chain_rule_closure* of a nonterminal is the set of all rules that begin derivation sequences that derive that nonterminal and contain only chain rules. The function can be extended to a set of nonterminals as follows:

$$chain\_rule\_closure(N_1) = \cup_{n \in N_1} chain\_rule\_closure(n)$$

Given a regular tree grammar in normal form and a subject tree, function *Match* in Figure 17.10 computes the rules that match at the root. For the time being, we ignore the costs of the rules.

The function computes the matching rules at the root of the subject tree by recursively computing matching rules and hence matching nonterminals at the children. This suggests a bottom-up strategy that computes matching rules and nonterminal sets at the children of a node before computing the sets at the node. Each such set can be thought of as a state. The computation that finds the matching rules at a node from the nonterminals that match at its children need not be performed at matching time, as all the sets under consideration are finite and can be precomputed and stored in the form of tables — one for each operator and terminal. This set of tables is actually an encoding of the transition function of a finite-state bottom-up tree-pattern-matching automaton, which computes the state at a node corresponding to an operator from the states of its children. For a terminal symbol $a \in T$, the table τ_a will contain just one set of rules. For an operator $op \in OP$, of arity k, the table τ_{op} will be a k-dimensional table. Each dimension is indexed with indices corresponding to the states described above. Assume that such tables are precomputed and stored as auxiliary information to be used during matching. The function *TableMatch* shown in Figure 17.11 will find the matching rules at the root of a subject tree.

function *Table Match(t)*
begin
if $t = a \in T$ **then**
    Table Match $= \tau_a$
else
    let $t = op(t_1, t_2, \ldots t_k)$ where *arity (op)* $= k$
    Table Match $= \tau_{op}(TableMatch)(t_1), TableMatch(t_2) \ldots TableMatch(t_k))$
end if
end function

FIGURE 17.11   Function that computes matches using precomputed tables.

The function *TableMatch* computes transitions in a bottom-up fashion, performing a constant amount of computation per node. Thus, matching time with precomputed tables is linear in the size of the subject tree. The size of the table for an operator of arity k is $O(2^{|N| \times maxarity})$. The table sizes computed in this manner are huge and can be reduced in the following way.

Assume that *States* is the set of states that is precomputed and indexed by integers from the set I. Let us call each element of *States* an *itemset*. Each such set consists of a set of rules, satisfying the property that there is some subject tree matched by exactly this set of rules. Let *itemset*(i) denote the set indexed by i. We define for each operator op and each dimension j, $1 \leq j \leq arity(op)$, an equivalence relation R_{op}^j as follows: for i_p and $i_q \in I$, $i_p R_{op}^j i_q$, if $nt(itemset[i_p]) \cap child\_NT(op, j) = nt(itemset[i_q]) \cap child\_NT(op, j)$. In other words, two indices are put into the same equivalence class of operator op in dimension j if their corresponding nonterminal sets project onto the same sets in the jth dimension of operator op. If i_p and i_q are in the same equivalence class of R_{op}^j, then it follows that for all $(i_1, i_2, \ldots, i_{j-1}, i_{j+1}, \ldots, i_k)$, $\tau_{op}(i_1, i_2, \ldots, i_{j-1}, i_p, i_{j+1}, \ldots, i_k) = \tau_{op}(i_1, i_2, \ldots, i_{j-1}, i_q, i_{j+1}, \ldots, i_k)$. For the case $k = 2$ this means that i_p and i_q correspond to the indices of identical rows or columns. This duplication can be avoided by storing just one copy. We therefore use index maps as follows. The mapping from the set of indices in I to the set of indices of equivalence classes of R_{op}^j denoted by I_{op}^j is denoted by μ_{op}^j. Thus, we have the mapping

$$\mu_{op}^j : I \mapsto I_{op}^j, 1 \leq j \leq k, arity(op) = k$$

The table τ_{op} is now indexed by elements of I_{op}^j in dimension j instead of those of I. At matching time one extra index table lookup is necessary in each dimension to obtain the resulting element of *States*. This is expressed by the following relation that will replace the table lookup statement of function *TableMatch*. $\tau_{op}(i_1, i_2, \ldots, i_k) = \theta_{op}(\mu_{op}^1(i_1), \mu_{op}^2(i_2), \ldots, \mu_{op}^k(i_k))$, where θ_{op} is the compressed table.

The next step is the direct generation of compressed tables. Informally, the algorithm works as follows. It begins by finding elements of *States* for all symbols of arity 0. It then finds elements of *States* that result from derivation trees of height increasing by one at each iteration until there is no change to the set *States*. At each iteration, elements of *States* corresponding to all operators that contribute to derivation trees of that height are computed. For each operator op and each dimension j of that operator, only nonterminals in $child\_NT(op, j)$ that are also members of a set in *States* computed so far will contribute to new sets associated with op. Such a collection of subsets for op in the jth dimension at iteration i is called *repset*(op, j, i). Thus, choices for a sequence of children of op are confined to elements drawn from a tuple of sets in the Cartesian product of collections at each iteration. Each such tuple is called a *repset_tuple*. Iteration is confined only to elements drawn from new tuples formed at the end of every iteration. At the end of each iteration, the new values of *repset*(op, j, i) are computed for the next iteration. The computation is complete when there is no change to *repset*(op, j, i) for all operators in all dimensions, for then no new tuples are generated. The procedure for precomputing compressed tables is given in Figure 17.12. We illustrate with the help of an example (adapted from [20]).

```
procedure MainNoCost( )
States = ∅
itemset = ∅
for each a ∈ T do
    mrules = mrules (a )
    mnonterminals = nt(mrules(a))
    match_rules = mrules ∪ chain_rule_closure(mnonterminals)
    itemset = match_rules
    States = States ∪ {itemset}
end for
generate child_NT (op, j) for each op ∈ OP and j, 1 ≤ j ≤ arity (op);
generate repset (op, j, 1) for each op ∈ OP and j, 1 ≤ j ≤ arity (op) and
update index maps;
i = 1; repset_product₀ = ∅
repeat
    for each op ∈ OP do
        let repset_productᵢ = ∏ⱼ₌₁...ₐᵣᵢₜᵧ₍ₒₚ₎ repset (op, j, i)
        for each repset_tuple = (S₁, S₂, . . . Sₖ) ∈ repset_productᵢ−repset_productᵢ₋₁
        do
            itemset = ∅
            for each (n₁, n₂ . . . nₖ) with nᵢ ∈ Sᵢ, 1 ≤ i ≤ k do
                mrules = {r : n → op (n₁, n₂ . . . nₖ) ∈ P }
                mnonterminals = nt(mrules)
                match_rules = mrules ∪ chain_rule_closure(mrules)
                itemset = itemset ∪ match_rules
            end for
            θₒₚ (S₁, S₂ . . . Sₖ) = itemset
            States = States ∪ {itemset}
        end for
    end for
    i = i + 1
    generate repset(op, j, i) for each op ∈ OP and j, 1 ≤ arity(op) and update index maps
until repset (op, j, i) = repset (op, j, i−1) ∀ₒₚ∀ j, 1 ≤ j ≤ arity(op)
end
```

FIGURE 17.12 Algorithm to precompute compressed tables without costs.

Example 17.6

Let the following be the rules of a regular tree gramar. The rules and nonterminals are numbered for convenience:

- $stmt \rightarrow := (addr, reg)$ [1]
- $addr \rightarrow +(reg, con)$ [0]
- $addr \rightarrow reg$ [0]
- $reg \rightarrow +(reg, con)$ [1]
- $reg \rightarrow con$ [1]
- $con \rightarrow CONST$ [0]

The nonterminals are numbered as follows: $stmt = 1$, $addr = 2$, $reg = 3$, and $con = 4$.

The operators are $:=$ and $+$ both of arity 2, and there is a single terminal $CONST$ of arity 0. Below are the results of the first iteration of the algorithm:

- There is only one symbol of arity 0, namely $CONST$.
  $mrules = \{con \rightarrow CONST\}$
  $mnonterminals = \{con\}$
  $match\_rules = \{con \rightarrow CONST, reg \rightarrow con, addr \rightarrow reg\}$

Thus, after processing symbols of arity 0 $States = \{con \rightarrow CONST, reg \rightarrow con, addr \rightarrow reg\}$.
Assume this set has index 1. Thus, $I = \{1\}$.
Referring to the state by its index, $nt(1) = \{con, reg, addr\}$.

- Consider the operator $+$.
  The set $child\_NT(+, 1) = \{reg\}$ and $child\_NT(+, 2) = \{con\}$.
  Thus, $repset(+, 1, 1) = child\_NT(+, 1) \cap nt(1) = \{\{reg\}\}$, $\mu_+^1(1) = 1$, $I_+^1 = \{1\}$. Here the projection onto the first dimension of operator $+$ gives the set containing a single set $\{reg\}$ assigned index 1. For ease of understanding, we will use the indices and the actual sets they represent interchangeably. $repset(+, 2, 1) = child\_NT(+, 2) \cap nt(1) = \{\{con\}\}$, $\mu_+^2 = 1$, $I_+^2 = \{1\}$.
  Thus, for $i = 1$ $repset\_product$ for $+ = \{\{\{reg\}\}, \{\{con\}\}\}$.
  For $repset\_tuple = (\{reg\}, \{con\}), i = 1,$
  $mrules = \{reg \rightarrow +(reg, con), addr \rightarrow +(reg, con)\}$
  $mnonterminals = \{reg, addr\}$
  $match\_rules = \{reg \rightarrow +(reg, con), addr \rightarrow +(reg, con), addr \rightarrow reg\}$
  This set $match\_rules$ is added as a new element of $States$ with index 2.
  Thus, $\theta_+(1, 1) = 2$.
  There are no more states added due to operator $+$ at iteration 1.

- Consider the operator $:=$.
  The set $child\_NT(1, :=) = \{addr\}$ and $child\_NT(2, :=) = \{reg\}$.
  Thus, $repset(:=, 1, 1) = \{\{addr\}\}$ and $\mu_{:=}^1 = 1$, $I_{:=}^1 = \{1\}$.
  $repset(:=, 2, 1) = \{\{reg\}\}$ and $\mu_{:=}^2 = 1$, $I_{:=}^1 = \{1\}$.
  Thus, for $i = 1$ $repset\_product$ for operator $:= = \{\{\{addr\}\}, \{\{reg\}\}\}$.
  For $repset\_tuple = (\{addr\}, \{reg\})$ and $i = 1,$
  $mrules = \{stmt \rightarrow := (addr, reg)\}$, $mnonterminals = \{stmt\}$
  $match\_rules = \{stmt \rightarrow := (addr, reg)\}$
  A new state corresponding to $match\_rules$ is added to $States$ with index 3. Thus $\theta_{:=}(1, 1) = 3$. There are no more states added due to operator $:=$ at iteration 1.

- At the end of the iteration for $i = 1$, $States = \{1, 2, 3\}$.

- It turns out that no more states can be added to $States$.

We next show how costs can be included in the states of the bottom-up tree-pattern-matching automaton. We want to capture the following information. Supposing we had a subject tree t and we computed all matching rules and nonterminals as well as minimal costs for each rule and each nonterminal that matched at a node. If we now compute the difference between the cost of each rule and that of the cheapest rule matching at the same node in the tree, we obtain the differential cost. If these differential costs are bounded, they can be precomputed and stored as part of the item in the itemset. Likewise, we can store differential costs with each nonterminal.

As before, let $match\_rules(t)$ be the set of rules matching at the root of a subject tree t. We now define the set of (rule, cost) pairs, $itemset$ matching the root of t.

$$itemset = \{(r, \Delta_r) | r \in match\_rules(t), \Delta_r = cost(r) - min\{cost(r') | r' \in match\_rules(t)\}\}$$

If the costs are bounded for all such pairs, we can precompute them by augmenting the procedure in Figure 17.12. The function that performs the computation for arity 0 symbols is given in Figure 17.13.

Given this procedure, we present the algorithm for precomputing tables with costs. We note that $repset(op, i, j)$ is a collection of sets whose elements are $< nonterminal, cost >$ pairs. The iterative procedure *IterativeMain* in Figure 17.14, for precomputation of itemsets, first calls *IterativeArityZeroTables* to create the tables for symbols of arity 0. It then iterates over patterns of increasing height until no further items are generated. The procedure *IterativeComputeTransitions* in Figure 17.15 creates the new states for each operator at each iteration and updates $States$.

We illustrate the procedure for the grammar of Example 17.6.

```
procedure Iterative Arity Zero Tables
States = ∅
for each a ∈ T do
    itemset = ∅
    mrules = rules(a)
    mnonterminals = nt(rules(a))
    match_rules = mrules ∪ chain_rule_closure (mnonterminals)
    match_NT = nt(match_rules)
    Δr = ∞, τ ∈ match_rules
    Dn = ∞, n ∈ match_NT
    COSTmin = min{rule_cost(r)|r ∈ mrules}
    for each r in mrules do
        Δr = COSTτ − COSTmin
    end for
    for each n in mnonterminals do
        Dn = min {Δr |∃r ∈ mrules, n ∈ nt(r)}
    end for
    repeat
        for each r: n → n1 such that r ∈ chain_rule_closure(mnonterminals))
        do
            Dn = min {Dn, Dn1 + rule_cost(ir)}
            Δr = min {Δr, Dn1 + rule_cost(r)}
        end for
    until no change to any Dn or Δr
    itemset = {(r, Δr)}|r ∈ match_rules}
    τa = itemset
    States = States ∪ {itemset}
end for
end procedure
```

FIGURE 17.13   Computation of arity 0 tables with static costs.

Example 17.7

The following steps are carried out for the only symbol *CONST* of arity 0:

- $mrules = \{6 : con \rightarrow CONST\}$, $mnonterminals = \{con\}$.
- $match\_rules = \{6 : con \rightarrow CONST, 5 : reg \rightarrow con, 3 : addr \rightarrow con\}$, $match\_NT = \{con, reg, addr\}$.
- $\Delta_6 = \infty, \Delta_5 = \infty, \Delta_3 = \infty$.
- $D_4 = \infty, D_3 = \infty, D_2 = \infty$.
- $COST_{min} = min\{rule\_cost(con \rightarrow CONST)\} = 0$.

```
procedure IterativeMain( )
Iterative Arity Zero Tables
generate repset(op, j, 1) for each op ∈ OP and j, 1 ≤ j ≤ arity (op)
i = 1; repset_product0 = ∅
repeat
    for each op ∈ OP do
        IterativeComputeTransition (op, i)
    end for
    i = i + 1
    generate repset (op, j, i) for each op ∈ OP and j, 1 ≤ j ≤ arity(op) and
    update index maps
until repset (op, j, i) = repset(op, j, i−1)∀op∀j, 1 ≤ j ≤ arity (op)
end
```

FIGURE 17.14   Procedure to precompute reduced tables with static costs.

procedure *IterativeComputeTransition* (op, i)
let *repset_product$_i$* $= \prod_{j=1\ldots arity_{(op)}}$ *repset* (op, j, i)
for each *repset_tuple* $= (S_1, S_2, \ldots S_k) \in$ *repset_product$_i$* -*repset_product$_{i--1}$*
do

  itemset $= \emptyset$; *mrules* $= \emptyset$;
  for each $(< n_1 \, D_{n1} >, < n_2, D_{n2} > \ldots < n_k, D_{n_k} >), < n_i \, D_i > \in S_i$
  do

    if $r : n \rightarrow op \, (n_1, n_2, \ldots n_k) \in P$ **then**
      $C_{rhs,r} = D_{n1} + D_{n2} + \cdots D_{n_k}$
      mrules $=$ *mrules* $\cup \{r\}$
    end if
  end for
  mnonterminals $=$ *nt*(*mrules*)
  match_rules $=$ *mrules* $\cup$ *chain_rule_closure*(*mrules*)
  match_NT $=$ *nt*(*match_rules*)
  $\Delta_r = \infty, r \in$ *match_rules*; $D_n = \infty, n \in$ *match_NT*
  for each r in *mrules* **do**
    $COST_r = C_{rhs,r} +$ *rule_cost*(r)
  end for
  $COST_{min} = min \, \{COST_r | r \in mrules\}$
  for each r in *mrules* **do**
    $\Delta_r = COST_r - COST_{min}$
  end for
  for each n in *mnonterminals* **do**
    $D_n = min \, \{\Delta_r | n \in nt(r)\}$
  end for
  repeat
    for each $r : n \rightarrow n_1$ such that $r \in$ *chain_rule_closure* (*match_NT*))
    do
      $D_n = min \, \{D_n, D_{n1} +$ *rule_cost*$(r)\}$
      $\Delta_r = min \, \{\Delta_r, D_{n1} +$ *rule.cost*$(r)\}$
    end for
  until no change to any D_n or Δ_n
  itemset $=$ *itemset* $\cup \, \{(r, \Delta_r)\} | r \in$ *match_rules*, $\Delta_r \leq \Delta_{r'}$ if $nt(r) = nt(r')\}$
  $\theta_{op} (S_1, S_2. \ldots S_k) =$ *itemset*
  States $=$ *States* $\cup \, \{itemset\}$
end for
end procedure

FIGURE 17.15   Procedure for computing transitions on operators.

- $\Delta_6 =$ *rule_cost*$(con \rightarrow CONST) - COST_{min} = 0$.
- $D_4 = 0$.
- After the first iteration of the repeat-until loop $D_4 = 0, D_3 = 1, D_2 = \infty, \Delta_6 = 0, \Delta_5 = 1, \Delta_3 = \infty$.
- After the second iteration of the repeat-until loop $D_4 = 0, D_3 = 1, D_2 = 1, \Delta_6 = 0, \Delta_5 = 1, \Delta_3 = 1$.
- There is no change at the next iteration, so *States* $= \{\{< con \rightarrow CONST, 0 >, < reg \rightarrow con, 1 >, < addr \rightarrow reg, 1 >\}\}$.

We next consider the operator $+$ of arity 2. *child_NT*$(+, 1) = \{reg\}$, *child_NT*$(+, 2) = \{con\}$. *repset* $(+, 1, 1) = \{\{< reg, 0 >\}\}$, *repset*$(+, 2, 1) = \{\{< con, 0 >\}\}$. The following steps are then carried out for the operator $+$ at the first iteration:

- *repset_product$_1$* $= \{\{< reg, 0 >\}\} \times \{\{< con, 0 >\}\}$.
- *repset_tuple* $= (\{< reg, 0 >\}, \{< con, 0 >\})$.
- *mrules* $= \{2 : addr \rightarrow + (reg \, con), 4 : reg \rightarrow + (reg \, con)\}$.

- $C_{rhs,2} = 0, C_{rhs,4} = 0$.
- $mnonterminals = \{addr, reg\}$.
- $match\_rules = \{2 : addr \rightarrow + (reg\ con), 4 : reg \rightarrow + (reg\ con), 3 : addr \rightarrow reg\}$.
- $match\_NT = \{addr, reg\}$.
- $\Delta_2 = \Delta_4 = \Delta_3 = \infty$.
- $D_3 = D_2 = \infty$.
- $COST_2 = 0 + 0 = 0, COST_4 = 0 + 1 = 1, COST_{min} = 0$.
- $\Delta_2 = 0, \Delta_4 = 1, D_2 = 0, D_3 = 1$.
- There is no change to these sets during the first iteration of the while loop, so the value of *itemset* after discarding more expensive rules for the same nonterminal is $itemset = \{<addr \rightarrow + (reg\ con), 0>, <reg \rightarrow + (reg\ con), 1>\}$.
- Thus, $States = \{\{<con \rightarrow CONST, 0>, <reg \rightarrow con, 1>, <addr \rightarrow reg, 1>\}\} \cup \{\{<addr \rightarrow + (reg\ con), 0>, <reg \rightarrow + (reg\ con), 1>\}\}$.

After processing the operator $:=$ in a similar manner, we get the following three itemsets in *States*:

- $\{<con \rightarrow CONST, 0>, <reg \rightarrow con, 1>, <addr \rightarrow reg, 1>\}$
- $\{<addr \rightarrow + (reg\ con), 0>, <reg \rightarrow + (reg\ con), 1>\}$
- $\{<stmt \rightarrow := (addr, reg), 0>\}$

17.4.2   A Worklist-Based Approach to Bottom-Up Code-Generator Generators

Proebsting [38] employs a worklist approach to the computation of itemsets; the presentation that follows is based on [38]. A state is implemented as a set of tuples, each tuple containing:

- A nonterminal that matches a node
- The normalized cost of this nonterminal
- The rule that generated this nonterminal at minimal cost

A tuple structured as above is called an *item*; a collection of such items is termed an *itemset*. Each itemset represents a state of the underlying cost-augmented tree-pattern-matching automaton whose set of states is *States*. Each itemset is represented as an array of (rule, cost) pairs indexed by nonterminals. Thus, $itemset[n].cost$ refers to the normalized cost of nonterminal n of the itemset, and $itemset[n].rule$ gives a rule that generates that nonterminal at minimal cost. A cost of ∞ in any position indicates that no rule derives the given nonterminal. The empty state $(\emptyset)$ has all costs equal to infinity.

The procedure *WorklistMain()* in Figure 17.16 manipulates a worklist that processes itemsets. Assume that *States* is a table that maintains a one-to-one mapping from itemsets to nonnegative integers. The routine *WorklistArityZeroTables* in Figure 17.17 computes the tables for all terminals in T.

The routine *WorklistComputeTransition* shown in Figure 17.18 augments the operator tables with a new transition computed from an itemset in the worklist. The itemset is projected in each dimension of each operator and combined with other represener sets for that operator to check if the combination leads to a new state. The closure is computed only if this is a new state. Finally, the itemset is added to the worklist and the set of states, and the appropriate transition table is updated.

Proebsting has shown that an optimization that he calls *state trimming* considerably reduces table sizes. We briefly explain one of the optimizations, called *triangle trimming*. Consider the two derivation trees shown in Figure 17.19. Both these have the same root and leaves except for a single leaf nonterminal. Both trees use different rules for the operator *op* to reduce to A. Let $r_1 : A \rightarrow op(X, Q)$ and $r_2 : B \rightarrow op(Y, R)$, with $A \rightarrow B, R \rightarrow Q$, and $Y \rightarrow Z$ being chain rules. Triangle trimming notes that both reductions to A involve different nonterminals for a left child state related to operator *op* that occur in the same state. Let that state be *state*. If $state[X].cost$ exceeds or equals $state[Z].cost$ in all contexts, we can eliminate nonterminal X from all such states. Considerable savings in storage have been reported using this optimization.

```
procedure WorklistMain( )
    States = ∅
    WorkList = ∅
    Worklist Arity Zero Tables
    while WorkList ≠ ∅ do
        itemset = next itemset from WorkList
        for op ∈ OP do
            WorklistComputeTransition (op, itemset)
        end for
    end while
end procedure
```

FIGURE 17.16   Worklist processing routine.

17.4.3   Hard Coded Bottom-Up Code-Generator Generators

Hard coded code-generator generators are exemplified in the work of Fraser et al. [20] and Emmelmann et al. [17]. They mirror their input specifications in the same way that recursive descent parsers mirror the structure of the underlying $LL(1)$ grammar. Examples of such tools are BEG [17] and iburg [20]. Code generators generated by such tools are easy to understand and debug, as the underlying logic is simple. The code generator that is output typically works in two passes on the subject tree. In a first bottom-up, left-to-right pass it labels each node with the set of nonterminals that match at the node. Then in a second top-down pass, it visits each node, performing appropriate semantic actions, such as generating code. The transition tables used in the techniques described earlier in this section are thus encoded in the flow of control of the code generator, with cost computations being performed dynamically.

17.4.4   The Code Generation Pass

Following the first pass, where all the nodes of the IR tree are labeled with a state, a second pass over the tree generates the optimal code. Each rule has associated with it a case number that specifies a set of actions to

```
procedure WorklistArityZeroTables
    for a ∈ T do
        itemset = ∅
        for each r ∈ rules (a) do
            itemset[nt(r)] = (r, rule_cost(r))
        end for
        // normalize costs
        for all n ∈ N do
            itemset[n].cost = itemset[n].cost − min_i {itemset[i].cost}
        end for
        // compute chain rule closure
        repeat
            for all r such that r : n → m is a chain rule do
                cost = rule_cost(r) + itemset[m].cost
                if cost < itemset[n].cost then
                    itemset[n] = (r, cost)
                end if
            end for
        until no changes to itemset
        Append itemset to WorkList
        States = States ∪ {itemset}
        τ_a = itemset
    end for
end procedure
```

FIGURE 17.17   The computation of tables of arity 0.

procedure *WorklistComputeTransition (op, itemset)*
for $i = 1$ to *arity(op)* **do**
   repstate $= \emptyset$
   for $n \in N$ **do**
      if *child_rules(n, i)* $\cup$ *rules (op)* $\neq \emptyset$ **then**
         repstate[n].cost = *itemset[n].cost*
      end if
   end for
   for all $n \in N$ **do**
      repstate[n].cost = *repstate[n].cost* $-$ *min$_i$* {repstate [i].cost}
   end for
   μ^i_{op} *(itemset)* = *repstate*
   if *repstate* $\in I^i_{op}$ **then**
      $I^i_{op} = I^i_{op} \cup \{repstate\}$
      for each *repset_tuple* = $(S_1, S_2, \ldots S_{i-1}, repstate\ S_{i+1}, \ldots S_k)$ where $S_j \in$ $I^i_{op}, j \neq i$ **do**
         newitemset $= \emptyset$
         for each rule r of the form $n \to op\ n_1 n_2 \ldots n_{arity(op)}$ in *rules (op)* **do**
            cost = *rule_cost(r)* + *repstate[n_i].cost* + $\Sigma_{j \neq i} S_j [n_j].cost$
            if *cost* < *newitemset[n]*.cost **then**
               newitemset[n] = $(r, cost)$
            end if
         end for
         for all $n \in N$ **do**
            newitemset[n].cost = *newitemset[n].cost*$-$*min$_i$* {newitemset[i].cost}
         end for
         if *newitemset* $\notin$ *States* **then**
            repeat
               for all r such that $r: n \to m$ is a chain rule **do**
                  cost = *rule_cost(r)* + *newitemset[m].cost*
                  if *cost* < *newitemset[n].cost* **then**
                     newitemset[n] = $(r, cost)$
                  end if
               end for
            until no changes to *newitemset*
            append *newitemset* to *WorkList*
            States = *States* $\cup$ {*newitemset*}
         end if
         $\theta_{op} (S_1, S_2, \ldots, S_{i-1}, repset, S_{i+1}, \ldots S_k)$ = *newitemset*
      end for
   end if
end for
end procedure

FIGURE 17.18 Procedure to compute transitions on operators.

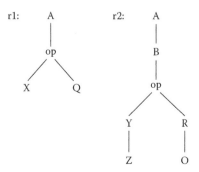

FIGURE 17.19 Two derivation trees to illustrate triangle trimming.

```
procedure GenerateCode
top_down_traverse(root, S)
execute actions in root.caselist in reverse order of the list
end procedure
```

FIGURE 17.20   Code generation routine.

be executed when the rule is matched. The actions could include allocation of a register, emission of code, or computation of some attribute. During this phase each node n will be assigned a list of case numbers stored in $n.caselist$ in the reverse order of execution. This is described in the procedures *GenerateCode* and *TopDownTraverse* in Figures 17.20 and 17.21, respectively.

17.5   Techniques Extending LR-Parsers

The idea of LR-based techniques for table-driven code generation had been proposed earlier by Graham and Glanville [23]. However, their approach cannot be applied in general to the problem of regular tree parsing for ambiguous tree grammars, as it does not carry forward all possible choices in order to be able to report all matches. The technique described here can be viewed as an extension of the $LR(0)$ parsing strategy and is based on the work reported in Shankar et al. [40] and Madhavan et al. [34]. Let G' be the context-free grammar obtained by replacing all right-hand sides of productions of G by postorder listings of the corresponding trees in $TREES(A \cup N)$. Note that G is a regular tree grammar whose associated language contains trees, whereas G' is a context-free grammar whose language contains strings with symbols from A. Of course, these strings are just the linear encodings of trees.

Let $post(t)$ denote the postorder listing of the nodes of a tree t. The following (rather obvious) claim underlies the algorithm:

```
procedure TopDownTraverse (node, nonterminal)
if node is a leaf then
    if node.state[nonterminal].rule is r : X → Y, Y ∈ N then
        append case number of r to node.caselist
        TopDownTravese(node, Y)
    else
        if node.state[nonterminal].rule is r : X → a, a ∈ T then
            append case number of r to node.caselist
            execute actions in node.caselist in reverse order
        end if
    end if
else
    if node.state [nonterminal].rule is r : X → Y, Y ∈ N then
        append case number of r to node.caselist
        TopDownTraverse (node, Y)
    else
        if node.state[nonterminal].rule is r : X → op (X₁, X₂, . . . Xₖ) then
            append case number of r to node.caselist
            for i = 1 to k do
                TopDownTraverse (child (i, node), Xᵢ)
            end for
        end if
        execute actions in node.caselist in reverse order
    end if
end if
end procedure
```

FIGURE 17.21   Top-down traversal for code generation.

A tree t is in $L(G)$ if and only if $post(t)$ is in $L(G')$. Also, any tree α in $TREES(A \cup N)$ that has an associated S-derivation tree in G has a unique sentential form $post(\alpha)$ of G' associated with it.

The problem of finding matches at any node of a subject tree t is equivalent to that of parsing the string corresponding to the postorder listing of the nodes of t. Assuming that a bottom-up parsing strategy is used, parsing corresponds to reducing the string to the start symbol, by a sequence of *shift* and *reduce* moves on the parsing stack, with a match of rule r being reported at node j whenever r is used for the reduction at the corresponding position in the string. Thus, in contrast with earlier methods that seek to construct *a tree automaton* to solve the problem, a *deterministic pushdown automaton* is constructed for the purpose.

17.5.1   Extension of the $LR(0)$ Parsing Algorithm

We assume that the reader is familiar with the notions of rightmost derivation sequences, handles, viable prefixes of right sentential forms, and items being valid for viable prefixes. Definitions can be found in [29]. The meaning of an *item* in this section corresponds to that understood in LR parsing theory. By a viable prefix *induced* by an input string is the stack contents that result from processing the input string during an *LR* parsing sequence. If the grammar is ambiguous, there may be several viable prefixes induced by an input string.

The key idea used in the algorithm is contained in the theorem below [40].

Theorem 17.1   *Let G' be a normal form context-free grammar derived from a regular tree grammar. Then all viable prefixes induced by an input string are of the same length.*

To apply the algorithm to the problem of tree-pattern matching, the notion of *matching* is refined to one of *matching in a left context*.

Definition 17.7   *Let n be any node in a tree t. A subtree t_i is said to be to the left of node n in the tree if the node m at which the subtree t_i is rooted occurs before n in a postorder listing of t. t_i is said to be a maximal subtree to the left of n if it is not a proper subtree of any subtree that is also to the left of n.*

Definition 17.8   *Let $G = (N, T, P, S)$ be a regular tree grammar in normal form and t be a subject tree. Then rule $X \to \beta$ matches at node j in left context α, $\alpha \in N^*$ if:*

- $X \to \beta$ *matches at node j or equivalently, $X \Rightarrow \beta \Rightarrow^* t'$, where t' is the subtree rooted at j.*
- *If α is not ϵ, then the sequence of maximal complete subtrees of t to the left of j, listed from left to right, is $t_1, t_2, \ldots, t_k$, with t_i having an X_i-derivation tree, $1 \le i \le k$, where $\alpha = X_1 X_2 \ldots X_k$.*
- *The string $X_1 X_2 \ldots X_k X$ is a prefix of the postorder listing of some tree in $TREES(A \cup N)$ with an S-derivation.*

Example 17.8

We reproduce the tree grammar of Example 17.6 as a context-free grammar below:

1. *stmt* $\to$ *addr reg* :=   [1]
2. *addr* $\to$ *reg con* +   [0]
3. *addr* $\to$ *reg*   [0]
4. *reg* $\to$ *reg con*+   [1]
5. *reg* $\to$ *con*   [1]
6. *con* $\to$ *CONST*   [0]

Consider the subject tree of Figure 17.22 and the derivation tree alongside. The rule *con* $\to$ *CONST* matches at node 2 in left context ϵ. The rule *con* $\to$ *CONST* matches at node 3 in left context *addr*. The rule *reg* $\to$ *reg con* + matches at node 5 in left context *addr*.

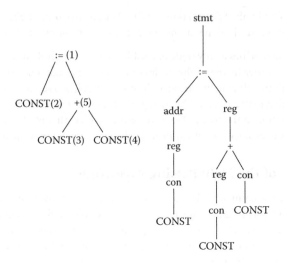

FIGURE 17.22   A derivation tree for a subject tree derived by the grammar of Example 17.8.

The following property forms the basis of the algorithm. Let t be a subject tree with postorder listing $a_1 \ldots a_j w, a_i \in A, w \in A^*$. Then rule $X \to \beta$ matches at node j in the left context α if and only if there is a rightmost derivation in the grammar G' of the form

$$S \Rightarrow^* \alpha X z \Rightarrow^* \alpha \, post(\beta)z \Rightarrow^* \alpha a_h \ldots a_j z \Rightarrow^* a_1 \ldots a_j z, z \in A^*$$

where $a_h \ldots a_j$ is the subtree rooted at node j.

Since there is a direct correspondence between obtaining rightmost derivation sequences in G' and finding matches of rules in G, the possibility of using an LR-like parsing strategy for tree parsing is obvious. Since all viable prefixes are of the same length, a deterministic finite automaton (DFA) can be constructed that recognizes *sets* of viable prefixes. We call this device the *auxiliary automaton*. The grammar is first augmented with the production $Z \to S\$$ to make it prefix free. Next, the auxiliary automaton is constructed; this plays the role that a DFA for a canonical set of LR items does in an LR parsing process. We first explain how this automaton is constructed without costs. The automaton M is defined as follows:

$$M = (Q, \Sigma, \delta, q_0, F)$$

where each state of Q contains a set of items of the grammar:

- $\Sigma = A \cup 2^N$.
- $q_0 \in Q$ is the start state.
- F is the state containing the item $Z \longrightarrow S\$$.
- $\delta : Q \times (A \cup 2^N) \mapsto Q$.

Transitions of the automaton are thus either on terminals or on sets of nonterminals. A set of nonterminals will label an edge iff all the nonterminals in the set match some subtree of a tree in the language generated by the regular tree grammar *in the same left context*. The precomputation of M is similar to the precomputation of the states of the DFA for canonical sets of $LR(0)$ items for a context-free grammar. However, there is one important difference. In the DFA for $LR(0)$ items, transitions on nonterminals are determined just by looking at the sets of items in any state. Here we have transitions on *sets* of nonterminals. These cannot be determined in advance, as we do not know a priori which rules are matched simultaneously when matching is begun from a given state. Therefore, transitions on sets of nonterminals are added as and when these sets are determined. Informally, at each step, we compute the set of items generated by making a transition on some element of A. Because the grammar is in normal form, each such transition leads

```
procedure TreeParser (a, M, matchpairs)
// The input string of length n + 1 including the end marker is in array a
// M is the DFA (constructed from the context free grammar) which
controls the parsing process with transition functions δ_A and δ_{LC}.
// matchpairs is a set of pairs (i, m) such that the set of rules in m matches
at node i in a left context induced by the sequence of complete subtrees to the left of i.
stack = q_0; matchpairs = ∅
current_state = q_0
for i = 1 to n do
    current_state: = δ_A (current_state, a[i]);
    match_rules = current_state.match_rules
    // The entry in the table δ_A directly gives the set of rules matched.
    pop(stack) arity (a|[i] + 1 times;
    current_state: = δ_{LC} (topstack, S_m);
    // S_m is the set of nonterminals matched after chain rule application
    match_rules = match_rules ∪ current_state.match_rules
    // add matching rules corresponding to chain rules that are matched
    matchpairs = matchpairs ∪ {(i, match_rules)}
    push(current_state)
end for
end procedure
```

FIGURE 17.23   Procedure for tree parsing using bottom-up context-free parsing approach.

to a state, termed a *matchset*, that calls for a reduction by one or more productions called *match_rules*. Since all productions corresponding to a given operator are of the same length (because operator arities are fixed and the grammar is in normal form), a reduction involves popping off a set of right-hand sides from the parsing stack and making a transition on a set of nonterminals corresponding to the left-hand sides of all productions by which we have performed reductions, from each state (called an *LCset*) that can be exposed on the stack after popping off the set of handles. This gives us, perhaps, a new state, which is then added to the collection if it is not present. Two tables encode the automaton. The first, δ_A, encodes the transitions on elements of A. Thus, it has as row indices, the indices of the *LCsets* and as columns, elements of A. The second, δ_{LC}, encodes the transitions of the automaton on sets of nonterminals. The rows are indexed by *LCsets* and the columns by indices of sets of nonterminals. The operation of the matcher, which is effectively a tree parser, is defined in Figure 17.23. Clearly, the algorithm is linear in the size of the subject tree. It remains to describe the precomputation of the auxiliary automaton coded by the tables δ_A and δ_{LC}.

17.5.2   Precomputation of Tables

The start state of the auxiliary automaton contains the same set of items as would the start state of the DFA for sets of $LR(0)$ items. From each state, for instance, q, identified to be a state of the auxiliary automaton, we find the state entered on a symbol of A, for instance, a. (This depends only on the set of items in the first state.) The second state, for instance, m (which we will refer to as a matchstate) will contain only complete items. We then set $\delta_A(q, a)$ to the pair $(match\_rules(m), S_m)$, where $match\_rules(m)$ is the set of rules that match at this point and S_m is the set of left-hand side nonterminals of the associated productions of the context-free grammar. Next we determine all states that have paths of length $arity(a) + 1$ to q. We refer to such states as *valid left context* states for q. These are the states that can be exposed on the stack while performing a reduction after the handle is popped off the stack. If p is such a state, we compute the state r corresponding to the itemset obtained by making transitions on elements of S_m augmented by all nonterminals that can be reduced to because of chain rules. These new itemsets are computed using the usual rules that are used for computing sets of $LR(0)$ items. Finally, the *closure* operation on resulting items completes the new itemset associated with r. The closure operation here is the conventional one used for constructing canonical sets of LR items [6].

function *Validlc(p, m)*
    if *NTSET (p, rhs(m)) = S_m* **then**
        Validlc := true
    else
        Validlc := false
    end if
end function

FIGURE 17.24   Function to compute valid left contexts.

Computing states that have paths of the appropriate length to a given state is expensive. A very good approximation is computed by the function *Validlc* in Figure 17.24. This function just examines the sets of items in a matchstate and a candidate left context state and decides whether the candidate is a valid left context state. For a matchstate m let $rhs(m)$ be the set of right-hand sides of productions corresponding to complete items in m.

For a matchstate m and a candidate left context state p, define

$$NTSET(p, rhs(m)) = \{B \mid B \rightarrow .\alpha \in itemset(p), \alpha \in rhs(m)\}$$

Then a necessary, but not a sufficient, condition for p to be a valid left context state for a matchstate corresponding to a matchset m is $NTSET(p, rhs[m]) = S_m$. (The condition is only necessary because there may be another production that always matches in this left context when the others do but that is not in the matchset.)

Before we describe the preprocessing algorithm, we have to define the costs that we will associate with items. The definitions are extensions of those used in Section 17.4 and involve keeping track of costs associated with rules partially matched (as that is what an item encodes) in addition to costs associated with rules fully matched.

Definition 17.9   *The absolute cost of a nonterminal X matching an input symbol a in left context ϵ is represented by $abscost(\epsilon, X, a)$. For a derivation sequence d represented by $X \Rightarrow X_1 \Rightarrow X_2 \ldots \Rightarrow X_n \Rightarrow a$, let $C_d = rulecost(X_n \rightarrow a) + \sum_{i=1}^{n-1} rulecost(X_i \rightarrow X_{i+1}) + rulecost(X \rightarrow X_1)$; then $abscost(\epsilon, X, a) = min_d(C_d)$.*

Definition 17.10   *The absolute cost of a nonterminal X matching a symbol a in left context α is defined as follows:*

$$abscost(\alpha, X, a) = abscost(\epsilon, X, a) \text{ if } X \text{ matches in left context } \alpha$$
$$abscost(\alpha, X, a) = \infty \text{ otherwise}$$

Definition 17.11   *The relative cost of a nonterminal X matching a symbol a in left context α is $cost(\alpha, X, a) = abscost(\alpha, X, a) - min_{y \in N}\{abscost(\alpha, Y, a)\}$.*

Having defined costs for trees of height 1, we next look at trees of height greater than 1. Let t be a tree of height greater than 1.

Definition 17.12   *The cost $abscost(\alpha, X, t) = \infty$ if X does not match t in left context α. If X matches t in left context α, let $t = a(t_1, t_2, \ldots, t_q)$ and $X \longrightarrow Y_1 Y_2 \ldots Y_q a$ where Y_i matches $t_i, 1 \leq i \leq q$. Let $abscost(\alpha, X \rightarrow Y_1 Y_2 \ldots Y_q a, t) = rulecost(X \rightarrow Y_1 \ldots Y_q a) + cost(\alpha, Y_1, t_1) + cost(\alpha Y_1, Y_2, t_2) + \ldots + cost(\alpha Y_1 Y_2 \ldots Y_{q-1}, Y_q, t_q)$. Hence, define*

$$abscost(\alpha, X, t) = min_{X \Rightarrow \beta \Rightarrow^* t}\{abscost(\alpha, X \Rightarrow \beta, t)\}$$

Definition 17.13   *The relative cost of a nonterminal X matching a tree t in left context α is $cost(\alpha, X, t) = abscost(\alpha, X, t) - min_{Y \Rightarrow^* t}\{abscost(\alpha, Y, t)\}$.*

function *Goto(itemset, a)*
$Goto = \{[A \rightarrow \alpha a., c] | [A \rightarrow \alpha.a, c'] \in intemset$ and
$c = c' + rule\_cost(A \rightarrow \alpha a) -$
$\min\{c'' + rule\_cost(B \rightarrow \beta a) | [B \rightarrow \beta.a, c''] \in itemset\}\}$
end function

FIGURE 17.25 The function to compute transitions on elements of A.

We now proceed to define a few functions that will be used by the algorithm. The function *Goto* in Figure 17.25 makes a transition from a state on a terminal symbol in A and computes normalized costs. Each such transition always reaches a matchstate, as the grammar is in normal form.

The *reduction* operation on a set of complete augmented items *itemset$_1$* with respect to another set of augmented items, *itemset$_2$*, is encoded in the function *Reduction* in Figure 17.26. The function *Closure* is displayed in Figure 17.27 and encodes the usual closure operation on sets of items. The function *ClosureReduction* is shown in Figure 17.28. Having defined these functions, we present the routine for precomputation in Figure 17.29.

The procedure *LRMain* will produce the auxiliary automaton for states with cost information included in the items. Equivalence relations that can be used to compress tables are described in [34]. We now look at an example with cost precomputation. The context-free grammar obtained by transforming the grammar of Example 17.2 is displayed in Example 17.9.

Example 17.9

$G = (V, B, G, a(2), b(0), P, V)$
P:
$V \rightarrow VBa\ [0]$
$V \rightarrow GVa\ [1]$
$V \rightarrow G\ \ \ \ [1]$
$G \rightarrow B\ \ \ \ [1]$
$V \rightarrow b\ \ \ \ [7]$
$B \rightarrow b\ \ \ \ [4]$

The automaton is shown in Figure 17.30.

function *Reduction(itemset$_2$, itemset$_1$)*
// First compute costs of nonterminals in matchsets
$S - S_{itemset1}$
$cost(X) = \min\{c_i | [X \rightarrow \alpha_{i.}, c_i] \in itemset_1\}$ if $X \in S\infty$ otherwise
// process chain rules and obtain updated costs of nonterminals
$temp = \cup\{[A \rightarrow B., c] | \exists [A \rightarrow .B, 0] \in itemset_2 \wedge [B \rightarrow \gamma., c_1] \in$
$itemset_1 \wedge c = c_1 + rule\_cost(A \rightarrow B)\}$
repeat
$\ \ \ \ S = S \cup \{X | [X \rightarrow Y., c] \in temp\}$
$\ \ \ \ $**for** $X \in S$ **do**
$\ \ \ \ \ \ \ \ cost(X) = min(cost(X), \min\{c_i | \exists [X \rightarrow Yi., c_i] \in temp\})$
$\ \ \ \ \ \ \ \ temp = \{(A \rightarrow B., c] | \exists [A \rightarrow .B, 0] \in itemset_2 \wedge [B \rightarrow Y., c_1] \in$
$\ \ \ \ \ \ \ \ temp \wedge c = c_1 + rule\_cost(A \rightarrow B)\}$
$\ \ \ \ $**end for**
until no change to *cost* array or *temp* $= \phi$
// Compute reduction
$Reduction = \cup\{[A \rightarrow \alpha B.\beta, c] | [A \rightarrow \alpha, B\beta, c_1] \in itemset_2 \wedge B \in S \wedge c = cost(B) + c_1$ if $\beta \neq \epsilon$ else
//This is a complete item corresponding to a chain rule
$c = rule\_cost(A \rightarrow B) - \min\{c_i | \exists [X \rightarrow .Y, 0] \in itemset_2, \wedge c_i = rule\_cost(X \rightarrow Y)\}$
end function

FIGURE 17.26 Function that performs reduction by a set of rules given the LCstate and the matchstate.

```
function Closure (itemset)
repeat
    itemset = itemset ∪ {[A → .α,0]|[B → .Aβ,c] ∈ itemset
until no change to itemset
Closure = itemset
end function
```

FIGURE 17.27   Function to compute the closure of a set of items.

```
function Closure Reduction(itemset)
Closure Reduction = Closure(Reduction(itemset))
end function
```

FIGURE 17.28   Function to compute *ClosureReduction* of a set of items.

```
procedure LRMain( )
lcsets : = ∅
matchsets : = ∅
list: = Closure ({[S → .α, 0] |S → α ∈ P})
while list is not empty do
    delete next element q from list and add it to lcsets
    for each a ∈ A such there is a transition on a from q do
        m: Goto(q, a)
        δ_A(q, a): = (match (m), S_m)
        if m is not in matchsets then
            matchsets: = matchsets ∪ {m}
            for each state r in lcsets do
                if Validle(r, m) then
                    p: = ClosureReduction (r, m)
                    δ_LC (r, S_m):= (match (p), p)
                    if p is not in list or lcsets then
                        append p to list
                    end if
                end if
            end for
        end if
    end for
    for each state t in matchsets do
        if Validlc(q, t) then
            s : = ClosureReduction (q, t)
            δ_LC(q, S_t): = (match (s), s)
            if s is not in list or lcsets then
                append s to list
            end if
        end if
    end for
end while
end procedure
```

FIGURE 17.29   Algorithm to construct the auxiliary automaton.

Example 17.10

Let us look at a typical step in the preprocessing algorithm. Let the starting state be q_0, that is, the first *LCset*.

$$q_0 = \{[S → .V\$,0], [V → .V B a, 0], [V → .G V a, 0], [V → .G, 0], [G → .B, 0],$$
$$[V → .b, 0], [B → .b, 0]\}$$

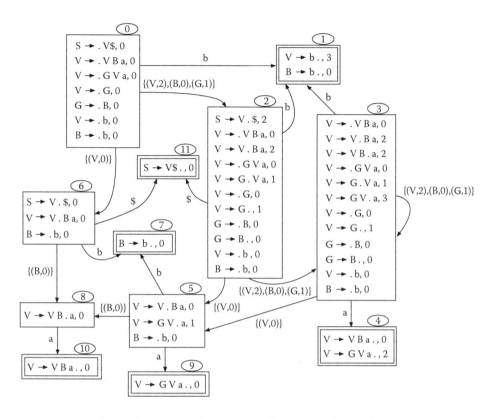

FIGURE 17.30 Auxiliary automaton for grammar of Example 17.9.

Using the definition of the *goto* operation, we can compute the matchset q_1 as

$$q_1 = Goto(q_0, b) = \{ [V \rightarrow b ., 3], [B \rightarrow b ., 0] \}$$

In the matchset, the set of matching nonterminals S_{q_1} is

$$S_{q_1} = \{V, B\}$$

with costs 3 and 0, respectively.

Now we can compute the set *ClosureReduction*(q_0, q_1). First we compute *Reduction*(q_0, q_1).
Initialization:

$$S = S_{q_1} = \{V, B\}$$
$$cost(V) = 3$$
$$cost(B) = 0$$
$$cost(G) = \infty$$
$$temp = \{[G \rightarrow B ., 1]\}$$

Processing chain rules:
Iteration 1:

$$S = S \cup \{G\} = \{V, B, G\}$$
$$cost(V) = 3$$
$$cost(B) = 0$$
$$cost(G) = 1$$
$$temp = \{[V \rightarrow G ., 2]\}$$

Iteration 2:

$$S = S \cup \{V\} = \{V, B, G\}$$
$$cost(V) = 2$$
$$cost(B) = 0$$
$$cost(G) = 1$$
$$temp = \phi$$

Computing reduction:

$$Reduction = \{[S \rightarrow V.\$, 2], [V \rightarrow V . B\, a,\, 2], [V \rightarrow G . V\, a,\, 1], [V \rightarrow G .,\, 1], [G \rightarrow B .,\, 0]\}$$

Once we have $Reduction(q_0, q_1)$, we can use the function *Closure* and compute *ClosureReduction*. Therefore,

$$q_2 = ClosureReduction(q_0, q_1)$$
$$= Closure(Reduction(q_0, q_1))$$
$$= Closure(\{[S \rightarrow V.\$, 2], [V \rightarrow V . B\, a,\, 2], [V \rightarrow G . V\, a,\, 1], [V \rightarrow G .,\, 1], [G \rightarrow B .,\, 0], \})$$
$$= \{[S \rightarrow V.\$, 2], [V \rightarrow . V\, B\, a,\, 0], \{[V \rightarrow V . B\, a,\, 2], [V \rightarrow . G\, V\, a,\, 0], [V \rightarrow G . V\, a,\, 1],$$
$$[V \rightarrow . G,\, 0], [V \rightarrow G .,\, 1], [G \rightarrow . B,\, 0],$$
$$[G \rightarrow B .,\, 0], [V \rightarrow . b,\, 0], [B \rightarrow . b,\, 0]\}$$

The auxiliary automaton for the grammar of Example 17.6 is shown in Figure 17.31. Though the number of states for this example exceeds that for the conventional bottom-up tree-pattern-matching automaton, it has been observed that for real machines, the tables tend to be smaller than those for conventional bottom-up tree-pattern-matching automata [34]. This is perhaps because separate tables need not be maintained for each operator. An advantage of this scheme is that it allows the machinery of attribute grammars to be used along with the parsing.

17.6 Related Issues

A question that arises when generating a specification for a particular target architecture is the following: Can a specification for a target machine produce code for every possible intermediate code tree produced by the front-end? (We assume here, of course, that the front-end generates a correct intermediate code tree.) This question has been addressed by Emmelmann [16], who refers to the property that is desired of the specification as the *completeness* property. The problem reduces to one of containment of the language $L(T)$ of all possible intermediate code trees in $L(G)$, the language of all possible trees generated by the regular tree grammar constituting the specification. Thus, the completeness test is the problem of testing the subset property of two regular tree grammars, which is decidable. An algorithm is given in [16].

A second question has to do with whether a code-generator generator designed to compute normalized costs statically terminates on a given input. If the grammar is such that relative costs diverge, the code-generator generator will not halt. A sufficient condition for ensuring that code-generator generators based on extensions of LR parsing techniques halt on an input specification is given in [34]. However, it is shown that there are specifications that can be handled by the tool but that fail the test.

An important issue is the generation of code for a directed acyclic graph (DAG) where shared nodes represent common subexpressions. The selection of optimal code for DAGs has been shown to be intractable [5], but there are heuristics that can be employed to generate code [6]. The labeling phase of a bottom-up tree parser can be modified to work with DAGs. One possibility is that the code generator could, in the top-down phase, perform code generation in the normal way but count visits for each node. For the first visit it could evaluate the shared subtree into a register and keep track of the register assigned. On

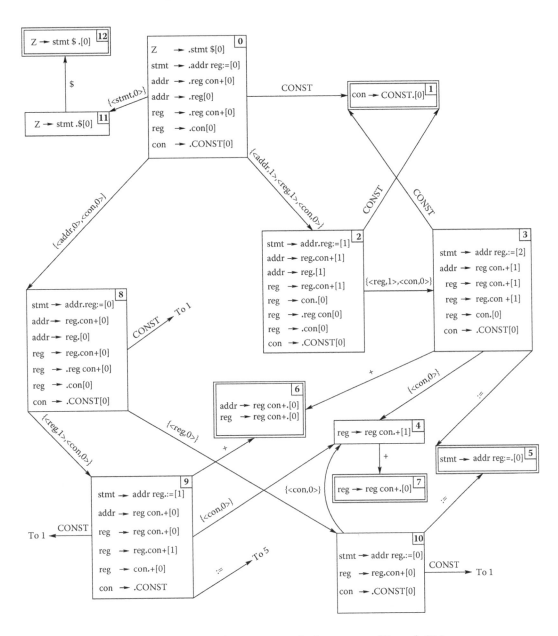

FIGURE 17.31 Auxiliary automaton for the grammar of Example 17.6.

subsequent visits to the node it could reuse the value stored in the register. However, assigning common subexpressions to registers is not always a good solution, especially when addressing modes in registers provide free computations associated with offsets and immediate operands. One solution to this problem involves adding a DAG operator to the intermediate language [10].

17.7 Conclusion and Future Work

We have described various techniques for the generation of instruction selectors from specifications in the form of tree grammars. Top-down, bottom-up, and LR-parser-based techniques have been described in detail. Instruction selection using the techniques described in this chapter is useful and practical, but the formalism is not powerful enough to capture features such as pipelines, clustered architectures, and so on. While it would be useful to have a single formalism for specification from which a complete code generator can be derived, no commonly accepted framework exists as yet. Instruction selection is important for complex instruction set computer architectures, but for reduced instruction set architectures there is a shift in the functional emphasis from code selection to instruction scheduling. In addition, because computation must be done in registers and because the ratio of memory access time to cycle time is high, some form of global register allocation is necessary. The interaction between instruction scheduling and register allocation is also important. Bradlee [11] has implemented a system that integrates instruction scheduling and global register allocation into a single tool. The advent of embedded processors with clustered architectures and VLIW instruction formats has added an extra dimension to the complexity of code-generator generators. The impact of compiler techniques on power consumption has been the subject of research only recently. Considerable work on retargetable compilers for embedded processors is already available, including MSSQ [35], RECORD [32], SPAM [8], CHESS [31, 43], CodeSyn [33], and AVIV [25]. Much more work is needed to address related problems on a sound theoretical basis.

Acknowledgments

The author gratefully acknowledges the assistance of J. Surya Kumari, Balvinder Singh, and V. Suresh in the preparation of the figures.

References

1. A. V. Aho and M. J. Corasick. 1975. Efficient string matching: An aid to bibliographic search. *Communications of the ACM* 18(6):333–40.
2. A. V. Aho and M. Ganapathi. 1985. Efficient tree pattern matching: An aid to code generation. In *Proceedings of the 12th ACM Symposium on Principles of Programming Languages*, 334–40. New York: ACM Press.
3. A. V. Aho, M. Ganapathi, and S. W. Tjiang. 1989. Code generation using tree matching and dynamic programming. *ACM Transactions on Programming Languages and Systems* 11(4):491–516.
4. A. V. Aho and S. C. Johnson. 1976. Optimal code generation for expression trees. *Journal of the ACM* 23(3):146–60.
5. A. V. Aho, S. C. Johnson, and J. D. Ullman. 1977. Code generation for expressions with common subexpressions. *Journal of the ACM* 24(1):21–28.
6. A. V. Aho, R. Sethi, and J. D. Ullman. 1986. *Compilers: Principles, techniques, and tools*. Reading MA: Addison Wesley.
7. A. W. Appel. 1987. Concise specifications of locally optimal code generators. Technical Report CS-TR-080-87, Department of Computer Science, Princeton University, Princeton, NJ.
8. G. Araujo, S. Malik, and M. Lee. 1996. Using register transfer paths in code generation for heterogenous memory-register architectures. In *Proceedings of the 33rd Design Automation Conference (DAC)*: IEEE CS Press.
9. A. Balachandran, D. M. Dhamdhere, and S. Biswas. 1990. Efficient retargetable code generation using bottom up tree pattern matching. *Computer Languages* 3(15):127–40.
10. John Boyland and H. Emmelmann. 1991. Discussion: Code generator specification techniques. In *Code-generation—Concepts, tools, techniques*. Workshops in computing series, ed. R. Giegerich and S. L. Graham, 66–69. Heidelberg, Germany: Springer-Verlag.
11. D. G. Bradlee. 1991. Retargetable instruction scheduling for pipelined processors. PhD thesis, University of Washington, Seattle, WA.
12. R. G. G. Cattell. 1980. Automatic derivation of code generators from machine descriptions. *ACM Transactions on Programming Languages and Systems* 2(2):173–90.

13. D. Chase. 1987. An improvement to bottom up tree pattern matching. In *Proceedings of the 14th ACM Symposium on Principles of Programming Languages*, 168–77. New York: ACM Press.

14. T. W. Christopher, P. J. Hatcher, and R. C. Kukuk. 1984. Using dynamic programming to generate optimised code in a Graham-Glanville style code generator. In *Proceedings of the ACM SIGPLAN 1984 Symposium on Compiler Construction*, 25–36. New York: ACM Press.

15. H. Emmelmann. 1991. Code selection by regularly controlled term rewriting. In *Code-generation — Concepts, tools, techniques*. Workshops in computing series, ed. R. Giegerich and S. L. Graham, 3–29. Heidelberg, Germany: Springer-Verlag.

16. H. Emmelmann. Testing completeness of code selector specifications. In *Proceedings of the 4th International Conference on Compiler Construction, CC '92*. Vol. 641 of Lecture Notes in Computer Science, 163–75. New York: Springer.

17. H. Emmelmann, F. Schroer, and R. Landwehr. 1987. BEG — A generator for efficient back ends. In *Proceedings of the SIGPLAN '89 Conference on Programming Language Design and Implementation*, 227–37. New York: ACM Press.

18. C. Ferdinand, H. Seidl, and R. Wilhelm. 1994. Tree automata for code selection. *Acta Informatica* 31:741–60.

19. C. Fraser. 1977. Automatic generation of code generators. PhD thesis, Yale University, New Haven, CT.

20. C. W. Fraser, D. R. Hanson, and T. A. Proebsting. 1992. Engineering a simple, efficient code-generator generator. *ACM Letters on Programming Languages and Systems* 1(3):213–26.

21. M. Ganapathi and C. W. Fischer. 1985. Affix grammar driven code generation. *ACM Transactions on Programming Languages and Systems* 7(4):560–99.

22. M. Ganapathi, C. N. Fischer, and J. L. Hennessy. 1982. Retargetable compiler code generation. *Computing Surveys* 14(4).

23. S. L. Graham, and R. S. Glanville. 1978. A new method for compiler code generation. In *Proceedings of the 5th ACM Symposium on Principles of Programming Languages*, 231–40. New York: ACM Press.

24. S. L. Graham, R. Henry, and R. A. Schulman. 1982. An experiment in table driven code generation. In *Proceedings of the SIGPLAN '82 Symposium on Compiler Construction*. New York: ACM Press.

25. S. Hanono and S. Devadas. 1998. Instruction selection, resource allocation and scheduling in the AVIV retargetable code generator. In *Proceedings of the 35th Design Automation Conference (DAC)*: IEEE CS Press.

26. P. Hatcher and T. Christopher. 1986. High-quality code generation via bottom-up tree pattern matching. In *Proceedings of the 13th ACM Symposium on Principles of Programming Languages*, 119–30. New York: ACM Press.

27. R. R. Henry and P. C. Damron. 1989. Performance of table driven generators using tree pattern matching. Technical Report 89-02-02, Computer Science Department, University of Washington, Seattle.

28. C. Hoffman and M. J. O'Donnell. 1982. Pattern matching in trees. *Journal of the ACM* 29(1):68–95.

29. J. E. Hopcroft and J. D. Ullman. 1979. *An introduction to automata theory, languages and computation*. Reading, MA: Addison Wesley.

30. R. Landwehr, H. St. Jansohn, and G. Goos. 1982. Experience with an automatic code generator generator. In *Proceedings of the SIGPLAN '82 Symposium on Compiler Construction*. New York: ACM Press.

31. D. Lanneer, J. Van Praet, A. Kifli, K. Schoofs, W. Geurts, F. Thoen, and G. Goossens. 1995. CHESS: Retargetable code generation for embedded DSP processors. In *Code generation for embedded processors*, ed. P. Marwedel, G. Goosens, Chapter 5.

32. R. Leupers and P. Marwedel. 1997. Retargetable generation of code selectors from HDL processor models. In *Proceedings of the European Design and Test Conference (ED & TC)*: IEEE CS Press.

33. C. Liem. 1997. *Retargetable compilers for embedded core processors*. Dordrecht, The Netherlands: Kluwer Academic.

34. Maya Madhavan, Priti Shankar, S. Rai, and U. Ramakrishna. 2000. Extending Graham-Glanville techniques for optimal code generation. *ACM Transactions on Programming Languages and Systems* 22(6):973–1001.

35. P. Marwedel. 1993. Tree-based mapping of algorithms to predefined structures. In *Proceedings of the International Conference on Computer-Aided Design (ICCAD)*. New York: ACM Press.

36. A. Nymeyer and J. P. Katoen. 1997. Code generation based on formal BURS theory and heuristic search. *Acta Informatica* 34(8):597–636.

37. E. Pelegri-Llopart and S. L. Graham. 1988. Optimal code generation for expression trees. In *Proceedings of the 15th ACM Symposium on Principles of Programming Languages*, 119–29. New York: ACM Press.

38. T. A. Proebsting. 1995. BURS automata generation. *ACM Transactions on Programming Languages and Systems* 17(3):461–86.

39. T. A. Proebsting and B. R. Whaley. 1996. One-pass optimal tree parsing — with or without trees. In *International Conference on Compiler Construction*, LNCS 1060, 294–308: Springer.

40. P. Shankar, A. Gantait, A. R. Yuvaraj, and M. Madhavan. 2000. A new algorithm for linear regular tree pattern matching. *Theoretical Computer Science* 242:125–42.

41. S. W. K. Tjiang. 1985. Twig reference manual. Computing Science Technical Report 120, AT&T Bell Labs, Murray Hills, NJ.

42. S. W. K. Tjiang. 1993. An olive twig. Technical report, Synopsys, Inc.

43. J. Van Praet, D. Lanneer, W. Geurts, and G. Goossens. 2001. Processor modeling and code selection for retargetable compilation. *ACM Transactions on Design Automation of Digital Systems* 6(3):277–307.

44. B. Weisgerber and R. Willhelm. 1988. Two tree pattern matchers for code selection. In *Compiler compilers and high speed compilation*. Vol. 371 of Lecture Notes in Computer Science, 215–29: Springer.

45. R. Wilhelm and D. Maurer. 1995. *Compiler design*. International Computer Science Series. Reading, MA: Addison Wesley.

18

A Retargetable Very Long Instruction Word Compiler Framework for Digital Signal Processors

Subramanian Rajagopalan
Synopsis (India) EDA Software Pvt. Ltd.,
Bangalore, India
Subramanian.Rajagopalan@synopsys.com

Sharad Malik
Department of Electrical Engineering,
Princeton University, Princeton, NJ
sharad@ee.princeton.edu

18.1 Introduction

Digital signal processors (DSPs) are used in a wide variety of embedded systems ranging from safety-critical flight navigation systems to common electronic items such as cameras, printers, and cellular phones. DSPs are also some of the more popular processing element cores available in the market today for use in system-on-a-chip (SOC)-based design for embedded processors. These systems not only have to meet the real-time constraints and power consumption requirements of the application domain, but also need to adapt to the fast-changing applications for which they are used. Thus, it is very important that the target processor be well matched to the particular application to meet the design goals. This in turn requires that DSP compilers produce good-quality code and be highly retargetable to enable a system designer to quickly evaluate different architectures for the application on hand.

Unlike their general-purpose counterparts, an important requirement for embedded system software is that it has to be sufficiently dense to fit within the limited quantity of silicon area, either random-access memory (RAM) or read-only memory (ROM), dedicated to program memory on the chip. This requirement arises because of the limited and expensive on-chip program memory. To achieve this goal, and at the same time not sacrifice dynamic performance, DSPs have often been designed with special architectural features such as address generation units, special addressing modes for better memory access, computation units optimized for digital signal processing, accumulator-based data-paths, and multiple memory banks. Hence, to produce good-quality code, it is necessary for DSP compilers to incorporate a large set of optimizations to support and exploit these special architectural features.

This chapter is organized as follows. DSP architectures are first presented and classified from a compiler developer's view in Section 18.2. An overview of the constraints imposed by DSP architectures is then provided in Section 18.3, using an illustrative example. The need for retargetable methodologies and previous efforts on this are discussed in Section 18.4. Some of the retargetable code generation and optimization techniques that have been developed to exploit the special DSP features are then presented in Section 18.5, followed by the summary in Section 18.6.

18.2   Digital Signal Processor Architectures

A close examination of DSP architectures and the requirements for a DSP compiler suggests that DSPs can be modeled as very long instruction word (VLIW) processors. A VLIW processor is a fully statically scheduled processor capable of issuing multiple operations per cycle. Figure 18.1 shows a simplified overview of the organization of instruction set architectures (ISAs) of VLIW processors.

The ISA of a VLIW processor is composed of multiple instruction templates that define the sets of operations that can be issued in parallel. Each slot in an instruction can be filled with one operation from a set of possible operations. Each operation consists of an opcode and a set of operands. The opcode of an operation defines the operation's resource usage when it executes. An operand in an operation can be a register, a memory address, or an immediate operand. The register operand in an operation can be a single machine word register, a part of a machine word register, or a set of registers. Although the memory address is shown as a single operand in Figure 18.1, in general, the address itself can be composed of multiple operands depending on the addressing mode used in the operation. DSPs have statically determined ILP and thus are specific instances of VLIW compilers, albeit with special constraints.

As compilers for VLIW processors are well studied, they serve as useful starting points for developing compilers for DSP processors. There is one major point of difference, though. Compilers for VLIW architectures need optimizations to exploit the application's instruction level parallelism (ILP) to primarily

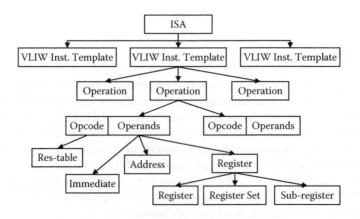

FIGURE 18.1   VLIW instruction set architecture.

obtain good dynamic performance, and this is often achieved at the expense of increased static code size. For DSPs, however, static code size is equally or even more important than the dynamic performance.

This section gives an overview of the features of different classes of DSP architectures from a compiler perspective. The basic architecture of DSPs is a Harvard architecture with separate data and program memories. Some of the architectural features of DSPs are motivated by the applications:

- Because DSP applications often work with multiple arrays of data, DSPs feature multiple data memory banks to extract ILP from memory operations.
- Because vector products and convolutions are common computations in signal processing, a fast single cycle multiply and multiply-accumulate-based data-paths are common.
- Multiple addressing modes, including special modes such as circular addressing, are used to optimize variable access.
- Fixed or floating point architectures are used depending on the application's requirements such as precision, cost, and scope.
- Hardware to support zero overhead loops is present for efficient stream-based iterative computation.

For the discussion in the rest of this chapter, a VLIW instruction in an ISA is defined as a set of operations allowed by the ISA to be issued in parallel in a single cycle. ISAs of DSPs can be classified based on the following attributes:

- Fixed operation width (FOW) ISA or variable operation width (VOW) ISA. A fixed operation width ISA is an ISA in which all the operations are encoded using a constant number of bits. A variable operation width ISA is any ISA that does not use a fixed operation width for all operations.
- Fixed operation issue ISA or variable operation issue ISA. A fixed operation issue ISA is an ISA in which the number of operations issued in each cycle remains a constant. A variable operation issue ISA can issue a varying number of operations in each cycle.
- Fixed instruction width ISA or variable instruction width ISA. A fixed instruction width ISA is an ISA in which the number of bits used to encode all the VLIW instructions is a constant. A variable instruction width ISA is any ISA whose instruction-encoding size is not a constant.

Of these three attributes, operation width and instruction width are more important because the issue width of a DSP is usually dependent on these two attributes. Hence, programmable DSP architectures can be broadly classified into four categories as shown in Figure 18.2:

- ISAs with FOW and fixed instruction width (Figure 18.2a).
- ISAs with FOW and variable instruction width (Figure 18.2b).

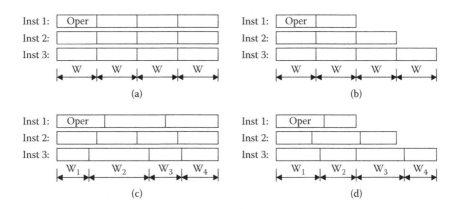

FIGURE 18.2 Different classes of DSP architectures: (a) Fixed operation width and fixed instruction width; (b) fixed operation width and variable instruction width; (c) variable operation width and fixed instruction width; (d) variable operation width and variable instruction width.

- ISAs with VOW and fixed instruction width (Figure 18.2c).
- ISAs with VOW and variable instruction width (Figure 18.2d).

Because the architectural features of the first two categories of DSPs are more due to the FOW attribute, the two categories can be combined into a single FOW category. Similarly, because compiling for architectures in the last category is similar to VOW and fixed instruction width architectures, they can also be combined into a single VOW architecture category.

18.2.1   Fixed Operation Width Digital Signal Processor Architectures

The ISA of an FOW architecture consists of reduced instruction set computer (RISC)-style operations, all of which are encoded using a constant number of bits. From Figure 18.2a and 18.2b, it can be seen that the issue width and instruction width of these processors are tied to one another. As the issue width is increased to get more ILP, the instruction width to be decoded also increases proportionally. Examples in this category include the Texas Instruments (TI) TIC6X series DSPs. Many high-end DSPs and, in particular, floating point DSP architectures fall into this category. Some of the features of these architectures are:

- These architectures are regular with fewer instruction level constraints on operands and operations when compared with the other categories of DSPs. The primary reason for this regularity is the encoding of the instructions. Not only is the encoding of each operation in an instruction independent, but the encoding of opcode and operands within an operation is also separate.
- A large general-purpose register set, which is a key to exploiting ILP, is an important feature of these architectures. This set may even be organized into multiple register files.
- The regularity and the large register set together make these DSPs more compiler friendly and easier to program than their VOW counterparts. This also stems from the fact that these architectures tend to have an orthogonal set of operands, operations, and instructions.
- Because of the general nature of operations and increased ILP, these processors are capable of handling a wider variety of applications, and the applications themselves can be large. Specialized functional units are sometimes used to improve a group of applications.
- To extract more ILP, there is a trend in these architectures to support speculation and predication, which are features in general-purpose processors. Because a discussion of speculation and predication is beyond the scope of this chapter, readers are referred to [7, 11].
- One of the drawbacks of fixed instruction width FOW DSPs is code size. For every VLIW slot in an instruction that the compiler is not able to fill with a useful operation, no-operations (NO-OPs) need to be filled in. This often leads to an increase in code size because the instruction width remains constant. To overcome this problem, some processors allow a compiler to signal the end of an instruction in each operation by turning on a flag in the last operation of an instruction. This prevents any unnecessary NO-OPs from being inserted, and it is the responsibility of the hardware to separate the instructions before or during issue. With this feature enabled, fixed instruction width FOW processors appear similar to variable instruction width FOW processors.

Because the features of these processors are very similar to VLIW processors, their compilers can leverage the work done in VLIW compilation. Thus, the DSP compiler techniques and optimizations described in this chapter may not be applicable or useful to most processors in this category.

18.2.2   Variable Operation Width Digital Signal Processor Architectures

The features of VOW architectures shown in Figures 18.2c and 18.2d are in sharp contrast to those of FOW-style architectures. These DSPs are used in applications such as cellular telephones. Hence, the binary code size of applications running on these processors needs to be extremely dense because of the limited and expensive on-chip program memory. Also, a conflicting requirement of good dynamic performance exists because these applications usually run on battery. Examples in this category include Fujitsu Hiperion, TI

TMS320C25, Motorola DSP56000, and so on. Most of the features of VOW DSPs described in this chapter reflect the decisions made by hardware architects to meet these conflicting requirements:

- To contain the code size of applications, but still exploit ILP, VOW DSP instructions are encoded such that the instruction width does not change with issue width. This leads to a situation where all the operations in an instruction are encoded together and operands within an operation are also tied together in an encoding that leads to irregular architectures with restricted ILP support.
- These DSPs feature only a small number of registers with many constraints on how each register can and cannot be used.
- The irregular architecture and the heterogeneous register set pose a challenge to DSP compilers that have to tackle the numerous constraints and yet meet the conflicting requirements. This is primarily due to the irregular operation and instruction encoding in VOW DSPs. This leads to nonorthogonal operands within operations, nonorthogonal operations within instructions, and a nonorthogonal instruction set.
- The DSPs in this class have traditionally been programmed in assembly. This drastically restricts the domain of applications to small programs that are usually kernels of larger applications that can be accelerated using DSPs.
- To compensate for the lack of a rich register set and to utilize the data memory banks effectively, these DSPs have an address register set and an address generation unit that enable address computations to be performed in parallel with other operations.

Section 18.3 discusses how some of these features affect the development of compilers for DSPs. For a more detailed discussion on DSP architectures, the readers are referred to [26].

18.3 Compilation for Digital Signal Processors

In this section, some of the constraints posed by VOW DSP architectures to compilers are described using an example from [49]. Some common DSP ISA constraints are shown in Figure 18.3, where the constraints between the different entries are represented in dotted lines. Figure 18.3a shows operation ILP constraints in the instruction template that restrict the set of operations that can be performed in parallel.

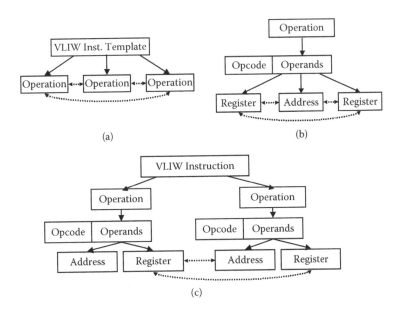

FIGURE 18.3 DSP constraints.

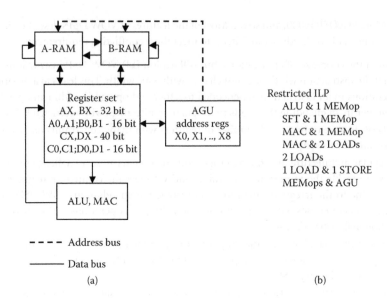

FIGURE 18.4   (a) Hiperion DSP data-path; (b) some operation ILP constraints [49].

Figure 18.3b shows operand constraints that exist between operands within an operation. This can be of different types such as register–register constraints, register-addressing mode or register–immediate size constraints, and so on. Finally, Figure 18.3c extends the operand constraints across operations within an instruction.

Figure 18.4, taken from [49], shows the data-path of Hiperion, a fixed-point DSP core. It is a fixed instruction width VOW DSP. Its data-path consists of dual memory banks, an arithmetic and logic unit (ALU) comprising a shift unit, an add unit, a multiply unit, and an address generation unit. There are eight address registers (ARs) and four accumulators, each with 32b and each of which can also be accessed as two general-purpose registers, each with 16b. In addition to instructions with parallel memory access operations, restricted ILP also exists between arithmetic and memory operations. Some of these constraints are now briefly described and explained with examples:

- Because addressing modes occupy more instruction bits, many restrictions exist on the addressing mode that a memory operation can use in different instructions. This includes the set of ARs that can be used by a particular addressing mode in a particular instruction.
- The fixed instruction width and the VOW together enforce several constraints on what set of operations can be performed in parallel.
- Operations in a single instruction share operands because of a lack of bits to encode the operations independently. This may or may not affect data flow.
- Operations in a single instruction have to split an already small number of instruction bits to encode their operands, leading to register allocation constraints that arise not from a lack of registers, but from a lack of bits to encode more registers.
- Large immediate operands require multiple instruction words to encode them, leading to larger code size and a potentially extra operation execution time.

The constraints described in this section are representative of constraints found in many commercial DSPs and are difficult to capture in both behavior description-based machine descriptions (MDs) (MDs based on ISA descriptions) and structural description-based MDs (MDs that extract ISA information from description of a processor data-path). It is also important to note that the internal data-paths of a processor are seldom transparent to the application developer; thus, the latter may not even be an option.

18.3.1   Operation Instruction Level Parallelism Constraints

These constraints describe what set of operations can and cannot be issued in parallel. Figure 18.4b shows an example set of ILP constraints. *MEMop* stands for either a *LOAD* or a *STORE* operation; *ALU* stands for operations that use the ALU such as *ADD*, *SUB*, and *Shift* operations; and *MAC* stands for all operations associated with the MAC unit such as *Multiply* and *Multiply-Accumulate*. Although two *LOAD* operations can be issued in parallel with a *MAC* operation, only one *LOAD* can be issued in parallel with an ALU operation. Hence, these constraints do not need to be limited by physical resources alone.

18.3.2   Operand Instruction Level Parallelism Constraints

These constraints describe how registers should be assigned to operands of operations issued in parallel. These constraints may or may not affect data flow.

Example 18.1

Constraint Affecting Data Flow: *MUL dest, source1, source2; LOAD dest1, [M1] ; LOAD dest2, [M2]* is a valid single-cycle schedule only if *dest1* and *dest2* are assigned the same registers as *source1* and *source2*. It is the responsibility of the compiler to ensure that the two sources of the multiply operation are not used after this instruction. Hence, these constraints are data-flow-affecting constraints. In addition, there can also be restrictions on what registers can be assigned to each operand in an operation or instruction.

Example 18.2

Constraint Not Affecting Data Flow: *ADD dest1, source1, source2; LOAD dest2, [M]* is a valid single-cycle schedule only if the following conditions are satisfied. If *dest1* is assigned the register *CX*, then:

1. Condition: *dest2* can be assigned a register only from the following set of registers {*A0,A1,B0,B1, D0,D1,CX,DX*}.
2. Condition: *source1* can be assigned only one of {*CX, A0*}.
3. Condition: *source2* must be assigned register *A1*.

The first constraint is an example of operand constraints across operations, and the other two are examples of operand constraints within an operation.

Most of these constraints can be attributed to the nonorthogonality property of VOW DSP architectures. Apart from the lack of orthogonality in operand, operation, and instruction encodings, this also includes difficulty in classifying arithmetic operations into similar classes, for example, an *ADD* and *SUB* may have completely different sets of constraints.

18.4   Retargetable Compilation

Section 18.2 described the architectural features of DSPs that were designed by hardware architects with the intent of meeting application demands, and Section 18.3 described the constraints posed by the architectures for compiler developers that are in addition to traditional compiler issues. While the compiler developers need a clean abstraction of the architecture that allows them to develop reusable or synthesizable compilers that do not need to know the actual architecture they are compiling for, the hardware architects need a configurable abstraction of the compiler that allows them to quickly evaluate the architectural feature that they have designed without much knowledge of how the feature is supported by the compiler. Traditionally, the compiler has been split into two phases, namely, the front end that converts the application program into a semantically equivalent common intermediate representation and the back end that takes the intermediate representation as input and emits the assembly code for the given target. This partially eases the job of the compiler developer and the hardware designer of having to match the application to the target.

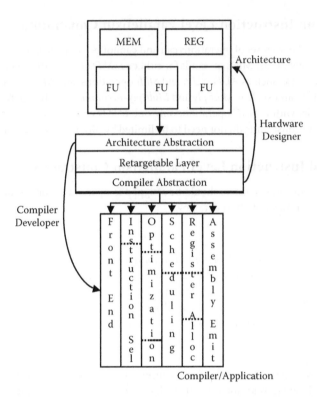

FIGURE 18.5   Retargetable interface.

However, this coarse-grain reusability is not sufficient, because the growing number of applications and the increasing number of processor designs both require an efficient fine-grain reusability of components that is often referred to as retargetability. This is shown in Figure 18.5, where a retargetable interface captures the hardware artifacts, exports them to the compiler developer, captures the ability to parameterize the compiler (masking the algorithms, complexity, etc.), and exports it to the architecture designer. In this design, each phase of the compiler may be individually configurable, or each phase itself may be divided into configurable components as shown by dotted lines within each phase in Figure 18.5. An interesting side effect of such a retargetable design is that the abstraction does not require the user to be an expert in both architectures and compilers.

For an SOC design, an architect may be faced with many options for the cores that meet the demands of the application on hand. The designer then needs a software tool-set such as a compiler and a simulator to quickly evaluate the set of DSP cores available, pick the best core, and reduce the time to market the design. This set of DSP cores includes programmable cores that may still have to be developed for a set of applications, therefore also requiring the development of the necessary tool-set, and off-the-shelf processor cores that are readily available with the tool-set. Although in the second case, the architectures are predefined and the necessary compilers for the specific processors may be readily available, in the former, the architecture itself is not defined and it is not practically possible to develop an architecture-specific compiler for every potential architecture solution. Hence, clearly, a need exists to design the compiler framework so that it is easy either to synthesize a new compiler for every potential architecture or to reuse parts of the compiler to the maximum extent possible by minimizing the amount of architecture-specific components. Based on this need for retargetability, different types of retargetability have been defined with respect to the extent of reuse [56].

18.4.1   Automatic Retargetability

An automatically retargetable compiler framework has built-in support for all potential architectures that meet the demands of the applications on hand. Compilers for specific architectures can then be generated by configuring a set of parameters in the framework. Although this is an attractive solution for retargetability, it is limited in scope to a regular set of parameterizable architectures and is not capable of supporting a wide variety of specialized architectural features found in DSPs.

18.4.2   User Retargetability

A user retargetable compiler framework relies on an architectural description to synthesize a compiler for a specific architecture. Although support exists for a wide variety of machine-independent optimizations, this type of framework also suffers from the drawback of not having the ability to automatically generate machine-dependent optimizations. This level of retargetability has been achieved only in the instruction selection phase of the back ends. Some examples in this category include Twig [1] and Iburg [16], both of which automatically generate instruction selectors from a given specification.

18.4.3   Developer Retargetability

A developer is an experienced compiler designer, and a developer retargetable framework is based on a library of parameterized optimizations for architectural features. A compiler for a specific architecture can be synthesized in this framework by stringing together the set of relevant optimizations from the developer's library with the correct set of parameters. If the library lacks an optimization that can be applied to a certain new architectural feature, then the developer is responsible for developing a parameterized optimization module, the associated procedure interface, and its insertion into the library. The new optimization module can then be used for designs in the future that have the new architectural feature. The success of this framework lies in the extent of code reuse or optimization module reuse. Hence, this type of retargetability is best suited for evaluation of architectures in a single family.

Figure 18.6 shows a potential retargetable compiler development methodology that can be used to design DSP compilers. The methodology starts either by hand-coding DSP kernels in assembly, or, if possible, by using a simple machine-specific compiler that produces correct code with no optimizations. For example, this compiler can be obtained from a simple modification of a compiler developed for a closely related processor, or it can be a first cut of the compiler or one of the publicly available retargetable research compilers that fits the domain of the applications. The developer then examines the assembly code for inefficiencies and suggests optimizations for the various regular and irregular parts of the DSP architecture. This includes efficient capture of the processor ISA and micro-architecture using machine descriptions and general-purpose performance optimizations for the regular parts. For the special architectural features, a parameterized optimization library is developed, and the optimizations are added to this library as and when they are developed. An optimizing compiler is then assembled using the different components, and the cycle is repeated until satisfactory code quality is achieved. Conventionally, retargetability has always been achieved with some kind of MD that describes the processor ISA and/or structure. This database is then queried by the different phases of compilation to generate code specific to the target processor.

The success of a retargetable methodology can be evaluated or measured based on several qualities such as *nature* (parameterizable or synthesizable components), *modularity* (fine- or coarse-grain capture of hardware artifacts), *efficiency* (quality of code generated for a set of processors and time taken to produce the code in the first place), and *extensibility* (ability to add new modules or optimizations and ability to extend the framework to other domains). These are not all independent or arranged in any order of preference.

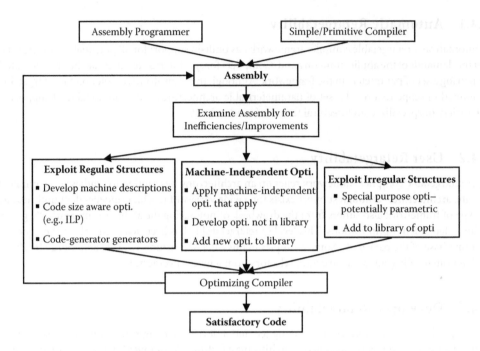

FIGURE 18.6   Retargetable compiler methodology.

18.5   Retargetable Digital Signal Processor Compiler Framework

In this section, the flow of a retargetable DSP compiler for an architecture with various constraints mentioned in Section 18.3 is described. The DSP code generator flow is shown in Figure 18.7. Because the compiler flow shown in Figure 18.7 is very similar to a general-purpose compiler flow, only the issue of

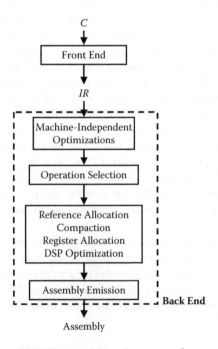

FIGURE 18.7   DSP code generator flow.

DSP-specific optimizations is addressed in this section. In general, the wide variety of machine-independent compiler optimizations used in compilers for general-purpose processors can also be applied to DSP compilers [2]. However, certain precautions need to be taken before directly applying an optimization because of the nature of the DSP architectures. For example, in most VOW-style DSPs, an extra instruction word is needed to store large constants if the operation uses one. Because this may also cause a one-cycle fetch penalty, constant propagation[1] must be applied selectively. Other examples that can have a negative effect include common subexpression elimination[2] and copy propagation.[3] These sets of optimizations tend to increase the register pressure by extending live ranges that can lead to negative results in architectures with small register sets. Some of the most beneficial optimizations for DSPs are the loop optimizations such as loop invariant code motion,[4] induction variable elimination,[5] strength reduction,[6] and so on. Optimizations such as loop unrolling,[7] however, increase static code size, and some loop optimizations also tend to increase register pressure. The trade-off in either case is the gain in reduction of execution time. Most DSP applications involve some kind of kernel loops that account for the majority of their execution time. Hence, optimizing loops is an important task in a successful DSP compiler framework. However, optimizations such as loop induction variable elimination can prevent efficient utilization of zero overhead looping features in DSPs.

In general, aggressive machine-independent optimizations can also destroy information needed by DSP optimizations such as array reference allocation and offset assignment. Similar to general-purpose compilation, DSP compilation also suffers from the interplay between the various optimizations that is commonly known as the *phase-ordering problem*. For example, the most commonly used example of the phase-ordering problem is that between scheduling and register allocation. If scheduling is done prior to register allocation, the variable live ranges can be long, which can increase register pressure and cause excessive spills (save/restore registers to/from memory). This may require another scheduling phase to be introduced after the allocation to take care of the spill code. However, if register allocation is performed first, several false dependencies are introduced that can lead to extended static schedules. The situation is no different in the case of DSP compilation. In the examples given in Sections 18.3.1 and 18.3.2, the constraints can be viewed as the problems of the scheduler, thereby creating unnecessary constraints for the register allocator, or they can be viewed as constraints for the register allocator, leading to bad schedules that do not exploit the ILP features of DSPs. Although the different phases in DSP compilation are listed in Figure 18.7, they may not necessarily be performed in the order shown.

Another concern about optimizations is in terms of the compilation time. Most optimizations described in this section have been shown to be in NP [17]. Given this, only a few options are available to a user, such as sacrificing optimality and converting the problem to one that can be solved in polynomial (preferably linear) time or attempting to solve smaller versions of the problem optimally or use good heuristics. Hence, the user often needs to make a time versus performance trade-off. This is an important issue in a retargetable design space exploration environment where faster techniques can be used to eliminate large portions of the search space and longer near-optimal solutions can be applied to arrive at the final choice. Because of the many constraints in DSP compilation, some of the techniques use methods such as *simulated annealing* and *linear programming* to solve the problems and more often attempt to solve multiple problems or phases at a time. This leads to an increase in complexity and a potential increase

---

[1] If there is an operation that moves a constant c to a virtual register u, then the uses of u may be replaced by c.

[2] If a program computes an expression more than once with exactly the same sources, it may be possible to remove redundant expressions.

[3] If there is an MOV operation that moves one virtual register v to another virtual register u, then the uses of u may be replaced by use of v.

[4] Moving loop invariant computations outside the loop.

[5] Removing the loop induction variable from inside the loop.

[6] Replacing expensive computations such as multiply and divide with simple operations such as shift, add, etc.

[7] Placing two or more copies of the loop body in a row to improve efficiency.

in compilation time. In such cases, the trade-off is between optimality with some generality within an architecture class and compilation time, complexity, and scalability. Heuristics can result in completely different behavior even for small changes in the constraints, whereas an exact method can produce robust solutions. This is another reason why a compiler framework and an MD need to be tied to one another.

This section briefly describes some of the common DSP optimizations performed to either exploit specialized architectural features or merely satisfy data-path constraints such as those mentioned in Section 18.3. For details on the conventional phases of general-purpose compilation such as scheduling and register allocation, the readers are referred to the corresponding chapters in this book.

18.5.1   Instruction Selection

Instruction selection corresponds to the task of selecting the sequence of appropriate target architecture opcodes from the intermediate representation of the application program. The output of most front ends of compilers is a forest of directed acyclic graphs (DAGs) semantically equivalent to the input program. It has been shown that the problem of covering the DAG with target opcodes is an NP-complete problem even for a single register machine [10, 17, 54]. However, optimal solutions exist when the intermediate representation is in the form of a sequence of expression trees (e.g., Iburg [16] and Twig [1]). Some of the DSP compilers use heuristics that convert a forest of DAGs to expression trees, potentially losing overall optimality, followed by locally (per tree) optimal pattern-matching algorithms to do instruction selection. Araujo proposed a heuristic to transform a DAG into a series of expression trees for acyclic architectures, classifying the ISA of DSPs as either cyclic or acyclic based on a register transfer graph (RTG) model [4].

The RTG model of an ISA describes how the different instructions utilize the transfer paths between the various storage locations to perform computations. The registers transparent to the developer through the processor ISA are divided into register classes, where each class has a specific function in the data-path. For example, a typical DSP data-path consists of an address register class, an accumulator class, and a general-purpose register class.

18.5.1.1   Register Transfer Graph

An RTG is a directed acyclic multi-graph where each node represents a register class, and an edge between nodes r_i and r_j is labeled with instructions in the ISA that take operands from location r_i and store the result in location r_j. Memory is assumed to be infinitely large and not represented in the RTG. However, an arrowhead is added to the register class nodes for memory transfer operations to indicate the direction of memory transfer.

Figure 18.8a shows a simple processor data-path; the corresponding RTG for this data-path is shown in Figure 18.8b. For clarity, the operations are not shown in the RTG. The accumulator (ACC), register set R_i, and register set R_j are the three register classes in the data-path and hence have a vertex each in the RTG. Because the data-path allows the accumulator to be both a source operand and a destination operand of some ALU operations, a self-edge exists around ACC in Figure 18.8b. Because some ALU operations have R_i and R_j as sources and ACC as the destination, a directed edge exists from both R_i and R_j to ACC. The remaining arrowheads in the RTG represent memory operations. Load operations can be performed only with R_i or R_j as destinations, and store operations can be performed using only the accumulator.

An architecture is said to be cyclic (acyclic) if its RTG has (no) cycles where a cycle is composed of at least two distinct vertices in the RTG. Acyclic ISAs have the property that any data-path cycle, which is not a self-loop, including a path between two nodes r_i and r_j in the RTG, goes through memory. Hence, the data-path shown in Figure 18.8a is acyclic because the RTG in Figure 18.8b does not have a cycle with two distinct vertices. Araujo and Malik show that spill-free code can be generated from expression trees for architectures with acyclic RTGs and provide a linear time algorithm for optimal scheduling for such architectures [5].

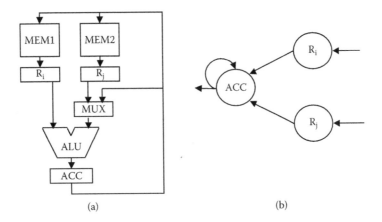

FIGURE 18.8 (a) Processor data-path; (b) RTG for the data-path.

To transform the DAG into a series of expression trees, a four-phase heuristic is used:

- Partial register allocation is done for operands of operations for which the ISA itself clearly specifies the allocation. For example, for the data-path in Figure 18.8a, the results of all the ALU operations are always written to the accumulator.
- The architectural information and the set of constraints imposed by the ISA are used to determine the set of edges in the DAG that can be broken without loss of optimality in the subsequent phases. For example, for the data-path in Figure 18.8a, the result of an ALU operation cannot be used as the left operand of a dependent operation. Such dependencies have to go through memory, and thus the corresponding DAG edge, called the natural edge, can be broken without any loss. Similarly, if an ALU operation is dependent on the results of two other ALU operations, at least one of the dependencies has to go through memory. Such edges are called pseudo-natural edges, and breaking such an edge does not guarantee optimality, but there is a reasonable probability that this is the case.
- Selective edges of the DAG are then marked and disconnected from the DAG to generate a forest of trees, while preserving the data flow constraints of the DAG.
- Optimal code is finally generated and scheduled for each expression tree.

Because the conversion of a DAG to expression trees potentially sacrifices optimality, Liao et al. proposed a technique using a combinatorial optimization algorithm called binate covering for instruction selection of DAGs. This can be solved either exactly or heuristically using branch and bound techniques [34]. This method, briefly summarized here, primarily involves the following steps for each basic block:

- All patterns that match each node in the subject program graph and their corresponding costs are first obtained. This phase does not take into account the data-path constraints and assumes that the associated data transfers necessary are for free.
- The covering matrix is then constructed, with a Boolean variable for each of the matched patterns as columns and the clauses and conditions that need to be satisfied for a legal cover of the program graph using the patterns as rows. This set of conditions is two-fold:
  - Each node must be covered by at least one pattern. This is represented as a set of clauses, one for each node in the program DAG, each of which is the inclusive OR of the Boolean variables that cover the node. For each row, or clause, an entry of 1 is made in the columns of Boolean variables included in the clause.
  - For each match, all the non-leaf inputs to the match must be outputs of other matches. This is represented by a set of implication clauses. For each non-leaf input, I, of each match, M, the set of all matches, M_I, that can generate I are determined. If M is chosen to cover a node or a

set of nodes in the program DAG, at least one of the matches from the set M_I must be chosen to obtain the input, I, to M. This needs to be true for each of the inputs of M and for all such matches chosen to cover the program graph. In the covering matrix, this is captured by entering a 0 in the column corresponding to M for all the implication clauses generated by M, and in each row corresponding to an implication clause of M, a 1 is entered in the columns of the variables included in the clause apart from M.

- A binate covering of the covering matrix that minimizes the cost and satisfies all the covering constraints — every row has either a 1 in the entry corresponding to a selected column or a 0 in the corresponding unselected column. The cost of a cover is the total cost of the selected columns. The main purpose of this covering is to generate complex instructions in the ISA from the program DAG that may not be possible by transforming to expression trees.
- The program graph is then modified into a new graph based on the obtained covering to reflect the complex instructions that have been generated.
- Additional clauses and costs corresponding to the irregular data-path constraints, such as register class constraints of the selected instructions, and the consequent data transfer constraints and costs are added.
- A new binate covering of the modified graph using the new set of clauses is then obtained. The two-phase approach is used to contain the size of the code generation problem by solving for a smaller number of clauses.

Leupers and Marwedel developed a two-phase linear programming–based method for DSPs with complex instructions [30, 32]. In the first phase, a tree-pattern matcher is used to obtain a cover on expression trees. In the second phase, complex instructions, which include a set of operations that can be issued in parallel, are generated from the cover (i.e., code scheduling is performed using linear programming). Leupers and Bashford developed a constraint logic programming–based code selection technique for DSPs and media processors [27]. This technique addresses two potential drawbacks of tree-based methods. The first is that splitting DAGs into trees can prevent generation of chained operations. For example, a multiply–accumulate operation cannot be generated if the result of the multiply is used by multiple operations, and covering expression trees with operation patterns does not take into account the available ILP in the ISA. Hence, the instruction selection algorithm takes the data flow graphs as input, keeps track of the set of all alternatives (both operation and instruction patterns that are possible), and then uses constraint logic programming to arrive at a cover. In addition, the authors incorporate support for single instruction multiple data (SIMD)-style operations where an SIMD operation is a set of independent identical operations operating on separate virtual subregisters of a register.

In CodeSyn, Paulin et al. use a hierarchical tree-like representation of the operations in the ISA to cover a control and data flow graph (CDFG) [43, 44]. If an operation in the tree does not match, the search space for the matching patterns is reduced by pruning the hierarchical tree at the operation node. Once all the matching tree patterns are found, dynamic programming is used to select the cover of the CDFG.

Van Praet et al. use their instruction set graph (ISG) model of the target architecture, along with the control and data flow graph representation of the program, in a branch-and-bound–based instruction selection algorithm in the CHESS compiler [25, 47]. The ISG is a mixed model of structural and behavioral models that captures the connectivity, ILP, and constraints of the target processor. CHESS also performs instruction selection on the program DAG by covering the DAG with patterns called bundles. Bundles are partial instructions in the target processor that are generated dynamically for a DAG from the ISG when required. This is a key difference between other techniques that statically determine all possible bundles. Initially nodes in the DAG may be covered by more than one bundle that is then reduced to the minimum number of bundles required to cover the DAG by a branch-and-bound strategy.

In addition, there are systems that solve multiple code generation problems in a combined manner. Hanono and Devadas provide a branch-and-bound–based solution for performing instruction selection and partial register allocation in a parallelism-aware manner on DAGs [21]. At the core of this method is the split-node DAG representation of the program DAG. The split-node DAG is used to represent the program DAG with the set of all alternatives available for the nodes in the program DAG along with the corresponding data transfers. The algorithm then searches for a low-cost solution with the maximal parallelism among nodes in the split-node DAG while looking for the minimal number of instructions that cover the program DAG, on which graph-coloring-based register allocation is then performed. The valid operations, instructions, and various constraints of the ISA are represented using the "ISDL" description.

Novack et al. developed the mutation scheduling framework, which is a unified method to perform instruction selection, register allocation, and scheduling [40, 41]. This is achieved by defining a *mutation set* for each expression generating a value. The mutation set keeps track of all the possible machine-dependent methods of implementing an expression, and this set is dynamically updated as scheduling is performed. For example, the references to a value depend on the register or memory assigned to it, and the expression that records this information is added into the corresponding mutation set. During scheduling, an expression used to implement the corresponding value may be changed to better adapt to the processor resource constraints by replacing it with another expression in the mutation set.

Wilson et al. provide a linear programming–based integrated solution that performs instruction selection, register allocation, and code scheduling on a DFG [65]. Wess provides another code generation technique based on trellis diagrams, an alternate form of representing instructions [63]. For more work in code generation, readers are referred to [9, 22, 24, 33, 36, 39, 46, 48, 51, 55].

18.5.2 Offset Assignment Problem

As stated in Section 18.2, DSPs provide special ARs, various addressing modes, and address generation units. The address generation units are used by auto-increment or auto-decrement arithmetic operations, which operate on ARs used in memory accesses. The offset assignment (OA) problem addresses the issue of finding an ordering of variables within a memory bank that can reduce the amount of address computation code to a minimum by optimally utilizing the auto-increment and auto-decrement feature. This is shown in the example in Figure 18.9 taken from [35, 58]. Figures 18.10b and 18.10d show the simple OA (SOA) optimized code sequence and the SOA unoptimized code sequence, respectively, for the piece of code in Figure 18.9a. Five additional address modification operations are needed by the unoptimized version that places variables in the order in which they are accessed in the code. Whereas SOA addresses the offset assignment problem with one AR and increments and decrements by 1, the multiple AR and the *l* increment and decrement variant are addressed by general offset assignment (GOA) and *l*-SOA, respectively.

The following sequence of steps summarizes the work done by Liao et al. to solve the SOA problem at the basic block level [35]:

- For the sequence of arithmetic computations shown in Figure 18.9a, the first step in SOA is to construct the access sequence shown in Figure 18.9b. This is done by adding the source variables in each operation, from left to right, to the access sequence followed by the destination operand to the access.
- The next step is the construction of the access graph shown in Figure 18.9c. Each vertex in this graph corresponds to a variable in the access sequence. An edge with weight *w* exists between two

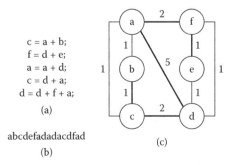

```
c = a + b;
f = d + e;
a = a + d;
c = d + a;
d = d + f + a;
```

(a)

abcdefadadacdfad

(b)

(c)

FIGURE 18.9 (a) Code sequence; (b) access sequence; (c) access graph [35].

```
                              MOV AR, &a                                         MOV AR, &a
                              LD [AR]        ; (a)                               LD [AR]++      ; (a)
                              AR -= 3                                            LD [AR]++      ; (b)
                              LD [AR]++      ; (b)                               c = a + b
                              c = a + b                                          ST [AR]++      ; (c)
                              ST [AR]++      ; (c)                               LD [AR]++      ; (d)
                              LD [AR]        ; (d)                               LD [AR]++      ; (e)
                              AR += 3                                            f = d + e
  b (0)                       LD [AR]--      ; (e)                               ST [AR]        ; (f)
                              f = d + e                    a (0) - AR            AR -= 5
  c (1)                       ST [AR]--      ; (f)                               LD [AR]        ; (a)
                              LD [AR]--      ; (a)         b (1)                 AR += 3
  d (2)                       LD [AR]++      ; (d)                               LD [AR]        ; (d)
                              a = a + d                    c (2)                 a = a + d
  a (3) - AR                  ST [AR]--      ; (a)                               AR -= 3
                              LD [AR]++      ; (d)         d (3)                 ST [AR]        ; (a)
  f (4)                       LD [AR]        ; (a)                               AR += 3
                              c = d + a                    e (4)                 LD [AR]        ; (d)
  e (5)                       AR -= 2                                            AR -= 3
                              ST [AR]++      ; (c)         f (5)                 LD [AR]        ; (a)
                              LD [AR]        ; (d)                               c = d + a
                              AR += 2                                            AR += 2
  (a)                         LD [AR]--      ; (f)         (c)                   ST [AR]++      ; (c)
                              d = d + f                                          LD [AR]        ; (d)
                              LD [AR]--      ; (a)                               AR += 2
                              d = d + a                                          LD [AR]        ; (f)
                              ST [AR]        ; (d)                               d = d + f
                                                                                 AR -= 5
                                                                                 LD [AR]        ; (a)
                                                                                 d = d + a
                                                                                 AR += 3
                                                                                 ST [AR]        ; (d)

                              (b)                                                (d)
```

FIGURE 18.10   (a) Optimized memory placement; (b) optimized code sequence; (c) unoptimized memory placement; (d) unoptimized code sequence.

vertices if and only if the corresponding two variables are adjacent to each other w times in the access sequence.

- From the access graph, the cost of an assignment is equal to the sum of the weights of all edges connecting pairs of vertices that are not assigned adjacent locations in memory. Hence, the variables should be assigned memory locations in such a way that the sum of the weights of all edges connecting variables not assigned contiguous locations in memory is minimized. Liao et al. have shown that this problem is equivalent to finding the maximum-weighted path covering (MWPC) of the access graph [35]. Because MWPC is NP-hard, heuristics are used to find a good assignment of variables to memory locations. In Figure 18.9c, the dark edges form an MWPC of the access graph with a cost of 4. Figure 18.10a shows the memory placement using the SOA path cover shown in Figure 18.9c and the consequent pseudo-assembly for a data-path like the one shown in Figure 18.8a. Additional optimizations such as memory propagation can be applied to the pseudo-assembly to prevent some unnecessary loads and stores. The heuristic given by Liao et al. to find the MWPC is a greedy approach that looks at the edges in the access graph in decreasing order of weight and adds an edge to the cover if it does not form a cycle with the edges already in the cover and if it does not increase the degree of a node in the cover to more than 2.

- A heuristic to solve the GOA problem with k address registers based on the SOA problem was also provided by Liao. This heuristic recursively partitions the accessed variables into two partitions, solves the simple assignment problem on each partition, and then decides to either partition further or return the current partition based on the costs of each assignment. The total number of partitions generated is dependent on the number of address registers k in the architecture.

In the presence of control flow, the exact access sequence in a procedure can be known only during execution time. To perform offset assignment optimization at the procedure level instead of at the basic block level, the access graphs of the basic blocks are merged with equal weighting and the variables connected by control flow edges. After including the increments and decrements for each basic block, a separate phase decides whether to include an increment or a decrement across basic blocks.

Bartley was the first to address the SOA problem [8]. Leupers and Marwedel have also addressed the SOA problem [31]. Their work includes an improvement to the MWPC heuristic by introducing a tie-breaking function when the MWPC heuristic is faced with edges of equal weights and another heuristic to partition the vertices of the access graph for the GOA problem.

Sudarsanam et al. address the issue of an l-SOA problem that corresponds to the case with one AR and an offset of +/- l and the l, k-GOA problem with k address registers and a maximum offset of l [57]. In this work, the authors note that with an offset of more than 1, some of the edges in the access graph but not in the MWPC cover do not contribute to the cost of the cover. To identify these edges, called induced edges, they define an induced $(l + 1)$ clique of a cover C of the access graph G. Intuitively, if there is a path P of length l in G, then using the free 1-l auto-increment and auto-decrement feature, it is possible to cover every edge that is a part of the complete subgraph induced by the $(l + 1)$ vertices of the path P in cover C on G. Hence, edges in G that are induced by a cover C are called induced edges; the sum of the weights of all edges in C and the edges induced by C is defined as the induced weight; the sum of the weights of all edges that are in G but are neither in C nor in the induced edges of C is defined as the l-induced cost of C. The l-SOA problem now reduces to finding a cover of an access graph G with the minimum l-induced cost. They define the (k, l)-GOA problem as given an access sequence L for a set of variables V and partition V into at most k partitions so that the total sum of the induced cost of each l-SOA and the associated setup costs is minimized. Other work in OA includes [29, 60, 64].

18.5.3   Reference Allocation

The DSP compiler phases of memory bank allocation and register allocation together constitute reference allocation. Memory bank allocation is performed to exploit the ILP between arithmetic operations and memory operations and between memory operations themselves. The register allocator is similar to the general-purpose register allocator that decides which variables should be assigned to registers and to which register each variable should be assigned. As described in Section 18.3, DSPs can have numerous operand and operation constraints that directly affect reference allocation. A general technique developed by Sudarsanam and Malik for reference allocation is presented here [58]. This is a simulated annealing-based technique that simultaneously performs memory bank allocation and register allocation phases while taking into account various constraints similar to those described in Section 18.3. The assumptions made by this technique include statically allocating all static and global variables to one memory bank and only compiling applications that are nonrecursive in nature. Also, the reference allocation technique described is performed after code compaction or scheduling.

The first step of the algorithm is to generate the constraint graph. The constraint graph has a vertex for each symbolic register and variable. The problem of reference allocation is then transformed to a graph-labeling problem where every vertex representing a symbolic register must be labeled with a physical register and every vertex representing a variable must be labeled with a memory bank. The different types of constraint-weighted edges, where the weight corresponds to the penalty of not satisfying the constraint, are:

- A *red edge* is added between two symbolic registers if and only if they are simultaneously live [2]. The only exception to adding this edge is between two symbolic registers that are accessed in the same instruction by two memory operations. In this case, another type of edge ensures correctness. This edge ensures that the two symbolic registers connected by the edge are not assigned the same physical register. The cost of this edge is the amount of spill code that needs to be inserted if this constraint is violated. In the algorithm, this cost is assumed to be a large constant that significantly reduces the chances of assigning two symbolic registers the same architectural register. Hence, the algorithm does not compute the spill cost for each symbolic register.
- A *green edge* is added between two symbolic registers accessed by parallel memory operations to take into account any constraints that may exist between them. For example, for the instruction

"MOV var_i, reg_j MOV var_k reg_l," a green edge would be added in the constraint graph between the vertices corresponding to reg_j and reg_l to take care of any register constraints between the two symbolic registers. If a restriction exists on the two variables var_i and var_k concerning the memory banks to which they can be allocated, this is captured by a pointer to var_i and var_k from reg_j and reg_l, respectively. The cost of this edge is the statically estimated instruction execution count because the two operations have to be issued in separate instructions if this constraint is not satisfied.

- Similar to green edges, *blue, brown,* and *yellow edges* are added to the constraint graph with appropriate costs to represent constraints between parallel memory and register transfer operations. Blue edges are added for instructions involving a register transfer and a load operation. Brown edges are added for instructions involving an immediate load and a register transfer. Yellow edges are added for instructions involving a register transfer and a store operation.

- A *black edge* is added to represent operand constraints within an operation. These edges are added between symbolic registers and global vertices, which correspond to the register set of the DSP architecture. A black edge prevents the assignment of a symbolic register to the architectural register that it connects, because of the encodings of the ISA. Each black edge has an infinite cost because an unsatisfied black edge constraint is not supported by the hardware. For example, the accumulator cannot be a destination operand of load operations in the processor data-path shown in Figure 18.9a, so black edges would be added between symbolic registers representing the destination operands of load operations and the global vertex representing the accumulator.

Once the constraint graph has been constructed, the reference allocation algorithm uses simulated annealing to arrive at a low-cost labeling of the constraint graph. The authors reported that a greedy solution implemented to solve the constraint graph produced results very close to those produced by the simulated annealing algorithm [58]. The problem, however, with this approach is that for more varieties of constraints between operands, more complex formulations and additional colored edges would be needed.

Saghir et al. developed an algorithm to exploit the dual memory banks in DSPs using compaction-based partitioning and partial data duplication [52, 53]. In this method, an interference graph is constructed for each basic block, with the variables accessed in the program as vertices, and an edge is added between every pair of memory operations that can be legally performed, based on both the ISA and the data flow in the program. The edges are labeled with a cost that signifies the performance penalty if the corresponding two variables are not accessed simultaneously. The interference graph is then partitioned into two sets corresponding to the two memory banks such that the overall cost is minimum. During the interference graph construction, memory operations accessing the same variables and locations may be marked for duplication; these are placed in both memory banks, and operations to preserve data integrity are inserted.

For more work on reference allocation, readers are referred to [23, 45, 62].

18.5.4   Register Allocation for Irregular Operand Constraints

Register allocation is an important phase in compilation for DSPs. Most compiler optimizations that work on the IR generate new temporaries assuming that they can be assigned to machine registers without any penalty. A poor register allocator can undo the effects of the prior optimization phases and weaken the performance of the compiled code, as register allocation can affect both code size and performance. Some of the challenges posed by nonorthogonal instruction sets were presented in Section 18.3. This section presents an adaptation of the integer linear programming–based register allocation technique developed by Appel and George to handle such irregular constraints [3].

While integer linear program is NP-complete, this technique has several advantages compared to the technique presented in Section 18.5.3. The integer linear programming–based works by Appel and George [3] and by George and Matthias [19] have been shown to work well under various constraints. The model provides for flexibility, that is, the integer linear programming part of the allocator can be used

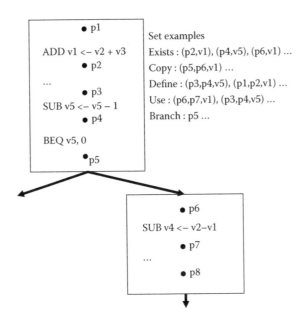

Set examples

Exists : (p2,v1), (p4,v5), (p6,v1) ...

Copy : (p5,p6,v1) ...

Define : (p3,p4,v5), (p1,p2,v1) ...

Use : (p6,p7,v1), (p3,p4,v5) ...

Branch : p5 ...

FIGURE 18.11   An integer linear program–based modeling example.

only to decide if a temporary should be in a register or memory at a program location, and it can decide which register the temporary should reside in. AMPL [15] is used to develop the integer linear program. AMPL provides a clean separation between the allocation model, which is highly architecture dependent, and the program data, which are dependent on the application program. This is a huge bonus in ease of implementation and reuse of models. The AMPL representation is compact and easy to read, debug, and retarget. The flexibility provided by "AMPL" to choose solvers is also an important feature.

The original allocator works in two parts, using integer linear programming to decide which temporaries are to reside in registers and which ones in memory at each program point, followed by a graph-coloring-based assignment. The integer linear program model primarily works out of two sets of data, namely, the set of all program points P and the set of all temporaries V. Figure 18.11 shows a code-snippet with basic blocks, the set of program points $p_1, p_2, \ldots, p_8$, and some temporaries $v_1, v_2, \ldots, v_5$. The set of register allocation constraints are captured as tuples in the model. In the set of *Exists* of tuples (p_j, v_i), $p_j \in P$, $v_i \in V$, each tuple (p_j, v_i) represents the fact that temporary v_j either is live at program point p_i or was defined in an operation just prior to program point p_j. *Copy*, a set of tuples of the form (p_j, p_k, v_i), $v_i \in V$; $p_j, p_k \in P$, where p_j and p_k are adjacent program points, specifies that v_i is unchanged from p_j to p_k. *Branch* is a set of all the program points $p_i \in P$ that are immediately after a branch operation. In addition, there are several sets of tuples of the form $(p_j, p_k, v_1, v_2, \ldots, v_n)$, $v_1, v_2, \ldots, v_n \in V$; $p_j, p_k \in P$. For example, the set of all tuples (p_j, p_k, v_i) denotes that v_i is used (similarly for defined) at an operation between points p_j and p_k.

A key point about this model is that at each program point a temporary can be live only in a register or memory, but not both. To capture this information, a set of four integer linear program variables is used for each v that *exists* at each p with the following interpretation, shown pictorially in Figure 18.12.

- $r_{(p,v)} = 1$ implies v is live in a register before and after p.
- $s_{(p,v)} = 1$ implies v is live in a register before p and spilled to memory between the operation before p and p.
- $l_{(p,v)} = 1$ implies v is live in memory before p and loaded into a machine register after p.
- $m_{(p,v)} = 1$ implies v is live in memory before and after p.

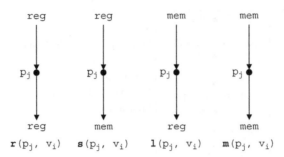

FIGURE 18.12   Storage assignment possibilities for v_i at program point p_j where it is live.

Given the definition of the variables and using the constraint sets, the set of equations can be written for register allocation as follows:

- $\forall(p, v) \in$ *Exists,*
  $l_{(p,v)} + s_{(p,v)} + r_{(p,v)} + m_{(p,v)} = 1$
- $\forall(p_i, p_j, v_i) \in$ *Copy,*
  $l_{(p_j,v_i)} + r_{(p_j,v_i)} = r_{(p_k,v_i)} + s_{(p_k,v_i)}$
  $\forall(p_j, p_k, v_i) \in$ *Copy,*
  $s_{(p_j,v_i)} + m_{(p_j,v_i)} = m_{(p_k,v_i)} + l_{(p_k,v_i)}$
- $\forall(p_j, p_k, v_i) \in$ *Defined,*
  $s_{(p_k,v_i)} + r_{(p_k,v_i)} = 1$
  Similar equations can be written for temporaries that are used, and so on. It must be noted that there can be no spill code immediately after branch operations, and hence there are equations enforcing that condition.
- Equations governing the number of machine registers at each program point. This can vary, for example, because of parameter passing registers or caller/callee save registers being used up at certain points in the program.

A simple objective that can be used for the integer linear program would be to minimize the number of loads and stores, but more complex functions can be easily incorporated. The solution of the integer linear program does not guarantee a valid coloring in the graph coloring phase since a temporary can interfere with more temporaries than the number of registers in the overall interference graph (though not at a program point). Appel and George use a combination of splitting live ranges, optimistic coalescing, and special $x86$ instructions to address this issue [3].

In the presence of irregular constraints, integer linear programming can also be used to assign the registers to temporaries by identifying the set of valid register assignments at each program point. Note that the set of valid registers can be different for a given temporary at different program points depending upon the context due to the nature of irregular operand constraints mentioned in Section 18.3.2. Other constraints such as register-pairs (for example, overlapping double and float registers) and temporary-pairs (constraints relating pairs of temporaries) can be incorporated into the model via simple extensions. The equations can be augmented to reflect the fact that a machine register can be assigned to only one temporary at each program point, and each temporary inturn can be assigned to only one machine register at each program point.

The primary restriction of the integer linear program model presented here is that a temporary can be in either register or memory at a program point, but not in both. This can be relaxed by including more variables at each program point. For example, by using six variables instead of four with the following interpretation, the quality of allocation can possibly be improved upon:

- $r_{(p,v)} = 1$ implies v is live only in a register before and after p.
- $s_{(p,v)} = 1$ implies v is live only in a register before p and only in memory after p.
- $l_{(p,v)} = 1$ implies v is live only in memory before p and live in both register and memory after p.
- $m_{(p,v)} = 1$ implies v is live only in memory before and after p.
- $k_{(p,v)} = 1$ implies v is live in register and memory before p and only in memory after p.
- $b_{(p,v)} = 1$ implies v is live in register and memory before and after p.

Similarly, more variables can be meaningfully incorporated into this model to potentially improve the quality of results, but at the cost of increased total number of variables for a program.

For more work on register allocation, readers are referred to Chapter 21.

18.5.5 Irregular ILP Constraints

Pressures to reduce code size and yet meet the power and execution time constraints of embedded applications often force designers to design processors, such as DSPs, with irregular architectures. As a consequence, DSPs are designed with small instruction widths and nonorthogonal opcode and operand encodings. This poses several problems to a compiler as illustrated in Section 18.3. In this section, the artificial resource allocation (ARA) algorithm for the operation ILP problem developed by Rajagopalan et al. is described [49, 50]. Conventional VLIW compilers use a reservation table–based scheduler that keeps track of the processor's resource usage as operations are scheduled. One of the highlights of the ARA algorithm is to allow compilers for irregular DSP architectures to use processor-independent table-based schedulers instead of writing processor-specific schedulers. This is achieved by transforming the set of irregular operation ILP constraints to a set of artificial regular constraints that can subsequently be used by conventional VLIW schedulers.

The ARA algorithm takes as input the set of all possible combinations of operations that can be issued in parallel and produces as output an augmented resource usage of the MD such that the following constraints are satisfied:

Constraint 1: Every pair of operations that cannot be issued in parallel must share an artificial resource.
Constraint 2: Every pair of operations that can be issued in parallel must not share a resource.
Constraint 3: The total number of artificial resources generated must be minimum. This condition is used to reduce the size of the reservation table and potentially the scheduler time.

The ARA algorithm is explained next with an example from the *Fujitsu Hiperion* DSP ISA, as shown in Figure 18.13 taken from [49]:

- The first step of the ARA algorithm is construction of a *compatibility graph G*. A vertex in the compatibility graph corresponds to an operation in the ISA. An edge exists between two vertices only if the two corresponding operations can be performed in parallel. For example, in Figure 18.13 there are five vertices in the compatibility graph, one each for Add, Shift, Multiply, and two Loads.

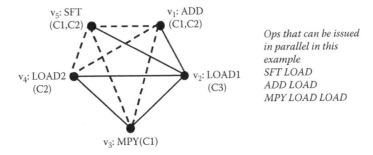

FIGURE 18.13 ARA algorithm applied to a subset of *Hiperion* ISA [49].

The parallel combinations shown in the right of Figure 18.13 are captured by the straight-line edges in the compatibility graph.

- The complement G' of the compatibility graph shown in dotted lines in Figure 18.13 is then constructed. An edge between two vertices in G' implies that the two operations cannot be performed in parallel. An immediate solution that follows is to assign an artificial resource to each edge in G'. Because this solution can cause the size of the reservation table to grow significantly, a better solution is obtained in the next step.
- The minimum number of artificial resources needed to satisfy the operation ILP constraints is obtained by performing the minimum edge clique cover on G' [50]. The problem of finding the minimum edge clique cover on G' is first converted to an equivalent problem of finding the minimum vertex clique cover on graph G_1. G_1 is obtained from G' as follows. A vertex in G_1 corresponds to an edge in G', and an edge exists between two vertices in G_1 if and only if the vertices connected by the corresponding edges in G' form a clique. Hence, a vertex clique cover of G_1 identifies larger cliques in G. This problem in turn is transformed to graph coloring [17] on the complement of G_1. In the Figure 18.13 example, three cliques, C_1, C_2, and C_3, are required to obtain the minimum edge clique cover.
- The final step in the ARA algorithm translates the result of the minimum edge clique cover algorithm into the resource usage section of the processor machine description.

The advantages of using this algorithm are that it is a highly retargetable solution to the operation ILP problem and it helps avoid processor-specific schedulers when such irregular constraints exist. Eichenberger and Davidson have provided techniques to compact the size of machine descriptions of general-purpose processors [14]. Gyllenhaal and Hwu provided methods to optimize not only the size of the MDs but also the improvement of the efficiency of queries [20]. For more work on MDs, readers are referred to the chapter 16 of this handbook.

18.5.6   Array Reference Allocation

The ISA of DSPs provides special address-modifying instructions that update address registers used by memory operations. Because many signal processing applications contain kernels that operate on arrays, one of the most important DSP optimizations is to assign ARs to array references so that a majority of the address computation operations can be replaced by auto-increment and auto-decrement address modifying operations that are free. This can significantly affect the static code size and the dynamic performance of the assembly code because address computation for successive memory operations is performed in parallel with current memory operations. In this section, the array reference allocation algorithm developed by Araujo and others is described [12, 42, 49].

Global reference allocation (GRA) is the problem of allocating ARs to array references such that the number of simultaneously live ARs is kept below the maximum number of such registers available in the processor, and the number of instructions required to update them is minimized. The local version of this problem, called *local reference allocation* (LRA), has all references restricted to basic block boundaries. There are known efficient graph-based solutions for LRA [18, 28]. Araujo et al. proposed a solution for optimizing array references within loops based on an index graph structure [6]. The index graph is constructed as follows. A vertex exists for each array access in the loop. An edge exists between two accesses only if the indexing distance between the two accesses is less than the limit of the auto-increment and auto-decrement limit in hardware. The array reference allocation problem then deals with finding the disjoint path cycle cover that minimizes the number of paths and cycles. This is similar to the offset assignment algorithm explained in Section 18.5.2.

Whereas general-purpose register allocation concerns itself with allocating a fixed set of general-purpose registers to a potentially much larger set of virtual registers in the program, reference allocation pertains to assigning a fixed set of address registers to the various memory access references in the program. When two virtual registers are assigned the same general register, it is the allocator's responsibility to insert the appropriate spill code (store to memory and load from memory). Similarly, when reference allocation combines two references using a single address register, it is the reference allocator's responsibility to insert appropriate update operations.

In the GRA algorithm developed by Cintra and Araujo called live range growth (LRG), to assign ARs to array references [12], the set of array reference ranges is first initialized to all the independent array references in the program. The next step involves reducing the number of array reference ranges in the program to the number of ARs in the processor. This involves combining array reference ranges and at the same time minimizing the number of reference update operations. To quantify the cost of combining reference ranges, the notion of an indexing distance is defined as follows:

Definition 18.1   *Let a and b be array references and s be the increment of the loop containing these references. Let* $index(a)$ *be a function that returns the subscript expression of reference a. The indexing distance between a and b is the positive integer:*

$$d(a,b) = \begin{cases} |index(b) - index(a)| & \text{if } a < b \\ |index(b) - index(a) + s| & \text{if } a > b \end{cases}$$

where a < b (a > b) if a (b) precedes b (a) in the schedule order.

An update operation is required when combining two reference ranges whenever the indexing distance is greater (less) than the maximum (minimum) allowed automatic increment (decrement) value. Hence, the cost of combining two ranges is the number of update instructions that need to be inserted. This is shown in the example in Figure 18.14b taken from [49], where the two live ranges R and S in Figure 18.14a, with an indexing distance of 1, have been merged into a single live range. To maintain correctness, necessary address register update operations have been inserted both within basic blocks and along the appropriate control flow edges.

To facilitate the computation of the indexing distances, an important requirement is that the references in the control flow graph are in single reference form [12], a variation of static single assignment (SSA) form [13].

Although GRA is a typical DSP optimization, it is implemented along with machine-independent optimizations. It is performed before most of the classical optimizations because techniques such as common subexpression elimination may destroy some opportunities for applying GRA.

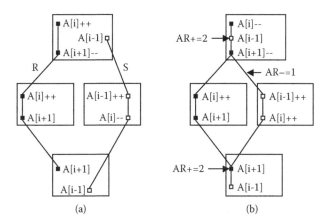

FIGURE 18.14   (a) Two array reference live ranges, R, S; (b) after merging live ranges R and S into a single live range [49].

Leupers, Basu, and Marwedel have proposed an optimal algorithm for AR allocation to array references in loops [28]. Liem, Paulin, and Jerraya have developed an array reference analysis tool that takes, as input, the C program with array references and the address resources of the hardware and produces an equivalent C program output in which the array references have been converted to optimized pointer references for the given hardware [37].

18.5.7 Addressing Modes

18.5.7.1 Paged Absolute Addressing

A consequence of reducing instruction width to reduce static code size, as noted in Section 18.2, is that memory operations using absolute addressing mode (i.e., using a physical address of memory location) to access data have to be encoded in more than one instruction word. To overcome this extra word penalty and potentially an extra cycle penalty, architectures such as the TI TMS320C25 [61] feature a paged absolute addressing mode. In these architectures, a single data memory space is partitioned into N nonoverlapping pages numbered 0 through $N - 1$, and a special page pointer register is dedicated to storing the ordinal of the page that is currently accessed. The absolute address can then be specified by loading the page pointer register with the appropriate page number and specifying the offset within the page in the memory operation. This can save instruction memory words and reduce execution time for a sequence of memory operations accessing the same page. In addition, because a machine register is used instead of ARs or general-purpose registers, a better register assignment of the program code is also possible. Because DSP programs are nonrecursive in nature, the automatic variables are statically allocated in memory to exploit absolute addressing and relieve address registers. In this section, some optimizations developed by Sudarsanam et al. to exploit absolute paged addressing modes are described [59].

The *LDPK* operation is used in the TI TMS320C25 DSP to load the page pointer register. To justify the use of absolute addressing for automatic variables, the *LDPK* operation overhead should be minimum. Hence, this algorithm tries to reduce the number of *LDPK* operations using data flow analysis [2]. The main steps in this algorithm are as follows:

- Assuming that code is generated on a per basic block basis, the algorithm conservatively assumes that the first operation to use absolute addressing in each basic block should be preceded by an *LDPK* operation.
- Another conservative assumption is that the value of the page pointer register is not preserved across procedure calls and hence must be restored after each procedure call.
- An *LDPK* operation can be suppressed before an operation using absolute addressing if the new value of the page pointer register is the same as the current value of the page pointer register.
- By using data flow analysis [2] on the assembly code, the set of unnecessary *LDPK* operations are determined across basic blocks. For each basic block B and page pointer register DP, three variables are used: IN(B), which holds the value of DP at the entry of B; OUT(B), which holds the value of DP at the exit of B; and LAST(B), which holds the ordinal of the last referenced page in B.
- A global data flow analysis described here is then performed to identify the redundant *LDPK* operations. While traversing B in reverse order, LAST(B) is assigned the value UNKNOWN after a procedure call and prior to an *LDPK* operation. LAST(B) is assigned the ordinal of the last *LDPK* operation prior to a procedure call while traversing B. Finally, LAST(B) is assigned PROPAGATE if neither an *LDPK* operation nor a procedure call is encountered after the traversal of B.
- For each basic block, B, IN(B) is computed as the intersection of all OUT sets of its predecessors. If the intersection is empty, IN(B) is assigned UNKNOWN. If LAST(B) is PROPAGATE, OUT(B) is set as IN(B); otherwise OUT(B) is set as LAST(B).
- The previous step is repeated when there is a change in OUT sets.
- After computing the data flow variables for each basic block, if the first *LDPK* operation of a basic block B is not preceded by a procedure call, then it can be removed if the ordinal of the *LDPK* operation is the same as IN(B).

An extension of this algorithm for interprocedural optimization was also provided by Sudarsanam et al. [59]. Lin provides a simple postpass algorithm that removes redundant *LDPK* instructions within basic blocks, assuming that an *LDPK* is generated for each access of a statically allocated variable [38].

18.6 Summary

DSPs are a significant fraction of the processor market and, in particular, low-cost embedded DSPs are in a wide variety of products and applications of daily use. Hence, it is important not only to design the necessary tools to program these processors but also to develop tools that can help in an automatic cooperative synthesis of both the architecture and the associated software tool-set for a given set of applications. A first step in this direction is the design of a highly reusable and retargetable compiler for DSPs. In this chapter, DSP architectures were first classified into two classes: FOW DSPs and VOW DSPs. Although DSPs in the former category are more regular with orthogonal instruction sets, the ones in the latter category are the low-end processors with highly optimized irregular architectures. VOW DSPs are also more cost sensitive and require special algorithms to exploit their hardware features and thus are the focus of this chapter. Sections 18.2 and 18.3 explain the consequences of a VOW-based design and point out the nature of constraints that a DSP compiler must solve. In Section 18.4, retargetable compilation is classified into three categories: automatic retargetability, user retargetability, and developer retargetability, depending on the level of automation, the level of user interaction in retargeting, and the level of architecture coverage provided by the compiler framework. Retargetability is a key factor in design automation and reduction of time to market. A potential approach for retargetable compiler development is also presented along with some qualitative measures.

Finally, in Section 18.5, some DSP architecture–specific issues that a compiler framework developer must solve to achieve a high level of retargetability and good efficiency were discussed. Some of the common DSP optimizations that have been developed were presented. In addition to the architecture constraints described, new features are available such as media operations that usually pack multiple identical operations into a single operation operating on smaller bit-width operands. Hence, DSP compiler frameworks also need to be extensible in that it should be easy to add new architectural features and new algorithms to exploit such features to the framework.

References

1. A. Aho, M. Ganapathi, and S. Tjiang. 1989. Code generation using tree matching and dynamic programming. *ACM Transactions on Programming Languages and Systems* 11(4):491–516.
2. A. V. Aho, R. Sethi, and J. D. Ullman. 1988. *Compilers: Principles, techniques, and tools.* Reading, MA: Addison-Wesley.
3. A. W. Appel and L. George. 2001. Optimal spilling for CISC machines with few registers. In *Proceedings of the ACM SIGPLAN Conference on Programming Language Design and Implementation,* 243–53.
4. G. Araujo. 1997. Code generation algorithms for digital signal processors. PhD thesis, Princeton University, Princeton, NJ.
5. G. Araujo and S. Malik. 1970. Code generation for fixed-point DSPs. *ACM Transactions on Design Automation of Electronic Systems* 3(2):136–61.
6. G. Araujo, A. Sudarsanam, and S. Malik. 1996. Instruction set design and optimizations for address computation in DSP processors. In *9th International Symposium on Systems Synthesis,* 31–37. Washington, DC: IEEE Computer Society Press.
7. D. I. August, D. A. Connors, S. A. Mahlke, J. W. Sias, K. M. Crozier, B.-C. Cheng, P. R. Eaton, Q. B. Olaniran, and W. W. Hwu. 1998. Integrated predicated and speculative execution in the IMPACT EPIC architecture. In *Proceedings of the 25th International Symposium on Computer Architecture.* Washington, DC: IEEE Computer Society Press.

8. D. H. Bartley. 1992. Optimizing stack frame accesses for processors with restricted addressing modes. *Software — Practice and Experience* 22(2), 101–110, John Wiley & Sons, Inc.

9. S. Bashford and R. Leupers. 1999. Phase-coupled mapping of data flow graphs to irregular data paths. *Design Automation for Embedded Systems* 4(2/3):119–65.

10. J. L. Bruno and R. Sethi. 1976. Code generation for one-register machine. *Journal of the ACM* 23(3):502–10.

11. P. P. Chang, S. A. Mahlke, W. Y. Chen, N. J. Warter, and W. W. Hwu. 1991. IMPACT: An architectural framework for multiple-instruction-issue processors. In *Proceedings of the 18th International Symposium on Computer Architecture*, 266–75. http://www.crhc.uiuc.edu/IMPACT.

12. M. Cintra and G. Araujo. 2000. Array reference allocation using SSA-Form and live range growth. In *Proceedings of the ACM SIGPLAN LCTES 2000*, 26–33.

13. R. Cytron, J. Ferrante, B. Rosen, M. Wegman, and F. Zadeck. 1989. An efficient method of computing static single assignment form. In *Proceedings of the ACM POPL'89*, 23–25. New York: ACM Press.

14. A. E. Eichenberger and E. S. Davidson. 1996. A reduced multipipeline machine description that preserves scheduling constraints. In *Proceedings of the ACM SIGPLAN Conference on Programming Language Design and Implementation*. New York: ACM Press.

15. Robert Fourer, David M. Gay, and Brian W. Kernighan. 1995. *AMPL: A mathematical language for mathematical programming*. Danvers, MA: Boyd and Fraser.

16. C. Fraser, D. Hanson, and T. Proebsting. 1992. Engineering a simple, efficient code-generator generator. *ACM Letters of Programming Languages and Systems* 1(3):213–26.

17. M. R. Garey and D. S. Johnson. 1979. *Computers and intractability: A guide to the theory of NP-completeness*. W.H. Freeman.

18. C. Gebotys. 1997. DSP address optimization using a minimum cost circulation technique. In *Proceedings of the International Conference on Computer-Aided Design*, 100–103. Washington, DC: IEEE Computer Society Press.

19. Lal George and Matthias Blume. 2003. Taming the IXP network processor. In *Proceedings of the ACM SIGPLAN'03 Conference on Programming Language Design and Implementation*, 26–37. New York: ACM Press.

20. J. C. Gyllenhaal and W. W. Hwu. 1996. Optimization of machine descriptions for efficient use. In *Proceedings of the 29th Annual ACM/IEEE International Symposium on Microarchitecture*, pp. 349–358.

21. S. Hanono and S. Devadas. 1998. Instruction selection, resource allocation, and scheduling in the AVIV retargetable code generator. In *Proceedings of the 35th Design Automation Conference*, 510–15. ACM/IEEE.

22. R. Hartmann. 1992. Combined scheduling and data routing for programmable ASIC systems. In *European Conference on Design Automation (EDAC)*, 486–90.

23. D. J. Kolson, A. Nicolau, N. Dutt, and K. Kennedy. 1995. Optimal register assignment to loops for embedded code generation. In *Proceedings of 8th International Symposium on System Synthesis*.

24. D. Lanneer, M. Cornero, G. Goossens, and H. D. Man. 1994. Data routing: A paradigm for efficient data-path synthesis and code generation. In *Proceedings of 7th IEEE/ACM International Symposium on High Level Synthesis*, 17–21.

25. D. Lanneer, J. V. Praet, A. Kifli, K. Schoofs, W. Geurts, F. Thoen, and G. Goossens. 1995. CHESS: Retargetable code generation for embedded DSP processors. In *Code generation for embedded processors*, chapter 5, 85–102. Dordrecht, Netherlands: Kluwer Academic Publishers.

26. P. Lapsley, J. Bier, A. Shoham, and E. A. Lee. 1997. *DSP processor fundamentals*. Washington, DC: IEEE Press.

27. R. Leupers and S. Bashford. 2000. Graph based code selection techniques for embedded processors. *ACM Transactions on Design Automation of Electronic Systems* 5(4), 794–814.

28. R. Leupers, A. Basu, and P. Marwedel. 1998. Optimized array index computation in DSP programs. In *Proceedings of the Asia and South Pacific Design Automation Conference*, pp. 87–92.

29. R. Leupers and F. David. 1998. A uniform optimization technique for offset assignment problems. In *Proceedings of the 11th International Symposium on System Synthesis*, pp. 3–8. Washington, DC: IEEE Computer Society.

30. R. Leupers and P. Marwedel. 1995. Time-constrained code compaction for DSPs. In *Proceedings of the 8th International Symposium on System Synthesis*, pp. 54–59. New York: ACM Press.

31. R. Leupers and P. Marwedel. 1996. Algorithms for address assignment in DSP code generation. In *Proceedings of International Conference on Computer Aided Design*, pp. 109–112. Washington, DC: IEEE Computer Society.

32. R. Leupers and P. Marwedel. 1996. Instruction selection for embedded DSPs with complex instructions. In *European Design and Automation Conference (EURO-DAC)*, pp. 200–205. Washington, DC: IEEE Computer Society.

33. R. Leupers and P. Marwedel. 1998. Retargetable code generation based on structural processor descriptions. *Design Automation for Embedded Systems* 3(1):1–36.

34. S. Liao, S. Devadas, K. Keutzer, and S. Tjiang. 1995. Instruction selection using binate covering for code size optimization. In *Proceedings of the International Conference on Computer-Aided Design*, 393–99.

35. S. Liao, S. Devadas, K. Keutzer, S. Tjiang, and A. Wang. 1996. Storage assignment to decrease code size. *ACM Transactions on Programming Languages and Systems* 18:235–53.

36. C. Liem. 1997. *Retargetable code generation for digital signal processors*. Dordrecht, The Netherlands: Kluwer Academic.

37. C. Liem, P. Paulin, and A. Jerraya. 1996. Address calculation for retargetable compilation and exploration of instruction-set architectures. In *Proceedings of the 33rd Design Automation Conference*, pp. 597–600. New York: ACM Press.

38. W. Lin. 1995. An optimizing compiler for the TMS320C25 DSP processor. Master's thesis, University of Toronto, Toronto, Canada.

39. P. Marwedel and G. Goossens, eds. *Code generation for embedded processors*. Dordrecht, The Netherlands: Kluwer Academic.

40. S. Novack and A. Nicolau. 1994. Mutation scheduling: A unified approach to compiling for fine-grain parallelism. In *Languages and compilers for parallel computing*, 16–30.

41. S. Novack, A. Nicolau, and N. Dutt. 1995. A unified code generation approach using mutation scheduling. In *Code generation for embedded processors*, 65–84. Dordrecht, Netherlands: Kluwer Academic.

42. G. Ottoni, S. Rigo, G. Araujo, S. Rajagopalan, and S. Malik. 2001. Optimal live range merge for address register allocation in embedded programs. In *Proceedings of the CC01 — International Conference on Compiler Construction*, pp. 274–288. Springer-Verlag.

43. P. Paulin, C. Liem, T. May, and S. Sutarwala. 1994. CodeSyn: A re-targetable code synthesis sytem. In *Proceedings of the 7th International High-Level Synthesis Workshop*, p. 94. Washington, DC: IEEE Computer Society.

44. P. Paulin, C. Liem, T. May, and S. Sutarwala. 1995. Flexware: A flexible firmware development environment for embedded systems. In *Code generation for embedded processors*, chapter 4, 65–84. Dordrecht, The Netherlands: Kluwer Academic.

45. D. Powell, E. Lee, and W. Newman. 1992. Direct synthesis of optimized DSP assembly code from signal flow block diagrams. In *Proceedings of the International Conference on Acoustics, Speech and Signal Processing*, vol. 5, 553–56.

46. J. V. Praet, G. Goossens, D. Lanneer, and H. D. Man. 1994. Instruction set definition and instruction selection for ASIPs. In *7th International Symposium on High Level Synthesis*, 11–16.

47. J. V. Praet, D. Lanneer, W. Geurts, and G. Goossens. 2001. Processor modeling and code selection for retargetable compilation. *ACM Transactions on Design Automation of Electronic Systems* 6(2):277–307.

48. J. V. Praet, D. Lanneer, G. Goossens, and W. G. D. Man. 1996. A graph based processor model for retargetable code generation. In *European Design and Test Conference*, p. 102. Washington, DC: IEEE Computer Society.

49. A. Rajagopalan, S. P. Rajan, S. Malik, S. Rigo, G. Araujo, and K. Takayama. 2001. A retargetable VLIW compiler framework for DSPs with instruction-level-parallelism. *IEEE Transactions on Computer Aided Design of Integrated Circuits and Systems* 20(11):1319–28.

50. S. Rajagopalan, M. Vachharajani, and S. Malik. 2000. Handling irregular ILP within conventional VLIW schedulers using artificial resource constraints. In *Proceedings of International Conference on Compilers Architecture and Synthesis for Embedded Systems*, pp. 157–164. New York: ACM Press.

51. K. Rimey and P. N. Hilfinger. 1988. Lazy data routing and greedy scheduling for application specific signal processors. In *21st Annual Workshop on Microprogramming and Microarchitecture (MICRO-21)*, 111–15.

52. M. A. R. Saghir, P. Chow, and C. G. Lee. 1995. Automatic data partitioning for HLL DSP compilers. In *Proceedings of the 6th International Conference on Signal Processing Applications and Technology*, 866–71.

53. M. A. R. Saghir, P. Chow, and C. G. Lee. 1996. Exploiting dual data-memory banks in digital signal processors. In *Proceedings of the 7th International Conference on Architectural Support for Programming Languages and Operating Systems*, 134–243.

54. R. Sethi. 1975. Complete register allocation problems. *SIAM Journal of Computing* 4(3):226–48.

55. M. Strik, J. van Meerbergen, A. Timmer, and J. Jess. 1995. Efficient code generation for in-house DSP cores. In *European Design and Test Conference*, 244–49.

56. A. Sudarsanam. 1998. Code generation libraries for retargetable compilation for embedded digital signal processors. PhD thesis, Princeton University, Princeton, NJ.

57. A. Sudarsanam, S. Liao, and S. Devadas. 1997. Analysis and evaluation of address arithmetic capabilities in custom DSP architectures. In *Proceedings of ACM/IEEE Design Automation Conference*, pp. 287–292. New York: ACM Press.

58. A. Sudarsanam and S. Malik. 1995. Memory bank and register allocation in software synthesis for ASIPs. In *Proceedings of the 1995 IEEE/ACM International Conference on Computer-Aided Design*, pp. 388–392. Washington, DC: IEEE Computer Society.

59. A. Sudarsanam, S. Malik, S. Tjiang, and S. Liao. 1997. Optimization of embedded DSP programs using post-pass data-flow analysis. In *Proceedings of 1997 International Conference on Acoustics, Speech, and Signal Processing*, p. 695. Washington, DC: IEEE Computer Society.

60. N. Sugino, H. Miyazaki, S. Iimure, and A. Nishihara. 1996. Improved code optimization method utilizing memory addressing operation and its application to DSP compiler. In *International Symposium on Circuits and Systems*, Vol. 2, pp. 249–252.

61. Texas Instruments. 1993. *TMS320C2x user's guide*, c ed.

62. B. Wess. 1991. Automatic code generation for integrated digital signal processors. In *Proceedings of the International Symposium on Circuits and Systems*, 33–36.

63. B. Wess. 1995. Code generation based on trellis diagrams. In *Code generation for embedded processors*, chapter 11, 188–202. Dordrecht, The Netherlands: Kluwer Academic.

64. B. Wess and M. Gotschlich. 1997. Constructing memory layouts for address generation units supporting offset 2 access. In *Proceedings of ICASSP*, p. 683. Washington, DC: IEEE Computer Society.

65. T. Wilson, G. Grewal, S. Henshall, and D. Banerji. 1995. An ILP based approach to code generation. In *Code generation for embedded processors*, chapter 6, 103–18. Dordrecht, The Netherlands: Kluwer Academic.

19
Instruction Scheduling

R. Govindarajan

Supercomputer Education and Research Centre, Department of Computer Science and Automation, Indian Institute of Science, Bangalore, India;
govind@csa.iisc.ernet.in

19.1 Introduction

Ever since the advent of reduced instruction set computers (RISC) [67, 122] and their pipelined execution of instructions, instruction scheduling techniques have gained importance as they rearrange instructions to "cover" the delay or latency that is required between an instruction and its dependent successor. Without such reordering, pipelines would stall, resulting in wasted processor cycles. Pipeline stalls would also occur while executing control transfer instructions, such as branch and jump instructions. In architectures that support *delayed branching*, where the control transfers are effected in a *delayed* manner [68], instruction reordering is again useful to cover the stall cycles with useful instructions. Instruction scheduling can be limited to a single *basic block* — a region of straight line code with a single point of entry and a single point of exit, separated by decision points and merge points [3, 73, 107] — or to multiple basic blocks. The former is referred as *basic block scheduling*, and the latter as *global scheduling* [3, 73, 107].

A significant amount of research conducted in industry and academia has resulted in processors that issue multiple instructions per cycle and hence exploit *instruction-level parallelism* (ILP) [132]. Exploiting ILP has lent itself as a viable approach for providing continuously increasing performance without having to rewrite applications. ILP processors have been classified into two broad categories, namely VLIW (very long instruction word) processors [32, 46, 136] and superscalar processors [77, 150, 151] depending on whether the parallelism is exposed at compile time or at runtime by dynamic instruction scheduling hardware. In VLIW machines, a compiler identifies independent instructions and communicates them to the hardware by packing them in a single long word instruction. At runtime, a long word instruction is fetched and decoded, and the independent instructions in it are executed in parallel in the multiple functional units available in the VLIW architecture. In a superscalar machine, however, a complex hardware identifies independent instructions and issues them in parallel at runtime. The explicitly parallel instruction computing (EPIC) architecture is a variant of the VLIW architecture where the compiler identifies and exposes independent instructions to the hardware [143, 145].

Multiple instruction issue per cycle has become a common feature in modern processors (see, e.g., [70, 82, 114, 145, 175]). The success of ILP processors has placed even more pressure on instruction scheduling methods, as exposing instruction-level parallelism is the key to the performance of ILP processors. Instruction scheduling can be done by hardware at runtime [68, 151] or by software at compile time [52, 66, 89, 132, 170]. In this discussion, we concentrate on compile-time instruction scheduling methods. Such compile-time instruction scheduling is solely responsible for exposing and exploiting the parallelism available in a program in a VLIW architecture. Thus, without the instruction scheduler, the (early) VLIW processors could not have achieved an ILP of 7 to 14 operations that they are capable of issuing in a cycle [32, 136].

In superscalar processors, instruction scheduling hardware determines at runtime the independent instructions that can be issued in parallel. However, the scope of the runtime scheduler is limited to a narrow window of 16 or 32 instructions [151]. Hence, compile-time techniques may be needed to expose parallelism beyond this window size. Furthermore, in a certain class of superscalar processors, namely the *in-order* issue processors [68, 151], instructions are issued in program order. Hence, when an instruction is stalled because of a data dependence, instructions beyond the stalled one are also stalled. Instruction scheduling can be beneficially applied for these in-order issue architectures to rearrange instructions and hence exploit higher ILP. Thus, even superscalar processors can benefit from the parallelism exposed by a compile-time scheduler.

Instruction scheduling methods for basic blocks may result in a moderate improvement (less than 5 to 10%) in performance, in terms of the execution time of a schedule, for simple pipelined RISC architectures [66]. However, the performance improvement achieved for multiple instruction issue processors could be significant (more than 20 to 50%) [102]. Instruction scheduling beyond basic blocks can achieve even higher performance improvement, ranging from 50 to 300% [73, 74, 100].

Instruction scheduling is typically performed after machine-independent optimizations, such as copy propagation, common subexpression elimination, loop-invariant code motion, constant folding, dead-code elimination, strength reduction, and control flow optimizations [3, 107]. Instruction scheduling is performed either on the target machine's assembly code or on a low-level code that is very close to the machine's assembly code. In certain implementations, instruction scheduling is performed after register allocation — another important compiler optimization that determines which variables are stored in registers and which remain in memory. When instruction scheduling follows register allocation, it is referred to as a *Postpass scheduling* approach [3, 107]. In the *Prepass scheduling*, or *prescheduling*, approach, instruction scheduling precedes register allocation. In Prepass scheduling, since register allocation is performed subsequently, any code introduced as a result of register spills is not scheduled by the scheduler. Hence, in Prepass scheduling, to handle the spill code, the instruction scheduler may be invoked again after register allocation. Instruction scheduling and register phases influence each other, so the ordering between the two phases in a compiler is an important issue. A number of methods integrate the two phases to produce efficient register-allocated instruction schedules [16, 19, 54, 106, 115, 126].

Early work on instruction scheduling related it to the problem of code compaction in microprogramming [49]. This relationship between microprogram compaction and instruction scheduling has been beneficially used in local or basic block scheduling. As pointed out in [132], this mindset remained a serious obstacle to achieving good performance in global scheduling (instruction scheduling beyond basic blocks) until trace scheduling [46] and other approaches were proposed. These latter approaches minimize the execution time of the most likely *trace* or control path, rather than obtaining a compact schedule. Furthermore, the similarities between job shop scheduling [30] and instruction scheduling were also well understood. Instruction scheduling borrows a number of concepts and algorithms from scheduling theory. Many of the theoretical results in instruction scheduling owe their origin to job scheduling.

It should be mentioned here that the objective of this chapter is more to provide an overview of instruction scheduling methods than to provide a comprehensive survey of all existing techniques. The following section presents the necessary background. Simple scheduling methods for covering pipeline delays are discussed in Section 19.3. Subsequently, we describe basic block instruction scheduling methods for VLIW and superscalar processors in Section 19.4. Section 19.5 deals with global scheduling techniques. In Section 19.6, phase ordering issues relating to instruction scheduling and register allocation are presented. Section 19.7 discusses recent research in instruction scheduling. Finally, we provide a concluding summary in Section 19.8.

19.2 Background

In this section we review the relevant background. The following subsection presents a number of definitions. In Section 19.2.2 we describe a representation for data dependences, used by instruction scheduling methods, and its construction. Last, we discuss various performance metrics for instruction scheduling methods in Section 19.2.3.

Before we proceed further, let us clarify the use of various notations in the programming examples. We use t1, t2, and so on to represent temporaries or symbolic registers, and r1, r2, and so on to represent (logical) registers assigned to temporaries. Symbols, such as x, y and a, b represent variables stored in memory locations.

19.2.1 Definitions

Two instructions i1 and i2 are said to be data dependent on each other if they share a common operand (register or memory operand), and the shared operand appears as a destination operand[1] in at least one of the instructions [3, 73, 107]. Consider the following sequence of instructions:

```
i1:   r1 ← load (r2)
i2:   r3 ← r1 + 4
i3:   r1 ← r4 - r5
```

Instruction i2 has r1 as one of its source operands, which is written by i1. This dependence from i1 to i2 is said to be a *flow dependence* or *true data dependence*. Thus, in any legal execution of the above sequence, the operand read of register r1 in instruction i2 must take place after the result value of instruction i1 is written. The dependence between instructions i2 and i3 due to register r1 is an *anti-dependence*. Here, instruction i3 must write the result value in r1 only after i2 has read its operand from r1. Last, there is an *output-dependence* between instructions i1 and i3, where the order in which they write to the destination must be the same as the program order (i.e., i1 before i3) for correct program behavior.

---

[1]If the shared operand appears as a source operand in both instructions, then there is an *input-dependence* between the two instructions. Since an input-dependence does not constrain the execution order, we do not consider this any further in our discussion.

A data dependence could also arise through a memory operand. Detecting such a data dependence accurately at compile time is hard, especially if the memory operands are accessed using indirect addressing modes. The problem becomes harder in the presence of *memory aliasing*, where two or more variables point to the same memory location. As a consequence, a conservative analysis of data dependence must assume a true dependence from each previous store to every subsequent load instruction. For the same reason, there is an anti-dependence from each load to every previous store instruction. Last, there is an output-dependence from each store to all subsequent stores.

Anti- and output-dependences together are referred to as *false dependences*. These dependences arise from the reuse of the same register variable or memory location. By appropriately *renaming* the destination register of i3, that is, using a different destination register, the anti- and output-dependences can be eliminated. If the dependences are analyzed before register allocation, then the code sequence uses only temporaries. Since there is no limit on the number of temporaries that can be used, anti- and output-dependences on the temporaries do not normally occur. However, anti- and output-dependences on memory variables accessed through load and store operations can still occur.

A *basic block* is a region of straight line code [3, 73, 107]. The execution control, also referred to as control flow, enters a basic block at the beginning, that is, the first instruction in the basic block, and exits at the end, that is, the last instruction. A control flow transfer or jump occurs from one program point to another because of control transfer instructions such as branch, procedure call, and return.

The control flow in a program is represented by a *control flow graph* whose nodes represent basic blocks. There is an arc between two blocks if a control transfer between them is possible. An instruction i is said to be control dependent on a conditional branch instruction b (or the predicate associated with it) if the outcome of the conditional branch determines whether or not instruction i is executed. For the sequence

```
b1:  if (t1 > 0) goto i2
i1:  t2 ← t3 + t4
i2:  t2 ← t3 - t4
```

instructions use *temporaries* or symbolic registers. Instruction i1 is executed only if the condition associated with b1 evaluates to false. Thus, instruction i1 is control dependent on b1, whereas instruction i2, which is executed irrespective of what b1 evaluates to, is *control independent*.

Last, we informally define the notion of the live range of a variable or a temporary that is used in register allocation. A variable or a temporary is said to be *defined* when it is assigned a value, that is, when the variable or temporary is the destination of an instruction. A variable is said to be *used* when it appears as a source operand in an instruction. The *last use* of a variable is a program point or instruction where the variable is used for the last time in the program or used for the last time before it is redefined at a subsequent program point. The *live range* of a variable starts from its definition point and ends at its last use. A variable is said to be *live* during its live range. For the example code shown in Figure 19.1a,

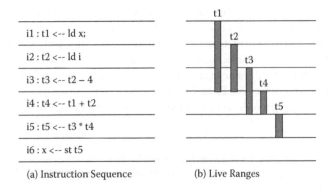

i1 : t1 <-- ld x;	
i2 : t2 <-- ld i	
i3 : t3 <-- t2 – 4	
i4 : t4 <-- t1 + t2	
i5 : t5 <-- t3 * t4	
i6 : x <-- st t5	
(a) Instruction Sequence	(b) Live Ranges

FIGURE 19.1   An example code sequence and live ranges.

the live ranges are depicted in Figure 19.1b. The temporary t2 is defined in instruction i2, and its last use is at i4. Thus, the live range of t2 is from i2 to i4. We follow the convention that the live range ends just before the last use so that the last use instruction can reuse the same register as the destination register.

When the live ranges of two variables do not overlap, they can share the same register. The number of variables that are live simultaneously indicates the number of registers that would be required. Informally, the number of simultaneously live variables is referred to as the *register pressure* at a given program point. For our example, assuming that no other variable is live into this basic block, the number of simultaneously live variables is three at instruction i3. When the number of available registers is less than the number of simultaneously live variables, the register allocation phase decides which variables or temporaries do not reside in registers. Load and store instructions are introduced, respectively, to load these temporaries from memory when necessary and spill them to memory locations subsequently. This is referred to as *register spills*, and the load and store instructions are referred to as *spill code*.

19.2.2 Directed Acyclic Graph

The data dependence among instructions in an instruction sequence can be represented by means of a *dependence graph*. The nodes of the dependence graph represent the instructions, and a directed edge between a pair of nodes represents a data dependence. The dependence graph for instructions in a basic block is acyclic. Such a dependence graph is termed a *directed acyclic graph* (DAG). In a DAG, node v is said to be a *successor* or *immediate successor* of u if there exists an edge (u, v). Similarly, node u is the predecessor (or immediate predecessor) of node v if there exists an edge (u, v) in the DAG. We use the term *descendents* to refer to all nodes that can be reached from a node.

Next, let us discuss DAG construction for a basic block. A DAG can be constructed either in a forward pass or in a backward pass of the basic block [152]. In a forward pass method, at each step a new node corresponding to an instruction in the sequence is added to the graph. By comparing against all previous instructions, the dependences among the instructions are determined, and appropriate dependence arcs between the corresponding nodes are added. This approach requires $O(n^2)$ steps, where n is the number of instructions. The dependences among instructions can also be determined using a table building approach where a list of definitions and current uses is maintained [152]. The dependences checked for could be through general-purpose registers (or temporaries), memory locations, and special-purpose registers such as condition code registers.

Consider the example code given in Figure 19.2a. The DAG for this example code sequence is shown in Figure 19.2b. (Often a DAG is also drawn in a bottom-up manner [3]. A DAG drawn in this manner is shown in Figure 19.7.) Since this code uses temporaries rather than register values, there are no anti- and output-dependences through register variables or temporaries. The dependence arc from node i2 to i9 is due to an anti-dependence on memory variable b. As the dependences due to memory locations could not be analyzed accurately, that is, the memory references could not be *disambiguated*, the anti-dependence arcs (i1,i8), (i1,i9), and (i2,i8) are also added. These anti-dependence edges are represented as broken lines in the figure. Finally, in the absence of memory disambiguation, there is also an output-dependence from i8 to i9. The output-dependence arc is indicated by a dash–dot line.

Each arc (u, v) in the dependence graph is associated with a weight that is the execution latency of u. In a DAG, a node that has no incoming arc is referred to as a *source node*. A node that has no outgoing arc is termed as a *sink node*. There could be multiple source nodes and sink nodes in a DAG. In our example DAG, nodes i1 and i2 are source nodes and i8 and i9 are sink nodes. For convenience, it is typical to add a fictitious source node, and edges from this node to every other node in the DAG are introduced. This fictitious node is henceforth referred to as the source node in the DAG. Similarly, there is a fictitious sink node, referred to as the sink node. Edges are added from each node in the DAG to the sink node. A weight 0 is associated with each of these newly introduced edges from the source node or to the sink node. To avoid clutter in the figure, we do not normally show the fictitious source and sink nodes and the associated edges.

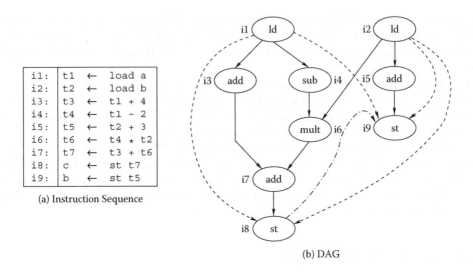

i1:	t1	←	load a
i2:	t2	←	load b
i3:	t3	←	t1 + 4
i4:	t4	←	t1 - 2
i5:	t5	←	t2 + 3
i6:	t6	←	t4 * t2
i7:	t7	←	t3 + t6
i8:	c	←	st t7
i9:	b	←	st t5

(a) Instruction Sequence

(b) DAG

FIGURE 19.2   An instruction sequence and its dependence graph.

A path in a DAG is said to be a *critical path* if the sum of the weights associated with the arcs in this path is (one of) the maximum among all paths. In our example, if the execution latency of each instruction is one cycle, then the path involving nodes i1, i4, i6, i7, and i8 is a critical path. Instructions in the critical path need to be given higher priority while scheduling, so that the execution time of the schedule is reduced.

19.2.3   Performance Metrics for Scheduling Methods

Typically, the objective of an instruction scheduling method is to reduce the execution time of a schedule, also referred to as the *schedule length*. Among the different scheduling methods, one that achieves the shortest schedule length is said to have the best performance or the best-quality schedule. A schedule with the shortest schedule length is referred to as an *optimal schedule* (in terms of schedule length). Note that the schedule length or execution time of a schedule is a static measure, as it refers to the execution of the schedule once. The overall execution time of a program is a dynamic measure that is also of interest.

Since obtaining an *optimal schedule* is an NP-complete problem [90, 120], the time taken to construct a schedule, referred to as the *schedule time*, is also an important performance metric. Scheduling methods that have unacceptably long schedule times could not be used in a production compiler. Often, a scheduling method in a production compiler is expected to produce a reasonable-quality schedule for a basic block within a few milliseconds. However, in certain application domains, such as digital signal processing, or in embedded applications, where the code is compiled once (at design time) and run subsequently, the schedule time is less of a constraint; in these applications, the compilation process itself may take several hours.

Apart from schedule length and schedule time, a performance metric that is of interest in an instruction scheduling method is the register pressure of the constructed schedule. Schedules with higher register pressure are likely to incur more register spills, which, in turn, may increase the schedule length. Thus, besides a lower execution time, a schedule with lower register pressure is often preferred. Many global scheduling methods, which schedule instructions beyond basic blocks, often cause an increase in the code size. Hence, the code size of the scheduled code is another metric that is of interest when comparing different schedules. Code size is especially important in embedded applications, where an increase in code size increases the on-chip program memory, which, in turn, increases the system cost.

Last, in embedded systems [123], power dissipated or energy consumed by the schedule is critical. Schedules which consume lower power without incurring significant performance penalty, in terms of execution time, are often preferred in embedded applications.

The initial sections of this chapter focus on instruction scheduling methods for high performance, where schedule length is the main performance metric. In Section 19.6 we discuss issues relating to register pressure. Finally, in Section 19.7 we discuss instruction scheduling methods for a specific application domain, namely digital signal processing, and for low-power embedded applications, where code size, power, and performance are important.

19.3   Instruction Scheduling for RISC Architectures

In this section we discuss early instruction scheduling methods proposed for handling pipeline stalls. First we present a simple, generic architecture model and the need for instruction scheduling in this architecture model. In Section 19.3.2 we present the instruction scheduling method developed by Gibbons and Muchnick [52] in detail. An alternative approach that combines register allocation and scheduling for pipeline stalls is discussed in Section 19.3.3. Last, a brief review of other instruction scheduling methods is presented in Section 19.3.4.

19.3.1   Architecture Model

In a simple RISC architecture instructions are executed in a pipelined manner. Instruction execution in a simple five-stage RISC pipeline is shown in Figure 19.3. Briefly, the instruction fetch (IF) stage is responsible for fetching instructions from memory. Instruction decode and operand fetch takes place in the decode (ID) stage. In the execute stage (EX), the specified operation is performed; for memory operations, such as load or store, address calculation takes place in this stage. The memory stage (MEM) is for load and store operations. Finally in the write-back stage (WB), the result of an arithmetic instruction or the value loaded from memory for a load instruction is stored in the destination register.

Let instruction (i+1) be dependent on instruction i; that is, (i+1) reads the result produced by instruction i. It can be seen that instruction i+1 may read the operand value (in the ID stage) before instruction i completes, that is, before i finishes the write-back in the destination register. This dependence should cause the execution of instruction (i+1) to stall until instruction i writes the result, to ensure correct program behavior. This is known as a *data hazard* [68].

Another situation that may warrant stalls in the pipeline is due to control hazards [68]. If instruction i is a control transfer instruction, such as a conditional branch, unconditional branch, subroutine call, or

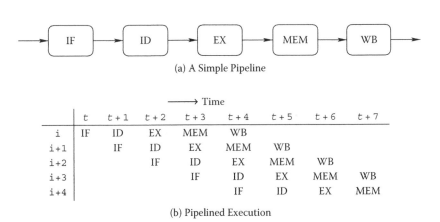

(a) A Simple Pipeline

	t	$t+1$	$t+2$	$t+3$	$t+4$	$t+5$	$t+6$	$t+7$
i	IF	ID	EX	MEM	WB			
i+1		IF	ID	EX	MEM	WB		
i+2			IF	ID	EX	MEM	WB	
i+3				IF	ID	EX	MEM	WB
i+4					IF	ID	EX	MEM

(b) Pipelined Execution

FIGURE 19.3   Instruction execution in a five-stage pipeline.

return instruction, then the subsequent instruction to be fetched may not be the instruction immediately following i. The location or *target address* from where the next instruction has to be fetched is not known until instruction i completes execution. Thus fetch, decode, and execution of subsequent instructions should be stalled until the control transfer instruction is complete.

In either situation, that is, when we have a data or control hazard, the pipeline needs to be stalled to ensure correct program execution. The pipeline may include hardware support, referred to as *pipeline interlock*, to detect such occurrences and stall the subsequent instructions appropriately [68]. With a pipeline interlock, the dependent instruction and all subsequent instructions are stalled for a few cycles. The number of stall cycles required depends on the latency of instruction i, pipeline implementation, as well as whether or not other hardware mechanisms, such as *result forwarding* or *bypassing*, exist [68]. Pipeline forwarding reduces the number of stalls required. In a typical pipeline only certain pairs of consecutive instructions cause such a stall, and typically these stalls are for one or two cycles, except in cases where there is a dependency from either a multi-cycle operation such as a floating point multiply or a memory load that causes a data cache miss.

In processors that do not have pipeline interlocks, for example, in the MIPS R2000 processor [78], either the compiler or the programmer has to explicitly introduce no-op (no operation) instruction(s) to ensure correct program execution. Alternatively, either the compiler or the programmer could reorder instructions, preserving data dependences, such that dependent instructions appear a few instructions apart. Such reordering would be useful for architectures with or without pipeline interlock hardware, as it avoids pipeline stalls. It is easy to see how instruction reordering is useful in avoiding stalls due to data dependence. For control hazard stalls, reordering is useful only if the pipeline supports *delayed branching*, a common feature in most of the RISC pipelines [68]. Under delayed branching, a few (typically one or two) instructions following the branch are executed irrespective of the control transfer. The instructions following a branch or control transfer instruction are said to occupy the branch *delay slots*. The instructions that appear in the delay slot must preserve both control and data dependences. If such instructions cannot be found, the delay slots should be filled with no-op instructions.

19.3.2   A Simple Instruction Scheduling Method

Optimal instruction scheduling to minimize the number of stalls under arbitrary pipeline constraints is known to be an NP-complete problem [65, 90, 120]. Several heuristic methods have been proposed in the literature [52, 65, 66, 120, 127, 170]. All these methods deal with instruction reordering within a basic block. We shall discuss two of the methods (developed by Gibbons and Muchnick [52] and Proebsting and Fischer [127]) in detail and compare the rest.

The method proposed by Gibbons and Muchnick [52] assumes that (a) there must be one cycle stall between a load and an immediately following dependent instruction (which uses the loaded value) and (b) the architecture has hardware interlocks. Thus, the goal of the scheduling method is to reduce the pipeline stalls as far as possible; it is neither mandatory to remove all the stalls nor necessary to insert no-ops where stalls could not be avoided.

The instruction reordering must preserve the data dependences present in the original instruction sequence. The dependences among instructions in a basic block are represented by a DAG.

Consider the evaluation of a simple statement:

$$d = (a + b) * (a + b - c)$$

The code sequence to compute the expression on a RISC architecture is shown in Figure 19.4a. The DAG for the sequence of instructions is shown in Figure 19.4b. The given instruction order incurs two stall cycles: one at instruction i3, as i3 immediately follows a dependent load i2, and another at instruction i5, which immediately follows the dependent load instruction i4. We shall now discuss how Gibbons and Muchnick's method obtains an instruction schedule with a reduced number of stalls while preserving program dependences.

If instructions in the basic block are scheduled in the topological order of the DAG, the dependences are preserved. An instruction is said to be *ready* if all its immediate predecessors in the DAG have been

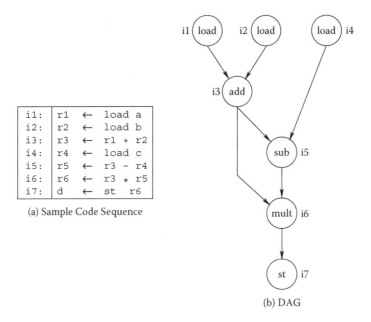

i1:	r1	←	load a
i2:	r2	←	load b
i3:	r3	←	r1 + r2
i4:	r4	←	load c
i5:	r5	←	r3 - r4
i6:	r6	←	r3 * r5
i7:	d	←	st r6

(a) Sample Code Sequence

(b) DAG

FIGURE 19.4 Instruction sequence for d = (a+b)*(a+b-c) and its DAG.

scheduled. Ready instructions are kept in a ReadyList. Among the instructions from the ReadyList, the "best" instruction is selected based on the following two guidelines:

1. An instruction that will not interlock with the one just scheduled.
2. An instruction that is most likely to cause interlock with the instructions after it.

While the first guideline tries to reduce the stalls, the second guideline attempts to schedule early those instructions that may cause stalls, so that there may be a wider choice of instructions to follow them. In addition to these guidelines, the method uses three heuristics, applied in the specified order, to select the next instruction to be scheduled:

1. Choose an instruction that causes a stall with any of its immediate successors in the DAG; this is somewhat similar to guideline 2 stated above.
2. Choose an instruction that has the largest number of immediate successors in the DAG, as this will potentially make many successor instructions ready.
3. Schedule an instruction that is farthest from the sink node in the DAG. This will enable the schedule process to be balanced among various paths toward the sink node.

After scheduling each instruction, the list of ready instructions is updated, including any successor instruction that has now become ready. The scheduling process proceeds in this way until all instructions in the basic block are scheduled. The scheduling algorithm is shown in Figure 19.5. The schedule generated by the above method for the example is depicted in Figure 19.6. This schedule incurs no stalls, as none of the load instructions are immediately followed by a dependent (arithmetic) instruction.

The worst-case complexity of the instruction scheduling method is $O(n^2)$. This happens in the degenerated case when all remaining instructions are in the ReadyList at each time step.

19.3.3 Combined Code Generation and Register Allocation Method

Proebsting and Fischer propose a linear time code scheduling algorithm, which integrates code scheduling and register allocation for a simple RISC architecture [127]. The scheduling method, known as the *delayed load scheduling* (DLS) method, assumes that the leaf nodes are memory loads. It produces *optimal code*,

```
Input:The DAG for the basic block
Output:The reordered instruction sequence.

From the ReadyList of instructions by including all source nodes
while (there are instructions to be scheduled in the DAG) do
{
    choose a ready instruction based on guidelines (i) and (ii)
      and also based on the static heuristics (a) -- (c);
    schedule the instruction, and remove it from ReadyList;
    add newly enabled instructions to ReadyList;
{
```

FIGURE 19.5   Gibbons–Muchnick scheduling method.

in terms of both the schedule length and the number of registers used, for a restricted architecture when the delay stall due to load instructions is one cycle and when the DAG is a tree. The method produces an efficient, near-optimal, schedule for the general case, i.e., when the delay is greater than one cycle or when the dependence graph is a DAG.

The DLS method is an adaptation of the Sethi–Ullman method [3, 144] for generating code (instruction sequence) for basic blocks from its DAG representation. To understand the DLS method, first we shall explain the Sethi–Ullman method, henceforth referred to as the SU method, with the help of an example. Unlike the RISC instruction scheduling method discussed in Section 19.3.2, the SU and DLS methods are applied at the time of code generation. These methods are applied to low-level intermediate form, typically the 3-address code, and can produce code sequence with register allocation. Hence, these methods can be considered integrated methods for code generation and register allocation.

19.3.3.1   The Sethi–Ullman Method

The objective of the SU method is to generate a minimum-length code sequence for an expression tree. The generated code sequence also requires the minimum number of temporaries. The SU method does not deal with pipeline stalls and hence may schedule a dependent instruction immediately after the load instruction on which it is dependent. Furthermore, the SU method considers basic blocks with no live-in registers; that is, all values are available in memory. Hence, the DAG representation of the basic block consists of leaf nodes that correspond to memory values, which must be loaded in registers (through load instructions) to perform any operation on them. The interior nodes of a DAG are all arithmetic operations.

Let us once again consider the basic block for the statement

$$d = (a + b) * (a + b - c)$$

The 3-address code for the basic block is shown in Figure 19.7a. This code sequence uses temporaries and is unoptimized. In the 3-address code, the evaluation of (a+b) takes place twice. This would be eliminated by common subexpression elimination optimization. Hence, we call the 3-address code given here unoptimized code. Without common subexpression elimination, the DAG for the basic block is a tree (as shown in Figure 19.7b).

i1:	r1	←	load a
i2:	r2	←	load b
i4:	r4	←	load c
i3:	r3	←	r1 + r2
i5:	r5	←	r3 − r4
i6:	r6	←	r3 * r5
i7:	d	←	st r6

FIGURE 19.6   A schedule for the instruction sequence in Figure 19.4.

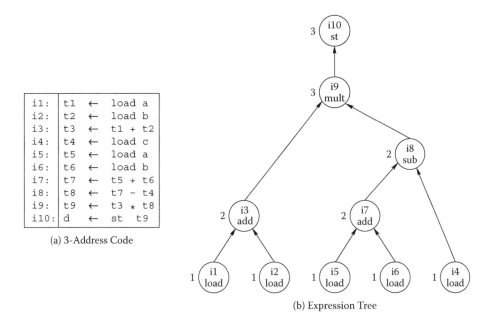

```
i1:  | t1  ←  load a
i2:  | t2  ←  load b
i3:  | t3  ←  t1 + t2
i4:  | t4  ←  load c
i5:  | t5  ←  load a
i6:  | t6  ←  load b
i7:  | t7  ←  t5 + t6
i8:  | t8  ←  t7 - t4
i9:  | t9  ←  t3 * t8
i10: | d   ←  st   t9
```

(a) 3-Address Code

(b) Expression Tree

FIGURE 19.7   3-Address code and expression tree for d = (a+b)*(a+b-c).

The code sequence shown in Figure 19.8a has a higher register pressure. A register-allocated code for this sequence requires four registers. If the architecture has fewer than four registers, a few values need to be spilled and reloaded at subsequent points in the computation. The spill loads and stores will increase the length of the code sequence. The SU method tries to minimize the length of the code sequence and the number of registers used when the dependence graph is a DAG. The method finds an optimal solution, optimal in terms of code length and number of registers used, when the dependence graph is a tree. An optimal solution for the DAG is shown in Figure 19.8b. Let us describe the method for the special case when the DAG is a tree.

The SU method has two phases: the first phase assigns a *label* for each node of the tree, and the second phase is a tree traversal that generates code as the nodes/subtrees of the DAG are visited. Intuitively, the label of a node represents the number of registers or temporaries that are needed to compute the subtree rooted under the node. In the tree traversal phase, a subtree is completely traversed before proceeding to other sibling subtrees. The traversal order among the siblings is based on the labels, and the node with a larger label is visited first. Thus, the method generates code first for the subtree that requires more registers.

```
i1:  | r1  ←  load a
i2:  | r2  ←  load b
i3:  | r1  ←  r1 + r2
i4:  | r2  ←  load c
i5:  | r3  ←  load a
i6:  | r4  ←  load b
i7:  | r3  ←  r3 + r4
i8:  | r2  ←  r3 - r2
i9:  | r1  ←  r1 * r2
i10: | d   ←  st   r1
```

(a) Code Sequence Using
4 Registers

```
i5:  | r1  ←  load a
i6:  | r2  ←  load b
i7:  | r1  ←  r1 + r2
i4:  | r2  ←  load c
i8:  | r1  ←  r1 - r2
i1:  | r2  ←  load a
i2:  | r3  ←  load b
i3:  | r2  ←  r2 + r3
i9:  | r1  ←  r1 * r2
i10: | d   ←  st   r1
```

(b) Optimal Code Sequence
with 3 Registers

FIGURE 19.8   Register-allocated code sequences.

The labeling phase traverses the tree in post-order, visiting all children before visiting a node. The label assigned for a node n corresponding to a binary operator is given by

$$label(n) = \begin{cases} \max(l_1, l_2) & \text{if } l_1 \neq l_2 \\ l_1 + 1 & \text{if } l_1 = l_2 \end{cases}$$

where l_1 and l_2 are the labels assigned to the left and right children of n. The label assignments for the nodes are shown in Figure 19.7b.

The label assigned to a nonbinary operator n, having k children, with $l_1, l_2, \ldots, l_k$ as the labels of the children nodes arranged in nonincreasing order, is

$$label(n) = \max_{1 \leq i \leq k} (l_i + i - 1)$$

Intuitively, when there are r children with the same label l_i (i.e., requiring the same number l_i registers for computing the subtrees rooted under them), then $l_i + r - 1$ registers are required for computing the subtree rooted under the parent node.

Each leaf node is assigned the label 1. For architectures supporting register–memory operands, only the leftmost child of each node that is a leaf node need be assigned the label 1. Other leaf nodes are assigned the label 0. The intuitive reasoning behind this is that the operand corresponding to the right child (in the case of a binary operator) that is a leaf node can be used directly as a memory operand (without requiring it to be loaded in a register) in the instruction corresponding to the parent node. Whereas for RISC architectures where all arithmetic operators have only register operands, all leaf nodes that are memory locations must first be loaded in a register. Hence, they are assigned a label 1.

Next we describe the second phase of the SU method, the generate code sequence phase. This phase traverses the labeled tree recursively, first generating the code for the child with the higher label. When the children have the same label value, they can be traversed in any order. The method maintains a register stack of the available registers. If the node visited is a leaf node (or a node with a label 1), then the code, a load instruction, is generated for the node. The register on the top of the stack is used as the destination register. For machines that support register–memory operands, a load instruction is generated only for the leftmost child that has a label value 1. The right child (assuming the parent node to be a binary operator) can be used directly as a memory operand when the code for the parent node is generated. Since we concentrate on RISC architectures with register–register operands in this discussion, we omit further details for complex instruction set computer (CISC) architectures. The interested reader is referred to [3, 144].

The code for an interior node is generated by emitting the corresponding instruction, with the source operands the same as the destination operand of the children nodes, and the destination register the same as that of the left child. In the special case when the DAG is a tree, while generating the code for the parent node, it is always possible to free the destination registers of the children nodes. This is because the value produced by a node is used only by its parent, and in a tree, there is at most one parent for each node. The destination registers of the right children can be freed and are pushed into the register stack so that they can be reused subsequently. The original algorithm also uses a swap operation on the register stack to ensure that a left child and its parents are evaluated on the same register. We leave out these details in our discussion. Last, since the code is generated by traversing the tree recursively, generating code for the children nodes before generating the code for a parent node, the dependences are easily satisfied. For the tree used in our motivating example, the instruction sequence generated by the SU method is shown in Figure 19.8b. This sequence uses three registers.

In the general case where the dependence graph is a DAG (and not a tree), the optimal code generation problem is known to be NP-complete [2]. A near-optimal solution is obtained by partitioning the DAG into trees. A node with more than one parent is termed a shared node. The partitioning is done in such a way that each root and/or shared node forms the root of the tree, with the maximal subtree that includes no shared nodes except as leaves. Shared nodes with more than one parent can be turned into as many leaves as necessary. More details can be found in [3, 144].

```
i5:   r1  ←  load a
i6:   r2  ←  load b
i7:   r1  ←  r1 + r2   % 1 stall
i4:   r2  ←  load c
i8:   r1  ←  r1 - r2   % 1 stall
i1:   r2  ←  load a
i2:   r3  ←  load b
i3:   r2  ←  r2 + r3   % 1 stall
i9:   r1  ←  r1 * r2
i10:  d   ←  st  r1
```

```
i5:   r1  ←  load a
i6:   r2  ←  load b
i4:   r3  ←  load c
i1:   r4  ←  load a
i7:   r1  ←  r1 + r2
i2:   r2  ←  load b
i8:   r1  ←  r1 - r3
i3:   r4  ←  r4 + r2
i9:   r1  ←  r1 * r4
i10:  d   ←  st  r1
```

(a) Stalls in Sethi–Ullman Sequence   (b) DLS Sequence with No Stalls

FIGURE 19.9   Instruction sequences with and without stalls.

In the expression used in our example, (a+b) is common to the left and right subtrees of the expression tree. More specifically, the subtrees rooted on i3 and i7 compute the same expression. On performing common subexpression elimination, one of these subtrees is eliminated, resulting in a DAG. By splitting the DAG into a set of subtrees, it is possible to obtain a code sequence using three registers.

19.3.3.2   DLS Method

Next we shall discuss the DLS method, which integrates code generation and register allocation for pipelined architectures that incur delays [127]. As before, we shall first discuss the method when the DAG is a tree. The main idea behind the DLS method is to produce a canonical form of the instruction sequence. Suppose the sequence consists of L memory load instructions (referred to as *loads*) and $(L - 1)$ arithmetic operations[2] (referred to as *operations*) and uses R registers. Then the canonical form consists of R load instructions followed by an alternating sequence of $(L - R)$ ⟨operation, load⟩ pairs, followed by $(R - 1)$ operations. The canonical form is generated from the sequence generated by the SU method.

The DLS method is a three-phase method, starting with a labeling phase where the nodes in the expression tree are assigned the *minReg* value. The *minReg* value is the label given by the labeling phase of the SU method. The second phase, *order*, generates the relative order of *loads* and *operations* in two separate data structures. The ordering is accomplished by recursively generating the order for the left and right subtrees, starting with the one that has the higher register requirement. The ordering phase is similar to that in the SU method discussed in the previous subsection. The schedule phase also assigns registers for all the instructions as discussed in the SU method. The number of registers R used is one more than the *minReg* value of the root of the expression tree.

Last, the *schedule* phase of the DLS method essentially generates the canonical order, listing R loads followed by an alternating sequence of ⟨operation,load⟩ pairs. The DLS method generates an optimal instruction sequence without any pipeline stall except in two special cases: (a) where the expression tree consists of a single *load* and (b) where the expression tree consists of a single *operation* and two leaf nodes (*load* instructions).

Let us consider the example expression introduced in Section 19.3.3.1. An optimal schedule for this code, considering no pipeline delays, is shown in Figure 19.8b. Now let us consider an architecture with delay $D = 1$ between a load and a dependent instruction. In this example, we have $L = 5$ loads, $(L - 1) = 4$ (binary) operations, and a unary operation (store). The sequence obtained from the SU method causes three stall cycles, at instructions i7, i8, and i3 as shown in Figure 19.9a. The SU schedule uses three registers. To avoid the stalls completely, the DLS sequence uses one more register than in SU schedules; hence $R = 3 + 1 = 4$. The sequence shown in Figure 19.9b is obtained from the DLS method, which

---

[2]There are $(L - 1)$ (binary) operations in a binary tree with L leaf nodes.

incurs no stall cycle. In this sequence, initially we have $R = 4$ loads, followed by an alternating sequence of $(L - R) = (5 - 4) = 1$ ⟨*operation, load*⟩ pair, followed by four arithmetic operations.[3] This sequence is optimal in terms of the number of execution cycles. It uses one more register than is used by the SU method but completely avoids all stall cycles. It should be noted here that among the sequences that incur the lowest execution cycles, this sequence uses the minimum number of registers; that is, no other instruction sequence incurs the minimum number of execution cycles and requires fewer registers.

The complexity of the DLS method is $O(n)$, as the labeling and ordering phases can be performed by traversing through the nodes once (bottom-up), and the schedule phase visits each node exactly once. Not only is this superior to the $O(n^2)$ complexity of the instruction scheduling method developed by Gibbons and Muchnick [52] (discussed in Section 19.3.2), but it performs scheduling and register allocation together in a single framework. The DLS method also serves as an excellent heuristic when the dependence graph is an arbitrary DAG or when the delay is greater than 1. Last, recall that the DLS method requires that the leaf nodes be memory load operations. This precludes register variables, or live-in registers, in the expression tree. An extension that relaxes this constraint is presented in [87, 166].

19.3.4   Other Pipeline Scheduling Methods

A heuristic pipeline scheduling method that is performed during the code generation was implemented in the PL-8 compiler [11]. A major advantage of this method is that it performs code scheduling before register allocation and hence avoids false dependences. Hennessy and Gross describe a heuristic method that is applied after code generation and register allocation [65, 66]. This method uses a dependence graph representation that eliminates false dependences. However, to accomplish this, their method needs to check for deadlocks in scheduling. It uses a look-ahead window to avoid deadlock. The worst-case running time of this method is $O(n^4)$.

Last, we briefly discuss filling the delay slot of a branch instruction. If a processor supports delayed branching [68], moving independent instructions in the branch delay slots helps reduce the number of control hazard stalls. Some of the instruction scheduling methods discussed in this section, e.g., the Gibbons–Muchnick method [52] and the Hennessy–Gross method [65], can be used to fill the branch delay slot with a useful instruction. It is best to fill the branch delay slot with an instruction from the basic block that the branch terminates. Otherwise, an instruction from either the target block (of the branch) or the fall-through block, whichever is most likely to be executed, is selected to be placed in the delay slot. The selected instruction either occurs as a source node in both (target and fall-through) basic blocks, has a destination register that is not live-in in the other block, or has a destination register that can be renamed.

The instruction scheduling methods discussed in this section do not consider (functional unit) resource constraints. They merely try to reorder instructions to reduce the no-op instructions or the pipeline stalls needed to ensure correct program behavior. In contrast, the instruction scheduling methods to be discussed in the following section for VLIW and superscalar architectures take into consideration the resource constraints.

19.4   Basic Block Scheduling

Instruction scheduling can be broadly classified based on whether the scheduling is for a single basic block, multiple basic blocks, or control flow loops involving single or multiple basic blocks [132]. Algorithms that schedule single basic blocks are termed *local scheduling* algorithms and are the topic of discussion in this section. Algorithms that deal with multiple basic blocks or basic blocks with cyclic control flow are

---

[3]In this sequence, because of the additional unary store operation, we have R, instead of $(R - 1)$, arithmetic operations at the end of the sequence.

termed *global scheduling* algorithms and are dealt with in Section 19.5. The term *cyclic scheduling* is used to refer to methods that schedule single or multiple basic blocks with cyclic control flow. Cyclic scheduling overlaps the execution of multiple instances of a static basic block corresponding to different iterations.

In this section we discuss local or basic block scheduling methods for VLIW and superscalar processors. First we present the necessary preliminaries in the following section. In Section 19.4.2 we present the basic list scheduling algorithm. Operation-based instruction scheduling methods are discussed in Section 19.4.3. An exact approach to obtain an optimal schedule using an integer linear programming formulation is presented in Section 19.4.4. Section 19.4.5 deals with resource usage models that are used in instruction scheduling methods. We present a few case studies in Section 19.4.6.

19.4.1 Preliminaries

With the advent of multiple instruction issue processors, namely superscalar processors [68,77,151] and VLIW architectures [47, 136], it has become important to expose ILP at compile time. Both VLIW and superscalar architectures have a number of functional units and are capable of executing multiple independent operations in a single cycle. Hence, instruction scheduling for these architectures must identify the instructions that can be executed in parallel in the same cycle. In a VLIW architecture the identification of independent instructions, and their reordering to expose ILP, must be done at compile time. Multiple independent instructions and operations that can be issued in the same cycle should be packed in a single long word instruction for a VLIW architecture.

Superscalar processors provide hardware mechanisms to detect dependences between instructions at runtime and to schedule multiple independent instructions in a single cycle. *In-order* issue superscalar processors are capable of issuing multiple independent instructions in each cycle; however, once they encounter an instruction for which the source operand(s) is (are) not yet ready (the instruction[s] producing the source operand[s] has [have] not completed execution), the dependent instruction as well as all future instructions are stalled until the dependent instruction becomes data ready. *Out-of-order* issue superscalar processors, however, are capable of issuing independent instructions, even beyond a dependent stalled instruction. In other words, they can issue instructions out of the order in which they appear in the program. Both in-order and out-of-order issue processors benefit by a runtime register-renaming mechanism [151] that helps eliminate false dependences (anti- and output-dependences). Because of the hardware support available in superscalar processors, instruction reordering to expose ILP is not mandatory, although such a reordering would certainly benefit both in-order and out-of-order superscalar processors. This is especially the case for in-order issue superscalar processors.

The instruction schedule constructed for a VLIW or a superscalar processor must satisfy both dependence constraints and resource constraints. Dependence constraints ensure that an instruction is not executed until all the instructions on which it is dependent are scheduled and their executions are complete. Once again, dependences among instructions are represented by means of a DAG. Since local instruction scheduling deals only with basic blocks, the dependence graph will be acyclic. In our discussion we shall assume that register allocation is performed after instruction scheduling and that the DAG is constructed from an instruction sequence that uses temporaries (rather than registers) and hence avoids all anti- and output-dependence arcs for all non-memory operations.

Resource constraints ensure that the constructed schedule does not require more resources (functional units) than available in the architecture. The resource usage model in a realistic instruction scheduling method needs to take into consideration the finite resources available in the architecture, the actual (non-unit) execution latencies incurred by some of the instructions, and simple or complex resource usage patterns. In a simple resource usage pattern, each instruction uses one resource for a single cycle, so multiple instructions scheduled on different cycles do not cause any structural hazard. With such a simple resource usage pattern, it is possible to schedule a new instruction in the functional unit in each cycle, whereas when the resource usage pattern is complex, a single resource could be used for multiple cycles.

The resource usage model specifies the usage pattern of resources for different classes of instructions (such as integer, load/store, multiply, floating point (FP) add, FP multiply, and FP divide instruction classes) as well as the available functional units. A simple representation for resource usage is a *reservation table*, which is an $r \times \ell$ matrix, where ℓ is the latency of the instruction and r is the number of stages in the functional unit [85]. An entry $R[r, t]$ is 1 if resource r is used t time steps after the initiation of an instruction in the functional unit and 0 otherwise. The resource usage pattern is simple when the functional unit is fully pipelined. A pipelined functional unit can initiate a new operation in every cycle. We defer discussion of complex resource usage and more sophisticated resource usage models to Section 19.4.5.

The resource requirements of a schedule can be modeled using a global resource reservation table (GRT), an $M \times T$ matrix, where M is the number of resources (including all stages of all functional units as well as other resources such as memory ports) whose contention must be explicitly modeled in the schedule, and T is an upper bound on the length of the schedule (i.e., the number of time steps taken by the schedule). An entry $GRT[r, t]$ is either 1 or 0, representing whether or not resource r is used at time step t in the current schedule. As a schedule is constructed, the GRT represents the resource requirements of the partial schedule of instructions that are already scheduled. Any new instruction scheduled should not cause a *resource contention*, that is, two or more instructions requiring the same resource at the same time step, with the partial schedule. Resource contention is checked by a contention query model that checks whether scheduling an instruction at time step t causes any conflicting resource requirements with instructions that are already scheduled. We shall return to the contention query model in greater detail in Section 19.4.5.

An instruction scheduling method assumes a fixed execution latency for each instruction. However, this does not cover variations in latency caused by events like cache misses. A schedule constructed with an underestimated or optimistic value of the latency may cause unnecessary stalls when a cache miss occurs. This would be the case even though there may be enough parallelism in the basic block to hide the latency. However, when a pessimistic latency value is used, the schedules are unnecessarily stretched, even for cache hit cases. Balanced scheduling [81] and improved balanced scheduling methods [99] generate schedules that can adapt more readily to uncertainties in memory latencies. These methods use a load latency estimate that is based on the number of independent instructions available in the basic block to mask the load latency. All other instruction scheduling methods assume an optimistic estimate for the execution latency, which will be followed in our discussion in the rest of this chapter.

19.4.2   List Scheduling Method

The list scheduling method schedules instructions from time step 0, starting with the source instructions in the basic block. At each time step t, it maintains a list of ready instructions (ReadyList) that are data ready, that is, instructions whose predecessors have already been scheduled and would produce the result value in the destination register by time t. List scheduling is a greedy heuristic method that always schedules a ready instruction in the current time step whenever there is no resource conflict.

A list scheduling algorithm is similar to Gibbons and Muchnick's instruction scheduling method discussed in Section 19.3.2, except that multiple instructions can be scheduled in the same step. The resource requirements of scheduled instructions are maintained in the GRT. At each time step, among the set of ready instructions from the ReadyList, instructions are scheduled one at a time based on a certain priority ordering of instructions. The priorities assigned to different instructions are decided by heuristics. List scheduling methods differ in the way they assign priorities to the instructions. We shall discuss some of the important heuristics that have been used in various instruction scheduling methods in Section 19.4.2.1. It should be noted here that the priorities assigned to instructions can be either static, that is, assigned once and remain constant throughout the instruction scheduling, or dynamic, that is, change during the instruction scheduling and hence require that the priorities of unscheduled instructions be recomputed after scheduling each instruction. Although the basic list scheduling algorithm discussed below assumes a static priority, it can easily be adapted for a heuristic that assigns dynamic priority values.

After scheduling all instructions in the ReadyList that do not cause a resource conflict, the time step is incremented by 1. Any instruction that has now become data ready is included in the ReadyList. The ReadyList is sorted on decreasing order of priority. The scheduling process continues in this way until all the instructions are scheduled. The complete algorithm is shown in Figure 19.10.

```
Input: DAG
Output: Instruction Schedule
AssignPriority (DAG) ; /* assigns priority to each instruction
                        in the DAG based on the priority policy */
ReadyList = source nodes in the DAG;
timestep = 0;
while (there exists an unscheduled instruction in the DAG) do
{
   Sort ReadyList in non-decreasing priority order;
   while (not all instructions in ReadyList are tried)
   {
     pick next instruction i from ReadyList;
     check for resource conflict;
     if (instruction can be scheduled)
     {
        update GRT (i, timestep);
        remove instruction i from ReadyList;
     }
   }
   increment timestep by 1;
   add instructions that have now become data ready in ReadyList;
}
```

FIGURE 19.10   A generic list scheduling algorithm.

We illustrate the list scheduling method with the help of a simple example. Consider the code sequence and its 3-address representation shown in Figures 19.11a and 19.11b, respectively. Its DAG is depicted in Figure 19.11c. Assume that the target architecture consists of two integer functional units, which can execute integer instructions as well as load/store instructions, and one multiply/divide unit. All functional units are fully pipelined. Also assume that the execution latencies of the add, mult, load, and store instructions are, respectively, one, three, two, and one cycles. These latency values also imply that there

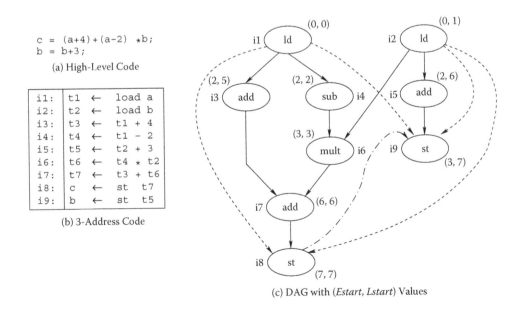

(a) High-Level Code

```
c = (a+4)+(a-2) *b;
b = b+3;
```

(b) 3-Address Code

i1:	t1	←	load a
i2:	t2	←	load b
i3:	t3	←	t1 + 4
i4:	t4	←	t1 - 2
i5:	t5	←	t2 + 3
i6:	t6	←	t4 * t2
i7:	t7	←	t3 + t6
i8:	c	←	st t7
i9:	b	←	st t5

(c) DAG with (*Estart*, *Lstart*) Values

FIGURE 19.11   An example code, 3-address representation, and DAG.

Time	Int. Unit 1	Int. Unit 2	Mult. Unit
0	t1 ← load a	t2 ← load b	
1			
2	t4 ← t1 - 2	t5 ← t2 + 3	
3	t3 ← t1 + 4	b ← st t5	t6 ← t4 * t2
4			
5			
6	t7 ← t6 + t3		
7	c ← st t7		

FIGURE 19.12 Schedule for the example code in Figure 19.11.

should be one stall cycle between a load and a dependent instruction and two stall cycles between a mult and a dependent instruction. There are no stall cycles between an add and a dependent instruction. Furthermore, the path i1 → i4 → i6 → i7 → i8 is the critical path in the DAG.

A list schedule for the example code is shown in Figure 19.12. Instructions on the critical path are scheduled at their earliest possible start time to achieve this schedule, whose length is 8. Note that if the two add instructions (i3 and i5) are scheduled ahead of the sub instruction in time step 2, it would have delayed the execution of instructions on the critical path, namely, i4, i6, i7, and i8 instructions, and hence would have increased the schedule length. To achieve schedules that require fewer execution cycles, the scheduling method should use an efficient heuristic that gives priorities to instructions on the critical path.

The schedule is presented as a parallel schedule, as shown in Figure 19.12, to a VLIW architecture, where multiple instructions that can be executed in the same cycle are packed into a single long word instruction. For a superscalar architecture, the parallel instructions in each cycle are linearized in a simple way, for instance, from left to right. It is shown in [147] that the linearizing order could have a performance impact on out-of-order issue superscalar processors [68, 151]. We defer a discussion of the linearization order to Section 19.6.3.3.

It has been shown that the worst-case performance of a list scheduling method is within twice the optimal schedule [90, 120]. That is, if T_{list} is the execution time of a schedule constructed by a list scheduler, and T_{opt} is the minimal execution time that would be required by any schedule for the given resource constraint, then T_{list}/T_{opt} is less than 2. The quality of list scheduling can degrade and approach the above bound as the number of resources increases and/or the maximum of the latencies of all instructions increases.

It should be noted here that the list scheduling method uses a greedy approach, trying to schedule instructions as soon as possible. If there are enough resources, in the list scheduling method each instruction would get scheduled at the earliest start time (*Estart*) possible. Furthermore, the list scheduling method described above makes a forward pass of the DAG, starting from a source node. It is possible to have a backward pass list scheduling method that schedules instructions starting from the sink node to the source node. While forward pass schedulers attempt to schedule an instruction at the earliest time possible, backward pass schedulers typically attempt to schedule each instruction as late as possible. The instruction scheduler of the GNU C Compiler (version 2) [158] and the local instruction scheduler of the Cydra 5 compiler [36] are backward pass schedulers, while the scheduler in the IBM XL compiler family makes a forward pass [170].

19.4.2.1 Heuristics Used

The list scheduling method uses heuristics to assign priorities to instructions. These priorities are used in selecting the ready instructions for scheduling in each time step. This section briefly discusses some of the heuristics used. An extensive survey and a classification of the various heuristics used in instruction scheduling methods are presented in [152].

A commonly used heuristic is based on the maximum distance of a node to the sink node. The maximum distance (*MaxDistance*) of the sink node to itself is 0. The MaxDistance of a node u is defined as

$$\text{MaxDistance}(u) = \max_{i=1\cdots k}(\text{MaxDistance}(v_i) + \text{weight}(u, v_i))$$

where $v_1, v_2, \ldots, v_k$ are the successors of node u in the DAG. MaxDistance is calculated using a backward pass on the DAG and is a static priority. Priority is given to nodes with larger MaxDistance. A variation of this heuristic is to consider the maximum distance to the sink node, where distance is measured in terms of the path length (number of edges in the path) and not as the sum of execution latencies of the instructions in the path.

Another heuristic used in the list scheduling method is to give priority to instructions that have larger execution latency. The maximum number of children heuristic gives priority to instructions that have more successors. A refinement of this is to consider only successors for which this instruction is the only parent. An alternative is to consider not only the immediate successors, but all the descendents of the instruction. In all these cases, giving higher priority to instructions with more descendents may enable more instructions to be added to the ReadyList. All these heuristics are static in nature and are computationally inexpensive compared to dynamic heuristics.

Many list scheduling methods give higher priority to instructions that have the smallest Estart. The Estart value of the fictitious source node is 0. The Estart value of an instruction v is defined as

$$\text{Estart}(v) = \max_{i=1\cdots k}(\text{Estart}(u_i) + \text{weight}(u_i, v))$$

where $u_1, u_2, \ldots, u_k$ are the predecessors of v. Similarly, priorities can be given to instructions with the smallest *Lstart* (latest start time), which is defined as

$$\text{Lstart}(u) = \min_{i=1\cdots k}(\text{Lstart}(v_i) - \text{weight}(u, v_i))$$

where $v_1, v_2, \ldots, v_k$ are the successors of u. The Lstart value of the sink node is set the same as its Estart value. Estart and Lstart are computed using a forward or a backward pass of the DAG. The Estart and Lstart values of the instructions in our example DAG are also shown in Figure 19.11c.

The difference between Lstart(u) and Estart(u), referred to as *slack* or *mobility*, can also be used to assign priorities to the nodes. Instructions with lower slack are given higher priority. Instructions on the critical path may have a slack 0 and hence get priority over instructions on the off-critical path. The instructions on the critical path of the DAG shown in Figure 19.11, namely i1, i4, i6, and i8, have a slack 0, indicating that there is no slack or freedom in scheduling them.

Many list scheduling methods use Estart and Lstart as static measures, although their values can be calculated after scheduling instructions in each step. Instructions that are scheduled in the current time step may affect the Estart (or Lstart) values of successor (or predecessor) nodes, so these values need to be recomputed in each time step. Slack can also be treated as a static or a dynamic measure. The list scheduling method described in [120] uses a combination of weighted path length and Lstart values.

A heuristic based on computing a *force* metric is used in scheduling data path operations in behavioral synthesis [124]. The *self force* of each instruction u at time step t reflects the effect of an attempted time step assignment t to instruction u on the overall instruction concurrency. The force is positive if the time step causes an increase in the concurrency and negative otherwise. The *predecessor* and *successor forces* refer to the effect of scheduling an instruction at a time step on its predecessors and successors, respectively. The *force* metric is a product, $K \cdot x$, where K represents the extent of concurrency of each type of instruction at a given time step t, and x is a function of the slack of the instruction u. Instructions with the lowest force are given the highest priority. For further details on calculating the force metric, the reader is referred to [124]. Although the objective of the force-directed scheduling method in behavioral synthesis is to minimize the resources while minimizing the execution time, it still obtains an efficient schedule for the instructions. In this sense, the force-directed heuristic may serve as a useful heuristic in an instruction scheduling method as well.

Last, priorities can be given to instructions that define fewer registers (or temporaries), that is, instructions that start fewer new live ranges. Intuitively, it is advantageous to defer the scheduling of an instruction that defines new temporaries to a later time step, as it would defer the increase in register pressure. Such a heuristic is typically used in Prepass scheduling methods. Likewise, it is advantageous to give higher priority to instructions that end the live range of a variable or temporary. Version 2 of the GNU C compiler uses this heuristic [158].

19.4.3   Operation Scheduling Method

While a list scheduling method schedules instructions on a cycle-by-cycle basis, an operation scheduling method attempts to schedule instructions one after another, trying to find the first time step at which each instruction can be scheduled. Operation-driven schedulers sort the instructions in the DAG in a topological order, giving priority to instructions on the critical path [139]. An operation scheduling method could be non-backtracking or backtracking. Here we discuss a backtracking method.

In a backtracking operation scheduling method, at each iteration, an instruction i is selected, based on a certain priority function. An attempt is made to schedule the instruction i at time t, which is between Estart (i) and Lstart (i) and does not cause a resource conflict. The scheduling of an instruction at time step t may affect the Estart and Lstart values of other unscheduled instructions. If dynamic priority is used to select the instruction, the priorities of unscheduled instructions are recomputed.

If there is no time step t between the Estart (i) and Lstart (i) at which the instruction can be scheduled, an already scheduled instruction j, which has conflicting resource requirements with this instruction, is de-scheduled, making room for this instruction. The de-scheduled instruction j is put back in the list of unscheduled instructions and is scheduled subsequently. For the method not to get into a loop where a set of instructions de-schedule each other, a threshold on the number of de-scheduled instructions is kept. When this threshold is exceeded, the partial schedule is discarded and new Lstart values for instructions are computed by increasing the Lstart value of the sink node.

19.4.4   An Optimal Instruction Scheduling Method

The resource-constrained instruction scheduling problem is known to be NP-complete [90, 120]. The instruction scheduling problem has been formulated as an integer linear programming problem [10, 24, 29]. Such an approach is attractive for the evaluation of (performance) bounds that can be achieved by any heuristic method. Also, more recently, Wilken, Liu, and Heffernan [174] have shown that optimal schedules can be obtained in a reasonable time even for large basic blocks, so such an optimal scheduling method can be applied even to production compilers.

In this section we illustrate an integer linear programming formulation for resource-constrained basic block instruction scheduling. We assume a simple resource model in which all functional units are fully pipelined. Altman et al. [8] present methods for modeling functional units with a complex resource usage pattern in an integer linear programming formulation, in the context of software pipelining — an instruction scheduling method for iterative computation [72, 88, 131–133]. Our discussion will consider an architecture consisting of functional units of different types, for example, integer arithmetic and logic unit (ALU), load/store unit, FP add unit, and FP mult/divide units, and the execution latency of instructions in these functional units can be different. We will assume that there are R_r instances in functional unit type r.

Let σ_i represent the time step at which instruction i is scheduled and $d_{(i,j)}$ represent the weight of edge (i, j). To satisfy dependence constraints, for each arc (i, j) in the DAG,

$$\sigma_j \geq \sigma_i + d_{(i,j)} \qquad (19.1)$$

To represent the schedule in a form that can be used in the integer linear programming formulation, a matrix K of size $n \times T$ is used, where n is the number of instructions or nodes in the DAG, and T is an estimate of the (worst-case) execution time of the schedule. Typically, T is the sum of the execution times of all the instructions in the basic block. Note that T is a constant and can be obtained from the DAG. An element of K, for instance, $K[i,t]$, is 1 if instruction i is scheduled at time step t and 0 otherwise. The schedule time σ_i of instruction i can be expressed using K as

$$\sigma_i = k_{i,0} \cdot 0 + k_{i,1} \cdot 1 + \cdots + k_{i,T-1} \cdot (T-1)$$

This can be written in matrix form for all σ_i's as

$$
\begin{bmatrix} \sigma_0 \\ \sigma_1 \\ \vdots \\ \sigma_{n-1} \end{bmatrix}
=
\begin{bmatrix}
k_{0,0} & k_{0,1}, & \cdots & k_{0,T-1} \\
k_{1,0} & k_{1,1} & \cdots & k_{1,T-1} \\
\vdots & \vdots & \vdots & \vdots \\
k_{n-1,0} & k_{n-1,1} & \cdots & k_{n-1,T-1}
\end{bmatrix}
*
\begin{bmatrix} 0 \\ 1 \\ \vdots \\ T-1 \end{bmatrix}
\tag{19.2}
$$

To express that each instruction is scheduled exactly once within the schedule, the constraint

$$\sum_t k_{i,t} = 1, \quad \forall i \tag{19.3}$$

is included in the formulation.

Last, the resource constraint that no more than R_r instructions are scheduled in any time step, where R_r is the number of functional units of type r, can be enforced through the equation

$$\sum_{i \in F(r)} k_{i,t} \leq R_r, \quad \forall t \text{ and } \forall r \tag{19.4}$$

where $F(r)$ represents the set of instructions that can be executed in functional unit type r.

The objective function is to minimize the execution time or schedule length. This can be represented as

$$\text{minimize}(\max_i (\sigma_i + d_{(i,j)}))$$

To express this in linear form, we introduce

$$z \geq \sigma_i + d_{(i,j)} \tag{19.5}$$

Now, the objective is to minimize z subject to Equations 19.1 to 19.5.

19.4.5 Resource Usage Models

This subsection deals with different resource usage models used in instruction scheduling. First we motivate the need for sophisticated resource usage models. In the subsequent subsection, we review some of the existing resource usage models.

19.4.5.1 Motivation

Modern processors implement very aggressive arithmetic and instruction pipelines. With an aggressive multiple instruction issue, structural hazard resolution in modern processors is expected to be more complex. Furthermore, in certain emerging application areas, such as mobile computing or space vehicle on-board computing, the size, weight, and power consumption may put tough requirements on the processor architecture design, which, in turn, may result in more resource sharing. All these lead to pipelines with more structural hazards. With such complex resource usage, the scheduling method must check and avoid any structural hazard, for example, contention for hardware resources by instructions. Such a check for resource contention is done by a *contention query module* [40].

The contention query module uses a resource usage model that specifies the resource usage patterns of various instructions in the target architecture. The contention query module answers the query, "Given a target machine and a partial schedule, can a new instruction be placed in time slot t without causing any resource conflicts?" Since a resource contention check needs to be performed before scheduling every instruction, and for each instruction several time steps of the schedule need to be examined, a significant part of the schedule time is spent in the contention query module. This is especially the case when the resource usage pattern is complex because of many structural hazards. Thus, an efficient resource usage model is critical for reducing the contention check time in instruction scheduling.

Portability and preciseness are two important aspects in choosing a resource model. Compilers designed to support a wide range of processors usually define the machine details to the scheduler in a form that can be easily modified when porting the compiler across different processors [62]. A portable model can only approximately model the complex execution constraints that are typical in modern-day superscalar and VLIW processors. Precise modeling of machine resources is important to avoid some of the stalls in the pipeline. Precise modeling of resource usages often involves a very low-level representation of the machine description that is generally coded directly into the compiler. As a result, it is tedious and time consuming to modify the code every time the compiler is targeted for a new processor.

19.4.5.2   Reservation Table Model

Traditionally, the resource usage pattern of an instruction i is represented using a reservation table. Instructions with identical resource usage patterns are said to belong to the same instruction class. The resource usage of any instruction in instruction class I has a single reservation table RT_I, which is an $m_I \times l_I$ bit matrix, where m_I is the number of resources needed by the instruction for its execution in the pipeline and l_I is the execution latency of the instruction [132]. An entry $RT_M[r, t] = 1$ indicates that the resource r is needed by this instruction t cycles after it is launched. Typically, each row of the reservation table is stored as a bit vector. The reservation tables for two instruction classes I_1 and I_2 are shown in Figure 19.13.

Apart from storing the reservation tables for each instruction class, the contention query module also maintains a GRT that is used to keep track of the machine state at every point in the schedule. The GRT is an $M \times T$ bit matrix, where M is the *total* number of resources in the target machine and T is an upper bound on the length of the schedule.

To answer the query, "Can an instruction of class I be scheduled in the current cycle?" the scheduler performs bit-wise AND operations of the nonzero bit vectors of RT_I with the corresponding bit vectors in the global reservation table, starting from the current cycle. If the results of the AND operations are all 0's, the instruction can be scheduled in the current cycle; otherwise it cannot be scheduled. On scheduling the instruction, similar bit-wise OR operations are performed on the GRT to reflect the resource usages of the scheduled instruction.

19.4.5.3   Reduced Reservation Table Model

The reduced reservation table (RRT) approach, developed by Eichenberger and Davidson [40], for answering a contention query is similar to the reservation table (RT) approach, except that the RRT approach uses a simplified reservation table. This simplified table is derived by eliminating much of the redundant information in the original reservation table. However, the scheduling constraints present in the original RT are preserved in the RRT. The resource usages are modeled using *logical* resources, unlike the RT model, wherein the actual resources of the target architecture are used. This offers a compact form of representing the RT, thus reducing the space required to store the tables and minimizing the time spent in contention queries.

Resources	Time Steps			
	0	1	2	3
r_0	1	0	0	0
r_1	0	1	1	0
r_2	0	0	0	1

(a) Reservation Table for I_1

Resources	Time Steps			
	0	1	2	3
r_0	1	0	0	0
r_3	0	1	0	0
r_4	0	0	1	1

(b) Reservation Table for I_2

FIGURE 19.13   Reservation table (RT) for the example machine.

Logical Resources	Time Steps	
	0	1
r'_0	1	0
r'_1	1	1

(a) RRT for I_1

Logical Resources	Time Steps	
	0	1
r'_0	1	0
r'_1	1	1

(b) RRT for I_2

FIGURE 19.14   Reduced reservation table (RRT) for the example machine.

First we define *forbidden latency*. For an ordered pair of instruction classes, A and B, a latency value f is said to be *forbidden* if two instructions, one belonging to instruction class A and another to B, when initiated on the respective functional unit types with a latency f between them cause a structural hazard. Such a structural hazard occurs if there exists at least one resource r and a time step t such that $RT_A[r,(t+f)]$ and $RT_B[r,t]$ are both 1. The forbidden latency set $F_{A,A}$ consists of the forbidden latencies between two initiations of the same instruction class. Similarly, the forbidden latency set $F_{A,B}$ consists of the forbidden latencies between two initiations of two different instruction classes. For example, for the RTs shown in Figure 19.13, latency 0 is in the forbidden set F_{I_1,I_2}, as two instructions, one each from these two instruction classes, initiated with a latency 0 (i.e., initiated in the same time step t), require the resource r_0 at time step t. Similarly, latency 1 is in F_{I_1,I_1} and latency 1 is in F_{I_2,I_2}, as resource r_1 for instruction class I_1 and r_4 for instruction class I_2 are required for two consecutive time steps.

Construction of the RRT is explained in detail in [40]. The RRT approach uses a set of logical resources to model all the forbidden latencies of the original resource usage. We shall illustrate the RRT approach using the example machine considered in Section 19.4.5.2 (refer to Figure 19.13). The RRTs for this machine are shown in Figure 19.14. Logical resources r'_0, r'_1, and r'_2 are used to model the resource usages. Note that the resource r'_0 models the forbidden latencies $0 \in F_{I_1,I_2}$ and $0 \in F_{I_2,I_1}$. Furthermore, the resource r'_1 models the forbidden latency $1 \in F_{I_1,I_1}$ and the resource r'_2 models the forbidden latency $1 \in F_{I_2,I_2}$. Last, forbidden latencies $0 \in F_{I_1,I_1}$ and $0 \in F_{I_2,I_2}$ are modeled by every resource. It can be verified that these RRTs model all and only those forbidden latencies that are present in the original reservation table. Compared to the RT in Figure 19.13, the RRT in Figure 19.14 is compact. The advantage of the RRT model is that it is likely to have fewer logical resources than physical resources. For the example machine, the number of logical resources is three, while the number of physical resources in the RT model is five. As a consequence, the space requirements of both the resource model and the contention check computation become efficient.

The contention query module of the instruction scheduler uses the RRT in the same manner as in the case of the original RT. The differences, however, are in the size of the GRTs and the individual RTs. The GRT in this case consists only of M' rows, where M' is the total number of logical resources in the machine. For the considered example, M' is 3, which is significantly less than the number of physical resources. As a consequence, the size of the GRT also reduces significantly.

19.4.5.4 Automaton-Based Approaches

The automaton approach models resource usage using a finite state automaton. This approach processes the reservation tables off-line to generate all possible legal initiation sequences in the architecture. The states of the automaton correspond to machine states at different points in the scheduling process. The automaton is constructed just once for the target architecture, and thereafter the compiler uses this during the instruction scheduling phase.

Müller's method constructs the automaton directly using the RTs [108]. Each state in this automaton is essentially a snapshot of the GRT (refer to Section 19.4.5.2) in the partial schedule. Proebsting and Fraser [128] improved upon Müller's technique by using *collision matrices* (to be defined later) for constructing the automaton. Bala and Rubin [12] extended Proebsting's technique to complex machines and introduced the notion of *factoring*. Although Proebsting's method directly produces the minimal

$$CM_{I_1} = \begin{matrix} I_1 \\ I_2 \end{matrix} \begin{bmatrix} 1 & 1 \\ 1 & 0 \end{bmatrix}; \quad CM_{I_2} = \begin{matrix} I_1 \\ I_2 \end{matrix} \begin{bmatrix} 1 & 0 \\ 1 & 1 \end{bmatrix}$$

FIGURE 19.15   Collision matrices for instruction classes I_1 and I_2.

automaton, it is still large for complex machines. Instead of building one large automaton, Bala and Rubin used a factoring scheme to create multiple smaller automatons, the sum total of whose states is less than the number of states in the original automaton. The factoring scheme is based on the observation that modern-day processors typically divide the instruction set into different *classes*, and each class is executed by a different functional unit. For instance, the integer and floating point units have different pipelines and use separate resources. As such, these resources can be divided into separate factors, and the automaton can be constructed separately for each of the factors.

Before we proceed with the construction of the automaton, we define *collision matrix* [85]. A *collision matrix* CM_I for the instruction class I is a bit matrix of size $n \times \ell'$, where n is the number of instruction classes and ℓ' is the longest repeat latency of an instruction class. The repeat latency of an instruction class is the minimum value such that any latency f greater than or equal to the repeat latency is permissible for the instruction class [85]. The collision matrix CM_I specifies whether or not a resource conflict will occur in initiating instructions of various classes, including itself, at different time steps. The rows of the collision matrix represent various instruction classes, and the columns represent different time steps. More specifically, the entry $CM_I[J, t] = 1$ if t is a forbidden latency between the instructions classes I and J, that is, $t \in F_{I,J}$. The collision matrices for the instruction classes considered in Section 19.4.5.2 are shown in Figure 19.15.

The construction of the finite state automaton proceeds as follows: each state F in the automaton is associated with a state matrix SM_F, which is an $n \times \ell'$ bit matrix. Given a state F and an instruction of class I, it is legal to issue I in the current cycle from state F, if $SM_F[I, 0]$ is 0. A legal issue causes a state transition $F \xrightarrow{I} F'$. The state matrix $SM_{F'}$ is computed by ORing the respective rows of SM_F with the collision matrix CM_I. The automaton for the motivating example is shown in Figure 19.16a. When the automaton reaches a state where all the entries in the first column are 1, it means no more instructions of any instruction class can be issued from the current cycle. State F_2 is an example of such a state. Hence, the state is *cycle advanced*, or the automaton moves to the next time step, which results in left-shifting the state matrix by one column [12]. This instruction class is marked as CA in Figure 19.16a. When the automaton-based resource model is used in conjunction with a simple list scheduler (such as the one discussed in Section 19.4.2), which schedules instructions on a cycle-by-cycle basis, it suffices to examine only transition latency zero, provided cycle advancing transitions are considered.

The automaton is represented in the form of a transition table that is used by the scheduler. The transition table is a two-dimensional matrix of size $N \times r$, where N is the number of states in the automaton and r is the number of instruction classes, including the pseudo-instruction class CA. The entries in the transition table are either state numbers or null (denoting illegal transitions). The transition table corresponding to the automaton for our motivating example is shown in Figure 19.16b. Thus, in the automaton-based approach, answering a contention query corresponds to a transition table lookup; updating the machine state on scheduling an instruction is changing to a new state. Both of these are constant time operations.

Two major concerns in using the automaton-based approach are the construction time of the automaton and the space requirements of the transition table. The construction of the automaton, though a one-time cost incurred at the time of compiler construction for this target architecture, could be significant, as the number of states in the automaton can grow very large, with an increase in either the number of instruction classes or the latencies of instructions. For example, for the DEC Alpha architecture, Bala and Rubin report 13,254 states when the automaton is constructed for integer and floating point instruction classes together.

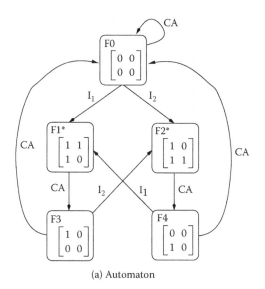

Current State	State Transition Upon		
	I_1	I_2	CA
F0	F1	F2	F0
F1*	-	-	F3
F2*	-	-	F4
F3	-	F2	F0
F4	F1	-	F0

(* Indicates Cycle Advancing States)

(a) Automaton (b) Transition Table

FIGURE 19.16 Automaton for the example machine.

When separate automatons are constructed, the number of states decreases to 237 and 232. For the Cydra 5 architecture [136], the number of states in the factored automaton is 1,127. Thus, the number of states is still large, which directly contributes to the increase in space requirements. This is because the transition table is an $N \times n$ matrix, where N is the number of states in the automaton and n is the number of instruction classes.

In [75, 76] the automaton model is further extended to a *group automaton model* that reduces the space requirements by observing and eliminating certain symmetry in the constructed automaton. It identifies instruction groups, sets of instruction classes, which exhibit this symmetric behavior. The work also proposes a resource model based on collision matrices, referred to as the *dynamic collision matrix* model. This resource model combines collision matrices with an RT-based approach to strike a good balance between the space and time requirements of the resource usage model. A classification of resource usage models is also presented in [76].

An automaton-based resource model for software pipelining has been proposed independently by Govindarajan et al. for instruction classes that do not share any resource [56]. It is subsequently extended to instruction classes that share resources in [58, 177]. The state diagram of the automaton, referred to as *modulo-scheduled (MS)-state diagram*, though similar to Bala and Rubin's automaton, has two important differences. First, the MS-state diagram is specialized for software pipelining or modulo scheduling, which take into account the repetitive scheduling of different instances of the same instruction, corresponding to different iterations. It is argued in [56, 58] that Bala and Rubin's automaton could not be directly used in a software pipelining method. Second, in the MS-state diagram, for each state, there is a state transition corresponding to each permissible latency, including latency 0, whereas in Bala and Rubin's automaton the state transitions in a state correspond only to latency value 0.

19.4.5.5 AND/OR-Tree Model

Gyllenhaal et al. proposed a two-tier model for resource usage representation [62]. It allows the user to specify the resource usage or machine description in a high-level language. The high-level language is designed to specify the machine description in an easy-to-understand, maintainable, and retargetable manner. The high-level description is translated into a low-level representation for efficient use by the instruction scheduler. The two-tier model helps to easily retarget the scheduler for different architectures.

Gyllenhaal et al. also proposed a new representation for a machine description based on the AND/OR-tree concept used in search algorithms, which is especially useful when a single instruction class has multiple *options* or resource usage patterns to choose from. This happens, for example, when there are two decode units, and an instruction can use either one of them in the decode stage. The new representation is an AND-tree of OR-trees. While the OR-trees encode the multiple options available within a stage, the AND-tree represents the usage of different stages. The AND/OR-tree model uses the short-circuit property of the AND/OR-trees to detect the resource conflicts quickly. This reduces the space complexity of the RT-based approaches and the computation time required to answer contention queries.

19.4.6 Case Study

In this section we shall review the instruction scheduling methods used in (a) the compiler for the Cydra 5 VLIW architecture [14, 36], (b) the GNU C compiler [158], and (c) the IBM XL compiler family [170] as case studies.

19.4.6.1 Instruction Scheduling in the Cydra 5 Compiler

The scheduler implemented in the Cydra 5 compiler is a backward pass list scheduler. It works bottom-up, scheduling from the sink node of the DAG. For each instruction i from the priority list, the scheduler attempts to schedule the instruction starting from the largest possible start time, based on its (already) scheduled successors. The priority algorithm ensures that all successor instructions are placed in the list ahead of an instruction. In addition, the priority algorithm can give greater priority either to instructions with the least slack or to instructions that reduce register lifetimes. The former heuristic results in schedules with low execution time but may increase the register pressure. This may cause register spills. The second heuristic is used to counter this effect. In the Cydra 5 compiler instruction scheduling is performed prior to register allocation.

19.4.6.2 Instruction Scheduling in the GNU C Compiler

The instruction scheduling method used in the GNU C compiler (version 2) also follows backward pass list scheduling [158]. The method orders instructions based on a priority algorithm that gives relatively higher priority to all successor instructions compared to the parent instruction. By placing instructions with higher priority later in the schedule than ones with lower priority, the scheduler preserves dependence constraints. Furthermore, instructions with larger execution time are also given higher priority, exposing instructions on the critical path.

The algorithm then starts scheduling by issuing the instruction with the highest priority, scheduling from the last instruction in the basic block to the first. Each time an instruction is scheduled, a check is performed on each predecessor instruction p to see if it has no more unscheduled successors. Such instructions are marked "ready" and added to the ReadyList in priority order. When all instructions are scheduled, the algorithm terminates. This scheduling method works well to produce good schedules but generally increases the register pressure and results in poor performance when the number of available registers is less than the required number. For this purpose, the list scheduling method also gives higher priority to instructions that end live ranges of variables.

19.4.6.3 Instruction Scheduling in the IBM XL Compiler Family

The XL family compiler of IBM uses a forward pass list scheduling method [170]. The primary priority heuristic used is based on the maximum distance to the sink node. It also uses a combination of smallest Estart value, minimum liveness, and greatest uncovering — corresponding to how many (ready) instructions will be added to the ReadyList — heuristics. The list scheduling algorithm starts scheduling from the source node, attempting to schedule instructions at time steps closer to its Estart time. The scheduling method is applied both as a Prepass method (before register allocation) and as a Postpass method (after register allocation). It takes care of a number of types of delay stalls and schedules fixed and floating point instructions in an alternating sequence for the RS/6000 processor.

19.5 Global Scheduling

Instruction scheduling within a basic block has limited scope, as the average size of a basic block is quite small, typically in the range of 5 to 20 instructions. Thus, even if the basic block scheduling method produces optimal schedules, the performance, in terms of the exploited ILP, is low. This is especially a serious concern in architectures that support greater ILP, for example, VLIW architectures with several functional units or superscalar architectures that can issue multiple instructions every cycle. The reason for the low ILP, especially near the beginning and end of basic blocks, is that basic block boundaries act like barriers, not allowing the movement of instructions past them.

Global instruction scheduling techniques, in contrast to local scheduling, schedule instructions beyond basic blocks, that is, overlapping the execution of instructions from successive basic blocks. These global scheduling methods are either for a set of basic blocks with acyclic control flow among them [36, 46, 74, 102] or for single or multiple basic blocks of a loop [25, 31, 49, 88, 133]. The former case is referred to as *global acyclic scheduling* and the latter as *cyclic scheduling*. First, we discuss a few global acyclic scheduling methods. Section 19.5.2 deals with cyclic scheduling.

19.5.1 Global Acyclic Scheduling

Early global scheduling methods performed local scheduling within each basic block and then tried to move instructions from one block to an empty slot in a neighboring block [25, 162]. However, these methods followed an ad hoc approach in moving instructions. Furthermore, the local compaction or scheduling that took place in each of the blocks resulted in several instruction movements (reorderings) that were done without understanding the opportunities available in neighboring blocks. Hence, some of these reorderings may have to be undone to get better performance. In contrast, global acyclic scheduling methods, such as *trace scheduling* [46], *percolation scheduling* [111], *superblock scheduling* [74], *hyperblock scheduling* [102], and *region scheduling* [63], take a global view in scheduling instructions from different basic blocks. In the following subsections we describe these approaches.

19.5.1.1 Trace Scheduling

Trace scheduling attempts to minimize the overall execution time of a program by identifying frequently executed *traces* — acyclic sequences of basic blocks in the control flow graph — and scheduling the instructions in each trace as if they were in a single basic block. The trace scheduling method identifies the most frequently executed trace, a single path in the control flow graph, by identifying the unscheduled basic block that has the highest execution frequency; the trace is then extended forward and backward along the most frequent edges. The frequency of edges and basic blocks are obtained by a linear combination of branch probabilities and loop trip counts obtained either through heuristics or through profiling [13]. Various profiling methods are discussed in greater detail in [60].

The instructions in the selected trace (including branch instructions) are then scheduled using a list scheduling method. The objective of the scheduling is to reduce the schedule length and hence the execution time of the instructions in the trace. During the scheduling, instructions can move above or below a branch instruction. Such movement of instructions may warrant compensation code to be inserted at the beginning or end of the trace. After scheduling the most frequently executed trace, the next trace (involving unscheduled basic blocks) is selected and scheduled. This process continues until all the basic blocks are considered.

Let us illustrate the trace scheduling method with the help of the example code shown in Figure 19.17a, adapted from [73]. The instruction sequence and the control flow graph for the code are shown in Figures 19.17b and 19.17c. Consider a simple two-way issue architecture with two integer units. Let us assume that the latency of an integer instruction, such as an `add`, `sub`, or `mov` instruction, is one cycle, and that of a `load` instruction is two cycles. Thus, there is a stall of one cycle between a `load` and a dependent instruction. For simplicity, we assume here that branch instructions do not require any stall cycles.

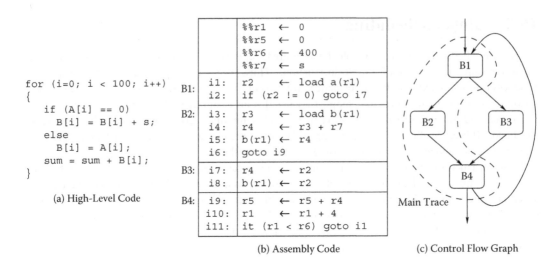

```
for (i=0; i < 100; i++)
{
    if (A[i] == 0)
        B[i] = B[i] + s;
    else
        B[i] = A[i];
    sum = sum + B[i];
}
```

(a) High-Level Code

	%%r1	← 0
	%%r5	← 0
	%%r6	← 400
	%%r7	← s
B1:	i1:	r2 ← load a(r1)
	i2:	if (r2 != 0) goto i7
B2:	i3:	r3 ← load b(r1)
	i4:	r4 ← r3 + r7
	i5:	b(r1) ← r4
	i6:	goto i9
B3:	i7:	r4 ← r2
	i8:	b(r1) ← r2
B4:	i9:	r5 ← r5 + r4
	i10:	r1 ← r1 + 4
	i11:	it (r1 < r6) goto i1

(b) Assembly Code

(c) Control Flow Graph

FIGURE 19.17   Multiple basic blocks example.

A basic block scheduling method achieves the schedule shown in Figure 19.18. As instructions cannot be moved beyond basic block boundaries, this is the best schedule that can be achieved for the given machine. The first column in the figure represents the time steps at which the instructions can be issued. The time steps shown in parentheses for instructions i9 to i11 correspond to the schedule time when the control flow is B1 → B3 → B4. Note that the extra cycle (time step 4) in the schedule (after instructions i7 and i8) in basic block B3 is due to the stall for the load instruction (i8) at the basic block boundary. It takes nine cycles for the path B1 → B2 → B4 and seven cycles for B1 → B3 → B4.

Assume that basic blocks *B*1, *B*2, and *B*4 are frequently executed, and they form the main trace. By allowing the instructions in basic block *B*2 to be moved above the split point in the control flow graph, a compact schedule for the most frequently executed trace can be obtained. For example, instruction i3 can be moved to block *B*1. Such movement of instructions above a conditional branch instruction is referred to as *speculative code motion*. Moving an instruction that could raise an exception, such as a memory load or a divide instruction, speculatively above a control split point, requires additional hardware support as discussed in [23]. This is to avoid raising unwanted exceptions.

The original program semantics must be ensured under speculative code motion. For this, the destination register of an instruction i should not be live on entry on alternative paths on which i is control dependent.

Time	Int. Unit 1		Int. Unit 2	
0	i1:	r2 ← load a(r1)		
1				
2	i2:	if (r2 != 0) goto i7		
3	i3:	r3 ← load b(r1)		
4				
5	i4:	r4 ← r3 + r7		
6	i5:	b(r1) ← r4	i6:	goto i9
3	i7:	r4 ← r2	i8:	b(r1) ← r2
4				
7(5)	i9:	r5 ← r5 + r4	i10:	r1 ← r1 + 4
8(6)	i11:	if (r1 < r6) goto i1		

FIGURE 19.18   Basic block schedule for the instruction sequence of Figure 19.17.

Time		Int. Unit 1		Int. Unit 2
0	i1:	r2 ← load a(r1)	i3:	r3 ← load b(r1)
1				
2	i2:	if (r2 != 0) goto i7	i4:	r4 ← r3 + r7
3	i5:	b(r1) ← r4		
4(5)	i9:	r5 ← r5 + r4	i10:	r1 ← r1 + 4
5(6)	i11:	if (r1 < r6) goto i1		

3	i7:	r4 ← r2	i8:	b(r1) ← r2
4	i12:	goto i9		

FIGURE 19.19   Trace schedule for the instruction sequence of Figure 19.17.

The reason is that when execution proceeds on an alternative path, instruction i, which was speculatively executed, would have modified the destination register, which is live on entry in this path. Suppose register r3 is live on entry for basic block B3 in our example; that is, there is some instruction j in B3 for which r3 is a source operand, and there are no instructions in B3, before j, that define r3. Then speculative motion of i3 from basic block B2 to B1 will destroy the live-in value of r3 for instruction j. To perform speculative code motion of an instruction whose destination register is live-in on an alternative path, the destination register must be renamed appropriately at compile time.

A schedule for the main trace is shown in Figure 19.19. In this schedule, the main trace consisting of basic blocks B1, B2, and B4 can be executed in six cycles, while the off-trace path involving B1, B3, and B4 can be executed in seven cycles. By scheduling instructions across basic blocks, the execution time of the main trace is reduced from nine to six cycles.

When execution goes through the less frequently executed path, the *off-trace path*, to preserve correct program execution, some of the instructions may be duplicated. The code inserted to ensure correct program behavior and thus compensate for the code movement is known as *compensation code*. For example, if an instruction from basic block B1 is moved down below the control split point to B2, then a compensation code has to be inserted in B3. Several other examples of code movement and the required compensation code are illustrated in [100].

The trace scheduling algorithm should maintain the need for introducing such compensation code at various program points. This is known as *bookkeeping*. The compensation code may increase the schedule length of other traces. Since the objective is to reduce the overall execution time, and since the trace that is scheduled first is the most frequently executed one, compacting the schedule of the instructions in this trace is desirable, even if this increases the schedule length of other traces. A key property of trace scheduling, as pointed out in [132], is that the decisions as to whether to move an instruction from one basic block to another, where to schedule it, and so on, are all made jointly in the same compiler phase. The trace scheduling method was implemented in the Bulldog compiler [43]. The work was later enhanced into a production-quality multiflow compiler [100].

19.5.1.2 Superblock Scheduling

Hwu et al. proposed a variant of trace scheduling called *superblock scheduling* [74] in the IMPACT project [23]. The motivation and the basic idea behind superblocks comes from the observation that the complexities involved in maintaining bookkeeping information in trace scheduling result from several incoming control flow edges at various points in a trace. The bookkeeping associated with these entrances, known as side entrances, can be avoided if the side entrances themselves are eliminated in the trace. For example, in Figure 19.17c, there is a side entrance to the trace at basic block B4. Thus, by eliminating the side entrances, the (control flow) join points, as well as the associated bookkeeping, can be eliminated in a superblock. To summarize, a superblock trace consists of a sequence of basic blocks with a single entry (at the beginning of the superblock) and multiple exits. Superblocks are formed in two steps:

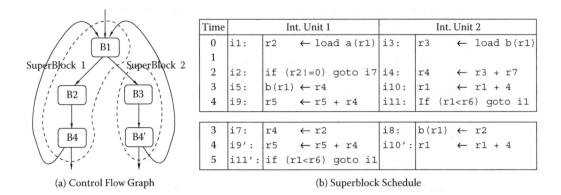

Time	Int. Unit 1			Int. Unit 2		
0	i1:	r2	← load a(r1)	i3:	r3	← load b(r1)
1						
2	i2:	if (r2!=0) goto i7		i4:	r4	← r3 + r7
3	i5:	b(r1) ← r4		i10:	r1	← r1 + 4
4	i9:	r5	← r5 + r4	i11:	If (r1<r6) goto i1	

Time	Int. Unit 1			Int. Unit 2		
3	i7:	r4	← r2	i8:	b(r1) ← r2	
4	i9':	r5	← r5 + r4	i10':	r1	← r1 + 4
5	i11':	if (r1<r6) goto i1				

(a) Control Flow Graph   (b) Superblock Schedule

FIGURE 19.20   Superblock formation and scheduling.

in the first step traces are identified using profile information; the second step performs *tail duplication* to eliminate side entrances.

We explain the construction of superblocks with the help of the example discussed in Section 19.5.1.1. Once again, let us assume that basic blocks B1, B2, and B4 constitute the main trace as shown in Figure 19.17c. The second trace in the control flow graph makes a control flow entry to $B4$ and hence a side entrance to the main trace. To form superblocks for this trace, the tail block $B4$ is replicated to eliminate the side entrance. The superblocks for the control flow graph are shown in Figure 19.20a.

Three optimizations to enlarge the size of a superblock have been proposed in [74] that, in turn, enhance the scope for exploiting higher ILP. Additional dependence-removing optimizations are subsequently performed to expose greater ILP. After these optimizations, the instructions in enlarged superblocks are scheduled using a list scheduling method. A schedule for the superblock is shown in Figure 19.20b. Although no bookkeeping code is needed in this example, avoiding the side entrance enables further compaction of the schedule for the main trace or superblock 1. It can be seen that superblock 1 can be executed in five cycles. When the control flows to superblock 2, six cycles are needed for the execution.

Both trace and superblock scheduling consider a linear sequence of basic blocks from a single control flow path. Both methods can move instructions from one basic block to another. However, store instructions that write into memory locations are not speculatively moved above branches, as this would modify a memory location whose old contents may be needed in an alternative path when execution proceeds on an off-trace path. Likewise, instructions that could cause an exception, such as load, store, integer divide, and floating point instructions, are typically not speculatively moved; otherwise, additional hardware support, in the form of non-trapping instructions, would be required [23].

19.5.1.3   Hyperblock Scheduling

Trace scheduling and superblock scheduling rely on the existence of a main trace, the most frequently executed path in the control flow graph. While this is likely in scientific computations, it may not be the case in control-intensive symbolic computing that dominates integer benchmark programs. To handle multiple control flow paths simultaneously, Mahlke et al. proposed *hyperblock scheduling* [102]. In this approach, the control flow graph is IF-converted [7] to eliminate conditional branches. IF-conversion replaces conditional branches with corresponding comparison instructions, each of which sets a predicate. Instructions that are control dependent on a branch are replaced by predicated instructions that are dependent on the corresponding predicate. For example, an instruction t1 ← t2+t3 that is control dependent on the condition (t4 == 0) is converted to

$$i : p1 \leftarrow (t4 == 0)$$
$$i' : t1 \leftarrow t2 + t3, if\, p1$$

Instruction i' is predicated on p1, and t1 ← t2+t3 is performed only if p1 is true. Thus, by using IF-conversion, a control dependence can be changed to a data dependence. In architectures supporting predicated execution [23, 80, 136], a predicated instruction is executed as a normal instruction if the predicate is true; it is treated as a no-op otherwise.

A hyperblock is a set of predicated basic blocks and, as with superblocks, has a single entry and multiple exits. However, unlike a superblock, which consists of instructions from only one path of control, a hyperblock may consist of instructions from multiple paths of control. The presence of multiple control flow paths in a hyperblock enables better scheduling for programs with heavily biased branches. The region of blocks chosen to form a hyperblock is from an innermost loop, although a hyperblock is not necessarily restricted only to loops. While conventional IF-conversion can predicate all basic blocks in an innermost loop, hyperblocks selectively predicate only those that would improve program performance. A heuristic based on the frequency of execution, size, and characteristics (such as whether or not they contain function calls) of basic blocks is used in selecting the blocks for predication. The reason for being selective in predication is that combining unnecessary basic blocks (from different control flow paths) results in wasting the available resources, leading to poor performance.

The selected set of basic blocks should (a) not have a side entrance and (b) not contain an inner loop. Tail duplication is done to eliminate side entrances in a hyperblock. Loop peeling is performed on a nested inner loop that iterates only a few times to enable inclusion of both inner and outer loops in the hyperblock. Last, when basic blocks from different control paths are included in a hyperblock, and when the execution times of the control paths are vastly different, *node splitting* is performed on nodes subsequent to the merge point (corresponding to the multiple control path). Node splitting duplicates the merge and its successor nodes.

Once the blocks are selected for a hyperblock, they are IF-converted. Then certain hyper-block-specific optimizations are performed [102]. Finally, the instructions in a hyperblock are scheduled using a list scheduling method. In hyperblock scheduling, two instructions that are in mutually exclusive control flow paths may be scheduled on the same resource. If the architecture does not support predicated execution, reverse If-conversion [172] is performed to regenerate the control flow paths.

Let us once again consider the control flow graph shown in Figure 19.17c. If basic blocks B2 and B3 are both equally likely to be executed, the superblock scheduling method can choose only one of the two basic blocks, whereas both can appear in a hyperblock (refer to Figure 19.21a). A new instruction i2' that sets a predicate is introduced in the code. Instructions i3, i4, and i5 are predicated on p1 while i7 and i8 are predicated on the complement of p1 (i.e., !p1). Since instructions i3 and i4 are now data dependent on i2', they can be scheduled only after time step 2. This results in a lengthier schedule. However, by identifying that these two instructions can be speculatively executed, *predicate promotion* can be performed on these instructions [102], and they can be scheduled earlier. The resulting schedule is shown in Figure 19.21b. Note, however, that the resulting hyperblock schedule takes six cycles and hence results in a performance degradation if the control flow path B1 → B2 → B4 is taken.

Tail duplication and node splitting performed to form hyperblocks result in duplication of code. This may increase the code size significantly. Another concern in hyperblock scheduling is that an aggressive selection for alternate control flow paths may unnecessarily increase the resource usage and hence may result in degenerated schedules. Hence, in order for hyperblock scheduling to be effective, both code duplication and the inclusion of alternate control flow paths must be kept under check.

19.5.1.4 Other Global Acyclic Scheduling Methods

In [103] global acyclic scheduling methods have been classified as either profile-driven or structure-driven approaches. The global acyclic methods discussed in the earlier subsections fall under the profile-driven approach. They identify the most frequently executed paths using profile information and coalesce them into an extended basic block. In contrast, the structure-driven approach identifies and attempts to increase parallelism along all execution paths by moving operations between basic blocks and considering program structure, without using profile information. Examples of structure-driven approaches are region scheduling [61], percolation scheduling [111], and global scheduling [15]. In this section we briefly review

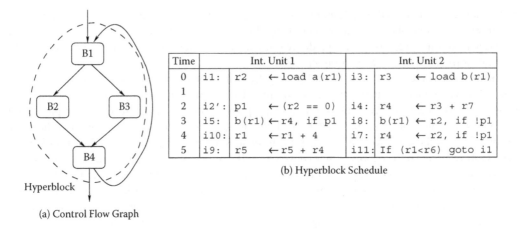

(a) Control Flow Graph

Time	Int. Unit 1			Int. Unit 2		
0	i1:	r2	← load a(r1)	i3:	r3	← load b(r1)
1						
2	i2′:	p1	← (r2 == 0)	i4:	r4	← r3 + r7
3	i5:	b(r1)	← r4, if p1	i8:	b(r1)	← r2, if !p1
4	i10:	r1	← r1 + 4	i7:	r4	← r2, if !p1
5	i9:	r5	← r5 + r4	i11:	If (r1<r6) goto i1	

(b) Hyperblock Schedule

FIGURE 19.21 Hyperblock formation and scheduling.

some of the structure-driven global scheduling methods, as well as a few other profile-driven scheduling methods.

Trace scheduling is generalized to deal with general control flow in *percolation scheduling* [111]. Percolation scheduling uses four transformations, namely delete, move, move conditional, and unify. For each node in the control flow graph, it tries to apply each of the four transformations and repeats the transformation until none can be applied. Three out of four of these transformations move operations upward in the control flow graph. Delete removes a node when it is empty. Percolation scheduling originally assumed unbounded resources. It is then extended to non-unit execution latencies, but still with unbounded resources, in [113]. *Trailblazing* [112] is another extension to percolation scheduling that exploits the structure of the hierarchical task graph representation to move instructions across large blocks of code in a nonincremental way, that is, without having to move instructions in a step-by-step manner through every block in the control flow path. This facilitates both efficient code motion and elimination of code explosion in certain cases.

Meld scheduling is a simple but effective instruction scheduling method across basic blocks that was used in the Cydra 5 compiler [36]. It follows a simple basic block scheduling approach, except that during the scheduling of a basic block B, if the predecessor (or the successor block), for instance, B', has already been scheduled, then the resource usage information at the end (or, respectively, the beginning) of the schedule for B' is used as the resource usage at the start (or end) of the the basic block B. The resource usage entering into the basic block should take into account the multiple basic blocks from which (to which) control flow could enter (leave) the block B. Taking into account the resource usage at the boundary of neighboring basic blocks and scheduling instructions from the current basic block allows the overlap of instructions across basic blocks. The work of Abraham et al. generalizes this idea and quantitatively evaluates the benefits of meld scheduling [1].

Another global code scheduling, called *region scheduling*, is discussed in [61]. This approach is based on an extended program dependence graph representation allowing code motion between regions consisting of control equivalent statements [45]. Regions are classified according to their parallelism content, which is used to drive a set of powerful code transformations. Golumbic and Rainish proposed several simple schemes for scheduling instructions beyond basic blocks [53].

A global instruction scheduling method, also based on a program dependence graph, has been implemented in the IBM XL family of compilers for the IBM RS/6000 systems [15]. The scheduling method proceeds by processing one basic block at a time. However, when scheduling instructions in a basic block B, instructions from control equivalent blocks as well as instructions from successors of B and successors of control equivalent blocks are also considered. While the latter two categories of instructions (from successor blocks) are considered *speculatively executed instructions*, instructions from control equivalent blocks are considered *useful* instructions. During a scheduling step, speculative instructions can be scheduled, provided they are data ready, resources for them are available, and they are schedulable across basic blocks. However, preference is given to useful instructions as opposed to speculatively executed ones. This is especially important in machines with a few functional units, such as the RS/6000 system.

Next we turn our attention to a few other profile-driven scheduling methods. Hank, Hwu, and Rau proposed *region-based compilation*, an approach that allows an arbitrary collection of basic blocks, possibly extending over multiple function bodies, to be considered as a compilation unit [63]. The region formation approach is a generalization of profile-based trace selection. A region can expand across more than one control path. Region formation considers aggressive function inlining to extend regions across function bodies. The region formation approach is proposed as a generalized technique that is applicable to the entire compilation process, including ILP optimizations, instruction scheduling, and register allocation. A global scheduling technique that operates over a restricted region, a single entry subgraph, is proposed in [101]. A region-based register allocation approach is discussed in [83].

Trace scheduling and superblock scheduling operate on linear sequences of basic blocks from a single control flow path and favor the current trace path at the expense of instructions in the off-trace trace. Hsu and Davidson [71] and, more recently, Havanaki et al. [64] proposed global scheduling methods that operate on a tree of basic blocks, possibly involving multiple control flow paths. A *treegion*, as the name suggests, is a tree in the control flow graph, where except for the root node, all other nodes (basic blocks) have a single incoming edge. Scheduling of instructions in the tree of basic blocks can benefit from profile information. Furthermore, a *treegion* [64] can be enlarged by performing tail duplication of merge nodes (and its successors). Compile-time register renaming is used to allow speculative code motion of instructions above their control-dependent branches.

The integrated global local scheduling (IGLS) approach [103] is a hybrid of profile-driven and structure-driven scheduling approaches. This method avoids the tail duplication and bookkeeping overheads of profile-driven approaches. IGLS orders the selection of blocks using profile information. However, in applying the code reordering transformation, it follows a structure-driven approach and does not necessarily restrict code reordering to any trace or extended block. Also, the selection of the appropriate code motion and the target block selection are made flexible and depend on the block's current properties such as the available parallelism within the block. The method has been implemented in the SGI MIPSpro compiler.

An important consequence of aggressive speculative scheduling of instructions in a global instruction scheduling method is that it may unduly delay some of the paths in the global region (such as a superblock) considered for the schedule. This happens especially when the resources (functional units) are limited. The reason for this is that the profile information is used only during the formation of the global region and not while scheduling the instructions. Fisher proposed the use of *speculative yield* — the probability that a speculatively scheduled instruction produces useful work — along with dependence height (similar to MaxDistance, defined in Section 19.4.2.1) in scheduling instructions in the global region [48]. *Successive retirement* is another profile-independent scheduling heuristic that attempts to retire each path (or exit) in order, as early as possible [27]. This heuristic, applied to superblock regions, minimizes speculation so that it only speculates when there are no non-speculative instructions available. The *speculative hedge* heuristic attempts to ensure that no path gets delayed unnecessarily by accounting for different processor resources while scheduling, and not just using a common scheduling priority function based on dependence height [37]. Last, the treegion scheduling method [64], by virtue of scheduling multiple paths in parallel, also avoids unduly penalizing the off-trace paths.

19.5.2 Cyclic Scheduling

To exploit higher ILP in loops, several cyclic scheduling methods have been proposed to overlap the execution of instructions from multiple basic blocks, where the multiple blocks could be multiple instances of the same static basic block corresponding to different iterations. Early cyclic scheduling methods unrolled the loop several times and performed some form of global scheduling on the unrolled loop [49]. While

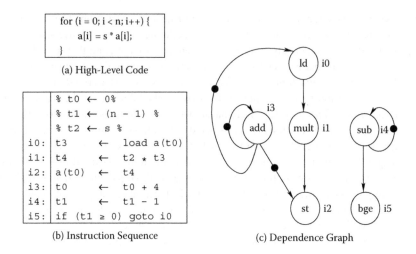

FIGURE 19.22   Software pipelining example.

this approach exploits greater ILP within the unrolled iteration, there is very little or no overlap across the iterations of the unrolled loop.

Software pipelining [6, 88, 132, 133] overlaps the execution of instructions from multiple iterations of a loop. The objective here is to sustain a high initiation rate, where the initiation of a subsequent iteration may start even before the previous iteration is complete. We discuss software pipelining only briefly here, as this is the topic of discussion of another chapter [5] in this book.

Let us first explain software pipelining with the help of the example (refer to Figure 19.22) adapted from [57, 135]. The dependences among the instructions in the loop are represented by means of a data dependence graph (DDG). The DDG, unlike the DAG used so far for acyclic scheduling, may be cyclic. In particular, a dependence from an instruction i to i' could be across iterations. That is, the value produced by i in the jth iteration could be used by i' in iteration $(j + d)$. Such a dependence is known as a *loop-carried dependence* with a *dependence distance* d. A loop-carried dependence is marked in the DDG by tokens on the dependence arc. The number of tokens present in an arc indicates the dependence distance.

For the instruction sequence shown in Figure 19.22b, the dependence graph is depicted in Figure 19.22c. In this graph, we assume that the possible dependence from store to load can be disambiguated and hence omitted. We shall assume an architecture with two Integer functional units and two floating point units. Let the latencies of instructions in these functional units be one and two cycles, respectively. Furthermore, we will assume that Load and Store instructions are executed by a Load/Store unit with execution times of two and one time units, respectively. An acyclic scheduling approach will be able to achieve a schedule in which the execution time of each iteration is five cycles. This corresponds to an initiation rate of one-fifth iteration per cycle.

Software pipelining overlaps successive iterations of the loop and hence can exploit higher ILP. Successive iterations of a loop are initiated with an initiation interval (II) and initiation rate (1/II). The minimum initiation interval (MII) achievable for a given loop is governed by resource constraints and recurrences or cyclic data dependences. The MII of a loop is the maximum of resource MII (ResMII) and recurrence MII (RecMII).

RecMII is determined from the dependence cycle(s) in the DDG [137]. Specifically,

$$RecMII = \max_{\forall cycles\ C} \left\lceil \frac{\text{sum of execution latencies of instructions in } C}{\text{sum of dependence distances in } C} \right\rceil$$

ResMII, for simple resource usage patterns (fully pipelined functional units), is given by

$$\text{ResMII} = \max_r \left\lceil \frac{N_r}{F_r} \right\rceil$$

where N_r is the number of instructions that can be executed in a functional unit of type r and F_r is the number of instances of a type r functional unit. For our target architecture,

$$\text{ResMII} = \max(\text{ResMII}_{Int}, \text{ResMII}_{FP}, \text{ResMII}_{Ld/St})$$

$$\text{ResMII} = \max \left(\frac{3}{2}, \frac{1}{2}, \frac{2}{1} \right) = 2$$

For the DDG in Figure 19.22b there are two self cycles on instructions i3 and i4. Hence, the RecMII for the loop is

$$\text{RecMII} = \max \left(\frac{1}{1}, \frac{1}{1} \right) = 1$$

Thus,

$$\text{MII} = \max(\text{RecMII}, \text{ResMII}) = \max(1, 2) = 2$$

In our discussion we consider periodic *linear schedules*, under which various instructions begin their execution at time steps given by a simple linear relationship. The jth iteration of an instruction i begins execution at time $II \cdot j + t_i$, where $t_i \geq 0$ is an integer offset and II is the initiation interval of the given schedule. It can be seen that t_i is also the schedule time of instruction i in iteration 0.

Figure 19.23 gives a resource-constrained schedule with $II = 2$ for our example loop. This schedule is obtained with $II = 2$, $t_{i0} = 0$, $t_{i1} = 2$, $t_{i2} = 5$, $t_{i3} = 3$, $t_{i4} = 4$, and $t_{i5} = 5$. The schedule has a prolog (from time step 0 to time step 3), a repetitive pattern (at time steps 4 and 5), and an epilog code starting from time step 6 to time step 9 as shown in the figure. Furthermore, at the first time step in the repetitive pattern

Time Step	Iter. 0	Iter. 1	Iter. 2	
0	i0 : ld			Prolog
1				
2	i1 : mult	i0 : ld		
3	i3 : add			
4	i4 : sub	i1 : mult	i0 : ld	Kernel
5	i2 : st i5 : bge	i3 : add		
6		i4 : sub	i1 : mult	Epilog
7		i2 : st i5 : bge	i3 : add	
8			i4 : sub	
9			i2 : st i5 : bge	

FIGURE 19.23   A software pipelined schedule with $II = 2$.

(time step 4), instructions i0, i1, and i4 are scheduled. Instructions i2, i3, and i5 are scheduled at the second time step (time step 5). The repetitive kernel is executed ($n - 2$) times in this case, and hence t1 must be appropriately set. The resource requirement in both cycles in the repetitive kernel is within the available resources. Furthermore, the schedule shown above is one of those resource-constrained schedules that achieves the lowest initiation interval (MII $= 2$). Hence, the schedule is a *rate-optimal* schedule.

Obtaining a rate-optimal resource-constrained software pipelined schedule is known to be NP-complete [88, 132]. Hence, many of the proposed methods for software pipelining attempt to obtain a near-optimal resource-constrained schedule. A number of heuristic methods for software pipelining have been proposed [36, 38, 51, 88, 105, 130, 131, 153, 168], starting with the work of Rau and Glaeser [133] and its application in the floating point systems (FPS) compiler and Cydra 5 compiler [35, 36]. Some of these algorithms backtrack some of the scheduling decisions to obtain efficient schedules.

A resource-constrained software pipelining method using list scheduling and hierarchical reduction of cyclic components has been proposed by Lam [88]. Her approach identifies *strongly connected components* — the maximal connected subgraph of the underlying undirected graph, where there is a path between every pair of nodes — and schedules the instructions in them. The strongly connected component is then treated as a single pseudo-operation with a complex resource usage pattern. Thus, the remaining DDG is reduced in a hierarchical way. Other heuristic-based scheduling methods have been proposed by Gasperoni and Schwiegelshohn [51], Wang and Eisenbeis [168], and Rau [131]. The problem of obtaining a rate-optimal resource-constrained software pipelined schedule is formulated as an integer linear programming problem in [44, 55, 57]. Altman et al. have extended their integer linear program formulation to handle complex resource usage patterns by unifying the scheduling and mapping problem in a single framework [8]. Efficient integer linear program formulation is proposed in [41] that makes use of a *structured 1-0 formulation* [26].

In [56, 58] a novel scheduling method, called co-scheduling, has been proposed that is a heuristic method that uses an MS-state diagram, an automaton-based resource usage model. The MS-state diagram model, proposed independently, is similar to the finite state automaton approach proposed by Bala and Rubin [12]; the main difference is that the former incorporates information about the initiation interval (II).

In addition to obtaining efficient schedules, in terms of low II, many software pipelining methods also attempt to reduce the register requirements of the constructed schedule. Huff's slack scheduling [72] is an iterative solution that gives priority to scheduling instructions with minimum *slack* (as defined in Section 19.4.2.1) and tries to schedule an instruction at a time step that minimizes register pressure. Stage scheduling constructs a schedule with lower register requirements from an already constructed resource-constrained software pipeline schedule either using a number of heuristics [39] or by solving a linear programming problem [42]. It should be noted here that the newly constructed schedule and the original schedule have the same repetitive kernel and t_i values (the schedule time of instruction i in iteration 0). The hypernode reduction modulo scheduling (HRMS) method [98], register-sensitive software pipelining [34], and swing modulo scheduling [97] are some of the other software pipelining methods that reduce the register requirements of the software pipelined schedule.

Register allocation of software pipelined schedules has been studied in [135]. A number of register allocation strategies were discussed and evaluated for architectures with and without specific hardware support. An important issue in software pipelining is that of handling the live ranges of the same variable corresponding to different iterations that overlap with themselves. For example, the value produced by instruction i1 at time step 2 in the schedule shown in Figure 19.23 is used by instruction i2 only at time step 5. However, another instance of i1 corresponding to the next iteration is executed (at time step 4) that could overwrite the destination register. *Modulo variable expansion* is a technique that unrolls the schedule a required number of times and renames the destination register appropriately to handle multiple simultaneously live values [88, 133]. Hardware support in the form of *rotating registers* was proposed in Cydra 5 [136] as a solution to this problem. With rotating registers, unrolling of loop schedules as in modulo variable expansion is not necessary. A software pipelining method that is sensitive to modulo variable expansion is proposed in [164]. This method first unrolls the loop an estimated number of times and schedules it in such a way as to avoid overlapping live ranges of the same variable.

If the register requirement is higher than the available registers, earlier approaches follow one of the following two simplistic approaches:

- Spill some of the variables to memory. In this approach, register spills require additional load and store operations, which need to be scheduled in the kernel. However, if the memory unit is saturated in the kernel such that the spill loads/stores could not be scheduled, then the II value needs to be increased and the loop must be rescheduled for the new II value.
- Reschedule the loop with a larger II but without inserting spills. Increased II in general reduces the register requirement of the schedule. Hence, with the new II, the register requirement of the constructed schedule may be lower than the number of available registers.

In general, increasing the II produces worse schedules than adding spill code. However, for some loops, the introduction of spill code may increase the II beyond what is achieved by the other method [4]. Thus, a hybrid method that adds spill code in some cases and increases the II in others can produce better results. The hybrid approaches [97, 140, 169, 176] first spill as many variables as required to have a register-constrained schedule without increasing the II; if this does not get the desired result, that is, even with the spill a schedule cannot be constructed with a register requirement less than the available register, then the II is increased. More recently, an on-the-fly spilling method has been proposed [4, 176] that takes into account register requirements during loop scheduling. If the register pressure of the partial schedule is high and greater than the available number of registers, the scheme inserts and schedules spill instructions on-the-fly. Further, if the spill code does not reduce register pressure sufficiently, then backtracking is used and selected instructions are unscheduled in order to reduce register pressure.

Loops consisting of multiple basic blocks with arbitrary acyclic control flow in the loop body pose another important challenge for software pipelining. Lam's hierarchical reduction approach, which schedules strongly connected components and reduces them as a single pseudo-operation, can handle conditionals as well [88]. In her approach, the two branches of a conditional are first scheduled independently. The entire conditional is then represented as a single node whose resource usage at each time step is the union of resource usages of the two branches, with the length of the schedule equal to the maximum of the lengths of the branches. After the entire loop is scheduled, the explicit control structure is regenerated by inserting conditionals. Another approach to handle conditionals in a loop body is by performing IF-conversion [7]. The IF-converted (or predicated) code can be scheduled [35] for architectures that support predicated execution [23, 80, 136] as if it were a single basic block. However, the resource usage for predicated code is the sum of the resource usages rather than their union.

The enhanced modulo scheduling method [171] follows an approach similar to software pipelining predicated code. However, it regenerates the explicit control structure as in the hierarchical reduction method [88]. This not only eliminates the disadvantage on resource requirements of predicated methods, but also does not require hardware support for predicated execution. In [173] a software pipelining method that uses multiple II values has been proposed. The scheduling procedure is reminiscent of trace scheduling; the most likely trace of execution is chosen and scheduled separately with the smallest possible II. The next trace is scheduled on top of this trace, filling in holes with an II that is a multiple of the smallest II and so on.

A comprehensive survey of various software pipelining methods can be found in [6, 132]. A survey of the recent advances in software pipelining can be found in [138].

19.6 Scheduling and Register Allocation

In this section we discuss the interaction between instruction scheduling and register allocation, another important phase in an optimizing compiler. Register allocation determines which frequently used variables are kept in registers to reduce memory references. Instruction scheduling and register allocation phases influence each other, so the ordering of these two phases in a compiler is an important issue.

19.6.1   Phase Ordering Issues

In many early compilers instruction scheduling and register allocation phases were performed separately, with each phase being ignorant of the requirements of the other, leading to degradation of performance. The performance degradation can be explained as follows. In *Postpass* scheduling where register allocation precedes instruction scheduling [52, 66], the register allocator, in an attempt to reduce the register requirements, may reuse the same register for different variables. This reuse of registers could result in anti- and output-dependences, which in turn will limit the scheduler's reordering opportunities. On aggressive multiple instruction issue processors, especially those that are statically scheduled, the parallelism lost may far outweigh any penalties incurred due to spill code.

However, in a *Prepass* method [11, 54, 170], instruction scheduling is performed before register allocation. This typically increases the lifetimes of registers, possibly leading to more spills and hence degrading performance. Furthermore, any spill code generated after the register allocation pass may go unscheduled, as scheduling was done before register allocation. This may even lead to illegal schedules in statically scheduled processors if the resources required for the spill code are not available. Therefore, it is customary that Prepass scheduling is followed by register allocation and Postpass scheduling.

19.6.2   Integrated Methods

A number of integrated techniques have been proposed in the literature to introduce some communication between the two phases [17, 19, 54, 126]. These integrated techniques increase the ILP exposed to the processor without drastically increasing the number of spills and hence improve performance considerably. We discuss two of these integrated methods, namely integrated prepass scheduling (IPS) [54] and the parallel interference graph method [126], in detail. A number of other integrated methods have also been proposed in the literature [16, 19, 106, 115], which are reviewed briefly.

19.6.2.1   Integrated Prepass Scheduling

In IPS [54], instruction scheduling precedes register allocation, but the scheduler is given a bound on the number of registers, which guides it to increase parallelism when the register pressure is low and to limit the parallelism otherwise.

The basic idea is to keep track of the number of available registers during the scheduling phase. Each issued instruction may create a new live register and terminate the lifetime of some registers. Hence, this method keeps track of the number of available registers at each scheduling step. The main algorithm switches between two schedulers. When there are enough registers, the scheduler uses CSP (*code scheduler to avoid pipeline delays*), which schedules instructions to avoid delays in pipelined machines. When the number of registers falls below a threshold, the scheduler switches to CSR (*code scheduling to minimize registers usage*), which essentially controls the use of registers.

Switching between CSP and CSR is driven by the number of available registers, AVLREG. AVLREG is increased when a live range ends and decreased when an instruction creates live registers. CSP is responsible for code scheduling most of the time. When AVLREG falls below a threshold (for instance, 1), CSR is invoked. The goal of CSR at this point is to find the next instruction that will not increase the number of live registers, or, if possible, decrease that number. That is, CSR tries to schedule an instruction that frees more registers than the number of live registers it creates. After AVLREG is restored to an acceptable value, CSP resumes scheduling. Thus, IPS performs Prepass scheduling without excessively increasing the register requirements of the schedule.

The scheduling phase is subsequently followed by the global register allocation phase. The IPS scheduler is similar to the list scheduler described in Section 19.4.2. The IPS scheduler uses the DDG of a basic block to perform the scheduling in each basic block.

19.6.2.2   Parallel Interference Graph Method

The integrated technique developed by Pinter is based on the coloring of a graph called the *parallel interference graph* [126]. The graph provides a single framework within which the considerations of both register allocation and instruction scheduling can be applied simultaneously. In this technique, the parallel interference graph — an interference graph that also takes into account scheduling constraints — is first constructed. Using this graph, register allocation is carried out, which is then followed by instruction scheduling. Hence, this is a Postpass method.

The parallel interference graph combines properties of the traditional interference graph and the scheduling graph. However, a simple combination of the two graphs is not possible because the vertices in the two graphs represent different things: the vertices in the interference graph stand for symbolic or virtual registers in the program, while the vertices in the scheduling graph correspond to instructions in the program. Likewise, an edge in the interference graph indicates an interference of live ranges of two symbolic registers, whereas an edge in the scheduling graph represents a precedence constraint between two instructions.

To see how the two graphs are combined consider the example code sequence shown in Figure 19.24a. The live ranges of variables are also shown in the same figure. Figure 19.24b shows the DAG for the code. The DAG gives the precedence constraints of the program. The transitive closure of this graph is generated and the edge directions are removed (refer to Figure 19.24c). The transitive closure edges are shown as dash–dot lines in the graph in Figure 19.24c. To this new graph all the machine-related dependencies that are not of precedence type are added. For example, consider a target machine with only one Integer unit and one Load/Store unit. Then, instructions i1, i2, and i6 that execute on the Load/Store unit form a group. Similarly, instructions i3, i4, and i5, which execute on the Integer unit, form another group. Any pair of instructions in the same group cannot be executed in parallel. This constraint is represented by adding an edge between each pair of instructions in a group. A machine constraint edge, shown as dashed line, is

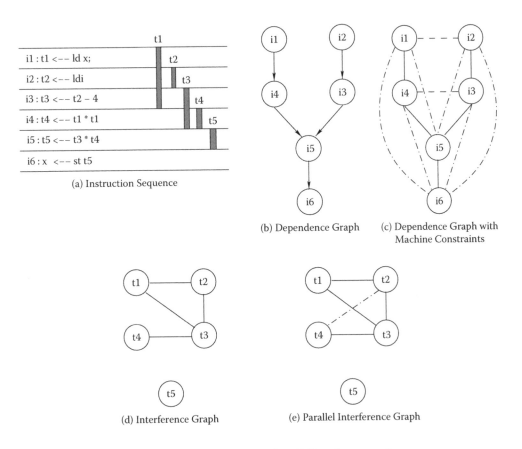

(a) Instruction Sequence

(b) Dependence Graph

(c) Dependence Graph with Machine Constraints

(d) Interference Graph

(e) Parallel Interference Graph

FIGURE 19.24 Construction of parallel interference graph.

added only if neither a dependency edge nor a transitive closure edge already exists between that pair of instructions. Figure 19.24c shows the graph after transitive closure and machine-related edges are added. For example, edges (i1,i2) and (i3,i4) are machine-related edges. The edges in the *complement* of this graph represent the actual parallelism available in the given program. The complement graph consists of only two edges, namely, (i1,i3) and (i2,i4). If we can ensure that the two definitions corresponding to each edge in the complement graph are given different registers, then no false dependence will be introduced by the register allocator. For example, the live ranges for t1 and t3, corresponding to the complement edge (i1, i3), should be given different registers to ensure that no false dependences are introduced between i1 and i3.

In the interference graph, nodes represent symbolic or virtual registers. An edge is added between a pair of nodes in the interference graph if their live ranges overlap. The interference graph for the code sequence is shown in Figure 19.24d. Now the parallel interference graph is built by adding edges from the complement graph to the interference graph, if they are not already present. In our example, the complement edge (i2,i4) should be added to the parallel interference graph. The resulting parallel interference graph is shown in Figure 19.24e. An optimal coloring of this graph will ensure that no false dependence will be introduced. While coloring the graph, if it is found that spill code has to be added, the scheduling edges in the interference graph are removed one at a time to avoid spilling, thus giving up some possible parallelism.

19.6.2.3   Other Integrated Methods

The unified resource allocator (URSA) method deals with function unit and register allocation simultaneously [16]. The method uses a three-phase *measure–reduce–assign* approach, where resource requirements are measured and program regions of excess requirements are identified in the first phase. The second phase reduces the requirements to what is available in the architecture, and the final phase carries out resource assignment. Norris and Pollock [115] proposed a cooperative scheduler-sensitive global register allocator, which is followed by local instruction scheduling. The scheduler-sensitive global register allocator is a graph coloring allocator that takes into consideration the scheduler's objectives throughout each phase of its allocation. The potential for code reordering is reflected in the construction of the interference graph. Scheduling constraints and possibilities are also taken into consideration when the allocator cannot find a coloring and decides to spill.

Bradlee et al. [19] developed an integrated approach called RASE, in which a prescheduling phase is run to calculate cost estimates for guiding the register allocator. A global register allocator then uses the cost estimates and spill costs to obtain an allocation and to determine a limit on the number of local registers for each block. A final scheduler is run using the register limit from allocation and inserting spill code as it schedules.

Motwani et al. studied a combined register allocation and instructions scheduling problem (CRISP) [106]. They formulated the problem as a single optimization problem and proposed an efficient heuristic algorithm, called the (α, β)-combined algorithm. The parameters α and β provide relative weightage for register pressure and ILP.

19.6.2.4   Evaluation of Integrated Methods

Several studies have compared the Prepass and Postpass scheduling methods with integrated techniques [17, 19, 22, 116]. In [19] Bradlee et al. compared three code generation strategies, namely, Postpass, integrated Prepass scheduling, and their own integrated technique called RASE. Their study, conducted for a statically scheduled in-order issue processor, demonstrated that while some level of integration is necessary to produce efficient schedules, the implementation and compilation expense of strategies that very closely couple the two phases is unnecessary. Chang et al. studied the importance of Prepass scheduling using the IMPACT compiler [23]. Their method applies both Prepass and Postpass scheduling to control-intensive nonscientific applications. Their study considers single-issue, superscalar, and superpipelined processors. Their evaluation also included superblock scheduling [74]. Their study reveals that Prepass scheduling does not improve the performance in control-intensive applications when a restricted percolation model is used. With a more general code motion, scheduling before register allocation is important to achieve good speedup, especially for machines with 48 or more registers.

In [116] Norris and Pollock describe a strategy for providing cooperation between register allocation and instruction scheduling. They considered both global and local instruction scheduling techniques. They experimentally compared their strategy with other cooperative and noncooperative techniques. Their results suggest that either a cooperative or noncooperative global instruction scheduling phase, followed by register allocation that is sensitive to the subsequent local instruction scheduling and local instruction scheduling yields good performance over noncooperative methods. Berson et al. [17] compared two previous integrated strategies [54, 115] with their strategy [16], which is based on register *reuse DAGs* for measuring the register pressure. They evaluated register spilling and register splitting methods for reducing register requirements. They studied the performance of the above methods on a six-issue VLIW architecture. Their results reveal that (a) the importance of integrated methods is more significant for programs with higher register pressure, (b) methods that use precomputed information (prior to instruction scheduling) on register demands perform better than the ones that compute register demands on-the-fly, for example, using register pressure as an index for register demands, and (c) live range splitting is more effective than live range spilling.

19.6.3 Phase Ordering in Out-of-Order Issue Processors

Many modern processors (e.g., MIPS R10000 [175], DEC Alpha 21264 [82], and the AMD K5 [148]) support out-of-order issue. In an out-of-order issue processor, instructions are scheduled dynamically with the help of complex hardware support mechanisms such as *register renaming* and *instruction window*. Register renaming is a technique by which logical registers are mapped to hardware physical registers or locations in the reorder buffer [151]. Such mapping removes anti- and output-dependences and hence exposes greater ILP in the program. Furthermore, the number of available physical registers is typically larger (roughly twice) than the number of logical registers visible to the register allocator. The instruction window holds the fetched and decoded instructions; the dynamic issue hardware selects data-ready instructions from the window and issues them. Instructions may be issued in an order different from the original program order. The register-renaming mechanism and the reorder buffer together remove anti- and output-dependences. This, in spirit, is similar to what the integrated register allocation and instruction scheduling techniques do at compile time. This makes the issues in phase ordering for out-of-order issue processors different from those for statically scheduled processors, namely, in-order issue and VLIW processors.

19.6.3.1 Evaluation of Phase Ordering in Out-of-Order Processors

The phase ordering problem in the context of out-of-order issue has been studied in [163, 165]. The study investigates (a) whether complex compile-time techniques do improve the overall performance and (b) whether a Prepass-like or a Postpass-like approach should be followed for out-of-order issue processors. The study observes an insignificant improvement in performance due to integrated methods when scheduling is limited to basic blocks. Furthermore, it advocates Postpass-like methods, as it is important to minimize register spills in out-of-order issue processors, even at the expense of obscuring some ILP [163, 165].

19.6.3.2 Minimum Register Instruction Sequencing

Recall the optimal code generation problem and a solution to it, the Sethi–Ullman method for integrated code generation and register allocation, discussed in Section 19.3.3.1. This problem is revisited in the context of out-of-order issue superscalar processors in [59]. The problem addressed in this work is that of obtaining an instruction sequence for a DAG that uses the minimum number of registers. This problem, termed minimum register instruction sequencing (MRIS), is motivated by the fact that in out-of-order issue processors it is important to reduce the number of register spills, even at the expense of not exposing ILP. It should be noted here that the MRIS problem and its solution take into account neither the resource constraints in the architecture nor the execution latencies of instructions. The emphasis here is to generate an instruction *sequence* rather than a *schedule*.

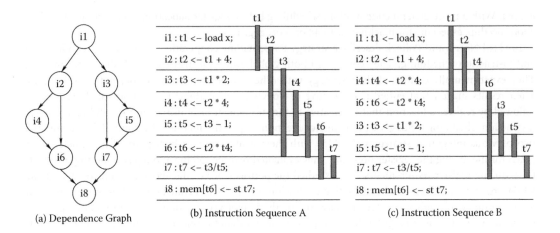

(a) Dependence Graph   (b) Instruction Sequence A   (c) Instruction Sequence B

FIGURE 19.25   Instruction sequences with different register requirements.

Let us motivate the MRIS problem with the help of an example. Consider the computation represented by the DAG shown in Figure 19.25. Two possible instruction sequences for this DAG are also shown in the figure along with the live ranges of the variables t1 to t7. For the instruction sequence A shown in Figure 19.25b, four variables are simultaneously live during instruction i5; therefore, four registers are required for sequence A. If the number of available registers is less than four, sequence A will result in spill loads/stores. However, for sequence B, shown in Figure 19.25c, only three variables are simultaneously live, so this sequence requires only three registers. In this example, the minimum register requirement is three. Hence, the sequence shown in Figure 19.25c is one of the minimum register sequences.

A solution to the MRIS problem proposed in [59] proceeds by identifying which instructions can share the same register in *any* legal instruction sequence. A complete answer to this question is known to be NP-hard [50]. The approach proposed in [59] uses the notion of an *instruction lineage*, which corresponds to a sequence of instructions that forms a path in the DAG, that is, a sequence of instructions {i1, i2, i3, ..., in} in the DAG, where i2 is the successor of i1, i3 is the successor of i2, and so on. When {i1, i2, ..., in} forms a *lineage*, the instructions in a lineage share the same register. That is, the register assigned to i1 is passed on to i2 (i1's *heir*), which is passed on to i3, and so on. Because of data dependence between pairs of instructions in the lineage, any legal sequence will order the instructions as i1, i2, ..., in. Hence, the instructions in a lineage can certainly share the same destination register.

When an instruction i1 has more than one successor, one of the successors, for instance, i2, is chosen as the legal heir. To make i2 the *last use* instruction of i1, and hence reuse the destination register, sequencing arcs are added from each successor of i1 to the chosen heir i2. For example, for the DAG shown in Figure 19.25a, L1 = [i1, i3, i7, i8) forms a lineage. Typically, the last node in a lineage either is a store node (in this case, i8) or is already in some other lineage. Thus, all instructions in a lineage except the last one share the same destination register. To emphasize that the last instruction in a lineage does not use the same destination register, a semi-open interval notation is used for a lineage, as in L1 = [i1, i3, i7, i8). Since instruction i3 is chosen as the heir of i1, a sequencing edge is added from i2 to i3. A simple but efficient heuristic based on the maximum distance (MaxDistance), measured in terms of the path length to the sink node, is used to select heirs. If the MaxDistance heuristic used is dynamic, that is, it is calculated after the introduction of each set of sequencing edges, then the introduction of sequencing edges does not introduce cycles in the DAG [59]. The remaining lineages for the DAG are L2 = [i2, i6, i8), L3 = [i5, i7), and L4 = [i4, i6).

To address the question of whether the live ranges of two lineages definitely overlap in any legal schedule, a sufficient condition is established in [59]. The sufficient condition tests whether there exists a path from the start node of lineage L1 to the end node of L2 and vice versa. If such paths exist, the live ranges of two lineages overlap in all legal sequences, and the lineages cannot share the same register. In our example, lineages L1 and L2 overlap. So do the pairs of lineages (L1,L3), (L1,L4), (L2,L3), and (L2,L4), but lineages L3 and L4 do not necessarily overlap in all sequences. Hence, they can be made to share the same register. Doing so would result in sequencing the execution of some of the instructions because of false dependences. However, in an out-of-order issue processor, these false dependences would be removed at runtime, and hence the parallelism will be exposed.

Based on the overlap relation, a lineage interference graph is constructed and colored using a traditional graph coloring algorithm. The number of colors required to color the graph is a heuristic lower bound on the minimum registers required. Using this lower bound as a guideline, a modified list scheduling method is used to generate a sequence that results in a near-optimal solution to the MRIS problem. This approach to the MRIS problem was found to be very effective in reducing the register pressure and in reducing the number of spill loads and stores in a number of SPEC benchmarks. Although the sequencing method does not take into consideration resource constraints and execution latencies, the execution time of the generated sequence was found to be comparable to that generated by a production-quality compiler.

19.6.3.3 Linearization of the Instruction Schedule

A superscalar processor expects a linear sequence of instructions. Hence, a parallel instruction schedule, such as the one shown in Figure 19.12, is presented to a superscalar processor by *linearizing* it in a simple way. The linearization method sequences the instructions in each cycle of the schedule in a left-to-right order. However, simple linearization methods are not aware of the register-renaming capabilities of out-of-order issue superscalar processors and may generate a sequence that would have higher register pressure. As a consequence, it may result in spill code. Furthermore, the linearization does not take into account the size of the instruction window [151], the register-renaming capabilities of superscalar architecture, or the in-order graduation mechanism. These may result in certain inefficiencies in the form of stall cycles. An efficient linearization method that is sensitive to register pressure and is aware of the architectural features of out-of-order issue superscalar processors has been proposed in [147].

The linearization proposed in [147] uses a set of *matching* conditions that ensure the ILP available in the given parallel schedule is not lost in the linearization process. The linearization method is an extension of the list scheduling method and adds instructions to the linear sequence in such a way that it reduces the register pressure without losing any parallelism compared to the given parallel schedule. The method was applied to basic blocks.

19.7 Recent Research in Instruction Scheduling

In this section we report some of the recent research work on instruction scheduling.

19.7.1 Instruction Scheduling for Application-Specific Processors

Instruction scheduling methods have been proposed for application-specific processors such as digital signal processing (DSP). Originally, most DSP applications, or their important kernels, were hand-coded in assembly language. The application programmer is required to perform the necessary instruction re-ordering to take full advantage of the parallelism that is available in these processors. With the increasing complexity of the processors and their programmability, programming in higher-level languages and compilation techniques to produce efficient code automatically are becoming increasingly important. A major challenge in applying existing instruction scheduling methods arises from the irregularities of DSP processors [92]. These irregularities include having special-purpose registers in the data path, heterogeneous registers, dedicated memory address generation units, chained instructions such as multiply-and-accumulate, saturation arithmetic, multi-stage functional units, and parallel memory banks. Furthermore, in most cases, code running on a DSP's processor also has to meet real-time constraints. Thus, instruction scheduling for DSP processors, which needs to take into account the irregularities of the architectures and the real-time constraints in resource usage, poses a major challenge.

Several instruction scheduling methods for DSP processors have been proposed. A DSP-specific code compaction technique was developed in [159] that considers both resource and timing constraints. Instruction scheduling for the TriMedia VLIW processor [125] is reported in [69]. The instruction scheduling problem is transformed into an integer linear program problem in [95]. Another integer linear program formulation for integrated instruction scheduling and register allocation is proposed in [20]. Methods for simultaneous register allocation and instruction scheduling for DSP processors (involving heterogeneous registers) are proposed in [28, 96]. An extensive survey of code generation for signal processing systems is presented in [18].

Certain DSP processors, for example, the Texas Instruments TMS320C series [157], support multiple *operating modes*, such as the sign-extension mode and product-shift mode, which provide slightly different execution semantics for instructions [9]. Multiple operating modes raise another interesting instruction scheduling problem. Here the objective is to schedule instructions, making use of the multiple modes, while reducing the number of mode setting instructions required, and hence the associated overhead cost.

Code size is another key concern in application-specific processors. As code size relates to on-chip program memory in these processors, code size can directly influence the cost of the system. Hence, compilation methods in general, and instruction scheduling methods in particular, optimize the code not only for performance, but also for code size. In the presence of code size constraints, the scope of global code scheduling methods is limited, as these methods are known for their code bloating problem.

The TMS C6x DSP processor, well known for its compiler-friendly architecture, has a cluster of functional units [157]. Each cluster has a register file of its own. Each functional unit in a cluster can access two read ports and one write port to its local register file. In addition, at most one functional unit in a cluster can access a register file in a cross cluster in any given time step. If more than one functional unit needs to access data across clusters, or multiple accesses to a cross register file are needed, it is desirable to explicitly copy this data to a local register.

In this type of architecture, known as *clustered architecture*, associated with the instruction scheduling is the problem of *mapping*, which assigns an instruction and its destination operand to a cluster. Methods that perform instruction scheduling before mapping [21] or mapping before scheduling [134] could result in poor schedules, as the first phase makes certain decisions without knowing their consequences on the subsequent phase. To take full advantage of the ILP that can be exploited in this architecture, a compiler needs to perform instruction scheduling along with instruction mapping. This requires that the two problems, assignment and scheduling, be solved in a unified framework as in [94, 118]. A modulo scheduling method for clustered architecture is discussed in [117, 141].

19.7.2   Instruction Scheduling for IA-64

The explicitly parallel instruction computing (EPIC) architecture is a synthesis of VLIW and superscalar design principles [143]. In this architecture, the compiler is responsible for exploiting the available ILP as in a VLIW architecture. However, the hardware routes instructions to appropriate functional units from a compact instruction encoding and inserts necessary stalls for data dependences to address the problem of increased code size in VLIW architecture. The Itanium (IA-64) architecture [145] is a statically scheduled architecture that follows the EPIC approach. The architecture also has several kinds of hardware support, for example, predicated execution and rotating registers [136], for exploiting higher ILP. In this architecture, the compiler exposes the ILP and packs independent instruction into instruction bundles. Certain restrictions on which types of instructions can be grouped into a bundle, along with which instructions are independent and what resources are available in a given cycle, make the problem of instruction scheduling and bundling an interesting problem. We discuss instruction scheduling in IA-64 in some detail in this section.

Each instruction in IA-64 is categorized into one of six instruction types: integer ALU (A-type), non-ALU integer (I-type), memory (M-type), floating point (F-type), branch (B-type), and extended (L/X-type). An instruction bundle contains three instructions. However, only certain groupings of instruction types are allowed. More specifically, only 12 specific bundle types are allowed. The bundle type of an instruction is specified in the 5-bit template field of the instruction. The template bits help in decoding and routing instructions and indicate the position of stops that mark the end of groups of instructions that can be executed in parallel. IA-64 can issue a maximum of two bundles, consisting of a maximum of six instructions per clock cycle. These six slots can contain at most two I-type, two M-type, two F-type, and three B-type instructions.

The problem of scheduling in IA-64 involves reordering of instructions, functional unit binding, and bundle allocation. Kastner and Winkel proposed an integer linear program–based approach for basic block instruction scheduling for IA-64 [79]. They addressed the problem by subdividing it into two phases, namely, macro-scheduling and bundling. The objective of the macro-scheduling phase is to find a minimum-length schedule, where instructions are grouped into instruction groups honoring the dependence constraints. The instructions in one group can execute simultaneously in a single cycle. The number of instructions in a particular instruction group is also bound by the number of instructions that can be issued in a single cycle, six in the case of IA-64. The macro-scheduling problem is formulated as an integer linear program.

The bundling phase transforms the sequence of instruction groups into an efficient sequence of feasible IA-64 bundles. The bundling phase splits the instruction group according to the bundle constraints. Such splitting is necessary since the macro-scheduling phase only ensures that there are at most six instructions in the instruction group. However, due to bundling restrictions, it may be necessary to generate more than two bundles. Second, empty slots in the bundle have to be filled with a no-op operation, and a no-op operation in IA-64 must be associated with an instruction type. A naive selection of the no-op type may result in further splits, causing more penalties. For example, if the instruction group contains two M-type instructions, and we need to associate a type to the no-op in the group, then associating M-type to the no-op will necessitate the instruction group to be split into more than two bundles. The bundling phase itself is divided into two subphases, namely, micro-scheduling and sequencing. Micro-scheduling generates a separate bundle sequence for each instruction group that is a partitioning problem, modeled again as an integer linear programming problem. The result of micro-scheduling is an optimal bundle sequence for each instruction group. These partial bundle sequences are combined by the sequencing phase in a space-efficient manner, minimizing the number of bundles, to form the final schedule.

The two-phase approach of macro-scheduling and bundling proposed by Kastner and Winkel may result in a suboptimal solution compared to an integrated approach that combines both the phases. However, the complexity of the integrated approach, especially in the context of an integer linear program problem that needs to solved, is very high and may negate the benefit.

19.7.3 Instruction Scheduling for Spatial Architectures

As the process technology used in modern processors continues to shrink to below 100 nanometers, wire delays — rather than gate delays — dominate and limit the frequency scalability of the architecture [119]. To overcome the impact of wire delays, heavily partitioned execution substrates (processing elements) have been proposed. Examples of these architectures include clustered VLIW architecture [21], Raw [156, 167], instruction-level distributed processing [84], Grid [110], TRIPS [142], and Wavescalar [155]. Of these, the Raw, Grid, TRIPS, and Wavesacalar architectures, known as spatial or tiled architectures, distribute their computing resources and require a compiler to explicitly partition instructions across the computing resources. Communication between distant resources can incur one or more cycles of delays. Furthermore, the communication across tiles is exposed to the compiler to make intelligent trade-offs between the amount of communication introduced and the parallelism exploited in the execution of the schedule. Thus, instruction scheduling becomes both a spatial mapping and a temporal ordering problem. We discuss the space–time scheduling methods for these architectures in this section. In particular, we discuss instruction scheduling for the TRIPS architecture [142].

TRIPS architecture consists of a grid of execution units connected by a thin operand routing network. The latency required to pass a value between ALUs is dependent on the distance traveled by the operand. Each ALU includes an integer unit, a floating unit, and an instruction buffer for holding instructions and their output values. Each block of instruction can have up to 128 instructions, and eight such blocks can be simultaneously in execution.

The TRIPS architecture executes the program as a sequence of multi-instruction blocks that are executed atomically. Within each block, the compiler encodes instruction dependences as a statically constructed data flow graph. Since these architectures expose the placement of instructions on a grid of ALUs to the compiler, a static placement and dynamic issue (SPDI) paradigm [109] is preferred, where the compiler selects the ALU in which the instruction is going to execute (static placement) and the hardware executes these instructions when their operands become ready (dynamic placement). This is in contrast to VLIW architectures, in which the compiler identifies both the time slot and the functional unit to which the instruction is issued, and superscalar processors, where both the issue and mapping of instructions to functional units are performed by the hardware.

A scheduler for such an architecture needs to map the data flow graph onto a grid of ALUs [109, 149]. A greedy list scheduling algorithm for TRIPS [109] tries to minimize the completion time by minimizing the static routing latency between a pair of instructions and at the same time exploiting ILP. The scheduler determines the number of instruction slots needed to schedule a group of instructions. Then a sorted list of instructions based on their priority is maintained. The priority is assigned based on the instruction's depth from the root node and its height from the leaf node in the data flow graph. The ready time includes the routing delay of the operand(s) to the ALU from other ALUs. More specifically, the time at which an instruction is ready to execute at a given slot is determined by the latest time at which the parent of the instruction completes added to the communication latency from the execution unit of the parent to the instruction slot selected. Then the instruction is assigned to a slot that minimizes the time at which the instruction is ready to execute. The decision on which slot to place an instruction is guided by a number of heuristics, including critical path ordering, load balancing, cache locality, and register output [109].

Spatial path scheduling [33] improves the performance when compared to the greedy scheduling method [109]. The basic algorithm computes criticality based on routing distances using all known anchor points, that is, fixed positions for operations in the block, such as register accesses. The basic algorithm is augmented with heuristics to model contention on the ALUs and network links, estimation of interblock critical paths, and lookahead planning of path routes. In addition, it uses a simulated annealing and a weight-based heuristic to obtain improved schedules.

The problem of instruction placement has significant impact on the performance of the spatial scheduling algorithms. An instruction placement performance model is developed in [104] that takes into consideration several factors that contribute to overall performance. Their model is composed of three separate components, namely, inter-instruction operand latency, data cache coherence overhead, and contention for computing resources (ALUs). They also propose a unified model that combines these components in proportion to their relative contribution to overall performance.

RAWCC [93], the compiler for the Raw architecture [156, 167], compiles general-purpose sequential programs to exploit ILP in the distributed Raw architecture. The Raw compiler handles the orchestration of computation and communication by performing spatial and temporal instruction scheduling, as well as data partitioning using a distributed on-chip memory model to exploit ILP. The central tasks of the orchestrater are the assignment and scheduling of instructions in a basic block. The Raw compiler performs assignments in three steps:

- **Clustering:** Clustering groups together instructions in such a way that the benefits obtained by executing the instructions within a cluster in parallel are overshadowed by the cost of communication across the tiles.
- **Merging:** Merging reduces the number of clusters down to the number of processing units by merging the clusters.
- **Placement:** Placement performs the mapping of merged clusters to the processing units (or tiles), taking into account the topology of the interconnect.

The assignment phase is followed by a scheduling phase that is performed with a traditional list scheduler. In [129] a convergent scheduler is proposed that is composed of independent passes, each implementing a heuristic that addresses a particular problem or constraint in the spatial and temporal instruction scheduling. Each pass provides spatial and temporal preferences for each instruction, which are not absolute. Furthermore, each pass expresses the confidence of its preferences, as well as preferences for multiple space and time slots. By applying a series of passes that address all the relevant constraints, the convergent scheduler can produce a schedule that satisfies all the important constraints. Passes can be run multiple times and in any order. The order in which the different passes are applied can vary with the architecture. Hence, using genetic programming, a machine learning technique, convergent scheduling automatically finds good ordering of the passes for a given architecture [129].

19.7.4 Instruction Scheduling for Low Power

Another area that has been receiving increasing attention is instruction scheduling for low power in embedded processors [123]. Embedded processors are used in many handheld devices, such as cell phones, pagers, digital cameras, toys, and so on, in which battery longevity and size are key factors that determine system cost. As the overall power dissipated or energy consumed directly relates to battery life, embedded processors generally have low power requirements. In these systems, it is quite common that an application is compiled more for power efficiency than for performance. By reordering instructions, the power dissipated can be decreased.

The transition or switching activities — toggling of signals from 0 to 1 or vice versa — that take place on the system bus, or more specifically on the instruction bus, can be reduced by instruction reordering methods, which in turn help reduce power. Su et al. [154] propose a technique that reorders instructions in such a way that the toggles between the encodings or machine codes of adjacent instructions are reduced. This is accomplished by a simple list scheduling method that uses a priority function that gives a higher priority to an instruction in the ReadyList that has the lowest power cost. The power cost of an instruction is estimated based on the last scheduled instruction in the partial schedule and a power cost table. The essential idea of this method is to reduce the amount of switching activity between adjacent instructions to reduce the power consumed.

Another scheduling method for reducing power consumption is presented by Tiwari et al. [160]. The goal of this work is to judiciously select instructions as opposed to reordering them to reduce power consumption. This approach uses a power table, which contains power consumed by individual instructions as well as certain commonly paired instructions. Using this power table, code is rescheduled to use instructions that result in less power consumption. In [161] a method to reduce the peak power dissipation is proposed. This method uses a predefined per cycle energy dissipation threshold and limits the number of instructions that can be scheduled in a given cycle based on this threshold. A method to reduce the power consumption on the instruction bus of a VLIW architecture is proposed in [91]. Each VLIW instruction, referred to as a long word instruction, consists of a number of instructions or operations. This method uses a greedy approach to reschedule operations within a long instruction word, as well as across long instruction words, but within a limited instruction window size, to reduce the switching activities among the instructions. When the operations are rescheduled across long word instructions, dependence constraints are preserved. The method attempts to obtain a schedule that consumes low power, without sacrificing performance. Several scheduling strategies that attempt to reduce energy consumption with and without sacrificing performance are evaluated in [121].

More recently, a number of scheduling methods [86, 146] have been proposed that deal with voltage and frequency scaling, an approach in which the operating voltage and/or the operating frequency are scaled down to reduce power consumption.

19.8 Summary

Instruction scheduling methods rearrange instructions in a code sequence to expose ILP for multiple instruction issue processors and to reduce the number of stall cycles incurred in single-issue, pipelined processors. Simple scheduling methods that cover pipeline stalls use information on the number of stall cycles required between dependent instructions. Basic block instruction scheduling methods are limited to rearranging instructions in a straight line code sequence with a single control flow entry and exit. In this chapter, we have reviewed several approaches to basic block instruction scheduling, including list scheduling, operation scheduling, and integer linear programming–based methods. The heuristics used in list scheduling methods, and resource models used for modeling complex resource usage patterns, have also been discussed.

Global scheduling refers to instruction scheduling that extends beyond instructions in a basic block. In the case of global acyclic scheduling, the control flow graph on which the scheduling method is applied is acyclic. Trace scheduling and superblock scheduling consider an acyclic control flow subgraph that consists of a single control flow path. In hyperblock scheduling and treegion scheduling, multiple control flow paths can be explored in a single trace. Software pipelining methods schedule multiple instances of either a single or multiple static basic block, corresponding to multiple iterations, of a cyclic control flow graph.

Register allocation is a closely related ILP compilation issue. Instruction scheduling and register allocation phases in an optimizing compiler influence each other. Issues related to the phase ordering of instruction scheduling and register allocation in both statically scheduled and dynamically scheduled multiple instruction issue processors have been discussed in this chapter. The need to reduce register spills, even at the expense of obscuring some ILP, has resulted in new approaches for instruction sequencing. Last, recent research on the application of instruction scheduling methods for application-specific processors and low-power embedded systems has been briefly discussed.

Acknowledgments

The author is grateful to Jose Nelson Amaral, Guang R. Gao, Matthew Jacob, and Hongbo Yang for their numerous suggestions on the various drafts of this chapter and to Bhama Sridharan and T. S. Subashree for proofreading an initial draft. The author acknowledges Jose Nelson Amaral, Erik R. Altman, A. Dani, N. S. S. Narasimha Rao, S. M. Sandhya, V. Janaki Ramanan, Hongbo Rong, Raul Silvera, Madhavi G. Valluri, Jian Wang, Hongbo Yang, and Chihong Zhang, with whom he has coauthored several papers on instruction scheduling and software pipelining that have directly and indirectly contributed to this chapter. Special thanks to Santosh G. Nagarakatte, who helped revise the chapter for the second edition. Last, the research funding received from the Department of Science and Technology, India, and from the Indian Institute of Science, Bangalore, which supported this work, is acknowledged.

References

1. S. G. Abraham, V. Kathail, and B. L. Deitrich. 1996. Meld scheduling: Relaxing scheduling constraints across region boundaries. In *Proceedings of the 29th Annual International Symposium on Microarchitecture*, 308–21. Washington, DC: IEEE Computer Society.
2. A. V. Aho, S. C. Johnson, and J. D. Ullman. 1977. Code generation for expressions with common subexpressions. *Journal of the ACM* 24(1):146–60.
3. A. V. Aho, R. Sethi, and J. D. Ullman. 1988. *Compilers — Principles, techniques, and tools*, corrected ed. Reading, MA: Addison-Wesley.
4. A. Aletà, J. M. Codina, A. González, and D. Kaeli. 2005. De-mystifying on-the-fly spill code. In *Proceedings of the 2005 ACM SIGPLAN Conference on Programming Language Design and Implementation*, 180–89. New York: ACM Press.
5. V. H. Allan and S. J. Allan. 2002. Software pipelining. In *The compiler design handbook: Optimization and machine code generation*, ed. Y. N. Srikant and P. Shankar. Boca Raton, FL: CRC Press.
6. V. H. Allan, R. Jones, R. Lee, and S. J. Allan. 1974. Software pipelining. *ACM Computing Surveys* 27(3):367–432.
7. J. R. Allen, K. Kennedy, C. Porterfield, and J. Warren. 1983. Conversion of control dependence to data dependence. In *Conference Record of the Tenth Annual ACM Symposium on Principles of Programming Languages*, 177–89.

8. E. R. Altman, R. Govindarajan, and G. R. Gao. 1995. Scheduling and mapping: Software pipelining in the presence of structural hazards. In *Proceedings of the ACM SIGPLAN '95 Conference on Programming Language Design and Implementation*, 139–50. New York: ACM Press.

9. G. Araujo, S. Devadas, K. Keutzer, S. Liao, S. Malik, A. Sudarsanam, S. Tjiang, and A. Wang. 1995. Challenges in code generation for embedded processors. In *Code generation for embedded processors*, ed. P. Marwedel and G. Goossens. Boston, MA: Kluwer Academic.

10. S. Arya. 1985. An optimal instruction-scheduling model for a class of vector processors. *IEEE Transactions on Computers* 34(11):981–95.

11. M. Auslander and M. Hopkins. 1982. An overview of the PL.8 compiler. In *Proceedings of the SIGPLAN '82 Symposium on Compiler Construction*, 22–31. New York: ACM Press.

12. V. Bala and N. Rubin. 1995. Efficient instruction scheduling using finite state automata. In *Proceedings of the 28th Annual International Symposium on Microarchitecture*, 46–56. Los Alamitos, CA: IEEE Computer Society Press.

13. T. Ball and J. R. Larus. 1994. Optimally profiling and tracing programs. *ACM Transactions on Programming Languages and Systems* 16(4):1319–60.

14. G. R. Beck, D. W. L. Yen, and T. L. Anderson. 1993. The Cdra 5 minisupercomputer: Architecture and implementation. *Journal of Supercomputing* 7:143–80.

15. D. Bernstein and M. Rodeh. 1991. Global instruction scheduling for superscalar machines. In *Proceedings of the ACM SIGPLAN '91 Conference on Programming Language Design and Implementation*, 241–55. New York: ACM Press.

16. D. A. Berson, R. Gupta, and M. L. Soffa. 1993. URSA: A unified resource allocator for registers and functional units in VLIW architectures. In *Proceedings of the IFIP WG 10.3 Working Conference on Architectures and Compilation Techniques for Fine and Medium Grain Parallelism*, ed. M. Cosnard, K. Ebcioğlu, and J.-L. Gaudiot, number A-23 in IFIP Transactions, 243–54. Amsterdam: North-Holland.

17. D. Berson, R. Gupta, and M. L. Soffa. 1998. An evaluation of integrated scheduling and register allocation techniques. In *Proceedings of the 11th International Workshop on Languages and Compilers for Parallel Computing*. Lecture Notes in Computer Science. Heidelberg, Germany: Springer-Verlag.

18. S. S. Bhattacharyya, R. Leupers, and P. Marwedel. 2000. Software synthesis and code generation for signal processing systems. *IEEE Transactions on Circuits and Systems II — Analog and Digital Signal Processing* 47(9):849–75.

19. D. G. Bradlee, S. J. Eggers, and R. R. Henry. 1991. Integrating register allocation and instruction scheduling for RISCs. In *Proceedings of the Fourth International Conference on Architectural Support for Programming Languages and Operating Systems*, 122–31. New York: ACM Press.

20. T. Bruggen and A. Ropers. 1999. Optimized code generation for digital signal processors. Technical report, Institute for Integrated Signal Processing Systems, 52074 Aachen, Germany.

21. A. Capitanio, N. Dutt, and A. Nicolau. 1992. Partitioned register files for VLIWs: A preliminary analysis of tradeoffs. In *Proceedings of the 25th Annual International Symposium on Microarchitecture*, 292–300. Los Alamitos, CA: IEEE Computer Society Press.

22. P. P. Chang, D. M. Lavery, S. A. Mahlke, W. Y. Chen, and W. W. Hwu. 1995. The importance of prepass code scheduling for superscalar and superpipelined processors. *IEEE Transactions on Computers* 44(3):353–70.

23. P. P. Chang, S. A. Mahlke, W. Y. Chen, N. J. Warter, and W. W. Hwu. 1991. IMPACT: An architectural framework for multiple-instruction-issue processors. In *Proceedings of the 18th Annual International Symposium on Computer Architecture*, 266–75. New York: ACM Press.

24. L.-F. Chao and E. H.-M. Sha. 1993. Rate-optimal static scheduling for DSP dataflow programs. In *Proceedings of 1993 Great Lakes Symposium on VLSI*, 80–84.

25. A. E. Charlesworth. 1981. An approach to scientific array processing: The architectural design of the AP-120B/FPS-164 family. *Computer* 14(9):18–27.

26. S. Chaudhuri, R. A. Walker, and J. E. Mitchell. 1994. Analyzing and exploiting the structure of the constraints in the ILP approach to the scheduling problem. *IEEE Transactions on Very Large Scale Integration (VLSI) Systems* 2(4):456–71.

27. C. Chekuri, R. Motwani, R. Johnson, B. Natarajan, B. R. Rau, and M. Schlansker. 1996. Profile-driven instruction level parallel scheduling with applications to super blocks. In *Proceedings of the 29th Annual International Symposium on Microarchitecture*, 58–67. Washington, DC: IEEE Computer Society.

28. W.-K. Cheng and Y.-L. Lin. 1999. Code generation of nested loops for DSP processors with heterogeneous registers and structural pipelining. *ACM Transactions on Design Automation of Electronic Systems (TODAES)* 4(3):231–56.

29. H.-C. Chou and C.-P. Chung. 1995. An optimal instruction scheduler for superscalar processor. *IEEE Transactions on Parallel and Distributed Systems* 6(3):303–13.

30. E. G. Coffman. 1976. *Computer and job-shop scheduling theory.* New York: John Wiley & Sons.

31. D. Cohen. 1977. A methodology for programming a pipeline array processor. In *Proceedings of the Tenth Annual Microprogramming Workshop*, 82–89. Piscataway, NJ: IEEE Press.

32. R. P. Colwell, R. P. Nix, J. J. O'Donnell, D. B. Papworth, and P. K. Rodman. 1988. A VLIW architecture for a trace scheduling compiler. *IEEE Transactions on Computers* 37(8).

33. K. E. Coons, X. Chen, D. Burger, K. S. McKinley, and S. K. Kushwaha. 2006. A spatial path scheduling algorithm for EDGE architectures. In *Proceedings of the 12th International Conference on Architectural Support for Programming Languages and Operating Systems*, 129–40. New York: ACM Press.

34. A. Dani, V. Janaki Ramanan, and R. Govindarajan. 1998. Register-sensitive software pipelining. In *Proceedings of the 12th International Parallel Processing Symposium and the 9th Symposium on Parallel and Distributed Processing.* Washington, DC: IEEE Computer Society.

35. J. C. Dehnert, P. Y.-T. Hsu, and J. P. Bratt. 1989. Overlapped loop support in the Cydra 5. In *Proceedings of the Third International Conference on Architectural Support for Programming Languages and Operating Systems*, 26–38: 967–979. New York: ACM Press.

36. J. C. Dehnert and R. A. Towle. 1993. Compiling for Cydra 5. *Journal of Supercomputing* 7:181–27.

37. B. L. Deitrich and W. W. Hwu. 1996. Speculative hedge: Regulating compile-time speculation against profile variations. In *Proceedings of the 29th Annual International Symposium on Microarchitecture*, 70–79. Washington, DC: IEEE Computer Society.

38. K. Ebcioglu and A. Nicolau. 1989. A global resource-constrained parallelization technique. In *Conference Proceedings, 1989 International Conference on Supercomputing*, 154–63. New York: ACM Press.

39. A. E. Eichenberger and E. S. Davidson. 1995. Stage scheduling: A technique to reduce the register requirements of a modulo schedule. In *Proceedings of the 28th Annual International Symposium on Microarchitecture*, 338–49. Los Alamitos, CA: IEEE Computer Society Press.

40. A. E. Eichenberger and E. S. Davidson. 1996. A reduced multipipeline machine description that preserves scheduling constraints. In *Proceedings of the ACM SIGPLAN '96 Conference on Programming Language Design and Implementation*, 12–22. New York: ACM Press.

41. A. E. Eichenberger and E. S. Davidson. 1997. Efficient formulation for optimal modulo schedulers. In *Proceedings of the ACM SIGPLAN '97 Conference on Programming Language Design and Implementation*, 194–205. New York: ACM Press.

42. A. E. Eichenberger, E. S. Davidson, and S. G. Abraham. 1994. Minimum register requirements for a modulo schedule. In *Proceedings of the 27th Annual International Symposium on Microarchitecture*, 75–84. New York: ACM Press.

43. J. R. Ellis. 1984. Bulldog: A compiler for VLIW architectures. Research Report YALEU/DCS/RR-364, Department of Computer Science, Yale University, New Haven, CT.

44. P. Feautrier. 1995. Fine-grain scheduling under resource constraints. In *Proceedings of the 7th International Workshop on Languages and Compilers for Parallel Computing*, ed. K. Pingali, U. Banerjee, D. Gelernter, A. Nicolau, and D. Padua, 1–15. Lecture Notes in Computer Science, vol. 892. Heidelberg, Germany: Springer-Verlag.

45. J. Ferrante, K. J. Ottenstein, and J. D. Warren. 1987. The program dependence graph and its use in optimization. *ACM Transactions on Programming Languages and Systems* 9(3):319–49.

46. J. A. Fisher. 1981. Trace scheduling: A technique for global microcode compaction. *IEEE Transactions on Computers* 30(7):478–90.

47. J. A. Fisher. 1983. Very long instruction word architectures and the ELI-512. In *Proceedings of the 10th Annual International Symposium on Computer Architecture*, 140–50. Los Alamitos, CA: IEEE Computer Society Press.

48. J. A. Fisher. 1993. Global code generation for instruction-level parallelism: Trace scheduling-2. Technical Report HPL-93-43, Hewlett-Packard Laboratory, Palo Alto, CA.

49. J. A. Fisher, D. Landskov, and B. D. Shriver. 1981. Microcode compaction: Looking backward and looking forward. In *Proceedings of the 1981 National Computer Conference*, 95–102.

50. M. R. Garey and D. S. Johnson. 1979. *Computers and intractability: A guide to the theory of NP-completeness.* New York: W. H. Freemann.

51. F. Gasperoni and U. Schwiegelshohn. 1991. Efficient algorithms for cyclic scheduling. Research Report RC 17068, IBM T. J. Watson Research Center, Yorktown Heights, NY.

52. P. B. Gibbons and S. S. Muchnick. 1986. Efficient instruction scheduling for a pipelined architecture. In *Proceedings of the SIGPLAN '86 Symposium on Compiler Construction*, 11–16. New York: ACM Press.

53. M. C. Golumbic and V. Rainish. 1990. Instruction scheduling beyond basic blocks. *IBM Journal of Research and Development* 34(1):93–97.

54. J. R. Goodman and W.-C. Hsu. 1988. Code scheduling and register allocation in large basic blocks. In *Conference Proceedings, 1988 International Conference on Supercomputing*, 442–52. New York: ACM Press.

55. R. Govindarajan, E. R. Altman, and G. R. Gao. 1994. Minimizing register requirements under resource-constrained rate-optimal software pipelining. In *Proceedings of the 27th Annual International Symposium on Microarchitecture*, 85–94. New York: ACM Press.

56. R. Govindarajan, E. R. Altman, and G. R. Gao. 1996. Co-scheduling hardware and software pipelines. In *Proceedings of the Second International Symposium on High-Performance Computer Architecture*, 52–61. Washington, DC: IEEE Computer Society.

57. R. Govindarajan, E. R. Altman, and G. R. Gao. 1996. A framework for resource-constrained rate-optimal software pipelining. *IEEE Transactions on Parallel and Distributed Systems* 7(11):1133–49.

58. R. Govindarajan, N. S. S. Narasimha Rao, E. R. Altman, and G. R. Gao. 2000. Enhanced co-scheduling: A software pipelining method using modulo-scheduled pipeline theory. *International Journal of Parallel Programming* 28(1):1–46.

59. R. Govindarajan, H. Yang, C. Zhang, J. Nelson Amaral, and G. R. Gao. 2001. Minimum register instruction sequence problem: Revisiting optimal code generation for dags. In *Proceedings of the International Parallel and Distributed Processing Symposium*.

60. R. Gupta, E. Mehofer, and Y. Zhang. 2002. Profile guided compiler optimizations. In *The compiler design handbook: Optimization and machine code generation*, ed. Y. N. Srikant and P. Shankar. Boca Raton, FL: CRC Press.

61. R. Gupta and M. L. Soffa. 1987. Region scheduling. In *Proceedings of the Second International Conference on Supercomputing '87*, ed. L. P. Kartashev and S. I. Kartashev, vol. III, 141–48.

62. J. C. Gyllenhaal, W. W. Hwu, and B. R. Rau. 1996. Optimization of machine descriptions for efficient use. In *Proceedings of the 29th Annual International Symposium on Microarchitecture*, 349–58. Washington, DC: IEEE Computer Society.

63. R. E. Hank, W. W. Hwu, and B. R. Rau. 1995. Region-based compilation: An introduction and motivation. In *Proceedings of the 28th Annual International Symposium on Microarchitecture*, 158–68. Los Alamitos, CA: IEEE Computer Society Press.

64. W. A. Havanki, S. Banerjia, and T. M. Conte. 1998. Treegion scheduling for wide issue processors. In *Proceedings of the Fourth International Symposium on High-Performance Computer Architecture*, 266–76. Washington, DC: IEEE Computer Society.

65. J. L. Hennessy and T. R. Gross. 1982. Code generation and reorganization in the presence of pipeline constraints. In *Conference Record of the Ninth Annual ACM Symposium on Principles of Programming Languages*, 120–27. New York: ACM Press.

66. J. L. Hennessy and T. Gross. 1983. Postpass code optimization of pipeline constraints. *ACM Transactions on Programming Languages and Systems* 5(3):422–48.

67. J. L. Hennessy, N. Jouppi, F. Baskett, T. Gross, and J. Gill. 1982. Hardware/software tradeoffs for increased performance. In *Proceedings of the Symposium on Architectural Support for Programming Languages and Operating Systems*, 2–11. Palo Alto, CA: ACM Press.

68. J. L. Hennessy and D. A. Patterson. 1996. *Computer architecture: A quantitative approach*, 2nd ed. San Francisco: Morgan Kaufmann.

69. J. Hoogerbrugge and L. Augusteijn. 1999. Instruction scheduling for TriMedia. *Journal of Instruction-Level Parallelism* 1(1).

70. P. Y.-T. Hsu. Design of the R-8000 microprocessor. Technical report, MIPS Technologies, Inc., Mountainview, CA.

71. P. Y.-T. Hsu and E. S. Davidson. 1986. Highly concurrent scalar processing. In *Proceedings of the 13th Annual International Symposium on Computer Architecture*, 386–95.

72. R. A. Huff. 1993. Lifetime-sensitive modulo scheduling. In *Proceedings of the ACM SIGPLAN '93 Conference on Programming Language Design and Implementation*, 258–67. New York: ACM Press.

73. W. W. Hwu, R. E. Hank, D. E. Gallagher, S. A. Mahlke, D. M. Lavery, G. E. Haab, J. C. Gyllenhaal, and D. I. August. 1995. Compiler technology for future microprocessors. *Proceedings of the IEEE* 83(12):1625–40.

74. W. W. Hwu, S. A. Mahlke, W. Y. Chen, P. P. Chang, N. H. Warter, R. G. Ouellette, R. E. Hank, T. Kyohara, G. E. Haab, J. G. Holm, and D. M. Lavery. 1993. The superblock: An effective technique for VLIW and superscalar compilation. *Journal of Supercomputing* 7:229–48.

75. V. Janaki Ramanan. 1999. Efficient resource usage modelling. M.Sc.(Engg) thesis, Indian Institute of Science, Supercomputer Education and Research Centre, Bangalore, India.

76. V. Janaki Ramanan and R. Govindarajan. 1999. Resource usage models for instruction scheduling: Two new models and a classification. In *Conference Proceedings of the 1999 International Conference on Supercomputing*. New York: ACM Press.

77. M. Johnson. 1991. *Superscalar microprocessor design*. Englewood Cliffs, NJ: Prentice-Hall.

78. G. Kane. 1987. *MIPS R2000 RISC architecture*. Englewood Cliffs, NJ: Prentice-Hall.

79. D. Kastner and S. Winkel. 2001. ILP-based instruction scheduling for IA-64. In *Proceedings of the Workshop on Languages, Compilers, and Tools for Embedded Systems (LCTES 2001)*, 145–54. New York: ACM Press.

80. V. Kathail, M. S. Schlansker, and B. R. Rau. 1994. HPL PlayDoh architecture specification: Version 1.0. Technical report, Hewlett-Packard Laboratory, Palo Alto, CA.

81. D. Kerns and S. Eggers. 1993. Balanced scheduling: Instruction scheduling when memory latency is uncertain. In *Proceedings of the ACM SIGPLAN '93 Conference on Programming Language Design and Implementation*, 278–89. New York: ACM Press.

82. R. E. Kessler. 1999. The Alpha 21264 microprocessor. *IEEE Micro* 19(2):24–36.

83. H. Kim, K. Gopinath, and V. Kathail. 1999. Region based register allocation for EPIC processors with predication. In *Proceedings of the International Conference on Parallel Computing* (ParCo '99), 38–44, Imperial College Press, London, U.K.

84. H.-S. Kim and J. E. Smith. 2002. An instruction set and microarchitecture for instruction level distributed processing. In *Proceedings of the 29th International Symposium on Computer Architecture (ISCA)*, 71–81. Anchorage, AK: IEEE Computer Society.

85. P. M. Kogge. 1981. *The architecture of pipelined computers*. New York: McGraw-Hill.

86. C. M. Krishna and Y.-H. Lee. 2000. Voltage-switching scheduling algorithms for low power in hard real-time systems. In *Proceedings of the IEEE Real-Time Technology and Application Symposium*. Washington, DC: IEEE Computer Society.

87. S. M. Kurlander, T. A. Proebsting, and C. N. Fischer. 1995. Efficient instruction scheduling for delayed-load architectures. *ACM Transactions on Programming Languages and Systems* 17(5): 740–76.

88. M. Lam. 1988. Software pipelining: An effective scheduling technique for VLIW machines. In *Proceedings of the SIGPLAN '88 Conference on Programming Language Design and Implementation*, 318–28. New York: ACM Press.

89. M. Lam. 1990. Instruction scheduling for superscalar architectures. *Annual Review of Computer Science* 4:173–201.

90. E. Lawler, J. K. Lenstra, C. Martel, B. Simons, and L. Stockmeyer. 1987. Pipeline scheduling: A survey. Technical report, IBM Research Division, San Jose, CA.

91. C. Lee, J. Lee, T. Hwang, and S. Tsai. 2000. Compiler optimization on instruction scheduling for low power. In *Proceedings of the 13th International Symposium on System Synthesis (ISSS'00)*. Washington, DC: IEEE Computer Society.

92. E. A. Lee. 1988. Programmable DSP architectures: Part I. In *IEEE ASSP*. Washington, DC: IEEE Computer Society Press.

93. W. Lee, R. Barua, M. Frank, D. Srikrishna, J. Babb, V. Sarkar, and S. Amarasinghe. 1998. Space-time scheduling of instruction-level parallelism on a Raw machine. In *Proceedings of the Eighth International Conference on Architectural Support for Programming Languages and Operating Systems*, 46–57. New York: ACM Press.

94. R. Leupers. 2000. Instruction scheduling for clustered VLIW DSPs. In *Proceedings of the 2000 International Conference on Parallel Architectures and Compilation Techniques*, 291–300. Washington, DC: IEEE Computer Society.

95. R. Leupers and P. Marwedel. 1997. Time-constrained code compaction for DSPs. *IEEE Transactions on Very Large Scale Integration (VLSI) Systems* 5(1):112–22.

96. S. Liao, S. Devadas, K. Keutzer, S. Tjiang, and A. Wang. 1995. Storage assignment to decrease code size. In *Proceedings of the ACM SIGPLAN '95 Conference on Programming Language Design and Implementation*, 186–95. New York: ACM Press.

97. J. Llosa, A. González, E. Ayguadé, and M. Valero. 1996. Swing modulo scheduling: A lifetime-sensitive approach. In *Proceedings of the 1996 Conference on Parallel Architectures and Compilation Techniques (PACT '96)*, 80–86. Washington, DC: IEEE Computer Society.

98. J. Llosa, M. Valero, E. Ayguadé, and A. González. 1995. Hypernode reduction modulo scheduling. In *Proceedings of the 28th Annual International Symposium on Microarchitecture*, 350–60. Los Alamitos, CA: IEEE Computer Society Press.

99. J. L. Lo and S. J. Eggers. 1995. Improving balanced scheduling with compiler optimizations that increase instruction-level parallelism. In *Proceedings of the ACM SIGPLAN '95 Conference on Programming Language Design and Implementation*, 151–62. New York: ACM Press.

100. P. G. Lowney, S. M. Freudeberger, T. J. Karzes, W. D. Lichtensein, R. P. Nix, J. S. O'Donnell, and J. C. Ruttenberg. 1993. The multiflow trace scheduling compiler. *Journal of Supercomputing* 7:51–142.

101. U. Mahadevan and S. Ramakrishnan. 1994. Instruction scheduling over regions: A framework for scheduling across basic blocks. In *Proceedings of the 5th International Conference on Compiler Construction, CC '94*, ed. P. A. Fritzson, 419–34. Lecture Notes in Computer Science, vol. 786. Heidelberg, Germany: Springer-Verlag.

102. S. A. Mahlke, D. C. Lin, W. Y. Chen, R. E. Hank, and R. A. Bringmann. 1992. Effective compiler support for predicated execution using the hyperblock. In *Proceedings of the 25th Annual International Symposium on Microarchitecture*, 45–54. Los Alamitos, CA: IEEE Computer Society Press.

103. S. Mantripragada, S. Jain, and J. Dehnert. 1998. A new framework for integrated global local scheduling. In *Proceedings of the International Conference on Parallel Architectures and Compilation Techniques (PACT '98)*, 167–74. Washington, DC: IEEE Computer Society.

104. M. Mercaldi, S. Swanson, A. Petersen, A. Putnam, A. Schwerin, M. Oskin, and S. J. Eggers. 2006. Instruction scheduling for a tiled dataflow architecture. In *Proceedings of the 12th International*

Conference on Architectural Support for Programming Languages and Operating Systems, 141–50. New York: ACM Press.

105. S.-M. Moon and K. Ebcioğlu. 1992. An efficient resource-constrained global scheduling technique for superscalar and VLIW processors. In *Proceedings of the 25th Annual International Symposium on Microarchitecture*, 55–71. Los Alamitos, CA: IEEE Computer Society Press.

106. R. Motwani, K. Palem, V. Sarkar, and S. Reyen. 1995. Combined instruction scheduling and register allocation. Technical Report TR 1995-698, New York University, Department of Computer Science, New York.

107. S. S. Muchnick. 1997. *Advanced compiler design and implementation*. San Francisco: Morgan Kaufmann.

108. T. Müller. 1993. Employing finite automata for resource scheduling. In *Proceedings of the 26th Annual International Symposium on Microarchitecture*, 12–20. Los Alamitos, CA: IEEE Computer Society Press.

109. R. Nagarajan, S. K. Kushwaha, D. Burger, K. S. McKinley, C. Lin, and S. W. Keckler. 2004. Static placement, dynamic issue (SPDI) scheduling for EDGE architectures. In *Proceedings of the 13th International Conference on Parallel Architectures and Compilation Techniques*, 74–84. Washington, DC: IEEE Computer Society.

110. R. Nagarajan, K. Sankaralingam, D. Burger, and S. Keckler. 2001. A design space evaluation of grid processor architectures. In *Proceedings of the 34th International Symposium on Microarchitecture (MICRO)*, 40–51. Washington, DC: IEEE Computer Society.

111. A. Nicolau. 1985. Percolation scheduling: A parallel compilation technique. Technical Report TR 85-678, Department of Computer Science, Cornell University, Ithaca, NY.

112. A. Nicolau and S. Novack. 1993. Trailblazing: A hierarchical approach to percolation scheduling. In *Proceedings of the 1993 International Conference on Parallel Processing*. Washington, DC: IEEE Computer Society.

113. A. Nicolau and R. Potasman. 1990. Realistic scheduling: Compaction for pipelined architectures. In *Proceedings of the 23rd Annual Workshop on Microprogramming and Microarchitecture*, 69–79. Los Alamitos, CA: IEEE Computer Society Press.

114. K. B. Normoyle, M. A. Csoppenszky, A. Tzeng, T. P. Johnson, C. D. Furman, and J. Mostoufi. 1998. UltraSPARC-II: Expanding the boundaries of a system on a chip. *IEEE Micro* 18(2):14–24.

115. C. Norris and L. L. Pollock. 1993. A scheduler-sensitive global register allocator. In *Proceedings of the 1993 Supercomputing Conference*. New York: ACM Press.

116. C. Norris and L. L. Pollock. 1995. An experimental study of several cooperative register allocation and instruction scheduling strategies. In *Proceedings of the 28th Annual International Symposium on Microarchitecture*, 169–79. Los Alamitos, CA: IEEE Computer Society Press.

117. E. Nystrom and A. E. Eichenberger. 1998. Effective cluster assignment for modulo scheduling. In *Proceedings of the 31st Annual International Symposium on Microarchitecture*, 103–14. Los Alamitos, CA: IEEE Computer Society Press.

118. E. Özer, S. Banerjia, and T. M. Conte. 1998. Unified assign and schedule: A new approach to scheduling for clustered register file microarchitectures. In *Proceedings of the 31st Annual International Symposium on Microarchitecture*, 308–15. Los Alamitos, CA: IEEE Computer Society Press.

119. S. Palacharla, N. Jouppi, and J. E. Smith. 1997. Complexity-effective superscalar processors. In *Proceedings of the 24th International Symposium on Computer Architecture*, 206–18. New York: ACM Press.

120. K. V. Palem and B. B. Simons. 1990. Scheduling time-critical instructions on RISC machines. In *Conference Record of the Seventeenth Annual ACM Symposium on Principles of Programming Languages*, 270–80. New York: ACM Press.

121. A. Parikh, M. Kandemir, N. Vijaykrishnan, and M. J. Irwin. 2000. Energy-aware instruction scheduling. In *Proceedings of the 7th International Conference on High Performance Computing*, 335–44. London, UK: Springer-Verlag.

122. D. A. Patterson and C. H. Sequin. 1981. RISC I: A reduced instruction set VLSI computer. In *Proceedings of the 8th Annual Symposium on Computer Architecture*, 443–57. Los Alamitos, CA: IEEE Computer Society Press.

123. P. G. Paulin, M. Cornero, and C. Liem. 1996. Trends in embedded system technology, an industrial perspective. In *Hardware/software co-design*, ed. M. Giovanni and M. Sami. Boston, MA: Kluwer Academic.

124. P. G. Paulin and J. P. Knight. 1989. Force-directed scheduling for the behavioral synthesis of ASIC's. *IEEE Transactions on Computer-Aided Design* 8(6):661–79.

125. Philips Semiconductors. http://www.trimedia.philips.com.

126. S. S. Pinter. 1993. Register allocation with instruction scheduling: A new approach. In *Proceedings of the ACM SIGPLAN '93 Conference on Programming Language Design and Implementation*, 248–57. New York: ACM Press.

127. T. A. Proebsting and C. N. Fischer. 1991. Linear-time, optimal code scheduling for delayed-load architectures. In *Proceedings of the ACM SIGPLAN '91 Conference on Programming Language Design and Implementation*, 256–67. New York: ACM Press.

128. T. A. Proebsting and C. W. Fraser. 1994. Detecting pipeline structural hazards quickly. In *Conference Record of the 21st ACM SIGPLAN-SIGACT Symposium on Principles of Programming Languages*, 280–86. New York: ACM Press.

129. D. Puppin, M. Stephenson, U.-M. O'Reilly, M. C. Martin, and S. Amarasinghe. 2003. Adapting convergent scheduling using machine-learning. In *Proceedings of the 16th International Workshop on Languages and Compilers for Parallel Computing (LCPC)*.

130. M. Rajagopalan and V. H. Allan. 1993. Efficient scheduling of fine grain parallelism in loops. In *Proceedings of the 26th Annual International Symposium on Microarchitecture*, 2–11. Los Alamitos, CA: IEEE Computer Society Press.

131. B. R. Rau. 1994. Iterative modulo scheduling: An algorithm for software pipelining loops. In *Proceedings of the 27th Annual International Symposium on Microarchitecture*, 63–74. New York: ACM Press.

132. B. R. Rau and J. A. Fisher. 1993. Instruction-level parallel processing: History, overview and perspective. *Journal of Supercomputing* 7:9–50.

133. B. R. Rau and C. D. Glaeser. 1981. Some scheduling techniques and an easily schedulable horizontal architecture for high performance scientific computing. In *Proceedings of the 14th Annual Microprogramming Workshop*, 183–98. Piscataway, NJ: IEEE Press.

134. B. R. Rau, V. Kathail, and S. Aditya. 1999. Machine-description driven compilers for EPIC and VLIW processors. *Design Automation for Embedded Systems* 4(2/3):71–118.

135. B. R. Rau, M. Lee, P. P. Tirumalai, and M. S. Schlansker. 1992. Register allocation for software pipelined loops. In *Proceedings of the ACM SIGPLAN '92 Conference on Programming Language Design and Implementation*, 283–99. New York: ACM Press.

136. B. R. Rau, D. W. L. Yen, W. Yen, and R. A. Towle. 1989. The Cydra 5 departmental supercomputer — Design philosophies, decisions, and trade-offs. *Computer* 22(1):12–35.

137. R. Reiter. 1968. Scheduling parallel computations. *Journal of the ACM* 15(4):590–99.

138. H. Rong and R. Govindarajan. 2007. Advances in software pipelining. Chapter 20, this volume.

139. J. C. Ruttenberg. 1985. Delayed-binding code generation for a VLIW supercomputer. Ph.D. thesis, Yale University, New Haven, CT.

140. J. Ruttenberg, G. R. Gao, A. Stoutchinin, and W. Lichtenstein. 1996. Software pipelining showdown: Optimal vs. heuristic methods in a production compiler. In *Proceedings of the ACM SIGPLAN 1996 Conference on Programming Language Design and Implementation*, 1–11. New York: ACM Press.

141. J. Sánchez and A. González. 2000. Modulo scheduling for a fully-distributed clustered VLIW architecture. In *Proceedings of the 33rd Annual International Symposium on Microarchitecture*, 124–33. New York: ACM Press.

142. K. Sankaralingam, R. Nagarajan, H. Liu, C. Kim, J. Huh, D. Burger, S. W. Keckler, and C. R. Moore. 2003. Exploiting ILP, TLP, and DLP with the polymorphous TRIPS architecture. In *Proceedings of the 30th International Symposium on Computer Architecture*, 422–33. New York: ACM Press.

143. M. Schlansker and B. Rau. 2000. EPIC: Explicitly parallel instruction computing. *IEEE Computer* 30(2):37–45.

144. R. Sethi and J. D. Ullman. 1970. The generation of optimal code for arithmetic expressions. *Journal of the ACM* 17(4):715–28.

145. H. Sharangpani and H. Arora. 2000. Itanium processor microarchitecture. *IEEE Micro* 20(5):24–43.

146. D. Shin, S. Lee, and J. Kim. 2001. Intra-task voltage scheduling for low-energy hard real-time applications. *IEEE Design & Test of Computers* 18(2):20–30.

147. R. Silvera, J. Wang, G. R. Gao, and R. Govindarajan. 1997. A register pressure sensitive instruction scheduler for dynamic issue processors. In *Proceedings of the 1997 International Conference on Parallel Architectures and Compilation Techniques*, 78–89. Washington, DC: IEEE Computer Society.

148. M. Slater. 1994. AMD's K5 designed to outrun Pentium. *Microprocessor Report*. Micro Design Resources, 8(4):1–11.

149. A. Smith, J. Gibson, B. Maher, N. Nethercote, B. Yoder, D. Burger, K. S. McKinley, and J. Burrill. 2006. Compiling for EDGE architectures. In *Proceedings of the International Symposium on Code Generation and Optimization*, 185–95. Washington, DC: IEEE Computer Society.

150. J. E. Smith, G. E. Dermer, B. D. Vanderwarn, S. D. Klinger, C. M. Rozewski, D. L. Fowler, K. R. Scidmore, and J. P. Laudon. 1987. The ZS-1 central processor. In *Proceedings of the Second International Conference on Architectural Support for Programming Languages and Operating Systems*, 199–204. Los Alamitos, CA: IEEE Computer Society Press.

151. J. E. Smith and G. S. Sohi. 1995. The microarchitecture of superscalar processors. *Proceedings of the IEEE* 83(12):1609–24.

152. M. Smotherman, S. Krishnamurthy, P. S. Aravind, and D. Hunnicutt. 1991. Efficient DAG construction and heuristic calculation for instruction scheduling. In *Proceedings of the 24th Annual International Symposium on Microarchitecture*, 93–102. New York: ACM Press.

153. B. Su, S. Ding, J. Wang, and J. Xia. 1987. Microcode compaction with timing constraints. In *Proceedings of the 20th Annual Workshop on Microprogramming*, 59–68. New York: ACM Press.

154. C.-L. Su, C.-Y. Tsui, and A. M. Despain. 1994. Low power architecture and compilation techniques for high-performance processors. In *Proceedings of the IEEE COMPCON*, 489–98. Washington, DC: IEEE Computer Society Press.

155. S. Swanson, K. Michelson, A. Schwerin, and M. Oskin. 2003. WaveScalar. In *Proceedings of the 36th Annual IEEE/ACM International Symposium on Microarchitecture*, 291–303. Washington, DC: IEEE Computer Society Press.

156. M. Taylor, J. Kim, J. Miller, D. Wentzlaff, F. Ghodrat, B. Greenwald, H. Hoffman, J.-W. Lee, P. Johnson, W. Lee, A. Ma, A. Saraf, M. Seneski, N. Shnidman, V. S. M. Frank, S. Amarasinghe, and A. Agarwal. 2002. The Raw microprocessor: A computational fabric for software circuits and general purpose programs. *IEEE Micro* 22(2):25–35.

157. Texas Instruments. 1998. *TMS320C62xx CPU and instruction set reference guide*. http://www. ti.com/sc/c6x.

158. M. D. Tiemann. 1989. The GNU instruction scheduler — cs343 course report. Technical report, Stanford University, Computer Science, Stanford, CA.

159. A. Timmer, M. Strik, J. van Meerbergen, and J. Jess. 1995. Conflict modelling and instruction scheduling in code generation for in-house DSP cores. In *Proceedings of the 32nd ACM/IEEE Design Automation Conference*, 593–98. New York: ACM Press.

160. V. Tiwari, S. Malik, and A. Wolfe. 1994. Power analysis of embedded software: A first step towards software power minimization. *IEEE Transactions on Very Large Scale Integration (VLSI) Systems* 2:437–45.

161. M. C. Toburen, T. M. Conte, and M. Reilly. 1998. Instruction scheduling for low power dissipation in high performance processors. In *Proceedings of the Power Driven Microarchitecture Workshop*. Hingham, MA: Kluwer Academic Publishers.

162. M. Tokoro, T. Takizuka, E. Tamura, and I. Yamamura. 1977. A technique for global optimization of microprograms. In *Proceedings of the Tenth Annual Microprogramming Workshop*, 41–50. Piscataway, NJ: IEEE Press.

163. M. G. Valluri. 1999. Evaluation of register allocation and instruction scheduling methods in multiple issue processors. M.Sc.(Engg) thesis, Indian Institute of Science, Supercomputer Education and Research Centre, Bangalore, India.

164. M. G. Valluri and R. Govindarajan. 1998. Modulo-variable expansion sensitive software pipelining. In *Proceedings of the 5th International Conference on High Performance Computing*, 334–41. Washington, DC: IEEE Computer Society Press.

165. M. G. Valluri and R. Govindarajan. 1999. Evaluating register allocation and instruction scheduling techniques in out-of-order issue processors. In *Proceedings of the 1999 International Conference on Parallel Architectures and Compilation Techniques*. Washington, DC: IEEE Computer Society Press.

166. R. Venugopal and Y. N. Srikant. 1995. Scheduling expression trees with reusable registers on delayed-load architectures. *Computer Languages* 21(1):49–65.

167. E. Waingold, M. Taylor, D. Srikrishna, V. Sarkar, W. Lee, V. Lee, J. Kim, M. Frank, P. Finch, R. Barua, J. Babb, S. Amarasinghe, and A. Agarwal. 1997. Baring it all to software: Raw machines. *IEEE Computer* 30(9):86–93.

168. J. Wang and C. Eisenbeis. 1993. Decomposed software pipelining: A new approach to exploit instruction level parallelism for loop programs. In *Proceedings of the IFIP WG 10.3 Working Conference on Architectures and Compilation Techniques for Fine and Medium Grain Parallelism*, ed. M. Cosnard, K. Ebcioğlu, and J.-L. Gaudiot, 3–14. Number A-23 in IFIP Transactions. Amsterdam: North-Holland.

169. J. Wang, A. Krall, M. A. Ertl, and C. Eisenbeis. 1994. Software pipelining with register allocation and spilling. In *Proceedings of the 27th Annual International Symposium on Microarchitecture*, 95–99.

170. H. S. Warren, Jr. 1990. Instruction scheduling for the IBM RISC system/6000 processor. *IBM Journal of Research and Development* 34(1):85–92. New York: ACM Press.

171. N. J. Warter, G. E. Haab, J. W. Bockhaus, and K. Subramanian. 1992. Enhanced modulo scheduling for loops with conditional branches. In *Proceedings of the 25th Annual International Symposium on Microarchitecture*, 170–79. Los Alamitos, CA: IEEE Computer Society Press.

172. N. J. Warter, S. A. Mahlke, W. W. Hwu, and B. R. Rau. 1993. Reverse if-conversion. In *Proceedings of the ACM SIGPLAN '93 Conference on Programming Language Design and Implementation*, 290–99. New York: ACM Press.

173. N. J. Warter-Perez and N. Partamian. 1995. Modulo scheduling with multiple initiation intervals. In *Proceedings of the 28th Annual International Symposium on Microarchitecture*, 111–18. Los Alamitos, CA: IEEE Computer Society Press.

174. K. Wilken, J. Liu, and M. Heffernan. 2000. Optimal instruction scheduling using integer programming. In *Proceedings of the ACM SIGPLAN '00 Conference on Programming Language Design and Implementation*, 121–33. New York: ACM Press.

175. K. C. Yeager. 1996. The MIPS R10000 superscalar microprocessor. *IEEE Micro* 16(2):28–40.

176. J. Zalamea, J. Llosa, E. Ayguadé, and M. Valero. 2000. Improved spill code generation for software pipelined loops. In *Proceedings of the ACM SIGPLAN '00 Conference on Programming Language Design and Implementation*, 134–44. New York: ACM Press.

177. C. Zhang, R. Govindarajan, S. Ryan, and G. R. Gao. 1999. Efficient state-diagram construction methods for software pipelining. CAPSL Technical Memo 28, Department of Electrical and Computer Engineering, University of Delaware, Newark, DE. ftp://ftp.capsl.udel.edu/pub/doc/memos.

20

Advances in Software Pipelining

Hongbo Rong

Microsoft Corporation,
Redmond, WA
hongbor@microsoft.com

R. Govindarajan

Supercomputer Education and
Research Centre, Department of
Computer Science and Automation,
Indian Institute of Science,
Bangalore, India;
govind@csa.iisc.ernet.in

20.1 Introduction

Many state-of-the-art processors have multiple functional units and execute several instructions simultaneously to exploit *instruction-level parallelism* (ILP) [97]. The ILP architectures include very long instruction word (VLIW) architectures [27, 43, 88], superscalar processors [56, 97, 98], and explicitly parallel instruction computing (EPIC) architectures [96]. For these architectures, it is important for the compiler to expose parallelism in the application to the underlying hardware. The compiler assists the hardware by statically scheduling independent instructions in the same time step, scheduling dependent instructions apart to satisfy dependences, and mapping the instructions onto appropriate hardware resources. This role is critical for VLIW and EPIC architectures, which fully rely on the compiler to expose parallelism. In these architectures, the compiler identifies and packs a set of independent instructions into a single long word instruction and communicates it to the hardware. The hardware fetches and decodes the long word instruction and executes the instructions in it in parallel, without being required to check the dependences between them. For superscalar processors, although the hardware dynamically identifies independent instructions and issues them to the resources for execution in parallel, the ability is limited by the size of the instruction window, from which instructions are scheduled, and its hardware complexity. With instructions scheduled according to their dependences and mapped to appropriate resources by the compiler, the chances of finding independent instructions at runtime are increased, while the chances of stalling instruction issue resulting from nonavailability of dependent values and/or resources are decreased.

Software pipelining is an instruction scheduling technique to expose ILP from loops. It is particularly important, as loops usually dominate the execution time of scientific applications. Software pipelining reduces the execution time of a loop by overlapping the execution of different iterations of the loop. Successive iterations are initiated at a (usually fixed) interval, called the *initiation interval* (*II*). The aim is to minimize the II and thus maximize the throughput.

Analogous to hardware pipelining, where instructions flow from resource to resource synchronously according to a pipeline schedule, in software pipelining, iterations flow from resource to resource synchronously [60]. Several iterations are concurrently active in different stages of the pipeline. When the pipeline is full, it enters a steady state. In this steady state, a kernel of length T time steps, T equal to II, repeats periodically. Every T time steps, effectively a single iteration gets completed and flushed out of the pipeline, and a new iteration is initiated and pushed into the pipeline. The software pipelined schedule must satisfy the dependences of the loop, including both intra- and inter-iteration dependences, and the resource constraints imposed by the architecture.

Numerous proposals on software pipelining for ILP architectures have been reported [1, 2, 17, 32, 33, 38, 40, 47, 50, 54, 59, 67, 83, 85, 104, 108]. The past decade has witnessed dramatic changes in processor architecture: from single-core to multi-core, from single-threaded to multi-threaded architectures, and from single VLIW to clustered VLIW architectures. The hardware resources have become increasingly abundant. In response to these changes, new constraints have been imposed on software pipelining to adapt it to these newer architectures. There is also a trend to go beyond the traditional limit that exposes ILP from the innermost loop and go to loop nests for more ILP [44, 59, 70, 79, 83, 92, 106].

This chapter surveys these advances in software pipelining. We focus on *modulo scheduling*, the most common approach found in the literature and product compilers. Throughout this chapter the terms *instruction* and *operation*, *time step* and *cycle*, and *loop nest* and *nested loop* will be used interchangeably. We use *single loop* to refer to a loop without any inner loop, or the innermost loop of a loop nest.

The rest of this section presents a historical perspective on software pipelining. Section 20.2 reviews traditional software pipelining of single loops, particularly modulo scheduling, including techniques for scheduling, register allocation, and code generation. Section 20.3 discusses recent advances in software pipelining of single loops, including power-aware and register pressure–aware methods, software pipelining for the new architectures that emerged in the past decade, and the fundamental link between software pipelining and hardware circuit retiming. In Section 20.4, we present the recent progress in software pipelining of nested loops, including a survey of several traditional and recent approaches and a detailed introduction to a new scheduling methodology, its register allocation, and code generation. Finally, we summarize the chapter and highlight future research directions in Section 20.5.

20.1.1   A Historical Perspective

ILP was originally exploited within a basic block by performing simple instruction scheduling via list scheduling methods, using an approach similar to microcode compaction [43]. The available ILP within a basic block, however, is limited [57, 71]. Trace scheduling breaks the limitation and schedules instructions across basic block boundaries [43]. However, it does not schedule instructions across the back edge of a loop, and therefore still cannot overlap different iterations of the loop. This weakness can be remedied to some extent by unrolling the loop by a certain factor k and applying trace scheduling to the unrolled loop body [42, 66]. Still, there is no overlap between the iterations of the unrolled loop. To achieve such overlapping, one can continue to unroll the loop until a repeating pattern is found in the schedule. The repeating pattern, known as the kernel, is rerolled to obtain a compact loop schedule [42, 100].

Software pipelining aims to identify such a pattern without unrolling. It is related to hardware pipeline scheduling, which was developed by Patel and Davidson to obtain optimal throughput [25, 60, 77, 84]. The objective in both hardware and software pipelining is to overlap the execution of successive instructions (iterations) in the case of hardware (software) pipelining and fully utilize the hardware resources. Early work constructed a software pipelined schedule by ad hoc hand coding [25, 26]. Rau and Glaeser [85] developed the first compiler to automatically generate software pipelines, drawing upon and generalizing the theory of hardware pipeline design [77]. A variety of software pipelining methods have been proposed since then [1, 2, 17, 32, 33, 38, 40, 45, 47, 50, 54, 59, 67, 83, 85, 100, 101, 104, 108]. Surveys on them can be found in [8, 9, 84].

The problem of constructing an optimal software pipeline that has the lowest possible initiation interval under the resource constraints imposed by the architecture is known to be NP-complete [59]. This can be shown by transforming the resource-constrained scheduling problem [46] to the software pipelining problem. The optimal resource-constrained software pipelining problem can be formulated as an integer linear programming problem [36, 40, 51], which will be discussed in detail in Section 20.2.4. Most software pipelining methods, however, employ heuristics to obtain a near-optimal schedule. Rau [87] classifies them into two broad categories: "move-then-schedule" and "schedule-then-move."

The move-then-schedule approach moves instructions, one by one, across the back edge of the loop, either in the forward or in the backward direction, and then schedules them to achieve a software pipelined schedule [31, 32, 47, 67]. The moving of an instruction through the back edge effectively moves the instruction from one iteration to another. Thus, the overlapping of different iterations is achieved. The subsequent scheduling compacts the loop body. Such a control flow transformation maintains the structure of the loop, thus naturally exposing a repeating pattern, which forms the new loop body.

The schedule-then-move approach focuses directly on the creation of a schedule that maximizes performance and subsequently discovers the code motions that are implicit in the schedule. There are two ways of doing this. The first, "unroll-while-scheduling," simultaneously unrolls and schedules the loop until it gets to a point where the rest of the schedule would be a repetition of an existing portion of the schedule [2].

The second approach in the schedule-then-move category, known as *modulo scheduling*, is particularly interesting [85]. Its objective is to derive a schedule for a single iteration of the loop, such that when the same schedule is repeated for the other iterations at an initiation interval, both intra- and inter-iteration dependences are satisfied, and no resource usage conflict arises among instructions of either the same or different iterations. It specifies a set of constraints for a legal schedule, which are respected during the scheduling process. This approach is interesting in that it is relatively simple and efficient. Empirically, it has a computational complexity of $O(w^2)$, where w is the number of instructions in the loop body [87]. It has also been shown to be able to produce optimal code, in terms of II in most cases [94]. For these reasons, almost every production compiler implementing software pipelining chooses this approach. As such, it is the main focus of this chapter.

Modulo scheduling is essentially an extension of list scheduling to loops. The instructions to be scheduled are ordered into a priority list. One difficulty in ordering is that the data dependence graph may have recurrences in it. Early work by Lam [59] addresses this by using a hierarchical approach; it first considers the strongly connected components (SCCs) in the underlying data dependence graph and schedules the instructions in them first. The schedule for each SCC is abstracted as a macro instruction. With this, the dependence graph becomes acyclic, and an acyclic scheduling algorithm based on list scheduling can be used. The hierarchical scheduling approach also enables software pipelining loops with conditionals (*if-then-else* constructs) and nested loops. Since then, a number of heuristic approaches have been proposed for constructing modulo schedules.

By overlapping iterations, software pipelining increases the register pressure in the constructed schedule. Thus, many software pipelining methods, in addition to obtaining efficient schedules by minimizing II, also attempt to reduce the register requirements [28, 34, 51, 54, 64, 65]. A discussion of register-constrained modulo schedules is presented in Section 20.3.1.

In addition to increased register pressure, software pipelining also introduces a new kind of register allocation problem, in which the lifetimes of the same variable from successive iterations overlap each other. This happens when a lifetime is longer than II time steps. Normally, this would result in a value being overwritten before its last use. One solution to overcome this is via modulo variable expansion [59], which unrolls the scheduled kernel a sufficient number of times, for instance, k times, such that no lifetime is longer than $k * II$ time steps. Then the overlapped lifetimes in the unrolled kernel get distinct registers allocated to them. However, unrolling the loop increases the code size of the schedule. Unrolling can be avoided by using a hardware support called *rotating registers*, proposed in the Cydra architecture [88]. Instead of unrolling the kernel and allocating different registers to the overlapping lifetimes at compile time, the kernel is kept untouched, but the lifetimes are automatically mapped to different physical registers at runtime by the hardware. Section 20.2.5.2 will introduce the register allocation solution for a software pipelined schedule in the presence of rotating registers [30, 86]. In case the register requirement is more than the available physical registers, some lifetimes are spilled. Spilling code can be inserted either during or after the scheduling process [6, 116].

Early software pipelining methods did not consider loops with conditional constructs. Lam's hierarchical reduction approach [59] schedules the two branches of a conditional individually and then reduces the entire conditional construct into a macro instruction whose scheduling constraints are the union of both branches. This does not require any special hardware support. Alternatively, for architectures supporting predicated execution, IF-conversion [10] can be applied to convert the conditional constructs into straight line code, and then software pipelining can be employed as usual [88]. The generated code, however, can also be converted back to have conditional constructs via reverse-IF conversion [109]. Then the code can be executed on machines without predication support as well. Warter and Partamian also proposed an interesting software pipelining method that uses multiple *II*s for scheduling the different control flow paths in the loop [110].

Last, software pipelining has traditionally been restricted to single (non-nested) loops or the innermost loop of a loop nest. Initial efforts to extend software pipelining to loop nests schedule the loops hierarchically, starting from the innermost loop to the outer loops [59, 70, 106]. Another direction is to apply the traditional hyperplane scheduling [61], which is usually used in large-array hardware structures such as systolic arrays, to a uniprocessor such that the hyperplanes repeat [44, 83]. This approach does not consider resource constraints, though. The other approach combines software pipelining with loop transformation [19, 79, 112], for example, applying unroll-and-jam to the loop nest, followed by software pipelining. More recently, a novel approach that modulo schedules an arbitrary loop level in a loop nest under resource constraints has been proposed [92]. We discuss these methods in greater detail in Section 20.4.

20.2   Software Pipelining for Single Loops

This section reviews the basic scheduling, register allocation, and code generation techniques for software pipelining of single loops. We focus on modulo scheduling for its simplicity and usefulness. We assume that the loop body does not contain any conditional or function call. While conditionals can be converted into linear code via IF-conversion [10] and predicated execution [88], function calls essentially prevent a loop from being software pipelined.

We first illustrate modulo scheduling with a simple example in Section 20.2.1 and present the required background knowledge in Section 20.2.2. For the resource-constrained modulo scheduling problem, we present a heuristic solution as a generic modulo scheduling framework in Section 20.2.3 and an optimal solution via integer linear programming in Section 20.2.4. For the obtained modulo schedule, the register allocation and code generation techniques, with and without hardware support, are then described in Section 20.2.5.

20.2.1   A Motivating Example

Figure 20.1a shows a single loop. Its intermediate representation is shown in Figure 20.1b, where a *temporary name* (TN) represents a variable. In this example, there are four variables, TN1 to TN4. We use the notation TN$\{d\}$ to refer to the TN value defined d iterations before. For example, TN2$\{1\}$ is the TN2 value defined in the previous loop iteration, and TN2$\{2\}$ is the TN2 value defined in the previous previous loop iteration.

The data dependence graph for the loop is shown in Figure 20.1c, where a node represents an instruction, and a directed arc between a pair of nodes represents a data dependence between the two instructions. Each arc is labeled with the nonnegative dependence distance. When the dependence distance is 0, the

```
real A[N];
int B[N];
for (i = 2; i < N; i++) {
    A[i] = (int)A[i−1]+B[i−2];
    B[i] = A[i];
}
```

(a)

```
TN3 = start address of array A
TN4 = start address of array B
TN3 = TN3+4                        //Address of A [1]
ld4 TN1 {1} = [TN3], 4             //Read A [1]. Increment the address
ld4 TN2 {2} = [TN4], 4            // Read B [0]. Increment the address
ld4 TN2 {1} = [TN4], 4            //Read B[1]. Increment the address
for (i =2; i < N; i++) {
    a: TN1 = TN1{1}+TN2 {2}      //TN1 = A[i−1]+B [i−2]
    b: st4 [TN4] = TN1            //B [i] = TN1
    c: ld4 TN2 = [TN4], 4        //Read B[i]. Increment the address
    d: fs4 [TN3] = TN1, 4        // A[i] = TN1. Increment the address
}
```

(b)

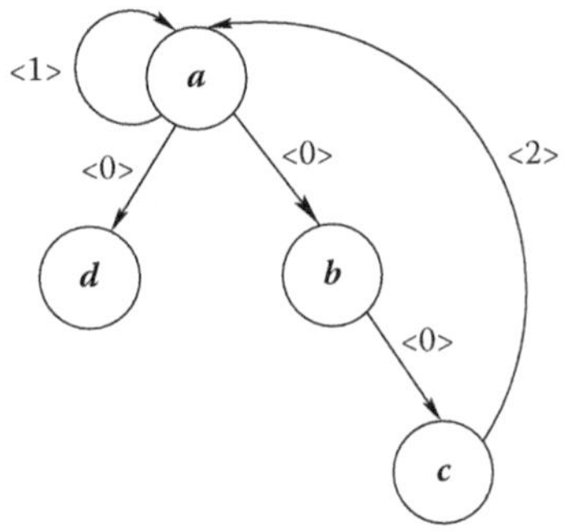

(c)

FIGURE 20.1   An example loop and its data dependence graph. (a) A single loop in C language. (b) The intermediate representation. (c) Data dependence graph (for the loop only).

dependence is within an iteration. Otherwise, the dependence is across loop iterations and is referred to as a *loop-carried dependence*. For example, there is a loop-carried dependence from instruction *c* to *a* and a loop-carried dependence from *a* to itself.

Suppose all the instructions are integer operations, except instruction *d*, which is a floating point (FP) operation. Consider a simple architecture with two Integer arithmetic and logic units (ALUs) and one FP ALU. Both are fully pipelined. The latency of the Integer units is one cycle, while that of the FP unit is two cycles.

In traditional instruction scheduling, the instructions in the loop body are scheduled to make it compact. Instructions are not moved across the loop back edge, and thus the iterations of the loop do not overlap. Such a nonoverlapping schedule is shown in Figure 20.2a. Note that the original lexical order between the instructions is not important in constructing the schedule. Their order in the schedule is determined only by the dependences between them. In this schedule, an iteration takes four time steps to complete, and then the next iteration is issued. In other words, the throughput of this schedule is one-fourth of an iteration per time step.

Time Step	Iteration $i = 2$		Iteration $i = 3$	
	Int. ALU	FP ALU	Int. ALU	FP ALU
0	a			
1		d		
2	b			
3	c			
4			a	
5				d
6			b	
7			c	

(a)

Time Step	Iteration i = 2		Iteration i = 3		Iteration i = 4		Iteration i = 5	
	Int. ALU	FP ALU	Int. ALU	FP ALU	Int. ALU	FP ALU	Int. ALU	FP ALU
0	a							
1		d						
2	b		a					
3	c			d				
4			b		a			
5			c			d		
6					b		a	
7					c			d

(b)

FIGURE 20.2   Comparison of two schedules for the motivating example. (a) A schedule without any overlap between iterations. (b) A software pipelined schedule with $II = 2$.

To achieve a higher iteration throughput, even before the current iteration is complete, a subsequent loop iteration can be allowed to initiate, provided the dependences to it are satisfied and the required resources are available. For such a schedule, the loop iterations are naturally overlapped (refer to Figure 20.2b). Every iteration of the loop has the same schedule, and successive iterations are initiated at an interval of two time steps. A repetitive pattern, called the *kernel*, appears in the schedule from time step 2. The kernel consists of b and c from one iteration and a and d from the next iteration. Before the kernel is the *prolog*, which consists of time steps 0 and 1. Likewise, there is an *epilog* at the end of the schedule, which is not shown in this figure.

Such a schedule is called a *modulo schedule*. In this schedule, a software pipeline is constructed for the loop iterations. Every two time steps, a new iteration is initiated into the pipeline. At the same time, a previous iteration finishes execution and exits the pipeline. For example, at time step 4, iteration 4 enters the pipeline, while iteration 2 exits from it. This leads to a throughput of half an iteration per time step. Compared to the schedule shown in Figure 20.2a, the modulo schedule gives better throughput by overlapping successive iterations of the loop.

To illustrate the code generated for a modulo schedule, we show the rewritten loop in a high-level language in Figure 20.3 for the modulo schedule depicted in Figure 20.2b. In the rewritten loop, $a(i)$ refers to the instance of instruction a in iteration i. To be brief, we may also refer to the instance simply as "instruction a in iteration i." The binding between the instructions and the functional units is not shown. This information is encoded in the very long instructional word (VLIW) and conveyed to the hardware. From now on, we would ignore such information for simplicity.

Figure 20.4 specifically shows the kernel. It has two time steps, which is equal to the initiation interval,[1] represented as $T = 2$. It has two *stages*, denoted $S = 2$. Each stage has instructions from a distinct iteration. Note that S and T are independent. Although they are equal here, it is just a coincidence.

---

[1]We use the notation T to represent the initiation interval of a schedule and II the initiation interval in general.

$$
\left.\begin{array}{l} a(2) \\ d(2) \end{array}\right\} \text{Prolog}
$$

$$
L' : \text{for}(i = 3; i < N; i + +)\{
$$

$$
\left.\begin{array}{cc} b(i-1) & a(i) \\ c(i-1) & d(i) \end{array}\right\} \text{Kernal}
$$

$$
\}
$$

$$
\left.\begin{array}{l} b(N-1) \\ c(N-1) \end{array}\right\} \text{Epilog}
$$

FIGURE 20.3 Loop rewriting with the software pipelined schedule.

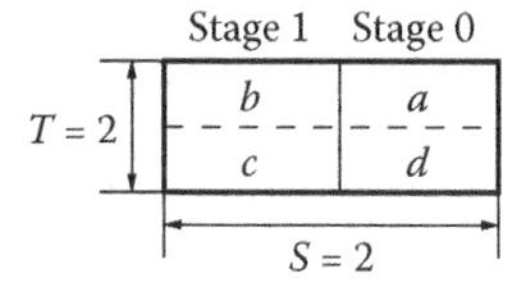

FIGURE 20.4 Kernel of the software pipelined schedule.

20.2.2 Background

Below we introduce the loop model and certain basic concepts relevant to modulo scheduling, including the data dependence graph, modulo scheduling constraints, the minimum distance between two instructions in a legal schedule, and the computation of the minimum initiation interval (MII).

20.2.2.1 Loop Model

We assume a single loop L that has an *index variable i* and a *trip count* $N > 1$. In our motivating example above, the index is not normalized. However, from now on, for convenience, we will assume that the index has been normalized to change from 0 to $N - 1$ with unit step. Then iterations 0, 1, and so on are the first, second, and so on iterations. The loop body contains no branch or function call.

20.2.2.2 Data Dependence

A data dependence from instruction a to instruction b is represented as $(a \rightarrow b, \delta, \boldsymbol{d})$, where δ is the *dependence latency* and $\boldsymbol{d}$ is the *dependence distance vector*. This notation will later be applied to nested loops as well. However, in a single (non-nested) loop, the dependence distance vector is one-dimensional. Let $\boldsymbol{d} = \langle d \rangle$. We can simply refer to d as the *dependence distance*. It indicates that for any i, instruction a in loop iteration i accesses (writes or reads) a value that is accessed by instruction b in loop iteration $i + d$. For example, in Figure 20.1c, the dependence from instruction a to instruction b can be represented as $(a \rightarrow b, 1, \langle 0 \rangle)$. It has a unit dependence latency, as the integer functional unit used to execute it has unit latency. The dependence distance is 0, which means the two dependent instructions are in the same iteration.

The data dependence $(a \rightarrow b, \delta, \boldsymbol{d})$ can be a flow, anti-, or output dependence [68]. Assume instruction a lexically appears before b in the program. We say it is a *flow dependence* if instruction a writes to a variable that is read by instruction b. It is an *anti-dependence* if instruction a reads a variable that instruction b writes to. It is an *output dependence* if both instructions write to the same variable.

A flow dependence is a true dependence, in that the order of the two instructions cannot be changed; the producer instruction a must be scheduled before the consumer instruction b at least δ time steps earlier, where the dependence latency δ is usually the latency of the functional unit that executes the producer instruction a. That is, the ith iteration of a must be executed at least δ time steps ahead of the $(i + d)$th iteration of b. The other two kinds of dependences, anti- and output dependences, are false dependences, as they can be removed simply via renaming the variable, which is accessed by the two instructions, to two different variables. So we do not consider false data dependences henceforth. Further, control dependences within the loop body have been converted into data dependences using IF-conversion [10] and predicated execution [88]. Therefore, in this chapter, a dependence refers to a (data) flow dependence by default.

For a variable, among all the flow dependences on it, the maximal dependence distance d implies that the variable requires d number of *live-in values* defined before the loop. For example, in Figure 20.1b, there is one dependence $c \to a$ for TN2 with a distance of 2. So before the loop, we need to prepare two live-in values for TN2. The two live-in values are the TN2{2} and TN2{1} loaded before the loop. They will be consumed by the first and second iterations of the loop, respectively.

All the flow dependences of the loop compose the *data dependence graph* (DDG). For an instruction a, we use $PREDS(a)$ to denote the set of its immediate predecessors in the dependence graph; that is, if $b \in PREDS(a)$, then there exists an edge $b \to a$ in the DDG. Then $PREDS^+(a)$, the transitive closure of $PREDS(a)$, represents the set of instructions that have a path to instruction a in the DDG. Similarly, we use $SUCCS(a)$ to represent the set of its immediate successors in the graph; that is, if $b \in SUCCS(a)$, then there exists an edge $a \to b$ in the DDG. $SUCCS^+(a)$ is the transitive closure of $SUCCS(a)$, which represents the set of instructions that have a path from instruction a in the DDG.

A path in the DDG is a *recurrence* or *cycle* if it starts and ends at the same instruction. It is a *simple recurrence* or *simple cycle* if no instruction appears more than once in it. In this chapter, when we use the terms *recurrence* or *cycle*, we refer to *simple* recurrence (cycle) by default.

A strongly connected component (SCC) in the DDG is a set of instructions such that for any two instructions a and b in the SCC, there is a path from a to b and a path from b to a. An SCC may contain more than one recurrence.

20.2.2.3   Modulo Scheduling Constraints

Let $\sigma(a, i)$ be the schedule time of $a(i)$, the instruction a in iteration i, and T be the initiation interval. Modulo scheduling specifies that the following set of constraints must be met to achieve a legal software pipelined schedule:

- **Modulo property:** Two instances of an instruction corresponding to two successive iterations are scheduled with an offset of T time steps. That is,

$$\sigma(a, i) + T = \sigma(a, i + 1) \tag{20.1}$$

- **Dependence constraints:** For every dependence $(a \to b, \delta, \langle d \rangle)$, the producer instruction a in the ith iteration must be scheduled at least δ time steps before the $(i + d)$th instance of the consumer instruction b. That is,

$$\sigma(a, i) + \delta \leq \sigma(b, i + d) \tag{20.2}$$

- **Resource constraints:** At any time step in the modulo scheduled kernel, no hardware resource is allocated to more than one instruction.

Because of the modulo property, we can take the schedule time of an instruction in the first iteration as a reference. Therefore, from now on, unless stated otherwise, by *schedule time* of an instruction a, we refer to its schedule time in the first iteration, that is, $\sigma(a, 0)$.

20.2.2.4 Minimum Distance between Two Instructions

According to the dependence constraints in Equation 20.2, for a dependence $(a \rightarrow b, \delta, \langle d \rangle)$,

$$\sigma(b, 0) - \sigma(a, 0) \geq \delta - d * T \tag{20.3}$$

The right-hand side is the minimum number of time steps by which instruction b must succeed instruction a in the same iteration in any feasible schedule. This is the *minimum distance* between the two instructions in the schedule, denoted as

$$MinDist(a, b) = \delta - d * T \tag{20.4}$$

For two instructions that are not directly connected in the dependence graph, their minimum distance is transitively calculated as follows:

$$MinDist(a, b) = \begin{cases} -\infty & \text{If there is no path} \\ & \text{from } a \text{ to } b \text{ in the} \\ & \text{dependency graph.} \\ \max_{\forall c \in PREDS(b)} \left(MinDist(a, c) + MinDist(c, b) \right) & \text{Otherwise} \end{cases} \tag{20.5}$$

An all-pairs shortest-path algorithm, such as the Bellman–Ford algorithm or the Gabow–Tarjan algorithm [62], can be used to perform the above computation of *MinDist*. The time complexity for the *MinDist* computation is $O(w^3)$, where w is number of instructions in the loop.

20.2.2.5 Computation of the Minimum Initiation Interval

The performance of a modulo schedule is determined mainly by the *II*. The smaller the *II* is, the higher is the throughput of the software pipeline, leading to better performance. The *II* has a lower bound governed by the recurrences in the dependence graph, termed *recurrence minimum initiation interval (RecMII)*, and a lower bound determined by the resource constraints of the underlying architecture, termed *resource minimum initiation interval (ResMII)*. The minimum II for any valid modulo schedule is then

$$MII = \max(RecMII, ResMII) \tag{20.6}$$

RecMII can be calculated by enumerating all the recurrences in the dependence graph.

$$RecMII = \max_{\forall recurrence\ R} \left\lceil \frac{\delta(R)}{d(R)} \right\rceil \tag{20.7}$$

where $\delta(R)$ is the sum of the dependence latencies of the instructions in recurrence R of the dependence graph, and $d(R)$ is the sum of the dependence distances around the recurrence [89]. Those recurrences with the maximum value of $\lceil \frac{\delta(R)}{d(R)} \rceil$ are known as *critical recurrences*. Although there could be many recurrences in a DDG, in practice, there are only a few [54].

An alternative approach to finding *RecMII* is to make use of a property of *MinDist*. It is interesting to apply the *MinDist* constraint to a recurrence. From Equations 20.3 and 20.4, we have

$$0 \geq MinDist(a, a) \tag{20.8}$$

This property still holds when the recurrence around instruction a spans over other instructions. Therefore, *for any instruction a, if the II is legal, then MinDist(a, a) $\leq$ 0.* Otherwise, that would indicate that the schedule time of the instruction is strictly greater than that of itself, which is a contradiction. Therefore, we can also compute *RecMII* by enumerating the values of *II* until $MinDist(a, a) \leq 0$ for all instruction a in any recurrence. The minimum value of *II* that meets the above constraint is *RecMII*.

The resource constraints also impose a lower bound on *II*. This bound, for pipelined functional units, is given by [59, 85]:

$$ResMII = \max_{\forall r} \left(\left\lceil \frac{N_r}{F_r} \right\rceil \right) \tag{20.9}$$

where N_r represents the number of instructions that execute on a functional unit of type r, and F_r is the number of functional units of type r. If the functional units are non-pipelined or have complex structural hazards, calculation of *ResMII* is based on the total number of time steps for which a functional unit of a type r is required. In these cases, *ResMII* is defined as

$$ResMII = \max_{\forall r} \left\lceil \frac{\sum_a N_{a,r}}{F_r} \right\rceil \tag{20.10}$$

where $N_{a,r}$ represents the maximum number of time steps for which instruction a uses any of the stages of a functional unit of type r. For example, for a non-pipelined functional unit, $N_{a,r}$ equals the latency of the functional unit.

Example 20.1

In our example dependence graph of Figure 20.1c, there are two recurrences. One is $a \to b \to c \to a$ with a latency of three cycles and dependence distance of 2. The other is $a \to a$ with a latency of one cycle and dependence distance of 1. Thus,

$$RecMII = \max \left(\left\lceil \frac{3}{2} \right\rceil, \left\lceil \frac{1}{1} \right\rceil \right) = 2$$

Clearly, the first recurrence is critical.

We have two different types of pipelined functional units, namely integer and FP ALUs. The number of functional units in these types is two and one, respectively. They are used by three and one instructions, respectively. Hence,

$$ResMII = \max \left(\left\lceil \frac{3}{2} \right\rceil, \left\lceil \frac{1}{1} \right\rceil \right) = 2$$

Therefore, $MII = \max(2, 2) = 2$.

In a production compiler, one is interested in *MII*, rather than *RecMII* or *ResMII*. One can start with an initial value of $II = ResMII$ and perform a binary search, and arrive at the lowest II under which $MinDist(a, a) \leq 0$ for any a that is in a recurrence cycle. This lowest II is *MII*. This approach avoids explicit computation of *RecMII*.

20.2.3   A Heuristic Modulo Scheduling Framework

20.2.3.1   Overview

Typically, modulo scheduling is an extension of list scheduling. It sorts the instructions into a priority-order list and attempts to schedule the instructions one by one according to the order, satisfying both resource constraints and dependence constraints. Modulo scheduling methods differ mainly in:

- How they construct the order.
- Within the feasible time range, which time step is chosen for scheduling an instruction.
- How to proceed when there is no feasible time step for scheduling an instruction.

Despite the variations in different heuristics, they all try to first schedule those instructions which in their assessment are hard to schedule. Usually, the instructions in a recurrence cycle, and more specifically those in a critical recurrence cycle, are given higher priorities than others.

A typical modulo scheduler starts by computing *MII* as introduced in Section 20.2.2.5. It chooses an initiation interval $T = MII$ and attempts to construct a module schedule under this *II*. Each instruction has a *feasible time range* within which it can be scheduled without violating any *dependence* constraint with the already scheduled instructions, as will be described in Section 20.2.3.2. However, some feasible time step may lead to *resource* conflict with an already scheduled instruction. Therefore, the scheduler inspects every feasible time step until one is found that does not result in a resource conflict. The resource usage is kept track of by a *modulo reservation table* (MRT), which will be introduced in Section 20.2.3.3. The instruction is then scheduled into this time step t, and the MRT is updated. The feasible time range of every unscheduled instruction that has a path from or to this instruction is updated accordingly, as will be discussed in Section 20.2.3.2.

If the instruction cannot be scheduled at any time step within the range, a time step t' (not necessarily in the feasible range) is chosen based on some heuristic, and one or more of the already scheduled instructions that have resource conflicts (and/or dependence violation when t' is not in the feasible range) in scheduling the current instruction at time step t' are evicted out of the schedule to accommodate the current instruction. These evicted instructions are inserted back into the list of unscheduled instructions according to their priority order. Their resource usages are removed from the MRT. The current instruction is scheduled at t', and the MRT is appropriately updated. Then, the feasible ranges for all the unscheduled instructions, including the just evicted ones, are recalculated. Consequently, the priorities of the unscheduled instructions are also changed. The scheduler proceeds in this way until all instructions in the loop are scheduled.

If scheduling and eviction of instructions happen frequently and finally go beyond a certain threshold, the scheduler increments the initiation interval T by 1 and starts all over again. Increasing the *II* should ease both the resource and dependence constraints, especially for instructions on critical recurrences. Therefore, it should increase the chances of a successful scheduling. If the scheduler fails repeatedly even after increasing the *II* a number of times, it gives up. Then simple instruction scheduling, instead of modulo scheduling, can be applied to the loop body, to expose parallelism within a single loop iteration.

The above process is shown in Algorithm 20.1 as a general modulo scheduling framework, where *ESTART(a)* and *LSTART(a)* are the lower and upper bounds, respectively, of the feasible time range for instruction a. The calculation of *ESTART* and *LSTART* is explained in Section 20.2.3.2. The algorithm produces the schedule time for each instruction. With these schedule times, it is trivial to derive the prolog, kernel, and epilog. Hence, these are not explicitly shown in the algorithm.

The computational complexity of the algorithm depends on the specific algorithms used for *MII* calculation, prioritization of instructions, the heuristic used for choosing the time step at which an instruction is scheduled, the threshold value that determines when to increment the *II*, and so on. As an example, Rau's iterative modulo scheduling method [87] is similar to the algorithm framework here. It has the worst-case complexity that is exponential in w, the total number of instructions in the loop. However, the empirical computational complexity is only $O(w^2)$.

Algorithm 20.1

A Generic Modulo Scheduling Framework

Input: *Dependence graph, resource usage table RT_a for each instruction a, a thresholdII as the maximum II to be tried, and a budget as the limit of total instructions to be attempted during scheduling under a particular II.*

Output: $\sigma(a, 0)$ *for each instruction a.*

1: *Compute MII;*
2: *T = MII;*
3: **while** T < *thresholdII* **do**
4:   *instructionsTried = 0;*
5:   *Reset MRT;*
6:   **for all** *instruction* a **do**
7:     *Compute ESTART(a) and LSTART(a) under the initiation interval* T;
8:     *Compute the priority for* a;

```
 9:        Scheduled(a) = false;
10:     end for
11:     while there are unscheduled instructions do
12:        Pick an unscheduled instruction a that has the highest priority;
13:        for t = ESTART(a) to LSTART(a) do
14:           if Instruction a can be scheduled at time t without resource conflict then
15:              σ(a, 0) = t
16:              Scheduled(a) = true;
17:              Update MRT with RT_a;
18:              Update ESTART(b) ∀b ∈ SUCCS⁺(a) and b is unscheduled;
19:              Update LSTART(b) ∀b ∈ PREDS⁺(a) and b is unscheduled;
20:              break;
21:           end if
22:        end for
23:        if t > LSTART(a) then
24:           Choose a time step t′ using some heuristics
25:           σ(a, 0) = t′
26:           Scheduled(a) = true;
27:           Evict any already scheduled instruction that has resource conflict with a;
28:           Update ESTART(b) ∀b ∈ SUCCS⁺(a);
29:           Update LSTART(b) ∀b ∈ PREDS⁺(a);
30:           Evict any already scheduled instruction b in SUCCS⁺(a) or PREDS⁺(a) whose schedule time is
              not within its updated feasible range [ESTART(b), LSTART(b)];
31:           Update MRT by removing the resource usage of all the evicted instructions;
32:           Update MRT with RT_a;
33:        end if
34:        Recompute the priorities of the unscheduled instructions;
35:        instructionsTried + +;
36:        if instructionsTried > budget then
37:           T + +;goto step 3;
38:        end if
39:     end while
40:     return SUCCESS
41:  end while
42:  return FAILURE
```

20.2.3.2   Feasible Time Ranges and Dependence Constraints

Now we explain how the feasible time ranges for the instructions are computed during scheduling. The feasible range is identified by a lower bound, the *earliest start (ESTART)* time, and an upper bound, the *latest start (LSTART)* time.

To calculate the ESTART and LSTART of an instruction, first, insert into the dependence graph a dummy start instruction *START* and a dummy stop instruction *STOP*. Then add a dependence from *START* to every instruction a that has no predecessor in the graph, and let $MinDist(START, a) = 0$. Similarly, add a dependence to *STOP* from every instruction b that has no successor in the graph, and let $MinDist(b, STOP) = 0$. Then the feasible range for any instruction a is calculated as follows [30, 54, 87]:

$$ESTART(a) = \begin{cases} 0 & \text{If } a \text{ is } START \\ \max_{b \in PREDS(a)}(ESTART(b) + MinDist(b, a)) & \text{Otherwise} \end{cases} \qquad (20.11)$$

and

$$LSTART(a) = \begin{cases} ESTART(STOP) & \text{If } a \text{ is } STOP \\ \min_{b \in SUCCS(a)}(LSTART(b) - MinDist(a,b)) & \text{Otherwise} \end{cases} \tag{20.12}$$

Note that because of the recurrences in the DDG, an instruction b may be within both $PREDS(a)$ and $SUCCS(a)$, and therefore, scheduling of it will affect both $LSTART$ and $ESTART$ of instruction a. The width of the time range,

$$LSTART(a) - ESTART(a) \tag{20.13}$$

is referred to as the *slack* of instruction a. The slack indicates how much freedom the scheduler has in scheduling the instruction.

Usually, given an II, the minimum distance ($MinDist$) for every instruction pair can be calculated beforehand. Using these, the feasible time range and the slack of each instruction can be derived.

Note that the minimum distance between two instructions is a static concept: it never changes. However, the feasible time range and slack of an instruction are dynamic; at the beginning of scheduling, when no instruction is scheduled yet, the initial values of the range and the slack for every instruction are computed according to Equations 20.11, 20.12, and 20.13. Once an instruction b is scheduled at a time step t, it affects the feasible ranges and slacks of the unscheduled instructions from or to which there is a path in the dependence graph. We may consider that $ESTART(b) = LSTART(b) = t$ and apply the new values to the above equations to update the ranges and slacks of these unscheduled instructions.

20.2.3.3 Modulo Reservation Table and Resource Constraints

During scheduling, the resource constraints are checked with the help of an MRT [59, 85, 87]. The MRT records the resource usage of the schedule as it is constructed. Because of the modulo property, the resource usage of a kernel represented by the MRT also represents the resource usage for the entire schedule.

The MRT is an $M \times T$ 0-1 matrix, where M is is the total number of resources in the architecture and T is the initiation interval. For each resource, the MRT records its usage in the T time steps in the kernel. Initially, all entries in the MRT are 0. If an instruction uses a resource r at time step t, then the entry $MRT(r, t \mod T)$ is set to 1.

Usually, an instruction a uses more than one hardware resource. For example, it may need the bus, issue slots, and various pipeline stages of an ALU. Thus, the instruction has its own resource usage pattern, represented by a reservation table RT_a, which is an $M \times d$ 0-1 matrix, where d is the latency of the instruction. In scheduling instruction a, the scheduler checks the availability of resources and updates the MRT by performing binary operations between MRT and RT_a.

To check whether there is any potential resource conflict in scheduling an instruction a at time step t, first RT_a is wrapped around the kernel with its first time step overlapped with time step ($t \mod T$) of the kernel, its second step with time step ($[t + 1] \mod T$), and so on. The scheduler performs non-destructive bit-wise AND operations between the corresponding entries of RT_a and MRT. By *non-destructive*, we mean that MRT is consulted but not modified. If any entry of the resulting matrix is 1, this means a resource that is requested by the instruction has already been taken by some other instruction scheduled in the kernel, and hence it is not feasible to schedule the instruction at the given time step. If the result of the bit-wise AND operation is a zero matrix, the instruction can be scheduled at t without resource conflict. The MRT is updated by performing bit-wise OR operations with RT_a [84].

Note that the latency of the instruction can be arbitrarily long, and therefore, it is possible that a single entry of the MRT overlaps with more than one entry of the instruction's reservation table, after it is wrapped around the kernel. These entries from the instruction's reservation table should not have any resource conflict between themselves. Thus, if a single resource is used by an instruction in two time steps that are separated by l cycles, then an initiation interval $II = l$ is not possible. Using that II will lead to violation of the resource constraints given in Section 20.2.2.3.

The number of bit-wise AND and OR operations performed for checking resource constraints can become a source of inefficiency, especially when the resource usage is complex or the number of instructions tried is large. Efficient resource modeling for general instruction scheduling and software pipelining has been proposed in [13, 30, 35, 51, 52, 81, 82].

20.2.3.4 Heuristics in Modulo Scheduling

Let us now look at a few heuristic methods and relate them to the generic modulo scheduling framework.

Lam [59] proposed a method that establishes several important points on resource-constrained modulo scheduling. First, modulo scheduling can be accomplished without special hardware support by unrolling the kernel by a small number of times. This is referred to as *modulo variable expansion*, which is explained in greater detail in Section 20.2.5. Second, the instructions in the SCCs in the dependence graph are more important than other instructions, because scheduling one instruction in an SCC constrains the scheduling of all the other instructions in the same SCC. Not all SCCs are equally important. Only SCCs that correspond to the critical recurrences, which strongly constrain the initiation interval, should be given highest priority in scheduling. Third, list scheduling for an acyclic dependence graph can be extended to handle a cyclic dependence graph as well. This is accomplished as follows. First, schedule each SCC in the order of priority. The schedule for an SCC is kept as a single macro instruction in the DDG, which represents the collective resource usage of the instructions in the SCC. When every SCC is replaced with a macro instruction, the DDG becomes acyclic. Then apply the acyclic scheduling algorithm to schedule the macro instructions and other instructions to arrive at a resource-constrained software pipelined schedule. Next, the control constructs in the loop body, if any, can be processed hierarchically, starting with the innermost control construct. After a construct is scheduled, the entire construct is reduced into a single macro instruction, representing all the scheduling constraints of its components with other constructs. This is called *hierarchical reduction*. An application of this technique is to handle the conditionals in the loop body. The two branches of a conditional are scheduled separately, and then the entire conditional construct is reduced into a macro instruction, with the union of the scheduling constraints of the two branches as its scheduling constraints. The same principle can be used to schedule nested loops as well. The work of Muthukumar and Doshi [70] and Wang and Gao [106] are two applications of this principle.

Rau's iterative modulo scheduling method [87] is in line with the generic modulo scheduling framework shown in Algorithm 20.1. It uses a height-based heuristic to prioritize the instructions. The height of an instruction a is equal to $MinDist(a, STOP)$, where $STOP$ is the dummy stop instruction introduced in the dependence graph, as described in Section 20.2.3.2. The larger the $MinDist$ is, the higher is the priority. Rau observes that (a) the height priority helps complete the scheduling process for most of the single loops in a single pass, and (b) it gives higher priority to instructions in SCCs that more strongly constrain the initiation interval. Iterative scheduling uses a fixed budget, $k * w$, where k is a small constant and w is the number of instructions in the loop. The budget also determines the number of different IIs to be tried, and how many times the eviction of instructions can happen under a particular II before the II is incremented.

Huff's slack scheduling method [54] is also similar to the generic modulo scheduling algorithm. Here, the instructions are prioritized according to their slacks. Instructions with smaller slacks are scheduled prior to those with higher slack values. All the above three methods [54, 59, 87] ensure that the instructions in SCCs are given higher priorities in scheduling, although they use different heuristics. Lam's method explicitly finds all the SCCs and schedules them first; Rau and Huff's methods implicitly favor the SCCs by prioritizing the instructions appropriately.

A number of methods, including Huff's slack scheduling, in addition to constructing a software pipeline, also minimize register pressure [28, 34, 37, 54, 64, 65, 107]. Many of them follow the general modulo scheduling framework shown in Algorithm 20.1 with the exception that the schedule time steps for instructions are chosen to reduce the register lifetimes.

Another unique approach to modulo scheduling, called *decomposed software pipelining* (DESP), decomposes the resource-constrained cyclic scheduling problem into a *cyclic* scheduling problem *without* resource constraints, followed by an *acyclic* scheduling problem *with* resource constraints. Both subproblems are simpler than the original problem. In other words, it first handles dependence constraints without considering resource constraints and then modifies the schedule to enforce resource constraints. The process is as follows. Assuming there are unlimited resources and considering the dependence constraints only, an initial schedule is found using classical graph theoretic algorithms. Then certain edges in the DDG can be removed to make the DDG acyclic. Hence, the scheduling problem now becomes resource-constrained scheduling of acyclic DDG. For this, traditional list scheduling can be applied, which adjusts the initial schedule under the resource constraints. This approach was proposed by Gasperoni and Schwiegelshohn [47]. Later, Wang et al. [105] proposed another perspective to DESP. They view the kernel as a two-dimensional matrix of instructions. Thus, the scheduling problem is to find the row and column numbers for each instruction. The row number corresponds to the time step within the kernel at which the instruction is scheduled. The column number relates to the stage number or iteration to which the instruction belongs. They find the row and column numbers in two steps, which is similar to the DESP approach. Calland et al. [17] improved DESP by applying circuit retiming, which was originally a hardware circuit design technique, to identify the edges to be removed in the DDG. This not only improves the efficiency of the method, but, more interestingly, establishes a link between software pipelining and circuit retiming. We will discuss this approach in more detail in Section 20.3.2.

20.2.4 An Optimal Modulo Scheduling Formulation

The resource-constrained optimal modulo scheduling problem, which is known to be NP-complete, is formulated as an optimization problem in [40, 51] as below.

Problem 20.1

Given a loop $\mathcal{L}$ and a machine architecture, construct a modulo schedule for a given II that achieves a certain optimal objective, subject to the modulo property, dependence constraints, and resource constraints as stated in Section 20.2.2.3.

Usually, the scheduler tries different *IIs* starting from *MII*, and solves the above optimization problem for each *II*. The first successful solution is the one that both minimizes *II* and achieves the optimal objective.

In this section, we present only the formulation for the modulo property and dependence and resource constraints, which stay the same for different optimal objectives. The objectives can be a minimization of resources such as the functional units or registers required by the schedule [11, 35, 51], minimization of power consumption, or power variation [113–115], and so on. We will show the formulation for minimizing power consumption and power variation in Section 20.3.3. If the optimality objective is null, that is, there is no objective function in the formulation, then the problem becomes a constraint solving problem.

In this section, we discuss the integer linear programming solution presented in [51] to the above problem. For simplicity, we assume an architecture with pipelined functional units. For architectures with non-pipelined functional units or functional units with complex resources, the formulation is presented in [11, 12].

Assume the loop consists of w instructions. For convenience, we number the instructions instruction $0, 1, \ldots, w - 1$. The architecture has F_r functional units of type r. For any instruction a, let $t_a = \sigma(a, 0)$ be the time step at which the instruction in the first iteration (iteration 0) is scheduled. Then, according to the modulo property, the instruction in iteration i is scheduled at time $\sigma(a, i) = t_a + i * T$, where T is the initiation interval.

First, consider the dependence constraints given in Equation 20.2. If there is a dependence $(a \rightarrow b, \delta, \langle d \rangle)$, then

$$\sigma(a, i) + \delta \leq \sigma(b, i + d)$$

This can be rewritten as

$$t_a + i * T + \delta \leq t_b + (i + d) * T$$

This, in turn, can be written as an integer constraint:

$$t_b - t_a \geq \delta - d * T \qquad \forall \text{ dependence } (a \rightarrow b, \delta, \langle d \rangle) \qquad (20.14)$$

Next, we represent the resource constraints in a linear form. Define a pair of values, k_a and o_a, for each t_a, such that

$$t_a = k_a * T + o_a$$

where $0 \leq o_a < T$. Thus, if $\mathcal{K}, \mathcal{O},$ and Γ are w-element row vectors

$$\Gamma = \begin{bmatrix} t_o \\ t_1 \\ \vdots \\ t_{w-1} \end{bmatrix}, \mathcal{K} = \begin{bmatrix} k_o \\ k_1 \\ \vdots \\ k_{w-1} \end{bmatrix}, \text{ and } \mathcal{O} = \begin{bmatrix} o_o \\ o_1 \\ \vdots \\ o_{w-1} \end{bmatrix}$$

then we can write Γ in the following way:

$$\Gamma = \mathcal{K} * T + \mathcal{O}$$

The repetitive kernel of the software pipelined loop is represented by a two-dimensional 0-1 matrix $\mathcal{M}$ of size $T \times w$. An entry $\mathcal{M}[t,a] = 1$ represents that instruction a is scheduled at time step t in the repetitive kernel; that is, $o_a = t$. The relation between $\mathcal{O}$ and $\mathcal{M}$ can be written as

$$\mathcal{O} = \begin{bmatrix} o_0 \\ o_1 \\ \vdots \\ o_{w-1} \end{bmatrix} = \begin{bmatrix} m_{0,0} & m_{0,1} & \cdots & m_{0,(w-1)} \\ m_{1,0} & m_{1,1} & \cdots & m_{1,(w-1)} \\ \vdots & \vdots & & \vdots \\ m_{(T-1),0} & m_{(T-1),1} & \cdots & m_{(T-1),(w-1)} \end{bmatrix}^{\text{Trans.}} \times \begin{bmatrix} 0 \\ 1 \\ \vdots \\ (T-1) \end{bmatrix}$$

Now Γ can be expressed as

$$\Gamma = \mathcal{K} * T + \mathcal{M}^{\text{Trans.}} \times \begin{bmatrix} 0 \\ 1 \\ \vdots \\ (T-1) \end{bmatrix} \qquad (20.15)$$

Each instruction must be scheduled exactly once in the repetitive kernel. Therefore,

$$\sum_{t=0}^{T-1} m_{t,a} = 1, \quad \forall a \in [0, w-1] \qquad (20.16)$$

The total requirement of the schedule for resource type r at time step t is given by

$$\sum_{a \in \mathcal{I}_r} m_{t,a},$$

where $\mathcal{I}_r$ represents the set of instructions that use functional unit type r. The resource constraints specify that the resource requirement of the schedule at any time step is less than or equal to the number of available resources. That is,

$$\sum_{a \in \mathcal{I}_r} m_{t,a} \leq F_r, \quad \forall t \in [0, T-1] \text{ and } \forall r \qquad (20.17)$$

We do not discuss the formulation of the objective function here. The resource-constrained optimal software pipelining problem is to solve Equations 20.14, 20.15, 20.16, and 20.17, subject to the objective function.

20.2.5 Register Allocation and Code Generation

Having discussed how to construct the modulo schedule, next we focus on register allocation and code generation for the constructed schedule.

First, let us introduce the concepts of *scalar lifetime* and *vector lifetime* [86]. A scalar lifetime is the lifetime of a loop variable for a given iteration of the loop. Each variable has one producer instruction that produces the value and one or more instructions that consume the value. The scalar lifetime starts when the producer is issued and ends when all of the consumers have finished.[2] The scalar lifetimes corresponding to the first, second, . . . , iterations of the loop, are referred to as the first scalar lifetime, the second scalar lifetime, . . . , respectively. The scalar lifetimes of a variable over all iterations of the loop compose the vector lifetime of the variable. In this chapter, the term *lifetime* refers to a vector lifetime by default, unless stated otherwise.

The lifetimes of the variables in a software pipelined schedule have unique features, which makes register allocation and code generation for it different from those for a traditional schedule. We illustrate the features with the help of an example. Consider the simple example loop in Figure 20.5a. In this example, there are two variables, TN1 and TN2. If a TN has d number of live-in values, correspondingly, it has d number of *live-in scalar lifetimes*. Note that these scalar lifetimes are defined outside of the loop and therefore appear earlier than the first scalar lifetime (of the loop). We assume that the value of TN1 produced in the last iteration is used after the loop, that is, TN1 has a *live-out scalar lifetime*.

Figure 20.5b shows a modulo schedule for the loop, assuming there are two general-purpose pipelined functional units that can execute any instruction and assuming that the latencies of instructions a, b, and c are, five, one, and one cycle(s), respectively. To illustrate the lifetimes of the variables, we unroll the schedule (see Figure 20.5c). The scalar lifetimes for variables TN1 and TN2 are shown beside the instructions that produce them.

The lifetimes have interesting features. It is unique to software pipelining that for each variable, its scalar lifetimes are produced regularly every *II* time steps, and all the scalar lifetimes have the same length, except for the live-in and live-out scalar lifetimes; if the lifetime is longer than *II*, when a new scalar lifetime is produced, the previous one is still live, and thus they are overlapped in time. Overlapping scalar lifetimes must be given different registers to keep them live at the same time. Such overlapping between the scalar lifetimes of the same variable is unique to a software pipelined schedule. In contrast, in a traditional schedule, overlapping happens only between the scalar lifetimes of different variables; for the same variable, a scalar lifetime of it ends before the next one is produced, and they never overlap.

To allocate registers efficiently for such a schedule, it is imperative to address the challenge of overlapping scalar lifetimes and make use of the repetition. There are two subproblems: first, how to address the overlap between the scalar lifetimes of the same variable and, second, how to address the overlap between those of different variables.

There are two approaches to solving the above problems. A software-only approach proposed by Lam, known as *modulo variable expansion* (MVE) [59], addresses the first subproblem and leaves the second one to the global register allocator in a later phase. It unrolls the kernel of the software pipelined schedule by a small number of times, such that a group of overlapping scalar lifetimes are given distinct variable names, and the names can be repetitively reused by other scalar lifetimes. The subsequent global register allocator will allocate distinct registers to these names. The unrolling leads to code size expansion, though. The other

---

[2]This definition is to support precise interrupt handling. Assume the following two instructions are executed at the same time step: $a: x = \cdots$ and $b : \cdots = y$. Since x is live from this time step, and y is live until instruction b is completed, x and y will not be allocated the same register. If during the execution, an interrupt happens, the hardware will automatically re-execute both instructions. It can do so because all the input registers still keep the original values, without being overwritten. For example, y is still kept in a register, which is different from that of x, and cannot be overwritten by x, even if x might have been produced. If precise interrupt handling is not necessary, then the definition of the scalar lifetime can be such that it starts after the producer has finished and ends when all the consumers have started.

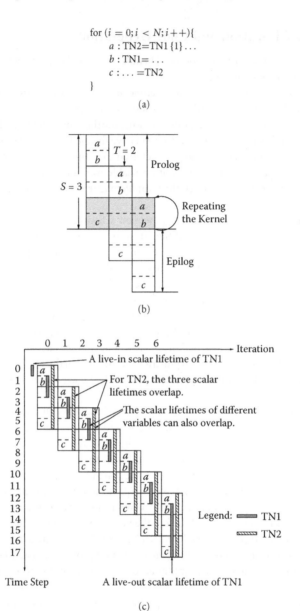

FIGURE 20.5   A modulo schedule and lifetimes of its variables. (a) An example single loop. (b) A possible modulo schedule with $T = 2$ and $S = 3$. The kernel is highlighted in darker color. (c) The unrolled schedule and the lifetimes of the variables. Source: Figures (a) and (b) are from [91], courtesy of ACM.

approach, developed first by Dehnert et al., uses special hardware support in the form of *rotating register files* [88]. With such support, the adjacent scalar lifetimes of a variable are automatically mapped to a set of consecutive registers during execution time. Under such an assumption, the lifetimes of the variables are represented elegantly, and both subproblems are solved in the same framework. With additional support for *predicated execution*, in code generation, only the kernel is generated, without requiring unrolling of the loop at all and without explicit prolog and epilog.

The two schemes are trade-offs in compiler and architecture functionality and in performance and code size. Basically, both schemes expand a single variable into an array of variables (for MVE) or registers (for the rotating registers approach) so that different instances of the same variable that are simultaneously live can be stored in distinct registers. MVE assumes no hardware support, while the other approach does. Below we describe the two approaches in detail.

20.2.5.1   Software-Only Approach: Modulo Variable Expansion

All the scalar lifetimes, except the live-in and live-out scalar lifetimes, have the same length. If the length is l, and the II is T time steps, then the variable has $\lceil \frac{l}{T} \rceil$ number of scalar lifetimes that are overlapping, that is, live simultaneously. The variable in these scalar lifetimes is renamed as distinct variables, each one of which can be allocated a different register by a global register allocator in a subsequent phase.

Consider the variable TN2 in Figure 20.5c as an example. The length of its lifetime is six time steps, and $T = 2$. Thus, it always has three scalar lifetimes overlapping during the pipelined kernel execution. We can expand TN2 to three variables, TN2, TN2′, and TN2″, as shown in Figure 20.6a. The three variables are repeatedly reused. Now no two scalar lifetimes of the same variable overlap. The length of a scalar lifetime of TN1 is two time steps, which is less than or equal to T. Hence, it does not require any MVE. Then the lifetimes of the variables in the software pipelined schedule can be allocated registers using a traditional global register allocator, for which there are many options available, including Chaitin's graph coloring approach [23].

Expansion of the variables implies unrolling the kernel and renaming the variables. In general, the kernel is unrolled u number of times, where

$$ u = \max_v \left(\left\lceil \frac{l_v}{T} \right\rceil \right) $$

where l_v is the length of a scalar lifetime for variable v, not considering live-in and live-out scalar lifetimes.

Note that in the unrolled kernel, a variable v has u number of instances now. These instances are renamed to x number of variables, where x is the smallest factor of u, and x is no less than $\lceil \frac{l_v}{T} \rceil$. The instances in the first x number of iterations are renamed to distinct names. Then these names are applied to the next x number of iterations in the unrolled kernel, and so on. As u is divisible by x, the u instances of the variable v will be renamed to these names exactly an integer number ($\frac{u}{x}$) of times.

For the motivating example, the lengths of a scalar lifetime of TN1 and TN2 are two and six time steps. Therefore,

$$ u = \max \left(\left\lceil \frac{2}{2} \right\rceil, \left\lceil \frac{6}{2} \right\rceil \right) = 3 $$

Thus, the kernel is unrolled three times. For TN1, the smallest factor of u no less than $\lceil \frac{2}{2} \rceil$ is 1. For TN2, the smallest factor of u no less than $\lceil \frac{6}{2} \rceil$ is 3. Therefore, the two variables are expanded to one and three variables, respectively. This is illustrated in Figure 20.6b.

Suppose the trip count of the original loop is N. Software pipelining the loop results in a kernel with S number of stages. Assume the kernel is unrolled u times for modulo variable expansion, and the unrolled kernel is executed p times. Since the prolog and epilog together contribute to $S - 1$ number of iterations, and the unrolled kernel contributes $u * p$ iterations, we have the following relationship:

$$ N = u * p + (S - 1) + q $$

where $0 \le q < u$ is the remaining number of loop iterations not covered by the software pipelined schedule. These loop iterations are executed in the original sequential form. This is known as *preconditioning*. In other words, for MVE, the generated code has two parts: one consisting of

$$ q = (N - (S - 1)) \mod u \tag{20.18} $$

number of non-software pipelined loop iterations, and the other consisting of all the other iterations in pipelined fashion. For the example shown in Figure 20.5, $S = 3$ and $u = 3$. Thus $q = (N - 2) \mod 3$. The code generation (loop rewriting) for this loop is shown in Figure 20.6b.

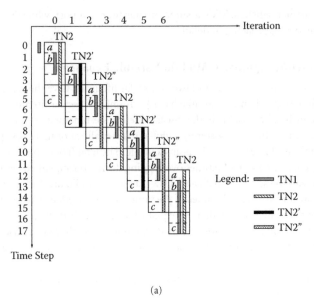

(a)

```
/* Preconditioned loop iterations */
q = (N − 2) mod 3;
for (i = 0; i < q; i++) {
    TN2 = TN1 {1} ...
    TN1 = ...
    ... = TN2
}
/* Below are the software pipelined loop iterations */
/* Prolog: */
    TN2 = TN1 {1} ...
    TN1 = ...
            TN2′ = TN1 {1} ...
            TN1= ...
/* In the loop body, the kernel is unrolled 3 times, and the variables are expanded: */
for (i = 0; i < (N − 2 − q) div 3; i++) {
                        TN2″ = TN1 {1} ...
    ... = TN2           TN1 = ...
                                    TN2 = TN1 {1} ...
        ... =TN2′                   TN1= ...
                                                TN2′ = TN1{1} ...
                    ... = TN2″                  TN1= ...
}
/* Epilog: */
                        ... =TN2
                                    ... =TN2′
```

(b)

FIGURE 20.6   Illustration of modulo variable expansion. (a) Expand the variable TN2 to three variables to remove overlapping of the scalar lifetimes of the same variable. TN2 is expanded to TN2, TN2′, and TN2″ repeatedly. (b) MVE code generation (loop rewriting).

In the above discussion, we have ignored the live-in and live-out scalar lifetimes. In practice, some of the iterations of the kernel need to be peeled off at the beginning and at the end of the loop. These iterations of the kernel are put into the prolog and epilog, respectively. This makes the prolog and epilog long enough to contain the live-in and live-out scalar lifetimes, such that in determining the unrolling factor u, above, we need only consider the length of a scalar lifetime that is not live-in or live-out. The scalar lifetimes in the prolog and epilog are renamed separately, honoring the constraints imposed by renaming for the unrolled kernel [86]. Our example happens to require no peeling of the iterations of the kernel.

20.2.5.2 Hardware-Supported Approach

In this subsection, we briefly review how to perform register allocation and code generation for a modulo scheduled loop when the target architecture provides support for software pipelining in the form of rotating register files and predicated execution [30, 86]. We first explain such hardware support and then discuss the solution in detail.

20.2.5.2.1 Hardware Support for Modulo Scheduling

A hardware support, called a rotating register file, has been proposed to address the problem of overlapping scalar lifetimes of the same variable. Such a support was originally provided in the Cydra architecture [88] and more recently in the IA-64 architecture [55].

A *rotating register file* is a set of physical registers organized as a circle. Let C be the circumference of the circle, that is, the total number of registers in the rotating register file. The problem of overlapping scalar lifetimes of the same variable is solved under this convention: if the first scalar lifetime of the variable is allocated to physical register r_x, then the second one is *automatically* allocated to the next physical register $r_{(x-1) \bmod C}$, the third to $r_{(x-2) \bmod C}$, and so on. Here we modulo the register indexes by the circumference C because of the cyclic nature of the rotating register file; the register index wraps around to the highest index when it becomes -1. The register allocator guarantees that when the register index wraps around, the scalar lifetime originally in the physical register with the highest index is no longer live.

The convention is realized by the cooperation between hardware and software. In the hardware side, there is a separate base register for the rotating register file. In the software side, the code generator replaces a variable with a *virtual register*, which is mapped to a set of *physical registers* at runtime with the help of the hardware. To distinguish between the virtual and physical register names, we use γ to denote the former and r to represent the latter. The correspondence between the two is achieved as follows. An access to a virtual register γ_x is automatically translated by the hardware into an access to a physical register $r_{(x+base) \bmod C}$, where *base* is the content of the base register.

Thus, in this approach, a variable is mapped to a single virtual register γ_x without any explicit expansion. In the code generated for the schedule, the base register is initialized to 0, which means γ_x is initially mapped to the physical register r_x. Then the kernel is repeatedly executed, once every II cycles. At the end of an execution of the kernel, a special instruction is issued to "rotate" the register file. The rotation is to decrement the content of the base register. The effect is such that with the repetitive execution of the kernel, the variable in the first, second, third, ... loop iterations is automatically mapped to physical registers r_x, $r_{(x-1) \bmod C}$, $r_{(x-2) \bmod C}$, ...

Register allocation for the loop variables for such a rotating register file happens before code generation. The impact of the above convention upon the register allocator is that it needs only to consider which *physical register* is to be allocated to the *first* scalar lifetime of a variable. Once that is done, all the other scalar lifetimes of the variable automatically have physical registers allocated, according to the convention. Suppose the physical registers allocated to the scalar lifetimes are r_x, $r_{(x-1) \bmod C}$, and so on. Then in code generation, a corresponding single *virtual register* γ_x is used to represent them. In execution of the code, the virtual register is mapped to r_x, $r_{(x-1) \bmod C}$, and so on, dynamically, just as the register allocator expects.

In the IA-64 architecture, there are three classes of rotating register files: integer, floating point, and predicate. Their base registers are initialized by a special instruction at the same time, and they are rotated simultaneously by the "rotation" instruction. The integer, floating point, and predicate variables are allocated to the three register files separately via the same register allocation method.

20.2.5.2.2   *Register Allocation*

In register allocation, the variables in the same register class, for example, integer registers, are considered together and allocated registers from the rotating register file of that class. The allocation method for each register class is the same. Hence, we discuss register allocation for only one register file.

Each variable has a vector lifetime in a software pipelined schedule. Now that the consecutive scalar lifetimes in it will be mapped to consecutive physical registers, we can represent the vector lifetime on a *space–time diagram*, where time is on the horizontal axis and the physical registers are on the vertical axis, assuming that there is an infinite number of registers. A vector lifetime is composed of a *leading blade* in case of live-in values, a *trailing blade* in case of live-out values, and a *wand* (the diagonal band) [86].

Because of the repetition of the scalar lifetimes, the first scalar lifetime is taken as a reference in representation of the whole vector lifetime. A vector lifetime is represented by a 4-tuple *(start, end, omega, alpha)* [86]. The *start* and *end* values refer to the start and end time steps of the first scalar lifetime. The scalar lifetime corresponding to iteration i starts at $(start + i * T)$ and ends at $(end + i * T)$, where T is the initiation interval. *Omega* is the number of live-in values for the loop variable. *Alpha* represents the number of live-out values for the loop variable. Figure 20.7a illustrates these ideas with an example.

Intuitively, register allocation packs the vector lifetimes on the space–time diagram as close as possible without any conflict. A physical register r_x is said to be *allocated* to a vector lifetime v if it is allocated to the first scalar lifetime of v. The next physical register $r_{(x-1) \bmod C}$ is then allocated to the second scalar

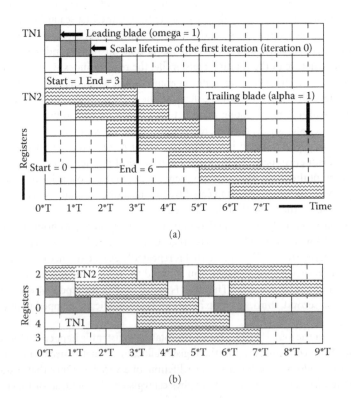

(a)

(b)

FIGURE 20.7 Register allocation for a modulo schedule with rotating register file support. (a) Space–time diagram for the example in Figure 20.5a with $N_1 = 7$. TN2 is made of a wand only. TN1 has a leading blade for it has a live-in value, and a trailing blade (assume it has a live-out value). The first scalar lifetime of TN1 starts at time step 1 and ends at 3, while that of TN2 starts at time step 0 and ends at 6. So the vector lifetimes of TN1 and TN2 can be represented as *(start, end, omega, alpha)* = (1,3, 1, 1) and (0, 6, 0, 0), respectively. (b) An optimal bin packing for the space–time diagram in Figure 20.7a. TN1 is allocated physical register 0, and TN2 physical register 2, with a circumference of 5 after the diagram is wrapped into a cylinder. Source: From [91], courtesy of ACM.

lifetime, and so on. *Conflict* refers to the fact that two scalar lifetimes that overlap in time are allocated to the same register.

To avoid any conflict, two vector lifetimes must have a certain *distance* between them in the space–time diagram. For two vector lifetimes A and B, let r_A and r_B be the physical registers allocated to A and B, respectively. The distance $DIST[A, B]$ is the legal range of $r_B - r_A$, within which A and B do not conflict in the space–time diagram. It is defined as $[d_3, +\infty]$, where d_3 is the lower bound value calculated as shown below. There is no upper bound (or the upper bound is $+\infty$).[3]

The distance $DIST[A,B]$ must meet the following conditions [86]. The wand of B must be to the right of the wand of A, and the leading/trailing blades of B must be above the leading/trailing blades of A. Formally,

$$start(B) + (r_B - r_A) * T \geq end(A)$$
$$r_B - r_A \geq omega(A) \text{ if } omega(B) > 0$$
$$r_B - r_A \geq alpha(A) \text{ if } alpha(A) > 0$$

That is,

$$d_1 = \left\lceil \frac{end(A) - start(B)}{T} \right\rceil \qquad (20.19)$$

$$d_2 = \begin{cases} d_1 & \text{if } omega(B) = 0 \\ max(d_1, omega(A)) & \text{otherwise} \end{cases} \qquad (20.20)$$

$$d_3 = \begin{cases} d_2 & \text{if } alpha(A) = 0 \\ max(d_2, alpha(A)) & \text{otherwise} \end{cases} \qquad (20.21)$$

where d_3 is the final value for the lower bound of the distance.

For the example in Figure 20.7a, TN1 and TN2 are represented as $(start, end, omega, alpha) = (1, 3, 1, 1)$ and $(0, 6, 0, 0)$, respectively. Thus, $DIST[TN2, TN1] = [3, +\infty]$, according to the above equations. This can also be seen intuitively from the figure. Similarly, we can calculate that $DIST[TN1, TN2] = [2, +\infty]$.

Because of the cyclic nature of the rotating register file, the space–time diagram can be wrapped up as a *cylinder* with time along the axis and registers on the circumference of the cylinder. Therefore, the register allocation problem consists of packing the vector lifetimes on the surface of the cylinder, such that there is no conflict and the circumference is minimized. Figure 20.7b illustrates the bin-packing concept.

In general, an optimal bin-packing problem is known to be NP-complete [86]. Therefore, heuristics are employed to sort the vector lifetimes and insert them one by one on the surface of the space–time cylinder without backtracking. Three sorting heuristics were proposed in [86]. In *start time ordering*, the earliest vector lifetime is inserted first. In *adjacency ordering*, the vector lifetime that minimizes the horizontal distance with the previously inserted lifetime is given higher priority. Last, *conflict ordering* follows an approach that is similar to what graph coloring [23] does for scalar lifetimes. The insertion of the chosen vector lifetime is then decided using one of the three strategies: best fit, first fit, and end fit. *Best fit* finds a register that minimizes the current register usage. *First fit* chooses the first compatible register starting from register 0. *End fit* starts from the register allocated to the vector lifetime inserted at the last step.

One might see the similarity between this process and scheduling. The definition of $DIST[A, B]$ is similar to *MinDist*, while the sorted ordering is similar to the priority list of instructions, and the insertion of the vector lifetimes is similar to scheduling instructions.

The above register allocation approach is very efficient. To get an indication of its efficiency, consider *MaxLive*, the maximum number of scalar lifetimes, of the same variable or different variables, that are simultaneously live. This is a lower bound on the number of registers required for the scalar lifetimes.

---

[3]The distance defined in [86] contains only the lower bound. Here we expand it to a range for clarity.

In traditional local register allocation for basic blocks, the lower bound *MaxLive* can always be achieved. In a software pipelined schedule, even with vector lifetimes, it was found that the tightness of this lower bound for the register requirement holds in many cases. That is, the register allocation approach, when combined with the best lifetime ordering and insertion heuristics, results in an actual register requirement that is close to the lower bound. Experiments show that the average register requirements vary from 1.0 to 1.3 times the *MaxLive* value [86]. This efficiency is reasonable, as the approach has fully considered the special features of the vector lifetimes in a software pipelined schedule and made use of them in the whole process, especially in lifetime representation, distance calculation, and ordering.

20.2.5.2.3 Code Generation

With rotating register files and predicated execution, it is possible to generate code for a software pipelined schedule that contains only the kernel. In such a schedule there is no explicit prolog or epilog.

As discussed in Section 20.2.1, a modulo scheduled kernel has S stages, where each stage has instructions from a distinct iteration. To control the execution of the stages during the prolog and epilog, the code generation scheme associates a predicate register with each stage of the kernel. This approach does not result in any code size increase.

We consider IA-64 [55] as the target architecture for the code generation. For the modulo schedule shown in Figure 20.5b and the register allocation in Figure 20.7b, the kernel-only code generation scheme results in the code depicted in Figure 20.8a. The code first initializes control registers and then executes the kernel, followed by a `br.ctop` instruction that branches back to the kernel.

The initialization consists of a `clrrrb` instruction in line 2, which resets the base registers corresponding to the integer, floating point, and predicate rotating register files. In lines 3 and 4, the *loop count register LC* is set to the original trip count minus 1, and the *epilog count register EC* is set to the total number of stages, $S = 3$.

The stages are controlled by S virtual predicate registers. For the IA-64 architecture,[4] $\rho16$, $\rho17$, ..., $\rho(16 + S - 1)$ are assigned to each stage in the kernel from right to left. The instruction in line 5 initializes $\rho16$ to 1 and all the other predicate rotating registers to 0. Line 6 sets up the live-in value for TN1.

Lines 9 and 10 compose the kernel. It consists of instructions a and b from iteration i and c from iteration $(i - 2)$. We defer a discussion on the register assignment for the variables to a later part of this section. The `Br.ctop` instruction in line 11 is a branch instruction in the IA-64 instruction set architecture (ISA), which rotates the three rotating register files simultaneously, decrements LC, and sets $\rho16$ to 1 if $LC > 0$; otherwise, that is, if $LC = 0$, it decrements EC and resets $\rho16$. Then it branches to the target. When finally both LC and EC become 0, it does not take the branch, but fails through.

To help us understand the kernel-only code more clearly, Figure 20.8b shows the dynamic execution process of the code. For simplicity, assume $N = 7$. The software pipeline is controlled by LC, EC, and the predicates of the three stages: $\rho16$ to $\rho18$. Their initial setting and the changes after every `br.ctop` instruction are listed at the right side. Initially, $LC = N - 1 = 6$, and $EC = S = 3$. After the kernel is executed once, the `br.ctop` instruction rotates the register files, decrements LC, and sets $\rho16$. The effect is that LC changes from 6 to 0, and at the same time, because of register rotation, all $\rho16$ to $\rho18$ gradually become 1. These three predicates are always associated with the stages from right to left, as annotated in the first kernel iterations. In the initial iteration, as $\rho18$ has a value 0, the instruction c is effectively not executed. Thus, the ineffective stages are automatically ignored in execution, as the corresponding predicates are 0. The ineffective stages are highlighted in darker color (the change of the corresponding predicates is shown in the triangle at the upper-right side). This dynamically forms the prolog.

After LC becomes 0, the `br.ctop` instruction rotates the register files, decrements EC, and resets $\rho16$ instead. The effect is that EC changes from 3 to 0, and in the same time, all $\rho16$ to $\rho18$ gradually become 0,

---

[4]In the IA-64 architecture, the predicate registers are named as $p16$, $p17$, ... To emphasize that they are actually virtual register names used by the code generator, we use the Greek letter ρ for predicate registers. Similarly, the Greek letter γ is used to denote integer virtual register names.

```
1        /* Intialization: */
2        clrrrb;;
3        LC = N − 1
4        EC = 3
5        mov pr.rot = 1 << 16
6        mov γ33 = live-in value of TN1;;
7        /* The software pipelined schedule: */
8    L':
9                                         a: (ρ16) TN2 (γ34) = TN1 {1} (γ33);;
10       c: (ρ 18) ... = TN2 (γ36)         b: (ρ16) TN1 (γ32) = ... ;;
11       br.ctop L';;
```

(a)

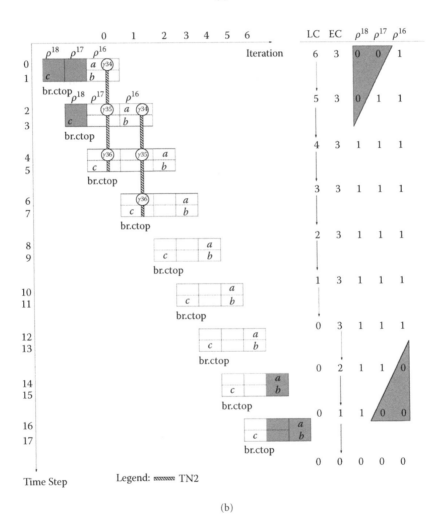

(b)

FIGURE 20.8 Code generation with rotating register files and predicated execution. (a) Kernel-only code for the software pipelined schedule in Figure 20.5b. (b) Execution of the kernel-only code (assume $N = 7$).

which makes the latter stages ineffective and prevents them from execution (see the triangle at the lower-right side). This dynamically forms the epilog.

After the kernel is executed for the last time, both LC and EC have been exactly decremented to 0. The `br.ctop` instruction fails through, and the loop finishes.

Now let us explain how the registers are assigned. As illustrated in Figure 20.7b, the two variables TN1 and TN2 are allocated physical registers 0 and 2, respectively. Assume they are integer variables. In IA-64, the integer rotating registers are from $\gamma 32$ to $\gamma 127$. Hence, two virtual registers with distance of 2, for instance, $\gamma 32$ and $\gamma 34$, are assigned to TN1 and TN2, respectively.

As mentioned earlier, at the end of the kernel, a `br.ctop` instruction rotates the rotating register files. Rotation of a register file means the base register corresponding to it is decremented. In other words, a physical register corresponding to virtual register $\gamma 32$ before the rotation corresponds to virtual register $\gamma 33$ after the rotation. Since a rotation occurs every *II* time steps, and TN1 is originally allocated $\gamma 32$, after one rotation, the value in the same physical register corresponds to $\gamma 33$. This explains why in line 6, we move the live-in value to $\gamma 33$, corresponding to TN1{1}. Similarly, TN2's value stays in a single physical register but is named $\gamma 34$, $\gamma 35$, and $\gamma 36$ with the rotations. That is why in line 10, in instruction *c*, TN2 uses $\gamma 36$ as a source register. This can be clearly seen from Figure 20.8b, where we show the name change of the physical register for TN2 in the first two iterations; it originally corresponds to $\gamma 34$. After one rotation by `br.ctop`, its corresponds to $\gamma 35$, and after another rotation, it corresponds to $\gamma 36$.

In the above example for kernel-only register allocation and code generation, a `br.ctop` instruction is scheduled after the kernel. To reduce the kernel execution time, such an instruction should be scheduled together with the instructions in the original loop body, as is done in the open research compiler [74]. It must be scheduled such that it is at the last time step of the kernel and executed after all the other instructions in the kernel.

20.3   Advances in Single-Loop Software Pipelining

Substantial progress has been made in software pipelining in the past decade. In the context of traditional uniprocessor architectures, efforts to construct software pipelined schedules that require no more than the available registers have been made. There has also been an interesting discovery on the relationship between software pipelining and circuit retiming [17], a hardware pipeline design technique.

The past decade has also seen fundamental changes in the direction of architectural design. Power consumption has become a major design constraint for high-performance and embedded systems. As aggressive ILP techniques applied to traditional architectures result in diminishing returns and technology constraints limit the scalability of the large monolithic hardware architecture, newer complexity-effective architectures, for example, clustered VLIW and multi-core architectures, have emerged. To meet the demands of these newer architectures and constraints, software pipelining has been adapted appropriately.

In this section, we survey these recent advances in software pipelining. First, we discuss the progress in the traditional uniprocessor context, namely register-constrained software pipelining (Section 20.3.1), the link between software pipelining and circuit retiming (Section 20.3.2), and power-aware software pipelining (Section 20.3.3). Section 20.3.4 deals with extensions to software pipelining for newer architectures.

20.3.1   Register-Constrained Software Pipelining

Although software pipelining and register allocation are coupled problems, they are usually addressed independently for simplicity. As a result, though a number of software pipelining methods considered reducing the register requirement as an additional objective [28, 34, 51, 54, 64], they still do not explicitly consider the number of available registers and limit the requirements of the constructed schedule to the available registers. This has resulted in scheduling being performed without considering register constraints, that is, assuming that an infinite number of registers are available, and then register allocation is attempted for the constructed schedule. In case the schedule requires more registers than are available from the target processor, earlier approaches follow one of the two options: *posteriori spilling* or *rescheduling with a larger II*. Recent advances, however, have considered register constraints during scheduling. We discuss these methods in some detail in this section.

20.3.1.1 Posteriori Spilling

Posteriori spilling selects some of the loop variables as spill candidates and adds the required spill loads and stores to the loop body. It reschedules the loop body with the *same II* and performs register allocation again [86]. The spilling is called posteriori because the spill code is generated after the initial scheduling. The spilling heuristic used in different methods varies in terms of the following aspects:

- Whether every *use* of a spilled candidate variable requires a spill load, or only certain *uses* of it require spill loads.
- The priority function used for selecting the spill candidates.
- The heuristic used to schedule the spill instructions.
- How many loop variables are spilled at a time before rescheduling the loop with the same *II*. As the loop is rescheduled after spilling, its register requirement may decrease. However, with the introduction of the spills, the memory units might be saturated and cannot accommodate all the loads and stores under the current *II*. In this case, the added spill code is removed, and the original loop body is rescheduled after incrementing *II* by 1.

Posteriori spilling was used in the MIPS compiler [94]. The number of variables selected for spilling increases exponentially. The first scheduling failure results in one variable being spilled, the second, two variables, the third, four variables, and so on. Spill candidates are selected by a heuristic, which computes the ratio of the number of time steps spanned by the scalar lifetime of a variable to the number of *uses* of the variable. The higher the ratio, the greater is the cost and, hence, the lower is the benefit to keep the lifetime in a register.

Another set of heuristics has been proposed for posteriori spilling in [116]. It spills candidates that are live at the "critical" time steps, the time steps at which the total number of scalar lifetimes equals *MaxLive*. That is, these time steps have the maximal register requirement. The number of candidates to be spilled is proportional to the difference between *MaxLive* and the number of available registers. It foresees the situation when it is better not to perform spilling, but to reschedule the loop with a larger *II* (without inserting spill code). This happens if the spill loads and stores to be inserted would overcommit the memory units under the current *II*.

20.3.1.2 Rescheduling with a Larger *II*

The other option, rescheduling the original loop with a larger *II* without inserting any spill code, results in a schedule with a lower throughput and possibly with fewer iterations overlapped. It presumes that the register requirement is somewhat proportional to the number of concurrently executing iterations. At the expense of reduced performance, a feasible register allocation may be found [86]. This option was used in the Cydra 5 compiler [30].

The two alternatives, posteriori scheduling and rescheduling with a larger *II*, were evaluated and compared by Llosa et al. [65]. They observed that, in general, rescheduling with an increased *II* performs poorly and might not find a schedule requiring fewer registers than available for some loops. Posteriori spilling always leads to a feasible schedule and, in many cases, results in more efficient schedules with a lower *II* than the other approach.

20.3.1.3 Register-Constrained Software Pipelining

While decoupling scheduling and register allocation does simplify the compilation process, it is beneficial to consider register constraints during scheduling, at least to some extent, to avoid failures in allocating registers for the constructed schedule.

Two approaches consider register requirement during scheduling [6, 28, 54, 64, 116]. One attempts to *reduce register requirement*, with no guarantee that the final schedule will require fewer registers than available [28, 54, 64]. The other approach *monitors register requirement and introduces spill code on-the-fly* during scheduling [6].

Reducing register requirement is based on the obvious fact that shortening the scalar lifetimes in the schedule lowers register pressure. The scalar lifetimes can be shortened by carefully prioritizing and scheduling the instructions such that for a variable, the producer instructions and the consumer instructions are scheduled closer to each other. In slack scheduling [54], the priority list is based on the slack of an instruction, which does not particularly favor register pressure. However, the decision of whether to schedule an instruction as early as possible (closer to the ESTART time) or as late as possible (closer to the LSTART time) is used for this purpose. In hypernode reduction modulo scheduling [64], the priority order of the instructions is such that only all the predecessors or all the successors of an instruction in the dependence graph, but not both, are ahead of the instruction in the priority list. During scheduling, the instruction is scheduled as early (or as late) as possible if its predecessors (or, respectively, successors) have already been scheduled.

The other approach, monitoring register requirement and introducing spilling code on-the-fly, is based on the observation that register requirement of a schedule is usually close to the lower bound, *MaxLive*. As mentioned in Section 20.2.5.2.2, *MaxLive* is a surprisingly tight lower bound for register requirement. Therefore, we can estimate register requirement with this lower bound, and monitor it during scheduling. In scheduling, if *MaxLive* in the current partial schedule is greater than the available number of registers, spill instructions can be inserted on-the-fly to keep the *MaxLive* of the partial schedule lower than the available registers. The final schedule would almost guarantee success in the subsequent register allocation phase [86]. Even if the spill code does not reduce register requirement sufficiently, backtracking can be used to reschedule some instructions for further reduction in register requirement.

A hybrid approach has been proposed in [6], which combines the benefits of on-the-fly and posteriori spilling. It makes two scheduling passes. The first pass tries to schedule the loop without adding spill code. If the *MaxLive* of the schedule is found to be less than the number of available registers, then register allocation is performed, since it is likely to succeed. Otherwise, another scheduling pass starts, which reschedules the loop with on-the-fly spill code generation. The same scheduling strategies are used in both passes, which makes the two schedules as similar as possible. Therefore, although the complete information of the lifetimes is not available until the rescheduling terminates, the *MaxLive* can be estimated by approximating the lifetimes of variables whose producer and/or consumers are not yet scheduled to those obtained in the first scheduling pass. Spill codes are inserted on-the-fly, guided by the information obtained from the schedule of the first pass. With respect to the first pass, the spilling is posteriori. With respect to the second pass, however, it is on-the-fly.

20.3.2 Modulo Scheduling and Retiming

Recent years have seen interesting advances relating modulo scheduling and *circuit retiming* [15–17, 21, 22, 24, 95, 99, 102]. The former is a technique for constructing software pipelines, while the latter is for optimizing hardware pipelines. Modulo scheduling has been applied to optimize synchronous circuits [15, 21, 99, 102], which was usually done by retiming. Conversely, retiming has been combined with modulo scheduling for scheduling loops [4, 16, 17, 95] for various objectives, such as maximizing throughput and minimizing power, register pressure, area, and so on.

What is the relationship between these two technologies? Why can they be combined? This section answers these questions based on the literatures [16, 17, 63]. We focus on a single loop with a single basic block of instructions.

20.3.2.1 Retiming

First, let us cite the concepts of synchronous circuit and retiming as defined by Leiserson and Saxe [63], with a minor change in notations. A circuit is modeled as a directed multi-graph $G = \langle V, E, \delta, w \rangle$, where V is the set of functional units, and E is the directed interconnections between them. A node $a \in V$ has a latency or propagation delay $\delta(a)$. An edge $e \in E$ is weighted with a register count $w(e)$. The weight means there are $w(e)$ number of registers located on the edge.

This graph is a *synchronous circuit* if the following conditions are met:

D1: The latency $\delta(a)$ is nonnegative for each node $a \in V$.

W1: The register count $w(e)$ is a nonnegative integer for all edges $e \in E$.

W2: In any directed cycle of G, there is at least one edge with a positive register count.

Condition W2 is to prevent a cycle in which every edge is zero weighted, which may lead to race conditions, oscillation, and so on. In other words, in a synchronous circuit, a feedback loop must have at least one register as a buffer. A cycle in a synchronous circuit is similar to a recurrence cycle in a DDG.

In the terminology of pipelining, a path between two registers, without any other register in between, is a pipeline stage. The length of the path, which is the total latencies of the functional units in the path, is the latency of the pipeline stage. The total weight of the path, which is the sum of the register counts of the edges in the path, is 0 by definition. Among all such zero-weighted paths, the longest one determines the *clock period*.

To maximize the throughput of the pipeline, it is necessary to minimize the clock period. This is done by retiming the synchronous circuit. Retiming reduces the length of the longest zero-weighted path by inserting or deleting registers, but without affecting the circuit structure otherwise [63].

Formally, a *retiming* of the circuit is an integer-valued vertex labeling function $r : V \rightarrow \mathbb{Z}$. It maps the original graph G to a new graph $G_r = \langle V, E, \delta, w_r \rangle$, where for each edge $e : a \rightarrow b \in E$, the new weight w_r is defined as $w_r(e) = w(e) + r(b) - r(a)$.

From the above definition, it is clear that retiming does not change the total weight of a cycle. That is, retiming does not violate condition W2. Of course, it does not change condition D1. Therefore, for a synchronous circuit, if a retiming guarantees that the retimed circuit still satisfies condition W1, then the retimed circuit is still a synchronous circuit, and the retiming is said to be *legal*.

20.3.2.2 Modulo Scheduling and Retiming

Several observations have been made about the relationship between retiming and software pipelining.

20.3.2.2.1 A DDG Is Equivalent to a Synchronous Circuit, Assuming No Resource Constraints

For each instruction, assume there is a functional unit exclusively assigned to it. Thus, an instruction in the DDG can be regarded as a functional unit in a circuit. A dependence between two instructions is a directed edge between two functional units, and the dependence distance is the edge's weight (register count). The latency of a functional unit is the maximal dependence latency among all the dependences for which the corresponding instruction is the producer. According to the discussion in Section 20.2.2.2, the DDG contains only flow dependences for software pipelining. Flow dependences' latencies must be positive. Therefore, the latency of every functional unit must be positive, which satisfies the above condition D1. The dependence distance is always a nonnegative integer, which respects condition W1. It is also impossible for the DDG to have a cycle in which every edge has a dependence distance of 0. This meets condition W2. In short, the DDG is a synchronous circuit.

20.3.2.2.2 Retiming Is a Mapping of Instructions to Pipeline Stages

From the perspective of pipelining, retiming groups into the same pipeline stage some functional units (i.e., instructions) that were originally in different stages by deleting the registers on the paths between them; retiming also divides into different pipelines stages some other functional units (instructions) that were originally in the same stage, by inserting registers on the paths between them. In other words, retiming is a mapping of instructions to pipeline stages.

20.3.2.2.3 Any Legal Modulo Schedule Has a Legally Retimed DDG

Modulo scheduling also maps the instructions to pipeline stages. It does so by retiming the DDG implicitly. According to the dependence constraints in Equation 20.2, for a dependence $(a \rightarrow b, \delta, \langle d \rangle)$, a valid modulo schedule must have the following:

$$\sigma(a, i) + \delta \leq \sigma(b, i + d) \tag{20.22}$$

for any loop iteration i. Assume the schedule times for the two instructions are as follows:

$$\sigma(a, 0) = s_a * T + o_a \quad \text{where } 0 \le o_a < T$$
$$\sigma(b, 0) = s_b * T + o_b \quad \text{where } 0 \le o_b < T$$

where T is the initiation interval. In other words, instructions a and b are scheduled into stages s_a and s_b, respectively. Thus, Equation 20.22 can be rewritten as

$$\sigma(b, i + d) - \sigma(a, i) = (d + s_b - s_a) * T + o_b - o_a \ge \delta$$

Dividing both sides with T, and taking the floor function, we get

$$\left\lfloor \frac{\sigma(b, i + d) - \sigma(a, i)}{T} \right\rfloor = d + s_b - s_a + \left\lfloor \frac{o_b - o_a}{T} \right\rfloor \ge \left\lfloor \frac{\delta}{T} \right\rfloor$$

Since $-T < o_b - o_a < T$ and $\delta \ge 0$, there must be $\left\lfloor \frac{o_b - o_a}{T} \right\rfloor = 0$, and $\left\lfloor \frac{\delta}{T} \right\rfloor \ge 0$. Therefore,

$$\left\lfloor \frac{\sigma(b, i + d) - \sigma(a, i)}{T} \right\rfloor = d + s_b - s_a \ge 0 \tag{20.23}$$

Remember, d is the dependence distance, which is the register count in terms of retiming. If we label the functional units corresponding to instructions a and b with their stage numbers s_a and s_b, respectively, then $d + s_b - s_a$ is the new register count of the edge between the two functional units, according to the definition of retiming. In the above relationship, $d + s_b - s_a \ge 0$ exactly says that condition W1 is guaranteed, and, therefore, this is a legal retiming.

The above relationship also answers this question: What does it mean by the new register count $d + s_b - s_a$? From the equation $\left\lfloor \frac{\sigma(b, i+d) - \sigma(a, i)}{T} \right\rfloor = d + s_b - s_a$, it is clear that it refers to the total number of stages, i.e., the total number of instances of the kernel, a value, produced by instruction a in loop iteration i, has to cross before it reaches the consumer instruction b in loop iteration $i + d$.

Finally, what does it mean by the original register count d? It means that before scheduling/retiming, all the instructions in the same iteration belong to a single pipeline stage ($s_b = s_a = 0$). Therefore, the value produced in loop iteration i must cross d number of stages to reach the consumer in loop iteration $i + d$. The number of stages d happens to equal the dependence distance in this case.

In summary, the above discussion helps with a deeper understanding of retiming. The labels attached to the functional units are the indexes of the pipeline stages where the corresponding instructions (in the first loop iteration) are scheduled. The register count of an edge is the total pipeline stages for a value to cross.

From Equation 20.23, we can further conclude:

$$\left\lfloor \frac{\sigma(b, i + d)}{T} \right\rfloor = \left\lfloor \frac{\sigma(a, i)}{T} \right\rfloor \iff d + s_b - s_a = 0 \tag{20.24}$$

The above relationship means that iff the new register count is 0, the producer $\sigma(a, i)$ and the consumer $\sigma(b, i + d)$ are in the same stage, that is, they are executed in the same instance of the kernel [16, 17].

Example 20.2

Let us illustrate these points with the help of the DDG for our motivating example discussed in Section 20.2.1. Before modulo scheduling, every instruction is in a single stage i.e., the same instance of the kernel. Therefore, we can imagine that every instruction in the DDG is labeled with stage number 0 (see Figure 20.9a).

After modulo scheduling, instructions a and d are scheduled in stage 0, and the others in stage 1, as shown by the kernel in Figure 20.4. In other words, there is a retiming r, which has $r(a) = r(d) = 0$, and $r(b) = r(c) = 1$. Label the instructions in the DDG with their stages and update the weight of every edge accordingly. We get Figure 20.9b.

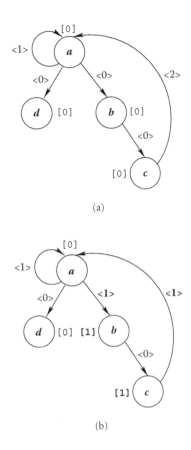

(a)

(b)

FIGURE 20.9 Illustration of retiming. The stage number of an instruction is enclosed in square brackets, [ ]. (a) The original DDG before retiming; (b) the DDG after retiming.

Compare the two figures. The weights of edges $a \rightarrow b$ and $c \rightarrow a$ have been changed from 0 to 1, and from 2 to 1, respectively. The weights of other edges are not changed. Let us take edge $c \rightarrow a$ to illustrate the meaning of the weights. According to the dependence, a result of instruction c in loop iteration 0 needs to be transferred to instruction a in iteration 2. Since every iteration is in a single stage before modulo scheduling, it needs to cross two stages. Therefore, the edge originally has a weight 2, which is equal to the dependence distance. In the modulo schedule in Figure 20.2b, however, the producer instruction c in iteration 0 and the consumer instruction a in iteration 2 are scheduled at time steps 3 and 4, respectively. One is in stage $\lfloor \frac{3}{T} \rfloor = 1$, the other in stage $\lfloor \frac{4}{T} \rfloor = 2$. Here $T = 2$ is the initiation interval. Therefore, the value needs to cross $2 - 1 = 1$ stage. That is why the weight of the edge $c \rightarrow a$ is changed to 1.

It should be noted that a retimed schedule is not necessarily a software pipelined schedule. This is because retiming finds only the stages for each instruction but does not consider the latencies of the instruction, nor their exact time steps in a stage, as can be seen from Equation 20.23. Therefore, given a retiming, there can be more than one schedule.

Under an infinite number of resources, a modulo schedule with $II = RecMII$ implies a retimed graph with the shortest clock period. In addition, the modulo schedule divides the clock period equally into II number of time steps.

In short, modulo scheduling implicitly retimes the DDG by allocating instructions to pipeline stages, but it is more than retiming. In this sense, all possible modulo schedules compose a subset of all possible retimed schedules (Figure 20.10).

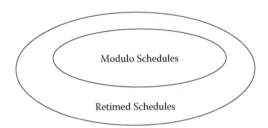

FIGURE 20.10   The space of modulo scheduling is a subspace of retiming.

20.3.2.2.4 *The Legally Retimed DDG May Be Further Retimed, Leading to a Change in the Modulo Schedule*

The retimed DDG may be optimized by a sequence of other retiming. Consequently, the modulo schedule is changed with each retiming.

20.3.2.2.5 *Resource-Constrained Modulo Scheduling Achieves Resource Sharing between Instructions, but Retiming Alone Usually Cannot Share Resources*

Retiming inherently assumes that one instruction is bound to one distinct functional unit. There is no sharing of the functional units between the instructions. Modulo scheduling allows different instructions to share the same functional unit by putting them into different time steps in the kernel. This can be used to optimize hardware circuit area; in compiling the modulo schedule into hardware, the input register to the functional unit is replaced with a sequence of input registers, one for each time step in the kernel. This is called *c-slowdown* in hardware terminology [99].

20.3.2.3 Decomposed Software Pipelining

Retiming assigns instructions to pipeline stages. Modulo scheduling is more than retiming. It specifies the exact time step in the kernel for each instruction. Therefore, to construct a modulo schedule, one can perform retiming to decide the stages of the instructions first, and then specify their time steps in the kernel. This approach is called decomposed software pipelining (DESP) [16, 17, 47, 105].

In this approach, first, the DDG is retimed to minimize the clock period. In this step, only dependence constraints are considered. This decides the stage for every instruction, assuming infinite resources. Second, under the resource constraints, the exact time step of each instruction in its stage is determined. The dependences whose register counts after retiming are 0 are kept as they are within the same instance of the kernel, according to the Equation 20.24, while all the other dependences are removed. Note that before and after retiming, the DDG is a synchronous circuit. According to condition W2 in the definition of a synchronous circuit, these zero-weighted dependences that remain in the DDG cannot form any cycle. In other words, the dependence graph becomes acyclic. Therefore, any acyclic scheduling method, for example, list scheduling, is then applied to arrange the instructions. It does not change the stages of the instructions, but only their time steps in the kernel. The height of the schedule is the initiation interval.

20.3.2.4 Improving Modulo Scheduling with Retiming

Retiming can be used before modulo scheduling to help achieve a better schedule than using modulo scheduling alone. Appropriate retiming reduces the total number of zero-weighted dependences. Therefore, retiming the dependence graph before modulo scheduling can provide modulo scheduling with less-constrained dependences [22, 99].

It is interesting to note that retiming, however, does not reduce *MII*, the lower bound of the *II*. This lower bound is determined by resources, and by the total latencies and total distances of a cycle in the DDG. Although retiming may change the weight of some dependences, the total weight of the cycle stays the same, as stated in Section 20.3.2.1. Retiming helps modulo scheduling not because it reduces *MII*, but because the retimed graph has fewer dependence constraints and thus may require less time to construct a schedule.

20.3.3 Power-Aware Software Pipelining

Power consumption has become one of the major design constraints in today's high-performance processors and embedded systems [69]. Excessive power consumption in small chip areas causes high chip temperature, affecting processor reliability adversely and increasing hardware packaging cost. Furthermore, for handheld systems, such as laptops and PDAs, reducing power or energy consumption also helps lengthen the battery life. Thus, it is of importance to reduce the power or energy consumed by the processor without significantly degrading its performance.

As software pipelining is the focus of this chapter, we concentrate only on recent compiler techniques that reduce the power or energy consumption of software pipelined schedules. General power-aware computing techniques are discussed in greater detail elsewhere [14]. First we present the necessary background on power and energy consumption [69]. This is followed by a discussion on power-aware software pipelining methods [113–115].

20.3.3.1 Background on Power/Energy Consumption

The power dissipated by a CMOS logic circuit can be categorized into three parts: *dynamic power, short-circuit power*, and *leakage power*. The dynamic power is due to the charging and discharging of the capacitive load on the output of each gate. The short-circuit power, as the name indicates, is due to the short-circuit current, which momentarily flows between the supply voltage and ground when the output of a CMOS gate switches. The leakage power is the power lost due to the leakage current regardless of the gate's state. Among the three, the dynamic power dominates, especially when a processor technology 90 nanometer or higher is used.

The dynamic power of a CMOS circuit scales quadratically with the supply voltage. In particular, if f is the clock frequency, C is the capacitance, and V is the supply voltage, then the dynamic power dissipated is given by

$$P_{dyn} \propto CV^2 f$$

It is easy to observe that a reduction in supply voltage can bring about a quadratic reduction in dynamic power. As a side effect, reduction in the supply voltage also lowers the operating frequency f. Thus, reducing the supply voltage has a significant effect on dynamic power consumption. That is why voltage scaling is performed in modern processors which reduces the supply voltage (and operating frequency) to decrease the power/energy consumption. This is accomplished either by changing the supply voltage and frequency of the processor dynamically or by providing multiple functional units that have different operating frequencies and voltages [80] or different architectural implementations [118]. It is the responsibility of the compiler to identify the appropriate functional units, or the operating frequency of the functional units, to save power.

In addition to the power dissipated, the power variation exhibited by the system across different cycles is also important. In high-performance embedded systems, the number of on-chip activities varies drastically on a cycle-by-cycle basis. This results in power supply voltage noise. Large power variation is usually correlated with high peak power and a large number of "hot spots." Both are detrimental to functional unit reliability. Ironically, such power variations are often exacerbated by commonly used power-saving circuit design techniques.

In an embedded system or handheld device, while the power dissipated directly concerns the input power supply, cooling, and thermal effects, the energy consumed concerns the battery capacity. Reducing the energy consumption is important, as it increases the battery's lifetime and reduces its cost and weight. The energy consumed is proportional to the product of power and the work time of the system:

$$E \propto P \cdot t$$

where P is the average power dissipated and t is the work time. In the context of software pipelining, since the execution time and energy consumption of the kernel dominate those of the loop, P can be defined as the sum of the power consumed at each time step in the kernel, and t as the total times the kernel is executed.

20.3.3.2   Power-Aware Software Pipelining

Power-aware software pipelining methods [113–115] reorder instructions at compile time based on an accurate estimate of power consumption at each time step, with the objective of minimizing either the overall power consumed or the power variation in the kernel. They are motivated by the following facts:

- Instructions usually have nonzero slacks. As introduced in Section 20.2.3.2, an instruction a has a feasible schedule time range $[ESTART(a), LSTART(a)]$, and its slack is $LSTART(a) - ESTART(a)$. When the slack is nonzero, the instruction has more than one time step to schedule without violating the dependence constraints with the already scheduled instructions and without degrading the performance of the schedule.
- Modern processor design tends to make functional units fully pipelined to reduce structural hazards. Thus, it is possible to delay the schedule time of certain noncritical instructions and obtain a schedule that uses fewer functional units without degrading the performance significantly.

20.3.3.2.1   Reducing Power Consumption

The method proposed in [113] attempts to reduce the power consumed while retaining the optimality of the software pipelined schedule. It assumes an architecture where functional units are pipelined. It minimizes the number of functional units that are active at a time step. A functional unit is said to be active at time step t if any stage of the functional unit is busy at that time step. Note that from the viewpoint of minimizing the power and energy, it is preferable to keep multiple stages of a functional unit active than to have multiple functional units, each with one stage active. The problem is formulated as follows:

Problem 20.2

Given a loop $\mathcal{L}$ and a machine architecture, construct a modulo schedule for a given II that consumes the minimal power, subject to the modulo property, dependence constraints, and resource constraints as stated in Section 20.2.2.3.

The resource-constrained modulo scheduling problem was formulated in Section 20.2.4. The dependence and resource constraints are expressed using Equations 20.14, 20.15, 20.16, and 20.17. Now we need to model the power consumed at each time step in the schedule as well. For this, we model the number of functional units that are active at each time step and minimize them.

To model the functional units that are active at a given time step, it is necessary to model the latency of each instruction, which specifies how long a functional unit is active. We assume that an instruction that takes s time steps is executed in a functional unit that has exactly s stages. At the first time step it is initiated, it is in the first stage of the functional unit; at the next time step, it is in the second stage, and so on. Therefore, an instruction a issued at time step t will be in stage s at time step $(t + s) \mod T$, where T is the initiation interval. To model the usage of the stages of different functional units, a three-dimensional array $\mathcal{U} = [u_{t,a,s}]$ is introduced as

$$u_{t,a,s} = 1 \iff \text{instruction } a \text{ is in stage } s \text{ at time step } t \text{ in the kernel}$$

The variable $u_{t,a,s}$ can be defined using matrix $\mathcal{M}$ as below:

$$u_{t,a,s} = \begin{cases} m_{(t+s) \bmod T, a} & \forall a \in [0, w-1], \forall t \in [0, T-1], \text{ and } \forall s \in [0, \delta(a) - 1] \\ 0 & \forall a \in [0, w-1], \forall t \in [0, T-1], \text{ and } \forall s \in [\delta(a), L_r - 1] \end{cases} \tag{20.25}$$

where $\delta(a)$ is the latency of instruction a, and L_r is the maximum latency of all instructions that can be executed on a functional unit of type r. The second case of Equation 20.25 is necessary, since different types of instructions (e.g., floating point divide and square root) executing in the same functional unit may have different execution latencies.

In a software pipelined schedule, the number of functional units of type r used at time step t is given by

$$F_{t,r} = \max_{s \in [0, L_r - 1]} \left(\sum_{a \in \mathcal{I}(r)} u_{t,a,s} \right)$$

The above equation can be expressed in a linear form as

$$F_{t,r} \geq \sum_{a \in \mathcal{I}(r)} u_{t,a,s} \quad \forall t \in [0, T-1], \forall s \in [0, L_r - 1], \text{ and } \forall r \tag{20.26}$$

Our objective is to minimize the power consumed by the active functional units. Let F_r represent the number of r-type functional units in the given architecture and Pd_r (Pl_r) be the power dissipated (leaked) by an active (inactive) functional unit of type r in a single time step. The total amount of power consumed at time step t by all functional units is given by

$$P_t = \sum_r (Pd_r * F_{t,r} + Pl_r * (F_r - F_{t,r}))$$

Then the power consumed by all the functional units in the software pipelined kernel is given by

$$P = \sum_{t=0}^{(T-1)} P_t \tag{20.27}$$

The objective function minimizes P, subject to Equations 20.14, 20.15, 20.16, 20.17, 20.25, 20.26, and 20.27.

20.3.3.2.2 *Reducing Power Variations*

The above formulation can be slightly modified to model and reduce power variations across different time steps, which is important from the viewpoint of the reliability of the system. Such a formulation is proposed in [114]. This formulation minimizes $\sum_t |P_t - P_{avg}|$, where P_{avg} denotes the average power consumed during the execution of loop L. This formulation also uses $\mathcal{U} = [u_{t,a,s}]$, the variable as defined in Equation 20.25. P_{avg} is

$$P_{avg} = \frac{\sum_{t=0}^{T-1} P_t}{T}$$

The software pipelined schedule desired is the one that has a smooth power profile, which means the actual power consumed at each time step is as close to the average power as possible. The difference between the actual power consumed at time step t and the average power is

$$\left| P_t - P_{avg} \right|$$

The objective function is to minimize the value of the following expression:

$$\sum_{t=0}^{T-1} \left| P_t - P_{avg} \right|$$

To formulate this as an integer linear programming problem, we introduce two inequalities as follows:

$$D_t \geq P_t - P_{avg} \tag{20.28}$$

$$D_t \geq P_{avg} - P_t \tag{20.29}$$

and define the objective function as

$$\min \sum_{t=0}^{T-1} D_t \qquad (20.30)$$

Although this does not explicitly specify an upper bound for D_t, the objective function (Equation 20.30) ensures that D_t for all t in $[0, T-1]$ will get its smallest permissible value. Consequently, D_t must be equal to the maximum of the right-hand sides of Equations 20.28 and 20.29.

Thus, a software pipelined schedule that has a balanced power profile can be achieved by solving the integer linear programming problem with the objective function (Equation 20.30) under the constraints given by Equations 20.14, 20.15, 20.16, 20.17, 20.25, 20.26, 20.28, and 20.29. The above formulation is shown to be effective in minimizing the peak power, number of hot spots, and power variation [114].

A similar effort to reduce peak power and power variation is proposed in [115]. This approach is based on the heuristic iterative modulo scheduling method [87]. The idea is to make the power consumed at each time step as close to the average power as possible. In other words, it avoids peaks in the power consumption in the kernel, making the power profile of the kernel as "flat" as possible in terms of power usage. Specifically, for an instruction, it chooses from its feasible time range a time step such that once the instruction is scheduled in this time step, the power profile of the constructed partial schedule is as flat as possible.

20.3.4   Software Pipelining for New Architectures

Software pipelining traditionally targets monolithic VLIW architectures. However, in conventional monolithic architectures, as the complexity grows and feature sizes decrease to tens of nanometers, wire delays tend to dominate gate delays, which limits the increase in processor clock frequencies [76]. *Clustering* has been proposed as an effective microarchitectural approach to mitigate the negative effect of wire delays. The main idea is to have a hierarchical organization of the interconnection wires such that units that communicate frequently are interconnected through short and fast wires, and those that communicate infrequently use longer and slower wires. More specifically, in a clustered VLIW architecture [18], the functional units are grouped into clusters and the clusters are interconnected. Examples include Texas Instruments' TMS320C6x [103], BOPS's ManArray [78], HP/ST's Lx [39], and Equator's MAP1000 [5, 48]. The cluster approach has also been followed in the superscalar architecture, with Alpha 21264 [58] as an important example. More recently, *multi-core architectures* [73] have become a commercially viable option that can effectively utilize the few 100 million transistors available in the chip. In both clustered VLIW and multi-core architectures, software pipelining remains a useful compiler technology to extract parallelism from loops.

20.3.4.1   Overview of Challenges and Solutions

In a clustered VLIW architecture, the front end of the microarchitecture is similar to that of a monolithic architecture. However, the functional units are grouped into clusters, with each cluster having its own register file. Enough datapaths are available for a functional unit to access the required operands, typically two source operands and one destination operand, from the register file of the same cluster, referred to as the local register file. However, only a few ports are available for all the functional units in a cluster to access a register file in a non-local cluster, referred to as remote register file. Therefore, when multiple functional units in a cluster try to access a remote register file, only a few (equal to the number of remote ports) of the accesses can go through in a single cycle, while others are stalled. To reduce the number of remote register file accesses, the value from a remote register can be copied to a local register through an explicit `copy` operation.

The clustered VLIW architecture introduces a few new challenges to software pipelining. First, it is no longer sufficient to just specify the time step for an instruction, but it is also required to specify to which cluster (and what functional unit) the instruction is scheduled. The latter, referred to as cluster assignment or mapping, adds an additional dimension to the software pipelining problem. Second, the latency of a dependence may vary because of potential access to a remote register file. Even among the remote clusters, the access time to different remote register files may be different because of the topology of the inter-cluster interconnect. That is, clustered machines are imbalanced in access time to the register banks. The nonhomogeneous latency in accessing remote register files needs to be taken into account while scheduling the instructions. Last, the value of a remote register may need to be copied via an *inter-cluster transfer/copy* instruction from the remote to the local register file. Such an instruction may be introduced into the dependence graph dynamically during the software pipelining process. Thus, the DDG used in the scheduling process itself gets modified during scheduling, depending on the instruction scheduled and the clusters to which they are assigned.

Thus, software pipelining for a clustered VLIW architecture needs to address three closely related key problems: cluster assignment, scheduling, and register spilling. Depending on the order of solving the problems, there have been several different approaches [3, 7, 18, 41, 53, 72, 117]:

- **Cluster assignment first:** In this approach, the instructions are first assigned to specific clusters. Once the assignment is done, edges in the dependence graph that span across two clusters are annotated with higher latencies. The subsequent scheduling phase, where any traditional software pipelining method can be employed, uses the modified DDG and determines the schedule time for each instruction and the functional unit within the cluster to which the instruction is assigned. Cluster assignment can be performed explicitly by graph partitioning [3, 7] or pre-scheduling [72], or implicitly by register partitioning [53].

- **Scheduling first:** In this approach, instructions are scheduled as if the target architecture is an idealized monolithic VLIW equivalent, where all functional units are within a single cluster. Then the interconnection in the cluster architecture is considered and the required inter-cluster transfer (copy) instructions are inserted into the schedule for the values flowing between clusters [18]. One major issue with this approach is that if the copy instructions happen to be inserted into a critical cycle in the dependence graph, then *II* is increased. Furthermore, there should be free remote ports available in the time step in which the copy instruction is to be scheduled.

- **Integrated scheduling with cluster assignment:** In this approach, the dependence graph is dynamically updated with the assignment and scheduling of the instructions and the insertion of new instructions. The new instructions inserted include the inter-cluster transfers (copy instructions) and the spill loads and stores if register spilling is also integrated [41, 117].

Software pipelining for multi-core architectures [49] has similar challenges and can be addressed similarly. However, functional units in such architectures are not lock-stepped as in clustered VLIWs. Hence, it is natural to expose coarse-grained parallelism in multi-core architecture. The concept of software pipelining for them needs to be extended accordingly; for example, an "instruction" may be a macro instruction that contains a set of machine instructions.

Below we describe the first and the last approaches in greater detail for clustered VLIW architectures and the second approach for multi-core architectures, which are less sensitive to the increase in *II*.

20.3.4.2 Cluster Assignment First Approach

In this approach, cluster assignment and scheduling are not coupled; that is, they are performed in two independent phases. We focus on how cluster assignment is done. Three approaches based on graph partitioning [7], pre-scheduling [72], and register partitioning [53] have been proposed. We review them briefly.

20.3.4.2.1 Cluster Assignment by Graph Partitioning

In the graph partitioning–based method [7], a preliminary cluster assignment is obtained by partitioning the dependence graph. This is followed by a scheduling phase that integrates register allocation and spill code generation. More specifically, the cluster assignment is translated into the following graph partitioning problem:

Problem 20.3

Given the data dependence graph of a loop and an II, partition the nodes of the DDG into C number of sets, where C is the number of clusters, such that the inter-cluster data transfers have minimum impact on the execution time of the loop.

The approach proposed in [7] partitions the graph such that no resource in a cluster is fully saturated. The DDG is partitioned using the following multilevel strategy:

- **Coarsening the graph:** Each dependence edge in the DDG is associated with a weight, which is determined using two factors: (a) the impact adding a delay to this edge has on the execution time and (b) the slack of the edge, defined as the number of delay cycles that could be added to this edge without affecting the execution time. The weight of the edge combines these two factors into a single metric. The graph is coarsened using a maximum weight-matching algorithm iteratively. Here, a *matching* refers to a set of edges in the DDG such that no two edges have a node in common. The weight of a matching is the sum of the weights of the edges in the matching. In each iteration, a matching with the maximum total weight is found. For each of the edges, the two nodes connected by it are combined into a single macro node, and the edge is removed. This process continues until the number of macro nodes in the graph equals the number of clusters. Each macro node is then assigned to a cluster.

- **Refining the partition:** This is a reverse step of coarsening. It tries to improve the partition by relocating some nodes to clusters other than their original ones. The nodes can be macro nodes or nodes in the original DDG. A refinement is said to improve a partition if the workload tends to be more balanced across clusters or if the impact of the inter-cluster edges upon the execution time is reduced.

  To achieve load balancing, the resources are considered from the most saturated to the least saturated. For each saturated resource, a node containing an instruction that uses this resource is moved to another cluster, if this resource in that cluster is not saturated, and the other resources that were previously saturated, but have been considered before, do not become saturated again.

  In addition to achieving load balancing, the refinement process relocates some nodes or swaps some pairs of nodes to reduce the impact of the inter-cluster edges upon the execution time. Such nodes are at the boundary of a cluster; that is, each of them is connected by a dependence edge that spans two clusters. All such possible relocations and swappings are sorted, and the one that results in the maximum reduction in execution time is performed. Ties are broken by giving preference to the refinement that results in a larger increase of the total slack of the inter-cluster edges or a larger reduction of the total number of inter-cluster edges.

20.3.4.2.2 Cluster Assignment by Pre-Scheduling

In its essence, cluster assignment is also a scheduling problem that schedules the instructions into the clusters. Pre-scheduling is therefore another way to perform cluster assignment. Once pre-scheduling is performed, the dependence edges that span across two clusters are marked as inter-cluster edges. For such an edge, a copy instruction can be explicitly inserted on the edge. More specifically, pre-scheduling mimics a modulo scheduling, which computes *MII*, sorts the instructions, schedules them one by one into the clusters, reserves resources beforehand for copy instructions if some values are predicted to be produced in a cluster and used in a different cluster, inserts copy instructions whenever necessary as the resources have already been reserved, or unreserves the resources if the values are actually produced and used in the same cluster, and reassigns some instructions or increases *II* if the assignment process cannot proceed [72].

In the context of traditional monolithic VLIW architectures, priority is given to instructions in the critical cycles during scheduling. In the context of clustered VLIW architectures, it is also important to avoid inserting copy instructions that increase *II*. To achieve this purpose, first, the nodes in the SCCs are assigned to clusters, avoiding splitting them across the clusters with copy instructions whenever possible. The SCCs themselves are also ordered. The most constraining SCC that results in the highest *RecMII* is ordered first, then the next most constraining SCC, and so on. Second, whenever possible, data-dependent nodes are assigned to the same cluster. For this purpose, a heuristic is borrowed from the hypernode reduction modulo scheduling [64], as discussed in Section 20.3.1.3. In this heuristic, a node is ordered after all its predecessor nodes or all its successor nodes. This increases the chance that the node is assigned to the same cluster as its predecessors or successors.

As the results of pre-scheduling, we find an *II* under which all instructions are assigned specific clusters, and copy instructions are introduced in the DDG. However, pre-scheduling does not determine the schedule time step for the instructions, which makes it different from a full modulo scheduling.

Once pre-scheduling is done, any traditional modulo scheduling method can use the new dependence graph and the cluster assignment results to arrive at a final schedule, under the *II* found by pre-scheduling. This scheduling, in addition to the cluster assignment, specifies the exact time step and the functional unit in which each instruction is scheduled. If the modulo scheduling fails, the *II* is incremented, and the whole process, pre-scheduling followed by scheduling, repeats.

Note that the purpose of pre-scheduling is different from the latter (modulo) scheduling phase. The former aims to produce a DDG with explicit copy instructions that have minimum impact on the execution time, whereas the latter modulo scheduling phase works on the new dependence graph and produces the complete schedule.

20.3.4.2.3 *Cluster Assignment by Register Partitioning*

It is good if an instruction accesses all its register operands from the local register file. In the heuristic proposed in [53], a *register component graph* is built, where each node is a register operand, and an edge connects two register operands of the same instruction. The nodes that are not connected are good candidates to be assigned to different register banks. Each edge has an associated weight that represents the benefit of assigning the two nodes connected by the edge to the same register bank. Basically, the benefit is higher if the instructions containing the two nodes are more critical. For example, if the instructions have less mobility (slack) or they are within a deeply nested loop, the weight of the edge is higher.

To be more precise, a pre-scheduling phase can be applied first to software pipeline the loop, assuming an equivalent monolithic architecture. Then a different kind of edge with a low weight is added in the register component graph between the destination operands of two instructions scheduled at the same time step in the pre-schedule. This edge indicates that it is beneficial not to allocate the two operands to the same register bank. Instead, distribute them to two different register banks. This increases the chance of the two instructions being scheduled in the same time step.

Note that the register component graph is different from the data dependence graph: its nodes represent register operands, not instructions. The graph needs to be partitioned into *C* sets, where *C* is the number of clusters. Any graph partitioning algorithm can be applied to partition the nodes. Once the clusters for the register operands are known, the latter scheduling phase can attempt to assign instructions to appropriate clusters, based on the clusters of their register operands.

20.3.4.3 Scheduling Integrated with Cluster Assignment

MIRS_C (*modulo scheduling with integrated register spilling and cluster assignment*) [117] performs modulo scheduling, spilling, and cluster assignment in a single framework. In this approach, the dependence graph is updated on-the-fly as a result of cluster assignment, spilling, or eviction of instructions out of the schedule. Below we briefly describe the scheduling process.

The instructions in the DDG are sorted into a priority list and scheduled in that order. The cluster for the current instruction is chosen with the twin objectives of load balancing and communication minimization. If an instruction requires a value produced by an instruction scheduled in another cluster, or it produces a value consumed by an instruction in a different cluster, a copy instruction is introduced in the dependence graph. This copy instruction is scheduled first, followed by the current instruction.

After scheduling an instruction, the register pressure of the resulting partial schedule is checked. If *MaxLive*, the maximum number of simultaneously live values in the partial schedule, exceeds the total available registers by a certain factor, then a *use* of a value is chosen and spilled. The chosen use is such that spilling it will lower *MaxLive* and such that it requires a minimum number of loads and stores on average. The loads and stores are then introduced into the dependence graph and the priority list.

Recall that heuristic scheduling methods often involve a backtracking step if the current instruction cannot be scheduled in any time step between its feasible time range. The same is done by MIRS_C: some time step is heuristically chosen, and the current instruction is scheduled in this time step. Consequently, some other already scheduled instructions are evicted from the partial schedule if they compete for the same resource, or violate dependence constraints, with this instruction. In the MIRS_C approach, if an ejected instruction is the predecessor or a unique successor of a previously introduced copy instruction, the copy is also ejected from the schedule and removed from the dependence graph as well. The above process repeats until all instructions are scheduled. Otherwise, when the number of reschedules is greater than a certain threshold, the *II* is incremented and the scheduling restarts again.

20.3.4.4 Software Pipelining for Multi-Core Architectures

Coarse-grained software pipelining has been used to exploit parallelism for streaming applications (image, video, DSP, etc.) [49]. These applications are naturally represented by a set of autonomous actors, referred to as *filters*, which communicate over explicit data channels. The filters are fired repeatedly in execution. To facilitate understanding, one may think of a filter as a macro instruction, and the filters (macro instructions) compose the body of a loop. Software pipelining is especially attractive when there are loop-carried dependences between the filters.

The approach proposed in [49] performs software pipelining followed by core assignment. First, the architecture is treated as a conventional single-core processor, without considering the interconnections between the cores, where each core is a functional unit. With this underlying assumption, software pipelining schedules the filters to the cores like the traditional software pipelining schedules instructions to functional units. The prolog is constructed to buffer enough data items such that the filters in the kernel are guaranteed to be independent. This allows each filter to execute completely independently during each iteration of the kernel, as they are reading and writing to buffers rather than communicating directly. The buffers could be stored in a variety of places, such as the local memory of the core, a hardware FIFO, a shared on-chip cache, or an off-chip DRAM.

In the second step, the core assignment is performed. As the filters are independent, any set of filters, contiguous or not, can be mapped to the same core. The mapping follows two criteria: load balancing and synchronization minimization. To achieve load balancing, filters are sorted in order of decreasing work (computation load), and then they are assigned to the cores in that order. A filter is assigned to the core that has the least amount of work so far. To reduce the synchronization cost, the load balancing algorithm is wrapped with a selective fusion pass: two adjacent filters in the data flow graph that have the smallest combined work are fused into a single filter. After each fusion step, the load balancing algorithm is re-executed. This fusion–load balancing cycle repeats until the core that is a bottleneck in the pipeline has its workload increased by more than a given threshold (10%). When this happens, the fusion is reversed and the process terminates.

Decoupled software pipelining (DSWP) [75] is another promising scheduling approach for multi-core architectures. It divides the functionality of the original loop into more than one loop, each of them being a thread and assigned to a separate core. This approach requires all the instructions in an SCC in the dependence graph to be within the same thread. Two threads communicate and synchronize through a queue, called a *synchronization array*.

The original loop can have complex control flow within it and can be a nested loop. Correspondingly, their dependence graph can have control dependences in it. Each SCC in the dependence graph is coalesced into a single node. After this, the dependence graph becomes acyclic. This graph, where a node represents an SCC, is partitioned into a number of subgraphs, using a heuristic algorithm.

The partitions are then assigned to the cores in the following way. Consider the cores one by one. From the nodes whose predecessors in the dependence graph have already been scheduled, choose a node with the longest estimated execution time and assign it to the current core. If the total estimated execution time of the nodes for the current core is close to the overall estimated execution time divided by the desired number of threads, then assignment to the current core is finished, and the next core becomes the current core.

Once the core assignment is completed, actual code is generated for each thread. Intuitively, all the instructions in the nodes (SCCs) that are assigned to the same core, along with their basic blocks and the control flow between the blocks, are extracted from the original loop as a thread. As such, a basic block can be duplicated in two different threads but may contain a different set of instructions. The last step of DSWP is to introduce a pair of *producer* and *consumer* instructions, for each inter-thread dependence, in the producer and the consumer threads, respectively. They access a queue, the synchronization array, to store and remove the value. This realizes both synchronization and communications between threads, without involving the operating system or communicating via shared memory, which are slow. However, for a memory dependence, the value is still stored into the memory, instead of into the queue. In this case, the producer and consumer instructions serve only as tokens to enforce memory instruction ordering constraints.

20.4 Software Pipelining for Nested Loops

In this section, we discuss various approaches for software pipelining of nested loops. Nested loops are important, as they contribute to most of the execution time of scientific applications.

Software pipelining of nested loops is more complicated than that of single loops, as the operation instances in the n-dimensional iteration space need to be distributed to distinct time steps, such that resources are not overcommitted at any time step, dependences are satisfied, the total number of time steps is minimized, and there is a repeating pattern(s) in the schedule to enable loop rewriting for code generation. A simple approach is to only software pipeline the innermost loop and execute the outer loops sequentially. Unfortunately, this approach may not always result in good performance for one or more of the following reasons:

- The innermost loop may have a tight recurrence, due to which the amount of ILP exploited is limited.
- The trip count of the innermost loop is low, due to which the kernel is repeated only a few times and the overheads of the prolog and epilog dominate.
- The data locality exhibited by the innermost loop may not exploit the cache architecture of the system efficiently, resulting in poor overall performance.

Thus, it is important to arrive at other techniques for software pipelining nested loops that can better exploit parallelism and data locality for higher performance.

This section introduces several approaches for such a purpose. We survey some of the traditional approaches, and study two new methods, unroll-and-squash [79], and single-dimension software pipelining [92], in more detail. The former combines a loop transformation with software pipelining and employs general techniques to reduce code size in a software pipelined schedule. The latter is interesting because of its generality and efficiency; it subsumes the classical modulo scheduling as a special case and generates a schedule with the shortest computation time compared to traditional modulo scheduling approaches under identical conditions. It also extends the traditional hyperplane scheduling [29, 61] to handle resource constraints. The approach has systematically addressed all of the following: scheduling, register allocation, and code generation, for software pipelining of loop nests.

20.4.1 Basic Concepts

First, let us establish some basic definitions. Formally, a nested loop is referred to as "a loop nest." An n-*deep loop nest* is composed of n loops, $L_1, L_2, \ldots, L_n$, from the outermost to the innermost level, with each level having exactly one loop. Each loop $L_x(1 \leq x \leq n)$ has an index variable i_x and a *trip count* $N_x > 1$. The index is normalized to change from 0 to $N_x - 1$ with a unit step.

The loop nest is a *perfect loop nest* if all the operations are within the innermost loop. In this case, an instance of an operation o has an n-dimensional *index vector* $\mathbf{I} = (i_1, i_2, \ldots, i_n)$, denoted by $o(\mathbf{I})$.

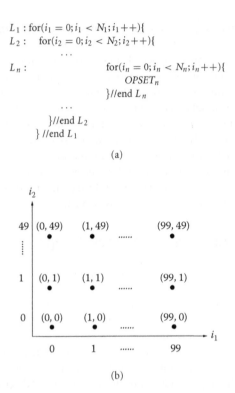

$L_1 :$ for$(i_1 = 0; i_1 < N_1; i_1++)\{$
$L_2 :$ for$(i_2 = 0; i_2 < N_2; i_2++)\{$
            $\cdots$

$L_n :$ for$(i_n = 0; i_n < N_n; i_n++)\{$
                                 $OPSET_n$
                                 $\}//$end L_n

      $\cdots$

      $\}//$end L_2
      $\} //$end L_1

(a)

(b)

FIGURE 20.11   The perfect loop nest model and an example iteration space. (a) The generic perfect loop nest model. (b) An example iteration space for a two-deep perfect loop nest (assume $N_1 = 100$ and $N_2 = 50$).

Figure 20.11a shows the generic perfect loop nest model, where $OPSET_n$ is the set of operations, which are in the innermost loop.

If there is any operation between two loop levels, the loop nest is said to be an *imperfect loop nest*. Suppose there is an operation o, which is in loop L_x but outside the inner loop L_{x+1}. An instance of it has an index vector $I = (i_1, i_2, \ldots, i_x)$. For uniformity, we can expand the index vector to be n-dimensional as well. If the operation is lexically before the inner loop L_{x+1}, the vector is expanded to $I = (i_1, i_2, \ldots, i_x, 0, \ldots, 0)$; otherwise, it is expanded to $I = (i_1, i_2, \ldots, i_x, N_{x+1} - 1, \ldots, N_n - 1)$. The operation instance is also denoted by $o(I)$.

The loop nest has an *iteration space*, where one point stands for the set of the operation instances that have the same index vector I. Such a point is called an *iteration point* and is identified by the index vector I. The iteration space for a two-deep perfect loop nest is illustrated in Figure 20.11b.

Note the difference between an iteration point and an iteration: an *iteration* of the L_x loop is one execution of the L_x loop body. Thus, the L_x loop has a total of N_x number of iterations, and each of the iterations contains $N_{x+1} * N_{x+2} * \cdots * N_n$ number of iteration points.

The iteration space is *rectangular* if its bounds $N_1, N_2, \ldots,$ and N_n, do not change during the execution of the loop nest, although they can change before and after it.

As before, a data dependence from operation a to operation b is represented as $(a \rightarrow b, \delta, d)$, where $\delta \geq 0$ is the dependence latency, and $d = \langle d_1, d_2, \ldots, d_n \rangle$ is the dependence distance vector, with d_1 the distance at the outermost level and d_n the innermost. The distance vector indicates that for any I, operation instance $a(I)$ produces a value that is consumed by operation instance $b(I + d)$.

The *sign of a vector* is that of its first nonzero element, either positive or negative. If all elements are 0, the vector is a *zero vector*.

20.4.2 Overview

Under resource constraints, it is an NP-complete problem to find the shortest schedule for a loop nest. In response to the complexity, the approaches to software pipelining nested loops can be broadly classified into two categories: (a) optimal approaches, which ensure dependence constraints are satisfied and the repetitive patterns are identified for compact code generation, but without considering resource constraints [44, 83], and (b) heuristic approaches, which address all three aspects (viz., resource constraints, dependence constraints, and repetitive patterns) simultaneously [59, 70, 79, 92, 106].

20.4.2.1 Optimal Approaches

The optimal approaches, which do not consider resource constraints, are extensions of the classical hyperplane scheduling [29, 61]. Hyperplane scheduling takes an iteration point as a unit and defines its schedule time as $I.\pi$, where $I = (i_1, i_2, \ldots, i_n)$ is the iteration point, π is a scheduling vector, and "$.$" is the inner product operator. It is extended to exploit ILP by giving a distinct scheduling vector and/or an offset to each operation in the loop body. The general form of the schedule time for an instance of operation o in an iteration point I is given by [29]

$$f(o, I) = I.\pi(o) + \mathit{offset}(o) \tag{20.31}$$

where $\pi(o)$ and $\mathit{offset}(o)$ are the scheduling vector and the offset, respectively. A special case arises when all the scheduling vectors for all the operations are the same:

$$f(o, I) = I.\pi + \mathit{offset}(o) \tag{20.32}$$

As an example, the r-periodic scheduling [44] defines

$$\pi = (\frac{T_1}{r}, \frac{T_2}{r}, \ldots, \frac{T_n}{r})^{Transpose} \tag{20.33}$$

and

$$\mathit{offset}(o) = \frac{A(o)}{r} \tag{20.34}$$

where $T_x (1 \leq x \leq n)$, $A(o)$, and r are positive integers. The schedule time function means that every T_x time steps, r number of iterations of loop L_x are issued. The scheduling problem is to find the scheduling vector and the offset for each operation such that the schedule time of the latest operation instance is minimized. If $f(o, I)$ is not an integer, then $\lfloor f(o, I) \rfloor$ is taken as the practical schedule time. Similarly, another optimal approach [83] defines

$$\pi = (a_1, a_2, \ldots, a_n)^{Transpose} \tag{20.35}$$

where a_x is a rational, and $\mathit{offset}(o)$ is also a rational.

20.4.2.2 Heuristic Approaches

Heuristic approaches essentially extend software pipelining methods for single (non-nested) loops. There have been several kinds of extensions: hierarchical reduction pipelines each loop in the loop nest, starting from the innermost one [59, 70, 106]; unroll-and-squash [79] overlaps the iterations of the outer loop in a two-deep loop nest and pipelines the inner loop; single-dimension software pipelining (SSP) overlaps and pipelines an arbitrary loop level, not necessarily the innermost one, in a loop nest. We defer a discussion of SSP and the register allocation and code generation for it to a subsequent subsection.

20.4.2.2.1 Hierarchical Reduction Methods

Hierarchical reduction methods [59, 70, 106] software pipeline the innermost loop first, then reduce the resulting loop (the loop that repeats the kernel) into a single macro operation. This macro operation represents the collective scheduling constraints of the entire kernel. The macro operation cannot be executed with any other operations. This is ensured by marking all the resources as consumed by the macro operation. The prolog and the epilog are outside the macro operation and are naturally merged with the other operations at the outer loop level. Then the outer loop can be software pipelined as a single loop. In this way, hierarchical reduction keeps the pipeline in a steady state by repeatedly executing the kernels and fills and drains the pipelines only at the beginning and the end of the whole execution. This is especially important from the viewpoint of performance when the inner loops have small trip counts.

One commonality across hierarchical approaches [59, 70, 106] is that they are innermost-loop centric, as these approaches schedule each loop level successively, starting with the innermost one. Among these works, the principle of hierarchical reduction was first proposed in [59], and then the work in [70] applied the same principle to the IA-64 architecture and worked out the details of the code generation, taking advantage of the hardware support for software pipelining in the IA-64 architecture, including predicated execution and rotating register files.

20.4.2.2.2 *Unroll-and-Squash*

Unroll-and-squash [79] was developed from unroll-and-jam [20]. It applies to a two-deep loop nest whose outer loop iterations are independent and can be executed in parallel. Given an unrolling factor u, unroll-and-jam unrolls the outer loop u times, and jams them together. The new inner loop is u times as big as the original one. Unroll-and-squash aims to achieve the same functionality of unroll-and-jam but reduces the code size of the inner loop such that it is nearly equal to that of the original inner loop. It does this by pipelining the inner loop and feeding the pipeline with data from the u different outer loop iterations in a round-robin fashion. The data are stored in rotating registers. One may regard this as a particular code size reduction technique for unroll-and-jam. However, the key to do this is the application of pipelining and rotating registers, which are very general. In fact, unroll-and-squash can be related to unroll-and-jam in just the way the kernel-only code generation scheme is related to modulo variable expansion (see Section 20.2.5). They use the same techniques to avoid code duplication. However, unroll-and-squash implicitly borrows the idea of hierarchical reduction in its process.

Let us illustrate unroll-and-squash with the help of the example loop nest shown in Figure 20.12a. Suppose we are given an unrolling factor $u = 3$. Unroll-and-jam unrolls the outer loop u times and jams them together. In the new loop nest, the operations both at the outer loop level and the inner loop level are u times as many as before. Among the u copies of the original operations, each copy is from a distinct outer loop iteration (refer to Figure 20.12b).

Since the inner loop body is composed of u copies of the same set of operations, it can be rearranged in a pipeline fashion, as shown in Figure 20.12c. Now the inner loop body is a short pipeline. If we simulate the execution of the inner loop, we can overlap the draining and filling parts of the pipelines of two successive inner loop iterations (see Figure 20.12d). This is the same idea as hierarchical reduction.

We can see that at each cycle, except for the beginning and the end of the execution, there is a steady state consisting of operations g and f. They take input operands from u original outer loop iterations alternatively. The same effect can be achieved by using a single copy of the operations g and f, but they take values from the different outer loop iterations in a round-robin fashion as shown in Figure 20.12e. Each variable is expanded into an array of u registers, corresponding to the u original outer loop iterations. The u registers are connected as a cycle, that is, as a rotating register file. It is rotated at the end of each execution of the current loop body. Executing the current loop body u times is equivalent to executing the unrolled loop body once. The rotation of the registers can be easily implemented in hardware, and in that case, the code size is reduced for the inner loop.

20.4.3 Single-Dimension Software Pipelining

SSP [90–93] is a resource-constrained scheduling method to pipeline a loop nest. In contrast to traditional innermost-loop-centric approaches [59, 70, 106], SSP identifies and pipelines the most profitable loop level from the entire loop nest. Here profitability can be measured in terms of ILP, data reuse, or any other

for ($i_1 = 0; i_1 < N_1; i_1{+}{+}$){
   TN1 = 0
   for ($i_2 = 0; i_2 < N_2; i_2{+}{+}$){
      TN2 = f(TN1)
      TN1 = g(TN2)
   }
}

(a)

for ($i_1 = 0; i_1 < N_1; i_1{+}{=} 3$){
   TN1=0   TN1' = 0   TN1'' = 0
   for ($i_2 = 0; i_2 < N_2; i_2{+}{+}$){
      TN2=f(TN1)   TN2' =f(TN1')   TN2'' =f(TN1'')
      TN1=g(TN2)   TN1' =g(TN2')   TN1'' =g(TN2'')
   }
}

(b)

for ($i_1 = 0; i_1 < N_1; i_1{+}{=} 3$){
   TN1=0   TN1' = 0   TN1'' = 0
   for ($i_2 = 0; i_2 < N_2; i_2{+}{+}$){
      TN2=f(TN1)
      TN1=g(TN2)   TN2' =f(TN1')   TN2'' =f(TN1'')
                  TN1' =g(TN2')   TN1'' =g(TN2'')
   }
}

(c)

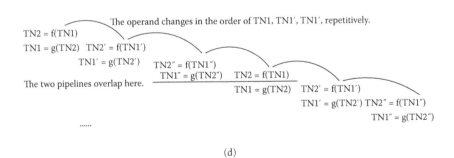

(d)

FIGURE 20.12 An illustration of unroll-and-squash. (a) Source loop nest. Here f and g are two operations. (b) The loop nest after unroll-and-jam. (c) Pipeline the inner loop. (d) Execution of the inner loop. The pipeline of the first iteration of the inner loop is overlapped with that of the second iteration. A variable repetitively takes the values from the three original outer loop iterations, as illustrated by the change of TN1. This suggests that the inner loop can be compacted by using rotating register files. (e) The loop nest after the inner loop is compacted. The inner loop contains a single set of the operations, but consumes the data from the three original outer loop iterations in a round-robin fashion. (Note: The inner loop trip count is roughly tripled.) A variable is expanded into a register file with three registers. Rotation is inserted at the end of the loop body. Here the rotation is simulated by moving registers, with the help of two temporary registers, *tmp*1 and *tmp*2. The moving operations can be eliminated if rotation is done by hardware automatically, and thus code size is reduced.

```
for  (i₁ = 0; i₁ < N₁; i₁ + = 3) {
        TN1 = 0              TN1' = 0              TN1" = 0

    /* Below is a compact equivalent to the inner loop */

        TN2 = f(TN1)
    for (i₂ = 0; i₂ < 3 * N₂ – 1; i₂ ++) {
        TN1 = g(TN2) TN2' = f(TN1')

    /* Rotate the values for TN1 and TN2* /

        tmp1 = TN1;        TN1= TN1';        TN1' = TN";        TN1" = tmp1;
        tmp2 = TN2;        TN2' = TN2';       TN2' = TN2";        TN2" = tmp2;
    }

                            TN1" = g(TN2")
}
```

(e)

FIGURE 20.12   (Continued)

optimization criteria. SSP retains the simplicity of the classical modulo scheduling technique for single loops and actually subsumes it as a special case.

Let us motivate the method with the help of a simple example. Figure 20.13a shows a perfect loop nest in C language. For simplicity, assume that each statement is an operation. Figure 20.13b is the data dependence graph, where each dependence edge is associated with the distance vector.

It can be seen that the inner loop has no parallelism because of the dependence cycle $a \rightarrow b \rightarrow a$ at this level. Thus, modulo scheduling of the inner loop would fail to find any parallelism for this example. The innermost-centric approaches expose extra parallelism by overlapping the filling and draining of the pipelines between successive outer loop iterations. Since modulo scheduling cannot find any parallelism, there is no filling or draining and therefore no overlapping. Thus, the innermost-centric approaches cannot find any parallelism either.

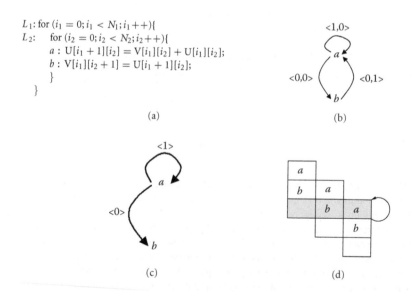

L_1: for $(i_1 = 0; i_1 < N_1; i_1++)$ {
L_2:   for $(i_2 = 0; i_2 < N_2; i_2++)$ {
      $a : U[i_1 + 1][i_2] = V[i_1][i_2] + U[i_1][i_2]$;
      $b : V[i_1][i_2 + 1] = U[i_1 + 1][i_2]$;
      }
    }

(a)   (b)   (c)   (d)

FIGURE 20.13   Illustration of the single-dimension software pipelining approach. (a) An example loop nest. (b) The DDG. (c) The simplified DDG of the L_1 loop. (d) One-dimensional schedule. (e) Software pipelined slices. (f) The final schedule. Source: From [92], courtesy of IEEE.

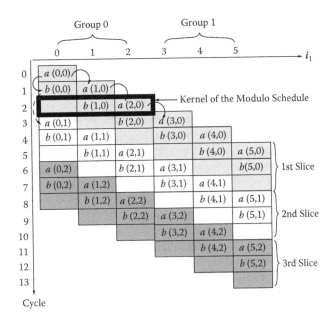

(e)

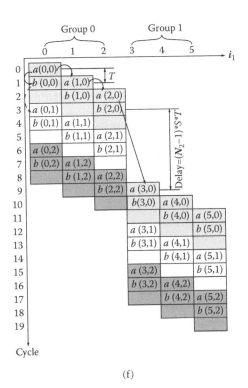

(f)

FIGURE 20.13 (Continued)

One may argue that a loop interchange transformation before software pipelining will solve this problem. Unfortunately, that may destroy the data reuse in the original loop nest; for large arrays each iteration point will introduce three cache misses for accessing $U[i_1 + 1][i_2]$, $V[i_1][i_2]$, and $U[i_1][i_2]$, as they are now accessed column-wise rather than row-wise. However, if N_2 is small, then even if the inner loop did not contain the tight recurrence, the overhead of the prolog and epilog would dominate, resulting in poor performance.

The above example shows the limitation of the traditional software pipelining: it cannot see the whole loop nest to better exploit parallelism, nor can it exploit the data reuse potential of the outer loop(s). This raises the question: Why not select a better loop to software pipeline, not necessarily the innermost one?

These arguments motivate identifying the most appropriate loop level in the loop nest and software pipelining it. SSP essentially accomplishes this using a novel approach. Before we proceed to the details, we illustrate the approach through the motivating example shown in Figure 20.13a.

Let us assume operations a and b in Figure 20.13a have a latency of one and two cycles, respectively. Further assume that we have two functional units, and both are pipelined and can perform any of the operations. The DDG is in Figure 20.13b. Suppose that the outer loop is selected for software pipelining. If only the dependences at this loop level are considered, a *simplified DDG*, shown in Figure 20.13c, can be obtained. We shall explain later why considering these dependences alone suffices. These dependences include $a \rightarrow a$ and $a \rightarrow b$, both having a distance vector in the form of $\langle d_1, 0 \rangle$ in the original DDG. In the simplified DDG, such a dependence distance vector has been simplified into $\langle d_1 \rangle$, as we software pipeline only the outer loop, and only the distance d_1 is useful.

Now that the DDG is 1-dimensional, from it and the resource constraints, a modulo schedule can be constructed using any modulo scheduling method, as if the outer loop were a single loop. An example modulo schedule is shown in Figure 20.13d, where the initiation interval $T = 1$, and there are $S = 3$ number of stages. This schedule is referred to as the *one-dimensional (1-D) schedule*.

Our task is to software pipeline the outer loop and hence overlap its iterations. Let each iteration of the outer loop run sequentially and successive iterations run in parallel. We consider that the operations belonging to iteration points $(i_1, 0)$, for all i_1, constitute the first *slice*, operations belonging to points $(i_1, 1)$ constitute the second slice, and so on. By applying the 1-D modulo schedule to each slice, one can obtain an *ideal schedule*, as shown in Figure 20.13e. In this ideal schedule, successive L_1 iterations are scheduled with the initiation interval $T = 1$ cycle. Here we assume $N_1 = 6$ and $N_2 = 3$ for simplicity.

Although the resource constraints are respected within each modulo scheduled slice, they are violated between slices, because a slice is issued greedily without waiting for the resources to be released by the previous slice. For example, at cycle 3, there are three operations, although there are only two functional units available. To remove the conflicts, we cut the slices into *groups*, with each group having $S = 3$ iterations of the outer loop. Each group, except the first one, is pushed down by $(N_2 - 1) * S * T$ cycles relative to its previous group. The delay is designed to ensure that repeating patterns definitely appear. This leads to the *final schedule* that maps each instance of an operation to its schedule time, as shown in Figure 20.13f. Not only dependence and resource constraints are respected, but the parallelism degree exploited in a modulo scheduled slice ($S = 3$) is still preserved, and the resources are fully used. A dependence is still respected after the pushing-down because that action either does not affect or only increases the time distance between the producer and consumer operations of the dependence, as illustrated by the dependences in Figure 20.13e before the pushing-down and in Figure 20.13f after that.

In the final schedule, one can identify some repeating patterns, if some ineffective operation instances are added. The final schedule with such operation instances is shown in Figure 20.14, where the added operation instances are highlighted by the shaded part. They are ineffective, as their first indexes are beyond the legal range of i_1, the outer loop index variable (the range is assumed to be $[0, 6)$ in our example). For target architectures with predication support like IA-64, predicate registers can be used to make them ineffective during execution of the final schedule, as illustrated later.

With the added ineffective operation instances, it is clear that the final schedule is composed of two repeating patterns, referred to as the *outermost loop pattern (OLP)* and the *inner loop execution segment*

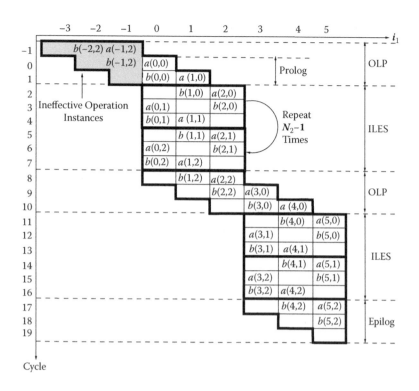

FIGURE 20.14 Repeating patterns in the final schedule. Source: From [93], courtesy of ACM.

(ILES). An OLP drains the pipeline of a previous group and fills the pipeline with a new group simultaneously. An ILES starts when the pipeline is filled with the new group and continues to run all the inner
loops of the new group until the group is going to drain. Then another OLP starts. Note that an ILES itself
is composed of $N_2 - 1 = 2$ number of a smaller pattern, as shown in the Figure 20.14. Apart from the OLP
and ILES, the final schedule also contains a prolog and an epilog. The first two cycles in the final schedule
are the prolog. As we can see from the figure, it is part of the first OLP. The last three cycles form the epilog.
Based on the above discussion, it is straightforward to rewrite the final schedule in a more compact form
(refer to Figure 20.15).

L_1': for $(i_1 = 0; i_1 < N_1; i_1 += 3)\{$

OLP
$$\begin{array}{llll} b(i_1 - 2, N_2 - 1) & a(i_1 - 1, N_2 - 1) & & \\ & b(i_1 - 1, N_2 - 1) & a(i_1, 0) & \\ & & b(i_1, 0) & a(i_1 + 1, 0) \end{array}$$

L_2': for $(i_2 = 1; i_2 < N_2; i_2++)\ \{$

ILES
$$\begin{array}{llll} & & b(i_1 + 1, i_2 - 1) & a(i_1 + 2, i_2 - 1) \\ a(i_1, i_2) & & & b(i_1 + 2, i_2 - 1) \\ b(i_1, i_2) & a(i_1 + 1, i_2) & \end{array}$$

$\}$

$\}$

Epilog
$$\begin{array}{ll} b(i_1 - 2, N_2 - 1) & a(i_1 - 1, N_2 - 1) \\ & b(i_1 - 1, N_2 - 1) \end{array}$$

FIGURE 20.15 Rewritten loop nest. Source: From [93], courtesy of ACM.

Below we will introduce the scheduling, register allocation, and code generation of the SSP approach in detail.

20.4.3.1   Scheduling

SSP consists of the following three steps:

- **Loop selection:** Select the appropriate loop level in the loop nest, based on certain criteria, such as the ILP or data locality exploited. The selected loop will be software pipelined by the next two steps, while its outer loops, if any, remain intact and may be parallelized using other methods. The loop level to be software pipelined must have a rectangular iteration space.
- **1-D schedule construction:** The n-dimensional ($n \geq 1$) DDG of the selected loop is reduced into a 1-D DDG, called a *simplified DDG*. Based on this DDG and the resource constraints, a *1-D schedule* is computed, represented by a *kernel*. No matter how many inner loops the selected loop has, it is scheduled as if it were a single loop. Any traditional modulo scheduling technique may be applied to construct this 1-D schedule.
- **Final schedule computation:** The 1-D schedule is mapped back to the n-dimensional iteration space to form the *final schedule*, which is semantically equivalent to the selected loop. In theory, this step computes a function that specifies the schedule time of each operation instance. In practice, this step translates into code generation for a target architecture, which will be discussed in Section 20.4.3.3. Register allocation can be performed right after 1-D schedule construction, and the allocation result can then be directly used in generating code. A discussion on register allocation for SSP will be presented in Section 20.4.3.2.

The above approach is referred to as *single-dimension software pipelining*, as the problem of multi-dimensional scheduling is simplified to 1-D scheduling and mapping.

The scheduling problem can be simplified because of the way the iterations of the selected loop are overlapped and the dependences are handled in this situation. For simplicity, let us assume that the loop selected for software pipelining is the outermost loop L_1. As illustrated in Figures 20.13e and 20.13f, the iterations run in parallel, while each of them runs sequentially.[5] For this, the set of iteration points $(i_1, 0, \ldots, 0, 0)$ for $i_1 \in [0, N_1)$ can be considered to be in the first slice, $(i_1, 0, \ldots, 0, 1)$ to be in the second slice, and so on. Successive slices are closely packed together. A 1-D (modulo) schedule is constructed for a slice. The same schedule is applied to all the other slices.

In terms of slices, there are three kinds of dependences: *zero*, *positive*, and *negative dependences*. Let the distance vector be $\langle d_1, d_2, \ldots, d_n \rangle$, where the first element d_1 must be nonnegative. The dependence is classified according to the sign of the sub-vector composed by the other elements, $\langle d_2, \ldots, d_n \rangle$. The dependence is said to be a zero, positive, or negative dependence if the sub-vector is a zero, positive, or negative vector, respectively.

A zero dependence occurs within the same slice. Hence, it has to be considered in constructing the 1-D schedule of the slice. In addition, only the distance d_1 is useful for software pipelining. That is, the dependence distance vector can be simplified as $\langle d_1 \rangle$ in pipelining. A positive dependence is from a slice to a next slice and is naturally resolved, as the two slices are executed sequentially. Thus, it can be ignored in constructing the 1-D schedule. Figure 20.16 illustrates the concepts. By keeping only the zero dependences and only their first elements in the distance vectors, we get the simplified DDG. This DDG, together with the resource constraints, is used in constructing the 1-D modulo schedule.

A negative dependence is from a slice to an earlier slice. Such a dependence cannot be directly handled by SSP. Henceforth, we assume that for the selected loop, we have only zero and positive dependences.

---

[5]By "sequential," we mean that the iteration points in the same iteration of the selected loop run in lexical order. This does not prevent the operation instances within the same individual iteration point from running in parallel.

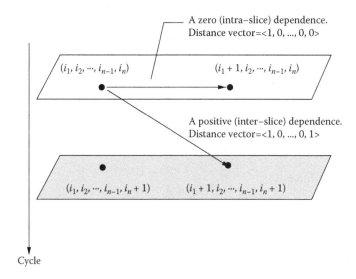

FIGURE 20.16 The dependences in an *n*-dimensional loop nest in two successive slices, where each parallelogram represents a slice, and each dot an iteration point. Although not shown, each slice is modulo scheduled. The outermost level L_1 is assumed to be the chosen loop. The intra-slice dependence must have all distance elements, except the first one, being 0. Source: From [92], courtesy of IEEE.

As illustrated in Figure 20.13f, to resolve resource conflicts between the slices, we cut them into groups, with each group having S number of L_1 iterations. Each group, except the first one, is delayed (pushed down) until the previous group drains and resources are available again. It is important to see that if a dependence is respected before the pushing-down, it is still respected after that. Therefore, we need only consider how to satisfy the dependences before the pushing-down.

In summary, in SSP, we consider only the zero dependences, which can be simplified to be 1-D, and we consider them only during the construction of the 1-D schedule (before the pushing-down). This enables us to reduce the *n*-dimensional scheduling problem to a 1-D scheduling problem. With this understanding, below we explain the three steps of SSP scheduling.

20.4.3.1.1 Loop Selection
To construct the most efficient software pipelined schedule, it is desirable to select the loop level with a lower initiation interval (higher parallelism) or a better data reuse potential (better cache effect) or both. All the loop levels that have rectangular iteration spaces are considered, and the initiation interval and data reuse are evaluated for each of them. The minimum initiation interval at loop level L_x is estimated as $\max(RecMII_x, ResMII)$, where $RecMII_x$ and $ResMII$ are the MIIs determined, respectively, by the recurrences in the simplified DDG of loop L_x and by the available hardware resources. $RecMII_x$ is calculated according to Equation 20.7, with the simplified DDG as the dependence graph. $ResMII$ is computed by Equations 20.9 and 20.10 for pipelined and non-pipelined resources, respectively.

Data reuse potential is estimated in the following way. As shown in Figure 20.13f, when we software pipeline a loop level, the iterations of the loop run in groups, with S number of iterations in a group. A tile containing $S * S$ number of iteration points is cut from any one of the groups, and the average number of memory accesses per iteration point within the tile is evaluated, based on a classical work [111]. If the estimated number of memory accesses is low, then the data reuse potential is high.

20.4.3.1.2 1-D Schedule Construction
Without loss of generality, assume L_1 is the selected level. The DDG is first reduced to obtain a simplified DDG, which consists of only zero dependences with 1-D distance vectors, as discussed earlier. Based on

the simplified DDG and the hardware resource constraints, a 1-D schedule is constructed. Since the DDG is 1-D, from the viewpoint of scheduling, L_1 is treated as if it were a single loop, without any inner loops. Any modulo scheduling method can be applied to schedule L_1 to obtain the 1-D schedule.

Let T be the initiation interval of the 1-D schedule and S be the number of stages. Let the schedule time for any operation instance $o(i_1)$ in the 1-D schedule be $\sigma(o, i_1)$, where $0 \le \sigma(o, i_1) < S * T$ when $i_1 = 0$. The 1-D schedule must satisfy the following properties:

- Modulo property:

$$\sigma(o, i_1) + T = \sigma(o, i_1 + 1) \tag{20.36}$$

- Dependence constraints:

$$\sigma(o_1, i_1) + \delta \le \sigma(o_2, i_1 + k) \tag{20.37}$$

  for every dependence arc $(o_1 \to o_2, \delta, \langle k \rangle)$ in the simplified DDG.
- Resource constraints: At any time step in the modulo scheduled kernel, no hardware resource is allocated to more than one operation.
- Sequential constraints: If $n > 1$, then

$$S * T - \sigma(o, 0) \ge \delta \tag{20.38}$$

  for every positive dependence with operation o as the producer operation, and δ as the dependence latency.

The first three constraints are exactly the same as those of the classical modulo scheduling, as presented in Section 20.2.2.3. The additional sequential constraints enforce sequential order between two successive iteration points in the same L_1 iteration. Consequently, the two successive slices containing the iteration points are executed sequentially. This ensures that all positive dependences are honored at runtime. The sequential constraints affect only loop nests with more than one loop.

20.4.3.1.3 *Final Schedule Computation*

For any operation o in the iteration point $\mathbf{I} = (i_1, i_2, \dots, i_n)$, the schedule time $f(o, \mathbf{I})$ is given by

$$\begin{aligned} f(o, \mathbf{I}) = {} & \sigma(o, i_1) \\ & + \sum_{2 \le x \le n} (i_x * (\prod_{x < y \le n+1} N_y) * S * T) \\ & + \left\lfloor \frac{i_1}{S} \right\rfloor * ((\prod_{2 \le x \le n+1} N_x) - 1) * S * T \end{aligned} \tag{20.39}$$

where $N_{n+1} = 1$.

The sum of the first two terms is the schedule time of $o(\mathbf{I})$ before pushing down the groups. The third term is the delay caused by the push-down. For example, consider the two-deep loop nest in Figure 20.13a. From the 1-D schedule in Figure 20.13d, we know that $S = 3$, $T = 1$, and $\sigma(a, i_1) = 0 + i_1 * T$. Thus, for any operation instance $a(i_1, i_2)$, we have the final schedule

$$f(a, (i_1, i_2)) = i_1 + i_2 * 3 + \left\lfloor \frac{i_1}{3} \right\rfloor * (N_2 - 1) * 3$$

For instance, when $N_2 = 3$, we have $f(a, (4, 1)) = 13$, as can be seen from Figure 20.13f.

The final schedule defined in Equation 20.39 has been shown to respect all the n-dimensional dependences in the original DDG and the resource constraints, although it considers only the 1-D simplified dependences during 1-D schedule construction [93]. In terms of performance, it has been shown that the SSP schedule is no worse than that of innermost-loop-centric modulo scheduling approaches [59, 70, 106] if (a) they use the same initiation interval T, (b) they have the same number of stages S, and (c) the trip count of the loop level selected for SSP is divisible by S. In other words, SSP has the shortest computation time among these approaches. Intuitively, this is because the final schedule produced by SSP always issues one iteration point every T cycles, without any hole, as can be seen from the example in Figure 20.13f.

If the loop nest is a single loop ($n = 1$), the sequential constraints are trivially satisfied. Other constraints are exactly the same as those of the classical modulo scheduling. The final schedule is $f(o, (i_1)) = \sigma(o, i_1)$. In this sense, classical modulo scheduling is subsumed by SSP as a special case.

The SSP final schedule can be related to the traditional hyperplane scheduling methods [29,61] and actually extends the latter with resource constraints. Rewrite the mapping function for the final schedule in Equation 20.39 as follows:

$$f(o, \boldsymbol{I}) = \boldsymbol{I}.\pi + \textit{offset}(o, \boldsymbol{I}) \tag{20.40}$$

where $\boldsymbol{I} = (i_1, i_2, \cdots, i_n)$, "." represents the inner product operator,

$$\pi = (T, (\prod_{2 < y \leq n+1} N_y) * S * T, \cdots, N_{n+1} * S * T)^{\textit{Transpose}} \tag{20.41}$$

and

$$\textit{offset}(o, \boldsymbol{I}) = \sigma(o, 0) + \left\lfloor \frac{i_1}{S} \right\rfloor * ((\prod_{2 \leq x \leq n+1} N_x) - 1) * S * T \tag{20.42}$$

This function consists of two parts. The first part, $\boldsymbol{I}.\pi$, corresponds to hyperplane scheduling, which determines how to allocate the iteration points to slices. The second part, $\textit{offset}(o, \boldsymbol{I})$, enforces dependences and resource constraints at the instruction level.

20.4.3.1.4 Extension to Imperfect Loop Nests

The above approach has been extended to imperfect loop nests. Figure 20.17a shows the general loop nest model, where $OPSET_x$ represents a set of non-branch operations between the beginnings of two adjacent loops. In this model, there is no operation between the ends of two adjacent loops. Assume the outermost loop is chosen for software pipelining. The corresponding kernel (1-D schedule) is shown in Figure 20.17b. All the loops have the same initiation interval T. The operations in the same loop L_x, including its inner loops, are scheduled into contiguous stages. The first such stage is referred to as f_x, and the last one, l_x. The total number of stages for loop L_x is termed S_x, which equals $l_x - f_x + 1$. In the current loop nest model, all loops will have the same last stage. That is, $l_1 = l_2 = \cdots = l_n$. Operations in the same stage must be from the same loop level. Figure 20.17c shows the general form of the final schedule.

The principle of overlapping the iterations of the selected loop is essentially the same as that for perfect loop nests. Intuitively, in the final schedule, the iterations of the selected loop are issued in parallel, whereas the inner loops within each of the iterations run sequentially. Let S_n be the total number of stages corresponding to the innermost loop in the kernel and T be the initiation interval of the kernel. Every T cycles, an iteration of the selected loop is issued, until the processor resources become insufficient to support any new iteration. Then a single group of S_n iterations, already issued, execute their inner loops in parallel. Until this group finishes execution and frees the resources, all the other iterations stall. Such a stalling period is an ILES.

Because of the regular stalls and issuing, repeating patterns naturally appear in the final schedule. The final schedule is composed of a prolog, the repetition of an OLP and an ILES, and an epilog (see Figure 20.17c).

Compared with the perfect loop nest case, the overlapping is different only in two minor points. First, a group has S_n iterations, in contrast with S iterations in the perfect case. Second, the prolog is no longer a part of the first OLP. It is a separate component of the final schedule now.

Because the principle of overlapping the iterations is basically unchanged, the principles of the three steps in SSP (loop selection and generating the 1-D and final schedules) for imperfect loop nests also stay the same as those for perfect loop nests.

L_1: for $(i_1 = 0; i_1 < N_1; i_1++)\{$
        $OPSET_1$
L_2:   for $(i_2 = 0; i_2 < N_2; i_2++)\{$
          $OPSET_2$
        . . .
L_n: for $(i_n = 0; i_n < N_n; i_n++)\{$
            $OPSET_n$
          $\}$ //end L_n
      . . .
    $\}$ //end L_2
   $\}$ //end L_1

(a)

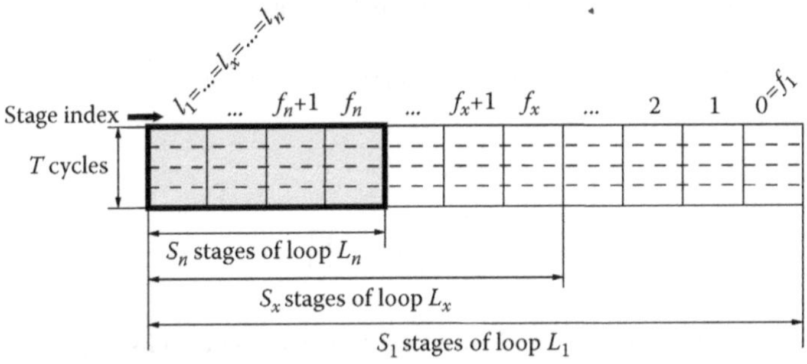

(b)

Prolog
L_1' : for$(i_1 = 0; i_1 < N_1; i_1+ = S_n)\{$
      OLP
      ILES
$\}$
Epilog

(c)

FIGURE 20.17 Generic loop nest scheduled by SSP. (a) A general loop nest model. (b) The general form of a kernel, assuming the outermost loop L_1 is chosen for scheduling. (c) The general form of a final schedule. Note that i_1 is incremented by S_n, which means a group contains S_n number of outermost loop iterations. Source: From [91], courtesy of ACM.

Example 20.3

Figure 20.18a shows a two-deep imperfect loop nest. Assume that L_1 is selected for scheduling, three functional units are available, and operations a, b, c, d, and e have latencies of two, five, one, four, and one cycles, respectively. A 1-D schedule can be found, as shown in Figure 20.18b. Based on the 1-D schedule, the final schedule in Figure 20.18c is formed as follows:

Initially, an L_1 iteration is issued every two cycles and executed sequentially. The issuing continues until the 10th cycle, when an ILES starts. If new iterations continue to be issued, there would be resource conflicts with the already running iterations. In the ILES, only the first group of $S_n = 3$ iterations continue executing their inner loops. Once done, the control goes back to L_1' and starts another OLP. To clarify the purpose of the OLP, we divide it into two parts by a vertical line. In the OLP, the left part drains the pipeline for previous group(s), which releases the processor resources, while at the same time, the right part fills the pipeline with the L_1 iterations of the following group(s), which get the resources.

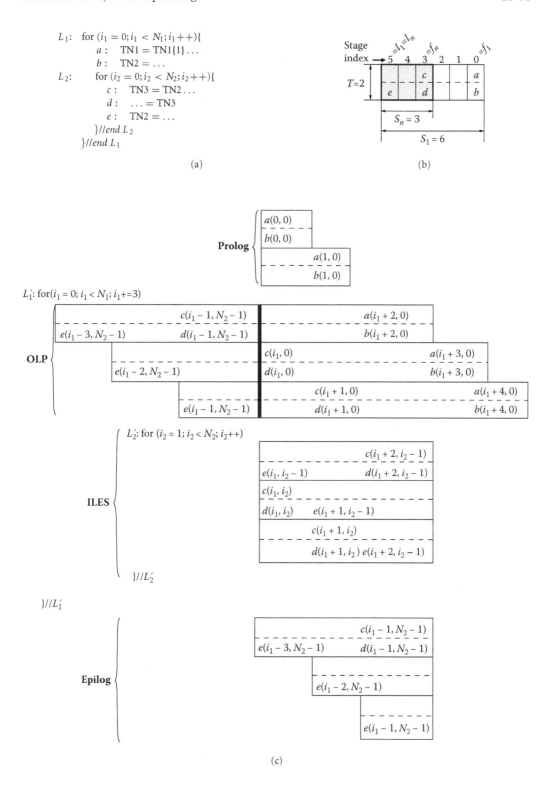

FIGURE 20.18 Illustration of software pipelining of an imperfect loop nest: (a) a two-deep loop nest; (b) a possible 1-D schedule (kernel); (c) final schedule in the form of a parallel loop nest. Source: From [91], courtesy of ACM.

To facilitate understanding, the final schedule is further illustrated cycle by cycle in Figure 20.19, with ineffective operation instances being masked from execution. To prepare for the introduction of register allocation below, we also show the scalar lifetimes of TN2 for the first six iterations. Each scalar lifetime is shown to the right of the iteration that produces it.

The extension to a more general imperfect loop nest model and constructing the 1-D schedule with multiple IIs for different loop levels can be found elsewhere [93].

20.4.3.2   Register Allocation

In this section, we show how register allocation is performed for the variables in an SSP final schedule, with the architectural support of a rotating register file, which was introduced in Section 20.2.5.2.1.

For simplicity, let us assume that the outermost loop L_1 is software pipelined. Unless stated otherwise, the term *iteration* always refers to an iteration of the outermost loop. A *loop variable* refers to a temporary name that has a definition in loop L_1, including its inner loops. For simplicity, we assume that at each loop level, the variable is either not defined or, if defined, defined only once. This does not prevent it from being defined at different loop levels.

First, let us briefly review how register allocation is done for single-loops. For single-loop software pipelining, the scalar lifetimes of a loop variable in successive iterations of the loop form a repetitive pattern. Therefore, the collection of the scalar lifetimes, the vector lifetime, can be represented with the first scalar lifetime as a reference. Based on the representation, the distances between all pairs of vector lifetimes can be computed. Then the vector lifetimes are packed on the surface of a space–time cylinder without violating the distances between them [30, 86]. This classical method was described in Section 20.2.5.2.

The software pipelined schedule of a loop nest is considerably more complex, and so are the vector lifetimes in it. As illustrated in Figure 20.19, there may be multiple *intervals* in a single scalar lifetime; pushing down a group may stretch some intervals and delay the others in it. Here by an "interval," we refer to a section of a scalar lifetime, whose start is a definition to the loop variable, and whose end is the last use of this definition. In addition, if a variable lives through a loop without being redefined, the corresponding interval of this variable in a scalar lifetime may be unknown at compile time; the length is dependent on the trip counts of this loop and its inner loops.

The key to the entire register allocation problem is to efficiently abstract the vector lifetimes, and based on the abstraction, accurately measure the distances of the vector lifetimes. Both tasks are based on a dynamic view of a vector lifetime, which exposes the different aspects of the vector lifetime step by step and thus enables incremental representation and distance calculation, and eventually a final solution is achieved. The dynamic view starts from the *simplest form* of the vector lifetime, and moves to the *ideal form*, and finally to the *final form*. The simplest form corresponds to the special case where the outermost loop is assumed to be a single loop; that is, the trip count of every inner loop equals 1. The ideal form corresponds to the ideal schedule where all iterations initiate at the constant II, without being delayed, and the intervals defined at inner loops are exposed, by considering the trip count of every inner loop bigger than 1. The final form corresponds to the complete final schedule, where the iterations are delayed in groups. For our example schedule in Figure 20.19, the three forms for TN2 are shown in Figure 20.21.

The concept of distance is expanded. For a pair of vector lifetimes, there is a *conservative distance* and an *aggressive distance*. The conservative distance does not allow for interleaving of the lifetimes, whereas the aggressive distance does. The aggressive distance enforces finer control upon the selection of a register for a vector lifetime. Figure 20.20 shows a register allocation for our example with and without interleaving.

The register allocation problem is still formulated as bin packing of the vector lifetimes on the surface of a cylinder, such that there is no conflict between any two scalar lifetimes and the circumference is minimized. The difference is the vector lifetimes themselves. The solution to the problem consists of the following steps:

- **Lifetime normalization:** First, the vector lifetimes are normalized such that any interval has a length known at compile time.
- **Lifetime representation:** The normalized vector lifetimes are then abstracted by a set of parameters.

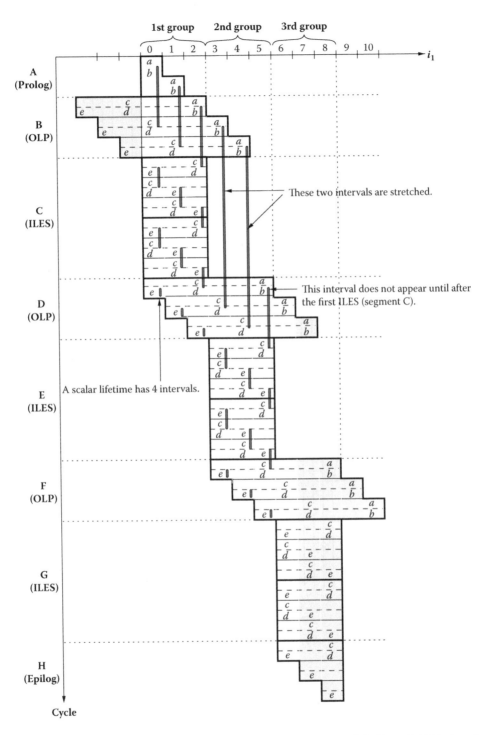

FIGURE 20.19 The unrolled final schedule (with $N_1 = 7$, $N_2 = 3$) for the example in Figure 20.18. To show the new features of the vector lifetimes, the scalar lifetimes of TN2 in the first two groups are illustrated. Each of them has four intervals. Due to the pushing-down of the second group, the first intervals defined in iterations 3 and 4 are stretched, and the first interval in iteration 5 is delayed. The ineffective operation instances are shaded. In order to show how the code generation technique masks them from execution, we intentionally assume $N_1 = 7$ so that the last group has two ineffective L_1 iterations. We will show the code generation later. Source: From [91], courtesy of ACM.

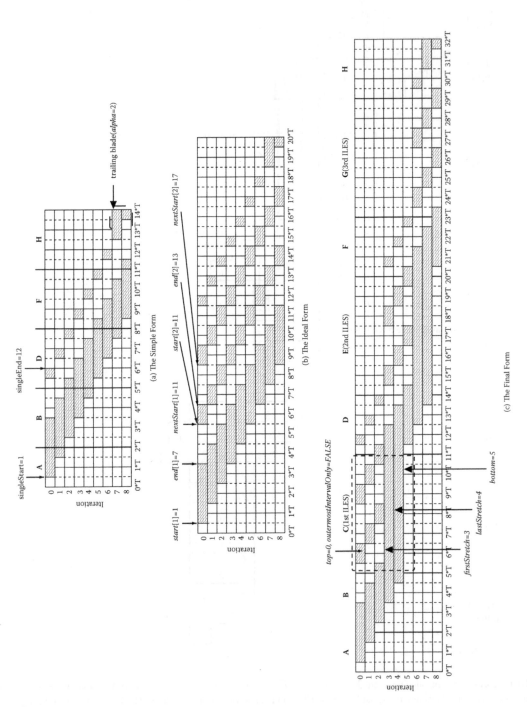

FIGURE 20.20   The vector lifetimes of TN2 in the final schedule in Figure 20.19. Assume TN2 has two live-out values (*alpha* = 2), and there is no live-in value (*omega* = 0).
Source: From [91], courtesy of ACM.

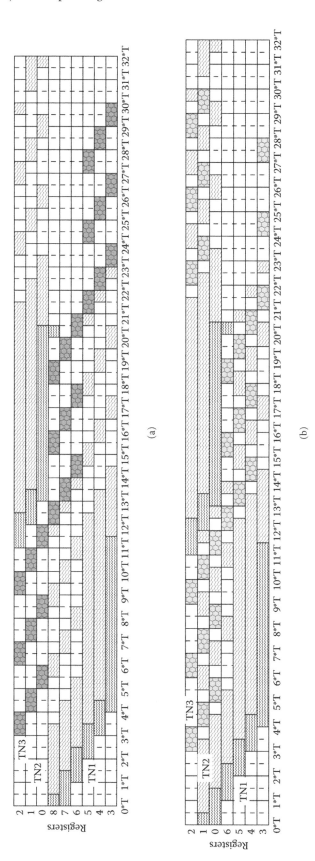

FIGURE 20.21 Register allocation examples: (a) register allocation using the conservative distance, circumference = 9; (b) register allocation using the aggressive distance, circumference = 7. The second scalar lifetime of TN3 is interleaved with the first scalar lifetime of TN2, and so on. Source: From [91], courtesy of ACM.

- **Distance calculation:** Using those parameters, a conservative distance and an aggressive distance between any two vector lifetimes, including the distances between a vector lifetime and itself, are computed. Then either the conservative or the aggressive distance can be used in the following steps.
- **Sorting:** The distance is used to order the vector lifetimes.
- **Bin packing:** The ordered vector lifetimes are inserted one by one onto the surface of the space–time cylinder with one of the strategies (best, first, and end fits), assuming the maximum circumference, that is, the maximal available registers in the architecture, with the distance between any pair of vector lifetimes respected.
- **Circumference minimization:** The last step minimizes the circumference of the cylinder so that the minimum number of registers are used.

This solution subsumes the classical register allocation for software pipelined single loops [30, 86] as a special case.

The first step, lifetime normalization, conceptually inserts a dummy copy operation TN = TN in the innermost loop if TN is live through any loop without being redefined. The introduction of the copy breaks an interval of TN in the middle such that TN is no longer live through the loop. The fourth step, sorting, can be done by a process similar to that in the traditional register allocation for single loops. Below we describe the other steps in more detail.

20.4.3.2.1   Lifetime Representation

For a loop variable, the simplest form is characterized in the same way as in the traditional register allocation for software pipelined single loops. The first scalar lifetime is taken as a point of reference. *SingleStart* and *singleEnd* are the start and end times of the first scalar lifetime. *Omega* and *alpha* are the total live-in values and total live-out values, respectively.

The ideal form exposes the intervals defined in the inner loops. Still, the first scalar lifetime is the reference. Also within it, for the interval defined at each inner loop level, the first instance of the interval is taken as a reference. For each inner loop L_x, the interval defined at this loop level and the hole following it are represented by $start[x]$, $end[x]$, and $nextStart[x]$, respectively. $End[x] - start[x]$ is the maximal possible size of the interval, and $nextStart[x] - end[x]$ is the smallest possible size of the hole.

Compared with the simple and ideal forms, the new thing in the final form is the ILESs. We take the first ILES as a reference. For the ILES, *top* is the iteration index of the intervals at the top of it, and *bottom* is one plus the iteration index of the intervals at the bottom of it. When there is no interval in it, *(top, bottom)* is set as $(+\infty, -\infty)$. The difference $bottom - top$ represents the vertical thickness of the vector lifetime in an ILES. *FirstStretch* and *lastStretch* are the iteration indexes of the first and last stretched intervals, respectively, that appear in the ILES. If there is no stretched interval at all, we set $(firstStretch, lastStretch) = (+\infty, -\infty)$. The boolean variable *outermostIntervalOnly* is true when the loop variable is defined only at the outermost loop level.

Lifetime representation is illustrated along with the three forms of the lifetime of TN2 in Figure 20.21.

20.4.3.2.2   Distance Calculation

Like the single-loop register allocation, a physical register r_A is said to be allocated to vector lifetime A if it is allocated to the first scalar lifetime. For any two vector lifetimes A and B, let r_A and r_B be the physical registers to be allocated to A and B, respectively. Each of the *conservative distance*, denoted $CONS[A, B]$, and the *aggressive distance*, denoted $AGGR[A, B]$, define a legal range of $r_B - r_A$, within which A and B do not conflict in the space–time diagram. The conservative distance does not allow the vector lifetimes to interleave, while the aggressive distance does. For simplicity, we introduce the computation of the conservative distance only.

$CONS[A, B]$ is computed in the same way as in the single-loop case introduced in Section 20.2.5.2.2. There are two additional conditions: first, $singleEnd(A)$ is adjusted. In the simplest form, an ILES is compressed into a time line. Any scalar lifetime that ends within the ILES should appear as if it ends at this time line (which is at time step $l_n * T$ in general). Second, in the final form, an ILES of B should be above the corresponding ILES of A. That is, let

$$adjustedSingleEnd(A) = \begin{cases} singleEnd(A) & \text{if } outermostIntervalOnly(A) \\ max(singleEnd(A), l_n * T) & \text{otherwise} \end{cases}$$

Then

$$singleStart(B) + (r_B - r_A) * T \geq adjustedSingleEnd(A),$$
$$r_B - r_A \geq omega(A) \qquad \text{if } omega(B) > 0$$
$$r_B - r_A \geq alpha(A) \qquad \text{if } alpha(A) > 0$$
$$r_B - bottom(B) \geq r_A - top(A)$$

The first inequality says that B's wand must be to the right of A's wand. The next two inequalities say that the leading and trailing blades of B must be above the leading and trailing blades of A. The last inequality says that an ILES of B is above the corresponding ILES of A. An illustration is given in Figure 20.22.

Solving these inequalities results in the following formulas:

$$d_1 = \left\lceil \frac{adjustedSingleEnd(A) - singleStart(B)}{T} \right\rceil \tag{20.43}$$

$$d_2 = \begin{cases} d_1 & \text{if } omega(B) = 0 \\ max(d_1, omega(A)) & \text{otherwise.} \end{cases} \tag{20.44}$$

$$d_3 = \begin{cases} d_2 & \text{if } alpha(A) = 0 \\ max(d_2, alpha(A)) & \text{otherwise.} \end{cases} \tag{20.45}$$

$$d_4 = max(d_3, bottom(B) - top(A)) \tag{20.46}$$

$$\implies CONS[A, B] = [d_4, +\infty] \tag{20.47}$$

When the loop nest is a single loop ($n = 1$), the above formulas become completely the same as those in Section 20.2.5.2.2. Although the aggressive distance is not introduced here, in this case, it is proved that $CONS[A, B] = AGGR[A, B]$. In this sense, the register allocation approach subsumes the classical register allocation for software pipelined single loops as a special case. Not just the distance calculation, but the whole solution is an extension of the classical approach.

20.4.3.2.3 Bin Packing on the Cylinder

The bin packing is guided by either the conservative or aggressive distance between the vector lifetimes, depending on the heuristics used. Once the distance between any two vector lifetimes is computed, the vector lifetimes are sorted and inserted one after the other on the surface of a space–time cylinder of a circumference equal to the maximum number of available registers, R. Let us assume that the physical registers are named $0, 1, \ldots, R - 1$, respectively.

The inserting of the vector lifetimes is similar to a scheduling process. For any vector lifetime B, find a physical register $r_B \in [0, R - 1]$ such that allocating r_B to B does not lead to conflict with any already inserted vector lifetime A.

Let r_A be the physical register allocated to A. Imagine we insert the vector lifetimes first to a space–time diagram and then wrap this diagram into a cylinder. On the diagram, the set of legal registers that B can

The first ILES (segment C) is compressed into this line.

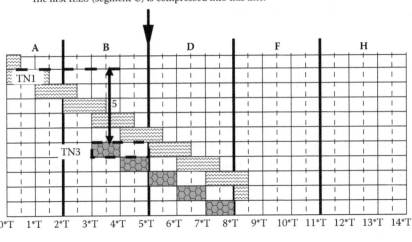

(a)

TN1's bottom is above TN3's top.

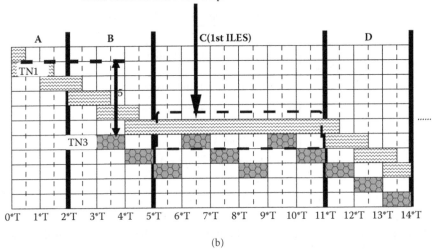

(b)

FIGURE 20.22   Illustration: Calculating the conservative distance $CONS[TN3, TN1]$. The simplest and final forms are used to show the intuition. The result is $CONS[TN3, TN1] = [5, +\infty]$. Note that the first scalar lifetime of $TN3$ has intervals in the first ILES, and in the simplest form they are imagined to be compressed into the timeline $5 * T$. Thus, the scalar lifetime is imagined to end at that line. We enclose this scalar lifetime in a dotted box in the first figure to illustrate this fact. Source: From [91], courtesy of ACM.

choose from is

$$(r_A + DIST[A, B]) \cup (r_A - DIST[B, A])$$

where $DIST$ is the given distance ($CONS$ or $AGGR$), and $r_A \pm set$ denotes a new set that is $\{r_A \pm i | \forall i \in set\}$. Any register not in the above set cannot be allocated to B. Therefore, after the diagram is wrapped into a cylinder, the illegal registers for B resulting from the conflict with A are

$$illegal(B, A, x) = \{i \bmod x | \forall i \in \mathbb{N}, i \notin (r_A + DIST[A, B]), \text{ and } i \notin (r_A - DIST[B, A])\}$$

where x is the circumference of the cylinder. Here it is R.

For all such already inserted vector lifetimes A, the illegal registers for B equal

$$\cup illegal(B, A, x)$$

Therefore, the bin-packing problem can be reinterpreted as follows:

Problem 20.4

Given a set of vector lifetimes, an order between them, the distance DIST (either CONS or AGGR) between any pair of them, and a strategy (e.g., first fit, last fit, best fit, etc.), find a mapping for each vector lifetime, one by one, according to the order, such that for the current vector lifetime B, it is mapped to a physical register r_B, where $r_B \in [0, R - 1]$, and $r_B \notin \cup illegal(B, A, R)$ for any vector lifetime A before B in this order.

This problem statement gives the solution as well. Note that the resulting mapping is dependent on whether the given distance is conservative or aggressive and what strategy is used.

20.4.3.2.4 *Circumference Minimization*

So far, the solution uses the maximum number of physical registers, R, as the circumference of the cylinder. The last step tries to decrease the actual number of registers required. This leads to a new circumference c.

First, the allocation that resulted from the bin-packing step is translated such that the smallest register allocated equals 0. That is, for each vector lifetime A whose physical register allocated is r_A, it is reallocated to $r_A - minReg$, where $minReg$ is the smallest register allocated. Note that we have been using a number to represent register.

Then the circumference c is initialized as the distance between the smallest and the biggest registers allocated. The circumference c is legal if there is no conflict between the vector lifetimes. A vector lifetime A may conflict with itself after wrapping around if the circumference is not within $DIST[A, A]$, where $DIST$ is the same as that in the bin-packing process, which is either $CONS$ or $AGGR$. It also conflicts with a different vector lifetime B, if B is allocated an illegal register with respect to A under the current circumference. In other words, c is legal if

$$c \in DIST[A, A] \qquad \forall \text{ vector lifetime } A$$

and

$$r_B \notin \cup illegal(B, A, c) \qquad \forall \text{ vector lifetimes } A \text{ and } B, A \neq B$$

where r_B is the register allocated to B.

The circumference c is incremented until it is legal.

20.4.3.3 Code Generation

Code generation for a software pipelined loop nest extends that for a software pipelined single loop introduced in Section 20.2.5.2.3. Again we take IA-64 as the target architecture, which has hardware support in the form of rotating register files and predicated execution. The new issue to be addressed here is how to control the hardware support, which was designed to support single-loop software pipelining only, such that it works for nested loop software pipelining as well.

In [90] the above issue is addressed by rotating registers only at the outermost (the selected) loop level. That is, perform register rotation only in the prolog, OLP, and epilog. In an ILES, do not rotate the registers, but access them statically. In addition, the control registers LC and EC are set up appropriately to ensure that exactly N_1 number of iterations are issued, and all the ineffective operation instances are masked from execution.

We explain the solution with an illustration for our example final schedule in Figure 20.18c. Assume we use aggressive distance during register allocation. Then, according to the allocation result in Figure 20.20b, TN1, TN2, and TN3 are allocated registers 0, 1, and 2, respectively. For IA-64, we assign $\gamma 32$, $\gamma 33$, and $\gamma 34$ to them.

Figure 20.23 shows the corresponding IA-64 assembly code. The initialization phase (lines 1 to 5) is similar to that in the software pipelined single loop illustrated in Figure 20.8a, so we do not repeat the explanation. The setting of the control registers LC and EC shown in the figure is general to any deep loop nest. We omit the derivation of the formulas here.

Starting from the prolog (line 6), the code is essentially a one-to-one translation of the operations in the abstract rewritten loop in Figure 20.18c to assembly, so we do not explain it in detail. There are several minor differences: the L'_1 is realized by a branch from line 29 back to line 12; before executing the epilog code (line 30), the control registers are asserted to have become 0, and then EC is reset to another value to drain the pipeline (line 31); after the epilog, the control registers must have become 0 again (line 41).

To be clear, we simulate the execution of the code, and show the process in Figure 20.24. In the prolog, each OLP, and the epilog, a rotation of the register files is performed via a `br.ctop` operation after each execution of a kernel (or a partial kernel). However, in an ILES, there is no rotation at all, and the rotating registers are accessed just like static registers. Take the first scalar lifetime of TN2 as an example. Before entering an ILES, the physical register containing a value of it is originally named $\gamma 33$, and renamed $\gamma 34$, $\gamma 35$, $\gamma 36$, and $\gamma 37$, in order, with the rotations. Then its name stays $\gamma 37$ throughout the whole ILES. We also illustrated the second scalar lifetime of TN2 to help understanding.

Next we explain how the loop is controlled. The stages of a kernel are predicated, from right to left, with the predicate registers $\rho 16$ to $\rho 21$, as illustrated by the first three (partial) kernels. Initially, only $\rho 16$ is 1. Step by step, all of them become 1. The stages predicated by 0 are automatically ignored in execution, as highlighted by shaded color. The change of the predicates is shown by the shaded triangle in the upper-right side.

With the execution, LC decrements step by step, until finally it becomes 0 in segment D, when EC starts to decrement, which gradually resets the predicate registers to 0, as shown by the shaded triangle in the lower-right side. After the ILES of the last group (segment G), another `br.ctop` is executed, and LC and EC become 0 exactly. This corresponds to the assertion in line 30 of Figure 20.23. Note that this detail is important for the correctness of the code generation: if either LC or EC is not 0, the `br.ctop` (or more exactly, `br.ctop` L'_1 in line 29 of Figure 20.23) will cause the control to transfer to line 12 of Figure 20.23, and another OLP will be executed, which is wrong. The initial setting of LC and EC in lines 2 and 3 make sure this erroneous transfer never occurs. When all the groups have been issued, the control will definitely transfer to the epilog.

In the epilog (segment H), since we still have some operation instances to execute, we reset EC to finish them. After that, it can be seen that LC and EC become 0 again. This corresponds to the assertion in line 41 of Figure 20.23.

20.5   Summary and Future Directions

Software pipelining originally emerged as a software approach to exploit ILP available in loops in uniprocessors. Numerous approaches to software pipelining have been proposed and studied in detail. Both optimal and heuristic approaches have been proposed. The techniques for scheduling, register allocation, and code generation have been extensively investigated. In this chapter, we have described the fundamental techniques in detail and surveyed various research directions.

As traditional architectures cease to scale well, distributed architectures such as clustered VLIW and multi-core processors become prevalent. Power has also become a first-order constraint in architectural design. Consequently, software pipelining has been adapted to these newer architectures, and to save power/energy consumption.

Although software pipelining is powerful in extracting fine-grain parallelism, this is also its limitation. Fine-grain parallelism becomes exploitable usually after the intermediate representation of a program is lowered to machine level during compilation. In such lowering, the high-level context has been largely lost. A comprehensive solution to speed up a loop should combine the power of loop transformation and threading at high levels and that of software pipelining at low levels, so that parallelism at different levels

```
 1  Initialization:
        clrrrb;;
 2      LC = N₁ − 1
 3      EC = S₁ − 1 + ((N₁ − 1)mod Sₙ)
 4      movpr.rot = 1 << 16
 5      mov γ33 = live-in value of TN1;;
 6  Prolog:                                                a:(ρ16)γ32 = γ33;;
 7                                                         b:(ρ16)γ33 = ...;;
 8      br.ctop next;;
 9    next:                                                a:(ρ16)γ32 = γ33;;
10                                                         b:(ρ16)γ33 = ...;;
11      br.ctop L'₁;;
12  L'₁:           c:(ρ19)γ37 = γ36...   a:(ρ16)γ32 = γ33;;   a:(ρ16)γ32 = γ33;;
13                 d:(ρ19)... = γ37      b:(ρ16)γ33 = ...;;   b:(ρ16)γ33 = ...;;
14      e:(ρ21)γ38 = ...
15      br.ctop next1;;
16    next1:       c:(ρ19)γ37 = γ36...   a:(ρ16)γ32 = γ33;;   a:(ρ16)γ32 = γ33;;
17        e:(ρ21)γ38 = ...  d:(ρ19)... = γ37   b:(ρ16)γ33 = ...;;   b:(ρ16)γ33 = ...;;
18      br.ctop next2;;
19    next2:       c:(ρ19)γ37 = γ36...   c:(ρ18)γ36 = γ35...
20                 d:(ρ19)... = γ37      d:(ρ18)... = γ36
21  L'₂; for(i₂ = 1; i₂ < N₂; i₂ + +){   e:(ρ19)γ36 = ...      e:(ρ18)γ35 = ...
22      e:(ρ20)γ37 = ...                 c:(ρ19)γ37 = γ36...
23                 c:(ρ20)γ38 = γ37...   d:(ρ19)... = γ37
24                 d:(ρ20)... = γ38      e:(ρ19)γ36 = ...
25                                       c:(ρ19)γ37 = γ36...
26                                       d:(ρ19)... = γ37
27  }//endL'₂
28
29  br.ctop L'₁
30  Epilog:
31      assert(LC = EC = 0);;
32      EC = Sₙ                          e:(ρ21)γ38 = ...      c:(ρ19)γ37 = γ36... ;;
33                                                             d:(ρ19)... = γ37;;
34      br.ctop next3;;
35    next3:                                                   e:(ρ21)γ38 = ...
36
37      br.ctop next4;;
38    next4:
39      e:(ρ21)γ38 = ...;;
40      br.ctop next5;;
41    next5:
        assert(LC = EC = 0);
```

FIGURE 20.23 Code for the software pipelined loop nest.

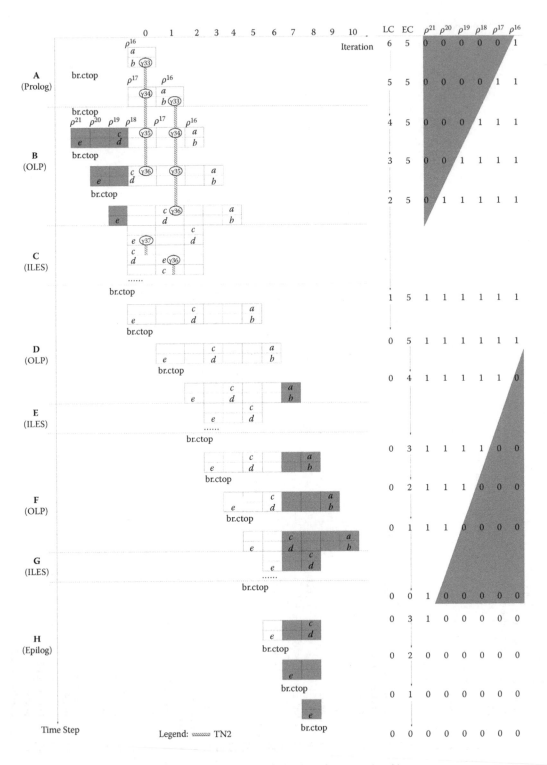

FIGURE 20.24 Execution of the code for the software pipelined loop nest.

is exploited systematically in the whole compilation process. Little work addresses this problem, despite its importance.

Acknowledgments

The authors acknowledge Erik R. Altman, A. Dani, Alban Douillet, Guang R. Gao, Zhizhong Tang N. S. S. Narasimha Rao, S. M. Sandhya, V. Janaki Ramanan, Raul Silvera, Madhavi G. Valluri, Jian Wang, Hongbo Yang, and Chihong Zhang, with whom they have coauthored several papers on instruction scheduling and software pipelining that have both directly and indirectly contributed to this chapter. The authors thank Aditya Thakur and Santosh G. Nagarakatte for their numerous suggestions and careful proofreading. The chapter has used excerpts from several papers in the references [90–93], courtesy of IEEE and ACM.

References

1. A. Aiken and A. Nicolau. 1988. Optimal loop parallelization. In *PLDI '88: Proceedings of the ACM SIGPLAN 1988 Conference on Programming Language Design and Implementation*, 308–17. New York: ACM Press.
2. A. Aiken and A. Nicolau. 1991. A realistic resource-constrained software pipelining algorithm. In *Advances in languages and compilers for parallel processing*, ed. D. Gelernter et al. 274–90. Cambridge, MA: MIT Press.
3. Cagdas Akturan and Margarida F. Jacome. 2001. Caliber: A software pipelining algorithm for clustered embedded VLIW processors. In *International Conference on Computer Aided Design (ICCAD)*, San Jose, CA. 112–18. New York: ACM Press.
4. Cagdas Akturan and Margarida F. Jacome. 2001. Rs-fdra: A register sensitive software pipelining algorithm for embedded VLIW processors. In *CODES '01: Proceedings of the Ninth International Symposium on Hardware/Software Codesign*, 67–72. New York: ACM Press.
5. Alex Aletà, Josep M. Codina, Antonio González, and David Kaeli. 2004. Removing communications in clustered microarchitectures through instruction replication. *ACM Trans. Archit. Code Optim.* 1(2):127–51.
6. Alex Aletà, Josep M. Codina, Antonio González, and David Kaeli. 2005. Demystifying on-the-fly spill code. In *PLDI '05: Proceedings of the 2005 ACM SIGPLAN Conference on Programming Language Design and Implementation*, 180–89. New York: ACM Press.
7. Alex Aletà, Josep M. Codina, Jesús Sánchez, and Antonio González. 2001. Graph-partitioning based instruction scheduling for clustered processors. In *MICRO 34: Proceedings of the 34th Annual ACM/IEEE International Symposium on Microarchitecture*, 150–59. Washington, DC: IEEE Computer Society Press.
8. Vicki H. Allan and Stephen J. Allan. 2002. Software pipelining. In *Compiler design handbook: Optimizations and machine code generation*, 689–738. Boca Raton, FL: CRC Press.
9. Vicki H. Allan, Reese B. Jones, Randall M. Lee, and Stephen J. Allan. 1995. Software pipelining. *ACM Comput. Surv.* 27(3):367–432.
10. J. R. Allen, Ken Kennedy, Carrie Porterfield, and Joe Warren. 1983. Conversion of control dependence to data dependence. In *Conference record of the 10th Annual ACM Symposium on Principles of Programming Languages*, 177–89. New York: ACM Press.
11. Erik R. Altman, R. Govindarajan, and Guang R. Gao. 1995. Scheduling and mapping: Software pipelining in the presence of structural hazards. In *PLDI '95: Proceedings of the ACM SIGPLAN 1995 Conference on Programming Language Design and Implementation*, 139–50. New York: ACM Press.
12. Erik R. Altman, R. Govindarajan, and Guang R. Gao. 1998. A unified framework for instruction scheduling and mapping for function units with structural hazards. *J. Parallel Distrib. Comput.* 49(2):259–93.
13. Vasanth Bala and Norman Rubin. 1995. Efficient instruction scheduling using finite state automata. In *MICRO 28: Proceedings of the 28th Annual International Symposium on Microarchitecture*, 46–56. Los Alamitos, CA: IEEE Computer Society Press.

14. Luca Benini, Alessandro Bogliolo, and Giovanni De Micheli. 2002. A survey of design techniques for system-level dynamic power management. In *Readings in hardware/software co-design*, 231–48. Norwell, MA: Kluwer Academic.

15. Francois R. Boyer, El Mostapha Aboulhamid, Yvon Savaria, and Michel Boyer. 2001. Optimal design of synchronous circuits using software pipelining techniques. *ACM Trans. Des. Autom. Electron. Syst.* 6(4):516–32.

16. Pierre-Yves Calland, Alain Darte, and Yves Robert. 1996. A new guaranteed heuristic for the software pipelining problem. In *ICS '96: Proceedings of the 10th International Conference on Supercomputing*, 261–69. New York: ACM Press.

17. Pierre-Yves Calland, Alain Darte, and Yves Robert. 1998. Circuit retiming applied to decomposed software pipelining. *IEEE Trans. Parallel Distrib. Syst.* 9(1):24–35.

18. Andrea Capitanio, Nikil Dutt, and Alexandru Nicolau. 1992. Partitioned register files for VLIWs: A preliminary analysis of tradeoffs. In *MICRO 25: Proceedings of the 25th Annual International Symposium on Microarchitecture*, 292–300. Los Alamitos, CA: IEEE Computer Society Press.

19. Steve Carr, Chen Ding, and Philip Sweany. 1996. Improving software pipelining with unroll-and-jam. In *HICSS '96: Proceedings of the 29th Hawaii International Conference on System Sciences (HICSS'96)*, Vol. 1, *Software technology and architecture*, 183. Washington, DC: IEEE Computer Society Press.

20. Steve Carr and Ken Kennedy. 1994. Improving the ratio of memory operations to floating-point operations in loops. *ACM Trans. Prog. Lang. Syst.* 16(6):1768–810.

21. Noureddine Chabini and Yvon Savaria. 2001. Methods for optimizing register placement in synchronous circuits derived using software pipelining techniques. In *ISSS '01: Proceedings of the 14th International Symposium on Systems Synthesis*, 209–14. New York: ACM Press.

22. Noureddine Chabini and Wayne Wolf. 2004. An approach for integrating basic retiming and software pipelining. In *EMSOFT '04: Proceedings of the 4th ACM International Conference on Embedded Software*, 287–96. New York: ACM Press.

23. G. J. Chaitin. 1982. Register allocation & spilling via graph coloring. In *CC'82*, 98–101. New York: ACM Press.

24. Liang-Fang Chao and Andrea LaPaugh. 1993. Rotation scheduling: A loop pipelining algorithm. In *DAC '93: Proceedings of the 30th International Conference on Design Automation*, 566–72. New York: ACM Press.

25. A. E. Charlesworth. 1981. An approach to scientific array processing: The architectural design of the ap-120b/fps-164 family. *Computer* 14(9):18–27.

26. Danny Cohen. 1978. A methodology for programming a pipeline array processor. In *MICRO 11: Proceedings of the 11th Annual Workshop on Microprogramming*, 82–89. Washington, DC: IEEE Press.

27. Robert P. Colwell, Robert P. Nix, John J. O'Donnell, David B. Papworth, and Paul K. Rodman. 1988. A VLIW architecure for a trace scheduling compiler. *International Parallel Processing Symposium on Parallel and Distributed Processing (IPPS/SPDP)* 37(8):967–79. Washington, DC: IEEE Computer Society.

28. Amod K. Dani, V. Janaki Ramanan, and Ramaswamy Govindarajan. 1998. Register-sensitive software pipelining. In *IPPS/SPDP*, 194–98.

29. Alain Darte and Yves Robert. Constructive methods for scheduling uniform loop nests. *IEEE Trans. Parallel Distrib. Syst.* 5(8):814–22.

30. James C. Dehnert and Ross A. Towle. 1993. Compiling for the Cydra 5. *J. Supercomput.* 7(1-2): 181–227.

31. K. Ebcioğlu and Toshio Nakatani. 1990. A new compilation technique for parallelizing loops with unpredictable branches on a VLIW architecture. In *Selected papers of the Second Workshop on Languages and Compilers for Parallel Computing*, 213–29. London: Pitman.

32. Kemal Ebcioğlu. 1987. A compilation technique for software pipelining of loops with conditional jumps. In *MICRO 20: Proceedings of the 20th Annual Workshop on Microprogramming*, 69–79. New York: ACM Press.

33. Kemal Ebcioglu and Alexandru Nicolau. 1989. A global resource-constrained parallelization technique. In *International Conference on Super Computing (ICS)*, Heraklion, Crete. 154–63. New York: ACM Press.

34. Alexandre E. Eichenberger and Edward S. Davidson. 1995. Stage scheduling: A technique to reduce the register requirements of a modulo schedule. In *MICRO 28: Proceedings of the 28th Annual International Symposium on Microarchitecture*, 338–49. Los Alamitos, CA: IEEE Computer Society Press.

35. Alexandre E. Eichenberger and Edward S. Davidson. 1996. A reduced multipipeline machine description that preserves scheduling constraints. In *PLDI '96: Proceedings of the ACM SIGPLAN 1996 Conference on Programming Language Design and Implementation*, 12–22. New York: ACM Press.

36. Alexandre E. Eichenberger and Edward S. Davidson. 1997. Efficient formulation for optimal modulo schedulers. In *SIGPLAN Conference on Programming Language Design and Implementation*, 194–205. New York: ACM Press.

37. Alexandre E. Eichenberger, Edward S. Davidson, and Santosh G. Abraham. 1994. Minimum register requirements for a modulo schedule. In *MICRO 27: Proceedings of the 27th Annual International Symposium on Microarchitecture*, 75–84. New York: ACM Press.

38. Alexandre E. Eichenberger, Edward S. Davidson, and Santosh G. Abraham. 1995. Optimum modulo schedules for minimum register requirements. In *ICS '95: Proceedings of the 9th International Conference on Supercomputing*, 31–40. New York: ACM Press.

39. Paolo Faraboschi, Geoffrey Brown, Joseph A. Fisher, Giuseppe Desoli, and Fred Homewood. 2000. Lx: A technology platform for customizable VLIW embedded processing. In *ISCA '00: Proceedings of the 27th Annual International Symposium on Computer Architecture*, 203–13. New York: ACM Press.

40. Paul Feautrier. 1994. Fine-grain scheduling under resource constraints. In *Proceedings of the 7th International Workshop on Languages and Compilers for Parallel Computing*, 1–15. London: Springer-Verlag.

41. Marcio Merino Fernandes, Josep Llosa, and Nigel Topham. 1999. Distributed modulo scheduling. In *HPCA '99: Proceedings of the 5th International Symposium on High Performance Computer Architecture*, 130. Washington, DC: IEEE Computer Society Press.

42. J. A. Fisher, D. Landskor, and B. D. Shriver. 1981. Microcode compaction: Looking backward and looking forward. In *Proceedings of the 1981 National Computer Conference*, 95–102. Arlington, VA: AFIPS Press.

43. Joseph A. Fisher. 1981. Trace scheduling: A technique for global microcode compaction. *IEEE Trans. Comput.* 30(7):478–90.

44. Guang R. Gao, Qi Ning, and Vincent Van Dongen. 1993. Software pipelining for nested loops. ACAPS Tech Memo 53, School of Computer Science, McGill University, Montréal, Québec.

45. Guang R. Gao, Yue-Bong Wong, and Qi Ning. 1991. A timed petri-net model for fine-grain loop scheduling. In *Proceedings of the ACM SIGPLAN 1991 Conference on Programming Language Design and Implementation*, 204–18. New York: ACM Press.

46. Michael R. Garey and David S. Johnson. 1979. *Computers and intractability: A guide to the theory of NP-completeness.* New York: Freeman.

47. Franco Gasperoni and Uwe Schwiegelshohn. 1992. Scheduling loops on parallel processors: A simple algorithm with close to optimum performance. In *CONPAR '92/VAPP V: Proceedings of the Second Joint International Conference on Vector and Parallel Processing*, 625–36. London: Springer-Verlag.

48. P. N. Glaskowsky. 1998. Map1000 unfolds at equator. *Microprocessor Report* 12.

49. Michael Gordon, William Thies, and Saman Amarasinghe. 2006. Exploiting coarse-grained task, data, and pipeline parallelism in stream programs. In *Proceedings of the International Conference on Architectural Support for Programming Languages and Operating Systems*, San Jose, CA, October 2006.

50. R. Govindarajan, Erik R. Altman, and Guang R. Gao. 1994. Minimizing register requirements under resource-constrained rate-optimal software pipelining. In *MICRO 27: Proceedings of the 27th Annual International Symposium on Microarchitecture*, 85–94. New York: ACM Press.

51. R. Govindarajan, Erik R. Altman, and Guang R. Gao. 1996. A framework for resource-constrained rate-optimal software pipelining. *IEEE Trans. Parallel Distrib. Syst.* 7(11):1133–49.

52. John C. Gyllenhaal, Wen Mei W. Hwu, and B. Ramabriohna Rau. 1996. Optimization of machine descriptions for efficient use. In *MICRO 29: Proceedings of the 29th Annual ACM/IEEE International Symposium on Microarchitecture*, 349–58. Washington, DC: IEEE Computer Society Press.

53. J. Hiser, S. Carr, P. Sweany, and S. J. Beaty. 2000. Register assignment for software pipelining with partitioned register banks. In *IPDPS '00: Proceedings of the 14th International Symposium on Parallel and Distributed Processing*, 211–18. Washington, DC: IEEE Computer Society Press.

54. Richard A. Huff. 1993. Lifetime-sensitive modulo scheduling. *SIGPLAN Notices* 28(6):258–67.

55. Intel. *Intel IA-64 architecture software developer's manual*, Vol. 1, *IA-64 application architecture*. Santa Clara, CA: Intel.

56. Mike Johnson. 1991. *Superscalar microprocessor design*. New York: Prentice-Hall.

57. N. P. Jouppi and D. W. Wall. 1989. Available instruction-level parallelism for superscalar and super-pipelined machines. In *Proceedings of the 3rd International Conference on Architectural Support for Programming Languages and Operating Systems (ASPLOS)*, Vol. 24, 272–82. New York: ACM Press.

58. R. E. Kessler. 1999. The alpha 21264 microprocessor. *IEEE Micro* 19(2):24–36.

59. Monica Lam. Software pipelining: An effective scheduling technique for VLIW machines. *SIGPLAN Notices* 23(7):318–28.

60. Monica S. Lam. 2004. Software pipelining: An effective scheduling technique for VLIW machines. *ACM SIGPLAN Notices* 39(4):244–56.

61. L. Lamport. 1974. The parallel execution of DO loops. *Commun. ACM* 17(2):83–93.

62. Eugene L. Lawler. 1976. *Combinatorial optimization: Networks and matroids*. New York: Holt, Rinehart, and Winston.

63. Charles E. Leiserson and James B. Saxe. 1991. Retiming synchronous circuitry. *Algorithmica* 6(1):5–35.

64. Josep Llosa, Mateo Valero, Eduard Ayguadé, and Antonio González. 1995. Hypernode reduction modulo scheduling. In *MICRO 28: Proceedings of the 28th Annual International Symposium on Microarchitecture*, 350–60. Los Alamitos, CA: IEEE Computer Society Press.

65. Josep Llosa, Mateo Valero, and Eduard Ayguadé. 1996. Heuristics for register-constrained software pipelining. In *MICRO 29: Proceedings of the 29th Annual ACM/IEEE International Symposium on Microarchitecture*, 250–61. Washington, DC: IEEE Computer Society Press.

66. P. Geoffrey Lowney, Stefan M. Freudenberger, Thomas J. Karzes, W. D. Lichtenstein, Robert P. Nix, John S. O'Donnell, and John C. Ruttenberg. 1993. The multiflow trace scheduling compiler. *J. Supercomput.* 7(1–2):51–142.

67. Soo-Mook Moon and Kemal Ebcioglu. 1992. An efficient resource-constrained global scheduling technique for superscalar and VLIW processors. In *MICRO 25: Proceedings of the 25th Annual International Symposium on Microarchitecture*, 55–71. Los Alamitos, CA: IEEE Computer Society Press.

68. Steven S. Muchnick. 1997. *Advanced compiler design and implementation*. San Francisco: Morgan Kaufmann.

69. Trevor Mudge. 2001. Power: A first-class architectural design constraint. *Computer* 34(4): 52–58.

70. Kalyan Muthukumar and Gautam Doshi. 2001. Software pipelining of nested loops. *Lecture Notes Comput. Sci.* 2027:165–81.

71. Alexandru Nicolau and Joseph A. Fisher. 1984. Measuring the parallelism available for very long instruction word architectures. *IEEE Trans. Comput.* C-33(11):968–76.

72. Erik Nystrom and Alexandre E. Eichenberger. 1998. Effective cluster assignment for modulo scheduling. In *MICRO 31: Proceedings of the 31st Annual ACM/IEEE International Symposium on Microarchitecture*, 103–14. Los Alamitos, CA: IEEE Computer Society Press.

73. Kunle Olukotun, Basem A. Nayfeh, Lance Hammond, Ken Wilson, and Kunyung Chang. 1996. The case for a single-chip multiprocessor. In *ASPLOS-VII: Proceedings of the Seventh International*

Conference on Architectural Support for Programming Languages and Operating Systems, 2–11. New York: ACM Press.

74. Open research compiler for Itanium processor family. http://ipf-orc.sourceforge.net/.

75. Guilherme Ottoni, Ram Rangan, Adam Stoler, and David I. August. 2005. Automatic thread extraction with decoupled software pipelining. In *MICRO 38: Proceedings of the 38th Annual IEEE/ACM International Symposium on Microarchitecture*, 105–18. Washington, DC: IEEE Computer Society Press.

76. Subbarao Palacharla, Norman P. Jouppi, and J. E. Smith. 1997. Complexity-effective superscalar processors. *SIGARCH Comput. Archit. News* 25(2):206–18.

77. Janak H. Patel and Edward S. Davidson. 1976. Improving the throughput of a pipeline by insertion of delays. In *ISCA '76: Proceedings of the 3rd Annual Symposium on Computer Architecture*, 159–64. New York: ACM Press.

78. Gerald G. Pechanek and Stamatis Vassiliadis. The ManArray embedded processor architecture. In *EUROMICRO* 26, 1348–55. Washington, DC: IEEE Computer Society Press.

79. D. Petkov, R. Harr, and S. Amarasinghe. 2002. Efficient pipelining of nested loops: Unroll-and-squash. In *16th International Parallel and Distributed Processing Symposium (IPDPS '02)*. Washington, DC: IEEE Computer Society Press.

80. R. Pyreddy and G. Tyson. 2001. Evaluating design tradeoffs in dual speed pipelines. In *Workshop on Complexity-Effective Design*, June 2001.

81. V. Janaki Ramanan and R. Govindarajan. 1991. Resource usage models for instruction scheduling: Two new models and a classification. In *ICS '99: Proceedings of the 13th International Conference on Supercomputing*, 417–24. New York: ACM Press.

82. V. Janaki Ramanan and Ramaswamy Govindarajan. 1999. Resource usage modelling for software pipelining. In *HiPC '99: Proceedings of the 6th International Conference on High Performance Computing*, 111–19. London: Springer-Verlag.

83. J. Ramanujam. 1994. Optimal software pipelining of nested loops. In *Proceedings of the 8th International Parallel Processing Symposium*, 335–42. Washington, DC: IEEE Computer Society Press.

84. B. R. Rau and J. A. Fisher. 1993. Instruction-level parallel processing: History, overview and perspective. *J. Supercomput.* 7:9–50.

85. B. R. Rau and C. D. Glaeser. 1981. Some scheduling techniques and an easily schedulable horizontal architecture for high performance scientific computing. In *MICRO 14: Proceedings of the 14th Annual Workshop on Microprogramming*, 183–98. Washington, DC: IEEE Computer Society Press.

86. B. R. Rau, M. Lee, P. P. Tirumalai, and M. S. Schlansker. 1992. Register allocation for software pipelined loops. In *Proceedings of the ACM SIGPLAN 1992 Conference on Programming Language Design and Implementation*, 283–99.

87. B. Ramakrishna Rau. 1994. Iterative modulo scheduling: An algorithm for software pipelining loops. In *Proceedings of the 27th Annual International Symposium on Microarchitecture*, 63–74. Washington, DC: IEEE Computer Society Press.

88. B. Ramakrishna Rau, David W. L. Yen, Wei Yen, and Ross A. Towie. 1989. The Cydra 5 departmental supercomputer: Design philosophies, decisions, and trade-offs. *Computer* 22(1):12–26, 28–30, 32–35.

89. Raymond Reiter. 1968. Scheduling parallel computations. *J. ACM* 15(4):590–99.

90. Hongbo Rong, Alban Douillet, R. Govindarajan, and Guang R. Gao. 2004. Code generation for single-dimension software pipelining of multi-dimensional loops. In *CGO '04: Proceedings of the International Symposium on Code Generation and Optimization*, 175–86. Washington, DC: IEEE Computer Society Press.

91. Hongbo Rong, Alban Douillet, and Guang R. Gao. 2005. Register allocation for software pipelined multi-dimensional loops. In *PLDI'05: Proceedings of the ACM SIGPLAN 2005 Conference on Programming Language Design and Implementation*. New York: ACM Press.

92. Hongbo Rong, Zhizhong Tang, R. Govindarajan, Alban Douillet, and Guang R. Gao. 2004. Single-dimension software pipelining for multi-dimensional loops. In *CGO '04: Proceedings of the*

International Symposium on Code Generation and Optimization, 163–74. Washington, DC: IEEE Computer Society Press.

93. Hongbo Rong, Zhizhong Tang, R. Govindarajan, Alban Douillet, and Guang R. Gao. 2007. Single-dimension software pipelining for multi-dimensional loops. *ACM Trans. Architecture and Code Optimization* 4(1).

94. John Ruttenberg, G. R. Gao, A. Stoutchinin, and W. Lichtenstein. 1996. Software pipelining showdown: Optimal vs. heuristic methods in a production compiler. In *PLDI '96: Proceedings of the ACM SIGPLAN 1996 Conference on Programming Language Design and Implementation*, 1–11. New York: ACM Press.

95. F. S'anchez and J. Cortadella. 1998. Reducing register pressure in software pipelining. *J. Inf. Sci. Eng.* Special Issue on Compiler Techniques for High-Performance Computing 14(1):265–79.

96. Michael S. Schlansker and B. Ramakrishna Rau. 2000. EPIC: Explicitly parallel instruction computing. *Computer* 33(2):37–45.

97. J. Smith and G. Sohi. 1995. The microarchitecture of superscalar processors. *Proc. IEEE.* 83(D): 1609–24.

98. J. E. Smith, G. E. Dermer, B. D. Vanderwarn, S. D. Klinger, C. M. Rozewski, D. L. Fowler, K. R. Scidmore, and J. P. Laudon. 1987. The ZS-1 central processor. In *Proceedings of the 2nd International Conference on Architectural Support for Programming Languages and Operating System (ASPLOS)*, Vol. 22, 199–204. New York: ACM Press.

99. Greg Snider. 2002. Performance-constrained pipelining of software loops onto reconfigurable hardware. In *FPGA '02: Proceedings of the 2002 ACM/SIGDA Tenth International Symposium on Field-Programmable Gate Arrays*, 177–86. New York: ACM Press.

100. B. Su, S. Ding, and J. Xia. 1986. Urpr—An extension of urcr for software pipelining. In *MICRO 19: Proceedings of the 19th Annual Workshop on Microprogramming*, 94–103. New York: ACM Press.

101. Bogong Su, Shiyuan Ding, Jian Wang, and Jinshi Xia. 1987. Gurpr—A method for global software pipelining. In *MICRO 20: Proceedings of the 20th Annual Workshop on Microprogramming*, 88–96. New York: ACM Press.

102. Welson Sun, Michael J. Wirthlin, and Stephen Neuendorffer. 2006. Combining module selection and resource sharing for efficient fpga pipeline synthesis. In *FPGA '06: Proceedings of the 2006 ACM/SIGDA 14th International Symposium on Field Programmable Gate Arrays*, 179–88. New York: ACM Press.

103. Texas Instruments, Inc. 1998. Tms320c62x/67x cpu and instruction set reference guide.

104. Vincent Van Dongen, Guang R. Gao, and Qi Ning. 1992. A polynomial time method for optimal software pipelining. In *CONPAR*, ed. Luc Bougé, Michel Cosnard, Yves Robert, and Denis Trystram, 613–24. Vol. 634 of Lecture Notes in Computer Science. Heidelberg, Germany: Springer-Verlag.

105. Jian Wang, Christine Eisenbeis, Martin Jourdan, and Bogong Su. 1994. Decomposed software pipelining: A new perspective and a new approach. *Int. J. Parallel Program.* 22(3):351–73.

106. Jian Wang and Guang R. Gao. 1996. Pipelining-dovetailing: A transformation to enhance software pipelining for nested loops. In *CC '96: Proceedings of the 6th International Conference on Compiler Construction*, 1–17. London: Springer-Verlag.

107. Jian Wang, Andreas Krall, M. Anton Ertl, and Christine Eisenbeis. 1994. Software pipelining with register allocation and spilling. In *MICRO 27: Proceedings of the 27th Annual International Symposium on Microarchitecture*, 95–99. New York: ACM Press.

108. Nancy J. Warter, Grant E. Haab, Krishna Subramanian, and John W. Bockhaus. 1992. Enhanced modulo scheduling for loops with conditional branches. In *MICRO 25: Proceedings of the 25th Annual International Symposium on Microarchitecture*, 170–79. Los Alamitos, CA: IEEE Computer Society Press.

109. Nancy J. Warter, Scott A. Mahlke, Wen-Mei W. Hwu, and B. Ramakrishna Rau. 1993. Reverse if-conversion. In *PLDI '93: Proceedings of the ACM SIGPLAN 1993 Conference on Programming Language Design and Implementation*, 290–99. New York: ACM Press.

110. Nancy J. Warter-Perez and Noubar Partamian. 1995. Modulo scheduling with multiple initiation intervals. In *MICRO 28: Proceedings of the 28th Annual International Symposium on Microarchitecture*, 111–19. Los Alamitos, CA: IEEE Computer Society Press.

111. Michael E. Wolf and Monica S. Lam. 1991. A data locality optimizing algorithm. *SIGPLAN Notices*, 26(6):30–44.

112. Michael E. Wolf, Dror E. Maydan, and Ding-Kai Chen. 1996. Combining loop transformations considering caches and scheduling. In *Proceedings of the 29th Annual International Symposium on Microarchitecture (MICRO 29)*, 274–86. Los Alamitos, CA: IEEE Computer Society Press.

113. Hongbo Yang, Guang R. Gao, and Clement Leung. 2002. On achieving balanced power consumption in software pipelined loops. In *CASES '02: Proceedings of the 2002 International Conference on Compilers, Architecture, and Synthesis for Embedded Systems*, 210–17, New York: ACM Press.

114. Hongbo Yang, R. Govindarajan, Guang R. Gao, George Cai, and Ziang Hu. 2002. Exploiting schedule slacks for rate-optimal power-minimum software pipelining. In *Proceedings of the 3rd Workshop on Compilers and Operating Systems for Low Power*, 5–1.

115. Han-Saem Yun and Jihong Kim. 2001. Power-aware modulo scheduling for high-performance vliw processors. In *ISLPED '01: Proceedings of the 2001 International Symposium on Low Power Electronics and Design*, 40–45. New York: ACM Press.

116. Javier Zalamea, Josep Llosa, Eduard Ayguadé, and Mateo Valero. 2000. Improved spill code generation for software pipelined loops. In *PLDI '00: Proceedings of the ACM SIGPLAN 2000 Conference on Programming Language Design and Implementation*, 134–44. New York: ACM Press.

117. Javier Zalamea, Josep Llosa, Eduard Ayguadé, and Mateo Valero. 2001. Modulo scheduling with integrated register spilling for clustered vliw architectures. In *MICRO 34: Proceedings of the 34th Annual ACM/IEEE International Symposium on Microarchitecture*, 160–69. Washington, DC: IEEE Computer Society Press.

118. W. Zhang, N. Vijaykrishnan, M. Kandemir, M. J. Irwin, D. Duarte, and Y-F. Tsai. 2001. Exploiting Vliw schedule slacks for dynamic and leakage energy reduction. In *MICRO 34: Proceedings of the 34th Annual ACM/IEEE International Symposium on Microarchitecture*, 102–13. Washington, DC: IEEE Computer Society Press.

21

Advances in Register Allocation Techniques

V. Krishna Nandivada
IBM India Research Labs,
New Delhi, India
nvkrishna@in.ibm.com

21.1 Introduction

Compilers introduce an unbounded number of temporary registers during different phases of compilation; these temporary registers arise from the translation of programmer declared variables, simplification of complex expressions, and introduction of temporaries during different optimization phases. However, the target hardware is constrained by the limited number of actual available registers. The task of the *register allocator* is to map these temporary registers to real registers and memory locations. In the text, we shall be abbreviating temporary registers as *pseudos* and actual registers as *registers*.

The register allocation problem has historically been studied under two subproblems: register assignment and spilling. Register assignment is the phase of assigning of machine registers to pseudos wherever possible. Spilling is the combined act of storing a currently used pseudo to memory and reloading it for the next use. Even though these two subproblems can be solved sequentially, such an approach leads to inefficient solutions. This has led researchers to integrate the two subproblems into one super problem [22]. The scope of the register allocation phase has also increased because of the interdependence of this phase on other subphases, such as coalescing, rematerialization, code scheduling, and so on. Researchers have proposed solutions to these issues by closely integrating the solutions to these subphases with the register allocation phase. Pointers to some of the classical solutions are presented in Section 21.1.1, and some of the later advances in this area are presented later. These advances have not yet been rigorously tested in industrial-strength compilers. Thus, our presentation is more in the nature of describing concepts and reporting on some recent experiences of the researchers in the area.

21.1.1   Classical Approaches

The register allocation problem has been studied in great detail [3, 9, 11, 12, 22, 46] for a wide variety of architectures [3, 13, 28]. The register allocation problem has been shown to be NP-complete [35, 52], and researchers have explored heuristic-based [9, 11, 12] as well as practically optimal solutions (for example, solutions based on genetic algorithms [18] and solutions based on integer linear programs [3, 22, 46]). In this chapter, we provide some of the latest advances in register allocation. As a reference, we provide a list of standard terminology and definitions here:

- *Basic block:* A sequence of branch and label-free instructions.
- *Live range:* A range of instructions, over which a pseudo is live.
- *Reaching definitions:* All the definitions that reach a program point.
- *Coalescing:* An add-on phase in register allocation that coalesces pseudos related by move instruction [38].
- *Rematerialization:* An add-on phase in register allocation that replaces a pseudo with the expression that computes the value in the pseudo [38].
- *Local allocation:* Register allocation done within one basic block.
- *Global allocation:* Register allocation for a function or procedure.
- *Interprocedural allocation:* Register allocation done across functions and procedures.
- *Interference graph:* An undirected graph $G = (V, E)$, where the set of vertices represent the set of live ranges and $v_1, v_2 \in V, (v_1, v_2) \in E$ iff the live ranges corresponding to v_1 and v_2 interfere. That is, the live ranges have common instructions.
- *Graph coloring:* Given an undirected graph, assign colors to the nodes of the graph, such that no two neighbors get the same color. The register allocation problem has long been studied as a variant of the graph coloring problem [12].
- *Spill cost:* The cost of introduced spill instructions in the code. It can be computed both dynamically and statically.
- *Chaitin–Briggs register allocation:* Chaitin proposed graph-coloring-based register allocator [12], and Briggs extended it with his optimistic register allocation algorithm [8]. The main phases of the two allocators are shown in Figure 21.1.
  A brief explanation of each of the phases is given below:
  - *Renumber:* A unique name is given to each live range.
  - *Build:* Builds an interference graph.
  - *Coalesce:* Remove trivial register–register copies. Build–coalesce phases are repeated until no more coalescing can be conducted.
  - *Spill cost:* Calculate the cost of spilling every node of the interference graph.
  - *Simplify:* Repeatedly find trivially colorable nodes (nodes with a degree less than the number of available registers). If no such node can be found, then based on a heuristic, spill a node.

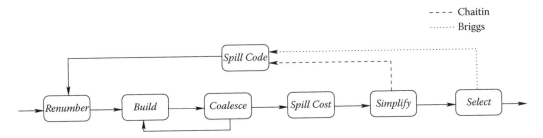

FIGURE 21.1 Overview of Chaitin and Briggs allocator. The dotted and dashed lines show the difference between the two, the dashed being part of Chaitin's allocation and the dotted part of Briggs'.

- *Select:* Assign colors (registers) to the nodes of the graph in the order determined by the phase *Simplify*.
- *Spill code:* Insert spill code for the use/def of the live ranges to be spilled (as decided by *Simplify*).
- *Iterated register coalescing:* George and Appel [21] proposed an aggressive coalescing scheme to eliminate copy instructions. Figure 21.2 presents an overview of their algorithm. A brief explanation of the phases that are different from Chaitin–Briggs register allocation is given below.
  - *Build:* Along with building an interference graph, this phase identifies nodes that are a part (source or destination) of move instructions (called *move-related* nodes).
  - *Simplify:* Like Chaitin–Briggs register allocation but removes non-move-related nodes only.
  - *Coalesce:* Similar to Chaitin–Briggs register allocation. Repeat *Build-Simplify-Coalesce* until there is no change.
  - *Freeze:* If there are still nodes to be colored and the previous phases have not succeeded in making the graph colorable, then pick one *move-related* node and make it non-move-related.
  - *Select:* Similar to Chaitin–Briggs register allocation.
  - *Potential/actual spill:* The *potential spill* phase notes the potential spill candidates, but the actual spilling is done in the *actual spill* phase after taking into consideration any changes made by the *Select* phase.
- *Linear scan register allocation:* A very fast register allocation algorithm that does the allocation by doing only one pass over the code. The algorithm does not perform very well compared to other register allocators in terms of the execution time of the generated code, but it does very well in terms of compilation time and the space utilization [49].
- *Three-address code:* A program in which every statement has at most three operands, two source operands, and one destination operand. For example, $x = y \ op \ z$ [1].

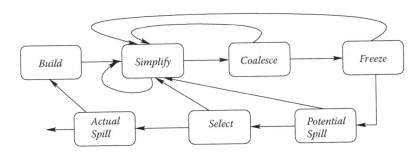

FIGURE 21.2 Overview of iterated register coalescing algorithm.

- *Static single assignment:* A program is considered to be in static single assignment (SSA) form if every pseudo has exactly one definition [16].
- *Split variables:* While translating a program into SSA form, multiple definitions to the same variable are replaced with different versions of the same variable. These instances are called the split variables.

Over the past few years, significant advances have been made in the area of register allocation. There have been new clarifications on the complexity of the register allocation problem, new results that improve the precision of the register allocation results (chordal graphs), new frontiers in register allocation (bitwidth-aware register allocation), new applications of register allocation (code size reduction), new extensions to register allocation (stack location allocation), and studies evaluating the correctness of register allocation. We will be exploring each of these in the following sections.

21.2 Language

Register allocation can be considered as a program translation phase, which takes as input a program with pseudos and returns as output a program where pseudos have been replaced with registers and memory locations. We define a simple three-address code language as our input language. The syntax for operands (pseudos and registers) in this language is given in Figure 21.3. The set P represents the set of pseudos, and set R represents the set of registers. The input program can contain statements that use constants and operands shown in Figure 21.3. The compiler during different stages of compilation can generate code containing registers (called precolored registers). Hence, registers are present as an operand in the grammar. The register allocator must respect the data flow constraints between these pre-colored registers in the input program (see Section 21.11).

After the completion of the register allocation phase, the program is still in the same format as the original code, with one difference: the pseudo to register mapping is encoded in the program, along with the new memory instructions added by the register allocator. That is, an operand v in the register allocated program can be either a register or a pseudo–register pair:

$$v \in (P \times R) + R$$

Note that this representation allows a pseudo to be mapped to different registers at different stages of the program. Some of the register allocators insert additional instructions such as move instructions, bit-wise operations, and so on.

The problem of register allocation is to translate an input program with operands in the language shown in Figure 21.3 to a program with the pseudo operands replaced with pseudo–register pairs. Figure 21.4 shows an example program and the register allocated program.

$$
\begin{array}{lll}
P & ::= & \{p_1, p_2, \cdots, p_n\} \\
R & ::= & \{\Re_1, \Re_2, \cdots, \Re_n\} \\
pr & \in PR & \subseteq (P + R)
\end{array}
$$

FIGURE 21.3   Grammar for the operands in the input program.

$$
\begin{array}{ll}
p_1 := \Re_0 + \Re_1 & (p_1, \Re_1) := \Re_0 + \Re_1 \\
p_2 := p_1 + 2 & (p_2, \Re 0) := (p_1, \Re_1) + \Re_0 \\
p_3 := p_1 + p_2 & (p_3, \Re_0) := (p_1, \Re_1) + (p_2, \Re_0)
\end{array}
$$

$$\text{(a)} \qquad\qquad\qquad\qquad \text{(b)}$$

FIGURE 21.4   Register allocation example. (a) Input program. (b) Register allocated program. The input program contains three pseudos and two pre-colored registers.

21.3   Advances in Graph Coloring Techniques

An observation that has been successfully used in recent studies is that many of the interference graphs induced by the live ranges of variables in programs are *1-perfect* [2]. A graph G is said to be 1-perfect if the chromatic number of G is equal to the size of the maximum clique in G. Andersson [2] shows that all the interference graphs present in the corpus of graphs studied by George and Appel [21] are 1-perfect. However, recognition of 1-perfect graphs is an NP-complete problem [43], so this observation cannot be used directly. A graph G is said to be *perfect* if, for each vertex-induced subgraph G′ of G, the graph G′ is 1-perfect. A vertex-induced subgraph (sometimes simply called an induced subgraph) consists of a subset of the vertices of a graph together with any edges whose endpoints are both in this subset. Grötschel et al. [23] show that the k-colorability problem can be solved in polynomial time for perfect graphs.

Gavril [20] shows that *chordal* graphs come with interesting properties, because of which problems such as minimum coloring, maximum clique, minimum covering by cliques, and maximum independent set, which are NP-complete in general, have polynomial time solutions in the context of chordal graphs. A graph is chordal if every cycle of four or more nodes in the graph has a *chord*, an edge joining two nodes that are not adjacent in the cycle (see below for further explanation). Chordal graphs are a subset of perfect graphs, so solving the minimum coloring problem in polynomial time leads to the register allocation problem being solved in polynomial time. This optimal coloring of chordal graphs can be undertaken in time linear in the number of edges and vertices.

Pereira and Palsberg [47] use the features of chordal graphs to do register allocation optimally for a large set of interference graphs. Hack et al. [24] and Brisk et al. [10] have independently proved the chordality of the interference graphs of programs in SSA form.

21.3.1   Chordal Graphs and Impact on Register Allocation

21.3.1.1   Chordal Graphs and Properties

Figures 21.5a and 21.5b present examples of chordal graphs. Figure 21.5a has no cycle of more than three nodes, and Figure 21.5b has a chord *bd* in the only possible cycle *abcd*. Figure 21.5c is another example of a chordal graph. However, removal of any of the six dashed edges would make the graph non-chordal.

A useful observation by Pereira and Palsberg [47] is that most of the graphs in the corpus of interference graphs of George and Appel [21] are chordal. It turns out that in a program in SSA form, without unstructured jumps (for example, with no arbitrary goto statements) (called *strict* programs), it will always generate chordal graphs. Three independent groups [6, 10, 24] came up with this key observation around the same time frame. This observation helps clearly decouple register assignment and spilling. For programs written in a structured language such as Java, where arbitrary goto statements do not appear, the interference graph is guaranteed to be chordal, provided the program is in SSA form. Thus, we can know precisely the optimal number of colors required to color the graph and hence can proceed to spilling without going through a coloring phase. In the absence of such decoupling, George and Appel [21] proposed an iterative algorithm to do register allocation, where the register allocator iteratively goes over phases that do coalescing, spilling, and coloring. However, this new optimality result helped Pereira and Palsberg [47] and Hack et al. [24] to design register allocators similar to the one shown in Figure 21.6.

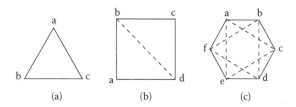

(a)   (b)   (c)

FIGURE 21.5   Example of chordal graphs. Graphs (a), (b), (c) are chordal graphs. However, removal of any of the dashed edges makes the graph non-chordal.

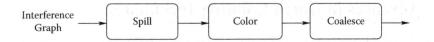

FIGURE 21.6 Optimal coloring helps decoupling of the phases in register allocation.

Function Greedy-color(V, *seo V*)
BEGIN
        while(true)
            v = next(*seo V*);
            if v == null then break;
            $C(v)$ = lowest color not used in the neighbors of v.
END

FIGURE 21.7 Sample greedy algorithm to color the nodes of an interference graph.

Function Maximum Cardinality Search (V, E)
BEGIN
        For all $v \in$ V, do $w(v) = 0$.
        ActiveSet = V
        For $i = 1..|V|$
            Let v' the node with maximum weight in ActiveSet.
            output v'
            Increase the weight of the neighbors of v' by 1.
            ActiveSet = ActiveSet - $\{v'\}$
END

FIGURE 21.8 Function maximum cardinality search generating a SEO.

In an undirected graph $G = (V, E)$, a vertex v is called *simplicial* if its neighborhood[1] in G is a clique. An ordering of the vertices of the graph G is called a *simplicial elimination ordering* (also called perfect elimination ordering) of G if every vertex $v_i \in V$ is a simplicial vertex in the subgraph induced by $\{v_1, v_2, \ldots, v_i\}$. In Figure 21.5b, the vertices a and c are simplicial. However b and d are not.

A useful property of chordal graphs is given by Dirac [17]:

> *An undirected graph without self-loops is chordal if and only if it has a simplicial elimination ordering.*

All three graphs in Figure 21.5 have simplicial elimination orderings: Figure 21.5a: a, b, c; Figure 21.5b: a, b, c, d; and Figure 21.5c: a, b, c, d, e, f.

Gavril [20] shows that a greedy approach can give optimal coloring if the order of vertices chosen for coloring is a simplicial elimination order. A sample greedy algorithm is shown in Figure 21.7.

The argument *seo V* is a simplicial elimination ordering (SEO) of the vertices of the graph under consideration, and V is the set of vertices. The function next(*seo V*) returns the next vertex in the order, if there is one present, or else returns null. The map $C(v)$ gives the color of the vertex v.

Given a graph G, an algorithm known as a *maximum cardinality search* [56] (Figure 21.8) recognizes and determines an SEO of a chordal graph in $O(|E| + |V|)$ time. The algorithm associates a weight w

---

[1] In a graph $G = (V, E)$, the neighborhood of a vertex $v \in V$ is the set of vertices V' such that $\forall v' \in V'\ (v, v') \in E$.

with each vertex v, which is initialized to 0. It keeps a set of vertices called ActiveSet (initialized to the set of vertices). At each iteration, it outputs a vertex v', which has the maximum weight among the vertices of the ActiveSet, increases the weight of the neighbors of v', and removes v' from the ActiveSet. The ties are resolved arbitrarily.

21.3.1.2 Implementation and Comments

Register assignment for strict programs in SSA form can be done optimally in polynomial time and thus allows for the decoupling of register assignment and spilling phases. Pereira and Palsberg exploited this to good effect and came up with a strategy that does near-optimal register allocation. A program in SSA form must also have to remove the ϕ nodes, and this inserts additional copy instructions. Pereira and Palsberg [47] and Hack et al. [24] employ an explicit approximate coalescing phase to get performance benefits. For the programs not in SSA form, the results generated by these approaches will not be optimal.

Pereira and Palsberg [47] use these algorithms and show that 95% of the methods of the Java 1.5 library have chordal interference graphs. Hack et al. [24] feed their heuristic-based coalescing solution to an integer linear program (ILP)-based coalescing solution and find significant speed-ups in the ILP-based coalescing method.

21.4 Register Allocation and Static Single Assignment

In Section 21.3 we showed that a program in SSA form has interesting properties such that the optimal register assignment can be done in polynomial time. In this section, we show some more issues that are specific to programs in SSA form.

In a program in SSA form, ϕ function nodes represent the control-flow-directed value renaming. Unlike regular function nodes, the arguments of a ϕ node may be contained in the same register as well. Another interesting point about ϕ nodes is that there is no specific ordering among the ϕ nodes present at the beginning of a basic block, and semantically, all the ϕ nodes present (at the beginning of a basic block) can be executed simultaneously.

The *classical* way to translate a ϕ function (*removal* of ϕ nodes) is by replacing it with a sequence of copy instructions along different control flow edges. For example, say we have a ϕ node in basic block b_0

$$\Re_1 = \phi(\Re_2, \Re_3)$$

representing the flow of value in $\Re_2$ across one edge (for instance, e_1, connecting a basic block b_1 and b_0) and in $\Re_3$ across another edge (for instance, e_2, connecting a basic block b_2 and b_0). The *classical* way to remove the ϕ function is to do the following:

- Create a new basic block b_{01} and add a copy instruction $\Re_1 := \Re_2$ in that. Remove the control flow edge between b_0 and b_1. Make b_{01} the successor of b_0 and b_1 the successor of b_{01}.
- Create a new basic block b_{02} and add a copy instruction $\Re_1 := \Re_3$ in that. Remove the control flow edge between b_0 and b_2. Make b_{02} the successor of b_0 and b_2 the successor of b_{02}.

These copy instructions introduce additional interference among the live ranges, and this increases the register pressure further. Pereira and Palsberg [48] prove that register allocation after classical SSA elimination is NP-complete. An interesting observation by Hack et al. [24] is that ϕ nodes can be removed in such a way the register demand does not increase; that is, register demand for the ϕ node never exceeds the number of variables the ϕ nodes define. In the following example,

$$\Re_3 := \phi(\Re_1, \Re_2)$$
$$\Re_4 := \phi(\Re_2, \Re_1)$$

the ϕ nodes pick the values of $\Re_1$ and $\Re_2$, or $\Re_2$ and $\Re_1$, into $\Re_3$ and $\Re_4$. In other words, we choose a permutation of the incoming values.

The set of copy instructions across any control flow edge can be thought of as a permutation of the incoming values.

$$(a_1, \ldots, a_n) = Perm(b_1, \ldots, b_n)$$

The operation *Perm* can be thought of as a "simultaneous" copy assigning b_1 to a_1, b_2 to a_2, and so on. Note that this may require some copies and some swap instructions. This *Perm* function can be implemented in various ways.

21.4.1 Swapping without Side Effects

Each permutation of size n can be written as a sequence of swaps and is thus implementable using n registers. Some architectures have instructions to swap two registers. Even architectures that do not support such an instruction can do a register swap by using three xor (exclusive or) instructions. (Note that register swap by the use of add or subtract operations is not *safe*; these operations can cause overflow and result in incorrect translation.) Let us say that at a join point, we have a set of ϕ instructions:

$$(v_1, \Re_1) := \phi(\Re_1, \Re_2)$$
$$(v_2, \Re_2) = \phi(\Re_3, \Re_1)$$

That is, we are getting values for v_1 in $\Re_1$ along edge e_1 and for $\Re_2$ along edge e_2. Similarly, we are getting values for v_2 in $\Re_1$ and $\Re_2$ along edges e_1 and e_2, respectively. Along edge e_1, we would need a copy (from $\Re_3$ to $\Re_1$). We can insert a swap $(\Re_1, \Re_2)$, along edge e_2. Let us take another example. Say we want to do the following permutation along some edge:

$$(\Re_1, \Re_2, \Re_3) \rightarrow (\Re_2, \Re_3, \Re_1)$$

This can be implemented by the following swap operations:

$$swap\ \Re_2, \Re_3$$
$$swap\ \Re_1, \Re_2$$

A simple algorithm can be used to generate swap operations required for a given permutation. First, we have to decompose the permutation into disjoint cycles. For example, in the permutation

$$(1, 2, 3, 4, 5, 6) \rightarrow (6, 3, 1, 4, 5, 2)$$

there are two disjoint cycles: (1 6 2 3) and (4 5). Each cycle (a b c d ...) can be implemented with a sequence of swaps: (a b) (b c) (c d) and so on.

21.4.2 Scratch Register

The permutation operation can also be undertaken using a scratch register, if that is available. This would require at most $n + 1$ number of moves, where n is the size of the source set. For the above example involving four registers shown in the context of "side-effect-less swap," we would insert the following set of move instructions:

$$\Re_X := \Re_1$$
$$\Re_1 := \Re_2$$
$$\Re_2 := \Re_3$$
$$\Re_3 := \Re_X$$

In this way, we can remove the ϕ functions without increasing the register pressure. Thus, it can be seen that assigning of registers to pseudos can be done in polynomial time if the program is in SSA form.

A program in non-SSA form can be converted to a program in SSA form in polynomial time. Since register assignment can be done in polynomial time for programs in SSA form with no arbitrary goto statements, does it imply that P = NP? The answer is no. The overall register allocation problem is still NP-complete. Intuitively, this hardness can be seen from two angles:

- If a normal program P requires K registers, and the SSA-converted program P_{ssa} requires K' registers, then definitely $K' \leq K$. (This follows directly from the fact that allocation for P_{ssa} can be done using exactly the same allocation as that of P. It can, however, use fewer registers owing to the reduced interference.) Thus, finding the optimal coloring for P requires more work than finding K'.
- If we consider programs only in SSA form, without arbitrary gotos, even then, the register allocation problem is still hard. Bouchez et al. [7] show that the hardness comes from pseudos that could not be put in registers — ones that are spilled and coalesced.

21.4.3 Linear Scan Register Allocation in the Context of the SSA Form

Linear scan register allocation has come as a good alternative to many traditional register allocations, when the time taken to do register allocation is an important consideration. A T linear scan algorithm computes live intervals of pseudos in a program and finds the overlaps in them by scanning them sequentially and allocating the same registers to nonoverlapping intervals. With the popularity of SSA as an intermediate form, the resulting impact of such a representation on different analyses cannot be ignored; linear scan register allocation is no exception.

The SSA form simplifies the data flow information in the sense that resulting intermediate code tends to have shorter live ranges than the intermediate form in non-SSA form. Figure 21.9 shows the snippet of a program in SSA form. The pseudo p has been split into two pseudos p_1 and p_2. A ϕ function is inserted at the merge point that computes p_3. If we study the live ranges of the pseudos p_1 and p_2, we see that the live ranges of p_1 and p_2 interfere and hence cannot be assigned the same register. However, the key point to note is that p_1 and p_2 can never be live together at the same time at the ϕ instruction; p_1 will be live if the path under consideration has basic block B1, and p_2 will be live if the path has basic block B2. To use this key point to good effect, we have to treat the ϕ function in a special way for the liveness analysis. Note that we cannot ignore the ϕ function altogether, as otherwise, the semantics of the data flow (the merger of the split variables in the ϕ function) will not be preserved.

Assigning the same register to all the SSA variables (p_1, p_2, etc.) generated from a particular variable p might not be beneficial in general. A simple example to this effect is presented in Figure 21.10.

Hence, we need a scheme that does more than blindly assign the same register to all the split variables of a variable and at the same time understands the special nature of the ϕ function.

21.4.3.1 Implementation

Mössenböck and Pfeiffer [37] present a scheme that understands the above-mentioned special semantics of the ϕ function. They present extensions to the linear scan register allocation algorithm for architectures with irregular register constraints.

For each split variable p_1 and p_2, being joined in a $\phi(p_1, p_2)$ node, they insert two move instructions in the basic blocks preceding the ϕ node; $p_4 := p_1$ in the basic block in which the definition of p_1 gets

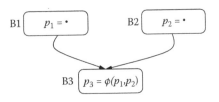

FIGURE 21.9 ϕ functions require special treatment.

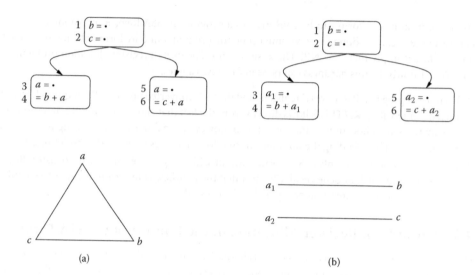

FIGURE 21.10   Assigning the same register to all the split values of the original pseudo might not give the optimal solution: (a) Shows the case where all the split copies of the variable a get the same register. In this case, the liveness interference graph shows that the live ranges of the variables a, b, and c interfere with each other. And we need three colors to color the graph. (b) Shows the case that by treating the split copies as different variables we get to see two disconnected graphs for the live range interference; the graph can be colored using two colors.

carried, and $p_5 := p_2$ in the other basic block. Also, they change the ϕ node to $\phi(p_4, p_5)$. Now the two original live ranges, corresponding to the variables p_1 and p_2, do not overlap, and the algorithm is at its discretion to assign them the same register. They use a heuristic to coalesce live ranges, so that all live ranges present in a coalesced live range get the same register. The heuristic is that two live ranges x and y can be coalesced if the register that is assigned to x is not assigned to any coalesced live range z, such that y overlaps with z.

Once these live ranges are coalesced, they present a variation to the classical linear scan algorithm, which understands these coalesced live ranges. This linear scan algorithm uses these coalesced live ranges rather than simple live ranges used by the original linear scan algorithm.

21.4.3.2   Complexity

One chief drawback of this approach is that compared to the original linear scan algorithm (linear in the number of live ranges), the complexity of this improvised algorithm is quadratic in the number of live ranges. For an algorithm that is aimed at reducing the register allocation time, having nonlinear complexity can be discouraging. However, in practice, the authors claim that the algorithm exhibits near-linear-time cost.

21.5   Bitwidth-Aware Register Allocation

Media and network applications have introduced new challenges and problems in research. In the context of register allocation, one of the issues that has added newer dimensions is the extensive use of subword data in these applications. In a classical view, the register allocation process maps pseudos to a complete register, not any subparts in it. An interesting point to note is that in case a pseudo needs only a few of the available bits in a register, then allocating all the bits of the register (full register, that is) to the pseudo can be considered wasteful. To be able to utilize this observation for register allocation and allocate a part of a register for each pseudo, we need a direct way to reference bit sections within the register. Architectures of modern embedded processors [19, 44, 57] allow this. The following example picks up an 8-bit value from $\Re_2$ and a 4-bit value from $\Re_3$ and stores them in a 12-bit destination in $\Re_1$.

$$\Re_1[4..15] := \Re_2[0..7] + \Re_3[2..5]$$

The goal is to take advantage of such instructions and reference multiple pseudos in one register simultaneously. For example, in the above example, we could use just one register to store/access all three values. Such an approach reduces the register requirements, which can help reduce the memory accesses (hence, improved execution time). Reduction in register requirements can also help in reducing the power consumption of the embedded device [29].

An operand v in the bitwidth-aware register allocated program can be represented by the following grammar:

$$\text{baR} \subseteq R \times N \times N$$
$$\text{PRPair} := (P \times \text{baR})$$
$$v \in (\text{PRPair} + \text{baR})$$

A bitwidth-aware register allocator replaces each pseudo p with a triple $(\Re, m, n)$, signifying that the bits $[m..n]$ of register $\Re$ are mapped to the pseudo p.

To take advantage of this observation during register allocation, we first need to compute the *bitwidth* information of the pseudos, which gives the sizes of the pseudos at different points of the program. Now we have to solve one more problem during the process of register allocation. Along with solving the register assignment and spilling problems, we have to solve the bin-packing problem to pack multiple pseudos into the same register. This second problem can be seen as a variation of register allocation. For the register allocation problem, we consider the interferences of the pseudos among each other and generate a pseudo to register map (PsR) in such a way that

$$\forall ps_1, ps_2 \in P, ps_1 \neq ps_2, ps_1 \text{ and } ps_2 \text{ interfere } \Rightarrow \text{PsR}(ps_1) \neq \text{PsR}(ps_2)$$

For the bitwidth-aware register allocation problem, we consider the interferences of pseudos among each other and then generate a pseudo to subregister map (PssR) in such a way that

$$\forall ps_1, ps_2 \in P, ps_1 \neq ps_2, ps_1 \text{ and } ps_2 \text{ interfere}$$
$$\Rightarrow$$
$$\text{PssR}(ps_1) \neq \text{PssR}(ps_2) \text{ OR bitSequence}(\text{PssR}(ps_1)) \cap \text{bitSequence}(\text{PssR}(ps_2)) = nil$$

For each pseudo ps, PssR returns a subregister in the form $(\Re_j, x, y)$, indicating that the bits $[x..y]$ of the register $\Re_j$ are mapped to ps. The function bitSequence$(\Re_j, x, y)$ returns a set $\{x, x + 1, \ldots, y - 1\}$.

The bitwidth-aware register allocator thus typically has two phases: bitwidth analysis and register coalescing.

21.5.1 Bitwidth Analysis

Even though the declared sizes of the pseudos give an upper limit on the sizes of the pseudos, it might not be a tight bound. For example, in the following code:

```
int a₁,n;
read n;
a₁ = n & 0xff;
...
```

the size of a_1 after the assignment is bound by 8 bits, even though it is declared as an integer (32 bits). Thus, to obtain precise bitwidth information, the compiler has to do some form of *bitwidth analysis*.

Let us look at another example, shown in Figure 21.11. In this figure, we have two variables of interest: x1 and x2. The maximum size of x1 is 16 bits and that of x2 is 8 bits. Assuming our registers are of size

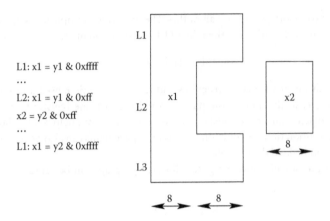

L1: x1 = y1 & 0xffff

...

L2: x1 = y1 & 0xff

x2 = y2 & 0xff

...

L1: x1 = y2 & 0xffff

FIGURE 21.11   A sample code sequence and its corresponding live ranges.

Operation	Details	Impact
$v>>t$	t is a compile time constant	t trailing bits of v are unused
$v<<t$	t is a compile time constant	t leading bits of v are unused
$v \& c$	c is a compile time constant	l leading and t leading bits of v are unused
	with l leading and t trailing zero bits	leading and t leading bits are unused

FIGURE 21.12   Arithmetic operations and their impact on the size.

16 bits, we cannot pack them in the same register if we do not take their lifetime into consideration. At any point in the program the combined size of x1 and x2 is always at most 16 bits and hence can be packed in one register of 16 bits. Besides the logical-and (&) operator, other operators impact the size of a pseudo. Figure 21.12 shows the different arithmetic operations and their corresponding impacts.

Thus, the challenge is to identify the minimal width of each live range at each program point and do so efficiently.

21.5.1.1   Implementation

Tallam and Gupta [55] introduced the problem of bitwidth-aware register allocation and provided an iterative data flow–based algorithm to compute the bitwidth information. To compute the bitwidth information, for each variable they track:

- *Dead bits:* If all the computations following a program point p that use the value of a variable v, at p, can be performed without explicitly referring to some bits in the representation of v, then these bits are called dead bits.
- *Live bits:* For each variable v, all the nondead bits are called live bits.

Hence, in the representation of the variable v, we will have a sequence of live bits, squeezed in between two sequences of dead bits. That is, there is a leading sequence of dead bits and a trailing sequence of dead bits to each sequence of live bits. The goal of the bitwidth analysis is to determine the live and dead bits for each of the variables at each program point.

Tallam and Gupta present a three-phase approach, as shown in Figure 21.13, for the computation of the live and dead bits:

- *Compute zero bit sections:* A forward analysis is carried out to compute a conservative estimate of the leading and trailing zero bit sections in each pseudo at each program point. This is computed using the information in Figure 21.12 and propagated along assignment statements. For example, in an assignment statement $x := y$, the zero bit sections of y are propagated to x.

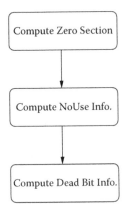

FIGURE 21.13 Computation of bitwidth-aware live ranges.

- *Compute NoUse information:* For each statement s_p that refers to the value of a pseudo p, $NoUse(s_p, v)$ returns a pair (l, t) such that the leading l bits and trailing t bits of v need not be explicitly referred to during the execution of s_p. Along with the rules inferred from Figure 21.12, the zero bit section information is used to generate NoUse information. NoUse information is propagated along the expressions using the following rules.

s_p	Details	Impact
$op\ (p, ..)$	p has at least l leading zero bits	l leading bits of p are not used
$|(p, ..)$	p has at least l leading zero bits and t trailing zero bits	l leading bits and t trailing bits of p are not used

- *Compute dead bit information:* A backward analysis is carried out to determine the leading and trailing dead bit sections for each pseudo. If the bit sections in $NoUse(s_p, p)$, are dead at the statement after s, then they are also dead immediately before s_p. If a statement s defines p and does not use p, then all bits of p are dead before s.

While iteratively computing zero bits, NoUse, and dead bit information, at any merge point, a conservative merge is done.

21.5.2 Variable Packing

Once the sizes of the pseudos (live bits) at different program points are known, the goal is to pack the pseudos, keeping the total size as the constraint. This packing of the pseudos can be seen as coalescing of the interfering pseudos. Formally, this problem can be stated as:

> **Live range coalescing:** *Given a set of live ranges L and two constants k and R, does there exist a live range coalescing that takes L as an input and outputs k coalesced live ranges such that the width of no coalesced live range exceeds R?*

Tallam and Gupta [55] show that live range coalescing is an NP-complete problem. Given an interference graph, we will present an iterative algorithm for this problem in this section. In each step a pair of interfering live ranges can be coalesced into one node, provided that at no program point does the *maximum interference width* (MIW) exceed the size of a register. Given a pair of live ranges lr_1 and lr_2, the MIW is defined as

$$MIW(lr_1, lr_2) = MAX_{\forall \text{program point } p}(width(lr_1, p) + width(lr_2, p))$$

where $width(lr_i, p)$ returns the width of the live range lr_i at program point p. Two live ranges lr_1 and lr_2 can be coalesced iff $MIW(lr_1, lr_2) \leq |R|$, where $|R|$ is the number of the bits in each register. This check requires $O(N)$ number of comparisons, where N is the number of program points.

Tallam and Gupta [55] and Barik and Sarkar [4] present heuristic-based algorithms to compute an estimated MIW (EMIW) in $O(1)$ time. Tallam and Gupta propose a scheme of edge labels to compute EMIW. Each edge (A, B) is labeled with a pair of values (A_b, B_a), A_b signifying the maximum width of A, while interfering with B and vice versa. EMIW (A, B) can be computed as $A_b + B_a$. During the coalescing phase, labels for the edges with coalesced nodes need to be updated.

If there is an edge between node C and node A, and node A has an edge to node B, then in the coalesced graph we will have an edge between AB and C. The edge label for the same would be (A_c, C_a). If node C is connected to both A and B, then we have three candidates for EMIW:

$$E_1 := max(A_b, A_c) + B_c + C_b$$
$$E_2 := max(B_a, B_c) + A_c + C_a$$
$$E_3 := max(C_a, C_b) + A_b + B_a$$

These three candidates signify the maximum interference of one node assuming the other two interfere maximally. Take, for example, E_1. The maximum A will contribute is given by $max(A_b, A_c)$. Hence, the EMIW for the three nodes will include the maximum of A_b and A_c and the EMIW for (B, C), which is obtained by E_1. Tallam and Gupta prove that for an EMIW to be safe, it has to be in the range of $E_{min} = min(E_1, E_2, E_3)$ and $E_{max} = max(E_1, E_2, E_3)$. They also show that picking E_{min} as the EMIW might not be safe, as under some circumstances $E_{min} < MIW$ and hence is not suitable.

There can be many possible EMIWs between E_{min} and E_{max}. Tallam and Gupta propose picking the middle one between E_1, E_2, and E_3. On choosing the middle one, the edge label for the edge (AB, C) is given by the following:

$$(AB_c, C_{ab}) = \begin{cases} (max(A_b, A_c) + B_c, C_b) & \text{if } E_{middle} = E_1 \\ (max(B_a, B_c) + A_c, C_a) & \text{if } E_{middle} = E_2 \\ (A_b + B_a, max(C_b, C_a)) & \text{if } E_{middle} = E_3 \end{cases}$$

Barik and Sarkar [4] generate multiple intermediate EMIWs and pick the least among them.

21.5.2.1 Handling Loops and Arrays

Besides handling the arithmetic and logical operations shown above, some high-level abstractions help in estimating the size of some more variables. For example, if we can identify loop induction variables, then we can easily place a bound on their sizes. In general, if we can get a closed-form expression for any pseudo, we can get a bound on their size.

Besides handling the scalar pseudos, we can use similar logic to handle arrays and help reduce the sizes of the arrays. In the simplest of the cases, we can compute the size of the array elements as the maximum size of any element in the array. This can be computed by going over the whole code and treating array accesses separately. This information can be used to pack multiple array elements into one word and consequentially reduce the number of memory accesses required to iterate over the array. See Barik and Sarkar [4] for details.

21.5.3 Comments

The works presented in this section give heuristics to do coalescing. How they perform in comparison with exact solutions needs to be tested.

21.6　Extensions to Register Allocation: Code Size Reduction

Register allocation has traditionally been used for code optimization targeted for execution time gains. In the past two decades attempts have been made to focus the goal on code size reduction as well. The reason for the shift of focus is the advent of new embedded processors with very limited memory, and they come with constraints on the code size.

Most of the research in this area has been target architecture specific. For example, Liao et al. [34], Leupers and Marwedel [33], and Rao and Pandey [50] target digital signal processor (DSP) type of processors and the auto-increment, auto-decrement addressing modes. Sudarsanam and Mailk [54] target architectures with two memory units and the parallel data access modes. Park et al. [46] and Naik and Palsberg [39] target architectures with more than one register bank and instruction with an RP (register pointer, pointing to the "current" memory bank)–relative addressing mode. Paek et al. [45] target architectures with multiple-load/multiple-store instructions and the generated spill code. In this section, we will present some of the key ideas behind the last two works.

The basic idea behind all of the above work on code size reduction has been modifying the data layout, such that the different instructions can be clubbed together. The clubbed instructions are the ones that access memory or memory addresses. Let us look at the example given below.

BEGIN
```
    int a, b, c, d, e;
    ...
    L1: d = a + c;
    b = d + e;
    L2:
```
END

Say at label L1, all the variables are in memory, and at label L2, all of the values must be back in memory. We first have to access variable a, then c, and then store the sum in d. Then we have to load variable e and then store the sum in b. Some architectures provide efficient mechanisms to access memory addresses next to each other (for example, auto-increment address mode can set up the base address being accessed). Thus, it would be advantageous to have a and c next to each other. That way, we can access variable a from some base address, and then we do not need to set up the base address again; use of auto-increment address mode would set the base address up automatically. Similarly, while storing b and d, if the architecture provides instructions to do multiple stores, it would be efficient (in terms of code size) to store them using one instruction. This will again require that variables b and d be next to each other, so instead of a syntax-driven layout (a, b, c, d, e), an efficient layout for these variables above would be a, c, b, d, e. In other words, depending on the architecture specifications of instructions, the input program changing the layout of the variables can result in efficient access of the variables.

21.6.1　Implementation

21.6.1.1　Z86E30 with Multiple Register Banks

The Z86E30 [58] architecture does not have explicit memory. All declared variables must be stored in registers. The processor has 16 banks of 16 registers each. Accessing any register requires the use of an RP. If the RP already points to the required bank, it need not be explicitly referenced. Otherwise, either the RP has to be set or the bank number has to be included in the instruction, resulting in *long* instructions (each long instruction requires one extra byte). Thus, accessing two variables present in two different banks would require instructions to set the RP or use a long instruction, so it would be advantageous to map variables into banks, depending on the locality. That is, variables that are accessed next to each other should be placed in the same bank, thus avoiding the code space required to access a different bank.

Naik and Palsberg [37] phrase the register allocation problem as an ILP problem, with an objective function to minimize the estimated size of the target code. The idea is that an instruction requiring register from a bank, different from the "current" bank (pointed by the RP), either has to use a long instruction (requires one extra byte) or has to use an instruction to change the current bank (requires two extra bytes). The goal of the ILP is to minimize the code size, taking into consideration these constraints.

```
Function = ComputeCodeSize
BEGIN
      codeSize = 0
      for i
        codeSize += spaceForRPReference (i)
        codeSize += spaceForRPManipulationInsts (i)
      return codeSize
END
```

For each instruction, the space required for explicitly referring the bank number and any RP manipulation (setting of RP, [re]storing RP) is summed up as the variable codeSize and fed to the ILP for minimization. The maps spaceForRPReference and spaceForRPManipulationInsts are set up by the ILP formulation.

21.6.1.2 Comments

Naik and Palsberg [37] claim that their approach results in code that is quite comparable to hand-generated code for their Z86E30 processor benchmarks. How their technique might fare for bigger benchmarks needs to be checked.

21.6.1.3 StrongARM Architectures with Load/Store Multiple Instructions

Processors like StrongARM provide special instructions to load/store multiple registers at the same time. Thus, multiple memory reference instructions can be replaced by one load/store multiple instruction (LDM/STM). However, constraints must be satisfied to be able to use an LDM/STM:

- *C1: Memory address constraint.* The sequence of memory locations from or to which m operands are loaded/stored must be contiguous, starting from the address specified by the content of a base register.
- *C2: Register sequence constraint.* The sequence of m registers corresponding to the m operands must be in an increasing sequence (need not be contiguous).
- *C3: Bound on number of operands.* The number of operands that can be used in an LDM/STM is bound by the limitation of the LDM/STM instruction. (For example, for the StrongARM LDM/STM instructions can have at most 16 operands.)

To be able to compact the multiple loads and stores scattered around the program, we must be able to:

- Group the loads (stores) into different sets. This would require moving load (store) instructions, and this must be done without altering the semantics of the program.
- Change the layout of the variables and possibly their register assignment such that the constraints *C1, C2,* and *C3* are satisfied.

Paek et al. [45] propose a three-step process to solve the problem:

- Construct a *load–store graph,* such that it satisfies constraint *C3.* A load–store graph is a multi-graph with nodes representing the variables used in the load/store instructions in the original program. Edges between the nodes indicate an overlap in the lifetime of the two variables (that is, they interfere).
- Generate temporary LDM/STM instructions, such that they satisfy constraint *C1.* This requires changing the layout of the variables to satisfy *C1.* Finding an optimal layout is an NP-complete problem [41]. Researchers have used approximate as well as exact methods (ILP based) to solve the problem.

- *Register assignment.* This step is required to satisfy the constraint *C2*. The generated temporary LDM/STM instructions might not satisfy *C2*. Paek et al. propose a conservative approach and break the LDM/STM instructions until *C2* is satisfied for all the LDM/STM instructions.

21.7 Register Allocation Super Optimizations

Register allocation has been long understood to be a key optimization, and its interdependence with many other phases of optimization is explored here. Some researchers have tried to come up with a super optimization that does register allocation and the dependent optimization together. In this section, we will present one such optimization.

21.7.1 Register Allocation + Stack Location Allocation

To offset the increasing gap between processor and memory speed, some of the modern architectures allow efficient accesses to RAM for some specific types of accesses. For example, in memories like SDRAM, accessing 64 bits of data requires the same amount of time as 32 bits. That is, it is more advantageous to access the SDRAM for 64 bits than 32 bits at a time. Architectures like StrongARM, which are part of popular platforms such as Intel IXPs [26] and Stargate [36], have instructions that allow multiple memory accesses in one instruction (load-multiple/store-multiple). A compiler can take advantage of these architecture and memory features to generate efficient spill code. Nandivada and Palsberg [41] explore this to generate efficient stack accesses for the local variables, by adding a phase after register allocation (SLA, stack location allocation). However, there is a phase ordering issue between SLA and register allocation. A unified method that takes into consideration both problems simultaneously could do better.

An important factor needed for an integrated solution is a unified metric to evaluate the impact of register allocation and SLA. Nandivada and Palsberg [42] present an ILP-based solution (SLA + RA [register allocation] = SARA) to this problem. Besides the constraints generated for the register allocation problem, they generate the constraints for SLA. They model the two-phase approach of register allocation followed by SLA in one problem. The register allocation tries to generate contiguous memory accesses next to each other wherever possible, and the SLA constraints try to merge these memory accesses into single load-multiple/store-multiple instructions. Each of these load-multiple/store-multiple instructions helps reduce the execution time, compared to the individual load/store instructions. The key point to note here is the unified metric to evaluate the solution; their ILP tries to minimize the memory access time needed because of the spill instructions inserted by the register allocator. The goal of the ILP is to minimize the cost function defined in Figure 21.14. MemAccessCost is the cost of accessing the individual words in memory. MemAccessCostReduction is the savings one gets by using load-multiple/store-multiple instructions. The map NumberOfSingleMemAccess(i) gives the number of single memory accesses at instruction i. The map NumberOfMultpleMemAccess(i) gives the number of multiple memory accesses at instruction i.

Function ComputeCost
BEGIN
    Cost = 0
    For each Instruction i
        Cost $+ =$ MemAccessCost $\times$ NumberOfSingleMemAccess(i)
        Cost $- =$ MemAccessCostReduction $\times$ NumberOfMultpleMemAccess(i).
    return Cost
END

FIGURE 21.14 Function to compute the unified cost for SARA.

21.7.1.1   Comments

The combined phase of RA+SLA adds significant compile time overhead, though it gives cognizable performance improvement (3 to 15%) over a sequential application of RA and SLA. However, the practicality of this method needs to be seen on bigger benchmarks (e.g., Spec2000, etc.).

21.8   Data Structures

Traditionally, graph-based register allocator designers have kept interference graphs as the main source of representation of the dependencies among the pseudos. Program semantics and architectures enforce additional constraints among the pseudos and registers. These constraints include requirements such as (a) architectural specification of some of the instructions mandating the use of specific registers (use of ST0 in *fmul* operations in x86, etc.), (b) compiler conventions forcing the behavior of some registers (caller save, callee save, return register, etc.), (c) architecture forcing certain structures in the operands of certain instructions (e.g., StrongARM requires that the load-multiple instruction have target registers in increasing order of their number; in IA-64 a coupled load requires a pair of even and odd registers), (d) architectural specification prohibiting the use of certain types of registers (e.g., use of floating point of registers in non-floating-point instructions, etc.), and (e) compiler optimization phases introducing dependencies among pseudos (e.g., while doing whole program register allocation, the compiler could decide on arbitrary registers to pass the function arguments).

Representing these constraints is fairly intuitive when represented as mathematical constraints, but representing and processing them in a graph is hard. The reasons are many:

- *Representation*: It is not clear how to represent all the different constraints at the same time.
- *Evaluation*: Given a graph with multiple constraints, processing them to arrive at a metric that can give an ordering of the vertices (pseudos) is hard.
- *Transitive constraints*: Graphs are convenient to represent constraints between pairs of nodes. Representing relations that involve multiple nodes (e.g., transitive relations) is not trivial.

There are advantages of including multiple constraints in one graph. It results in a unified analysis that takes into consideration all the factors at the same time. To be able to get a unified picture, one needs a common metric to measure and compare different constraints and then come up with a unified metric.

21.8.1   Implementation

Koseki et al. [32] present a scheme to represent multiple constraints as part of the interference graph, and they use it for the register assignment process. Their scheme consists of two key data structures: register preference graph and coloring precedence graph. They use these two graphs to do the register assignment.

21.8.1.1   Register Preference Graph

A register preference graph (RPG) is a directed graph with nodes representing live ranges, registers, and register classes and edges representing the binary relation between the two nodes. The preferences can be prioritized according to the benefit the preference would derive. The benefit is calculated as a weighted metric (weighted by the execution frequency or program structure) estimating the performance differential between when the preference is honored and when the node is located in memory; this metric gives the *strength* of the preference. One way to calculate the strengths is by estimating the difference in the number of processor cycles required to access a variable when it is in its preferred register and when it is located in memory. Figure 21.15 represents a sample RPG. Pseudo p_1 prefers a floating point register, with a strength of w_1; pseudo p_2 and p_3 prefer integer registers with a strength of w_2. All the pseudos prefer caller save registers with a strength of w_3 and callee save registers with a strength of w_4. Pseudo p_2 can be coalesced with pseudo p_3 with a strength of w_5 when p_3 is assigned a caller save register and a strength of w_6 when p_3 is assigned a callee save register.

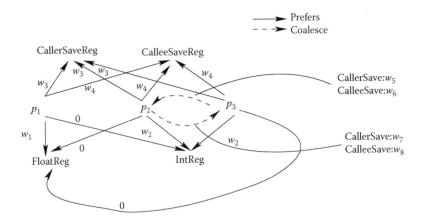

FIGURE 21.15 Sample Register Preference Graph

Koseki et al. [32] track many preference types, for example, coalesce (source node tries to use the same register as the destination), sequential+ (source node tries to use the previous register relative to the destination), sequential− (source node tries to use the next register relative to the destination), and prefers (source tries to use one of the registers included in the destination). These preferences add their strengths to the edges of the graph. The key point here is to ensure that in the presence of diverse constraints, the evaluation metric still needs to be uniform.

21.8.1.2 Coloring Precedence Graph

A coloring presence graph (CPG) is a directed graph with nodes representing live ranges and edges representing the precedence of register selection between the two connected nodes. One can pick any precedence as the criterion for setting an order among the nodes. Koseki et al. [32] use the degree of the node as the precedence criterion, so the nodes with lowest degrees have higher precedence. The precedence is represented by directed edges.

21.8.1.3 Register Assignment

In this phase, the two graphs, RPG and CPG, are consulted, and register assignment is done. Once we have these two preference graphs, the register assignment would proceed in standard fashion. First, the CPG is consulted and nodes are picked in a certain order of traversal. These nodes are then checked in the RPG to compute their preference and accordingly assigned registers, as per the availability.

21.8.1.4 Comments

Koseki et al. [32] present a useful data structure to represent multiple constraints. Their representation assumes that each variable resides either in memory or in register for its whole lifetime. It would be useful to be able to relax this constraint. It would also be interesting to see how different subproblems of register allocation, besides register assignment, can be framed using similar data structures.

21.9 Validating Register Allocation

Given a register allocation algorithm, it is a nontrivial task to determine if the algorithm does the register allocation correctly. The problem is undecidable, as the problem in general reduces to program verification. However, given a register allocation (the output of a register allocation algorithm for a given input program), we can conservatively validate the allocation.

A simple idea is that register allocation preserves the data flow of the original program. That is, if a variable is defined at label L_1, and used in L_2, without any other definitions in between, then the

$$L_1 : p_1 = n$$
$$L_2 : p_2 = 2$$
$$L_3 : p_3 = p_1 + p_2$$

$$L_1 : (p_1, \Re_j) := (n, \Re_j)$$
$$L_2 : (p_2, \Re_k) := 2$$
$$L_3 : (p_3, \Re_k) := (p_1, \Re_j) + (p_2, \Re_l)$$

FIGURE 21.16   Register allocation validation. Register allocation is valid for p_1, but not for p_2.

definition and the use must utilize the same register (or memory location). Or in other words, all the def-use relations must be preserved. Let us look at the program shown in Figure 21.16 and the corresponding register allocation.

Variable p_1 is defined at label L_1 and used at L_3. Similarly, variable p_2 is defined at label L_2 and used at L_3. A valid register allocation algorithm will ensure that the register assigned to p_1 at the *def* instruction labeled L_1 will be the same as the register for p_1 at the *use* instruction labeled L_2 (unless the register is copied to some other register). The variable p_1 is defined in register $\Re_j$ and used from $\Re_j$ and hence is valid.

However, variable p_2 defined in register $\Re_k$ at label L_2 is being used from $\Re_l$ at label L_3 and hence is an invalid allocation.

Two types of instructions provide interesting features for the validation process:

- *Copy instructions:* The register allocator might insert copy/move instructions. These copy/move instructions break some old def-use pairs and create new ones. While checking the validity of the assigned registers, these updated def-uses should be taken into consideration. For example, in the above example, say there is a new copy instruction ($\Re_l := \Re_k$) between L_2 and L_3. Then the register allocation for the variable p_2 would also be considered valid.

- *Branch join points:* If two definitions of a pseudo reach at a join point in two different registers, it would result in a conflicting situation. To avoid such scenarios, for any *use* of a pseudo with multiple reaching definitions, the defined values must reach the use point in the same register or memory location.

21.9.1   Implementation

Huang et al. [25] present an iterative data flow method to validate the register allocation. Their process consists of three phases: model extraction, constraint generation, and constraint solving.

21.9.1.1   Model Extraction

Given the input program and *output* register allocated program, two maps are generated:

- A map from each of the non-spill output instructions to an instruction in the input program. The map has no entries for the spill instructions or any copy instructions inserted by the register allocator.
- A map from each of the output operands (registers and memory locations) to the input pseudos.

These two maps are used in the next two phases to do validation. The basic idea of the map generation is to build a one–one mapping between the output program and the input program. This one–one mapping is an important requirement of such an analysis (see Section 21.9.2).

21.9.1.2   Constraint Generation

In this phase data flow constraints are generated. These equations control the flow of values in the output program. For each program point, three types of information are generated:

- *Active values:* These are the values (of pseudos) that are currently held in the operands (registers and memory locations).
- *Evicted values:* These are the old values (of pseudos) that were present in the operands, but now the same operands hold active values for some other pseudos.

$$L_1 : (a, \mathfrak{R}_1) := 0$$
$$\text{store } \mathfrak{R}_1, a$$
$$L_2 : (b, \mathfrak{R}_1) := 1$$
$$\cdots$$

FIGURE 21.17 A snippet of register allocated program. Variables a and b both are assigned the register $\mathfrak{R}_1$. Register $\mathfrak{R}_1$ gets spilled after instruction L_1.

- *Stale values:* These are the old values (of pseudos) that were present in the operands, but now the same operands hold new active values for the same pseudos.

For each statement, this information is kept in three sets: active set, stale set, and evicted set. Let us look at the example in Figure 21.17. In this example, after L_2 is processed, the current value of b becomes the active value in $\mathfrak{R}_1$, the value of a in $\mathfrak{R}_1$ is evicted, and any old value of b in any other location becomes stale.

To generate the data flow equations, we start with initial empty values for these sets. These sets are populated at each instruction as per the def-use information of that instruction and the maps generated in phase 1. At any join point, the evicted and stale sets get merged. The intersection of the elements of the active sets (A_1 and A_2) is used as the new active set. The elements of $[A_1 - (A_1 \cap A_2) \cup A_2 - (A_1 \cap A_2)]$ are added to the stale set.

21.9.1.3   Constraint Solving and Error Checking

In this phase a pass is made over the input program, and then for each instruction, it checks the pseudos in the original program and operands in the output program. If any of the mappings of pseudos (to operands) at any instruction is an element of the stale set or evicted set, then an error is flagged. If no error is flagged during this pass, the register allocation is considered validated.

21.9.2   Limitations of Data Flow–Based Validation

There are limitations to the data flow–based validation process:

- *Control flow:* This data flow–based validation process has an underlying assumption that the control flow has not changed significantly. It is based on the assumption that after register allocation is done, there is a direct mapping from the instruction and pseudos of the original program to the instructions and pseudo–register pairs in the new program. The register allocator may add new instructions (for spilling, coalescing, etc.), and these instructions can still be mapped as an intermediate computation between two instructions of the original program. However, this might not be true if the register allocator is allowed to modify the control flow of the program. For example, if the register allocator is allowed to delete instructions, (re)move definitions or uses of pseudos, or reschedule the original code, then a data flow–based validation framework, like the one suggested above, would find it hard to map the new instructions and pseudos to the original ones and might not be able to validate the allocation.
- *Computations:* The register allocator might insert instructions that do arithmetic or some other computation. It would be very hard for the validator to interpret arbitrary computations and validate the output.
- *Validation not verification:* An instance of register allocation validated by a process similar to the above does not give any guarantees about the correctness of the register allocator.

21.10   Other Interesting Ideas

In this section, we present some more interesting ideas that have been proposed and extended in recent years but not categorized under any head in this chapter.

21.10.1   Live Range Splitting

A naive approach to spilling a live range is by spilling the entire live range. That is, for every *def* in the live range, a store is inserted and a load is inserted for every *use*. Such an approach can lead to significant performance degradation if the uses/defs of the live range are present in a frequently executed part of the program.

The point to note is that the standard interference graphs give a skewed notion of interference; if a live range interferes with K other live ranges, it gives an impression that all of them interfere simultaneously. In reality, the degree of interference varies at different program points. Hence, a live range may be assigned to a register at some program points and spilled at others. Also, based on the execution frequency, the cost of spilling a live range is not the same at all program points; spill instructions inserted in frequently executed code are more expensive than those inserted in infrequently executed code. Some researchers [8, 31] have proposed splitting of live ranges into subranges before the register assignment phase, and others [5, 14, 15, 27] have proposed it during the register assignment phase and assigning registers to some of these subranges. Live range splitting done before register allocation may increase the register pressure, because of insertion of copy instructions, and since spilling is done as another separate pass after register assignment, these approaches may miss out on some opportunities to find the best spilling candidates.

Nandivada and Palsberg [42] propose a scheme of fine grain splitting during the combined phase of register assignment and spilling. A live range can be split at any program point, irrespective of whether there is a use-def of the corresponding variable at the program point. This splitting is solely driven by the execution time penalty of the overall spill code.

Nakaike et al. [40] propose a scheme (employed before register assignment) to split the live ranges at each join and fork point of basic blocks. To overcome the penalty of copy instructions, they coalesce subranges in the frequently executed code based on profile information. Nakaike et al. restrict their splitting to subranges that have interference exceeding the number of available registers. They remove additional copies present even after coalescing, by spilling some of the subranges. They show that their method is scalable and gives improvements over other live range splitting algorithms.

21.10.2   Progressive Register Allocation

Traditional register allocators work with a notion of 0/1 completion. That is, they either give the designated solution or not. There are fast algorithms that give inferior solutions, and there are slow algorithms that produce superior solutions. It would be conceivable to have an algorithm that produces an initial solution and gradually improves the solution based on time or other resources. Typical candidates for such an arrangement include solutions based on ILPs genetic algorithms, and other exact solutions. With some amount of engineering (providing a simple initial solution), these algorithms, which incrementally find better solutions, can be guaranteed to give a valid solution at any point during the running of the algorithm. Thus, a compiler could interrupt the register allocator during the running of the allocator, and the allocator would return a valid solution.

Koes and Goldstein [30] present a global progressive register allocator that quickly finds an initial solution and progressively finds better solutions until a provably optimal solution is obtained or a set time limit is reached. Their algorithm is derived from a model of register allocation based on multicommodity network flows that explicitly represent register allocation subproblems such as spill code optimization, register preferences, copy insertion, and constant rematerialization.

The register allocation problem can be intuitively represented as a multicommodity network flow model. The edge costs are used to model the cost of spilling and the cost of register preference. Each node in the network represents an allocation class (register, constant, or memory). The variables are represented as commodities, with each variable definition represented as a source and the last use of each variable shown as a sink. Thus, valid allocations can be represented as a valid flow (from source to sink), and cost of flow is the cost of the allocation.

At each iteration, Koes and Goldstein compute a lower bound by using the theory of Lagrangian relaxation, which guides the optimal solution and gives a hint of the quality of the current solution. The authors show that, even though there is no guarantee of finding an optimal solution, compared to the standard iterated graph coloring approach, their approach gives good improvements in code size reduction.

It would be interesting to see the extensions to these ideas to optimize for execution time and evaluate the impact. Also, it would be interesting to encode properties such as register coalescing and more complex features such as bitwidth information in the flow graph.

21.10.3 Register Assignment and Hardware Features

Smith et al. [53] extend the standard graph-coloring-based register assignment algorithms to take into consideration two common hardware register features: register classes (a single register may be part of multiple register classes) and register aliases (multiple register names may be aliases for a single hardware register).

In traditional graph-coloring-based register allocation algorithms (Chaitin–Briggs), two nodes are considered to be interfering if there is a connecting edge between them. If in the *Simplify* phase of the register allocation, a node is found to have a degree less than K (K being the number of available registers), then it is considered to be trivially colorable. Smith et al. extend this notion to take into consideration classes and aliasing. If a node n has m neighbors, and m is less than the number of registers available in the class of registers designated for n (C_n), then n is trivially colorable.

To handle aliasing information, Smith et al. [53] first introduce the notion of an *alias* map. For each register r, $alias(r)$ returns the set of register name with which it aliases. This definition is extended to sets: $alias(S)$ is the union of the aliases of all the members of S. Similar to the notion of degree of the node, Smith et al. propose a measure called $squeeze^*(n)$, defined as the maximum set of registers that could be denied to n and are part of C_n. It can be computed as

$$squeeze^*(n) = \max_{S \in \text{colorings of } n\text{'s neighbors}} |C_n \cap alias(S)|$$

If $squeeze^*(n) < |C_n|$, then the node is trivially colorable.

This measure *squeeze*∗ requires enumerating of the colors of all the neighbors and can be very expensive. The authors instead propose a heuristic to compute it by taking into consideration the worst-case interfering scenario in the context of C_n that can be computed as a table and looked up during compilation.

Smith et al. show that their heuristic does well to color SPECFP 2000 floating point benchmarks. However, it needs to be seen how their work compares against a scheme like that of Koseki et al. [32] and extends to other register allocation preferences.

21.11 Common Misconceptions

In this section, a few of the common misconceptions that are prevalent in the community are addressed. Some of these misconceptions may seem trivial, but the author has found many contrasting experiences during his interactions with different researchers to merit their inclusion:

- *Register assignment:* Looking at the vast research work on the register assignment problem, one might get a deceptive notion that register assignment is the most important part of the register allocation process. However, spill code generation and coalescing, to some extent, have more impact on the overall performance. Irrespective of how optimal the register assignment process is, if the spill code generation is not done well (for instance, too many spill instructions inside frequently executed code) or if too many unnecessary register–register copy instructions remain in the code, then the runtime penalty can be very high.

- *Registers used:* Another interesting misconception is about the number of used registers. Besides the theoretical importance of finding out the minimum number of required registers, the practical significance of this result is minimal. In an architecture with N number of registers, the more important question is the following: *Can register allocation be done for a given program with N number of registers?*

The machine code generated by two different register allocation algorithms that use m_1 ($<N$) and m_2 ($<N$) will have exactly the same execution time, provided both the allocators spill alike.

- *0/1 assignment:* Many register allocation techniques are based on the premise that a pseudo either gets a register or is placed in memory. That is, a pseudo either gets a register for the whole of its lifetime or a memory location, and every time a pseudo is accessed, which has been mapped to a memory location, we need a free register. Thus, programs in three-address code form might require up to two free registers at all program points; this can lead to a considerable increase in program execution time (arising out of the increased spill code owing to two fewer registers). There are other drawbacks of such an approach as well. Many times it might be useful to keep the pseudo in register for some part of the program and then place it in memory for some other part. An intuition for such an approach comes from the experience that at different program points, different pseudos might be considered *hot* and thus be needed in registers for efficient program execution. In such a scenario, permanently assigning registers and memory locations to a pseudo for all of its lifetime might not be a good idea.

- *Caller and callee save:* Two register allocations using exactly the same number of registers and spilling the matching variables (live ranges) can result in machine codes that take different amounts of time to execute. This might happen because of the difference in the types of available registers. The registers that are considered caller save, if used in a function, must be saved and restored at the beginning and end of the function. Depending on the number of times the function is called during execution, the impact of these stores and restores can become significant.

- *Pre-colored registers:* The input program given to a register allocator invariably uses both pseudos and registers. Typically, pre-colored registers are used to respect calling conventions (passing parameters, return values). A register allocator must respect the data flow semantics of these registers and not interfere with the pre-colored registers. However, not using these registers in the whole program might not be an efficient way to handle them. The register allocator can compute the liveness information for these pre-colored registers, and depending on the liveness information (the program points where the pre-colored registers are not live), the pre-colored registers can be used for mapping to pseudos.

- *Evaluation metric:* Register allocation is one of the most important optimization phases in the optimization sequence of any compiler. Measuring the number of registers used or the number of live ranges spilled or the number of spill instructions generated does not show the actual impact. However, many recent papers limit themselves to some of these static numbers and draw conclusions based upon them. The most reliable metric to compare and contrast two register allocation algorithms is to compare the final output, for example, (a) execution times of the code generated by the two allocators if the goal is to optimize for speed, (b) final code size of the two allocators if the code size reduction is the goal, and (c) power consumption during execution if power saving is the goal.

- *Interdependence:* To be able to understand and estimate the real impact of a register allocation algorithm, one needs to measure the impact in the presence of other dependent optimizations. Register allocation has an impact on many optimizations, so measuring its impact in isolation might not be very useful. A way out is to implement the register allocator as a module in an optimizing compiler and then measure the execution time/code size/power consumption of different benchmark programs in the presence of other optimizations.

21.12   Conclusion and Future Directions

The research work presented in this chapter is indicative of the interest in the area of register allocation. Though this problem has been studied for a long time, the problem continues to generate challenging problems and interesting solutions.

An interesting observation by Sarkar and Barik [51] is that even though the graph-based register allocation algorithms are faster than approaches depending on exact methods, they still take a lot of time. Efficient and scalable approaches to register allocation algorithms will continue to be areas of interest, owing to the increasing sizes of applications under consideration.

Because of the close interaction between the register allocator module and the target architecture, different evolving architectures open new frontiers for register allocation. Specialized register allocation techniques in the context of multi-core architectures (with many available registers) and embedded systems (with very few registers) will be two key areas of research in the future.

Processors constrained by power present a different dimension to the register allocation problem. Specialized register access mechanisms set up in these processors open up new goals for register allocation.

Irregular architectures (e.g., IA32, x86) come with features such as register pairing (a value can be stored in two adjacent registers), register classes, subword register access, and so on. Fast and scalable register allocation techniques understanding these features will go a long way.

References

1. A. Aho, R. Sethi, and J. Ullman. 1986. *Compilers: Principles, techniques, and tools.* Reading, MA: Addison-Wesley.
2. Christian Andersson. 2003. Register allocation by optimal graph coloring. In *12th Conference on Compiler Construction*, Warsaw, Poland 34–45. Springer.
3. Andrew W. Appel and Lal George. 2001. Optimal spilling for CISC machines with few registers. In *SIGPLAN'01 Conference on Programming Language Design and Implementation*, Snowbird, UT, 243–53. New York: ACM Press.
4. Rajkishore Barik and Vivek Sarkar. 2006. Enhanced bitwidth-aware register allocation. In *International Conference on Compiler Construction*, Vienna, Austria 263–76. Springer.
5. Peter Bergner, Peter Dahl, David Engebretsen, and Matthew T. O'Keefe. 1997. Spill code minimization via interference region spilling. In *SIGPLAN Conference on Programming Language Design and Implementation*, Las Vegas, NV 287–95. New York: ACM Press.
6. Florent Bouchez, Alain Darte, Christophe Guillon, and Fabrice Rastello. 2005. Register allocation and spill complexity under ssa. Technical Report RR2005-33, LIP, ENS-Lyon, France.
7. Florent Bouchez, Alain Darte, Christophe Guillon, and Fabrice Rastello. 2006. Register allocation: What does the NP-completeness proof of Chaitin et al. really prove? In *Fifth Annual Workshop on Duplicating, Deconstructing and Debunking.*
8. Preston Briggs. 1992. Register allocation via graph coloring. PhD thesis, Rice University, Houston, TX.
9. Preston Briggs, Keith D. Cooper, and Linda Torczon. 1994. Improvements to graph coloring register allocation. *ACM Transactions on Programming Languages and Systems* 16(3):428–55.
10. Philip Brisk, Foad Dabiri, Jamie Macbeth, and Majid Sarrafzadeh. 2005. Polynomial time graph coloring register allocation. In *International Workshop on Logic and Synthesis.*
11. D. Callahan and B. Koblenz. 1991. Register allocation via hierarchical graph coloring. In *Proceedings of the ACM SIGPLAN '91 Conference on Programming Language Design and Implementation*, Vol. 26, 192–203. New York: ACM Press.
12. G. J. Chaitin. 1982. Register allocation and spilling via graph coloring. *SIGPLAN Notices* 17(6):98–105.
13. G. J. Chaitin, M. A. Auslander, A. K. Chandra, J. Cocke, M. E. Hopkins, and P. W. Markstein. 1981. Register allocation via coloring. *Computer Languages* 6(1): 47–57.
14. Frederick Chow and John Hennessy. 1984. Register allocation by priority-based coloring. In *Proceedings of the SIGPLAN Symposium on Compiler Construction*, Montreal, Canada 222–32. New York: ACM Press.
15. Keith D. Cooper and L. Taylor Simpson. 1998. Live range splitting in a graph coloring register allocator. In *Proceedings of the International Conference on Compiler Construction*, Lisbon, Portugal, 174–87. Springer.
16. Ron Cytron, Jeanne Ferrante, Barry K. Rosen, Mark N. Wegman, and F. Kenneth Zadeck. 1991. Efficiently computing static single assignment form and the control dependence graph. *ACM Transactions on Programming Languages and Systems* 13(4):451–90.

17. G. A. Dirac. 1961. On rigid circuit graphs. *Abhandlungen aus dem Mathematischen Seminar der Universiat Hamburg* 25:71–75.

18. K. M. Elleithy and E. G. Abd-El-Fattah. 1999. A genetic algorithm for register allocation. In *Ninth Great Lakes Symposium on VLSI*, Ann Arbor, MI 226. Washington, DC: IEEE Computer Society.

19. J. Fridman. 1999. Data alignment for sub-word parallelism in DSP. In *IEEE Workshop on Signal Processing Systems (SiPS)*, Taipei, Taiwan 251–60. IEEE.

20. Fanica Gavril. 1972. Algorithms for minimum coloring, maximum clique, minimum covering by cliques, and maximum independent set of a chordal graph. *SIAM Journal of Computing* 1(2):180–87.

21. Lal George and Andrew W. Appel. 1996. Iterated register coalescing. *ACM Transactions on Programming Languages and Systems* 18(3):300–24.

22. David W. Goodwin and Kent D. Wilken. 1996. Optimal and near-optimal global register allocations using 0-1 integer programming. *Software — Practice & Experience* 26(8):929–68.

23. Martin Grötschel, László Lovász, and Alexander Schrijver. 1988. *Geometric algorithms and combinatorial optimization.* Heidelberg, Germany: Springer-Verlag.

24. Sebastian Hack, Daniel Grund, and Gerhard Goos. 2006. Register allocation for programs in SSA-form. In *International Conference on Compiler Construction*, Vienna, Austria 247–62. Springer.

25. Yuqiang Huang, Bruce R. Childers, and Mary Lou Soffa. 2006. Catching and identifying bugs in register allocation. In *International Static Analysis Symposium*, Seoul, Korea 281–300. Springer.

26. Intel(r) IXP1200 network processor. http://www.intel.com/design/network/products/npfamily/ixp1200.htm.

27. H. Kim. 2001. Region-based register allocation for EPIC architectures. PhD thesis, Department of Computer Science, New York University.

28. H. Kim, K. Gopinath, V. Kathail, and B. Narahari. 1999. Fine-grained register allocation for EPIC processors with predication. In *International Conference on Parallel and Distributed Processing Techniques and Applications*, Las Vegas, NV, 2760–66. CSREA Press.

29. Johnson Kin, Munish Gupta, and William H. Mangione-Smith. 1997. The filter cache: An energy efficient memory structure. In *International Symposium on Microarchitecture*, Research Triangle Park, NC, 184–93. New York: ACM Press.

30. David Ryan Koes and Seth Copen Goldstein. 2006. A global progressive register allocator. In *Proceedings of the Conference on Programming Language Design and Implementation*, Ottawa, Canada, 204–15. New York: ACM Press.

31. Priyadarshan Kolte and Mary Jean Harrold. 1993. Load/store range analysis for global register allocation. In *Conference on Programming Language Design and Implementation*, Albuquerque, NM, 268–77. New York: ACM Press.

32. Akira Koseki, Hideaki Komatsu, and Toshio Nakatani. 2002. Preference-directed graph coloring. In *Proceedings of the ACM SIGPLAN Conference on Programming Language Design and Implementation*, 33–44. New York: ACM Press.

33. Rainer Leupers and Peter Marwedel. 1996. Algorithms for address assignment in DSP code generation. In *Proceedings of the IEEE International Conference on Computer Aided Design*. Washington, DC: IEEE Computer Society Press.

34. Stan Liao, Srinivas Devadas, Kurt Keutzer, Steven Tjiang, and Albert Wang. 1996. Storage assignment to decrease code size. *ACM Transactions on Programming Languages and Systems* 18(3):235–53.

35. Vincenzo Liberatore, Martin Farach-Colton, and Ulrich Kremer. 1999. Evaluation of algorithms for local register allocation. In *Compiler Construction, 8th International Conference, CC'99.* Vol. 1575 of Lecture Notes in Computer Science. Heidelberg, Germany: Springer-Verlag.

36. Low-power, small-size, 400mhz, linux single board computer. http://www.xbow.com/Products/XScale.htm.

37. Hanspeter Mössenböck and Michael Pfeiffer. 2002. Linear scan register allocation in the context of SSA form and register constraints. *Lecture Notes in Computer Science* 2304, 229–246: Springer.

38. Steven S. Muchnick. 1997. *Advanced compiler design and implementation.* San Francisco: Morgan Kaufmann.

39. Mayur Naik and Jens Palsberg. 2004. Compiling with code-size constraints. *Transactions on Embedded Computing Systems* 3(1):163–81.

40. Takuya Nakaike, Tatsushi Inagaki, Hideaki Komatsu, and Toshio Nakatani. 2006. Profile-based global live-range splitting. In *Conference on Programming Language Design and Implementation*, Ottawa, Canada, 216–27. New York: ACM Press.

41. V. Krishna Nandivada and Jens Palsberg. 2003. Efficient spill code for SDRAM. In *Proceedings of the International Conference on Compilers, Architecture and Synthesis for Embedded Systems*, San Jose, CA, 24–31. New York: ACM Press.

42. V. Krishna Nandivada and Jens Palsberg. 2006. SARA: Combining stack allocation and register allocation. In *International Conference on Compiler Construction*: Vienna, Austria, Springer.

43. Assaf Natanzon, Ron Shamir, and Roded Sharan. 1999. Complexity classification of some edge modification problems. In *Proceedings of the 25th International Workshop on Graph-Theoretic Concepts in Computer Science*, 65–77. London: Springer-Verlag.

44. Xiaoning Nie, Lajos Gazsi, Frank Engel, and Gerhard Fettweis. 1999. A new network processor architecture for high-speed communications. In *Proceedings of the IEEE Workshop on SIGNAL Processing Systems (SiPS'99)*, Taipei, Taiwan, 548–57. IEEE.

45. Yunheung Paek, Minwook Ahn, Doosan Cho, and Taehwan Kim. 2007. Efficient embedded code generation with multiple load/store instructions. *Software — Practice & Experience*. In Press.

46. Jinpyo Park, Je-Hyung Lee, and Soo-Mook Moon. 2001. Register allocation for banked register file. In *Proceedings of Workshop on Languages, Compilers and Tools for Embedded Systems*, Snowbird, UT, 39–47. New York: ACM Press.

47. Fernando M. Q. Pereira and Jens Palsberg. 2005. Register allocation via coloring of chordal graphs. In *Proceedings of the 3rd Asian Symposium on Programming Languages and Systems*, Tsukuba, Japan: Springer.

48. Fernando M. Q. Pereira and Jens Palsberg. 2006. Register allocation after classical SSA elimination is NP-complete. In *Foundations of Software Science and Computation Structures (FoSSaCS)* Vienna, Austria, 79–93. Springer.

49. Massimiliano Poletto and Vivek Sarkar. 1999. Linear scan register allocation. *ACM Transactions on Programming Languages and Systems* 21(5):895–913.

50. A. Rao and S. Pande. 1999. Storage assignment optimizations to generate compact and efficient code on embedded DSPs. *In Proceedings of the ACM SIGPLAN'95 Conference on Programming Language Design and Implementation*, 128–38. New York: ACM Press.

51. Vivek Sarkar and Rajkishore Barik. 2007. Extended linear scan: An alternate foundation for global register allocation. In *International Conference on Compiler Construction*, Braga, Portugal: Springer.

52. Ravi Sethi. 1973. Complete register allocation problems. In *Proceedings of the Fifth Annual ACM Symposium on Theory of Computing*, 182–95. New York: ACM Press.

53. Michael D. Smith, Norman Ramsey, and Glenn Holloway. 2004. A generalized algorithm for graph-coloring register allocation. In *Proceedings of the Conference on Programming Language Design and Implementation*, Washington, DC, 277–88. New York: ACM Press.

54. A. Sudarsanam and S. Malik. 2000. Memory bank and register allocation in software synthesis for ASIPs. *ACM Transactions on Design Automation of Electronic System* 5(2):388–92.

55. Sriraman Tallam and Rajiv Gupta. 2003. Bitwidth aware global register allocation. In *Proceedings of the 30th Symposium on Principles of Programming Languages*, New Orleans, LA, 85–96. New York: ACM Press.

56. Robert Endre Tarjan and Mihalis Yannakakis. 1985. Simple linear-time algorithms to test chordality of graphs, test acyclicity of hypergraphs, and selectively reduce acyclic hypergraphs. *SIAM Journal of Computing* 14(1):254–55.

57. Jens Wagner and Rainer Leupers. 2001. C compiler design for an industrial network processor. In *LCTES/OM*, Snowbird, UT, 155–64. New York: ACM Press.

58. Zilog z8 microcontroller user manual. http:// www.zilog.com.

Index

T - #0299 - 101024 - C0 - 254/178/42 [44] - CB - 9781420043822 - Gloss Lamination

Q

Qualifier
  checking, **9**-16–17
  implemented in clarity, **9**-18
  inference, **9**-17
  programming discipline, **9**-17
  refine existing types, **9**-12
  specific typing rules, **9**-17–18
  type soundness theorem, **9**-18

R

RADL. *see* Rockwell Architecture Description Language
  (RADL)
RAM. *see* Random-access memory (RAM)
Random-access memory (RAM), **18**-2
Rapid type analysis (RTA), **13**-15–16, **13**-17, **13**-19
RASE, **19**-40
Rate-optimal schedule, **19**-36
RAWCC, **19**-46
Raw compiler, **19**-46–47
RBAC-based trust management (RT), **2**-14
RCCFG. *see* Reverse concurrent control flow graph
  (RCCFG)
RCV. *see* Redundancy class variable (RCV)
RDG. *see* Reduced dependence graph (RDG)
Reachability, **6**-2, **12**-3, **12**-12
Reachability graph, **6**-2
Reachable object, **6**-24
Reaching definitions, **21**-2
Read-only memory (ROM), **18**-2
Read Ports (RP), **7**-14–15
Real-time constraints, **10**-14
Real-valued and Boolean-valued operators, **8**-19
Receive-compute-send execution pattern, **15**-28
Receive node, **14**-20
RecMII. *see* Recurrence minimum initiation interval
  (RecMII)
Reconstruction phase, **2**-22
Recording
  age of objects, **6**-18, **6**-19–20
  history using predicates, **12**-35
  intergenerational pointers, **6**-20–21
Rectangular tiling, **15**-24
Recurrence minimum initiation interval (RecMII), **20**-9
Recursive-partitioning-based regression techniques, **8**-11
Red edge, **18**-17
Redistribution phase, **2**-22
Reduced bit-width instruction set architecture (rISA),
  3-7
Reduced dependence graph (RDG)
  basic and refined, **15**-18–19
Reduced instruction set computer (RISC), **16**-8, **18**-4,
  19-1
  architecture model, **19**-7–8
  instruction scheduling, **19**-7–13
Reduced reservation table (RRT), **19**-22–23, **19**-23
Reducing IR trees
  procedure, **17**-11

Reducing power consumption, **20**-34–35
Reducing power variations, **20**-35–36
Reduction, **15**-12
  loop, **15**-36
  operation, **17**-29
  simplification, **15**-12–15
Redundancy class variable (RCV), **11**-39
Reference affinity, **5**-23
Reference allocation, **18**-17–18
Reference count
  collectors, **6**-6–8
  object state, **6**-7
Refinement types for ML language, **9**-19
Region-based compilation, **19**-33
  algorithm, **7**-4–5
Regions
  code instrumentation, **7**-8
  computation, **7**-7–8
  example, **7**-7
  rules for combining, **7**-6
  sample values, **7**-9
  scheduling, **19**-32
Register access, **1**-37
Register allocation, **8**-2, **8**-4, **20**-17–18, **20**-22, **20**-56–60
  bitwidth-aware, **21**-10–14
  chordal graphs, **21**-5–6
  code sequences, **19**-11
  code size reduction, **21**-15–16
  data structures, **21**-18–19
  extensions, **21**-15–16
  future directions, **21**-24
  graph coloring technique, **21**-5–6
  instruction scheduling, **19**-37–42
  interaction, **8**-3
  irregular operand constraints, **18**-18–19
  language, **21**-4
  live ranges, **8**-5
  method, **19**-9–12
  misconceptions, **21**-23
  modulo schedule, **20**-22
  program, **21**-21
  SSP, **20**-56
  stack location, **21**-17
  static single assignment, **21**-7–9
  super optimizations, **21**-17–18
  techniques, **21**-1–25
  validating, **21**-19–20
Register assignment
  comments, **21**-7
  hardware features, **21**-23
  misconceptions, **21**-23
Register component graph, **20**-39
Register-constrained software pipelining, **20**-26,
  20-27–28
Register partitioning, **20**-39
Register preference graph (RPG), **21**-18
Register pressure, **19**-5
Register sequence constraint, **21**-16
Register spills, **19**-5
Registers used, **21**-23
Register transfer graph, **18**-12–15